TOPIC	*SYMBOL*	*MEANING*	
FUNCTIONS	$f(a)$	value of the function f at a	54
	$f: A \to B$	function from A to B	54
	$f_1 + f_2$	sum of the functions f_1 and f_2	56
	$f_1 f_2$	product of the functions f_1 and f_2	56
	$f(S)$	image of the set S under f	56
	$i_A(s)$	identity function on A	59
	$f^{-1}(x)$	inverse of f	60
	$f \circ g$	composition of f and g	61
	$\lfloor x \rfloor$	floor function of x	64
	$\lceil x \rceil$	ceiling function of x	64
	a_n	term of $\{a_i\}$ with subscript n	67
	$\sum_{i=1}^{n} a_n$	sum of $a_1, a_2, \ldots, a_n$	68
	$\sum_{\alpha \in S} a_\alpha$	sum of a_α over $\alpha \in S$	70
	$\prod_{i=1}^{n} a_n$	product of $a_1, a_2, \ldots, a_n$	73
	$f(x) = O(g(x))$	$f(x)$ is big-O of$g(x)$	74
	$n!$	n factorial	78
	$\min(x, y)$	minimum of x and y	111
	$\sim$	approximately equal to	253
	$\max(x, y)$	maximum of x and y	112
INTEGERS	$a \mid b$	a divides b	106
	$a \nmid b$	a does not divide b	106
	$gcd(a, b)$	greatest common divisor of a and b	110
	$lcm(a, b)$	least common multiple of a and b	112
	$a \mathbf{\ mod\ } b$	remainder when a is divided by b	112
	$a \not\equiv b \pmod{m}$	a is not congruent to b modulo m	113
	$a \equiv b \pmod{m}$	a is congruent to b modulo m	113
	$(a_k a_{k-1} \cdots a_1 a_0)_b$	base b representation	123
MATRICES	$[a_{ij}]$	matrix with entries a_{ij}	145
	$\mathbf{A} + \mathbf{B}$	matrix sum of $\mathbf{A}$ and $\mathbf{B}$	145
	$\mathbf{AB}$	matrix product of $\mathbf{A}$ and $\mathbf{B}$	146
	$\mathbf{I}_n$	identity matrix of order n	149
	$\mathbf{A}^t$	transpose of $\mathbf{A}$	150
	$\mathbf{A} \vee \mathbf{B}$	join of $\mathbf{A}$ and $\mathbf{B}$	151
	$\mathbf{A} \wedge \mathbf{B}$	the meet of $\mathbf{A}$ and $\mathbf{B}$	151
	$\mathbf{A} \odot \mathbf{B}$	Boolean product of $\mathbf{A}$ and $\mathbf{B}$	151
	$\mathbf{A}^{[n]}$	nth Boolean power of $\mathbf{A}$	153

Discrete Mathematics and Its Applications

Also Available from McGraw-Hill

Schaum's Outline Series in Mathematics & Statistics

Most outlines include basic theory, definitions, and hundreds of solved problems and supplementary problems with answers.

Titles on the Current List Include:

Advanced Calculus
Advanced Mathematics
Analytic Geometry
Beginning Calculus
Boolean Algebra
Calculus, 3d edition
Calculus of Finite Differences & Difference Equations
College Algebra
Complex Variables
Descriptive Geometry
Differential Equations
Differential Geometry
Discrete Math
Elementary Algebra
Review of Elementary Mathematics
Essential Computer Math
Finite Mathematics
First Year College Mathematics
Fourier Analysis
General Topology
Geometry, 2d edition
Group Theory
Laplace Transforms
Linear Algebra, 2d edition
Mathematical Handbook of Formulas & Tables
Matrices
Matrix Operations
Modern Algebra
Modern Elementary Algebra
Modern Introductory Differential Equations
Numerical Analysis, 2d edition
Partial Differential Equations
Plane Geometry
Probability
Probability & Statistics
Projective Geometry
Real Variables
Set Theory & Related Topics
Statistics, 2d edition
Technical Mathematics
Tensor Calculus
Trigonometry, 2d edition
Vector Analysis

Schaum's Solved Problems Books

Each title in this series is a complete and expert source of solved problems containing thousands of problems with worked out solutions.

Titles on the Current List Include:

3000 Solved Problems in Calculus
2500 Solved Problems in College Algebra and Trigonometry
2500 Solved Problems in Differential Equations
2000 Solved Problems in Discrete Mathematics
3000 Solved Problems in Linear Algebra
2000 Solved Problems in Numerical Analysis
3000 Solved Problems in Precalculus
2000 Solved Problems in Statistics

Discrete Mathematics and Its Applications

Second Edition

Kenneth H. Rosen
AT&T Bell Laboratories

McGraw-Hill, Inc.
New York St. Louis San Francisco Auckland Bogotá
Caracas Hamburg Lisbon London Madrid Mexico Milan Montreal
New Delhi Paris San Juan São Paulo Singapore Sydney Tokyo Toronto

Discrete Mathematics and Its Applications

1 2 3 4 5 6 7 8 9 0 DOC DOC 9 0 9 8 7 6 5 4 3 2 1

ISBN 0-07-053744-5

This book was set in Times Roman by Publication Services.
The editors were Richard Wallis and John M. Morriss;
the production supervisor was Friederich W. Schulte.
The cover was designed by Joseph Gillians.
Project supervision was done by Publication Services.
R. R. Donnelley & Sons Company was printer and binder.

Cover painting: "Between the Clock and the Bed." © Jasper Johns / VAGA New York 1990.

Library of Congress Cataloging-in-Publication Data

Rosen, Kenneth H.
Discrete mathematics and its applications / Kenneth H. Rosen. — 2nd ed.
p. cm.
Includes bibliographical references (p.) and index.
ISBN 0-07-053744-5
1. Mathematics. 2. Computer science—Mathematics. I. Title.
QA39.2.R654 1991
511—dc20 90-26129

Contents

Preface xi

To the Student xxi

1
The Foundations: Logic, Sets, and Functions 1

1.1	Logic	2
1.2	Propositional Equivalences	12
1.3	Predicates and Quantifiers	19
1.4	Sets	32
1.5	Set Operations	41
1.6	Functions	54
1.7	Sequences and Summations	66
1.8	The Growth of Functions	74

Key Terms and Results 84
Review Questions 87
Supplementary Exercises 88
Computer Projects 90

2
The Fundamentals: Algorithms, the Integers, and Matrices 91

2.1	Algorithms	92
2.2	Complexity of Algorithms	98
2.3	The Integers and Division	105
2.4	Integers and Algorithms	120
2.5	Applications of Number Theory	129
2.6	Matrices	144

Key Terms and Results 156
Review Questions 157
Supplementary Exercises 159
Computer Projects 160

3
Mathematical Reasoning 161

3.1 Methods of Proof 162
3.2 Mathematical Induction 175
3.3 Recursive Definitions 193
3.4 Recursive Algorithms 202
3.5 Program Correctness 208

Key Terms and Results 215
Review Questions 216
Supplementary Exercises 217
Computer Projects 220

4
Counting 222

4.1 The Basics of Counting 223
4.2 The Pigeonhole Principle 233
4.3 Permutations and Combinations 240
4.4 Discrete Probability 250
4.5 Probability Theory 257
4.6 Generalized Permutations and Combinations 271
4.7 Generating Permutations and Combinations 281

Key Terms and Results 287
Review Questions 288
Supplementary Exercises 289
Computer Projects 293

5
Advanced Counting Techniques 294

5.1 Recurrence Relations 295
5.2 Solving Recurrence Relations 305
5.3 Divide-and-Conquer Relations 312
5.4 Inclusion-Exclusion 318
5.5 Applications of Inclusion-Exclusion 326

Key Terms and Results 335
Review Questions 336

Supplementary Exercises 337
Computer Projects 339

6
Relations 340

6.1 Relations and Their Properties 341
6.2 *n*-ary Relations and Their Applications 351
6.3 Representing Relations 358
6.4 Closures of Relations 365
6.5 Equivalence Relations 378
6.6 Partial Orderings 386

Key Terms and Results 401
Review Questions 403
Supplementary Exercises 404
Computer Projects 407

7
Graphs 408

7.1 Introduction to Graphs 409
7.2 Graph Terminology 416
7.3 Representing Graphs and Graph Isomorphism 429
7.4 Connectivity 442
7.5 Euler and Hamilton Paths 451
7.6 Shortest Path Problems 466
7.7 Planar Graphs 476
7.8 Graph Coloring 486

Key Terms and Results 496
Review Questions 497
Supplementary Exercises 499
Computer Projects 502

8
Trees 504

8.1 Introduction to Trees 505
8.2 Applications of Trees 519
8.3 Tree Traversal 526
8.4 Trees and Sorting 543
8.5 Spanning Trees 552
8.6 Minimal Spanning Trees 563
Key Terms and Results 571

Review Questions 572
Supplementary Exercises 573
Computer Projects 576

9

Boolean Algebra 577

9.1	Boolean Functions	578
9.2	Representing Boolean Functions	584
9.3	Logic Gates	588
9.4	Minimization of Circuits	596

Key Terms and Results 609
Review Questions 610
Supplementary Exercises 611
Computer Projects 613

10

Modeling Computation 614

10.1	Languages and Grammars	615
10.2	Finite-State Machines with Output	626
10.3	Finite-State Machines with No Output	635
10.4	Language Recognition	644

Key Terms and Results 656
Review Questions 657
Supplementary Exercises 658
Computer Projects 660

Appendixes A–0

A.1	Exponential and Logarithmic Functions	A–0
A.2	Pseudocode	A–3
A.3	Generating Functions	A–9

Suggested Readings B–0

Answers to Odd-Numbered Exercises S–0

Index of Biographies I–0

Index I–1

Preface

In writing the first edition of this book and revising it for the second edition, I have been guided by two purposes that have resulted from my longstanding experience and interest in teaching discrete mathematics. For the student, my purpose was to write in a precise, readable manner, with the concepts and techniques of discrete mathematics clearly presented and demonstrated. For the instructor, my purpose was carefully to design a flexible, comprehensive teaching tool that uses proven pedagogical techniques in mathematics.

I have been extremely gratified by the tremendous success of the first edition. The many improvements in the second edition have been made possible by the generous feedback of scores of instructors and hundreds of students at many of the more than 200 schools where this book has been successfully used.

This text is designed for a one- or two-term introductory discrete mathematics course to be taken by students in a wide variety of majors, including mathematics, computer science, and engineering. College algebra is the only prerequisite.

Goals of a Discrete Mathematics Course

A discrete mathematics course has more than one purpose. Students should learn a particular set of mathematical facts and how to apply them; but more importantly, such a course should teach students how to think mathematically. To achieve these goals, this text stresses mathematical reasoning and the different ways problems are solved. Five important themes are interwoven in this text: mathematical reasoning, combinatorial analysis, discrete structures, applications and modeling, and algorithmic thinking. A successful discrete mathematics course should blend and carefully balance all five of these themes.

1. *Mathematical Reasoning:* Students must understand mathematical reasoning in order to read, comprehend, and construct mathematical arguments. This

text starts with a discussion of mathematical logic, which serves as the foundation for the subsequent discussions of methods of proof. The technique of mathematical induction is stressed through many different types of examples of such proofs and a careful explanation of why mathematical induction is a valid proof technique.

2. *Combinatorial Analysis:* An important problem-solving skill is the ability to count or enumerate objects. The discussion of enumeration in this book begins with the basic techniques of counting. The stress is on performing combinatorial analysis to solve counting problems, not on applying formulae.
3. *Discrete Structures:* A course in discrete mathematics should teach students how to work with discrete structures, which are the abstract mathematical structures used to represent discrete objects and relationships between these objects. These discrete structures include sets, permutations, relations, graphs, trees, and finite-state machines.
4. *Applications and Modeling:* Discrete mathematics has applications to almost every conceivable area of study. There are many applications to computer science in this text, as well as applications to such diverse areas as chemistry, botany, zoology, linguistics, geography, and business. These applications are natural and important uses of discrete mathematics and are not contrived. Modeling with discrete mathematics is an extremely important problem-solving skill, which students have the opportunity to develop by constructing their own models in some of the exercises in the book.
5. *Algorithmic Thinking:* Certain classes of problems are solved by the specification of an algorithm. After an algorithm has been described, a computer program can be constructed implementing it. The mathematical portions of this activity, which include the specification of the algorithm, the verification that it works properly, and the analysis of the computer memory and time required to perform it, are all covered in this text. Algorithms are described using both English and an easily understood form of pseudocode.

Features

ACCESSIBILITY: There are no mathematical prerequisites beyond college algebra for this text. The few places in the book where calculus is referred to are explicitly noted. Most students should easily understand the pseudocode used in the text to express algorithms, regardless of whether they have formally studied programming languages. There is no formal computer science prerequisite.

Each chapter begins at an easily understood and accessible level. Once basic mathematical concepts have been carefully developed, more difficult material and applications to other areas of study are presented.

FLEXIBILITY: This text has been carefully designed for flexible use. The dependence of chapters on previous material has been minimized. Each chapter is divided into sections of approximately the same length, and each section is divided into subsections that form natural blocks of material for teaching. Instructor's can easily pace their lectures using these blocks.

WRITING STYLE: The writing style in this book is direct and pragmatic. Precise mathematical language is used without excessive formalism and abstraction. Notation is introduced and used when appropriate. Care has been taken to balance the mix of notation and words in mathematical statements.

EXTENSIVE CLASSROOM USE: This book has been used at over 250 schools and more than 200 have used it more than once. The feedback from instructors and students at many of the schools has helped make the second edition an even more successful teaching tool than the first edition.

MATHEMATICAL RIGOR AND PRECISION: All definitions and theorems in this text are stated extremely carefully so that students will appreciate the precision of language and rigor needed in mathematics. Proofs are motivated and developed slowly; their steps are all carefully justified. Recursive definitions are explained and used extensively.

FIGURES AND TABLES: This text contains more than 500 figures. The figures are designed to illustrate key concepts and steps of proofs. Color has been carefully used in figures to illustrate important points. Whenever possible, tables have been used to summarize key points and illuminate quantitative relationships.

WORKED EXAMPLES: Over 600 examples are used to illustrate concepts, relate different topics, and introduce applications. In the examples, a question is first posed, then its solution is presented with the appropriate amount of detail.

APPLICATIONS: The applications included in this text demonstrate the utility of discrete mathematics in the solution of real-world problems. Applications to a wide variety of areas including computer science, psychology, chemistry, engineering, linguistics, biology, business, and many other areas are included in this text.

ALGORITHMS: Results in discrete mathematics are often expressed in terms of algorithms; hence, key algorithms are introduced in each chapter of the book. These algorithms are expressed in words and in an easily understood form of structured pseudocode, which is described and specified in Appendix 2. The computational complexity of the algorithms in the text are also analyzed at an elementary level.

HISTORICAL INFORMATION: The background of many topics is succinctly described in the text. Brief biographies of nearly 50 mathematicians and computer scientists are included as footnotes. These biographies include information about the lives, careers, and accomplishments of these important contributors to discrete mathematics. In addition, numerous historical footnotes are included that supplement the historical information in the main body of the text.

KEY TERMS AND RESULTS: A list of key terms and results follows each chapter. The key terms include only the most important that students should learn, not every term defined in the chapter.

EXERCISES: There are over 2400 exercises in the text. There are many different types of questions posed. There is an ample supply of straightforward exercises that develop basic skills, a large number of intermediate exercises, and a good supply of challenging exercises. Exercises are stated clearly and unambiguously, and all are carefully graded for level of difficulty. Exercise sets contain special discussions, with exercises, that develop new concepts not covered in the text, permitting students to discover new ideas through their own work. Exercises that are somewhat more difficult than average are marked with a single star; those that are much more challenging are marked with two stars. Exercises whose solutions require calculus are explicitly noted. Exercises that develop results used in the text are clearly identified with the symbol ☞. Solutions to all odd-numbered exercises are provided at the back of the text. The solutions include proofs in which most of the steps are clearly spelled out.

REVIEW QUESTIONS: A set of review questions is provided at the end of each chapter. These questions are designed to help students focus their study on the most important concepts and techniques of that chapter. To answer these questions students need to write long answers, rather than just perform calculations or give short replies.

SUPPLEMENTARY EXERCISE SETS: Each chapter is followed by a rich and varied set of supplementary exercises. These exercises are generally more difficult than those in the exercise sets following the sections. The supplementary exercises reinforce the concepts of the chapter and integrate different topics more effectively.

COMPUTER PROJECTS: Each chapter is followed by a set of computer projects. The 135 computer projects tie together what students may have learned in computing and in discrete mathematics. Computer projects that are more difficult than average, from both a mathematical and a programming point of view, are marked with a star, and those that are extremely challenging are marked with two stars.

APPENDIXES: There are three appendixes to the text. The first covers exponential and logarithmic functions, reviewing some basic material used heavily in the course; the second specifies the pseudocode used to describe algorithms in this text; and the third discusses generating functions.

SUGGESTED READING: A list of suggested readings for each chapter is provided in a section at the end of the text. These suggested readings include books at or below the level of this text, more difficult books, expository articles, and articles in which discoveries in discrete mathematics were originally published.

Changes in the Second Edition

This new edition includes a wide range of improvements and additions that make the book more readable, teachable, flexible, and interesting. However, the basic features of the first edition, as well as the tone and emphases, have been retained. The following are the most noteworthy changes in the second edition.

ENHANCED COVERAGE OF KEY TOPICS: Many topics troublesome to students have been given expanded coverage. For example, the material on quantifiers has been expanded, including the addition of examples from Lewis Carroll; material on big-O estimates has been expanded and clarified; the treatment of mathematical induction has been enhanced; the material on the recursive definition of sets has been improved; and the coverage of graph theory, including bipartite graphs and graph isomorphism, has been enhanced.

NEW TOPICS: The second edition of the book contains two new optional sections requested by many instructors. A section on probability theory has been added to Chapter 4, supplementing the material on discrete probability. This section covers probability functions, expected values, conditional probability, and Bernoulli trials; it also applies this material to computing the average-case complexity of algorithms. A section on number theory has been added to Chapter 2; it covers such topics as the Chinese Remainder Theorem, arithmetic with large integers, Fermat's Little Theorem, pseudoprimes, and public key cryptography.

NEW APPLICATIONS: Many applications have been added. These include pseudorandom number generation, hashing functions, computer arithmetic with large integers, public key cryptography, and interconnection networks for parallel processors.

NEW EXAMPLES: More than 50 examples have been added, including many that help clarify difficult points. Examples in the first edition identified as less than effective have been replaced with better examples.

NEW AND IMPROVED EXERCISES: More than 500 new exercises have been added. These include routine exercises (especially where lacking in the first edition), many intermediate-level exercises, and selected challenging exercises. Exercises that were unclear or ambiguous have been clarified or deleted. In the second edition exercises of a particular type occur both as odd-numbered and even-numbered exercises. The grading of exercises has been reviewed and revised.

REVIEW QUESTIONS: Review questions are now provided at the end of each chapter. These questions are designed to help students think through the most important ideas in each chapter. To answer these questions students must write out complete answers, rather than just provide short responses.

BIOGRAPHICAL AND HISTORICAL FOOTNOTES: More than 50 brief biographies covering ancient, classical, and modern mathematicians and computer scientists have been added as footnotes. In addition, interesting historical notes are included throughout the book as footnotes, supplementing the historical information included in the main body of the text.

IMPROVED DESIGN: The design of the book has been enhanced for easier reading. For example, all definitions are now screened in color so that they are clearly highlighted.

NEW AND IMPROVED ANCILLARIES: A supplemental volume *Applications of Discrete Mathematics,* is now available for use by instructors in conjunction with the text. The **Student Solutions Guide** for the second edition contains material that helps students answer review questions and provides sample tests and crib sheets for each chapter to help students prepare for exams. Material designed to help instructors teaching a writing-across-the-curriculum-style course using this text and an accompanying Student Workbook will also be available.

How to Use This Book

This text has been carefully written and constructed to support discrete mathematics courses at several levels. The following table identifies the core and optional sections. An introductory one-term course in discrete mathematics at the sophomore level can be based on the core sections of the text, with other sections covered at the discretion of the instructor. A two-term introductory course could include all the optional mathematics sections in addition to the core sections. A course with a strong computer science emphasis can be taught by covering some or all of the optional computer science sections.

Chapter	*Core Sections*	*Optional Computer Science Sections*	*Optional Mathematics Sections*
1	1.1–1.9 (as needed)		
2	2.1–2.3, 2.6 (as needed)	2.4	2.5
3	3.1–3.3	3.4, 3.5	
4	4.1-4.4	4.7	4.5, 4.6
5	5.1, 5.4	5.3	5.2, 5.5
6	6.1, 6.3, 6.5	6.2	6.4, 6.6
7	7.1–7.5		7.6–7.8
8	8.1	8.2, 8.3, 8.4	8.5, 8.6
9		9.1–9.4	
10		10.1–10.4	

Instructors using this book can adjust the level of difficulty of their course by omitting the more challenging examples at the end of sections as well as the more challenging exercises. The dependence of chapters on earlier chapters is shown in the following chart.

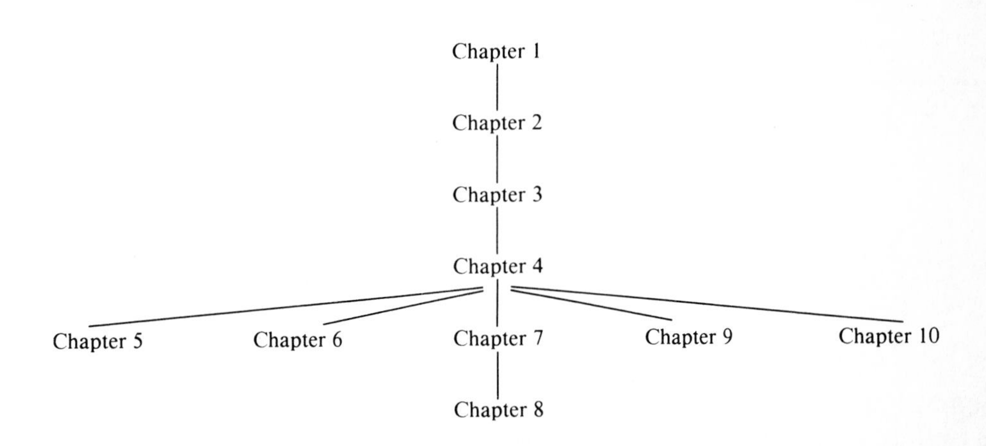

Ancillaries

STUDENT SOLUTIONS GUIDE: This student manual, available separately, contains full solutions to all the odd-numbered problems in the exercise sets. These solutions explain why a particular method is used and why it works. For some exercises, one or two other possible approaches are described to show that a problem may be solved in several different ways. The guide also includes sample tests for each chapter and a sample crib sheet for each chapter, both designed to help students prepare for exams. Students have found this guide extremely useful.

INSTRUCTOR'S RESOURCE GUIDE: This manual contains full solutions to even-numbered exercises in the text. It also provides suggestions on how to teach the material in each chapter of the book, including the points to stress in each section and how to put the material into perspective. Furthermore, the manual contains sample examination questions for each chapter; the solutions to these sample questions are provided as well. Finally, sample syllabi are presented.

COMPUTER PROJECTS SOLUTIONS GUIDE: A manual containing solutions to the computer projects is available to instructors who adopt the text. This manual gives the code in Pascal for these projects, including sample input and output. The programs are available on a disc that will run on a DOS PC.

APPLICATIONS OF DISCRETE MATHEMATICS: This ancillary is a separate text that can be used either in conjunction with the text or independently. It contains more than 20 chapters (each with its own set of exercises) written by instructors who have used the text. Following a common format similar to that of the text, the chapters in this book can be used as a text for a separate course, for a student seminar, or for a student doing independent study. Subsequent editions of this ancillary are planned that will broaden the range of applications covered. Instructors are invited to submit additional applications for possible inclusion in later versions.

Acknowledgments

I would like to thank the many instructors and students at many different schools who have used this book and provided me with their valuable feedback and helpful suggestions. Their input has made this a much better book than it would have been otherwise. In particular, I would like to thank my students at Monmouth College for their help in the preparation of the second edition. I especially want to thank Jerrold Grossman and John Michaels for their technical reviews of the second edition and their "eagle eyes," which have helped ensure the accuracy of this book.

I thank the many, many reviewers of both the first and second editions. These reviewers have provided much helpful criticism and encouragement to me. I hope this edition lives up to their high expectations.

Eric Allender,
Rutgers University
Stephen Andrilli,
La Salle University
Jack R. Barone,
Baruch College
Ken W. Bosworth,
University of Maryland
Russell Campbell,
University of Northern Iowa
Tim Carroll,
Bloomsburg University
Peter Collinge,
Monroe Community College
Ron Davis,
Millersville University
Thomas Dowling,
Ohio State University
Ladnor Geissinger,
University of North Carolina
Paul Gormley,
Villanova University
Jerrold Grossman,
Oakland University
Donald Hutchison,
Clackamas Community College
Kenneth Johnson,
North Dakota State University
David Jonah,
Wayne State University
W. Thomas Kiley,
George Mason University
Gary Klatt,
University of Wisconsin
Nicholas Krier,
Colorado State University
Robert Lavelle,
Iona College
Ernie Manes,
University of Massachusetts
Francis Masat,
Glassboro State College
Robert McGuigan,
Westfield State College
J. M. Metzger,
University of North Dakota
John Michaels,
SUNY,
Brockport
D. R. Morrison,
University of New Mexico

Jeffrey Nunemacher,
Ohio Wesleyan University
Charles Parry,
Virginia Polytechnic Institute and State University
Thomas W. Parsons,
Hofstra University
Mary K. Prisco,
University of Wisconsin–Green Bay
Harold Reiter,
University of North Carolina
Amy L. Rocha,
San Jose State University
Janet Roll,
University of Findlay
Douglas Shier,
University of Maryland
Wallace Terwilligen,
Bowling Green State University
Lisa Townsley-Kulich,
Illinois Benedictine College
David S. Tucker,
Midwestern State University
Thomas Upson,
Rochester Institute of Technology
Roman Voronka,
New Jersey Institute of Technology
James Walker,
University of South Carolina

I want to thank the executives of McGraw-Hill for their support of this project, especially Seibert Adams and Eric Munson. Thanks go to Richard Wallis, the sponsoring editor, for his enthusiastic support, to Mary Verville, who has been a strong advocate in her role as Marketing Manager, and to Maggie Lanzillo, Editorial Assistant. I would like to offer my continuing gratitude to the original editor of this book, Wayne Yuhasz (now of CRC Press), who helped make the original edition of this book as successful as it was and who has continued to provide me with useful advice. Also, I am grateful to Jeff Pepper (of Osborne McGraw-Hill) for his suggestions and editorial review. I offer thanks to the staff of McGraw-Hill for coordinating the production of this book, including John Morriss, Editing Manager, and to the staff of Publication Services, including Melissa G. Madsen and John Fancher, who served as Managing Editors, for their fine job of producing this edition.

As always, I am grateful for the support given to me by my management at AT&T Bell Laboratories, including Len Stier, Dick Rosinski, Allan Breithaupt, Ken Huber, Tony Barrese, Randy Pilc, Jim Day, Bill O'Shea, and Joe Timko. They have provided an environment that has helped me develop professionally and provided me with the resources needed to make this book a success.

Kenneth H. Rosen

To the Student

What is discrete mathematics? Discrete mathematics is the part of mathematics devoted to the study of discrete objects. (Here *discrete* means consisting of distinct or unconnected elements.) The kind of problems solved using discrete mathematics include: How many ways are there to choose a valid password on a computer system? What is the probability of winning a lottery? Is there a link between two computers in a network? What is the shortest path between two cities using a transportation system? How can a list of integers be sorted so that the integers are in increasing order? How many steps are required to do such a sorting? How can a circuit be designed that adds two integers? You will learn the discrete structures and techniques needed to solve problems such as these.

More generally, discrete mathematics is used whenever objects are counted, when relationships between finite sets are studied, and when processes involving a finite number of steps are analyzed. A key reason for the growth in the importance of discrete mathematics is that information is stored and manipulated by computing machines in a discrete fashion.

There are several important reasons for studying discrete mathematics. First, through this course you can develop your mathematical maturity, that is, your ability to understand and create mathematical arguments. You will not get very far in your studies in the mathematical sciences without these skills. Second, discrete mathematics is the gateway to more advanced courses in all parts of the mathematical sciences. Math courses based on the material studied in discrete mathematics include logic, set theory, number theory, linear algebra, abstract algebra, combinatorics, graph theory, and probability theory (the discrete part of the subject). Discrete mathematics provides the mathematical foundations for many computer science courses, including data structures, algorithms, data base theory, automata theory, formal languages, compiler theory, computer security, and operating systems. Students find these courses much more difficult when they have not had the appropriate mathematical foundations from discrete math. Also, discrete mathematics contains the necessary mathematical background for solving problems in operations research, including many discrete optimization techniques, chemistry, engineering, biology, and so on. In the text, we will study applications to some of these areas.

Finally, I would like to offer some helpful advice to students about how best to learn discrete mathematics. You will learn the most by working exercises. I suggest you do as many as you possibly can, including both the exercises at the end of each section of the text and the supplementary exercises at the end of each chapter. Always attempt exercises yourself before consulting the answers at the end of the book or the Student Solutions Guide. Only once you have put together a solution, or you find yourself at an impasse, should you look up the suggested solution. At that point you will find the discussions in the Student Solutions Guide most helpful. When doing exercises, keep in mind that the more difficult exercises are marked as described in the following table.

Key to the Exercises

No marking	A routine exercise
*	A difficult exercise
**	An extremely challenging exercise
☞	An exercise containing a result used in the text
(requires calculus)	An exercise whose solution requires the use of limits

Discrete Mathematics and Its Applications

1

The Foundations: Logic, Sets, and Functions

This chapter reviews the foundations of discrete mathematics. Three important topics are covered: logic, sets, and functions. The rules of logic specify the precise meaning of mathematical statements. For instance, these rules tell us what statements such as, "There exists an integer that is greater than 100 that is a power of 2," and, "For every integer n the sum of the positive integers not exceeding n is $n(n+1)/2$," mean. Furthermore, logic is the basis of all mathematical reasoning. Also, logic has practical applications to the design of computing machines.

Much of discrete mathematics is devoted to the study of discrete structures, which are used to represent discrete objects. All discrete structures are built up from sets, which are collections of objects. Examples of discrete structures built up from sets include combinations, which are unordered collections of objects used extensively in counting; relations, which are sets of ordered pairs that represent relationships between objects; graphs, which are sets of vertices and edges that connect vertices; and finite state machines, which are use to model computing machines.

The concept of a function is extremely important in discrete mathematics. A function assigns to each element of a set precisely one element of a set. Such useful structures as sequences and strings are special types of functions. Functions are used to represent the number of steps a procedure uses to solve a problem. The analysis of algorithms uses terminology and concepts related to the growth of functions. Recursive functions, defined by specifying their values at positive integers in terms of their values at smaller positive integers, are used to solve many counting problems.

1.1 Logic

INTRODUCTION

The rules of logic give precise meaning to mathematical statements. These rules are used to distinguish between valid and invalid mathematical arguments. Since a major goal of this book is to teach the reader how to understand and how to construct correct mathematical arguments, we begin our study of discrete mathematics with an introduction to logic.

In addition to its importance in understanding mathematical reasoning, logic has numerous applications to computer science. These rules are used in the design of computer circuits, the construction of computer programs, the verification of the correctness of programs, and in many other ways. We will discuss each of these applications in the following chapters.

PROPOSITIONS

Our discussion begins with an introduction to the basic building blocks of logic, propositions. A **proposition** is a statement that is either true or false, but not both.

EXAMPLE 1 All the following statements are propositions.

1. Washington, D.C. is the capital of the United States of America.

2. Toronto is the capital of Canada.

3. $1 + 1 = 2$.

4. $2 + 2 = 3$.

Propositions 1 and 3 are true, whereas 2 and 4 are false. ■

Some sentences that are not propositions are given in the next example.

EXAMPLE 2 Consider the following sentences.

1. What time is it?

2. Read this carefully.

3. $x + 1 = 2$.

4. $x + y = z$.

Sentences 1 and 2 are not propositions because they are not statements. Sentences 3 and 4 are not propositions because they are neither true nor false, since the variables in these sentences have not been assigned values. Various ways to form propositions from sentences of this type will be discussed in Section 3 of this chapter. ■

Letters are used to denote propositions, just as letters are used to denote variables. The conventional letters used for this purpose are $p, q, r, s, \ldots$. The **truth value** of a proposition is true, denoted by T, if it is a true proposition and false, denoted by F, if it is a false proposition.

We now turn our attention to methods for producing new propositions from those that we already have. These methods were discussed by the English mathematician George Boole in 1854 in his book *The Laws of Thought*. Many mathematical statements are constructed by combining one or more propositions. New propositions, called **compound propositions**, are formed from existing propositions using logical operators.

DEFINITION 1. Let p be a proposition. The statement

"It is not the case that p."

is another proposition, called the *negation* of p. The negation of p is denoted by $\neg p$. The proposition $\neg p$ is read "not p."

EXAMPLE 3 Find the negation of the proposition

"Today is Friday."

Solution: The negation is

"It is not the case that today is Friday."

This negation can be more simply expressed by

"Today is not Friday." ■

George Boole (1815–1864). George Boole, the son of a cobbler, was born in Lincoln, England in November 1815. Because of his family's difficult financial situation, Boole had to struggle to educate himself while supporting his family. Nevertheless, he became one of the most important mathematicians of the 1800s. Although he considered a career as a clergyman, he decided instead to go into teaching and soon afterward opened a school of his own. In his preparation for teaching mathematics, Boole, unsatisfied with textbooks of his day, decided to read the works of the great mathematicians. While reading papers of the great French mathematician Lagrange, Boole made discoveries in the calculus of variations, the branch of analysis dealing with finding curves and surfaces optimizing certain parameters.

In 1848 Boole published *The Mathematical Analysis of Logic*, the first of his contributions to symbolic logic. In 1849 he was appointed professor of mathematics at Queen's College in Cork, Ireland. In 1854 he published *The Laws of Thought*, his most famous work. In this book Boole introduced what is now called Boolean algebra in his honor. Boole wrote textbooks on differential equations and on difference equations that were used in Great Britain until the end of the 19th century. Boole married in 1855; his wife was the niece of the professor of Greek at Queen's College. In 1864 Boole died from pneumonia, which he contracted as a result of keeping a lecture engagement even though he was soaking wet.

TABLE 1 The Truth Table for the Negation of a Proposition.

p	$\neg p$
T	F
F	T

A **truth table** displays the relationships between the truth values of propositions. Truth tables are especially valuable in the determination of the truth values of propositions constructed from simpler propositions. Table 1 displays the truth values of a proposition and its negation.

The negation of a proposition can also be considered the result of the operation of the **negation operator** on a proposition. The negation operator constructs a new proposition from a single existing proposition. We will now introduce the logical operators that are used to form new propositions from two or more existing propositions. These logical operators are also called **connectives**.

DEFINITION 2. Let p and q be propositions. The proposition "p and q," denoted by $p \wedge q$, is true when both p and q are true and is false otherwise. The proposition $p \wedge q$ is called the *conjunction* of p and q.

The truth table for $p \wedge q$ is shown in Table 2. Note that there are four rows in this truth table, one row for each possible combination of truth values of the propositions p and q.

EXAMPLE 4 Find the conjunction of the propositions p and q where p is the proposition "Today is Friday" and q is the proposition "It is raining today."

Solution: The conjunction of these propositions, $p \wedge q$, is the proposition "Today is Friday and it is raining today." This proposition is true on rainy Fridays and is false on any day that is not a Friday and on Fridays when it does not rain. ■

DEFINITION 3. Let p and q be propositions. The proposition "p or q," denoted by $p \vee q$, is the proposition that is false when p and q are both false and true otherwise. The proposition $p \vee q$ is called the *disjunction* of p and q.

The truth table for $p \vee q$ is shown in Table 3.

TABLE 2 The Truth Table for the Conjunction of Two Propositions.

p	q	$p \wedge q$
T	T	T
T	F	F
F	T	F
F	F	F

TABLE 3 The Truth Table for the Disjunction of Two Propositions.

p	q	$p \vee q$
T	T	T
T	F	T
F	T	T
F	F	F

The use of the connective "or" in a disjunction corresponds to one of the two ways the word "or" is used in English, namely, in an inclusive way. A disjunction is true when either of the two propositions in it is true or when both are true. For instance, the "inclusive or" is being used in the statement

"Students who have taken calculus or computer science can take this class."

Here, we mean that students who have taken both calculus and computer science can take the class, as well as the students who have taken just one of the two subjects. On the other hand, we are using the "exclusive or" when we say

"Students who have taken calculus or computer science, but not both, can enroll in this class."

Here, we mean that students who have taken both calculus and a computer science course cannot take the class. Only those who have taken exactly one of the two courses can take the class.

Similarly, when a menu at a restaurant states, "Soup or salad comes with an entree," the restaurant almost always means that customers can have either soup or salad, but not both. Hence, this is an exclusive, rather than an inclusive, or.

EXAMPLE 5 What is the disjunction of the propositions p and q where p and q are the same propositions as in Example 4?

Solution: The disjunction of p and q, $p \vee q$, is the proposition

"Today is Friday or it is raining today."

This proposition is true on any day that is either a Friday or a rainy day (including rainy Fridays). It is only false on days that are not Fridays when it also does not rain. ■

As was previously remarked, the use of the connective "or" in a disjunction corresponds to one of the two ways the word "or" is used in English, namely, in an inclusive way. Thus, a disjunction is true when either of the two propositions in it is true or when both are true. Sometimes, we use "or" in an exclusive sense. When the exclusive or is used to connect the propositions p and q, the proposition "p or q (but not both)," is obtained. This proposition is true when p is true and q is false or vice versa, and it is false when both p and q are false and when both are true.

DEFINITION 4. Let p and q be propositions. The *exclusive or* of p and q, denoted by $p \oplus q$, is the proposition that is true when exactly one of p and q is true and is false otherwise.

The truth table for the "exclusive or" of two propositions is displayed in Table 4.

We will discuss several other important ways that propositions may be combined.

TABLE 4 The Truth Table for the Exclusive Or of Two Propositions.

p	q	$p \oplus q$
T	T	F
T	F	T
F	T	T
F	F	F

TABLE 5 The Truth Table for the Implication $p \rightarrow q$.

p	q	$p \rightarrow q$
T	T	T
T	F	F
F	T	T
F	F	T

DEFINITION 5. Let p and q be propositions. The *implication* $p \rightarrow q$ is the proposition that is false when p is true and q is false and true otherwise. In this implication p is called the *hypothesis* (or *antecedent* or *premise*) and q is called the *conclusion* (or *consequence*).

The truth table for the implication $p \rightarrow q$ is shown in Table 5.

Because implications arise in many places in mathematical reasoning, a wide variety of terminology is used to express $p \rightarrow q$. Some of the more common ways of expressing this implication are:

- "if p, then q"
- "p implies q"
- "if p, q"
- "p only if q"
- "p is sufficient for q"
- "q if p"
- "q whenever p"
- "q is necessary for p."

Note that $p \rightarrow q$ is false only in the case that p is true but q is false, so that it is true when both p and q are true, and when p is false (no matter what truth value q has). The way we have defined implications is more general than the meaning attached to implications in the English language. For instance, the implication

"If it is sunny today, then we will go to the beach."

is an implication used in normal language, since there is a relationship between the hypothesis and the conclusion. Further, this implication is considered valid unless it is indeed sunny today, but we do not go to the beach. On the other hand, the implication

"If today is Friday, then $2 + 3 = 5$."

is true from the definition of implication, since its conclusion is true. (The truth value of the hypothesis does not matter then.) The implication

"If today is Friday, then $2 + 3 = 6$."

is true every day except Friday, even though $2 + 3 = 6$ is false.

We would not use these last two implications in natural language, since there is no relationship between the hypothesis and the conclusion in either implication. In mathematical reasoning we consider implications of a more general sort than we use in English. The mathematical concept of an implication is independent of a cause-and-effect relationship between hypothesis and conclusion.

Unfortunately, the if-then construction used in many programming languages is different from that used in logic. Most programming languages contain statements such as **if** p **then** S, where p is a proposition and S is a program segment (one or more statements to be executed). When execution of a program encounters such a statement, S is executed if p is true, whereas S is not executed if p is false. This is illustrated in the following example.

EXAMPLE 6 What is the value of the variable x after the statement

if $2 + 2 = 4$ **then** $x := x + 1$

if $x = 0$ before this statement is encountered? (Here the symbol $:=$ stands for assignment. The statement $x := x + 1$ represents the assignment of the value of $x + 1$ to x.)

Solution: Since $2 + 2 = 4$ is true, the assignment statement $x := x + 1$ is executed. Hence, x has the value $0 + 1 = 1$ after this statement is encountered. ■

We can build up compound propositions using the negation operator and the different connectives defined so far. Parentheses are used to specify the order in which the various logical operators in a compound proposition are applied. In particular, the logical operators in the innermost parentheses are applied first. For instance, $(p \vee q) \wedge (\neg r)$ is the conjunction of $p \vee q$ and $\neg r$. To cut down on the number of parentheses needed, we specify that the negation operator is applied before all other logical operators. This means that $\neg p \wedge q$ is the conjunction of $\neg p$ and q, namely $(\neg p) \wedge q$, not the negation of the conjunction of p and q, namely $\neg(p \wedge q)$.

There are some related implications that can be formed from $p \rightarrow q$. The proposition $q \rightarrow p$ is called the **converse** of $p \rightarrow q$. The **contrapositive** of $p \rightarrow q$ is the proposition $\neg q \rightarrow \neg p$.

EXAMPLE 7 Find the converse and the contrapositive of the implication

"If today is Thursday, then I have a test today."

Solution: The converse is

"If I have a test today, then today is Thursday."

TABLE 6 **The Truth Table for the Biconditional $p \leftrightarrow q$.**

p	q	$p \leftrightarrow q$
T	T	T
T	F	F
F	T	F
F	F	T

And the contrapositive of this implication is

"If I do not have a test today, then today is not Thursday." ■

We now introduce another way to combine propositions.

DEFINITION 6. Let p and q be propositions. The *biconditional* $p \leftrightarrow q$ is the proposition that is true when p and q have the same truth values and is false otherwise.

The truth table for $p \leftrightarrow q$ is shown in Table 6. Note that the biconditional $p \leftrightarrow q$ is true precisely when both the implications $p \rightarrow q$ and $q \rightarrow p$ are true. Because of this, the terminology

"p if and only if q"

is used for this biconditional. Other common ways of expressing the proposition $p \leftrightarrow q$ are: "p is necessary and sufficient for q" and "if p then q, and conversely."

LOGIC AND BIT OPERATIONS

Computers represent information using bits. A **bit** has two possible values, namely, zero and one. This meaning of the word bit comes from *b*inary dig*it*, since zeros and ones are the digits used in binary representations of numbers. The well-known statistician John Tukey introduced this terminology in 1946. A bit can be used to represent a truth value, since there are two truth values, namely, *true* and *false*. As is customarily done, we will use a one bit to represent *true* and a zero bit to represent *false*. A variable is called a **Boolean variable** if its value is either *true* or *false*. Consequently, a Boolean variable can be represented using a bit.

Computer **bit operations** correspond to the logical connectives. By replacing *true* by a one and *false* by a zero in the truth tables for the operators $\wedge$, $\vee$, and $\oplus$, the tables shown in Table 7 for the corresponding bit operations are obtained. We will

TABLE 7 **Tables for the Bit Operators *OR*, *AND*, and *XOR*.**

$\vee$	0	1	$\wedge$	0	1	$\oplus$	0	1
0	0	1	0	0	0	0	0	1
1	1	1	1	0	1	1	1	0

Historical Note: There were several other suggested words for a binary digit, including *binit* and *bigit*, that never were widely accepted. The adoption of the word *bit* may be due to its meaning as a common English word. For an account of Tukey's coining of the word bit, see the April 1984 issue of *Annals of the History of Computing*.

also use the notation *OR*, *AND*, and *XOR* for the operators $\vee$, $\wedge$, and $\oplus$, as is done in various programming languages.

Information is often represented using bit strings, which are sequences of zeros and ones. When this is done, operations on the bit strings can be used to manipulate this information.

DEFINITION 7. A *bit string* is a sequence of zero or more bits. The *length* of this string is the number of bits in the string.

EXAMPLE 8 101010011 is a bit string of length nine. ■

We can extend bit operations to bit strings. We define the **bitwise *OR*, bitwise *AND***, and **bitwise *XOR*** of two strings of the same length to be the strings that have as their bits the *OR, AND,* and *XOR* of the corresponding bits in the two strings, respectively. We use the symbols $\vee$, $\wedge$, and $\oplus$ to represent the bitwise *OR*, bitwise *AND*, and bitwise *XOR* operations, respectively. We illustrate bitwise operations on bit strings with the following example.

EXAMPLE 9 Find the bitwise *OR*, bitwise *AND*, and bitwise *XOR* of the bit strings 01101 10110 and 11000 11101. (Here, and throughout this book, bit strings will be split into blocks of five bits to make them easier to read.)

Solution: The bitwise *OR*, bitwise *AND*, and bitwise *XOR* of these strings are obtained by taking the *OR, AND*, and *XOR* of the corresponding bits, respectively. This gives us

```
01101 10110
11000 11101
-----------
11101 11111   bitwise OR
01000 10100   bitwise AND
10101 01011   bitwise XOR
```
■

Exercises

1. Which of the following sentences are propositions? What are the truth values of those that are propositions?
   a) Boston is the capital of Massachusetts.
   b) Miami is the capital of Florida.
   c) $2 + 3 = 5$.
   d) $5 + 7 = 10$.
   e) $x + 2 = 11$.
   f) Answer this question.
   g) $x + y = y + x$ for every pair of real numbers x and y.
2. Which of the following are propositions? What are the truth values of those that are propositions?
   a) Do not pass go.
   b) What time is it?
   c) There are no black flies in Maine.

d) $4 + x = 5$.
e) $x + 1 = 5$ if $x = 1$.
f) $x + y = y + z$ if $x = z$.

3. What is the negation of each of the following propositions?
a) Today is Thursday.
b) There is no pollution in New Jersey.
c) $2 + 1 = 3$.
d) The summer in Maine is hot and sunny.

4. Let p and q be the propositions

p: I bought a lottery ticket this week.
q: I won the million dollar jackpot on Friday.

Express each of the following propositions as an English sentence.
a) $\neg p$ **b)** $p \vee q$
c) $p \to q$ **d)** $p \wedge q$
e) $p \leftrightarrow q$ **f)** $\neg p \to \neg q$
g) $\neg p \wedge \neg q$ **h)** $\neg p \vee (p \wedge q)$

5. Let p and q be the propositions

p: It is below freezing.
q: It is snowing.

Write the following propositions using p and q and logical connectives.
a) It is below freezing and snowing.
b) It is below freezing but not snowing.
c) It is not below freezing and it is not snowing.
d) It is either snowing or below freezing (or both).
e) If it is below freezing, it is also snowing.
f) It is either below freezing or it is snowing, but it is not snowing if it is below freezing.
g) That it is below freezing is necessary and sufficient for it to be snowing.

6. Let p, q, and r be the propositions

p: You have the flu.
q: You miss the final examination.
r: You pass the course.

Express each of the following propositions as an English sentence.
a) $p \to q$ **b)** $\neg q \leftrightarrow r$
c) $q \to \neg r$ **d)** $p \vee q \vee r$
e) $(p \to \neg r) \vee (q \to \neg r)$
f) $(p \wedge q) \vee (\neg q \wedge r)$

7. Let p and q be the propositions

p: You drive over 65 miles per hour.
q: You get a speeding ticket.

Write the following propositions using p and q and logical connectives.
a) You do not drive over 65 miles per hour.
b) You drive over 65 miles per hour, but you do not get a speeding ticket.
c) You will get a speeding ticket if you drive over 65 miles per hour.
d) If you do not drive over 65 miles per hour, then you will not get a speeding ticket.
e) Driving over 65 miles per hour is sufficient for getting a speeding ticket.
f) You get a speeding ticket but you do not drive over 65 miles per hour.
g) Whenever you get a speeding ticket, you are driving over 65 miles per hour.

8. Let p, q, and r be the propositions

p: You get an A on the final exam.
q: You do every exercise in this book.
r: You get an A in this class.

Write the following propositions using p, q, and r and logical connectives.
a) You get an A in this class, but you do not do every exercise in this book.
b) You get an A on the final, you do every exercise in this book, and you get an A in this class.
c) To get an A in this class, it is necessary for you to get an A on the final.
d) You get an A on the final, but you don't do every exercise in this book; nevertheless, you get an A in this class.
e) Getting an A on the final and doing every exercise in this book is sufficient for getting an A in this class.
f) You will get an A in this class if and only if you either do every exercise in this book or you get an A on the final.

9. For each of the following sentences, state what the sentence means if the "or" is an inclusive or (that is, a disjunction) versus an exclusive or. Which of these meanings of "or" do you think is intended?
a) To take discrete mathematics, you must have taken calculus or a course in computer science.
b) When you buy a new car from Acme Motor Company, you get \$2000 back in cash or a 2% car loan.
c) Dinner for two includes two items from column A or three items from column B.
d) School is closed if more than two feet of snow fall or if the wind chill is below -100.

10. An explorer is captured by a group of cannibals. There are two types of cannibals, those that always tell the truth and those that always lie. The cannibals will barbecue the explorer unless he can determine whether a particular cannibal always lies or always tells the truth. He is allowed to ask the cannibal exactly one question.

a) Explain why the question "Are you a liar?" does not work.

b) Find a question that the explorer can use to determine whether the cannibal always lies or always tells the truth.

11. Write each of the following statements in the form "if p, then q." (*Hint:* Refer to the list of common ways to express implications listed in this section.)

a) It snows whenever the wind blows from the northeast.

b) The apple trees will bloom if it stays warm for a week.

c) That the Pistons win the championship implies that they beat the Lakers.

d) It is necessary to walk eight miles to get to the top of Longs Peak.

e) To get tenure as a professor, it is sufficient to be world-famous.

f) If you drive more than 400 miles, you will need to buy gasoline.

g) Your guarantee is good only if you bought your CD player less than 90 days ago.

12. Write each of the following propositions in the form "p if and only if q."

a) For you to get an A in this course, it is necessary and sufficient that you learn how to solve discrete mathematics problems.

b) If you read the newspaper every day, you will be informed, and conversely.

c) It rains if it is a weekend day, and it is a weekend day if it rains.

d) You can see the wizard only if the wizard is not in, and the wizard is not in only if you can see him.

13. State the converse and contrapositive of each of the following implications.

a) If it snows today, I will ski tomorrow.

b) I come to class whenever there is going to be a quiz.

c) A positive integer is a prime only if it has no divisors other than 1 and itself.

14. State the converse and contrapositive of each of the following implications.

a) If it snows tonight, then I will stay at home.

b) I go to the beach whenever it is a sunny summer day.

c) When I stay up late, it is necessary that I sleep until noon.

15. Construct a truth table for each of the following compound propositions.

a) $p \wedge \neg p$

b) $p \vee \neg p$

c) $(p \vee \neg q) \rightarrow q$

d) $(p \vee q) \rightarrow (p \wedge q)$

e) $(p \rightarrow q) \leftrightarrow (\neg q \rightarrow \neg p)$

f) $(p \rightarrow q) \rightarrow (q \rightarrow p)$

16. Construct a truth table for each of the following compound propositions.

a) $p \oplus p$ **b)** $p \oplus \neg p$

c) $p \oplus \neg q$ **d)** $\neg p \oplus \neg q$

e) $(p \oplus q) \vee (p \oplus \neg q)$ **f)** $(p \oplus q) \wedge (p \oplus \neg q)$

17. Construct a truth table for each of the following compound propositions.

a) $p \rightarrow \neg q$

b) $\neg p \leftrightarrow q$

c) $(p \rightarrow q) \vee (\neg p \rightarrow q)$

d) $(p \rightarrow q) \wedge (\neg p \rightarrow q)$

e) $(p \leftrightarrow q) \vee (\neg p \leftrightarrow q)$

f) $(\neg p \leftrightarrow \neg q) \leftrightarrow (p \leftrightarrow q)$

18. What is the value of x after each of the following statements is encountered in a computer program, if $x = 1$ before the statement is reached?

a) **if** $1 + 2 = 3$ **then** $x := x + 1$

b) **if** $(1+1 = 3)$ *OR* $(2+2 = 3)$ **then** $x := x+1$

c) **if** $(2+3 = 5)$ *AND* $(3+4 = 7)$ **then** $x := x+1$

d) **if** $(1+1 = 2)$ *XOR* $(1+2 = 3)$ **then** $x := x+1$

e) **if** $x < 2$ **then** $x := x + 1$

19. Find the bitwise *OR*, bitwise *AND*, and bitwise *XOR* of each of the following pairs of bit strings.

a) 10 11110, 01 00001

b) 111 10000, 101 01010

c) 00011 10001, 10010 01000

d) 11111 11111, 00000 00000

20. Evaluate each of the following expressions.

a) $11000 \wedge (01011 \vee 11011)$

b) $(01111 \wedge 10101) \vee 01000$

c) $(01010 \oplus 11011) \oplus 01000$

d) $(11011 \vee 01010) \wedge (10001 \vee 11011)$

Fuzzy logic is used in artificial intelligence. In fuzzy logic, a proposition has a truth value that is a number between 0 and 1, inclusive. A proposition with a truth value of 0 is false and one with a truth value of 1 is true. Truth values that are between 0 and 1 indicate varying degrees of truth. For instance, the truth value 0.8 can be assigned to the statement "Fred is happy," since Fred is happy most of the time, and the truth value 0.4 can be assigned to the statement "John is happy," since John is happy slightly less than half the time.

21. The truth value of the negation of a proposition in fuzzy logic is 1 minus the truth value of the proposition. What are the truth values of the statements "Fred is not happy" and "John is not happy"?

22. The truth value of the conjunction of two propositions in fuzzy logic is the minimum of the truth values of the two propositions. What are the truth values of the statements "Fred and John are happy" and "Neither Fred nor John is happy"?

23. The truth value of the disjunction of two propositions in fuzzy logic is the maximum of the truth values of the two propositions. What are the truth values of the statements "Fred is happy, or John is happy" and "Fred is not happy, or John is not happy"?

★24. Is the assertion "This statement is false" a proposition?

1.2 Propositional Equivalences

INTRODUCTION

An important type of step used in a mathematical argument is the replacement of a statement with another statement with the same truth value. Because of this, methods that produce propositions with the same truth value as a given compound proposition are used extensively in the construction of mathematical arguments.

We begin our discussion with a classification of compound propositions according to their possible truth values.

DEFINITION 1. A compound proposition that is always true, no matter what the truth values of the propositions that occur in it, is called a *tautology*. A compound proposition that is always false is called a *contradiction*. Finally, a proposition that is neither a tautology nor a contradiction is called a *contingency*.

Tautologies and contradictions are often important in mathematical reasoning. We give an example to illustrate these types of propositions.

EXAMPLE 1 We can construct examples of tautologies and contradictions using just one proposition. Consider the truth tables of $p \vee \neg p$ and $p \wedge \neg p$, shown in Table 1. Since $p \vee \neg p$ is always true, it is a tautology. Since $p \wedge \neg p$ is always false, it is a contradiction. ■

TABLE 1 **Examples of a Tautology and a Contradiction.**

p	$\neg p$	$p \vee \neg p$	$p \wedge \neg p$
T	F	T	F
F	T	T	F

LOGICAL EQUIVALENCES

Compound propositions that always have the same truth value are called **logically equivalent**. We can also define this notion as follows.

DEFINITION 2. The propositions p and q are called *logically equivalent* if $p \leftrightarrow q$ is a tautology. The notation $p \Longleftrightarrow q$ denotes that p and q are logically equivalent.

One way to determine whether two propositions are equivalent is to use a truth table. In particular, the propositions p and q are equivalent if and only if the columns giving their truth values agree. The following example illustrates this method.

EXAMPLE 2 Show that $\neg(p \vee q)$ and $\neg p \wedge \neg q$ are logically equivalent. This equivalence is one of *De Morgan's laws* for propositions, named after the English mathematician Augustus De Morgan of the mid-19th century.

Solution: The truth tables for these propositions are displayed in Table 2. Since the truth values of the propositions $\neg(p \vee q)$ and $\neg p \wedge \neg q$ agree for all possible combinations of the truth values of p and q, it follows that these propositions are logically equivalent. ■

Augustus De Morgan (1806–1871). Augustus De Morgan was born in India, where his father was a colonel in the Indian army. De Morgan's family moved to England when he was seven months old. He attended private schools where he developed a strong interest in mathematics in his early teens. De Morgan studied at Trinity College, Cambridge, graduating in 1827. Although he considered entering medicine or law, he decided on a career in mathematics. He won a position at University College, London, in 1828 but resigned when the college dismissed a fellow professor without giving reasons. However, he resumed this position in 1836 when his successor died, staying there until 1866.

De Morgan was a noted teacher who stressed principles over techniques. His students included many famous mathematicians, including Ada Augusta, Countess of Lovelace, who was Charles Babbage's collaborator in his work on computing machines (see page 16 for biographical notes on Ada Augusta). (De Morgan cautioned the countess against studying too much mathematics, since it might interfere with her childbearing abilities!)

De Morgan was an extremely prolific writer. He wrote more than 1000 articles for more than 15 periodicals. De Morgan also wrote textbooks on many subjects, including logic, probability, calculus, and algebra. In 1838 he presented what was perhaps the first clear explanation of an important proof technique known as mathematical induction (discussed in Section 3.2 of this text), a term he coined. In the 1840s De Morgan made fundamental contributions to the development of symbolic logic. He invented notations that helped him prove propositional equivalences, such as the laws that are named after him. In 1842 De Morgan presented what was perhaps the first precise definition of a limit and developed some tests for convergence of infinite series. De Morgan was also interested in the history of mathematics and wrote biographies of Newton and Halley.

In 1837 De Morgan married Sophia Frend, who wrote his biography in 1882. De Morgan's research, writing, and teaching left little time for his family or social life. Nevertheless, he was noted for his kindness, humor, and wide range of knowledge.

TABLE 2 **Truth Tables for $\neg(p \vee q)$ and $\neg p \wedge \neg q$.**

p	q	$p \vee q$	$\neg(p \vee q)$	$\neg p$	$\neg q$	$\neg p \wedge \neg q$
T	T	T	F	F	F	F
T	F	T	F	F	T	F
F	T	T	F	T	F	F
F	F	F	T	T	T	T

TABLE 3 **Truth Tables for $\neg p \vee q$ and $p \rightarrow q$.**

p	q	$\neg p$	$\neg p \vee q$	$p \rightarrow q$
T	T	F	T	T
T	F	F	F	F
F	T	T	T	T
F	F	T	T	T

EXAMPLE 3 Show that the propositions $p \rightarrow q$ and $\neg p \vee q$ are logically equivalent.

Solution: We construct the truth table for these propositions in Table 3. Since the truth values of $\neg p \vee q$ and $p \rightarrow q$ agree, these propositions are logically equivalent. ■

EXAMPLE 4 Show that the propositions $p \vee (q \wedge r)$ and $(p \vee q) \wedge (p \vee r)$ are logically equivalent. This is the *distributive law* of disjunction over conjunction.

Solution: We construct the truth table for these propositions in Table 4. Since the truth values of $p \vee (q \wedge r)$ and $(p \vee q) \wedge (p \vee r)$ agree, these propositions are logically equivalent. ■

TABLE 4 **A Demonstration that $p \vee (q \wedge r)$ and $(p \vee q) \wedge (p \vee r)$ Are Logically Equivalent.**

p	q	r	$q \wedge r$	$p \vee (q \wedge r)$	$p \vee q$	$p \vee r$	$(p \vee q) \wedge (p \vee r)$
T	T	T	T	T	T	T	T
T	T	F	F	T	T	T	T
T	F	T	F	T	T	T	T
T	F	F	F	T	T	T	T
F	T	T	T	T	T	T	T
F	T	F	F	F	T	F	F
F	F	T	F	F	F	T	F
F	F	F	F	F	F	F	F

Remark: A truth table of a compound proposition involving three different propositions requires eight rows, one for each possible combination of truth values of the three propositions. In general 2^n rows are required if a compound proposition involves n propositions.

Table 5 contains some important equivalences. In these equivalences, **T** denotes any proposition that is true, and **F** denotes any proposition that is false. The reader is asked to verify these equivalences in the exercises at the end of the section.

The associative law for disjunction shows that the expression $p \vee q \vee r$ is well defined, in the sense that it does not matter whether we first take the disjunction of p and q and then the disjunction of $p \vee q$ with r, or if we first take the disjunction of q and r and then take the disjunction of p and $q \vee r$. Similarly, the expression $p \wedge q \wedge r$ is well defined. By extending this reasoning, it follows that $p_1 \vee p_2 \vee \cdots \vee p_n$ and

TABLE 5 **Logical Equivalences.**

Equivalence	*Name*
$p \wedge \mathbf{T} \Leftrightarrow p$ $p \vee \mathbf{F} \Leftrightarrow p$	Identity laws
$p \vee \mathbf{T} \Leftrightarrow \mathbf{T}$ $p \wedge \mathbf{F} \Leftrightarrow \mathbf{F}$	Domination laws
$p \vee p \Leftrightarrow p$ $p \wedge p \Leftrightarrow p$	Idempotent laws
$\neg(\neg p) \Leftrightarrow p$	Double negation law
$p \vee q \Leftrightarrow q \vee p$ $p \wedge q \Leftrightarrow q \wedge p$	Commutative laws
$(p \vee q) \vee r \Leftrightarrow p \vee (q \vee r)$ $(p \wedge q) \wedge r \Leftrightarrow p \wedge (q \wedge r)$	Associative laws
$p \vee (q \wedge r) \Leftrightarrow (p \vee q) \wedge (p \vee r)$ $p \wedge (q \vee r) \Leftrightarrow (p \wedge q) \vee (p \wedge r)$	Distributive laws
$\neg(p \wedge q) \Leftrightarrow \neg p \vee \neg q$ $\neg(p \vee q) \Leftrightarrow \neg p \wedge \neg q$	De Morgan's laws

TABLE 6 Some Useful Logical Equivalences.

$p \vee \neg p \Leftrightarrow \mathbf{T}$
$p \wedge \neg p \Leftrightarrow \mathbf{F}$
$(p \to q) \Leftrightarrow (\neg p \vee q)$

$p_1 \wedge p_2 \wedge \cdots \wedge p_n$ are well defined whenever $p_1, p_2, \ldots, p_n$ are propositions. Also, note that De Morgan's laws extend to

$$\neg(p_1 \vee p_2 \cdots \vee p_n) \Leftrightarrow (\neg p_1 \wedge \neg p_2 \wedge \cdots \wedge \neg p_n)$$

and

$$\neg(p_1 \wedge p_2 \cdots \wedge p_n) \Leftrightarrow (\neg p_1 \vee \neg p_2 \vee \cdots \vee \neg p_n).$$

(Methods for proving these identities will be given in Chapter 3.)

The logical equivalences in Table 5, as well as any others that have been established (such as those shown in Table 6), can be used to construct additional logical equivalences. The reason for this is that a proposition in a compound proposition can be replaced by one that is logically equivalent to it without changing the truth value of the compound proposition. This technique is illustrated in Examples 5 and 6, where we also use the fact that if p and q are logically equivalent and q and r are logically equivalent, then p and r are logically equivalent (see Exercise 36).

EXAMPLE 5 Show that $\neg(p \vee (\neg p \wedge q))$ and $\neg p \wedge \neg q$ are logically equivalent.

Solution: We could use a truth table to show these compound propositions are equivalent. Instead, we will establish this equivalence by developing a series of logical equivalences, using one of the equivalences in Table 5 at a time, starting with $\neg(p \vee (\neg p \wedge q))$ and ending with $\neg p \wedge \neg q$. We have the following equivalences.

Ada Augusta, Countess of Lovelace (1815–1852). Ada Augusta was the only child from the marriage of the famous poet Lord Byron and Annabella Millbanke, who separated when she was one month old. She was raised by her mother, who encouraged her intellectual talents. She was taught by the mathematicians William Frend and Augustus De Morgan. In 1838 she married Lord King, later elevated to Earl of Lovelace. Together they had three children.

Ada Augusta continued her mathematical studies after her marriage, assisting Charles Babbage in his work on an early computing machine, called the "Analytic Engine." The most complete accounts of this machine are found in her writings. After 1845 she and Babbage worked toward the development of a system to predict horse races. Unfortunately, their system did not work well, leaving Ada heavily in debt at the time of her death. The programming language Ada is named in honor of the Countess of Lovelace.

$\neg(p \vee(\neg p \wedge q))$	$\Leftrightarrow \neg p \wedge \neg(\neg p \wedge q)$	from the second De Morgan's law
	$\Leftrightarrow \neg p \wedge[\neg(\neg p) \vee \neg q]$	from the first De Morgan's law
	$\Leftrightarrow \neg p \wedge(p \vee \neg q)$	from the double negation law
	$\Leftrightarrow (\neg p \wedge p) \vee(\neg p \wedge \neg q)$	from the distributive law
	$\Leftrightarrow \mathbf{F} \vee(\neg p \wedge \neg q)$	since $\neg p \wedge p \Leftrightarrow \mathbf{F}$
	$\Leftrightarrow (\neg p \wedge \neg q) \vee \mathbf{F}$	from the commutative law for disjunction
	$\Leftrightarrow \neg p \wedge \neg q$	from the identity law for **F**

Consequently $\neg(p \vee(\neg p \wedge q))$ and $\neg p \wedge \neg q$ are logically equivalent. ■

EXAMPLE 6 Show that $(p \wedge q) \rightarrow (p \vee q)$ is a tautology.

Solution: To show this statement is a tautology, we will use logical equivalences to demonstrate that it is logically equivalent to **T**. (*Note:* This could also be done using a truth table.)

$(p \wedge q) \rightarrow (p \vee q)$	$\Leftrightarrow \neg(p \wedge q) \vee(p \vee q)$	by Example 3
	$\Leftrightarrow (\neg p \vee \neg q) \vee(p \vee q)$	by the first De Morgan's law
	$\Leftrightarrow (\neg p \vee p) \vee(\neg q \vee q)$	by the associative and commutative laws for disjunction
	$\Leftrightarrow \mathbf{T} \vee \mathbf{T}$	by Example 1 and the commutative law for disjunction
	$\Leftrightarrow \mathbf{T}$	by the domination law ■

Exercises

1. Use truth tables to verify the following equivalences.
   a) $p \wedge \mathbf{T} \Leftrightarrow p$ b) $p \vee \mathbf{F} \Leftrightarrow p$
   c) $p \wedge \mathbf{F} \Leftrightarrow \mathbf{F}$ d) $p \vee \mathbf{T} \Leftrightarrow \mathbf{T}$
   e) $p \vee p \Leftrightarrow p$ f) $p \wedge p \Leftrightarrow p$
2. Show that $\neg(\neg p)$ and p are logically equivalent.
3. Use truth tables to verify the commutative laws
   a) $p \vee q \Leftrightarrow q \vee p$ b) $p \wedge q \Leftrightarrow q \wedge p$
4. Use truth tables to verify the associative laws
   a) $(p \vee q) \vee r \Leftrightarrow p \vee(q \vee r)$
   b) $(p \wedge q) \wedge r \Leftrightarrow p \wedge(q \wedge r)$
5. Use truth tables to verify the distributive law $p \wedge(q \vee r) \Leftrightarrow (p \wedge q) \vee(p \wedge r)$.
6. Use a truth table to verify the equivalence $\neg(p \wedge q) \Leftrightarrow \neg p \vee \neg q$.
7. Show that each of the following implications is a tautology by using truth tables.
   a) $(p \wedge q) \rightarrow p$ b) $p \rightarrow (p \vee q)$
   c) $\neg p \rightarrow (p \rightarrow q)$ d) $(p \wedge q) \rightarrow (p \rightarrow q)$
   e) $\neg(p \rightarrow q) \rightarrow p$ f) $\neg(p \rightarrow q) \rightarrow \neg q$
8. Show that each of the following implications is a tautology by using truth tables.
   a) $[\neg p \wedge(p \vee q)] \rightarrow q$
   b) $[(p \rightarrow q) \wedge(q \rightarrow r)] \rightarrow (p \rightarrow r)$
   c) $[p \wedge(p \rightarrow q)] \rightarrow q$
   d) $[(p \vee q) \wedge(p \rightarrow r) \wedge(q \rightarrow r)] \rightarrow r$
9. Show that each implication in Exercise 7 is a tautology without using truth tables.
10. Show that each implication in Exercise 8 is a tautology without using truth tables.
11. Verify the following equivalences, which are known as the **absorption laws**.
    a) $[p \vee(p \wedge q)] \Leftrightarrow p$
    b) $[p \wedge(p \vee q)] \Leftrightarrow p$
12. Determine whether $(\neg p \wedge(p \rightarrow q)) \rightarrow \neg q$ is a tautology.
13. Determine whether $(\neg q \wedge(p \rightarrow q)) \rightarrow \neg p$ is a tautology.

14. Show that $p \leftrightarrow q$ and $(p \wedge q) \vee (\neg p \wedge \neg q)$ are equivalent.

15. Show that $(p \rightarrow q) \rightarrow r$ and $p \rightarrow (q \rightarrow r)$ are not equivalent.

The **dual** of a compound proposition that contains only the logical operators $\vee$, $\wedge$, and $\neg$ is the proposition obtained by replacing each $\vee$ by $\wedge$, each $\wedge$ by $\vee$, each **T** by **F**, and each **F** by **T**. The dual of s is denoted by s^*.

16. Find the dual of each of the following propositions.
a) $p \wedge \neg q \wedge \neg r$ **b)** $(p \wedge q \wedge r) \vee s$
c) $(p \vee \mathbf{F}) \wedge (q \vee \mathbf{T})$

17. Show that $(s^*)^* = s$.

18. Show that the logical equivalences in Table 5, except for the double negation law, come in pairs, where each pair contains propositions that are duals of each other.

★★**19.** Why are the duals of two equivalent compound propositions also equivalent, where these compound propositions contain only the operators $\wedge, \vee$, and $\neg$?

20. Find a compound proposition involving the propositions p, q, and r that is true when p and q are true and r is false, but is false otherwise. (*Hint:* Use a conjunction of each proposition or its negation.)

21. Find a compound proposition involving the propositions p, q, and r that is true when exactly two of p, q, and r are true and is false otherwise. (*Hint:* Form a disjunction of conjunctions. Include a conjunction for each combination of values for which the proposition is true. Each conjunction should include each of the three propositions or their negations.)

22. Suppose that a truth table in n propositional variables is specified. Show that a compound proposition with this truth table can be formed by taking the disjunction of conjunctions of the variables or their negations, with one conjunction included for each combination of values for which the compound proposition is true. The resulting compound proposition is said to be in **disjunctive normal form**.

A collection of logical operators is called **functionally complete** if every compound proposition is logically equivalent to a compound proposition involving only these logical operators.

23. Show that $\neg$, $\wedge$, and $\vee$ form a functionally complete collection of logical operators. (*Hint:* Use the fact that every proposition is logically equivalent to one in disjunctive normal form, as shown in Exercise 22.)

★**24.** Show that $\neg$ and $\wedge$ form a functionally complete collection of logical operators. (*Hint:* First use De Morgan's law to show that $p \vee q$ is equivalent to $\neg(\neg p \wedge \neg q)$.)

★**25.** Show that $\neg$ and $\vee$ form a functionally complete collection of logical operators.

The following exercises involve the logical operators *NAND* and *NOR*. The proposition p *NAND* q is true when either p or q, or both, are false, and it is false when both p and q are true. The proposition p *NOR* q is true when both p and q are false, and is false otherwise. The propositions p *NAND* q and p *NOR* q are denoted by $p \mid q$ and $p \downarrow q$, respectively.

26. Construct a truth table for the logical operator *NAND*.

27. Show that $p \mid q$ is logically equivalent to $\neg(p \wedge q)$.

28. Construct a truth table for the logical operator *NOR*.

29. Show that $p \downarrow q$ is logically equivalent to $\neg(p \vee q)$.

30. In this exercise we will show that $\{\downarrow\}$ is a functionally complete collection of logical operators.
a) Show that $p \downarrow p$ is logically equivalent to $\neg p$.
b) Show that $(p \downarrow q) \downarrow (p \downarrow q)$ is logically equivalent to $p \vee q$.
c) Conclude from parts (a) and (b), and Exercise 25, that $\{\downarrow\}$ is a functionally complete collection of logical operators.

★**31.** Find a proposition equivalent to $p \rightarrow q$ using only the logical operator $\downarrow$.

32. Show that $\{\mid\}$ is a functionally complete collection of logical operators.

33. Show that $p \mid q$ and $q \mid p$ are equivalent.

34. Show that $p \mid (q \mid r)$ and $(p \mid q) \mid r$ are not equivalent, so that the logical operator $\mid$ is not associative.

★**35.** How many different truth tables of compound propositions are there that involve the propositions p and q?

36. Show that if p, q, and r are compound propositions such that p and q are logically equivalent and q and r are logically equivalent, then p and r are logically equivalent.

☞ **37.** Show that $p \rightarrow q$ and $\neg q \rightarrow \neg p$ are logically equivalent.

1.3 Predicates and Quantifiers

INTRODUCTION

Statements involving variables, such as

"$x > 3$," "$x = y + 3$," and "$x + y = z$,"

are often found in mathematical assertions and in computer programs. These statements are neither true nor false when the values of the variables are not specified. In this section we will discuss the ways that propositions can be produced from such statements.

The statement "x is greater than three" has two parts. The first part, the variable x, is the subject of the statement. The second part, the **predicate** "is greater than three," refers to a property that the subject of the statement can have. We can denote the statement "x is greater than three" by $P(x)$, where P denotes the predicate "is greater than three" and x is the variable. The statement $P(x)$ is also said to be the value of the **propositional function** P at x. Once a value has been assigned to the variable x, the statement $P(x)$ has a truth value. Consider the following example.

EXAMPLE 1 Let $P(x)$ denote the statement "$x > 3$." What are the truth values of $P(4)$ and $P(2)$?

Solution: The statement $P(4)$ is obtained by setting $x = 4$ in the statement "$x > 3$." Hence, $P(4)$, which is the statement "$4 > 3$," is true. However, $P(2)$, which is the statement "$2 > 3$," is false. ■

We can also have statements that involve more than one variable. For instance, consider the statement "$x = y + 3$." We can denote this statement by $Q(x, y)$, where x and y are variables and Q is the predicate. When values are assigned to the variables x and y, the statement $Q(x, y)$ has a truth value.

EXAMPLE 2 Let $Q(x, y)$ denote the statement "$x = y + 3$." What are the truth values of the propositions $Q(1, 2)$ and $Q(3, 0)$?

Solution: To obtain $Q(1, 2)$, set $x = 1$ and $y = 2$ in the statement $Q(x, y)$. Hence, $Q(1, 2)$ is the statement "$1 = 2 + 3$," which is false. The statement $Q(3, 0)$ is the proposition "$3 = 0 + 3$," which is true. ■

Similarly, we can let $R(x, y, z)$ denote the statement "$x + y = z$." When values are assigned to the variables x, y, and z, this statement has a truth value.

EXAMPLE 3 What are the truth values of the propositions $R(1, 2, 3)$ and $R(0, 0, 1)$?

Solution: The proposition $R(1, 2, 3)$ is obtained by setting $x = 1$, $y = 2$, and $z = 3$ in the statement $R(x, y, z)$. We see that $R(1, 2, 3)$ is the statement "$1 + 2 = 3$," which is true. Also note that $R(0, 0, 1)$, which is the statement "$0 + 0 = 1$," is false. ■

In general, a statement involving the n variables $x_1, x_2, \ldots, x_n$ can be denoted by

$$P(x_1, x_2, \ldots, x_n).$$

A statement of the form $P(x_1, x_2, \ldots, x_n)$ is the value of the **propositional function** P at the n-tuple $(x_1, x_2, \ldots, x_n)$, and P is also called a predicate.

Propositional functions occur in computer programs, as the following example demonstrates.

EXAMPLE 4 Consider the statement

if $x > 0$ **then** $x := x + 1$.

When this statement is encountered in a program, the value of the variable x at that point in the execution of the program is inserted into $P(x)$, which is "$x > 0$." If $P(x)$ is true for this value of x, the assignment statement $x := x + 1$ is executed, so that the value of x is increased by one. If $P(x)$ is false for this value of x, the assignment statement is not executed, so that the value of x is not changed. ■

QUANTIFIERS

When all the variables in a propositional function are assigned values, the resulting statement has a truth value. However, there is another important way to change propositional functions into propositions, called **quantification**. Two types of quantification will be discussed here, namely, universal quantification and existential quantification.

Many mathematical statements assert that a property is true for all values of a variable. Such a statement is expressed using a universal quantification. The universal quantification of a proposition forms the proposition that is true if and only if $P(x)$ is true for all values of x in the **universe of discourse**. The universe of discourse specifies the possible values of the variable x.

DEFINITION 3. *The universal quantification* of $P(x)$ is the proposition

"$P(x)$ is true for all values of x in the universe of discourse."

The notation

$$\forall x\ P(x)$$

denotes the universal quantification of $P(x)$. The proposition $\forall x\ P(x)$ is also expressed as

"For all x $P(x)$"

or

"For every x $P(x)$."

EXAMPLE 5 Express the statement

"Every student in this class has studied calculus."

as a universal quantification.

Solution: Let $P(x)$ denote the statement

"x has studied calculus."

Then the statement "Every student in this class has studied calculus." can be written as $\forall x\ P(x)$, where the universe of discourse consists of the students in this class.

This statement can also be expressed as

$$\forall x\,(S(x) \rightarrow P(x))$$

where $S(x)$ is the statement

"x is in this class."

$P(x)$ is as before, and the universe of discourse is the set of all students. ■

Example 5 illustrates that there is often more than one good way to express a quantification.

EXAMPLE 6 Let $P(x)$ be the statement "$x + 1 > x$." What is the truth value of the quantification $\forall x\, P(x)$, where the universe of discourse is the set of real numbers?

Solution: Since $P(x)$ is true for all real numbers x, the quantification

$$\forall x\, P(x)$$

is true. ■

EXAMPLE 7 Let $Q(x)$ be the statement "$x < 2$." What is the truth value of the quantification $\forall x\, Q(x)$, where the universe of discourse is the set of real numbers?

Solution: $Q(x)$ is not true for all real numbers x, since, for instance, $Q(3)$ is false. Consequently,

$$\forall x\, Q(x)$$

is false. ■

When all the elements in the universe of discourse can be listed, say $x_1, x_2, \ldots, x_n$, it follows that the universal quantification $\forall x\, P(x)$ is the same as the conjunction

$$P(x_1) \wedge P(x_2) \wedge \cdots \wedge P(x_n),$$

since this conjunction is true if and only if $P(x_1), P(x_2), \ldots, P(x_n)$ are all true.

EXAMPLE 8 What is the truth value of $\forall x\, P(x)$, where $P(x)$ is the statement "$x^2 < 10$," and the universe of discourse consists of the positive integers not exceeding 4?

Solution: The statement $\forall x\, P(x)$ is the same as the conjunction

$$P(1) \wedge P(2) \wedge P(3) \wedge P(4),$$

since the universe of discourse consists of the integers 1, 2, 3, and 4. Since $P(4)$, which is the statement "$4^2 < 10$," is false, it follows that $\forall x\, P(x)$ is false. ■

Many mathematical statements assert that there is an element with a certain property. Such statements are expressed using existential quantification. With existential quantification, we form the proposition that is true if and only if $P(x)$ is true for at least one value of x in the universe of discourse.

DEFINITION 3. The *existential quantification* of $P(x)$ is the proposition

"There exists an element x in the universe of discourse such that $P(x)$ is true."

We use the notation

$$\exists x\, P(x)$$

for the existential quantification of $P(x)$. The existential quantification $\exists x\, P(x)$ is also expressed as

"There is an x such that $P(x)$,"

"There is at least one x such that $P(x)$,"

or

"For some x $P(x)$."

EXAMPLE 9 Let $P(x)$ denote the statement "$x > 3$." What is the truth value of the quantification $\exists x\, P(x)$, where the universe of discourse is the set of real numbers?

Solution: Since "$x > 3$" is true, for instance, when $x = 4$, the existential quantification of $P(x)$, $\exists x\, P(x)$, is true. ■

EXAMPLE 10 Let $Q(x)$ denote the statement "$x = x + 1$." What is the truth value of the quantification $\exists x\, Q(x)$, where the universe of discourse is the set of real numbers?

Solution: Since $Q(x)$ is false for every real number x, the existential quantification of $Q(x)$, $\exists x\, Q(x)$, is false. ■

When all the elements in the universe of discourse can be listed, say $x_1, x_2, \ldots, x_n$, the existential quantification $\exists x\, P(x)$ is the same as the disjunction

$$P(x_1) \vee P(x_2) \vee \cdots \vee P(x_n),$$

since this disjunction is true if and only if at least one of $P(x_1), P(x_2), \ldots, P(x_n)$ is true.

EXAMPLE 11 What is the truth value of $\exists x\, P(x)$ where $P(x)$ is the statement "$x^2 > 10$," and the universe of discourse consists of the positive integers not exceeding 4?

Solution: Since the universe of discourse is $\{1, 2, 3, 4\}$, the proposition $\exists x\, P(x)$ is the same as the disjunction

$$P(1) \vee P(2) \vee P(3) \vee P(4).$$

Since $P(4)$, which is the statement "$4^2 > 10$," is true, it follows that $\exists x\, P(x)$ is true. ■

Table 1 summarizes the meaning of the universal and the existential quantifiers.

It is sometimes helpful to think in terms of looping and searching when determining the truth value of a quantification. Suppose that there are n objects in the universe of discourse for the variable x. To determine whether $\forall x\, P(x)$ is true, we can loop through all n values of x to see if $P(x)$ is always true. If we encounter a value x for which $P(x)$ is false, then we have shown that $\forall x\, P(x)$ is false. Otherwise $\forall x\, P(x)$ is true. To see whether $\exists x\, P(x)$ is true, we loop through the n values of x searching for a value for which $P(x)$ is true. If we find one, then $\exists x\, P(x)$ is true. If we never find such an x, we have determined that $\exists x\, P(x)$ is false. (Note that this

TABLE 1 **Quantifiers.**

Statement	**When true?**	**When false?**
$\forall x\, P(x)$	$P(x)$ is true for every x	There is an x for which $P(x)$ is false
$\exists x\, P(x)$	There is an x for which $P(x)$ is true	$P(x)$ is false for every x

searching procedure does not apply if there are infinitely many values in the universe of discourse. However, it is still a useful way of thinking about the truth values of quantifications.)

EXAMPLES FROM LEWIS CARROLL (optional)

Lewis Carroll (really C. L. Dodgson writing under a pseudonym), the author of *Alice in Wonderland*, is also the author of several works on symbolic logic. His books contain many examples of reasoning using quantifiers. The next two examples come from his book *Symbolic Logic*; other examples from this book are given in the exercise set at the end of this section. These examples illustrate how quantifiers are used to express various types of statements.

EXAMPLE 12 Consider the following statements. The first two are called *premises* and the third is called the *conclusion*. The entire set is called an *argument*.

"All lions are fierce."
"Some lions do not drink coffee."
"Some fierce creatures do not drink coffee."

(In Section 3.1 we will discuss the issue of determining whether the conclusion is a valid consequence of the premises. In this example, it is.) Let $P(x)$, $Q(x)$, and $R(x)$ be the statements "x is a lion," "x is fierce," and "x drinks coffee," respectively. Assuming that the universe of discourse is the set of all creatures, express the statements in the argument using quantifiers and $P(x)$, $Q(x)$, and $R(x)$.

Solution: We can express these statements as:

$$\forall x(P(x) \rightarrow Q(x))$$
$$\exists x(P(x) \wedge \neg R(x))$$
$$\exists x(Q(x) \wedge \neg R(x))$$

Charles Lutwidge Dodgson (1832–1898). We know Charles Dodgson as Lewis Carroll—the pseudonym he used in his writings on logic. Dodgson, the son of a clergyman, was the third of 11 children, all of whom stuttered. He was uncomfortable in the company of adults and is said to have spoken without stuttering only to young girls, many of whom he entertained, corresponded with, and photographed (often in the nude). Although attracted to young girls, he was extremely puritanical and religious. His friendship with the three young daughters of Dean Liddell led to his writing *Alice in Wonderland*, which brought him money and fame.

Dodgson graduated from Oxford in 1854 and obtained his master of arts degree in 1857. He was appointed lecturer in mathematics at Christ Church College, Oxford, in 1855. He was ordained in the Church of England in 1861 but never practiced his ministry. His writings include articles and books on geometry, determinants, and the mathematics of tournaments and elections. He used the pseudonym Lewis Carroll for his many works on recreational logic.

(Note: The second statement cannot be written as

$$\exists x(P(x) \rightarrow \neg R(x)).$$

The reason is that $P(x) \rightarrow \neg R(x)$ is true whenever x is not a lion, so that $\exists x(P(x) \rightarrow \neg R(x))$ is true as long as there is at least one creature that is not a lion, even if every lion drinks coffee. Similarly, the third statement cannot be written as

$$\exists x(Q(x) \rightarrow \neg R(x)).)$$

■

EXAMPLE 13 Consider the following statements, of which the first three are premises and the fourth is a valid conclusion.

"All hummingbirds are richly colored."
"No large birds live on honey."
"Birds that do not live on honey are dull in color."
"Hummingbirds are small."

Let $P(x)$, $Q(x)$, $R(x)$, and $S(x)$ be the statements "x is a hummingbird," "x is large," "x lives on honey," and "x is richly colored," respectively. Assuming that the universe of discourse is the set of all birds, express the statements in the argument using quantifiers and $P(x)$, $Q(x)$, $R(x)$, and $S(x)$.

Solution: We can express the statements in the argument as:

$$\forall x(P(x) \rightarrow S(x))$$
$$\neg \exists x(Q(x) \wedge R(x))$$
$$\forall x(\neg R(x) \rightarrow \neg S(x))$$
$$\forall x(P(x) \rightarrow \neg Q(x))$$

(Note we have assumed that "small" is the same as "not large" and that "dull in color" is the same as "not richly colored." To show that the fourth statement is a valid conclusion of the first three, we need to use rules of inference that will be discussed in Section 3.1.) ■

BINDING VARIABLES

When a quantifier is used on the variable x, or when we assign a value to this variable, we say that this occurrence of the variable is **bound**. An occurrence of a variable that is not bound by a quantifier or set equal to a particular value is said to be **free**. All the variables that occur in a propositional function must be bound to turn it into a proposition. This can be done using a combination of universal quantifiers, existential quantifiers, and value assignments.

Many mathematical statements involve multiple quantifications of propositional functions involving more than one variable. It is important to note that the order of the quantifiers is important, unless all the quantifiers are universal quantifiers or all are existential quantifiers. These remarks are illustrated by Examples 14, 15, and 16. In each of these examples the universe of discourse for each variable is the set of real numbers.

EXAMPLE 14 Let $P(x, y)$ be the statement "$x + y = y + x$." What is the truth value of the quantification $\forall x \forall y P(x, y)$?

Solution: The quantification

$$\forall x \forall y P(x, y)$$

denotes the proposition

"For all real numbers x and for all real numbers y it is true that $x + y = y + x$."

Since $P(x, y)$ is true for all real numbers x and y, the proposition $\forall x \forall y P(x, y)$ is true. ■

EXAMPLE 15 Let $Q(x, y)$ denote "$x + y = 0$." What are the truth values of the quantifications $\exists y \forall x Q(x, y)$ and $\forall x \exists y Q(x, y)$?

Solution: The quantification

$$\exists y \forall x Q(x, y)$$

denotes the proposition

"There is a real number y such that for every real number x $Q(x, y)$ is true."

No matter what value of y is chosen, there is only one value of x for which $x + y = 0$. Since there is no real number y such that $x + y = 0$ for all real numbers x, the statement $\exists y \forall x Q(x, y)$ is false.

The quantification

$$\forall x \exists y Q(x, y)$$

denotes the proposition

"For every real number x there is a real number y such that $Q(x, y)$ is true."

Given a real number x, there is a real number y such that $x + y = 0$, namely, $y = -x$. Hence, the statement $\forall x \exists y Q(x, y)$ is true. ■

Example 15 illustrates that the order in which quantifiers appear makes a difference. The statement $\exists x \forall y P(x, y)$ and $\forall y \exists x P(x, y)$ are not logically equivalent. The statement $\exists x \forall y P(x, y)$ is true if and only if there is an x that makes $P(x, y)$ true for every y. So, for this statement to be true, there must be a particular value of x for which $P(x, y)$ is true regardless of the choice of y. On the other hand, the statement $\forall y \exists x P(x, y)$ is true if and only if for every value of y, there is a value of x for which $P(x, y)$ is true. So, for this statement to be true, no matter which y you choose, there must be a value of x (possibly depending on the y you choose) for which $P(x, y)$ is true. In other words, in the second case y can depend on x, whereas in the first case, x is a constant independent of y.

From these observations, it follows that if $\exists x \, \forall y \, P(x, y)$ is true, then $\forall y \, \exists x P(x, y)$ must also be true. However, if $\forall y \, \exists x \, P(x, y)$ is true, it is not necessary for $\exists x \, \forall y \, P(x, y)$ to be true. (See Supplementary Exercises 8 and 10 at the end of this chapter.)

In working with quantifications of more than one variable, it is sometimes helpful to think in terms of nested loops. (Of course, if there are infinitely many elements in the universe of discourse of some variable, we cannot actually loop through all values. Nevertheless, this way of thinking is helpful in understanding nested quantifiers.) For example, to see whether $\forall x \, \forall y \, P(x, y)$ is true, we loop through the values for x, and for each x we loop through the values for y. If we find that $P(x, y)$ is true for all values for x and y, we have determined that $\forall x \, \forall y \, P(x, y)$ is true. If we ever hit a value x for which we hit a value y for which $P(x, y)$ is false, then we have shown that $\forall x \, \forall y \, P(x, y)$ is false.

Similarly, to determine whether $\forall x \, \exists y \, P(x, y)$ is true, we loop through the values for x. For each x we loop through the values for y until we find a y for which $P(x, y)$ is true. If for all x we hit such a y, then $\forall x \, \exists y \, P(x, y)$ is true; if for some x we never hit such a y, then $\forall x \, \exists y \, P(x, y)$ is false.

To see whether $\exists x \, \forall y \, P(x, y)$ is true, we loop through the values for x until we find an x for which $P(x, y)$ is always true when we loop through all values for y. Once we find such an x, we know that $\exists x \, \forall y \, P(x, y)$ is true. If we never hit such an x, then we know that $\exists x \, \forall y \, P(x, y)$ is false.

Finally, to see whether $\exists x \, \exists y \, P(x, y)$ is true, we loop through the values for x, where for each x we loop through the values for y until we hit an x for which we hit a y for which $P(x, y)$ is true. The statement $\exists x \, \exists y \, P(x, y)$ is false only if we never hit an x for which we hit a y such that $P(x, y)$ is true.

Table 2 summarizes the meanings of the different possible quantifications involving two variables.

TABLE 2 **Quantifications of Two Variables.**

Statement	*When true?*	*When false?*
$\forall x \, \forall y \, P(x, y)$ $\forall y \, \forall x \, P(x, y)$	$P(x, y)$ is true for every pair x, y	There is a pair x, y for which $P(x, y)$ is false
$\forall x \, \exists y \, P(x, y)$	For every x there is a y for which $P(x, y)$ is true	There is an x such that $P(x, y)$ is false for every y
$\exists x \, \forall y \, P(x, y)$	There is an x for which $P(x, y)$ is true for every y	For every x there is a y for which $P(x, y)$ is false
$\exists x \, \exists y \, P(x, y)$ $\exists y \, \exists x \, P(x, y)$	There is a pair x, y for which $P(x, y)$ is true	$P(x, y)$ is false for every pair x, y

Quantifications of more than two variables are also common, as Example 16 illustrates.

EXAMPLE 16 Let $Q(x, y, z)$ be the statement "$x + y = z$." What are the truth values of the statements $\forall x \forall y \exists z\, Q(x, y, z)$ and $\exists z \forall x \forall y\, Q(x, y, z)$?

Solution: Suppose that x and y are assigned values. Then, there exists a real number z such that $x + y = z$. Consequently, the quantification

$$\forall x \forall y \exists z\, Q(x, y, z),$$

which is the statement

"For all real numbers x and for all real numbers y there is a real number z such that $x + y = z$,"

is true. The order of the quantification here is important, since the quantification

$$\exists z \forall x \forall y\, Q(x, y, z),$$

which is the statement

"There is a real number z such that for all real numbers x and for all real numbers y it is true that $x + y = z$,"

is false, since there is no value of z that satisfies the equation $x + y = z$ for all values of x and y. ■

The following example illustrates how quantifiers can be used to express statements involving several variables. As the example illustrates, there usually is more than one way to use quantifiers to express a statement.

EXAMPLE 17 Use quantifiers to express the statement "There is a woman who has taken a flight on every airline in the world."

Solution: Let $P(w, f)$ be "w has taken f" and $Q(f, a)$ be "f is a flight on a." We can express the statement as

$$\exists w \forall a \exists f (P(w, f) \wedge Q(f, a))$$

where the universes of discourse for w, f, and a consist of all the women in the world, all airplane flights, and all airlines, respectively.

The statement could also be expressed as

$$\exists w \forall a \exists f\, R(w, f, a)$$

where $R(w, f, a)$ is "w has taken f on a." Although this is more compact, it somewhat obscures the relationship between the variables. Consequently, the first solution is usually preferable. ■

Quantifiers are often used in the definition of mathematical concepts. One example that you may be familiar with is the concept of limit, which is important in calculus.

EXAMPLE 18 (Calculus required) Express the definition of a limit using quantifiers.

Solution: Recall that the definition of the statement

$$\lim_{x \to a} f(x) = L$$

is: For every real number $\epsilon > 0$ there exists a real number $\delta > 0$ such that $|f(x) - L| < \epsilon$ whenever $0 < |x - a| < \delta$. This definition of a limit can be phrased in terms of quantifiers by

$$\forall \epsilon\, \exists \delta\, \forall x (0 < |x - a| < \delta \rightarrow |f(x) - L| < \epsilon),$$

where the universe of discourse for the variables δ and ϵ is the set of positive real numbers and for x is the set of real numbers. ■

We will often want to consider the negation of a quantified expression. For instance, consider the negation of the statement

"Every student in the class has taken a course in calculus."

This statement is a universal quantification, namely,

$$\forall x\, P(x),$$

where $P(x)$ is the statement "x has taken a course in calculus." The negation of this statement is, "It is not the case that every student in the class has taken a course in calculus." This is equivalent to, "There is a student in the class who has not taken a course in calculus." And this is simply the existential quantification of the negation of the original propositional function, namely,

$$\exists x \neg P(x).$$

This example illustrates the following equivalence:

$$\neg \forall x\, P(x) \Longleftrightarrow \exists x \neg P(x).$$

Suppose we wish to negate an existential quantification. For instance, consider the proposition "There is a student in this class who has taken a course in calculus." This is the existential quantification

$$\exists x\, Q(x),$$

where $Q(x)$ is the statement "x has taken a course in calculus." The negation of this statement is the proposition "It is not the case that there is a student in this class who has taken a course in calculus." This is equivalent to, "Every student in this class has not taken calculus," which is just the universal quantification of the negation of the original propositional function, or, phrased in the language of quantifiers,

$$\forall x \neg Q(x).$$

TABLE 3 **Negating Quantifiers.**

Negation	*Equivalent Statement*	*When is Negation true?*	*When false?*
$\neg\exists x\, P(x)$	$\forall x \neg P(x)$	$P(x)$ is false for every x	There is an x for which $P(x)$ is true
$\neg\forall x\, P(x)$	$\exists x \neg P(x)$	There is an x for which $P(x)$ is false	$P(x)$ is true for every x

This example illustrates the equivalence

$$\neg\exists x\, Q(x) \Leftrightarrow \forall x \neg Q(x).$$

Negations of quantifiers are summarized in Table 3.

Exercises

1. Let $P(x)$ denote the statement "$x \leq 4$." What are the truth values of the following?
   a) $P(0)$ b) $P(4)$ c) $P(6)$
2. Let $P(x)$ be the statement "the word x contains the letter a." What are the truth values of the following?
   a) P(orange) b) P(lemon)
   c) P(true) d) P(false)
3. Let $Q(x, y)$ denote the statement "x is the capital of y." What are the truth values of the following?
   a) Q(Denver, Colorado)
   b) Q(Detroit, Michigan)
   c) Q(Massachusetts, Boston)
   d) Q(New York, New York)
4. State the value of x after the statement **if** $P(x)$ **then** $x := 1$ is executed, where $P(x)$ is the statement "$x > 1$," if the value of x when this statement is reached is
   a) $x = 0$ b) $x = 1$ c) $x = 2$
5. Let $P(x)$ be the statement "x spends more than five hours every weekday in class," where the universe of discourse for x is the set of students. Express each of the following quantifications in English.
   a) $\exists x\, P(x)$ b) $\forall x\, P(x)$
   c) $\exists x \neg P(x)$ d) $\forall x \neg P(x)$
6. Let $P(x, y)$ be the statement "x has taken class y," where the universe of discourse for x is the set of all students in your class and for y is the set of all computer science courses at your school. Express each of the following quantifications in English.
   a) $\exists x \exists y\, P(x, y)$ b) $\exists x \forall y\, P(x, y)$
   c) $\forall x \exists y\, P(x, y)$ d) $\exists y \forall x\, P(x, y)$
   e) $\forall y \exists x\, P(x, y)$ f) $\forall x \forall y\, P(x, y)$
7. Use quantifiers to express the statements:
   a) Every computer science student needs a course in discrete mathematics.
   b) There is a student in this class who owns a personal computer.
   c) Every student in this class has taken at least one computer science course.
   d) There is a student in this class who has taken at least one course in computer science.
   e) Every student in this class has been in every building on campus.
   f) There is a student in this class who has been in every room of at least one building on campus.
   g) Every student in this class has been in at least one room of every building on campus.
8. A discrete mathematics class contains 1 mathematics major who is a freshman, 12 mathematics majors who are sophomores, 15 computer science majors who are sophomores, 2 mathematics majors who are juniors, 2 computer science majors who are juniors, and 1 computer science major who is a senior. Express each of the following statements in terms of quantifiers and then determine its truth value.
   a) There is a student in the class who is a junior.
   b) Every student in the class is a computer science major.

c) There is a student in the class who is neither a mathematics major nor a junior.
d) Every student in the class is either a sophomore or a computer science major.
e) There is a major such that there is a student in the class in every year of study with that major.

9. Let $P(x)$ be the statement "$x = x^2$." If the universe of discourse is the set of integers, what are the truth values of the following?
a) $P(0)$ b) $P(1)$ c) $P(2)$
d) $P(-1)$ e) $\exists x\, P(x)$ f) $\forall x\, P(x)$

10. Let $Q(x, y)$ be the statement "$x + y = x - y$." If the universe of discourse for both variables is the set of integers, what are the truth values of the following?
a) $Q(1, 1)$ b) $Q(2, 0)$
c) $\forall y\, Q(1, y)$ d) $\exists x\, Q(x, 2)$
e) $\exists x\, \exists y\, Q(x, y)$ f) $\forall x\, \exists y\, Q(x, y)$
g) $\exists y\, \forall x\, Q(x, y)$ h) $\forall y\, \exists x\, Q(x, y)$
i) $\forall x\, \forall y\, Q(x, y)$

11. Suppose the universe of discourse of the propositional function $P(x, y)$ consists of pairs x and y where x is 1, 2, or 3 and y is 1, 2, or 3. Write out the following propositions using disjunctions and conjunctions.
a) $\exists x\, P(x, 3)$ b) $\forall y\, P(1, y)$
c) $\forall x\, \forall y\, P(x, y)$ d) $\exists x\, \exists y\, P(x, y)$
e) $\exists x\, \forall y\, P(x, y)$ f) $\forall y\, \exists x\, P(x, y)$

12. Express the negations of the following propositions using quantifiers. Also, express these negations in English.
a) Every student in this class likes mathematics.
b) There is a student in this class who has never seen a computer.
c) There is a student in this class who has taken every mathematics course offered at this school.
d) There is a student in this class who has been in at least one room of every building on campus.

Exercises 13–16 are based on questions found in the book *Symbolic Logic* by Lewis Carroll.

13. Let $P(x)$, $Q(x)$, and $R(x)$ be the statements "x is a professor," "x is ignorant," and "x is vain," respectively. Express each of the following statements using quantifiers, logical connectives, and $P(x)$, $Q(x)$, and $R(x)$, where the universe of discourse is the set of all people.
a) No professors are ignorant.
b) All ignorant people are vain.
c) No professors are vain.
★d) Does (c) follow from (a) and (b)? If not, is there a correct conclusion?

14. Let $P(x)$, $Q(x)$, and $R(x)$ be the statements "x is a clear explanation," "x is satisfactory," and "x is an excuse," respectively. Suppose that the universe of discourse for x is the set of all English text. Express each of the following statements using quantifiers, logical connectives, and $P(x)$, $Q(x)$, and $R(x)$.
a) All clear explanations are satisfactory.
b) Some excuses are unsatisfactory.
c) Some excuses are not clear explanations.
★d) Does (c) follow from (a) and (b)? If not, is there a correct conclusion?

15. Let $P(x)$, $Q(x)$, $R(x)$, and $S(x)$ be the statements "x is a baby," "x is logical," "x is able to manage a crocodile," and "x is despised," respectively. Suppose that the universe of discourse is the set of all people. Express each of the following statements using quantifiers, logical connectives, and $P(x)$, $Q(x)$, $R(x)$, and $S(x)$.
a) Babies are illogical.
b) Nobody is despised who can manage a crocodile.
c) Illogical persons are despised.
d) Babies cannot manage crocodiles.
★e) Does (d) follow from (a), (b), and (c)? If not, is there a correct conclusion?

16. Let $P(x)$, $Q(x)$, $R(x)$, and $S(x)$ be the statements "x is a duck," "x is one of my poultry," "x is an officer," and "x is willing to waltz," respectively. Express each of the following statements using quantifiers, logical connectives, and $P(x)$, $Q(x)$, $R(x)$, and $S(x)$.
a) No ducks are willing to waltz.
b) No officers ever decline to waltz.
c) All my poultry are ducks.
d) My poultry are not officers.
★e) Does (d) follow from (a), (b), and (c)? If not, is there a correct conclusion?

17. Show that the statements $\neg \exists x\, \forall y\, P(x, y)$ and $\forall x\, \exists y\, \neg P(x, y)$ have the same truth value.

18. Show that $\forall x\, (P(x) \wedge Q(x))$ and $\forall x\, P(x) \wedge \forall x\, Q(x)$ have the same truth value.

19. Show that $\exists x\, (P(x) \vee Q(x))$ and $\exists x\, P(x) \vee \exists x\, Q(x)$ have the same truth value.

20. Establish the following logical equivalences, where A is a proposition not involving any quantifiers.
a) $(\forall x\; P(x)) \vee A \Leftrightarrow \forall x (P(x) \vee A)$
b) $(\exists x\; P(x)) \vee A \Leftrightarrow \exists x (P(x) \vee A)$

21. Establish the following logical equivalences, where A is a proposition not involving any quantifiers.

a) $(\forall x\, P(x)) \wedge A \Leftrightarrow \forall x (P(x) \wedge A)$

b) $(\exists x\, P(x)) \wedge A \Leftrightarrow \exists x (P(x) \wedge A)$

22. Show that $\forall x\, P(x) \vee \forall x\, Q(x)$ and $\forall x\, (P(x) \vee Q(x))$ are not logically equivalent.

23. Show that $\exists x\, P(x) \wedge \exists x\, Q(x)$ and $\exists x (P(x) \wedge Q(x))$ are not logically equivalent.

★24. Show that $\forall x\, P(x) \vee \forall x\, Q(x)$ and $\forall x\, \forall y\, (P(x) \vee Q(y))$ are logically equivalent. (The new variable y is used to combine the quantifications correctly.)

★25. **a)** Show that $\forall x\, P(x) \wedge \exists x\, Q(x)$ and $\forall x\, \exists y\, (P(x) \wedge Q(y))$ are equivalent.

b) Show that $\forall x P(x) \vee \exists x Q(x)$ and $\forall x \exists y\, (P(x) \vee Q(y))$ are equivalent.

26. The notation $\exists! x\, P(x)$ denotes the proposition

"There exists a unique x such that $P(x)$ is true."

If the universe of discourse is the set of integers, what are the truth values of

a) $\exists! x (x > 1)$ **b)** $\exists! x (x^2 = 1)$

c) $\exists! x (x + 3 = 2x)$ **d)** $\exists! x (x = x + 1)$

27. What are the truth values of the following statements?

a) $\exists! x\, P(x) \rightarrow \exists x\, P(x)$

b) $\forall x\, P(x) \rightarrow \exists! x\, P(x)$

c) $\exists! x\, \neg P(x) \rightarrow \neg \forall x\, P(x)$

28. Write out the quantification $\exists! x\, P(x)$, where the universe of discourse consists of the integers 1, 2, and 3, in terms of negations, conjunctions, and disjunctions.

★29. Express the quantification $\exists! x\, P(x)$ using universal quantifications, existential quantifications, and logical operators.

A statement is in *prenex normal form (PNF)* if and only if it is of the form

$$Q_1 x_1 Q_2 x_2 \cdots Q_k x_k P(x_1, x_2, \ldots, x_k)$$

where each Q_i, $i = 1, 2, \ldots, k$ is either the existential quantifier or the universal quantifier, and $P(x_1, \ldots, x_k)$ is a predicate involving no quantifiers. For example, $\exists x\, \forall y\, (P(x, y) \wedge Q(y))$ is in prenex normal form, whereas $\exists x\, P(x) \vee \forall x\, Q(x)$ is not (since the quantifiers do not all occur first).

Every statement formed from propositional variables, predicates, **T**, and **F** using logical connectives and quantifiers is equivalent to a statement in prenex normal form. Exercise 31 asks for a proof of this fact.

★30. Put the following statements in prenex normal form. (*Hint:* Use logical equivalence from Tables 5 and 6 in Section 1.2, Table 2 in this section, and Exercises 18–21 and 24–25 in this section.)

a) $\exists x\, P(x) \vee \exists x\, Q(x) \vee A$, where A is a proposition not involving any quantifiers.

b) $\neg(\forall x\, P(x) \vee \forall x\, Q(x))$

c) $\exists x\, P(x) \rightarrow \exists x\, Q(x)$

★★31. Show how to transform an arbitrary statement to a statement in prenex normal form that is equivalent to the given statement.

1.4 Sets

INTRODUCTION

We will study a wide variety of discrete structures in this book. These include relations, which consist of ordered pairs of elements; combinations, which are unordered collections of elements; and graphs, which are sets of vertices and edges connecting vertices. Moreover, we will illustrate how these and other discrete structures are used in modeling and problem solving. In particular, many examples of the use of discrete structures in the storage, communication, and manipulation of data will be described. In this section we study the fundamental discrete structure upon which all other discrete structures are built, namely, the set.

Sets are used to group objects together. Often, the objects in a set have similar properties. For instance, all the students who are currently enrolled in your school make up a set. Likewise, all the students currently taking a course in discrete mathematics at any school make up a set. In addition, those students who are enrolled in your school who are taking a course in discrete mathematics form a set that can be obtained by taking the elements common to the first two collections. The language of sets is a means to study such collections in an organized fashion.

Note that the term *object* has been used without specifying what an object is. This description of a set as a collection of objects, based on the intuitive notion of an object, was first stated by the German mathematician Georg Cantor in 1895. The theory that results from this intuitive definition of a set leads to **paradoxes,** or logical inconsistencies, as the English philosopher Bertrand Russell showed in 1902 (see Exercise 24 for a description of one of these paradoxes). These logical inconsistencies can be avoided by building set theory starting with basic assumptions, called **axioms**. We will use Cantor's original version of set theory, known as **naive set theory**, without developing an axiomatic version of set theory, since all sets considered in this book can be treated consistently using Cantor's original theory.

We now proceed with our discussion of sets.

Georg Cantor (1845–1918). Georg Cantor was born in St. Petersburg, Russia, where his father was a successful merchant. Cantor developed his interest in mathematics in his teens. He began his university studies in Zurich in 1862, but when his father died he left Zurich. He continued his university studies at the University of Berlin in 1863, where he studied under the eminent mathematicians Weierstrass, Kummer, and Kronecker. He received his doctor's degree in 1867 after having written a dissertation on number theory. Cantor assumed a position at the University of Halle in 1869, where he continued working until his death.

Cantor is considered the founder of set theory. His contributions in this area include the discovery that the set of real numbers is uncountable. He is also noted for his many important contributions to analysis. Cantor also was interested in philosophy and wrote papers relating his theory of sets with metaphysics.

Cantor married in 1874 and had five children. His melancholy temperament was balanced by his wife's happy disposition. Although he received a large inheritance from his father, he was poorly paid as a professor. To mitigate this, he tried to obtain a better-paying position at the University of Berlin. His appointment there was blocked by Kronecker, who did not agree with Cantor's views on set theory. Cantor suffered from mental illness throughout the later years of his life. He died in 1918 in a psychiatric clinic.

Bertrand Russell (1872–1970). Bertrand Russell was born into a prominent English family active in the progressive movement and with a strong commitment to liberty. He became an orphan at an early age and was placed in the care of his father's parents, who had him educated at home. He entered Trinity College, Cambridge in 1890, where he excelled in mathematics and in moral science. He won a fellowship on the basis of his work on the foundations of geometry. In 1910 Trinity College appointed him to a lectureship in logic and the philosophy of mathematics.

Russell fought for progressive causes throughout his life. He held strong pacifist views, and his protests against World War I led to dismissal from his position at Trinity College. He was imprisoned for six months in 1918 because of an article he wrote that was branded seditious. Russell fought for women's suffrage in Great Britain. In 1961, at the age of 89, he was imprisoned for the second time for his protests for nuclear disarmament.

Russell's greatest work was in his development of principles that could be used as a foundation for all of mathematics. His most famous work is *Principia Mathematica*, written with Alfred North Whitehead, which attempts to deduce all of mathematics using a set of primitive axioms. He wrote many books on philosophy, physics, and his political ideas. Russell won the Nobel Prize for literature in 1950.

DEFINITION 1. The objects in a set are also called the *elements* or *members* of the set. A set is said to *contain* its elements.

There are several ways to describe sets. One way is to list all the members of a set, when this is possible. We use a notation where all members of the set are listed between brackets. For example, the notation $\{a, b, c, d\}$ represents the set with the four elements a, b, c, and d.

EXAMPLE 1 The set V of all vowels in the English alphabet can written as $V = \{a, e, i, o, u\}$. ■

EXAMPLE 2 The set O of odd positive integers less than 10 can be expressed by $O = \{1, 3, 5, 7, 9\}$. ■

EXAMPLE 3 Although sets are usually used to group together elements with common properties, there is nothing that prevents a set from having seemingly unrelated elements. For instance, $\{a, 2, \text{Fred}, \text{New Jersey}\}$ is the set containing the four elements a, 2, Fred, and New Jersey. ■

Uppercase letters are usually used to denote sets. The boldface letters **N**, **Z**, and **R** will be reserved to represent the set of natural numbers $\{0, 1, 2, 3, \ldots\}$, the set of integers $\{\ldots, -2, -1, 0, 1, 2, \ldots\}$, and the set of real numbers, respectively. We will occasionally use the notation $\mathbf{Z}^+$ to denote the set of positive integers.

Sometimes the brackets notation is used to describe a set without listing all its members. Some members of the set are listed, and then *ellipses* (...) are used when the general pattern of the elements is obvious.

EXAMPLE 4 The set of positive integers less than 100 can be denoted by $\{1, 2, 3, \ldots, 99\}$. ■

Since many mathematical statements assert that two differently specified collections of objects are really the same set, we need to understand what it means for two sets to be equal.

DEFINITION 2. Two sets are *equal* if and only if they have the same elements.

EXAMPLE 5 The sets $\{1, 3, 5\}$ and $\{3, 5, 1\}$ are equal, since they have the same elements. Note that the order in which the elements of a set are listed does not matter. Also, note that it does not matter if an element of a set is listed more than once, so that $\{1, 3, 3, 3, 5, 5, 5, 5\}$ is the same as the set $\{1, 3, 5\}$ since they have the same elements. ■

Another way to describe a set is to use **set builder** notation. We characterize all those elements in the set by stating the property or properties they must have to be members. For instance, the set O of all odd positive integers less than 10 can be written as

$$O = \{x \mid x \text{ is an odd positive integer less than } 10\}.$$

We often use this type of notation to describe sets when it is impossible to list all the elements of the set. For instance, the set of all real numbers can be written as

$$\mathbf{R} = \{x \mid x \text{ is a real number}\}.$$

Sets can also be represented graphically using Venn diagrams, named after the English mathematician John Venn, who introduced their use in 1881. In Venn diagrams the **universal set** U, which contains all the objects under consideration, is represented by a rectangle. Inside this rectangle, circles or other geometrical figures are used to represent sets. Sometimes points are used to represent the particular elements of the set. Venn diagrams are often used to indicate the relationships between sets. We show how a Venn diagram is used to represent a set in the following example.

EXAMPLE 6 Draw a Venn diagram that represents V, the set of vowels in the English alphabet.

Solution: We draw a rectangle to indicate the universal set U, which is the set of the 26 letters of the English alphabet. Inside this rectangle we draw a circle to represent V. Inside this circle we indicate the elements of V with points (see Figure 1). ■

We will now introduce some notation used to describe membership in sets. We write $a \in A$ to denote that a is an element of the set A. The notation $a \notin A$ denotes that A is not a member of the set A. Note that lowercase letters are usually used to denote elements of sets.

There is a special set that has no elements. This set is called the **empty set** or **null set** and is denoted by $\emptyset$. The empty set can also be denoted by $\{\,\}$ (that is, we

---

John Venn (1834–1923). John Venn was born into a London suburban family noted for its philanthropy. He attended London schools and got his mathematics degree from Caius College, Cambridge in 1857. He was elected a fellow of this college and held his fellowship there until his death. He took holy orders in 1859 and, after a brief stint of religious work, returned to Cambridge, where he developed programs in the moral sciences. Besides his mathematical work, Venn had an interest in history and wrote extensively about his college and family.

Venn's book *Symbolic Logic* clarifies ideas originally presented by Boole. In this book, Venn presents a systematic development of a method that uses geometric figures to analyze logical arguments. The diagrams used by Venn are known today as *Venn diagrams* and are primarily used to illustrate relationships between sets. In addition to his work on symbolic logic, he made contributions to probability theory described in his widely used textbook on that subject.

represent the empty set with a pair of brackets that enclose all the elements in this set). Often, a set of elements with certain properties turns out to be the null set. For instance, the set of all positive integers that are greater than their squares is the null set.

DEFINITION 3. The set A is said to be a *subset* of B if and only if every element of A is also an element of B. We use the notation $A \subseteq B$ to indicate that A is a subset of the set B.

We see that $A \subseteq B$ if and only if the quantification

$$\forall x(x \in A \rightarrow x \in B)$$

is true. For instance, the set of all odd positive integers less than 10 is a subset of the set of all positive integers less than 10. The set of all computer science majors at your school is a subset of the set of all students at your school.

The null set is a subset of every set, that is,

$$\emptyset \subseteq S$$

whenever S is a set. To establish that the null set is a subset of S, we must show that every element of the null set is also in S. In other words, we must show that the implication "if $x \in \emptyset$, then $x \in S$" is always true. We need only note that the hypothesis of this implication, namely, "$x \in \emptyset$," is always false to see that this implication is always true. Hence, the empty set is a subset of every set. Furthermore, note that every set is a subset of itself (the reader should verify this). Consequently, if P is a set, we know that $\emptyset \subseteq P$ and $P \subseteq P$.

When we wish to emphasize that a set A is a subset of the set B, but $A \neq B$, we write $A \subset B$ and say that A is a **proper subset** of B. Venn diagrams can be used to show that a set A is a subset of a set B. We draw the universal set U as a rectangle. Within this rectangle we draw a circle for B. Since A is a subset of B, we draw the circle for A within the circle for B. This relationship is shown in Figure 2.

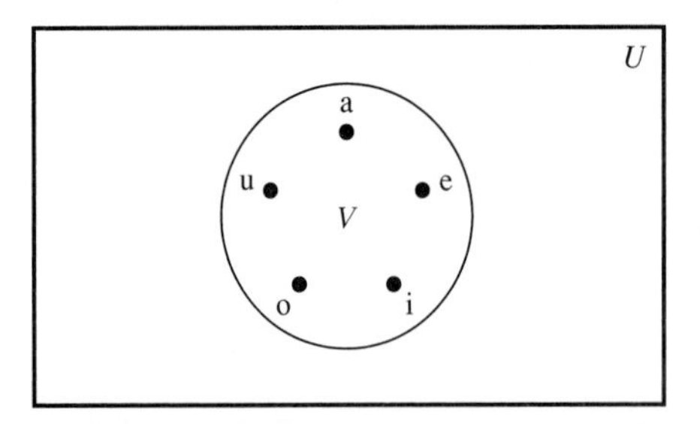

FIGURE 1 Venn Diagram for the Set of Vowels.

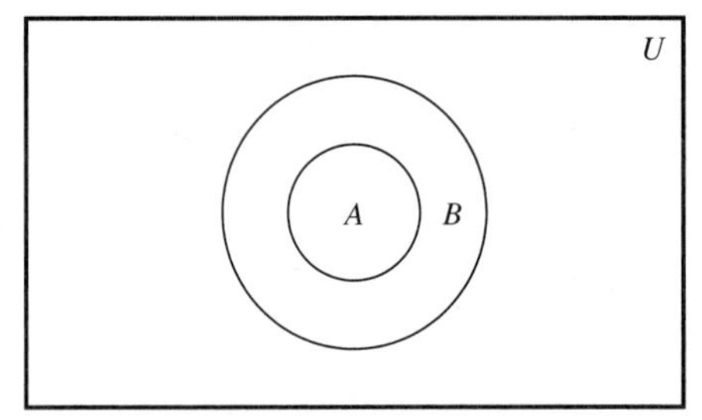

FIGURE 2 Venn Diagram Showing that A is a Subset of B.

One way to show that two sets have the same elements is to show that each set is a subset of the other. In other words, we can show that if A and B are sets with $A \subseteq B$ and $B \subseteq A$, then $A = B$. This turns out to be a useful way to show that two sets are equal.

Sets may have other sets as members. For instance, we have the sets

$$\{\emptyset, \{a\}, \{b\}, \{a, b\}\}$$

and

$$\{x \mid x \text{ is a subset of the set } \{a, b\}\}.$$

Note that these two sets are equal.

Sets are used extensively in counting problems. For these applications we need to discuss the size of sets.

DEFINITION 4. Let S be a set. If there are exactly n distinct elements in S where n is a nonnegative integer, we say that S is a *finite set* and that n is the *cardinality* of S. The cardinality of S is denoted by $|S|$.

EXAMPLE 7 Let A be the set of odd positive integers less than 10. Then $|A| = 5$. ■

EXAMPLE 8 Let S be the set of letters in the English alphabet. Then $|S| = 26$. ■

EXAMPLE 9 Since the null set has no elements, it follows that $|\emptyset| = 0$. ■

We will also be interested in sets that are not finite.

DEFINITION 5. A set is said to be *infinite* if it is not finite.

EXAMPLE 10 The set of positive integers is infinite. ■

It is possible to develop a theory of cardinality of infinite sets. This will be discussed in Section 1.7.

THE POWER SET

Many problems involve testing all combinations of elements of a set to see if they satisfy some property. To consider all such combinations of elements of a set S, we build a new set that has as its members all the subsets of S.

DEFINITION 6. Given a set S, the *power set* of S is the set of all subsets of the set S. The power set of S is denoted by $P(S)$.

EXAMPLE 11 What is the power set of the set $\{0, 1, 2\}$?

Solution: The power set $P(\{0, 1, 2\})$ is the set of all subsets of $\{0, 1, 2\}$. Hence,

$$P(\{0, 1, 2\}) = \{\emptyset, \{0\}, \{1\}, \{2\}, \{0, 1\}, \{0, 2\}, \{1, 2\}, \{0, 1, 2\}\}.$$

Note that the empty set and the set itself are members of this set of subsets. ■

EXAMPLE 12 What is the power set of the empty set? What is the power set of the set $\{\emptyset\}$?

Solution: The empty set has exactly one subset, namely, itself. Consequently,

$$P(\emptyset) = \{\emptyset\}.$$

The set $\{\emptyset\}$ has exactly two subsets, namely, $\emptyset$ and the set $\{\emptyset\}$ itself. Therefore,

$$P(\{\emptyset\}) = \{\emptyset, \{\emptyset\}\}.$$ ■

If a set has n elements, then its power set has 2^n elements. We will demonstrate this fact in several ways in subsequent sections of the text.

CARTESIAN PRODUCTS

The order of elements in a collection is often important. Since sets are unordered, a different structure is needed to represent ordered collections. This is provided by **ordered *n*-tuples**.

DEFINITION 7. The *ordered n-tuple* $(a_1, a_2, \ldots, a_n)$ is the ordered collection that has a_1 as its first element, a_2 as its second element, ..., and a_n as its nth element.

We say that two ordered n-tuples are equal if and only if each corresponding pair of their elements is equal. In other words, $(a_1, a_2, \ldots, a_n) = (b_1, b_2, \ldots, b_n)$ if and only if $a_i = b_i$, for $i = 1, 2, \ldots, n$. In particular, 2-tuples are called **ordered pairs**. The ordered pairs (a, b) and (c, d) are equal if and only if $a = c$ and $b = d$. Note that (a, b) and (b, a) are not equal unless $a = b$.

Many of the discrete structures we will study in later chapters are based on the notion of the *Cartesian product* of sets (named after René Descartes). We first define the Cartesian product of two sets.

DEFINITION 8. Let A and B be sets. The *Cartesian product* of A and B, denoted by $A \times B$, is the set of all ordered pairs (a, b) where $a \in A$ and $b \in B$. Hence,

$$A \times B = \{(a, b) \mid a \in A \wedge b \in B\}.$$

EXAMPLE 13 Let A represent the set of all students at a university, and let B represent the set of all courses offered at the university. What is the Cartesian product $A \times B$?

Solution: The Cartesian product $A \times B$ consists of all the ordered pairs of the form (a, b) where a is a student at the university and b is a course offered at the university. The set $A \times B$ can be used to represent all possible enrollments of students in courses at the university. ■

René Descartes (1596–1650). René Descartes was born into a noble family near Tours, France, about 200 miles southwest of Paris. He was the third child of his father's first wife, who died several days after his birth. Because of René's poor health, his father, a provincial judge, let his son's formal lessons slide, until at the age of eight René entered the Jesuit college at La Flèche. The rector of the school took a liking to him and permitted him to stay in bed until late in the morning because of his frail health. From then on, Descartes spent his mornings in bed; he considered these times his most productive hours for thinking.

Descartes left school in 1612, moving to Paris, where he spent two years studying mathematics. He earned a law degree in 1616 from the University of Poitiers. At 18 Descartes became disgusted with studying and decided to see the world. He moved to Paris and became a successful gambler. However, he grew tired of bawdy living and moved to the suburb of Saint-Germain, where he devoted himself to mathematical study. When his gambling friends found him, he decided to leave France and undertake a military career. However, he never did any fighting. One day, while escaping the cold in an overheated room at a military encampment, he had several feverish dreams, which revealed his future career as a mathematician and philosopher.

After ending his military career, he traveled throughout Europe. He then spent several years in Paris, where he studied mathematics and philosophy and constructed optical instruments. Descartes decided to move to Holland, where he spent 20 years wandering around the country, accomplishing his most important work. During this time he wrote several books, including the *Discours*, which contains his contributions to analytic geometry, for which he is best known. He also made fundamental contributions to philosophy.

In 1649 Descartes was invited by Queen Christina to visit her court in Sweden to tutor her in philosophy. Although he was reluctant to live in what he called "the land of bears amongst rocks and ice," he finally accepted the invitation and moved to Sweden. Unfortunately, the winter of 1649–1650 was extremely bitter. Descartes caught pneumonia and died in mid-February.

EXAMPLE 14 What is the Cartesian product of $A = \{1, 2\}$ and $B = \{a, b, c\}$?

Solution: The Cartesian product $A \times B$ is

$$A \times B = \{(1, a), (1, b), (1, c), (2, a), (2, b), (2, c)\}.$$

■

The Cartesian products $A \times B$ and $B \times A$ are not equal, unless $A = \emptyset$ or $B = \emptyset$, so that $A \times B = \emptyset$, or unless $A = B$ (see Exercise 22 at the end of this section). This is illustrated in the following example.

EXAMPLE 15 Show that the Cartesian product $B \times A$ is not equal to the Cartesian product $A \times B$, where A and B are as in Example 14.

Solution: The Cartesian product $B \times A$ is

$$B \times A = \{(a, 1), (a, 2), (b, 1), (b, 2), (c, 1), (c, 2)\}.$$

This is not equal to $A \times B$, which was found in Example 14. ■

The Cartesian product of more than two sets can also be defined.

DEFINITION 9. The *Cartesian product* of the sets $A_1, A_2, \ldots, A_n$, denoted by $A_1 \times A_2 \times \cdots \times A_n$, is the set of ordered n-tuples $(a_1, a_2, \ldots, a_n)$ where a_i belongs to A_i for $i = 1, 2, \ldots, n$. In other words

$$A_1 \times A_2 \times \cdots \times A_n = \{(a_1, a_2, \ldots, a_n) \mid a_i \in A_i \text{ for } i = 1, 2, \ldots, n\}.$$

EXAMPLE 16 What is the Cartesian product $A \times B \times C$ where $A = \{0, 1\}$, $B = \{1, 2\}$, and $C = \{0, 1, 2\}$?

Solution: The Cartesian product $A \times B \times C$ consists of all ordered triples (a, b, c) where $a \in A$, $b \in B$, and $c \in C$. Hence,

$$A \times B \times C = \{(0,1,0), (0,1,1), (0,1,2), (0,2,0), (0,2,1), (0,2,2), (1,1,0), (1,1,1), (1,1,2), (1,2,0), (1,2,1), (1,2,2)\}.$$

■

Exercises

1. List the members of the following sets.
   a) $\{x \mid x \text{ is a real number such that } x^2 = 1\}$
   b) $\{x \mid x \text{ is a positive integer less than } 12\}$
   c) $\{x \mid x \text{ is the square of an integer and } x < 100\}$
   d) $\{x \mid x \text{ is an integer such that } x^2 = 2\}$

2. Use set builder notation to give a description of each of the following sets.
   a) $\{0, 3, 6, 9, 12\}$
   b) $\{-3, -2, -1, 0, 1, 2, 3\}$
   c) $\{m, n, o, p\}$

3. Determine whether each of the following pairs of sets are equal.
   a) $\{1, 3, 3, 3, 5, 5, 5, 5, 5\}, \{5, 3, 1\}$
   b) $\{\{1\}\}, \{1, \{1\}\}$
   c) $\emptyset, \{\emptyset\}$
4. Suppose that $A = \{2, 4, 6\}$, $B = \{2, 6\}$, $C = \{4, 6\}$, and $D = \{4, 6, 8\}$. Determine which of these sets are subsets of which other of these sets.
5. Determine whether each of the following statements is true or false.
   a) $x \in \{x\}$ b) $\{x\} \subseteq \{x\}$ c) $\{x\} \in \{x\}$
   d) $\{x\} \in \{\{x\}\}$ e) $\emptyset \subseteq \{x\}$ f) $\emptyset \in \{x\}$
6. Use a Venn diagram to illustrate the relationship $A \subseteq B$ and $B \subseteq C$.
7. Suppose that A, B, and C are sets such that $A \subseteq B$ and $B \subseteq C$. Show that $A \subseteq C$.
8. Find two sets A and B such that $A \in B$ and $A \subseteq B$.
9. What is the cardinality of each of the following sets?
   a) $\{a\}$ b) $\{\{a\}\}$
   c) $\{a, \{a\}\}$ d) $\{a, \{a\}, \{a, \{a\}\}\}$
10. What is the cardinality of each of the following sets?
    a) $\emptyset$ b) $\{\emptyset\}$
    c) $\{\emptyset, \{\emptyset\}\}$ d) $\{\emptyset, \{\emptyset\}, \{\emptyset, \{\emptyset\}\}\}$
11. Find the power set of each of the following sets.
    a) $\{a\}$ b) $\{a, b\}$ c) $\{\emptyset, \{\emptyset\}\}$
12. Can you conclude that $A = B$ if A and B are two sets with the same power set?
13. How many elements do each of the following sets have?
    a) $P(\{a, b, \{a, b\}\})$ b) $P(\{\emptyset, a, \{a\}, \{\{a\}\}\})$
    c) $P(P(\emptyset))$
14. Determine whether each of the following sets is the power set of a set.
    a) $\emptyset$ b) $\{\emptyset, \{a\}\}$
    c) $\{\emptyset, \{a\}, \{\emptyset, a\}\}$ d) $\{\emptyset, \{a\}, \{b\}, \{a, b\}\}$
15. Let $A = \{a, b, c, d\}$ and $B = \{y, z\}$. Find
    a) $A \times B$ b) $B \times A$
16. What is the Cartesian product $A \times B$ where A is the set of courses offered by the mathematics department at a university and B is the set of mathematics professors at this university?
17. What is the Cartesian product $A \times B \times C$ where A is the set of all airlines and B and C are both the set of all cities in the United States?
18. Suppose that $A \times B = \emptyset$ where A and B are sets. What can you conclude?
19. Let A be a set. Show that $\emptyset \times A = A \times \emptyset = \emptyset$.
20. Let $A = \{a, b, c\}$, $B = \{x, y\}$, and $C = \{0, 1\}$. Find
    a) $A \times B \times C$ b) $C \times B \times A$
    c) $C \times A \times B$ d) $B \times B \times B$
21. How many different elements does $A \times B$ have if A has m elements and B has n elements?
22. Show that $A \times B \neq B \times A$ when A and B are nonempty unless $A = B$.

★23. Show that the ordered pair (a, b) can be defined in terms of sets as $\{\{a\}, \{a, b\}\}$. (*Hint:* First show that $\{\{a\}, \{a, b\}\} = \{\{c\}, \{c, d\}\}$ if and only if $a = c$ and $b = d$.)

★24. In this exercise **Russell's paradox** is presented. Let S be the set that contains a set x if the set x does not belong to itself, so that $S = \{x \mid x \notin x\}$.
   a) Show that the assumption that S is a member of S leads to a contradiction.
   b) Show that the assumption that S is not a member of S leads to a contradiction.

   From parts (a) and (b) it follows that the set S cannot be defined as it was. This paradox can be avoided by restricting the types of elements sets can have.

★25. Describe a procedure for listing all the subsets of a finite set.

1.5 Set Operations

INTRODUCTION

Two sets can be combined in many different ways. For instance, starting with the set of mathematics majors and the set of computer science majors at your school, we can form the set of students who are mathematics majors or computer science majors, the set of students who are joint majors in mathematics and computer science, the set of all students not majoring in mathematics, and so on.

DEFINITION 1. Let A and B be sets. The *union* of the sets A and B, denoted by $A \cup B$, is the set that contains those elements that are either in A or in B, or in both.

An element x belongs to the union of the sets A and B if and only if x belongs to A or x belongs to B. This tells us that

$$A \cup B = \{x \mid x \in A \vee x \in B\}.$$

The Venn diagram shown in Figure 1 represents the union of two sets A and B. The shaded area within the circle representing A or the circle representing B is the area that represents the union of A and B.

We will give some examples of the union of sets.

EXAMPLE 1 The union of the sets $\{1, 3, 5\}$ and $\{1, 2, 3\}$ is the set $\{1, 2, 3, 5\}$, that is, $\{1, 3, 5\} \cup \{1, 2, 3\} = \{1, 2, 3, 5\}$. ■

EXAMPLE 2 The union of the set of all computer science majors at your school and the set of all mathematics majors at your school is the set of students at your school who are majoring either in mathematics or in computer science (or in both). ■

DEFINITION 2. Let A and B be sets. The *intersection* of the sets A and B, denoted by $A \cap B$, is the set containing those elements in both A and B.

An element x belongs to the intersection of the sets A and B if and only if x belongs to A and x belongs to B. This tells us that

$$A \cap B = \{x \mid x \in A \wedge x \in B\}.$$

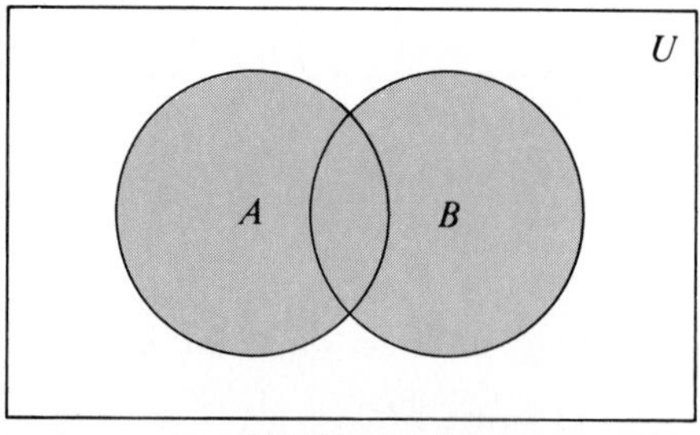

$A \cup B$ is shaded.

FIGURE 1 **Venn Diagram Representing the Union of A and B.**

The Venn diagram shown in Figure 2 represents the intersection of two sets A and B. The shaded area that is within both the circles representing the sets A and B is the area that represents the intersection of A and B.

We give some examples of the intersection of sets.

EXAMPLE 3 The intersection of the sets $\{1, 3, 5\}$ and $\{1, 2, 3\}$ is the set $\{1, 3\}$, that is, $\{1, 3, 5\} \cap \{1, 2, 3\} = \{1, 3\}$. ■

EXAMPLE 4 The intersection of the set of all computer science majors at your school and the set of all mathematics majors is the set of all students who are joint majors in mathematics and computer science. ■

DEFINITION 3. Two sets are called *disjoint* if their intersection is the empty set.

EXAMPLE 5 Let $A = \{1, 3, 5, 7, 9\}$ and $B = \{2, 4, 6, 8, 10\}$. Since $A \cap B = \emptyset$, A and B are disjoint. ■

We often are interested in finding the cardinality of the union of sets. To find the number of elements in the union of two finite sets A and B, note that $|A| + |B|$ counts each element that is in A but not in B or in B but not in A exactly once, and each element that is in both A and B exactly twice. Thus, if the number of elements that are in both A and B is subtracted from $|A| + |B|$, elements in $A \cap B$ will be counted only once. Hence,

$$|A \cup B| = |A| + |B| - |A \cap B|.$$

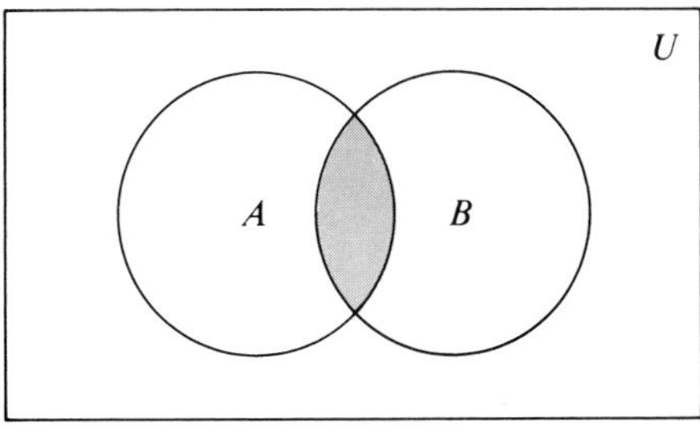

$A \cap B$ is shaded.

FIGURE 2 **Venn Diagram Representing the Intersection of A and B.**

The generalization of this result to unions of an arbitrary number of sets is called the **principle of inclusion-exclusion**. The principle of inclusion-exclusion is an important technique used in the art of enumeration. We will discuss this principle and other counting techniques in detail in Chapters 4 and 5.

There are other important ways to combine sets.

DEFINITION 4. Let A and B be sets. The *difference* of A and B, denoted by $A-B$, is the set containing those elements that are in A but not in B. The difference of A and B is also called the *complement of B with respect to A.*

An element x belongs to the difference of A and B if and only if $x \in A$ and $x \notin B$. This tells us that

$$A - B = \{x \mid x \in A \wedge x \notin B\}.$$

The Venn diagram shown in Figure 3 represents the difference of the sets A and B. The shaded area inside the circle that represents A and outside the circle that represents B is the area that represents $A - B$.

We give some examples of differences of sets.

EXAMPLE 6 The difference of $\{1, 3, 5\}$ and $\{1, 2, 3\}$ is the set $\{5\}$, that is, $\{1, 3, 5\} - \{1, 2, 3\} = \{5\}$. This is different from the difference of $\{1, 2, 3\}$ and $\{1, 3, 5\}$, which is the set $\{2\}$. ■

EXAMPLE 7 The difference of the set of computer science majors at your school and the set of mathematics majors at your school is the set of all computer science majors at your school who are not also mathematics majors. ■

Once the universal set U has been specified, the **complement** of a set can be defined.

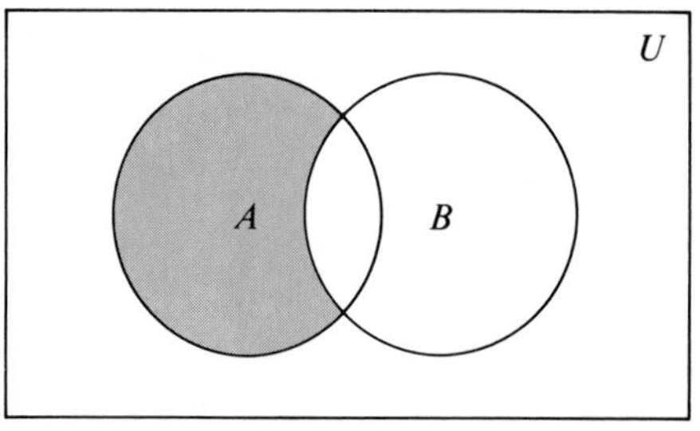

$A - B$ is shaded.

FIGURE 3 **Venn Diagram for the Difference of A and B.**

DEFINITION 5. Let U be the universal set. The *complement* of the set A, denoted by $\overline{A}$, is the complement of A with respect to U. In other words, the complement of the set A is $U - A$.

An element belongs to $\overline{A}$ if and only if $x \notin A$. This tells us that

$$\overline{A} = \{x \mid x \notin A\}.$$

In Figure 4 the shaded area outside of the circle that represents A is the area representing $\overline{A}$.

We give some examples of the complement of a set.

EXAMPLE 8 Let $A = \{a, e, i, o, u\}$ (where the universal set is the set of letters of the English alphabet). Then $\overline{A} = \{b, c, d, f, g, h, j, k, l, m, n, p, q, r, s, t, v, w, x, y, z\}$. ■

EXAMPLE 9 Let A be the set of positive integers greater than 10 (with universal set the set of all positive integers). Then $\overline{A} = \{1, 2, 3, 4, 5, 6, 7, 8, 9, 10\}$. ■

SET IDENTITIES

Table 1 lists the most important set identities. We will prove several of these identities here, using three different methods. These methods are presented to illustrate that there are often many different approaches to the solution of a problem. The proofs of the remaining identities will be left as exercises. The reader should note the similarity between these set identities and the logical equivalences discussed in Section 1.2. In fact, the set identities given can be proved directly from the corresponding logical equivalences. Furthermore, both are special cases of identities that hold for Boolean algebra (discussed in Chapter 9).

One way to prove that two sets are equal is to show that one of the sets is a subset of the other and vice versa. We illustrate this type of proof by establishing the second of De Morgan's laws.

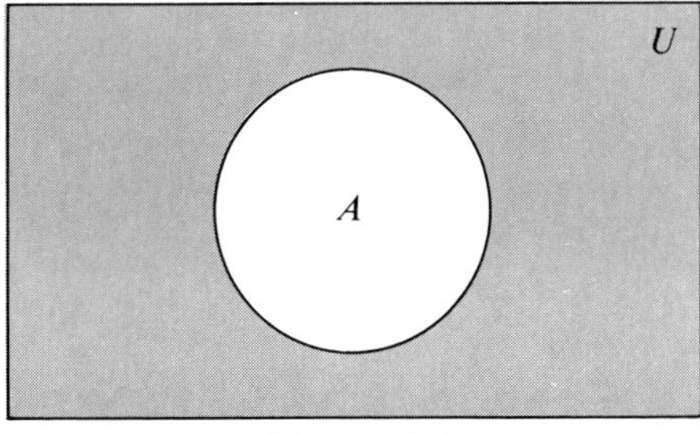

$\overline{A}$ is shaded.

FIGURE 4 **Venn Diagram for the Complement of the Set A.**

TABLE 1 Set Identities.

Identity	*Name*
$A \cup \varnothing = A$ $A \cap U = A$	Identity laws
$A \cup U = U$ $A \cap \varnothing = \varnothing$	Domination laws
$A \cup A = A$ $A \cap A = A$	Idempotent laws
$\overline{(\overline{A})} = A$	Complementation law
$A \cup B = B \cup A$ $A \cap B = B \cap A$	Commutative laws
$A \cup (B \cup C) = (A \cup B) \cup C$ $A \cap (B \cap C) = (A \cap B) \cap C$	Associative laws
$A \cap (B \cup C) = (A \cap B) \cup (A \cap C)$ $A \cup (B \cap C) = (A \cup B) \cap (A \cup C)$	Distributive laws
$\overline{A \cup B} = \overline{A} \cap \overline{B}$ $\overline{A \cap B} = \overline{A} \cup \overline{B}$	De Morgan's laws

EXAMPLE 10 Prove that $\overline{A \cap B} = \overline{A} \cup \overline{B}$ by showing that each set is a subset of the other.

Solution: First, suppose that $x \in \overline{A \cap B}$. It follows that $x \notin A \cap B$. This implies that $x \notin A$ or $x \notin B$. Hence, $x \in \overline{A}$ or $x \in \overline{B}$. Thus, $x \in \overline{A} \cup \overline{B}$. This shows that $\overline{A \cap B} \subseteq \overline{A} \cup \overline{B}$.

Now suppose that $x \in \overline{A} \cup \overline{B}$. Then $x \in \overline{A}$ or $x \in \overline{B}$. It follows that $x \notin A$ or $x \notin B$. Hence, $x \notin A \cap B$. Therefore, $x \in \overline{A \cap B}$. This demonstrates that $\overline{A} \cup \overline{B} \subseteq \overline{A \cap B}$. Since we have demonstrated that each set is a subset of the other, these two sets must be equal and the identity is proved. ■

Another way to verify set identities is to use set builder notation and the rules of logic. Consider the following proof of the second of De Morgan's laws.

EXAMPLE 11 Use set builder notation and logical equivalences to show that $\overline{A \cap B} = \overline{A} \cup \overline{B}$.

Solution: The following chain of equalities provides a demonstration of this identity:

$$\begin{aligned}\overline{A \cap B} &= \{x \mid x \notin A \cap B\}\\ &= \{x \mid \neg(x \in (A \cap B))\}\\ &= \{x \mid \neg(x \in A \wedge x \in B)\}\\ &= \{x \mid x \notin A \vee x \notin B\}\\ &= \{x \mid x \in \overline{A} \vee x \in \overline{B}\}\\ &= \{x \mid x \in \overline{A} \cup \overline{B}\}.\end{aligned}$$

Note that the second De Morgan's law for logical equivalences was used in the fourth equality of this chain. ■

Set identities can also be proved using **membership tables.** We consider each combination of sets that an element can belong to and verify that elements in the same combinations of sets belong to both the sets in the identity. To indicate that an element is in a set a 1 is used, and to indicate that an element is not in a set a 0 is used. (The reader should note the similarity between membership tables and truth tables.)

EXAMPLE 12 Use a membership table to show that $A \cap (B \cup C) = (A \cap B) \cup (A \cap C)$.

Solution: The membership table for these combinations of sets is shown in Table 2. This table has eight rows. Since the columns for $A \cap (B \cup C)$ and $(A \cap B) \cup (A \cap C)$ are the same, the identity is valid. ■

TABLE 2 **A Membership Table for the Distributive Property.**

A	B	C	$B \cup C$	$A \cap (B \cup C)$	$A \cap B$	$A \cap C$	$(A \cap B) \cup (A \cap C)$
1	1	1	1	1	1	1	1
1	1	0	1	1	1	0	1
1	0	1	1	1	0	1	1
1	0	0	0	0	0	0	0
0	1	1	1	0	0	0	0
0	1	0	1	0	0	0	0
0	0	1	1	0	0	0	0
0	0	0	0	0	0	0	0

Additional set identities can be established using those that we have already proved. Consider the following example.

EXAMPLE 13 Let A, B, and C be sets. Show that

$$\overline{A \cup (B \cap C)} = (\overline{C} \cup \overline{B}) \cap \overline{A}.$$

Solution: We have

$$\begin{aligned}
\overline{A \cup (B \cap C)} &= \overline{A} \cap \overline{(B \cap C)} && \text{by the first De Morgan's law} \\
&= \overline{A} \cap (\overline{B} \cup \overline{C}) && \text{by the second De Morgan's law} \\
&= (\overline{B} \cup \overline{C}) \cap \overline{A} && \text{by the commutative law for intersections} \\
&= (\overline{C} \cup \overline{B}) \cap \overline{A} && \text{by the commutative law for unions}
\end{aligned}$$

■

GENERALIZED UNIONS AND INTERSECTIONS

Since unions and intersections of sets satisfy associative laws, the sets $A \cup B \cup C$ and $A \cap B \cap C$ are well defined when A, B, and C are sets. Note that $A \cup B \cup C$ contains those elements that are in at least one of the sets A, B, and C, and that $A \cap B \cap C$ contains those elements that are in all of A, B, and C. These combinations of the three sets, A, B, and C, are shown in Figure 5.

EXAMPLE 14 Let $A = \{0, 2, 4, 6, 8\}$, $B = \{0, 1, 2, 3, 4\}$, and $C = \{0, 3, 6, 9\}$. What are $A \cup B \cup C$ and $A \cap B \cap C$?

Solution: The set $A \cup B \cup C$ contains those elements in at least one of A, B, and C. Hence,

$$A \cup B \cup C = \{0, 1, 2, 3, 4, 6, 8, 9\}.$$

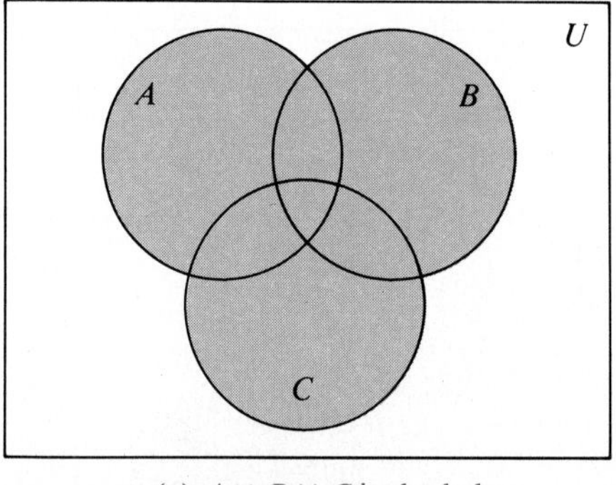

(a) $A \cup B \cup C$ is shaded.

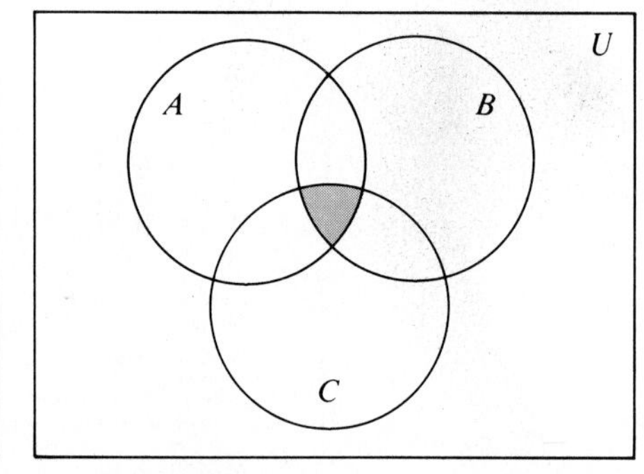

(b) $A \cap B \cap C$ is shaded.

FIGURE 5 **The Union and Intersection of A, B, and C.**

The set $A \cap B \cap C$ contains those elements in all three of A, B, and C. Thus,

$$A \cap B \cap C = \{0\}.$$ ■

We can also consider unions and intersections of an arbitrary number of sets. We use the following definitions.

DEFINITION 6. The *union* of a collection of sets is the set that contains those elements that are members of at least one set in the collection.

We use the notation

$$A_1 \cup A_2 \cup \cdots \cup A_n = \bigcup_{i=1}^{n} A_i$$

to denote the union of the sets $A_1, A_2, \ldots, A_n$.

DEFINITION 7. The *intersection* of a collection of sets is the set that contains those elements that are members of all the sets in the collection.

We use the notation

$$A_1 \cap A_2 \cap \cdots \cap A_n = \bigcap_{i=1}^{n} A_i$$

to denote the intersection of the sets $A_1, A_2, \ldots, A_n$. We illustrate generalized unions and intersections with the following example.

EXAMPLE 15 Let $A_i = \{i, i+1, i+2, \ldots\}$. Then

$$\bigcup_{i=1}^{n} A_i = \bigcup_{i=1}^{n} \{i, i+1, i+2, \ldots\} = \{1, 2, 3, \ldots\} = \mathbf{N},$$

and

$$\bigcap_{i=1}^{n} A_i = \bigcap_{i=1}^{n} \{i, i+1, i+2, \ldots\} = \{n, n+1, n+2, \ldots\}.$$ ■

COMPUTER REPRESENTATION OF SETS

There are various ways to represent sets using a computer. One method is to store the elements of the set in an unordered fashion. However, if this is done, the operations of computing the union, intersection, or difference of two sets would be time consuming, since each of these operations would require a large amount of searching for elements. We will present a method for storing elements using an arbitrary ordering of the elements of the universal set. This method of representing sets makes computing combinations of sets easy.

Assume that the universal set U is finite (and of reasonable size so that the number of elements of U is not larger than the memory size of the computer being used). First, specify an arbitrary ordering of the elements of U, for instance $a_1, a_2, \ldots, a_n$. Represent a subset A of U with the bit string of length n where the ith bit in this string is 1 if a_i belongs to A, and is 0 if a_i does not belong to A. The following example illustrates this technique.

EXAMPLE 16 Let $U = \{1, 2, 3, 4, 5, 6, 7, 8, 9, 10\}$, and the ordering of elements of U has the elements in increasing order, i.e., $a_i = i$. What bit strings represent the subset of all odd integers in U, the subset of all even integers in U, and the subset of integers not exceeding five in U?

Solution: The bit string that represents the set of odd integers in U, namely, $\{1, 3, 5, 7, 9\}$, has a one bit in the first, third, fifth, seventh, and ninth positions, and a zero elsewhere. It is

10101 01010.

(We have split this bit string of length 10 into two blocks of length five for easy reading since long bit strings are difficult to read.) Similarly, we represent the subset of all even integers in U, namely, $\{2, 4, 6, 8, 10\}$, by the string

01010 10101.

The set of all integers in U that do not exceed five, namely, $\{1, 2, 3, 4, 5\}$, is represented by the string

11111 00000. ■

Using bit strings to represent sets, it is easy to find complements of sets and unions, intersections, and differences of sets. To find the bit string for the complement of a set from the bit string for that set, we simply change each 1 to a 0 and each 0 to 1, since $x \in A$ if and only if $x \notin \overline{A}$. Note that this operation corresponds to taking the negation of each bit when we associate a bit with a truth value, with 1 representing true and 0 false.

EXAMPLE 17 We have seen that the bit string for the set $\{1, 3, 5, 7, 9\}$ (with universal set $\{1, 2, 3, 4, 5, 6, 7, 8, 9, 10\}$) is

10101 01010,

What is the bit string for the complement of this set?

Solution: The bit string for the complement of this set is obtained by replacing 0s with 1s and vice versa. This yields the string

$$01010\ 10101,$$

which corresponds to the set $\{2, 4, 6, 8, 10\}$. ■

To obtain the bit string for the union and intersection of two sets we perform bitwise Boolean operations on the bit strings representing the two sets. The bit in the ith position of the bit string of the union is 1 if either of the bits in the ith position in the two strings is 1 (or both are 1) and is 0 when both bits are 0. Hence, the bit string for the union is the bitwise *OR* of the bit strings for the two sets. The bit in the ith position of the bit string of the intersection is 1 when the bits in the corresponding position in the two strings are both 1 and is 0 when either of the two bits is 0 (or both are). Hence, the bit string for the intersection is the bitwise *AND* of the bit strings for the two sets.

EXAMPLE 18 The bit strings for the sets $\{1, 2, 3, 4, 5\}$ and $\{1, 3, 5, 7, 9\}$ are 11111 00000 and 10101 01010, respectively. Use bit strings to find the union and intersection of these sets.

Solution: The bit string for the union of these sets is

$$11111\ 00000 \vee 10101\ 01010 = 11111\ 01010,$$

which corresponds to the set $\{1, 2, 3, 4, 5, 7, 9\}$. The bit string for the intersection of these sets is

$$11111\ 00000 \wedge 10101\ 01010 = 10101\ 00000$$

which corresponds to the set $\{1, 3, 5\}$. ■

Exercises

1. Let A be the set of students who live within one mile of school and let B be the set of students who walk to classes. Describe the students in each of the following sets.
  a) $A \cap B$ **b)** $A \cup B$
  c) $A - B$ **d)** $B - A$

2. Suppose that A is the set of sophomores at your school and B is the set of students in discrete mathematics at your school. Express each of the following sets in terms of A and B.
  a) the set of sophomores taking discrete mathematics in your school
  b) the set of sophomores at your school who are not taking discrete mathematics
  c) the set of students at your school who either are sophomores or are taking discrete mathematics
  d) the set of students at your school who either are not sophomores or are not taking discrete mathematics

3. Let $A = \{1, 2, 3, 4, 5\}$ and $B = \{0, 3, 6\}$. Find
  a) $A \cup B$ **b)** $A \cap B$
  c) $A - B$ **d)** $B - A$

4. Let $A = \{a, b, c, d, e\}$ and $B = \{a, b, c, d, e, f, g, h\}$. Find

a) $A \cup B$ b) $A \cap B$
c) $A - B$ d) $B - A$

5. Let A be a set. Show that $\overline{\overline{A}} = A$.

6. Let A be a set. Show that
a) $A \cup \emptyset = A$ b) $A \cap \emptyset = \emptyset$
c) $A \cup A = A$ d) $A \cap A = A$
e) $A - \emptyset = A$ f) $A \cup U = U$
g) $A \cap U = A$ h) $\emptyset - A = \emptyset$

7. Let A and B be sets. Show that
a) $A \cup B = B \cup A$ b) $A \cap B = B \cap A$

8. Find the sets A and B if $A - B = \{1, 5, 7, 8\}$, $B - A = \{2, 10\}$, and $A \cap B = \{3, 6, 9\}$.

9. Show that if A and B are sets, then $\overline{A \cup B} = \overline{A} \cap \overline{B}$.
a) by showing each side is a subset of the other side
b) using a membership table

10. Let A and B be sets. Show that
a) $(A \cap B) \subseteq A$
b) $A \subseteq (A \cup B)$
c) $A - B \subseteq A$
d) $A \cap (B - A) = \emptyset$
e) $A \cup (B - A) = A \cup B$

11. Show that if A and B are sets, then $A - B = A \cap \overline{B}$.

12. Show that if A and B are sets, then $(A \cap B) \cup (A \cap \overline{B}) = A$.

13. Let A, B, and C be sets. Show that
a) $A \cup (B \cup C) = (A \cup B) \cup C$
b) $A \cap (B \cap C) = (A \cap B) \cap C$
c) $A \cup (B \cap C) = (A \cup B) \cap (A \cup C)$

14. Let A, B, and C be sets, Show that $(A - B) - C = (A - C) - (B - C)$.

15. Let $A = \{0, 2, 4, 6, 8, 10\}$, $B = \{0, 1, 2, 3, 4, 5, 6\}$, and $C = \{4, 5, 6, 7, 8, 9, 10\}$. Find
a) $A \cap B \cap C$ b) $A \cup B \cup C$
c) $(A \cup B) \cap C$ d) $(A \cap B) \cup C$

16. Draw the Venn diagrams for each of the following combinations of the sets A, B, and C.
a) $A \cap (B \cup C)$
b) $\overline{A} \cap \overline{B} \cap \overline{C}$
c) $(A - B) \cup (A - C) \cup (B - C)$

17. What can you say about the sets A and B if the following are true?
a) $A \cup B = A$ b) $A \cap B = A$
c) $A - B = A$ d) $A \cap B = B \cap A$
e) $A - B = B - A$

18. Can you conclude that $A = B$ if A, B, and C are sets such that
a) $A \cup C = B \cup C$?
b) $A \cap C = B \cap C$?

19. Let A and B be subsets of a universal set U. Show that $A \subseteq B$ if and only if $\overline{B} \subseteq \overline{A}$.

The **symmetric difference** of A and B, denoted by $A \oplus B$, is the set containing those elements in either A or B, but not in both A and B.

20. Find the symmetric difference of $\{1, 3, 5\}$ and $\{1, 2, 3\}$.

21. Find the symmetric difference of the set of computer science majors at a school and the set of mathematics majors at this school.

22. Draw a Venn diagram for the symmetric difference of the sets A and B.

23. Show that $A \oplus B = (A \cup B) - (A \cap B)$.

☞ **24.** Show that $A \oplus B = (A - B) \cup (B - A)$.

25. Show that if A is a subset of a universal set U, then
a) $A \oplus A = \emptyset$ b) $A \oplus \emptyset = A$
c) $A \oplus U = \overline{A}$ d) $A \oplus \overline{A} = U$

26. Show that if A and B are sets, then
a) $A \oplus B = B \oplus A$ b) $(A \oplus B) \oplus B = A$

27. What can you say about the sets A and B if $A \oplus B = A$?

★**28.** Determine whether the symmetric difference is associative; that is, if A, B, and C are sets, does it follow that

$$A \oplus (B \oplus C) = (A \oplus B) \oplus C?$$

★**29.** Suppose that A, B, and C are sets such that $A \oplus C = B \oplus C$. Must it be the case that $A = B$?

★**30.** Show that if A, B, and C are sets, then

$$|A \cup B \cup C| = |A| + |B| + |C| - |A \cap B| - |A \cap C| - |B \cap C| + |A \cap B \cap C|.$$

(This is a special case of the inclusion-exclusion principle, which will be studied in Chapter 5.)

31. Let $A_i = \{1, 2, 3, \ldots, i\}$ for $i = 1, 2, 3, \ldots$. Find
a) $\bigcup_{i=1}^{n} A_i$ b) $\bigcap_{i=1}^{n} A_i$

32. Let $A_i = \{i, i + 1, i + 2, \ldots\}$. Find
a) $\bigcup_{i=1}^{n} A_i$ b) $\bigcap_{i=1}^{n} A_i$

33. Let A_i be the set of all bit strings of length not exceeding i. Find
a) $\bigcup_{i=1}^{n} A_i$ b) $\bigcap_{i=1}^{n} A_i$

34. Suppose that the universal set is $U = \{1, 2, 3, 4, 5, 6, 7, 8, 9, 10\}$. Express each of the following sets with bit strings where the ith bit in the string is 1 if i is in the set and 0 otherwise.

a) $\{3, 4, 5\}$ **b)** $\{1, 3, 6, 10\}$
c) $\{2, 3, 4, 7, 8, 9\}$

35. Using the same universal set as in the last problem, find the set specified by each of the following bit strings.
a) 11110 01111 **b)** 01011 11000
c) 10000 00001.

36. What subsets of a finite universal set do the following bit strings represent?
a) the string with all zeros
b) the string with all ones

37. What is the bit string corresponding to the difference of two sets?

38. What is the bit string corresponding to the symmetric difference of two sets?

39. Show how bitwise operations on bit strings can be used to find the following combinations of $A = \{a, b, c, d, e\}$, $B = \{b, c, d, g, p, t, v\}$, $C = \{c, e, i, o, u, x, y, z\}$, and $D = \{d, e, h, i, n, o, t, u, x, y\}$.
a) $A \cup B$ **b)** $A \cap B$
c) $(A \cup D) \cap (B \cup C)$ **d)** $A \cup B \cup C \cup D$

40. How can the union and intersection of n sets that all are subsets of the universal set U be found using bit strings?

41. The **successor** of the set A is the set $A \cup \{A\}$. Find the successors of the following sets.
a) $\{1, 2, 3\}$ **b)** $\emptyset$ **c)** $\{\emptyset\}$ **d)** $\{\emptyset, \{\emptyset\}\}$

42. How many elements does the successor of a set with n elements have?

Sometimes the number of times an element occurs in an unordered collection matters. **Multisets** are unordered collections of elements where an element can occur as a member more than once. The notation $\{m_1 \cdot a_1, m_2 \cdot a_2, \ldots, m_r \cdot a_r\}$ denotes the multiset with element a_1 occurring m_1 times, element a_2 occurring m_2 times, and so on. The numbers m_i, $i = 1, 2, \ldots, r$ are called the **multiplicities** of the elements a_i, $i = 1, 2, \ldots, r$.

Let P and Q be multisets. The **union** of the multisets P and Q is the multiset where the multiplicity of an element is the maximum of its multiplicities in P and Q. The **intersection** of P and Q is the multiset where the multiplicity of an element is the minimum of its multiplicities in P and Q. The **difference** of P and Q is the multiset where the multiplicity of an element is the multiplicity of the element in P less its multiplicity in Q unless this difference is negative, in which case the multiplicity is 0. The **sum** of P and Q is the multiset where the multiplicity of an element is the sum of multiplicities in P and Q. The union, intersection, and difference of P and Q are denoted by $P \cup Q$, $P \cap Q$, and $P - Q$, respectively (where these operations should not be confused with the analogous operations for sets). The sum of P and Q is denoted by $P + Q$.

43. Let A and B be the multisets $\{3 \cdot a, 2 \cdot b, 1 \cdot c\}$ and $\{2 \cdot a, 3 \cdot b, 4 \cdot d\}$, respectively. Find
a) $A \cup B$ **b)** $A \cap B$ **c)** $A - B$
d) $B - A$ **e)** $A + B$

44. Suppose that A is the multiset that has as its elements the types of computer equipment needed by one department of a university where the multiplicities are the number of pieces of each type needed, and B is the analogous multiset for a second department of the university. For instance A could be the multiset $\{107 \cdot$ personal computers, $44 \cdot$ modems, $6 \cdot$ minicomputers$\}$ and B could be the multiset $\{14 \cdot$ personal computers, $6 \cdot$ modems, $2 \cdot$ mainframes$\}$.
a) What combination of A and B represents the equipment the university should buy assuming both departments use the same equipment?
b) What combination of A and B represents the equipment that will be used by both departments if both departments use the same equipment?
c) What combination of A and B represents the equipment that the second department uses, but the first department does not, if both departments use the same equipment?
d) What combination of A and B represents the equipment that the university should purchase if the departments do not share equipment?

Fuzzy sets are used in artificial intelligence. Each element in the universal set U has a **degree of membership**, which is a real number between 0 and 1 (including 0 and 1), in a fuzzy set S. The fuzzy set S is denoted by listing the elements with their degrees of membership (elements with zero degree of membership are not listed). For instance, we write {0.6 Alice, 0.9 Brian, 0.4 Fred, 0.1 Oscar, 0.5 Rita} for the set F (of famous people) to indicate that Alice has a 0.6 degree of membership in F, Brian has a 0.9 degree of membership in F, Fred has a 0.4 degree of membership in F, Oscar has a 0.1 degree of membership in F, and Rita has a 0.5 degree of membership in F (so that Brian is the most famous and Oscar is the least famous of these people). Also, suppose that R is the set of rich people with $R =$ {0.4 Alice, 0.8 Brian, 0.2 Fred, 0.9 Oscar, 0.7 Rita}.

45. The **complement** of a fuzzy set S is the set $\overline{S}$, with the degree of the membership of an element in S equal to 1 minus the degree of membership of this element in S. Find $\overline{F}$ (the fuzzy set of people who are not famous) and $\overline{R}$ (the fuzzy set of people who are not rich).
46. The **union** of two fuzzy sets S and T is the fuzzy set $S \cup T$, where the degree of membership of an element in $S \cup T$ is the maximum of the degrees of membership of this element in S and in T. Find the fuzzy set $F \cup R$ of rich or famous people.
47. The **intersection** of two fuzzy sets S and T is the fuzzy set $S \cap T$, where the degree of membership of an element in $S \cap T$ is the minimum of the degrees of membership of this element is S and in T. Find the fuzzy set $F \cap R$ of rich and famous people.

1.6 Functions

INTRODUCTION

In many instances we assign to each element of a set a particular element of a second set (which may be the same as the first). For instance, suppose that each student in a discrete mathematics class is assigned a letter grade from the set $\{A, B, C, D, F\}$. For instance, the grades may be A for Adams, C for Chou, B for Goodfriend, A for Rodriguez, and F for Stevens. This assignment of grades is illustrated in Figure 1.

This assignment is an example of a function. The concept of a function is extremely important in discrete mathematics. Functions are used in the definition of such discrete structures as sequences and strings. Functions are also used to represent how long it takes a computer to solve problems of a given size. Recursive functions, which are functions defined in terms of themselves, are used throughout computer science; they will be studied in Chapter 3. This section reviews the basic concepts involving functions needed in discrete mathematics.

DEFINITION 1. Let A and B be sets. A *function* f from A to B is an assignment of a unique element of B to each element of A. We write $f(a) = b$ if b is the unique element of B assigned by the function f to the element a of A. If f is a function from A to B, we write $f : A \to B$.

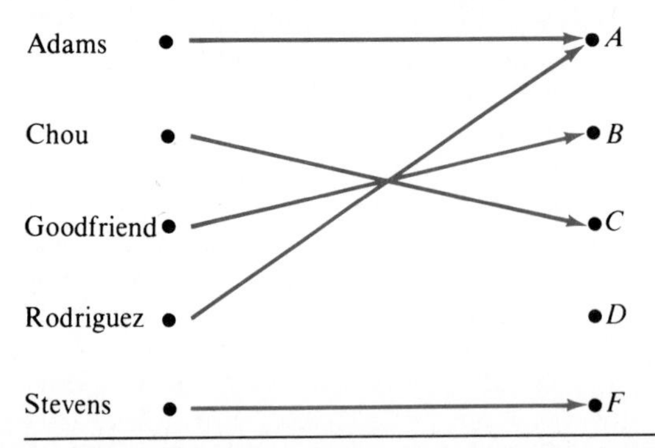

FIGURE 1 **Assignment of Grades in a Discrete Mathematics Class.**

Functions are specified in many different ways. Sometimes we explicitly state the assignments. Often we give a formula, such as $f(x) = x + 1$, to define a function. Other times we use a computer program to specify a function.

DEFINITION 2. If f is a function from A to B, we say that A is the *domain* of f and B is the *codomain* of f. If $f(a) = b$, we say that b is the *image* of a and a is a *pre-image* of b. The *range* of f is the set of all images of elements of A. Also, if f is a function from A to B, we say that f *maps* A to B.

Figure 2 represents a function f from A to B.

Consider the example that began this section. Let G be the function that assigns a grade to a student in our discrete mathematics class. Note that G(Adams) $= A$, for instance. The domain of G is the set {Adams, Chou, Goodfriend, Rodriguez, Stevens}, and the codomain is the set $\{A, B, C, D, F\}$. The range of G is the set $\{A, B, C, F\}$, because there are students who are assigned each grade except D. Also consider the following examples.

EXAMPLE 1 Let f be the function that assigns the last two bits of a bit string of length two or greater to that string. Then, the domain of f is the set of all bit strings of length two or greater, and both the codomain and range are the set $\{00, 01, 10, 11\}$. ■

EXAMPLE 2 Let f be the function from $\mathbf{Z}$ to $\mathbf{Z}$ that assigns the square of an integer to this integer. Then, $f(x) = x^2$, where the domain of f is the set of all integers, the codomain of f can be chosen to be the set of all integers, and the range of f is the set of all nonnegative integers that are perfect squares, namely, $\{0, 1, 4, 9, \dots\}$. ■

EXAMPLE 3 (For students familiar with Pascal) The domain and codomain of functions are often specified in programming languages. For instance, the Pascal statement

function *floor*(x: real): integer

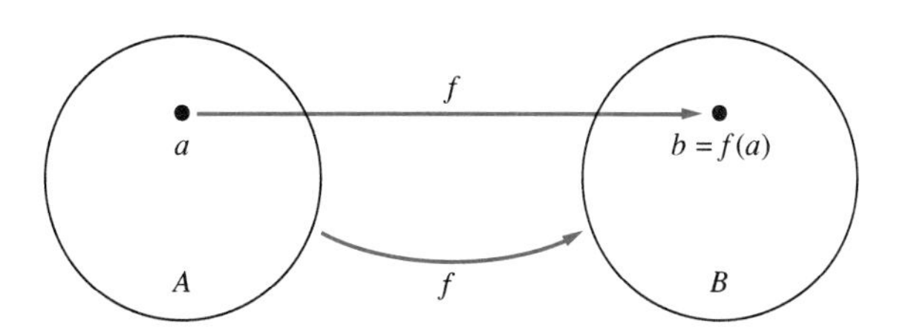

FIGURE 2 **The Function f Maps A to B.**

states that the domain of the floor function is the set of real numbers and its codomain is the set of integers. ■

Two real-valued functions with the same domain can be added and multiplied.

DEFINITION 3. Let f_1 and f_2 be functions from A to $\mathbf{R}$. Then $f_1 + f_2$ and $f_1 f_2$ are also functions from A to $\mathbf{R}$ defined by

$$(f_1 + f_2)(x) = f_1(x) + f_2(x),$$
$$(f_1 f_2)(x) = f_1(x) f_2(x).$$

Note that the functions $f_1 + f_2$ and $f_1 f_2$ have been defined by specifying their values at x in terms of the values of f_1 and f_2 at x.

EXAMPLE 4 Let f_1 and f_2 be functions from $\mathbf{R}$ to $\mathbf{R}$ such that $f_1(x) = x^2$ and $f_2(x) = x - x^2$. What are the functions $f_1 + f_2$ and $f_1 f_2$?

Solution: From the definition of the sum and product of functions, it follows that

$$(f_1 + f_2)(x) = f_1(x) + f_2(x) = x^2 + (x - x^2) = x$$

and

$$(f_1 f_2)(x) = x^2(x - x^2) = x^3 - x^4.$$ ■

When f is a function from a set A to a set B, the image of a subset of A can also be defined.

DEFINITION 4. Let f be a function from the set A to the set B and let S be a subset of A. The *image* of S is the subset of B that consists of the images of the elements of S. We denote the image of S by $f(S)$, so that

$$f(S) = \{f(s) \mid s \in S\}.$$

EXAMPLE 5 Let $A = \{a, b, c, d, e\}$ and $B = \{1, 2, 3, 4\}$ with $f(a) = 2$, $f(b) = 1$, $f(c) = 4$, $f(d) = 1$, and $f(e) = 1$. The image of the subset $S = \{b, c, d\}$ is the set $f(S) = \{1, 4\}$. ■

ONE-TO-ONE AND ONTO FUNCTIONS

Some functions have distinct images at distinct members of their domain. These functions are said to be **one-to-one**.

DEFINITION 5. A function f is said to be *one-to-one*, or *injective*, if and only if $f(x) = f(y)$ implies that $x = y$ for all x and y in the domain of f. A function is said to be an *injection* if it is one-to-one.

Remark: A function f is one-to-one if and only if $f(x) \neq f(y)$ whenever $x \neq y$. This way of expressing that f is one-to-one is obtained by taking the contrapositive of the implication in the definition.

We illustrate this concept by giving examples of functions that are one-to-one and other functions that are not one-to-one.

EXAMPLE 6 Determine whether the function f from $\{a, b, c, d\}$ to $\{1, 2, 3, 4, 5\}$ with $f(a) = 4$, $f(b) = 5$, $f(c) = 1$, and $f(d) = 3$ is one-to-one.

Solution: The function f is one-to-one since f takes on different values at the four elements of its domain. This is illustrated in Figure 3. ■

EXAMPLE 7 Determine whether the function $f(x) = x^2$ from the set of integers to the set of integers is one-to-one.

Solution: The function $f(x) = x^2$ is not one-to-one because, for instance, $f(1) = f(-1) = 1$, but $1 \neq -1$. ■

EXAMPLE 8 Determine whether the function $f(x) = x + 1$ is one-to-one.

Solution: The function $f(x) = x + 1$ is a one-to-one function. To demonstrate this, note that $x + 1 \neq y + 1$ when $x \neq y$. ■

We now give some conditions that guarantee that a function is one-to-one.

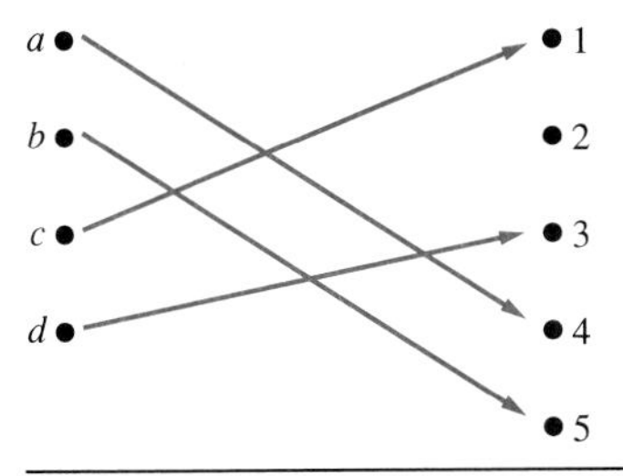

FIGURE 3 **A One-to-One Function.**

DEFINITION 6. A function f whose domain and codomain are subsets of the set of real numbers is called *strictly increasing* if $f(x) < f(y)$ whenever $x < y$ and x and y are in the domain of f. Similarly, f is called *strictly decreasing* if $f(x) > f(y)$ whenever $x < y$ and x and y are in the domain of f.

From these definitions, we see that a function that is either strictly increasing or strictly decreasing must be one-to-one.

For some functions the range and the codomain are equal. That is, every member of the codomain is the image of some element of the domain. Functions with this property are called **onto** functions.

DEFINITION 7. A function f from A to B is called *onto*, or *surjective*, if and only if for every element $b \in B$ there is an element $a \in A$ with $f(a) = b$. A function f is called a *surjection* if it is onto.

We give examples of onto functions and functions that are not onto.

EXAMPLE 9 Let f be the function from $\{a, b, c, d\}$ to $\{1, 2, 3\}$ defined by $f(a) = 3$, $f(b) = 2$, $f(c) = 1$, and $f(d) = 3$. Is f an onto function?

Solution: Since all three elements of the codomain are images of elements in the domain, we see that f is onto. This is illustrated in Figure 4. ■

EXAMPLE 10 Is the function $f(x) = x^2$ from the set of integers to the set of integers onto?

Solution: The function f is not onto since there is no integer x with $x^2 = -1$, for instance. ■

EXAMPLE 11 Is the function $f(x) = x + 1$ from the set of integers to the set of integers onto?

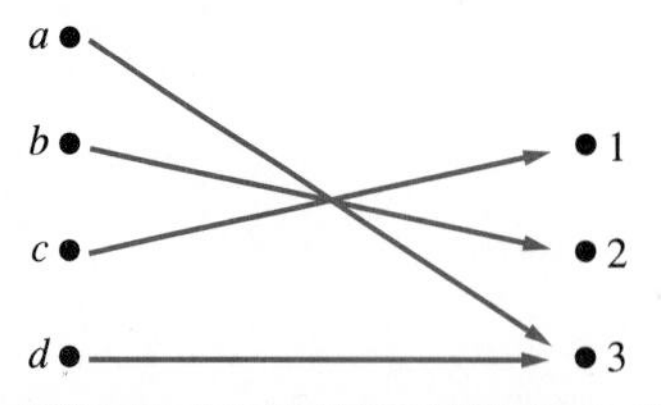

FIGURE 4 **An Onto Function.**

Solution: This function is onto, since for every integer y there is an integer x such that $f(x) = y$. To see this, note that $f(x) = y$ if and only if $x + 1 = y$, which holds if and only if $x = y - 1$. ∎

DEFINITION 8. The function f is a *one-to-one correspondence* or a *bijection* if it is both one-to-one and onto.

The following examples illustrate the concept of a bijection.

EXAMPLE 12 Let f be the function from $\{a, b, c, d\}$ to $\{1, 2, 3, 4\}$ with $f(a) = 4$, $f(b) = 2$, $f(c) = 1$, and $f(d) = 3$. Is f a bijection?

Solution: The function f is one-to-one and onto. It is one-to-one since the function takes on distinct values. It is onto since all four elements of the codomain are images of elements in the domain. Hence, f is a bijection. ∎

Figure 5 displays four functions where the first is one-to-one but not onto, the second is onto but not one-to-one, the third is both one-to-one and onto, and the fourth is neither one-to-one nor onto. The fifth correspondence in Figure 5 is not a function, since it sends an element to two different elements.

Suppose that f is a function from a set A to itself. If A is finite, then f is one-to-one if and only if it is onto. (This follows from the result in Exercise 34.) This is not necessarily the case if A is infinite (as will be shown in Section 1.7).

EXAMPLE 13 Let A be a set. The *identity function* on A is the function $\iota_A : A \to A$ where

$$\iota_A(x) = x$$

where $x \in A$. In other words, the identity function ι_A is the function that assigns each element to itself. The function ι_A is one-to-one and onto, so that it is a bijection. ∎

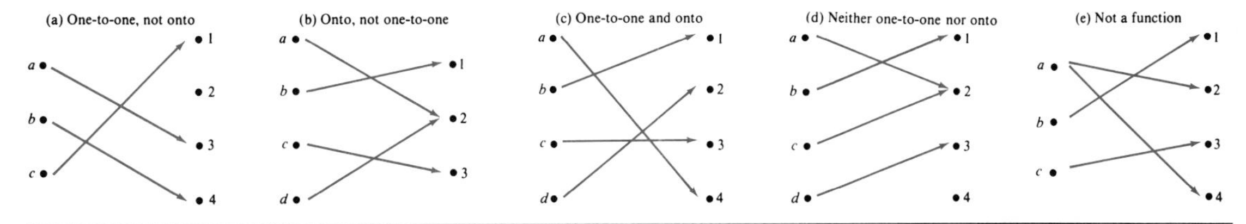

FIGURE 5 Examples of Different Types of Correspondences.

INVERSE FUNCTIONS AND COMPOSITIONS OF FUNCTIONS

Now consider a one-to-one correspondence f from the set A to the set B. Since f is an onto function, every element of B is the image of some element in A. Furthermore, because f is also a one-to-one function, every element of B is the image of a *unique* element of A. Consequently, we can define a new function from B to A that reverses the correspondence given by f. This leads to the following definition.

DEFINITION 9. Let f be a one-to-one correspondence from the set A to the set B. The *inverse function* of f is the function that assigns to an element b belonging to B the unique element a in A such that $f(a) = b$. The inverse function of f is denoted by f^{-1}. Hence, $f^{-1}(b) = a$ when $f(a) = b$.

Figure 6 illustrates the concept of a an inverse function.

If a function f is not a one-to-one correspondence, we cannot define an inverse function of f. When f is not a one-to-one correspondence, either it is not one-to-one or it is not onto. If f is not one-to-one, some element b in the codomain is the image of more than one element in the domain. If f is not onto, for some element b in the codomain, no element a in the domain exists for which $f(a) = b$. Consequently, if f is not a one-to-one correspondence, we cannot assign to each element b in the codomain a unique element a in the domain such that $f(a) = b$ (because for some b there is either more than one such a or no such a).

A one-to-one correspondence is called **invertible** since we can define an inverse of this function. A function is **not invertible** if it is not a one-to-one correspondence, since the inverse of such a function does not exist.

EXAMPLE 14 Let f be the function from $\{a, b, c\}$ to $\{1, 2, 3\}$ such that $f(a) = 2$, $f(b) = 3$, and $f(c) = 1$. Is f invertible, and if it is, what is its inverse?

Solution: The function f is invertible since it is a one-to-one correspondence. The inverse function f^{-1} reverses the correspondence given by f, so that $f^{-1}(1) = c$, $f^{-1}(2) = a$, and $f^{-1}(3) = b$. ■

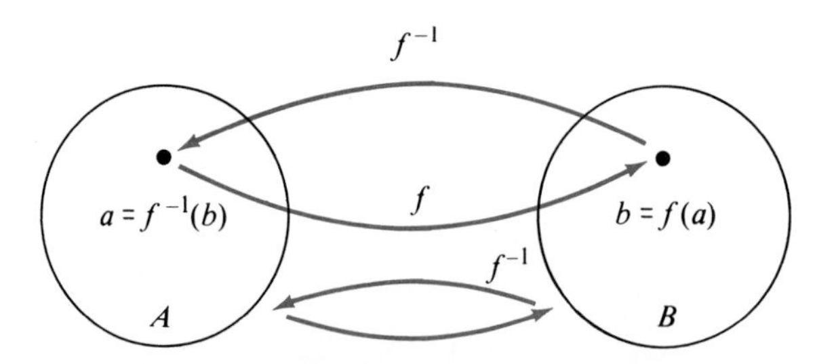

FIGURE 6 **The Function f^{-1} Is the Inverse of the Function f.**

EXAMPLE 15 Let f be the function from the set of integers to the set of integers such that $f(x) = x + 1$. Is f invertible, and if it is, what is its inverse?

Solution: The function f has an inverse since it is a one-to-one correspondence, as we have shown. To reverse the correspondence, suppose that y is the image of x, so that $y = x + 1$. Then $x = y - 1$. This means that $y - 1$ is the unique element of **Z** that is sent to y by f. Consequently, $f^{-1}(y) = y - 1$. ■

EXAMPLE 16 Let f be the function from **Z** to **Z** with $f(x) = x^2$. Is f invertible?

Solution: Since $f(-1) = f(1) = 1$, f is not one-to-one. If an inverse function were defined, it would have to assign two elements to 1. Hence, f is not invertible. ■

DEFINITION 10. Let g be a function from the set A to the set B and let f be a function from the set B to the set C. The *composition* of the functions f and g, denoted by $f \circ g$ is defined by

$$(f \circ g)(a) = f(g(a)).$$

In other words, $f \circ g$ is the function that assigns to the element a of A the element assigned by f to $g(a)$. Note that the composition $f \circ g$ cannot be defined unless the range of g is a subset of the domain of f. In Figure 7 the composition of functions is shown.

EXAMPLE 17 Let g be the function from the set $\{a, b, c\}$ to itself such that $g(a) = b$, $g(b) = c$, and $g(c) = a$. Let f be the function from the set $\{a, b, c\}$ to the set $\{1, 2, 3\}$ such that

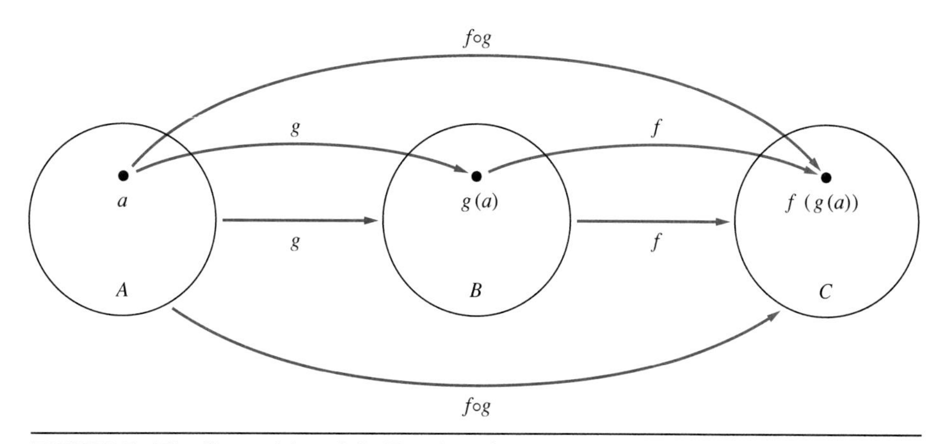

FIGURE 7 **The Composition of the Functions f and g.**

$f(a) = 3$, $f(b) = 2$, and $f(c) = 1$. What is the composition of f and g, and what is the composition of g and f?

Solution: The composition $f \circ g$ is defined by $(f \circ g)(a) = f(g(a)) = f(b) = 2$, $(f \circ g)(b) = f(g(b)) = f(c) = 1$, and $(f \circ g)(c) = f(g(c)) = f(a) = 3$.

Note that $g \circ f$ is not defined, because the range of f is not a subset of the domain of g. ■

EXAMPLE 18 Let f and g be the functions from the set of integers to the set of integers defined by $f(x) = 2x + 3$ and $g(x) = 3x + 2$. What is the composition of f and g? What is the composition of g and f?

Solution: Both the compositions $f \circ g$ and $g \circ f$ are defined. Moreover,

$$(f \circ g)(x) = f(g(x)) = f(3x + 2) = 2(3x + 2) + 3 = 6x + 7$$

and

$$(g \circ f)(x) = g(f(x)) = g(2x + 3) = 3(2x + 3) + 2 = 6x + 11$$

■

Remark: Note that even though $f \circ g$ and $g \circ f$ are defined for the functions f and g in Example 18, $f \circ g$ and $g \circ f$ are not equal. In other words, the commutative law does not hold for the composition of functions.

When the composition of a function and its inverse is formed, in either order, an identity function is obtained. To see this, suppose that f is a one-to-one correspondence from the set A to the set B. Then the inverse function f^{-1} exists and is a one-to-one correspondence from B to A. The inverse function reverses the correspondence of the original function, so that $f^{-1}(b) = a$ when $f(a) = b$, and $f(a) = b$ when $f^{-1}(b) = a$. Hence,

$$(f^{-1} \circ f)(a) = f^{-1}(f(a)) = f^{-1}(b) = a,$$

and

$$(f \circ f^{-1})(b) = f(f^{-1}(b)) = f(a) = b.$$

Consequently $f^{-1} \circ f = \iota_A$ and $f \circ f^{-1} = \iota_B$, where ι_A and ι_B are the identity functions on the sets A and B, respectively. That is, $(f^{-1})^{-1} = f$.

THE GRAPHS OF FUNCTIONS

We can associate a set of pairs in $A \times B$ to each function from A to B. This set of pairs is called the **graph** of the function and is often displayed pictorially to aid in understanding the behavior of the function.

DEFINITION 11. Let f be a function from the set A to the set B. The *graph* of the function f is the set of ordered pairs $\{(a, b) \mid a \in A \text{ and } f(a) = b\}$.

From the definition, the graph of a function f from A to B is the subset of $A \times B$ containing the ordered pairs with the second entry equal to the element of B assigned by f to the first entry.

EXAMPLE 19 Display the graph of the function $f(n) = 2n + 1$ from the set of integers to the set of integers.

Solution: The graph of f is the set of ordered pairs of the form $(n, 2n + 1)$ where n is an integer. This graph is displayed in Figure 8. ■

EXAMPLE 20 Display the graph of the function $f(x) = x^2$ from the set of integers to the set of integers.

Solution: The graph of f is the set of ordered pairs of the form $(x, f(x)) = (x, x^2)$ where x is an integer. This graph is displayed in Figure 9. ■

SOME IMPORTANT FUNCTIONS

Next, we introduce two important functions in discrete mathematics, namely, the floor and ceiling functions. Let x be a real number. The floor function rounds x down to the closest integer less than or equal to x, and the ceiling function rounds x up to the

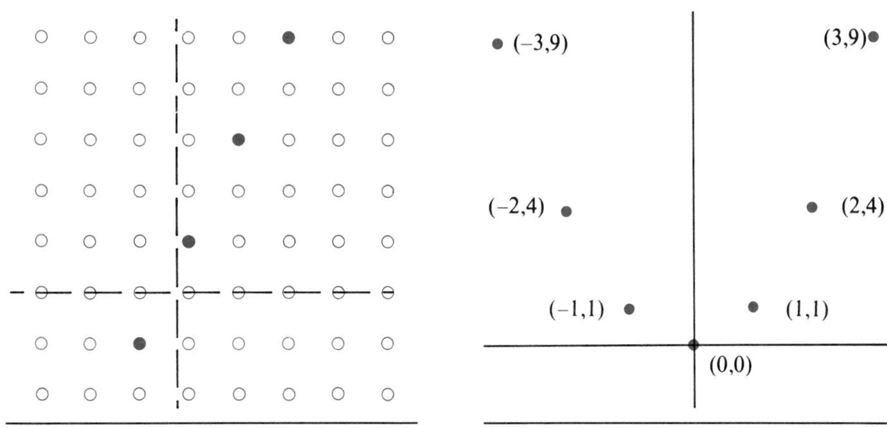

FIGURE 8 **The Graph of the Function $f(n) = 2n + 1$ from Z to Z.**

FIGURE 9 **The Graph of $f(x) = x^2$ from Z to Z.**

closest integer greater than or equal to x. These functions are often used when objects are counted. They play an important role in the analysis of the number of steps used by procedures to solve problems of a particular size.

DEFINITION 12. The *floor function* assigns to the real number x the largest integer that is less than or equal to x. The value of the floor function at x is denoted by $\lfloor x \rfloor$. The *ceiling function* assigns to the real number x the smallest integer that is greater than or equal to x. The value of the ceiling function at x is denoted by $\lceil x \rceil$.

Remark: The floor function is often also called the *greatest integer function*. It is often denoted by $[x]$.

EXAMPLE 21 We display some values of the floor and ceiling functions:

$$\lfloor \tfrac{1}{2} \rfloor = 0, \lceil \tfrac{1}{2} \rceil = 1, \lfloor -\tfrac{1}{2} \rfloor = -1, \lceil -\tfrac{1}{2} \rceil = 0, \lfloor 3.1 \rfloor = 3, \lceil 3.1 \rceil = 4, \lfloor 7 \rfloor = 7, \lceil 7 \rceil = 7.$$

■

We display the graphs of the floor and ceiling functions in Figure 10.

There are certain types of functions that will be used throughout the text. These include polynomial, logarithmic, and exponential functions. A brief review of the properties of these functions needed in this text is given in Appendix 1. In this book the notation $\log x$ will be used to denote the logarithm to the base two of x, since two is the base that we will usually use for logarithms. We will denote logarithms to the base b, where b is any real number greater than one, by $\log_b x$.

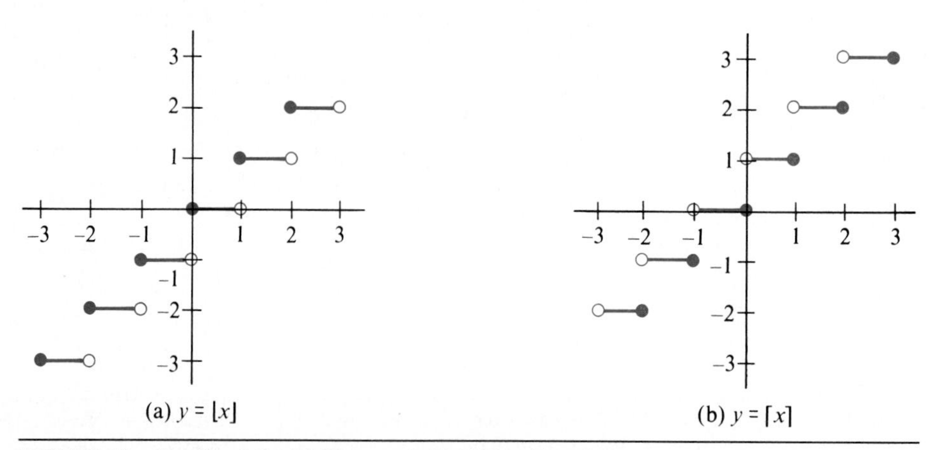

FIGURE 10 **Graphs of the (a) Floor and (b) Ceiling Functions.**

Exercises

1. Why is f not a function from **R** to **R** in the following equations?
   a) $f(x) = 1/x$
   b) $f(x) = \sqrt{x}$
   c) $f(x) = \pm\sqrt{(x^2 + 1)}$
2. Determine whether f is a function from **Z** to **R** if
   a) $f(n) = \pm n$
   b) $f(n) = \sqrt{n^2 + 1}$
   c) $f(n) = 1/(n^2 - 4)$
3. Determine whether f is a function from the set of all bit strings to the set of integers if
   a) $f(S)$ is the position of a zero bit in S.
   b) $f(S)$ is the number of one bits in S.
   c) $f(S)$ is the smallest integer i such that the ith bit of S is 1 and $f(S) = 0$ when S is the empty string, the string with no bits.
4. Find the domain and range of the following functions.
   a) the function that assigns to each nonnegative integer its last digit
   b) the function that assigns the next largest integer to a positive integer
   c) the function that assigns to a bit string the number of one bits in the string
   d) the function that assigns to a bit string the number of bits in the string
5. Find the following values.
   a) $\lceil \frac{3}{4} \rceil$
   b) $\lfloor \frac{7}{8} \rfloor$
   c) $\lceil -\frac{3}{4} \rceil$
   d) $\lfloor -\frac{7}{8} \rfloor$
   e) $\lceil 3 \rceil$
   f) $\lfloor -1 \rfloor$
6. Determine whether each of the following functions from $\{a, b, c, d\}$ to itself is one-to-one.
   a) $f(a) = b,\ f(b) = a,\ f(c) = c,\ f(d) = d$
   b) $f(a) = b,\ f(b) = b,\ f(c) = d,\ f(d) = c$
   c) $f(a) = d,\ f(b) = b,\ f(c) = c,\ f(d) = d$
7. Which functions in Exercise 6 are onto?
8. Determine whether each of the following functions from **Z** to **Z** is one-to-one.
   a) $f(n) = n - 1$
   b) $f(n) = n^2 + 1$
   c) $f(n) = n^3$
   d) $f(n) = \lceil n/2 \rceil$
9. Which functions in Exercise 8 are onto?
10. Give an example of a function from **N** to **N** that is
    a) one-to-one but not onto.
    b) onto but not one-to-one.
    c) both onto and one-to-one (but different from the identity function).
    d) neither one-to-one nor onto.
11. Determine whether each of the following functions is a bijection from **R** to **R**.
    a) $f(x) = 2x + 1$
    b) $f(x) = x^2 + 1$
    c) $f(x) = x^3$
    d) $f(x) = (x^2 + 1)/(x^2 + 2)$
12. Let $f(x) = 2x$. What is
    a) $f(\mathbf{Z})$?
    b) $f(\mathbf{N})$?
    c) $f(\mathbf{R})$?
13. Suppose that g is a function from A to B and f is a function from B to C.
    a) Show that if both f and g are one-to-one functions, then $f \circ g$ is also one-to-one.
    b) Show that if both f and g are onto functions, then $f \circ g$ is also onto.

★14. If f and $f \circ g$ are one-to-one, does it follow that g is one-to-one? Justify your answer.

★15. If f and $f \circ g$ are onto, does it follow that g is onto? Justify your answer.

16. Find $f \circ g$ and $g \circ f$ where $f(x) = x^2 + 1$ and $g(x) = x + 2$ are functions from **R** to **R**.
17. Find $f + g$ and fg for the functions f and g given in Exercise 16.
18. Let $f(x) = ax + b$ and $g(x) = cx + d$ where a,b,c, and d are constants. Determine for which constants a, b, c, and d it is true that $f \circ g = g \circ f$.
19. Show that the function $f(x) = ax + b$ from **R** to **R** is invertible, where a and b are constants, with $a \neq 0$, and find the inverse of f.
20. Let f be a function from the set A to the set B. Let S and T be subsets of A. Show that
    a) $f(S \cup T) = f(S) \cup f(T)$
    b) $f(S \cap T) \subseteq f(S) \cap f(T)$
21. Give an example to show that the inclusion in part (b) in Exercise 20 may be proper.

Let f be a function from the set A to the set B. Let S be a subset of B. We define the **inverse image** of S to be the subset of A containing all pre-images of all elements of S. We denote the inverse image of S by $f^{-1}(S)$, so that $f^{-1}(S) = \{a \in A \mid f(a) \in S\}$.

22. Let f be a function from **R** to **R** defined by $f(x) = x^2$. Find
    a) $f^{-1}(\{1\})$
    b) $f^{-1}(\{x \mid 0 < x < 1\})$
    c) $f^{-1}(\{x \mid x > 4\})$
23. Let $g(x) = \lfloor x \rfloor$. Find
    a) $g^{-1}(\{0\})$
    b) $g^{-1}(\{-1, 0, 1\})$
    c) $g^{-1}(\{x \mid 0 < x < 1\})$
24. Let f be a function from A to B. Let S and T be subsets of B. Show that
    a) $f^{-1}(S \cup T) = f^{-1}(S) \cup f^{-1}(T)$
    b) $f^{-1}(S \cap T) = f^{-1}(S) \cap f^{-1}(T)$

25. Let f be a function from A to B. Let S be a subset of B. Show that $f^{-1}(\overline{S}) = \overline{f^{-1}(S)}$.

26. Show that $\lceil x \rceil = -\lfloor -x \rfloor$.

27. Let x be a real number. Show that $\lfloor 2x \rfloor = \lfloor x \rfloor + \lfloor x + \frac{1}{2} \rfloor$.

28. Draw the graph of the function $f(n) = 1 - n^2$ from **Z** to **Z**.

29. Draw the graph of the function $f(x) = \lfloor 2x \rfloor$ from **R** to **R**.

30. Draw the graph of the function $f(x) = \lfloor x/2 \rfloor$ from **R** to **R**.

31. Find the inverse function of $f(x) = x^3 + 1$.

32. Suppose that f is an invertible function from Y to Z and g is an invertible function from X to Y. Show that the inverse of the composition $f \circ g$ is given by $(f \circ g)^{-1} = g^{-1} \circ f^{-1}$.

33. Let S be a subset of a universal set U. The **characteristic function** f_s of S is the function from U to the set $\{0, 1\}$ such that $f_S(x) = 1$ if x belongs to S and $f_S(x) = 0$ if x does not belong to S. Let A and B be sets. Show that for all x

a) $f_{A \cap B}(x) = f_A(x) \cdot f_B(x)$

b) $f_{A \cup B}(x) = f_A(x) + f_B(x) - f_A(x) \cdot f_B(x)$

c) $f_{\overline{A}}(x) = 1 - f_A(x)$

d) $f_{A \oplus B}(x) = f_A(x) + f_B(x) - 2f_A(x)f_B(x)$

34. Suppose that f is a function from A to B, where A and B are finite sets with $|A| = |B|$. Show that f is one-to-one if and only if it is onto.

A program designed to evaluate a function may not produce the correct value of the function for all elements in the domain of this function. For example, a program may not produce a correct value because evaluating the function may lead to an infinite loop or an overflow.

To study such situations, we use the concept of a partial function. A **partial function** f from a set A to a set B is an assignment to each element a in a subset of A, called the **domain of definition** of f, of a unique element b in B. The sets A and B are called the **domain** and **codomain** of f, respectively. We say that f is **undefined** for elements in A that are not in the domain of definition of f. We write $f : A \to B$ to denote that f is a partial function from A to B. (This is the same notation as is used for functions. The context in which the notation is used determines whether f is a partial function or a total function.) When the domain of definition of f equals A, we say that f is a **total function**.

35. For each of the following partial functions, determine its domain, codomain, domain of definition, and the set of values for which it is undefined. Also, determine whether it is a total function.

a) $f : \mathbf{Z} \to \mathbf{R},\ f(n) = 1/n$

b) $f : \mathbf{Z} \to \mathbf{Z},\ f(n) = \lceil n/2 \rceil$

c) $f : \mathbf{Z} \times \mathbf{Z} \to \mathbf{Q},\ f(m, n) = m/n$

d) $f : \mathbf{Z} \times \mathbf{Z} \to \mathbf{Z},\ f(m, n) = mn$

e) $f : \mathbf{Z} \times \mathbf{Z} \to \mathbf{Z},\ f(m, n) = m - n$ if $m > n$

36. a) Show that a partial function from A to B can be viewed as a function f^* from A to $B \cup \{u\}$ where u is not an element of B and

$$f^*(a) = \begin{cases} f(a) & \text{if } a \text{ belongs to the domain of definition of } f \\ u & \text{if } f \text{ is undefined at } a \end{cases}$$

b) Using the construction in (a), find the function f^* corresponding to each partial function in Exercise 35.

1.7 Sequences and Summations

INTRODUCTION

Sequences are used to represent ordered lists of elements. Sequences are used in discrete mathematics in many ways. They can be used to represent solutions to certain counting problems, as we will see in Chapter 5. They are also an important data structure in computer science. This section contains a review of the concept of a function, as well as the notation used to represent sequences and sums of terms of sequences.

When the elements of an infinite set can be listed, the set is called countable. We will conclude this section with a discussion of both countable and uncountable sets.

SEQUENCES

A sequence is a discrete structure used to represent an ordered list.

DEFINITION 1. A *sequence* is a function from a subset of the set of integers (usually either the set $\{0, 1, 2, \ldots\}$ or the set $\{1, 2, 3, \ldots\}$) to a set S. We use the notation a_n to denote the image of the integer n. We call a_n a *term* of the sequence.

We use the notation $\{a_n\}$ to describe the sequence. (Note that a_n represents an individual term of the sequence $\{a_n\}$. Also note that the notation $\{a_n\}$ for a sequence conflicts with the notation for a set. However, the context in which we use this notation will always make it clear when we are dealing with sets and when we are dealing with sequences.)

We describe sequences by listing the terms of the sequence in order of increasing subscripts.

EXAMPLE 1 Consider the sequence $\{a_n\}$, where

$$a_n = 1/n.$$

The list of the terms of this sequence, beginning with a_1, namely,

$$a_1, a_2, a_3, a_4, \ldots$$

starts with

$$1, \tfrac{1}{2}, \tfrac{1}{3}, \tfrac{1}{4}, \ldots.$$

■

EXAMPLE 2 Consider the sequence $\{b_n\}$ with $b_n = (-1)^n$. The list of the terms of this sequence, $b_0, b_1, b_2, b_3, \ldots$, begins with

$$1, -1, 1, -1, 1, \ldots.$$

■

EXAMPLE 3 Consider the sequence $c_n = 5^n$. The list of the terms of the sequence $c_0, c_1, c_2, c_3, c_4, c_5, \ldots$ begins with

$$1, 5, 25, 125, 625, 3125, \ldots.$$

■

Sequences of the form

$$a_1, a_2, \ldots, a_n$$

are often used in computer science. These finite sequences are also called **strings**. This string is also denoted by $a_1 a_2 \cdots a_n$. (Recall that bit strings, which are finite sequences of bits, were introduced in Section 1.1.) The **length** of the string S is the number of terms in this string. The **empty string** is the string that has no terms. The empty string has length zero.

EXAMPLE 4 The string *abcd* is a string of length four. ■

SUMMATIONS

Next, we introduce **summation notation**. We begin by describing the notation used to express the sum of the terms

$$a_m, a_{m+1}, \ldots, a_n$$

from the sequence $\{a_n\}$. We use the notation

$$\sum_{j=m}^{n} a_j$$

to represent

$$a_m + a_{m+1} + \cdots + a_n.$$

Here, the variable j is called the **index of summation**, and the choice of the letter j as the variable is arbitrary; that is, we could have used any other letter, such as i or k. Or, in other words,

$$\sum_{j=m}^{n} a_j = \sum_{i=m}^{n} a_i = \sum_{k=m}^{n} a_k.$$

Here, the index of summation runs through all integers starting with its **lower limit** m and ending with its **upper limit** n. The uppercase Greek letter sigma, Σ, is used to denote summation. We give some examples of summation notation.

EXAMPLE 5 Express the sum of the first 100 terms of the sequence $\{a_n\}$, where $a_n = 1/n$ for $n = 1, 2, 3, \ldots$.

Solution: The lower limit for the index of summation is 1, and the upper limit is 100. We write this sum as

$$\sum_{j=1}^{100} (1/j).$$

■

EXAMPLE 6 What is the value of $\sum_{j=1}^{5} j^2$?

Solution: We have

$$\begin{aligned}\sum_{j=1}^{5} j^2 &= 1^2 + 2^2 + 3^2 + 4^2 + 5^2 \\ &= 1 + 4 + 9 + 16 + 25 \\ &= 55.\end{aligned}$$

■

EXAMPLE 7 What is the value of $\sum_{k=4}^{8} (-1)^k$?

Solution: We have

$$\begin{aligned}\sum_{k=4}^{8} (-1)^k &= (-1)^4 + (-1)^5 + (-1)^6 + (-1)^7 + (-1)^8 \\ &= 1 + (-1) + 1 + (-1) + 1 \\ &= 1.\end{aligned}$$

■

EXAMPLE 8 A *geometric progression* is a sequence of the form

$$a, ar, ar^2, ar^3, \ldots, ar^k$$

where a, the initial term, and r, the common ratio, are real numbers. Sums of terms of geometric progressions commonly arise; such sums are called *geometric series*. We will find a formula for S, the sum of the first $n + 1$ terms of a geometric progression with initial term a and common nonzero ratio r, that is,

$$S = \sum_{j=0}^{n} ar^j.$$

To compute S, first multiply both sides of the equality by r and then manipulate the resulting sum as follows:

$$\begin{aligned}rS &= r\sum_{j=0}^{n} ar^j \\ &= \sum_{j=0}^{n} ar^{j+1} \\ &= \sum_{k=1}^{n+1} ar^k \quad \text{(This equality is obtained by shifting the index of summation, setting } k = j + 1.\text{)} \\ &= \sum_{k=0}^{n} ar^k + (ar^{n+1} - a) \\ &= S + (ar^{n+1} - a)\end{aligned}$$

From these equalities, we see that

$$rS = S + (ar^{n+1} - a).$$

Solving for S shows that if $r \neq 1$

$$S = \frac{ar^{n+1} - a}{r - 1}.$$

If $r = 1$, then clearly the sum equals $(n + 1)a$. ■

EXAMPLE 9 Double summations arise in many contexts. An example of a double summation is

$$\sum_{i=1}^{4} \sum_{j=1}^{3} ij.$$

To evaluate the double sum, first expand the inner summation and then continue by computing the outer summation:

$$\begin{aligned}\sum_{i=1}^{4} \sum_{j=1}^{3} ij &= \sum_{i=1}^{4} (i + 2i + 3i) \\ &= \sum_{i=1}^{4} 6i \\ &= 6 + 12 + 18 + 24 = 60.\end{aligned}$$ ■

We can also use summation notation to add all values of a function, or terms of an indexed set, where the index of summation runs over all values in a set. That is, we write

$$\sum_{s \in S} f(s)$$

to represent the sum of the values $f(s)$, for all members s of S.

EXAMPLE 10 What is the value of $\sum_{s \in \{0,2,4\}} s$?

Solution: Since $\sum_{s \in \{0,2,4\}} s$ represents the sum of the values of s for all the members of the set $\{0,2,4\}$, it follows that

$$\sum_{s \in \{0,2,4\}} s = 0 + 2 + 4 = 6.$$ ■

CARDINALITY (optional)

Recall that in Section 1.4, the cardinality of a finite set was defined to be the number of elements in the set. It is possible to extend the concept of cardinality to all sets, both finite and infinite, with the following definition.

DEFINITION 2. The sets A and B have the same *cardinality* if and only if there is a one-to-one correspondence from A to B.

To see that this definition agrees with the previous definition of the cardinality of a finite set as the number of elements in that set, note that there is a one-to-one correspondence between any two finite sets with n elements, where n is a nonnegative integer.

We will now split infinite sets into two groups, those with the same cardinality as the set of natural numbers, and those with different cardinality.

DEFINITION 3. A set that is either finite or has the same cardinality as the set of natural numbers is called *countable*. A set that is not countable is called *uncountable*.

We now give examples of countable and uncountable sets.

EXAMPLE 11 Show that the set of odd positive integers is a countable set.

Solution: To show that the set of odd positive integers is countable, we will exhibit a one-to-one correspondence between this set and the set of natural numbers. Consider the function

$$f(n) = 2n - 1$$

from **N** to the set of odd positive integers. We show that f is a one-to-one correspondence by showing it is both one-to-one and onto. To see that it is one-to-one, suppose that $f(n) = f(m)$. Then $2n - 1 = 2m - 1$, so that $n = m$. To see that it is onto, suppose that t is an odd positive integer. Then t is 1 less than an even integer $2k$, where k is a natural number. Hence $t = 2k - 1 = f(k)$. We display this one-to-one correspondence in Figure 1. ■

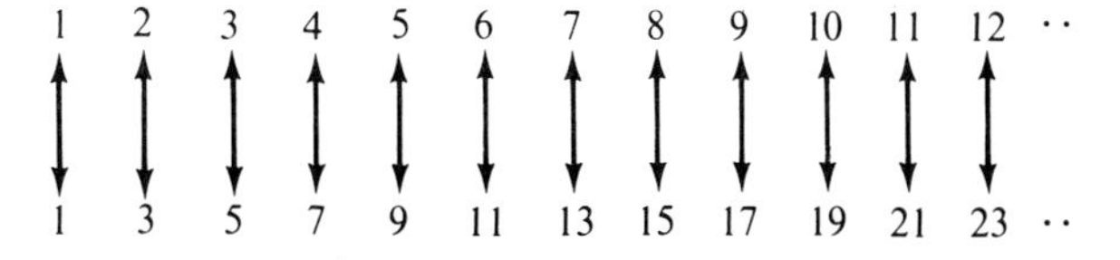

FIGURE 1 **A One-to-One Correspondence between N and the Set of Odd Positive Integers.**

An infinite set is countable if and only if it is possible to list the elements of the set in a sequence (indexed by the natural numbers). The reason for this is that a one-to-one correspondence f from the set of natural numbers to a set S can be expressed in terms of a sequence $a_1, a_2, \ldots, a_n, \ldots$ where $a_1 = f(1), a_2 = f(2), \ldots, a_n = f(n), \ldots$. For instance, the set of odd integers can be listed in a sequence $a_1, a_2, \ldots, a_n, \ldots$ where $a_n = 2n - 1$.

We now give an example of an uncountable set.

EXAMPLE 12 Show that the set of real numbers is an uncountable set.

Solution: To show that the set of real numbers is uncountable, we suppose that the set of real numbers is countable and arrive at a contradiction. Then, the subset of all real numbers that fall between 0 and 1 would also be countable (since any subset of a countable set is also countable; see Exercise 20 at the end of the section). Under this assumption, the real numbers between 0 and 1 can be listed in some order, say $r_1, r_2, r_3, \ldots$. Let the decimal representation of these real numbers be

$$\begin{aligned}
r_1 &= 0.d_{11}d_{12}d_{13}d_{14}\ldots \\
r_2 &= 0.d_{21}d_{22}d_{23}d_{24}\ldots \\
r_3 &= 0.d_{31}d_{32}d_{33}d_{34}\ldots \\
r_4 &= 0.d_{41}d_{42}d_{43}d_{44}\ldots \\
&\;\;\vdots
\end{aligned}$$

where $d_{ij} \in \{0, 1, 2, 3, 4, 5, 6, 7, 8, 9\}$. (For example, if $r_1 = 0.23794102\ldots$, we have $d_{11} = 2$, $d_{12} = 3$, $d_{13} = 7$, and so on.) Then, form a new real number with decimal expansion $r = 0.d_1d_2d_3d_4\ldots$, where the decimal digits are determined by the following rule:

$$d_i = \begin{cases} 4 \text{ if } d_{ii} \neq 4 \\ 5 \text{ if } d_{ii} = 4. \end{cases}$$

(As an example, suppose that $r_1 = 0.23794102\ldots$, $r_2 = 0.44590138\ldots$, $r_3 = 0.09118764\ldots$, $r_4 = 0.80553900\ldots$, and so on. Then we have $r = 0.d_1d_2d_3d_4\ldots = 0.4544\ldots$, where $d_1 = 4$ since $d_{11} \neq 4$, $d_2 = 5$ since $d_{22} = 4$, $d_3 = 4$ since $d_{33} \neq 4$, $d_4 = 4$ since $d_{44} \neq 4$, and so on.)

Every real number has a unique decimal expansion (when the possibility that the expansion has a tail end that consists entirely of the digit 9 is excluded). Then, the real number r is not equal to any of $r_1, r_2, \ldots,$ since the decimal expansion of r differs from the decimal expansion of r_i in the ith place to the right of the decimal point, for each i.

Since there is a real number r between 0 and 1 that is not in the list, the assumption that all the real numbers between 0 and 1 could be listed must be false. Therefore, all the real numbers between 0 and 1 cannot be listed, so that the set of real numbers between 0 and 1 is uncountable. Any set with an uncountable subset is uncountable (see Exercise 21 at the end of this section). Hence, the set of real numbers is uncountable. ■

Exercises

1. Find the following terms of the sequence $\{a_n\}$ where $a_n = 2 \cdot (-3)^n + 5^n$.

a) a_0 **b)** a_1 **c)** a_4 **d)** a_5

2. What is the term a_8 of the sequence $\{a_n\}$ if a_n equals

a) 2^{n-1}? **b)** 7?
c) $1 + (-1)^n$? **d)** $-(-2)^n$?

3. What are the terms a_0, a_1, a_2, and a_3 of the sequence $\{a_n\}$ where a_n equals

a) $2^n + 1$? **b)** $(n+1)^{n+1}$?
c) $\lfloor n/2 \rfloor$? **d)** $\lfloor n/2 \rfloor + \lceil n/2 \rceil$?

4. What are the terms a_0, a_1, a_2, and a_3 of the sequence $\{a_n\}$ where a_n equals

a) $(-2)^n$? **b)** 3?
c) $7 + 4^n$? **d)** $2^n + (-2)^n$?

5. What are the values of the following sums?

a) $\sum_{k=1}^{5}(k+1)$ **b)** $\sum_{j=0}^{4}(-2)^j$

c) $\sum_{i=1}^{10} 3$ **d)** $\sum_{j=0}^{8}(2^{j+1} - 2^j)$

6. What are the values of the following sums, where $S = \{1, 3, 5, 7\}$?

a) $\sum_{j \in S} j$ **b)** $\sum_{j \in S} j^2$

c) $\sum_{j \in S} (1/j)$ **d)** $\sum_{j \in S} 1$

7. What is the value of each of the following sums of terms of a geometric progression?

a) $\sum_{j=0}^{8} 3 \cdot 2^j$ **b)** $\sum_{j=1}^{8} 2^j$

c) $\sum_{j=2}^{8} (-3)^j$ **d)** $\sum_{j=0}^{8} 2 \cdot (-3)^j$

8. Find the value of each of the following sums.

a) $\sum_{j=0}^{8}(1 + (-1)^j)$ **b)** $\sum_{j=0}^{8}(3^j - 2^j)$

c) $\sum_{j=0}^{8}(2 \cdot 3^j + 3 \cdot 2^j)$ **d)** $\sum_{j=0}^{8}(2^{j+1} - 2^j)$

9. Compute each of the following double sums.

a) $\sum_{i=1}^{2}\sum_{j=1}^{3}(i+j)$ **b)** $\sum_{i=0}^{2}\sum_{j=0}^{3}(2i+3j)$

c) $\sum_{i=1}^{3}\sum_{j=0}^{2} i$ **d)** $\sum_{i=0}^{2}\sum_{j=1}^{3} ij$

10. Compute each of the following double sums.

a) $\sum_{i=1}^{3}\sum_{j=1}^{2}(i-j)$ **b)** $\sum_{i=0}^{3}\sum_{j=0}^{2}(3i+2j)$

c) $\sum_{i=1}^{3}\sum_{j=0}^{2} j$ **d)** $\sum_{i=0}^{2}\sum_{j=0}^{3} i^2 j^3$

11. Show that $\sum_{j=1}^{n}(a_j - a_{j-1}) = a_n - a_0$ where $a_0, a_1, \ldots, a_n$ is a sequence of real numbers. This type of sum is called **telescoping**.

12. Use the identity $1/(k(k+1)) = 1/k - 1/(k+1)$ and Exercise 11 to compute $\sum_{k=1}^{n} 1/(k(k+1))$.

13. Sum both sides of the identity $k^2 - (k-1)^2 = 2k - 1$ from $k = 1$ to $k = n$ and use Exercise 11 to find

a) a formula for $\sum_{k=1}^{n}(2k-1)$ (the sum of the first n odd natural numbers).

b) a formula for $\sum_{k=1}^{n} k$.

★**14.** Use the technique given in Exercise 11, together with the result of Exercise 13b, to find a formula for $\sum_{k=1}^{n} k^2$.

There is also a special notation for products. The product of $a_m, a_{m+1}, \ldots, a_n$ is represented by

$$\prod_{j=m}^{n} a_j.$$

15. What are the values of the following products?

a) $\prod_{i=0}^{10} i$ **b)** $\prod_{i=5}^{8} i$
c) $\prod_{i=1}^{100}(-1)^i$ **d)** $\prod_{i=1}^{10} 2$

The value of the **factorial function** at a positive integer n, denoted by $n!$, is the product of the positive integers from 1 to n, inclusive. Also, we specify that $0! = 1$.

16. Express $n!$ using product notation.

17. Find $\sum_{j=0}^{4} j!$.

18. Find $\prod_{j=0}^{4} j!$.

19. Determine whether each of the following sets is countable or uncountable. For those that are countable, exhibit a one-to-one correspondence between the set of natural numbers and that set.

a) the negative integers
b) the even integers
c) the real numbers between 0 and $\frac{1}{2}$
d) integers that are multiples of 7

20. Show that a subset of a countable set is also countable.

☞ **21.** Show that if A is an uncountable set and $A \subseteq B$, then B is uncountable.

★**22.** Show that the union of two countable sets is countable.

★★**23.** Show that the union of a countable number of countable sets is countable.

★**24.** A real number is called **rational** if it can be written as the quotient of two integers. Show that the set of rational numbers between 0 and 1 is countable. (*Hint:* List the elements of this set in order of increasing $p + q$ where p is the numerator and q is the denominator of a fraction p/q in lowest terms.)

★**25.** Show that the set of all bit strings is countable.

★**26.** Show that the set of real numbers that are solutions of quadratic equations $ax^2 + bx + c = 0$ where a, b, and c are integers, is countable.

★**27.** Show that the set of all computer programs in a particular programming language is countable. (*Hint:* A computer program written in a programming language can be thought of as a string of symbols from a finite alphabet.)

★**28.** Show that the set of functions from the positive integers to the set $\{0, 1, 2, 3, 4, 5, 6, 7, 8, 9\}$ is uncountable. (*Hint:* First set up a one-to-one correspondence between the set of real numbers between 0 and 1 and a subset of these functions. Do this by associating to the real number $0.d_1d_2 \ldots d_n \ldots$ the function f with $f(n) = d_n$.)

★**29.** We say that a function is **computable** if there is a computer program that finds the values of this function. Use Exercises 27 and 28 to show that there are functions that are not computable.

1.8 The Growth of Functions

INTRODUCTION

Suppose that a computer program reorders any list of n integers into a list where the integers are in increasing order. One important consideration concerning the practicality of this program is how long a computer takes to solve this problem. An analysis may show that the time used to reorder a list of n integers (where these integers are less than some specified size) is less than $f(n)$ microseconds, where $f(n) = 100n \log n + 25n + 9$. To analyze the practicality of the program, we need to understand how quickly this function grows as n grows. This section reviews some important methods used in estimating the growth of functions. We will introduce the notation most commonly used in the analysis of the growth of functions, namely, **big-*O*** notation. We will develop some useful results about the growth of functions using this notation.

BIG-*O* NOTATION

The growth of functions is often described using a special notation. The following definition describes this notation.

DEFINITION 1. Let f and g be functions from the set of integers or the set of real numbers to the set of real numbers. We say that $f(x) = O(g(x))$ if there are constants C and k such that

$$|f(x)| \leq C|g(x)|$$

whenever $x > k$. We also say that $f(x)$ is $O(g(x))$. (This is read as $f(x)$ is "big-oh" of $g(x)$.)

Remark: To show $f(x) = O(g(x))$, we need only find *one* pair of constants C and k such that $|f(x)| \leq C|g(x)|$ if $x > k$. However, a pair C,k that satisfies the definition is *never* unique. Moreover, if one such pair exists, there are *infinitely many* such pairs. A simple way to see this is to note that if C,k is one such pair, any pair $C'k'$ with $C < C'$ and $k < k'$ also satisfies the definition, since $|f(x)| \leq C|g(x)| \leq C'|g(x)|$ whenever $x > k' > k$.

EXAMPLE 1 Show that $f(x) = x^2 + 2x + 1$ is $O(x^2)$.

Solution: Since

$$0 \leq x^2 + 2x + 1 \leq x^2 + 2x^2 + x^2 = 4x^2$$

whenever $x > 1$, it follows that $f(x) = O(x^2)$. (To apply the definition of big-O notation here, take $C = 4$ and $k = 1$. It is not necessary to use absolute values here since all functions in these equalities are positive when x is positive.)

Another approach is to note that when $x > 2$, it follows that $2x \leq x^2$. Consequently, if $x > 2$, we see that

$$0 \leq x^2 + 2x + 1 \leq x^2 + x^2 + x^2 = 3x^2.$$

(We apply the definition with $C = 3$ and $k = 2$.)

Observe that in the relationship $f(x) = O(x^2)$, x^2 can be replaced by any function with larger values than x^2, for example, $f(x) = O(x^3)$, $f(x) = O(x^2+2x+7)$, and so on. It is also true that x^2 is $O(x^2 + 2x + 1)$, since $x^2 < x^2 + 2x + 1$ whenever $x \geq 1$.

Figure 1 illustrates the relationship $x^2 + 2x + 1 = O(x^2)$. ■

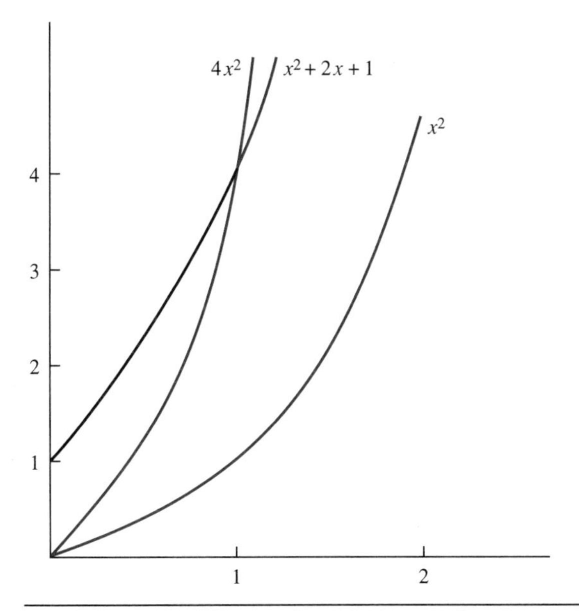

FIGURE 1 The Function $x^2 + 2x + 1 = O(x^2)$.

Note that in Example 1 we have two functions, $f(x) = x^2 + 2x + 1$ and $g(x) = x^2$, such that $f(x) = O(g(x))$ and $g(x) = O(f(x))$, the latter fact following from the inequality $x^2 \leq x^2 + 2x + 1$, which holds for all nonnegative real numbers x. We say that two functions $f(x)$ and $g(x)$ that satisfy both of these big-O relationships are of the *same order*. (See Exercises 20 and 21.)

Remark: Note that the equals sign in the notation $f(x) = O(g(x))$ does *not* represent a genuine equality. Rather, this notation tells us that an inequality holds relating the values of the functions f and g for sufficiently large numbers in the domains of these functions.

Big-O notation has been used in mathematics for almost a century. In computer science it is widely used in the analysis of algorithms, as will be seen in Chapter 2. The German mathematician Paul Bachmann first introduced big-O notation in 1892 in an important book on number theory. The big-O symbol is sometimes called a **Landau symbol** after the German mathematician Edmund Landau, who used this notation throughout his work.

When $f(x) = O(g(x))$, and $h(x)$ is a function that has larger absolute values than $g(x)$ does for sufficiently large values of x, it follows that $f(x) = O(h(x))$. In other words, the function $g(x)$ in the relationship $f(x) = O(g(x))$ can be replaced by a function with larger absolute values. To see this, note that if

$$|f(x)| \leq C|g(x)| \qquad \text{if } x > k,$$

and if $|h(x)| > |g(x)|$ for all $x > k$, then

$$|f(x)| \leq C|h(x)| \qquad \text{if } x > k.$$

Hence, $f(x) = O(h(x))$.

Paul Gustav Heinrich Bachmann (1837–1920). Paul Bachmann, the the son of a Lutheran pastor, shared his father's pious lifestyle and love of music. His mathematical talent was discovered by one of his teachers, even though he had difficulties with some of his early mathematical studies. After recuperating from tuberculosis in Switzerland, Bachmann studied mathematics, first at the University of Berlin and later at Göttingen, where he attended lectures presented by the famous number theorist Dirichlet. He received his doctorate under the German number theorist Kummer in 1862; his thesis was on group theory. Bachmann was a professor at Breslau and later at Münster. After he retired from his professorship, he continued his mathematical writing, played the piano, and served as a music critic for newspapers. Bachmann's mathematical writings include a five-volume survey of results and methods in number theory, a two-volume work on elementary number theory, a book on irrational numbers, and a book on the famous conjecture known as Fermat's Last Theorem. He introduced big-O notation in his 1892 book *Analytische Zahlentheorie*.

Edmund Landau (1877–1938). Edmund Landau, the son of a Berlin gynecologist, attended high school and university in Berlin. He received his doctorate in 1899, under the direction of Frobenius. Landau first taught at the University of Berlin and then moved to Göttingen, where he was a full professor until the Nazis forced him to stop teaching. Landau's main contributions to mathematics were in the field of analytic number theory. In particular, he established several important results concerning the distribution of primes. He is the author of a three-volume exposition on number theory, as well as other books on number theory and mathematical analysis.

When big-O notation is used, the function g in the relationship $f(x) = O(g(x))$ is chosen to be as small as possible (sometimes from a set of reference functions, such as functions of the form x^n where n is a positive integer).

In subsequent discussions, we will almost always deal with functions that take on only positive values. All references to absolute values can be dropped when working with big-O estimates for such functions. Figure 2 illustrates the relationship $f(x) = O(g(x))$.

The following example illustrates how big-O notation is used to estimate the growth of functions.

EXAMPLE 2 Show that $7x^2$ is $O(x^3)$.

Solution: The inequality $7x^2 < x^3$ holds whenever $x > 7$. (We see this by dividing both sides of this inequality by x^2.) Hence, $7x^2$ is $O(x^3)$, taking $C = 1$ and $k = 7$ in the definition of big-O notation. ■

EXAMPLE 3 Example 2 shows that $7x^2$ is $O(x^3)$. Is it also true that x^3 is $O(7x^2)$?

Solution: To determine whether x^3 is $O(7x^2)$, it is necessary to determine whether there are constants C and k such that $x^3 \leq C(7x^2)$ whenever $x > k$. This inequality is equivalent to the inequality $x < 7C$, which is obtained by dividing both sides by x^2. No such C can exist since x can be made arbitrarily large. Hence x^3 is *not* $O(7x^2)$. ■

We now give some examples involving functions that have the set of positive integers as their domains.

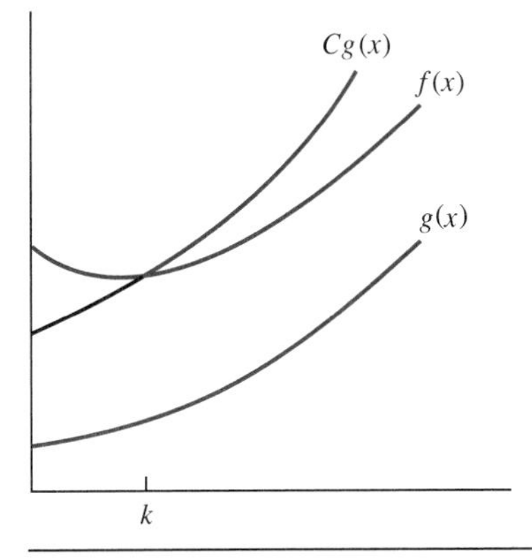

FIGURE 2 The Function $f(x) = O(g(x))$.

EXAMPLE 4 How can big-O notation be used to estimate the sum of the first n positive integers?

Solution: Since each of the integers in the sum of the first n positive integers does not exceed n, it follows that

$$1 + 2 + \cdots + n \leq n + n + \cdots + n = n^2.$$

From this inequality it follows that $1 + 2 + 3 + \cdots + n = O(n^2)$, taking $C = 1$ and $k = 1$ in the definition of big-O notation. (In this example the domains of the functions in the big-O relationship are the set of positive integers.) ■

In the next example big-O estimates will be developed for the factorial function and its logarithm. These estimates will be important in the analysis of the number of steps used in sorting procedures.

EXAMPLE 5 Give big-O estimates for the factorial function and the logarithm of the factorial function, where the factorial function $f(n) = n!$ is defined by

$$n! = 1 \cdot 2 \cdot 3 \cdot \cdots \cdot n$$

whenever n is a positive integer, and $0! = 1$. For example,

$$1! = 1, \qquad 2! = 1 \cdot 2 = 2, \qquad 3! = 1 \cdot 2 \cdot 3 = 6,$$
$$4! = 1 \cdot 2 \cdot 3 \cdot 4 = 24.$$

Note that the function $n!$ grows rapidly. For instance,

$$20! = 2{,}432{,}902{,}008{,}176{,}640{,}000.$$

Solution: A big-O estimate for $n!$ can be obtained by noting that each term in the product does not exceed n. Hence,

$$\begin{aligned} n! &= 1 \cdot 2 \cdot 3 \cdot \cdots \cdot n \\ &\leq n \cdot n \cdot n \cdot \cdots \cdot n \\ &= n^n. \end{aligned}$$

This inequality shows that $n! = O(n^n)$. Taking logarithms of both sides of the inequality established for $n!$, we obtain

$$\log n! \leq \log n^n = n \log n.$$

This implies that

$$\log n! = O(n \log n).$$ ■

EXAMPLE 6 In Section 3.2 we will show that

$$n < 2^n$$

whenever n is a positive integer. Using this inequality we can conclude that

$$n = O(2^n).$$

(Take $k = C = 1$ in the definition of big-O notation.) Since the logarithm function is increasing, taking logarithms (base 2) of both sides of this inequality shows that

$$\log n < n.$$

It follows that

$$\log n = O(n).$$

(Again we take $C = k = 1$ in the definition of big-O notation.)

If we have logarithms to a base b, where b is different from 2, we still have $\log_b n = O(n)$ since

$$\log_b n = \frac{\log n}{\log_b 2} < \frac{n}{\log_b 2}$$

whenever n is a positive integer. (We have used Theorem 3 in Appendix 1 to see that $\log_b n = \log n / \log_b 2$.) ■

THE GROWTH OF COMBINATIONS OF FUNCTIONS

Many algorithms are made up of two or more separate subprocedures. The number of steps used by a computer to solve a problem with input of a specified size using such an algorithm is the sum of the number of steps used by these subprocedures. To give a big-O estimate for the number of steps needed, it is necessary to find big-O estimates for the number of steps used by each subprocedure and then combine these estimates.

Big-O estimates of combinations of functions can be provided if care is taken when different big-O estimates are combined. In particular, it is often necessary to estimate the growth of the sum and the product of two functions. What can be said if big-O estimates for each of two functions are known? To see what sort of estimates hold for the sum and the product of two functions, suppose that $f_1(x) = O(g_1(x))$ and $f_2(x) = O(g_2(x))$.

From the definition of big-O notation, there are constants C_1, C_2, k_1, and k_2 such that

$$|f_1(x)| \leq C_1|g_1(x)|$$

when $x > k_1$, and

$$|f_2(x)| \leq C_2|g_2(x)|$$

when $x > k_2$. To estimate the sum of $f_1(x)$ and $f_2(x)$, note that

$$\begin{aligned}|(f_1 + f_2)(x)| &= |f_1(x) + f_2(x)| \\ &\le |f_1(x) + f_2(x)| \quad \text{(using the triangle inequality } |a + b| \le |a| + |b|\text{)}.\end{aligned}$$

When x is greater that both k_1 and k_2, it follows from the inequalities for $|f_1(x)|$ and $|f_2(x)|$ that

$$\begin{aligned}|f_1(x)| + |f_2(x)| &< C_1|g_1(x)| + C_2|g_2(x)| \\ &\le C_1|g(x)| + C_2|g(x)| \\ &= (C_1 + C_2)|g(x)| \\ &= C|g(x)|,\end{aligned}$$

where $C = C_1 + C_2$ and $g(x) = \max(|g_1(x)|, |g_2(x)|)$. (Here $\max(a, b)$ denotes the maximum, or larger of a and b.)

This inequality shows that $|(f_1 + f_2)(x)| \le C|g(x)|$ whenever $x > k$, where $k = \max(k_1, k_2)$. We state this useful result as the following theorem.

THEOREM 1 Suppose that $f_1(x) = O(g_1(x))$ and $f_2(x) = O(g_2(x))$. Then

$$(f_1 + f_2)(x) = O(\max(g_1(x), g_2(x))).$$

We often have big-O estimates for f_1 and f_2 in terms of the same function g. In this situation, Theorem 1 can be used to show that $(f_1 + f_2)(x)$ is also $O(g(x))$, since $\max(g(x), g(x)) = g(x)$. This result is stated in the following corollary.

COROLLARY 1 Suppose that $f_1(x)$ and $f_2(x)$ are both $O(g(x))$. Then

$$(f_1 + f_2)(x) = O(g(x)).$$

In a similar way big-O estimates can be derived for the product of the functions f_1 and f_2. When x is greater than $\max(k_1, k_2)$ it follows that

$$\begin{aligned}|(f_1 f_2)(x)| &= |f_1(x)||f_2(x)| \\ &\le C_1|g_1(x)|C_2|g_2(x)| \\ &\le C_1C_2|(g_1g_2)(x)| \\ &\le C|(g_1g_2)(x)|,\end{aligned}$$

where $C = C_1C_2$. From this inequality, it follows that $f_1(x)f_2(x)$ is $O(g_1g_2)$, since there are constants C and k, namely, $C = C_1C_2$ and $k = \max(k_1, k_2)$, such that $|(f_1f_2)(x)| \le C|g_1(x)g_2(x)|$ whenever $x > k$. This result is stated in the following theorem.

THEOREM 2 Suppose that $f_1(x) = O(g_1(x))$ and $f_2(x) = O(g_2(x))$. Then

$$(f_1f_2)(x) = O(g_1(x)g_2(x)).$$

The goal in using big-O notation to estimate functions is to choose a function $g(x)$ that grows relatively slowly with $f(x) = O(g(x))$. The following examples illustrate how to use Theorems 1 and 2 to do this. The type of analysis given in these examples is often used in the analysis of the time used to solve problems using computer programs.

EXAMPLE 7 Give a big-O estimate for $f(n) = 3n \log(n!) + (n^2 + 3) \log n$ where n is a positive integer.

Solution: First, the product $3n \log(n!)$ will be estimated. From Example 5 we know that $\log(n!)$ is $O(n \log n)$. Using this estimate, and the fact that $3n$ is $O(n)$, Theorem 2 gives the estimate

$$3n \log(n!) = O(n^2 \log n).$$

Next, the product $(n^2 + 3) \log n$ will be estimated. Since $(n^2 + 3) < 2n^2$ when $n > 2$, it follows that $n^2 + 3$ is $O(n^2)$. Thus, from Theorem 2 it follows that $(n^2 + 3) \log n = O(n^2 \log n)$. Using Theorem 1 to combine the two big-O estimates for the products show that

$$f(n) = 3n \log(n!) + n^2 \log n = O(n^2 \log n).$$ ■

EXAMPLE 8 Give a big-O estimate for $f(x) = (x + 1) \log (x^2 + 1) + 3x^2$.

Solution: First, a big-O estimate for $(x + 1) \log (x^2 + 1)$ will be found. Note that $(x + 1)$ is $O(x)$. Furthermore, $x^2 + 1 \leq 2x^2$ when $x > 1$. Hence,

$$\log(x^2 + 1) \leq \log(2x^2) = \log 2 + \log x^2 = \log 2 + 2 \log x \leq 3 \log x,$$

if $x > 2$. This shows that $\log (x^2 + 1)$ is $O(\log x)$.

From Theorem 2 it follows that

$$(x + 1) \log(x^2 + 1) = O(x \log x).$$

Since $3x^2$ is $O(x^2)$, Theorem 1 tells us that

$$\begin{aligned} f(x) &= O(\max(x \log x, x^2)) \\ &= O(x^2), \end{aligned}$$

where we have used the inequality $x \log x \leq x^2$, for $x > 1$. ■

Polynomials can often be used to estimate the growth of functions. Instead of analyzing the growth of polynomials each time they occur, we would like a result that can always be used to estimate the growth of a polynomial. The following theorem does this. It shows that the leading term of a polynomial dominates its growth by asserting that a polynomial of degree n or less is $O(x^n)$.

THEOREM 3 Let $f(x) = a_n x^n + a_{n-1} x^{n-1} + \cdots + a_1 x + a_0$, where $a_0, a_1, \ldots, a_{n-1}, a_n$ are real numbers. Then $f(x) = O(x^n)$.

Proof: Using the triangle inequality, if $x > 1$ we have

$$\begin{aligned} |f(x)| &= |a_n x^n + a_{n-1} x^{n-1} + \cdots + a_1 x + a_0| \\ &\leq |a_n| x^n + |a_{n-1}| x^{n-1} + \cdots + |a_1| x + |a_0| \\ &= x^n (|a_n| + |a_{n-1}|/x + \cdots + |a_1|/x^{n-1} + |a_0|/x^n) \\ &\leq x^n (|a_n| + |a_{n-1}| + \cdots + |a_1| + |a_0|) \end{aligned}$$

when $x > 1$. This shows that

$$|f(x)| < Cx^n$$

where $C = |a_n| + |a_{n-1}| + \cdots + |a_0|$ whenever $x > 1$. Hence, $f(x) = O(x^n)$. ■

As mentioned before, big-O notation is used to estimate the number of operations needed to solve a problem using a specified procedure or algorithm. The functions used in these estimates often include the following:

$$1, \log n, n, n \log n, n^2, 2^n, n!$$

Using calculus it can be shown that each function in the list is smaller than the succeeding function, in the sense that the ratio of a function and the succeeding function tends to zero as n grows without bound. Figure 3 displays the graphs of these functions, using a scale for the values of the functions that doubles for each successive marking on the graph.

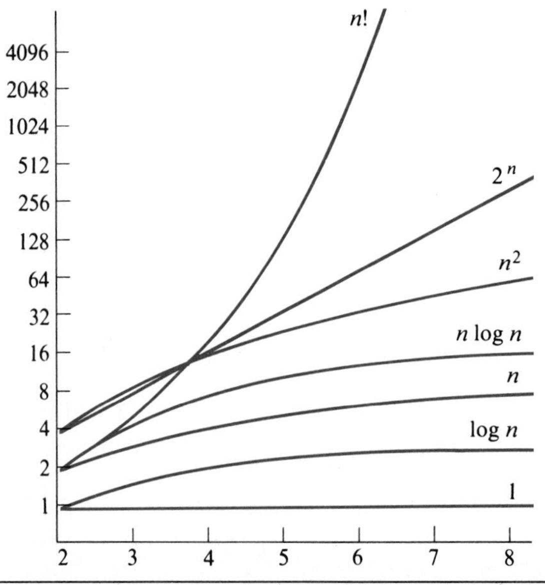

FIGURE 3 A Display of the Growth of Functions Commonly Used in Big-O Estimates

Exercises

1. Determine whether each of the following functions is $O(x)$.

a) $f(x) = 10$ **b)** $f(x) = 3x + 7$
c) $f(x) = x^2 + x + 1$ **d)** $f(x) = 5 \log x$
e) $f(x) = \lfloor x \rfloor$ **f)** $f(x) = \lceil x/2 \rceil$

2. Determine whether each of the following functions is $O(x^2)$.

a) $f(x) = 17x + 11$ **b)** $f(x) = x^2 + 1000$
c) $f(x) = x \log x$ **d)** $f(x) = x^4/2$
e) $f(x) = 2^x$ **f)** $f(x) = \lfloor x \rfloor \cdot \lceil x \rceil$

3. Use the definition of $f(x) = O(g(x))$ to show that $x^4 + 9x^3 + 4x + 7$ is $O(x^4)$.

4. Use the definition of $f(x) = O(g(x))$ to show that $2^x + 17$ is $O(3^x)$.

5. Show that $(x^2 + 1)/(x + 1)$ is $O(x)$.

6. Show that $(x^3 + 2x)/(2x + 1)$ is $O(x^2)$.

7. Show that $x^2 + 4x + 17$ is $O(x^3)$ but that x^3 is not $O(x^2 + 4x + 17)$.

8. Show that x^3 is $O(x^4)$ but that x^4 is not $O(x^3)$.

9. Show that $3x^4 + 1$ is $O(x^4/2)$ and $x^4/2$ is $O(3x^4 + 1)$.

10. Show that $x \log x$ is $O(x^2)$ but that x^2 is not $O(x \log x)$.

11. Show that 2^n is $O(3^n)$ but that 3^n is not $O(2^n)$.

12. Is $x^3 = O(g(x))$, if g is the given function? (For example, if $g(x) = x + 1$, this question asks whether $x^3 = O(x + 1)$.)

a) $g(x) = x^2$ **b)** $g(x) = x^3$
c) $g(x) = x^2 + x^3$ **d)** $g(x) = x^2 + x^4$
e) $g(x) = 3^x$ **f)** $g(x) = x^3/2$

13. Explain what it means for a function to be $O(1)$.

14. Show that if $f(x) = O(x)$, then $f(x) = O(x^2)$.

15. Suppose that $f(x)$, $g(x)$, and $h(x)$ are functions such that $f(x)$ is $O(g(x))$ and $g(x)$ is $O(h(x))$. Show that $f(x)$ is $O(h(x))$.

16. Let k be a positive integer. Show that $1^k + 2^k + \cdots + n^k = O(n^{k+1})$.

17. Give as good a big-O estimate as possible for each of the following functions.

a) $(n^2 + 8)(n + 1)$
b) $(n \log n + n^2)(n^3 + 2)$
c) $(n! + 2^n)(n^3 + \log(n^2 + 1))$

18. Give a big-O estimate for each of the following functions. For the function g in your estimate $f = O(g)$ use a simple function g of smallest order.

a) $(n^3 + n^2 \log n)(\log n + 1) + (17 \log n + 19)(n^3 + 2)$
b) $(2^n + n^2)(n^3 + 3^n)$
c) $(n^n + n2^n + 5^n)(n! + 5^n)$

19. Give a big-O estimate for each of the following functions. For the function g in your estimate $f = O(g)$ use a simple function g of the smallest order.

a) $n \log(n^2 + 1) + n^2 \log n$
b) $(n \log n + 1)^2 + (\log n + 1)(n^2 + 1)$
c) $n^{2^n} + n^{n^2}$

Let $f(x)$ and $g(x)$ be functions from the set of real numbers or the set of positive integers to the set of real numbers. We write $f(x) = \Theta(g(x))$ when there are positive real numbers C_1 and C_2 and a positive integer k such that

$$C_1|g(x)| \leq |f(x)| \leq C_2|g(x)|$$

whenever $x > k$.

20. a) Show that $3x + 7$ is $\Theta(x)$.
b) Show that $2x^2 + x - 7$ is $\Theta(x^2)$.
c) Show that $\lfloor x + 1/2 \rfloor$ is $\Theta(x)$.
d) Show that $\log(x^2 + 1)$ is $\Theta(\log_2 x)$.
e) Show that $\log_{10} x$ is $\Theta(\log_2 x)$.

21. Show that $f(x) = \Theta(g(x))$ if and only if $f(x) = O(g(x))$ and $g(x) = O(f(x))$.

22. a) Show that $3x^2 + x + 1 = \Theta(3x^2)$.
b) Express the relationship in part (a) using a picture showing the functions $3x^2 + x + 1$, $C_1 \cdot 3x^2$, and $C_2 \cdot 3x^2$, and the constant k on the x-axis, where C_1, C_2 and k are the constants you found in part (a) to show that $3x^2 + x + 1 = \Theta(3x^2)$.

23. Express the relationship $f(x) = \Theta(g(x))$ using a picture. Show the graphs of the functions $f(x)$, $C_1|g(x)|$, and $C_2|g(x)|$, as well as the constant k on the x-axis.

24. Give a big-O estimate of the product of the first n odd positive integers.

25. Show that if f and g are real-valued functions such that $f(x) = O(g(x))$, then $f^k(x) = O(g^k(x))$. (Note that $f^k(x) = f(x)^k$.)

26. Show that if $f(x)$ is $O(\log_b x)$ where $b > 1$, then $f(x)$ is $O(\log_a x)$ where $a > 1$.

27. Suppose that $f(x)$ is $O(g(x))$ where f and g are increasing and unbounded functions. Show that $\log |f(x)|$ is $O(\log |g(x)|)$.

28. Suppose that $f(x)$ is $O(g(x))$. Does it follow that $2^{f(x)} = O(2^{g(x)})$?

The following problems deal with another type of asymptotic notation, called **little-*o*** notation. Because little-*o* notation is based on the concept of limits, a knowledge of calculus is needed for these problems. We say that $f(x) = o(g(x))$, or $f(x)$ is $o(g(x))$ (read $f(x)$ is "little-oh" of $g(x)$), when

$$\lim_{x\to\infty} \frac{f(x)}{g(x)} = 0.$$

29. (Calculus required) Show that
a) x^2 is $o(x^3)$ **b)** $x \log x$ is $o(x^2)$
c) x^2 is $o(2^x)$ **d)** $x^2 + x + 1$ is not $o(x^2)$

30. (Calculus required)
a) Show that if $f(x)$ and $g(x)$ are functions such that $f(x) = o(g(x))$ and c is a constant, then $cf(x) = o(g(x))$ where $(cf)(x) = cf(x)$.
b) Show that if $f_1(x)$, $f_2(x)$, and $g(x)$ are functions such that $f_1(x) = o(g(x))$ and $f_2(x) = o(g(x))$, then $(f_1 + f_2)(x) = o(g(x))$, where $(f_1 + f_2)(x) = f_1(x) + f_2(x)$.

31. (Calculus required) Represent pictorially that $x \log x$ is $o(x^2)$ by graphing $x \log x$, x^2, and $x \log x / x^2$. Explain how this picture shows that $x \log x = o(x^2)$.

32. (Calculus required) Express the relationship $f(x) = o(g(x))$ using a picture. Show the graphs of $f(x)$, $g(x)$, and $f(x)/g(x)$.

★33. (Calculus required) Suppose that $f(x) = o(g(x))$. Does it follow that $2^{f(x)} = o(2^{g(x)})$?

★34. (Calculus required) Suppose that $f(x) = o(g(x))$. Does it follow that $\log|f(x)| = o(\log|g(x)|)$?

35. (Calculus required) The two parts of this problem describe the relationship between little-*o* and big-*O* notation.
a) Show that if $f(x)$ and $g(x)$ are functions such that $f(x) = o(g(x))$, then $f(x) = O(g(x))$.
b) Show that if $f(x)$ and $g(x)$ are functions such that $f(x) = O(g(x))$, then it does not necessarily follow that $f(x) = o(g(x))$.

36. (Calculus required) Show that if $f(x)$ is a polynomial of degree n and $g(x)$ is a polynomial of degree m where $m > n$, then $f(x) = o(g(x))$.

37. (Calculus required) Show that if $f_1(x) = O(g(x))$ and $f_2(x) = o(g(x))$, then $f_1(x) + f_2(x) = O(g(x))$.

38. (Calculus required) Let H_n be the nth **harmonic number**

$$H_n = 1 + \frac{1}{2} + \frac{1}{3} + \cdots + \frac{1}{n}.$$

Show that $H_n = O(\log n)$. (*Hint:* First establish the inequality

$$\sum_{j=2}^{n} \frac{1}{j} < \int_1^n \frac{1}{x}dx$$

by showing that the sum of the areas of the rectangles of height $1/j$ with base from $j-1$ to j, for $j = 2, 3, \ldots, n$ is less than the area under the curve $y = 1/x$ from 2 to n.)

★39. Show that $n \log n$ is $O(\log n!)$.

40. Determine whether $\log(n!) = O(n \log n)$. Justify your answer.

Key Terms and Results

LOGIC (SECTIONS 1–3):

TERMS

proposition: a statement that is true or false
truth value: true or false
$\neg$p (negation of p): the proposition with truth value opposite to the truth value of p
logical operators: operators used to combine propositions
compound proposition: a proposition constructed by combining propositions using logical operators
truth table: a table displaying the truth values of propositions
$p \vee q$ (disjunction of p and q): the proposition that is true unless both p and q are false

$p \wedge q$ (conjunction of p and q): the proposition that is true only when both p and q are true
$p \oplus q$ (exclusive or of p and q): the proposition that is true when exactly one of p and q is true
$p \rightarrow q$ (p implies q): the proposition that is false only when p is true and q is false
converse of $p \rightarrow q$: the implication $q \rightarrow p$
contrapositive of $p \rightarrow q$: the implication $\neg q \rightarrow \neg p$
$p \leftrightarrow q$ (biconditional): the proposition that is true only when p and q have the same truth value
bit: either a zero or a one
Boolean variable: a variable that has a value of zero or one
bit operation: an operation on a bit or bits
bit string: a list of bits
bitwise operations: operations on bit strings that operate on each bit in one string and the corresponding bit in the other string
tautology: a compound proposition that is always true
contradiction: a compound proposition that is always false
contingency: a compound proposition that is sometimes true and sometimes false
logical equivalence: compound propositions are logically equivalent if they always have the same truth values
propositional function: the combination of a variable and a predicate
universe of discourse: the domain of the variable in a propositional function
$\exists x P(x)$ (existential quantification of $P(x)$): the proposition that is true if and only if there exists an x in the universe of discourse such that $P(x)$ is true
$\forall x P(x)$ (universal quantification of $P(x)$): the proposition that is true if and only if $P(x)$ is true for all x in the universe of discourse
free variable: a variable not bound in a propositional function

RESULTS

The logical equivalences given in Tables 5 and 6 in Section 2

SETS (SECTIONS 4–5):

TERMS

set: a collection of distinct objects
axiom: a basic assumption of a theory
paradox: a logical inconsistency
element, member of a set: an object in a set
$\emptyset$ (empty set, null set): the set with no members
universal set: the set containing all objects under consideration
Venn diagram: a graphical representation of a set or sets
$S = T$ (set equality): S and T have the same elements
$S \subseteq T$ (S is a subset of T): every element of S is also an element of T
$S \subset T$ (S is a proper subset of T): S is a subset of T and $S \neq T$
finite set: a set with n elements where n is a nonnegative integer
infinite set: a set that is not finite
$|S|$ (the cardinality of S): the number of elements in S
$P(S)$ (the power set of S): the set of all subsets of S

$A \cup B$ (the union of A and B): the set containing those elements that are in at least one of A and B

$A \cap B$ (the intersection of A and B): the set containing those elements that are in both A and B.

$A - B$ (the difference of A and B): the set containing those elements that are in A but not in B

$\overline{A}$ (the complement of A): the set of elements in the universal set that are not in A

$A \oplus B$ (the symmetric difference of A and B): the set containing those elements in exactly one of A and B.

membership table: a table displaying the membership of elements in sets

RESULTS

The set identities given in Table 1 in Section 5.

FUNCTIONS (SECTIONS 6–8):

TERMS

function from A to B: an assignment of an element of B to each element of A

domain of f: the set A where f is a function from A to B

codomain of f: the set B where f is a function from A to B

b is the image of a under f: $b = f(a)$

a is a pre-image of b under f: $f(a) = b$

range of f: the set of images of f

onto function, surjection: a function from A to B such that every element of B is the image of some element in A

one-to-one function, injection: a function such that the images of elements in its domain are all different

one-to-one correspondence, bijection: a function that is both one-to-one and onto

inverse of f: the function that reverses the correspondence given by f (when f is a bijection)

$f \circ g$ (composition of f and g): the function that assigns $f(g(x))$ to x

$\lfloor x \rfloor$ (floor function): the largest integer not exceeding x

$\lceil x \rceil$ (ceiling function): the smallest integer greater than or equal to x

sequence: a function with domain that is a subset of the set of integers

string: a finite sequence

$\sum_{i=1}^{n} a_i$: the sum $a_1 + a_2 + \cdots + a_n$

$\prod_{i=1}^{n} a_i$: the product $a_1 a_2 \cdots a_n$

countable set: a set that is either finite or that can be placed in one-to-one correspondence with the set of positive integers

uncountable set: a set that is not countable

RESULTS

The set of real numbers is uncountable.

$\log n! = O(n \log n)$

If $f_1(x) = O(g_1(x))$ and $f_2(x) = O(g_2(x))$, then $(f_1 + f_2)(x) = O(\max(g_1(x), g_2(x)))$ and $(f_1 f_2)(x) = O(g_1(x) g_2(x))$.

If $a_0, a_1, \ldots, a_n$ are real numbers, then $a_n x^n + a_{n-1} x^{n-1} + \cdots + a_1 x + a_0$ is $O(x^n)$.

Review Questions

1. a) Define the negation of a proposition.
   b) What is the negation of "This is a boring course"?
2. a) Define (using truth tables) the disjunction, conjunction, exclusive or, conditional, and biconditional of the propositions p and q.
   b) What are the disjunction, conjunction, exclusive or, implication, and biconditional of the propositions "I'll go to the movies tonight" and "I'll finish my discrete mathematics homework"?
3. a) Describe at least five different ways to write the implication $p \rightarrow q$ in English.
   b) Define the converse and contrapositive of an implication.
   c) State the converse and the contrapositive of the implication "If it is sunny tomorrow, then I will go for a walk in the woods."
4. a) What does it mean for two propositions to be logically equivalent?
   b) Describe the different ways to show that two compound propositions are logically equivalent.
   c) Show in at least two different ways that the compound propositions $\neg p \vee (r \rightarrow \neg q)$ and $\neg p \vee \neg q \vee \neg r$ are equivalent.
5. *(Depends on the Exercise Set in Section 1.2)*
   a) Given a truth table, explain how to use disjunctive normal form to construct a compound proposition with this truth table.
   b) Explain why part (a) shows that the operators $\wedge, \vee$, and $\neg$ are functionally complete.
   c) Is there an operator such that the set containing just this operator is functionally complete?
6. What are the universal and existential quantifications of a predicate $P(x)$? What are their negations?
7. a) What is the difference between the quantification $\exists x \forall y P(x, y)$ and $\forall y \exists x P(x, y)$ where $P(x, y)$ is a predicate?
   b) Give an example of a predicate $P(x, y)$ such that $\exists x \forall y P(x, y)$ and $\forall y \exists x P(x, y)$ have different truth values.
8. a) Define the union, intersection, difference, and symmetric difference of two sets.
   b) What are the union, intersection, difference, and symmetric difference of the set of positive integers and the set of odd integers?
9. a) Define what it means for two sets to be equal.
   b) Describe the ways to show that two sets are equal.
   c) Show in at least two different ways that the sets $A - (B \cap C)$ and $(A - B) \cup (A - C)$ are equal.
10. Explain the relationship between logical equivalences and set identities.
11. a) Define $|S|$, the cardinality of the set S.
    b) Give a formula for $|A \cup B|$ where A and B are sets.
12. a) Define the power set of a set S.
    b) When is the empty set in the power set of a set S?
    c) How many elements does the power set of a set S with n elements have?
13. a) Define the domain, codomain, and the range of a function.
    b) Let $f(n)$ be the function from the set of integers to the set of integers such that $f(n) = n^2 + 1$. What is the domain, codomain, and range of this function?
14. a) Define what it means for a function from the set of positive integers to the set of positive integers to be one-to-one.
    b) Define what it means for a function from the set of positive integers to the set of positive integers to be onto.
    c) Give an example of a function from the set of positive integers to the set of positive integers that is both one-to-one and onto.
    d) Give an example of a function from the set of positive integers to the set of positive integers that is one-to-one but not onto.
    e) Give an example of a function from the set of positive integers to the set of positive integers that is not one-to-one but is onto.
    f) Give an example of a function from the set of positive integers to the set of positive integers that is neither one-to-one nor onto.
15. a) Define the inverse of a function.
    b) When does a function have an inverse?
    c) Does the function $f(n) = 10 - n$ from the set of integers to the set of integers have an inverse? If so, what is it?
16. a) Define the floor and ceiling functions from the set of real numbers to the set of integers.
    b) For which real numbers x is it true that $\lfloor x \rfloor = \lceil x \rceil$?
17. a) Use summation notation to express the sum of the powers of 2 from 2^0 to 2^n.
    b) What is the value of the sum in part (a)?

18. **a)** What does it mean for a set to be countable? Give a precise definition.
b) Is the set of negative integers countable? Why or why not?
c) Is the set of rational numbers with denominators greater than 3 countable? Why or why not?
d) Is the set of real numbers between 2 and 3 countable? Why or why not?

19. **a)** State the definition of the statement $f(n) = O(g(n))$ where $f(n)$ and $g(n)$ are functions from the set of positive integers to the set of real numbers.
b) Use the definition of $f(n) = O(g(n))$ directly to prove or disprove that $n^2 + 18n + 107 = O(n^3)$.
c) Use the definition of $f(n) = O(g(n))$ directly to prove or disprove that $n^3 = O(n^2 + 18n + 107)$.

20. **a)** How can you produce a big-O estimate for a function that is the sum of different terms where each term is the product of several functions?
b) Give a big-O estimate for the function $f(n) = (n! + 1)(2^n + 1) + (n^{n-2} + 8n^{n-3})(n^3 + 2^n)$. For the function g in your estimate $f = O(g)$ use a simple function of smallest possible order.

Supplementary Exercises

1. Let p be the proposition "I will do every exercise in this book" and q be the proposition "I will get an 'A' in this course." Express each of the following as a combination of p and q.
a) I will get an "A" in this couse only if I do every exercise in this book.
b) I will get an "A" in this course and I will do every exercise in this book.
c) Either I will not get an "A" in this course or I will not do every exercise in this book.
d) For me to get an "A" in this course it is necessary and sufficient that I do every exercise in this book.

2. Find the truth table of the compound proposition $(p \vee q) \rightarrow (p \wedge \neg r)$.

3. Show that the following propositions are tautologies.
a) $(\neg q \wedge (p \rightarrow q)) \rightarrow \neg p$
b) $((p \vee q) \wedge \neg p) \rightarrow q$

4. Give the converse and the contrapositive of the following implications.
a) If it rains today, then I will drive to work.
b) If $|x| = x$, then $x \geq 0$.
c) If n is greater than 3, then n^2 is greater than 9.

5. Find a compound proposition involving the propositional variables p, q, r, and s that is true when exactly three of these propositional variables are true and is false otherwise.

6. Let $P(x)$ be the statement "student x knows calculus" and let $Q(y)$ be the statement "class y contains a student who knows calculus." Express each of the following as quantifications of $P(x)$ and $Q(y)$.
a) Some students know calculus.
b) Not every student knows calculus.
c) Every class has a student in it who knows calculus.
d) Every student in every class knows calculus.
e) There is at least one class with no students who know calculus.

7. Let $P(m, n)$ be the statement "m divides n" where the universe of discourse for both variables is the set of positive integers. Determine the truth values of each of the following propositions.
a) $P(4, 5)$
b) $P(2, 4)$
c) $\forall m \forall n P(m, n)$
d) $\exists m \forall n P(m, n)$
e) $\exists n \forall m P(m, n)$
f) $\forall n P(1, n)$

8. Let $P(x,y)$ be a propositional function. Show that the implication $\exists x \forall y P(x, y) \rightarrow \forall y \exists x P(x, y)$ is a tautology.

9. Let $P(x)$ and $Q(x)$ be propositional functions. Show that $\exists x (P(x) \rightarrow Q(x))$ and $\forall x P(x) \rightarrow \exists x Q(x)$ always have the same truth value.

10. If $\forall y \exists x P(x, y)$ is true, does it necessarily follow that $\exists x \forall y P(x, y)$ is true?

11. If $\forall x \exists y P(x, y)$ is true, does it necessarily follow that $\exists x \forall y P(x, y)$ is true?

12. Find the negations of the following statements.
a) If it snows today, then I will go skiing tomorrow.
b) Every person in this class understands mathematical induction.

c) Some students in this class do not like discrete mathematics.

d) In every mathematics class there is some student who falls asleep during lectures.

13. Express the following statement using quantifiers: "Every student in this class has taken some course in every department in the school of mathematical sciences."

14. Express the following statement using quantifiers: "There is a building on the campus of some college in the United States in which every room is painted white."

15. Let A be the set of English words that contain the letter x and let B be the set of English words that contain the letter q. Express each of the following sets as a combination of A and B.

a) The set of English words that do not contain the letter x.

b) The set of English words that contain both an x and a q.

c) The set of English words that contain an x but not a q.

d) The set of English words that do not contain either an x or a q.

e) The set of English words that contain an x or a q, but not both.

16. Show that if A is a subset of B, then the power set of A is a subset of the power set of B.

17. Suppose that A and B are sets such that the power set of A is a subset of the power set of B. Does it follow that A is a subset of B?

18. Let **E** denote the set of even integers and **O** denote the set of odd integers. As usual, let **Z** denote the set of all integers. Determine each of the following.

a) $\mathbf{E} \cup \mathbf{O}$ **b)** $\mathbf{E} \cap \mathbf{O}$

c) $\mathbf{Z} - \mathbf{E}$ **d)** $\mathbf{Z} - \mathbf{O}$

19. Show that if A is a set and U is the universal set, then

a) $A \cap \overline{A} = \varnothing$ **b)** $A \cup \overline{A} = U$

20. Show that if A and B are sets, then

a) $A = A \cap (A \cup B)$

b) $A = A \cup (A \cap B)$

21. Show that if A and B are sets, then $A - (A - B) = A \cap B$.

22. Let A and B be sets. Show that $A \subseteq B$ if and only if $A \cap B = A$.

23. Let A, B, and C be sets. Show that $(A - B) - C$ is not necessarily equal to $A - (B - C)$.

24. Suppose that A, B, and C are sets. Prove or disprove that $(A - B) - C = (A - C) - B$.

25. Suppose that A, B, C, and D are sets. Prove or disprove that $(A - B) - (C - D) = (A - C) - (B - D)$.

26. Show that if A and B are finite sets, then $|A \cap B| \leq |A \cup B|$. Determine when this relationship is an equality.

27. Let A and B be sets in a finite universal set U. List the following in order of increasing size.

a) $|A|, |A \cup B|, |A \cap B|, |U|, |\varnothing|$

b) $|A - B|, |A \oplus B|, |A| + |B|, |A \cup B|, |\varnothing|$

28. Let A and B be subsets of the finite universal set U. Show that $|\overline{A} \cap \overline{B}| = |U| - |A| - |B| + |A \cap B|$.

29. Let f and g be functions from $\{1, 2, 3, 4\}$ to $\{a, b, c, d\}$ and from $\{a, b, c, d\}$ to $\{1, 2, 3, 4\}$, respectively, such that $f(1) = d$, $f(2) = c$, $f(3) = a$, $f(4) = b$, and $g(a) = 2$, $g(b) = 1$, $g(c) = 3$, $g(d) = 2$.

a) Is f one-to-one? Is g one-to-one?

b) Is f onto? Is g onto?

c) Does either f or g have an inverse? If so, find this inverse.

30. Let f be a one-to-one function from the set A to the set B. Let S and T be subsets of A. Show that $f(S \cap T) = f(S) \cap f(T)$.

31. Give an example to show that the equality in Exercise 30 may not hold if f is not one-to-one.

32. Show that $\lceil x + 1 \rceil = \lceil x \rceil + 1$ whenever x is a real number.

33. Find the value of the following quantities.

a) $\sum_{i=0}^{3} \left(\sum_{j=0}^{4} ij \right)$ **b)** $\prod_{j=1}^{4} \left(\sum_{i=0}^{3} j \right)$

c) $\sum_{i=1}^{5} \left(\sum_{j=0}^{i} 1 \right)$ **d)** $\prod_{i=1}^{3} \left(\prod_{j=0}^{i} j \right)$

34. Is the set of irrational numbers between 0 and 1 countable? Justify your answer.

★★35. A real number is called **algebraic** if it is the root of a polynomial with integer coefficients. Show that there are a countable number of algebraic numbers. (*Hint:* Use the fact that a polynomial of degree n has at most n distinct roots.)

36. Show that $8x^3 + 12x + 100 \log x$ is $O(x^3)$.

37. Give a big-O estimate for $(x^2 + x(\log x)^3)(2^x + x^3)$.

38. Find a big-O estimate for $\sum_{j=1}^{n} j(j + 1)$.

★39. Show that $n!$ is not $O(2^n)$.

★40. Show that n^n is not $O(n!)$.

Computer Projects

WRITE PROGRAMS WITH THE SPECIFIED INPUT AND OUTPUT.

1. Given the truth values of the propositions p and q, find the truth values of the conjunction, disjunction, exclusive or, implication, and biconditional of these propositions.
2. Given two bit strings of length n, find the bitwise *AND*, bitwise *OR*, and bitwise *XOR* of these strings.
3. Given the truth values of the propositions p and q in fuzzy logic, find the truth value of the disjunction and the conjunction of p and q (see Exercises 21–23 of Section 1.1).
4. Given subsets A and B of a set with n elements, use bit strings to find $\overline{A}$, $A \cup B$, $A \cap B$, $A - B$, and $A \oplus B$.
5. Given multisets A and B from the same universal set, find $A \cup B$, $A \cap B$, $A - B$, and $A + B$ (see preamble to Exercise 43 of Section 1.5).
6. Given fuzzy sets A and B, find $\overline{A}$, $A \cup B$, and $A \cap B$ (see preamble to Exercise 45 of Section 1.5).
7. Given a function f from $\{1, 2, \ldots, n\}$ to the set of integers, determine whether f is one-to-one.
8. Given a function f from $\{1, 2, \ldots, n\}$ to itself, determine whether f is onto.
9. Given a bijection f from the set $\{1, 2, \ldots, n\}$ to itself, find f^{-1}.
10. Given the terms of a sequence $a_1, a_2, \ldots, a_n$, find $\sum_{j=1}^{n} a_j$ and $\prod_{j=1}^{n} a_j$.

2

The Fundamentals: Algorithms, the Integers, and Matrices

Many problems can be solved by considering them as special cases of general problems. For instance, consider the problem of locating the largest integer in the sequence 101, 12, 144, 212, 98. This is a specific case of the problem of locating the largest integer in a sequence of integers. To solve this general problem we must give an algorithm, which specifies a sequence of steps used to solve this general problem. We will study algorithms for solving many different types of problems in this book. For instance, algorithms will be developed for finding the greatest common divisor of two integers, for generating all the orderings of a finite set, for searching a list, and for finding the shortest path between two vertices in a network. One important consideration concerning an algorithm is its computational complexity. That is, what are the computer resources needed to use this algorithm to solve a problem of a specified size? We will illustrate the analysis of the complexity of algorithms in this chapter.

The set of integers plays a fundamental role in discrete mathematics. In particular, the concept of division of integers is fundamental to computer arithmetic. We will briefly review some of the important concepts of number theory, the study of integers and their properties. Some important algorithms involving integers will be studied, including the Euclidean algorithm for computing greatest common divisors, which was first described thousands of years ago. Integers can be represented using any positive integer greater than one as a base. Binary expansions, which are used throughout computer science, are representations with two as the base. In this chapter we discuss base b representations of integers and give an algorithm for finding them. Algorithms for integer arithmetic, which were the first procedures called algorithms, will also be discussed. This chapter also introduces several important applications of number theory. For example, in this chapter we will use number theory to make messages secret, to generate pseudorandom numbers, and to assign memory locations to computer files.

Matrices are used in discrete mathematics to represent a variety of discrete structures. We review the basic material about matrices and matrix arithmetic needed to represent relations and graphs. Matrix arithmetic will be used in numerous algorithms involving these structures.

2.1 Algorithms

INTRODUCTION

There are many general classes of problems that arise in discrete mathematics. For instance: given a sequence of integers, find the largest one; given a set, list all of its subsets; given a set of integers, put them in increasing order; given a network, find the shortest path between two vertices. When presented with such a problem, the first thing to do is to construct a model that translates the problem into a mathematical context. Discrete structures used in such models include sets, sequences, and functions—structures discussed in Chapter 1—as well as such other structures as permutations, relations, graphs, trees, networks, and finite state machines—concepts that will be discussed in later chapters.

Setting up the appropriate mathematical model is only part of the solution. To complete the solution, a method is needed that will solve the general problem using the model. Ideally, what is required is a procedure that follows a sequence of steps that leads to the desired answer. Such a sequence of steps is called an **algorithm**.

DEFINITION 1. An *algorithm* is a definite procedure for solving a problem using a finite number of steps.

The term *algorithm* is a corruption of the name *al-Khowarizmi*, an Arabian mathematician of the ninth century, whose book on Hindu numerals is the basis of modern decimal notation. Originally, the word *algorism* was used for the rules for performing arithmetic using decimal notation. *Algorism* evolved into the word *algorithm* by the 18th century. With the growing interest in computing machines, the concept of an algorithm was given a more general meaning, to include all definite procedures for solving problems, not just the procedures for performing arithmetic. (We will discuss algorithms for performing arithmetic with integers in Section 2.4.)

In this book, we will discuss algorithms that solve a wide variety of problems. In this section we will use the problem of finding the largest integer in a finite sequence

---

Abu Ja'far Mohammed ibn Musa al-Khowarizmi (born before 800, died after 847). al-Khowarizmi, an astronomer and mathematician, was a member of the House of Wisdom, an academy of scientists in Baghdad. The name al-Khowarizmi means "from the town of Kowarzizm," which is now called Khiva and is part of Uzbekistan in the USSR. al-Khowarizmi wrote books on mathematics, astronomy, and geography. Western Europeans first learned about algebra from his works. The word *algebra* comes from al-jabr, part of the title of his book *Kitab al-jabr w'al muquabala*. This book was translated into Latin and was a widely used textbook. His book on the use of Hindu numerals describes procedures for arithmetic operations using these numerals. European authors used a Latin corruption of his name, which later evolved to the word algorithm to describe the subject of arithmetic with Hindu numerals.

of integers to illustrate the concept of an algorithm and the properties algorithms have. Also, we will describe algorithms for locating a particular element in a finite set. In subsequent sections, procedures for finding the greatest common divisor of two integers, for finding the shortest path between two points in a network, for multiplying matrices, and so on, will be discussed.

EXAMPLE 1 Describe an algorithm for finding the largest element in a finite sequence of integers.

Even though the problem of finding the maximum element in a sequence is relatively trivial, it provides a good illustration of the concept of an algorithm. Also, there are many instances where the largest integer in a finite sequence of integers is required. For instance, a university may need to find the highest score on a competitive exam taken by thousands of students. Or a sports organization may want to identify the member with the highest rating each month. We want to develop an algorithm that can be used whenever the problem of finding the largest element in a finite sequence of integers arises.

We can specify a procedure for solving this problem in several ways. One method is simply to use the English language to describe the sequence of steps used. We now provide such a solution.

Solution of Example 1: We perform the following steps.

1. Set the temporary maximum equal to the first integer in the sequence. (The temporary maximum will be the largest integer examined at any stage of the procedure.)
2. Compare the next integer in the sequence to the temporary maximum, and if it is larger than the temporary maximum, set the temporary maximum equal to this integer.
3. Repeat the previous step if there are more integers in the sequence.
4. Stop when there are no integers left in the sequence. The temporary maximum at this point is the largest integer in the sequence. ■

An algorithm can also be described using a computer language. However, when this is done, only those instructions permitted in this language can be used. This often leads to a description of the algorithm that is complicated and difficult to understand. Furthermore, since many programming languages are in common use, it would be undesirable to choose one particular language. So, instead of using a particular computer language to specify algorithms, a form of **pseudocode** will be used in this book. (All algorithms will also be described using the English language.) Pseudocode provides an intermediate step between an English language description of an algorithm and an implementation of this algorithm in a programming language. The steps of the algorithm are specified using instructions resembling those used in programming languages. However, in pseudocode, the instructions used can include any well-defined operations or statements. A computer program can be produced in any computer language using the pseudocode description as a starting point.

The pseudocode used in this book is loosely based on the programming language Pascal. However, the syntax of Pascal, or that of other programming languages, will not be followed. Furthermore, any well-defined instruction can be used in this pseudocode. The details of the pseudocode used in the text are given in Appendix 2. The reader should refer to this appendix whenever the need arises.

A pseudocode description of the algorithm for finding the maximum element in a finite sequence follows.

ALGORITHM 1 Finding the Maximum Element in a Finite Sequence.

procedure $max(a_1, a_2, \ldots, a_n$: integers)
$max := a_1$
for $i := 2$ **to** n
  if $max < a_i$ **then** $max := a_i$
{*max* is the largest element}

This algorithm first assigns the initial term of the sequence, a_1, to the variable *max*. The "for" loop is used to successively examine terms of the sequence. If a term is greater than the current value of *max*, it is assigned to be the new value of *max*.

There are several properties that algorithms generally share. They are useful to keep in mind when algorithms are described. These properties are:

- *Input*. An algorithm has input values from a specified set.
- *Output*. From each set of input values an algorithm produces output values from a specified set. The output values comprise the solution to the problem.
- *Definiteness*. The steps of an algorithm must be defined precisely.
- *Finiteness*. An algorithm should produce the desired output after a finite (but perhaps large) number of steps for any input in the set.
- *Effectiveness*. It must be possible to perform each step of an algorithm exactly and in a finite amount of time.
- *Generality*. The procedure should be applicable for all problems of the desired form, not just for a particular set of input values.

EXAMPLE 2 Show that Algorithm 1 for finding the maximum element in a finite sequence of integers has all the properties listed.

Solution: The input to Algorithm 1 is a sequence of integers. The output is the largest integer in the sequence. Each step of the algorithm is precisely defined, since only assignments, a finite loop, and conditional statements occur. The algorithm uses a finite number of steps, since it terminates after all the integers in the sequence have been examined. The algorithm can be carried out in a finite amount of time since each step is either a comparison or an assignment. Finally, Algorithm 1 is general, since it can be used to find the maximum of any finite sequence of integers. ■

SEARCHING ALGORITHMS

The problem of locating an element in an ordered list occurs in many contexts. For instance, a program that checks the spelling of words looks for words in a dictionary, which is just an ordered list of words. Problems of this kind are called **searching problems.** We will discuss several algorithms for searching in this section. We will study the number of steps used by each of these algorithms in Section 2.2.

The general searching problem can be described as follows: Locate an element x in a list of distinct elements $a_1, a_2, \ldots, a_n$, or determine that it is not in the list. The solution to this search problem is the location of the term in the list that equals x (that is, i is the solution if $x = a_i$) and is 0 if x is not in the list.

The first algorithm that we will present is called the **linear search** or **sequential search** algorithm. The linear search algorithm begins by comparing x and a_1. When $x = a_1$, the solution is the location of a_1, namely 1. When $x \neq a_1$, compare x with a_2. If $x = a_2$, the solution is the location of a_2, namely 2. When $x \neq a_2$, compare x with a_3. Continue this process, comparing x successively with each term of the list until a match is found, where the solution is the location of that term, unless no match occurs. If the entire list has been searched without locating x, the solution is 0. The pseudocode for the linear search algorithm is displayed as Algorithm 2.

ALGORITHM 2 The Linear Search Algorithm.

procedure *linear search*(x: integer, $a_1, a_2, \ldots, a_n$: distinct integers)
$i := 1$
while ($i \leq n$ and $x \neq a_i$)
  $i := i + 1$
if $i \leq n$ **then** *location* $:= i$
else *location* $:= 0$
{*location* is the subscript of term that equals x, or is 0 if x is not found}

We will now consider another searching algorithm. This algorithm can be used when the list has terms occurring in order of increasing size (for instance, if the terms are numbers, they are listed from smallest to largest; if they are words, they are listed in lexicographic, or alphabetic, order). This second algorithm is called the **binary search algorithm.** It proceeds by comparing the element to be located to the middle term of the list. The list is then split into two smaller sublists of the same size, or where one of these smaller lists has one fewer term than the other. The search continues by restricting the search to the appropriate sublist based on the comparison of the element to be located and the middle term. In the next section, it will be shown that the binary search algorithm is much more efficient than the linear search algorithm. The following example demonstrates how a binary search works.

EXAMPLE 3 To search for 19 in the list

1 2 3 5 6 7 8 10 12 13 15 16 18 19 20 22,

first split this list, which has 16 terms, into two smaller lists with eight terms each, namely,

1 2 3 5 6 7 8 10   12 13 15 16 18 19 20 22.

Then, compare 19 and the largest term in the first list. Since $10 < 19$, the search for 19 can be restricted to the list containing the 9th through the 16th terms of the original list. Next, split this list, which has eight terms, into the two smaller lists of four terms each, namely,

12 13 15 16   18 19 20 22.

Since $16 < 19$ (comparing 19 with the largest term of the first list) the search is restricted to the second of these lists, which contains the 13th through the 16th terms of the original list. The list 18 19 20 22 is split into two lists, namely,

18 19   20 22.

Since 19 is not greater than the largest term of the first of these two lists, which is also 19, the search is restricted to the first list: 18 19, which contains the 13th and 14th terms of the original list. Next, this list of two terms is split into two lists of one term each: 18 and 19. Since $18 < 19$, the search is restricted to the second list: the list containing the 14th term of the list: 19. Now that the search has been narrowed down to one term, a comparison is made, and 19 is located as the 14th term in the original list. ■

We now specify the steps of the binary search algorithm. To search for the integer x in the list $a_1, a_2, \ldots, a_n$, where $a_1 < a_2 < \cdots < a_n$, begin by comparing x with the middle term of the sequence, a_m, where $m = \lfloor (n+1)/2 \rfloor$. (Recall that $\lfloor x \rfloor$ is the greatest integer not exceeding x.) If $x > a_m$, the search for x can be restricted to the second half of the sequence, which is $a_{m+1}, a_{m+2}, \ldots, a_n$. If x is not greater than a_m, the search for x can be restricted to the first half of the sequence, which is $a_1, a_2, \ldots, a_m$.

The search has now been restricted to a list with no more than $\lceil n/2 \rceil$ elements. Using the same procedure, compare x to the middle term of the restricted list. Then restrict the search to the first or second half of the list. Repeat this process until a list with one term is obtained. Then determine whether this term is x. Pseudocode for the binary search algorithm is displayed as Algorithm 3.

ALGORITHM 3 The Binary Search Algorithm

```
procedure binary search(x: integer, a_1, a_2, ..., a_n: increasing integers)
i := 1  {i is left endpoint of search interval}
j := n  {j is right endpoint of search interval}
while i < j
begin
    m := ⌊(i + j)/2⌋
    if x > a_m then i := m + 1
    else j := m
end
if x = a_i then location := i
else location := 0
{location is the subscript of term equal to x, or 0 if x is not found}
```

Algorithm 3 proceeds by successively narrowing down the part of the sequence being searched. At any given stage only the terms beginning with a_i and ending with a_j are under consideration. In other words, i and j are the smallest and largest subscripts of the remaining terms, respectively. Algorithm 3 continues narrowing the part of the sequence being searched until only one term of the sequence remains. When this is done, a comparison is made to see whether this term equals x.

Exercises

1. List all the steps used by Algorithm 1 to find the maximum of the list 1, 8, 12, 9, 11, 2, 14, 5, 10, 4.

2. Determine which characteristics of an algorithm the following procedures have and which they lack.

a)
```
procedure double(n: positive integer)
while n > 0
    n := 2n
```

b)
```
procedure divide(n: positive integer)
while n ≥ 0
begin
    m := 1/n
    n := n − 1
end
```

c)
```
procedure sum(n: positive integer)
sum := 0
while i < 10
    sum := sum + i
```

d)
```
procedure choose(a,b: integers)
x := either a or b
```

3. Devise an algorithm that finds the sum of all the integers in a list.

4. Devise an algorithm to compute x^n, where x is a real number and n is an integer. (*Hint:* First give a procedure for computing x^n when n is nonnegative by successive multiplication by x, starting with 1. Then extend this procedure, and use the fact that $x^{-n} = 1/x^n$ to compute x^n when n is negative.)

5. Describe an algorithm that interchanges the values of the variables x and y, using only assignments. What is the minimum number of assignment statements needed to do this?

6. Describe an algorithm that uses only assignment statements that replaces the triple (x, y, z) with (y, z, x). What is the minimum number of assignment statements needed?

7. List all the steps used to search for 9 in the sequence 1, 3, 4, 5, 6, 8, 9, 11 using

a) a linear search. **b)** a binary search.

8. List all the steps used to search for 7 in the sequence given in Exercise 7.

9. Describe an algorithm that inserts an integer x in the appropriate position into the list $a_1, a_2, \ldots, a_n$ of integers that are in increasing order.

10. Describe an algorithm for finding the smallest integer in a finite sequence of natural numbers.

11. Describe an algorithm that locates the first occurrence of the largest element in a list of integers, where the integers in the list are not necessarily distinct.

12. Describe an algorithm that locates the last occurrence of the smallest element in a list of integers, where the integers in the list are not necessarily distinct.

13. Describe an algorithm that produces the maximum, median, mean, and minimum of a set of three integers. (The **median** of a set of integers is the middle element in the list when these integers are listed in order of increasing size. The **mean** of a set of integers is the sum of the integers divided by the number of integers in the set.)

14. Describe an algorithm for finding both the largest and the smallest integers in a finite sequence of integers.

15. Describe an algorithm that puts the first three terms of a sequence of integers of arbitrary length in increasing order.

16. Describe an algorithm to find the longest word in an English sentence (where a word is a string of letters and a sentence is a list of words, separated by blanks).

17. Describe an algorithm that determines whether a function from a finite set to another finite set is onto.

18. Describe an algorithm that determines whether a function from a finite set to another finite set is one-to-one.

19. Describe an algorithm that will count the number of ones in a bit string by examining each bit of the string to determine whether it is a one bit.

20. Change Algorithm 3 so that the binary search procedure compares x to a_m at each stage of the algorithm, with the algorithm terminating if $x = a_m$. What advantage does this version of the algorithm have?

21. The **ternary search algorithm** locates an element in a list of increasing integers by successively splitting the list into three sublists of equal (or as close to equal as possible) size, and restricting the search to the appropriate piece. Specify the steps of this algorithm.

22. Specify the steps of an algorithm that locates an element in a list of increasing integers by successively splitting the list into four sublists of equal (or as close to equal as possible) size, and restricting the search to the appropriate piece.

23. A **mode** of a list of integers is an element that occurs at least as often as each of the other elements. Devise an algorithm that finds a mode in a list of nondecreasing integers.

24. Devise an algorithm that finds all modes (defined in Exercise 23) in a list of nondecreasing integers.

2.2 Complexity of Algorithms

INTRODUCTION

When does an algorithm provide a satisfactory solution to a problem? First, it must always produce the correct answer. How this can be demonstrated will be discussed in Chapter 3. Second, it should be efficient. The efficiency of algorithms will be discussed in this section.

How can the efficiency of an algorithm be analyzed? One measure of efficiency is the time used by a computer to solve the problem using this algorithm, when the input values are of a specified size. A second measure is the amount of computer memory required to implement the algorithm when the input values are of a specified size.

Questions such as these involve the **computational complexity** of the algorithm. An analysis of the time required to solve a problem of a particular size involves the **time complexity** of the algorithm. An analysis of the computer memory required involves the **space complexity** of the algorithm. Considerations of the time and space complexity of an algorithm are essential when algorithms are implemented. It is obviously important to know whether an algorithm will produce an answer in a microsecond, a minute, or a billion years. Likewise, the required memory must be available to solve a problem, so that space complexity must be taken into account.

Considerations of space complexity are tied in with the particular data structures used to implement the algorithm. Because data structures are not dealt with in detail in this book, space complexity will not be considered. We will restrict our attention to time complexity.

The time complexity of an algorithm can be expressed in terms of the number of operations used by the algorithm when the input has a particular size. The operations used to measure the time complexity of algorithms may be the comparison of integers, the addition of integers, the multiplication of integers, the division of integers, or any other basic operation.

Time complexity is described in terms of the number of operations required instead of actual computer time because of the difference in time needed for different computers to perform basic operations. Moreover, it is quite complicated to break all operations down to the basic bit operations that a computer uses. Furthermore, the fastest computers in existence can perform basic bit operations (for instance, adding, multiplying, comparing, or exchanging two bits) in 10^{-9} seconds (1 nanosecond), but personal computers may require 10^{-6} seconds (1 microsecond), which is 1000 times as long, to do the same operations.

We illustrate how to analyze the time complexity of an algorithm by considering Algorithm 1 of Section 2.1, which finds the maximum of a finite set of integers.

EXAMPLE 1 Describe the time complexity of Algorithm 1 of Section 2.1 for finding the maximum element in a set.

Solution: The number of comparisons will be used as the measure of the time complexity of the algorithm, since comparisons are the basic operations used.

To find the maximum element of a set with n elements, listed in an arbitrary order, the temporary maximum is first set equal to the initial term in the list. Then, after a comparison has been done to determine that the end of the list has not yet been reached, the temporary maximum and second term are compared, updating the temporary maximum to the value of the second term if it is larger. This procedure is continued, using two additional comparisons for each term of the list—one to determine that the end of the list has not been reached and another to determine whether to update the temporary maximum. Since two comparisons are used for each of the second through the nth elements and one more comparison is used to exit the loop when $i = n + 1$, exactly $2(n - 1) + 1 = 2n - 1$ comparisons are used whenever this algorithm is applied. Hence, the algorithm for finding the maximum of a set of n elements has time complexity

$$O(n),$$

measured in terms of the number of comparisons used. ■

Next, the time complexity of searching algorithms will be analyzed.

EXAMPLE 2 Describe the time complexity of the linear search algorithm.

Solution: The number of comparisons used by the algorithm will be taken as the measure of the time complexity. At each step of the loop in the algorithm, two comparisons are performed—one to see whether the end of the list has been reached and one to compare the element x with a term of the list. Finally, one more comparison is made outside the loop. Consequently, if $x = a_i$, $2i + 1$ comparisons are used. The most comparisons, $2n$, are required when the element is not in the list. In this case, $2n$ comparisons are used to determine that x is not a_i, for $i = 1, 2, \ldots n$, an additional comparison is used to exit the loop, and one comparison is made outside the loop. So when x is not in the list, a total of $2n + 2$ comparisons are used. Hence, a linear search requires at most $O(n)$ comparisons. ■

The type of complexity analysis done in Example 2 is a **worst-case** analysis. By the worst-case performance of an algorithm, we mean the largest number of operations needed to solve the given problem using this algorithm on input of specified size. Worst-case analysis tells us how many operations an algorithm requires to guarantee that it will produce a solution.

EXAMPLE 3 Describe the time complexity of the binary search algorithm.

Solution: For simplicity, assume there are $n = 2^k$ elements in the list $a_1, a_2, \ldots, a_n$ where k is a nonnegative integer. Note that $k = \log n$. (If n, the number of elements in the list, is not a power of 2, the list can be considered part of a larger list with 2^{k+1} elements, where $2^k < n < 2^{k+1}$. Here 2^{k+1} is the smallest power of 2 larger than n.)

At each stage of the algorithm, i and j, the locations of the first term and the last term of the restricted list at that stage, are compared to see whether the restricted list has more than one term. If $i < j$, a comparison is done to determine whether x is greater than the middle term of the restricted list.

At the first stage the search is restricted to a list with 2^{k-1} terms. So far, two comparisons have been used. This procedure is continued, using two comparisons at each stage to restrict the search to a list with half as many terms. In other words, two comparisons are used at the first stage of the algorithm when the list has 2^k elements, two more when the search has been reduced to a list with 2^{k-1} elements, two more when the search has been reduced to a list with 2^{k-2} elements, and so on, until two comparisons are used when the search has been reduced to a list with $2^1 = 2$ elements. Finally, when one term is left in the list, one comparison tells us that there are no additional terms left, and one more comparison is used to determine if this term is x.

Hence, at most $2k + 2 = 2\log n + 2$ comparisons are required to perform a binary search when the list being searched has 2^k elements. (If n is not a power of 2, the original list is expanded to a list with 2^{k+1} terms, where $k = \lfloor \log n \rfloor$, and the search requires at most $2\lfloor \log n \rfloor + 2$ comparisons.) Consequently, a binary search requires at most

$$O(\log n)$$

comparisons. From this analysis it follows that the binary search algorithm is more efficient, in the worst case, than a linear search. ■

Another important type of complexity analysis, besides worst-case analysis, is called **average-case** analysis. The average number of operations used to solve the problem over all inputs of a given size is found in this type of analysis. Average-case time complexity analysis is usually much more complicated than worst-case analysis. However, the average-case analysis for the linear search algorithm can be done without difficulty, as shown in Example 4.

EXAMPLE 4 Describe the average-case performance of the linear search algorithm, assuming that the element x is in the list.

Solution: There are n types of possible inputs when x is known to be in the list. If x is the first term of the list, three comparisons are needed, one to determine whether the end of the list has been reached, one to compare x and the first term, and one outside the loop. If x is the second term of the list, two more comparisons are needed, so that a total of five comparisons are used. In general, if x is the ith term of the list, two comparisons will be used at each of the i steps of the loop, and one outside the loop, so that a total of $2i + 1$ comparisons are needed. Hence, the average number of comparisons used equals

$$\frac{3 + 5 + 7 + \cdots + (2n + 1)}{n} = \frac{2(1 + 2 + 3 + \cdots + n) + n}{n}.$$

In Section 3.2 we will show that

$$1 + 2 + 3 + \cdots + n = \frac{n(n + 1)}{2}.$$

Hence, the average number of comparisons used by the linear search algorithm (when x is known to be in the list) is

$$\frac{2[n(n+1)/2]}{n} + 1 = n + 2 = O(n).$$

Remark: In this analysis it has been assumed that x is in the list being searched and it is equally likely that x is in any position. It is also possible to do an average-case analysis of this algorithm when x may not be in the list (see Exercise 13 at the end of this section). ■

Table 1 displays some common terminology used to describe the time complexity of algorithms. For instance, an algorithm is said to have **exponential complexity** if it has time complexity $O(b^n)$, where $b > 1$, measured in terms of some specified type of operation. Similarly, an algorithm with time complexity $O(n^b)$ is said to have **polynomial complexity.** The linear search algorithm has **linear** (worst- or average-case) **complexity** and the binary search algorithm has **logarithmic** (worst-case) **complexity**, measured in terms of the number of comparisons used.

A big-O estimate of the time complexity of an algorithm expresses how the time required to solve the problem changes as the input grows in size. In practice, the best estimate (that is, with the smallest reference function) that can be shown is used. However, big-O estimates of time complexity cannot be directly translated into the actual amount of computer time used. One reason for this is that a big-O estimate $f(n) = O(g(n))$, where $f(n)$ is the time complexity of an algorithm and $g(n)$ is a reference function, means that $f(n) < Cg(n)$ when $n > k$, where C and k are constants. So without knowing the constants C and k in the inequality, this estimate cannot be used to determine an upper bound on the number of operations used. Moreover, as remarked before, the time required for an operation depends on the type of operation and the computer being used.

However, the time required for an algorithm to solve a problem of a specified size can be determined if all operations can be reduced to the bit operations used by

TABLE 1 Commonly Used Terminology for the Complexity of Algorithms.

Complexity	*Terminology*
$O(1)$	Constant complexity
$O(\log n)$	Logarithmic complexity
$O(n)$	Linear complexity
$O(n \log n)$	$n \log n$ complexity
$O(n^b)$	Polynomial complexity
$O(b^n)$, where $b > 1$	Exponential complexity
$O(n!)$	Factorial complexity

TABLE 2 The Computer Time Used by Algorithms.

Problem Size	*Bit Operations Used*					
n	$\log n$	n	$n \log n$	n^2	2^n	$n!$
10	3×10^{-9} sec	10^{-8} sec	3×10^{-8} sec	10^{-7} sec	10^{-6} sec	3×10^{-3} sec
10^2	7×10^{-9} sec	10^{-7} sec	7×10^{-7} sec	10^{-5} sec	4×10^{13} yr	*
10^3	1.0×10^{-8} sec	10^{-6} sec	1×10^{-5} sec	10^{-3} sec	*	*
10^4	1.3×10^{-8} sec	10^{-5} sec	1×10^{-4} sec	10^{-1} sec	*	*
10^5	1.7×10^{-8} sec	10^{-4} sec	2×10^{-3} sec	10 sec	*	*
10^6	2×10^{-8} sec	10^{-3} sec	2×10^{-2} sec	17 min	*	*

the computer. Table 2 displays the time needed to solve problems of various sizes with an algorithm using the indicated number of bit operations. Times of more than 10^{100} years are indicated with an asterisk. (In Section 2.4 the number of bit operations used to add and multiply two integers will be discussed.) In constructing this table, each bit operation is assumed to take 10^{-9} seconds, which is the time required for a bit operation using the fastest computers today. In the future, these times will decrease as faster computers are developed.

It is important to know how long a computer will need to solve a problem. For instance, if an algorithm requires ten hours, it may be worthwhile to spend the computer time (and money) required to solve this problem. But, if an algorithm requires ten billion years to solve a problem, it would be unreasonable to use resources to implement this algorithm. One of the most interesting phenomena of modern technology is the tremendous increase in the speed and memory space of computers. Another important factor that decreases the time needed to solve problems on computers is **parallel processing,** which is the technique of performing sequences of operations simultaneously. Because of the increased speed of computation, increases in computer memory, and the use of algorithms that take advantage of parallel processing, problems that were considered impossible to solve five years ago are now routinely solved, and certainly five years from now this statement will still be true.

Exercises

1. How many comparisons are used by the algorithm given in Exercise 10 of Section 2.1 to find the smallest natural number in a sequence of n natural numbers?

2. Write the algorithm that puts the first four terms of a list of arbitrary length in increasing order. Show that this algorithm has time complexity $O(1)$ in terms of the number of comparisons used.

3. Suppose that an element is known to be among the first four elements in a list of 32 elements. Would a linear search or a binary search locate this element more rapidly?

4. Determine the number of multiplications used to find x^{2^k} starting with x and successively squaring (to find x^2, x^4, and so on). Is this a more efficient way to find x^{2^k} than by multiplying x by itself the appropriate number of times?

5. Give a big-O estimate for the number of comparisons used by the algorithm that determines the number of ones in a bit string by examining each bit of the string to determine whether it is a one bit (see Exercise 19 of Section 2.1).

★6. **a)** Show that the following algorithm determines the number of one bits in the bit string S.

procedure *bit count*(S: bit string)
count := 0
while $S \neq 0$
begin
  count := *count* +1
  $S := S \wedge (S - 1)$
end {*count* is the number of ones in S}

Here $S - 1$ is the bit string obtained by changing the rightmost one bit of S to a zero and all the zero bits to the right of this to ones. (Recall that $S \wedge (S - 1)$ is the bitwise *AND* of S and $S - 1$.)

b) How many bitwise *AND* operations are needed to find the number of one bits in a string S?

7. The conventional algorithm for evaluating a polynomial $a_n x^n + a_{n-1}x^{n-1} + \cdots + a_1 x + a_0$ at $x = c$ can be expressed in pseudocode by

procedure *polynomial*($c, a_0, a_1, \ldots, a_n$: real numbers)
power := 1
$y := a_0$
for i := 1 **to** n
begin
  power := *power* $*$ c
  $y := y + a_i *$ *power*
end $\{y = a_n c^n + a_{n-1}c^{n-1} + \cdots + a_1 c + a_0\}$

where the final value of y is the value of the polynomial at $x = c$.

a) Evaluate $3x^2+x+1$ at $x = 2$ by working through each step of the algorithm.

b) Exactly how many multiplications and additions are used to evaluate a polynomial of degree n at $x = c$? (Do not count additions used to increment the loop variable.)

8. There is a more efficient algorithm (in terms of the number of multiplications and additions used) for evaluating polynomials than the conventional algorithm described in the previous exercise. This algorithm is called **Horner's method.** The following pseudocode shows how to use this method to find the value of $a_n x^n + a_{n-1}x^{n-1} + \cdots + a_1 x + a_0$ at $x = c$.

procedure *Horner*($c, a_0, a_1, a_2, \ldots, a_n$: real numbers)
$y := a_n$
for i := 1 **to** n
  $y := y * c + a_{n-i}$
$\{y = a_n c^n + a_{n-1}c^{n-1} + \cdots + a_1 c + a_0\}$

a) Evaluate $3x^2+x+1$ at $x = 2$ by working through each step of the algorithm.

b) Exactly how many multiplications and additions are used by this algorithm to evaluate a polynomial of degree n at $x = c$? (Do not count additions used to increment the loop variable.)

9. How large a problem can be solved in one second using an algorithm that requires $f(n)$ bit operations, where each bit operation is carried out in 10^{-9} seconds, with the following values for $f(n)$?

a) $\log n$ **b)** n **c)** $n \log n$
d) n^2 **e)** 2^n **f)** $n!$

10. How much time does an algorithm take to solve a problem of size n if this algorithm uses $2n^2 + 2^n$ bit operations, each requiring 10^{-9} seconds, with the following values of n?

a) 10 **b)** 20 **c)** 50 **d)** 100

11. How much time does an algorithm using 2^{50} bit operations need if each bit operation takes the following number of seconds?

a) 10^{-6} seconds
b) 10^{-9} seconds
c) 10^{-12} seconds

12. Determine the least number of comparisons, or best-case performance,

a) required to find the maximum of a sequence of n integers, using Algorithm 1 of Section 2.1.

b) used to locate an element in a list of n terms with a linear search.

c) used to locate an element in a list of n terms using a binary search.

13. Analyze the average-case performance of the linear search algorithm, if exactly half the time element x

is not in the list and if x is in the list it is equally likely to be in any position.

14. An algorithm is called **optimal** for the solution of a problem with respect to a specified operation if there is no algorithm for solving this problem using fewer operations.

a) Show that Algorithm 1 in Section 2.1 is an optimal algorithm with respect to the number of comparisons of integers. (*Note:* Comparisons used for bookkeeping in the loop are not of concern here.)

b) Is the linear search algorithm optimal with respect to the number of comparisons of integers (not including comparisons used for bookkeeping in the loop)?

15. Describe the worst-case time complexity, measured in terms of comparisons, of the ternary search algorithm described in Exercise 21 of Section 2.1.

16. Describe the worst-case time complexity, measured in terms of comparisons, of the search algorithm described in Exercise 22 of Section 2.1.

17. Analyze the worst-case time complexity of the algorithm you devised in Exercise 23 of Section 2.1 for locating a mode in a list of nondecreasing integers.

18. Analyze the worst-case time complexity of the algorithm you devised in Exercise 24 of Section 2.1 for locating all modes in a list of nondecreasing integers.

2.3 The Integers and Division

INTRODUCTION

The part of discrete mathematics involving the integers and their properties belongs to the branch of mathematics called **number theory.** This section is the beginning of a three-section introduction to number theory. In this section we will review some basic concepts of number theory, including divisibility, greatest common divisors, and modular arithmetic. In Section 2.4 we will describe several important algorithms from number theory, tying together the material in Sections 2.1 and 2.2 on algorithms and their complexity with the notions introduced in this section. For example, we will introduce algorithms for finding the greatest common divisor of two positive integers and for performing computer arithmetic using binary expansions. Finally, in Section 2.5 we will continue our study of number theory by introducing some important results and their applications to computer arithmetic and cryptology, the study of secret messages.

The ideas that we will develop in this section are based on the notion of divisibility. One important concept based on divisibility is that of a prime number. A prime is an integer greater than 1 that is divisible only by 1 and by itself. Determining whether an integer is prime is important in applications to cryptology. An important theorem from number theory, the Fundamental Theorem of Arithmetic, asserts that every positive integer can be written uniquely as the product of prime numbers. Factoring integers into their prime factors is important in cryptology. Division of an integer by a positive integer produces a quotient and a remainder. Working with these remainders leads to modular arithmetic, which is used throughout computer science. We will discuss three applications of modular arithmetic in this section: generating pseudorandom numbers, assigning computer memory locations to files, and encrypting and decrypting messages.

DIVISION

When one integer is divided by a second, nonzero integer, the quotient may or may not be an integer. For example, $12/3 = 4$ is an integer, whereas $11/4 = 2.75$ is not. This leads to the following definition.

DEFINITION 1. If a and b are integers with $a \neq 0$, we say that *a divides b* if there is an integer c such that $b = ac$. When a divides b we say that a is a *factor* of b and that b is a *multiple* of a. The notation $a \mid b$ denotes that a divides b. We write $a \nmid b$ when a does not divide b.

In Figure 1 a number line indicates which integers are divisible by the positive integer d.

EXAMPLE 1 Determine whether $3 \mid 7$ and whether $3 \mid 12$.

Solution: It follows that $3 \nmid 7$, since $7/3$ is not an integer. On the other hand, $3 \mid 12$ since $12/3 = 4$. ■

EXAMPLE 2 Let n and d be positive integers. How many positive integers not exceeding n are divisible by d?

Solution: The positive integers divisible by d are all the integers of the form dk, where k is a positive integer. Hence, the number of positive integers divisible by d that do not exceed n equals the number of integers k with $0 < dk \leq n$, or with $0 < k \leq n/d$. Therefore, there are $\lfloor n/d \rfloor$ positive integers not exceeding n that are divisible by d. ■

Some of the basic properties of divisibility of integers are given in Theorem 1.

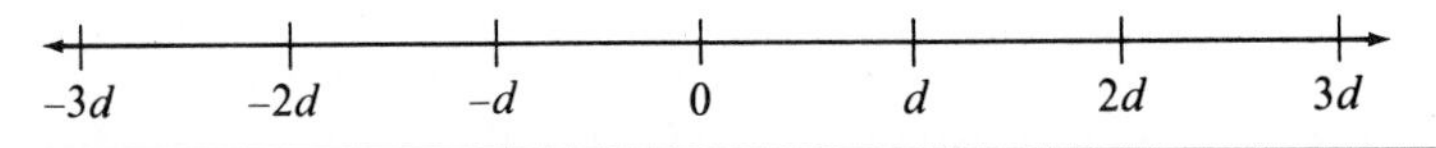

FIGURE 1 **Integers Divisible by the Positive Integer d.**

THEOREM 1 Let a, b, and c be integers. Then

1. if $a \mid b$ and $a \mid c$, then $a \mid (b + c)$;
2. if $a \mid b$, then $a \mid bc$ for all integers c;
3. if $a \mid b$ and $b \mid c$, then $a \mid c$.

Proof: Suppose that $a \mid b$ and $a \mid c$. Then, from the definition of divisibility, it follows that there are integers s and t with $b = as$ and $c = at$. Hence,

$$b + c = as + at = a(s + t).$$

Therefore, a divides $b + c$. This establishes part (1) of the theorem. The proofs of parts (2) and (3) are left as exercises for the reader. ■

Every positive integer greater than 1 is divisible by at least two integers, since a positive integer is divisible by 1 and by itself. Integers that have exactly two different positive integer factors are called **primes.**

DEFINITION 2. A positive integer p greater than 1 is called *prime* if the only positive factors of p are 1 and p. A positive integer that is greater than 1 and is not prime is called *composite*.

EXAMPLE 3 The integer 7 is prime since its only positive factors are 1 and 7, whereas the integer 9 is composite since it is divisible by 3. ■

The primes are the building blocks of positive integers, as the Fundamental Theorem of Arithmetic shows. The proof will be omitted.

THEOREM 2 **THE FUNDAMENTAL THEOREM OF ARITHMETIC** Every positive integer can be written uniquely as the product of primes, where the prime factors are written in order of increasing size. (Here, a product can have zero, one, or more than one prime factor.)

The next example gives some prime factorizations of integers.

EXAMPLE 4 The prime factorizations of 100, 641, 999, and 1024 are given by

$$100 = 2 \cdot 2 \cdot 5 \cdot 5 = 2^2 5^2,$$
$$641 = 641,$$
$$999 = 3 \cdot 3 \cdot 3 \cdot 37 = 3^3 \cdot 37,$$

and

$$1024 = 2 \cdot 2 \cdot 2 \cdot 2 \cdot 2 \cdot 2 \cdot 2 \cdot 2 \cdot 2 \cdot 2 = 2^{10}.$$ ■

It is often important to show that a given integer is prime. For instance, in cryptology large primes are used in some methods for making messages secret. One procedure for showing that an integer is prime is based on the following observation.

THEOREM 3 If n is a composite integer, then n has a prime divisor less than or equal to $\sqrt{n}$.

Proof: If n is composite, it has a factor a with $1 < a < n$. Hence, $n = ab$, where both a and b are positive integers greater than 1. We see that $a \leq \sqrt{n}$ or $b \leq \sqrt{n}$, since otherwise $ab > \sqrt{n} \cdot \sqrt{n} = n$. Hence, n has a positive divisor not exceeding $\sqrt{n}$. This divisor is either prime or, by the Fundamental Theorem of Arithmetic, has a prime divisor. In either case, n has a prime divisor less than or equal to $\sqrt{n}$. ■

From Theorem 3, it follows that an integer is prime if it is not divisible by any prime less than or equal to its square root. In the following example this observation is used to show that 101 is prime.

EXAMPLE 5 Show that 101 is prime.

Solution: The only primes not exceeding $\sqrt{101}$ are 2, 3, 5, and 7. Since 101 is not divisible by 2, 3, 5, or 7 (the quotient of 101 and each of these integers is not an integer), it follows that 101 is prime. ■

Since every integer has a prime factorization, it would be useful to have a procedure for finding this prime factorization. Consider the problem of finding the prime factorization of n. Begin by dividing n by successive primes, starting with the smallest prime, 2. If n has a prime factor, then by Theorem 3 a prime factor p not exceeding $\sqrt{n}$ will be found. So, if no prime factor not exceeding $\sqrt{n}$ is found, then n is prime. Otherwise, if a prime factor p is found, continue by factoring n/p. Note that n/p has no prime factors less than p. Again, if n/p has no prime factor greater than or equal to p and not exceeding its square root, then it is prime. Otherwise, if it has a prime factor q, continue by factoring $n/(pq)$. This procedure is continued until the factorization has been reduced to a prime. This procedure is illustrated in the following example.

EXAMPLE 6 Find the prime factorization of 7007.

Solution: To find the prime factorization of 7007, first perform divisions of 7007 by successive primes, beginning with 2. None of the primes 2, 3, and 5 divides 7007. However, 7 divides 7007, with $7007/7 = 1001$. Next, divide 1001 by successive primes, beginning with 7. It is immediately seen that 7 also divides 1001, since $1001/7 = 143$. Continue by dividing 143 by successive primes, beginning with 7. Although 7 does not divide 143, 11 does divide 143, and $143/11 = 13$. Since 13 is prime, the procedure is completed. It follows that the prime factorization of 7007 is $7 \cdot 7 \cdot 11 \cdot 13 = 7^2 \cdot 11 \cdot 13$. ■

Using trial division together with Theorem 3 gives procedures for factoring and for primality testing. However, these procedures are not the most efficient algorithms known for these tasks. Recently, factoring and primality testing have become important in applications of number theory to cryptology. This has led to a large interest in developing efficient algorithms for both tasks. The reader should consult the suggested readings listed at the end of the book for discussions of various algorithms for primality testing and factoring.

THE DIVISION ALGORITHM

We have seen that an integer may or may not be divisible by another. However, when an integer is divided by a positive integer, there always is a quotient and a remainder, as the division algorithm shows.

THEOREM 4 **THE DIVISION ALGORITHM** Let a be an integer and d a positive integer. Then there are unique integers q and r, with $0 \leq r < d$, such that $a = dq + r$. ■

Remark: Theorem 4 is not really an algorithm. (Why not?) Nevertheless, we use its traditional name.

DEFINITION 3. In the equality given in the division algorithm, d is called the *divisor*, a is called the *dividend*, q is called the *quotient*, and r is called the *remainder*.

The following two examples illustrate the division algorithm.

EXAMPLE 7 What are the quotient and remainder when 101 is divided by 11?

Solution: We have

$$101 = 11 \cdot 9 + 2.$$

Hence, the quotient when 101 is divided by 11 is 9, and the remainder is 2. ■

EXAMPLE 8 What are the quotient and remainder when -11 is divided by 3?

Solution: We have

$$-11 = 3(-4) + 1.$$

Hence, the quotient when -11 is divided by 3 is -4, and the remainder is 1.

Note that the remainder cannot be negative. Consequently, the remainder is *not* -2, even though

$$-11 = 3(-3) - 2,$$

since $r = -2$ does not satisfy $0 \leq r < 3$. ■

Note that the integer a is divisible by the integer d if and only if the remainder is zero when a is divided by d.

GREATEST COMMON DIVISORS AND LEAST COMMON MULTIPLES

The largest integer that divides both of two integers is called the **greatest common divisor** of these integers.

DEFINITION 4. Let a and b be integers, not both zero. The largest integer d such that $d \mid a$ and $d \mid b$ is called the *greatest common divisor* of a and b. The greatest common divisor of a and b is denoted by $gcd(a, b)$.

The greatest common divisor of two integers, not both zero, exists because the set of common divisors of these integers is finite. One way to find the greatest common divisor of two integers is to find all the positive common divisors of both integers and then take the largest divisor. This is done in the following examples. Later, a more efficient method of finding greatest common divisors will be given.

EXAMPLE 9 What is the greatest common divisor of 24 and 36?

Solution: The positive common divisors of 24 and 36 are 1, 2, 3, 4, 6, and 12. Hence, $gcd(24, 36) = 12$. ■

EXAMPLE 10 What is the greatest common divisor of 17 and 22?

Solution: The integers 17 and 22 have no positive common divisors other than 1, so that $gcd(17, 22) = 1$. ■

Since it is often important to specify that two integers have no common positive divisor other than 1, we have the following definition.

DEFINITION 5. The integers a and b are *relatively prime* if their greatest common divisor is 1.

EXAMPLE 11 From Example 10 it follows that 17 and 22 are relatively prime, since $gcd(17, 22) = 1$. ■

Since we often need to specify that no two integers in a set of integers have a common positive divisor greater than 1, we make the following definition.

DEFINITION 6. The integers $a_1, a_2, \ldots, a_n$ are *pairwise relatively prime* if $gcd(a_i, a_j) = 1$ whenever $1 \leq i < j \leq n$.

EXAMPLE 12 Determine whether the integers 10, 17, and 21 are pairwise relatively prime and whether the integers 10, 19, and 24 are pairwise relatively prime.

Solution: Since $gcd(10, 17) = 1$, $gcd(10, 21) = 1$ and $gcd(17, 21) = 1$, we conclude that 10, 17, and 21 are pairwise relatively prime.

Since $gcd(10, 24) = 2 > 1$, we see that 10, 19, and 24 are not pairwise relatively prime. ■

Another way to find the greatest common divisor of two integers is to use the prime factorizations of these integers. Suppose that the prime factorizations of the integers a and b, neither equal to zero, are

$$a = p_1^{a_1} p_2^{a_2} \cdots p_n^{a_n}, \quad b = p_1^{b_1} p_2^{b_2} \cdots p_n^{b_n},$$

where each exponent is a nonnegative integer, and where all primes occurring in the prime factorization of either a or b are included in both factorizations, with zero exponents if necessary. Then $gcd(a, b)$ is given by

$$gcd(a, b) = p_1^{\min(a_1, b_1)} p_2^{\min(a_2, b_2)} \cdots p_n^{\min(a_n, b_n)},$$

where $\min(x,y)$ represents the minimum of the two numbers x and y. To show that this formula for $gcd(a, b)$ is valid, we must show that the integer on the right-hand side divides both a and b, and that no larger integer also does. This integer does divide both a and b, since the power of each prime in the factorization does not exceed the power of this prime in either the factorization of a or that of b. Further, no larger integer can divide both a and b, because the exponents of the primes in this factorization cannot be increased, and no other primes can be included.

EXAMPLE 13 Since the prime factorizations of 120 and 500 are $120 = 2^3 \cdot 3 \cdot 5$ and $500 = 2^2 \cdot 5^3$, the greatest common divisor is

$$gcd(120, 500) = 2^{\min(3,2)} 3^{\min(1,0)} 5^{\min(1,3)} = 2^2 3^0 5^1 = 20.$$

■

Prime factorizations can also be used to find the **least common multiple** of two integers.

DEFINITION 7. The *least common multiple* of the positive integers a and b is the smallest positive integer that is divisible by both a and b. The least common multiple of a and b is denoted by $lcm(a, b)$.

The least common multiple exists because the set of integers divisible by both a and b is nonempty, and every nonempty set of positive integers has a least element (by the well-ordering property, which will be discussed in Chapter 3). Suppose that the prime factorizations of a and b are as before. Then the least common multiple of a and b is given by

$$lcm(a, b) = p_1^{\max(a_1,b_1)} p_2^{\max(a_2,b_2)} \cdots p_n^{\max(a_n,b_n)}$$

where $\max(x, y)$ denotes the maximum of the two numbers x and y. This formula is valid since a common multiple of a and b has at least $\max(a_i, b_i)$ factors of p_i in its prime factorization, and the least common multiple has no other prime factors besides those in a and b.

EXAMPLE 14 What is the least common multiple of $2^3 3^5 7^2$ and $2^4 3^3$?

Solution: We have

$$lcm(2^3 3^5 7^2, 2^4 3^3) = 2^{\max(3,4)} 3^{\max(5,3)} 7^{\max(2,0)} = 2^4 3^5 7^2.$$ ■

The following theorem gives the relationship between the greatest common divisor and least common multiple of two integers. It can be proved using the formulae we have derived for these quantities. The proof of this theorem is left as an exercise for the reader.

THEOREM 5 Let a and b be positive integers. Then

$$ab = gcd(a, b) lcm(a, b).$$

MODULAR ARITHMETIC

In some situations we care only about the remainder of an integer when it is divided by some specified positive integer. For instance, when we ask what time it will be (on a 24-hour clock) 50 hours from now, we care only about the remainder when 50 plus the current hour is divided by 24. Since we are often interested only in remainders, we have special notations for them.

DEFINITION 8. Let a be an integer and m be a positive integer. We denote by a **mod** m the remainder when a is divided by m.

It follows from the definition of remainder that a **mod** m is the integer r such that $a = qm + r$ and $0 \leq r < m$.

EXAMPLE 15 We see that 17 **mod** 5 = 2, −133 **mod** 9 = 2, and 2001 **mod** 101 = 82. ■

We also have a notation to indicate that two integers have the same remainder when they are divided by the positive integer m.

DEFINITION 9. If a and b are integers and m is a positive integer, then a is *congruent to b modulo m* if m divides $a - b$. We use the notation $a \equiv b \pmod{m}$ to indicate that a is congruent to b modulo m. If a and b are not congruent modulo m, we write $a \not\equiv b \pmod{m}$.

Note that $a \equiv b \pmod{m}$ if and only if a **mod** $m = b$ **mod** m.

EXAMPLE 16 Determine whether 17 is congruent to 5 modulo 6 and whether 24 and 14 are congruent modulo 6.

Solution: Since 6 divides $17 - 5 = 12$, we see that $17 \equiv 5 \pmod{6}$. However, since $24 - 14 = 10$ is not divisible by 6, we see that $24 \not\equiv 14 \pmod{6}$. ■

The great German mathematician Carl Friedrich Gauss developed the concept of congruences at the end of the 19th century. The notion of congruences has played an

---

Karl Friedrich Gauss (1777–1855) Karl Friedrich Gauss, the son of a bricklayer, was a child prodigy. He demonstrated his potential at the age of 10, when he quickly solved a problem assigned by a teacher to keep the class busy. The teacher asked the students to find the sum of the first 100 positive integers. Gauss realized that this sum could be found by forming 50 pairs, each with the sum 101: $1 + 100, 2 + 99, \ldots, 50 + 51$. This brilliance attracted the sponsorship of patrons, including Duke Ferdinand of Brunswick, who made it possible for Gauss to attend Caroline College and the University of Göttingen. While a student, he invented the method of least squares, which is used to estimate the most likely value of a variable from experimental results. In 1796 Gauss made a fundamental discovery in geometry, advancing a subject that had not advanced since ancient times. He showed that a 17-sided regular polygon could be drawn using just a ruler and compass.

In 1799 Gauss presented the first rigorous proof of the Fundamental Theorem of Arithmetic, which states that a polynomial of degree n has exactly n roots (counting multiplicities). Gauss achieved worldwide fame when he successfully calculated the orbit of the first asteroid discovered, Ceres, using scanty data.

Gauss was called the Prince of Mathematics by his contemporary mathematicians. Although Gauss is noted for his many discoveries in geometry, algebra, analysis, astronomy, and physics, he had a special interest in number theory, which can be seen from his statement "Mathematics is the queen of the sciences, and the theory of numbers is the queen of mathematics." Gauss laid the foundations for modern number theory with the publication of his book *Disquisitiones Arithmeticae* in 1801.

important role in the development of number theory. The following theorem provides a useful way to work with congruences.

THEOREM 6 Let m be a positive integer. The integers a and b are congruent modulo m if and only if there is an integer k such that $a = b + km$.

Proof: If $a \equiv b \pmod{m}$, then $m \mid (a - b)$. This means that there is an integer k such that $a - b = km$, so that $a = b + km$. Conversely, if there is an integer k such that $a = b + km$, then $km = a - b$. Hence, m divides $a - b$, so that $a \equiv b \pmod{m}$. ■

The following theorem shows how congruences work with respect to addition and multiplication.

THEOREM 7 Let m be a positive integer. If $a \equiv b \pmod{m}$ and $c \equiv d \pmod{m}$, then

$$a + c \equiv b + d \pmod{m}$$

and

$$ac \equiv bd \pmod{m}.$$

Proof: Since $a \equiv b \pmod{m}$ and $c \equiv d \pmod{m}$, there are integers s and t with $b = a + sm$ and $d = c + tm$. Hence,

$$b + d = (a + sm) + (c + tm) = (a + c) + m(s + t)$$

and

$$bd = (a + sm)(c + tm) = ac + m(at + cs + stm).$$

Hence,

$$a + c \equiv b + d \pmod{m}$$

and

$$ac \equiv bd \pmod{m}.$$ ■

EXAMPLE 17 Since $7 \equiv 2 \pmod{5}$ and $11 \equiv 1 \pmod{5}$, it follows from Theorem 7 that

$$18 = 7 + 11 \equiv 2 + 1 = 3 \pmod{5}$$

and that

$$77 = 7 \cdot 11 \equiv 2 \cdot 1 = 2 \pmod{5}.$$ ■

APPLICATIONS OF CONGRUENCES

Number theory has applications to a wide range of areas. We will introduce three applications in this section: the use of congruences to assign memory locations to computer files, the generation of pseudorandom numbers, and cryptosystems based on modular arithmetic.

EXAMPLE 18 Hashing Functions The central computer at your school maintains records for each student. How can memory locations be assigned so that student records can be retrieved quickly? The solution to this problem is to use a suitably chosen **hashing function.** Records are identified using a **key**, which uniquely identifies each student's records. For instance, student records are often identified using the social security number of the student as the key. A hashing function h assigns memory location $h(k)$ to the record that has k as its key.

In practice, many different hashing functions are used. One of the most common is the function

$$h(k) = k \textbf{ mod } m$$

where m is the number of available memory locations.

Hashing functions should be easily evaluated so that files can be quickly located. The hashing function $h(k) = k \textbf{ mod } m$ meets this requirement; to find $h(k)$, we need only compute the remainder when k is divided by m. Furthermore, the hashing function should be onto, so that all memory locations are possible. The function $h(k) = k \textbf{ mod } m$ also satisfies this property.

For example, when $m = 111$, the record of the student with social security number 064212848 is assigned to memory location 14, since

$$h(064212848) = 064212848 \textbf{ mod } 111 = 14.$$

Similarly, since

$$h(037149212) = 037149212 \textbf{ mod } 111 = 65,$$

the record of the student with social security number 037149212 is assigned to memory location 65.

Since a hashing function is not one-to-one (since there are more possible keys than memory locations), more than one file may be assigned to a memory location. When this happens, we say that a **collision** occurs. One way to resolve a collision is to assign the first free location following the occupied memory location assigned by the hashing function. For example, after making the two earlier assignments, we assign location 15 to the record of the student with the social security number 107405723. To see this, first note that $h(k)$ maps this social security number to location 14, since

$$h(107405723) = 107405723 \textbf{ mod } 111 = 14,$$

but this location is already occupied (by the file of the student with social security number 064212848). However, memory location 15, the first location following memory location 14, is free.

There are many more sophisticated ways to resolve collisions that are more efficient than the simple method we have described. These are discussed in the references on hashing functions given at the end of the book. ■

EXAMPLE 19 Pseudorandom Numbers Randomly chosen numbers are often needed for computer simulations. Different methods have been devised for generating numbers that have properties of randomly chosen numbers. Because numbers generated by systematic methods are not truly random, they are called **pseudorandom numbers.**

The most commonly used procedure for generating pseudorandom numbers is the **linear congruential method.** We choose four integers: the **modulus** m, **multiplier** a, **increment** c, and **seed** x_0, with $2 \leq a < m, 0 \leq c < m$, and $0 \leq x_0 < m$. We generate a sequence of pseudorandom numbers $\{x_n\}$, with $0 \leq x_n < m$ for all n, by successively using the congruence

$$x_{n+1} = (ax_n + c) \textbf{ mod } m.$$

(This is an example of a recursive definition, discussed in Section 3.3. In that section we will show that such sequences are well defined.)

Many computer experiments require the generation of pseudorandom numbers between 0 and 1. To generate such numbers, we divide numbers generated with a linear congruential generator by the modulus: that is, we use the numbers x_n/m.

For instance, the sequence of pseudorandom numbers generated by choosing $m = 9, a = 7, c = 4$, and $x_0 = 3$, can be found as follows:

$$\begin{aligned}
x_1 &= 7x_0 + 4 = 7 \cdot 3 + 4 = 25 \textbf{ mod } 9 = 7,\\
x_2 &= 7x_1 + 4 = 7 \cdot 7 + 4 = 53 \textbf{ mod } 9 = 8,\\
x_3 &= 7x_2 + 4 = 7 \cdot 8 + 4 = 60 \textbf{ mod } 9 = 6,\\
x_4 &= 7x_3 + 4 = 7 \cdot 6 + 4 = 46 \textbf{ mod } 9 = 1,\\
x_5 &= 7x_4 + 4 = 7 \cdot 1 + 4 = 11 \textbf{ mod } 9 = 2,\\
x_6 &= 7x_5 + 4 = 7 \cdot 2 + 4 = 18 \textbf{ mod } 9 = 0,\\
x_7 &= 7x_6 + 4 = 7 \cdot 0 + 4 = 4 \textbf{ mod } 9 = 4,\\
x_8 &= 7x_7 + 4 = 7 \cdot 4 + 4 = 32 \textbf{ mod } 9 = 5,\\
x_9 &= 7x_8 + 4 = 7 \cdot 5 + 4 = 39 \textbf{ mod } 9 = 3.
\end{aligned}$$

Since $x_9 = x_0$ and each term depends only on the previous term, the following sequence is generated:

$$3, 7, 8, 6, 1, 2, 0, 4, 5, 3, 7, 8, 6, 1, 2, 0, 4, 5, 3, \ldots.$$

This sequence contains nine different numbers before repeating.

Most computers actually do use linear congruential generators to generate pseudorandom numbers. Often, a linear congruential generator with increment $c = 0$ is used. Such a generator is called a **pure multiplicative generator.** For example, the pure multiplicative generator with modulus $2^{31} - 1$ and multiplier $7^5 = 16{,}807$ is widely used. With these values, it can be shown that $2^{31} - 2$ numbers are generated before repetition begins. ■

CRYPTOLOGY

Congruences have many applications to discrete mathematics and computer science. Discussions of these applications can be found in the suggested readings given at the end of the book. One of the most important applications of congruences involves **cryptology,** which is the study of secret messages. One of the earliest known uses of cryptology was by Julius Caesar. He made messages secret by shifting each letter three letters forward in the alphabet (sending the last three letters of the alphabet to the first three). For instance, using this scheme the letter B is sent to E and the letter X is sent to A. This is an example of **encryption,** that is, the process of making a message secret.

To express Caesar's encryption process mathematically, first replace each letter by an integer from 0 to 25, based on its position in the alphabet. For example, replace A by 0, K by 10, and Z by 25. Caesar's encryption method can be represented by the function f that assigns to the nonnegative integer p, $p \leq 25$, the integer $f(p)$ in the set $\{0, 1, 2, \ldots, 25\}$ with

$$f(p) = (p + 3) \textbf{ mod } 26.$$

In the encrypted version of the message, the letter represented by p is replaced with the letter represented by $(p + 3)$ **mod** 26.

EXAMPLE 20 What is the secret message produced from the message "MEET YOU IN THE PARK" using the Caesar cipher?

Solution: First replace the letters in the message with numbers. This produces

12 4 4 19   24 14 20   8 13   19 7 4   15 0 17 10.

Now replace each of these numbers p by $f(p) = (p + 3)$ **mod** 26. This gives

15 7 7 22   1 17 23   11 16   22 10 7   18 3 20 13.

Translating this back to letters produces the encrypted message "PHHW BRX LQ WKH SDUN." ■

To recover the original message from a secret message encrypted by the Caesar cipher, the function f^{-1}, the inverse of f, is used. Note that the function f^{-1} sends an integer p from $\{0, 1, 2, \ldots, 25\}$ to $f^{-1}(p) = (p - 3)$ **mod** 26. In other words, to find the original message, each letter is shifted back three letters in the alphabet, with the first three letters sent to the last three letters of the alphabet. The process of determining the original message from the encrypted message is called **decryption.**

There are various ways to generalize the Caesar cipher. For example, instead of shifting each letter by 3, we can shift each letter by k, so that

$$f(p) = (p + k) \textbf{ mod } 26.$$

Such a cipher is called a **shift cipher.** Note that decryption can be carried out using

$$f^{-1}(p) = (p - k) \textbf{ mod } 26.$$

Obviously, Caesar's method and shift ciphers do not provide a high level of security. There are various ways to enhance this method. One approach that slightly enhances the security is to use a function of the form

$$f(p) = (ap + b) \textbf{ mod } 26,$$

where a and b are integers, chosen such that f is a bijection. (Such a mapping is called an *affine transformation.*) This provides a number of possible encryption systems. The use of one of these systems is illustrated in the following example.

EXAMPLE 21 What letter replaces the letter K when the function $f(p) = (7p + 3)$ **mod** 26 is used for encryption?

Solution: First, note that 10 represents K. Then, using the encryption function specified, it follows that $f(10) = (7 \cdot 10 + 3)$ **mod** $26 = 21$. Since 21 represents V, K is replaced by V in the encrypted message. ■

Caesar's encryption method, and the generalization of this method, proceed by replacing each letter of the alphabet by another letter in the alphabet. Encryption methods of this kind are vulnerable to attacks based on the frequency of occurrence of letters in the message. More sophisticated encryption methods are based on replacing blocks of letters with other blocks of letters. There are a number of techniques based on modular arithmetic for encrypting blocks of letters. A discussion of these may be found in the suggested readings listed at the end of the book.

Exercises

1. Does 17 divide each of the following numbers?
a) 68 **b)** 84 **c)** 357 **d)** 1001

2. Show that if a is an integer other than 0, then
a) 1 divides a. **b)** a divides 0.

3. Show that part (2) of Theorem 1 is true.

4. Show that part (3) of Theorem 1 is true.

5. Show that if $a \mid b$ and $b \mid a$, where a and b are integers, then $a = b$ or $a = -b$.

6. Show that if a, b, c, and d are integers such that $a \mid c$ and $b \mid d$, then $ab \mid cd$.

7. Show that if a, b, and c are integers such that $ac \mid bc$, then $a \mid b$.

8. Are the following integers primes?
a) 19 **b)** 27 **c)** 93
d) 101 **e)** 107 **f)** 113

9. In each of the following cases, what are the quotient and remainder?
a) 19 is divided by 7
b) -111 is divided by 11
c) 789 is divided by 23
d) 1001 is divided by 13
e) 0 is divided by 19
f) 3 is divided by 5
g) -1 is divided by 3
h) 4 is divided by 1

10. Find the prime factorization of each of the following.
a) 39 **b)** 81 **c)** 101
d) 143 **e)** 289 **f)** 899

11. Find the prime factorization of 10!.

★12. How many zeros are there at the end of 100!?

★13. An **irrational number** is a real number x that cannot be written as the ratio of two integers. Show that $\log_2 3$ is an irrational number.

14. Which positive integers less than 12 are relatively prime to 12?

15. Determine whether the following sets of integers are pairwise relatively prime.
a) (11,15,19) **b)** (14,15,21)
c) (12,17,31,37) **d)** (7,8,9,11)

16. We call a positive integer **perfect** if it equals the sum of its divisors other than itself.

a) Show that 6 and 28 are perfect.
b) Show that $2^{p-1}(2^p - 1)$ is a perfect number when $2^p - 1$ is prime.

17. Let m be a positive integer. Show that $a \equiv b \pmod{m}$ if a **mod** $m = b$ **mod** m.

18. Let m be a positive integer. Show that a **mod** $m = b$ **mod** m if $a \equiv b \pmod{m}$.

19. Show that if $2^n - 1$ is prime, then n is prime. (*Hint:* Use the identity $2^{ab} - 1 = (2^a - 1)(2^{a(b-1)} + 2^{a(b-2)} + \cdots + 2^a + 1)$).

20. Determine whether each of the following integers is prime.
a) $2^7 - 1$ b) $2^9 - 1$
c) $2^{11} - 1$ d) $2^{13} - 1$

21. The value of the **Euler ϕ-function** at the positive integer n is defined to be the number of positive integers less than or equal to n that are relatively prime to n. (*Note:* ϕ is the Greek letter phi.) Find
a) $\phi(4)$ b) $\phi(10)$ c) $\phi(13)$

22. Show that n is prime if and only if $\phi(n) = n - 1$.

23. What is the value of $\phi(p^k)$ when p is prime and k is a positive integer?

24. What are the greatest common divisors of the following pairs of integers?
a) $2^2 \cdot 3^3 \cdot 5^5,\ 2^5 \cdot 3^3 \cdot 5^2$
b) $2 \cdot 3 \cdot 5 \cdot 7 \cdot 11 \cdot 13,\ 2^{11} \cdot 3^9 \cdot 11 \cdot 17^{14}$
c) $17,\ 17^{17}$
d) $2^2 \cdot 7,\ 5^3 \cdot 13$
e) $0,\ 5$
f) $2 \cdot 3 \cdot 5 \cdot 7,\ 2 \cdot 3 \cdot 5 \cdot 7$

★25. Show that if n and k are positive integers, then $\lceil n/k \rceil = \lfloor (n-1)/k \rfloor + 1$.

26. Show that if a is an integer and d is a positive integer greater than 1, then the quotient and remainder obtained when a is divided by d are $\lfloor a/d \rfloor$ and $a - d\lfloor a/d \rfloor$, respectively.

27. Evaluate the following quantities.
a) 13 **mod** 3 b) −97 **mod** 11
c) 155 **mod** 19 d) −221 **mod** 23

28. List five integers that are congruent to 4 modulo 12.

29. Decide whether each of the following integers is congruent to 5 modulo 17.
a) 80 b) 103 c) −29 d) −122

30. If the product of two integers is $2^7 3^8 5^2 7^{11}$ and their greatest common divisor is $2^3 3^4 5$, what is their least common multiple?

31. Show that if a and b are positive integers then $ab = gcd(a, b)\ lcm(a, b)$. (*Hint:* Use the prime factorizations of a and b and the formulae for $gcd(a, b)$ and $lcm(a, b)$ in terms of these factorizations.)

32. Show that if $a \equiv b \pmod{m}$ and $c \equiv d \pmod{m}$, where a, b, c, d, and m are integers with $m \geq 2$, then $a - c \equiv b - d \pmod{m}$.

33. Show that if $n \mid m$, where n and m are positive integers greater than 1, and if $a \equiv b \pmod{m}$, where a and b are integers, then $a \equiv b \pmod{n}$.

34. Show that if a, b, c, and m are integers such that $m \geq 2$, $c > 0$, and $a \equiv b \pmod{m}$, then $ac \equiv bc \pmod{mc}$.

35. Show that $ac \equiv bc \pmod{m}$, where a, b, c, and m are integers with $m \geq 2$, does not necessarily imply that $a \equiv b \pmod{m}$.

36. Show that if a, b, and m are integers such that $m \geq 2$ and $a \equiv b \pmod{m}$, then $gcd(a, m) = gcd(b, m)$.

37. Show that if a, b, k, and m are integers such that $k \geq 1$, $m \geq 2$, and $a \equiv b \pmod{m}$, then $a^k \equiv b^k \pmod{m}$ whenever k is a positive integer.

38. Which memory locations are assigned by the hashing function $h(k) = k$ **mod** 101 to the records of students with the following social security numbers?
a) 104578690 b) 432222187
c) 372201919 d) 501338753

39. A parking lot has 31 visitor spaces, numbered from 0 to 30. Visitors are assigned parking spaces using the hashing function $h(k) = k$ **mod** 31, where k is the number formed from the first three digits on a visitor's license plate.
a) Which spaces are assigned by the hashing function to cars that have the following first three digits on their license plates?

317, 918, 007, 100, 111, 310

b) Describe a procedure visitors should follow to find a free parking space, when the space they are assigned is occupied.

40. What sequence of pseudorandom numbers is generated using the linear congruential generator $x_{n+1} = (4x_n + 1)$ **mod** 7 with seed $x_0 = 3$?

41. What sequence of pseudorandom numbers is generated using the pure multiplicative generator $x_{n+1} = 3x_n$ **mod** 11 with seed $x_0 = 2$?

42. Write an algorithm in pseudocode for generating a sequence of pseudorandom numbers using a linear congruential generator.

43. Encrypt the message "DO NOT PASS GO" by translating the letters into numbers, applying the encryption function given, and then translating the numbers back into letters.

a) $f(p) = (p + 3)$ **mod** 26 (the Caesar cipher)
b) $f(p) = (p + 13)$ **mod** 26
c) $f(p) = (3p + 7)$ **mod** 26

44. Decrypt the following messages encrypted using the Caesar cipher.

a) EOXH MHDQV
b) WHVW WRGDB
c) HDW GLP VXP

2.4 Integers and Algorithms

INTRODUCTION

As mentioned in Section 2.1, the term *algorithm* originally referred to procedures for performing arithmetic operations using the decimal representations of integers. These algorithms, adapted for use with binary representations, are the basis for computer arithmetic. They provide good illustrations of the concept of an algorithm and the complexity of algorithms. For these reasons, they will be discussed in this section.

There are many important algorithms involving integers besides those used in arithmetic. We will begin our discussion of integers and algorithms with the Euclidean algorithm. It is one of the most useful algorithms, and perhaps the oldest algorithm in mathematics. We will also describe an algorithm for finding the base b expansion of a positive integer for any base b.

THE EUCLIDEAN ALGORITHM

The method described in Section 2.3 for computing the greatest common divisor of two integers, using the prime factorizations of these integers, is inefficient. The reason is that it is time-consuming to find prime factorizations. We will give a more efficient method of finding the greatest common divisor, called the **Euclidean algorithm.** This algorithm has been known since ancient times. It is named after the ancient Greek mathematician Euclid, who included a description of this algorithm in his *Elements*.

Before describing the Euclidean algorithm, we will show how it is used to find $gcd(91, 287)$. First, divide 287, the larger of the two integers, by 91, the smaller, to obtain

$$287 = 91 \cdot 3 + 14.$$

Any divisor of 91 and 287 must also be a divisor of $287 - 91 \cdot 3 = 14$. Also, any divisor of 91 and 14 must also be a divisor of $287 = 91 \cdot 3 + 14$. Hence, the greatest

Euclid (c. 350 B.C.E.). Euclid was the author of the most successful mathematics book ever written, the *Elements,* which appeared in over 1000 different editions from ancient to modern times. Little is known about Euclid's life, other than that he taught at the famous academy at Alexandria. Apparently, Euclid did not stress applications. When a student asked what he would get by learning geometry, Euclid explained that knowledge was worth acquiring for its own sake and told his servant to give the student a coin "since he must make a profit from what he learns."

common divisor of 91 and 287 is the same as the greatest common divisor of 91 and 14. This means that the problem of finding $gcd(91, 287)$ has been reduced to the problem of finding $gcd(91, 14)$.

Next, divide 91 by 14 to obtain

$$91 = 14 \cdot 6 + 7.$$

Since any common divisor of 91 and 14 also divides $91 - 14 \cdot 6 = 7$, and any common divisor of 14 and 7 divides 91, it follows that $gcd(91, 14) = gcd(14, 7)$.

Continue by dividing 14 by 7, to obtain

$$14 = 7 \cdot 2.$$

Since 7 divides 14, it follows that $gcd(14, 7) = 7$. Since $gcd(287, 91) = gcd(91, 14) = gcd(14, 7) = 7$, the original problem has been solved.

We now describe how the Euclidean algorithm works in generality. We will use successive divisions to reduce the problem of finding the greatest common divisor of two positive integers to the same problem with smaller integers, until one of the integers is zero.

The Euclidean algorithm is based on the following result about greatest common divisors and the division algorithm.

LEMMA 1 Let $a = bq + r$, where a, b, q, and r are integers. Then $gcd(a, b) = gcd(b, r)$.

Proof: If we can show that the common divisors of a and b are the same as the common divisors of b and r, we will have shown that $gcd(a, b) = gcd(b, r)$, since both pairs must have the same *greatest* common divisor.

So suppose that d divides both a and b. Then it follows that d also divides $a - bq = r$ (from Theorem 1 of Section 2.3). Hence, any common divisor of a and b is also a common divisor of b and r.

Likewise, suppose that d divides both b and r. Then d also divides $bq + r = a$. Hence, any common divisor of b and r is also a common divisor of a and b.

Consequently, $gcd(a, b) = gcd(b, r)$. ■

Suppose that a and b are positive integers with $a \geq b$. Let $r_0 = a$ and $r_1 = b$. When we successively apply the division algorithm, we obtain

$$\begin{aligned}
r_0 &= r_1 q_1 + r_2 && 0 \leq r_2 < r_1, \\
r_1 &= r_2 q_2 + r_3 && 0 \leq r_3 < r_2, \\
&\vdots \\
r_{n-2} &= r_{n-1} q_{n-1} + r_n && 0 \leq r_n < r_{n-1}, \\
r_{n-1} &= r_n q_n.
\end{aligned}$$

Eventually a remainder of zero occurs in this sequence of successive divisions, since the sequence of remainders $a = r_0 > r_1 > r_2 > \cdots \geq 0$ cannot contain more than a terms. Furthermore, it follows from Lemma 1 that

$$\begin{aligned}
gcd(a, b) &= gcd(r_0, r_1) = gcd(r_1, r_2) = \cdots = gcd(r_{n-2}, r_{n-1}) \\
&= gcd(r_{n-1}, r_n) = gcd(r_n, 0) = r_n.
\end{aligned}$$

Hence, the greatest common divisor is the last nonzero remainder in the sequence of divisions.

EXAMPLE 1 Find the greatest common divisor of 414 and 662 using the Euclidean algorithm.

Solution: Successive uses of the division algorithm give:

$$\begin{aligned} 662 &= 414 \cdot 1 + 248 \\ 414 &= 248 \cdot 1 + 166 \\ 248 &= 166 \cdot 1 + 82 \\ 166 &= 82 \cdot 2 + 2 \\ 82 &= 2 \cdot 41. \end{aligned}$$

Hence, $gcd(414, 662) = 2$, since 2 is the last nonzero remainder. ■

The Euclidean algorithm is expressed in pseudocode in Algorithm 1.

ALGORITHM 1 The Euclidean Algorithm.

```
procedure gcd(a, b: positive integers)
x := a
y := b
while y ≠ 0
begin
  r := x mod y
  x := y
  y := r
end {gcd(a, b) is x}
```

In Algorithm 1, the initial values of x and y are a and b, respectively. At each stage of the procedure, x is replaced by y, and y is replaced by x **mod** y, which is the remainder when x is divided by y. This process is repeated as long as $y \neq 0$. The algorithm terminates when $y = 0$, and the value of x at that point, the last nonzero remainder in the procedure, is the greatest common divisor of a and b.

We will study the time complexity of the Euclidean algorithm in Section 3 of Chapter 3, where we will show that the number of divisions required to find the greatest common divisor of a and b, where $a \geq b$, is $O(\log b)$.

REPRESENTATIONS OF INTEGERS

In everyday life we use decimal notation to express integers. For example, 965 is used to denote $9 \cdot 10^2 + 6 \cdot 10 + 5$. However, it is often convenient to use bases other than 10. In particular, computers usually use binary notation (with 2 as the base) when carrying out arithmetic, and octal (base 8) or hexadecimal (base 16) notation when expressing characters, such as letters or digits. In fact, we can use any positive integer greater than 1 as the base when expressing integers. This is stated in the following theorem.

THEOREM 1 Let b be a positive integer greater than 1. Then if n is a positive integer, it can be expressed uniquely in the form

$$n = a_k b^k + a_{k-1} b^{k-1} + \cdots + a_1 b + a_0,$$

where k is a nonnegative integer, $a_0, a_1, \ldots, a_k$ are nonnegative integers less than b, and $a_k \neq 0$.

The proof of this theorem may be found in the suggested readings referred to at the end of the book. The representation of n given in Theorem 1 is called the **base *b* expansion of *n*.** The base b expansion of n is denoted by $(a_k a_{k-1} \cdots a_1 a_0)_b$. For instance, $(245)_8$ represents $2 \cdot 8^2 + 4 \cdot 8 + 5 = 165$.

Choosing 2 as the base gives **binary expansions** of integers. In binary notation each digit is either a 0 or a 1. In other words, the binary expansion of an integer is just a bit string. Binary expansions (and related expansions that are variants of binary expansions) are used by computers to represent and do arithmetic with integers.

EXAMPLE 2 What is the decimal expansion of the integer that has $(101011111)_2$ as its binary expansion?

Solution: We have

$$(101011111)_2 = 2^8 + 2^6 + 2^4 + 2^3 + 2^2 + 2 + 1 = 351.$$ ■

Sixteen is another base used in computer science. The base 16 expansion of an integer is called its **hexadecimal** expansion. Sixteen different digits are required for such expansions. Usually, the hexadecimal digits used are 0, 1, 2, 3, 4, 5, 6, 7, 8, 9, A, B, C, D, E, and F, where the letters A through F represent the digits corresponding to the numbers 10 through 15 (in decimal notation).

EXAMPLE 3 What is the decimal expansion of the hexadecimal expansion of $(2AE0B)_{16}$?

Solution: We have

$$(2AE0B)_{16} = 2 \cdot 16^4 + 10 \cdot 16^3 + 14 \cdot 16^2 + 0 \cdot 16 + 11 = (175627)_{10}.$$ ■

Since a hexadecimal digit is represented using four bits, **bytes,** which are bit strings of length eight, can be represented by two hexadecimal digits. For instance, $(11100101)_2 = (E5)_{16}$ since $(1110)_2 = (E)_{16}$ and $(0101)_2 = (5)_{16}$.

We will now describe an algorithm for constructing the base b expansion of an integer n. First, divide n by b to obtain a quotient and remainder, that is

$$n = bq_0 + a_0, \quad 0 \le a_0 < b.$$

The remainder, a_0, is the rightmost digit in the base b expansion of n. Next, divide q_0 by b to obtain

$$q_0 = bq_1 + a_1, \quad 0 \le a_1 < b.$$

We see that a_1 is the second digit from the right in the base b expansion of n. Continue this process, successively dividing the quotients by b, obtaining additional base b digits as the remainders. This process terminates when we obtain a quotient equal to zero.

EXAMPLE 4 Find the base 8 expansion of $(12345)_{10}$.

Solution: First, divide 12345 by 8 to obtain

$$12345 = 8 \cdot 1543 + 1.$$

Successively dividing quotients by 8 gives

$$\begin{aligned} 1543 &= 8 \cdot 192 + 7 \\ 192 &= 8 \cdot 24 + 0 \\ 24 &= 8 \cdot 3 + 0 \\ 3 &= 8 \cdot 0 + 3. \end{aligned}$$

Since the remainders are the digits of the base 8 expansion of 12345, it follows that

$$(12345)_{10} = (30071)_8.$$ ■

The pseudocode given in Algorithm 2 finds the base b expansion $(a_{k-1} \cdots a_1 a_0)_b$ of the integer n.

ALGORITHM 2 Constructing Base b Expansions.

procedure *base b expansion*(n: positive integer)
$q := n$
$k := 0$
while $q \ne 0$
begin
  $a_k := q$ **mod** b
  $q := \lfloor q/b \rfloor$
  $k := k + 1$
end {the base b expansion of n is $(a_{k-1} \cdots a_1 a_0)_b$}

In Algorithm 2, q represents the quotient obtained by successive divisions by b, starting with $q = n$. The digits in the base b expansion are the remainders of these divisions and are given by q **mod** b. The algorithm terminates when a quotient $q = 0$ is reached.

ALGORITHMS FOR INTEGER OPERATIONS

The algorithms for performing operations with integers using their binary expansions are extremely important in computer arithmetic. We will describe algorithms for the addition and the multiplication of two integers expressed in binary notation. We will also analyze the computational complexity of these algorithms, in terms of the actual number of bit operations used. Throughout this discussion, suppose that the binary expansions of a and b are

$$a = (a_{n-1}a_{n-2}\cdots a_1a_0)_2, \quad b = (b_{n-1}b_{n-2}\cdots b_1b_0)_2,$$

so that a and b each have n bits (putting bits equal to 0 at the beginning of one of these expansions if necessary).

Consider the problem of adding two integers in binary notation. A procedure to perform addition can be based on the usual method for adding numbers with pencil and paper. This method proceeds by adding pairs of binary digits together with carries, when they occur, to compute the sum of two integers. This procedure will now be specified in detail.

To add a and b, first add their rightmost bits. This gives

$$a_0 + b_0 = c_0 \cdot 2 + s_0,$$

where s_0 is the rightmost bit in the binary expansion of $a + b$, and c_0 is the **carry**, which is either 0 or 1. Then add the next pair of bits and the carry,

$$a_1 + b_1 + c_0 = c_1 \cdot 2 + s_1,$$

where s_1 is the next bit (from the right) in the binary expansion of $a + b$, and c_1 is the carry. Continue this process, adding the corresponding bits in the two binary expansions and the carry, to determine the next bit from the right in the binary expansion of $a + b$. At the last stage, add a_{n-1}, b_{n-1}, and c_{n-2} to obtain $c_{n-1} \cdot 2 + s_{n-1}$. The leading bit of the sum is $s_n = c_{n-1}$. This procedure produces the binary expansion of the sum, namely, $a + b = (s_n s_{n-1} s_{n-2} \cdots s_1 s_0)_2$.

EXAMPLE 5 Add $a = (1110)_2$ and $b = (1011)_2$.

Solution: Following the procedure specified in the algorithm, first note that

$$a_0 + b_0 = 0 + 1 = 0 \cdot 2 + 1,$$

so that $c_0 = 0$ and $s_0 = 1$. Then, since

$$a_1 + b_1 + c_0 = 1 + 1 + 0 = 1 \cdot 2 + 0,$$

```
  1 1
  1 1 1 0
  1 0 1 1
1 1 0 0 1
```

FIGURE 1 Adding $(1110)_2$ and $(1011)_2$.

it follows that $c_1 = 1$ and $s_1 = 0$. Continuing,

$$a_2 + b_2 + c_1 = 1 + 0 + 1 = 1 \cdot 2 + 0,$$

so that $c_2 = 1$ and $s_2 = 0$. Finally, since

$$a_3 + b_3 + c_2 = 1 + 1 + 1 = 1 \cdot 2 + 1,$$

it follows that $c_3 = 1$ and $s_3 = 1$. This means that $s_4 = c_3 = 1$. Therefore, $s = a + b = (11001)_2$. This addition is displayed in Figure 1. ■

The algorithm for addition can be described using pseudocode as follows.

ALGORITHM 3 Addition of Integers.

procedure *add*(a,b: positive integers)
{the binary expansions of a and b are $(a_{n-1}a_{n-2} \cdots a_1a_0)_2$
  and $(b_{n-1}b_{n-2} \cdots b_1b_0)_2$, respectively}
$c := 0$
for $j := 0$ **to** $n - 1$
begin
  $d := \lfloor (a_j + b_j + c)/2 \rfloor$
  $s_j := a_j + b_j + c - 2d$
  $c := d$
end
$s_n := c$
{the binary expansion of the sum is $(s_n s_{n-1} \cdots s_0)_2$}

Next, the number of additions of bits used by Algorithm 3 will be analyzed.

EXAMPLE 6 How many additions of bits are required to use Algorithm 3 to add two integers with n bits (or less) in their binary representations?

Solution: Two integers are added by successively adding pairs of bits and, when it occurs, a carry. Adding each pair of bits and the carry requires three or fewer additions of bits. Thus, the total number of additions of bits used is less than three times the number of bits in the expansion. Hence, the number of additions of bits used by Algorithm 3 to add two n-bit integers is $O(n)$. ■

Next, consider the multiplication of two n-bit integers a and b. The conventional algorithm (used when multiplying with pencil and paper) works as follows. Using the distributive law, we see that

$$ab = a\sum_{j=0}^{n-1} b_j 2^j = \sum_{j=0}^{n-1} a(b_j 2^j).$$

We can compute *ab* using this equation. We first note that $ab_j = a$ if $b_j = 1$ and $ab_j = 0$ if $b_j = 0$. Each time we multiply a term by 2, we shift its binary expansion one place to the left and add a zero at the tail end of the expansion. Consequently, we can obtain $(ab_j)2^j$ by **shifting** the binary expansion of ab_j j places to the left, adding j zero bits at the tail end of this binary expansion. Finally, we obtain *ab* by adding the *n* integers $ab_j 2^j$, $j = 0, 1, 2, \ldots, n-1$.

The following example illustrates the use of this algorithm.

EXAMPLE 7 Find the product of $a = (110)_2$ and $b = (101)_2$.

Solution: First note that

$$ab_0 \cdot 2^0 = (110)_2 \cdot 1 \cdot 2^0 = (110)_2,$$
$$ab_1 \cdot 2^1 = (110)_2 \cdot 0 \cdot 2^1 = (0000)_2,$$

and

$$ab_2 \cdot 2^2 = (110)_2 \cdot 1 \cdot 2^2 = (11000)_2.$$

To find the product, add $(110)_2$, $(0000)_2$, and $(11000)_2$. Carrying out these additions (using Algorithm 3, including initial zero bits when necessary) shows that $ab = (11110)_2$. This multiplication is displayed in Figure 2. ■

```
    1 1 0
    1 0 1
    -----
    1 1 0
  0 0 0
1 1 0
---------
1 1 1 1 0
```

FIGURE 2 Multiplying $(110)_2$ and $(101)_2$.

This procedure for multiplication can be described using pseudocode as follows.

ALGORITHM 4 Multiplying Integers.

procedure *multiply*(*a*,*b*: positive integers)
{the binary expansions of a and b are $(a_{n-1}a_{n-2}\cdots a_1a_0)_2$
 and $(b_{n-1}b_{n-2}\cdots b_1b_0)_2$, respectively}
for $j := 0$ **to** $n-1$
begin
 if $b_j = 1$ **then** $c_j :=$ a shifted j places
 else $c_j := 0$
end
{$c_0, c_1, \ldots, c_{n-1}$ are the partial products}
$p := 0$
for $j := 0$ **to** $n-1$
 $p := p + c_j$
{p is the value ofab}

Next, we determine the number of additions of bits and shifts of bits used by Algorithm 4 to multiply two integers.

EXAMPLE 8 How many additions of bits and shifts of bits are used to multiply a and b using Algorithm 4?

Solution: Algorithm 4 computes the products of a and b by adding the partial products $c_0, c_1, c_2, \ldots,$ and c_{n-1}. When $b_j = 1$, we compute the partial product c_j by shifting the binary expansion of aj bits. When $b_j = 0$, no shifts are required since $c_j = 0$. Hence, to find all n of the integers $ab_j 2^j$, $j = 0, 1, \ldots, n-1$, requires at most

$$0 + 1 + 2 + \cdots + n - 1 = O(n^2)$$

shifts (from Example 4 in Section 1.8).

To add the integers ab_j from $j = 0$ to $j = n - 1$ requires the addition of an n-bit integer, an $(n+1)$-bit integer, . . . , and a $(2n)$-bit integer. We know from Example 8 that each of these additions requires $O(n)$ additions of bits. Consequently, a total of $O(n^2)$ additions of bits are required for all n additions. ■

Surprisingly, there are more efficient algorithms than the conventional algorithm for multiplying integers. One such algorithm, which uses $O(n^{1.585})$ bit operations to multiply n-bit numbers, will be described in Chapter 5.

Exercises

1. Use the Euclidean algorithm to find
 a) $gcd(12, 18)$ **b)** $gcd(111, 201)$
 c) $gcd(1001, 1331)$ **d)** $gcd(12345, 54321)$

2. Use the Euclidean algorithm to find
 a) $gcd(1, 5)$ **b)** $gcd(100, 101)$
 c) $gcd(123, 277)$ **d)** $gcd(1529, 14039)$
 e) $gcd(1529, 14038)$ **f)** $gcd(11111, 111111)$

3. How many divisions are required to find $gcd(21, 34)$ using the Euclidean algorithm?

4. How many divisions are required to find $gcd(34, 55)$ using the Euclidean algorithm?

5. Convert the following integers from decimal notation to binary notation.
 a) 231 **b)** 4532 **c)** 97644

6. Convert the following integers from decimal notation to binary notation.
 a) 321 **b)** 1023 **c)** 100632

7. Convert the following integers from binary notation to decimal notation.
 a) 11111 **b)** 10000 00001
 c) 10101 0101 **d)** 11010 01000 10000

8. Convert the following integers from binary notation to decimal notation.
 a) 11011 **b)** 10101 10101
 c) 11101 11110 **d)** 11111 00000 11111

9. Devise a simple method for converting from hexadecimal notation to binary notation.

10. Devise a simple method for converting from binary notation to hexadecimal notation.

11. Convert each of the following integers from hexadecimal notation to binary notation.
 a) 80E **b)** 135AB
 c) ABBA **d)** DEFACED

12. Convert each of the following integers from binary notation to hexadecimal notation.
 a) 111 10111 **b)** 10 10101 01010
 c) 11101 11011 10111

13. Show that every positive integer can be represented uniquely as the sum of distinct powers of 2. (*Hint:* Consider binary expansions of integers.)

14. It can be shown that every integer can be uniquely represented in the form

$$e_k 3^k + e_{k-1} 3^{k-1} + \cdots + e_1 3 + e_0,$$

where $e_j = -1, 0,$ or 1 for $j = 0, 1, 2, \ldots, k$. Expansions of this type are called **balanced ternary expansions.** Find the balanced ternary expansions of

a) 5. **b)** 13. **c)** 37. **d)** 79.

15. Show that a positive integer is divisible by 3 if and only if the sum of its decimal digits is divisible by 3.

One's complement representations of integers are used to simplify computer arithmetic. To represent positive and negative integers with absolute value less than 2^n, a total of $n+1$ bits are used. The leftmost bit is used to represent the sign. A 0 bit in this position is used for positive integers, and a 1 bit in this position is used for negative integers. For positive integers the remaining bits are identical to the binary expansion of the integer. For negative integers, the remaining bits are obtained by first finding the binary expansion of the absolute value of the integer, and then taking the complement of each of these bits, where the complement of a 1 is a 0 and the complement of a 0 is a 1.

16. Find the one's complement representations, using bit strings of length six, of the following integers.

a) 22 **b)** 31 **c)** −7 **d)** −19

17. What integer does each of the following one's complement representations of length five represent?

a) 11001 **b)** 01101 **c)** 10001
d) 11111

18. How is the one's complement representation of $-m$ obtained from the one's complement of m, when bit strings of length n are used?

19. How is the one's complement representation of the sum of two integers obtained from the one's complement representations of these integers?

20. How is the one's complement representation of the difference of two integers obtained from the one's complement representations of these integers?

21. Sometimes integers are encoded by using four-digit binary expansions to represent each decimal digit. This produces the **binary coded decimal** form of the integer. For instance, 791 is encoded in this way by 011110010001. How many bits are required to represent a number with n decimal digits using this type of encoding?

A **Cantor expansion** is a sum of the form

$$a_n n! + a_{n-1}(n-1)! + \cdots + a_2 2! + a_1 1!,$$

where a_i is an integer with $0 \le a_i \le i$ for $i = 1, 2, \ldots, n$.

22. Find the Cantor expansions of

a) 2 **b)** 7 **c)** 19
d) 87 **e)** 1000 **f)** 1,000,000

★23. Describe an algorithm that finds the Cantor expansion of an integer.

★24. Describe an algorithm to add two integers from their Cantor expansions.

25. Add $(10111)_2$ and $(11010)_2$ by working through each step of the algorithm for addition given in the text.

26. Multiply $(1110)_2$ and $(1010)_2$ by working through each step of the algorithm for multiplication given in the text.

27. Describe an algorithm for finding the difference of two binary expansions.

28. Estimate the number of bit operations used to subtract two binary expansions.

29. Devise an algorithm that, given the binary expansions of the integers a and b, determines whether $a > b$, $a = b$, or $a < b$.

30. How many bit operations does the comparison algorithm from Exercise 29 use when the larger of a and b has n bits in its binary expansion?

31. Estimate the complexity of Algorithm 2 for finding the base b expansion of an integer n in terms of the number of divisions used.

2.5 Applications of Number Theory

INTRODUCTION

Number theory has many applications, especially to computer science. In Section 2.3 we described several of these applications, including hashing functions, the generation of pseudorandom numbers, and shift ciphers. This section continues our introduction to number theory, developing some key results and presenting two important applications: a method for performing arithmetic with large integers and a recently invented type of cryptosystem, called a *public key system*. In such a cryptosystem, we do not have to keep encryption keys secret, since knowledge of an encryption key does not help someone decrypt messages in a realistic amount of time. Privately held decryption keys are used to decrypt messages.

Before developing these applications, we will introduce some key results that play a central role in number theory and its applications. For example, we will show how to solve systems of linear congruences modulo pairwise relatively prime integers using the Chinese Remainder Theorem, and then show how to use this result as a basis for performing arithmetic with large integers. We will introduce Fermat's Little Theorem and the concept of a pseudoprime, and show how to use these concepts to develop a public key cryptosystem.

SOME USEFUL RESULTS

An important result we will use throughout this section is that the greatest common divisor of two integers a and b can be expressed in the form

$$sa + tb,$$

where s and t are integers. In other words, $gcd(a, b)$ can be expressed as a **linear combination** with integer coefficients of a and b. For example, $gcd(6, 14) = 2$, and $2 = (-2) \cdot 6 + 1 \cdot 14$. We state this fact as Theorem 1.

THEOREM 1 If a and b are positive integers, then there exist integers s and t such that $gcd(a, b) = sa + tb$.

We will not give a formal proof of Theorem 1, but we will provide an example of a method for finding a linear combination of two integers equal to their greatest common divisor. (In this section, we will assume that a linear combination has integer coefficients.) The method proceeds by working backward through the divisions of the Euclidean algorithm.

EXAMPLE 1 Express $gcd(252, 198) = 18$ as a linear combination of 252 and 198.

Solution: To show that $gcd(252, 198) = 18$, the Euclidean algorithm uses the following divisions:

$$\begin{aligned} 252 &= 1 \cdot 198 + 54 \\ 198 &= 3 \cdot 54 + 36 \\ 54 &= 1 \cdot 36 + 18 \\ 36 &= 2 \cdot 18. \end{aligned}$$

Using the next-to-last division (the third division), we can express $gcd(252, 198) = 18$ as a linear combination of 54 and 36. We find that

$$18 = 54 - 1 \cdot 36.$$

The second division tells us that

$$36 = 198 - 3 \cdot 54.$$

Substituting this expression for 36 into the previous equation, we can express 18 as a linear combination of 54 and 198. We have

$$18 = 54 - 1 \cdot 36 = 54 - 1 \cdot (198 - 3 \cdot 54) = 4 \cdot 54 - 1 \cdot 198.$$

The first division tells us that

$$54 = 252 - 1 \cdot 198.$$

Substituting this expression for 54 into the previous equation, we can express 18 as a linear combination of 252 and 198. We conclude that

$$18 = 4 \cdot (252 - 1 \cdot 198) - 1 \cdot 198 = 4 \cdot 252 - 5 \cdot 198,$$

completing the solution. ■

The method used in Example 1 works for any pair of positive integers. (There are more efficient methods for expressing $gcd(a, b)$ as a linear combination of a and b; consult the references mentioned at the end of the book to learn more about these methods.)

We will use Theorem 1 to develop several useful results. One of our goals will be to prove the part of the Fundamental Theorem of Arithmetic asserting that a positive integer has at most one prime factorization. We will show that if a positive integer has a factorization into primes, where the primes are written in nondecreasing order, then this factorization is unique.

First, we need to develop some results about divisibility.

LEMMA 1 If a, b, and c are positive integers such that $gcd(a, b) = 1$ and $a \mid bc$, then $a \mid c$.

Proof: Since $gcd(a, b) = 1$, by Theorem 1 there are integers s and t such that

$$sa + tb = 1.$$

Multiplying both sides of this equation by c, we obtain

$$sac + tbc = c.$$

Using Theorem 1 of Section 2.3, we can use this last equation to show that $a \mid c$. By part 2 of that theorem, $a \mid tbc$. Since $a \mid sac$ and $a \mid tbc$, by part 1 of that theorem, we conclude that a divides $sac + tbc$, and hence $a \mid c$. This finishes the proof. ■

We will use the following generalization of Lemma 1 in the proof of uniqueness of prime factorizations. (The proof of Lemma 2 is left as an exercise in Section 3.2, since it can be most easily carried out using the method of mathematical induction, which will be covered in that section.)

LEMMA 2 If p is a prime and $p \mid a_1a_2 \cdots a_n$ where each a_i is an integer, then $p \mid a_i$ for some i.

We can now show that a factorization of an integer into primes is unique. That is, we will show that every integer can be written as the product of primes in nondecreasing order in at most one way. This is part of the Fundamental Theorem of Arithmetic. We will prove the other part, that every integer has a factorization into primes, in Section 3.2.

Proof: Suppose that the positive integer n can be written as the product of primes in two different ways, say $n = p_1p_2 \cdots p_s$ and $n = q_1q_2 \cdots q_t$, each p_i and q_j are primes such that $p_1 \leq p_2 \leq \cdots \leq p_s$ and $q_1 \leq q_2 \leq \cdots \leq q_t$.

When we remove all common primes from the two factorizations, we have

$$p_{i_1}p_{i_2} \cdots p_{i_u} = q_{j_1}q_{j_2} \cdots q_{j_v},$$

where no prime occurs on both sides of this equation and u and v are positive integers. By Lemma 2 it follows that p_{i_1} divides q_{j_k} for some k. Since no prime divides another prime, this is impossible. Consequently, there can be at most one factorization of n into primes in nondecreasing order. ■

Lemma 1 can also be used to prove a result about dividing both sides of a congruence by the same integer. We have shown (Theorem 7 in Section 2.3) that we can multiply both sides of a congruence by the same integer. However, dividing both sides of a congruence by an integer does not always produce a valid congruence, as the following example shows.

EXAMPLE 2 The congruence $14 \equiv 8 \pmod 6$ holds, but both sides of this congruence cannot be divided by 2 since $14/2 = 7$ and $8/2 = 4$, but $7 \not\equiv 4 \pmod 6$. ■

However, using Lemma 1, we can show that we can divide both sides of a congruence by an integer relatively prime to the modulus. This is stated as Theorem 2.

THEOREM 2 Let m be a positive integer and let a, b, and c be integers. If $ac \equiv bc \pmod m$ and $gcd(c, m) = 1$, then $a \equiv b \pmod m$.

Proof: Since $ac \equiv bc \pmod m$, $m \mid ac - bc = c(a - b)$. By Lemma 1, since $gcd(c, m) = 1$, it follows that $m \mid a - b$. We conclude that $a \equiv b \pmod m$. ■

LINEAR CONGRUENCES

A congruence of the form

$$ax \equiv b \pmod{m}$$

where m is a positive integer, a and b are integers, and x is a variable, is called a **linear congruence.** Such congruences arise throughout number theory and its applications.

How can we solve the linear congruence $ax \equiv b \pmod{m}$, that is, find all integers x that satisfy this congruence? One method that we will describe uses an integer $\overline{a}$ such that $\overline{a}a \equiv 1 \pmod{m}$, if such an integer exists. Such an integer $\overline{a}$ is said to be an **inverse** of a modulo m. Theorem 3 guarantees that an inverse of a modulo m exists whenever a and m are relatively prime.

THEOREM 3 If a and m are relatively prime integers and $m > 1$, then an inverse of a modulo m exists. Furthermore, this inverse is unique modulo m. (That is, there is a unique positive integer $\overline{a}$ less than m that is an inverse of a modulo m and every other inverse of a modulo m is congruent to $\overline{a}$ modulo m.)

Proof: By Theorem 1, since $gcd(a, m) = 1$, there are integers s and t such that

$$sa + tm = 1.$$

This implies that

$$sa + tm \equiv 1 \pmod{m}.$$

Since $tm \equiv 0 \pmod{m}$, it follows that

$$sa \equiv 1 \pmod{m}.$$

Consequently, s is an inverse of a modulo m. That this inverse is unique modulo m is left as Exercise 9 at the end of this section. ■

The proof of Theorem 3 describes a method for finding the inverse of a modulo m when a and m are relatively prime: find a linear combination of a and m that equals 1 (which can be done by working backward through the steps of the Euclidean algorithm); the coefficient of a in this linear combination is an inverse of a modulo m. We illustrate this procedure in Example 3.

EXAMPLE 3 Find an inverse of 3 modulo 7.

Solution: Since $gcd(3, 7) = 1$, Theorem 2 tells us that an inverse of 3 modulo 7 exists. The Euclidean algorithm ends quickly when used to find the greatest common divisor of 3 and 7:

$$7 = 2 \cdot 3 + 1.$$

From this equation we see that

$$-2 \cdot 3 + 1 \cdot 7 = 1.$$

This shows that -2 is an inverse of 3 modulo 7. (Note that every integer congruent to -2 modulo 7 is also an inverse of 3, such as 5, -9, 12, and so on.) ■

When we have an inverse $\overline{a}$ of a modulo m, we can easily solve the congruence $ax \equiv b \pmod{m}$ by multiplying both sides of the linear congruence by $\overline{a}$, as Example 4 illustrates.

EXAMPLE 4 What are the solutions of the linear congruence $3x \equiv 4 \pmod{7}$?

Solution: By Example 3 we know that -2 is an inverse of 3 modulo 7. Multiplying both sides of the congruence by -2 shows that

$$-2 \cdot 3x \equiv -2 \cdot 4 \pmod{7}.$$

Since $-6 \equiv 1 \pmod{7}$ and $-8 \equiv 6 \pmod{7}$, it follows that if x is a solution, then $x \equiv -8 \equiv 6 \pmod{7}$.

We need to determine whether every x with $x \equiv 6 \pmod{7}$ is a solution. Assume that $x \equiv 6 \pmod{7}$. Then, by Theorem 7 of Section 2.3, it follows that

$$3x \equiv 3 \cdot 6 = 18 \equiv 4 \pmod{7},$$

which shows that all such x satisfy the congruence. We conclude that the solutions to the congruence are the integers x such that $x \equiv 6 \pmod{7}$, namely, 6, 13, 20, ... and $-1, -8, -15, \ldots$. ■

THE CHINESE REMAINDER THEOREM

Systems of linear congruences arise in many contexts. For example, as we will see later, they are the basis for a method that can be used to perform arithmetic with large integers. Such systems can even be found as word puzzles in the writings of ancient Chinese and Hindu mathematicians, such as that given in Example 5.

EXAMPLE 5 In the first century, the Chinese mathematician Sun-Tsu asked:

> There are certain things whose number is unknown. When divided by 3, the remainder is 2; when divided by 5, the remainder is 3; and when divided by 7, the remainder is 2. What will be the number of things?

This puzzle can be translated into the following question: What are the solutions of the systems of congruences

$$x \equiv 2 \pmod{3},$$
$$x \equiv 3 \pmod{5},$$
$$x \equiv 2 \pmod{7}?$$

We will solve this system, and with it Sun-Tsu's puzzle, later in this section. ■

The *Chinese Remainder Theorem*, named after the Chinese heritage of problems involving systems of linear congruences, states that when the moduli of a system of linear congruences are pairwise relatively prime, there is a unique solution of the system modulo the product of the moduli.

THEOREM 4 **THE CHINESE REMAINDER THEOREM.** Let $m_1, m_2, \ldots, m_n$ be pairwise relatively prime positive integers. The system

$$x \equiv a_1 \pmod{m_1},$$
$$x \equiv a_2 \pmod{m_2},$$
$$\vdots$$
$$x \equiv a_n \pmod{m_n}$$

has a unique solution modulo $m = m_1 m_2 \cdots m_n$. (That is, there is a solution x with $0 \leq x < m$, and all other solutions are congruent modulo m to this solution.)

Proof: To establish this theorem, we need to show that a solution exists and that it is unique modulo m. We will show that a solution exists by describing a way to construct this solution; showing that the solution is unique modulo m is left as Exercise 20 at the end of this section.

To construct a simultaneous solution, first let

$$M_k = m/m_k$$

for $k = 1, 2, \ldots, n$. That is, M_k is the product of the moduli except for m_k. Since m_i and m_k have no common factors greater than 1 when $i \neq k$, it follows that $gcd(m_k, M_k) = 1$. Consequently, by Theorem 3, we know that there is an integer y_k, an inverse of M_k modulo m_k, such that

$$M_k y_k \equiv 1 \pmod{m_k}.$$

To construct a simultaneous solution, form the sum

$$x = a_1 M_1 y_1 + a_2 M_2 y_2 + \cdots + a_n M_n y_n.$$

We will now show that x is a simultaneous solution. First, note that since $M_j \equiv 0 \pmod{m_k}$ whenever $j \neq k$, all terms except the kth term in this sum are congruent to 0 modulo m_k. Since $M_k y_k \equiv 1 \pmod{m_k}$ we see that

$$x \equiv a_k M_k y_k \equiv a_k \pmod{m_k},$$

for $k = 1, 2, \ldots, n$. We have shown that x is a simultaneous solution to the n congruences. ■

The following example illustrates how to use the construction given in the proof of Theorem 4 to solve a system of congruences. We will solve the system given in Example 5, arising in Sun-Tsu's puzzle.

EXAMPLE 6 To solve the system of congruences in Example 5, first let $m = 3 \cdot 5 \cdot 7 = 105$, $M_1 = m/3 = 35$, $M_2 = m/5 = 21$, and $M_3 = m/7 = 15$. We see that 2 is an inverse of $M_1 = 35$ modulo 3, since $35 \equiv 2 \pmod 3$; 1 is an inverse of $M_2 = 21$ modulo 5, since $21 \equiv 1 \pmod 5$; and 1 is an inverse of $M_3 = 15 \pmod 7$, since $15 \equiv 1 \pmod 7$. The solutions to this system are those x such that

$$\begin{aligned} x \equiv a_1 M_1 y_1 + a_2 M_2 y_2 + a_3 M_3 y_3 &= 2 \cdot 35 \cdot 2 + 3 \cdot 21 \cdot 1 + 2 \cdot 15 \cdot 1 \pmod{105} \\ &= 233 \equiv 23 \pmod{105}. \end{aligned}$$

It follows that 23 is the smallest positive integer that is a simultaneous solution. We conclude that 23 is the smallest positive integer that leaves a remainder of 2 when divided by 3, a remainder of 3 when divided by 5, and a remainder of 2 when divided by 7. ■

COMPUTER ARITHMETIC WITH LARGE INTEGERS

Suppose that $m_1, m_2, \ldots, m_n$ are pairwise relatively prime integers greater than or equal to 2 and let m be their product. By the Chinese Remainder Theorem, we can show (see Exercise 18) that an integer a with $0 \le a < m$ can be uniquely represented by the n-tuple consisting of its remainders upon division by m_i, $i = 1, 2, \ldots, n$. That is, we can uniquely represent a by

$$(a \textbf{ mod } m_1, a \textbf{ mod } m_2, \ldots, a \textbf{ mod } m_n).$$

EXAMPLE 7 What are the pairs used to represent the nonnegative integers less than 12 when they are represented by the ordered pair where the first component is the remainder of the integer upon division by 3 and the second component is the remainder of the integer upon division by 4?

Solution: We have the following representations, obtained by finding the remainder of each integer when it is divided by 3 and by 4:

$$\begin{array}{lll} 0 = (0,0) & 4 = (1,0) & 8 = (2,0) \\ 1 = (1,1) & 5 = (2,1) & 9 = (0,1) \\ 2 = (2,2) & 6 = (0,2) & 10 = (1,2) \\ 3 = (0,3) & 7 = (1,3) & 11 = (2,3). \end{array}$$

■

To perform arithmetic with large integers, we select moduli $m_1, m_2, \ldots, m_n$ where each m_i is an integer greater than 2, $gcd(m_i, m_j) = 1$ whenever $i \neq j$, and

$m = m_1 m_2 \cdots m_n$ is greater than the result of the arithmetic operations we want to carry out.

Once we have selected our moduli, we carry out arithmetic operations with large integers by performing componentwise operations on the n-tuples representing these integers using their remainders upon division by $m_i, i = 1, 2, \ldots, n$. Once we have computed the value of each component in the result, we recover its value by solving a system of n congruences modulo $m_i, i = 1, 2, \ldots, n$. This method of performing arithmetic with large integers has several valuable features. First, it can be used to perform arithmetic with integers larger than can ordinarily be carried out on a computer. Second, computations with respect to the different moduli can be done in parallel, speeding up the arithmetic.

EXAMPLE 8 Suppose that performing arithmetic with integers less than 100 on a certain processor is much quicker than doing arithmetic with larger integers. We can restrict almost all our computations to integers less than 100 if we represent integers using their remainders modulo pairwise relatively prime integers less than 100. For example, we can use the moduli of 99, 98, 97, and 95. (These integers are relatively prime pairwise, since no two have a common factor greater than 1.)

By the Chinese Remainder Theorem, every nonnegative integer less than $99 \cdot 98 \cdot 97 \cdot 95 = 89{,}403{,}930$ can be represented uniquely by its remainders when divided by these four moduli. For example, we represent 123,684 as (33, 8, 9, 89), since 123,684 **mod** 99 = 33, 123,684 **mod** 98 = 8, 123,684 **mod** 97 = 9, and 123,684 **mod** 95 = 89. Similarly, we represent 413,456 as (32, 92, 42, 16).

To find the sum of 123,684 and 413,456, we work with these 4-tuples instead of these two integers directly. We add the 4-tuples componentwise and reduce each component with respect to the appropriate modulus. This yields

$$\begin{aligned}(33, 8, 9, 89) &+ (32, 92, 42, 16)\\ &= (65 \textbf{ mod } 99, 100 \textbf{ mod } 98, 51 \textbf{ mod } 97, 105 \textbf{ mod } 95)\\ &= (65, 2, 51, 10).\end{aligned}$$

To find the sum, that is, the integer represented by (65, 2, 51, 10), we need to solve the system of congruences

$$\begin{aligned}x &\equiv 65 \pmod{99}\\ x &\equiv 2 \pmod{98}\\ x &\equiv 51 \pmod{97}\\ x &\equiv 10 \pmod{95}\end{aligned}$$

It can be shown (see Exercise 27) that 537,140 is the unique nonnegative solution of this system less than 89,403,930. Consequently, 537,140 is the sum. Note that it is only when we have to recover the integer represented by (65,2,51,10) that we have to do arithmetic with integers larger than 100. ■

Particularly good choices for moduli for arithmetic with large integers are sets of integers of the form $2^k - 1$, where k is a positive integer, since it is easy to do binary arithmetic modulo such integers, and since it is easy to find sets of such integers that are pairwise relatively prime. (The second reason is a consequence of the fact that $gcd(2^a - 1, 2^b - 1) = 2^{gcd(a,b)} - 1$, as Exercise 29 shows.) Suppose, for instance, that we can do arithmetic with integers less than 2^{35} easily on our computer, but that working with larger integers requires special procedures. We can use pairwise relatively prime moduli less than 2^{35} to perform arithmetic with integers as large as their product. For example, as Exercise 30 shows, the integers $2^{35} - 1, 2^{34} - 1, 2^{33} - 1, 2^{31} - 1, 2^{29} - 1$, and $2^{23} - 1$ are pairwise relatively prime. Since the product of these six moduli exceeds 2^{184}, we can perform arithmetic with integers as large as 2^{184} (as long as the results do not exceed this number) by doing arithmetic modulo each of these six moduli, none of which exceeds 2^{35}.

PSEUDOPRIMES

In Section 2.3 we showed that an integer n is prime when it is not divisible by any prime p with $p \leq \sqrt{n}$. Unfortunately, using this criterion to show that a given integer is prime is inefficient. It requires that we find all primes not exceeding $\sqrt{n}$ and that we carry out trial division by each such prime to see whether it divides n.

Are there more efficient ways to determine whether an integer is prime? Ancient Chinese mathematicians believed that n was prime if and only if

$$2^{n-1} \equiv 1 \pmod{n}.$$

If this were true, it would provide an efficient primality test. Why did they believe this congruence could be used to determine whether an integer is prime? First, they observed that the congruence holds whenever n is prime. For example, 5 is prime and

$$2^{5-1} = 2^4 = 16 \equiv 1 \pmod{5}.$$

Second, they never found a composite integer n for which the congruence holds. The ancient Chinese were only partially correct. They were correct in thinking that the congruence holds whenever n is prime, but they were incorrect in concluding that n is necessarily prime if the congruence holds.

The great French mathematician Fermat showed that the congruence holds when n is prime. He proved the following, more general result.

Pierre de Fermat (1601–1665). Pierre de Fermat, one of the most important mathematicians of the 17th century, was a lawyer by profession. He is the most famous amateur mathematician in history. Fermat published little of his mathematical discoveries. It is through his correspondence with other mathematicians that we know of his work. Fermat was one of the inventors of analytic geometry and developed some of the fundamental ideas of calculus. Fermat, along with Pascal, gave probability theory a mathematical basis. Fermat formulated what is now the most famous unsolved problem in mathematics. He asserted that the equation $x^n + y^n = z^n$ has no nontrivial positive integer solutions when n is an integer greater than 2. In more than 300 years, no proof (or counterexample) has been found. In his copy of the works of the ancient Greek mathematician Diophantus, Fermat wrote that he had a proof but that it would not fit in the margin. Because no one has been able to find a proof, most people think that Fermat thought he had a proof, but it was incorrect. However, he may have been tempting others to look for a proof, not being able to find one himself.

THEOREM 5 **FERMAT'S LITTLE THEOREM** If p is prime and a is an integer not divisible by p, then

$$a^{p-1} \equiv 1 \pmod{p}.$$

Furthermore, for every integer a we have

$$a^p \equiv a \pmod{p}.$$

The proof of Theorem 5 is outlined in Exercise 17 at the end of this section.

Unfortunately, there are composite integers n such that $2^{n-1} \equiv 1 \pmod{n}$. Such integers are called **pseudoprimes.**

EXAMPLE 9 The integer 341 is a pseudoprime since it is composite ($341 = 11 \cdot 31$) and as Exercise 23 shows

$$2^{340} \equiv 1 \pmod{341}.$$

■

Although the ancient Chinese were wrong, pseudoprimes are relatively rare. Their scarcity—and the even greater scarcity of integers that pass more delicate tests that begin by determining whether an integer is a pseudoprime—can be used as the basis for efficient **probabilistic primality tests.** Such tests can be used to quickly show that it almost certainly is the case that a given integer is prime. (More precisely, these tests show that the probability that an integer that passes a series of tests is prime is close to 1; see Chapter 4 for a discussion of probability.) These probabilistic primality tests can be used, and are used, to find large primes extremely rapidly on computers.

PUBLIC KEY CRYPTOGRAPHY

In Section 2.3 we introduced methods for encrypting messages based on congruences. When these encryption methods are used, messages, which are strings of characters, are translated into numbers. Then the number for each character is transformed into another number, either using a shift or an affine transformation modulo 26. These methods are examples of **private key cryptosystems.** Knowing the encryption key lets you quickly find the decryption key. For example, when a shift cipher is used with encryption key k, a number p representing a letter is sent to

$$c = (p + k) \textbf{ mod } 26.$$

Decryption is carried out by shifting by $-k$, that is,

$$p = (c - k) \textbf{ mod } 26.$$

When a private key cryptosystem is used, a pair of people who wish to communicate in secret must have a separate key. Since anyone knowing this key can both encrypt and decrypt messages easily, these two people need to securely exchange the key.

In the mid-1970s, cryptologists introduced the concept of **public key cryptosystems.** When such cryptosystems are used, knowing how to send someone a message does not help you decrypt messages sent to this person. In such a system, every person can have a publicly known encryption key. Only the decryption keys are kept secret, and only the intended recipient of a message can decrypt it, since the encryption key does not let someone find the decryption key without an extraordinary amount of work (such as 2 billion years of computer time).

In 1976, three researchers at M.I.T.—Ron Rivest, Adi Shamir, and Len Adleman—introduced a public key cryptosystem, known as the **RSA system**, from the initials of its inventors. The RSA cryptosystem is based on modular exponentiation modulo the product of two large primes. Each individual has an encryption key consisting of a modulus $n = pq$, where p and q are large primes, say with 200 digits each, and an exponent e that is relatively prime to $(p-1)(q-1)$. To produce a usable key, two large primes must be found. This can be done quickly on a computer using probabilistic primality tests, referred to earlier in this section. However, the product of these primes $n = pq$, with approximately 400 digits, cannot be factored in a reasonable length of time. As we will see, this is an important reason why decryption cannot be done quickly without a separate decryption key.

RSA ENCRYPTION

In the RSA encryption method, messages are translated into sequences of integers. This can be done by translating each letter into an integer, as is done with the Caesar cipher. These integers are grouped together to form larger integers, each representing a block of letters. The encryption proceeds by transforming the integer M, representing the plaintext (the original message), to an integer C, representing the ciphertext (the encrypted message), using the function

$$C = M^e \ \mathbf{mod}\ n.$$

(To perform the encryption, we use an algorithm for fast modular multiplication, such as that described in Exercise 14 in the Supplementary Exercises at the end of this chapter.) We leave the encrypted message as blocks of numbers and send these to the intended recipient.

Example 10 illustrates how RSA encryption is performed. For practical reasons we use small primes p and q in this example, rather than primes with 100 or more digits. Although the cipher described in this example is not secure, it does illustrate the techniques used in the RSA cipher.

EXAMPLE 10 Encrypt the message STOP using the RSA cryptosystem with $p = 43$ and $q = 59$, so that $n = 43 \cdot 59 = 2537$, and with $e = 13$. Note that

$$gcd(e, (p-1)(q-1)) = gcd(13, 42 \cdot 58) = 1.$$

Solution: We translate the letters in STOP into their numerical equivalents and then group the numbers into blocks of four. We obtain

1819 1415.

We encrypt each block using the mapping

$$C = M^{13} \textbf{ mod } 2537.$$

Computations using fast modular multiplication show that $1819^{13} \textbf{ mod } 2537 = 2081$ and $1415^{13} \textbf{ mod } 2537 = 2182$. The encrypted message is 2081 2182. ■

RSA DECRYPTION

The plaintext message can be quickly recovered when the decryption key d, an inverse of e modulo $(p - 1)(q - 1)$, is known. (Such an inverse exists since $gcd(e, (p - 1)(q - 1)) = 1$.) To see this, note that if $de \equiv 1 \pmod{(p-1)(q-1)}$, there is an integer k such that $de = 1 + k(p - 1)(q - 1)$. It follows that

$$C^d = (M^e)^d = M^{de} = M^{1+k(p-1)(q-1)}.$$

By Fermat's Little Theorem (assuming that $gcd(M, p) = gcd(M, q) = 1$, which holds except in rare cases), it follows that $M^{p-1} \equiv 1 \pmod{p}$ and $M^{q-1} \equiv 1 \pmod{q}$. Consequently,

$$C^d \equiv M \cdot (M^{p-1})^{k(q-1)} \equiv M \cdot 1 \equiv M \pmod{p}$$

and

$$C^d \equiv M \cdot (M^{q-1})^{k(p-1)} \equiv M \cdot 1 \equiv M \pmod{q}.$$

Since $gcd(p, q) = 1$, it follows by the Chinese Remainder Theorem that

$$C^d \equiv M \pmod{pq}.$$

Example 11 illustrates how to decrypt messages sent using the RSA cryptosystem.

EXAMPLE 11 We receive the encrypted message 0981 0461. What is the decrypted message if it was encrypted using the RSA cipher from Example 10?

Solution: The message was encrypted using the RSA cryptosystem with $n = 43 \cdot 59$ and exponent 13. As Exercise 4 shows, $d = 937$ is an inverse of 13 modulo $42 \cdot 58 = 2436$. We use 937 as our decryption exponent. Consequently, to decrypt a block C, we compute

$$P = C^{937} \textbf{ mod } 2537.$$

To decrypt the message, we use the fast modular exponentiation algorithm to compute $0981^{937} \textbf{ mod } 2537 = 0704$ and $0461^{937} \textbf{ mod } 2537 = 1115$. Consequently, the numerical version of the original message is 0704 1115. Translating this back to English letters, we see that the message is HELP. ■

RSA AS A PUBLIC KEY SYSTEM

Why is the RSA cryptosystem suitable for public key cryptography? When we know the factorization of the modulus n, that is, when we know p and q, we can use the Euclidean algorithm to quickly find an exponent d inverse to e modulo $(p-1)(q-1)$. This lets us decrypt messages sent using our key. However, no method is known to decrypt messages that is not based on finding a factorization of n, or that does not also lead to the factorization of n. The most efficient factorization methods known (as of 1990) require billions of years to factor 400-digit integers. Consequently, when p and q are 200-digit primes, messages encrypted using $n = pq$ as the modulus cannot be found in a reasonable time unless the primes p and q are known.

Active research is underway to find new ways to efficiently factor integers. Integers thought, as recently as several years ago, to be far too large to be factored in a reasonable amount of time can now be factored routinely. Integers with more than 100 digits, as well as some with more than 150 digits, have been factored using team efforts. When new factorization techniques are found, it will be necessary to use larger primes to ensure secrecy of messages. Unfortunately, messages that were considered secure earlier can be saved and subsequently decrypted by unintended recipients when it becomes feasible to factor the $n = pq$ in the key used for RSA encryption.

The RSA method has been implemented and is used for some particularly sensitive applications. However, the most commonly used cryptosystem is a private key system known as DES (an acronym for the Data Encryption Standard). When DES is used, encryption and decryption can be performed extremely rapidly on a computer. Although some people believe that messages encrypted using DES can be broken by experts, DES is considered sufficiently secure in most situations. The use of public key cryptography, via the RSA system, is growing, but when the RSA system is used, encryption and decryption are too slow (using the current generation of computers) for many applications. However, there are applications that use both private key and public key systems. For example, a public key cryptosystem, such as RSA, can be used to distribute private keys to pairs of individuals when they wish to communicate. These people then use a private key system, such as DES, for encryption and decryption of messages.

Exercises

1. Express the greatest common divisor of each of the following pairs of integers as a linear combination of these integers.

a) 10,11 **b)** 21,44 **c)** 36,48 **d)** 34,55 **e)** 117,213 **f)** 0,223 **g)** 123,2347 **h)** 3454,4666 **i)** 9999,11111

2. Express the greatest common divisor of each of the following pairs of integers as a linear combination of these integers.

a) 9,11 **b)** 33,44
c) 35,78 **d)** 21,55
e) 101,203 **f)** 124,323
g) 2002,2339 **h)** 3457,4669
i) 10001,13422

3. Show that 15 is an inverse of 7 modulo 26.

4. Show that 937 is an inverse of 13 modulo 2436.

5. Find an inverse of 4 modulo 9.

6. Find an inverse of 2 modulo 17.

7. Find an inverse of 19 modulo 141.

8. Find an inverse of 144 modulo 233.

★9. Show that if a and m are relatively prime positive integers, then the inverse of a modulo m is unique modulo m. (*Hint:* Assume that there are two solutions b and c of the congruence $ax \equiv 1 \pmod{m}$. Use Theorem 2 to show that $b \equiv c \pmod{m}$.)

10. Show that an inverse of a modulo m does not exist if $gcd(a, m) > 1$.

11. Solve the congruence $4x \equiv 5 \pmod{9}$.

12. Solve the congruence $2x \equiv 7 \pmod{17}$.

★13. Show that if m is a positive integer greater than 1 and $ac \equiv bc \pmod{m}$, then $a \equiv b \pmod{m/gcd(c, m)}$.

14. **a)** Show that the positive integers less than 11, except 1 and 10, can be split into pairs of integers such that each pair consists of integers that are inverses of each other modulo 11.

b) Use part (a) to show that $10! \equiv -1 \pmod{11}$.

15. Show that if p is prime, the only solutions of $x^2 \equiv 1 \pmod{p}$ are integers x such that $x \equiv 1 \pmod{p}$ and $x \equiv -1 \pmod{p}$.

★16. **a)** Generalize the result in part (a) of Exercise 14; that is, show that if p is a prime, the positive integers less than p, except 1 and $p-1$, can be split into $(p-3)/2$ pairs of integers such that each pair consists of integers that are inverses of each other. (*Hint:* Use the result of Exercise 15.)

b) From part (a) conclude that $(p-1)! \equiv -1 \pmod{p}$ whenever p is prime. This result is known as **Wilson's theorem.**

c) What can we conclude if n is a positive integer such that $(n-1)! \not\equiv -1 \pmod{n}$?

★17. This exercise outlines a proof of Fermat's Little Theorem.

a) Suppose that a is not divisible by the prime p. Show that no two of the integers $1 \cdot a, 2 \cdot a, \ldots, (p-1)a$ are congruent modulo p.

b) Conclude from part (a) that the product of $1, 2, \ldots, p-1$ is congruent modulo p to the product of $a, 2a, \ldots, (p-1)a$. Use this to show that

$$(p-1)! \equiv a^{p-1}(p-1)! \pmod{p}.$$

c) Use Wilson's theorem (proved in Exercise 16) to show that $a^{p-1} \equiv 1 \pmod{p}$ if $p \nmid a$.

d) Use part (c) to show that $a^p \equiv a \pmod{p}$ for all integers a.

18. Use the Chinese Remainder Theorem to show that an integer a, with $0 \le a < m = m_1 m_2 \cdots m_n$ where the integers $m_1, m_2, \ldots, m_n$ are pairwise relatively prime, can be represented uniquely by the n-tuple (a **mod** m_1, a **mod** $m_2, \ldots,$ a **mod** m_n).

★19. Let $m_1, m_2, \ldots, m_n$ be pairwise relatively prime integers greater than or equal to 2. Show that if $a \equiv b \pmod{m_i}$ for $i = 1, 2, \ldots, n$, then $a \equiv b \pmod{m}$, where $m = m_1 m_2 \cdots m_n$.

★20. Complete the proof of the Chinese Remainder Theorem by showing that the simultaneous solution of a system of linear congruences modulo pairwise relatively prime integers is unique modulo the product of these moduli. (*Hint:* Assume that x and y are two simultaneous solutions. Show that $m_i \mid x - y$ for all i. Using Exercise 19, conclude that $m = m_1 m_2 \cdots m_n \mid x - y$.)

21. Which integers leave a remainder of 1 when divided by 2 and also leave a remainder of 1 when divided by 3?

22. Which integers are divisible by 5 but leave a remainder of 1 when divided by 3?

23. **a)** Show that $2^{340} \equiv 1 \pmod{11}$ by Fermat's Little Theorem and noting that $2^{340} = (2^{10})^{34}$.

b) Show that $2^{340} \equiv 1 \pmod{31}$ using the fact that $2^{340} = (2^5)^{68} = 32^{68}$.

c) Conclude from parts (a) and (b) that $2^{340} \equiv 1 \pmod{341}$.

24. Find the nonnegative integer a less than 28 represented by each of the following pairs, where each pair represents (a **mod** 4, a **mod** 7).

a) (0,0) **b)** (1,0) **c)** (1,1)
d) (2,1) **e)** (2,2) **f)** (0,3)
g) (2,0) **h)** (3,5) **i)** (3,6)

25. Express each nonnegative integer a less than 15 using the pair (a **mod** 3, a **mod** 5).

26. Explain how to use the pairs found in Exercise 25 to add 4 and 7.

27. Solve the system of congruences that arises in Example 8.

★**28.** Show that if a and b are positive integers, then $(2^a - 1) \mathbf{mod} (2^b - 1) = 2^{a \mathbf{mod} b} - 1$.

★★**29.** Use Exercise 28 to show that if a and b are positive integers, then $gcd(2^a - 1, 2^b - 1) = 2^{gcd(a,b)} - 1$. (*Hint:* Show that the remainders obtained when the Euclidean algorithm is used to compute $gcd(2^a - 1, 2^b - 1)$ are of the form $2^r - 1$, where r is a remainder arising when the Euclidean algorithm is used to find $gcd(a, b)$.)

30. Use Exercise 29 to show that the integers $2^{35} - 1, 2^{34} - 1, 2^{33} - 1, 2^{31} - 1, 2^{29} - 1$, and $2^{23} - 1$ are pairwise relatively prime.

★**31.** Show that we can easily factor n when we know that n is the product of two primes, p and q, and we know the value of $(p - 1)(q - 1)$.

32. Encrypt the message ATTACK using the RSA system with $n = 43 \cdot 59$ and $e = 13$, translating each letter into integers and grouping together pairs of integers, as done in Example 10.

33. What is the original message encrypted using the RSA system with $n = 43 \cdot 59$ and $e = 13$ if the encrypted message is 0667 1947 0671? (Note: Some computational aid is needed to do this in a realistic amount of time.)

2.6 Matrices

INTRODUCTION

Matrices are used throughout discrete mathematics to express relationships between elements in sets. In subsequent chapters we will use matrices in a wide variety of models. For instance, matrices will be used in models of communications networks and transportation systems. Many algorithms will be developed that use these matrix models. This section contains a review of matrix arithmetic that will be used in these algorithms.

DEFINITION 1. A *matrix* is a rectangular array of numbers. A matrix with m rows and n columns is called an $m \times n$ matrix. The plural of matrix is *matrices*. A matrix with the same number of rows as columns is called *square*.

EXAMPLE 1 The matrix

$$\begin{bmatrix} 1 & 1 \\ 0 & 2 \\ 1 & 3 \end{bmatrix}$$

is a 3×2 matrix. ■

We now introduce some terminology about matrices. Boldface uppercase letters will be used to represent matrices.

DEFINITION 2. Let

$$\mathbf{A} = \begin{bmatrix} a_{11} & a_{12} & \cdots & a_{1n} \\ a_{21} & a_{22} & \cdots & a_{2n} \\ \cdot & \cdot & & \cdot \\ \cdot & \cdot & & \cdot \\ \cdot & \cdot & & \cdot \\ a_{n1} & a_{n2} & \cdots & a_{nn} \end{bmatrix}.$$

The ith *row* of $\mathbf{A}$ is the $1 \times n$ matrix $[a_{i1}, a_{i2}, \ldots, a_{in}]$. The jth *column* of $\mathbf{A}$ is the $n \times 1$ matrix

$$\begin{bmatrix} a_{1j} \\ a_{2j} \\ \cdot \\ \cdot \\ \cdot \\ a_{nj} \end{bmatrix}.$$

The (i,j)th *element* or *entry* of $\mathbf{A}$ is the element a_{ij}, that is, the number in the ith row and jth column of $\mathbf{A}$. A convenient shorthand notation for expressing the matrix $\mathbf{A}$ is to write $\mathbf{A} = [a_{ij}]$, which indicates that $\mathbf{A}$ is the matrix with its (i,j)th element equal to a_{ij}.

MATRIX ARITHMETIC

The basic operations of matrix arithmetic will now be discussed, beginning with a definition of matrix addition.

DEFINITION 3. Let $\mathbf{A} = [a_{ij}]$ and $\mathbf{B} = [b_{ij}]$ be $m \times n$ matrices. The *sum* of $\mathbf{A}$ and $\mathbf{B}$, denoted by $\mathbf{A} + \mathbf{B}$, is the $m \times n$ matrix that has $a_{ij} + b_{ij}$ as its (i,j)th element. In other words, $\mathbf{A} + \mathbf{B} = [a_{ij} + b_{ij}]$.

The sum of two matrices of the same size is obtained by adding elements in the corresponding positions. Matrices of different sizes cannot be added, since the sum of two matrices is defined only when both matrices have the same number of rows and the same number of columns.

EXAMPLE 2 We have

$$\begin{bmatrix} 1 & 0 & -1 \\ 2 & 2 & -3 \\ 3 & 4 & 0 \end{bmatrix} + \begin{bmatrix} 3 & 4 & -1 \\ 1 & -3 & 0 \\ -1 & 1 & 2 \end{bmatrix} = \begin{bmatrix} 4 & 4 & -2 \\ 3 & -1 & -3 \\ 2 & 5 & 2 \end{bmatrix}.$$

■

We now discuss matrix products. A product of two matrices is defined only when the number of columns in the first matrix equals the number of rows of the second matrix.

DEFINITION 4. Let **A** be an $m \times k$ matrix and **B** be a $k \times n$ matrix. The *product* of **A** and **B**, denoted by **AB**, is the $m \times n$ matrix with (i,j)th entry equal to the sum of the products of the corresponding elements from the ith row of **A** and the jth column of **B**. In other words, if $\mathbf{AB} = [c_{ij}]$, then

$$c_{ij} = a_{i1}b_{1j} + a_{i2}b_{2j} + \cdots + a_{ik}b_{kj} = \sum_{t=1}^{k} a_{it}b_{tj}.$$

In Figure 1 the colored row of **A** and the colored column of **B** are used to compute the element c_{ij} of **AB**. The product of two matrices is not defined when the number of columns in the first matrix and the number of rows in the second matrix are not the same.

We now give some examples of matrix products.

EXAMPLE 3 Let

$$\mathbf{A} = \begin{bmatrix} 1 & 0 & 4 \\ 2 & 1 & 1 \\ 3 & 1 & 0 \\ 0 & 2 & 2 \end{bmatrix}$$

and

$$\mathbf{B} = \begin{bmatrix} 2 & 4 \\ 1 & 1 \\ 3 & 0 \end{bmatrix}.$$

Find **AB** if it is defined.

$$\begin{bmatrix} a_{11} & a_{12} & \cdots & a_{1k} \\ a_{21} & a_{22} & \cdots & a_{2k} \\ \vdots & \vdots & & \vdots \\ a_{i1} & a_{i2} & \cdots & a_{ik} \\ \vdots & \vdots & & \vdots \\ a_{m1} & a_{m2} & \cdots & a_{mk} \end{bmatrix} \begin{bmatrix} b_{11} & b_{12} & \cdots & b_{1j} & \cdots & b_{1n} \\ b_{21} & b_{22} & \cdots & b_{2j} & \cdots & b_{2n} \\ \vdots & & & \vdots & & \vdots \\ b_{k1} & b_{k2} & \cdots & b_{kj} & \cdots & b_{kn} \end{bmatrix} = \begin{bmatrix} c_{11} & c_{12} & \cdots & c_{1n} \\ c_{21} & c_{22} & \cdots & c_{2n} \\ \vdots & & c_{ij} & \vdots \\ c_{m1} & c_{m2} & \cdots & c_{mn} \end{bmatrix}$$

FIGURE 1 The Product of $A = [a_{ij}]$ and $B = [b_{ij}]$.

Solution: Since **A** is a 4×3 matrix and **B** is a 3×2 matrix, the product **AB** is defined and is a 4×2 matrix. To find the elements of **AB**, the corresponding elements of the rows of **A** and the columns of **B** are first multiplied and then these products are added. For instance, the element in the (3,1)th position of **AB** is the sum of the products of the corresponding elements of the third row of **A** and the first column of **B**, namely, $3 \cdot 2 + 1 \cdot 1 + 0 \cdot 3 = 7$. When all the elements of **AB** are computed, we see that

$$\mathbf{AB} = \begin{bmatrix} 14 & 4 \\ 8 & 9 \\ 7 & 13 \\ 8 & 2 \end{bmatrix}.$$

■

Matrix multiplication is *not* commutative. That is, if **A** and **B** are two matrices, it is not necessarily true that **AB** and **BA** are the same. In fact, it may be that only one of these two products is defined. For instance, if **A** is 2×3 and **B** is 3×4, then **AB** is defined and is 2×4; however, **BA** is not defined, since it is impossible to multiply a 3×4 matrix and a 2×3 matrix.

In general, suppose that **A** is an $m \times n$ matrix and **B** is an $r \times s$ matrix. Then **AB** is defined only when $n = r$ and **BA** is defined only when $s = m$. Moreover, even when **AB** and **BA** are both defined, they will not be the same size unless $m = n = r = s$. Hence, if both **AB** and **BA** are defined and are the same size, then both **A** and **B** must be square and of the same size. Furthermore, even with **A** and **B** both $n \times n$ matrices, **AB** and **BA** are not necessarily equal, as the following example demonstrates.

EXAMPLE 4 Let

$$\mathbf{A} = \begin{bmatrix} 1 & 1 \\ 2 & 1 \end{bmatrix} \quad \text{and} \quad \mathbf{B} = \begin{bmatrix} 2 & 1 \\ 1 & 1 \end{bmatrix}.$$

Does $\mathbf{AB} = \mathbf{BA}$?

Solution: We find that

$$\mathbf{AB} = \begin{bmatrix} 3 & 2 \\ 5 & 3 \end{bmatrix} \quad \text{and} \quad \mathbf{BA} = \begin{bmatrix} 4 & 3 \\ 3 & 2 \end{bmatrix}.$$

Hence, $\mathbf{AB} \neq \mathbf{BA}$. ■

ALGORITHMS FOR MATRIX MULTIPLICATION

The definition of the product of two matrices leads to an algorithm that computes the product of two matrices. Suppose that $\mathbf{C} = [c_{ij}]$ is the $m \times n$ matrix that is the product of the $m \times k$ matrix $\mathbf{A} = [a_{ij}]$ and the $k \times n$ matrix $\mathbf{B} = [b_{ij}]$. The algorithm based on the definition of the matrix product is expressed in pseudocode in Algorithm 1.

ALGORITHM 1 Matrix Multiplication.

```
procedure matrix multiplication(A,B: matrices)
for i := 1 to m
begin
  for j := 1 to n
  begin
    c_ij := 0
    for q := 1 to k
      c_ij := c_ij + a_iq b_qj
  end
end {C = [c_ij] is the product of A and B}
```

We can determine the complexity of this algorithm in terms of the number of additions and multiplications used.

EXAMPLE 5 How many additions of integers and multiplications of integers are used by Algorithm 1 to multiply two $n \times n$ matrices with integer entries?

Solution: There are n^2 entries in the product of **A** and **B**. To find each entry requires a total of n multiplications and n additions. Hence, a total of n^3 multiplications and n^3 additions are used. ■

Surprisingly, there are more efficient algorithms for matrix multiplication than that given in Algorithm 1. As Example 5 shows, multiplying two $n \times n$ matrices directly from the definition requires $O(n^3)$ multiplications and additions. Using other algorithms, two $n \times n$ matrices can be multiplied using $O(n^{\sqrt{7}})$ multiplications and additions. (Details of such algorithms can be found in the references given in the suggested readings at the end of the book.)

There is another important problem involving the complexity of the multiplication of matrices. How should the product $\mathbf{A}_1\mathbf{A}_2 \cdots \mathbf{A}_n$ be computed using the fewest multiplications of integers, where $\mathbf{A}_1, \mathbf{A}_2, \ldots, \mathbf{A}_n$ are $m_1 \times m_2, m_2 \times m_3, \ldots, m_n \times m_{n+1}$ matrices, respectively, and each has integers as entries? (Since matrix multiplication is associative, as shown in Exercise 13 at the end of this section, the order of the multiplication used does not matter.) Before studying this problem note that $m_1 m_2 m_3$ multiplications of integers are performed to multiply an $m_1 \times m_2$ matrix and an $m_2 \times m_3$ matrix using Algorithm 1 (see Exercise 23 at the end of this section). The following example illustrates this complexity problem.

EXAMPLE 6 In which order should the matrices $\mathbf{A}_1$, $\mathbf{A}_2$, and $\mathbf{A}_3$ where $\mathbf{A}_1$ is 30×20, $\mathbf{A}_2$ is 20×40, and $\mathbf{A}_3$ is 40×10, all with integer entries, be multiplied to use the least number of multiplications of integers?

Solution: There are two possible ways to compute $\mathbf{A}_1\mathbf{A}_2\mathbf{A}_3$. These are $\mathbf{A}_1(\mathbf{A}_2\mathbf{A}_3)$ and $(\mathbf{A}_1\mathbf{A}_2)\mathbf{A}_3$.

If $\mathbf{A}_2$ and $\mathbf{A}_3$ are first multiplied, a total of $20\cdot40\cdot10 = 8000$ multiplications of integers are used to obtain the 20×10 matrix $\mathbf{A}_2\mathbf{A}_3$. Then, to multiply $\mathbf{A}_1$ and $\mathbf{A}_2\mathbf{A}_3$ requires $30\cdot20\cdot10 = 6000$ multiplications. Hence, a total of

$$8000 + 6000 = 14{,}000$$

multiplications are used. On the other hand, if $\mathbf{A}_1$ and $\mathbf{A}_2$ are first multiplied, then $30\cdot20\cdot40 = 24,000$ multiplications are used to obtain the 30×40 matrix $\mathbf{A}_1\mathbf{A}_2$. Then, to multiply $\mathbf{A}_1\mathbf{A}_2$ and $\mathbf{A}_3$ requires $30\times40\times10 = 12,000$ multiplication. Hence, a total of

$$24{,}000 + 12{,}000 = 36{,}000$$

multiplications are used.

Clearly, the first method is more efficient. ■

Algorithms for determining the most efficient way to multiply n matrices are discussed in the suggested readings listed at the end of the book.

TRANSPOSES AND POWERS OF MATRICES

We now introduce an important matrix with entries that are zeros and ones.

DEFINITION 5. The *identity matrix of order n* is the $n\times n$ matrix $\mathbf{I}_n = [\delta_{ij}]$, where $\delta_{ij} = 1$ if $i = j$ and $\delta_{ij} = 0$ if $i \neq j$. Hence

$$\mathbf{I}_n = \begin{bmatrix} 1 & 0 & \cdots & 0 \\ 0 & 1 & \cdots & 0 \\ \cdot & \cdot & & \cdot \\ \cdot & \cdot & & \cdot \\ \cdot & \cdot & & \cdot \\ 0 & 0 & \cdots & 1 \end{bmatrix}.$$

Multiplying a matrix by an appropriately sized identity matrix does not change this matrix. In other words, when $\mathbf{A}$ is an $m\times n$ matrix, we have

$$\mathbf{A}\mathbf{I}_n = \mathbf{I}_m\mathbf{A} = \mathbf{A}.$$

Powers of square matrices can be defined. When $\mathbf{A}$ is an $n\times n$ matrix, we have

$$\mathbf{A}^0 = \mathbf{I}_n, \qquad \mathbf{A}^r = \underbrace{\mathbf{AAA}\cdots\mathbf{A}.}_{r\text{ times}}$$

The operation of interchanging the rows and columns of a square matrix is used in many algorithms.

DEFINITION 6. Let $\mathbf{A} = [a_{ij}]$ be an $m \times n$ matrix. The *transpose* of $\mathbf{A}$, denoted by $\mathbf{A}^t$, is the $n \times m$ matrix obtained by interchanging the rows and columns of $\mathbf{A}$. In other words, if $\mathbf{A}^t = [b_{ij}]$, then $b_{ij} = a_{ji}$ for $i = 1, 2, \ldots, n$ and $j = 1, 2, \ldots, m$.

EXAMPLE 7 The transpose of the matrix

$$\begin{bmatrix} 1 & 2 & 3 \\ 4 & 5 & 6 \end{bmatrix}$$

is the matrix

$$\begin{bmatrix} 1 & 4 \\ 2 & 5 \\ 3 & 6 \end{bmatrix}.$$

■

Matrices that do not change when their rows and columns are interchanged are often important.

DEFINITION 7. A square matrix $\mathbf{A}$ is called *symmetric* if $\mathbf{A} = \mathbf{A}^t$. Thus $\mathbf{A} = [a_{ij}]$ is symmetric if $a_{ij} = a_{ji}$ for all i and j with $1 \leq i \leq n$ and $1 \leq j \leq n$.

Note that a matrix is symmetric if and only if it is square and it is symmetric with respect to its main diagonal (which consists of entries that are in the ith row and ith column for some i). This symmetry is displayed in Figure 2.

EXAMPLE 8 The matrix

$$\begin{bmatrix} 1 & 1 & 0 \\ 1 & 0 & 1 \\ 0 & 1 & 0 \end{bmatrix}$$

is symmetric. ■

ZERO-ONE MATRICES

A matrix with entries that are either 0 or 1 is called a **zero-one matrix.** Zero-one matrices are often used to represent discrete structures, as we will see in Chapters 6 and 7. Algorithms using these structures are based on Boolean arithmetic with zero-

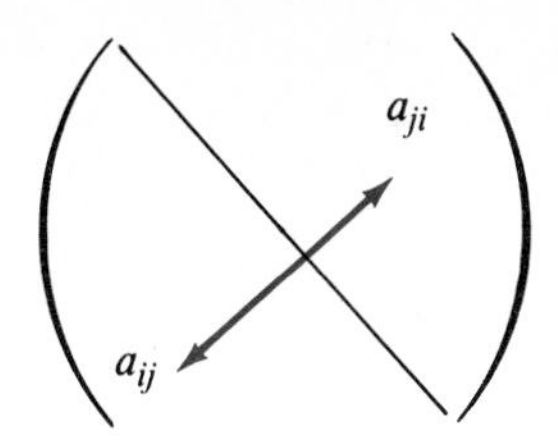

FIGURE 2 A Symmetric Matrix.

one matrices. This arithmetic is based on the Boolean operations $\vee$ and $\wedge$, which operate on pairs of bits, defined by

$$b_1 \wedge b_2 = \begin{cases} 1 & \text{if } b_1 = b_2 = 1 \\ 0 & \text{otherwise,} \end{cases}$$

$$b_1 \vee b_2 = \begin{cases} 1 & \text{if } b_1 = 1 \text{ or } b_2 = 1 \\ 0 & \text{otherwise.} \end{cases}$$

DEFINITION 8. Let $\mathbf{A} = [a_{ij}]$ and $\mathbf{B} = [b_{ij}]$ be $m \times n$ zero-one matrices. Then the *join* of $\mathbf{A}$ and $\mathbf{B}$ is the zero-one matrix with (i,j)th entry $a_{ij} \vee b_{ij}$. The join of $\mathbf{A}$ and $\mathbf{B}$ is denoted by $\mathbf{A} \vee \mathbf{B}$. The *meet* of $\mathbf{A}$ and $\mathbf{B}$ is the zero-one matrix with (i,j)th entry $a_{ij} \wedge b_{ij}$. The meet of $\mathbf{A}$ and $\mathbf{B}$ is denoted by $\mathbf{A} \wedge \mathbf{B}$.

EXAMPLE 9 Find the join and meet of the zero-one matrices

$$\mathbf{A} = \begin{bmatrix} 1 & 0 & 1 \\ 0 & 1 & 0 \end{bmatrix} \qquad \mathbf{B} = \begin{bmatrix} 0 & 1 & 0 \\ 1 & 1 & 0 \end{bmatrix}.$$

Solution: We find that the join of $\mathbf{A}$ and $\mathbf{B}$ is

$$\mathbf{A} \vee \mathbf{B} = \begin{bmatrix} 1 \vee 0 & 0 \vee 1 & 1 \vee 0 \\ 0 \vee 1 & 1 \vee 1 & 0 \vee 0 \end{bmatrix} = \begin{bmatrix} 1 & 1 & 1 \\ 1 & 1 & 0 \end{bmatrix}.$$

The meet of $\mathbf{A}$ and $\mathbf{B}$ is

$$\mathbf{A} \wedge \mathbf{B} = \begin{bmatrix} 1 \wedge 0 & 0 \wedge 1 & 1 \wedge 0 \\ 0 \wedge 1 & 1 \wedge 1 & 0 \wedge 0 \end{bmatrix} = \begin{bmatrix} 0 & 0 & 0 \\ 0 & 1 & 0 \end{bmatrix}.$$

■

We now define the **Boolean product** of two matrices.

DEFINITION 9. Let $\mathbf{A} = [a_{ij}]$ be an $m \times k$ zero-one matrix and $\mathbf{B} = [b_{ij}]$ be a $k \times n$ zero-one matrix. Then the *Boolean product* of $\mathbf{A}$ and $\mathbf{B}$, denoted by $\mathbf{A} \odot \mathbf{B}$, is the $m \times n$ matrix with (i,j)th entry $[c_{ij}]$ where

$$c_{ij} = (a_{i1} \wedge b_{1j}) \vee (a_{i2} \wedge b_{2j}) \vee \cdots \vee (a_{ik} \wedge b_{kj}).$$

Note that the Boolean product of **A** and **B** is obtained in an analogous way to the ordinary product of these matrices, but with addition replaced with the operation $\vee$ and with multiplication replaced with the operation $\wedge$. We give an example of the Boolean products of matrices.

EXAMPLE 10 Find the Boolean product of **A** and **B**, where

$$\mathbf{A} = \begin{bmatrix} 1 & 0 \\ 0 & 1 \\ 1 & 0 \end{bmatrix}, \quad \mathbf{B} = \begin{bmatrix} 1 & 1 & 0 \\ 0 & 1 & 1 \end{bmatrix}.$$

Solution: The Boolean product $\mathbf{A} \odot \mathbf{B}$ is given by

$$\mathbf{A} \odot \mathbf{B} = \begin{bmatrix} (1 \wedge 1) \vee (0 \wedge 0) & (1 \wedge 1) \vee (0 \wedge 1) & (1 \wedge 0) \vee (0 \wedge 1) \\ (0 \wedge 1) \vee (1 \wedge 0) & (0 \wedge 1) \vee (1 \wedge 1) & (0 \wedge 0) \vee (1 \wedge 1) \\ (1 \wedge 1) \vee (0 \wedge 0) & (1 \wedge 1) \vee (0 \wedge 1) & (1 \wedge 0) \vee (0 \wedge 1) \end{bmatrix}$$

$$= \begin{bmatrix} 1 \vee 0 & 1 \vee 0 & 0 \vee 0 \\ 0 \vee 0 & 0 \vee 1 & 0 \vee 1 \\ 1 \vee 0 & 1 \vee 0 & 0 \vee 0 \end{bmatrix}$$

$$= \begin{bmatrix} 1 & 1 & 0 \\ 0 & 1 & 1 \\ 1 & 1 & 0 \end{bmatrix}.$$

■

Algorithm 2 displays pseudocode for computing the Boolean product of two matrices.

ALGORITHM 2 The Boolean Product.

```
procedure Boolean product(A,B: zero-one matrices)
for i := 1 to m
begin
  for j := 1 to n
  begin
    c_ij := 0
    for q := 1 to k
      c_ij := c_ij ∨ (a_iq ∧ b_qj)
  end
end {C = [c_ij] is the Boolean product of A and B}
```

We can also define the Boolean powers of a square zero-one matrix. These powers will be used in our subsequent studies of paths in graphs, which are used to model such things as communications paths in computer networks.

DEFINITION 10. Let $\mathbf{A}$ be a square zero-one matrix and let r be a positive integer. The rth *Boolean power* of $\mathbf{A}$ is the Boolean product of r factors of $\mathbf{A}$. The rth Boolean product of $\mathbf{A}$ is denoted by $\mathbf{A}^{[r]}$. Hence

$$\mathbf{A}^{[r]} = \underbrace{\mathbf{A} \odot \mathbf{A} \odot \mathbf{A} \odot \cdots \odot \mathbf{A}}_{r \text{ times}}.$$

(This is well defined since the Boolean product of matrices is associative.) We also define $\mathbf{A}^{[0]}$ to be $\mathbf{I}_n$.

EXAMPLE 11 Let

$$\mathbf{A} = \begin{bmatrix} 0 & 0 & 1 \\ 1 & 0 & 0 \\ 1 & 1 & 0 \end{bmatrix}.$$

Find $\mathbf{A}^{[n]}$ for all positive integers n.

Solution: We find that

$$\mathbf{A}^{[2]} = \mathbf{A} \odot \mathbf{A} = \begin{bmatrix} 1 & 1 & 0 \\ 0 & 0 & 1 \\ 1 & 0 & 1 \end{bmatrix}.$$

We also find that

$$\mathbf{A}^{[3]} = \mathbf{A}^{[2]} \odot \mathbf{A} = \begin{bmatrix} 1 & 0 & 1 \\ 1 & 1 & 0 \\ 1 & 1 & 1 \end{bmatrix},$$

$$\mathbf{A}^{[4]} = \mathbf{A}^{[3]} \odot \mathbf{A} = \begin{bmatrix} 1 & 1 & 1 \\ 1 & 0 & 1 \\ 1 & 1 & 1 \end{bmatrix}.$$

Additional computation shows that

$$\mathbf{A}^{[5]} = \begin{bmatrix} 1 & 1 & 1 \\ 1 & 1 & 1 \\ 1 & 1 & 1 \end{bmatrix}.$$

The reader can now see that $\mathbf{A}^{[n]} = \mathbf{A}^{[5]}$ for all positive integers n with $n \geq 5$. ■

The number of bit operations used to find the Boolean product of two $n \times n$ matrices can be easily determined.

EXAMPLE 12 How many bit operations are used to find $\mathbf{A} \odot \mathbf{B}$, where $\mathbf{A}$ and $\mathbf{B}$ are $n \times n$ zero-one matrices?

Solution: There are n^2 entries in $\mathbf{A} \odot \mathbf{B}$. Using Algorithm 2, a total of n *OR*s and n *AND*s are used to find an entry of $\mathbf{A} \odot \mathbf{B}$. Hence, $2n$ bit operations are used to find each entry. Therefore, $2n^3$ bit operations are required to compute $\mathbf{A} \odot \mathbf{B}$ using Algorithm 2. ■

Exercises

1. Let

$$\mathbf{A} = \begin{bmatrix} 1 & 1 & 1 & 3 \\ 2 & 0 & 4 & 6 \\ 1 & 1 & 3 & 7 \end{bmatrix}.$$

a) What size is **A**?
b) What is the third column of **A**?
c) What is the second row of **A**?
d) What is the element of **A** in the (3,2)th position?
e) What is $\mathbf{A}^t$?

2. Find **A** + **B**, where

a) $\mathbf{A} = \begin{bmatrix} 1 & 0 & 4 \\ -1 & 2 & 2 \\ 0 & -2 & -3 \end{bmatrix}, \mathbf{B} = \begin{bmatrix} -1 & 3 & 5 \\ 2 & 2 & -3 \\ 2 & -3 & 0 \end{bmatrix}.$

b) $\mathbf{A} = \begin{bmatrix} -1 & 0 & 5 & 6 \\ -4 & -3 & 5 & -2 \end{bmatrix},$

$\mathbf{B} = \begin{bmatrix} -3 & 9 & -3 & 4 \\ 0 & -2 & -1 & 2 \end{bmatrix}.$

3. Find **AB** if

a) $\mathbf{A} = \begin{bmatrix} 2 & 1 \\ 3 & 2 \end{bmatrix}, \quad \mathbf{B} = \begin{bmatrix} 0 & 4 \\ 1 & 3 \end{bmatrix}.$

b) $\mathbf{A} = \begin{bmatrix} 1 & -1 \\ 0 & 1 \\ 2 & 3 \end{bmatrix}, \quad \mathbf{B} = \begin{bmatrix} 3 & -2 & -1 \\ 1 & 0 & 2 \end{bmatrix}.$

4. Find the product **AB**, where

a) $\mathbf{A} = \begin{bmatrix} 1 & 0 & 1 \\ 0 & -1 & -1 \\ -1 & 1 & 0 \end{bmatrix}, \mathbf{B} = \begin{bmatrix} 0 & 1 & -1 \\ 1 & -1 & 0 \\ -1 & 0 & 1 \end{bmatrix}.$

b) $\mathbf{A} = \begin{bmatrix} 1 & -3 & 0 \\ 1 & 2 & 2 \\ 2 & 1 & -1 \end{bmatrix}, \mathbf{B} = \begin{bmatrix} 1 & -1 & 2 & 3 \\ -1 & 0 & 3 & -1 \\ -3 & -2 & 0 & 2 \end{bmatrix}.$

5. Find a matrix **A** such that

$$\begin{bmatrix} 2 & 3 \\ 1 & 4 \end{bmatrix} \mathbf{A} = \begin{bmatrix} 3 & 0 \\ 1 & 2 \end{bmatrix}.$$

(*Hint:* Finding **A** requires that you solve systems of linear equations.)

6. Find a matrix **A** such that

$$\begin{bmatrix} 1 & 3 & 2 \\ 2 & 1 & 1 \\ 4 & 0 & 3 \end{bmatrix} \mathbf{A} = \begin{bmatrix} 7 & 1 & 3 \\ 1 & 0 & 3 \\ -1 & -3 & 7 \end{bmatrix}$$

7. Let **A** be an $m \times n$ matrix and let **0** be the $m \times n$ matrix that has all entries equal to zero. Show that $\mathbf{A} = \mathbf{0} + \mathbf{A} = \mathbf{A} + \mathbf{0}$.

8. Show that matrix addition is commutative; that is, show that if **A** and **B** are both $m \times n$ matrices, then $\mathbf{A} + \mathbf{B} = \mathbf{B} + \mathbf{A}$.

9. Show that matrix addition is associative; that is, show that if **A**, **B**, and **C** are all $m \times n$ matrices, then $\mathbf{A} + (\mathbf{B} + \mathbf{C}) = (\mathbf{A} + \mathbf{B}) + \mathbf{C}$.

10. Let **A** be a 3×4 matrix, **B** be a 4×5 matrix, and **C** be a 4×4 matrix. Determine which of the following products are defined and find the size of those that are defined.
a) **AB** **b)** **BA** **c)** **AC**
d) **CA** **e)** **BC** **f)** **CB**

11. What do we know about the sizes of the matrices **A** and **B** if both of the products **AB** and **BA** are defined?

12. In this exercise we show that matrix multiplication is distributive over matrix addition.
a) Suppose that **A** and **B** are $m \times k$ matrices and that **C** is a $k \times n$ matrix. Show that $(\mathbf{A} + \mathbf{B})\mathbf{C} = \mathbf{AC} + \mathbf{BC}$
b) Suppose that **C** is an $m \times k$ matrix and that **A** and **B** are $k \times n$ matrices. Show that $\mathbf{C}(\mathbf{A} + \mathbf{B}) = \mathbf{CA} + \mathbf{CB}$.

13. In this exercise we show that matrix multiplication is associative. Suppose that **A** is an $m \times p$ matrix, **B** is a $p \times k$ matrix, and **C** is a $k \times n$ matrix. Show that $\mathbf{A}(\mathbf{BC}) = (\mathbf{AB})\mathbf{C}$.

14. The $n \times n$ matrix $\mathbf{A} = [a_{ij}]$ is called a **diagonal matrix** if $a_{ij} = 0$ when $i \neq j$. Show that the product of two $n \times n$ diagonal matrices is again a diagonal matrix. Give a simple rule for determining this product.

15. Let

$$\mathbf{A} = \begin{bmatrix} 1 & 1 \\ 0 & 1 \end{bmatrix}.$$

Find a formula for $\mathbf{A}^n$, whenever n is a positive integer.

16. Show that $(\mathbf{A}^t)^t = \mathbf{A}$.

17. Let **A** and **B** be two $n \times n$ matrices. Show that

a) $(\mathbf{A} + \mathbf{B})^t = \mathbf{A}^t + \mathbf{B}^t$. **b)** $(\mathbf{AB})^t = \mathbf{B}^t\mathbf{A}^t$.

If **A** and **B** are $n \times n$ matrices with $\mathbf{AB} = \mathbf{BA} = \mathbf{I}_n$, then **B** is called the **inverse** of **A** (this terminology is appropriate since such a matrix **B** is unique) and **A** is said to be **invertible.** The notation $\mathbf{B} = \mathbf{A}^{-1}$ denotes that **B** is the inverse of **A**.

18. Show that

$$\begin{bmatrix} 2 & 3 & -1 \\ 1 & 2 & 1 \\ -1 & -1 & 3 \end{bmatrix}$$

is the inverse of

$$\begin{bmatrix} 7 & -8 & 5 \\ -4 & 5 & -3 \\ 1 & -1 & 1 \end{bmatrix}.$$

19. Let **A** be a 2×2 matrix with

$$\mathbf{A} = \begin{bmatrix} a & b \\ c & d \end{bmatrix}.$$

Show that if $ad - bc \neq 0$, then

$$\mathbf{A}^{-1} = \begin{bmatrix} \dfrac{d}{ad - bc} & \dfrac{-b}{ad - bc} \\ \dfrac{-c}{ad - bc} & \dfrac{a}{ad - bc} \end{bmatrix}.$$

20. Let

$$\mathbf{A} = \begin{bmatrix} -1 & 2 \\ 1 & 3 \end{bmatrix}.$$

a) Find $\mathbf{A}^{-1}$. (*Hint:* Use Exercise 19.)

b) Find $\mathbf{A}^3$.

c) Find $(\mathbf{A}^{-1})^3$.

d) Use your answers to (b) and (c) to show that $(\mathbf{A}^{-1})^3$ is the inverse of $\mathbf{A}^3$.

21. Let **A** be an invertible matrix. Show that $(\mathbf{A}^n)^{-1} = (\mathbf{A}^{-1})^n$ whenever n is a positive integer.

22. Let **A** be a matrix. Show that the matrix $\mathbf{AA}^t$ is symmetric. (*Hint:* Show that this matrix equals its transpose with the help of Exercise 17b.)

23. Show that the conventional algorithm uses $m_1m_2m_3$ multiplications to compute the product of the $m_1 \times m_2$ matrix **A** and the $m_2 \times m_3$ matrix **B**.

24. What is the most efficient way to multiply the matrices $\mathbf{A}_1$, $\mathbf{A}_2$, and $\mathbf{A}_3$ with sizes

a) 20×50, 50×10, 10×40?

b) 10×5, 5×50, 50×1?

25. What is the most efficient way to multiply the matrices $\mathbf{A}_1$, $\mathbf{A}_2$, $\mathbf{A}_3$, and $\mathbf{A}_4$ if the dimensions of these matrices are 10×2, 2×5, 5×20, and 20×3, respectively?

26. a) Show that the system of simultaneous linear equations

$$\begin{aligned} a_{11}x_1 + a_{12}x_2 + \cdots + a_{1n}x_n &= b_1 \\ a_{21}x_1 + a_{22}x_2 + \cdots + a_{2n}x_n &= b_2 \\ &\ \ \vdots \\ a_{n1}x_1 + a_{n2}x_2 + \cdots + a_{nn}x_n &= b_n \end{aligned}$$

in the variables $x_1, x_2, \ldots, x_n$ can be expressed as $\mathbf{AX} = \mathbf{B}$, where $\mathbf{A} = [a_{ij}]$, **X** is an $n \times 1$ matrix with x_i the entry in its ith row, and **B** is an $n \times 1$ matrix with b_i the entry in its ith row.

b) Show that if the matrix $\mathbf{A} = [a_{ij}]$ is invertible (as defined in the preamble to Exercise 18), then the solution of the system in part (a) can be found using the equation $\mathbf{X} = \mathbf{A}^{-1}\mathbf{B}$.

27. Use Exercises 18 and 26 to solve the system

$$\begin{aligned} 7x_1 - 8x_2 + 5x_3 &= 5 \\ -4x_1 + 5x_2 - 3x_3 &= -3 \\ x_1 - x_2 + x_3 &= 0 \end{aligned}$$

28. Let

$$\mathbf{A} = \begin{bmatrix} 1 & 1 \\ 0 & 1 \end{bmatrix} \quad \text{and} \quad \mathbf{B} = \begin{bmatrix} 0 & 1 \\ 1 & 0 \end{bmatrix}.$$

Find

a) $\mathbf{A} \vee \mathbf{B}$. **b)** $\mathbf{A} \wedge \mathbf{B}$. **c)** $\mathbf{A} \odot \mathbf{B}$.

29. Let

$$\mathbf{A} = \begin{bmatrix} 1 & 0 & 1 \\ 1 & 1 & 0 \\ 0 & 0 & 1 \end{bmatrix} \quad \text{and} \quad \mathbf{B} = \begin{bmatrix} 0 & 1 & 1 \\ 1 & 0 & 1 \\ 1 & 0 & 1 \end{bmatrix}.$$

Find

a) $\mathbf{A} \vee \mathbf{B}$. **b)** $\mathbf{A} \wedge \mathbf{B}$. **c)** $\mathbf{A} \odot \mathbf{B}$.

30. Find the Boolean product of **A** and **B**, where

$$\mathbf{A} = \begin{bmatrix} 1 & 0 & 0 & 1 \\ 0 & 1 & 0 & 1 \\ 1 & 1 & 1 & 1 \end{bmatrix} \quad \text{and} \quad \mathbf{B} = \begin{bmatrix} 1 & 0 \\ 0 & 1 \\ 1 & 1 \\ 1 & 0 \end{bmatrix}.$$

31. Let

$$\mathbf{A} = \begin{bmatrix} 1 & 0 & 0 \\ 1 & 0 & 1 \\ 0 & 1 & 0 \end{bmatrix}.$$

Find

a) $\mathbf{A}^{[2]}$. b) $\mathbf{A}^{[3]}$. c) $\mathbf{A} \vee \mathbf{A}^{[2]} \vee \mathbf{A}^{[3]}$.

32. Let **A** be a zero-one matrix. Show that

a) $\mathbf{A} \vee \mathbf{A} = \mathbf{A}$. b) $\mathbf{A} \wedge \mathbf{A} = \mathbf{A}$.

33. In this exercise we show that the meet and join operations are commutative. Let **A** and **B** be $m \times n$ zero-one matrices. Show that

a) $\mathbf{A} \vee \mathbf{B} = \mathbf{B} \vee \mathbf{A}$. b) $\mathbf{B} \wedge \mathbf{A} = \mathbf{A} \wedge \mathbf{B}$.

34. In this exercise we show that the meet and join operations are associative. Let **A**, **B**, and **C** be $m \times n$ zero-one matrices. Show that

a) $(\mathbf{A} \vee \mathbf{B}) \vee \mathbf{C} = \mathbf{A} \vee (\mathbf{B} \vee \mathbf{C})$.

b) $(\mathbf{A} \wedge \mathbf{B}) \wedge \mathbf{C} = \mathbf{A} \wedge (\mathbf{B} \wedge \mathbf{C})$.

35. We will establish distributive laws of the meet over the join operation in this exercise. Let **A**, **B**, and **C** be $m \times n$ zero-one matrices. Show that

a) $\mathbf{A} \vee (\mathbf{B} \wedge \mathbf{C}) = (\mathbf{A} \vee \mathbf{B}) \wedge (\mathbf{A} \vee \mathbf{C})$.

b) $\mathbf{A} \wedge (\mathbf{B} \vee \mathbf{C}) = (\mathbf{A} \wedge \mathbf{B}) \vee (\mathbf{A} \wedge \mathbf{C})$.

36. Let **A** be an $n \times n$ zero-one matrix. Let **I** be the $n \times n$ identity matrix. Show that $\mathbf{A} \odot \mathbf{I} = \mathbf{I} \odot \mathbf{A} = \mathbf{A}$.

37. In this exercise we will show that the Boolean product of zero-one matrices is associative. Assume that **A** is an $m \times p$ zero-one matrix, **B** is a $p \times k$ zero-one matrix, and **C** is a $k \times n$ zero-one matrix. Show that $\mathbf{A} \odot (\mathbf{B} \odot \mathbf{C}) = (\mathbf{A} \odot \mathbf{B}) \odot \mathbf{C}$.

Key Terms and Results

TERMS

algorithm: a definite procedure for solving a problem using a finite number of steps

searching algorithm: the problem of locating an element in a list

linear search algorithm: a procedure for searching a list element by element

binary search algorithm: a procedure for searching an ordered list by successively splitting the list in half

time complexity: the amount of time required for an algorithm to solve a problem

space complexity: the amount of storage space required for an algorithm to solve a problem

worst-case time complexity: the greatest amount of time required for an algorithm to solve a problem of a given size

average-case time complexity: the average amount of time required for an algorithm to solve a problem of a given size

$a \mid b$ (a divides b): there is an integer c such that $b = ac$

prime: a positive integer greater than 1 with exactly two positive integer divisors

composite: a positive integer greater than 1 that is not prime

$gcd(a,b)$ (greatest common divisor of a and b): the largest integer that divides both a and b

relatively prime integers: integers a and b such that $gcd(a, b) = 1$

pairwise relatively prime integers: a set of integers with the property that every pair of these integers is relatively prime

$lcm(a, b)$ (least common multiple of a and b): the smallest positive integer that is divisible by both a and b

a mod b: the remainder when the integer a is divided by the positive integer b

$a \equiv b \pmod{m}$ (a is congruent to b modulo m): $a - b$ is divisible by m

encryption: the process of making a message secret

decryption: the process of returning a secret message to its original form

$n = (a_k a_{k-1} \cdots a_1 a_0)_b$: the base b representation of n

binary representation: the base 2 representation of an integer

hexadecimal representation: the base 16 representation of an integer

linear combination of a and b with integer coefficients: a number of the form $sa + tb$ where s and t are integers

inverse of a modulo m: an integer $\overline{a}$ such that $\overline{a}a \equiv 1 \pmod{m}$
linear congruence: a congruence of the form $ax \equiv b \pmod{m}$ where x is a variable
pseudoprime to the base 2: a composite integer n such that $2^{n-1} \equiv 1 \pmod{n}$
private key encryption: encryption where both encryption keys and decryption keys must be kept secret
public key encryption: encryption where encryption keys are public knowledge, but decryption keys are kept secret
matrix: a rectangular array of numbers
matrix addition: see page 145
matrix multiplication: see page 146
$\mathbf{I}_n$ (identity matrix of order n): the $n \times n$ matrix that has entries equal to one on its diagonal and zeros elsewhere
$\mathbf{A}^t$ (transpose of A): the matrix obtained from **A** by interchanging the rows and columns
symmetric: a matrix is symmetric if it equals its transpose
zero-one matrix: a matrix with each entry equal to either 0 or 1
$\mathbf{A} \vee \mathbf{B}$ (the join of A and B): see page 151
$\mathbf{A} \wedge \mathbf{B}$ (the meet of A and B): see page 151
$\mathbf{A} \odot \mathbf{B}$ (the Boolean product of A and B): see page 151

RESULTS

The linear and binary search algorithms (given in Section 2.1).
The Fundamental Theorem of Arithmetic: Every positive integer can be written uniquely as the product of primes, where the prime factors are written in order of increasing size.
The division algorithm: Let a and d be integers with d positive. Then there are unique integers q and r with $0 \le r < d$ such that $a = dq + r$.
If a and b are positive integers, then $ab = gcd(a, b)lcm(a, b)$.
The Euclidean algorithm for finding greatest common divisors (see Algorithm 1 in Section 2.4).
Let b be a positive integer greater than 1. Then if n is a positive integer, it can be expressed uniquely in the form $n = a_k b^k + a_{k-1} b^{k-1} + \cdots + a_1 b + a_0$.
The algorithm for finding the base b expansion of an integer (see Algorithm 2 in Section 2.4).
The conventional algorithms for addition and multiplication of integers (given in Section 2.4).
The greatest common divisor of two integers can be expressed as a linear combination with integer coefficients of these integers.
If m is a positive integer and $gcd(a, m) = 1$, then a has a unique inverse modulo m.
The Chinese Remainder Theorem: A system of linear congruences modulo pairwise relatively prime integers has a unique solution modulo the product of these moduli.
Fermat's Little Theorem: If p is prime and $p \nmid a$, then $a^{p-1} \equiv 1 \pmod{p}$.

Review Questions

1. **a)** Define the term *algorithm*.
b) What are the different ways to describe algorithms?
c) What is the difference between an algorithm for solving a problem and a computer program that solves this problem?

2. **a)** Describe, using English, an algorithm for finding the largest, second largest, and third largest integers in a list of n integers.

b) Express this algorithm in pseudocode.

c) How many comparisons does the algorithm use?

3. a) Define what the worst-case time complexity, average-case time complexity, and best-case time complexity (in terms of comparisons) mean for an algorithm that finds the smallest integer in a list of n integers.

b) What are the worst-case, average-case, and best-case time complexities, in terms of comparisons, of the algorithm that finds the smallest integer in a list of n integers by comparing each of the integers with the smallest integer found so far?

4. a) Describe the linear search and binary search algorithm for finding an integer in a list of integers in increasing order.

b) Compare the worst-case time complexities of these two algorithms.

c) Is one of these algorithms always faster than the other (measured in terms of comparisons)?

5. State the Fundamental Theorem of Arithmetic.

6. a) Describe a procedure for finding the prime factorization of an integer.

b) Use this procedure to find the prime factorization of 80,707.

7. a) Define the greatest common divisor of two integers.

b) Describe at least three different ways to find the greatest common divisor of two integers. When does each method work best?

c) Find the greatest common divisor of 1,234,567 and 7,654,321.

d) Find the greatest common divisor of $2^3 3^5 5^7 7^9 11$ and $2^9 3^7 5^5 7^3 13$.

8. a) Define what it means for a and b to be congruent modulo 7.

b) Which pairs of the integers $-11, -8, -7, -1, 0, 3$, and 17 are congruent modulo 7?

c) Show that if a and b are congruent modulo 7, then $10a + 13$ and $-4b + 20$ are also congruent modulo 7.

9. Describe a procedure for converting decimal (base 10) expansions of integers into hexadecimal expansions.

10. a) How can you find a linear combination (with integer coefficients) of two integers that equals their greatest common divisor?

b) Express $gcd(84, 119)$ as a linear combination of 84 and 119.

11. a) What does it mean for $\overline{a}$ to be an inverse of a modulo m?

b) How can you find an inverse of a modulo m when m is a positive integer and $gcd(a, m) = 1$?

c) Find an inverse of 7 modulo 19.

12. a) How can an inverse of a modulo m be used to solve the linear congruence $ax \equiv b \pmod{m}$ when $gcd(a, m) = 1$?

b) Solve the linear congruence $7x \equiv 13 \pmod{19}$.

13. a) State the Chinese Remainder Theorem.

b) Find the solutions to the system $x \equiv 1 \pmod{4}$, $x \equiv 2 \pmod{5}$, and $x \equiv 3 \pmod{7}$.

14. Suppose that $2^{n-1} \equiv 1 \pmod{n}$. Is n necessarily prime?

15. a) What is the difference between a public key and a private key cryptosystem?

b) Explain why using shift ciphers is a private key system.

c) Explain why the RSA cipher system is a public key system.

16. Define the product of two matrices **A** and **B**. When is this product defined?

17. a) How many different ways are there to evaluate the product $\mathbf{A}_1\mathbf{A}_2\mathbf{A}_3\mathbf{A}_4$ by successively multiplying pairs of matrices, when this product is defined?

b) Suppose that $\mathbf{A}_1$, $\mathbf{A}_2$, $\mathbf{A}_3$, and $\mathbf{A}_4$ are 10×20, 20×5, 5×10, and 10×5 matrices, respectively. How should $\mathbf{A}_1\mathbf{A}_2\mathbf{A}_3\mathbf{A}_4$ be computed to use the least number of multiplications of entries?

Supplementary Exercises

1. **a)** Describe an algorithm for locating the last occurrence of the largest number in a list of integers.
b) Estimate the number of comparisons used.

2. **a)** Describe an algorithm for finding the first and second largest elements in a list of integers.
b) Estimate the number of comparisons used.

3. **a)** Give an algorithm to determine whether a bit string contains a pair of consecutive zeros.
b) How many comparisons does the algorithm use?

4. **a)** Suppose that a list contains integers that are in order of largest to smallest and an integer can appear repeatedly in this list. Devise an algorithm that locates all occurrences of an integer x in the list.
b) Estimate the number of comparisons used.

5. Find four numbers that are congruent to 5 modulo 17.

6. Show that if a and d are positive integers, then there are integers q and r such that $a = dq + r$ where $-d/2 < r \leq d/2$.

★7. Show that if $ac \equiv bc \pmod{m}$, then $a \equiv b \pmod{m/d}$ where $d = gcd(m, c)$.

★8. How many zeros are at the end of the binary expansion of $100_{10}!$?

9. Use the Euclidean algorithm to find the greatest common divisor of 10,223 and 33,341.

10. How many divisions are required to find $gcd(144, 233)$ using the Euclidean algorithm?

11. Find $gcd(2n + 1, 3n + 2)$, where n is a positive integer. (*Hint:* Use the Euclidean algorithm.)

12. **a)** Show that if a and b are positive integers with $a \geq b$, then $gcd(a, b) = a$ if $a = b$, $gcd(a, b) = 2gcd(a/2, b/2)$ if a and b are even, $gcd(a, b) = gcd(a/2, b)$ if a is even and b is odd, and $gcd(a, b) = gcd(a - b, b)$ if both a and b are odd.
b) Explain how to use (a) to construct an algorithm for computing the greatest common divisor of two positive integers that uses only comparisons, subtractions, and shifts of binary expansions, without using any divisions.
c) Find $gcd(1202, 4848)$ using this algorithm.

13. Show that an integer is divisible by 9 if and only if the sum of its decimal digits is divisible by 9.

14. **a)** Devise an algorithm for computing x^n **mod** m, where x is an integer and m and n are positive integers, using the binary expansion of n. (*Hint:* Perform successive squarings to obtain x **mod** m, x^2 **mod** m, x^4 **mod** m, and so on. Then multiply the appropriate powers of the form x^{2^k} **mod** m to obtain x^n **mod** m.)
b) Estimate the number of multiplications used by this algorithm.

A set of integers is called **mutually relatively prime** if the greatest common divisor of these integers is 1.

15. Determine whether the following sets of integers are mutually relatively prime.
a) 8, 10, 12 **b)** 12, 15, 25
c) 15, 21, 28 **d)** 21, 24, 28, 32

16. Find a set of four mutually relatively prime integers such that no two of them are relatively prime.

17. **a)** Suppose that messages are encrypted using the function $f(p) = (ap + b)$ **mod** 26 such that $gcd(a, 26) = 1$. Determine a function that can be used to decrypt messages.
b) The encrypted version of a message is LJMKG MGMXF QEXMW. If it was encrypted using the function $f(p) = (7p + 10)$ **mod** 26, what was the original message?

18. Show that the system of congruences $x \equiv 2 \pmod{6}$ and $x \equiv 3 \pmod{9}$ has no solutions.

19. Find all solutions of the system of congruences $x \equiv 4 \pmod{6}$ and $x \equiv 13 \pmod{15}$.

★20. **a)** Show that the system of congruences $x \equiv a_1 \pmod{m_1}$ and $x \equiv a_2 \pmod{m_2}$ has a solution if and only if $gcd(m_1, m_2) \mid a_1 - a_2$.
b) Show that the solution in part (a) is unique modulo $lcm(m_1, m_2)$.

21. Find $\mathbf{A}^n$ if $\mathbf{A}$ is

$$\begin{bmatrix} 0 & 1 \\ -1 & 0 \end{bmatrix}.$$

22. Show that if $\mathbf{A} = c\mathbf{I}$, where c is a real number and $\mathbf{I}$ is the $n \times n$ identity matrix, then $\mathbf{AB} = \mathbf{BA}$ whenever $\mathbf{B}$ is an $n \times n$ matrix.

23. Show that if $\mathbf{A}$ is a 2×2 matrix such that $\mathbf{AB} = \mathbf{BA}$ whenever $\mathbf{B}$ is a 2×2 matrix, then $\mathbf{A} = c\mathbf{I}$, where c is a real number and $\mathbf{I}$ is the 2×2 identity matrix.

An $n \times n$ matrix is called **upper triangular** if $a_{ij} = 0$ whenever $i > j$.

24. From the definition of the matrix product, devise an algorithm for computing the product of two upper triangular matrices that ignores those products in the computation that are automatically equal to zero.

25. Give a pseudocode description of the algorithm in Exercise 24 for multiplying two upper triangular matrices.

26. How many multiplications of entries are used by the algorithm found in Exercise 25 for multiplying two $n \times n$ upper triangular matrices?

27. Show that if **A** and **B** are invertible matrices and **AB** exists, then $(\mathbf{AB})^{-1} = \mathbf{B}^{-1}\mathbf{A}^{-1}$.

28. What is the best order to form the product **ABCD** if **A, B, C,** and **D** are matrices with dimensions 30×10, 10×40, 40×50, and 50×30, respectively? Assume that the number of multiplications of entries used to multiply a $p \times q$ matrix and a $q \times r$ matrix is pqr.

29. Let **A** be an $n \times n$ matrix and let **0** be the $n \times n$ matrix all of whose entries are zero. Show that the following are true.

a) $\mathbf{A} \odot \mathbf{0} = \mathbf{0} \odot \mathbf{A} = \mathbf{0}$

b) $\mathbf{A} \vee \mathbf{0} = \mathbf{0} \vee \mathbf{A} = \mathbf{A}$

c) $\mathbf{A} \wedge \mathbf{0} = \mathbf{0} \wedge \mathbf{A} = \mathbf{0}$

Computer Projects

WRITE PROGRAMS WITH THE FOLLOWING INPUT AND OUTPUT.

1. Given a list of n integers, find the largest integer in the list.

2. Given a list of n integers, find the first and last occurrences of the largest integer in the list.

3. Given a list of n distinct integers, determine the position of an integer in the list using a linear search.

4. Given an ordered list of n distinct integers, determine the position of an integer in the list using a binary search.

5. Given an ordered list of n integers and an integer x, find the number of comparisons used to determine the position of an integer in the list using a linear search and using a binary search.

6. Given a positive integer, determine whether it is prime.

7. Given a message, encrypt this message using the Caesar cipher; and given a message encrypted using the Caesar cipher, decrypt this message.

8. Given two positive integers, find their greatest common divisor using the Euclidean algorithm.

9. Given two positive integers, find their least common multiple.

★10. Given a positive integer, find the prime factorization of this integer.

11. Given a positive integer and a positive integer b greater than 1, find the base b expansion of this integer.

12. Given a positive integer, find the Cantor expansion of this integer (see Exercise 21 of Section 2.4).

13. Given a positive integer n, a modulus m, multiplier a, increment c, and seed x_0, with $0 \leq a < m$, $0 \leq c < m$, and $0 \leq x_0 < m$, generate the sequence of n pseudorandom numbers using the linear congruential generator $x_{n+1} = (ax_n + c) \mathbf{~mod~} m$.

14. Given positive integers a and b, find integers s and t such that $sa + tb = gcd(a, b)$.

15. Given n linear congruences modulo pairwise relatively prime moduli, find the simultaneous solution of these congruences modulo the product of these moduli.

16. Given an $m \times k$ matrix **A** and a $k \times n$ matrix **B**, find **AB**.

17. Given a square matrix **A** and a positive integer n, find $\mathbf{A}^n$.

18. Given a square matrix, determine whether it is symmetric.

19. Given an $n_1 \times n_2$ matrix **A**, an $n_2 \times n_3$ matrix **B**, an $n_3 \times n_4$ matrix **C**, and an $n_4 \times n_5$ matrix **D**, all with integer entries, determine the most efficient order to multiply these matrices (in terms of the number of multiplications and additions of integers).

20. Given two $m \times n$ Boolean matrices, find their meet and join.

21. Given an $m \times k$ Boolean matrix **A** and a $k \times n$ Boolean matrix **B**, find the Boolean product of **A** and **B**.

22. Given a square Boolean matrix **A** and a positive integer n, find $\mathbf{A}^{[n]}$.

3

Mathematical Reasoning

To understand written mathematics, we must understand what makes up a correct mathematical argument, that is, a proof. To learn mathematics, a person needs to construct mathematical arguments and not just read exposition. Obviously, this requires an understanding of the techniques used to build proofs. The goals of this chapter are to teach what makes up a correct mathematical argument and to give the student the necessary tools to construct these arguments.

Many mathematical statements assert that a property is true for all positive integers. Examples of such statements are that for every positive integer n: $n! \leq n^n$, $n^3 - n$ is divisible by 3, and the sum of the first n positive integers is $n(n+1)/2$. A major goal of this chapter, and the book, is to give the student a thorough understanding of mathematical induction, which is used to prove results of this kind.

In previous chapters we explicitly defined sets, sequences, and functions. That is, we described sets by listing their elements or by giving some property that characterizes these elements. We gave formulae for the terms of sequences and the values of functions. There is another important way to define such objects, based on mathematical induction. To define sequences and functions, some initial terms are specified, and a rule is given for finding subsequent values from values already known. For instance, we can define the sequence $\{2^n\}$ by specifying that $a_1 = 2$ and that $a_{n+1} = 2a_n$ for $n = 1, 2, 3, \ldots$. Sets can be defined by listing some of their elements and giving rules for constructing elements from those already known to be in the set. Such definitions, called recursive definitions, are used throughout discrete mathematics and computer science.

When a procedure is specified for solving a problem, this procedure *always* solves the problem correctly. Just testing to see that the correct result is obtained for a set of input values does not show that the procedure always works correctly. The correctness of a procedure can be guaranteed only by proving that it always yields the correct result. The final section of this chapter contains an introduction to the techniques of program verification. This is a formal technique to verify that procedures are correct. Program verification serves as the basis for attempts underway to prove in a mechanical fashion that programs are correct.

3.1 Methods of Proof

INTRODUCTION

Two important questions that arise in the study of mathematics are: (1) When is a mathematical argument correct? (2) What methods can be used to construct mathematical arguments? This section helps answer these questions by describing various forms of correct and incorrect mathematical arguments.

A **theorem** is a statement that can be shown to be true. We demonstrate that a theorem is true with a sequence of statements that form an argument, called a **proof**. To construct proofs, methods are needed to derive new statements from old ones. The statements used in a proof can include **axioms** or **postulates,** which are the underlying assumptions about mathematical structures, the hypotheses of the theorem to be proved, and previously proved theorems. The **rules of inference,** which are the means used to draw conclusions from other assertions, tie together the steps of a proof.

In this section rules of inference will be discussed. This will help clarify what makes up a correct proof. Some common forms of incorrect reasoning, called **fallacies,** will also be described. Then various methods commonly used to prove theorems will be introduced.

Remark: The terms *lemma* and *corollary* are used for certain types of theorems. A **lemma** (plural **lemmas** or **lemmata**) is a simple theorem used in the proof of other theorems. (For instance, Lemma 1 in Section 2.4 was used to prove the theorem that the Euclidean algorithm produces the greatest common divisor of two integers.) Complicated proofs are usually easier to understand when they are proved using a series of lemmas, where each lemma is proved individually. A **corollary** is a proposition that can be established directly from a theorem that has been proved.

RULES OF INFERENCE

The tautology $(p \wedge (p \rightarrow q)) \rightarrow q$ is the basis of the rule of inference called **modus ponens,** or the **law of detachment.** This tautology is written in the following way:

$$\begin{array}{l} p \\ \underline{p \rightarrow q} \\ \therefore q \end{array}$$

Using this notation, the hypotheses are written in a column and the conclusion below a bar. (The symbol $\therefore$ denotes "therefore.") Modus ponens states that if both an implication and its hypothesis are known to be true, then the conclusion of this implication is true.

EXAMPLE 1 Suppose that the implication "if it snows today, then we will go skiing" and its hypothesis, "it is snowing today," are true. Then, by modus ponens, it follows that the conclusion of the implication, "we will go skiing," is true. ■

EXAMPLE 2 The implication "if n is divisible by 3, then n^2 is divisible by 9," is true. Consequently, if n is divisible by 3, then by modus ponens, it follows that n^2 is divisible by 9. ■

Table 1 lists some important rules of inference. The verifications of these rules of inference can be found as exercises in Section 1.2. Here are some examples of arguments using these rules of inference.

EXAMPLE 3 State which rule of inference is the basis of the following argument: "It is below freezing now. Therefore, it is either below freezing or raining now."

TABLE 1 Rules of Inference.

Rule of Inference	*Tautology*	*Name*
$\dfrac{p}{\therefore p \vee q}$	$p \rightarrow (p \vee q)$	Addition
$\dfrac{p \wedge q}{\therefore p}$	$(p \wedge q) \rightarrow p$	Simplification
p $\underline{p \rightarrow q}$ $\therefore q$	$[p \wedge (p \rightarrow q)] \rightarrow q$	Modus ponens
$\neg q$ $\underline{p \rightarrow q}$ $\therefore \neg p$	$[\neg q \wedge (p \rightarrow q)] \rightarrow \neg p$	Modus tollens
$p \rightarrow q$ $\underline{q \rightarrow r}$ $\therefore p \rightarrow r$	$[(p \rightarrow q) \wedge (q \rightarrow r)] \rightarrow (p \rightarrow r)$	Hypothetical syllogism
$p \vee q$ $\underline{\neg p}$ $\therefore q$	$[(p \vee q) \wedge \neg p] \rightarrow q$	Disjunctive syllogism

Solution: Let p be the proposition "It is below freezing now" and q the proposition "It is raining now." Then this argument is of the form

$$\frac{p}{\therefore p \vee q}$$

This is an argument that uses the addition rule. ■

EXAMPLE 4 State which rule of inference is the basis of the following argument: "It is below freezing and raining now. Therefore, it is below freezing now."

Solution: Let p be the proposition "It is below freezing now," and let q be the proposition "It is raining now." This argument is of the form

$$\frac{p \wedge q}{\therefore p}$$

This argument uses the simplification rule. ■

EXAMPLE 5 State which rule of inference is used in the argument:

> If it rains today, then we will not have a barbecue today. If we do not have a barbecue today, then we will have a barbecue tomorrow. Therefore, if it rains today, then we will have a barbecue tomorrow.

Solution: Let p be the proposition "It is raining today," let q be the proposition "We will not have a barbecue today," and let r be the proposition "We will have a barbecue tomorrow." Then this argument is of the form

$$\begin{array}{c} p \rightarrow q \\ \underline{q \rightarrow r} \\ \therefore p \rightarrow r \end{array}$$

Hence, this argument is a hypothetical syllogism. ■

An argument built using rules of inference is said to be **valid.** When all propositions used in a valid argument are true, it leads to a correct conclusion. However, a valid argument can lead to an incorrect conclusion if one or more false propositions are used within the argument. For example,

> "If 101 is divisible by 3, then 101^2 is divisible by 9. 101 is divisible by 3. Consequently, 101^2 is divisible by 9."

is a valid argument based on modus ponens. However, the conclusion of this argument is false, since 9 does not divide $101^2 = 10,201$. The false proposition "101 is divisible by 3" has been used in the argument, which means that the conclusion of the argument may be false.

FALLACIES

There are several common fallacies that arise in incorrect arguments. These fallacies resemble rules of inference but are based on contingencies rather than tautologies. These are discussed here to show the distinction between correct and incorrect reasoning.

The proposition $[(p \rightarrow q) \wedge q] \rightarrow p$ is not a tautology, since it is false when p is false and q is true. However, there are many incorrect arguments that treat this as a tautology. This type of incorrect reasoning is called the **fallacy of affirming the conclusion.**

EXAMPLE 6 Is the following argument valid?

> If you do every problem in this book, then you will learn discrete mathematics.
> You learned discrete mathematics.
> Therefore, you did every problem in this book.

Solution: Let p be the proposition "You did every problem in this book." Let q be the proposition "You learned discrete mathematics." Then this argument is of the form: if $p \rightarrow q$ and q, then p. This is an example of an incorrect argument using the fallacy of affirming the conclusion. Indeed, it is possible for you to learn discrete mathematics in some way other than by doing every problem in this book. (You may learn discrete mathematics by reading, listening to lectures, doing some but not all the problems in this book, and so on.) ■

EXAMPLE 7 Let p be the proposition "$n \equiv 1 \pmod 3$," and let q be the proposition "$n^2 \equiv 1 \pmod 3$." The implication $p \rightarrow q$, which is "if $n \equiv 1 \pmod 3$, then $n^2 \equiv 1 \pmod 3$," is true. If q is true, so that $n^2 \equiv 1 \pmod 3$, does it follow that p is true, namely, that $n \equiv 1 \pmod 3$?

Solution: It would be incorrect to conclude that p is true, since it is possible that $n \equiv 2 \pmod 3$. If the incorrect conclusion that p is true is made, this would be an example of the fallacy of affirming the conclusion. ■

The proposition $[(p \rightarrow q) \wedge \neg p] \rightarrow \neg q$ is not a tautology, since it is false when p is false and q is true. There are many incorrect arguments that use this incorrectly as a rule of inference. This type of incorrect reasoning is called the **fallacy of denying the hypothesis.**

EXAMPLE 8 Let p and q be as in Example 6. If the implication $p \to q$ is true, and $\neg p$ is true, is it correct to conclude that $\neg q$ is true? In other words, is it correct to assume that you did not learn discrete mathematics if you did not do every problem in the book, assuming that if you do every problem in this book, then you will learn discrete mathematics?

Solution: It is possible that you learned discrete mathematics even if you did not do every problem in this book. This incorrect argument is of the form $p \to q$ and $\neg p$ imply $\neg q$, which is an example of the fallacy of denying the hypothesis. ■

EXAMPLE 9 Let p and q be as in Example 7. Is it correct to assume that if $\neg p$ is true then $\neg q$ is true, using the fact that $p \to q$ is true? In other words, is it correct to conclude that $n^2 \not\equiv 1 \pmod 3$ if $n \not\equiv 1 \pmod 3$, using the implication: if $n \equiv 1 \pmod 3$, then $n^2 \equiv 1 \pmod 3$?

Solution: It is incorrect to conclude that $n^2 \not\equiv 1 \pmod 3$ if $n \not\equiv 1 \pmod 3$, since $n^2 \equiv 1 \pmod 3$ when $n \equiv 2 \pmod 3$. This incorrect argument is another example of the fallacy of denying the hypothesis. ■

Many incorrect arguments are based on a fallacy called **begging the question.** This fallacy occurs when one or more steps of a proof are based on the truth of the statement being proved. In other words, this fallacy arises when a statement is proved using itself, or a statement equivalent to it. That is why this fallacy is also called **circular reasoning.**

EXAMPLE 10 Is the following argument correct? It supposedly shows that n is an even integer whenever n^2 is an even integer.

> Suppose that n^2 is even. Then $n^2 = 2k$ for some integer k. Let $n = 2l$ for some integer l. This shows that n is even.

Solution: This argument is incorrect. The statement "let $n = 2l$ for some integer l" occurs in the proof. No argument has been given to show that it is true. This is circular reasoning since this statement is equivalent to the statement being proved, namely, "n is even." Of course, the result itself is correct; only the method of proof is wrong. ■

METHODS OF PROVING THEOREMS

We proved several theorems in Chapters 1 and 2. Let us now be more explicit about the methodology of constructing proofs. We will describe how different types of statements are proved.

Because many theorems are implications, the techniques for proving implications are important. Recall that $p \to q$ is true unless p is true but q is false. Note that when the statement $p \to q$ is proved, it need only be shown that q is true if p is true; it is *not* usually the case that q is proved to be true. We will give the most common techniques for proving implications in the following discussion.

Suppose that the hypothesis p of an implication $p \to q$ is false. Then the implication $p \to q$ is true, because the statement has the form $\mathbf{F} \to \mathbf{T}$ or $\mathbf{F} \to \mathbf{F}$, and hence is true. Consequently, if it can be shown that p is false, then a proof, called a **vacuous proof,** of the implication $p \to q$ can be given. Vacuous proofs are often used to establish special cases of theorems that state that an implication is true for all positive integers (i.e., a theorem of the kind $\forall n\, P(n)$ where $P(n)$ is a propositional function). Proof techniques for theorems of this kind will be discussed in Section 3.2.

EXAMPLE 11 Show that the proposition $P(0)$ is true where $P(n)$ is the propositional function "If $n > 1$, then $n^2 > n$."

Solution: Note that the proposition $P(0)$ is the implication "If $0 > 1$, then $0^2 > 0$." Since the hypothesis $0 > 1$ is false, the implication $P(0)$ is automatically true.

Remark: The fact that the conclusion of this implication, $0^2 > 0$, is false is irrelevant to the truth value of the implication, because an implication with a false hypothesis is guaranteed to be true. ■

Suppose that the conclusion q of an implication $p \to q$ is true. Then $p \to q$ is true, since the statement has the form $\mathbf{T} \to \mathbf{T}$ or $\mathbf{F} \to \mathbf{T}$, which are true. Hence, if it can be shown that q is true, then a proof, called a **trivial proof,** of $p \to q$ can be given. Trivial proofs are often important when special cases of theorems are proved (see the discussion of proof by cases) and in mathematical induction, which is a proof technique discussed in Section 3.2.

EXAMPLE 12 Let $P(n)$ be the proposition "If a and b are positive integers with $a \geq b$, then $a^n \geq b^n$." Show that the proposition $P(0)$ is true.

Solution: The proposition $P(0)$ is "If $a \geq b$, then $a^0 \geq b^0$." Since $a^0 = b^0 = 1$, the conclusion of $P(0)$ is true. Hence, $P(0)$ is true. This is an example of a trivial proof. Note that the hypothesis, which is the statement "$a \geq b$," was not needed in this proof. ■

The implication $p \to q$ can be proved by showing that if p is true, then q must also be true. This shows that the combination p true and q false never occurs. A proof of this kind is called a **direct proof.** To carry out such a proof, assume that p is true and use rules of inference and theorems already proved to show that q must also be true.

EXAMPLE 13 Give a direct proof of the theorem "If n is odd, then n^2 is odd."

Solution: Assume that the hypothesis of this implication is true, namely, suppose that n is odd. Then $n = 2k + 1$, where k is an integer. It follows that $n^2 = (2k + 1)^2 = 4k^2 + 4k + 1 = 2(2k^2 + 2k) + 1$. Therefore, n^2 is odd (it is 1 more than twice an integer). ■

Since the implication $p \rightarrow q$ is equivalent to its contrapositive, $\neg q \rightarrow \neg p$, the implication $p \rightarrow q$ can be proved by showing that its contrapositive, $\neg q \rightarrow \neg p$, is true. This related implication is usually proved directly, but any proof technique can be used. An argument of this type is called an **indirect proof.**

EXAMPLE 14 Give an indirect proof of the theorem "If $3n + 2$ is odd, then n is odd."

Solution: Assume that the conclusion of this implication is false; namely, assume that n is even. Then $n = 2k$ for some integer k. It follows that $3n + 2 = 3(2k) + 2 = 6k + 2 = 2(3k + 1)$, so that $3n + 2$ is even (since it is a multiple of 2). Since the negation of the conclusion of the implication implies that the hypothesis is false, the original implication is true. ■

Suppose that a contradiction q can be found so that $\neg p \rightarrow q$ is true, that is, $\neg p \rightarrow \mathbf{F}$ is true. Then the proposition $\neg p$ must be false. Consequently, p must be true. This technique can be used when a contradiction, such as $r \wedge \neg r$, can be found so that it is possible to show that the implication $\neg p \rightarrow (r \wedge \neg r)$ is true. An argument of this type is called a **proof by contradiction.**

EXAMPLE 15 Prove that $\sqrt{2}$ is irrational by giving a proof by contradiction.

Solution: Let p be the proposition: "$\sqrt{2}$ is irrational." Suppose that $\neg p$ is true. Then $\sqrt{2}$ is rational. We will show that this leads to a contradiction. Under the assumption that $\sqrt{2}$ is rational, there exist integers a and b with $\sqrt{2} = a/b$, where a and b have no common factors (so that the fraction a/b is in lowest terms). Since $\sqrt{2} = a/b$, when both sides of this equation are squared, it follows that

$$2 = a^2/b^2.$$

Hence,

$$2b^2 = a^2.$$

This means that a^2 is even, implying that a is even. Furthermore, since a is even, $a = 2c$ for some integer c. Thus

$$2b^2 = 4c^2,$$

so that

$$b^2 = 2c^2.$$

This means that b^2 is even. Hence, b must be even as well.

It has been shown that $\neg p$ implies that $\sqrt{2} = a/b$, where a and b have no common factors, and 2 divides a and b. This is a contradiction. Hence, $\neg p$ is false, so that p: "$\sqrt{2}$ is irrational" is true. ■

An indirect proof of an implication can be rewritten as a proof by contradiction. In an indirect proof we show that $p \rightarrow q$ is true by using a direct proof to show that $\neg q \rightarrow \neg p$ is true. That is, in an indirect proof of $p \rightarrow q$ we assume that $\neg q$ is true and show that $\neg p$ must also be true. To rewrite an indirect proof of $p \rightarrow q$ as a proof by contradiction, we suppose that both p and $\neg q$ are true. Then we use the steps from the direct proof of $\neg q \rightarrow \neg p$ to show that $\neg p$ must also be true. This leads to the contradiction $p \wedge \neg p$, completing the proof by contradiction. Example 16 illustrates how an indirect proof of an implication can be rewritten as a proof by contradiction.

EXAMPLE 16 Give a proof by contradiction of the theorem "If $3n + 2$ is odd, then n is odd."

Solution: We assume that $3n + 2$ is odd and that n is not odd, so that n is even. Following the same steps as in the solution of Example 14 (an indirect proof of this theorem), we can show that if n is even, then $3n + 2$ is even. This contradicts the assumption that $3n + 2$ is odd, completing the proof. ■

To prove an implication of the form

$$(p_1 \vee p_2 \vee \cdots \vee p_n) \rightarrow q$$

the tautology

$$[(p_1 \vee p_2 \vee \cdots \vee p_n) \rightarrow q] \leftrightarrow [(p_1 \rightarrow q) \wedge (p_2 \rightarrow q) \wedge \cdots \wedge (p_n \rightarrow q)]$$

can be used as a rule of inference. This shows that the original implication with a hypothesis made up of a disjunction of the propositions $p_1, p_2, \ldots, p_n$ can be proved by proving each of the n implications $p_i \rightarrow q$, $i = 1, 2, \ldots, n$, individually. Such an argument is called a **proof by cases.** Sometimes to prove that an implication $p \rightarrow q$ is true, it is convenient to use a disjunction $p_1 \vee p_2 \vee \cdots \vee p_n$ instead of p as the hypothesis of the implication, where p and $p_1 \vee p_2 \vee \cdots \vee p_n$ are equivalent. Consider the following example.

EXAMPLE 17 Prove the implication "If n is an integer not divisible by 3, then $n^2 \equiv 1 \pmod{3}$."

Solution: Let p be the proposition "n is not divisible by 3," and let q be the proposition "$n^2 \equiv 1 \pmod{3}$." Then p is equivalent to $p_1 \vee p_2$ where p_1 is "$n \equiv 1 \pmod{3}$" and p_2 is "$n \equiv 2 \pmod{3}$." Hence, to show that $p \rightarrow q$ it can be shown that $p_1 \rightarrow q$ and $p_2 \rightarrow q$. It is easy to give direct proofs of these two implications.

First, suppose that p_1 is true. Then $n \equiv 1 \pmod{3}$, so that $n = 3k + 1$ for some integer k. Thus,

$$n^2 = 9k^2 + 6k + 1 = 3(3k^2 + 2k) + 1.$$

It follows that $n^2 \equiv 1 \pmod{3}$. Hence, the implication $p_1 \rightarrow q$ is true. Next, suppose that p_2 is true. Then $n \equiv 2 \pmod{3}$, so that $n = 3k + 2$ for some integer k. Thus,

$$n^2 = 9k^2 + 12k + 4 = 3(3k^2 + 4k + 1) + 1.$$

Hence, $n^2 \equiv 1 \pmod{3}$, so the implication $p_2 \rightarrow q$ is true.

Since it has been shown that both $p_1 \rightarrow q$ and $p_2 \rightarrow q$ are true, it can be concluded that $(p_1 \vee p_2) \rightarrow q$ is true. Moreover, since p is equivalent to $p_1 \vee p_2$, it follows that $p \rightarrow q$ is true. ■

To prove a theorem that is an equivalence, that is, one that is a statement of the form $p \leftrightarrow q$ where p and q are propositions, the tautology

$$(p \leftrightarrow q) \leftrightarrow [(p \rightarrow q) \wedge (q \rightarrow p)]$$

can be used. That is, the proposition "p if and only if q" can be proved if both the implications "if p, then q" and "if q, then p" are proved.

EXAMPLE 18 Prove the theorem "The integer n is odd if and only if n^2 is odd."

Solution: This theorem has the form "p if and only if q," where p is "n is odd" and q is "n^2 is odd." To prove this theorem, we need to show that $p \rightarrow q$ and $q \rightarrow p$ are true.

We have already shown (in Example 13) that $p \rightarrow q$ is true. We will use an indirect proof to prove that $q \rightarrow p$. Assume that its conclusion is false, namely, that n is even. Then $n = 2k$ for some integer k. Then $n^2 = 4k^2 = 2(2k^2)$, so that n^2 is even (since it is a multiple of 2). This completes the indirect proof of $q \rightarrow p$.

Since we have shown that both $p \rightarrow q$ and $q \rightarrow p$ are true, we have shown that the theorem is true. ■

Sometimes a theorem states that several propositions are equivalent. Such a theorem states that propositions $p_1, p_2, p_3, \ldots, p_n$ are equivalent. This can be written as

$$p_1 \leftrightarrow p_2 \leftrightarrow \cdots \leftrightarrow p_n,$$

which states that all n propositions have the same truth values. One way to prove these mutually equivalent is to use the tautology

$$[p_1 \leftrightarrow p_2 \leftrightarrow \cdots \leftrightarrow p_n] \leftrightarrow [(p_1 \rightarrow p_2) \wedge (p_2 \rightarrow p_3) \wedge \cdots \wedge (p_n \rightarrow p_1)].$$

This shows that if the implications $p_1 \rightarrow p_2, p_2 \rightarrow p_3, \ldots, p_n \rightarrow p_1$ can be shown to be true, then the propositions $p_1, p_2, \ldots, p_n$ are all equivalent.

EXAMPLE 19 Prove that when n is an integer, the following three statements are equivalent.

p_1: n **mod** $3 = 1$ or n **mod** $3 = 2$
p_2: n is not divisible by 3
p_3: $n^2 \equiv 1 \pmod{3}$

Solution: To show that the statements are equivalent, we can prove that the implications $p_1 \rightarrow p_2$, $p_2 \rightarrow p_3$, and $p_3 \rightarrow p_1$ are true.

We will use a direct proof to show that $p_1 \rightarrow p_2$ is true. Assume that n **mod** $3 = 1$ or 2. By the division algorithm, $n = 3q + r$ where $0 \leq r < 3$. By the definition of **mod**, we have $r = n$ **mod** 3. Since n is divisible by 3 if and only if $r = 0$, the assumption that n **mod** $3 = 1$ or 2 implies that n is not divisible by 3. This completes the proof that $p_1 \rightarrow p_2$ is true.

We have already shown that $p_2 \rightarrow p_3$ is true in Example 17.

We will use an indirect proof to show that $p_3 \rightarrow p_1$ is true. We assume that the conclusion of this implication is false, namely, that n **mod** 3 is neither 1 nor 2. Since n **mod** 3 equals 0, 1, or 2, we see that n **mod** $3 = 0$. This means that $3 \mid n$, so that $n = 3k$ for some integer k. This implies that $n^2 = 9k^2 = 3(3k^2)$, which shows that $n^2 \equiv 0 \pmod{3}$, so that p_3 is false. This completes the indirect proof that $p_3 \rightarrow p_1$, and it also completes the proof of the theorem. ■

THEOREMS AND QUANTIFIERS

Many theorems are stated as propositions that involve quantifiers. There are a variety of methods for proving theorems that are quantifications. We will describe some of the most important of these here.

Many theorems are assertions that objects of a particular type exist. A theorem of this type is a proposition of the form $\exists x\, P(x)$, where P is a predicate. A proof of a proposition of the form $\exists x\, P(x)$ is called an **existence proof.** There are several ways to prove a theorem of this type. Sometimes an existence proof of $\exists x\, P(x)$ can be given by finding an element a such that $P(a)$ is true. Such an existence proof is called **constructive.** It is also possible to give an existence proof that is **nonconstructive;** that is, we do not find an element a such that $P(a)$ is true, but rather prove that $\exists x\, P(x)$ is true in some other way. One common method of giving a nonconstructive existence proof is to use proof by contradiction and show that the negation of the existential quantification implies a contradiction. The concept of a constructive existence proof is illustrated by the following example.

EXAMPLE 20 A Constructive Existence Proof Show that there are n consecutive composite positive integers for every positive integer n. Note that this asks for proof of the quantification: $\forall n\, \exists x\, (x + i$ is composite for $i = 1, 2, \ldots, n)$.

Solution: Let

$$x = (n + 1)! + 1.$$

Consider the integers

$$x + 1, x + 2, \ldots, x + n.$$

Note that $i + 1$ divides $x + i = (n + 1)! + (i + 1)$ for $i = 1, 2, \ldots, n$. Hence, n consecutive composite positive integers have been given. Note that in the solution a number x such that $x + i$ is composite for $i = 1, 2, \ldots, n$ has been produced. Hence, this is an example of a constructive existence proof. ■

Remark: The proof in Example 20 can be found in the works of the ancient Greek mathematician Euclid.

An example of a nonconstructive existence proof is given next.

EXAMPLE 21 A Nonconstructive Existence Proof Show that for every positive integer n there is a prime greater than n. This problem asks for a proof of an existential quantification, namely, $\exists x\, Q(x)$, where $Q(x)$ is the proposition "x is prime and x is greater than n," and the universe of discourse is the set of positive integers.

Solution: Let n be a positive integer. To show that there is a prime greater than n, consider the integer $n! + 1$. Since every integer has a prime factor, there is at least one prime dividing $n! + 1$. (One possibility is that $n! + 1$ is already prime.) Note that when $n! + 1$ is divided by an integer less than or equal to n, the remainder equals 1. Hence, any prime factor of this integer must be greater than n. This proves the result. This argument is a nonconstructive existence proof because a prime larger than n has not been produced. It has simply been shown that one must exist. ■

Suppose a statement of the form $\forall x\, P(x)$ is false. How can we show this? Recall that the propositions $\neg\forall x\, P(x)$ and $\exists x \neg P(x)$ are equivalent. This means that if we find an element a such that $P(a)$ is false, then we have shown that $\exists x \neg P(x)$ is true, which means that $\forall x\, P(x)$ is false. An element a for which $P(a)$ is false is called a **counterexample**. Note that only one counterexample needs to be found to show that $\forall x\, P(x)$ is false.

EXAMPLE 22 Show that the assertion "All primes are odd" is false.

Solution: The statement "All primes are odd" is a universal quantification, namely,

$$\forall x\, O(x),$$

where $O(x)$ is the proposition "x is odd," and the universe of discourse is the set of primes. Note that $x = 2$ is a counterexample, since 2 is a prime number that is even. Hence, the statement "All prime numbers are odd" is false. ■

SOME COMMENTS ON PROOFS

We have described a variety of methods for proving theorems. The reader may have observed that no algorithm for proving theorems has been given here. Such a procedure does not exist.

There are many theorems whose proofs are easy to find by directly working through the hypotheses and definitions of the terms in the theorem. However, it is often difficult to prove a theorem without resorting to a clever use of an indirect proof, a proof by contradiction, or some other proof technique. Constructing proofs is an art that can be learned only by trying various lines of attack.

Moreover, many statements that appear to be theorems have resisted the persistent efforts of mathematicians for hundreds of years. For instance, as simple a statement as "every even positive integer greater than 4 is the sum of two primes" has not yet been shown to be true, and no counterexample has been found. This statement is known as **Goldbach's conjecture** and is one of many assertions in mathematics that is simple to state, with a truth value that is unknown.

Exercises

1. What rule of inference is used in each of the following arguments?

a) Alice is a mathematics major. Therefore, Alice is either a mathematics major or a computer science major.

b) Jerry is a mathematics major and a computer science major. Therefore, Jerry is a mathematics major.

c) If it is rainy, then the pool will be closed. It is rainy. Therefore, the pool is closed.

d) If it snows today, the university will close. The university is not closed today. Therefore, it did not snow today.

e) If I go swimming, then I will stay in the sun too long. If I stay in the sun too long, then I will sunburn. Therefore, if I go swimming, then I will sunburn.

2. What rule of inference is used in each of the following arguments?

a) Kangaroos live in Australia and are marsupials. Therefore, kangaroos are marsupials.

b) It is either hotter than 100 degrees today or the pollution is dangerous. It is less than 100 degrees outside today. Therefore, the pollution is dangerous.

c) Linda is an excellent swimmer. If Linda is an excellent swimmer, then she can work as a lifeguard. Therefore, Linda can work as a lifeguard.

d) Steve will work at a computer company this summer. Therefore, this summer Steve will

---

Christian Goldbach (1690–1764). Christian Goldbach was born in Königsberg, Prussia, the city noted for its famous bridge problem (which will be studied in Section 7.5). He became professor of mathematics at the Academy in St. Petersburg in 1725. In 1728 Goldbach went to Moscow to tutor the son of the tsar. He entered the world of politics when, in 1742, he became a staff member in the Russian Ministry of Foreign Affairs. Goldbach is best known for his correspondence with eminent mathematicians, including Leonhard Euler and Daniel Bernoulli, for his famous conjectures in number theory, and for several contributions to analysis.

work at a computer company or he will be a beach bum.

e) If I work all night on this homework, then I can answer all the exercises. If I answer all the exercises, I will understand the material. Therefore, if I work all night on this homework, then I will understand the material.

3. Determine whether each of the following arguments is valid. If an argument is correct, what rule of inference is being used? If it is not, what fallacy occurs?

a) If n is a real number such that $n > 1$, then $n^2 > 1$. Suppose that $n^2 > 1$. Then $n > 1$.

b) The number $\log_2 3$ is irrational if it is not the ratio of two integers. Therefore, since $\log_2 3$ cannot be written in the form a/b where a and b are integers, it is irrational.

c) If n is a real number with $n > 3$, then $n^2 > 9$. Suppose that $n^2 \leq 9$. Then $n \leq 3$.

d) A positive integer is either a perfect square or it has an even number of positive integer divisors. Suppose that n is a positive integer that has an odd number of positive integer divisors. Then n is a perfect square.

e) If n is a real number with $n > 2$, then $n^2 > 4$. Suppose that $n \leq 2$. Then $n^2 \leq 4$.

4. The following argument is an incorrect proof of the theorem "If n^2 is not divisible by 3, then n is not divisible by 3." The reason it is incorrect is that circular reasoning has been used. Where has the error in reasoning been made?

> If n^2 is not divisible by 3, then n^2 does not equal $3k$ for some integer k. Hence, n does not equal $3l$ for some integer l. Therefore, n is not divisible by 3.

5. Prove the proposition $P(0)$, where $P(n)$ is the proposition "If n is a positive integer greater than 1, then $n^2 > n$." What kind of proof did you use?

6. Prove the proposition $P(1)$, where $P(n)$ is the proposition "If n is a positive integer, then $n^2 \geq n$." What kind of proof did you use?

7. Let $P(n)$ be the proposition "If a and b are positive real numbers, then $(a + b)^n \geq a^n + b^n$." Prove that $P(1)$ is true. What kind of proof did you use?

8. Prove that the square of an even number is an even number using

a) a direct proof.

b) an indirect proof.

c) a proof by contradiction.

9. Prove that the sum of two odd integers is even.

10. Prove that the sum of two rational numbers is rational.

11. Prove that the sum of an irrational number and a rational number is irrational using a proof by contradiction.

12. Prove that the product of two rational numbers is rational.

13. Prove or disprove that the product of two irrational numbers is irrational.

14. Prove or disprove that the product of a nonzero rational number and an irrational number is irrational.

★15. Prove or disprove that $n^2 - n + 41$ is prime whenever n is a positive integer.

16. Prove or disprove that $2^n + 1$ is prime for all nonnegative integers n.

17. Show that $\sqrt[3]{3}$ is irrational.

★18. Show that $\sqrt{n}$ is irrational if n is a positive integer that is not a perfect square.

19. Prove that if x and y are real numbers, then $\max(x, y) + \min(x, y) = x + y$. (*Hint:* Use a proof by cases, with the two cases corresponding to $x \geq y$ and $x < y$, respectively.)

20. Prove that the square of an integer not divisible by 5 leaves a remainder of 1 or 4 when divided by 5. (*Hint:* Use a proof by cases, where the cases correspond to the possible remainders for the integer when it is divided by 5.)

21. Prove that if x and y are real numbers, then $|x| + |y| \geq |x + y|$ (where $|x|$ represents the absolute value of x, which equals x if $x \geq 0$ and equals $-x$ if $x \leq 0$).

22. Prove that if n is a positive integer, then n is even if and only if $7n + 4$ is even.

23. Prove that if n is a positive integer, then n is odd if and only if $5n + 6$ is odd.

24. Prove that $m^2 = n^2$ if and only if $m = n$ or $m = -n$.

★25. Let p be prime. Prove that $a^2 \equiv b^2 \pmod{p}$ if and only if $a \equiv b \pmod{p}$ or $a \equiv -b \pmod{p}$.

26. Prove or disprove that $n^2 - 1$ is composite whenever n is a positive integer greater than 1.

27. Prove or disprove that if m and n are integers such that $mn = 1$, then either $m = 1$ and $n = 1$, or else $m = -1$ and $n = -1$.

28. Prove or disprove that a **mod** $m + b$ **mod** $m = (a + b)$ **mod** m whenever m is a positive integer.

29. Prove or disprove that every positive integer can be written as the sum of the squares of two integers.

30. Prove that if n is a positive integer such that the sum of its divisors is $n + 1$, then n is prime. What kind of proof did you use?

31. Prove that at least one of the real numbers $a_1, a_2, \ldots, a_n$ is greater than or equal to the average of these numbers. What kind of proof did you use?

⋆32. Use Exercise 31 to show that if the first 10 positive integers are placed around a circle, in any order, there exist three integers in consecutive locations around the circle that have a sum greater than or equal to 17.

33. Prove that if n is an integer, the following four statements are equivalent: (i) n is even, (ii) $n + 1$ is odd, (iii) $3n + 1$ is odd, (iv) $3n$ is even.

34. Prove that if n is an integer, the following three statements are equivalent: (i) 5 divides n, (ii) 5 divides n^2, (iii) $n^2 \not\equiv \pm 1 \pmod{5}$.

35. Prove or disprove that there are three consecutive odd positive integers that are primes, that is, odd primes of the form p, $p + 2$, and $p + 4$.

36. Prove or disprove that given a positive integer n, there are n consecutive odd positive integers that are primes.

37. Which rules of inference are used to establish the conclusion of Lewis Carroll's argument described in Example 12 of Section 1.3?

38. Which rules of inference are used to establish the conclusion of Lewis Carroll's argument described in Example 13 of Section 1.3?

39. Give a constructive proof of the proposition: "For every positive integer n there is an integer divisible by more than n primes."

40. Find a counterexample to the proposition: "For every prime number n, $n + 2$ is prime."

⋆41. Prove that there are infinitely many primes congruent to 3 modulo 4. Is your proof constructive or nonconstructive? (*Hint:* One approach is to assume that there are only finitely many such primes $p_1, p_2, \ldots, p_n$. Let $q = 4p_1p_2\cdots p_n + 3$. Show that q must have a prime factor congruent to 3 modulo 4 not among the n primes $p_1, p_2, \ldots, p_n$.)

42. Prove or disprove that if $p_1, p_2, \ldots, p_n$ are the n smallest primes, then $p_1p_2\cdots p_n + 1$ is prime.

43. Show that the propositions p_1, p_2, p_3, p_4, and p_5 can be shown to be equivalent by proving that the implications $p_1 \to p_4$, $p_3 \to p_1$, $p_4 \to p_2$, $p_2 \to p_5$, and $p_5 \to p_3$ are true.

3.2 Mathematical Induction

INTRODUCTION

What is a formula for the sum of the first n positive odd integers? The sums of the first n positive odd integers for $n = 1, 2, 3, 4, 5$ are

$$\begin{aligned} 1 &= 1, \\ 1 + 3 &= 4, \\ 1 + 3 + 5 &= 9, \\ 1 + 3 + 5 + 7 &= 16, \\ 1 + 3 + 5 + 7 + 9 &= 25. \end{aligned}$$

From these values it is reasonable to guess that the sum of the first n positive odd integers is n^2. We need a method to *prove* that this *guess* is correct, if in fact it is.

Mathematical induction is an extremely important proof technique that can be used to prove assertions of this type. As we will see in this section and in subsequent chapters, mathematical induction is used extensively to prove results about a large variety of discrete objects. For example, it is used to prove results about the complexity of algorithms, the correctness of certain types of computer programs, theorems about graphs and trees, as well as a wide range of identities and inequalities.

In this section we will describe how mathematical induction can be used and why it is a valid proof technique. It is extremely important to note that mathematical induction can be used only to prove results obtained in some other way. It is *not* a tool for discovering formulae or theorems.

THE WELL-ORDERING PROPERTY

The validity of mathematical induction follows from the following fundamental axiom about the set of integers.

THE WELL-ORDERING PROPERTY Every nonempty set of nonnegative integers has a least element.

The well-ordering property can often be used directly in proofs.

EXAMPLE 1 Use the well-ordering property to prove the division algorithm. Recall that the division algorithm states that if a is an integer and d is a positive integer, then there are unique integers q and r with $0 \leq r < d$ and $a = dq + r$.

Solution: Let S be the set of nonnegative integers of the form $a - dq$ where q is an integer. This set is nonempty since $-dq$ can be made as large as desired (taking q to be a negative integer with large absolute value). By the well-ordering property S has a least element $r = a - dq_0$.

The integer r is nonnegative. It is also the case that $r < d$. If it were not, then there would be a smaller nonnegative element in S, namely, $a - d(q_0 + 1)$. To see this, suppose that $r \geq d$. Since $a = dq_0 + r$, it follows that $a - d(q_0 + 1) = (a - dq_0) - d = r - d \geq 0$. Consequently, there are integers q and r with $0 \leq r < d$. The proof that q and r are unique is left as an exercise for the reader. ■

MATHEMATICAL INDUCTION

Many theorems state that $P(n)$ is true for all positive integers n, where $P(n)$ is a propositional function. Mathematical induction is a technique for proving theorems of this kind. In other words, mathematical induction is used to prove propositions of the form $\forall n\, P(n)$, where the universe of discourse is the set of positive integers.

A proof by mathematical induction that $P(n)$ is true for every positive integer n consists of two steps:

1. *Basis step*. The proposition $P(1)$ is shown to be true.
2. *Inductive step*. The implication $P(n) \rightarrow P(n+1)$ is shown to be true for every positive integer n.

Here, $P(n)$ is called the **inductive hypothesis.** When we complete both steps of a proof by mathematical induction, we have proved that $P(n)$ is true for all positive integers n; that is, we have shown that $\forall n\, P(n)$ is true.

Expressed as a rule of inference, this proof technique can be stated as

$$[P(1) \wedge \forall n\, (P(n) \rightarrow P(n+1))] \rightarrow \forall n\, P(n).$$

Since mathematical induction is such an important technique, it is worthwhile to explain in detail the steps of a proof using this technique. The first thing we do to prove that $P(n)$ is true for all positive integers n is to show that $P(1)$ is true. This amounts to showing that the particular statement obtained when n is replaced by 1 in $P(n)$ is true. Then we must show that $P(n) \rightarrow P(n+1)$ is true for every positive integer n. To prove that this implication is true for every positive integer n, we need to show that $P(n+1)$ cannot be false when $P(n)$ is true. This can be accomplished by assuming that $P(n)$ is true and showing that *under this hypothesis* $P(n+1)$ must also be true.

Remark: In a proof by mathematical induction it is *not* assumed that $P(n)$ is true for all positive integers! It is only shown that *if it is assumed* that $P(n)$ is true, then $P(n+1)$ is also true. Thus, a proof by mathematical induction is not a case of begging the question, or circular reasoning.

When we use mathematical induction to prove a theorem, we first show that $P(1)$ is true. Then we know that $P(2)$ is true, since $P(1)$ implies $P(2)$. Further, we know that $P(3)$ is true, since $P(2)$ implies $P(3)$. Continuing along these lines, we see that $P(k)$ is true, for any positive integer k.

There are several useful illustrations of mathematical induction that can help you remember how this principle works. One of these involves a line of people, person 1, person 2, and so on. A secret is told to person 1, and each person tells the secret to the next person in line, if the former person hears it. Let $P(n)$ be the proposition that person n knows the secret. Then $P(1)$ is true, since the secret is told to person 1; $P(2)$ is true, since person 1 tells person 2 the secret; $P(3)$ is true, since person 2 tells person 3 the secret; and so on. By the principle of mathematical induction, every person in line learns the secret. This is illustrated in Figure 1. (Of course, it has been assumed that each person relays the secret in an unchanged manner to the next person, which is usually not true in real life.)

Historical note: The first known use of mathematical induction is in the work of the 16th-century mathematician Francesco Maurolico (1494–1575). Maurolico wrote extensively on the works of classical mathematics and made many contributions to geometry and optics. In his book *Arithmeticorum Libri Duo* Maurolico presented a variety of properties of the integers together with proofs of these properties. To prove some of these properties he devised the method of mathematical induction. His first use of mathematical induction in this book was to prove that the sum of the first n odd positive integers equals n^2.

FIGURE 1 **People Telling Secrets.**

Another way to illustrate the principle of mathematical induction is to consider an infinite row of dominoes, labeled $1, 2, 3, \ldots, n, \ldots$ where each domino is standing up. Let $P(n)$ be the proposition that domino n is knocked over. If the first domino is knocked over—i.e., if $P(1)$ is true—and if, whenever the nth domino is knocked over, it also knocks the $(n + 1)$th domino over—i.e., if $P(n) \rightarrow P(n + 1)$ is true—then all the dominoes are knocked over. This is illustrated in Figure 2.

Why Mathematical Induction is Valid Why is mathematical induction a valid proof technique? The reason comes from the well-ordering property. Suppose we know that $P(1)$ is true and that the proposition $P(n) \rightarrow P(n + 1)$ is true for all positive integers n. To show that $P(n)$ must be true for all positive integers, assume that there is at least one positive integer for which $P(n)$ is false. Then the set S of positive integers

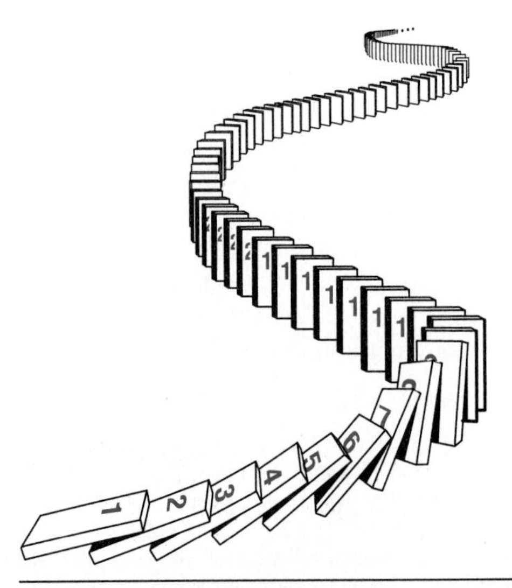

FIGURE 2 **Illustrating How Mathematical Induction Works Using Dominoes.**

for which $P(n)$ is false is nonempty. Thus, by the well-ordering property, S has a least element, which will be denoted by k. We know that k cannot be 1, since $P(1)$ is true. Since k is positive and greater than 1, $k - 1$ is a positive integer. Furthermore, since $k - 1$ is less than k, it is not in S, so $P(k - 1)$ must be true. Since the implication $P(k - 1) \to P(k)$ is also true, it must be the case that $P(k)$ is true. This contradicts the choice of k. Hence, $P(n)$ must be true for every positive integer n.

EXAMPLES OF PROOFS BY MATHEMATICAL INDUCTION

We will use a variety of examples to illustrate how theorems are proved using mathematical induction. We began by proving a formula for the sum of the first n odd positive integers. (Many theorems proved in this section via mathematical induction can be proved using different methods. However, it is worthwhile to try to prove a theorem in more than one way, since one method of attack may succeed while another approach may not.)

EXAMPLE 2 Use mathematical induction to prove that the sum of the first n odd positive integers is n^2.

Solution: Let $P(n)$ denote the proposition that the sum of the first n odd positive integers is n^2. We must first complete the basis step; that is, we must show that $P(1)$ is true. Then we must carry out the inductive step; that is, we must show that $P(n+1)$ is true when $P(n)$ is assumed to be true.

BASIS STEP: $P(1)$ states that the sum of the first one odd positive integers is 1^2. This is true since the sum of the first odd positive integer is 1.

INDUCTIVE STEP: To complete the inductive step we must show that the proposition $P(n) \to P(n + 1)$ is true for every positive integer n. To do this, suppose that $P(n)$ is true for a positive integer n; that is,

$$1 + 3 + 5 + \cdots + (2n - 1) = n^2.$$

(Note that the nth odd positive integer is $(2n - 1)$, since this integer is obtained by adding 2 a total of $n - 1$ times to 1.) We must show that $P(n + 1)$ is true, assuming that $P(n)$ is true. Note that $P(n + 1)$ is the statement that

$$1 + 3 + 5 + \cdots + (2n - 1) + (2n + 1) = (n + 1)^2.$$

So, assuming that $P(n)$ is true, it follows that

$$\begin{aligned} 1 + 3 + 5 + \cdots + (2n - 1) + (2n + 1) &= [1 + 3 + \cdots + (2n - 1)] + (2n + 1) \\ &= n^2 + (2n + 1) \\ &= n^2 + 2n + 1 \\ &= (n + 1)^2. \end{aligned}$$

This shows that $P(n+1)$ follows from $P(n)$. Note that we used the inductive hypothesis $P(n)$ in the second equality to replace the sum of the first n odd positive integers by n^2.

Since $P(1)$ is true and the implication $P(n) \rightarrow P(n+1)$ is true for all positive integers n, the principle of mathematical induction shows that $P(n)$ is true for all positive integers n. ■

The next example uses the principle of mathematical induction to prove an inequality.

EXAMPLE 3 Use mathematical induction to prove the inequality

$$n < 2^n$$

for all positive integers n.

Solution: Let $P(n)$ be the proposition "$n < 2^n$".

BASIS STEP: $P(1)$ is true, since $1 < 2^1 = 2$.

INDUCTIVE STEP: Assume that $P(n)$ is true for the positive integer n. That is, assume that $n < 2^n$. We need to show that $P(n+1)$ is true. That is, we need to show that $n + 1 < 2^{n+1}$. Adding 1 to both sides of $n < 2^n$, and then noting that $1 \leq 2^n$, gives

$$n + 1 < 2^n + 1 \leq 2^n + 2^n = 2^{n+1}.$$

We have shown that $P(n+1)$ is true, namely, that $n + 1 < 2^{n+1}$, based on the assumption that $P(n)$ is true. The induction step is complete.

Therefore, by the principle of mathematical induction, it has been shown that $n < 2^n$ is true for all positive integers n. ■

We will now show how to use mathematical induction to prove a theorem involving divisibility of integers.

EXAMPLE 4 Use mathematical induction to prove that $n^3 - n$ is divisible by 3 whenever n is a positive integer.

Solution: To construct the proof, let $P(n)$ denote the proposition: "$n^3 - n$ is divisible by 3".

BASIS STEP: $P(1)$ is true, since $1^3 - 1 = 0$ is divisible by 3.

INDUCTIVE STEP: Assume that $P(n)$ is true; that is, $n^3 - n$ is divisible by 3. We must show that $P(n+1)$ is true. That is, we must show that $(n+1)^3 - (n+1)$ is divisible by 3. Note that

$$\begin{aligned}(n + 1)^3 - (n + 1) &= (n^3 + 3n^2 + 3n + 1) - (n + 1)\\ &= (n^3 - n) + 3(n^2 + n).\end{aligned}$$

Since both terms in this sum are divisible by 3 (the first by the assumption of the inductive step, and the second because it is 3 times an integer), it follows that $(n + 1)^3 - (n + 1)$ is also divisible by 3. This completes the induction step. Thus, by the principle of mathematical induction, $n^3 - n$ is divisible by 3 whenever n is a positive integer. ■

Sometimes we need to show that $P(n)$ is true for $n = k, k + 1, k + 2, \ldots,$ where k is an integer other than 1. We can use mathematical induction to accomplish this as long as we change the basis step. For instance, consider Example 5, which proves that a summation formula is valid for all nonnegative integers, so that we need to prove that $P(n)$ is true for $n = 0, 1, 2, \ldots$.

EXAMPLE 5 Use mathematical induction to show that

$$1 + 2 + 2^2 + \cdots + 2^n = 2^{n+1} - 1$$

for all nonnegative integers n.

Solution: Let $P(n)$ be the proposition that this formula is correct for the integer n.

BASIS STEP: $P(0)$ is true since $2^0 = 1 = 2^1 - 1$.

INDUCTIVE STEP: Assume that $P(n)$ is true. To carry out the inductive step using this assumption, it must be shown that $P(n + 1)$ is true, namely,

$$1 + 2 + 2^2 + \cdots + 2^n + 2^{n+1} = 2^{(n+1)+1} - 1 = 2^{n+2} - 1.$$

Using the inductive hypothesis $P(n)$, it follows that

$$\begin{aligned} 1 + 2 + 2^2 + \cdots + 2^n + 2^{n+1} &= (1 + 2 + 2^2 + \cdots + 2^n) + 2^{n+1} \\ &= (2^{n+1} - 1) + 2^{n+1} \\ &= 2 \cdot 2^{n+1} - 1 \\ &= 2^{n+2} - 1. \end{aligned}$$

This finishes the inductive step, which completes the proof. ■

As Example 5 demonstrates, to use mathematical induction to show that $P(n)$ is true for $n = k, k + 1, k + 2, \ldots,$ where k is an integer other than 1, we show that $P(k)$ is true (the basis step) and then show that the implication $P(n) \rightarrow P(n + 1)$ is true for $n = k, k + 1, k + 2, \ldots$ (the inductive step). Note that k can be negative, zero, or positive. We leave it to the reader to show that this form of induction is valid (see Exercise 58).

The formula given in Example 5 is a special case of a general result for the sum of the terms of a **geometric progression,** which is a sequence of the form $a, ar, ar^2, \ldots, ar^n, \ldots,$ where a and r are real numbers. For instance the sequence in Example 5 is a geometric progression with $a = 1$ and $r = 2$. Likewise, the sequence $3, 15, 75, \ldots, 3 \cdot 5^n, \ldots$ is a geometric progression with $a = 3$ and $r = 5$.

The next example gives a formula for the sum of the first $n + 1$ terms of such a sequence. The proof of this general formula will use mathematical induction.

EXAMPLE 6 Sums of Geometric Progressions Use mathematical induction to prove the following formula for the sum of a finite number of terms of a geometric progression:

$$\sum_{j=0}^{n} ar^j = a + ar + ar^2 + \cdots + ar^n = \frac{ar^{n+1} - a}{r - 1},$$

when $r \neq 1$.

Solution: To prove this formula using mathematical induction, let $P(n)$ be the proposition that the sum of the first $n + 1$ terms of a geometric progression in this formula is correct.

BASIS STEP: $P(0)$ is true, since

$$a = \frac{ar - a}{r - 1}.$$

INDUCTIVE STEP: Assume that $P(n)$ is true. That is, assume

$$a + ar + ar^2 + \cdots + ar^n = \frac{ar^{n+1} - a}{r - 1}.$$

To show that this implies that $P(n+1)$ is true, add ar^{n+1} to both sides of this equation to obtain

$$a + ar + ar^2 + \cdots + ar^n + ar^{n+1} = \frac{ar^{n+1} - a}{r - 1} + ar^{n+1}.$$

Rewriting the right-hand side of this equation shows that

$$\frac{ar^{n+1} - a}{r - 1} + ar^{n+1} = \frac{ar^{n+1} - a}{r - 1} + \frac{ar^{n+2} - ar^{n+1}}{r - 1}$$

$$= \frac{ar^{n+2} - a}{r - 1}.$$

Combining these equations gives

$$a + ar + ar^2 + \cdots + ar^n + ar^{n+1} = \frac{ar^{n+2} - a}{r - 1}.$$

This shows that if $P(n)$ is true, then $P(n + 1)$ must also be true. This completes the inductive argument and shows that the formula for the sum of the terms of a geometric series is correct. ■

As previously mentioned, the formula in Example 5 is the case of the formula in Example 6 with $a = 1$ and $r = 2$. The reader should verify that putting these values for a and r in the general formula gives the same formula as in Example 5.

An important inequality for the sum of the reciprocals of a set of positive integers will be proved in the next example.

EXAMPLE 7 An Inequality for Harmonic Numbers The *harmonic numbers* $H_k, k = 1, 2, 3, \ldots,$ are defined by

$$H_k = 1 + \frac{1}{2} + \frac{1}{3} + \cdots + \frac{1}{k}.$$

For instance,

$$H_4 = 1 + \frac{1}{2} + \frac{1}{3} + \frac{1}{4} = \frac{25}{12}.$$

Use mathematical induction to show that

$$H_{2^n} \geq 1 + \frac{n}{2},$$

whenever n is a nonnegative integer.

Solution: To carry out the proof, let $P(n)$ be the proposition that $H_{2^n} \geq 1 + n/2$.

BASIS STEP: $P(0)$ is true, since $H_{2^0} = H_1 = 1 \geq 1 + 0/2$.

INDUCTIVE STEP: Assume that $P(n)$ is true, so that $H_{2^n} \geq 1 + n/2$. It must be shown that $P(n+1)$, which states that $H_{2^{n+1}} \geq 1 + (n+1)/2$, must also be true under this assumption. This can be done since

$$\begin{aligned}
H_{2^{n+1}} &= 1 + \frac{1}{2} + \frac{1}{3} + \cdots + \frac{1}{2^n} + \frac{1}{2^n + 1} + \cdots + \frac{1}{2^{n+1}} \\
&= H_{2^n} + \frac{1}{2^n + 1} + \cdots + \frac{1}{2^{n+1}} \\
&\geq \left(1 + \frac{n}{2}\right) + \frac{1}{2^n + 1} + \cdots + \frac{1}{2^{n+1}} \qquad \text{(by the inductive hypothesis)} \\
&\geq \left(1 + \frac{n}{2}\right) + 2^n \cdot \frac{1}{2^{n+1}} \qquad \text{(since there are } 2^n \text{ terms each not less than } 1/2^{n+1}\text{)} \\
&\geq \left(1 + \frac{n}{2}\right) + \frac{1}{2} \\
&= 1 + \frac{n+1}{2}.
\end{aligned}$$

This establishes the inductive step of the proof. Thus, the inequality for the harmonic numbers is valid for all nonnegative integers n.

Remark: The inequality established here can be used to show that the *harmonic series*

$$1 + \frac{1}{2} + \frac{1}{3} + \cdots + \frac{1}{n} + \cdots$$

is a divergent infinite series. This is an important example in the study of infinite series. ■

The next example shows how mathematical induction can be used to verify a formula for the number of subsets of a finite set.

EXAMPLE 8 The Number of Subsets of a Finite Set Use mathematical induction to show that if S is a finite set with n elements, then S has 2^n subsets. (We will prove this result directly in several ways in Chapter 4.)

Solution: Let $P(n)$ be the proposition that a set with n elements has 2^n subsets.

BASIS STEP: $P(0)$ is true, since a set with zero elements, the empty set, has exactly $2^0 = 1$ subsets, since it has one subset, namely, itself.

INDUCTIVE STEP: Assume that $P(n)$ is true, that is, that every set with n elements has 2^n subsets. It must be shown that under this assumption $P(n+1)$, which is the statement that every set with $n+1$ elements has 2^{n+1} subsets, must also be true. To show this, let T be a set with $n+1$ elements. Then, it is possible to write $T = S \cup \{a\}$ where a is one of the elements of T and $S = T - \{a\}$. The subsets of T can be obtained in the following way. For each subset X of S there are exactly two subsets of T, namely, X and $X \cup \{a\}$. (This is illustrated in Figure 3.) These constitute all the subsets of T and are all distinct. Since there are 2^n subsets of S, there are $2 \cdot 2^n = 2^{n+1}$ subsets of T. This finishes the induction argument. ■

EXAMPLE 9 Show that if n is a positive integer,

$$1 + 2 + \cdots + n = n(n+1)/2.$$

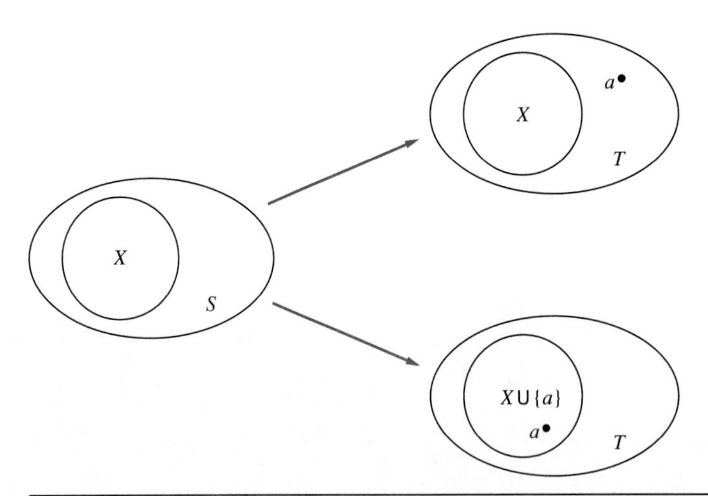

FIGURE 3 **Generating Subsets of a Set with $n+1$ Elements. Here $T = S \cup \{a\}$.**

Solution: Let $P(n)$ be the proposition that the sum of the first n positive integers is $n(n+1)/2$. We must do two things to prove that $P(n)$ is true for $n = 1, 2, 3, \ldots$. Namely, we must show that $P(1)$ is true and that the implication $P(n)$ implies $P(n+1)$ is true for $n = 1, 2, 3, \ldots$.

BASIS STEP: $P(1)$ is true, since $1 = 1(1+1)/2$.

INDUCTIVE STEP: Assume that $P(n)$ holds, so that

$$1 + 2 + \cdots + n = n(n+1)/2.$$

Under this assumption, it must be shown that $P(n+1)$ is true, namely, that

$$1 + 2 + \cdots + n + n + 1 = (n+1)[(n+1)+1]/2 = (n+1)(n+2)/2$$

is also true. Add $n+1$ to both sides of the equation in $P(n)$ to obtain

$$\begin{aligned} 1 + 2 + \cdots + n + (n+1) &= n(n+1)/2 + (n+1) \\ &= [(n/2) + 1](n+1) \\ &= (n+1)(n+2)/2. \end{aligned}$$

This last equation shows that $P(n+1)$ is true. This completes the inductive step and completes the proof. ■

EXAMPLE 10 Use mathematical induction to prove that $2^n < n!$ for every positive integer n with $n \geq 4$.

Solution: Let $P(n)$ be the proposition that $2^n < n!$.

BASIS STEP: To prove the inequality for $n \geq 4$ requires that the basis step be $P(4)$. Note that $P(4)$ is true, since $2^4 = 16 < 4! = 24$.

INDUCTIVE STEP: Assume that $P(n)$ is true. That is, assume that $2^n < n!$. We must show that $P(n+1)$ is true. That is, we must show that $2^{n+1} < (n+1)!$. Multiplying both sides of the inequality $2^n < n!$ by 2, it follows that

$$\begin{aligned} 2 \cdot 2^n &< 2 \cdot n! \\ &< (n+1) \cdot n! \\ &= (n+1)!. \end{aligned}$$

This shows that $P(n+1)$ is true when $P(n)$ is true. This completes the inductive step of the proof. Hence, it follows that $2^n < n!$ is true for all integers n with $n \geq 4$. ■

EXAMPLE 11 Use mathematical induction to prove the following generalization of one of De Morgan's laws:

$$\overline{\bigcap_{k=1}^{n} A_k} = \bigcup_{k=1}^{n} \overline{A_k},$$

whenever $A_1, A_2, \ldots, A_n$ are subsets of a universal set U, and $n \geq 2$.

Solution: Let $P(n)$ be the identity for n sets.

BASIS STEP: The statement $P(2)$ asserts that $\overline{A_1 \cap A_2} = \overline{A_1} \cup \overline{A_2}$. This is one of De Morgan's laws; it was proved in Section 1.5.

INDUCTIVE STEP: Assume that $P(n)$ is true, that is

$$\overline{\bigcap_{k=1}^{n} A_k} = \bigcup_{k=1}^{n} \overline{A_k}$$

whenever $A_1, A_2, \ldots, A_n$ are subsets of the universal set U. To carry out the inductive step it must be shown that if this equality holds for any n subsets of U, it must also be valid for any $n + 1$ subsets of U. Suppose that $A_1, A_2, \ldots, A_n, A_{n+1}$ are subsets of U. When the inductive hypothesis is assumed to hold, it follows that

$$\begin{aligned}
\overline{\bigcap_{k=1}^{n+1} A_k} &= \overline{\left(\bigcap_{k=1}^{n} A_k\right) \cap A_{n+1}} \\
&= \left(\overline{\bigcap_{k=1}^{n} A_k}\right) \cup \overline{A_{n+1}} && \text{(by De Morgan's law)} \\
&= \left(\bigcup_{k=1}^{n} \overline{A_k}\right) \cup \overline{A_{n+1}} && \text{(by the inductive hypothesis)} \\
&= \bigcup_{k=1}^{n+1} \overline{A_k}.
\end{aligned}$$

This completes the proof by induction. ■

The next example illustrates how mathematical induction can be used to prove a result about covering chessboards with pieces shaped like the letter L.

EXAMPLE 12 Let n be a positive integer. Show that any $2^n \times 2^n$ chessboard with one square removed can be tiled using L-shaped pieces, where these pieces cover three squares at a time, as shown in Figure 4.

FIGURE 4 **An L-shaped Piece.**

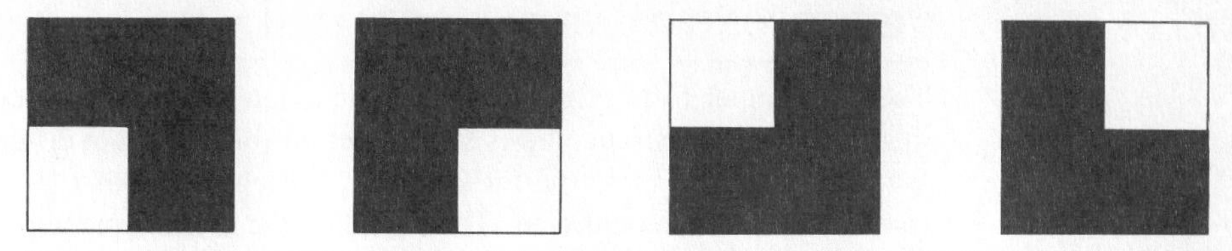

FIGURE 5 **Tiling 2×2 Chessboards with One Square Removed.**

Solution: Let $P(n)$ be the proposition that any $2^n \times 2^n$ chessboard with one square removed can be tiled using L-shaped pieces. We can use mathematical induction to prove that $P(n)$ is true for all positive integers n.

BASIS STEP: $P(1)$ is true, since any of the four 2×2 chessboards with one square removed can be tiled using one L-shaped piece, as shown in Figure 5.

INDUCTIVE STEP: Assume that $P(n)$ is true; that is, assume that any $2^n \times 2^n$ chessboard with one square removed can be tiled using L-shaped pieces. It must be shown that under this assumption $P(n+1)$ must also be true; that is, any $2^{n+1} \times 2^{n+1}$ chessboard with one square removed can be tiled using L-shaped pieces.

To see this, consider a $2^{n+1} \times 2^{n+1}$ chessboard with one square removed. Split this chessboard into four chessboards of size $2^n \times 2^n$, by dividing it in half in both directions. This is illustrated in Figure 6. No square has been removed from three of these four chessboards. The fourth $2^n \times 2^n$ chessboard has one square removed, so by the inductive hypothesis, it can be covered by L-shaped pieces. Now temporarily remove the square from each of the other three $2^n \times 2^n$ chessboards that has the center of the original, larger chessboard as one of its corners, as shown in Figure 7. By the inductive hypothesis, each of these three $2^n \times 2^n$ chessboards with a square removed can be tiled by L-shaped pieces. Furthermore, the three squares that were temporarily removed can be covered by one L-shaped piece. Hence, the entire $2^{n+1} \times 2^{n+1}$ chessboard can be tiled with L-shaped pieces. This completes the proof. ■

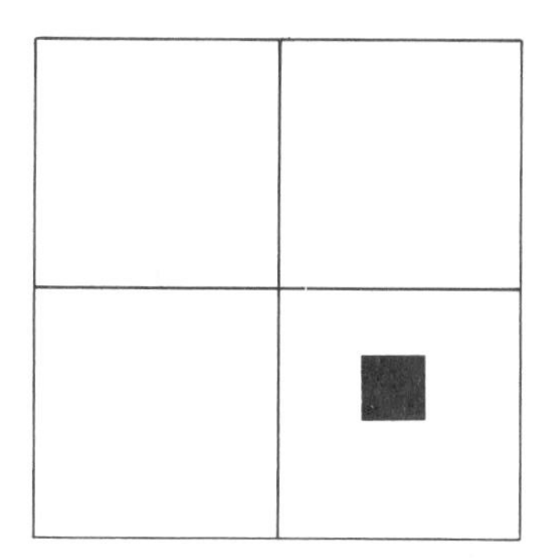

FIGURE 6 **Dividing a $2^{n+1} \times 2^{n+1}$ Chessboard into Four $2^n \times 2^n$ Chessboards.**

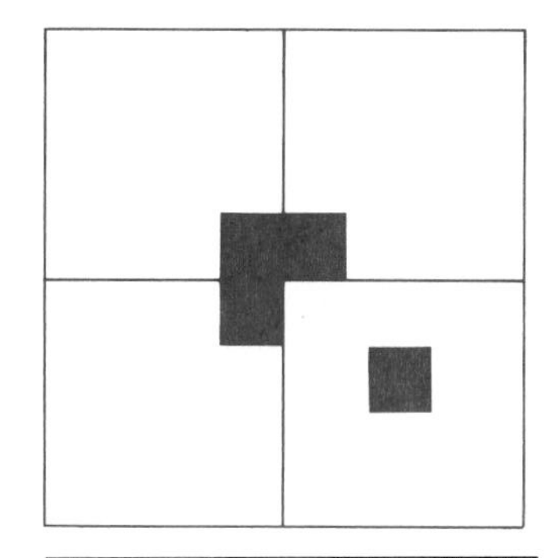

FIGURE 7 **Tiling the $2^{n+1} \times 2^{n+1}$ Chessboard with One Square Removed.**

THE SECOND PRINCIPLE OF MATHEMATICAL INDUCTION

There is another form of mathematical induction that is often useful in proofs. With this form we use the same basis step as before, but we use a different inductive step. We assume that $P(k)$ is true for $k = 1, \ldots, n$ and show that $P(n + 1)$ must also be true based on this assumption. This is called the **second principle of mathematical induction.** We summarize the two steps used to show that $P(n)$ is true for all positive integers n:

1. *Basis step*. The proposition $P(1)$ is shown to be true.
2. *Inductive step*. It is shown that $[P(1) \wedge P(2) \wedge \cdots \wedge P(n)] \rightarrow P(n + 1)$ is true for every positive integer n.

The two forms of mathematical induction are equivalent; that is, each can be shown to be a valid proof technique assuming the other. We leave it as an exercise for the reader to show this. We now give an example that shows how the second principle of mathematical induction is used.

EXAMPLE 13 Show that if n is an integer greater than 1, then n can be written as the product of primes.

Solution: Let $P(n)$ be the proposition that n can be written as the product of primes.

BASIS STEP: $P(2)$ is true, since 2 can be written as the product of one prime, itself.

INDUCTIVE STEP: Assume that $P(k)$ is true for all positive integers k with $k \leq n$. To complete the inductive step, it must be shown that $P(n + 1)$ is true under this assumption.

There are two cases to consider, namely, when $n + 1$ is prime and when $n + 1$ is composite. If $n + 1$ is prime, we immediately see that $P(n + 1)$ is true. Otherwise, $n + 1$ is composite and can be written as the product of two positive integers a and b with $2 \leq a \leq b < n + 1$. By the induction hypothesis, both a and b can be written as the product of primes. Thus, if $n + 1$ is composite, it can be written as the product of primes, namely, those primes in the factorization of a and those in the factorization of b. ■

Remark: Since 1 is a product of primes, namely, the *empty* product of no primes, we could have started the proof in Example 13 with $P(1)$ as the basis step. We chose not to since many people find this confusing.

Note that Example 13 completes the proof of the Fundamental Theorem of Arithmetic, which asserts that every nonnegative integer can be written uniquely as the product of primes in nondecreasing order. We showed in Section 2.5 that an integer has at most one such factorization into primes. Example 13 shows there is at least one such factorization.

Using the principle of mathematical induction, instead of the second principle of mathematical induction, to prove the result in Example 13 is difficult. However, as Example 14 shows, some results can be readily proved using either the principle of mathematical induction or the second principle of mathematical induction.

EXAMPLE 14 Prove that every amount of postage of 12 cents or more can be formed using only 4-cent and 5-cent stamps.

Solution: We will prove this result using the principle of mathematical induction. Then we will present a proof using the second principle of mathematical induction. Let $P(n)$ be the statement that postage of n cents can be formed using 4-cent and 5-cent stamps.

We begin by using the principle of mathematical induction.

BASIS STEP: Postage of 12 cents can be formed using three 4-cent stamps.

INDUCTIVE STEP: Assume that $P(n)$ is true, so that postage of n cents can be formed using 4-cent and 5-cent stamps. If at least one 4-cent stamp was used, replace it with a 5-cent stamp to form postage of $n + 1$ cents. If no 4-cent stamps were used, postage of n cents was formed using just 5-cent stamps. Since $n \geq 12$, at least three 5-cent stamps were used. So, replace three 5-cent stamps with four 4-cent stamps to form postage of $n + 1$ cents. This completes the inductive step, as well as the proof by the principle of mathematical induction.

Next, we will use the second principle of mathematical induction. We will show that postage of 12, 13, 14, and 15 cents can be formed and then show how to get postage of $n + 1$ cents for $n \geq 15$ from postage of $n - 3$ cents.

BASIS STEP: We can form postage of 12, 13, 14, and 15 cents using three 4-cent stamps, two 4-cent stamps and one 5-cent stamp, one 4-cent stamp and two 5-cent stamps, and three 5-cent stamps, respectively.

INDUCTIVE STEP: Let $n \geq 15$. Assume that we can form postage of k cents, where $12 \leq k \leq n$. To form postage of $n + 1$ cents, use the stamps that form postage of $n - 3$ cents together with a 4-cent stamp. This completes the inductive step, as well as the proof by the second principle of mathematical induction.

(There are other ways to approach this problem besides those described here. Can you find a solution that does not use mathematical induction?) ■

Remark: Example 14 shows how we can adapt the second principle of mathematical induction to handle cases where the inductive step is valid only for sufficiently large values of n. In particular, to show that $P(n)$ is true for $n = k, k+1, k+2, \ldots,$ where k is an integer, we first show that $P(k), P(k+1), P(k+2), \ldots, P(l)$ are true (the basis step), and then we show that $[P(k) \wedge P(k+1) \wedge P(k+2) \wedge \cdots \wedge P(n)] \rightarrow P(n+1)$ is true for every integer $n \geq l$ (the inductive step). For example, the basis step of the second proof in the solution of Example 14 shows that $P(12)$, $P(13)$, $P(14)$, and

$P(15)$ are true. We need to prove these cases separately since the inductive step, which shows that $[P(12) \wedge P(13) \wedge \cdots \wedge P(n)] \rightarrow P(n+1)$, only holds when $n \geq 15$.

We will discuss two important applications of mathematical induction in the following sections. The first involves the definition of sequences without giving explicit formulae for their terms. The second involves proving that computer programs are correct.

Exercises

1. Find a formula for the sum of the first n even positive integers.

2. Use mathematical induction to prove the formula that you found in Exercise 1.

3. Use mathematical induction to prove that $3 + 3 \cdot 5 + 3 \cdot 5^2 + \cdots + 3 \cdot 5^n = 3(5^{n+1} - 1)/4$ whenever n is a nonnegative integer.

4. Find a formula for

$$\frac{1}{1 \cdot 2} + \frac{1}{2 \cdot 3} + \cdots + \frac{1}{n(n+1)}$$

by examining the values of this expression for small values of n. Use mathematical induction to prove your result.

5. Show that $1^2 + 2^2 + \cdots + n^2 = n(n+1)(2n+1)/6$ whenever n is a positive integer.

6. Show that $1^3 + 2^3 + \cdots + n^3 = [n(n+1)/2]^2$ whenever n is a positive integer.

7. Prove that $1^2 + 3^2 + 5^2 + \cdots + (2n+1)^2 = (n+1)(2n+1)(2n+3)/3$ whenever n is a nonnegative integer.

8. Prove that $1 \cdot 1! + 2 \cdot 2! + \cdots + n \cdot n! = (n+1)! - 1$ whenever n is a positive integer.

★9. Show by mathematical induction that if $h > -1$, then $1 + nh \leq (1+h)^n$ for all nonnegative integers n. This is called **Bernoulli's inequality.**

10. Prove that $3^n < n!$ whenever n is a positive integer greater than 6.

11. Show that $2^n > n^2$ whenever n is an integer greater than 4.

12. Use mathematical induction to prove that $n! < n^n$ whenever n is a positive integer greater than 1.

13. Prove using mathematical induction that

$$1 \cdot 2 + 2 \cdot 3 + \cdots + n(n+1) = n(n+1)(n+2)/3$$

whenever n is a positive integer.

14. Use mathematical induction to prove that

$$1 \cdot 2 \cdot 3 + 2 \cdot 3 \cdot 4 + \cdots + n(n+1)(n+2) = n(n+1)(n+2)(n+3)/4.$$

15. Show that $1^2 - 2^2 + 3^2 - \cdots + (-1)^{n-1}n^2 = (-1)^{n-1}n(n+1)/2$ whenever n is a positive integer.

16. Prove that

$$1 + \frac{1}{4} + \frac{1}{9} + \cdots + \frac{1}{n^2} < 2 - \frac{1}{n}$$

whenever n is a positive integer greater than 1.

17. Show that any postage that is a positive integer number of cents greater than 7 cents can be formed using just 3-cent stamps and 5-cent stamps.

18. Use mathematical induction to show that 3 divides $n^3 + 2n$ whenever n is a nonnegative integer.

19. Use mathematical induction to show that 5 divides $n^5 - n$ whenever n is a nonnegative integer.

20. Use mathematical induction to show that 6 divides $n^3 - n$ whenever n is a nonnegative integer.

★21. Use mathematical induction to show that $n^2 - 1$ is divisible by 8 whenever n is an odd positive integer.

22. Use mathematical induction to show that $n^2 - 7n + 12$ is nonnegative whenever n is an integer greater than 3.

23. Use mathematical induction to prove that a set with n elements has $n(n-1)/2$ subsets containing exactly two elements whenever n is an integer greater than or equal to 2.

★**24.** Use mathematical induction to prove that a set with n elements has $n(n-1)(n-2)/6$ subsets containing exactly three elements whenever n is an integer greater than or equal to 3.

25. Use mathematical induction to prove that $\sum_{j=1}^{n} j^4 = n(n+1)(2n+1)(3n^2+3n-1)/30$ whenever n is a positive integer.

26. For which nonnegative integers n is $n^2 \leq n!$? Prove your answer using mathematical induction.

27. For which nonnegative integers n is $2n+3 \leq 2^n$? Prove your answer using mathematical induction.

28. Use mathematical induction to show that $1/(2n) \leq [1 \cdot 3 \cdot 5 \cdot \cdots \cdot (2n-1)]/(2 \cdot 4 \cdot \cdots \cdot 2n)$ whenever n is a positive integer.

29. **a)** Determine which amounts of postage can be formed using 5-cent and 6-cent stamps.

b) Prove your answer to (a) using the principle of mathematical induction.

c) Prove your answer to (a) using the second principle of mathematical induction.

30. Which amounts of money can be formed using just dimes and quarters? Prove your answer using a form of mathematical induction.

31. An automatic teller machine has only \$20 bills and \$50 bills. Which amounts of money can the machine dispense, assuming the machine has a limitless supply of these two denominations of bills? Prove your answer using a form of mathematical induction.

32. Suppose that

$$\mathbf{A} = \begin{bmatrix} a & 0 \\ 0 & b \end{bmatrix}$$

where a and b are real numbers. Show that

$$\mathbf{A}^n = \begin{bmatrix} a^n & 0 \\ 0 & b^n \end{bmatrix}$$

for every positive integer n.

33. Suppose that $\mathbf{A}$ and $\mathbf{B}$ are square matrices with the property $\mathbf{AB} = \mathbf{BA}$. Show that $\mathbf{AB}^n = \mathbf{B}^n\mathbf{A}$ for every positive integer n.

34. Suppose that m is a positive integer. Use mathematical induction to prove that if a and b are integers with $a \equiv b$ (mod m), then $a^k \equiv b^k \pmod{m}$ whenever k is a nonnegative integer.

35. Use mathematical induction to show that if $A_1, A_2, \ldots, A_n$ and B are sets, then

$$(A_1 \cup A_2 \cup \cdots \cup A_n) \cap B = (A_1 \cap B) \cup (A_2 \cap B) \cup \cdots \cup (A_n \cap B).$$

36. Prove that if $A_1, A_2, \ldots, A_n$ and $B_1, B_2, \ldots, B_n$ are sets such that $A_k \subseteq B_k$ for $k = 1, 2, \ldots, n$, then

a) $\bigcup_{k=1}^{n} A_k \subseteq \bigcup_{k=1}^{n} B_k$. **b)** $\bigcap_{k=1}^{n} A_k \subseteq \bigcap_{k=1}^{n} B_k$.

37. Use mathematical induction to prove that if $A_1, A_2, \ldots, A_n$ are subsets of a universal set U, then

$$\overline{\bigcup_{k=1}^{n} A_k} = \bigcap_{k=1}^{n} \overline{A_k}.$$

38. Use mathematical induction to show that $\neg(p_1 \vee p_2 \vee \cdots \vee p_n)$ is equivalent to $\neg p_1 \wedge \neg p_2 \wedge \cdots \wedge \neg p_n$ whenever $p_1, p_2, \ldots, p_n$ are propositions.

★**39.** Show that

$$[(p_1 \rightarrow p_2) \wedge (p_2 \rightarrow p_3) \wedge \cdots \wedge (p_{n-1} \rightarrow p_n)] \rightarrow [(p_1 \wedge p_2 \wedge \cdots \wedge p_{n-1}) \rightarrow p_n]$$

is a tautology whenever $p_1, p_2, \ldots, p_n$ are propositions.

40. Use the formula for the sum of the terms of a geometric progression to evaluate the following sums.

a) $4 + 4 \cdot 3 + 4 \cdot 3^2 + \cdots + 4 \cdot 3^8$

b) $3 + 3 \cdot 2^2 + 3 \cdot 2^4 + \cdots + 3 \cdot 2^{10}$

c) $1 - 2 + 2^2 - 2^3 + \cdots + (-1)^n 2^n$

41. What is wrong with the following "proof" that all horses are the same color?

> Let $P(n)$ be the proposition that all the horses in a set of n horses are the same color. Clearly, $P(1)$ is true. Now assume that $P(n)$ is true, so that all the horses in any set of n horses are the same color. Consider any $n+1$ horses; number these as horses $1, 2, 3, \ldots, n, n+1$. Now the first n of these horses all must have the same color, and the last n of these must also have the same color. Since the set of the first n horses and the set of the last n horses overlap, all $n+1$ must be the same color. This shows that $P(n+1)$ is true and finishes the proof by induction.

★**42.** Find the flaw with the following "proof" that $a^n = 1$ for all nonnegative integers n, whenever a is a nonzero real number.

BASIS STEP: $a^0 = 1$ is true by the definition of a^0.

INDUCTIVE STEP: Assume that $a^k = 1$ for all nonnegative integers k with $k \leq n$. Then note that

$$a^{n+1} = \frac{a^n \cdot a^n}{a^{n-1}} = \frac{1 \cdot 1}{1} = 1.$$

★43. Show that the second form of mathematical induction is a valid method of proof by showing that it follows from the well-ordering property.

★44. Show that the following form of mathematical induction is a valid method to prove that $P(n)$ is true for all positive integers n.

BASIS STEP: $P(1)$ and $P(2)$ are true.

INDUCTIVE STEP: For each positive integer n, if $P(n)$ and $P(n+1)$ are both true, then $P(n+2)$ is true.

In Exercises 45 and 46, H_n denotes the nth harmonic number.

★45. Use mathematical induction to show that $H_{2^n} \leq 1 + n$ whenever n is a nonnegative integer.

★46. Use mathematical induction to prove that

$$H_1 + H_2 + \cdots + H_n = (n+1)H_n - n.$$

★47. Prove that

$$1 + \frac{1}{\sqrt{2}} + \frac{1}{\sqrt{3}} + \cdots + \frac{1}{\sqrt{n}} > 2(\sqrt{n+1} - 1).$$

★48. Show that n lines separate the plane into $(n^2 + n + 2)/2$ regions if no two of these lines are parallel and no three pass through a common point.

★★49. Let $a_1, a_2, \ldots, a_n$ be positive real numbers. The **arithmetic mean** of these numbers is defined by

$$A = (a_1 + a_2 + \cdots + a_n)/n,$$

and the **geometric mean** of these numbers is defined by

$$G = (a_1 a_2 \cdots a_n)^{1/n}.$$

Use mathematical induction to prove that $A \geq G$.

★50. Use mathematical induction to show that 21 divides $4^{n+1} + 5^{2n-1}$ whenever n is a positive integer.

51. Use mathematical induction to prove Lemma 2 of Section 2.5, which states that if p is a prime and $p \mid a_1 a_2 \cdots a_n$, where a_i is an integer for $i = 1, 2, 3, \ldots, n$, then $p \mid a_i$ for some integer i.

★52. The well-ordering property can be used to show that there is a unique greatest common divisor of two positive integers. Let a and b be positive integers, and let S be the set of positive integers of the form $as + bt$, where s and t are integers.

a) Show that S is nonempty.

b) Use the well-ordering property to show that S has a smallest element c.

c) Show that if d is a common divisor of a and b, then d is a divisor of c.

d) Show that $c \mid a$ and $c \mid b$. (*Hint:* First, assume that $c \nmid a$. Then $a = qc + r$, where $0 < r < c$. Show that $r \in S$, contradicting the choice of c.)

e) Conclude from (c) and (d) that the greatest common divisor of a and b exists. Finish the proof by showing that this greatest common divisor of two positive integers is unique.

★53. Show that if $a_1, a_2, \ldots, a_n$ are n distinct real numbers, exactly $n - 1$ multiplications are used to compute the product of these n numbers no matter how parentheses are inserted into their product. (*Hint:* Use the second principle of mathematical induction and consider the last multiplication.)

54. Construct a tiling using L-shaped pieces of the 4×4 chessboard with the square in the upper left corner removed.

55. Construct a tiling using L-shaped pieces of the 8×8 chessboard with the square in the upper left corner removed.

56. Prove or disprove that all chessboards of the following shapes can be completely covered using L-shaped pieces whenever n is a positive integer.

a) 3×2^n **b)** 6×2^n

c) $3^n \times 3^n$ **d)** $6^n \times 6^n$

57. Let a be an integer and d be a positive integer. Show that the integers q and r with $a = dq + r$ and $0 \leq r < d$, which were shown to exist in Example 1, are unique.

☞ **58.** Use the principle of mathematical induction to show that $P(n)$ is true for $n = k, k+1, k+2, \ldots,$ where k is an integer, if $P(k)$ is true and the implication $P(n) \rightarrow P(n+1)$ is true for all positive integers n with $n \geq k$.

3.3 Recursive Definitions

INTRODUCTION

Sometimes it is difficult to define an object explicitly. However, it may be easy to define this object in terms of itself. This process is called **recursion.** For instance, the picture shown in Figure 1 is produced recursively. First, an original picture is given. Then a process of successively superimposing centered smaller pictures on top of the previous pictures is carried out.

We can use recursion to define sequences, functions, and sets. In previous discussions, we specified the terms of a sequence using an explicit formula. For instance, the sequence of powers of 2 is given by $a_n = 2^n$ for $n = 0, 1, 2, \ldots$. However, this sequence can also be defined by giving the first term of the sequence, namely, $a_0 = 1$, and a rule for finding a term of the sequence from the previous one, namely, $a_{n+1} = 2a_n$ for $n = 0, 1, 2, \ldots$.

FIGURE 1 **A Recursively Defined Picture.**

RECURSIVELY DEFINED FUNCTIONS

We can define a function with the set of nonnegative integers as its domain by

(*i*) specifying the value of the function at zero, and

(*ii*) giving a rule for finding its value at an integer from its values at smaller integers.

Such a definition is called a **recursive** or **inductive definition.**

EXAMPLE 1 Suppose that f is defined recursively by

$$f(0) = 3,$$
$$f(n+1) = 2f(n) + 3.$$

Find $f(1)$, $f(2)$, $f(3)$, and $f(4)$.

Solution: From the recursive definition it follows that

$$\begin{aligned} f(1) &= 2f(0) + 3 = 2 \cdot 3 + 3 = 9, \\ f(2) &= 2f(1) + 3 = 2 \cdot 9 + 3 = 21, \\ f(3) &= 2f(2) + 3 = 2 \cdot 21 + 3 = 45, \\ f(4) &= 2f(3) + 3 = 2 \cdot 45 + 3 = 93. \end{aligned}$$

■

Many functions can be studied using their recursive definitions. The factorial function is one such example.

EXAMPLE 2 Give an inductive definition of the factorial function $F(n) = n!$.

Solution: We can define the factorial function by specifying the initial value of this function, namely, $F(0) = 1$, and giving a rule for finding $F(n+1)$ from $F(n)$. This is obtained by noting that $(n+1)!$ is computed from $n!$ by multiplying by $n+1$. Hence, the desired rule is

$$F(n+1) = (n+1)F(n).$$

■

To determine a value of the factorial function, such as $F(5) = 5!$, from the recursive definition found in Example 2, it is necessary to use the rule that shows how to express $F(n+1)$ in terms of $F(n)$ several times:

$$\begin{aligned} F(5) &= 5F(4) = 5 \cdot 4F(3) = 5 \cdot 4 \cdot 3F(2) = 5 \cdot 4 \cdot 3 \cdot 2F(1) \\ &= 5 \cdot 4 \cdot 3 \cdot 2 \cdot 1F(0) = 5 \cdot 4 \cdot 3 \cdot 2 \cdot 1 \cdot 1 = 120. \end{aligned}$$

Once $F(0)$ is the only value of the function that occurs, no more reductions are necessary. The only thing left to do is to insert the value of $F(0)$ into the formula.

Recursively defined functions are well defined. This is a consequence of the principle of mathematical induction. (See Exercise 40 at the end of this section.) Additional examples of recursive definitions are given in the following examples.

EXAMPLE 3 Give a recursive definition of a^n where a is a nonzero real number and n is a nonnegative integer.

Solution: The recursive definition contains two parts. First a^0 is specified, namely, $a^0 = 1$. Then the rule for finding a^{n+1} from a^n, namely, $a^{n+1} = a \cdot a^n$, for $n = 0, 1, 2, 3, \ldots$, is given. These two equations uniquely define a^n for all nonnegative integers n. ■

EXAMPLE 4 Give a recursive definition of

$$\sum_{k=0}^{n} a_k.$$

Solution: The first part of the recursive definition is

$$\sum_{k=0}^{0} a_k = a_0.$$

The second part is

$$\sum_{k=0}^{n+1} a_k = \left(\sum_{k=0}^{n} a_k\right) + a_{n+1}.$$

■

In some recursive definitions of functions, the values of the function at the first k positive integers are specified, and a rule is given for determining the value of the function at larger integers from its values at some or all of the preceding k integers. That such definitions produce well-defined functions follows from the second principle of mathematical induction (see Exercise 41 at the end of this section).

EXAMPLE 5 The *Fibonacci numbers*, $f_0, f_1, f_2, \ldots$, are defined by the equations $f_0 = 0$, $f_1 = 1$, and

$$f_n = f_{n-1} + f_{n-2}$$

for $n = 2, 3, 4, \ldots$. What are the Fibonacci numbers f_2, f_3, f_4, f_5, f_6?

Fibonacci (c.1180–1228). Fibonacci (short for *filius Bonacci*, or "son of Bonacci") was also known as Leonardo of Pisa. He was born in the Italian commercial center of Pisa. Fibonacci was a merchant who traveled extensively throughout the Mideast, where he came into contact with Arabian mathematics. In his book *Liber Abaci*, Fibonacci introduced the European world to Arabic notation for numerals and algorithms for arithmetic. It was in this book that his famous rabbit problem (described in Section 5.1) appeared. Fibonacci also wrote books on geometry and trigonometry and on Diophantine equations, which involve finding integer solutions to equations.

Solution: Since the first part of the definition states that $f_0 = 0$ and $f_1 = 1$, it follows from the second part of the definition that

$$\begin{aligned} f_2 &= f_1 + f_0 = 1 + 0 = 1, \\ f_3 &= f_2 + f_1 = 1 + 1 = 2, \\ f_4 &= f_3 + f_2 = 2 + 1 = 3, \\ f_5 &= f_4 + f_3 = 3 + 2 = 5, \\ f_6 &= f_5 + f_4 = 5 + 3 = 8. \end{aligned}$$

■

We can use the recursive definition of the Fibonacci numbers to prove many properties of these numbers. We give one such property in the next example.

EXAMPLE 6 Show that $f_n > \alpha^{n-2}$, where $\alpha = (1 + \sqrt{5})/2$, whenever $n \geq 3$.

Solution: We can use the second principle of mathematical induction to prove this inequality. Let $P(n)$ be the statement $f_n > \alpha^{n-2}$. We want to show that $P(n)$ is true whenever n is an integer greater than or equal to 3.

First, note that

$$\alpha < 2 = f_3, \qquad \alpha^2 = (3 + \sqrt{5})/2 < 3 = f_4,$$

so that $P(3)$ and $P(4)$ are true. Now assume that $P(k)$ is true, namely, that $f_k > \alpha^{k-2}$, for all integers k with $3 \leq k \leq n$, where $n \geq 4$. We must show that $P(n+1)$ is true, that is that $f_{n+1} > \alpha^{n-1}$. Since α is a solution of $x^2 - x - 1 = 0$ (as the quadratic formula shows), it follows that $\alpha^2 = \alpha + 1$. Therefore,

$$\alpha^{n-1} = \alpha^2 \cdot \alpha^{n-3} = (\alpha + 1)\alpha^{n-3} = \alpha \cdot \alpha^{n-3} + 1 \cdot \alpha^{n-3} = \alpha^{n-2} + \alpha^{n-3}.$$

By the inductive hypothesis, if $n \geq 5$, it follows that

$$f_{n-1} > \alpha^{n-3}, \qquad f_n > \alpha^{n-2}.$$

Therefore, we have

$$f_{n+1} = f_n + f_{n-1} > \alpha^{n-2} + \alpha^{n-3} = \alpha^{n-1}.$$

It follows that $P(n+1)$ is true. This completes the proof.

Remark: The inductive step shows that whenever $n \geq 4$, $P(n+1)$ follows from the assumption that $P(k)$ is true for $3 \leq k \leq n$. Hence, the inductive step does *not* show that $P(3) \rightarrow P(4)$. Therefore, we had to show that $P(4)$ is true separately. ■

We can now show that the Euclidean algorithm uses $O(\log b)$ divisions to find the greatest common divisor of the positive integers a and b, where $a \geq b$.

THEOREM 1 **LAMÉ'S THEOREM** Let a and b be positive integers with $a \geq b$. Then the number of divisions used by the Euclidean algorithm to find $gcd(a, b)$ is less than five times the number of decimal digits in b.

Proof: Recall that when the Euclidean algorithm is applied to find $gcd(a, b)$ with $a \geq b$, the following sequence of equations (where $a = r_0$ and $b = r_1$) is obtained.

$$
\begin{aligned}
r_0 &= r_1 q_1 + r_2 & 0 \leq r_2 < r_1 \\
r_1 &= r_2 q_2 + r_3 & 0 \leq r_3 < r_2 \\
&\vdots \\
r_{n-2} &= r_{n-1} q_{n-1} + r_n & 0 \leq r_n < r_{n-1} \\
r_{n-1} &= r_n q_n.
\end{aligned}
$$

Here n divisions have been used to find $r_n = gcd(a, b)$. Note that the quotients $q_1, q_2, \ldots, q_{n-1}$ are all at least 1. Moreover, $q_n \geq 2$, since $r_n < r_{n-1}$. This implies that

$$
\begin{aligned}
r_n &\geq 1 = f_2, \\
r_{n-1} &\geq 2r_n \geq 2f_2 = f_3, \\
r_{n-2} &\geq r_{n-1} + r_n \geq f_3 + f_2 = f_4, \\
&\vdots \\
r_2 &\geq r_3 + r_4 \geq f_{n-1} + f_{n-2} = f_n, \\
b &= r_1 \geq r_2 + r_3 \geq f_n + f_{n-1} = f_{n+1}.
\end{aligned}
$$

It follows that if n divisions are used by the Euclidean algorithm to find $gcd(a, b)$ with $a \geq b$, then $b \geq f_{n+1}$. From Example 6 we know that $f_{n+1} > \alpha^{n-1}$ for $n > 2$, where $\alpha = (1 + \sqrt{5})/2$. Therefore, it follows that $b > \alpha^{n-1}$. Furthermore, since $\log_{10} \alpha \sim 0.208 > 1/5$, we see that

---

Gabriel Lamé (1795–1870). Gabriel Lamé entered the École Polytechnique in 1813, graduating in 1817. He continued his education at the École des Mines, graduating in 1820.

In 1820 Lamé went to Russia, where was appointed director of the Schools of Highways and Transportation in St. Petersburg. Not only did he teach, but he also planned roads and bridges while in Russia. He returned to Paris in 1832, where he helped set up an engineering firm. However, he soon left the firm, accepting the chair of physics at the École Polytechnique, a position he held until 1844. While holding this position, he was active in outside consulting as an engineer, serving as chief engineer of mines and participating in the building of railways.

Lamé contributed original work to number theory, applied mathematics, and thermodynamics. His best-known work involves the introduction of curvilinear coordinates. His work on number theory includes proving Fermat's Last Theorem for $n = 5$ as well as providing the upper bound for the number of divisions used by the Euclidean algorithm given in this text.

In the opinion of Gauss, one of the most important mathematicians of all time, Lamé was the foremost French mathematician of his time. However, French mathematicians considered him to be too practical, while French scientists considered him to be too theoretical.

$$\log_{10} b > (n - 1)\log_{10} \alpha > (n - 1)/5.$$

Hence, $n - 1 < 5 \cdot \log_{10} b$. Now suppose that b has k decimal digits. Then $b < 10^k$ and $\log_{10} b < k$. It follows that $n - 1 < 5k$, and since k is an integer, it follows that $n \leq 5k$. This finishes the proof. ■

Since the number of decimal digits in b, which equals $\lfloor \log_{10} b \rfloor + 1$, is less than or equal to $\log_{10} b + 1$, Theorem 1 tells us that the number of divisions required to find $gcd\,(a,b)$ with $a > b$ is less than $5(\log_{10} b + 1)$. Since $5(\log_{10} b + 1) = O(\log b)$, we see that $O(\log b)$ divisions are used by the Euclidean algorithm to find $gcd\,(a, b)$ whenever $a > b$.

RECURSIVELY DEFINED SETS

Recursive definitions are often used to define sets. When this is done, an initial collection of elements is given. Then the rules used to construct elements of the set from other elements already known to be in the set are given. Sets described in this way are well defined, and theorems about them can be proved using their recursive definitions. Some examples of recursive definitions of sets follow.

EXAMPLE 7 Let S be defined recursively by

$$3 \in S;$$
$$x + y \in S \text{ if } x \in S \text{ and } y \in S.$$

Show that S is the set of positive integers divisible by 3. (Note that implicit in this definition is the assumption that nothing belongs to S unless it can be generated using the two statements in the recursive definition of S.)

Solution: Let A be the set of all positive integers divisible by 3. To prove that $A = S$, we must show that A is a subset of S and that S is a subset of A. To prove that A is a subset of S, we must show that every positive integer divisible by 3 is in S. We will use mathematical induction to prove this.

Let $P(n)$ be the statement that $3n$ belongs to S. The basis step holds since, by the first part of the recursive definition of S, $3 \cdot 1 = 3$ is in S. To establish the inductive step, assume that $P(n)$ is true, namely, that $3n$ is in S. Since $3n$ is in S and since 3 is in S, it follows from the second part of the recursive definition of S that $3n + 3 = 3(n + 1)$ is also in S.

To prove that S is a subset of A, we use the recursive definition of S. First, the basis step of the definition specifies that 3 is in S. Since $3 = 3 \cdot 1$, all elements specified to be in S in this step are divisible by 3. To finish the proof, we must show that all integers in S generated using the second part of the recursive definition are in A. This consists of showing that $x + y$ is in A whenever x and y are elements of S also assumed to be in A. Now if x and y are both in A, it follows that $3 \mid x$ and $3 \mid y$. By Theorem 1 of Section 2.3, it follows that $3 \mid x + y$, completing the proof. ■

The recursive definition of a set in Example 7 is typical. First, an initial set of elements is given. Second, a rule is provided for generating new elements from those already known to be in the set. Implicit in the definition is that no element belongs to the set unless it is listed in the initial set of elements or it can be generated using the rule given for constructing new elements.

One of the most common uses of recursive definitions for sets is to define **well-formed formulae** in various systems. This is illustrated in the following examples.

EXAMPLE 8 The well-formed formulae of variables, numerals, and operators from $\{+, -, *, /, \uparrow\}$ (where $*$ denotes multiplication and $\uparrow$ denotes exponentiation) are defined by

> x is a well-formed formula if x is a numeral or a variable;
> $(f+g)$, $(f-g)$, $(f*g)$, (f/g), and $(f \uparrow g)$ are well-formed formulae if f and g are.

For instance, from this definition, since x and 3 are well-formed formulae, $(x+3)$, $(x-3)$, $(x*3)$, $(x/3)$, and $(x \uparrow 3)$ are well-formed formulae. Continuing, since y also is a well-formed formula, so are $((x+3)+y)$, $(y-(x*3))$, and so on. (Note that $(3/0)$ is a well-formed formula, since only syntax matters here.) ■

EXAMPLE 9 The well-formed formulae for compound propositions involving **T**, **F**, propositional variables, and operators $\{\neg, \wedge, \vee, \rightarrow, \leftrightarrow\}$ are defined by

> **T**, **F**, and p, where p is a propositional variable, are well-formed formulae;
> $(\neg p)$, $(p \vee q)$, $(p \wedge q)$, $(p \rightarrow q)$, $(p \leftrightarrow q)$ are well-formed formulae if p and q are well-formed formulae.

For example, if p, q, and r are propositional variables, then repeatedly using the recursive definition shows that $(p \vee q)$, $(r \wedge \mathbf{T})$, and $((p \vee q) \rightarrow (r \wedge \mathbf{T}))$ are well-formed formulae. ■

Recursive definitions are often used in the study of strings. Recall from Chapter 1 that a **string** over an alphabet Σ is a finite sequence of symbols from Σ. The set of strings over Σ is denoted by Σ^*. Two strings can be combined via the operation of **concatenation.** The concatenation of the strings x and y, denoted by xy, is the string x followed by the string y. For instance, the concatenation of $x = abra$ and $y = cadabra$ is $xy = abracadabra$. The following recursive definition is often used when results about strings are proved.

EXAMPLE 10 A Recursive Definition of the Set of Strings The set Σ^* of strings over the alphabet Σ can be defined recursively by $\lambda \in \Sigma^*$, where λ is the *empty string* containing no symbols, and $wx \in \Sigma^*$ whenever $w \in \Sigma^*$ and $x \in \Sigma$.

The first part of this definition says that the empty string belongs to Σ^*. The second part states that new strings are produced by concatenating strings in Σ^* with symbols from Σ. ■

The **length** of a string, which is the number of symbols in the string, can also be defined recursively.

EXAMPLE 11 Give a recursive definition of $l(w)$, the length of the string w.

Solution: The length of a string can be defined by

$$l(\lambda) = 0;$$
$$l(wx) = l(w) + 1 \text{ if } w \in \Sigma^* \text{ and } x \in \Sigma.$$ ■

The following example illustrates how the recursive definition of strings can be used in proofs.

EXAMPLE 12 Use mathematical induction to prove that $l(xy) = l(x) + l(y)$, where x and y belong to Σ^*, the set of strings over the alphabet Σ.

Solution: We will base our proof on the recursive definition of the set Σ^* given in Example 10. Let $P(y)$ be the statement that $l(xy) = l(x) + l(y)$ whenever x belongs to Σ^*.

BASIS STEP: To complete the basis step, we must show that $P(\lambda)$ is true. That is, we must show that $l(x\lambda) = l(x) + l(\lambda)$ for all $x \in \Sigma^*$. Since $l(x\lambda) = l(x) = l(x) + 0 = l(x) + l(\lambda)$ for every string x, it follows that $P(\lambda)$ is true.

INDUCTIVE STEP: To complete the inductive step, we assume that $P(y)$ is true and show that this implies that $P(ya)$ is true whenever $a \in \Sigma$. What we need to show is that $l(xya) = l(x) + l(ya)$ for every $a \in \Sigma$. To show this, note that by the recursive definition of $l(w)$ (given in Example 11), we have $l(xya) = l(xy) + 1$ and $l(ya) = l(y) + 1$. And, by the inductive hypothesis, $l(xy = l(x) + l(y)$. We conclude that $l(xya) = l(x) + l(y) + 1 = l(x) + l(ya)$. ■

Exercises

1. Find $f(1)$, $f(2)$, $f(3)$, and $f(4)$ if $f(n)$ is defined recursively by $f(0) = 1$ and for $n = 0, 1, 2, \ldots$
   a) $f(n+1) = f(n) + 2$.
   b) $f(n+1) = 3f(n)$.
   c) $f(n+1) = 2^{f(n)}$.
   d) $f(n+1) = f(n)^2 + f(n) + 1$.
2. Find $f(2)$, $f(3)$, $f(4)$, and $f(5)$ if f is defined recursively by $f(0) = f(1) = 1$ and for $n = 1, 2, \ldots$

a) $f(n+1) = f(n) - f(n-1)$.
b) $f(n+1) = f(n)f(n-1)$.
c) $f(n+1) = f(n)^2 + f(n-1)^3$.
d) $f(n+1) = f(n)/f(n-1)$.

3. Give a recursive definition of the sequence $\{a_n\}$, $n = 1, 2, 3, \ldots$ if
a) $a_n = 6n$. b) $a_n = 2n + 1$.
c) $a_n = 10^n$. d) $a_n = 5$.

4. Give a recursive definition of the sequence $\{a_n\}$, $n = 1, 2, 3, \ldots$ if
a) $a_n = 4n - 2$. b) $a_n = 1 + (-1)^n$.
c) $a_n = n(n+1)$. d) $a_n = n^2$.

5. Let F be the function such that $F(n)$ is the sum of the first n positive integers. Give a recursive definition of $F(n)$.

6. Give a recursive definition of $S_m(n)$, the sum of the integer m and the nonnegative integer n.

7. Give a recursive definition of $P_m(n)$, the product of the integer m and the nonnegative integer n.

In Exercises 8–13 f_n is the nth Fibonacci number.

8. Prove that $f_1^2 + f_2^2 + \cdots + f_n^2 = f_n f_{n+1}$ whenever n is a positive integer.

9. Prove that $f_1 + f_3 + \cdots + f_{2n-1} = f_{2n}$ whenever n is a positive integer.

★10. Show that $f_{n+1}f_{n-1} - f_n^2 = (-1)^n$ whenever n is a positive integer.

11. Determine the number of divisions used by the Euclidean algorithm to find the greatest common divisor of the Fibonacci numbers f_n and f_{n+1} where n is a nonnegative integer. Verify your answer using mathematical induction.

12. Let

$$\mathbf{A} = \begin{bmatrix} 1 & 1 \\ 1 & 0 \end{bmatrix}$$

Show that

$$\mathbf{A}^n = \begin{bmatrix} f_{n+1} & f_n \\ f_n & f_{n-1} \end{bmatrix}$$

whenever n is a positive integer.

13. (This exercise depends on the notion of the determinant of a 2×2 matrix.) By taking determinants of both sides of the equation in Exercise 12, prove the identity given in Exercise 10.

★14. Give a recursive definition of the functions max and min so that $\max(a_1, a_2, \ldots, a_n)$ and $\min(a_1, a_2, \ldots, a_n)$ are the maximum and minimum of the n numbers $a_1, a_2, \ldots, a_n$, respectively.

★15. Let $a_1, a_2 \ldots, a_n$, and $b_1, b_2, \ldots, b_n$ be real numbers. Use the recursive definitions that you gave in Exercise 14 to prove the following.
a) $\max(-a_1, -a_2, \ldots, -a_n)$
$= -\min(a_1, a_2, \ldots, a_n)$
b) $\max(a_1 + b_1, a_2 + b_2, \ldots, a_n + b_n)$
$\leq \max(a_1, a_2, \ldots, a_n) + \max(b_1, b_2, \ldots, b_n)$
c) $\min(a_1 + b_1, a_2 + b_2, \ldots, a_n + b_n)$
$\geq \min(a_1, a_2, \ldots, a_n) + \min(b_1, b_2, \ldots, b_n)$

16. Show that the set S defined by $1 \in S$ and $s + t \in S$ whenever $s \in S$ and $t \in S$ is the set of positive integers.

17. Give a recursive definition of the set of positive integers that are multiples of 5.

18. Give a recursive definition of
a) the set of odd positive integers.
b) the set of positive integer powers of 3.
c) the set of polynomials with integer coefficients.

19. Give a recursive definition of
a) the set of even integers.
b) the set of positive integers congruent to 2 modulo 3.
c) the set of positive integers not divisible by 5.

20. Show that any well-formed formula of numerals, variables, and operators from $\{+, -, *, /, \uparrow\}$ contains the same number of right and left parentheses.

21. Define well-formed formulae of sets, variables representing sets, and operators from $\{\,\bar{}\,, \cup, \cap, -\}$.

The **reversal** of a string is the string consisting of the symbols of the string in reverse order. The reversal of the string w is denoted by w^R.

22. Find the reversal of the following bit strings.
a) 0101 b) 11011
c) 10001 00101 11

23. Give a recursive definition of the reversal of a string. (*Hint:* First define the reversal of the empty string. Then write a string w of length $n + 1$ as xy, where x is a string of length n, and express the reverse of w in terms of x^R and y.)

★24. Give a recursive proof that $(w_1 w_2)^R = w_2^R w_1^R$.

25. Give a recursive definition of w^i where w is a string and i is a nonnegative integer. (Here w^i represents the concatenation of i copies of the string w.)

★26. Give a recursive definition of the set of bit strings that are palindromes.

27. When does a string belong to the set A of bit strings defined recursively by

$$\lambda \in A$$
$$0x1 \in A \text{ if } x \in A,$$

where λ is the empty string?

★28. Recursively define the set of bit strings that have more 0s than 1s.

29. Use Exercise 25 and mathematical induction to show that $l(w^i) = i \cdot l(w)$, where w is a string and i is a nonnegative integer.

★30. Show that $(w^R)^i = (w^i)^R$ whenever w is a string and i is a nonnegative integer; that is, show that the ith power of the reversal of a string is the reversal of the ith power of the string.

★31. A **partition** of a positive integer n is a way to write n as a sum of positive integers. For instance, $7 = 3+2+1+1$ is a partition of 7. Let P_m equal the number of different partitions of m, where the order of terms in the sum does not matter, and let P_{mn} be the number of different ways to express m as the sum of positive integers not exceeding n.

a) Show that $P_{mm} = P_m$.

b) Show that the following recursive definition for P_{mn} is correct:

$$P_{mn} = \begin{cases} 1 & \text{if } m = 1 \\ 1 & \text{if } n = 1 \\ P_{mn} & \text{if } m < n \\ 1 + P_{m,m-1} & \text{if } m = n > 1 \\ P_{m,n-1} + P_{m-n,n} & \text{if } m > n > 1 \end{cases}$$

c) Find the number of partitions of 5 and of 6 using this recursive definition.

Consider the following inductive definition of a version of **Ackermann's function.** This function was named after Wilhelm Ackermann, a German mathematician who was a student of the great mathematician David Hilbert. Ackermann's function plays an important role in the theory of recursive functions and in the study of the complexity of certain algorithms involving set unions. (There are several different variants of this function. All are called Ackermann's function and have similar properties, even though their values do not always agree.)

$$A(m, n) = \begin{cases} 2n & \text{if } m = 0 \\ 0 & \text{if } m \geq 1 \text{ and } n = 0 \\ 2 & \text{if } m \geq 1 \text{ and } n = 1 \\ A(m-1, A(m, n-1)) & \text{if } m \geq 1 \text{ and } n \geq 2 \end{cases}$$

Exercises 32 to 39 involve this version of Ackermann's function.

32. Find the following values of Ackermann's function.

a) $A(1, 0)$ **b)** $A(0, 1)$

c) $A(1, 1)$ **d)** $A(2, 2)$

33. Show that $A(m, 2) = 4$ whenever $m \geq 1$.

34. Show that $A(1, n) = 2^n$ whenever $n \geq 1$.

35. Find the following values of Ackermann's function.

a) $A(2, 3)$ **★b)** $A(3, 3)$

★36. Find $A(3, 4)$.

★★37. Prove that $A(m, n+1) > A(m, n)$ whenever m and n are nonnegative integers.

★38. Prove that $A(m+1, n) \geq A(m, n)$ whenever m and n are nonnegative integers.

39. Prove that $A(i, j) \geq j$ whenever i and j are nonnegative integers.

☞ **40.** Use mathematical induction to prove that a function F defined by specifying $F(0)$ and a rule for obtaining $F(n+1)$ from $F(n)$ is well defined.

☞ **41.** Use the second principle of mathematical induction to prove that a function F defined by specifying $F(0)$ and a rule for obtaining $F(n+1)$ from the values $F(k)$ for $k = 0, 1, 2, \ldots, n$ is well defined.

3.4 Recursive Algorithms

INTRODUCTION

Sometimes we can reduce the solution to a problem with a particular set of input to the solution of the same problem with smaller input values. For instance, the problem of finding the greatest common divisor of two positive integers a and b where $b > a$ can be reduced to finding the greatest common divisor of a pair of smaller integers, namely, b **mod** a and a, since $gcd(b \textbf{ mod } a, a) = gcd(a, b)$. When such a reduction can be done, the solution to the original problem can be found with a sequence of reductions, until the problem has been reduced to some initial case for which the solution is known. For instance, for finding the greatest common divisor, the reduction continues until the smaller of the two numbers is zero, since $gcd(a, 0) = a$ when $a > 0$.

We will see that algorithms that successively reduce a problem to the same problem with smaller input are used to solve a wide variety of problems.

DEFINITION 1. An algorithm is called *recursive* if it solves a problem by reducing it to an instance of the same problem with smaller input.

We will describe several different recursive algorithms in the following examples. The first example shows how a recursive algorithm can be constructed to evaluate a function from its recursive definition.

EXAMPLE 1 Give a recursive algorithm for computing a^n where a is a real number and n is a nonnegative integer.

Solution: We can base a recursive algorithm on the recursive definition of a^n. This definition states that $a^{n+1} = a \cdot a^n$ for $n > 0$, and the initial condition $a^0 = 1$. To find a^n, successively use the recursive condition to reduce the exponent until it becomes zero. We give this procedure in Algorithm 1. ■

ALGORITHM 1 A Recursive Algorithm for Computing a^n.

procedure *power*(a: nonzero real number, n: nonnegative integer)
if $n = 0$ **then** $power(a, n) := 1$
else $power(a, n) := a * power(a, n - 1)$

Next we give a recursive algorithm for finding greatest common divisors.

EXAMPLE 2 Give a recursive algorithm for computing the greatest common divisor of two nonnegative integers a and b with $a < b$.

Solution: We can base a recursive algorithm on the reduction $gcd(a, b) = gcd(b \textbf{ mod } a, a)$ and the condition $gcd(0, b) = b$ when $b > 0$. This produces the procedure in Algorithm 2. ■

ALGORITHM 2 A Recursive Algorithm for Computing $gcd(a, b)$.

procedure $gcd(a, b$: nonnegative integers with $a < b)$
if $a = 0$ **then** $gcd(0, b) := b$
else $gcd(a, b) := gcd(b$ **mod** $a, a)$

We will now give recursive versions of searching algorithms.

EXAMPLE 3 Express the linear search algorithm as a recursive procedure.

Solution: To *search* for x in the search sequence $a_1, a_2, \ldots, a_n$, at the ith step of the algorithm x and a_i are compared. If x equals a_i then i is the location of x. Otherwise, the search for x is reduced to a search in a sequence with one fewer element, namely, the sequence $a_{i+1}, \ldots, a_n$. We can now give a recursive procedure.

Let search(i, j, x) be the procedure that searches for x in the sequence $a_i, a_{i+1}, \ldots, a_j$. The input to the procedure consists of the triple $(1, n, x)$. The procedure terminates at a step if the first term of the remaining sequence is x or if there is only one term of the sequence and this is not x. If x is not the first term, but there are additional terms, the same procedure is carried out but with a search sequence of one fewer term, obtained by deleting the first term of the search sequence, which is not x. ■

ALGORITHM 3 A Recursive Sequential Search Algorithm.

procedure $search(i, j, x)$
if $a_i = x$ **then**
  $location := i$
else if $i = j$ **then**
  $location := 0$
else
  $search(i + 1, j, x)$

EXAMPLE 4 Construct a recursive version of a binary search algorithm.

Solution: Suppose we want to locate x in the sequence $a_1, a_2, \ldots, a_n$. To perform a binary search, we begin by comparing x with the middle term, $a_{\lfloor (n+1)/2 \rfloor}$. Our algorithm will terminate if x equals this term. Otherwise, we reduce the search to a smaller search sequence, namely, the first half of the sequence if x is smaller than the middle term of the original sequence, and the second half otherwise. We have reduced the solution of the search problem to the solution of the same problem with a sequence approximately half as long. We express this recursive version of a binary search algorithm as Algorithm 4. ■

ALGORITHM 4 A Recursive Binary Search Algorithm.

procedure *binary search*(x, i, j)
$m := \lfloor (i + j)/2 \rfloor$
if $x = a_m$ **then**
  location $:= m$
else if $(x < a_m$ and $i < m)$ **then**
  binary search$(x, i, m - 1)$
else if $(x > a_m$ and $j > m)$ **then**
  binary search$(x, m + 1, j)$
else *location* $:= 0$

RECURSION AND ITERATION

A recursive definition expresses the value of a function at a positive integer in terms of the values of the function at smaller integers. This means that we can devise a recursive algorithm to evaluate a recursively defined function at a positive integer.

EXAMPLE 5 The following recursive procedure gives the value of $n!$ when the input is a positive integer n. ■

ALGORITHM 5 A Recursive Procedure for Factorials.

procedure *factorial*(n: positive integer)
if $n = 1$ **then**
  factorial$(n) := 1$
else
  factorial$(n) := n * $*factorial*$(n - 1)$

There is another way to evaluate the factorial function at an integer from its recursive definition. Instead of successively reducing the computation to the evaluation of the function at smaller integers, we can start with the value of the function at 1 and successively apply the recursive definition to find the values of the function at successive larger integers. Such a procedure is called **iterative.** In other words, to find $n!$ using an iterative procedure, we start with 1, the value of the factorial function at 1, and multiply successively by each positive integer less than or equal to n. This procedure is shown in Algorithm 6.

ALGORITHM 6 An Iterative Procedure for Factorials.

```
procedure iterative factorial(n: positive integer)
x := 1
for i := 1 to n
   x := i * x
{x is n!}
```

After this code has been executed, the value of the variable x is $n!$. For instance, going through the loop six times gives $6! = 1 \cdot 2 \cdot 3 \cdot 4 \cdot 5 \cdot 6 = 720$.

Often an iterative approach for the evaluation of a recursively defined sequence requires much less computation than a procedure using recursion (unless special-purpose recursive machines are used). This is illustrated by the iterative and recursive procedures for finding the nth Fibonacci number. The recursive procedure is given first.

ALGORITHM 7 A Recursive Algorithm for Fibonacci Numbers.

```
procedure fibonacci(n: nonnegative integer)
if n = 0 then fibonacci(0) := 0
else if n = 1 then fibonacci(1) := 1
else fibonacci(n) := fibonacci(n − 1) + fibonacci(n − 2)
```

When we use a recursive procedure to find f_n, we first express f_n as $f_{n-1}+f_{n-2}$. Then we replace both of these Fibonacci numbers by the sum of two previous Fibonacci numbers, and so on. When f_1 or f_0 arises, it is replaced by its value.

Note that at each stage of the recursion, until f_1 or f_0 is obtained, the number of Fibonacci numbers to be evaluated has doubled. For instance, when we find f_4 using this recursive algorithm, we must carry out all the computations illustrated in the tree diagram in Figure 1. This tree consists of a root labeled with f_4, and branches

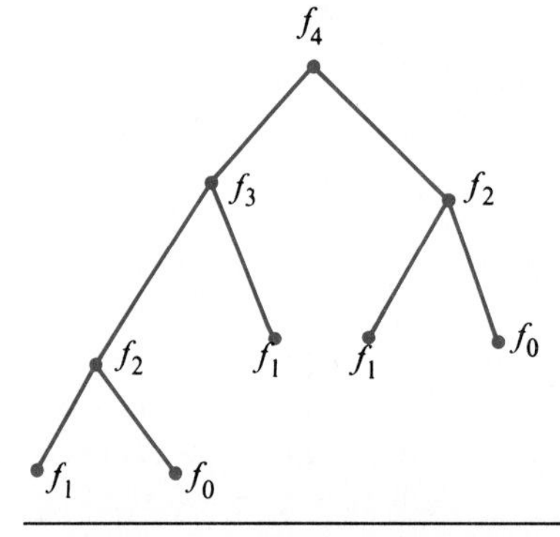

FIGURE 1 Evaluating f_4 Recursively.

from the root to vertices labeled with the two Fibonacci numbers f_3 and f_2 that occur in the reduction of the computation of f_4. Each subsequent reduction produces two branches in the tree. This branching ends when f_0 and f_1 are reached. The reader can verify that this algorithm requires $f_{n+1} - 1$ additions to find f_n.

Now consider the amount of computation required to find f_n using the following iterative approach.

ALGORITHM 8 An Iterative Algorithm for Computing Fibonacci Numbers.

```
procedure iterative fibonacci(n: nonnegative integer)
if n = 0 then y := 0
else
begin
  x := 0
  y := 1
  for i := 1 to n − 1
  begin
    z := x + y
    x := y
    y := z
  end
end
{y is the nth Fibonacci number}
```

This procedure initializes x as $f_0 = 0$ and y as $f_1 = 1$. When the loop is traversed, the sum of x and y is assigned to the auxiliary variable z. Then x is assigned the value of y and then y is assigned the value of the auxiliary variable z. Therefore, after going through the loop the first time, it follows that x equals f_1 and y equals $f_0 + f_1 = f_2$. Furthermore, after going through the loop $n - 1$ times, x equals f_{n-1} and y equals f_n (the reader should verify this statement). Only $n - 1$ additions have been used to find f_n with this iterative approach when $n > 1$. Consequently, this algorithm requires far less computation than does the recursive algorithm.

We have shown that a recursive algorithm may require far more computation than an iterative one when a recursively defined function is evaluated. It is sometimes preferable to use a recursive procedure even if it is less efficient than the iterative procedure. In particular, this is true when the recursive approach is easily implemented and the iterative approach is not. (Also, machines designed to handle recursion may be available that eliminate the advantage of using iteration.)

Exercises

1. Give a recursive algorithm for computing nx whenever n is a positive integer and x is an integer.

2. Give a recursive algorithm for finding the sum of the first n positive integers.

3. Give a recursive algorithm for finding the sum of the first n odd positive integers.
4. Give a recursive algorithm for finding the maximum of a finite set of integers.
5. Give a recursive algorithm for finding the minimum of a finite set of integers.
6. Devise a recursive algorithm for finding x^n **mod** m whenever n, x, and m are positive integers.
7. Give a recursive algorithm for finding $n!$ **mod** m whenever n and m are positive integers.
8. Give a recursive algorithm for finding a **mode** of a list of integers. (A **mode** is an element in the list that occurs at least as often as every other element.)
9. Devise a recursive algorithm for computing the greatest common divisor of two nonnegative integers a and b with $a < b$ using the fact that $gcd(a, b) = gcd(a, b - a)$.
10. Devise a recursive algorithm to find a^{2^n} where a is a real number and n is a positive integer. (*Hint:* Use the equality $a^{2^{n+1}} = (a^{2^n})^2$.)
11. How does the number of multiplications used by the algorithm in Exercise 10 compare to the number of multiplications used by Algorithm 1 to evaluate a^{2^n}?

★12. Use the algorithm in Exercise 10 to devise an algorithm for evaluating a^n when n is a nonnegative integer. (*Hint:* Use the binary expansion of n.)

★13. How does the number of multiplications used by the algorithm in Exercise 12 compare to the number of multiplications used by Algorithm 1 to evaluate a^n?

14. How many additions are used by the recursive and iterative algorithms given in Algorithms 7 and 8, respectively, to find the Fibonacci number f_7?
15. Devise a recursive algorithm to find the nth term of the sequence defined by $a_0 = 1$, $a_1 = 2$, and $a_n = a_{n-1} \cdot a_{n-2}$, for $n = 2, 3, 4, \ldots$.
16. Devise an iterative algorithm to find the nth term of the sequence defined in Exercise 15.
17. Is the recursive algorithm or the iterative algorithm for finding the sequence in Exercise 15 more efficient?
18. Give a recursive algorithm to find the number of partitions of a positive integer based on the recursive definition given in Exercise 31 in Section 3.3.
19. Give a recursive algorithm for finding the reversal of a bit string. (See the definition of the reversal of a bit string in the preamble of Exercise 22 in Section 3.3 on page 201.)
20. Give a recursive algorithm for finding the string w^i, the concatenation of i copies of w, when w is a bit string.
21. Give a recursive algorithm for computing values of the Ackermann function. (*Hint:* See Exercise 32 in Section 3.3.)

3.5 Program Correctness

INTRODUCTION

Suppose that we have designed an algorithm to solve a problem and have written a program to implement it. How can we be sure that the program always produces the correct answer? After all the bugs have been removed, so that the syntax is correct, we can test the program with sample input. It is not correct if an incorrect result is produced for any sample input. However, even if the program gives the correct answer for all sample input, it may not always produce the correct answer (unless all possible input has been tested). We need a proof to show that the program *always* gives the correct output.

Program verification, the proof of correctness of programs, uses the rules of inference and proof techniques described in this chapter, including mathematical induction. Since an incorrect program can lead to disastrous effects, a large amount of methodology has been constructed for verifying programs. Efforts have been devoted to automating program verification so that it can be carried out using a computer.

However, only limited progress has been made toward this goal. Indeed, some mathematicians and theoretical computer scientists feel it will never be realistic to mechanize the proof of correctness of complex programs.

Some of the concepts and methods used to prove that programs are correct will be introduced in this section. However, a complete methodology for program verification will not be developed in this book. This section is meant to be a brief introduction to the area of program verification, which ties together the rules of logic, proof techniques, and the concept of an algorithm.

PROGRAM VERIFICATION

A program is said to be **correct** if it produces the correct output for every possible input. A proof that a program is correct consists of two parts. The first part shows that the correct answer is obtained if the program terminates. This part of the proof establishes the **partial correctness** of the program. The second part of the proof shows that the program always terminates.

To specify what it means for a program to produce the correct output, two propositions are used. The first is the **initial assertion,** which gives the properties that the input values must have. The second is the **final assertion,** which gives the properties the output of the program should have, if the program did what was intended. The appropriate initial and final assertions must be provided when a program is checked.

DEFINITION 1. A program, or program segment, S is said to be *partially correct with respect to* the initial assertion p and the final assertion q, if whenever p is true for the input values of S and S terminates, then q is true for the output values of S. The notation $p\{S\}q$ indicates that the programs, or program segment, S is partially correct with respect to the initial assertion p and the final assertion q.

Note that the notion of partial correctness has nothing to do with whether a program terminates; it focuses only on whether the program does what it is expected to do if it terminates.

A simple example illustrates the concepts of initial and final assertions.

EXAMPLE 1 Show that the program segment

$$y := 2$$
$$z := x + y$$

is correct with respect to the initial assertion p: $x = 1$ and the final assertion q: $z = 3$.

Solution: Suppose that p is true, so that $x = 1$ as the program begins. Then y is assigned the value 2, and z is assigned the sum of the values of x and y, which is 3. Hence, S is correct with respect to the initial assertion p and the final assertion q. Thus, $p\{S\}q$ is true. ■

RULES OF INFERENCE

We will give some rules of inference for the proof of program correctness. The first of these is used to prove that a program is correct by splitting the program into a series of subprograms and then showing that each subprogram is correct.

Suppose that the program S is split into subprograms S_1 and S_2. Write $S = S_1; S_2$ to indicate that S is made up of S_1 followed by S_2. Suppose that the correctness of S_1 with respect to the initial assertion p and final assertion q, and the correctness of S_2 with respect to the initial assertion q and the final assertion r, have been established. It follows that if p is true and S_1 is executed and terminates, then q is true; and if q is true, and S_2 executes and terminates, then r is true. Thus, if p is true and $S = S_1; S_2$ is executed and terminates, then r is true. This rule of inference, called the **composition rule,** can be stated as

$$\begin{array}{l} \quad p\{S_1\}q \\ \quad \underline{q\{S_2\}r} \\ \therefore p\{S_1; S_2\}r \end{array}$$

This rule of inference will be used later in this section.

Next, some rules of inference for program segments involving conditional statements and loops will be given. Since programs can be split into segments for proofs of correctness, this will let us verify many different programs.

CONDITIONAL STATEMENTS

First, rules of inference for conditional statements will be given. Suppose that a program segment has the form

```
if condition then
  S
```

where S is a block of statements. Then S is executed if *condition* is true, and it is not executed when *condition* is false. To verify that this segment is correct with respect to the initial assertion p and final assertion q, two things must be done. First, it must be shown that when p is true and *condition* is also true, then q is true after S terminates. Second, it must be shown that when p is true and *condition* is false, then q is true (since in this case S does not execute).

This leads to the following rule of inference:

$$\begin{array}{l} \quad (p \wedge condition)\{S\}q \\ \quad \underline{(p \wedge \neg condition) \rightarrow q} \\ \therefore p\{\textbf{if } condition \textbf{ then } S\}q. \end{array}$$

The following example illustrates how this rule of inference is used.

EXAMPLE 2 Verify that the program segment

```
if x > y then
  y := x
```

is correct with respect to the initial assertion **T** and the final assertion $y \geq x$.

Solution: When the initial assertion is true and $x > y$, the assignment $y := x$ is carried out. Hence, the final assertion, which asserts that $y \geq x$, is true in this case. Moreover, when the initial assertion is true and $x > y$ is false, so that $x \leq y$, the final assertion is again true. Hence, using the rule of inference for program segments of this type, this program is correct with respect to the given initial and final assertions. ■

Similarly, suppose that a program has a statement of the form

```
if condition then
  S1
else
  S2
```

If *condition* is true, then S_1 executes; but if *condition* is false, then S_2 executes. So to verify that this program segment is correct with respect to the initial assertion p and the final assertion q, two things must be done. First, it must be shown that when p is true and *condition* is true, then q is true after S_1 terminates. Second, it must be shown that when p is true and *condition* is false, then q is true after S_2 terminates. This leads to the following rule of inference:

$$\begin{array}{l} (p \land \mathit{condition})\{S_1\}q \\ (p \land \neg \mathit{condition})\{S_2\}q \\ \hline \therefore p\{\textbf{if } \mathit{condition} \textbf{ then } S_1 \textbf{ else } S_2\}q. \end{array}$$

The following example illustrates how this rule of inference is used.

EXAMPLE 3 Verify that the program segment

```
if x < 0 then
  abs := −x
else
  abs := x
```

is correct with respect to the initial assertion **T** and the final assertion $abs = |x|$.

Solution: Two things must be demonstrated. First, it must be shown that if the initial assertion is true and $x < 0$, then $abs = |x|$. This is correct, since when $x < 0$ the assignment statement $abs := -x$ sets $abs = -x$, which is $|x|$ by definition when $x < 0$. Second, it must be shown that if the initial assertion is true and $x < 0$ is false, so that $x \geq 0$, then $abs = |x|$. This is also correct, since in this case the program uses the assignment statement $abs := x$, and x is $|x|$ by definition when $x \geq 0$, so that $abs := x$. Hence, using the rule of inference for program segments of this type, this segment is correct with respect to the given initial and final assertions. ∎

LOOP INVARIANTS

Next, proofs of correctness of **while** loops will be described. To develop a rule of inference for program segments of the type

```
while condition
  S
```

note that S is repeatedly executed until *condition* becomes false. An assertion that remains true each time S is executed must be chosen. Such an assertion is called a **loop invariant.** In other words, p is a loop invariant if $(p \wedge condition)\{S\}p$ is true.

Suppose that p is a loop invariant. It follows that if p is true before the program segment is executed, p and $\neg condition$ are true after termination, if it occurs. This rule of inference is

$$\frac{(p \wedge condition)\{S\}p}{\therefore p\{\textbf{while } condition\ S\}(\neg condition \wedge p).}$$

The use of a loop invariant is illustrated in the following example.

EXAMPLE 4 A loop invariant is needed to verify that the program segment

```
i := 1
factorial := 1
while i < n
begin
    i := i + 1
    factorial := factorial * i
end
```

terminates with *factorial* $= n!$ when n is a positive integer. Let p be the proposition "*factorial* $:= i!$ and $i \leq n$." We will prove that p is a loop invariant using mathematical induction. First, note that p is true before the loop is entered, since at that point *factorial* $= 1 = 1!$ and $1 \leq n$. Now assume p is true and $i < n$ after an execution of the loop. Assume that the **while** loop is executed again. First, i is incremented by 1. Thus, i is still less than or equal to n, since by the inductive hypothesis $i < n$ before the loop was entered, and i and n are positive integers. Furthermore, *factorial,* which was $(i-1)!$ by the inductive hypothesis, is set equal to $(i-1)! \cdot i = i!$. Hence, p remains true. Therefore p is a loop invariant. In other words, the assertion $[p \wedge (i < n)]\{S\}p$ is true. It follows that the assertion $p\{$**while** $i < n\ S\}[(i \geq n) \wedge p]$ is also true.

Furthermore, the loop terminates after $n - 1$ traversals with $i = n$, since i is assigned the value 1 at the beginning of the program, 1 is added to i at each traversal, and the loop terminates when $i \geq n$. Consequently, at termination *factorial* $= n!$. ■

A final example will be given to show how the various rules of inference can be used to verify the correctness of a longer program.

EXAMPLE 5 We will outline how to verify the correctness of the program S for computing the product of two integers.

```
procedure multiply(m,n: integers)
     { if n < 0 then a := -n
  S1 { else a := n

     { k := 0
  S2 { x := 0

     { while k < a
     { begin
  S3 {    x := x + m
     {    k := k + 1
     { end

     { if n < 0 then product := -x
  S4 { else product := x
```

The goal is to prove that after S is executed, *product* has the value mn. The proof of correctness can be carried out by splitting S into four segments, with $S = S_1; S_2; S_3; S_4$, as shown in the listing of S. The rule of composition can be used to build the correctness proof. Here is how the argument proceeds. The details will be left as an exercise for the reader.

Let p be the initial assertion that m and n are integers. Then, it can be shown that $p\{S_1\}q$ is true, when q is the proposition $p \wedge (a = |n|)$. Next, let r be the proposition $q \wedge (k = 0) \wedge (x = 0)$. It is easily verified that $q\{S_2\}r$ is true. It can be shown that "$x = mk$ and $k \leq a$" is an invariant for the loop in S_3. Furthermore, it is easy to see that the loop terminates after a iterations, with $k = a$, so that $x = ma$ at this point. Since r implies that $x = m \cdot 0$ and $0 \leq a$, the loop invariant is true before the loop is entered. Since the loop terminates with $k = a$, it follows that $r\{S_3\}s$ is true where s is the proposition "$x = ma$ and $a = |n|$." Finally, it can be shown that S_4 is correct with respect to the initial assertion s and final assertion t, where t is the proposition "$product = mn$".

Putting all this together, since $p\{S_1\}q$, $q\{S_2\}r$, $r\{S_3\}s$, and $s\{S_4\}t$ are all true, it follows from the rule of composition that $p\{S\}t$ is true. Furthermore, since all four segments terminate, S does terminate. This verifies the correctness of the program. ■

Exercises

1. Prove that the program segment

```
y := 1
z := x + y
```

is correct with respect to the initial assertion $x = 0$ and the final assertion $z = 1$.

2. Verify that the program segment

```
if x < 0 then x := 0
```

is correct with respect to the initial assertion **T** and the final assertion $x \geq 0$.

3. Verify that the program segment

```
x := 2
z := x + y
if y > 0 then
    z := z + 1
else
    z := 0
```

is correct with respect to the initial assertion $y = 3$ and the final assertion $z = 6$.

4. Verify that the program segment

```
if x < y then
    min := x
else
    min := y
```

is correct with respect to the initial assertion **T** and the final assertion $(x \leq y \wedge min = x) \vee (x > y \wedge min = y)$.

★5. Devise a rule of inference for verification of partial correctness of statements of the form

```
if condition 1 then
    S1
else if condition 2 then
    S2
    ⋮
else
    Sn
```

where $S_1, S_2, \ldots, S_n$ are blocks.

6. Use the rule of inference developed in Exercise 5 to verify that the program

```
if x < 0 then
  y := −2|x|/x
else if x > 0 then
  y := 2|x|/x
else if x = 0 then
  y := 2
```

is correct with respect to the initial assertion **T** and the final assertion $y = 2$.

7. Use a loop invariant to prove that the following program segment for computing the nth power, where n is a positive integer, of a real number x is correct.

```
power := 1
i := 1
while i ≤ n
begin
  power := power * x
  i := i + 1
end
```

★8. Prove that the iterative program for finding f_n given in Section 3.4 is correct.

9. Provide all the details in the proof of correctness given in Example 5.

10. Suppose that both the implication $p_0 \rightarrow p_1$ and the program assertion $p_1\{S\}q$ are true. Show that $p_0\{S\}q$ also must be true.

11. Suppose that both the program assertion $p\{S\}q_0$ and the implication $q_0 \rightarrow q_1$ are true. Show that $p\{S\}q_1$ also must be true.

12. The following program computes quotients and remainders.

```
r := a
q := 0
while r ≥ d
begin
  r := r − d
  q := q + 1
end
```

Verify that it is partially correct with respect to the initial assertion "a and d are positive integers" and the final assertion "q and r are integers such that $a = dq + r$ and $0 \leq r < d$."

13. Use a loop invariant to verify that the Euclidean algorithm (Algorithm 1 in Section 2.4) is partially correct with respect to the initial assertion "a and b are positive integers" and the final assertion $x = gcd(a, b)$.

Key Terms and Results

TERMS

theorem: a mathematical assertion that can be shown to be true

proof: a demonstration that a theorem is true

lemma: a simple theorem used to prove other theorems

corollary: a proposition that can be proved as a consequence of a theorem that has just been proved

rule of inference: an implication that is a tautology that is used to draw conclusions from known assertions

fallacy: an implication that is a contingency that is often incorrectly used to draw conclusions

circular reasoning or **begging the question:** reasoning where one or more steps are based on the truth of the statement being proved

vacuous proof: a proof that the implication $p \rightarrow q$ is true based on the fact that p is false

trivial proof: a proof that the implication $p \rightarrow q$ is true based on the fact that q is true

direct proof: a proof that the implication $p \rightarrow q$ is true that proceeds by showing that q must be true when p is true

indirect proof: a proof that the implication $p \rightarrow q$ is true that proceeds by showing that p must be false when q is false

proof by contradiction: a proof that a proposition p is true based on the truth of the implication $\neg p \rightarrow q$ where q is a contradiction

proof by cases: a proof of an implication where the hypothesis is a disjunction of propositions that shows that each hypothesis separately implies the conclusion

counterexample: an element x such that $P(x)$ is false

mathematical induction: a proof technique for statements of the form $\forall n \in \mathbf{N} P(n)$ consisting of a basis step and an inductive step

basis step: the proof of $P(1)$ in a proof of $\forall n \in \mathbf{N} P(n)$ by mathematical induction

inductive step: the proof of $P(n) \rightarrow P(n+1)$ in a proof of $\forall n \in \mathbf{N} P(n)$ by mathematical induction

recursive definition of a function: a definition of a function that specifies an initial set of values and a rule for obtaining values of this function at integers from its values at smaller integers

recursive definition of a set: a definition of a set that specifies an initial set of elements in the set and a rule for obtaining other elements from those in the set

recursive algorithm: an algorithm that proceeds by reducing a problem to the same problem with smaller input

iteration: a procedure based on the repeated use of operations in a loop

program correctness: verification that a procedure always produces the correct result

loop invariant: a property that remains true during every traversal of a loop

initial assertion: the statement specifying the properties of the input values of a program

final assertion: the statement specifying the properties the output values should have if the program worked correctly

RESULTS

The well-ordering property: Every nonempty set of nonnegative integers has a least element.

Principle of mathematical induction: The statement $\forall n P(n)$ is true if $P(1)$ is true and $\forall n[P(n) \rightarrow P(n+1)]$ is true.

Second principle of mathematical induction: The statement $\forall n P(n)$ is true if $P(1)$ is true and $\forall n[(P(1) \wedge \cdots \wedge P(n)) \rightarrow P(n+1)]$ is true.

Review Questions

1. a) Describe what is meant by a direct proof, an indirect proof, and a proof by contradiction of an implication $p \rightarrow q$.

b) Give a direct proof, an indirect proof, and a proof by contradiction of the statement: "If n is even, then $n+4$ is even."

2. a) Describe one way to prove the biconditional $p \leftrightarrow q$.

b) Prove the statement: "The integer $3n+2$ is odd if and only if the integer $9n+5$ is even, where n is an integer."

3. To prove that the statements p_1, p_2, p_3, and p_4 are equivalent, is it sufficient to show that the implications $p_4 \rightarrow p_2$, $p_3 \rightarrow p_1$, and $p_1 \rightarrow p_2$ are valid? If not, provide another set of implications that can be used to show that the four statements are equivalent.

4. a) Suppose that a statement of the form $\forall x P(x)$ is false. How can this be proved?

b) Show that the statement "For every positive integer n, the number n^2+1 is prime" is false.

5. a) What is the difference between a constructive existence proof and a nonconstructive existence proof?

b) Show that for every integer n there is an integer greater than n that is not divisible by either

3 or 5. Is your existence proof constructive or nonconstructive?

6. a) State the well-ordering property for the set of positive integers.

b) Use this property to show that every positive integer can be written as the product of primes.

7. a) Can you use the principle of mathematical induction to find a formula for the sum of the first n terms of a sequence?

b) Can you use the principle of mathematical induction to determine whether a given formula for the sum of the first n terms of a sequence is correct?

c) Find a formula for the sum of the first n even positive integers, and prove it using mathematical induction.

8. a) For which positive integers n is it true that $11n + 17 \leq 2^n$?

b) Prove the conjecture you made in part (a) using mathematical induction.

9. a) Which amounts of postage can be formed using only 5-cent and 9-cent stamps?

b) Prove the conjecture you made using mathematical induction.

c) Prove the conjecture you made using the second principle of mathematical induction.

d) Find a proof of your conjecture different from the ones you gave in (b) and (c).

10. Give three different examples of proofs that use the second principle of mathematical induction.

11. a) Explain why a function is well defined if it is defined recursively by specifying $f(1)$ and a rule for finding $f(n)$ from $f(n-1)$.

b) Provide a recursive definition of the function $f(n) = (n+1)!$.

12. a) Give a recursive definition of the Fibonacci numbers.

b) Show that $f_n > \alpha^{n-2}$ whenever $n \geq 3$ where f_n is the nth term of the Fibonacci sequence and $\alpha = (1 + \sqrt{5})/2$.

13. a) Explain why a sequence a_n is well defined if it is defined recursively by specifying a_1 and a_2 and a rule for finding a_n from $a_1, a_2, \ldots, a_{n-1}$ for $n = 3, 4, 5, \ldots$.

b) Find the value of a_n if $a_1 = 1$, $a_2 = 2$, and $a_n = a_{n-1} + a_{n-2} + \cdots + a_1$, for $n = 3, 4, 5, \ldots$.

14. Give two examples of how well-formed formulae are defined recursively for different sets of elements and operators.

15. a) Give a recursive definition of the length of a string.

b) Use the recursive definition from part (a) to prove that $l(xy) = l(x) + l(y)$.

16. a) What is a recursive algorithm?

b) Describe a recursive algorithm for computing the sum of n numbers in a sequence.

17. Describe a recursive algorithm for computing the greatest common divisor of two positive integers.

18. a) Does testing a computer program to see whether it produces the correct output for certain input values verify that the program always produces the correct output?

b) Does showing that a computer program is partially correct with respect to an initial assertion and a final assertion verify that the program always produces the correct output? If not, what else is needed?

19. What techniques can you use to show that a long computer program is partially correct with respect to an initial assertion and a final assertion?

20. What is a loop invariant? How is a loop invariant used?

Supplementary Exercises

1. Prove that the product of two odd numbers is odd.

2. Prove that $\sqrt{5}$ is irrational.

3. Prove or disprove that the sum of two irrational numbers is irrational.

4. Prove or disprove that n^2+n+1 is prime whenever n is a positive integer.

5. Determine whether the following is a valid argument. If n is greater than 5, then n^2 is greater than 25. Therefore, if n is an integer with n^2 greater than 25, it follows that n is greater than 5.

6. Prove that $n^4 - 1$ is divisible by 5 when n is not divisible by 5. Use a proof by cases, with four different cases—one for each of the nonzero remainders that an integer not divisible by 5 can have when you divide it by 5.

7. Prove that $|xy| = |x||y|$ by cases.

★8. We define the **Ulam numbers** by setting $u_1 = 1$ and $u_2 = 2$. Furthermore, after determining whether the integers less than n are Ulam numbers, we set n equal to the next Ulam number if it can be written uniquely as the sum of two different Ulam numbers. Note that $u_3 = 3$, $u_4 = 4$, $u_5 = 6$, and $u_6 = 8$.

a) Find the first 20 Ulam numbers.

b) Prove that there are infinitely many Ulam numbers.

9. Give a constructive proof that there is a polynomial $P(x)$ such that $P(x_1) = y_1, P(x_2) = y_2, \ldots, P(x_n) = y_n$, where $x_1, \ldots, x_n, y_1, \ldots, y_n$ are real numbers. (*Hint:* Let

$$P(x) = \sum_{i=1}^{n} \left(\prod_{i \neq j} \frac{x - x_j}{x_i - x_j} \right) y_i.)$$

10. Show that $1^3 + 3^3 + 5^3 + \cdots + (2n + 1)^3 = (n + 1)^2(2n^2 + 4n + 1)$ whenever n is a positive integer.

11. Show that $1 \cdot 2^0 + 2 \cdot 2^1 + 3 \cdot 2^2 + \cdots + n \cdot 2^{n-1} = (n - 1) \cdot 2^n + 1$ whenever n is a positive integer.

12. Show that

$$\frac{1}{1 \cdot 3} + \frac{1}{3 \cdot 5} + \cdots + \frac{1}{(2n - 1)(2n + 1)} = \frac{n}{2n + 1}$$

whenever n is a positive integer.

13. Show that

$$\frac{1}{1 \cdot 4} + \frac{1}{4 \cdot 7} + \cdots + \frac{1}{(3n - 2)(3n + 1)} = \frac{n}{3n + 1}$$

whenever n is a positive integer.

14. Use mathematical induction to show that $2^n > n^2 + n$ whenever n is an integer greater than 4.

15. Use mathematical induction to show that $2^n > n^3$ whenever n is an integer greater than 9.

16. Find an integer N such that $2^n > n^4$ whenever n is greater than N. Prove that your result is correct using mathematical induction.

17. Use mathematical induction to prove that $a - b$ is a factor of $a^n - b^n$ whenever n is a positive integer.

18. Use mathematical induction to prove that 9 divides $n^3 + (n+1)^3 + (n+2)^3$ whenever n is a nonnegative integer.

19. An **arithmetic progression** is a sequence of the form $a, a+d, a+2d, \ldots, a+nd$ where a and d are real numbers. Use mathematical induction to prove that the sum of these terms of an arithmetic progression is given by

$$a + (a+d) + \cdots + (a+nd) = (n+1)(2a+nd)/2.$$

20. Suppose that $a_j \equiv b_j \pmod{m}$ for $j = 1, 2, \ldots, n$. Use mathematical induction to prove that

a) $\sum_{j=1}^{n} a_j \equiv \sum_{j=1}^{n} b_j \pmod{m}$

b) $\prod_{j=1}^{n} a_j \equiv \prod_{j=1}^{n} b_j \pmod{m}$

★21. Determine which Fibonacci numbers are even, and use a form of mathematical induction to prove your conjecture.

★22. Determine which Fibonacci numbers are divisible by 3. Use a form of mathematical induction to prove your conjecture.

★23. Prove that $f_k f_n + f_{k+1} f_{n+1} = f_{n+k+1}$ for all nonnegative integers n, where k is a nonnegative integer and f_i denotes the ith Fibonacci number.

The sequence of **Lucas numbers** is defined by $l_0 = 2$, $l_1 = 1$, and $l_n = l_{n-1} + l_{n-2}$ for $n = 2, 3, 4, \ldots$.

24. Show that $f_n + f_{n+2} = l_{n+1}$ whenever n is a positive integer, where f_i and l_i are the ith Fibonacci number and ith Lucas number, respectively.

25. Show that $l_0^2 + l_1^2 + \cdots + l_n^2 = l_n l_{n+1} + 2$ whenever n is a nonnegative integer and l_i is the ith Lucas number.

★26. Use mathematical induction to show that the product of any n consecutive positive integers is divisible by $n!$. (*Hint:* Use the identity $m(m + 1)\cdots(m+n-1)/n! = (m-1)m(m+1)\cdots(m + n - 2)/n! + m(m + 1)\cdots(m + n - 2)/(n - 1)!$.)

27. Use mathematical induction to show that $(\cos x + i \sin x)^n = \cos nx + i \sin nx$ whenever n is a positive integer. (*Hint:* Use the identities $\cos(a + b) = \cos a \cos b - \sin a \sin b$ and $\sin(a + b) = \sin a \cos b + \cos a \sin b$.)

★**28.** Use mathematical induction to show that $\sum_{j=1}^{n} \cos jx = \cos[(n+1)x/2] \sin(nx/2)/\sin(x/2)$ whenever n is a positive integer and $\sin(x/2) \neq 0$.

The **McCarthy 91** function is defined using the rule

$$M(n) = \begin{cases} n - 10 & \text{if } n > 100 \\ M(M(n+11)) & \text{if } n \leq 100 \end{cases}$$

for all positive integers n.

29. By successively using the defining rule for $M(n)$, find

a) $M(102)$. **b)** $M(101)$. **c)** $M(99)$.
d) $M(97)$. **e)** $M(87)$. **f)** $M(76)$.

★★**30.** Show that the function $M(n)$ is a well-defined function from the set of positive integers to the set of positive integers. (*Hint:* Prove that $M(n) = 91$ for all positive integers n with $n \leq 101$.)

31. Is the following proof that

$$\frac{1}{1 \cdot 2} + \frac{1}{2 \cdot 3} + \cdots + \frac{1}{(n-1)n} = \frac{3}{2} - \frac{1}{n},$$

whenever n is a positive integer, correct? Justify your answer.

BASIS STEP: The result is true when $n = 1$ since

$$\frac{1}{1 \cdot 2} = \frac{3}{2} - \frac{1}{1}.$$

INDUCTIVE STEP: Assume that the result is true for n. Then

$$\frac{1}{1 \cdot 2} + \frac{1}{2 \cdot 3} + \cdots + \frac{1}{(n-1)n} + \frac{1}{n(n+1)} = \frac{3}{2} - \frac{1}{n} + \left(\frac{1}{n} - \frac{1}{n+1}\right) = \frac{3}{2} - \frac{1}{n+1}.$$

Hence, the result is true for $n + 1$ if it is true for n. This completes the proof.

★**32.** A jigsaw puzzle is put together by successively joining pieces that fit together into blocks. A move is made each time a piece is added to a block, or when two blocks are joined. Use the second form of mathematical induction to prove that no matter how the moves are carried out, exactly $n - 1$ moves are required to assemble a puzzle with n pieces.

★**33.** Show that n circles divide the plane into $n^2 - n + 2$ regions if every two circles intersect in exactly two points and no three circles contain a common point.

★**34.** Show that n planes divide three-dimensional space into $(n^3 + 5n + 6)/6$ regions if any three of these planes have a point in common and no four contain a common point.

★**35.** Use the well-ordering property to show that $\sqrt{2}$ is irrational. (*Hint:* Assume that $\sqrt{2}$ is rational. Show that the set of positive integers of the form $b\sqrt{2}$ has a least element a. Then show that $a\sqrt{2} - a$ is a smaller positive integer of this form.)

36. A set is **well ordered** is every nonempty subset of this set has a least element. Determine whether each of the following sets is well ordered.

a) the set of integers
b) the set of integers greater than -100
c) the set of positive rationals
d) the set of positive rationals with denominator less than 100

★**37.** Show that the well-ordering property can be proved when the principle of mathematical induction is taken as an axiom.

★**38.** Show that the first and second principles of mathematical induction are equivalent; that is, each can be shown to be valid from the other.

39. a) Show that if $a_1, a_2, \ldots, a_n$ are positive integers, then $gcd(a_1, a_2, \ldots, a_{n-1}, a_n) = gcd(a_1, a_2, \ldots, a_{n-2}, gcd(a_{n-1}, a_n))$.

b) Use part (a), together with the Euclidean algorithm, to develop a recursive algorithm for computing the greatest common divisor of a set of n positive integers.

★**40.** Describe a recursive algorithm for writing the greatest common divisor of n positive integers as a linear combination of these integers.

41. Find an explicit formula for $f(n)$ if $f(1) = 1$ and $f(n) = f(n-1) + 2n - 1$ for $n \geq 2$. Prove your result using mathematical induction.

★★**42.** Give a recursive definition of the set of bit strings that contain twice as many 0s as 1s.

43. Let S be the set of bit strings defined recursively by $\lambda \in S$ and $0x \in S$, $x1 \in S$ if $x \in S$, where λ is the empty string.

a) Find all strings in S of length not exceeding five.

b) Give an explicit description of the elements of S.

44. Let S be the set of strings defined recursively by $abc \in S$, $bac \in S$, $acb \in S$, and $abcx \in S$; $abxc \in S$, $axbc \in S$, $xabc \in S$ if $x \in S$.

a) Find all elements of S of length eight or less.

b) Show that every element of S has length divisible by three.

The set B of all **balanced strings of parentheses** is defined recursively by $\lambda \in B$, where λ is the empty string; $(x) \in B$, $xy \in B$ if $x, y \in B$.

45. Find all balanced strings of parentheses with four or fewer symbols.

46. Use induction to show that if x is a balanced string of parentheses, then the number of left parentheses equals the number of right parentheses in x.

Define the function N on the set of strings of parentheses by

$$N(\lambda) = 0, N(() = 1, N()) = -1,$$
$$N(uv) = N(u) + N(v),$$

where λ is the empty string, and u and v are strings. It can be shown that N is well defined.

47. Find

a) $N(())$ b) $N()))())(()$

c) $N((()(())$ d) $N(()((()))(()))$.

★★48. Show that a string w of parentheses is balanced if and only if $N(w) = 0$ and $N(u) \geq 0$ whenever u is a prefix of w, that is, $w = uv$.

★49. Give a recursive algorithm for finding all balanced strings of parentheses containing n or fewer symbols.

50. Give a recursive algorithm for finding the greatest common divisor of two nonnegative integers a and b with $a \leq b$, based on the fact that $gcd(a, b) = a$ if $a = b$, $gcd(a, b) = 2\,gcd(a/2, b/2)$ if a and b are even, $gcd(a, b) = (a/2, b)$ if a is even and b is odd, and $gcd(a, b) = gcd(b - a, b)$ if a and b are odd.

51. Verify the program segment

```
if x > y then
    x := y
```

with respect to the initial assertion **T** and the final assertion $x \leq y$.

★52. Develop a rule of inference for verifying recursive programs and use it to verify the recursive program for computing factorials given in Section 3.4.

Computer Projects

WRITE PROGRAMS WITH THE FOLLOWING INPUT AND OUTPUT.

1. Given a geometric progression $a, ar, ar^2, \ldots, ar^n$, find the sum of its terms.

2. Given a nonnegative integer n, find the sum of the n smallest positive integers.

★★3. Given a $2^n \times 2^n$ chessboard with one square missing, construct a tiling of this chessboard using L-shaped pieces.

★★4. Generate all well-formed formulae for expressions involving the variables x, y, and z and the operators $\{+, *, /, -\}$ with n or fewer symbols.

★★5. Generate all well-formed formulae for propositions with n or fewer symbols where each symbol is **T**, **F**, one of the propositional variables p and q, or an operator from $\{\neg, \vee, \wedge, \rightarrow, \leftrightarrow\}$.

6. Given a string, find its reversal.

7. Given a real number a and a nonnegative integer n, find a^n using recursion.

8. Given a real number a and a nonnegative integer n, find a^{2^n} using recursion.

★9. Given a real number a and a nonnegative integer n, find a^n using the binary expansion of n and a recursive algorithm for computing a^{2^k}.

10. Given two integers not both zero, find their greatest common divisor using recursion.

11. Given a list of integers and an element x, locate x in this list using a recursive implementation of a linear search.
12. Given a list of integers and an element x, locate x in this list using a recursive implementation of a binary search.
13. Given a nonnegative integer n, find the nth Fibonacci number using iteration.
14. Given a nonnegative integer n, find the nth Fibonacci number using recursion.
15. Given a positive integer, find the number of partitions of this integer. (See Exercise 31 of Section 3.3.)
16. Given positive integers m and n, find $A(m, n)$, the value of Ackermann's function at the pair (m,n). (See the preamble to Exercise 32 of Section 3.3.)

4

Counting

Combinatorics, the study of arrangements of objects, is an important part of discrete mathematics. This subject was studied as long ago as the 17th century, when combinatorial questions arose in the study of gambling games. Enumeration, the counting of objects with certain properties, is an important part of combinatorics. We must count objects to solve many different types of problems. For instance, counting is used to determine the complexity of algorithms. Furthermore, counting techniques are used extensively when probabilities of events are computed.

The basic rules of counting, which we will study in Section 4.1, can solve a tremendous variety of problems. For instance, we can use these rules to enumerate the different phone numbers possible in the United States, the allowable passwords on a computer system, and the different orders in which the runners in a race can finish. Another important combinatorial tool is the pigeonhole principle, which we will study in Section 4.2. This states that when objects are placed in boxes and there are more objects than boxes, then there is a box containing at least two objects. For instance, we can use this principle to show that among a set of 15 or more students, at least three were born on the same day of the week.

We can phrase many counting problems in terms of ordered or unordered arrangements of the objects of a set. These arrangements, called permutations and combinations, are used in many counting problems. For instance, suppose the 100 top finishers on a competitive exam taken by 2000 students are invited to a banquet. We can enumerate the possible sets of 100 students that will be invited, as well as the ways the top 10 prizes can be awarded.

We can analyze gambling games, such as poker, using counting techniques. We can also use these techniques to determine the probabilities of winning lotteries, such as the probability a person will win a lottery where 6 numbers are chosen from the first 48 positive integers.

Another problem in combinatorics involves generating all the arrangements of a specified kind. This is often important in computer simulations. We will devise algorithms to generate arrangements of various types.

4.1 The Basics of Counting

INTRODUCTION

A password on a computer system consists of six, seven, or eight characters. Each of these characters must be a digit or a letter of the alphabet. Each password must contain at least one digit. How many such passwords are there? The techniques needed to answer this question and a wide variety of other counting problems will be introduced in this section.

Counting problems arise throughout mathematics and computer science. For example, we must count the successful outcomes of experiments and all the possible outcomes of these experiments to determine probabilities of discrete events. We need to count the number of operations used by an algorithm to study its time complexity.

We will introduce the basic techniques of counting in this section. These methods serve as the basis for almost all counting techniques.

BASIC COUNTING PRINCIPLES

We will present two basic counting principles. Then we will show how they can be used to solve many different counting problems.

THE SUM RULE If a first task can be done in n_1 ways and a second task in n_2 ways, and if these tasks cannot be done at the same time, then there are $n_1 + n_2$ ways to do either task.

The following example illustrates how the sum rule is used.

EXAMPLE 1 Suppose either a member of the mathematics faculty or a student who is a mathematics major is chosen as a representative to a university committee. How many different choices are there for this representative if there are 37 members of the mathematics faculty and 83 mathematics majors?

Solution: The first task, choosing a member of the mathematics faculty, can be done in 37 ways. The second task, choosing a mathematics major, can be done in 83 ways. From the sum rule it follows that there are $37 + 83 = 120$ possible ways to pick this representative. ■

We can extend the sum rule to more than two tasks. Suppose that the tasks $T_1, T_2, \ldots, T_m$ can be done in $n_1, n_2, \ldots, n_m$ ways, respectively, and no two of these tasks can be done at the same time. Then the number of ways to do one of these tasks is $n_1 + n_2 + \cdots + n_m$. This extended version of the sum rule is often useful in counting problems, as Examples 2 and 3 show. This version of the sum rule can be proved using mathematical induction from the sum rule for two sets. (This is Exercise 45 at the end of the section.)

EXAMPLE 2 A student can choose a computer project from one of three lists. The three lists contain 23, 15, and 19 possible projects, respectively. How many possible projects are there to choose from?

Solution: The student can choose a project from the first list in 23 ways, from the second list in 15 ways, and from the third list in 19 ways. Hence, there are $23 + 15 + 19 = 57$ projects to choose from. ■

EXAMPLE 3 What is the value of k after the following code has been executed?

```
k := 0
for i_1 := 1 to n_1
  k := k + 1
for i_2 := 1 to n_2
  k := k + 1
  .
  .
  .
for i_m := 1 to n_m
  k := k + 1
```

Solution: The initial value of k is zero. This block of code is made up of m different loops. Each time a loop is traversed, 1 is added to k. Let T_i be the task of traversing the ith loop. The task T_i can be done in n_i ways, since the ith loop is traversed n_i times. Since no two of these tasks can be done at the same time, the sum rule shows that the final value of k, which is the number of ways to do one of the tasks T_i, $i = 1, 2, \ldots, m$, is $n_1 + n_2 + \cdots + n_m$. ■

The sum rule can be phrased in terms of sets as follows: If $A_1, A_2, \ldots, A_m$ are disjoint sets, then the number of elements in the union of these sets is the sum of the numbers of elements in them. To relate this to our statement of the sum rule, let T_i be the task of choosing an element from A_i for $i = 1, 2, \ldots, m$. There are $|A_i|$ ways to do T_i. From the sum rule, since no two of the tasks can be done at the same time, the number of ways to choose an element from one of the sets, which is the number of elements in the union, is

$$|A_1 \cup A_2 \cup \cdots \cup A_m| = |A_1| + |A_2| + \cdots + |A_m|.$$

This equality applies only when the sets in question are disjoint. The situation is much more complicated when these sets have elements in common. That situation will be briefly discussed later in this section and discussed in more depth in Chapter 5.

The product rule applies when a procedure is made up of separate tasks.

THE PRODUCT RULE Suppose that a procedure can be broken down into two tasks. If there are n_1 ways to do the first task and n_2 ways to do the second task after the first task has been done, then there are $n_1 n_2$ ways to do the procedure.

The following examples show how the product rule is used.

EXAMPLE 4 The chairs of an auditorium are to be labeled with a letter and a positive integer not exceeding 100. What is the largest number of chairs that can be labeled differently?

Solution: The procedure of labeling a chair consists of two tasks, namely, assigning one of the 26 letters and then assigning one of the 100 possible integers to the seat. The product rule shows that there are $26 \cdot 100 = 2600$ different ways that a chair can be labeled. Therefore, the largest number of chairs that can be labeled differently is 2600. ■

EXAMPLE 5 There are 32 microcomputers in a computer center. Each microcomputer has 24 ports. How many different ports to a microcomputer in the center are there?

Solution: The procedure of choosing a port consists of two tasks, first picking a microcomputer and then picking a port on this microcomputer. Since there are 32 ways to choose the microcomputer and 24 ways to choose the port no matter which microcomputer has been selected, the product rule shows that there are 768 ports. ■

An extended version of the product rule is often useful. Suppose that a procedure is carried out by performing the tasks $T_1, T_2, \ldots, T_m$. If task T_i can be done in n_i ways after tasks $T_1, T_2, \ldots,$ and T_{i-1} have been done, then there are $n_1 \cdot n_2 \cdot \cdots \cdot n_m$ ways to carry out the procedure. This version of the product rule can be proved by mathematical induction from the product rule for two tasks (see Exercise 46 at the end of the section).

EXAMPLE 6 How many different bit strings are there of length seven?

Solution: Each of the seven bits can be chosen in two ways, since each bit is either zero or one. Therefore, the product rule shows there are a total of $2^7 = 128$ different bit strings of length seven. ■

EXAMPLE 7 How many different license plates are available if each plate contains a sequence of three letters followed by three digits (and no sequences of letters are prohibited, even if they are obscene)?

Solution: There are 26 choices for each of the three letters and 10 choices for each of the three digits. Hence, by the product rule there are a total of $26 \cdot 26 \cdot 26 \cdot 10 \cdot 10 \cdot 10 =$ 17,576,000 possible license plates. ■

EXAMPLE 8 Counting Functions How many functions are there from a set with m elements to one with n elements?

Solution: A function corresponds to a choice of one of the n elements in the codomain for each of the m elements in the domain. Hence, by the product rule there are $n \cdot n \cdot \cdots \cdot n = n^m$ functions from a set with m elements to one with n elements. ■

EXAMPLE 9 Counting One-to-One Functions How many one-to-one functions are there from a set with m elements to one with n elements?

Solution: First note when $m > n$ there are no one-to-one functions from a set with m elements to a set with n elements. Now let $m \leq n$. Suppose the elements in the domain are $a_1, a_2, \ldots, a_m$. There are n ways to choose the value of the function at a_1. Since the function is one-to-one, the value of the function at a_2 can be picked in $n - 1$ ways (since the value used for a_1 cannot be used again). In general, the value of the function at a_k can be chosen in $n - k + 1$ ways. By the product rule, there are $n(n-1)(n-2)\cdots(n-m+1)$ one-to-one functions from a set with m elements to one with n elements. ■

EXAMPLE 10 The Telephone Numbering Plan The format of telephone numbers in North America is specified by a *numbering plan*. A telephone number consists of ten digits, which are split into a three-digit area code, a three-digit office code, and a four-digit station code. Because of signaling considerations, there are certain restrictions on some of these digits. To specify the allowable format, let X denote a digit that can take any of the values of 0 through 9, let N denote a digit that can take any of the values of 2 through 9, and let Y denote a digit that must be a 0 or a 1. Two numbering plans, which will be called the old plan and the new plan, will be discussed. (The old plan was in use in the 1960s, and the new plan will ultimately be in use everywhere in North America.) As will be shown, the new plan allows the use of more numbers.

In the old plan, the formats of the area code, office code, and station code are *NYX*, *NNX*, and *XXXX*, respectively. In the new plan, the formats of these codes are *NXX*, *NXX*, and *XXXX*, respectively. How many different North American telephone numbers are possible under the old plan and under the new plan?

Solution: By the product rule, there are $8 \cdot 2 \cdot 10 = 160$ area codes with format *NYX* and $8 \cdot 10 \cdot 10 = 800$ area codes with format *NXX*. Similarly, by the product rule, there are $8 \cdot 8 \cdot 10 = 640$ office codes with format *NNX* and $8 \cdot 10 \cdot 10 = 800$ with format *NXX*. The product rule also shows that there are $10 \cdot 10 \cdot 10 \cdot 10 = 10{,}000$ station codes with format *XXXX*.

Consequently, applying the product rule again, it follows that under the old plan there are

$$160 \cdot 640 \cdot 10{,}000 = 1{,}024{,}000{,}000$$

different numbers available in North America. Under the new plan there are

$$800 \cdot 800 \cdot 10{,}000 = 6{,}400{,}000{,}000$$

different numbers available. ■

EXAMPLE 11 What is the value of k after the following code has been executed?

```
k := 0
for i_1 := 1 to n_1
  for i_2 := 1 to n_2
    .
    .
    .
      for i_m := 1 to n_m
        k := k + 1
```

Solution: The initial value of k is zero. Each time the nested loop is traversed, 1 is added to k. Let T_i be the task of traversing the ith loop. Then the number of times the loop is traversed is the number of ways to do the tasks $T_1, T_2, \ldots, T_m$. The number of ways to carry out the task T_j, $j = 1, 2, \ldots, m$, is n_j, since the jth loop is traversed once for each integer i_j with $1 \leq i_j \leq n_j$. By the product rule, it follows that the nested loop is traversed $n_1 n_2 \cdots n_m$ times. Hence, the final value of k is $n_1 n_2 \cdots n_m$. ■

EXAMPLE 12 Counting Subsets of a Finite Set Use the product rule to show that the number of different subsets of a finite set S is $2^{|S|}$.

Solution: Let S be a finite set. List the elements of S in arbitrary order. Recall that there is a one-to-one correspondence between subsets of S and bit strings of length $|S|$. Namely, a subset of S is associated with the bit string with a 1 in the ith position if the ith element in the list is in the subset, and a 0 in this position otherwise. By the product rule, there are $2^{|S|}$ bit strings of length $|S|$. Hence, $|P(S)| = 2^{|S|}$. ■

The product rule is often phrased in terms of sets in the following way: If $A_1, A_2, \ldots, A_m$ are finite sets, then the number of elements in the Cartesian product of these sets is the product of the number of elements in each set. To relate this to the product rule, note that the task of choosing an element in the Cartesian product

$A_1 \times A_2 \times \cdots \times A_m$ is done by choosing an element in A_1, an element in $A_2, \ldots,$ and an element in A_m. From the product rule it follows that

$$|A_1 \times A_2 \times \cdots \times A_m| = |A_1| \cdot |A_2| \cdot \cdots \cdot |A_m|.$$

More Complex Counting Problems Many counting problems cannot be solved using just the sum rule or just the product rule. However, many complicated counting problems can be solved using both of these rules.

EXAMPLE 13 In a version of the computer language BASIC, the name of a variable is a string of one or two alphanumeric characters, where uppercase and lowercase letters are not distinguished. (An *alphanumeric* character is either one of the 26 English letters or one of the 10 digits.) Moreover, a variable name must begin with a letter and must be different from the five strings of two characters that are reserved for programming use. How many different variable names are there in this version of BASIC?

Solution: Let V equal the number of different variable names in this version of BASIC. Let V_1 be the number of these that are one character long and V_2 be the number of these that are two characters long. Then by the sum rule, $V = V_1 + V_2$. Note that $V_1 = 26$, since a one-character variable name must be a letter. Furthermore, by the product rule there are $26 \cdot 36$ strings of length two that begin with a letter and end with an alphanumeric character. However, five of these are excluded, so that $V_2 = 26 \cdot 36 - 5 = 931$. Hence, there are $V = V_1 + V_2 = 26 + 931 = 957$ different names for variables in this version of BASIC. ■

EXAMPLE 14 Each user on a computer system has a password, which is six to eight characters long, where each character is an uppercase letter or a digit. Each password must contain at least one digit. How many possible passwords are there?

Solution: Let P be the total number of possible passwords, and let P_6, P_7, and P_8 denote the number of possible passwords of length 6, 7, and 8, respectively. By the sum rule, $P = P_6 + P_7 + P_8$. We will now find P_6, P_7, and P_8. Finding P_6 directly is difficult. To find P_6 it is easier to find the number of strings of uppercase letters and digits that are six characters long, including those with no digits, and subtract from this the number of strings with no digits. By the product rule, the number of strings of six characters is 36^6, and the number of strings with no digits is 26^6. Hence,

$$P_6 = 36^6 - 26^6 = 2{,}176{,}782{,}336 - 308{,}915{,}776 = 1{,}867{,}866{,}560.$$

Similarly, it can be shown that

$$P_7 = 36^7 - 26^7 = 78{,}364{,}164{,}096 - 8{,}031{,}810{,}176 = 70{,}332{,}353{,}920$$

and

$$P_8 = 36^8 - 26^8 = 2{,}821{,}109{,}907{,}456 - 208{,}827{,}064{,}576$$
$$= 2{,}612{,}282{,}842{,}880.$$

Consequently,

$$P = P_6 + P_7 + P_8 = 2{,}684{,}483{,}063{,}360.$$

■

THE INCLUSION-EXCLUSION PRINCIPLE

When two tasks can be done at the same time, we cannot use the sum rule to count the number of ways to do one of the two tasks. Adding the number of ways to do each task leads to an overcount, since the ways to do both tasks are counted twice. To correctly count the number of ways to do one of the two tasks, we add the number of ways to do each of the two tasks and then subtract the number of ways to do both tasks. This technique is called the **principle of inclusion-exclusion.** Example 15 illustrates how we can solve counting problems using this principle.

EXAMPLE 15 How many bit strings of length eight either start with a 1 bit or end with the two bits 00?

Solution: The first task, constructing a bit string of length eight beginning with a 1 bit, can be done in $2^7 = 128$ ways. This follows by the product rule, since the first bit can be chosen in only one way and each of the other seven bits can be chosen in two ways.

The second task, constructing a bit string of length eight ending with the two bits 00, can be done in $2^6 = 64$ ways. This follows by the product rule, since each of the first six bits can be chosen in two ways and the last two bits can be chosen in only one way.

Both tasks, constructing a bit string of length eight that begins with a 1 and ends with 00, can be done in $2^5 = 32$ ways. This follows by the product rule, since the first bit can be chosen in only one way, each of the second through the sixth bits can be chosen in two ways, and the last two bits can be chosen in one way. Consequently, the number of bit strings of length eight that begin with a 1 and end with a 00, which equals the number of ways to do either the first task or the second task, equals $128 + 64 - 32 = 160$. ■

We can phrase this counting principle in terms of sets. Let A_1 and A_2 be sets. Let T_1 be the task of choosing an element from A_1 and T_2 the task of choosing an element from A_2. There are $|A_1|$ ways to do T_1 and $|A_2|$ ways to do T_2. The number of ways to do either T_1 or T_2 is the sum of the number of ways to do T_1 and the number of ways to do T_2, minus the number of ways to do both T_1 and T_2. Since there are $|A_1 \cup A_2|$ ways to do either T_1 or T_2 and $|A_1 \cap A_2|$ ways to do both T_1 and T_2, we have

$$|A_1 \cup A_2| = |A_1| + |A_2| - |A_1 \cap A_2|.$$

This is the formula given in Section 1.5 for the number of elements in the union of two sets.

The principle of inclusion-exclusion can be generalized to find the number of ways to do one of n different tasks or, equivalently, to find the number of elements in the union of n sets, whenever n is a positive integer. We will study the inclusion-exclusion principle and some of its many applications in Chapter 5.

TREE DIAGRAMS

Counting problems can be solved using **tree diagrams.** A tree consists of a root, a number of branches leaving the root, and possible additional branches leaving the endpoints of other branches. (We will study trees in detail in Chapter 8.) To use trees in counting, we use a branch to represent each possible choice. We represent the possible outcomes by the leaves, which are the endpoints of branches not having other branches starting at them.

EXAMPLE 16 How many bit strings of length four do not have two consecutive 1s?

Solution: The tree diagram in Figure 1 displays all bit strings of length four without two consecutive 1s. We see that there are eight bit strings of length four without two consecutive 1s. ■

EXAMPLE 17 A playoff between two teams consists of at most five games. The first team that wins three games wins the playoff. In how many different ways can the playoff occur?

Solution: The tree diagram in Figure 2 displays all the ways the playoff can proceed, with the winner of each game shown. We see that there are 20 different ways for the playoff to occur. ■

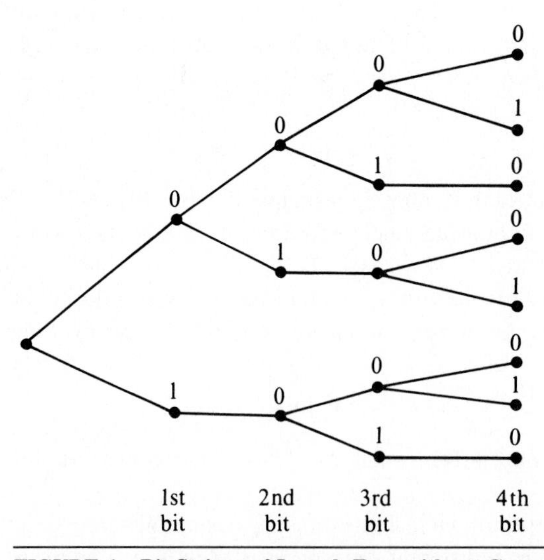

FIGURE 1 Bit Strings of Length Four without Consecutive 1s.

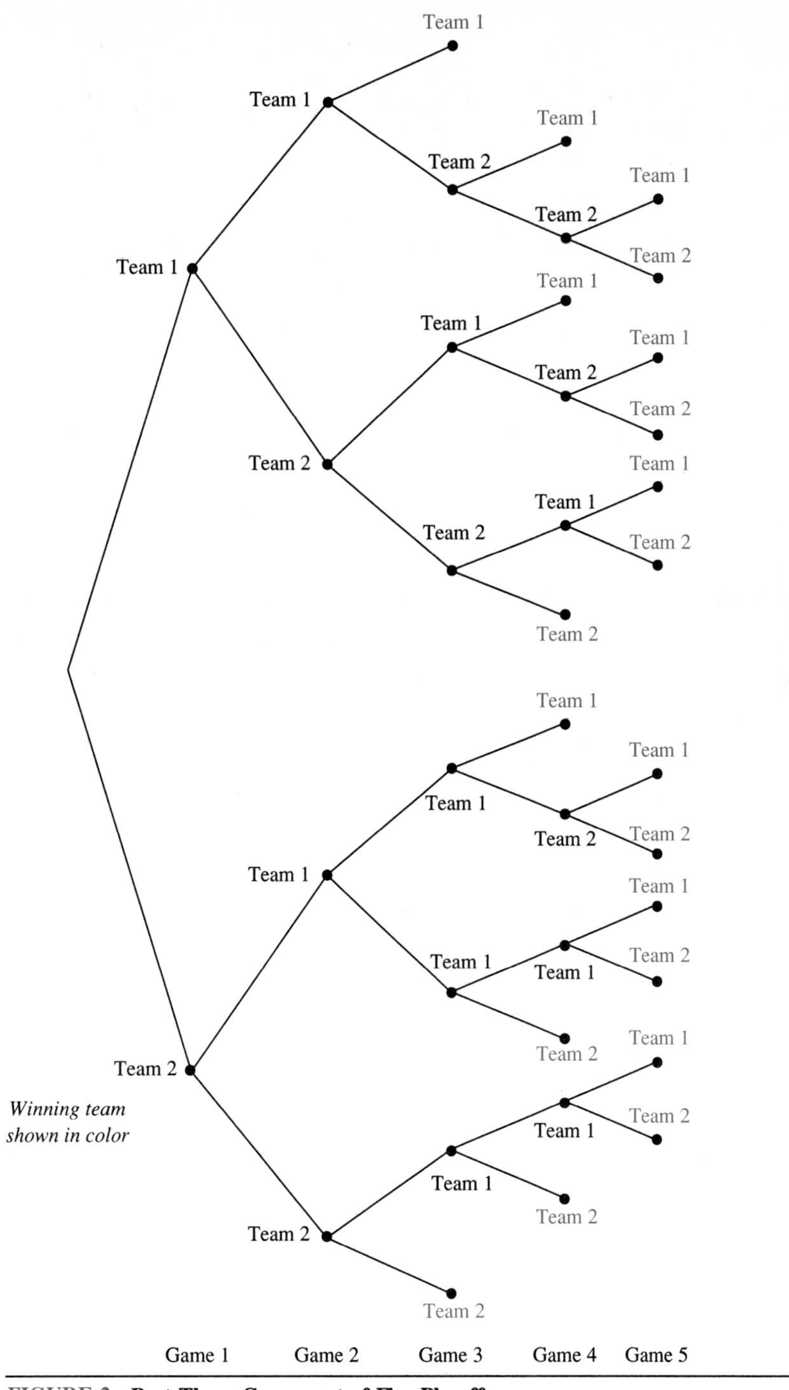

FIGURE 2 Best Three Games out of Five Playoffs.

Exercises

1. There are 18 mathematics majors and 325 computer science majors at a college.
   a) How many ways are there to pick two representatives, so that one is a mathematics major and the other is a computer science major?
   b) How many ways are there to pick one representative who is either a mathematics major or a computer science major?
2. An office building contains 27 floors and has 37 offices on each floor. How many offices are in the building?
3. A multiple-choice test contains 10 questions. There are four possible answers for each question.
   a) How many ways can a student answer the questions on the test if every question is answered?
   b) How many ways can a student answer the questions on the test if the student can leave answers blank?
4. A particular brand of shirt comes in 12 colors, has a male version and a female version, and comes in three sizes for each sex. How many different types of this shirt are made?
5. There are six different airlines that fly from New York to Denver and seven that fly from Denver to San Francisco. How many different possibilities are there for a trip from New York to San Francisco via Denver, when an airline is picked for the flight to Denver and an airline is picked for the continuation flight to San Francisco?
6. There are four major auto routes from Boston to Detroit and six from Detroit to Los Angeles. How many major auto routes are there from Boston to Los Angeles via Detroit?
7. How many different three-letter initials can people have?
8. How many different three-letter initials with none of the letters repeated can people have?
9. How many different three-letter initials are there that begin with an A?
10. How many bit strings are there of length eight?
11. How many bit strings of length 10 begin and end with a 1?
12. How many bit strings are there of length six or less?
13. How many bit strings with length not exceeding n, where n is a positive integer, consist entirely of 1s?
14. How many bit strings of length n, where n is a positive integer, start and end with 1s?
15. How many strings are there of lowercase letters of length four or less?
16. How many strings are there of four lowercase letters that have the letter x in them?
17. How many positive integers with exactly three decimal digits
    a) are divisible by 7?
    b) are odd?
    c) have the same three decimal digits?
    d) are not divisible by 4?
    e) are divisible by 3 or 4?
    f) are not divisible by either 3 or 4?
    g) are divisible by 3 but not by 4?
    h) are divisible by 3 and 4?
18. How many strings of four decimal digits
    a) do not contain the same digit twice?
    b) end with an even digit?
    c) have exactly three digits that are 9s?
19. A committee is formed containing either the governor or one of the two senators of each of the fifty states. How many ways are there to form this committee?
20. How many license plates can be made using either three digits followed by three letters or three letters followed by three digits?
21. How many license plates can be made using either two letters followed by four digits or two digits followed by four letters?
22. How many license plates can be made using either three letters followed by three digits or four letters followed by two digits?
23. How many license plates can be made using either two or three letters followed by either two or three digits?
24. How many different functions are there from a set with 10 elements to sets with the following numbers of elements?
    a) 2 b) 3 c) 4 d) 5
25. How many one-to-one functions are there from a set with five elements to sets with the following number of elements?
    a) 4 b) 5 c) 6 d) 7
26. How many functions are there from the set $\{1, 2, \ldots, n\}$, where n is a positive integer, to the set $\{0, 1\}$?

27. How many functions are there from the set $\{1, 2, \ldots, n\}$, where n is a positive integer, to the set $\{0, 1\}$
  a) that are one-to-one?
  b) that assign 0 to both 1 and n?
  c) that assign 1 to exactly one of the positive integers less than n?

28. How many subsets of a set with 100 elements have more than one element?

29. A **palindrome** is a string whose reversal is identical to the string. How many bit strings of length n are palindromes?

30. In how many ways can a photographer at a wedding arrange 6 people in a row from a group of 10 people, including the bride and groom, if
  a) the bride must be in the picture?
  b) both the bride and groom must be in the picture?
  c) exactly one of the bride and the groom is in the picture?

31. In how many ways can a photographer at a wedding arrange six people in a row, including the bride and groom, if
  a) the bride must be next to the groom?
  b) the bride is not next to the groom?
  c) the bride is positioned somewhere to the left of the groom?

32. How many bit strings of length seven either begin with two 0s or end with three 1s?

33. How many bit strings of length ten either begin with three 0s or end with two 0s?

★34. How many bit strings of length ten contain either five consecutive 0s or five consecutive 1s?

★★35. How many bit strings of length eight contain either three consecutive 0s or four consecutive 1s?

36. Every student in a discrete mathematics class is either a computer science or a mathematics major or is a joint major in these two subjects. How many students are in the class if there are 38 computer science majors (including joint majors), 23 mathematics majors (including joint majors), and 7 joint majors?

37. How many positive integers not exceeding 100 are divisible either by 4 or by 6?

38. The name of a variable in the C programming language is a string that can contain uppercase letters, lowercase letters, digits, or underscores. Further, the first character in the string must be a letter, either uppercase or lowercase, or an underscore. If the name of a variable is determined by its first eight characters, how many different variables can be named in C? (Note that the name of a variable may contain fewer than eight characters.)

39. Suppose that at some future time every telephone in the world is assigned a number that contains a country code 1 to 3 digits long, that is, of the form X, XX, or XXX, followed by a 10-digit telephone number of the form NXX–NXX–XXXX (as described in Example 10). How many different telephone numbers would be available worldwide under this numbering plan?

40. Use a tree diagram to find the number of bit strings of length four with no three consecutive 0s.

41. How many ways are there to arrange the letters a, b, c, and d such that a is not followed immediately by b?

42. Use a tree diagram to find the number of ways that the World Series can occur, where the first team that wins four games out of seven wins the series.

43. Use a tree diagram to determine the number of subsets of $\{3, 7, 9, 11, 24\}$ with the property that the sum of the elements in the subset is less than 28.

★44. Use the product rule to show that there are 2^{2^n} different truth tables for n propositions.

45. Use mathematical induction to prove the sum rule for m tasks from the sum rule for two tasks.

46. Use mathematical induction to prove the product rule for m tasks from the product rule for two tasks.

4.2 The Pigeonhole Principle

INTRODUCTION

Suppose that a flock of pigeons flies into a set of pigeonholes to roost. The **pigeonhole principle** states that if there are more pigeons than pigeonholes, then there must be

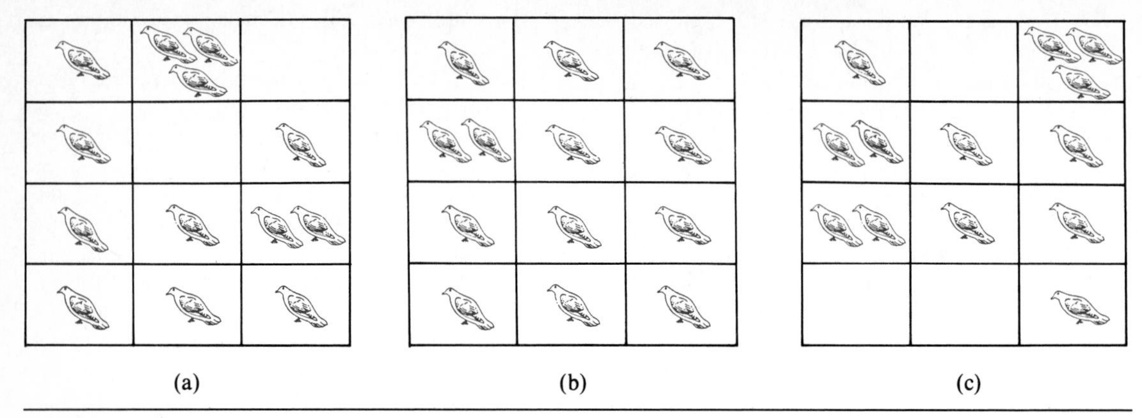

FIGURE 1 **There Are More Pigeons than Pigeonholes.**

at least one pigeonhole with at least two pigeons in it (see Figure 1). Of course, this principle applies to other objects besides pigeons and pigeonholes.

THEOREM 1 **THE PIGEONHOLE PRINCIPLE** If $k + 1$ or more objects are placed into k boxes, then there is at least one box containing two or more of the objects.

Proof: Suppose that none of the k boxes contains more than one object. Then the total number of objects would be at most k. This is a contradiction, since there are at least $k + 1$ objects. ■

The pigeonhole principle is also called the **Dirichlet drawer principle,** after the 19th-century German mathematician Dirichlet, who often used this principle in his work. The following examples show how the pigeonhole principle is used.

EXAMPLE 1 Among any group of 367 people, there must be at least two with the same birthday, because there are only 366 possible birthdays. ■

G. Lejeune Dirichlet (1805–1859). G. Lejeune Dirichlet was born into a French family living near Cologne, Germany. He studied at the University of Paris and held positions at the University of Breslau and the University of Berlin. In 1855 he was chosen to succeed Gauss at the University of Göttingen. Dirichlet is said to be the first person to master Gauss's *Disquisitiones Arithmeticae*, which appeared 20 years earlier. He is said to have kept a copy at his side even when he traveled. Dirichlet made many important discoveries in number theory, including the theorem that there are infinitely many primes in arithmetical progressions $an + b$ when a and b are relatively prime. He proved the $n = 5$ case of Fermat's Last Theorem, that there are no nontrivial solutions in integers to $x^5 + y^5 = z^5$. Dirichlet also made many contributions to analysis.

EXAMPLE 2 In any group of 27 English words, there must be at least two that begin with the same letter, since there are 26 letters in the English alphabet. ■

EXAMPLE 3 How many students must be in a class to guarantee that at least two students receive the same score on the final exam, if the exam is graded on a scale from 0 to 100 points?

Solution: There are 101 possible scores on the final. The pigeonhole principle shows that among any 102 students there must be at least 2 students with the same score. ■

THE GENERALIZED PIGEONHOLE PRINCIPLE

The pigeonhole principle states that there must be at least two objects in the same box when there are more objects than boxes. However, even more can be said when the number of objects exceeds a multiple of the number of boxes. For instance, among any set of 21 decimal digits there must be 3 that are the same. This follows because when 21 objects are distributed into 10 boxes, one box must have more than 2 objects.

THEOREM 2 **THE GENERALIZED PIGEONHOLE PRINCIPLE** If N objects are placed into k boxes, then there is at least one box containing at least $\lceil N/k \rceil$ objects.

Proof: Suppose that none of the boxes contains more than $\lceil N/k \rceil - 1$ objects. Then, the total number of objects is at most

$$k(\lceil N/k \rceil - 1) < k(((N/k) + 1) - 1) = N,$$

where the inequality $\lceil N/k \rceil < (N/k) + 1$ has been used. This is a contradiction since there are a total of N objects. ■

The following examples illustrate how the generalized pigeonhole principle is applied.

EXAMPLE 4 Among 100 people there are at least $\lceil 100/12 \rceil = 9$ who were born in the same month. ■

EXAMPLE 5 What is the minimum number of students required in a discrete mathematics class to be sure that at least six will receive the same grade, if there are five possible grades, A, B, C, D, and F?

Solution: The minimum number of students needed to guarantee that at least six students receive the same grade is the smallest integer N such that $\lceil N/5 \rceil = 6$. The smallest such integer is $N = 5 \cdot 5 + 1 = 26$. Thus, 26 is the minimum number of students needed to be sure that at least 6 students will receive the same grade. ■

EXAMPLE 6 What is the least number of area codes needed to guarantee that the 25 million phones in a state have distinct 10-digit telephone numbers? (Assume that telephone numbers are of the form *NXX–NXX–XXXX*, where the first three digits form the area code, N represents a digit from 2 to 9 inclusive, and X represents any digit.)

Solution: There are 8 million different phone numbers of the form *NXX–XXXX* (as shown in Example 10 of Section 4.1). Hence, by the generalized pigeonhole principle, among 25 million telephones, at least $\lceil 25{,}000{,}000/8{,}000{,}000 \rceil$ of them must have identical phone numbers. Hence, at least four area codes are required to ensure that all 10-digit numbers are different. ■

SOME ELEGANT APPLICATIONS OF THE PIGEONHOLE PRINCIPLE

In many interesting applications of the pigeonhole principle, the objects to be placed in boxes must be chosen in a clever way. A few such applications will be described here.

EXAMPLE 7 During a month with 30 days a baseball team plays at least 1 game a day, but no more than 45 games. Show that there must be a period of some number of consecutive days during which the team must play exactly 14 games.

Solution: Let a_j be the number of games played on or before the jth day of the month. Then $a_1, a_2, \ldots, a_{30}$ is an increasing sequence of distinct positive integers, with $1 \leq a_j \leq 45$. Moreover, $a_1 + 14, a_2 + 14, \ldots, a_{30} + 14$ is also an increasing sequence of distinct positive integers, with $15 \leq a_j + 14 \leq 59$.

The 60 positive integers $a_1, a_2, \ldots, a_{30}, a_1 + 14, a_2 + 14, \ldots, a_{30} + 14$ are all less than or equal to 59. Hence, by the pigeonhole principle two of these integers are equal. Since the integers a_j, $j = 1, 2, \ldots, 30$ are all distinct and the integers $a_j + 14$, $j = 1, 2, \ldots, 30$ are all distinct, there must be indices i and j with $a_i = a_j + 14$. This means that exactly 14 games were played from day $j + 1$ to day i. ■

EXAMPLE 8 Show that among any $n + 1$ positive integers not exceeding $2n$ there must be an integer that divides one of the other integers.

Solution: Write each of the $n + 1$ integers $a_1, a_2, \ldots, a_{n+1}$ as a power of 2 times an odd integer. In other words, let $a_j = 2^{k_j} q_j$ for $j = 1, 2, \ldots, n + 1$, where k_j

is a nonnegative integer and q_j is odd. The integers $q_1, q_2, \ldots, q_{n+1}$ are all odd positive integers less than $2n$. Since there are only n odd positive integers less than $2n$, it follows from the pigeonhole principle that two of the integers $q_1, q_2, \ldots, q_{n+1}$ must be equal. Therefore, there are integers i and j such that $q_i = q_j$. Let q be the common value of q_i and q_j. Then, $a_i = 2^{k_i}q$ and $a_j = 2^{k_j}q$. It follows that if $k_i < k_j$, then a_i divides a_j; while if $k_i > k_j$, then a_j divides a_i. ■

A clever application of the pigeonhole principle shows the existence of an increasing or a decreasing subsequence of a certain length in a sequence of distinct integers. Some definitions will be reviewed before this application is presented. Suppose that $a_1, a_2, \ldots, a_N$ is a sequence of real numbers. A **subsequence** of this sequence is a sequence of the form $a_{i_1}, a_{i_2}, \ldots, a_{i_m}$, where $1 \leq i_1 < i_2 < \cdots < i_m \leq N$. Hence, a subsequence is a sequence obtained from the original sequence by including some of the terms of the original sequence in their original order, and perhaps not including other terms. A sequence is called **strictly increasing** if each term is larger than the one that precedes it, and it is called **strictly decreasing** if each term is smaller than the one that precedes it.

THEOREM 3 Every sequence of $n^2 + 1$ distinct real numbers contains a subsequence of length $n + 1$ that is either strictly increasing or strictly decreasing.

The following example will be given before the theorem is proved.

EXAMPLE 9 The sequence 8, 11, 9, 1, 4, 6, 12, 10, 5, 7 contains 10 terms. Note that $10 = 3^2 + 1$. There are four increasing subsequences of length four, namely 1, 4, 6, 12; 1, 4, 6, 7; 1, 4, 6, 10; and 1, 4, 5, 7. There is also a decreasing subsequence of length four, namely, 11, 9, 6, 5. ■

The proof of the theorem will now be given.

Proof: Let $a_1, a_2, \ldots, a_{n^2+1}$ be a sequence of $n^2 + 1$ distinct real numbers. Associate an ordered pair with each term of the sequence, namely, associate (i_k, d_k) to the term a_k, where i_k is the length of the longest increasing subsequence starting at a_k, and d_k is the length of the longest decreasing subsequence starting at a_k.

Suppose that there are no increasing or decreasing subsequences of length $n + 1$. Then i_k and d_k are both positive integers less than or equal to n, for $k = 1, 2, \ldots, n^2 + 1$. Hence, by the product rule there are n^2 possible ordered pairs for (i_k, d_k). By the pigeonhole principle, two of these $n^2 + 1$ ordered pairs are equal. In other words, there exist terms a_s and a_t, with $s < t$ such that $i_s = i_t$ and $d_s = d_t$. We will show that this is impossible. Because the terms of the sequence are distinct, either $a_s < a_t$ or $a_s > a_t$. If $a_s < a_t$, then, since $i_s = i_t$, an increasing subsequence of length $i_t + 1$ can be built starting at a_s, by taking a_s followed by an increasing subsequence of length i_t beginning at a_t. This is a contradiction. Similarly, if $a_s > a_t$, it can be shown that d_s must be greater than d_t, which is a contradiction. ■

The final example shows how the generalized pigeonhole principle can be applied to an important part of combinatorics called **Ramsey theory,** after the English mathematician F. P. Ramsey. In general, Ramsey theory deals with the distribution of subsets of elements of sets.

EXAMPLE 10 Assume that in a group of six people, each pair of individuals consists of two friends or two enemies. Show that there are either three mutual friends or there are three mutual enemies in the group.

Solution: Let A be one of the six people. Of the five other people in the group, there are either three or more who are friends of A, or three or more who are enemies of A. This follows from the generalized pigeonhole principle, since when five objects are divided into two sets, one of the sets has at least $\lceil 5/2 \rceil = 3$ elements. In the former case, suppose that B, C, and D are friends of A. If any two of these three individuals are friends, then these two and A form a group of three mutual friends. Otherwise, B, C, and D form a set of three mutual enemies. The proof in the latter case, when there are three or more enemies of A, proceeds in a similar manner. ■

Exercises

1. Show that in any set of six classes there must be two that meet on the same day, assuming that no classes are held on weekends.
2. Show that if there are 30 students in a class, then at least 2 have last names that begin with the same letter.
3. A drawer contains a dozen brown socks and a dozen black socks, all unmatched. A man takes socks out at random in the dark. How many socks must he take out to be sure that he has at least two socks of the same color?
4. Let d be a positive integer. Show that among any group of $d+1$ (not necessarily consecutive) integers there are two with exactly the same remainder when they are divided by d.
5. Let n be a positive integer. Show that in any set of n consecutive integers there is exactly one divisible by n.
6. Show that if f is a function from S to T where S and T are finite sets with $|S| > |T|$, then there are elements s_1 and s_2 in S such that $f(s_1) = f(s_2)$, or in other words, f is not one-to-one.
7. How many students, each of whom comes from one of the 50 states, must be enrolled in a university to guarantee that there are at least 100 who come from the same state?
8. ★ Let (x_i, y_i), $i = 1, 2, 3, 4, 5$, be a set of five distinct points with integer coordinates in the xy plane. Show that the midpoint of the line joining at least one pair of these points has integer coordinates.

Frank Plumpton Ramsey (1903–1930). Frank Plumpton Ramsey, son of the president of Magdalene College, Cambridge, was educated at Winchester and Trinity Colleges. After graduating in 1923, he was elected as a fellow of King's College, Cambridge, where he spent the remainder of his life. Ramsey made important contributions to mathematical logic. What we now call Ramsey theory began with his clever combinatorial arguments, published in the paper "On a Problem of Formal Logic." Ramsey also made contributions to the mathematical theory of economics. He was noted as an excellent lecturer on the foundations of mathematics. His death at the age of 26 deprived the mathematical community and Cambridge University of a brilliant young scholar.

★**9.** Let (x_i, y_i, z_i), $i = 1, 2, 3, 4, 5, 6, 7, 8, 9$, be a set of nine distinct points with integer coordinates in xyz space. Show that the midpoint of at least one pair of these points has integer coordinates.

10. How many ordered pairs of integers (a, b) are needed to guarantee that there are two ordered pairs (a_1, b_1) and (a_2, b_2) such that a_1 **mod** $5 = a_2$ **mod** 5 and b_1 **mod** $5 = b_2$ **mod** 5?

11. **a)** Show that if five integers are selected from the first eight positive integers, there must be a pair of these integers with sum equal to 9.
b) Is the conclusion in part (a) true if four integers are selected rather than five?

12. **a)** Show that if seven integers are selected from the first 10 positive integers, there must be at least two pairs of these integers with sum 11.
b) Is the conclusion in part (a) true if six integers are selected rather than seven?

13. A company stores products in a warehouse. Storage bins in this warehouse are specified by their aisle, location in the aisle, and shelf. There are 50 aisles, 85 horizontal locations in each aisle, and 5 shelves throughout the warehouse. What is the least number of products the company can have so that at least two products must be stored in the same bin?

14. There are 51 houses on a street. Each house has an address between 1000 and 1099, inclusive. Show that at least two houses have addresses that are consecutive integers.

★**15.** Let x be an irrational number. Show that the absolute value of the difference between jx and the nearest integer to jx is less than $1/n$ for some positive integer j not exceeding n.

16. Find an increasing subsequence of maximal length and a decreasing subsequence of maximal length in the sequence 22, 5, 7, 2, 23, 10, 15, 21, 3, 17.

17. Construct a sequence of 16 positive integers that has no increasing or decreasing subsequence of 5 terms.

18. Show that if there are 101 people of different heights standing in a line, it is possible to find 11 people in the order they are standing in the line with heights that are either increasing or decreasing.

★**19.** Describe an algorithm in pseudocode for producing the largest increasing or decreasing subsequence of a sequence of distinct integers.

20. Show that in a group of five people (where any two people are either friends or enemies), there are not necessarily three mutual friends or three mutual enemies.

21. Show that in a group of 10 people (where any 2 people are either friends or enemies), there are either 3 friends or 4 mutual enemies, and there are either 3 mutual enemies or 4 mutual friends.

22. Use Exercise 21 to show that among any group of 20 people (where any 2 people are either friends or enemies), there are either 4 mutual friends or 4 mutual enemies.

23. Show that there are at least four people in California (population: 25 million) with the same three initials who were born on the same day of the year (but not necessarily in the same year).

24. Show that if there are 100,000,000 wage earners in the United States who earn less than 1,000,000 dollars, then there are two who earned exactly the same amount of money, to the penny, last year.

25. There are 38 different time periods during which classes at a university can be scheduled. If there are 677 different classes, how many different rooms will be needed?

26. A computer network consists of six computers. Each computer is directly connected to at least one of the other computers. Show that there are at least two computers in the network that are directly connected to the same number of other computers.

27. A computer network consists of six computers. Each computer is directly connected to zero or more of the other computers. Show that there are at least two computers in the network that are directly connected to the same number of other computers.

★**28.** Prove that at a party where there are at least two people, there are two people who know the same number of other people there.

29. An arm wrestler is the champion for a period of 75 hours. The arm wrestler had at least one match an hour, but no more than 125 total matches. Show that there is a period of consecutive hours during which the arm wrestler had exactly 24 matches.

★**30.** Is the statement in Exercise 29 true if 24 is replaced by
a) 2? **b)** 23? **c)** 25? **d)** 30?

31. Show that if f is a function from S to T where S and T are finite sets and $m = \lceil |S|/|T| \rceil$, then there are elements $s_1, s_2, \ldots, s_m$ of S such that $f(s_1) = f(s_2) = \cdots = f(s_m)$.

32. Suppose there are nine students in a discrete mathematics class at a small college.

a) Show that the class must have at least five male students or at least five female students.
b) Show that the class must have at least three male students or at least seven female students.

33. Suppose that every student in a discrete mathematics class of 25 students is a freshman, a sophomore, or a junior.
a) Show that there are at least 9 freshman, at least 9 sophomores, or at least 9 juniors in the class.
b) Show that there are either at least 3 freshman, at least 19 sophomores, or at least 5 juniors in the class.

34. Let $n_1, n_2, \ldots, n_t$ be positive integers. Show that if $n_1 + n_2 + \cdots + n_t - t + 1$ objects are placed into t boxes, then for some i, $i = 1, 2, \ldots, t$, the ith box contains at least n_i objects.

★35. A proof of Theorem 3 based on the generalized pigeonhole principle is outlined in this problem. The notation used is the same as that used in the proof in the text.
a) Assume that $i_k \leq n$ for $k = 1, 2, \ldots, n^2 + 1$. Use the generalized pigeonhole principle to show that there are $n + 1$ terms $a_{k_1}, a_{k_2}, \ldots, a_{k_{n+1}}$ with $i_{k_1} = i_{k_2} = \cdots = i_{k_{n+1}}$, where $1 \leq k_1 < k_2 < \cdots < k_{n+1}$.
b) Show that $a_{k_j} > a_{k_{j+1}}$ for $j = 1, 2, \ldots, n$. (*Hint:* Assume that $a_{k_j} < a_{k_{j+1}}$, and show that this implies that $i_{k_j} > i_{k_{j+1}}$, which is a contradiction).
c) Use parts (a) and (b) to show that if there is no increasing subsequence of length $n+1$, then there must be a decreasing subsequence of this length.

4.3 Permutations and Combinations

INTRODUCTION

Suppose a tennis team has 10 members. The coach has to select 5 players to make the trip to a match at another school. In addition, the coach has to prepare an ordered list of 4 players to play the 4 singles matches. In this section, methods will be developed to count the different unordered collections of the 5 players that are selected to make the trip and the different lists of 4 players to play the 4 singles matches. More generally, techniques will be introduced for counting the unordered selections of distinct objects and the ordered arrangements of objects of a finite set.

PERMUTATIONS

A **permutation** of a set of distinct objects is an ordered arrangement of these objects. We also are interested in ordered arrangements of some of the elements of a set. An ordered arrangement of r elements of a set is called an ***r*-permutation.**

EXAMPLE 1 Let $S = \{1, 2, 3\}$. The arrangement 3, 1, 2 is a permutation of S. The arrangement 3, 2 is a 2-permutation of S. ■

The number of r-permutations of a set with n elements is denoted by $P(n, r)$. We can find $P(n, r)$ using the product rule.

THEOREM 1 The number of r-permutations of a set with n distinct elements is

$$P(n, r) = n(n-1)(n-2)\cdots(n-r+1).$$

Proof: The first element of the permutation can be chosen in n ways, since there are n elements in the set. There are $n-1$ ways to choose the second element of the permutation, since there are $n-1$ elements left in the set after using the element picked for the first position. Similarly, there are $n-2$ ways to choose the third element, and so on, until there are exactly $n-r+1$ ways to choose the rth element. Consequently, by the product rule, there are

$$n(n-1)(n-2)\cdots(n-r+1)$$

r-permutations of the set. ■

From Theorem 1 it follows that

$$P(n, r) = n(n-1)(n-2)\cdots(n-r+1) = n!/(n-r)!$$

In particular, note that $P(n, n) = n!$. We will illustrate this result with some examples.

EXAMPLE 2 How many different ways are there to select 4 different players from 10 players on a team to play four tennis matches, where the matches are ordered?

Solution: The answer is given by the number of 4-permutations of a set with 10 elements. By Theorem 1, this is $P(10, 4) = 10 \cdot 9 \cdot 8 \cdot 7 = 5{,}040$. ■

EXAMPLE 3 Suppose that there are eight runners in a race. The winner receives a gold medal, the second-place finisher receives a silver medal, and the third-place finisher receives a bronze medal. How many different ways are there to award these medals, if all possible outcomes of the race can occur?

Solution: The number of different ways to award the medals is the number of 3-permutations of a set with 8 elements. Hence, there are $P(8, 3) = 8 \cdot 7 \cdot 6 = 336$ possible ways to award the medals. ■

EXAMPLE 4 Suppose that a saleswoman has to visit eight different cities. She must begin her trip in a specified city, but she can visit the other seven cities in any order she wishes. How many possible orders can the saleswoman use when visiting these cities?

Solution: The number of possible paths between the cities is the number of permutations of seven elements, since the first city is determined, but the remaining seven can be ordered arbitrarily. Consequently, there are $7! = 7 \cdot 6 \cdot 5 \cdot 4 \cdot 3 \cdot 2 \cdot 1 = 5040$ ways for the saleswoman to choose her tour. If, for instance, the saleswoman wishes to find the path between the cities with minimum distance, and she computes the total distance for each possible path, she must consider a total of 5040 paths! ■

COMBINATIONS

An ***r*-combination** of elements of a set is an unordered selection of r elements from the set. Thus, an r-combination is simply a subset of the set with r elements.

EXAMPLE 5 Let S be the set $\{1, 2, 3, 4\}$. Then $\{1, 3, 4\}$ is a 3-combination from S. ■

The number of r-combinations of a set with n distinct elements is denoted by $C(n, r)$.

EXAMPLE 6 We see that $C(4, 2) = 6$, since the 2-combinations of $\{a,b,c,d\}$ are the six subsets $\{a,b\}$, $\{a,c\}$, $\{a,d\}$, $\{b,c\}$, $\{b,d\}$, and $\{c,d\}$. ■

We can determine the number of r-combinations of a set with n elements using the formula for the number of r-permutations of a set. To do this, note that the r-permutations of a set can be obtained by first forming r-combinations and then ordering the elements in these combinations. The proof of the following theorem, which gives the value of $C(n,r)$, is based on this observation.

THEOREM 2 The number of r-combinations of a set with n elements, where n is a positive integer and r is an integer with $0 \leq r \leq n$, equals

$$C(n,r) = \frac{n!}{r!(n-r)!}.$$

Proof: The r-permutations of the set can be obtained by forming the $C(n,r)$ r-combinations of the set, and then ordering the elements in each r-combination, which can be done in $P(r,r)$ ways. Consequently,

$$P(n,r) = C(n,r) \cdot P(r,r).$$

This implies that

$$C(n,r) = \frac{P(n,r)}{P(r,r)} = \frac{n!/(n-r)!}{r!/(r-r)!} = \frac{n!}{r!(n-r)!}.$$ ■

The following corollary is helpful in computing the number of r-combinations of a set.

COROLLARY 1 Let n and r be nonnegative integers with $r \leq n$. Then $C(n,r) = C(n,n-r)$.

Proof: From Theorem 2 it follows that

$$C(n,r) = \frac{n!}{r!(n-r)!}$$

and

$$C(n,n-r) = \frac{n!}{(n-r)!\,[n-(n-r)]!} = \frac{n!}{(n-r)!\,r!}.$$

Hence, $C(n,r) = C(n,n-r)$. ■

There is another common notation for the number of r-combinations from a set with n elements, namely,

$$\binom{n}{r}.$$

This number is also called a **binomial coefficient.** The name *binomial coefficient* is used because these numbers occur as coefficients in the expansion of powers of binomial expressions such as $(a+b)^n$. We will discuss the **binomial theorem,** which expresses a power of a binomial expression as a sum of terms involving binomial coefficients, later in this section.

EXAMPLE 7 How many ways are there to select 5 players from a 10-member tennis team to make a trip to a match at another school?

Solution: The answer is given by the number of 5-combinations of a set with 10 elements. By Theorem 2, the number of such combinations is

$$C(10,5) = \frac{10!}{5!\,5!} = 252.$$ ■

BINOMIAL COEFFICIENTS

Some of the more important properties of the binomial coefficients will be discussed here. The first property to be discussed is an important identity.

THEOREM 3 PASCAL'S IDENTITY Let n and k be positive integers with $n \geq k$. Then

$$C(n+1,k) = C(n,k-1) + C(n,k).$$

Proof: Suppose that T is a set containing $n+1$ elements. Let a be an element in T, and let $S = T - \{a\}$. Note that there are $C(n+1,k)$ subsets of T containing k elements. However, a subset of T with k elements either contains a together with $k-1$ elements of S, or contains k elements of S and does not contain a. Since there are $C(n,k-1)$ subsets of $k-1$ elements of S, there are $C(n,k-1)$ subsets of k elements of T that contain a. And there are $C(n,k)$ subsets of k elements of T that do not contain a, since there are $C(n,k)$ subsets of k elements of S. Consequently,

$$C(n+1,k) = C(n,k-1) + C(n,k).$$ ■

$$\binom{0}{0}$$
$$\binom{1}{0} \binom{1}{1}$$
$$\binom{2}{0} \binom{2}{1} \binom{2}{2}$$
$$\binom{3}{0} \binom{3}{1} \binom{3}{2} \binom{3}{3}$$
$$\binom{4}{0} \binom{4}{1} \binom{4}{2} \binom{4}{3} \binom{4}{4}$$
$$\binom{5}{0} \binom{5}{1} \binom{5}{2} \binom{5}{3} \binom{5}{4} \binom{5}{5}$$
$$\binom{6}{0} \binom{6}{1} \binom{6}{2} \binom{6}{3} \binom{6}{4} \binom{6}{5} \binom{6}{6}$$
$$\binom{7}{0} \binom{7}{1} \binom{7}{2} \binom{7}{3} \binom{7}{4} \binom{7}{5} \binom{7}{6} \binom{7}{7}$$
$$\binom{8}{0} \binom{8}{1} \binom{8}{2} \binom{8}{3} \binom{8}{4} \binom{8}{5} \binom{8}{6} \binom{8}{7} \binom{8}{8}$$

...

(a)

By Pascal's identity:

$$\binom{6}{4} + \binom{6}{5} = \binom{7}{5}$$

1
1 1
1 2 1
1 3 3 1
1 4 6 4 1
1 5 10 10 5 1
1 6 15 20 15 6 1
1 7 21 35 35 21 7 1
1 8 28 56 70 56 28 8 1

...

(b)

FIGURE 1 **Pascal's Triangle.**

Remark: A combinatorial proof of Pascal's identity has been given. It is also possible to prove this identity by algebraic manipulation from the formula for $C(n, r)$ (see Exercise 47 at the end of this section).

Pascal's identity is the basis for a geometric arrangement of the binomial coefficients in a triangle, as shown in Figure 1.

The nth row in the triangle consists of the binomial coefficients

$$\binom{n}{k}, \quad k = 0, 1, \ldots, n.$$

This triangle is known as **Pascal's triangle.** Pascal's identity shows that when two adjacent binomial coefficients in this triangle are added, the binomial coefficient in the next row between these two coefficients is produced.

Blaise Pascal (1623–1662). Blaise Pascal exhibited his talents at an early age, although his father, who had made discoveries in analytic geometry, kept mathematics books away from him to encourage other interests. At 16 Pascal discovered an important result concerning conic sections. At 18 he designed a calculating machine, which he had built and sold. Pascal, along with Fermat, laid the foundations for the modern theory of probability. In this work he made new discoveries concerning what is now called Pascal's triangle. In 1654, Pascal abandoned his mathematical pursuits to devote himself to theology. After this, he returned to mathematics only once. One night, distracted by a severe toothache, he sought comfort by studying the mathematical properties of the cycloid. Miraculously, his pain subsided, which he took as a sign of divine approval of the study of mathematics.

The binomial coefficients enjoy many other identities besides Pascal's identity. Two other identities will be stated here. Combinatorial proofs will be given. Others may be found in the exercises at the end of the section.

THEOREM 4 Let n be a positive integer. Then

$$\sum_{k=0}^{n} C(n, k) = 2^n.$$

Proof: A set with n elements has a total of 2^n different subsets. Each subset has either zero elements, one element, two elements, . . . , or n elements in it. There are $C(n,0)$ subsets with zero elements, $C(n,1)$ subsets with one element, $C(n,2)$ subsets with two elements, . . . , and $C(n,n)$ subsets with n elements. Therefore,

$$\sum_{k=0}^{n} C(n,k)$$

counts the total number of subsets of a set with n elements. This shows that

$$\sum_{k=0}^{n} C(n,k) = 2^n.$$

■

THEOREM 5 **VANDERMONDE'S IDENTITY** Let m, n, and r be nonnegative integers with r not exceeding either m or n. Then

$$C(m + n, r) = \sum_{k=0}^{r} C(m, r - k)C(n, k).$$

Remark: This identity was discovered by mathematician Alexandre-Théophile Vandermonde in the 18th century.

Proof: Suppose that there are m items in one set and n items in a second set. Then the total number of ways to pick r elements from the union of these sets is $C(m + n, r)$. Another way to pick r elements from the union is to pick k elements from the first set and then $r - k$ elements from the second set, where k is an integer with $0 \leq k \leq r$. This can be done in $C(m, k)C(n, r - k)$ ways, using the product rule. Hence, the total number of ways to pick r elements from the union also equals

$$C(m + n, r) = \sum_{k=0}^{r} C(m, r - k)C(n, k).$$

This proves Vandermonde's identity. ■

Alexandre-Théophile Vandermonde (1735–1796). Because Alexandre-Théophile Vandermonde was a sickly child, his physician father directed him to a career in music. However, he later developed an interest in mathematics. His complete mathematical work consists of four papers published in 1771–1772. These papers include fundamental contributions on the roots of equations, on the theory of determinants, and on the knight's tour problem (introduced in the exercises in Section 7.5). Vandermonde's interest in mathematics lasted for only two years. Afterward, he published papers on harmony, on experiments with cold, and on the manufacture of steel. He also became interested in politics, joining the cause of the French revolution and holding several different positions in government.

THE BINOMIAL THEOREM

The binomial theorem gives the coefficients of the expansion of powers of binomial expressions. A **binomial** expression is simply the sum of two terms, such as $x + y$. (The terms can be products of constants and variables, but that does not concern us here.) The following example illustrates why this theorem holds.

EXAMPLE 8 The expansion of $(x + y)^3$ can be found using combinatorial reasoning instead of multiplying the three terms out. When $(x+y)^3 = (x+y)(x+y)(x+y)$ is expanded, all products of a term in the first sum, a term in the second sum, and a term in the third sum are added. Terms of the form x^3, x^2y, xy^2, and y^3 arise. To obtain a term of the form x^3, an x must be chosen in each of the sums, and this can be done in only one way. Thus, the x^3 term in the product has a coefficient of 1. To obtain a term of the form x^2y, an x must be chosen in two of the three sums (and consequently a y in the other sum). Hence, the number of such terms is the number of 2-combinations of three objects, namely, $C(3, 2)$. Similarly, the number of terms of the form xy^2 is the number of ways to pick one of the three sums to obtain an x (and consequently take a y from each of the other two terms). This can be done in $C(3, 1)$ ways. Finally, the only way to obtain a y^3 term is to choose the y for each of the three sums in the product, and this can be done in exactly one way. Consequently, it follows that

$$(x + y)^3 = x^3 + 3x^2y + 3xy^2 + y^3.$$ ■

The binomial theorem will now be stated.

THEOREM 6 **THE BINOMIAL THEOREM** Let x and y be variables, and let n be a positive integer. Then

$$(x + y)^n = \sum_{j=0}^{n} C(n, j)x^{n-j}y^j.$$

Proof: A combinatorial proof of the theorem will be given. The terms in the product when it is expanded are of the form $x^{n-j}y^j$ for $j = 0, 1, 2, \ldots, n$. To count the number of terms of the form $x^{n-j}y^j$, note that to obtain such a term it is necessary to choose $n - j$ xs from the n sums (so that the other j terms in the product are ys). Therefore, the coefficient of $x^{n-j}y^j$ is $C(n, n-j) = C(n, j)$. This proves the theorem. ■

The use of the binomial theorem is illustrated by the following examples.

EXAMPLE 9 What is the expansion of $(x + y)^4$?

Solution: From the binomial theorem it follows that

$$
\begin{aligned}
(x+y)^4 &= \sum_{j=0}^{4} C(4,j)x^{4-j}y^j \\
&= C(4,0)x^4 + C(4,1)x^3y + C(4,2)x^2y^2 + C(4,3)xy^3 + C(4,4)y^4 \\
&= x^4 + 4x^3y + 6x^2y^2 + 4xy^3 + y^4.
\end{aligned}
$$
■

EXAMPLE 10 What is the coefficient of $x^{12}y^{13}$ in the expansion of $(x+y)^{25}$?

Solution: From the binomial theorem it follows that this coefficient is

$$C(25,13) = \frac{25!}{13!\,12!} = 5{,}200{,}300.$$
■

EXAMPLE 11 What is the coefficient of $x^{12}y^{13}$ in the expansion of $(2x-3y)^{25}$?

Solution: First, note that this expression equals $(2x+(-3y))^{25}$. By the binomial theorem, we have

$$(2x+(-3y))^{25} = \sum_{j=0}^{25} C(25,j)(2x)^{25-j}(-3y)^j.$$

Consequently, the coefficient of $x^{12}y^{13}$ in the expansion is obtained when $j=13$, namely,

$$C(25,13)2^{12}(-3)^{13} = -\frac{25!}{13!\,12!}2^{12}3^{13}.$$
■

The binomial theorem can be used to give another proof of Theorem 4. Recall that this theorem states that $\sum_{k=0}^{n} C(n,k) = 2^n$ whenever n is a positive integer.

Proof: Using the binomial theorem we see that

$$2^n = (1+1)^n = \sum_{k=0}^{n} C(n,k)1^k1^{n-k} = \sum_{k=0}^{n} C(n,k).$$

This is the desired result. ■

The binomial theorem can also be used to prove the following identity.

THEOREM 7 Let n be a positive integer. Then

$$\sum_{k=0}^{n} (-1)^k C(n,k) = 0.$$

Proof: From the binomial theorem it follows that

$$0 = ((-1)+1)^n = \sum_{k=0}^{n} C(n,k)(-1)^k1^{n-k} = \sum_{k=0}^{n} C(n,k)(-1)^k.$$

This proves the theorem. ■

Exercises

1. List all the permutations of $\{a, b, c\}$.

2. How many permutations are there of the set $\{a, b, c, d, e, f, g\}$?

3. How many permutations of $\{a, b, c, d, e, f, g\}$ end with a?

4. Let $S = \{1, 2, 3, 4, 5\}$.

a) List all the 3-permutations of S.

b) List all the 3-combinations of S.

5. Find the value of each of the following quantities.

a) $P(6, 3)$ **b)** $P(6, 5)$ **c)** $P(8, 1)$
d) $P(8, 5)$ **e)** $P(8, 8)$ **f)** $P(10, 9)$

6. Find the value of each of the following quantities.

a) $C(5, 1)$ **b)** $C(5, 3)$ **c)** $C(8, 4)$
d) $C(8, 8)$ **e)** $C(8, 0)$ **f)** $C(12, 6)$

7. Find the number of 5-permutations of a set with nine elements.

8. In how many different orders can five runners finish a race?

9. How many possibilities are there for the win, place, and show (first, second, and third) positions in a horse race with 12 horses if all orders of finish are possible?

10. There are six different candidates for governor of a state. In how many different orders can the names of the candidates be printed on a ballot?

11. A group contains n men and n women. How many ways are there to arrange these people in a row if the men and women alternate?

12. In how many ways can a set of two positive integers less than 100 be chosen?

13. In how many ways can a set of five letters be selected from the English alphabet?

14. How many subsets with an odd number of elements does a set with 10 elements have?

15. How many subsets with more than two elements does a set with 100 elements have?

16. How many bit strings of length 10 have

a) exactly three 0s?

b) the same number of 0s as 1s?

c) at least seven 1s?

d) at least three 1s?

17. One hundred tickets, numbered $1, 2, 3, \ldots, 100$, are sold to 100 different people for a drawing. Four different prizes are awarded, including a grand prize (a trip to Tahiti).

a) How many ways are there to award the prizes?

b) How many ways are there to award the prizes if the person holding ticket 47 wins the grand prize?

c) How many ways are there to award the prizes if the person holding ticket 47 wins one of the prizes?

d) How many ways are there to award the prizes if the person holding ticket 47 does not win a prize?

e) How many ways are there to award the prizes if the people holding tickets 19 and 47 both win prizes?

f) How many ways are there to award the prizes if the people holding tickets 19, 47, and 73 all win prizes?

g) How many ways are there to award the prizes if the people holding tickets 19, 47, 73, and 97 all win prizes?

h) How many ways are there to award the prizes if none of the people holding tickets 19, 47, 73, and 97 wins a prize?

i) How many ways are there to award the prizes if the grand prize winner is a person holding ticket 19, 47, 73, or 97?

j) How many ways are there to award the prizes if the people holding tickets 19 and 47 win prizes, but the people holding tickets 73 and 97 do not win prizes?

18. Thirteen people on a softball team show up for a game.

a) How many ways are there to choose 10 players to take the field?

b) How many ways are there to assign the 10 positions to 10 of the 13 people who show up?

c) Of the 13 people who show up, 3 are women. How many ways are there to choose 10 players to take the field if at least one of these players must be a woman?

19. A club has 25 members.

a) How many ways are there to choose 4 members of the club to serve on an executive committee?

b) How many ways are there to choose a president, vice president, secretary, and treasurer of the club?

20. A professor writes 40 discrete mathematics true/false questions. Of the statements in these questions, 17 are true. If the questions can be positioned in any order, how many different answer keys are possible?

21. How many 4-permutations of the positive integers not exceeding 100 contain three consecutive integers in the correct order?

22. Seven women and nine men are on the faculty in the mathematics department at a school.

a) How many ways are there to select a committee of five members of the department if at least one woman must be on the committee?

b) How many ways are there to select a committee of five members of the department if at least one woman and at least one man must be on the committee?

23. The English alphabet contains 21 consonants and 5 vowels. How many strings of 6 lowercase letters of the English alphabet contain

a) exactly 1 vowel? **b)** exactly 2 vowels?
c) at least 1 vowel? **d)** at least 2 vowels?

24. How many strings of six lowercase letters from the English alphabet contain

a) the letter a?

b) the letters a and b?

c) the letters a and b in consecutive positions with a preceding b, with all the letters distinct?

d) the letters a and b, where a is somewhere to the left of b in the string, with all the letters distinct?

25. Suppose that a department contains 10 men and 15 women. How many ways are there to form a committee with 6 members if it must have the same number of men and women?

26. Suppose that a department contains 10 men and 15 women. How many ways are there to form a committee with 6 members if it must have more women than men?

27. How many bit strings contain exactly eight 0s and ten 1s if every 0 must be immediately followed by a 1?

28. How many bit strings contain exactly five 0s and fourteen 1s if every 0 must be immediately followed by two 1s?

29. How many bit strings of length 10 contain at least three 1s and at least three 0s?

30. How many ways are there to select 12 countries in the United Nations to serve on a council if 3 are selected from a block of 45, 4 are selected from a block of 57, and the others are selected from the remaining 69 countries?

31. How many license plates consisting of three letters followed by three digits contain no letter or digit twice?

32. How many ways are there to seat six people around a circular table, where seatings are considered to be the same if they can be obtained from each other by rotating the table?

33. Show that if n and k are positive integers, then

$$C(n+1,k) = (n+1)C(n,k-1)/k.$$

Use this identity to construct an inductive definition of the binomial coefficients.

34. Show that if p is a prime and k is an integer such that $1 \leq k \leq p-1$, then p divides $C(p,k)$.

35. Find the expansion of $(x+y)^5$.

36. Find the coefficient of x^5y^8 in $(x+y)^{13}$.

37. How many terms are there in the expansion of $(x+y)^{100}$?

38. What is the coefficient of x^7 in $(1+x)^{11}$?

39. What is the coefficient of x^9 in $(2-x)^{19}$?

40. What is the coefficient of x^8y^9 in the expansion of $(3x+2y)^{17}$?

41. What is the coefficient of $x^{101}y^{99}$ in the expansion of $(2x-3y)^{200}$?

★42. Give a formula for the coefficient of x^k in the expansion of $(x+1/x)^{100}$, where k is an integer.

★43. Give a formula for the coefficient of x^k in the expansion of $(x^2-1/x)^{100}$, where k is an integer.

44. The row of Pascal's triangle containing the binomial coefficients $C(10,k)$, $0 \leq k \leq 10$, is:

1 10 45 120 210 252 210 120 45 10 1

Use Pascal's identity to produce the row immediately following this row in Pascal's triangle.

45. What is the row of Pascal's triangle containing the binomial coefficients $C(9,k)$, $0 \leq k \leq 9$?

★46. Let n be a positive integer. What is the largest binomial coefficient $C(n,r)$, where r is a nonnegative integer less than or equal to n? Prove your answer is correct.

47. Prove Pascal's identity, using the formula for $C(n,r)$.

48. Prove the identity $C(n,r)C(r,k) = C(n,k) \cdot C(n-k, r-k)$, whenever n, r, and k are nonnegative integers with $r \leq n$ and $k \leq r$,

a) using a combinatorial argument.

b) using an argument based on the formula for the number of r-combinations of a set with n elements.

★49. Prove that

$$\sum_{k=0}^{r} C(n+k,k) = C(n+r+1,r),$$

whenever n and r are positive integers,

a) using a combinatorial argument.
b) using Pascal's identity.

50. Show that if n is a nonnegative integer, then $C(2n, 2) = 2C(n, 2) + n^2$
a) using a combinatorial argument.
b) by algebraic manipulation.

★51. Give a combinatorial proof that $\sum_{k=1}^{n} kC(n, k) = n2^{n-1}$. (*Hint:* Count in two ways the number of ways to select a committee and to then select a leader of the committee.)

★52. Give a combinatorial proof that $\sum_{k=1}^{n} kC(n, k)^2 = nC(2n - 1, n - 1)$. (*Hint:* Count in two ways the number of ways to select a committee, with a chairman, with n members from a group of n mathematics professors and n computer science professors, such that the chairman of the committee is a mathematics professor.)

53. Show that a set has the same number of subsets with an odd number of elements as it does subsets with an even number of elements.

★54. Prove the binomial theorem using mathematical induction.

55. In this exercise we will count the number of paths in the xy plane between the origin (0,0) and point (m,n) such that each path is made up of a series of steps, where each step is a move one unit to the right or a move one unit upward. (No moves to the left or downward are allowed.) Two such paths from $(0, 0)$ to $(5, 3)$ are illustrated here.

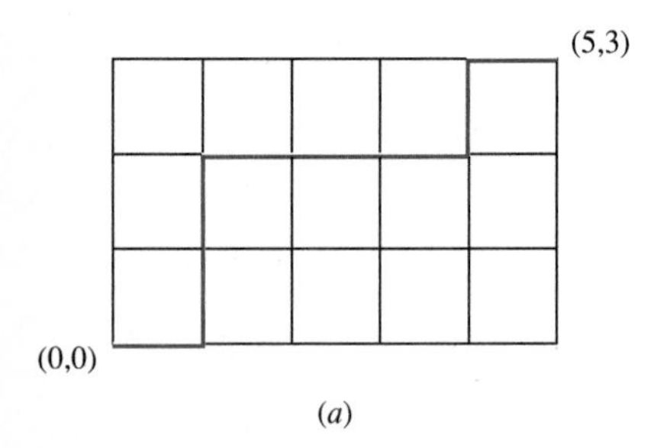

(*a*)

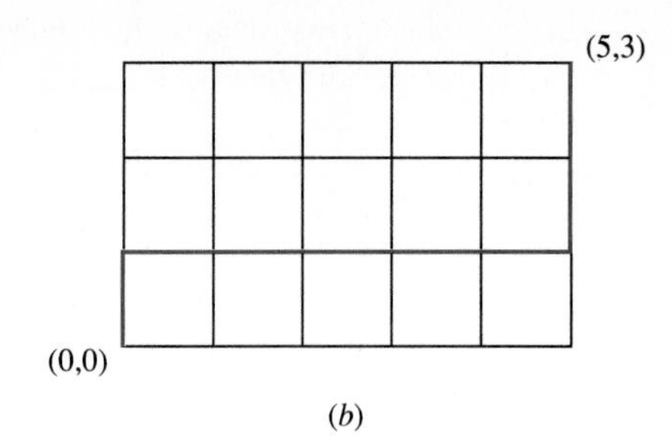

(*b*)

a) Show that each path of the type described can be represented by a bit string consisting of m 0s and n 1s, where a 0 represents a move one unit to the right and a 1 represents a move one unit upward.
b) Conclude from part (a) that there are $C(m + n, n)$ paths of the desired type.

56. Use Exercise 55 to prove that $C(n, k) = C(n, n - k)$ whenever k is an integer with $0 \leq k \leq n$. (*Hint:* Consider the number of paths of the type described in Exercise 55 from $(0, 0)$ to $(n - k, k)$ and from $(0, 0)$ to $(k, n - k)$.)

57. Use Exercise 55 to prove Theorem 4. (*Hint:* Count the number of paths with n steps of the type described in Exercise 55. Every such path must end at one of the points $(n - k, k)$ for $k = 0, 1, 2, \ldots, n$.)

58. Use Exercise 55 to prove Pascal's identity. (*Hint:* Show that a path of the type described in Exercise 55 from $(0, 0)$ to $(n+1-k, k)$ passes through either $(n + 1 - k, k - 1)$ or $(n - k, k)$, but not through both.)

59. Prove the identity in Exercise 49 using Exercise 55. (*Hint:* First, note that the number of paths from $(0, 0)$ to $(n + 1, r)$ equals $C(n + 1 + r, r)$. Second, count the number of paths by summing the number of these paths that start by going k units upward for $k = 0, 1, 2, \ldots, r$.)

4.4 Discrete Probability

INTRODUCTION

Combinatorics and probability theory share common origins. The theory of probability was first developed in the 17th century when certain gambling games were analyzed by the French mathematician Blaise Pascal. It was in these studies that Pascal discovered properties of the binomial coefficients. In the 18th century, the French mathematician Laplace, who also studied gambling, gave a definition of the probability of an event as the number of successful outcomes divided by the number of possible outcomes. For instance, the probability a die comes up an odd number when it is rolled is the number of successful outcomes—namely, the number of ways it can come up odd—divided by the number of possible outcomes—namely, the number of different ways the die can come up. There are a total of six possible outcomes—namely, 1, 2, 3, 4, 5, and 6—and exactly three of these are successful outcomes—namely, 1, 3, and 5. Hence, the probability that the die comes up an odd number is $3/6 = 1/2$. (Note that it has been assumed that all possible outcomes are equally likely, or, in other words, that the die is fair.)

In this section we will restrict ourselves to experiments that have finitely many, equally likely, outcomes. This permits us to use Laplace's definition of the probability of an event. We will continue our study of probability in Section 4.5, where we will study experiments with finitely many outcomes that are not necessarily equally likely. In Section 4.5 we will also introduce some key concepts in probability theory, including conditional probability, independence of events, random variables, and expected values.

FINITE PROBABILITY

An **experiment** is a procedure that yields one of a given set of possible outcomes. The **sample space** of the experiment is the set of possible outcomes. An **event** is a subset of the sample space. Laplace's definition of the probability of an event with finitely many possible outcomes will now be stated.

DEFINITION 1. The *probability* of an event E, which is a subset of a finite sample space S of equally likely outcomes, is $p(E) = |E|/|S|$.

Some additional examples are given here.

EXAMPLE 1 An urn contains four blue balls and five red balls. What is the probability that a ball chosen from the urn is blue?

Pierre Simon Laplace (1749–1827). Pierre Simon Laplace came from humble origins in Normandy. In his childhood he was educated in a school run by the Benedictines. At 16 he entered the University of Caen intending to study theology. However, he soon realized his true interests were with mathematics. After completing his studies he was named a provisional professor at Caen, and in 1769 he became professor of mathematics at the Paris Military School.

Laplace is best known for his contributions to celestial mechanics, the study of the motions of heavenly bodies. His work *Traité du Mécanique Céleste* is considered one of the greatest scientific works of the early 19th century. Laplace was one of the founders of probability theory and made many contributions to mathematical statistics. His work in this area is documented in his book *Thèorie Analytique des Probabilitées*, where he defined the probability of an event as the ratio of the number of favorable outcomes to the total number of outcomes of an experiment.

Laplace was famous for his political flexibility. He was loyal, in succession, to the French Republic, Napoleon, and King Louis XVIII. This permitted him to be productive before, during, and after the French revolution.

Solution: To calculate the probability, note that there are nine possible outcomes, and four of these possible outcomes produce a blue ball. Hence, the probability that a blue ball is chosen is $4/9$. ■

EXAMPLE 2 What is the probability that when two dice are rolled, the sum of the numbers on the two dice is 7?

Solution: There are a total of 36 possible outcomes when two dice are rolled. (The product rule can be used to see this; since each die has 6 possible outcomes, the total number of outcomes when two dice are rolled is $6^2 = 36$.) There are 6 successful outcomes, namely, $(1, 6)$, $(2, 5)$, $(3, 4)$, $(4, 3)$, $(5, 2)$, and $(6, 1)$, where the values of the first and second dice are represented by an ordered pair. Hence, the probability that a 7 comes up when two fair dice are rolled is $6/36 = 1/6$. ■

Lotteries have become extremely popular recently. We can easily compute the odds of winning different types of lotteries.

EXAMPLE 3 In a lottery, players win a large prize when they pick four digits that match, in the correct order, four digits selected by a random mechanical process. A smaller prize is won if only three digits are matched. What is the probability that a player wins the large prize? What is the probability that a player wins the small prize?

Solution: There is only one way to choose all four digits correctly. By the product rule, there are $10^4 = 10{,}000$ ways to choose four digits. Hence, the probability that a player wins the large prize is $1/10{,}000 = 0.0001$.

Players win the smaller prize when they correctly choose exactly three of the four digits. Exactly one digit must be wrong to get three digits correct, but not all four correct. By the sum rule, the number of ways to choose exactly three digits correctly can be obtained by adding the number of ways to choose four digits matching the digits picked in all but the ith position, for $i = 1, 2, 3, 4$. To count the number of successes with the first digit incorrect, note that there are nine possible choices for the first digit (all but the one correct digit), and one choice for each of the other digits, namely, the correct digits for these slots. Hence, there are nine ways to choose four digits where the first digit is incorrect, but the last three are correct. Similarly, there are nine ways to choose four digits where the second digit is incorrect, nine with the third digit incorrect, and nine with the fourth digit incorrect. Hence, there is a total of 36 ways to choose four digits with exactly three of the four digits correct. Thus, the probability that a player wins the smaller prize is $36/10{,}000 = 9/2500 = 0.0036$. ■

EXAMPLE 4 There are many lotteries now that award enormous prizes to people who correctly choose a set of six numbers out of the first n positive integers, where n is usually between 30 and 50. What is the probability that a person picks the correct six numbers out of 40?

Solution: There is only one winning combination. The total number of ways to choose six numbers out of 40 is

$$C(40, 6) = \frac{40!}{34!\,6!} = 3{,}838{,}380.$$

Consequently, the probability of picking a winning combination is $1/3{,}838{,}380 \sim 0.00000026$. (Here the symbol $\sim$ means approximately equal to.) ■

We can find the probability of hands in card games using the techniques developed so far. A deck of cards contains 52 cards. There are 13 different kinds of cards, with 4 cards of each kind. These kinds are twos, threes, fours, fives, sixes, sevens, eights, nines, tens, jacks, queens, kings, and aces. There are also four suits, spades, clubs, hearts, and diamonds, each containing 13 cards, with one card of each kind in a suit.

EXAMPLE 5 How many different hands of 5 cards from the deck of 52 are there?

Solution: There are $C(52, 5) = 2{,}598{,}960$ different hands with five cards. ■

EXAMPLE 6 Find the probability that a hand of five cards in poker contains four cards of one kind.

Solution: By the product rule, the number of hands of five cards with four cards of one kind is the product of the number of ways to pick one kind, the number of ways to pick the four of this kind out of the four in the deck of this kind, and the number of ways to pick the fifth card. This is

$$C(13, 1)C(4, 4)C(48, 1).$$

Since there is a total of $C(52, 5)$ different hands of five cards, the probability that a hand contains four cards of one kind is

$$\frac{C(13, 1)C(4, 4)C(48, 1)}{C(52, 5)} = \frac{13 \cdot 1 \cdot 48}{2{,}598{,}960} \sim 0.00024.$$ ■

EXAMPLE 7 What is the probability that a poker hand contains a full house, that is, three of one kind and two of another kind?

Solution: By the product rule, the number of hands containing a full house is the product of the number of ways to pick two kinds in order, the number of ways to pick three out of four for the first kind, and the number of ways to pick two out of four for the second kind. (Note that the order of the two kinds matters, since, for instance, three queens and two aces is different than three aces and two queens.) We see that the number of hands containing a full house is

$$P(13, 2)C(4, 3)C(4, 2) = 13 \cdot 12 \cdot 4 \cdot 6 = 3744.$$

Since there are 2,598,960 poker hands, the probability of a full house is

$$\frac{3744}{2{,}598{,}960} \sim 0.0014.$$

■

THE PROBABILITY OF COMBINATIONS OF EVENTS

We can use counting techniques to find the probability of events derived from other events.

THEOREM 1 Let E be an event in a sample space S. The probability of the event $\overline{E}$, the complementary event of E, is given by

$$p(\overline{E}) = 1 - p(E).$$

Proof: To find the probability of the event $\overline{E}$, note that $|\overline{E}| = |S| - |E|$. Hence,

$$p(\overline{E}) = \frac{|S| - |E|}{|S|} = 1 - \frac{|E|}{|S|} = 1 - p(E).$$

■

There is an alternate strategy for finding the probability of an event when a direct approach does not work well. Instead of determining the probability of the event, the probability of its complement can be found. This is often easier to do, as the following example shows.

EXAMPLE 8 A sequence of 10 bits is randomly generated. What is the probability that at least one of these bits is 0?

Solution: Let E be the event that at least one of the 10 bits is 0. Then $\overline{E}$ is the event that all the bits are 1s. Since the sample space S is the set of all bit strings of length 10, it follows that

$$\begin{aligned} p(E) &= 1 - p(\overline{E}) \\ &= 1 - \frac{|\overline{E}|}{|S|} \\ &= 1 - \frac{1}{2^{10}} \\ &= 1 - \frac{1}{1024} \\ &= \frac{1023}{1024}. \end{aligned}$$

Hence, the probability that the bit string will contain at least one 0 bit is 1023/1024. It is quite difficult to find this probability directly without using Theorem 1. ■

We can also find the probability of the union of two events.

THEOREM 2 Let E_1 and E_2 be events in the sample space S. Then

$$p(E_1 \cup E_2) = p(E_1) + p(E_2) - p(E_1 \cap E_2).$$

Proof: Using the formula given in Section 1.4 for the number of elements in the union of two sets, it follows that

$$|E_1 \cup E_2| = |E_1| + |E_2| - |E_1 \cap E_2|.$$

Hence,

$$\begin{aligned} p(E_1 \cup E_2) &= \frac{|E_1 \cup E_2|}{|S|} \\ &= \frac{|E_1| + |E_2| - |E_1 \cap E_2|}{|S|} \\ &= \frac{|E_1|}{|S|} + \frac{|E_2|}{|S|} - \frac{|E_1 \cap E_2|}{|S|} \\ &= p(E_1) + p(E_2) - p(E_1 \cap E_2). \end{aligned}$$

■

EXAMPLE 9 What is the probability that a positive integer selected at random from the set of positive integers not exceeding 100 is divisible by either 2 or 5?

Solution: Let E_1 be the event that the integer selected is divisible by 2, and let E_2 be the event that it is divisible by 5. Then $E_1 \cup E_2$ is the event that it is divisible by either 2 or 5. Also, $E_1 \cap E_2$ is the event that it is divisible by both 2 and 5, or equivalently, that it is divisible by 10. Since $|E_1| = 50$, $|E_2| = 20$, and $|E_1 \cap E_2| = 10$, it follows that

$$\begin{aligned} p(E_1 \cup E_2) &= p(E_1) + p(E_2) - p(E_1 \cap E_2) \\ &= \frac{50}{100} + \frac{20}{100} - \frac{10}{100} \\ &= \frac{3}{5}. \end{aligned}$$

■

Exercises

1. What is the probability that a card selected from a deck is an ace?

2. What is the probability that a die comes up 6 when it is rolled?

3. What is the probability that a randomly selected integer chosen from the first 100 positive integers is odd?

4. What is the probability that a randomly selected day of the year (from the 366 possible days) is in April?

5. What is the probability that the sum of the numbers on two dice is even when they are rolled?

6. What is the probability that a card selected from a deck is an ace or a heart?

7. What is the probability that a coin lands heads up six times in a row?

8. What is the probability that a five-card poker hand contains the ace of hearts?

9. What is the probability that a five-card poker hand does not contain the queen of hearts?

10. What is the probability that a five-card poker hand contains the two of diamonds and the three of spades?

11. What is the probability that a five-card poker hand contains the two of diamonds, the three of spades, the six of hearts, the ten of clubs, and the king of hearts?

12. What is the probability that a five-card poker hand contains exactly one ace?

13. What is the probability that a five-card poker hand contains at least one ace?

14. What is the probability that a five-card poker hand contains of five different kinds?

15. What is the probability that a five-card poker hand contains two pairs (that is, two of each of two different kinds and a fifth card of a third kind)?

16. What is the probability that a five-card poker hand contains a flush, that is, five cards of the same suit?

17. What is the probability that a five-card poker hand contains a straight, that is, five cards that have consecutive kinds? (Note that an ace can be considered either the lowest card of an A-2-3-4-5 straight or the highest card of a 10-J-Q-K-A straight.)

18. What is the probability that a five-card poker hand contains a straight flush, that is, five cards of the same suit of consecutive kinds?

★19. What is the probability that a five-card poker hand contains cards of five different kinds and does not contain a flush or a straight?

20. What is the probability that a five-card poker hand contains a royal flush, that is, the 10, jack, queen, king, and ace of one suit?

21. What is the probability that a die never comes up an even number when it is rolled six times?

22. What is the probability that a positive integer not exceeding 100 selected at random is divisible by 3?

23. What is the probability that a positive integer not exceeding 100 selected at random is divisible by 5 or 7?

24. Find the probability of winning the lottery by selecting the correct six integers from the positive integers not exceeding

a) 30. **b)** 36. **c)** 42. **d)** 48.

25. Find the probability of winning the lottery by selecting the correct six integers from the positive integers not exceeding

a) 50. **b)** 52. **c)** 56. **d)** 60.

26. To play the Pennsylvania superlottery, a player selects 7 numbers out of the first 80 positive integers. What is the probability that a person wins the grand prize by picking 7 numbers that are among the 11 numbers selected by the Pennsylvania lottery commission?

27. In a superlottery, players win a fortune if they choose the eight numbers selected by a computer from the positive integers not exceeding 100. What is the probability that a player wins this superlottery?

28. What is the probability that a player wins the prize offered for correctly choosing five (but not six) numbers out of six integers chosen between 1 and 40, inclusive, by a computer?

29. In roulette, a wheel with 38 numbers is spun. Of these, 18 are red, and 18 are black. The other 2 numbers, which are neither black nor red, are 0 and 00. The probability that when the wheel is spun it lands on any particular number is $1/38$.

a) What is the probability that the wheel lands on a red number?

b) What is the probability that the wheel lands on a black number twice in a row?

c) What is the probability that the wheel lands on 0 or 00?

d) What is the probability that the wheel does not land on 0 or 00 five times in a row?

e) What is the probability that the wheel lands on a number between 1 and 6, inclusive, on one spin, but does not land between them on the next spin?

30. Two events E_1 and E_2 are called **independent** if $p(E_1 \cap E_2) = p(E_1)p(E_2)$. For each of the following pairs of events, which are subsets of the set of all possible outcomes when a coin is tossed three times, determine whether or not they are independent.

a) E_1: the first coin comes up tails; E_2: the second coin comes up heads.

b) E_1: the first coin comes up tails; E_2: two, and not three, heads come up in a row.

c) E_1: the second coin comes up tails; E_2: two, and not three, heads come up in a row.

(We will study independence of events in more depth in Section 4.5.)

4.5 Probability Theory

INTRODUCTION

In Section 4.4 we introduced the notion of the probability of an event. (Recall that an event is a subset of the possible outcomes of an experiment.) We defined the probability of an event E as Laplace did, that is,

$$p(E) = \frac{|E|}{|S|},$$

the number of outcomes in E divided by the total number of outcomes. This definition assumes that all outcomes are equally likely. However, many experiments have outcomes that are not equally likely. For instance, a coin may be biased so that it comes up heads twice as often as tails. Similarly, the likelihood that the input of a linear search is a particular element in a list, or is not in the list, depends on how the input is generated. How can we model the likelihood of events in such situations? In this section we will show how to define probabilities of outcomes to study probabilities of experiments where outcomes may not be equally likely.

Suppose that a fair coin is flipped four times, and the first time it comes up heads. Given this information, what is the probability that heads comes up three times? To answer this and similar questions, we will introduce the concept of *conditional probability*. Does knowing that the first flip comes up heads change the probability that heads comes up three times? If not, these two events are called *independent*, a concept studied later in this section.

Many questions address a particular numerical value associated with the outcome of an experiment. For instance, when we flip a coin 100 times, what is the probability that exactly 40 heads appear? How many heads should we expect to appear? In this section we will study *random variables*, which are functions that associate numerical values to the outcomes of experiments, and their weighted averages, called *expected values*.

ASSIGNING PROBABILITIES

Suppose that the sample space S of an experiment consists of n possible outcomes, $x_1, x_2, \ldots, x_n$. We assign a probability $p(x_i)$ to each outcome x_i, $i = 1, 2, \ldots, n$. We require that two conditions be met:

$$(i) \quad 0 \leq p(x_i) \leq 1, \text{ for each } i$$

and

$$(ii) \quad \sum_{i=1}^{n} p(x_i) = 1.$$

Condition (i) states that the probability of each outcome is a nonnegative real number no greater than 1. Condition (ii) states that the sum of the probabilities of all possible outcomes should be 1; that is, when we do the experiment, it is a certainty that one of these outcomes occurs. This is a generalization of Laplace's definition in which each of the n outcomes is assigned probability $1/n$. Indeed, conditions (i) and (ii) are met when Laplace's definition of probabilities of equally likely outcomes is used. (See Exercise 4.)

To model an experiment, the probability $p(x)$ assigned to an outcome x should equal the limit of the number of times x occurs divided by the number of times the experiment is performed, as this number grows without bound. (We will assume that all experiments discussed have outcomes that are predictable on the average, so that this limit exists. We also assume that the outcomes of successive trials of an experiment do not depend on past results.)

Remark: In this section we will require that the number of possible outcomes be finite. A countably infinite number of outcomes can be handled similarly using infinite series, as illustrated in Exercises 35–37 at the end of this section. We will not discuss probabilities of events when the set of outcomes is not discrete, such as when the outcome of an experiment can be any real number. In such cases integral calculus is usually required for the study of the probabilities of events.

We can model experiments in which outcomes are either equally likely or not equally likely by choosing the appropriate function $p(x)$, as Example 1 illustrates.

EXAMPLE 1 What probabilities should we assign to the outcomes H (heads) and T (tails) when a fair coin is flipped? What probabilities should be assigned to these events when the coin is biased so that heads comes up twice as often as tails?

Solution: For a fair coin, the probability that heads comes up when the coin is flipped equals the probability that tails comes up, so that the events are equally likely. Consequently, we assign the probability $1/2$ to each of the two possible outcomes, that is, $p(H) = p(T) = 1/2$.

For the biased coin we have

$$p(H) = 2p(T).$$

Since

$$p(H) + p(T) = 1,$$

it follows that

$$2p(T) + p(T) = 3p(T) = 1.$$

We conclude that $p(T) = 1/3$ and $p(H) = 2/3$. ■

We now define the probability of an event as the sum of the probabilities of the outcomes in this event.

DEFINITION 1. The *probability* of the event E is the sum of the probabilities of the outcomes in E. That is, if $E = \{a_1, a_2, \ldots, a_m\}$, then

$$p(E) = \sum_{i=1}^{m} p(a_i).$$

EXAMPLE 2 Suppose that a die is biased (or loaded) so that 3 appears twice as often as each other number but that the other five outcomes are equally likely. What is the probability that an odd number appears when we roll this die?

Solution: We want to find the probability of the event $E = \{1, 3, 5\}$. By Exercise 2 at the end of this section, we have

$$p(1) = p(2) = p(4) = p(5) = p(6) = 1/7; p(3) = 2/7.$$

It follows that

$$p(E) = p(1) + p(3) + p(5) = 1/7 + 2/7 + 1/7 = 4/7.$$ ■

When events are equally likely, the definition of the probability of an event given in this section (Definition 1) agrees with Laplace's definition (Definition 1 of Section 4.4). To see this, suppose that there are n equally likely outcomes; each possible outcome has probability $1/n$, since the sum of their probabilities is 1. Suppose the event E contains m outcomes. According to Definition 1,

$$p(E) = \sum_{i=1}^{m} \frac{1}{n} = \frac{m}{n}.$$

Since $|E| = m$ and $|S| = n$, it follows that

$$p(E) = \frac{m}{n} = \frac{|E|}{|S|}.$$

This is Laplace's definition of the probability of the event E.

COMBINATIONS OF EVENTS

The formulae for probabilities of combinations of events in Section 4.4 continue to hold when we use Definition 1 to define the probability of an event. For example, Theorem 1 of Section 4.4 asserts that

$$p(\overline{E}) = 1 - p(E)$$

where $\overline{E}$ is the complementary event of the event E. This equality also holds when Definition 1 is used. To see this, note that since the sum of the probabilities of the n possible outcomes is 1, and each outcome is either in E or in $\overline{E}$, but not in both, we have

$$\sum_{i=1}^{n} p(x_i) = 1 = p(E) + p(\overline{E}).$$

Hence, $p(\overline{E}) = 1 - p(E)$.

Under Laplace's definition, by Theorem 2 in Section 4.4, we have

$$p(E_1 \cup E_2) = p(E_1) + p(E_2) - p(E_1 \cap E_2)$$

whenever E_1 and E_2 are events in a sample space S. This also holds when we define the probability of an event as we do in this section. To see this, note that $p(E_1 \cup E_2)$ is the sum of the probabilities of the outcomes in $E_1 \cup E_2$. When an outcome x is in one, but not both, of E_1 and E_2, $p(x)$ occurs in exactly one of the sums for $p(E_1)$ and $p(E_2)$. When an outcome x is in both E_1 and E_2, $p(x)$ occurs in the sum for $p(E_1)$, in the sum for $p(E_2)$, and in the sum for $p(E_1 \cap E_2)$, so that it occurs $1 + 1 - 1 = 1$ time on the right-hand side. Consequently, the left-hand side and right-hand side are equal.

CONDITIONAL PROBABILITY

Suppose that we flip a coin three times, and all eight possibilities are equally likely. Moreover, suppose we know that the event F, that the first flip comes up tails, occurs. Given this information, what is the probability of the event E, that an odd number of tails appears? Since the first flip comes up tails, there are only four possible outcomes: TTT, TTH, THT, and THH, where H and T represent heads and tails, respectively. An odd number of tails appears only for the outcomes TTT and THH. Since the eight outcomes have equal probability, each of the four possible outcomes, given that F occurs, should also have an equal probability of $1/4$. This suggests that we should assign the probability of $2/4 = 1/2$ to E, given that F occurs. This probability is called the **conditional probability** of E given F.

In general, to find the conditional probability of E given F, we use F as the sample space. For an outcome from E to occur, this outcome must also belong to $E \cap F$. With this motivation, we make the following definition.

DEFINITION 2. Let E and F be events with $p(F) > 0$. The *conditional probability* of E given F, denoted by $p(E \mid F)$, is defined as

$$p(E \mid F) = \frac{p(E \cap F)}{p(F)}.$$

EXAMPLE 3 What is the probability that a bit string of length four, generated at random so that each of the 16 bit strings of length four is equally likely, contains at least two consecutive 0s, given that its first bit is a 0? (We assume that 0 bits and 1 bits are equally likely.)

Solution: Let E be the event that a bit string of length four contains at least two consecutive 0s, and let F be the event that the first bit of a bit string of length four is a 0. The probability that a bit string of length four has at least two consecutive 0s, given that its first bit is a 0, equals

$$p(E \mid F) = \frac{p(E \cap F)}{p(F)}.$$

Since $E \cap F = \{0000, 0001, 0010, 0011, 0100\}$, we see that $p(E \cap F) = 5/16$. Since there are eight bit strings of length four that start with a 0, we have $p(F) = 8/16 = 1/2$. Consequently,

$$p(E \mid F) = \frac{5/16}{1/2} = \frac{5}{8}.$$

■

EXAMPLE 4 What is the conditional probability that a family with two children has two boys, given they have at least one boy? Assume that each of the possibilities *BB*, *BG*, *GB*, and *GG* is equally likely, where *B* represents a boy and *G* represents a girl.

Solution: Let E be the event that a family with two children has two boys, and let F be the event that a family with two children has at least one boy. It follows that $E = \{BB\}$, $F = \{BB, BG, GB\}$, and $E \cap F = \{BB\}$. Since the four possibilities are equally likely, it follows that $p(F) = 3/4$ and $p(E \cap F) = 1/4$. We conclude that

$$p(E \mid F) = \frac{p(E \cap F)}{p(F)} = \frac{1/4}{3/4} = \frac{1}{3}.$$

■

INDEPENDENCE

Suppose a coin is flipped four times, as described in the introduction to our discussion of conditional probability. Does knowing that the first flip comes up tails, (event F), alter the probability that tails comes up an odd number of times (event E)? In other words, is it the case that $p(E \mid F) = p(E)$? This equality is valid for the events E and F, since $p(E \mid F) = 1/2$ and $p(E) = 1/2$. Because this equality holds, we say that E and F are **independent events.**

Since $p(E \mid F) = p(E \cap F)/p(F)$, asking whether $p(E \mid F) = p(E)$ is the same as asking whether $p(E \cap F) = p(E)p(F)$. This leads to the following definition.

DEFINITION 3. The events E and F are said to be *independent* if and only if

$$p(E \cap F) = p(E)p(F).$$

EXAMPLE 5 Suppose E is the event that a randomly generated bit string of length four begins with a 1 and F is the event that a randomly generated bit string contains an even number of 0s. Are E and F independent, if the 16 bit strings of length four are equally likely?

Solution: There are eight bit strings of length four that begin with a 1: 1000, 1001, 1010, 1011, 1100, 1101, 1110, and 1111. There are also eight bit strings of length four that contain an even number of 1s: 0000, 0011, 0101, 0110, 1001, 1010, 1100, 1111. Since there are 16 bit strings of length four, it follows that

$$p(E) = p(F) = 8/16 = 1/2.$$

Because $E \cap F = \{1111, 1100, 1010, 1001\}$, we see that

$$p(E \cap F) = 4/16 = 1/4.$$

Since

$$p(E \cap F) = 1/4 = (1/2)(1/2) = p(E)p(F),$$

we conclude that E and F are independent. ■

EXAMPLE 6 Assume, as in Example 4, that each of the four ways a family can have two children is equally likely. Are the events E, that a family with two children has two boys, and F, that a family with two children has at least one boy, independent?

Solution: Since $E = \{BB\}$, we have $p(E) = 1/4$. In Example 4 we showed that $p(F) = 3/4$ and that $p(E \cap F) = 1/4$. Since $p(E \cap F) = 1/4 \neq 3/16 = (1/4)(3/4) = p(E)p(F)$, events E and F are not independent. ■

EXAMPLE 7 Are the events E, that a family with three children has children of both sexes, and F, that a family with three children has at most one boy, independent? Assume that the eight ways a family can have three children are equally likely.

Solution: By assumption, each of the eight ways a family can have three children, BBB, BBG, BGB, BGG, GBB, GBG, GGB, and GGG, has a probability of $1/8$. Since $E = \{BBG, BGB, BGG, GBB, GBG, GGB\}$, $F = \{BGG, GBG, GGB, GGG\}$, and $E \cap F = \{BGG, GBG, GGB\}$, it follows that $p(E) = 6/8 = 3/4$, $p(F) = 4/8 = 1/2$, and $p(E \cap F) = 3/8$. Since

$$p(E \cap F) = \frac{3}{8} = \frac{3}{4} \cdot \frac{1}{2} = p(E)p(F),$$

we conclude that E and F are independent. (This conclusion may seem surprising. Indeed, if we change the number of children, the conclusion may no longer hold. See Exercise 17 at the end of this section.) ■

BERNOULLI TRIALS AND THE BINOMIAL DISTRIBUTION

Suppose that an experiment can have only two possible outcomes. For instance, when a bit is generated at random, the possible outcomes are 0 and 1. When a coin is flipped, the possible outcomes are heads and tails. Each performance of an experiment with two possible outcomes is called a **Bernoulli trial,** after James Bernoulli, who made important contributions to probability theory. In general, a possible outcome of a Bernoulli trial is called a **success** or a **failure.** If p is the probability of a success and q is the probability of a failure, it follows that $p + q = 1$.

Many problems can be solved by determining the probability of k successes when an experiment consists of n independent Bernoulli trials. Consider the following example.

EXAMPLE 8 A coin is biased so that the probability of heads is $2/3$. What is the probability that exactly four heads come up when the coin is flipped seven times, assuming that the flips are independent?

Solution: There are $2^7 = 128$ possible outcomes when a coin is flipped seven times. The number of ways four of the seven flips can be heads is $C(7, 4)$. Since the seven flips are independent, the probability of each of these outcomes is $(2/3)^4(1/3)^3$. Consequently, the probability that exactly four heads appear is

$$
\begin{aligned}
C(7,4)(2/3)^4(1/3)^3 &= \frac{35 \cdot 16}{3^7} \\
&= \frac{560}{2187}.
\end{aligned}
$$

■

Following the same reasoning as used in Example 8, we can establish the following theorem, which tells us the probability of k successes in n independent Bernoulli trials.

James Bernoulli (1654–1705). James Bernoulli (also known as Jacob I), was born in Basel, Switzerland. He is one of the eight prominent mathematicians in the Bernoulli family (see Section 8.1 for the Bernoulli family tree of mathematicians). Following his father's wish, James studied theology and entered the ministry. But contrary to the desires of his parents, he also studied mathematics and astronomy. He traveled throughout Europe from 1676 to 1682, learning about the latest discoveries in mathematics and the sciences. Upon returning to Basel in 1682, he founded a school for mathematics and the sciences. He was appointed professor of mathematics at the University of Basel in 1687, remaining in this position for the rest of his life.

James Bernoulli is best known for the work the *Ars Conectandi*, published eight years after his death. In this work, he described the known results in probability theory and in enumeration, often providing alternative proofs of known results. This work also includes the application of probability theory to games of chance and his introduction of the theorem known as the **law of large numbers**. This law states that if $\epsilon > 0$, as n becomes arbitrarily large the probability approaches 1 that the number of times an event E occurs during n trials is within ϵ of $p(E)$.

THEOREM 1 **PROBABILITY OF k SUCCESSES IN n INDEPENDENT BERNOULLI TRIALS** The probability of k successes in n independent Bernoulli trials, with probability of success p and probability of failure $q = 1 - p$, is

$$C(n, k)p^k q^{n-k}.$$

Proof: When n Bernoulli trials are carried out, the outcome is an n-tuple $(t_1, t_2, \ldots, t_n)$, where $t_i = S$ (for success) or $t_i = F$ (for failure) for $i = 1, 2, \ldots, n$. Since the n trials are independent, the probability of each outcome of n trials consisting of k successes and $n - k$ failures (in any order) is $p^k q^{n-k}$. Since there are $C(n, k)$ n-tuples of Ss and Fs that contain k Ss, the probability of k successes is

$$C(n, k)p^k q^{n-k}.$$

■

We denote by $b(k; n, p)$ the probability of k successes in n independent Bernoulli trials with probability of success p and probability of failure $q = 1 - p$. Considered as a function of k, we call this function the **binomial distribution.** Theorem 1 tells us that $b(k; n, p) = C(n, k)p^k q^{n-k}$.

EXAMPLE 9 What is the probability that exactly eight 0 bits are generated when 10 bits are generated with the probability that a 0 bit is generated is 0.9, the probability that a 1 bit is generated is 0.1, and the bits are generated independently?

Solution: By Theorem 1, the probability that exactly eight 0 bits are generated is

$$b(8; 10, 0.9) = C(10, 8)(0.9)^8(0.1)^2 = 0.1937102445.$$

■

Note that the sum of the probabilities that there are k successes when n independent Bernoulli trials are carried out, for $k = 0, 1, 2, \ldots, n$, equals

$$\begin{aligned}\sum_{k=0}^{n} C(n, k)p^k q^{n-k} &= (p + q)^n \\ &= 1,\end{aligned}$$

as should be the case. The first equality in this string of equalities is a consequence of the binomial theorem. The second equality follows since $q = 1 - p$.

RANDOM VARIABLES

Many problems are concerned with a numerical value associated with the outcome of an experiment. For instance, we may want to know the probability that there are nine 1 bits generated when 10 bits are randomly generated, or we may want to know the probability that a coin comes up tails 11 times when it is flipped 20 times. To study problems of this type we introduce the concept of a random variable.

DEFINITION 4. A **random variable** is a function from the sample space of an experiment to the set of real numbers. That is, a random variable assigns a real number to each possible outcome.

Remark: Note that a random variable is a function. It is not a variable, and it is not random!

EXAMPLE 10 Suppose that a coin is flipped three times. Let $X(t)$ be the number of heads that appear when t is the outcome. Then the random variable $X(t)$ takes on the following values:

$$\begin{aligned}
X(HHH) &= 3,\\
X(HHT) &= X(HTH) = X(THH) = 2,\\
X(TTH) &= X(THT) = X(HTT) = 1,\\
X(TTT) &= 0.
\end{aligned}$$

■

EXAMPLE 11 Let X be the sum of the numbers that appear when a pair of dice is rolled. What are the values of this random variable for the 36 possible outcomes (i,j), where i and j are the numbers that appear on the first die and the second die, respectively, when these two dice are rolled?

Solution: The random variable X takes on the following values:

$$\begin{aligned}
X((1,1)) &= 2,\\
X((1,2)) &= X((2,1)) = 3,\\
X((1,3)) &= X((2,2)) = X((3,1)) = 4,\\
X((1,4)) &= X((2,3)) = X((3,2)) = X((4,1)) = 5,\\
X((1,5)) &= X((2,4)) = X((3,3)) = X((4,2)) = X((5,1)) = 6,\\
X((1,6)) &= X((2,5)) = X((3,4)) = X((4,3)) = X((5,2)) = X((6,1)) = 7,\\
X((2,6)) &= X((3,5)) = X((4,4)) = X((5,3)) = X((6,2)) = 8,\\
X((3,6)) &= X((4,5)) = X((5,4)) = X((6,3)) = 9,\\
X((4,6)) &= X((5,5)) = X((6,4)) = 10,\\
X((5,6)) &= X((6,5)) = 11,\\
X((6,6)) &= 12.
\end{aligned}$$

■

EXPECTED VALUES

Many questions can be formulated in terms of the value we expect a random variable to take, or more precisely, the average value of a random variable when an experiment is performed a large number of times. Questions of this kind include: How many heads are expected to appear when a coin is flipped 100 times? What is the expected number of comparisons used to find an element in a list using a linear search? To study such questions we introduce the concept of the expected value of a random variable.

DEFINITION 5. The *expected value* (or *expectation*) of the random variable $X(s)$ on the sample space $S = \{s_1, s_2, \ldots, s_n\}$ is equal to

$$E(X) = \sum_{i=1}^{n} p(s_i)X(s_i).$$

EXAMPLE 12 A fair coin is flipped three times. Let S be the sample space of the eight possible outcomes, and let X be the random variable that assigns to an outcome the number of heads in this outcome. What is the expected value of X?

Solution: In Example 10 we listed the values of X for the eight possible outcomes when a coin is flipped three times. Since the coin is fair and the flips are independent, the probability of each outcome is $1/8$. Consequently,

$$\begin{aligned} E(X) &= \frac{1}{8}(X(HHH) + X(HHT) + X(HTH) + X(THH) + X(TTH) \\ &\quad + X(THT) + X(HTT) + X(TTT)) \\ &= \frac{1}{8}(3 + 2 + 2 + 2 + 1 + 1 + 1 + 0) \\ &= \frac{12}{8} \\ &= \frac{3}{2}. \end{aligned}$$

■

When an experiment has relatively few outcomes, we can compute the expected value of a random variable directly from its definition, as was done in Example 12. However, when an experiment has a large number of outcomes, it may be inconvenient to compute the expected value of a random variable directly from its definition. Instead, we can find the expected value of a random variable by grouping together all outcomes assigned the same value by the random variable. In particular, suppose that X is a random variable with range $\{r_1, r_2, \ldots, r_m\}$, and let $p(X = r_j)$ be the probability that the random variable X takes the value r_j. Consequently, $p(X = r_j)$ is the sum of the probabilities of the outcomes x_i such that $X(x_i) = r_j$. It follows that

$$E(X) = \sum_{j=1}^{m} p(X = r_j)r_j.$$

Examples 13 and 14 illustrate the use of this formula. In Example 13 we will find the expected value of the sum of the numbers that appear on two fair dice when they are rolled. In Example 14 we will find the expected value of the number of successes when n Bernoulli trials are performed.

EXAMPLE 13 What is the expected value of the sum of the numbers that appear when a pair of fair dice is rolled?

Solution: Let X be the random variable equal to the sum of the numbers that appear when a pair of dice is rolled. In Example 11 we listed the value of X for the 36 outcomes of this experiment. The range of X is $\{2, 3, 4, 5, 6, 7, 8, 9, 10, 11, 12\}$. By Example 11 we see that

$$\begin{aligned}
p(X = 2) &= p(X = 12) = 1/36, \\
p(X = 3) &= p(X = 11) = 2/36 = 1/18, \\
p(X = 4) &= p(X = 10) = 3/36 = 1/12, \\
p(X = 5) &= p(X = 9) = 4/36 = 1/9, \\
p(X = 6) &= p(X = 8) = 5/36, \\
p(X = 7) &= 6/36 = 1/6.
\end{aligned}$$

Substituting these values in the formula, we have

$$\begin{aligned}
E(X) &= 2 \cdot \frac{1}{36} + 3 \cdot \frac{1}{18} + 4 \cdot \frac{1}{12} + 5 \cdot \frac{1}{9} + 6 \cdot \frac{5}{36} + 7 \cdot \frac{1}{6} \\
&\quad + 8 \cdot \frac{5}{36} + 9 \cdot \frac{1}{9} + 10 \cdot \frac{1}{12} + 11 \cdot \frac{1}{18} \cdot 12 \cdot \frac{1}{36} \\
&= 7.
\end{aligned}$$

■

EXAMPLE 14 What is the expected value of the number of successes when n Bernoulli trials are performed, where p is the probability of success on each trial?

Solution: Let X be the random variable equal to the number of successes in n trials. By Theorem 1 we see that $p(X = k) = C(n, k)p^k q^{n-k}$. Hence, from the formula for the expected value of a random variable that groups together outcomes assigned the same value by the random variable, we have

$$\begin{aligned}
E(X) &= \sum_{k=1}^{n} k\, p(X = k) \\
&= \sum_{k=1}^{n} kC(n, k)p^k q^{n-k} \\
&= \sum_{k=1}^{n} nC(n-1, k-1)p^k q^{n-k} \\
&= np \sum_{k=1}^{n} C(n-1, k-1)p^{k-1} q^{n-k} \\
&= np \sum_{j=0}^{n-1} C(n-1, j)p^j q^{n-1-j} \\
&= np(p + q)^{n-1} \\
&= np.
\end{aligned}$$

The third equality follows from the fact that $C(n, k) = nC(n-1, k-1)/k$, which follows from Exercise 33 of Section 4.3. The fifth equality is obtained by shifting the index of summation with $j = k - 1$ so that j ranges from 0 to $n - 1$ when k ranges from 1 to n. The sixth equality follows from the Binomial Theorem. The seventh equality follows since $p + q = 1$. From this computation, we conclude that the expected value of X equals np. This means that the expected number of successes in n Bernoulli trials is np. ■

AVERAGE-CASE COMPUTATIONAL COMPLEXITY

Computing the average-case computational complexity of an algorithm can be interpreted as computing the expected value of a random variable. Let the sample space of an experiment be the set of possible inputs a_j, $j = 1, 2, \ldots, n$, and let the random variable X assign to a_j the number of operations used by the algorithm when given a_j as input. Based on our knowledge of the input, we assign a probability $p(a_j)$ to each possible input value a_j. Then, the average-case complexity of the algorithm is

$$E(X) = \sum_{j=1}^{n} p(a_j)X(a_j).$$

This is the expected value of X.

In Example 15 we will illustrate how to find the average-case computational complexity of the linear search algorithm under different assumptions concerning the probability that the element for which we search is an element of the list.

EXAMPLE 15 Average-Case Computational Complexity of the Linear Search Algorithm. We are given an element x and a list of n distinct real numbers. The linear search algorithm, described in Section 2.1, locates x by successively comparing it to each element in the list, terminating when x is located or when all the elements have been examined and it has been determined that x is not in the list. What is the average-case computational complexity of the linear search algorithm if the probability that x is in the list is p and it is equally likely that x is any of the n elements in the list? (There are $n + 1$ possible types of input: the n numbers in the list and a number not in the list, which we treat as a single input.)

Solution: In Example 4 of Section 2.2 we showed that $2i + 1$ comparisons are used if x equals the ith element of the list and, in Example 2 of Section 2.2, that $2n + 2$ comparisons are used if x is not in the list. The probability that x equals a_i, the ith element in the list, is p/n, and the probability that x is not in the list is $q = 1 - p$. It follows that the average-case computational complexity of the linear search algorithm is

$$\begin{aligned} E &= 3p/n + 5p/n + \cdots + (2n + 1)p/n + (2n + 2)q \\ &= \frac{p}{n}(3 + 5 + \cdots + (2n + 1)) + (2n + 2)q \\ &= \frac{p}{n}((n + 1)^2 - 1) + (2n + 2)q \\ &= p(n + 2) + (2n + 2)q. \end{aligned}$$

(The third equality follows from Example 2 of Section 3.2.) For instance, when x is guaranteed to be in the list, we have $p = 1$ (so that the probability that $x = a_i$ is $1/n$ for each i) and $q = 0$. Then $E = n + 2$, as we showed in Example 4 in Section 2.2.

When p, the probability that x is in the list, is 1/2, it follows that $q = 1 - p = 1/2$, so that $E = (n+2)/2 + n + 1 = (3n+4)/2$. Similarly, if the probability that x is in the list is 3/4, we have $p = 3/4$ and $q = 1/4$, so that $E = 3(n+2)/4 + (n+1)/2 = (5n+8)/4$.

Finally, when x is guaranteed not to be in the list, we have $p = 0$ and $q = 1$. It follows that $E = 2n + 2$, which is not surprising since we have to search the entire list. ■

Exercises

1. What probability should be assigned to the outcome of heads when a biased coin is tossed, if heads is three times as likely to come up as tails? What probability should be assigned to the outcome of tails?

2. Find the probability of each outcome when a loaded die is rolled, if a 3 is twice as likely to appear as each of the other five numbers on the die.

3. Find the probability of each outcome when a biased die is rolled, if rolling a 2 or rolling a 4 is three times as likely to appear as rolling each of the other four numbers on the die.

4. Show that conditions (i) and (ii) are met under Laplace's definition of probability, when outcomes are equally likely.

5. A pair of dice is loaded. The probability that a 4 appears on the first die is $2/7$, and the probability that a 3 appears on the second die is $2/7$. Other outcomes for each die appear with probability $1/7$. What is the probability of 7 appearing as the sum of the numbers when the two dice are rolled?

6. Suppose that E and F are events such that $p(E) = 0.8$ and $p(F) = 0.6$. Show that $p(E \cap F) \geq 0.4$.

7. Show that if E and F are events, then $p(E \cap F) \geq p(E) + p(F) - 1$. This is known as **Bonferroni's inequality.**

8. Use mathematical induction to prove the following generalization of Bonferroni's inequality:

$$p(E_1 \cap E_2 \cap \cdots \cap E_n) \geq p(E_1) + p(E_2) + \cdots + p(E_n) - (n-1),$$

where $E_1, E_2, \ldots, E_n$ are n events.

9. Show that if $E_1, E_2, \ldots, E_n$ are events from a finite sample space, then

$$p(E_1 \cup E_2 \cup \cdots \cup E_n) \leq p(E_1) + p(E_2) + \cdots + p(E_n).$$

This is known as **Boole's inequality.**

Exercises 10–12 involve the famous problem about the probability that at least two people in a group have the same birthday.

10. What is the probability that two people have the same birthday? For this problem assume that all the 366 dates are equally likely as birthdays.

★11. **a)** What is the probability that in a group of n people, there are at least two with the same birthday? For this problem assume that all of the 366 dates are equally likely as birthdays. (*Hint:* Find the probability that in a group of n people, the birthdays of all the people are different.)

b) How many people are needed to make the probability greater than $1/2$ that at least two people have the same birthday?

★12. February 29 occurs only in leap years. Years divisible by 4, but not by 100, are always leap years. Years divisible by 100, but not by 400, are not leap years, but years divisible by 400 are leap years.

a) What probability distribution for birthdays should be used to reflect how often February 29 occurs?

b) Answer the question asked in part (a) of Exercise 11 using this probability distribution.

13. What is the conditional probability that exactly four heads appear when a fair coin is flipped five times, given that the first flip came up heads?

14. What is the conditional probability that exactly four heads appear when a fair coin is flipped five times, given that the first flip came up tails?

15. What is the conditional probability that a randomly generated bit string of length four contains at least two consecutive 0s, given that the first bit is a 1? (Assume the probabilities of a 0 and a 1 are the same.)

16. Let E be the event that a randomly generated bit string of length three contains an odd number of 1s, and let F be the event that the string starts with 1. Are E and F independent?

17. Let E and F be the events that a family of n children has children of both sexes and has at most one boy, respectively. Are E and F independent if

a) $n = 2$? **b)** $n = 4$? **c)** $n = 5$?

18. Assume that the probability a child is a boy is 0.51 and that the sexes of children born into a family are independent. What is the probability that a family of five children has

a) exactly three boys?
b) at least one boy?
c) at least one girl?
d) all children of the same sex?

19. A group of six people play the game of "odd person out" to determine who will buy refreshments. Each person flips a fair coin. If there is a person whose outcome is not the same as that of any other member of the group, this person has to buy the refreshments. What is the probability that there is an odd person out after the coins are flipped once?

20. Find the probability that a randomly generated bit string of length 10 does not contain a 0 if bits are independent and if

a) a 0 bit and a 1 bit are equally likely.
b) the probability that a bit is a 1 is 0.6.
c) the probability that the ith bit is a 1 is $1/2^i$ for $i = 1, 2, 3, \ldots, 10$.

21. Find the probability that a family with five children does not have a boy, if the sexes of children are independent and if

a) a boy and a girl are equally likely.
b) the probability of a boy is 0.51.
c) the probability that the ith child is a boy is $0.51 - (i/100)$.

22. Find the probability that a randomly generated bit string of length 10 begins with a 1 or ends with a 00 for the same conditions as in parts (a), (b), and (c) of Exercise 20, if bits are generated independently.

23. Find the probability that the first child of a family with five children is a boy or that the last two children of the family are girls, for the same conditions as in parts (a), (b), and (c) of Exercise 21.

24. Find each of the following probabilities when n independent Bernoulli trials are carried out with probability of success p.

a) the probability of no successes
b) the probability of at least one success
c) the probability of at most one success
d) the probability of at least two successes

25. Find each of the following probabilities when n independent Bernoulli trials are carried out with probability of success p.

a) the probability of no failures
b) the probability of at least one failure
c) the probability of at most one failure
d) the probability of at least two failures

26. What is the expected number of heads that come up when a fair coin is flipped 10 times?

27. What is the expected number of times a 6 appears when a fair die is rolled 10 times?

28. A coin is biased so that the probability a head comes up when it is flipped is 0.6. What is the expected number of heads that come up when it is flipped 10 times?

29. What is the expected sum of the numbers that appear on two dice, each biased so that a 3 comes up twice as often as each other number?

30. Show that the expected value of the sum of two random variables is the sum of their expected values. That is, show that if X and Y are random variables defined on a sample space S, then $E(X + Y) = E(X) + E(Y)$.

31. The final exam of a discrete mathematics course consists of 50 true/false questions, each worth 2 points, and 25 multiple-choice questions, each worth 4 points. The probability that Linda answers a true/false question correctly is 0.9, and the probability that she answers a multiple-choice question correctly is 0.8. What is her expected score on the final?

32. What is the expected sum of the numbers that appear when three fair dice are rolled?

33. Suppose that the probability that x is in a list of n distinct integers is $2/3$ and that it is equally likely that x equals any element in the list. Find the average number of comparisons used by the linear search algorithm to find x or to determine that it is not in the list.

★**34.** Suppose that the probability that x is the ith element in a list of n distinct integers is $i/[n(n+1)]$. Find the average number of comparisons used by the linear search algorithm to find x or to determine that it is not in the list.

In this section we have studied experiments with finitely many outcomes. In Exercises 35–37 we study an experiment with countably many outcomes. A coin is flipped until it comes up tails. The sample space of the experiment is $\{T, HT, HHT, HHHT, HHHHT, \ldots\}$. The probability the coin comes up tails is p.

35. What is the probability that the experiment ends after n flips, that is, the outcome consists of $n-1$ heads and a tail?

36. Show that the sum of the probabilities of the possible outcomes equals 1.

37. What is the probability that at most n flips are required for the experiment to end?

4.6 Generalized Permutations and Combinations

INTRODUCTION

In many counting problems, elements may be used repeatedly. For instance, a letter or digit may be used more than once on a license plate. When a dozen donuts are selected, each variety can be chosen repeatedly.

Also, some counting problems involve elements that are indistinguishable. For instance, to count the number of ways the letters of the word *SUCCESS* can be rearranged, the placement of identical letters must be considered. Other counting problems involve counting the ways to place distinguishable elements in boxes, such as dealing poker hands to four players.

In this section we will develop techniques to solve counting problems such as those just described.

PERMUTATIONS WITH REPETITION

Consider the following example of a counting problem when repetition is allowed.

EXAMPLE 1 How many strings of length n can be formed from the English alphabet?

Solution: By the product rule, since there are 26 letters, and since each letter can be used repeatedly, we see that there are 26^n strings of length n. ■

The following question involving probability also involves permutations with repetition.

EXAMPLE 2 What is the probability of drawing three red balls in a row from an urn containing five red balls and seven blue balls if a ball is put back into the urn after it is drawn?

Solution: By the product rule the number of successful outcomes—that is, the number of ways to draw three red balls—is 5^3, since for each drawing there are five red balls in the urn. The total number of outcomes is 12^3, since for each drawing there are 12 balls in the urn. Thus, the desired probability is $5^3/12^3 = 125/1728$. This is an example of **sampling with replacement.** ■

The number of r-permutations of a set with n elements when repetition is allowed is given in the following theorem.

THEOREM 1 The number of r-permutations of a set of n objects with repetition allowed is n^r.

Proof: There are n ways to select an element of the set for each of the r positions in the r-permutation when repetition is allowed, since for each choice all n objects are available. Hence, by the product rule there are n^r r-permutations when repetition is allowed. ■

COMBINATIONS WITH REPETITION

Consider the following examples of combinations with repetition of elements allowed.

EXAMPLE 3 How many ways are there to select four pieces of fruit from a bowl containing apples, oranges, and pears if the order in which the pieces are selected does not matter, only the type of fruit and not the individual piece matters, and there are at least four pieces of each type of fruit in the bowl?

Solution: To solve this problem we list all the ways possible to select the fruit. There are 15 ways:

4 apples	4 oranges	4 pears
3 apples, 1 orange	3 apples, 1 pear	3 oranges, 1 apple
3 oranges, 1 pear	3 pears, 1 apple	3 pears, 1 orange
2 apples, 2 oranges	2 apples, 2 pears	2 oranges, 2 pears
2 apples, 1 orange, 1 pear	2 oranges, 1 apple, 1 pear	2 pears, 1 apple, 1 orange

The solution is the number of 4-combinations with repetition allowed from a three-element set, $\{apple, orange, pear\}$. ■

To solve more complex counting problems of this type, we need a general method for counting the r-combinations of an n-element set. In Example 4 we will illustrate such a method.

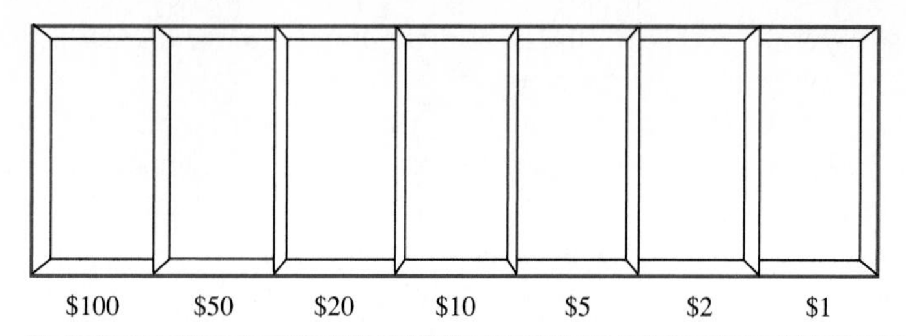

FIGURE 1 **Cash Box with Seven Types of Bills.**

EXAMPLE 4 How many ways are there to select five bills from a money bag containing \$1 bills, \$2 bills, \$5 bills, \$10 bills, \$20 bills, \$50 bills, and \$100 bills? Assume that the order in which the bills are chosen does not matter, that the bills of each denomination are indistinguishable, and that there are at least five bills of each type.

Solution: Since the order in which the bills are selected does not matter and seven different types of bills can be selected as many as five times, this problem involves counting 5-combinations with repetition allowed from a set with seven elements. Listing all possibilities would be tedious, since there are a large number of solutions. Instead, we will illustrate the use of a technique for counting combinations with repetition allowed.

Suppose that a cash box has seven compartments, one to hold each type of bill, as illustrated in Figure 1. These bins are separated by six dividers, as shown in the picture. The choice of five bills corresponds to placing five markers in the compartments holding different types of bills. Figure 2 illustrates this correspondence

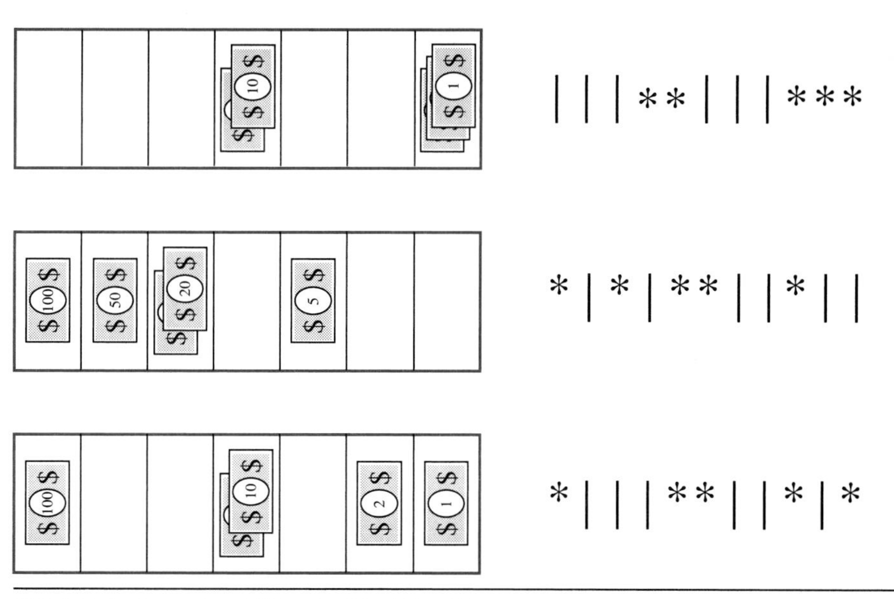

FIGURE 2 **Examples of Ways to Select Five Bills.**

for three different ways to select five bills, where the six dividers are represented by bars and the five bills by stars.

The number of ways to select five bills corresponds to the number of ways to arrange six bars and five stars. Consequently, the number of ways to select the five bills is the number of ways to select the positions of the five stars, from 11 possible positions. This corresponds to the number of unordered selections of 5 objects from a set of 11 objects, which can be done in $C(11, 5)$ ways. Consequently, there are

$$C(11, 5) = \frac{11!}{5!\,6!} = 462$$

ways to choose five bills from the bag with seven types of bills. ■

The following theorem generalizes this discussion.

THEOREM 2 There are $C(n + r - 1, r)$ r-combinations from a set with n elements when repetition of elements is allowed.

Proof: Each r-combination of a set with n elements when repetition is allowed can be represented by a list of $n - 1$ bars and r stars. The $n - 1$ bars are used to mark off n different cells, with the ith cell containing a star for each time the ith element of the set occurs in the combination. For instance, a 6-combination of a set with four elements is represented with three bars and six stars. Here

$$** \mid * \mid \mid ***$$

represents the combination containing exactly two of the first element, one of the second element, none of the third element, and three of the fourth element of the set.

As we have seen, each different list containing $n - 1$ bars and r stars corresponds to an r-combination of the set with n elements, when repetition is allowed. The number of such lists is $C(n - 1 + r, r)$, since each list corresponds to a choice of the r positions to place the r stars from the $n - 1 + r$ positions that contain r stars and $n - 1$ bars. ■

The following examples show how Theorem 2 is applied.

EXAMPLE 5 Suppose that a cookie shop has four different kinds of cookies. How many different ways can six cookies be chosen? Assume that only the type of cookie, and not the individual cookies or the order in which they are chosen, matters.

Solution: The number of ways to choose six cookies is the number of 6-combinations of a set with four elements. From Theorem 2 this equals $C(4 + 6 - 1, 6) = C(9, 6)$. Since

$$C(9, 6) = C(9, 3) = \frac{9 \cdot 8 \cdot 7}{1 \cdot 2 \cdot 3} = 84,$$

there are 84 different ways to choose the six cookies. ■

Theorem 2 can also be used to find the number of solutions of certain linear equations where the variables are integers subject to constraints. This is illustrated by the following example.

EXAMPLE 6 How many solutions does the equation

$$x_1 + x_2 + x_3 = 11$$

have, where x_1, x_2, and x_3 are nonnegative integers?

Solution: To count the number of solutions, we note that a solution corresponds to a way of selecting 11 items from a set with three elements, so that x_1 items of type 1, x_2 items of type 2, and x_3 items of type 3 are chosen. Hence, the number of solutions is equal to the number of 11-combinations with repetition allowed from a set with three elements. From Theorem 2 it follows that there are

$$C(3 + 11 - 1, 11) = C(13, 11) = C(13, 2) = \frac{13 \cdot 12}{1 \cdot 2} = 78$$

solutions.

The number of solutions of this equation can also be found when the variables are subject to constraints. For instance, we can find the number of solutions where the variables are integers with $x_1 \geq 1$, $x_2 \geq 2$, and $x_3 \geq 3$. A solution to the equation subject to these constraints corresponds to a selection of 11 items with x_1 items of type 1, x_2 items of type 2, and x_3 items of type 3, where, in addition, there is at least one item of type 1, two items of type 2, and three items of type 3. So, choose one item of type 1, two of type 2, and three of type 3. Then select five additional items. By Theorem 2 this can be done in

$$C(3 + 5 - 1, 5) = C(7, 5) = C(7, 2) = \frac{7 \cdot 6}{1 \cdot 2} = 21$$

ways. Thus, there are 21 solutions of the equation subject to the given constraints. ■

The following example shows how counting the number of combinations with repetition allowed arises in determining the value of a variable that is incremented each time a certain type of nested loop is traversed.

EXAMPLE 7 What is the value of k after the following pseudocode has been executed?

```
k := 0
for i_1 := 1 to n
  for i_2 := 1 to i_1
      .
      .
      .
    for i_m := 1 to i_{m-1}
      k := k + 1
```

Solution: Note that the initial value of k is 0 and that 1 is added to k each time the nested loop is traversed with a set of integers $i_1, i_2, \ldots, i_m$ such that

$$1 \leq i_m \leq i_{m-1} \leq \cdots \leq i_1 \leq n.$$

The number of such sets of integers is the number of ways to choose m integers from $\{1, 2, \ldots, n\}$, with repetition allowed. (To see this, note that once such a set has been selected, if we order the integers in the set in nondecreasing order, this uniquely defines an assignment of $i_m, i_{m-1}, \ldots, i_1$. Conversely, every such assignment corresponds to a unique unordered set.) Hence, from Theorem 2, it follows that $k = C(n+m-1, m)$ after this code has been executed. ■

The formulae for the numbers of ordered and unordered selections of r elements, chosen with and without repetition allowed from a set with n elements, are shown in Table 1.

TABLE 1 Combinations and Permutations with and without Repetition

Type	*Repetition Allowed?*	*Formula*
r-permutations	No	$\frac{n!}{(n-r)!}$
r-combinations	No	$\frac{n!}{r!(n-r)!}$
r-permutations	Yes	n^r
r-combinations	Yes	$\frac{(n+r-1)!}{r!(n-1)!}$

PERMUTATIONS OF SETS WITH INDISTINGUISHABLE OBJECTS

Some elements may be indistinguishable in counting problems. When this is the case care must be taken to avoid counting things more than once. Consider the following example.

EXAMPLE 8 How many different strings can be made by reordering the letters of the word

SUCCESS?

Solution: Because some of the letters of *SUCCESS* are the same, the answer is *not* given by the number of permutations of seven letters. This word contains three *S*s, two *C*s, one *U*, and one *E*. To determine the number of different strings that can be made by reordering the letters, first note that the three *S*s can be placed among the seven positions in $C(7, 3)$ different ways, leaving four positions free. Then the two *C*s can be placed in $C(4, 2)$ ways, leaving two free positions. The *U* can be placed in $C(2, 1)$ ways, leaving just one position free. Hence *E* can be placed in $C(1, 1)$ way. Consequently, from the product rule, the number of different strings that can be made is

$$\begin{aligned} C(7,3)C(4,2)C(2,1)C(1,1) &= \frac{7!}{3!4!} \cdot \frac{4!}{2!2!} \cdot \frac{2!}{1!1!} \cdot \frac{1!}{1!0!} \\ &= \frac{7!}{3!2!1!1!} \\ &= 420. \end{aligned}$$

■

Using the same sort of reasoning as in the previous example, the following theorem can be proved.

THEOREM 3 The number of different permutations of n objects, where there are n_1 indistinguishable objects of type 1, n_2 indistinguishable objects of type 2, . . . , and n_k indistinguishable objects of type k is

$$\frac{n!}{n_1!\,n_2!\cdots n_k!}.$$

Proof: To determine the number of permutations, first note that the n_1 objects of type 1 can be placed among the n positions in $C(n, n_1)$ ways, leaving $n - n_1$ positions free. Then the objects of type 2 can be placed in $C(n - n_1, n_2)$ ways, leaving $n - n_1 - n_2$ positions free. Continue placing the objects of type 3, . . . , type $k - 1$, until at the last stage n_k objects of type k can be placed in $C(n - n_1 - n_2 - \cdots - n_{k-1}, n_k)$ ways. Hence, by the product rule, the total number of different permutations is

$$\begin{aligned} &C(n, n_1)C(n - n_1, n_2)\cdots C(n - n_1 - \cdots - n_{k-1}, n_k) \\ &\qquad = \frac{n!}{n_1!(n-n_1)!}\,\frac{(n-n_1)!}{n_2!(n-n_1-n_2)!}\cdots\cdots\frac{(n-n_1-\cdots-n_{k-1})!}{n_k!0!} \\ &\qquad = \frac{n!}{n_1!\,n_2!\cdots n_k!}. \end{aligned}$$

■

DISTRIBUTING OBJECTS INTO BOXES

Some counting problems can be solved by enumerating the ways distinguishable objects can be placed into distinguishable boxes. Consider the following example in which the objects are cards and the "boxes" are hands of players.

EXAMPLE 9 How many ways are there to distribute hands of 5 cards to each of four players from the standard deck of 52 cards?

Solution: We will use the product rule to solve this problem. To begin, note that the first player can be dealt 5 cards in $C(52, 5)$ ways. The second player can be dealt 5 cards in $C(47, 5)$ ways, since only 47 cards are left. The third player can be dealt 5 cards in $C(42, 5)$ ways. Finally, the fourth player can be dealt 5 cards in $C(37, 5)$ ways. Hence, the total number of ways to deal four players 5 cards each is

$$C(52, 5)C(47, 5)C(42, 5)C(37, 5) = \frac{52!}{47!\,5!} \cdot \frac{47!}{42!\,5!} \cdot \frac{42!}{37!\,5!} \cdot \frac{37!}{32!\,5!} = \frac{52!}{5!\,5!\,5!\,5!\,32!}.$$

■

Remark: The solution to Example 9 equals the number of permutations of 52 objects, with 5 indistinguishable objects of each of four different types, and 32 objects of a fifth type. This equality can be seen by defining a one-to-one correspondence between permutations of this type and distributions of cards to the players. To define this correspondence, first order the cards from 1 to 52. Then cards dealt to the first player correspond to the cards in the positions assigned to objects of the first type in the permutation. Similarly, cards dealt to the second, third, and fourth players, respectively, correspond to cards in the positions assigned to objects of the second, third, and fourth type, respectively. The cards not dealt to any player correspond to cards in the positions assigned to objects of the fifth type. The reader should verify that this a one-to-one correspondence.

Example 9 is a typical problem that involves distributing distinguishable objects into distinguishable boxes. The distinguishable objects are the 52 cards, and the five distinguishable boxes are the hands of the four players and the rest of the deck. Counting problems that involve distributing distinguishable objects into boxes can be solved using the following theorem.

THEOREM 4 The number of ways to distribute n distinguishable objects into k distinguishable boxes, so that n_i objects are placed into box i, $i = 1, 2, \ldots, k$ equals

$$\frac{n!}{n_1!\,n_2!\cdots n_k!}.$$

The proof of Theorem 4 is left for the reader (see Exercises 43 and 44).

Exercises

1. In how many different ways can five elements be selected in order from a set with three elements when repetition is allowed?

2. In how many different ways can five elements be selected in order from a set with five elements when repetition is allowed?

3. How many strings of six letters are there?

4. Every day a student randomly chooses a sandwich for lunch from a pile of wrapped sandwiches. If there are six kinds of sandwiches, how many different ways are there for the student to choose sandwiches for the seven days of a week?

5. How many ways are there to assign three jobs to five employees if each employee can be given more than one job?

6. How many ways are there to select five unordered elements from a set with three elements when repetition is allowed?

7. How many ways are there to select three unordered elements from a set with five elements when repetition is allowed?

8. How many different ways are there to choose a dozen donuts from the 21 varieties at a donut shop?

9. A bagel shop has onion bagels, poppy seed bagels, egg bagels, salty bagels, pumpernickel bagels, sesame seed bagels, raisin bagels, and plain bagels. How many ways are there to choose

a) six bagels?
b) a dozen bagels?
c) two dozen bagels?
d) a dozen bagels with at least one of each kind?
e) a dozen bagels with at least three egg bagels and no more than two salty bagels?

10. A croissant shop has plain croissants, cherry croissants, chocolate croissants, almond croissants, apple croissants, and broccoli croissants. How many ways are there to choose

a) a dozen croissants?
b) three dozen croissants?
c) two dozen croissants with at least two of each kind?
d) two dozen croissants with no more than two broccoli croissants?
e) two dozen croissants with at least five chocolate croissants and at least three almond croissants?
f) two dozen croissants with at least one plain croissant, at least two cherry croissants, at least three chocolate croissants, at least one almond croissant, at least two apple croissants, and no more than three broccoli croissants?

11. How many ways are there to choose eight coins from a piggy bank containing 100 identical pennies and 80 identical nickels?

12. How many different combinations of pennies, nickels, dimes, quarters, and half dollars can a piggy bank contain if it has 20 coins in it?

13. A book publisher has 3000 copies of a discrete mathematics book. How many ways are there to store these books in their three warehouses if the copies of the book are indistinguishable?

14. How many solutions are there to the equation

$$x_1 + x_2 + x_3 + x_4 = 17$$

where x_1, x_2, x_3, and x_4 are nonnegative integers?

15. How many solutions are there to the equation

$$x_1 + x_2 + x_3 + x_4 + x_5 = 21$$

where x_i, $i = 1, 2, 3, 4, 5$, is a nonnegative integer such that

a) $x_1 \geq 1$?
b) $x_i \geq 2$ for $i = 1, 2, 3, 4, 5$?
c) $0 \leq x_1 \leq 10$?
d) $0 \leq x_1 \leq 3$, $1 \leq x_2 < 4$, and $x_3 \geq 15$?

16. How many solutions are there to the equation

$$x_1 + x_2 + x_3 + x_4 + x_5 + x_6 = 29$$

where x_i, $i = 1, 2, 3, 4, 5, 6$, is a nonnegative integer such that

a) $x_i > 1$ for $i = 1, 2, 3, 4, 5, 6$?
b) $x_1 \geq 1$, $x_2 \geq 2$, $x_3 \geq 3$, $x_4 \geq 4$, $x_5 > 5$, and $x_6 \geq 6$?
c) $x_1 \leq 5$?
d) $x_1 < 8$ and $x_2 > 8$?

17. How many strings of 10 ternary digits (0, 1, or 2) are there that contain exactly two 0s, three 1s, and five 2s?

18. How many strings of 20 decimal digits are there that contain two 0s, four 1s, three 2s, one 3, two 4s, three 5s, two 7s, and three 9s?

19. Suppose that a large family has 14 children, including two sets of identical triplets, three sets of identical twins, and two individual children. How many ways are there to seat these children in a row of chairs if the identical triplets or twins cannot be distinguished from one another?

20. How many solutions are there to the inequality

$$x_1 + x_2 + x_3 \leq 11$$

where x_1, x_2, and x_3 are nonnegative integers? (*Hint:* Introduce an auxiliary variable x_4 so that $x_1 + x_2 + x_3 + x_4 = 11$.)

21. How many positive integers less than 1,000,000 have the sum of their digits equal to 19?

22. How many positive integers less than 1,000,000 have exactly one digit equal to 9 and have a sum of digits equal to 13?

23. There are 10 questions on a discrete mathematics final exam. How many ways are there to assign scores to the problems if the sum of the scores is 100 and each question is worth at least 5 points?

24. Show that there are $C(n + r - q_1 - q_2 - \cdots - q_r - 1, n - q_1 - q_2 - \cdots - q_r)$ different unordered selections of n objects of r different types that include at least q_1 objects of type 1, q_2 objects of type $2, \ldots,$ and q_r objects of type r.

25. How many different bit strings can be transmitted if the string must begin with a 1 bit, must include three additional 1 bits (so that a total of four 1 bits is sent), must include a total of twelve 0 bits, and must have at least two 0 bits following each 1 bit?

26. How many different strings can be made from the letters in *MISSISSIPPI*, using all the letters?

27. How many different strings can be made from the letters in *ABRACADABRA*, using all the letters?

28. How many different strings can be made from the letters in *AARDVARK*, using all the letters, if all three *A*s must be consecutive?

29. How many different strings can be made from the letters in *ORONO*, using some or all of the letters?

30. How many strings with five or more characters can be formed from the letters in *SEERESS*?

31. How many strings with seven or more characters can be formed from the letters in *EVERGREEN*?

32. How many different bit strings can be formed using six 1s and eight 0s?

33. A student has three mangos, two papayas, and two kiwi fruits. If the student eats one piece of fruit each day, and only the type of fruit matters, in how many different ways can these fruits be consumed?

34. A professor packs her collection of 40 issues of a mathematics journal in four boxes with 10 issues per box. How many ways can she distribute the journals if

a) each box is numbered, so that they are distinguishable?

b) the boxes are identical, so that they cannot be distinguished?

35. How many ways are there to travel in *xyz* space from the origin $(0, 0, 0)$ to the point $(4, 3, 5)$ by taking steps one unit in the positive x direction, one unit in the positive y direction, or one unit in the positive z direction? (Moving in the negative x, y, or z direction is prohibited, so that no backtracking is allowed.)

36. How many ways are there to travel in *xyzw* space from the origin $(0, 0, 0, 0)$ to the point $(4, 3, 5, 4)$ by taking steps one unit in the positive x, positive y, positive z, or positive w direction?

37. How many ways are there to deal hands of 7 cards to each of five players from a standard deck of 52 cards?

38. In bridge, the 52 cards of a standard deck are dealt to four players. How many different ways are there to deal bridge hands to four players?

39. What is the probability that each player has a hand containing an ace when the 52 cards of a standard deck are dealt to four players?

40. In how many ways can a dozen books be placed on four distinguishable shelves

a) if the books are indistinguishable copies of the same title?

b) if no two books are the same, and the positions of the books on the shelves matter? (*Hint:* Break this into 12 tasks, placing each book separately. Start with the sequence 1, 2, 3, 4 to represent the shelves. Represent the books by b_i, $i = 1, 2, \ldots, 12$. Place b_1 to the right of one of the terms in 1, 2, 3, 4. Then successively place $b_2, b_3, \ldots,$ and b_{12}.)

41. How many ways can n books be placed on k distinguishable shelves

a) if the books are indistinguishable copies of the same title?

b) if no two books are the same, and the positions of the books on the shelves matter?

42. A shelf holds 12 books in a row. How many ways are there to choose 5 books so that no two adjacent books are chosen? (*Hint:* Represent the books that are chosen by bars and the books not chosen by stars. Count the number of sequences of five bars and seven stars so that no two bars are adjacent.)

★43. Use the product rule to prove Theorem 4, by first placing objects in the first box, then placing objects in the second box, and so on.

★**44.** Prove Theorem 4 by first setting up a one-to-one correspondence between permutations of n objects with n_i indistinguishable objects of type i, $i = 1, 2, 3, \ldots, k$, and the distributions of n objects in k boxes such that n_i objects are placed in box i, $i = 1, 2, 3, \ldots, k$ and then applying Theorem 3.

★**45.** In this exercise we will prove Theorem 2 by setting up a one-to-one correspondence between the set of r-combinations with repetition allowed of $S = \{1, 2, 3, \ldots, n\}$ and the set of r-combinations of the set $T = \{1, 2, 3, \ldots, n + r - 1\}$.

a) Arrange the elements in an r-combination, with repetition allowed, of S into an increasing sequence $x_1 \leq x_2 \leq \cdots \leq x_r$. Show that the sequence formed by adding $k - 1$ to the kth term is strictly increasing. Conclude that this sequence is made up of r distinct elements from T.

b) Show that the procedure described in (a) defines a one-to-one correspondence between the set of r-combinations, with repetition allowed, of S and the r-combinations of T. (*Hint:* Show the correspondence can be reversed by associating to the r-combination $\{x_1, x_2, \ldots, x_r\}$ of T, with $1 \leq x_1 < x_2 < \cdots < x_r \leq n + r - 1$, the r-combination with repetition allowed from S, formed by subtracting $k - 1$ from the kth element.)

c) Conclude that there are $C(n + r - 1, r)$ r-combinations with repetition allowed from a set with n elements.

46. How many ways are there to distribute five distinguishable objects into three indistinguishable boxes?

47. How many ways are there to distribute five indistinguishable objects into three indistinguishable boxes?

48. How many different terms are there in the expansion of $(x_1+x_2+\cdots+x_m)^n$ after all terms with identical sets of exponents are added?

★**49.** Prove the **multinomial theorem:** If n is a positive integer, then

$$(x_1 + x_2 + \cdots + x_m)^n = \sum_{n_1+n_2+\cdots+n_m=n} C(n; n_1, n_2, \ldots, n_m) x_1^{n_1} x_2^{n_2} \ldots x_n^{n_m},$$

where

$$C(n; n_1, n_2, \ldots, n_m) = \frac{n!}{n_1!\, n_2! \cdots n_m!}$$

is a **multinomial coefficient.**

50. Find the expansion of $(x + y + z)^4$.

51. Find the coefficient of $x^3y^2z^5$ in $(x + y + z)^{10}$.

52. How many terms are there in the expansion of $(x + y + z)^{100}$?

4.7 Generating Permutations and Combinations

INTRODUCTION

Methods for counting various types of permutations and combinations were described in the previous sections of this chapter. But sometimes permutations or combinations need to be generated, not just counted. Consider the following three problems. First, suppose that a salesman must visit six different cities. In which order should these cities be visited to minimize total travel time? One way to determine the best order is to determine the travel time for each of the $6! = 720$ different orders in which the cities can be visited and choose the one with the smallest travel time. Second, suppose some numbers from a set of six numbers have 100 as their sum. One way to find these numbers is to generate all the $2^6 = 64$ subsets and check the sum of their terms. Third, suppose a laboratory has 95 employees. A group of 12 of these employees with a particular set of 25 skills is needed for a project. (Each employee can have one or more of these skills.) One way to find such a set of employees is to generate all sets of 12 of these employees and check whether they have the desired skills. These examples show that it is often necessary to generate permutations and combinations to solve problems.

GENERATING PERMUTATIONS

Any set with n elements can be placed in one-to-one correspondence with the set $\{1, 2, 3, \ldots, n\}$. We can list the permutations of any set of n elements by generating the permutations of the n smallest positive integers and then replacing these integers with the corresponding elements. Many different algorithms have been developed to generate the $n!$ permutations of this set. We will describe one of these that is based on the **lexicographic ordering** of the set of permutations of $\{1, 2, 3, \ldots, n\}$. In this ordering, the permutation $a_1 a_2 \cdots a_n$ precedes the permutation of $b_1 b_2 \cdots b_n$, if for some k, with $1 \leq k \leq n$, $a_1 = b_1$, $a_2 = b_2, \ldots, a_{k-1} = b_{k-1}$, and $a_k < b_k$. In other words, a permutation of the set of the n smallest positive integers precedes (in lexicographic order) a second permutation if the number in this permutation in the first position where the two permutations disagree is smaller than the number in that position in the second permutation.

EXAMPLE 1 The permutation 23415 of the set $\{1, 2, 3, 4, 5\}$ precedes the permutation 23514, since these permutations agree in the first two positions, but the number in the third position in the first permutation, 4, is smaller than the number in the third position in the second permutation, 5. Similarly, the permutation 41532 precedes 52143. ■

An algorithm for generating the permutations of $\{1, 2, \ldots, n\}$ can be based on a procedure that constructs the next permutation in lexicographic order following a given permutation $a_1 a_2 \cdots a_n$. We will show how this can be done. First, suppose that $a_{n-1} < a_n$. Interchange a_{n-1} and a_n to obtain a larger permutation. No other permutation is both larger than the original permutation and smaller than the permutation obtained by interchanging a_{n-1} and a_n. For instance, the next largest permutation after 234156 is 234165. On the other hand, if $a_{n-1} > a_n$, then a larger permutation cannot be obtained by interchanging these last two terms in the permutation. Look at the last three integers in the permutation. If $a_{n-2} < a_{n-1}$, then the last three integers in the permutation can be rearranged to obtain the next largest permutation. Put the smaller of the two integers a_{n-1} and a_n that is greater than a_{n-1} in position $n - 2$. Then, place the remaining integer and a_{n-2} into the last two positions in increasing order. For instance, the next largest permutation after 234165 is 234516.

On the other hand, if $a_{n-2} > a_{n-1}$ (and $a_{n-1} > a_n$), then a larger permutation cannot be obtained by permuting the last three terms in the permutation. Based on these observations, a general method can be described for producing the next largest permutation in increasing order following a given permutation $a_1a_2 \cdots a_n$. First, find the integers a_j and a_{j+1} with $a_j < a_{j+1}$ and

$$a_{j+1} > a_{j+2} > \cdots > a_n,$$

that is, the last pair of adjacent integers in the permutation where the first integer in the pair is smaller than the second. Then, the next largest permutation in lexicographic order is obtained by putting in the *j*th position the least integer among $a_{j+1}, a_{j+2}, \ldots,$ and a_n that is greater than a_j, and listing in increasing order the rest of the integers $a_j, a_{j+1}, \ldots, a_n$ in positions $j+1$ to n. It is easy to see that there is no other permutation larger than the permutation $a_1a_2 \cdots a_n$, but smaller than the new permutation produced. (The verification of this fact is left as an exercise for the reader.)

EXAMPLE 2 What is the next largest permutation in lexicographic order after 362541?

Solution: The last pair of integers a_j and a_{j+1} where $a_j < a_{j+1}$ is $a_3 = 2$ and $a_4 = 5$. The least integer to the right of 2 that is greater than 2 in the permutation is $a_5 = 4$. Hence, 4 is placed in the third position. Then the integers 2, 5, and 1 are placed in order in the last three positions, giving 125 as the last three positions of the permutation. Hence, the next permutation is 364125. ■

To produce the $n!$ permutations of the integers $1, 2, 3, \ldots, n$, begin with the smallest permutation in lexicographic order, namely, $123 \cdots n$, and successively apply the procedure described for producing the next largest permutation of $n! - 1$ times. This yields all the permutations of the n smallest integers in lexicographic order.

EXAMPLE 3 Generate the permutations of the integers 1, 2, 3 in lexicographic order.

Solution: Begin with 123. The next permutation is obtained by interchanging 3 and 2 to obtain 132. Next, since $3 > 2$ and $1 < 3$, permute the three integers in 132. Put the smaller of 3 and 2 in the first position, and then put 1 and 3 in increasing order in positions 2 and 3 to obtain 213. This is followed by 231, obtained by interchanging 1 and 3, since $1 < 3$. the next largest permutation has 3 in the first position, followed by 1 and 2 in increasing order, namely, 312. Finally, interchange 1 and 2 to obtain the last permutation, 321. ■

Algorithm 1 displays the procedure for finding the next largest permutation in lexicographic order after a permutation that is not $n\ n-1\ n-2\ \cdots\ 2\ 1$, which is the largest permutation.

ALGORITHM 1 Generating the Next Largest Permutation in Lexicographic Order.

procedure *next permutation*($a_1 a_2 \cdots a_n$: permutation of
  $\{1, 2, \ldots, n\}$ not equal to $n \ n-1 \cdots \ 2 \ 1$)
$j := n - 1$
while $a_j > a_{j+1}$
  $j := j - 1$
$\{j$ is the largest subscript with $a_j < a_{j+1}\}$

$k := n$
while $a_j > a_k$
  $k := k - 1$
$\{a_k$ is the smallest integer greater than a_j to the right of $a_j\}$

interchange a_j and a_k
$r := n$
$s := j + 1$
while $r > s$
begin
  interchange a_r and a_s
  $r := r - 1$
  $s := s + 1$
end
{this puts the tail end of the permutation after the jth position in
  increasing order}

GENERATING COMBINATIONS

How can we generate all the combinations of the elements of a finite set? Since a combination is just a subset, we can use the correspondence between subsets of $\{a_1, a_2, \ldots, a_n\}$ and bit strings of length n.

Recall that the bit string corresponding to a subset has a 1 in position k if a_k is in the subset, and has a 0 in this position if a_k is not in the subset. If all the bit strings of length n can be listed, then by the correspondence between subsets and bit strings, a list of all the subsets is obtained.

Recall that a bit string of length n is also the binary expansion of an integer between 0 and $2^n - 1$. The 2^n bit strings can be listed in order of their increasing size as integers in their binary expansions. To produce all binary expansions of length n, start with the bit string $000 \ldots 00$, with n zeros. Then, successively find the next largest expansion until the bit string $111 \ldots 11$ is obtained. At each stage the next largest binary expansion is found by locating the first position, from the right, that is not a 1, then changing all the 1s to the right of this position to 0s and making this first 0 (from the right) a 1.

EXAMPLE 4 Find the next largest bit string after 10001 00111.

Solution: The first bit from the right that is not a 1 is the fourth bit from the right. Change this bit to a 1 and change all the following bits to 0s. This produces the next largest bit string, 10001 01000. ■

The procedure for producing the next largest bit string after $b_{n-1}b_{n-2}\cdots b_1b_0$ is given as Algorithm 2.

ALGORITHM 2 Generating the Next Largest Bit String.

procedure *next bit string*($b_{n-1}b_{n-2}\cdots b_1b_0$: bit string not equal to 11...11)
$i := 0$
while $b_i = 1$
begin
  $b_i := 0$
  $i := i + 1$
end
$b_i := 1$

Next, an algorithm for generating the r-combinations of the set $\{1, 2, 3, \ldots, n\}$ will be given. An r-combination can be represented by a sequence containing the elements in the subset in increasing order. The r-combinations can be listed using lexicographic order on these sequences. The next combinations after $a_1a_2\cdots a_r$ can be obtained in the following way: First, locate the last element a_i in the sequence such that $a_i \neq n - r + i$. Then, replace a_i with $a_i + 1$ and a_j with $a_i + j - i + 1$, for $j = i + 1, i + 2, \ldots, r$. It is left for the reader to show that this produces the next largest combination in lexicographic order. This procedure is illustrated with the following example.

EXAMPLE 5 Find the next largest 4-combination of the set $\{1, 2, 3, 4, 5, 6\}$ after $\{1, 2, 5, 6\}$.

Solution: The last term among the terms a_i with $a_1 = 1$, $a_2 = 2$, $a_3 = 5$, and $a_4 = 6$ such that $a_i \neq 6 - 4 + i$ is $a_2 = 2$. To obtain the next largest 4-combination, increment a_2 by 1 to obtain $a_2 = 3$. Then set $a_3 = 3 + 1 = 4$ and $a_4 = 3 + 2 = 5$. Hence the next largest 4-combination is $\{1, 3, 4, 5\}$. ■

Algorithm 3 gives this procedure in pseudocode.

ALGORITHM 3 Generating the Next r-Combination in Lexicographic Order.

procedure *next r-combination*($\{a_1, a_2, \ldots, a_r\}$: proper subset of $\{1, 2, \ldots, n\}$ not equal to $\{n - r + 1, \ldots, n\}$ with $a_1 < a_2 < \cdots < a_r$)
$i := r$
while $a_i = n - r + i$
    $i := i - 1$
$a_i := a_i + 1$
for $j := i + 1$ **to** r
    $a_j := a_i + j - i$

Exercises

1. Find the next largest permutation in lexicographic order after each of the following permutations.
a) 1432 **b)** 54123 **c)** 12453
d) 45231 **e)** 6714235 **f)** 31528764

2. Place the following permutations of {1,2,3,4,5,6} in lexicographic order: 234561, 231456, 165432, 156423, 543216, 541236, 231465, 314562, 432561, 654321, 654312, 435612.

3. Use Algorithm 1 to generate the 24 permutations of the first four positive integers in lexicographic order.

4. Use Algorithm 2 to list all the subsets of the set $\{1, 2, 3, 4\}$.

5. Use Algorithm 3 to list all the 3-combinations of $\{1, 2, 3, 4, 5\}$.

6. Show that Algorithm 1 produces the next largest permutation in lexicographic order.

7. Show that Algorithm 3 produces the next largest r-combination in lexicographic order after a given r-combination.

8. Develop an algorithm for generating the r-permutations of a set of n elements.

9. List all 3-permutations of $\{1, 2, 3, 4, 5\}$.

The remaining exercises in this section develop another algorithm for generating the permutations of $\{1, 2, 3, \ldots, n\}$. This algorithm is based on Cantor expansions of integers. Every nonnegative integer less than $n!$ has a unique Cantor expansion

$$a_1 1! + a_2 2! + \cdots + a_{n-1}(n - 1)!$$

where a_i is a nonnegative integer not exceeding i, for $i = 1, 2, \ldots, n - 1$. The integers $a_1, a_2, \ldots, a_{n-1}$ are called the **Cantor digits** of this integer.

Given a permutation of $\{1, 2, \ldots, n\}$, let a_{k-1}, $k = 2, 3, \ldots, n$, be the number of integers less than k that follow k in the permutation. For instance, in the permutation 43215, a_1 is the number of integers less than 2 that follow 2, so that $a_1 = 1$. Similarly, for this example $a_2 = 2$, $a_3 = 3$, and $a_4 = 0$. Consider the function from the set of permutations $\{1, 2, 3, \ldots, n\}$ to the set of nonnegative integers less than $n!$ that sends a permutation to the integer that has $a_1, a_2, \ldots, a_{n-1}$, defined in this way, as its Cantor digits.

10. Find the integers that correspond to the following permutations.
a) 246531 **b)** 12345 **c)** 654321

★11. Show that the correspondence described here is a bijection between the set of permutations of $\{1, 2, 3, \ldots, n\}$ and the nonnegative integers less than $n!$.

12. Find the permutations of $\{1, 2, 3, 4, 5\}$ that correspond to the following integers with respect to the correspondence between Cantor expansions and permutations as described before Exercise 10.
a) 3 **b)** 89 **c)** 111

13. Develop an algorithm for producing all permutations of a set of n elements based on the correspondence described in the preamble to Exercise 10.

Key Terms and Concepts

TERMS

combinatorics: the study of arrangements of objects
enumeration: the counting of arrangements of objects
tree diagram: a diagram made up of a root, branches leaving the root, and other branches leaving some of the endpoints of branches
permutation: an ordered arrangement of the elements of a set
***r*-permutation:** an ordered arrangement of r elements of a set
***P*(*n*, *r*):** the number of r-permutations of a set with n elements
***r*-combination:** an unordered selection of r elements of a set
***C*(*n*, *r*):** the number of r-combinations of a set with n elements
$\binom{n}{r}$ **(binomial coefficient):** also the number of r-combinations of a set with n elements
Pascal's triangle: a representation of the binomial coefficients where the ith row of the triangle contains $C(i, j)$ for $j = 0, 1, 2, \ldots, i$
probability of an event: the number of successful outcomes of this event divided by the number of possible outcomes
***p*(*E* | *F*) (conditional probability of *E* given *F*):** $p(E \cap F)/p(F)$
independent events: events E and F such that $p(E \cap F) = p(E)p(F)$
random variable: a function that assigns a real number to each outcome of an experiment
expected value of a random variable: the weighted average of a random variable, with values of the random variable weighted by the probability of outcomes
Bernoulli trial: an experiment with two possible outcomes

RESULTS

The sum rule: a basic counting technique that states that the number of ways to do a task in one of two ways is the sum of the number of ways to do these tasks if they cannot be done simultaneously
The product rule: a basic counting technique that states that the number of ways to do a procedure that consists of two subtasks is the product of the number of ways to do the first task and the number of ways to do the second task after the first task has been done
The pigeonhole principle: When more than k objects are placed in k boxes, there must be a box containing more than one object.
The generalized pigeonhole principle: When N objects are placed in k boxes, there must be a box containing at least $\lceil N/k \rceil$ objects.

$$P(n, r) = \frac{n!}{(n-r)!}$$

$$C(n, r) = \frac{n!}{r!(n-r)!}$$

Pascal's identity: $C(n+1, k) = C(n, k-1) + C(n, k)$

$$C(n, 0) + C(n, 1) + \cdots + C(n, n) = 2^n$$

The binomial theorem: $(x+y)^n = \sum_{k=0}^{n} C(n, k)x^{n-k}y^k$

The probability of k successes when n independent Bernoulli trials are carried out equals $C(n, k)p^k q^{n-k}$, where p is the probability of success and $q = 1 - p$ is the probability of failure.

There are n^r r-permutations of a set with n elements when repetition is allowed.

There are $C(n+r-1, r)$ r-combinations of a set with n elements when repetition is allowed.
There are $n!/(n_1! n_2! \cdots n_k!)$ permutations of n objects where there are n_i indistinguishable objects of type i for $i = 1, 2, 3, \ldots, k$.
The algorithm for generating the permutations of the set $\{1, 2, \ldots, n\}$.

Review Questions

1. Explain how the sum and product rules can be used to find the number of bit strings of length not exceeding 10.

2. Explain how to find the number of bit strings of length not exceeding 10 that have at least one 0 bit.

3. **a)** How can the product rule be used to find the number of functions from a set with m elements to a set with n elements?
b) How many functions are there from a set with 5 elements to a set with 10 elements?
c) How can the product rule be used to find the number of one-to-one functions from a set with m elements to a set with n elements?
d) How many one-to-one functions are there from a set with 5 elements to a set with 10 elements?
e) How many onto functions are there from a set with 5 elements to a set with 10 elements?

4. How can you find the number of possible outcomes of a playoff between two teams where the first team that wins four games wins the playoff?

5. How can you find the number of bit strings of length 10 that either begin with 101 or end with 010?

6. **a)** State the pigeonhole principle.
b) Explain how the pigeonhole principle can be used to show that among any 11 integers, at least 2 must have the same last digit.

7. **a)** State the generalized pigeonhole principle.
b) Explain how the generalized pigeonhole principle can be used to show that among any 91 integers, there are at least 10 that end with the same digit.

8. **a)** What is the difference between an r-combination and an r-permutation of a set with n elements?
b) Derive an equation that relates the number of r-combinations and the number of r-permutations of a set with n elements.
c) How many ways are there to select 6 students from a class of 25 to serve on a committee?
d) How many ways are there to select 6 students from a class of 25 to hold six different executive positions on a committee?

9. **a)** What is Pascal's triangle?
b) How can a row of Pascal's triangle be produced from the one above it?

10. What is meant by a combinatorial proof of an identity? How is such a proof different from an algebraic one?

11. Explain how to prove Pascal's identity using a combinatorial argument.

12. **a)** State the binomial theorem.
b) Explain how to prove the binomial theorem using a combinatorial argument.
c) Find the coefficient of $x^{100}y^{101}$ in the expansion of $(2x+5y)^{201}$.

13. **a)** Define the probability of an event when all outcomes are equally likely.
b) What is the probability that you select the 6 winning numbers in a lottery if the 6 different winning numbers are selected from the first 50 positive integers?

14. **a)** What conditions should be met by the probabilities assigned to the outcomes from a finite sample space?
b) What probabilities should be assigned to the outcome of heads and the outcome of tails if heads comes up three times as often as tails?

15. **a)** Define the conditional probability of an event E given an event F.
b) Suppose E is the event that when a die is rolled it comes up an even number, and F is the event that when a die is rolled it comes up 1, 2, or 3. What is the probability of F given E?

16. **a)** When are two events E and F independent?
b) Suppose E is the event that an even number appears when a fair die is rolled, and F is the event that a 5 or 6 comes up. Are E and F independent?

17. **a)** What is a random variable?
b) What are the values assigned by the random variable X that assigns to a roll of two dice the larger number that appears on the two dice?

18. **a)** Define the expected value of a random variable X.

b) What is the expected value of the random variable X that assigns to a roll of two dice the larger number that appears on the two dice?

19. a) Explain how the average-case computational complexity of an algorithm, with finitely many possible input values, can be interpreted as an expected value.

b) What is the average-case computational complexity of the linear search algorithm, if the probability that the element for which we search is in the list is $1/3$, and it is equally likely that this element is any of the n elements in the list?

20. a) What is meant by a Bernoulli trial?

b) What is the probability of k successes in n independent Bernoulli trials?

c) What is the expected value of the number of successes in n independent Bernoulli trials?

21. a) Explain how to find a formula for the number of ways to select r objects from n objects when repetition is allowed and order does not matter.

b) How many ways are there to select a dozen objects from among objects of five different types if objects of the same type are indistinguishable?

c) How many ways are there to select a dozen objects from these five different types if there must be at least three objects of the first type?

d) How many ways are there to select a dozen objects from these five different types if there cannot be more than four objects of the first type?

e) How many ways are there to select a dozen objects from these five different types if there must be at least two objects of the first type but no more than three objects of the second type?

22. a) Let n and r be positive integers. Explain why the number of solutions of the equation $x_1 + x_2 + \cdots + x_n = r$, where x_i is a nonnegative integer for $i = 1, 2, 3, \ldots, n$, equals the number of r-combinations of a set with n elements.

b) How many solutions in nonnegative integers are there to the equation $x_1 + x_2 + x_3 + x_4 = 17$?

c) How many solutions in positive integers are there to the equation in part (b)?

23. a) Derive a formula for the number of permutations of n objects of k different types where there are n_1 indistinguishable objects of type 1, n_2 indistinguishable objects of type $2, \ldots,$ and n_k indistinguishable objects of type k.

b) How many ways are there to order the letters of the word *INDISCREETNESS*?

24. Describe an algorithm for generating all the permutations of the set of the n smallest positive integers.

25. a) How many ways are there to deal hands of 5 cards to six players from a standard 52-card deck?

b) How many ways are there to distribute n distinguishable objects into k distinguishable boxes so that n_i objects are placed in box i?

26. Describe an algorithm for generating all the combinations of the set of the n smallest positive integers.

Supplementary Exercises

1. How many ways are there to choose 6 items from 10 distinct items when

a) the items in the choices are ordered and repetition is not allowed?

b) the items in the choices are ordered and repetition is allowed?

c) the items in the choices are unordered and repetition is not allowed?

d) the items in the choices are unordered and repetition is allowed?

2. How many ways are there to choose 10 items from 6 distinct items when

a) the items in the choices are ordered and repetition is not allowed?

b) the items in the choices are ordered and repetition is allowed?

c) the items in the choices are unordered and repetition is not allowed?

d) the items in the choices are unordered and repetition is allowed?

3. A test contains 100 true/false questions. How many different ways can a student answer the questions on the test, if answers may be left blank?

4. How many bit strings of length 10 either start with 000 or end with 1111?

5. How many bit strings of length 10 over the alphabet $\{a, b, c\}$ have either exactly three as or exactly four bs?

6. The internal telephone numbers in the phone system on a campus consist of five digits, with the first digit not equal to zero. How many different numbers can be assigned in this system?

7. An ice cream parlor has 28 different flavors, 8 different kinds of sauce, and 12 toppings.

a) In how many different ways can a dish of three scoops of ice cream be made where each flavor can be used more than once and the order of the scoops does not matter?

b) How many different kinds of small sundaes are there if a small sundae contains one scoop of ice cream, a sauce, and a topping?

c) How many different kinds of large sundaes are there if a large sundae contains three scoops of ice cream, where each flavor can be used more than once and the order of the scoops does not matter; two kinds of sauce, where each sauce can be used only once and the order of the sauces does not matter; and three toppings, where each topping can be used only once and the order of the toppings does not matter?

8. How many positive integers less than 1000

a) have exactly three decimal digits?

b) have an odd number of decimal digits?

c) have at least one decimal digit equal to 9?

d) have no odd decimal digits?

e) have two consecutive decimal digits equal to 5?

f) are palindromes (that is, read the same forward and backward)?

9. When the numbers from 1 to 1000 are written out in decimal notation, how many of the following digits are used?

a) 0 **b)** 1 **c)** 2 **d)** 9

10. There are 12 signs of the zodiac. How many people are needed to guarantee that at least six of these people have the same sign?

11. A fortune cookie company makes 213 different fortunes. A student eats at a restaurant that uses fortunes from this company. What is the largest possible number of times that the student can eat at the restaurant without getting the same fortune four times?

12. How many people are needed to guarantee that at least two were born on the same day of the week and in the same month (perhaps in different years)?

13. Show that there are at least two different five-element subsets of a set of 10 positive integers not exceeding 50 that have the same sum.

14. A package of baseball cards contains 20 cards. How many packages must be purchased to ensure that two cards in these packages are identical if there are a total of 550 different cards?

15. a) How many cards must be chosen from a deck to guarantee that at least two aces are chosen?

b) How many cards must be chosen from a deck to guarantee that at least two aces and two kinds are chosen?

c) How many cards must be chosen from a deck to guarantee that there are at least two cards of the same kind?

d) How many cards must be chosen from a deck to guarantee that there are at least two cards of two different kinds?

★16. Show that in any set of $n + 1$ positive integers not exceeding $2n$ there must be two that are relatively prime.

★17. Show that in a sequence of m integers there exists one or more consecutive terms with sum divisible by m.

18. Show that if five points are picked in the interior of a square with side length 2, then at least two of these points are no farther than $\sqrt{2}$ apart.

19. Show that the decimal expansion of a rational number must repeat itself from some point onward.

20. How many diagonals does a regular polygon with n sides have, where n is a positive integer with $n \geq 3$?

21. How many ways are there to choose a dozen donuts from 20 varieties

a) if there are no two donuts of the same variety?

b) if all donuts are of the same variety?

c) if there are no restrictions?

d) if there are at least two varieties?

e) if there must be at least six blueberry-filled donuts?

f) if there can be no more than six blueberry-filled donuts?

22. What is the probability that six consecutive numbers will be chosen as the winning numbers in a lottery where each number chosen is between 1 and 40 (inclusive)?

23. What is the probability that a hand of 13 cards contains no pairs?

24. Find n if

a) $P(n, 2) = 110$.

b) $P(n, n) = 5040$.

c) $P(n, 4) = 12P(n, 2)$.

25. Find n if

a) $C(n, 2) = 45$.

b) $C(n, 3) = P(n, 2)$.

c) $C(n, 5) = C(n, 2)$.

26. Show that if n and r are nonnegative integers and $n \geq r$, then

$$P(n+1, r) = P(n, r)(n+1)/(n+1-r).$$

27. Give a combinatorial proof that $C(n, r) = C(n, n-r)$.

28. Give a combinatorial proof of Theorem 7 of Section 4.3 by setting up a correspondence between the subsets of a set with an even number of elements and the subsets of this set with an odd number of elements. (*Hint:* Take an element a in the set. Set up the correspondence by putting a in the subset if it is not already in it and taking it out if it is in the subset.)

29. Let n and r be nonnegative integers with $r < n$. Show that

$$C(n, r-1) = C(n+2, r+1) - 2C(n+1, r+1) + C(n, r+1).$$

30. Use mathematical induction to prove that $\sum_{j=2}^{n} C(j, 2) = C(n+1, 3)$ whenever n is an integer greater than 1.

31. Use the binomial theorem to prove that $3^n = \sum_{k=0}^{n} C(n, k)2^k$. (*Hint:* Let $x = 1$ and $y = 2$ in the statement of the theorem.)

32. In this exercise we will derive a formula for the sum of the squares of the n smallest positive integers. We will count the number of triples (i, j, k) such that i, j, and k are integers such that $0 \leq i < k$, $0 \leq j < k$, and $1 \leq k \leq n$ in two ways.

a) Show that there are k^2 such triples with a fixed k. Conclude that there are $\sum_{k=1}^{n} k^2$ such triples.

b) Show that the number of such triples with $0 \leq i < j < k$ and the number of such triples with $0 \leq j < i < k$ both equal $C(n+1, 3)$.

c) Show that the number of such triples with $0 \leq i = j < k$ equals $C(n+1, 2)$.

d) Combining part (a) with parts (b) and (c), conclude that

$$\sum_{k=1}^{n} k^2 = 2C(n+1, 3) + C(n+1, 2) = n(n+1)(2n+1)/6.$$

★33. How many bit strings of length n, where $n \geq 4$, contain exactly two occurrences of 01?

34. What is the probability that a seven-card poker hand contains

a) four cards of one kind and three cards of a second kind?

b) three cards of one kind and pairs of each of two different kinds?

c) pairs of each of three different kinds and a single card of a fourth kind?

d) pairs of each of two different kinds and three cards of a third, fourth, and fifth kind?

e) cards of seven different kinds?

f) a seven-card flush?

g) a seven-card straight?

h) a seven-card straight flush?

35. What is the probability that a 13-card bridge hand contains

a) all 13 hearts?

b) 13 cards of the same suit?

c) 7 spades and 6 clubs?

d) 7 cards of one suit and 6 cards of a second suit?

e) 4 diamonds, 6 hearts, 2 spades, and 1 club?

f) 4 cards of one suit, 6 cards of a second suit, 2 cards of a third suit, and 1 card of the fourth suit?

36. Suppose that p and q are primes and $n = pq$. What is the probability that a randomly chosen positive integer less than n is not divisible by either p or q?

★37. Suppose that m and n are positive integers. What is the probability that a randomly chosen positive integer less than mn is not divisible by either m or n?

38. Suppose that $E_1, E_2, \ldots, E_n$ are n events with $p(E_i) > 0$ for $i = 1, 2, \ldots, n$. Show that

$$p(E_1 \cap E_2 \cap \cdots \cap E_n) = p(E_1)p(E_2 \mid E_1)p(E_3 \mid E_1 \cap E_2) \cdots p(E_n \mid E_1 \cap E_2 \cap \cdots \cap E_{n-1}).$$

39. We say that the events $E_1, E_2, \ldots, E_n$ are **mutually independent** if

$$p(E_{i_1} \cap E_{i_2} \cap \cdots \cap E_{i_m}) = p(E_{i_1})p(E_{i_2}) \cdots p(E_{i_m})$$

whenever i_j, $j = 1, 2, \ldots, m$, are integers with $1 \leq i_1 < i_2 < \cdots < i_m \leq n$, and $m \geq 2$.

a) Write out the conditions required for three events E_1, E_2, E_3 to be mutually independent.

b) Let E_1, E_2, and E_3 be the events that the first flip comes up heads, that the second flip comes up tails, and that the third flip comes up tails, respectively, when a fair coin is flipped three times. Are E_1, E_2, and E_3 mutually independent?

c) Let E_1, E_2, and E_3 be the events that the first flip comes up heads, that the third flip comes up heads, and that an even number of heads come up, respectively, when a fair coin is flipped three times. Are E_1, E_2, and E_3 mutually independent?

d) How many conditions must be checked to show that n events are mutually independent?

★40. Suppose that E and F are events with $p(F) \neq 0$. Show that the probability of E is the weighted average of the probability of E given F and the probability of E given the complement of F, $\overline{F}$, where the weights are the probabilities of F and $\overline{F}$, respectively. That is,

$$p(E) = p(E \mid F)p(F) + p(E \mid \overline{F})p(\overline{F}).$$

(*Hint:* Use the fact that $E = (E \cap F) \cup (E \cap \overline{F})$.)

★41. Suppose that E is an event from a sample space S and that $F_1, F_2, \ldots, F_n$ are mutually exclusive events such that $\bigcup_{i=1}^{n} F_i = S$. Assume that $p(E) \neq 0$ and $p(F_i) \neq 0$ for $i = 1, 2, \ldots, n$. Show that

$$p(F_j \mid E) = \frac{p(E \mid F_j)p(F_j)}{\sum_{i=1}^{n} p(E \mid F_i)p(F_i)}.$$

(*Hint:* Use the fact that $E = \bigcup_{i=1}^{n}(E \cap F_i)$.) This result is known as **Bayes' formula,** since it was developed by the English philosopher Thomas Bayes.

★42. A space probe near Neptune communicates with Earth using bit strings. Suppose that in its transmissions it sends a 1 one-third of the time and a 0 two-thirds of the time. When a 0 is sent, the probability it is received correctly is 0.9, and the probability it is received incorrectly (as a 1) is 0.1. When a 1 is sent, the probability it is received correctly is 0.8, and the probability it is received incorrectly (as a 0) is 0.2.

a) Use Exercise 40 to find the probability that a 0 is received.

b) Use Bayes' formula, given in Exercise 41, to find the probability that a 0 was transmitted, given that a 0 was received.

43. A professor writes 20 multiple-choice questions, each with possible answer a, b, c, or d, for a discrete mathematics test. If the number of questions with a, b, c, and d as their answer is 8, 3, 4, and 5, respectively, how many different answer keys are possible, if the questions can be placed in any order?

44. How many different arrangements are there of eight people seated at a round table, where two arrangements are considered the same if one can be obtained from the other by a rotation?

45. How many ways are there to assign 24 students to five faculty advisors?

46. How many ways are there to choose a dozen apples from a bushel containing 20 indistinguishable Delicious apples, 20 indistinguishable Macintosh apples, and 20 indistinguishable Granny Smith apples, if at least three of each kind must be chosen?

47. How many solutions are there to the equation $x_1 + x_2 + x_3 = 17$, where x_1, x_2, and x_3 are nonnegative integers with

a) $x_1 > 1, x_2 > 2$, and $x_3 > 3$?

b) $x_1 < 6$ and $x_3 > 5$?

c) $x_1 < 4, x_2 < 3$, and $x_3 > 5$?

48. a) How many different strings can be made from the word *PEPPERCORN* when all the letters are used?

b) How many of these strings start and end with the letter *P*?

c) In how many of these strings are the three letter *P*s consecutive?

49. How many subsets of a set with 10 elements

a) have fewer than 5 elements?

b) have more than 7 elements?

c) have an odd number of elements?

50. A witness to a hit-and-run accident tells the police that the license plate of the car in the accident, which contains three letters followed by three digits, starts with the letters AS and contains both the digits 1 and 2. How many different license plates can fit this description?

51. How many ways are there to put n identical objects into m distinct containers so that no container is empty?

52. How many ways are there to seat six boys and eight girls in a row of chairs so that no two boys are seated next to each other?

53. Devise an algorithm for generating all the r-permutations of a finite set when repetition is allowed.

54. Devise an algorithm for generating all the r-combinations of a finite set when repetition is allowed.

Computer Projects

WRITE PROGRAMS WITH THE FOLLOWING INPUT AND OUTPUT.

1. Given a positive integer n and a nonnegative integer not exceeding n, find the number of r-permutations and r-combinations of a set with n elements.

2. Given positive integers n and r, find the number of r-permutations when repetition is allowed and r-combinations when repetition is allowed of a set with n elements.

3. Given a positive integer n, find the probability of selecting the six integers from the set $\{1, 2, \ldots, n\}$ that were mechanically selected in a lottery.

4. Given a sequence of positive integers, find the longest increasing and the longest decreasing subsequence of the sequence.

★5. Given an equation $x_1 + x_2 + \cdots + x_n = C$, where C is a constant, and $x_1, x_2, \ldots, x_n$ are nonnegative integers, list all the solutions.

6. Given a positive integer n, list all the permutations of the set $\{1, 2, 3, \ldots, n\}$ in lexicographic order.

7. Given a positive integer n and a nonnegative integer r not exceeding n, list all the r-combinations of the set $\{1, 2, 3, \ldots, n\}$ in lexicographic order.

8. Given a positive integer n and a nonnegative integer r not exceeding n, list all the r-permutations of the set $\{1, 2, 3, \ldots, n\}$ in lexicographic order.

9. Given a positive integer n, list all the combinations of the set $\{1, 2, 3, \ldots, n\}$.

10. Given positive integers n and r, list all the r-permutations, with repetition allowed, of the set $\{1, 2, 3, \ldots, n\}$.

11. Given positive integers n and r, list all the r-combinations, with repetition allowed, of the set $\{1, 2, 3, \ldots, n\}$.

5

Advanced Counting Techniques

Many counting problems cannot be solved easily using methods discussed in Chapter 4. One such problem is: How many bit strings of length n do not contain two consecutive zeros? To solve this problem, let a_n be the number of such strings of length n. An argument can be given that shows $a_{n+1} = a_n + a_{n-1}$. This equation and the initial conditions $a_1 = 2$ and $a_2 = 3$ determine the sequence $\{a_n\}$. Moreover, an explicit formula can be found for a_n from the equation relating the terms of the sequence. As we will see, a similar technique can be used to solve many different types of counting problems.

There are many other kinds of counting problems that cannot be solved using the techniques discussed in Chapter 4, such as: How many ways are there to assign seven jobs to three employees so that each employee is assigned at least one job? How many primes are there less than 1000? Both of these problems can be solved by counting the number of elements in the union of sets. We will develop a technique, called the principle of inclusion-exclusion, that counts the number of elements in unions of sets, and we will show how this principle can be used to solve counting problems.

The techniques studied in this chapter, together with the basic techniques of Chapter 4, can be used to solve many counting problems. We will briefly consider another counting technique—namely, generating functions—in Appendix 3.

5.1

Recurrence Relations

INTRODUCTION

The number of bacteria in a colony doubles every hour. If a colony begins with five bacteria, how many will be present in n hours? To solve this problem, let a_n be the number of bacteria at the end of n hours. Since the number of bacteria doubles every hour, the relationship $a_n = 2a_{n-1}$ holds whenever n is a positive integer. This relationship, together with the initial condition $a_0 = 5$, uniquely determines a_n for all nonnegative integers n. We can find a formula for a_n from this information.

We cannot solve many types of counting problems using the techniques discussed in Chapter 4. However, some of these can be solved by finding relationships, called recurrence relations, between the terms of a sequence, as was done in the problem involving bacteria. We will study a variety of counting problems that can be modeled using recurrence relations. We will develop methods in this section and in the following section for finding explicit formulae for the terms of sequences that satisfy certain types of recurrence relations.

RECURRENCE RELATIONS

In Chapter 3 we discussed how sequences can be defined recursively. Recall that a recursive definition of a sequence specifies one or more initial terms and a rule for determining subsequent terms from those that precede them. Recursive definitions can be used to solve counting problems. When they are, the rule for finding terms from those that precede them is called a **recurrence relation.**

DEFINITION 1. A *recurrence relation* for the sequence $\{a_n\}$ is a formula that expresses a_n in terms of one or more of the previous terms of the sequence, namely, $a_0, a_1, \ldots, a_{n-1}$, for all integers n with $n \geq n_0$, where n_0 is a nonnegative integer. A sequence is called a *solution* of a recurrence relation if its terms satisfy the recurrence relation.

EXAMPLE 1 Let $\{a_n\}$ be a sequence that satisfies the recurrence relation $a_n = a_{n-1} - a_{n-2}$ for $n = 2, 3, 4, \ldots$, and suppose that $a_0 = 3$ and $a_1 = 5$. What are a_2 and a_3?

Solution: We see from the recurrence relation that $a_2 = a_1 - a_0 = 5 - 3 = 2$ and $a_3 = a_2 - a_1 = 2 - 5 = -3$. ■

EXAMPLE 2 Determine whether the sequence $\{a_n\}$ is a solution of the recurrence relation $a_n = 2a_{n-1} - a_{n-2}$ for $n = 2, 3, 4, \ldots$, where $a_n = 3n$ for every nonnegative integer n. Answer the same question where $a_n = 2^n$ and where $a_n = 5$.

Solution: Suppose that $a_n = 3n$ for every nonnegative integer n. Then, for $n \geq 2$, we see that $a_n = 2a_{n-1} - a_{n-2} = 2[3(n-1)] - 3(n-2) = 3n$. Therefore, $\{a_n\}$, where $a_n = 3n$, is a solution of the recurrence relation.

Suppose that $a_n = 2^n$ for every nonnegative integer n. Note that $a_0 = 1$, $a_1 = 2$, and $a_2 = 4$. Since $a_2 \neq 2a_1 - a_0 = 2 \cdot 2 - 1 = 3$, we see that $\{a_n\}$, where $a_n = 2^n$, is not a solution of the recurrence relation.

Suppose that $a_n = 5$ for every nonnegative integer n. Then for $n \geq 2$, we see that $a_n = 2a_{n-1} - a_{n-2} = 2 \cdot 5 - 5 = 5$. Therefore, $\{a_n\}$, where $a_n = 5$, is a solution of the recurrence relation. ■

The **initial conditions** for a sequence specify the terms that precede the first term where the recurrence relation takes effect. For instance, in Example 1, $a_0 = 3$ and $a_1 = 5$ are the initial conditions. The recurrence relation and initial conditions uniquely determine a sequence. This is the case since a recurrence relation, together with initial conditions, provide a recursive definition of the sequence. Any term of the sequence can be found from the initial conditions using the recurrence relation a sufficient number of times. However, there are better ways for computing the terms of certain classes of sequences defined by recurrence relations and initial conditions. We will discuss these methods in this section and in the following section.

MODELING WITH RECURRENCE RELATIONS

We can use recurrence relations to model a wide variety of problems, such as finding compound interest, counting rabbits on an island, determining the number of moves in the tower of Hanoi puzzle, and counting bit strings with certain properties.

EXAMPLE 3 Compound Interest Suppose that a person deposits $10,000 in a savings account at a bank yielding 11% per year with interest compounded annually. How much will be in the account after 30 years?

Solution: To solve this problem, let P_n denote the amount in the account after n years. Since the amount in the account after n years equals the amount in the account after $n-1$ years plus interest for the nth year, we see that the sequence $\{P_n\}$ satisfies the recurrence relation

$$P_n = P_{n-1} + 0.11P_{n-1} = (1.11)P_{n-1}.$$

The initial condition is $P_0 = 10{,}000$.

We can use an iterative approach to find a formula for P_n. Note that

$$\begin{aligned}
P_1 &= (1.11)P_0 \\
P_2 &= (1.11)P_1 = (1.11)^2 P_0 \\
P_3 &= (1.11)P_2 = (1.11)^3 P_0 \\
&\ \ \vdots \\
P_n &= (1.11)P_{n-1} = (1.11)^n P_0.
\end{aligned}$$

When we insert the initial condition $P_0 = 10{,}000$, the formula $P_n = (1.11)^n 10{,}000$ is obtained. We can use mathematical induction to establish its validity. That the formula is valid for $n = 0$ is a consequence of the initial condition. Now assume that $P_n = (1.11)^n 10{,}000$. Then, from the recurrence relation and the induction hypothesis,

$$P_{n+1} = (1.11)P_n = (1.11)(1.11)^n 10{,}000 = (1.11)^{n+1} 10{,}000.$$

This shows that the explicit formula for P_n is valid.

Inserting $n = 30$ into the formula $P_n = (1.11)^n 10{,}000$ shows that after 30 years the account contains

$$P_{30} = (1.11)^{30} 10{,}000 = \$228{,}922.97.$$ ■

The next example shows how the population of rabbits on an island can be modeled using a recurrence relation.

EXAMPLE 4 Rabbits and the Fibonacci Numbers Consider the following problem, which was originally posed by Leonardo di Pisa, also known as Fibonacci, in the 13th century in his book *Liber abaci*. A young pair of rabbits (one of each sex) is placed on an island. A pair of rabbits does not breed until they are two months old. After they are two months old, each pair of rabbits produces another pair each month, as shown in Figure 1. Find a recurrence relation for the number of pairs of rabbits on the island after n months, assuming that no rabbits ever die.

Solution: Denote by f_n the number of pairs of rabbits after n months. We will show that f_n, $n = 1, 2, 3, \ldots$, are the terms of the Fibonacci sequence.

The rabbit population can be modeled using a recurrence relation. At the end of the first month, the number of pairs of rabbits on the island is $f_1 = 1$. Since this pair does not breed during the second month, $f_2 = 1$ also. To find the number of pairs after n months, add the number on the island the previous month, f_{n-1}, and the number of newborn pairs, which equals f_{n-2}, since each newborn pair comes from a pair at least two months old.

Consequently, the sequence $\{f_n\}$ satisfies the recurrence relation

$$f_n = f_{n-1} + f_{n-2}$$

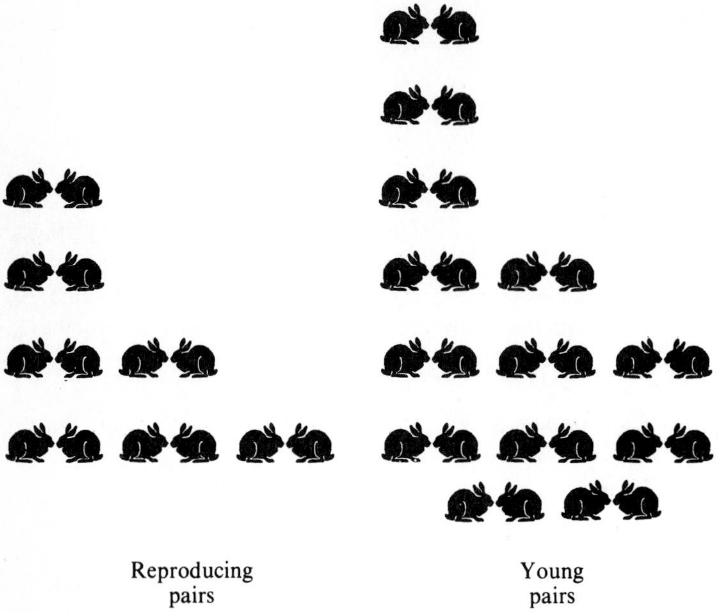

Month	Reproducing Pairs	Young Pairs	Total Pairs
1	0	1	1
2	0	1	1
3	1	1	2
4	1	2	3
5	2	3	5
6	3	5	8

Reproducing pairs

Young pairs

FIGURE 1 **Rabbits on an Island.**

for $n \geq 3$ together with the initial conditions $f_1 = 1$ and $f_2 = 1$. Since this recurrence relation and the initial conditions uniquely determine this sequence, the number of pairs of rabbits on the island after n months is given by the nth Fibonacci number. ■

The next example involves a famous puzzle.

EXAMPLE 5 The Tower of Hanoi A popular puzzle of the late 19th century, called the Tower of Hanoi, consists of three pegs mounted on a board together with disks of different sizes. Initially these disks are placed on the first peg in order of size, with the largest on the bottom (as shown in Figure 2). The rules of the puzzle allow disks to be moved one at a time from one peg to another as long as a disk is never placed on top of a smaller disk. The goal of the puzzle is to have all the disks on the second peg in order of size, with the largest on the bottom.

Let H_n denote the number of moves needed to solve the Tower of Hanoi problem with n disks. Set up a recurrence relation for the sequence $\{H_n\}$.

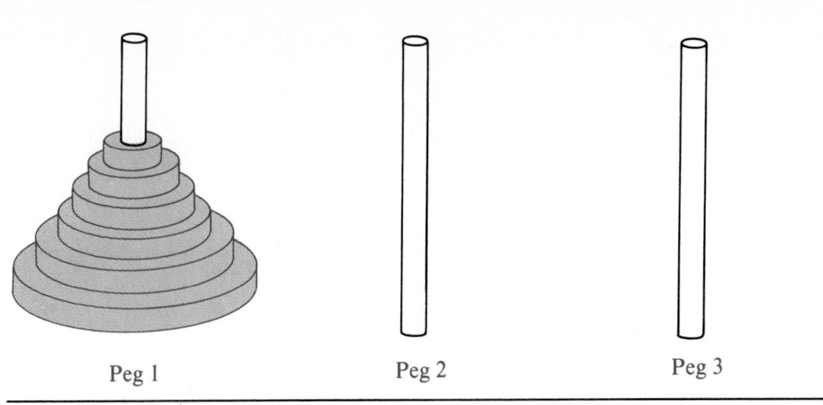

FIGURE 2 **The Initial Position in the Tower of Hanoi.**

Solution: Begin with n disks on peg 1. We can transfer the top $n-1$ disks, following the rules of the puzzle, to peg 3 using H_{n-1} moves (see Figure 3 for an illustration of the pegs and disks at this point). We keep the largest disk fixed during these moves. Then, we use one move to transfer the largest disk to the second peg. We can transfer the $n-1$ disks on peg 3 to peg 2, using H_{n-1} additional moves, placing them on top of the largest disk, which always stays fixed on the bottom of peg 2. Moreover, it is easy to see that the puzzle cannot be solved using fewer steps. This shows that

$$H_n = 2H_{n-1} + 1.$$

The initial condition is $H_1 = 1$, since one disk can be transferred from peg 1 to peg 2, according to the rules of the puzzle, in one move.

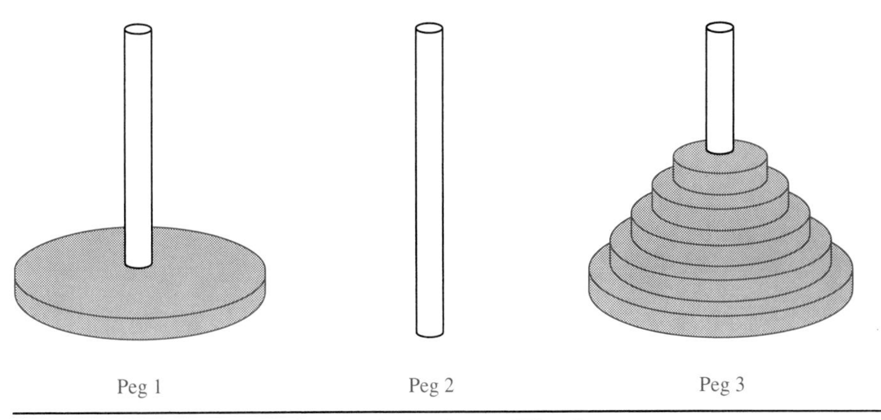

FIGURE 3 **An Intermediate Position in the Tower of Hanoi.**

We can use an iterative approach to solve this recurrence relation. Note that

$$\begin{aligned}
H_n &= 2H_{n-1} + 1 \\
&= 2(2H_{n-2} + 1) + 1 = 2^2 H_{n-2} + 2 + 1 \\
&= 2^2(2H_{n-3} + 1) + 2 + 1 = 2^3 H_{n-3} + 2^2 + 2 + 1 \\
&\vdots \\
&= 2^{n-1} H_1 + 2^{n-2} + 2^{n-3} + \cdots + 2 + 1 \\
&= 2^{n-1} + 2^{n-2} + \cdots + 2 + 1 \\
&= 2^n - 1.
\end{aligned}$$

We have used the recurrence relation repeatedly to express H_n in terms of previous terms of the sequence. In the next to last equality, the initial condition $H_1 = 1$ has been used. The last equality is based on the formula for the sum of the terms of a geometric series, which can be found in Example 5 in Section 3.2.

The iterative approach has produced the solution to the recurrence relation $H_n = 2H_{n-1} + 1$ with the initial condition $H_1 = 1$. This formula can be proved using mathematical induction. This is left as an exercise for the reader at the end of the section.

An ancient myth tells us that there is a tower in Hanoi where monks are transferring 64 gold disks from one peg to another, according to the rules of the puzzle. They take one second to move a disk. The myth says that the world will end when they finish the puzzle. How long after the monks started will the world end?

From the explicit formula, the monks require

$$2^{64} - 1 = 18{,}446{,}744{,}073{,}709{,}551{,}615$$

moves to transfer the disks. Making one move per second, it will take them more than 500 billion years to solve the puzzle, so the world should survive a while longer than it already has. ■

Example 6 illustrates how recurrence relations can be used to count bit strings of a specified length that have a certain property.

EXAMPLE 6 Find a recurrence relation and give initial conditions for the number of bit strings of length n that do not have two consecutive 0s. How many such bit strings are there of length five?

Solution: Let a_n denote the number of bit strings of length n that do not have two consecutive 0s. To obtain a recurrence relation for $\{a_n\}$, note that by the sum rule, the number of bit strings of length n that do have two consecutive 0s equals the number of such bit strings ending with a 0 plus the number of such bit strings ending with a 1. We will assume that $n \geq 3$, so that the bit string has at least three bits.

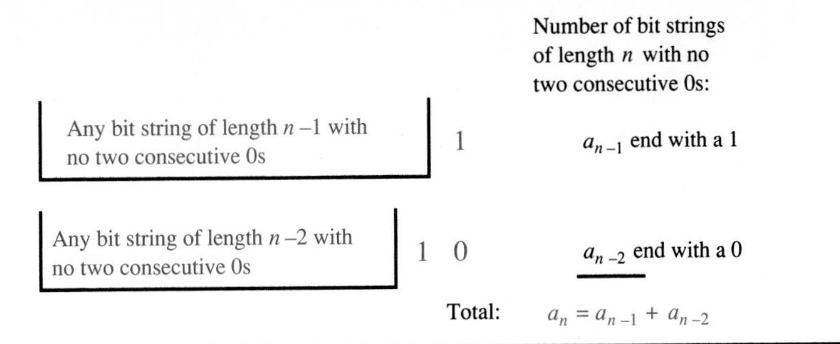

FIGURE 4 **Counting Bit Strings of Length n with No Two Consecutive 0s.**

The bit strings of length n ending with 1 that do not have two consecutive 0s are precisely the bit strings of length $n - 1$ with no two consecutive 0s with a 1 added at the end. Consequently, there are a_{n-1} such bit strings.

Bit strings of length n ending with a 0 that do not have two consecutive 0s must have 1 as their $(n - 1)$st bit; otherwise they would end with a pair of 0s. It follows that the bit strings of length n ending with a 0 that have no two consecutive 0s are precisely the bit strings of length $n - 2$ with no two consecutive 0s with 10 added at the end. Consequently, there are a_{n-2} such bit strings.

We conclude, as illustrated in Figure 4, that

$$a_n = a_{n-1} + a_{n-2}$$

for $n \geq 3$.

The initial conditions are $a_1 = 2$, since both bit strings of length one, 0 and 1, do not have consecutive 0s, and $a_2 = 3$, since the valid bit strings of length two are 01, 10, and 11. To obtain a_5, we use the recurrence relation three times to find that

$$a_3 = a_2 + a_1 = 3 + 2 = 5$$
$$a_4 = a_3 + a_2 = 5 + 3 = 8$$
$$a_5 = a_4 + a_3 = 8 + 5 = 13.$$

Remark: Note that $\{a_n\}$ satisfies the same recurrence relation as the Fibonacci sequence. Since $a_1 = f_3$ and $a_2 = f_4$ it follows that $a_n = f_{n+2}$. ■

The next example shows how a recurrence relation can be used to model the number of codewords that are allowable using certain validity checks.

EXAMPLE 7 Codeword Enumeration A computer system considers a string of decimal digits a valid codeword if it contains an even number of 0 digits. For instance, 1230407869 is valid, whereas 120987045608 is not valid. Let a_n be the number of valid n-digit codewords. Find a recurrence relation for a_n.

Solution: Note that $a_1 = 9$ since there are 10 one-digit strings, and only one, namely, the string 0, is not valid. A recurrence relation can be derived for this sequence by considering how a valid n-digit string can be obtained from strings of $n - 1$ digits. There are two ways to form a valid string with n digits from a string with one fewer digit.

First, a valid string of n digits can be obtained by appending a valid string of $n - 1$ digits with a digit other than 0. This appending can be done in nine ways. Hence, a valid string with n digits can be formed in this manner in $9a_{n-1}$ ways.

Second, a valid string of n digits can be obtained by appending a 0 to a string of length $n - 1$ that is not valid. (This produces a string with an even number of 0 digits since the invalid string of length $n - 1$ has an odd number of 0 digits.) The number of ways that this can be done equals the number of invalid $(n - 1)$-digit strings. Since there are 10^{n-1} strings of length $n - 1$, and a_{n-1} are valid, there are $10^{n-1} - a_{n-1}$ valid n-digit strings obtained by appending an invalid string of length $n - 1$ with a 0.

Since all valid strings of length n are produced in one of these two ways, it follows that there are

$$\begin{aligned} a_n &= 9a_{n-1} + (10^{n-1} - a_{n-1}) \\ &= 8a_{n-1} + 10^{n-1} \end{aligned}$$

valid strings of length n. ■

Exercises

1. Find the first five terms of the sequence defined by each of the following recurrence relations and initial conditions.

a) $a_n = 6a_{n-1}$, $a_0 = 2$
b) $a_n = a_{n-1}^2$, $a_1 = 2$
c) $a_n = a_{n-1} + 3a_{n-2}$, $a_0 = 1$, $a_1 = 2$
d) $a_n = na_{n-1} + n^2a_{n-2}$, $a_0 = 1$, $a_1 = 1$
e) $a_n = a_{n-1} + a_{n-3}$, $a_0 = 1$, $a_1 = 2$, $a_2 = 0$

2. Show that the sequence $\{a_n\}$ is a solution of the recurrence relation $a_n = -3a_{n-1} + 4a_{n-2}$ if

a) $a_n = 0$. **b)** $a_n = 1$.
c) $a_n = (-4)^n$. **d)** $a_n = 2(-4)^n + 3$.

3. Is the sequence $\{a_n\}$ a solution of the recurrence relation $a_n = 8a_{n-1} - 16a_{n-2}$ if

a) $a_n = 0$? **b)** $a_n = 1$?
c) $a_n = 2^n$? **d)** $a_n = 4^n$?
e) $a_n = n4^n$? **f)** $a_n = 2 \cdot 4^n + 3n4^n$?
g) $a_n = (-4)^n$? **h)** $a_n = n^24^n$?

4. For each of the following sequences find a recurrence relation satisfied by this sequence. (The answers are not unique since there are infinitely many different recurrence relations satisfied by any sequence.)

a) $a_n = 3$ **b)** $a_n = 2n$
c) $a_n = 2n + 3$ **d)** $a_n = 5^n$
e) $a_n = n^2$ **f)** $a_n = n^2 + n$
g) $a_n = n + (-1)^n$ **h)** $a_n = n!$

5. Find the solution to each of the following recurrence relations and initial conditions. Use an iterative approach, such as that used in Example 5.

a) $a_n = 3a_{n-1}$, $a_0 = 2$
b) $a_n = a_{n-1} + 2$, $a_0 = 3$
c) $a_n = a_{n-1} + n$, $a_0 = 1$
d) $a_n = a_{n-1} + 2n + 3$, $a_0 = 4$
e) $a_n = 2a_{n-1} - 1$, $a_0 = 1$
f) $a_n = 3a_{n-1} + 1$, $a_0 = 1$
g) $a_n = na_{n-1}$, $a_0 = 5$
h) $a_n = 2na_{n-1}$, $a_0 = 1$

6. A person deposits \$1000 in an account that yields 9% interest compounded yearly.

a) Set up a recurrence relation for the amount in the account at the end of n years.

b) Find an explicit formula for the amount in the account at the end of n years.

c) How much money will the account contain after 100 years?

7. Suppose that the number of bacteria in a colony triples every hour.

a) Set up a recurrence relation for the number of bacteria after n hours have elapsed.

b) If 100 bacteria are used to begin a new colony, how many bacteria will be in the colony in 10 hours?

8. Assume that the population of the world in 1988 is 6 billion and is growing 3% a year.

a) Set up a recurrence relation for the population of the world n years after 1988.

b) Find an explicit formula for the population of the world n years after 1988.

c) What will the population of the world be in 2001?

9. A factory makes custom sports cars at an increasing rate. In the first month only one car is made, in the second month two cars are made, and so on, with n cars made in the nth month.

a) Set up a recurrence relation for the number of cars produced in the first n months by this factory.

b) How many cars are produced in the first year?

c) Find an explicit formula for the number of cars produced in the first n months by this factory.

10. An employee joined a company in 1987 with a starting salary of \$50,000. Every year this employee receives a raise of \$1000 plus 5% of the salary of the previous year.

a) Set up a recurrence relation for the salary of this employee n years after 1987.

b) What is the salary of this employee in 1995?

c) Find an explicit formula for the salary of this employee n years after 1987.

11. Use mathematical induction to verify the formula derived in Example 5 for the number of moves required to complete the Tower of Hanoi puzzle.

12. a) Find a recurrence relation for the number of permutations of a set with n elements.

b) Use this recurrence relation to find the number of permutations of a set with n elements using iteration.

13. A vending machine dispensing books of stamps accepts only Susan B. Anthony dollar coins, \$1 bills, and \$5 bills.

a) Find a recurrence relation for the number of ways to deposit n dollars in the vending machine, where the order in which the coins and bills are deposited matters.

b) What are the initial conditions?

c) How many ways are there to deposit \$10 for a book of stamps?

14. A country uses as currency coins with values of 1 peso, 2 pesos, 5 pesos, and 10 pesos and bills with values of 5 pesos, 10 pesos, 20 pesos, 50 pesos, and 100 pesos. Find a recurrence relation for the number of ways to pay a bill of n pesos if the order in which the coins and bills are paid matters.

15. a) Find a recurrence relation for the number of bit strings of length n that contain a pair of consecutive 0s.

b) What are the initial conditions?

c) How many bit strings of length seven contain two consecutive 0s?

16. a) Find a recurrence relation for the number of bit strings of length n that contain three consecutive 0s.

b) What are the initial conditions?

c) How many bit strings of length seven contain three consecutive 0s?

17. a) Find a recurrence relation for the number of bit strings of length n that do not contain three consecutive 0s.

b) What are the initial conditions?

c) How many bit strings of length seven do not contain three consecutive 0s?

★18. a) Find a recurrence relation for the number of bit strings that contain the string 01.

b) What are the initial conditions?

c) How many bit strings of length seven contain the string 01?

19. a) Find a recurrence relation for the number of ways to climb n stairs if the person climbing the stairs can take one stair or two stairs at a time.

b) What are the initial conditions?

c) How many ways can this person climb a flight of stairs?

20. a) Find a recurrence relation for the number of ways to climb n stairs if the person climbing the stairs can take one, two, or three stairs at a time.

b) What are the initial conditions?

c) How many ways can this person climb a flight of eight stairs?

A string that contains only 0s, 1s, and 2s is called a **ternary string.**

21. a) Find a recurrence relation for the number of ternary strings that do not contain two consecutive 0s.
b) What are the initial conditions?
c) How many ternary strings of length six do not contain two consecutive 0s?

22. a) Find a recurrence relation for the number of ternary strings that contain two consecutive 0s.
b) What are the initial conditions?
c) How many ternary strings of length six contain two consecutive 0s?

★23. a) Find a recurrence relation for the number of ternary strings that do not contain two consecutive 0s or two consecutive 1s.
b) What are the initial conditions?
c) How many ternary strings of length six do not contain two consecutive 0s or two consecutive 1s?

★24. a) Find a recurrence relation for the number of ternary strings that contain either two consecutive 0s or two consecutive 1s.
b) What are the initial conditions?
c) How many ternary strings of length six contain either two consecutive 0s or two consecutive 1s?

★25. a) Find a recurrence relation for the number of ternary strings that do not contain consecutive symbols that are the same.
b) What are the initial conditions?
c) How many ternary strings of length six do not contain consecutive symbols that are the same?

★★26. a) Find a recurrence relation for the number of ternary strings that contain two consecutive symbols that are the same.
b) What are the initial conditions?
c) How many ternary strings of length six contain consecutive symbols that are the same?

27. Messages are transmitted over a communications channel using two signals. The transmittal of one signal requires 1 microsecond, and the transmittal of the other signal requires 2 microseconds.
a) Find a recurrence relation for the number of different messages consisting of sequences of these two signals, where each signal in the message is immediately followed by the next signal, that can be sent in n microseconds.
b) What are the initial conditions?
c) How many different messages can be sent in 10 microseconds using these two signals?

28. A bus driver pays all tolls, using only nickels and dimes, by throwing one coin at a time into the mechanical toll collector.
a) Find a recurrence relation for the number of different ways the bus driver can pay a toll of n cents (where the order in which the coins are used matters).
b) In how many different ways can the driver pay a toll of 45 cents?

29. a) Find the recurrence relation satisfied by R_n, where R_n is the number of regions that a plane is divided into by n lines, if no two of the lines are parallel and no three of the lines go through the same point.
b) Find R_n using iteration.

30. Find a recurrence relation for the number of bit sequences of length n with an even number of 0s.

31. How many bit sequences of length seven contain an even number of 0s?

32. Show that the Fibonacci numbers satisfy the recurrence relation $f_n = 5f_{n-4} + 3f_{n-5}$ for $n = 5, 6, 7, \ldots,$ together with the initial conditions $f_0 = 0$, $f_1 = 1$, $f_2 = 1$, $f_3 = 2$, and $f_4 = 3$. Use this recurrence relation to show that f_{5n} is divisible by 5, for $n = 1, 2, 3, \ldots$.

★33. Let $S(m, n)$ denote the number of onto functions from a set with m elements to a set with n elements. Show that $S(m, n)$ satisfies the recurrence relation

$$S(m, n) = n^m - \sum_{k=1}^{n-1} C(n, k)S(m, k)$$

whenever $m \geq n$ and $n > 1$, with the initial condition $S(m, 1) = 1$.

Let $\{a_n\}$ be a sequence of real numbers. The **backward differences** of this sequence are defined recursively as follows. The **first difference** ∇a_n is

$$\nabla a_n = a_n - a_{n-1}.$$

The **$(k+1)$th difference** $\nabla^{k+1} a_n$ is obtained from $\nabla^k a_n$ by

$$\nabla^{k+1} a_n = \nabla^k a_n - \nabla^k a_{n-1}.$$

34. Find ∇a_n for the sequence $\{a_n\}$ where
a) $a_n = 4$. **b)** $a_n = 2n$.
c) $a_n = n^2$. **d)** $a_n = 2^n$.

35. Find $\nabla^2 a_n$ for the sequences in Exercise 34.

36. Show that $a_{n-1} = a_n - \nabla a_n$.

37. Show that $a_{n-2} = a_n - 2\nabla a_n + \nabla^2 a_n$.

★**38.** Prove that a_{n-k} can be expressed in terms of $a_n, \nabla a_n, \nabla^2 a_n, \ldots, \nabla^k a_n$.

39. Express the recurrence relation $a_n = a_{n-1} + a_{n-2}$ in terms of a_n, ∇a_n, and $\nabla^2 a_n$.

40. Show that any recurrence relation for the sequence $\{a_n\}$ can be written in terms of $a_n, \nabla a_n, \nabla^2 a_n, \ldots$. The resulting equation involving the sequences and its differences is called a **difference equation.**

5.2 Solving Recurrence Relations

INTRODUCTION

A wide variety of recurrence relations occur in models. Some of these recurrence relations can be solved using iteration or some other ad hoc technique. However, there is an important class of recurrence relations that can be explicitly solved in a systematic way. These are recurrence relations that express the terms of a sequence as linear combinations of previous terms.

DEFINITION 1. A *linear homogeneous recurrence relation of degree* k *with constant coefficients* is a recurrence relation of the form

$$a_n = c_1 a_{n-1} + c_2 a_{n-2} + \cdots + c_k a_{n-k}$$

where $c_1, c_2, \ldots, c_k$ are real numbers, and $c_k \neq 0$.

The recurrence relation in the definition is **linear** since the right-hand side is a sum of multiples of the previous terms of the sequence. The recurrence relation is **homogeneous** since no terms occur that are not multiples of the a_js. The coefficients of the terms of the sequence are all **constants,** rather than functions that depend on n. The **degree** is k because a_n is expressed in terms of the previous k terms of the sequence.

It is a consequence of the second principle of mathematical induction that a sequence satisfying the recurrence relation in the definition is uniquely determined by this recurrence relation and the k initial conditions

$$a_0 = C_0, a_1 = C_1, \ldots, a_{k-1} = C_{k-1}.$$

EXAMPLE 1 The recurrence relation $P_n = (1.11)P_{n-1}$ is a linear homogeneous recurrence relation of degree one. The recurrence relation $f_n = f_{n-1} + f_{n-2}$ is a linear homogeneous recurrence relation of degree two. The recurrence relation $a_n = a_{n-5}$ is a linear homogeneous recurrence relation of degree five. ■

Some examples of recurrence relations that are not linear homogeneous recurrence relations with constant coefficients follow.

EXAMPLE 2 The recurrence relation $a_n = a_{n-1} + a_{n-2}^2$ is not linear. The recurrence relation $H_n = 2H_{n-1} + 1$ is not homogeneous. The recurrence relation $B_n = nB_{n-1}$ does not have constant coefficients. ■

Linear homogeneous recurrence relations are studied for two reasons. First, they often occur in modeling of problems. Second, they can be systematically solved.

SOLVING LINEAR HOMOGENEOUS RECURRENCE RELATIONS WITH CONSTANT COEFFICIENTS

The basic approach for solving linear homogeneous recurrence relations is to look for solutions of the form $a_n = r^n$, where r is a constant. Note that $a_n = r^n$ is a solution of the recurrence relation $a_n = c_1a_{n-1} + c_2a_{n-2} + \cdots + c_ka_{n-k}$ if and only if

$$r^n = c_1r^{n-1} + c_2r^{n-2} + \cdots + c_kr^{n-k}.$$

When both sides of this equation are divided by r^{n-k} and the right-hand side is subtracted from the left, we obtain the equivalent equation

$$r^k - c_1r^{k-1} - c_2r^{k-2} - \cdots - c_{k-1}r - c_k = 0.$$

Consequently, the sequence $\{a_n\}$ with $a_n = r^n$ is a solution if and only if r is a solution of this last equation, which is called the **characteristic equation** of the recurrence relation. The solutions of this equation are called the **characteristic roots** of the recurrence relation. As we will see, these characteristic roots can be used to give an explicit formula for all the solutions of the recurrence relation.

We will first develop results that deal with linear homogeneous recurrence relations with constant coefficients of degree two. Then corresponding general results when the degree may be greater than two will be stated. Because the proofs needed to establish the results in the general case are more complicated, they will not be given in the text.

We now turn our attention to linear homogeneous recurrence relations of degree two. First, consider the case when there are two distinct characteristic roots.

THEOREM 1 Let c_1 and c_2 be real numbers. Suppose that $r^2 - c_1r - c_2 = 0$ has two distinct roots r_1 and r_2. Then the sequence $\{a_n\}$ is a solution of the recurrence relation $a_n = c_1a_{n-1} + c_2a_{n-2}$ if and only if $a_n = \alpha_1r_1^n + \alpha_2r_2^n$ for $n = 0, 1, 2, \ldots$, where α_1 and α_2 are constants.

Proof: We must do two things to prove the theorem. First, it must be shown that if r_1 and r_2 are the roots of the characteristic equation, and α_1 and α_2 are constants, then the sequence $\{a_n\}$ with $a_n = \alpha_1 r_1^n + \alpha_2 r_2^n$ is a solution of the recurrence relation. Second, it must be shown that if the sequence $\{a_n\}$ is a solution, then $a_n = \alpha_1 r_1^n + \alpha_2 r_2^n$ for some constants α_1 and α_2.

Now we will show that if $a_n = \alpha_1 r_1^n + \alpha_2 r_2^n$, then the sequence $\{a_n\}$ is a solution of the recurrence relation. Since r_1 and r_2 are roots of $r^2 - c_1 r - c_2 = 0$, it follows that $r_1^2 = c_1 r_1 + c_2$, $r_2^2 = c_1 r_2 + c_2$.

From these equations, we see that

$$\begin{aligned} c_1 a_{n-1} + c_2 a_{n-2} &= c_1(\alpha_1 r_1^{n-1} + \alpha_2 r_2^{n-1}) + c_2(\alpha_1 r_1^{n-2} + \alpha_2 r_2^{n-2}) \\ &= \alpha_1 r_1^{n-2}(c_1 r_1 + c_2) + \alpha_2 r_2^{n-2}(c_1 r_2 + c_2) \\ &= \alpha_1 r_1^{n-2} r_1^2 + \alpha_2 r_2^{n-2} r_2^2 \\ &= \alpha_1 r_1^n + \alpha_2 r_2^n \\ &= a_n. \end{aligned}$$

This shows that the sequence $\{a_n\}$ with $a_n = \alpha_1 r_1^n + \alpha_2 r_2^n$ is a solution of the recurrence relation.

To show that every solution $\{a_n\}$ of the recurrence relation $a_n = c_1 a_{n-1} + c_2 a_{n-2}$ has $a_n = \alpha_1 r_1^n + \alpha_2 r_2^n$ for $n = 0, 1, 2, \ldots$, for some constants α_1 and α_2, suppose that $\{a_n\}$ is a solution of the recurrence relation, and the initial conditions $a_0 = C_0$ and $a_1 = C_1$ hold. It will be shown that there are constants α_1 and α_2 so that the sequence $\{a_n\}$ with $a_n = \alpha_1 r_1^n + \alpha_2 r_2^n$ satisfies these same initial conditions. This requires that

$$\begin{aligned} a_0 &= C_0 = \alpha_1 + \alpha_2 \\ a_1 &= C_1 = \alpha_1 r_1 + \alpha_2 r_2. \end{aligned}$$

We can solve these two equations for α_1 and α_2. From the first equation it follows that $\alpha_2 = C_0 - \alpha_1$. Inserting this expression into the second equation gives

$$C_1 = \alpha_1 r_1 + (C_0 - \alpha_1) r_2.$$

Hence,

$$C_1 = \alpha_1(r_1 - r_2) + C_0 r_2.$$

This shows that

$$\alpha_1 = \frac{(C_1 - C_0 r_2)}{r_1 - r_2}$$

and

$$\alpha_2 = C_0 - \alpha_1 = C_0 - \frac{(C_1 - C_0 r_2)}{r_1 - r_2} = \frac{C_0 r_1 - C_1}{r_1 - r_2},$$

where these expressions for α_1 and α_2 depend on the fact that $r_1 \neq r_2$. (When $r_1 = r_2$, this theorem is not true.) Hence, with these values for α_1 and α_2, the sequence $\{a_n\}$ with $\alpha_1 r_1^n + \alpha_2 r_2^n$ satisfies the two initial conditions. Since this recurrence relation and these initial conditions uniquely determine the sequence, it follows that $a_n = \alpha_1 r_1^n + \alpha_2 r_2^n$. ∎

The characteristic roots of a linear homogeneous recurrence relation with constant coefficients may be complex numbers. Theorem 1 (and also subsequent theorems in this section) still applies in this case. Recurrence relations with complex characteristic roots will not be discussed in the text. Readers familiar with complex numbers may wish to solve Exercises 22 and 23 at the end of this section.

The following examples illustrate the usefulness of the explicit formula given in Theorem 1.

EXAMPLE 3 What is the solution of the recurrence relation

$$a_n = a_{n-1} + 2a_{n-2}$$

with $a_0 = 2$ and $a_1 = 7$?

Solution: Theorem 1 can be used to solve this problem. The characteristic equation of the recurrence relation is $r^2 - r - 2 = 0$. Its roots are $r = 2$ and $r = -1$. Hence, the sequence $\{a_n\}$ is a solution to the recurrence relation if and only if

$$a_n = \alpha_1 2^n + \alpha_2(-1)^n,$$

for some constants α_1 and α_2. From the initial conditions, it follows that

$$\begin{aligned} a_0 &= 2 = \alpha_1 + \alpha_2 \\ a_1 &= 7 = \alpha_1 \cdot 2 + \alpha_2 \cdot (-1). \end{aligned}$$

Solving these two equations shows that $\alpha_1 = 3$ and $\alpha_2 = -1$. Hence, the solution to the recurrence relation and initial conditions is the sequence $\{a_n\}$ with

$$a_n = 3 \cdot 2^n - (-1)^n.$$

■

EXAMPLE 4 Find an explicit formula for the Fibonacci numbers.

Solution: Recall that the sequence of Fibonacci numbers satisfies the recurrence relation $f_n = f_{n-1} + f_{n-2}$ and also satisfies the initial conditions $f_0 = 0$ and $f_1 = 1$. The roots of the characteristic equation $r^2 - r - 1 = 0$ are $r_1 = (1 + \sqrt{5})/2$ and $r_2 = (1 - \sqrt{5})/2$. Therefore, from Theorem 1 it follows that the Fibonacci numbers are given by

$$f_n = \alpha_1 \left(\frac{1 + \sqrt{5}}{2}\right)^n + \alpha_2 \left(\frac{1 - \sqrt{5}}{2}\right)^n,$$

for some constants α_1 and α_2. The initial conditions $f_0 = 0$ and $f_1 = 1$ can be used to find these constants. We have

$$\begin{aligned} f_0 &= \alpha_1 + \alpha_2 = 0 \\ f_1 &= \alpha_1 \left(\frac{1 + \sqrt{5}}{2}\right) + \alpha_2 \left(\frac{1 - \sqrt{5}}{2}\right) = 1. \end{aligned}$$

The solution to these simultaneous equations for α_1 and α_2 is

$$\alpha_1 = 1/\sqrt{5}, \qquad \alpha_2 = -1/\sqrt{5}.$$

Consequently, the Fibonacci numbers are given by

$$f_n = \frac{1}{\sqrt{5}}\left(\frac{1+\sqrt{5}}{2}\right)^n - \frac{1}{\sqrt{5}}\left(\frac{1-\sqrt{5}}{2}\right)^n.$$

■

Theorem 1 does not apply when there is one characteristic root of multiplicity two. This case can be handled using the following theorem.

THEOREM 2 Let c_1 and c_2 be real numbers with $c_2 \neq 0$. Suppose that $r^2 - c_1 r - c_2 = 0$ has only one root r_0. A sequence $\{a_n\}$ is a solution of the recurrence relation $a_n = c_1 a_{n-1} + c_2 a_{n-2}$ if and only if $a_n = \alpha_1 r_0^n + \alpha_2 n r_0^n$, for $n = 0, 1, 2, \ldots$, where α_1 and α_2 are constants.

The proof of Theorem 2 is left as an exercise at the end of the section. The following example illustrates the use of this theorem.

EXAMPLE 5 What is the solution of the recurrence relation

$$a_n = 6a_{n-1} - 9a_{n-2}$$

with initial conditions $a_0 = 1$ and $a_1 = 6$?

Solution: The only root of $r^2 - 6r + 9 = 0$ is $r = 3$. Hence, the solution to this recurrence relation is

$$a_n = \alpha_1 3^n + \alpha_2 n 3^n$$

for some constants α_1 and α_2. Using the initial conditions, it follows that

$$a_0 = 1 = \alpha_1$$
$$a_1 = 6 = \alpha_1 \cdot 3 + \alpha_2 \cdot 3.$$

Solving these two equations shows that $\alpha_1 = 1$ and $\alpha_2 = 1$. Consequently, the solution to this recurrence relation and the initial conditions is

$$a_n = 3^n + n3^n.$$

■

We will now state the general result about the solution of linear homogeneous recurrence relations with constant coefficients, where the degree may be greater than two, under the assumption that the characteristic equation has distinct roots. The proof of this result will be left as an exercise for the reader.

THEOREM 3 Let $c_1, c_2, \ldots, c_k$ be real numbers. Suppose that the characteristic equation

$$r^k - c_1 r^{k-1} - \cdots - c_k = 0$$

has k distinct roots $r_1, r_2, \ldots, r_k$. Then a sequence $\{a_n\}$ is a solution of the recurrence relation

$$a_n = c_1 a_{n-1} + c_2 a_{n-2} + \cdots + c_k a_{n-k}$$

if and only if

$$a_n = \alpha_1 r_1^n + \alpha_2 r_2^n + \cdots + \alpha_k r_k^n$$

for $n = 0, 1, 2, \ldots,$ where $\alpha_1, \alpha_2, \ldots, \alpha_k$ are constants.

We illustrate the use of the theorem with an example.

EXAMPLE 6 Find the solution to the recurrence relation

$$a_n = 6a_{n-1} - 11a_{n-2} + 6a_{n-3}$$

with the initial conditions $a_0 = 2$, $a_1 = 5$, and $a_2 = 15$.

Solution: The characteristic polynomial of this recurrence relation is

$$r^3 - 6r^2 + 11r - 6.$$

The characteristic roots are $r = 1$, $r = 2$, and $r = 3$, since $r^3 - 6r^2 + 11r - 6 = (r-1)(r-2)(r-3)$. Hence, the solutions to this recurrence relation are of the form

$$a_n = \alpha_1 \cdot 1^n + \alpha_2 \cdot 2^n + \alpha_3 \cdot 3^n.$$

To find the constants α_1, α_2, and α_3, use the initial conditions. This gives

$$\begin{aligned} a_0 &= 2 = \alpha_1 + \alpha_2 + \alpha_3 \\ a_1 &= 5 = \alpha_1 + \alpha_2 \cdot 2 + \alpha_3 \cdot 3 \\ a_2 &= 15 = \alpha_1 + \alpha_2 \cdot 4 + \alpha_3 \cdot 9. \end{aligned}$$

When these three simultaneous equations are solved for α_1, α_2, and α_3, we find that $\alpha_1 = 1$, $\alpha_2 = -1$, and $\alpha_3 = 2$. Hence, the unique solution to this recurrence relation and the given initial conditions is the sequence $\{a_n\}$ with

$$a_n = 1 - 2^n + 2 \cdot 3^n.$$ ■

Exercises

1. Determine which of the following are linear homogeneous recurrence relations with constant coefficients. Also, find the degree of those that are.

a) $a_n = 3a_{n-1} + 4a_{n-2} + 5a_{n-3}$
b) $a_n = 2na_{n-1} + a_{n-2}$
c) $a_n = a_{n-1} + a_{n-4}$
d) $a_n = a_{n-1} + 2$
e) $a_n = a_{n-1}^2 + a_{n-2}$
f) $a_n = a_{n-2}$

2. Determine which of the following are linear homogeneous recurrence relations with coefficients. Also, find the degree of those that are.

a) $a_n = 3a_{n-2}$ b) $a_n = 3$
c) $a_n = a_{n-1}^2$ d) $a_n = a_{n-1} + 2a_{n-3}$
e) $a_n = a_{n-1}/n$
f) $a_n = 4a_{n-2} + 5a_{n-4} + 9a_{n-7}$

3. Solve the following recurrence relations together with the initial conditions given.
a) $a_n = 2a_{n-1}$ for $n \geq 1$, $a_0 = 3$
b) $a_n = a_{n-1}$ for $n \geq 1$, $a_0 = 2$
c) $a_n = 5a_{n-1} - 6a_{n-2}$ for $n \geq 2$, $a_0 = 1$, $a_1 = 0$
d) $a_n = 4a_{n-1} - 4a_{n-2}$ for $n \geq 2$, $a_0 = 6$, $a_1 = 8$
e) $a_n = -4a_{n-1} - 4a_{n-2}$ for $n \geq 2$, $a_0 = 0$, $a_1 = 1$
f) $a_n = 4a_{n-2}$ for $n \geq 2$, $a_0 = 0$, $a_1 = 4$
g) $a_n = a_{n-2}/4$ for $n \geq 2$, $a_0 = 1$, $a_1 = 0$

4. Solve the following recurrence relations together with the initial conditions given.
a) $a_n = a_{n-1} + 6a_{n-2}$ for $n \geq 2$, $a_0 = 3$, $a_1 = 6$
b) $a_n = 7a_{n-1} - 10a_{n-2}$ for $n \geq 2$, $a_0 = 2$, $a_1 = 1$
c) $a_n = 6a_{n-1} - 8a_{n-2}$ for $n \geq 2$, $a_0 = 4$, $a_1 = 10$
d) $a_n = 2a_{n-1} - a_{n-2}$ for $n \geq 2$, $a_0 = 4$, $a_1 = 1$
e) $a_n = a_{n-2}$ for $n \geq 2$, $a_0 = 5$, $a_1 = -1$
f) $a_n = -6a_{n-1} - 9a_{n-2}$ for $n \geq 2$, $a_0 = 3$, $a_1 = -3$
g) $a_{n+2} = -4a_{n+1} + 5a_n$ for $n \geq 0$, $a_0 = 2$, $a_1 = 8$

5. How many different messages can be transmitted in n microseconds using the two signals described in Exercise 27 of Section 5.1?

6. How many different messages can be transmitted in n microseconds using three different signals if one signal requires 1 microsecond for transmittal, the other two signals require 2 microseconds each for transmittal, and a signal in a message is followed immediately by the next signal?

7. In how many ways can a $2 \times n$ rectangular board be tiled using 1×2 and 2×2 pieces?

8. A model for the number of lobsters caught per year is based on the assumption that the number of lobsters caught in a year is the average of the number caught in the two previous years.
a) Find a recurrence relation for $\{L_n\}$, where L_n is the number of lobsters caught in year n, under the assumption for this model.
b) Find L_n if 100,000 lobsters were caught in year 1 and 300,000 were caught in year 2.

9. A deposit of \$100,000 is made to an investment fund at the beginning of a year. On the last day of each year two dividends are awarded. The first dividend is 20% of the amount in the account during that year. The second dividend is 45% of the amount in the account during the previous year.
a) Find a recurrence relation for $\{P_n\}$, where P_n is the amount in the account at the end of n years if no money is ever withdrawn.
b) How much is in the account after n years if no money has been withdrawn?

★10. Prove Theorem 2.

11. The **Lucas numbers** satisfy the recurrence relation

$$L_n = L_{n-1} + L_{n-2},$$

and the initial conditions $L_0 = 2$ and $L_1 = 1$.
a) Show that $L_n = f_{n-1} + f_{n+1}$ for $n = 2, 3, \ldots$, where f_n is the nth Fibonacci number.
b) Find an explicit formula for the Lucas numbers.

12. Find the solution to $a_n = 2a_{n-1} + a_{n-2} - 2a_{n-3}$ for $n = 3, 4, 5, \ldots$, with $a_0 = 3$, $a_1 = 6$, and $a_2 = 0$.

13. Find the solution to $a_n = 7a_{n-2} + 6a_{n-3}$ with $a_0 = 9$, $a_1 = 10$, and $a_2 = 32$.

14. Find the solution to $a_n = 5a_{n-2} - 4a_{n-4}$ with $a_0 = 3$, $a_1 = 2$, $a_2 = 6$, and $a_3 = 8$.

15. Find the solution to $a_n = 2a_{n-1} + 5a_{n-2} - 6a_{n-3}$ with $a_0 = 7$, $a_1 = -4$, and $a_2 = 8$.

★16. Prove Theorem 3.

17. Prove the following identity relating the Fibonacci numbers and the binomial coefficients:

$$f_{n+1} = C(n, 0) + C(n-1, 1) + \cdots + C(n-k, k)$$

where n is a positive integer and $k = \lfloor n/2 \rfloor$. (*Hint:* Let $a_n = C(n, 0) + C(n-1, 1) + \cdots + C(n-k, k)$. Show that the sequence $\{a_n\}$ satisfies the same recurrence relation and initial conditions satisfied by the sequence of Fibonacci numbers.)

18. A linear **inhomogeneous** recurrence relation with constant coefficients of degree k is a recurrence relation of the form

$$a_n = c_1 a_{n-1} + c_2 a_{n-2} + \cdots + c_k a_{n-k} + F(n).$$

Show that if $\{p_n\}$ is a solution of this inhomogeneous linear recurrence relation, then every solution is of the form $\{p_n + h_n\}$ where h_n is a solution of the associated linear homogeneous recurrence relation $a_n = c_1 a_{n-1} + c_2 a_{n-2} + \cdots + c_k a_{n-k}$. (*Hint:* Show that if $\{q_n\}$ is another solution, then $\{q_n - p_n\}$ is a solution of the associated linear homogeneous recurrence relation.)

19. Consider the inhomogeneous linear recurrence relation $a_n = 3a_{n-1} + 2^n$.

a) Show that $a_n = -2^{n+1}$ is a solution of this recurrence relation.

b) Use Exercise 18 to find all solutions of this recurrence relation.

c) Find the solution with $a_0 = 1$.

20. Consider the inhomogeneous linear recurrence relation $a_n = 2a_{n-1} + 2^n$.

a) Show that $a_n = n2^n$ is a solution of this recurrence relation.

b) Use Exercise 18 to find all solutions of this recurrence relation.

c) Find the solution with $a_0 = 2$.

21. a) Determine values of the constants A and B so that $a_n = An + B$ is a solution of the recurrence relation $a_n = 2a_{n-1} + n + 5$.

b) Use Exercise 18 to find all the solutions of this recurrence relation.

c) Find the solution of this recurrence relation with $a_0 = 4$.

22. a) Find the characteristic roots of the linear homogeneous recurrence relation $a_n = 2a_{n-1} - 2a_{n-2}$. (*Note:* These are complex numbers.)

b) Find the solution of the recurrence relation in part (a) with $a_0 = 1$ and $a_1 = 2$.

★23. a) Find the characteristic roots of the linear homogeneous recurrence relation $a_n = a_{n-4}$. (*Note:* These include complex numbers.)

b) Find the solution of the recurrence relation in part (a) with $a_0 = 1$, $a_1 = 0$, $a_2 = -1$, and $a_3 = 1$.

★24. Solve the simultaneous recurrence relations

$$a_n = 3a_{n-1} + 2b_{n-1}$$
$$b_n = a_{n-1} + 2b_{n-1}$$

with $a_0 = 1$ and $b_0 = 2$.

5.3 Divide-and-Conquer Relations

INTRODUCTION

Many recursive algorithms take a problem with given input and divide it into one or more smaller problems. This reduction is successively applied until the solutions of the smaller problems can be found quickly. For instance, we perform a binary search by reducing the search for an element in a list to the search for this element in a list half as long. We successively apply this reduction until one element is left. Another example of this type of recursive algorithm is a procedure for multiplying integers that reduces the problem of the multiplication of two integers to three multiplications of pairs of integers with half as many bits. This reduction is successively applied until integers with one bit are obtained. These procedures are called **divide-and-conquer** algorithms. In this section the recurrence relations that arise in the analysis of the complexity of these algorithms will be studied.

DIVIDE-AND-CONQUER RELATIONS

Suppose that an algorithm splits a problem of size n into a subproblems, where each subproblem is of size n/b (for simplicity, suppose that b divides n; in reality, the smaller problems are often of size equal to the nearest integer either less than or equal to, or greater than or equal to, n/b). Also, suppose that a total of $g(n)$ extra operations are required when this split of a problem of size n into smaller problems is made.

Then, if $f(n)$ represents the number of operations required to solve the problem, it follows that f satisfies the recurrence relation

$$f(n) = af(n/b) + g(n).$$

This is called a **divide-and-conquer** recurrence relation.

EXAMPLE 1 We introduced a binary search algorithm in Section 2.1. This binary search algorithm reduces the search for an element in a search sequence of size n to the binary search for this element in a search sequence of size $n/2$, when n is even. (Hence, the problem of size n has been reduced to *one* problem of size $n/2$). Two comparisons are needed to implement this reduction (one to determine which half of the list to use and the other to determine whether any terms of the list remain). Hence, if $f(n)$ is the number of comparisons required to search for an element in a search sequence of size n, then $f(n) = f(n/2) + 2$ when n is even. ■

EXAMPLE 2 Consider the following algorithm for locating the minimum and maximum elements of a sequence $a_1, a_2, \ldots, a_n$. If $n = 1$, then a_1 is the maximum and the minimum. If $n > 1$, split the sequence into two sequences, either where both have the same number of elements or where one of the sets has one element more than the other. The problem is reduced to finding the maximum and minimum of each of the two smaller sequences. The solution to the original problem results from the comparison of the separate maxima and minima of the two smaller sets to obtain the overall maximum and minimum.

Let $f(n)$ be the total number of comparisons needed to find the minimum and maximum elements of the set with n elements. We have shown that a problem of size n can be reduced into two problems of size $n/2$, when n is even, using two comparisons, one to compare the minima of the two sets and the other to compare the maxima of the two sets. This gives the recurrence relation $f(n) = 2f(n/2) + 2$ when n is even. ■

EXAMPLE 3 Surprisingly, there are more efficient algorithms than the conventional algorithm (described in Section 2.4) for multiplying integers. One of these algorithms, which uses a divide-and-conquer technique, will be described here. This fast multiplication algorithm proceeds by splitting each of two $2n$-bit integers into two blocks each with n bits. Then, the original multiplication is reduced from the multiplication of two $2n$-bit integers to three multiplications of n-bit integers, plus shifts and additions.

Suppose that a and b are integers with binary expansions of length $2n$ (add initial bits of zero in these expansions if necessary to make them the same length). Let

$$a = (a_{2n-1}a_{2n-2}\cdots a_1a_0)_2$$

and

$$b = (b_{2n-1}b_{2n-2}\cdots b_1b_0)_2.$$

Let

$$a = 2^n A_1 + A_0, \qquad b = 2^n B_1 + B_0,$$

where

$$A_1 = (a_{2n-1} \cdots a_{n+1} a_n)_2, \qquad A_0 = (a_{n-1} \cdots a_1 a_0)_2,$$
$$B_1 = (b_{2n-1} \cdots b_{n+1} b_n)_2, \qquad B_0 = (b_{n-1} \cdots b_1 b_0)_2.$$

The algorithm for fast multiplication of integers is based on the identity

$$ab = (2^{2n} + 2^n)A_1B_1 + 2^n(A_1 - A_0)(B_0 - B_1) + (2^n + 1)A_0B_0.$$

The important fact about this identity is that it shows that the multiplication of two $2n$-bit integers can be carried out using three multiplications of n-bit integers, together with additions, subtractions, and shifts. This shows that if $f(n)$ is the total number of bit operations needed to multiply two n-bit integers, then

$$f(2n) = 3f(n) + Cn.$$

The reasoning behind this equation is as follows. The three multiplications of n-bit integers are carried out using $3f(n)$-bit operations. Each of the additions, subtractions, and shifts uses a constant multiple of n-bit operations, and Cn represents the total number of bit operations used by these operations. ■

EXAMPLE 4 There are algorithms that multiply two $n \times n$ matrices, when n is even, using seven multiplications each of two $(n/2) \times (n/2)$ matrices and 15 additions of $(n/2) \times (n/2)$ matrices. Hence, if $f(n)$ is the number of operations (multiplications and additions) used, it follows that

$$f(n) = 7f(n/2) + 15n^2/4$$

when n is even. ■

As Examples 1–4 show, recurrence relations of the form $f(n) = af(n/b) + g(n)$ arise in many different situations. It is possible to derive estimates of the size of functions that satisfy such recurrence relations. Suppose that f satisfies this recurrence relation whenever n is divisible by b. Let $n = b^k$, where k is a positive integer. Then

$$\begin{aligned} f(n) &= af(n/b) + g(n) \\ &= a^2 f(n/b^2) + ag(n/b) + g(n) \\ &= a^3 f(n/b^3) + a^2 g(n/b^2) + ag(n/b) + g(n) \\ &\;\;\vdots \\ &= a^k f(n/b^k) + \sum_{j=0}^{k-1} a^j g(n/b^j). \end{aligned}$$

Since $n/b^k = 1$, it follows that

$$f(n) = a^k f(1) + \sum_{j=0}^{k-1} a^j g(n/b^j).$$

We can use this equation for $f(n)$ to estimate the size of functions that satisfy divide-and-conquer relations.

THEOREM 1 Let f be an increasing function that satisfies the recurrence relation

$$f(n) = af(n/b) + c$$

whenever n is divisible by b, where $a \geq 1$, b is an integer greater than 1, and c is a positive real number. Then

$$f(n) = \begin{cases} O(n^{\log_b a}) & \text{if } a > 1 \\ O(\log n) & \text{if } a = 1. \end{cases}$$

Proof: First let $n = b^k$. From the expression for $f(n)$ obtained in the discussion preceding the theorem, with $g(n) = c$, we have

$$f(n) = a^k f(1) + \sum_{j=0}^{k-1} a^j c = a^k f(1) + c \sum_{j=0}^{k-1} a^j.$$

First consider the case when $a = 1$. Then

$$f(n) = f(1) + ck.$$

Since $n = b^k$, we have $k = \log_b n$. Hence

$$f(n) = f(1) + c \log_b n.$$

When n is not a power of b, we have $b^k < n < b^{k+1}$, for a positive integer k. Since f is increasing, it follows that $f(n) \leq f(b^{k+1}) = f(1)+c(k+1) = (f(1)+c)+ck \leq (f(1) + c) + c\log_b n$. Therefore, in both cases, $f(n) = O(\log n)$ when $a = 1$.

Now suppose that $a > 1$. First assume that $n = b^k$ where k is a positive integer. From the formula for the sum of terms of a geometric progression (Example 6 of Section 3.2), it follows that

$$\begin{aligned} f(n) &= a^k f(1) + c(a^k - 1)/(a - 1) \\ &= a^k[f(1) + c/(a - 1)] - c/(a - 1) \\ &= C_1 n^{\log_b a} + C_2, \end{aligned}$$

since $a^k = a^{\log_b n} = n^{\log_b a}$ (see Exercise 4 in Appendix 1), where $C_1 = [f(1)+ c/(a - 1)]$ and $C_2 = -c/(a - 1)$.

Now suppose that n is not a power of b. Then $b^k < n < b^{k+1}$ where k is a nonnegative integer. Since f is increasing,

$$\begin{aligned} f(n) &\le f(b^{k+1}) = C_1 a^{k+1} + C_2 \\ &\le (C_1 a) a^{\log_b n} + C_2 \\ &\le (C_1 a) n^{\log_b a} + C_2, \end{aligned}$$

since $k \le \log_b n < k + 1$.

Hence, we have $f(n) = O(n^{\log_b a})$. ■

Remark: This proof gives an explicit formula for $f(n)$ where $n = b^k$.

The following examples illustrate how Theorem 1 is used.

EXAMPLE 5 Let $f(n) = 5f(n/2) + 3$ and $f(1) = 7$. Find $f(2^k)$ where k is a positive integer. Also, estimate $f(n)$ if f is an increasing function.

Solution: From the proof of Theorem 1, with $a = 5$, $b = 2$, and $c = 3$, we see that if $n = 2^k$, then

$$\begin{aligned} f(n) &= a^k[f(1) + c/(a-1)] + [-c/(a-1)] \\ &= 5^k[7 + (3/4)] - 3/4 \\ &= 5^k(31/4) - 3/4. \end{aligned}$$

Also, if $f(n)$ is increasing, Theorem 1 shows that $f(n) = O(n^{\log_b a}) = O(n^{\log 5})$. ■

We can use Theorem 1 to estimate the computational complexity of the binary search algorithm and the algorithm given in Example 2 for locating the minimum and maximum of a sequence.

EXAMPLE 6 Estimate the number of comparisons used by a binary search.

Solution: In Example 1 it was shown that $f(n) = f(n/2) + 2$ when n is even, where f is the number of comparisons required to perform a binary search on a sequence of size n. Hence, from Theorem 1, it follows that $f(n) = O(\log n)$. ■

EXAMPLE 7 Estimate the number of comparisons used to locate the maximum and minimum elements in a sequence using the algorithm given in Example 2.

Solution: In Example 2 we showed that $f(n) = 2f(n/2) + 2$, when n is even, where f is the number of comparisons needed by this algorithm. Hence, from Theorem 1, it follows that $f(n) = O(n^{\log 2}) = O(n)$. ■

We will now state a more general, and more complicated, theorem that is useful in analyzing the complexity of divide-and-conquer algorithms.

THEOREM 2 Let f be an increasing function that satisfies the recurrence relation

$$f(n) = af(n/b) + cn^d$$

whenever $n = b^k$, where k is a positive integer, $a \geq 1$, b is an integer greater than 1, and c and d are positive real numbers. Then

$$f(n) = \begin{cases} O(n^d) & \text{if } a < b^d \\ O(n^d \log n) & \text{if } a = b^d \\ O(n^{\log_b a}) & \text{if } a > b^d. \end{cases}$$

The proof of Theorem 2 is left for the reader as Exercises 17–21 at the end of this section.

EXAMPLE 8 Estimate the number of bit operations needed to multiply two n-bit integers using the fast multiplication algorithm.

Solution: Example 3 shows that $f(n) = 3f(n/2) + Cn$, when n is even, where $f(n)$ is the number of bit operations required to multiply two n-bit integers using the fast multiplication algorithm. Hence, from Theorem 2 it follows that $f(n) = O(n^{\log 3})$. Note that $\log 3 \sim 1.6$. Since the conventional algorithm for multiplication uses $O(n^2)$ bit operations, the fast multiplication algorithm is a substantial improvement over the conventional algorithm in terms of time complexity for sufficiently large integers. ■

EXAMPLE 9 Estimate the number of multiplications and additions required to multiply two $n \times n$ matrices using the matrix multiplication algorithm referred to in Example 4.

Solution: Let $f(n)$ denote the number of additions and multiplications used by the algorithm mentioned in Example 4 to multiply two $n \times n$ matrices. We have $f(n) = 7f(n/2) + 15n^2/4$, when n is even. Hence, from Theorem 2 it follows that $f(n) = O(n^{\log 7})$. Note that $\log 7 \sim 2.8$. Since the conventional algorithm for multiplying two $n \times n$ matrices uses $O(n^3)$ additions and multiplications, it follows that for sufficiently large integers n, this algorithm is substantially more efficient in time complexity than the conventional algorithm. ■

Exercises

1. How many comparisons are needed for a binary search in a set of 64 elements?
2. How many comparisons are needed to locate the maximum and minimum elements in a sequence with 128 elements using the algorithm in Example 2?
3. Multiply $(1110)_2$ and $(1010)_2$ using the fast multiplication algorithm.
4. Express the fast multiplication algorithm in pseudocode.
5. Determine a value for the constant C in Example 3 and use it to estimate the number of bit operations needed to multiply two 64-bit integers using the fast multiplication algorithm.
6. How many operations are needed to multiply two 32×32 matrices using the algorithm referred to in Example 4?
7. Suppose that $f(n) = f(n/3) + 1$ when n is divisible by 3, and $f(1) = 1$. Find
   a) $f(3)$. **b)** $f(27)$. **c)** $f(729)$.
8. Suppose that $f(n) = 2f(n/2) + 3$ when n is even, and $f(1) = 5$. Find
   a) $f(2)$. **b)** $f(8)$. **c)** $f(64)$.
   d) $f(1024)$.
9. Suppose that $f(n) = f(n/5) + 3n^2$ when n is divisible by 5, and $f(1) = 4$. Find
   a) $f(5)$. **b)** $f(125)$. **c)** $f(3125)$.
10. Find $f(n)$ when $n = 2^k$, where f satisfies the recurrence relation $f(n) = f(n/2) + 1$ with $f(1) = 1$.
11. Estimate the size of f in Exercise 10 if f is an increasing function.
12. Find $f(n)$ when $n = 3^k$, where f satisfies the recurrence relation $f(n) = 2f(n/3) + 4$ with $f(1) = 1$.
13. Estimate the size of f in Exercise 12 if f is an increasing function.
14. Suppose that there are $n = 2^k$ teams in an elimination tournament, where there are $n/2$ games in the first round, with the $n/2 = 2^{k-1}$ winners playing in the second round, and so on. Develop a recurrence relation for the number of rounds in the tournament.
15. How many rounds are there in the elimination tournament described in Exercise 14 when there are 32 teams?
16. Solve the recurrence relation for the number of rounds in the elimination tournament described in Exercise 14.

In Exercises 17–21, assume that f is an increasing function satisfying the recurrence relation $f(n) = af(n/b) + cn^d$, where $a \geq 1$, b is an integer greater than 1, and c and d are positive real numbers. These exercises supply a proof of Theorem 2.

★17. Show that if $a = b^d$ and n is a power of b, then $f(n) = f(1)n^d + cn^d \log_b n$.

18. Use Exercise 17 to show that if $a = b^d$, then $f(n) = O(n^d \log n)$.

★19. Show that if $a \neq b^d$ and n is a power of b, then $f(n) = C_1 n^d + c_2 n^{\log_b a}$, where $C_1 = b^d c/(b^d - a)$ and $C_2 = f(1) + b^d c/(a - b^d)$.

20. Use Exercise 19 to show that if $a < b^d$, then $f(n) = O(n^d)$.
21. Use Exercise 19 to show that if $a > b^d$, then $f(n) = O(n^{\log_b a})$.
22. Find $f(n)$ when $n = 4^k$, where f satisfies the recurrence relation $f(n) = 5f(n/4) + 6n$, with $f(1) = 1$.
23. Estimate the size of f in Exercise 22 if f is an increasing function.
24. Find $f(n)$ when $n = 2^k$, where f satisfies the recurrence relation $f(n) = 8f(n/2) + n^2$ with $f(1) = 1$.
25. Estimate the size of f in Exercise 24 if f is an increasing function.

5.4 Inclusion-Exclusion

INTRODUCTION

A discrete mathematics class contains 30 women and 50 sophomores. How many students in the class are either women or sophomores? This question cannot be answered unless more information is provided. Adding the number of women in the class and the number of sophomores probably does not give the correct answer, because women sophomores are counted twice. This observation shows that the number of students in the class that are either sophomores or women is the sum of the number of women and the number of sophomores in the class minus the number of women sophomores. A technique for solving such counting problems was introduced in Section 4.1. In this section we will generalize the ideas introduced in that section to solve a wider range of counting problems.

THE PRINCIPLE OF INCLUSION-EXCLUSION

How many elements are in the union of two finite sets? In Section 1.5 it was shown that the number of elements in the union of the two sets A and B is the sum of the numbers of elements in the sets minus the number of elements in their intersection. That is,

$$|A \cup B| = |A| + |B| - |A \cap B|.$$

As we showed in Section 4.1, the formula for the number of elements in the union of two sets is useful in counting problems. The following examples provide additional illustrations of the usefulness of this formula.

EXAMPLE 1 A discrete mathematics class contains 25 students majoring in computer science, 13 students majoring in mathematics, and 8 joint mathematics and computer science majors. How many students are in this class, if every student is majoring in mathematics, computer science, or both mathematics and computer science?

Solution: Let A be the set of students in the class majoring in computer science and B be the set of students in the class majoring in mathematics. Then $A \cap B$ is the set of students in the class that are joint mathematics and computer science majors. Since every student in the class is majoring in either computer science or mathematics (or both), it follows that the number of students in the class is $|A \cup B|$. Therefore,

$A \cup B| = |A| + |B| - |A \cap B| = 25 + 13 - 8 = 30$

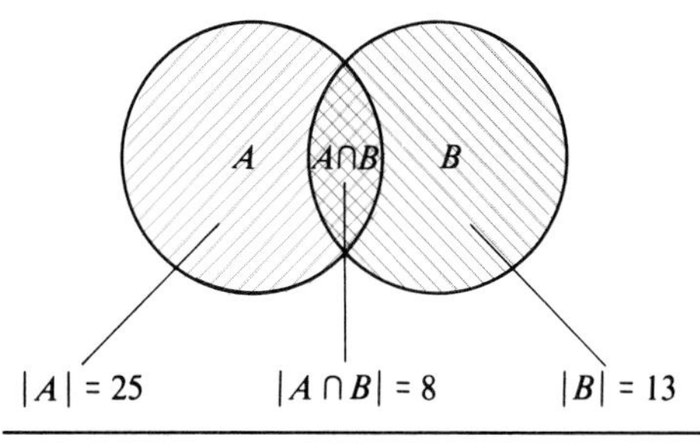

FIGURE 1 **The Set of Students in a Discrete Mathematics Class.**

$$\begin{aligned} |A \cup B| &= |A| + |B| - |A \cap B| \\ &= 25 + 13 - 8 \\ &= 30. \end{aligned}$$

Therefore, there are 30 students in the class. This computation is illustrated in Figure 1. ■

EXAMPLE 2 How many positive integers not exceeding 1000 are divisible by 7 or 11?

Solution: Let A be the set of positive integers not exceeding 1000 that are divisible by 7, and let B be the set of positive integers not exceeding 1000 that are divisible by 11. Then $A \cup B$ is the set of integers not exceeding 1000 that are divisible by either 7 or 11, and $A \cap B$ is the set of integers not exceeding 1000 that are divisible by both 7 and 11. From Example 2 of Section 2.3, we know that among the positive integers not exceeding 1000 there are $\lfloor 1000/7 \rfloor$ integers divisible by 7 and $\lfloor 1000/11 \rfloor$ divisible by 11. Since 7 and 11 are relatively prime, the integers divisible by both 7 and 11 are those divisible by $7 \cdot 11$. Consequently, there are $\lfloor 1000/(11 \cdot 7) \rfloor$ positive integers not exceeding 1000 that are divisible by both 7 and 11. It follows that there are

$$\begin{aligned} |A \cup B| &= |A| + |B| - |A \cap B| \\ &= \left\lfloor \frac{1000}{7} \right\rfloor + \left\lfloor \frac{1000}{11} \right\rfloor - \left\lfloor \frac{1000}{7 \cdot 11} \right\rfloor \\ &= 142 + 90 - 12 \\ &= 220 \end{aligned}$$

positive integers not exceeding 1000 that are divisible by either 7 or 11. This computation is illustrated in Figure 2.

The next example shows how to find the number of elements in a finite universal set that are outside the union of two sets.

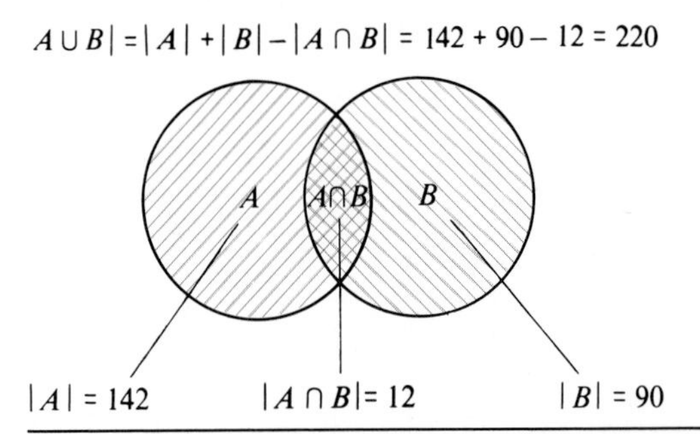

FIGURE 2 The Set of Positive Integers Not Exceeding 1000 Divisible by Either 7 or 11.

EXAMPLE 3 Suppose there are 1807 freshmen at your school. Of these, 453 are taking a course in computer science, 567 are taking a course in mathematics, and 299 are taking courses in both computer science and mathematics. How many freshmen are not taking a course in either computer science or in mathematics?

Solution: To find the number of freshmen who are not taking a course in either mathematics or computer science, subtract the number that are taking a course in either of these subjects from the total number of freshmen. Let A be the set of all freshmen taking a course in computer science, and let B be the set of all freshmen taking a course in mathematics. It follows that $|A| = 453$, $|B| = 567$, and $|A \cap B| = 299$. The number of freshmen taking a course in either computer science or mathematics is

$$|A \cup B| = |A| + |B| - |A \cap B| = 453 + 567 - 299 = 721.$$

Consequently, there are $1807 - 721 = 1086$ freshmen who are not taking a course in computer science or mathematics. This is illustrated in Figure 3. ■

Later in this section it will be shown how the number of elements in the union of a finite number of sets can be found. The result that will be developed is called the **principle of inclusion-exclusion.** Before considering unions of n sets, where n is any positive integer, a formula for the number of elements in the union of three sets A, B, and C will be derived. To construct this formula note that $|A| + |B| + |C|$ counts each element that is in exactly one of the three sets once, elements that are in exactly two of the sets twice, and elements in all three sets three times. This is illustrated in the first panel in Figure 4.

To remove the overcount of elements in more than one of the sets, subtract the number of elements in the intersections of all pairs of the three sets, obtaining

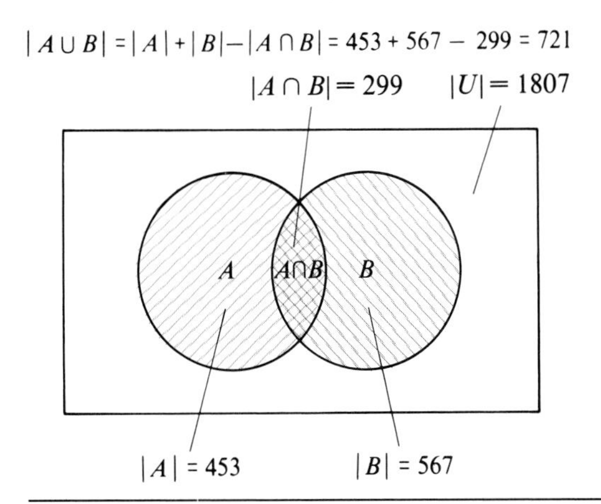

FIGURE 3 **The Set of Freshmen Not Taking a Course in Either Computer Science or Mathematics.**

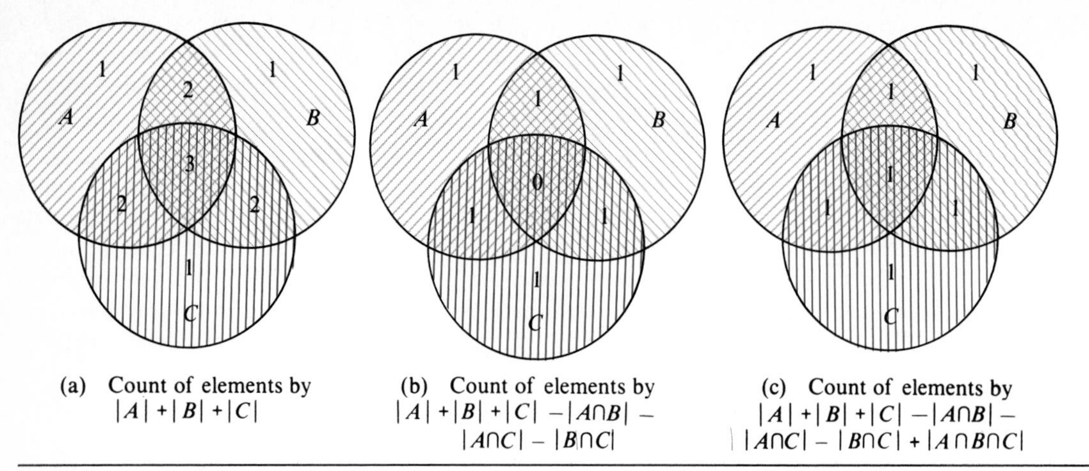

FIGURE 4 **Finding a Formula for the Number of Elements in the Union of Three Sets.**

$$|A| + |B| + |C| - |A \cap B| - |A \cap C| - |B \cap C|.$$

This expression still counts elements that occur in exactly one of the sets once. An element that occurs in exactly two of the sets is also counted exactly once, since this element will occur in one of the three intersections of sets taken two at a time. However, those elements that occur in all three sets will be counted zero times by this expression, since they occur in all three intersections of sets taken two at a time. This is illustrated in the second panel in Figure 4.

To remedy this undercount, add the number of elements in the intersection of all three sets. This final expression counts each element once, whether it is in one, two, or three of the sets. Thus,

$$|A \cup B \cup C| = |A| + |B| + |C| - |A \cap B| - |A \cap C| - |B \cap C| + |A \cap B \cap C|.$$

This formula is illustrated in the third panel of Figure 4.

The following examples illustrate how this formula can be used.

EXAMPLE 4 A total of 1232 students have taken a course in Spanish, 879 have taken a course in French, and 114 have taken a course in Russian. Further, 103 have taken courses in both Spanish and French, 23 have taken courses in both Spanish and Russian, and 14 have taken courses in both French and Russian. If 2092 students have taken at least one of Spanish, French, and Russian, how many students have taken a course in all three languages?

Solution: Let S be the set of students who have taken a course in Spanish, F the set of students who have taken a course in French, and R the set of students who have taken a course in Russian. Then

$$|S| = 1232, \qquad |F| = 879, \qquad |R| = 114,$$
$$|S \cap F| = 103, \qquad |S \cap R| = 23, \qquad |F \cap R| = 14,$$

and

$$|S \cup F \cup R| = 2092.$$

Inserting these quantities into the equation

$$|S \cup F \cup R| = |S| + |F| + |R| - |S \cap F| - |S \cap R| - |F \cap R| + |S \cap F \cap R|$$

gives

$$2092 = 1232 + 879 + 114 - 103 - 23 - 14 + |S \cap F \cap R|.$$

Solving for $|S \cap F \cap R|$ shows that $|S \cap F \cap R| = 7$. Therefore, there are seven students who have taken courses in Spanish, French, and Russian. This is illustrated in Figure 5. ■

We will now state and prove the inclusion-exclusion principle, which tells us how many elements are in the union of a finite number of finite sets.

THEOREM 1 **THE PRINCIPLE OF INCLUSION-EXCLUSION** Let $A_1, A_2, \ldots, A_n$ be finite sets. Then

$$|A_1 \cup A_2 \cup \cdots \cup A_n| = \sum_{1 \leq i \leq n} |A_i| - \sum_{1 \leq i < j \leq n} |A_i \cap A_j| + \sum_{1 \leq i < j < k \leq n} |A_i \cap A_j \cap A_k| - \cdots + (-1)^{n+1} |A_1 \cap A_2 \cap \cdots \cap A_n|.$$

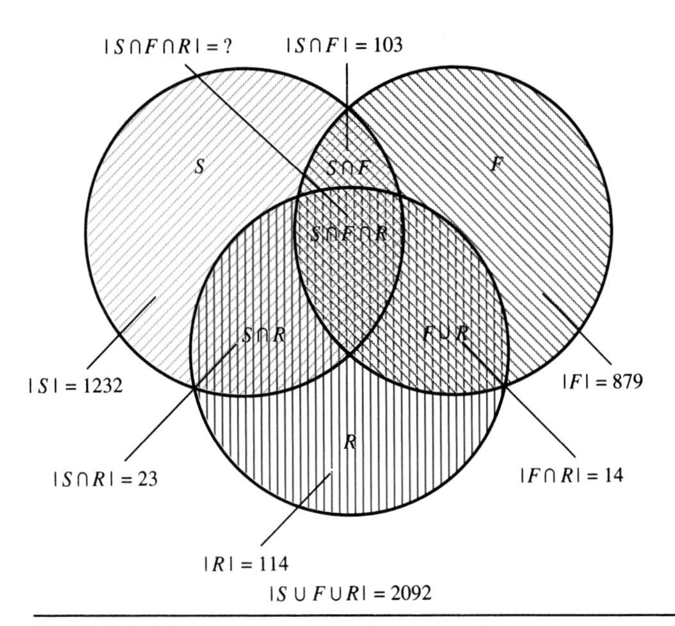

FIGURE 5 **The Set of Students Who Have Taken Courses in Spanish, French, and Russian.**

Proof: We will prove the formula by showing that an element in the union is counted exactly once by the right-hand side of the equation. Suppose that a is a member of exactly r of the sets $A_1, A_2, \ldots, A_n$ where $1 \leq r \leq n$. This element is counted $C(r, 1)$ times by $\Sigma|A_i|$. It is counted $C(r, 2)$ times by $\Sigma|A_i \cap A_j|$. In general, it is counted $C(r, m)$ times by the summation involving m of the sets A_i. Thus, this element is counted exactly

$$C(r, 1) - C(r, 2) + C(r, 3) - \cdots + (-1)^{r+1}C(r, r)$$

times by the expression on the right-hand side of this equation. Our goal is to evaluate this quantity. From Theorem 7 of Section 4.3, we have

$$C(r, 0) - C(r, 1) + C(r, 2) - \cdots + (-1)^{r}C(r, r) = 0.$$

Hence,

$$1 = C(r, 0) = C(r, 1) - C(r, 2) + \cdots + (-1)^{r+1}C(r, r).$$

Therefore, each element in the union is counted exactly once by the expression on the right-hand side of the equation. This proves the principle of inclusion-exclusion. ■

The inclusion-exclusion principle gives a formula for the number of elements in the union of n sets for every positive integer n. There are terms in this formula for the number of elements in the intersection of every nonempty subset of the collection of the n sets. Hence, there are $2^n - 1$ terms in this formula.

EXAMPLE 5 Give a formula for the number of elements in the union of four sets.

Solution: The inclusion-exclusion principle shows that

$$\begin{aligned}|A_1 \cup A_2 \cup A_3 \cup A_4| = {} & |A_1| + |A_2| + |A_3| + |A_4| \\ & - |A_1 \cap A_2| - |A_1 \cap A_3| - |A_1 \cap A_4| - |A_2 \cap A_3| - |A_2 \cap A_4| - |A_3 \cap A_4| \\ & + |A_1 \cap A_2 \cap A_3| + |A_1 \cap A_2 \cap A_4| + |A_1 \cap A_3 \cap A_4| + |A_2 \cap A_3 \cap A_4| \\ & - |A_1 \cap A_2 \cap A_3 \cap A_4|.\end{aligned}$$

Note that this formula contains 15 different terms, one for each nonempty subset of $\{A_1, A_2, A_3, A_4\}$. ■

Exercises

1. How many elements are in $A_1 \cup A_2$ if there are 12 elements in A_1, 18 elements in A_2, and
a) $A_1 \cap A_2 = \emptyset$?
b) $|A_1 \cap A_2| = 1$?
c) $|A_1 \cap A_2| = 6$?
d) $A_1 \subseteq A_2$?

2. There are 345 students at a college who have taken a course in calculus, 212 who have taken a course in discrete mathematics, and 188 who have taken courses in both calculus and discrete mathematics. How many students have taken a course in either calculus or discrete mathematics?

3. A survey of households in the United States reveals that 96% have at least one television set, 98% have telephone service, and 95% have telephone service and at least one television set. What percentage of households in the United States have neither telephone service nor a television set?

4. A marketing report concerning personal computers states that 650,000 owners will buy a modem for their machines next year and 1,250,000 will buy at least one software package. If the report states that 1,450,000 owners will buy either a modem or at least one software package, how many will buy both a modem and at least one software package?

5. Find the number of elements in $A_1 \cup A_2 \cup A_3$ if there are 100 elements in each set if

a) the sets are pairwise disjoint.

b) there are 50 common elements in each pair of sets and no elements in all three sets.

c) there are 50 common elements in each pair of sets and 25 elements in all three sets.

d) the sets are equal.

6. Find the number of elements in $A_1 \cup A_2 \cup A_3$ if there are 100 elements in A_1, 1000 in A_2, and 10,000 in A_3 if

a) $A_1 \subseteq A_2$ and $A_2 \subseteq A_3$.

b) the sets are pairwise disjoint.

c) there are two elements common to each pair of sets and one element in all three sets.

7. There are 2504 computer science students at a school. Of these, 1876 have taken a course in Pascal, 999 have taken a course in Fortran, and 345 have taken a course in C. Further, 876 have taken courses in both Pascal and Fortran, 231 have taken courses in both Fortran and C, and 290 have taken courses in both Pascal and C. If 189 of these students have taken courses in Fortran, Pascal, and C, how many of these 2504 students have not taken a course in any of these three programming languages?

8. In a survey of 270 college students, it is found that 64 like brussels sprouts, 94 like broccoli, 58 like cauliflower, 26 like both brussels sprouts and broccoli, 28 like both brussels sprouts and cauliflower, 22 like both broccoli and cauliflower, and 14 like all three vegetables. How many of the 270 students do not like any of these vegetables?

9. How many students are enrolled in a course either in calculus, discrete mathematics, data structures, or programming languages at a school if there are 507, 292, 312, and 344 students in these courses, respectively, 14 in both calculus and data structures, 213 in both calculus and programming languages, 211 in both discrete mathematics and data structures, 43 in both discrete mathematics and programming languages, and no student may take calculus and discrete mathematics, or data structures and programming languages, concurrently?

10. Find the number of positive integers not exceeding 100 that are not divisible by 5 or by 7.

11. Find the number of positive integers not exceeding 100 that are either odd or the square of an integer.

12. Find the number of positive integers not exceeding 1000 that are either the square or the cube of an integer.

13. How many bit strings of length eight do not contain six consecutive 0s?

★14. How many permutations of the 26 letters of the English alphabet do not contain any of the strings *fish*, *rat*, or *bird*?

15. How many permutations of the 10 digits either begin with the 3 digits 987, contain the digits 45 in the fifth and sixth positions, or end with the 3 digits 123?

16. How many elements are in the union of four sets if each of the sets has 100 elements, each pair of the sets shares 50 elements, each three of the sets share 25 elements, and there are 5 elements in all four sets?

17. How many elements are in the union of four sets if the sets have 50, 60, 70, and 80 elements, respectively, each pair of the sets has 5 elements in common, each triple of the sets has 1 common element, and no element is in all four sets?

18. How many terms are there in the formula for the number of elements in the union of 10 sets given by the principle of inclusion-exclusion?

19. Write out the explicit formula given by the principle of inclusion-exclusion for the number of elements in the union of five sets.

20. How many elements are in the union of five sets if the sets contain 10,000 elements each, each pair of sets has 1000 common elements, each triple of sets has 100 common elements, every four of the sets have 10 common elements, and there is 1 element in all five sets?

21. Write out the explicit formula given by the principle

THE SIEVE OF ERATOSTHENES

The principle of inclusion-exclusion can be used to find the number of primes not exceeding a specified positive integer. Recall that a composite integer is divisible by a prime not exceeding its square root. So, to find the number of primes not exceeding 100, first note that composite integers not exceeding 100 must have a prime factor not exceeding 10. Because the only primes less than 10 are 2, 3, 5, and 7, the primes not exceeding 100 are these four primes and those positive integers greater than 1 and not exceeding 100 that are divisible by none of 2, 3, 5, or 7. To apply the principle of inclusion-exclusion, let P_1 be the property that an integer is divisible by 2, let P_2 be the property that an integer is divisible by 3, let P_3 be the property that an integer is divisible by 5, and let P_4 be the property that an integer is divisible by 7. Thus, the number of primes not exceeding 100 is

$$4 + N(P_1'P_2'P_3'P_4').$$

Since there are 99 positive integers greater than 1 and not exceeding 100, the principle of inclusion-exclusion shows that

$$\begin{aligned} N(P_1'P_2'P_3'P_4') = 99 &- N(P_1) - N(P_2) - N(P_3) - N(P_4) \\ &+ N(P_1P_2) + N(P_1P_3) + N(P_1P_4) + N(P_2P_3) \\ &+ N(P_2P_4) + N(P_3P_4) \\ &- N(P_1P_2P_3) - N(P_1P_2P_4) - N(P_1P_3P_4) - N(P_2P_3P_4) \\ &+ N(P_1P_2P_3P_4). \end{aligned}$$

The number of integers not exceeding 100 (and greater than 1) that are divisible by all the primes in a subset of $\{2, 3, 5, 7\}$ is $\lfloor 100/N \rfloor$, where N is the product of the primes in this subset. (This follows since any two of these primes have no common factor.) Consequently,

$$\begin{aligned} N(P_1'P_2'P_3'P_4') &= 99 - \left\lfloor \frac{100}{2} \right\rfloor - \left\lfloor \frac{100}{3} \right\rfloor - \left\lfloor \frac{100}{5} \right\rfloor - \left\lfloor \frac{100}{7} \right\rfloor \\ &\quad + \left\lfloor \frac{100}{2\cdot 3} \right\rfloor + \left\lfloor \frac{100}{2\cdot 5} \right\rfloor + \left\lfloor \frac{100}{2\cdot 7} \right\rfloor + \left\lfloor \frac{100}{3\cdot 5} \right\rfloor \\ &\quad + \left\lfloor \frac{100}{3\cdot 7} \right\rfloor + \left\lfloor \frac{100}{5\cdot 7} \right\rfloor \\ &\quad - \left\lfloor \frac{100}{2\cdot 3\cdot 5} \right\rfloor - \left\lfloor \frac{100}{2\cdot 3\cdot 7} \right\rfloor - \left\lfloor \frac{100}{2\cdot 5\cdot 7} \right\rfloor - \left\lfloor \frac{100}{3\cdot 5\cdot 7} \right\rfloor \\ &\quad + \left\lfloor \frac{100}{2\cdot 3\cdot 5\cdot 7} \right\rfloor \\ &= 99 - 50 - 33 - 20 - 14 + 16 + 10 + 7 + 6 + 4 + 2 \\ &\quad - 3 - 2 - 1 - 0 + 0 \\ &= 21. \end{aligned}$$

Hence, there are $4 + 21 = 25$ primes not exceeding 100.

The **sieve of Eratosthenes** is used to find all primes not exceeding a specified positive integer. For instance, the following procedure is used to find the primes not exceeding 100. First the integers that are divisible by 2, other than 2, are deleted. Since 3 is the first integer greater than 2 that is left, all those integers divisible by 3, other than 3, are deleted. Since 5 is the next integer left after 3, those integers divisible by 5, other than 5, are deleted. The next integer left is 7, so those integers divisible by 7, other than 7, are deleted. Since all composite integers not exceeding 100 are divisible by 2, 3, 5, or 7, all remaining integers except 1 are prime. In Table 1, the panels display those integers deleted at each stage, where each integer divisible by 2, other than 2, is underlined in the first panel, each integer divisible by 3, other than 3, is underlined in the second panel, each integer divisible by 5, other than 5, is underlined in the third panel, and each integer divisible by 7, other than 7, is underlined in the fourth panel. The integers not underlined are the primes not exceeding 100.

THE NUMBER OF ONTO FUNCTIONS

The principle of inclusion-exclusion can also be used to determine the number of onto functions from a set with m elements to a set with n elements. First consider the following example.

EXAMPLE 2 How many onto functions are there from a set with six elements to a set with three elements?

Solution: Suppose that the elements in the codomain are b_1, b_2, and b_3. Let P_1, P_2, and P_3 be the properties that b_1, b_2, and b_3 are not in the range of the function, respectively. Note that a function is onto if and only if it has none of the properties P_1, P_2, or P_3. By the inclusion-exclusion principle it follows that the number of onto functions from a set with six elements to a set with three elements is

$$N(P_1'P_2'P_3') = N - [N(P_1) + N(P_2) + N(P_3)] \\ +[N(P_1P_2) + N(P_1P_3) + N(P_2P_3)] - N(P_1P_2P_3),$$

Eratosthenes (276–194 B.C.E.). It is known that Eratosthenes, born in Cyrene, a Greek colony west of Egypt, spent time studying at Plato's academy in Athens. We also know that King Ptolemy II invited Eratosthenes to Alexandria to tutor his son and that later Eratosthenes became chief librarian at the famous library at Alexandria, a central repository of ancient wisdom. Eratosthenes was an extremely versatile scholar, writing on mathematics, geography, astronomy, history, philosophy, and literary criticism. Besides his work in mathematics, he is most noted for his chronology of ancient history and for his famous measurement of the size of the earth.

TABLE 1 The Sieve of Eratosthenes.

Integers Divisible by 2 Other than 2 Receive an Underline

1	2	3	4	5	6	7	8	9	10
11	12	13	14	15	16	17	18	19	20
21	22	23	24	25	26	27	28	29	30
31	32	33	34	35	36	37	38	39	40
41	42	43	44	45	46	47	48	49	50
51	52	53	54	55	56	57	58	59	60
61	62	63	64	65	66	67	68	69	70
71	72	73	74	75	76	77	78	79	80
81	82	83	84	85	86	87	88	89	90
91	92	93	94	95	96	97	98	99	100

Integers Divisible by 3 Other than 3 Receive an Underline

1	2	3	4	5	6	7	8	9	10
11	12	13	14	15	16	17	18	19	20
21	22	23	24	25	26	27	28	29	30
31	32	33	34	35	36	37	38	39	40
41	42	43	44	45	46	47	48	49	50
51	52	53	54	55	56	57	58	59	60
61	62	63	64	65	66	67	68	69	70
71	72	73	74	75	76	77	78	79	80
81	82	83	84	85	86	87	88	89	90
91	92	93	94	95	96	97	98	99	100

Integers Divisible by 5 Other than 5 Receive an Underline

1	2	3	4	5	6	7	8	9	10
11	12	13	14	15	16	17	18	19	20
21	22	23	24	25	26	27	28	29	30
31	32	33	34	35	36	37	38	39	40
41	42	43	44	45	46	47	48	49	50
51	52	53	54	55	56	57	58	59	60
61	62	63	64	65	66	67	68	69	70
71	72	73	74	75	76	77	78	79	80
81	82	83	84	85	86	87	88	89	90
91	92	93	94	95	96	97	98	99	100

Integers Divisible by 7 Other than 7 Receive an Underline; Integers in Color Are Prime

1	2	3	4	5	6	7	8	9	10
11	12	13	14	15	16	17	18	19	20
21	22	23	24	25	26	27	28	29	30
31	32	33	34	35	36	37	38	39	40
41	42	43	44	45	46	47	48	49	50
51	52	53	54	55	56	57	58	59	60
61	62	63	64	65	66	67	68	69	70
71	72	73	74	75	76	77	78	79	80
81	82	83	84	85	86	87	88	89	90
91	92	93	94	95	96	97	98	99	100

where N is the total number of functions from a set with six elements to one with three elements. We will evaluate each of the terms on the right-hand side of this equation.

From Example 8 of Section 4.1, it follows that $N = 3^6$. Note that $N(P_i)$ is the number of functions that do not have b_i in their range. Hence, there are two choices for the value of the function at each element of the domain. Therefore, $N(P_i) = 2^6$. Furthermore, there are $C(3, 1)$ terms of this kind. Note that $N(P_iP_j)$ is the number of functions that do not have b_i and b_j in their range. Hence, there is only one choice for the value of the function at each element of the domain. Therefore, $N(P_iP_j) = 1^6 = 1$. Furthermore, there are $C(3, 2)$ terms of this kind. Also, note that $N(P_1P_2P_3) = 0$, since this term is the number of functions that have none of b_1, b_2, and b_3 in their range. Clearly, there are no such functions. Therefore, the number of onto functions from a set with six elements to one with three elements is

$$3^6 - C(3, 1)2^6 + C(3, 2)1^6 = 729 - 192 + 3 = 540.$$ ■

The general result that tells us how many onto functions there are from a set with m elements to one with n elements will now be stated. The proof of this result is left as an exercise for the reader.

THEOREM 1 Let m and n be positive integers with $m \geq n$. Then, there are

$$n^m - C(n, 1)(n-1)^m + C(n, 2)(n-2)^m - \cdots + (-1)^{n-1}C(n, n-1) \cdot 1^m$$

onto functions from a set with m elements to a set with n elements.

One of the many different applications of Theorem 1 will now be described.

EXAMPLE 3 How many ways are there to assign five different jobs to four different employees if every employee is assigned at least one job?

Solution: Consider the assignment of jobs as a function from the set of five jobs to the set of four employees. An assignment where every employee gets at least one job is the same as an onto function from the set of jobs to the set of employees. Hence, by Theorem 1 it follows that there are

$$4^5 - C(4, 1)3^5 + C(4, 2)2^5 - C(4, 3)1^5 = 1024 - 972 + 192 - 4 = 240$$

ways to assign the jobs so that each employee is assigned at least one job. ■

DERANGEMENTS

The principle of inclusion-exclusion will be used to count the permutations of n objects that leave no objects in their original positions. Consider the following example.

EXAMPLE 4 The Hatcheck Problem A new employee checks the hats of n people at a restaurant, forgetting to put claim check numbers on the hats. When customers return for their hats, the checker gives them back hats chosen at random from the remaining hats. What is the probability that no one receives the correct hat? ■

Remark: The answer is the number of ways the hats can be arranged so that there is no hat in its original position divided by $n!$, the number of permutations of n hats. We will return to this example after we find the number of permutations of n objects that leave no objects in their original position.

A **derangement** is a permutation of objects that leaves no object in its original position. To solve the problem posed in Example 4 we will need to determine the number of derangements of a set of n objects.

EXAMPLE 5 The permutation 21453 is a derangement of 12345 because no number is left in its original position. However, 21543 is not a derangement of 12345, because this permutation leaves 4 fixed. ■

Let D_n denote the number of derangements of n objects. For instance $D_3 = 2$ since the derangements of 123 are 231 and 312. We will evaluate D_n, for all positive integers n, using the principle of inclusion-exclusion.

THEOREM 2 The number of derangements of a set with n elements is

$$D_n = n!\left[1 - \frac{1}{1!} + \frac{1}{2!} - \frac{1}{3!} + \cdots + (-1)^n \frac{1}{n!}\right].$$

Proof: Let a permutation have property P_i if it fixes element i. The number of derangements is the number of permutations having none of the properties P_i for $i = 1, 2, \ldots, n$, or

$$D_n = N(P_1'P_2' \cdots P_n').$$

Using the principle of inclusion-exclusion, it follows that

$$D_n = N - \sum_i N(P_i) + \sum_{i<j} N(P_iP_j) - \sum_{i<j<k} N(P_iP_jP_k) + \cdots + (-1)^n N(P_1P_2 \cdots P_n),$$

where N is the number of permutations of n elements. This equation states that the number of permutations that fix no elements equals the total number of permutations, less the number that fix at least one element, plus the number that fix at least two elements, less the number that fix at least three elements, and so on. All the quantities that occur on the right-hand side of this equation will now be found.

Historical Note: In **rencontres** (matches), an Old French card game, the 52 cards in a deck are laid out in a row. The cards of a second deck are laid out with one card of the second deck on top of each card of the first deck. The score is determined by counting the number of matching cards in the two decks. In 1708 Pierre Raymond de Montmort (1678–1719) posed **le problème de recontres:** What is the probability that no matches take place in the game of rencontres? The solution to Montmort's problem is the probability that a randomly selected permutation of 52 objects is a derangement, namely $D_{52}/52!$, which, as we will see, is approximately $1/e$.

First, note that $N = n!$, since N is just the total number of permutations of n elements. Also, $N(P_i) = (n-1)!$. This follows from the product rule, since $N(P_i)$ is the number of permutations that fix element i, so that the ith position of the permutation is determined, but each of the remaining positions can be filled arbitrarily. Similarly,

$$N(P_iP_j) = (n-2)!,$$

since this is the number of permutations that fix elements i and j, but where the other $n-2$ elements can be arranged arbitrarily. In general, note that

$$N(P_{i_1}P_{i_2}\cdots P_{i_m}) = (n-m)!,$$

because this is the number of permutations that fix elements $i_1, i_2, \ldots, i_m$, but where the other $n-m$ elements can be arranged arbitrarily. Because there are $C(n, m)$ ways to choose m elements from n, it follows that

$$\sum_{1\le i\le n} N(P_i) = C(n,1)(n-1)!,$$

$$\sum_{1\le i<j\le n} N(P_iP_j) = C(n,2)(n-2)!,$$

and in general,

$$\sum_{1\le i_1<i_2<\cdots<i_m\le n} N(P_{i_1}P_{i_2}\cdots P_{i_m}) = C(n,m)(n-m)!.$$

Consequently, inserting these quantities into our formula for D_n gives

$$\begin{aligned} D_n &= n! - C(n,1)(n-1)! + C(n,2)(n-2)! - \cdots + (-1)^n C(n,n)(n-n)! \\ &= n! - \frac{n!}{1!(n-1)!}(n-1)! + \frac{n!}{2!(n-2)!}(n-2)! - \cdots + (-1)^n \frac{n!}{n!0!}0!. \end{aligned}$$

Simplifying this expression gives

$$D_n = n!\left[1 - \frac{1}{1!} + \frac{1}{2!} - \cdots + (-1)^n \frac{1}{n!}\right].$$ ■

It is now simple to find D_n for a given positive integer n. For instance, using Theorem 2, it follows that

$$D_3 = 3!\left[1 - \frac{1}{1!} + \frac{1}{2!} - \frac{1}{3!}\right] = 6\left(1 - 1 + \frac{1}{2} - \frac{1}{6}\right) = 2,$$

as we have previously remarked.

The solution of the problem in Example 4 can now be given.

Solution: The probability that no one receives the correct hat is $D_n/n!$. By Theorem 2, this probability is

$$\frac{D_n}{n!} = 1 - \frac{1}{1!} + \frac{1}{2!} - \cdots (-1)^n \frac{1}{n!}.$$

TABLE 2 **The Probability of a Derangement**

n	2	3	4	5	6	7
$D_n/n!$	0.50000	0.33333	0.37500	0.36667	0.36806	0.36786

The values of this probability for $2 \leq n \leq 7$ are displayed in Table 2.

Using methods from calculus it can be shown that

$$e^{-1} = 1 - \frac{1}{1!} + \frac{1}{2!} - \cdots + (-1)^n \frac{1}{n!} + \cdots \sim 0.368.$$

Since this is an alternating series with terms tending to zero, it follows that as n grows without bound, the probability that no one receives the correct hat converges to $e^{-1} \sim 0.368$. In fact, this probability can be shown to be within $1/(n+1)!$ of e^{-1}.

Exercises

1. Suppose that in a bushel of 100 apples there are 20 that have worms in them and 15 that have bruises. Only those apples with neither worms nor bruises can be sold. If there are 10 bruised apples that have worms in them, how many of the 100 apples can be sold?
2. Of 1000 applicants for a mountain climbing trip in the Himalayas, 450 get altitude sickness, 622 are not in good enough shape, and 30 have allergies. An applicant qualifies if and only if this applicant does not get altitude sickness, is in good shape, and does not have allergies. If there are 111 applicants who get altitude sickness and are not in good enough shape, 14 who get altitude sickness and have allergies, 18 who are not in good enough shape and have allergies, and 9 who get altitude sickness, are not in good enough shape, and have allergies, how many applicants qualify?
3. How many solutions does the equation $x_1 + x_2 + x_3 = 13$ have where x_1, x_2, and x_3 are nonnegative integers less than 6?
4. Find the number of solutions of the equation $x_1 + x_2 + x_3 + x_4 = 17$ where x_i, $i = 1, 2, 3, 4$, are nonnegative integers such that $x_1 \leq 3$, $x_2 \leq 4$, $x_3 \leq 5$, and $x_4 \leq 8$.
5. Find the number of primes less than 200 using the principle of inclusion-exclusion.
6. An integer is called **squarefree** if it is not divisible by the square of a positive integer greater than 1. Find the number of squarefree positive integers less than 100.
7. How many positive integers less than 10,000 are not the second or higher power of an integer?
8. How many onto functions are there from a set with seven elements to one with five elements?
9. How many ways are there to distribute six different toys to three different children such that each child gets at least one toy?
10. In how many ways can eight distinct balls be distributed into three distinct urns if each urn must contain at least one ball?
11. In how many ways can seven different jobs be assigned to four different employees so that each employee is assigned at least one job and the most difficult job is assigned to the best employee?
12. List all the derangements of $\{1, 2, 3, 4\}$.
13. How many derangements are there of a set with seven elements?
14. What is the probability that none of 10 people receives the correct hat if a hatcheck person hands their hats back randomly?
15. A machine that inserts letters into envelopes goes haywire and inserts letters randomly into envelopes. What is the probability that in a group of 100 letters

a) no letter is put into the correct envelope?
b) exactly 1 letter is put into the correct envelope?
c) exactly 98 letters are put into the correct envelope?
d) exactly 99 letters are put into the correct envelope?
e) all letters are put into the correct envelope?

16. A group of n students is assigned seats for each of two classes in the same classroom. How many ways can these seats be assigned if no student is assigned the same seat for both classes?

★17. How many ways can the digits 0, 1, 2, 3, 4, 5, 6, 7, 8, 9 be arranged so that no even digit is in its original position?

★18. Use a combinatorial argument to show that the sequence $\{D_n\}$, where D_n denotes the number of derangements of n objects, satisfies the recurrence relation

$$D_n = (n-1)(D_{n-1} + D_{n-2})$$

for $n \geq 2$.

★19. Use Exercise 18 to show that

$$D_n = nD_{n-1} + (-1)^n$$

for $n \geq 1$.

20. Use Exercise 19 to find an explicit formula for D_n.

21. For which positive integers n is D_n, the number of derangements of n objects, even?

22. Suppose that p and q are distinct primes. Use the principle of inclusion-exclusion to find $\phi(pq)$, the number of integers not exceeding pq that are relatively prime to pq.

★23. Use the principle of inclusion-exclusion to derive a formula for $\phi(n)$ when the prime factorization of n is

$$n = p_1^{a_1} p_2^{a_2} \cdots p_m^{a_m}.$$

★24. Show that if n is a positive integer, then

$$n! = C(n,0)D_n + C(n,1)D_{n-1} + \cdots + C(n,n-1)D_1 + C(n,n)D_0,$$

where D_k is the number of derangements of k objects.

25. How many derangements of $\{1,2,3,4,5,6\}$ begin with the integers 1, 2, and 3, in some order?

26. How many derangements of $\{1,2,3,4,5,6\}$ end with the integers 1, 2, and 3, in some order?

27. Prove Theorem 1.

Key Terms and Results

TERMS

recurrence relation: a formula expressing terms of a sequence, except for some initial terms, as a function of one or more previous terms of the sequence

initial conditions for a recurrence relation: the values of the terms of a sequence satisfying the recurrence relation before this relation takes effect

linear homogeneous recurrence relation with constant coefficients: a recurrence relation that expresses the terms of a sequence, except for the first few, as a linear combination of previous terms

characteristic roots of a linear homogeneous recurrence relation with constant coefficients: the roots of the polynomial associated with a linear homogeneous recurrence relation with constant coefficients

divide-and-conquer algorithm: an algorithm that solves a problem recursively by splitting it into a fixed number of smaller problems of the same type

sieve of Eratosthenes: a procedure for finding the primes less than a specified positive integer

derangement: a permutation of objects such that no object is in its original place

RESULTS

The formula for the number of elements in the union of two finite sets:

$$|A \cup B| = |A| + |B| - |A \cap B|$$

The formula for the number of elements in the union of three finite sets:

$$|A \cup B \cup C| = |A| + |B| + |C| - |A \cap B| - |A \cap C| - |B \cap C| + |A \cap B \cap C|$$

The principle of inclusion-exclusion:

$$|A_1 \cup A_2 \cup \cdots \cup A_n| = \sum_{1 \le i \le n} |A_i| - \sum_{1 \le i < j \le n} |A_i \cap A_j| + \sum_{1 \le i < j < k \le n} |A_i \cap A_j \cap A_k| - \cdots + (-1)^{n+1}|A_1 \cap A_2 \cap \cdots \cap A_n|$$

The number of onto functions from a set with m elements to a set with n elements:

$$n^m - C(n, 1)(n-1)^m + C(n, 2)(n-2)^m - \cdots + (-1)^{n-1}C(n, n-1) \cdot 1^m$$

The number of derangements of n objects:

$$D_n = n!\left[1 - \frac{1}{1!} + \frac{1}{2!} - \cdots + (-1)^n \frac{1}{n!}\right]$$

Review Questions

1. a) What is a recurrence relation?
   b) Find a recurrence relation for the amount of money that will be in an account after n years if \$1,000,000 is deposited in an account yielding 9%.
2. Explain how the Fibonacci numbers are used to solve Fibonacci's problem about rabbits.
3. a) Find a recurrence relation for the number of steps needed to solve the Tower of Hanoi puzzle.
   b) Show how this recurrence relation can be solved using iteration.
4. a) Explain how to find a recurrence relation for the number of bit strings of length n not containing two consecutive 1s.
   b) Describe another counting problem that has a solution satisfying the same recurrence relation.
5. Define a linear homogeneous recurrence relation of degree k.
6. a) Explain how to solve linear homogeneous recurrence relations of degree two.
   b) Solve the recurrence relation $a_n = 13a_{n-1} - 22a_{n-2}$ for $n \ge 2$ if $a_0 = 3$ and $a_1 = 15$.
   c) Solve the recurrence relation $a_n = 14a_{n-1} - 49a_{n-2}$ for $n \ge 2$ if $a_0 = 3$ and $a_1 = 35$.
7. a) Explain how to find $f(b^k)$ where k is a positive integer if $f(n)$ satisfies the divide-and-conquer recurrence relation $f(n) = af(n/b) + g(n)$ whenever b divides the positive integer n.
   b) Find $f(256)$ if $f(n) = 3f(n/4) + 5n/4$ and $f(1) = 7$.
8. a) Derive a divide-and-conquer recurrence relation for the number of comparisons used to find a number in a list using a binary search.
   b) Give a big-O estimate for the number of comparisons used by a binary search from the divide-and-conquer recurrence relation you gave in (a) using Theorem 1 in Section 5.3.
9. a) Give a formula for the number of elements in the union of three sets.
   b) Explain why this formula is valid.
   c) Explain how to use the formula from (a) to find the number of integers not exceeding 1000 that are divisible by 6, 10, or 15.
   d) Explain how to use the formula from (a) to find the number of solutions in nonnegative integers to the equation $x_1 + x_2 + x_3 + x_4 = 22$ with $x_1 < 8$, $x_2 < 6$, and $x_3 < 5$.
10. a) Give a formula for the number of elements in the union of four sets and explain why it is valid.
    b) Suppose the sets A_1, A_2, A_3, and A_4 each contain 25 elements, the intersection of any two of these sets contains 5 elements, the intersection of any three of these sets contains 2 elements, and 1 element is in all four of the sets. How many elements are in the union of the four sets?
11. a) State the principle of inclusion-exclusion.
    b) Outline a proof of this principle.
12. Explain how the principle of inclusion-exclusion can be used to count the number of onto functions from a set with m elements to a set with n elements.

13. **a)** How can you count the number of ways to assign m jobs to n employees so that each employee is assigned at least one job?
b) How many ways are there to assign seven jobs to three employees so that each employee is assigned at least one job?

14. Explain how the inclusion-exclusion principle can be used to count the number of primes not exceeding the positive integer n.

15. **a)** Define a derangement.
b) Why is counting the number of ways a hatcheck person can return hats to n people, so that no one receives the correct hat, the same as counting the number of derangements of n objects?
c) Explain how to count the number of derangements of n objects.

Supplementary Exercises

1. A group of 10 people begin a chain letter, with each person sending the letter to 4 other people. Each of these people sends the letter to 4 additional people.
a) Find a recurrence relation for the number of letters sent at the nth stage of this chain letter, if no person ever receives more than one letter.
b) What are the initial conditions for the recurrence relation in part (a)?
c) How many letters are sent at the nth stage of the chain letter?

2. Every hour 1% of a radioactive isotope decays.
a) Set up a recurrence relation for the amount of this isotope left after n hours.
c) Solve this recurrence relation.

3. Every hour the U.S. government prints 10,000 more \$1 bills, 4000 more \$5 bills, 3000 more \$10 bills, 2500 more \$20 bills, 1000 more \$50 bills, and the same number of \$100 bills as it did the previous hour. In the initial hour 1000 of each bill were produced.
a) Set up a recurrence relation for the amount of money produced in the nth hour.
b) What are the initial conditions for the recurrence relation in part (a)?
c) Solve the recurrence relation for the amount of money produced in the nth hour.
d) Set up a recurrence relation for the total amount of money produced in the first n hours.
e) Solve the recurrence relation for the total amount of money produced in the first n hours.

4. Suppose every hour there are two new bacteria in a colony for each bacterium that was present the previous hour, and that all bacteria two hours old die. The colony starts with 100 new bacteria.
a) Set up a recurrence relation for the number of bacteria present after n hours.
b) What is the solution of this recurrence relation?
c) When will the colony contain more than one million bacteria?

5. Messages are sent over a communications channel using two different signals. One signal requires two microseconds for transmittal, and the other signal requires three microseconds for transmittal. Each signal of a message is followed immediately by the next signal.
a) Find a recurrence relation for the number of different signals that can be sent in n microseconds.
b) What are the initial conditions of the recurrence relation in part (a)?
c) How many different messages can be sent in 12 microseconds?

6. A small post office has only 4-cent stamps, 6-cent stamps, and 10-cent stamps. Find a recurrence relation for the number of ways to form postage of n cents with these stamps if the order that the stamps are used matters. What are the initial conditions for this recurrence relation?

7. How many ways are there to form the following postages using the rules described in Exercise 6?
a) 12 cents **b)** 14 cents
c) 18 cents **d)** 22 cents

8. Find the solutions of the simultaneous system of congruences

$$\begin{aligned} a_n &= a_{n-1} + b_{n-1} \\ b_n &= a_{n-1} - b_{n-1} \end{aligned}$$

with $a_0 = 1$ and $b_0 = 2$.

9. Solve the recurrence relation $a_n = a_{n-1}^2/a_{n-2}$ if $a_0 = 1$ and $a_1 = 2$. (*Hint:* Take logarithms of both sides to obtain a recurrence relation for the sequence $\log a_n$, $n = 0, 1, 2, \ldots$.)

★10. Solve the recurrence relation $a_n = a_{n-1}^3 a_{n-2}^2$ if $a_0 = 2$ and $a_1 = 2$. (See the hint for Exercise 9.)

11. Suppose that the characteristic equation of a linear homogeneous recurrence relation with constant coefficients has one or more multiple roots. Give a general result that expresses the solutions of the recurrence relation in terms of the roots of the characteristic equation.

12. Use Exercise 11 to find the solution of the recurrence relation $a_n = 3a_{n-1} - 3a_{n-2} + a_{n-3}$ if $a_0 = 2$, $a_1 = 2$, and $a_2 = 4$.

★13. Suppose that in Example 4 of Section 5.1 a pair of rabbits leaves the island after reproducing twice. Find a recurrence relation for the number of rabbits on the island in the middle of the nth month.

14. Find the solution to the recurrence relation $f(n) = 3f(n/5) + 2n^4$, when n is divisible by 5, for $n = 5^k$, where k is a positive integer and $f(1) = 1$.

15. Estimate the size of f in Exercise 14 if f is an increasing function.

16. Find a recurrence relation that describes the number of comparisons used by the following algorithm: Find the largest and second largest elements of a sequence of n numbers recursively by splitting the sequence into two subsequences with an equal number of terms, or where there is one more term in one subsequence than in the other, at each stage. Stop when subsequences with two terms are reached.

17. Estimate the number of comparisons used by the algorithm described in Exercise 16.

Let $\{a_n\}$ be a sequence of real numbers. The **forward differences** of this sequence are defined recursively as follows. The **first forward difference** is $\Delta a_n = a_{n+1} - a_n$. The **(k+1)th forward difference** $\Delta^{k+1} a_n$ is obtained from $\Delta^k a_n$ by $\Delta^{k+1} a_n = \Delta^k a_{n+1} - \Delta^k a_n$.

18. Find Δa_n where
a) $a_n = 3$. **b)** $a_n = 4n + 7$.
c) $a_n = n^2 + n + 1$.

19. Let $a_n = 3n^3 + n + 2$. Find $\Delta^k a_n$ where k equals
a) 2. **b)** 3. **c)** 4.

★20. Suppose that $a_n = P(n)$ where P is a polynomial of degree d. Prove that $\Delta^{d+1} a_n = 0$ for all nonnegative integers n.

21. Let $\{a_n\}$ and $\{b_n\}$ be sequences of real numbers. Show that

$$\Delta(a_n b_n) = a_{n+1}(\Delta b_n) + b_n(\Delta a_n).$$

22. Suppose that 14 students get an A on the first exam in a discrete mathematics class, and 18 get an A on the second exam. If 22 students received an A on either the first exam or the second exam, how many students received an A on both exams?

23. There are 323 farms in Monmouth County that have at least one of horses, cows, and sheep. If 224 have horses, 85 have cows, 57 have sheep, and 18 farms have all three types of animals, how many farms have exactly two of these three types of animals?

24. Queries to a data base of student records at a college produced the following data: There are 2175 students at the college, 1675 of these are not freshmen, 1074 students have taken a course in calculus, 444 students have taken a course in discrete mathematics, 607 students are not freshmen and have taken calculus, 350 students have taken calculus and discrete mathematics, 201 students are not freshmen and have taken discrete mathematics, and 143 students are not freshmen and have taken both calculus and discrete mathematics. Can all the responses to the queries be correct?

25. Students in the school of mathematics at a university major in one or more of the following four areas: applied mathematics (AM), pure mathematics (PM), operations research (OR), and computer science (CS). How many students are in this school if (including joint majors) there are 23 students majoring in AM; 17 students majoring in PM; 44 in OR; 63 in CS; 5 in AM and PM; 8 in AM and CS; 4 in AM and OR; 6 in PM and CS; 5 in PM and OR; 14 in OR and CS; 2 in PM, OR, and CS; 1 in PM, AM, and OR; 1 in PM, AM, and CS; and 1 in all four fields.

26. How many terms are needed when the inclusion-exclusion principle is used to express the number of elements in the union of seven sets if no more than five of these sets have a common element?

27. How many solutions in positive integers are there to the equation $x_1 + x_2 + x_3 = 20$ with $2 < x_1 < 6$, $6 < x_2 < 10$, and $0 < x_3 < 5$?

28. How many positive integers less than 1,000,000 are
  a) divisible by 2, 3, or 5?
  b) not divisible by 7, 11, or 13?
  c) divisible by 3 but not by 7?

29. How many positive integers less than 200 are
  a) second or higher powers of integers?
  b) either second or higher powers of integers or primes?
  c) not divisible by the square of an integer greater than 1?
  d) not divisible by the cube of an integer greater than 1?
  e) not divisible by three or more primes?

★30. How many ways are there to assign six different jobs to three different employees if the hardest job is assigned to the most experienced employee and the easiest job is assigned to the least experienced employee?

31. What is the probability that exactly one person is given back the correct hat by a hatcheck person who gives n people their hats back at random?

32. How many bit strings of length six do not contain four consecutive 1s?

33. What is the probability that a bit string of length six contains at least four 1s?

Computer Projects

WRITE PROGRAMS WITH THE FOLLOWING INPUT AND OUTPUT.

1. Given a positive integer n, list all the moves required in the Tower of Hanoi puzzle to move n disks from one peg to another according to the rules of the puzzle.

2. Given a positive integer n, list all the bit sequences of length n that do not have a pair of consecutive 0s.

3. Given a recurrence relation $a_n = c_1 a_{n-1} + c_2 a_{n-2}$ where c_1 and c_2 are real numbers, initial conditions $a_0 = C_0$ and $a_1 = C_1$, and a positive integer k, find a_k using iteration.

4. Given a recurrence relation $a_n = c_1 a_{n-1} + c_2 a_{n-2}$ and initial conditions $a_0 = C_0$ and $a_1 = C_1$, determine the unique solution.

5. Given a recurrence relation of the form $f(n) = af(n/b) + c$, where a is a real number, b is a positive integer, and c is a real number, and a positive integer k, find $f(b^k)$ using iteration.

6. Given the number of elements in the intersection of three sets, the number of elements in each pairwise intersection of these sets, and the number of elements in each set, find the number of elements in their union.

7. Given a positive integer n, produce the formula for the number of elements in the union of n sets.

8. Given positive integers m and n, find the number of onto functions from a set with m elements to a set with n elements.

9. Given a positive integer n, list all the derangements of the set $\{1, 2, 3, \ldots, n\}$.

6

Relations

Relationships between elements of sets occur in many contexts. Every day we deal with relationships such as those between a business and its telephone number, an employee and his or her salary, a person and a relative, and so on. In mathematics we study relationships such as those between a positive integer and one that it divides, an integer and one that it is congruent to modulo 5, a real number and one that is larger than it, and so on. Relationships such as that between a program and a variable it uses and that between a computer language and a valid statement in this language often arise in computer science.

Relationships between elements of sets are represented using the structure called a relation. Relations can be used to solve problems such as determining which pairs of cities are linked by airline flights in a network, finding a viable order for the different phases of a complicated project, or producing a useful way to store information in computer data bases.

6.1 Relations and Their Properties

INTRODUCTION

The most direct way to express a relationship between elements of two sets is to use ordered pairs made up of two related elements. For this reason, sets of ordered pairs are called binary relations. In this section we introduce the basic terminology used to describe binary relations. Later in this chapter we will use relations to solve problems involving communications networks, project scheduling, and identifying elements in sets with common properties.

DEFINITION 1. Let A and B be sets. A *binary relation from A to B* is a subset of $A \times B$.

In other words, a binary relation from A to B is a set of ordered pairs where the first element of each ordered pair comes from A and the second element comes from B. We use the notation $a\, R\, b$ to denote that $(a, b) \in R$ and $a \not R\, b$ to denote that $(a, b) \notin R$. Moreover, when (a, b) belongs to R, a is said to be **related to** b by R.

Binary relations represent relationships between the elements of two sets. We will introduce n-ary relations, which express relationships among elements of more than two sets, later in this chapter. We will omit the word *binary* when there is no danger of confusion.

The following are examples of relations:

EXAMPLE 1 Let A be the set of students in your school, and let B be the set of courses. Let R be the relation that consists of those pairs (a, b) where a is a student enrolled in course b. For instance, if Jason Goodfriend and Deborah Sherman are enrolled in CS518, which is Discrete Mathematics, the pairs (Jason Goodfriend, CS518) and (Deborah Sherman, CS518) belong to R. If Jason Goodfriend is also enrolled in CS510, which is Data Structures, then the pair (Jason Goodfriend, CS510) is also in R. However, if Deborah Sherman is not enrolled in CS510, then the pair (Deborah Sherman, CS510) is not in R. ■

EXAMPLE 2 Let A be the set of all cities, and let B be the set of the 50 states in the United States of America. Define the relation R by specifying that (a, b) belongs to R if city a is in state b. For instance, (Boulder, Colorado), (Bangor, Maine), (Ann Arbor, Michigan), (Cupertino, California), and (Red Bank, New Jersey) are in R. ■

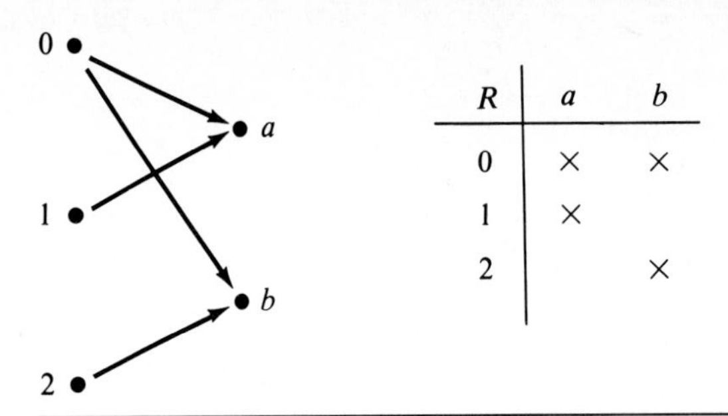

R	a	b
0	×	×
1	×	
2		×

FIGURE 1 **Displaying the Ordered Pairs in the Relation R from Example 3.**

EXAMPLE 3 Let $A = \{0, 1, 2\}$ and $B = \{a, b\}$. Then $\{(0, a), (0, b), (1, a), (2, b)\}$ is a relation from A to B. This means, for instance, that $0\,R\,a$, but that $1\,\not R\,b$. Relations can be represented graphically, as shown in Figure 1, using arrows to represent ordered pairs. Another way to represent this relation is to use a table, which is also done in Figure 1. We will discuss representations of relations in more detail in Section 6.3. ■

FUNCTIONS AS RELATIONS

Recall that a function f from a set A to a set B (as defined in Section 1.6) assigns a unique element of B to each element of A. The graph of f is the set of ordered pairs (a, b) such that $b = f(a)$. Since the graph of f is a subset of $A \times B$, it is a relation from A to B. Moreover, the graph of a function has the property that every element of A is the first element of exactly one ordered pair of the graph.

Conversely, if R is a relation from A to B such that every element in A is the first element of exactly one ordered pair of R, then a function can be defined with R as its graph. This can be done by assigning to an element a of A the unique element $b \in B$ such that $(a, b) \in R$.

A relation can be used to express a one-to-many relationship between the elements of the sets A and B, where an element of A may be related to more than one element of B. A function represents a relation where exactly one element of B is related to each element of A.

RELATIONS ON A SET

Relations from a set A to itself are of special interest.

DEFINITION 2. A *relation on* the set A is a relation from A to A.

In other words, a relation on a set A is a subset of $A \times A$.

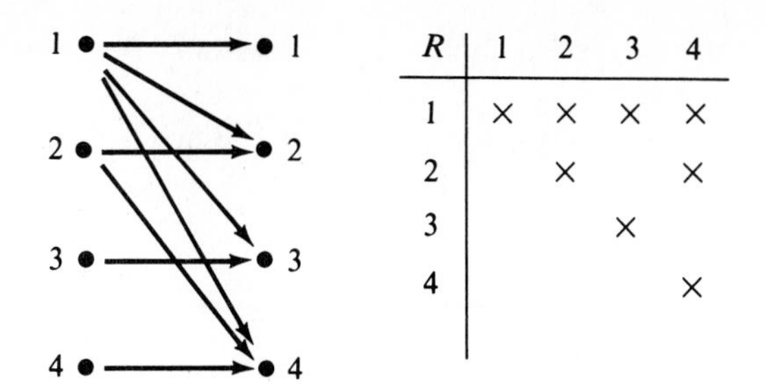

R	1	2	3	4
1	×	×	×	×
2		×		×
3			×	
4				×

FIGURE 2 **Displaying the Ordered Pairs in the Relation R from Example 4.**

EXAMPLE 4 Let A be the set $\{1, 2, 3, 4\}$. Which ordered pairs are in the relation $R = \{(a, b) \mid a$ divides $b\}$?

Solution: Since (a, b) is in R if and only if a and b are positive integers not exceeding 4 such that a divides b, we see that

$$R = \{(1, 1), (1, 2), (1, 3), (1, 4), (2, 2), (2, 4), (3, 3), (4, 4)\}.$$

The pairs in this relation are displayed both graphically and in tabular form in Figure 2. ■

Next, some examples of relations on the set of integers will be given.

EXAMPLE 5 Consider the following relations on the set of integers:

$$\begin{aligned}
R_1 &= \{(a, b) \mid a \leq b\} \\
R_2 &= \{(a, b) \mid a > b\} \\
R_3 &= \{(a, b) \mid a = b \text{ or } a = -b\} \\
R_4 &= \{(a, b) \mid a = b\} \\
R_5 &= \{(a, b) \mid a = b + 1\} \\
R_6 &= \{(a, b) \mid a + b \leq 3\}
\end{aligned}$$

Which of these relations contain each of the pairs $(1, 1)$, $(1, 2)$, $(2, 1)$, $(1, -1)$, and $(2, 2)$?

Remark: Unlike the relations in Examples 1–4, these are relations on an infinite set.

Solution: The pair $(1, 1)$ is in R_1, R_3, R_4, and R_6; $(1, 2)$ is in R_1 and R_6; $(2, 1)$ is in R_2, R_5, and R_6; $(1, -1)$ is in R_2, R_3, and R_6; and finally, $(2, 2)$ is in R_1, R_3, and R_4. ■

It is not hard to determine the number of relations on a finite set, since a relation on a set A is simply a subset of $A \times A$.

EXAMPLE 6 How many relations are there on a set with n elements?

Solution: A relation on a set A is a subset of $A \times A$. Since $A \times A$ has n^2 elements when A has n elements, and a set with m elements has 2^m subsets, there are 2^{n^2} subsets of $A \times A$. Thus, there are 2^{n^2} relations on a set with n elements. ■

PROPERTIES OF RELATIONS

There are several properties that are used to classify relations on a set. We will introduce the most important of these here.

In some relations an element is always related to itself. For instance, let R be the relation on the set of all people consisting of pairs (x, y) where x and y have the same mother and the same father. Then $x \, R \, x$ for every person x.

DEFINITION 3. A relation R on a set A is called *reflexive* if $(a, a) \in R$ for every element $a \in A$.

We see that a relation on A is reflexive if every element of A is related to itself. The following examples illustrate the concept of a reflexive relation.

EXAMPLE 7 Consider the following relations on $\{1, 2, 3, 4\}$:

$$
\begin{aligned}
R_1 &= \{(1,1), (1,2), (2,1), (2,2), (3,4), (4,1), (4,4)\} \\
R_2 &= \{(1,1), (1,2), (2,1)\} \\
R_3 &= \{(1,1), (1,2), (1,4), (2,1), (2,2), (3,3), (4,1), (4,4)\} \\
R_4 &= \{(2,1), (3,1), (3,2), (4,1), (4,2), (4,3)\} \\
R_5 &= \{(1,1), (1,2), (1,3), (1,4), (2,2), (2,3), (2,4), (3,3), (3,4), (4,4)\} \\
R_6 &= \{(3,4)\}
\end{aligned}
$$

Which of these relations are reflexive?

Solution: The relations R_3 and R_5 are reflexive since they both contain all pairs of the form (a, a), namely, $(1, 1)$, $(2, 2)$, $(3, 3)$, and $(4, 4)$. The other relations are not reflexive since they do not contain all of these ordered pairs. In particular, R_1, R_2, R_4, and R_6 are not reflexive since $(3, 3)$ is not in any of these relations. ■

EXAMPLE 8 Which of the relations from Example 5 are reflexive?

Solution: The reflexive relations from this example are R_1 (since $a \leq a$ for every integer a), R_3, and R_4. For each of the other relations in this example it is easy to find a pair of the form (a, a) that is not in the relation. (This is left as an exercise for the reader.) ■

EXAMPLE 9 Is the "divides" relation on the set of positive integers reflexive?

Solution: Since $a \mid a$ whenever a is a positive integer, the "divides" relation is reflexive. ■

In some relations an element is related to a second element if and only if the second element is also related to the first element. The relation consisting of pairs (x, y) where x and y are students at your school with at least one common class has this property. Other relations have the property that if an element is related to a second element, then this second element is not related to the first. The relation consisting of the pairs (x, y) where x and y are students at your school where x has a higher grade point average than y has this property.

DEFINITION 4. A relation R on a set A is called *symmetric* if $(b, a) \in R$ whenever $(a, b) \in R$, for $a, b \in A$. A relation R on a set A such that $(a, b) \in R$ and $(b, a) \in R$ only if $a = b$, for $a, b \in A$, is called *antisymmetric*.

That is, a relation is symmetric if and only if a is related to b implies that b is related to a. A relation is antisymmetric if and only if there are no pairs of distinct elements a and b with a related to b and b related to a. The terms *symmetric* and *antisymmetric* are not opposites, since a relation can have both of these properties or may lack both of them (see Exercise 6 at the end of this section). A relation cannot be both symmetric and antisymmetric if it contains some pair of the form (a, b) where $a \neq b$.

EXAMPLE 10 Which of the relations from Example 7 are symmetric and which are antisymmetric?

Solution: The relations R_2 and R_3 are symmetric, since in each case (b, a) belongs to the relation whenever (a, b) does. For R_2 the only thing to check is that both $(2, 1)$ and $(1, 2)$ are in the relation. For R_3, it is necessary to check that both $(1, 2)$ and $(2, 1)$ belong to the relation, and $(1, 4)$ and $(4, 1)$ belong to the relation. The reader should verify that none of the other relations is symmetric. This is done by finding a pair (a, b) so that it is in the relation but (b, a) is not.

R_4, R_5, and R_6 are all antisymmetric. For each of these relations there is no pair of elements a and b with $a \neq b$ such that both (a, b) and (b, a) belong to the relation. The reader should verify that none of the other relations is antisymmetric. This is done by finding a pair (a, b) with $a \neq b$ so that (a, b) and (b, a) are both in the relation. ■

EXAMPLE 11 Which of the relations from Example 5 are symmetric and which are antisymmetric?

Solution: The relations R_3, R_4, and R_6 are symmetric. R_3 is symmetric, for if $a = b$ or $a = -b$, then $b = a$ or $b = -a$. R_4 is symmetric since $a = b$ implies that $b = a$. R_6 is symmetric since $a + b \leq 3$ implies that $b + a \leq 3$. The reader should verify that none of the other relations is symmetric.

The relations R_1, R_2, R_4, and R_5 are antisymmetric. R_1 is antisymmetric because the inequalities $a \leq b$ and $b \leq a$ imply that $a = b$. R_2 is antisymmetric since it is impossible for $a > b$ and $b > a$. R_4 is antisymmetric, since two elements are related with respect to R_4 if and only if they are equal. R_5 is antisymmetric since it is impossible that $a = b + 1$ and $b = a + 1$. The reader should verify that none of the other relations is antisymmetric. ■

EXAMPLE 12 Is the "divides" relation on the set of positive integers symmetric? Is it antisymmetric?

Solution: This relation is not symmetric since $1 \mid 2$, but $2 \nmid 1$. It is antisymmetric, for if a and b are positive integers with $a \mid b$ and $b \mid a$, then $a = b$ (the verification of this is left as an exercise for the reader). ■

Let R be the relation consisting of all pairs (x, y) of students at your school where x has taken more credits than y. Suppose that x is related to y and y is related to z. This means that x has taken more credits than y and y has taken more credits than z. We can conclude that x has taken more credits than z, so that x is related to z. What we have shown is that R has the transitive property, which is defined as follows.

DEFINITION 5. A relation R on a set A is called *transitive* if whenever $(a, b) \in R$ and $(b, c) \in R$ then $(a, c) \in R$, for $a, b, c \in A$.

EXAMPLE 13 Which of the relations in Example 7 are transitive?

Solution: R_4, R_5, and R_6 are transitive. For each of these relations, we can show that it is transitive by verifying that if (a, b) and (b, c) belong to this relation, then (a, c) also does. For instance, R_4 is transitive, since $(3, 2)$ and $(2, 1)$, $(4, 2)$ and $(2, 1)$, $(4, 3)$ and $(3, 1)$, and $(4, 3)$ and $(3, 2)$ are the only such sets of pairs, and $(3, 1)$, $(4, 1)$, and $(4, 2)$ belong to R_4. The reader should verify that R_5 and R_6 are transitive.

R_1 is not transitive since $(3, 4)$ and $(4, 1)$ belong to R_1, but $(3, 1)$ does not. R_2 is not transitive since $(2, 1)$ and $(1, 2)$ belong to R_2, but $(2, 2)$ does not. R_3 is not transitive since $(4, 1)$ and $(1, 2)$ belong to R_3, but $(4, 2)$ does not. ■

EXAMPLE 14 Which of the relations in Example 5 are transitive?

Solution: The relations R_1, R_2, R_3, and R_4 are transitive. R_1 is transitive since $a \leq b$ and $b \leq c$ imply that $a \leq c$. R_2 is transitive since $a > b$ and $b > c$ imply that $a > c$. R_3 is transitive since $a = \pm b$ and $b = \pm c$ imply that $a = \pm c$. R_4 is clearly transitive, as the reader should verify. R_5 is not transitive since $(2, 1)$ and $(1, 0)$ belong to R_5, but $(2, 0)$ does not. R_6 is not transitive since $(2, 1)$ and $(1, 2)$ belong to R_6, but $(2, 2)$ does not. ■

EXAMPLE 15 Is the "divides" relation on the set of positive integers transitive?

Solution: Suppose that a divides b and b divides c. Then there are positive integers k and l such that $b = ak$ and $c = bl$. Hence, $c = akl$, so that a divides c. It follows that this relation is transitive. ■

The next example shows how to count the number of relations with a specified property.

EXAMPLE 16 How many reflexive relations are there on a set with n elements?

Solution: A relation R on a set A is a subset of $A \times A$. Consequently, a relation is determined by specifying whether each of the n^2 ordered pairs in $A \times A$ is in R. However, if R is reflexive, each of the n ordered pairs (a, a) for $a \in A$ must be in R. Each of the other $n(n-1)$ ordered pairs of the form (a, b) where $a \neq b$ may or may not be in R. Hence, by the product rule for counting, there are $2^{n(n-1)}$ reflexive relations (this is the number of ways to choose whether each element (a, b) with $a \neq b$ belongs to R). ■

The number of symmetric relations and the number of antisymmetric relations on a set with n elements can be found using reasoning similar to that in Example 16. (See Exercise 25 at the end of this section.) Counting the transitive relations on a set with n elements is a problem beyond the scope of this book.

COMBINING RELATIONS

Since relations from A to B are subsets of $A \times B$, two relations from A to B can be combined in any way two sets can be combined. Consider the following examples.

EXAMPLE 17 Let $A = \{1, 2, 3\}$ and $B = \{1, 2, 3, 4\}$. The relations $R_1 = \{(1, 1), (2, 2), (3, 3)\}$ and $R_2 = \{(1, 1), (1, 2), (1, 3), (1, 4)\}$ can be combined to obtain

$$R_1 \cup R_2 = \{(1, 1), (1, 2), (1, 3), (1, 4), (2, 2), (3, 3)\},$$
$$R_1 \cap R_2 = \{(1, 1)\},$$
$$R_1 - R_2 = \{(2, 2), (3, 3)\},$$
$$R_2 - R_1 = \{(1, 2), (1, 3), (1, 4)\}.$$

■

EXAMPLE 18 Let A and B be the set of all students and the set of all courses at a school, respectively. Suppose that R_1 consists of all ordered pairs (a, b) where a is a student who has taken course b, and R_2 consists of all ordered pairs (a, b) where a is a student who requires course b to graduate. What are the relations $R_1 \cup R_2$, $R_1 \cap R_2$, $R_1 \oplus R_2$, $R_1 - R_2$, and $R_2 - R_1$?

Solution: The relation $R_1 \cup R_2$ consists of all ordered pairs (a, b) where a is a student who either has taken course b or needs course b to graduate, and $R_1 \cap R_2$ is the set of all ordered pairs (a, b) where a is a student who has taken course b and needs this course to graduate. Also, $R_1 \oplus R_2$ consists of all ordered pairs (a, b) where student a has taken course b but does not need it to graduate, or needs course b to graduate but has not taken it. $R_1 - R_2$ is the set of ordered pairs (a, b) where a has taken course b but does not need it to graduate; that is, b is an elective course that a has taken. $R_2 - R_1$ is the set of all ordered pairs (a, b) where b is a course that a needs to graduate but has not taken. ■

There is another way that relations are combined that is analogous to the composition of functions.

DEFINITION 6. Let R be a relation from a set A to a set B and S a relation from B to a set C. The *composite* of R and S is the relation consisting of ordered pairs (a, c) where $a \in A$, $c \in C$, and for which there exists an element $b \in B$ such that $(a, b) \in R$ and $(b, c) \in S$. We denote the composite of R and S by $S \circ R$.

The following example illustrates how composites of relations are formed.

EXAMPLE 19 What is the composite of the relations R and S where R is the relation from $\{1, 2, 3\}$ to $\{1, 2, 3, 4\}$ with $R = \{(1, 1), (1, 4), (2, 3), (3, 1), (3, 4)\}$ and S is the relation from $\{1, 2, 3, 4\}$ to $\{0, 1, 2\}$ with $S = \{(1, 0), (2, 0), (3, 1), (3, 2), (4, 1)\}$?

Solution: $S \circ R$ is constructed using all ordered pairs in R and ordered pairs in S where the second element of the ordered pair in R agrees with the first element of the ordered pair in S. For example, the ordered pairs $(2, 3)$ in R and $(3, 1)$ in S produce the ordered pair (2, 1) in $S \circ R$. Computing all the ordered pairs in the composite, we find

$$S \circ R = \{(1, 0), (1, 1), (2, 1), (2, 2), (3, 0), (3, 1)\}.$$

■

The powers of a relation R can be inductively defined from the definition of a composite of two relations.

DEFINITION 7. Let R be a relation on the set A. The powers R^n, $n = 1, 2, 3, \ldots$, are defined inductively by

$R^1 = R$ and $R^{n+1} = R^n \circ R$.

The definition shows that $R^2 = R \circ R$, $R^3 = R^2 \circ R = (R \circ R) \circ R$, and so on.

EXAMPLE 20 Let $R = \{(1, 1), (2, 1), (3, 2), (4, 3)\}$. Find the powers R^n, $n = 2, 3, 4, \ldots$.

Solution: Since $R^2 = R \circ R$, we find that $R^2 = \{(1, 1), (2, 1), (3, 1), (4, 2)\}$. Furthermore, since $R^3 = R^2 \circ R$, $R^3 = \{(1, 1), (2, 1), (3, 1), (4, 1)\}$. Additional computation shows that R^4 is the same as R^3, so that $R^4 = \{(1, 1), (2, 1), (3, 1), (4, 1)\}$. It also follows that $R^n = R^3$ for $n = 5, 6, 7, \ldots$. The reader should verify this. ■

The following theorem shows that the powers of a transitive relation are subsets of this relation. It will be used in Section 6.4.

THEOREM 1 Let R be a transitive relation on a set A. Then $R^n \subseteq R$ for $n = 1, 2, 3, \ldots$.

Proof: We will use mathematical induction to prove this theorem. The theorem is trivially true for $n = 1$.

Assume that $R^n \subseteq R$ where n is a positive integer. This is the inductive hypothesis. To complete the inductive step we must show that this implies that R^{n+1} is also a subset of R. To show this, assume that $(a, b) \in R^{n+1}$. Then, since $R^{n+1} = R^n \circ R$, there is an element x with $x \in A$ such that $(a, x) \in R$ and $(x, b) \in R^n$. The inductive hypothesis, namely, that $R^n \subseteq R$, implies that $(x, b) \in R$. Furthermore, since R is transitive, and $(a, x) \in R$ and $(x, b) \in R$, it follows that $(a, b) \in R$. This shows that $R^{n+1} \subseteq R$, completing the proof. ■

Exercises

1. List the ordered pairs in the relation R from $A = \{0, 1, 2, 3, 4\}$ to $B = \{0, 1, 2, 3\}$ where $(a, b) \in R$ if and only if

a) $a = b$.
b) $a + b = 4$.
c) $a > b$.
d) $a \mid b$.
e) $gcd(a, b) = 1$.
f) $lcm(a, b) = 2$.

2. a) List all the ordered pairs in the relation $R = \{(a, b) \mid a \text{ divides } b\}$ on the set $\{1, 2, 3, 4, 5, 6\}$.
   b) Display this relation graphically, as was done in Example 4.
   c) Display this relation in tabular form, as was done in Example 4.
3. For each of the following relations on the set $\{1, 2, 3, 4\}$, decide whether it is reflexive, whether it is symmetric, whether it is antisymmetric, and whether it is transitive.
   a) $\{(2, 2), (2, 3), (2, 4), (3, 2), (3, 3), (3, 4)\}$
   b) $\{(1, 1), (1, 2), (2, 1), (2, 2), (3, 3), (4, 4)\}$
   c) $\{(2, 4), (4, 2)\}$
   d) $\{(1, 2), (2, 3), (3, 4)\}$
   e) $\{(1, 1), (2, 2), (3, 3), (4, 4)\}$
   f) $\{(1, 3), (1, 4), (2, 3), (2, 4), (3, 1), (3, 4)\}$
4. Determine whether the relation R on the set of all people is reflexive, symmetric, antisymmetric, and/or transitive, where $(a, b) \in R$ if and only if
   a) a is taller than b.
   b) a and b were born on the same day.
   c) a has the same first name as b.
   d) a and b have a common grandparent.
5. Determine whether the relation R on the set of all integers is reflexive, symmetric, antisymmetric, and/or transitive, where $(x, y) \in R$ if and only if
   a) $x \neq y$.
   b) $xy \geq 1$.
   c) $x = y + 1$ or $x = y - 1$.
   d) $x \equiv y \pmod 7$.
   e) x is a multiple of y.
   f) x and y are both negative or both nonnegative.
   g) $x = y^2$.
   h) $x \geq y^2$.
6. Give an example of a relation on a set that is
   a) symmetric and antisymmetric.
   b) neither symmetric nor antisymmetric.

☞ A relation R on the set A is **irreflexive** if for every $a \in A, (a, a) \notin R$. That is, R is irreflexive if no element in A is related to itself.

☞7. Which relations in Exercise 3 are irreflexive?
8. Which relations in Exercise 4 are irreflexive?
9. Can a relation on a set be neither reflexive nor irreflexive?

A relation R is called **asymmetric** if $(a, b) \in R$ implies that $(b, a) \notin R$.

10. Which relations in Exercise 3 are asymmetric?
11. Which relations in Exercise 4 are asymmetric?
12. Must an asymmetric relation also be antisymmetric? Must an antisymmetric relation be asymmetric? Give reasons for your answers.
13. How many different relations are there from a set with m elements to a set with n elements?

☞ Let R be a relation from a set A to a set B. The **inverse relation** from B to A, denoted by R^{-1}, is the set of ordered pairs $\{(b, a) \mid (a, b) \in R\}$. The **complementary relation** $\overline{R}$ is the set of ordered pairs $\{(a, b) \mid (a, b) \notin R\}$.

14. Let R be the relation $R = \{(a, b) \mid a < b\}$ on the set of integers. Find
    a) R^{-1}. b) $\overline{R}$.
15. Let R be the relation $R = \{(a, b) \mid a \text{ divides } b\}$ on the set of positive integers. Find
    a) R^{-1}. b) $\overline{R}$.
16. Let R be the relation on the set of all states in the United States consisting of pairs (a, b) where state a borders state b. Find
    a) R^{-1}. b) $\overline{R}$.
17. Suppose that the function f from A to B is a one-to-one correspondence. Let R be the relation that equals the graph of f. That is, $R = \{(a, f(a)) \mid a \in A\}$. What is the inverse relation R^{-1}?
18. Let $R_1 = \{(1, 2), (2, 3), (3, 4)\}$ and $R_2 = \{(1, 1), (1, 2), (2, 1), (2, 2), (2, 3), (3, 1), (3, 2), (3, 3), (3, 4)\}$ be relations from $\{1, 2, 3\}$ to $\{1, 2, 3, 4\}$. Find
    a) $R_1 \cup R_2$ b) $R_1 \cap R_2$ c) $R_1 - R_2$
    d) $R_2 - R_1$
19. Let A be the set of students at your school and B the set of books in the school library. Let R_1 and R_2 be the relations consisting of all ordered pairs (a,b) where student a is required to read book b in a course, and where student a has read book b, respectively. Describe the ordered pairs in each of the following relations.
    a) $R_1 \cup R_2$ b) $R_1 \cap R_2$ c) $R_1 \oplus R_2$
    d) $R_1 - R_2$ e) $R_2 - R_1$
20. Let R be the relation $\{(1, 2), (1, 3), (2, 3), (2, 4), (3, 1)\}$, and let S be the relation $\{(2, 1), (3, 1), (3, 2), (4, 2)\}$. Find $S \circ R$.
21. Let R be the relation on the set of people consisting of pairs (a,b) where a is a parent of b. Let S be the relation on the set of people consisting of pairs (a,b) where a and b are siblings (brothers or sisters). What are $S \circ R$ and $R \circ S$?
22. List the 16 different relations on the set $\{0, 1\}$.
23. How many of the 16 different relations on $\{0, 1\}$ contain the pair $(0, 1)$?

24. Which of the 16 relations on $\{0, 1\}$, which you listed in Exercise 22, are

a) reflexive? **b)** irreflexive?
c) symmetric? **d)** antisymmetric?
e) asymmetric? **f)** transitive?

★25. How many relations are there on a set with n elements that are

a) symmetric?
b) antisymmetric?
c) asymmetric?
d) irreflexive?
e) reflexive and symmetric?
f) neither reflexive nor irreflexive?

★26. How many transitive relations are there on a set with n elements if

a) $n = 1$? **b)** $n = 2$? **c)** $n = 3$?

27. Find the error in the "proof" of the following "theorem."

THEOREM Let R be a relation on a set A that is symmetric and transitive. Then R is reflexive.

Proof: Let $a \in A$. Take an element $b \in A$ such that $(a, b) \in R$. Since R is symmetric, we also have $(b, a) \in R$. Now using the transitive property, we can conclude that $(a, a) \in R$ since $(a, b) \in R$ and $(b, a) \in R$.

28. Suppose that R and S are reflexive relations on a set A. Prove or disprove each of the following statements.

a) $R \cup S$ is reflexive.
b) $R \cap S$ is reflexive.
c) $R \oplus S$ is irreflexive.
d) $R - S$ is irreflexive.
e) $S \circ R$ is reflexive.

29. Show that the relation R on a set A is symmetric if and only if $R = R^{-1}$ where R^{-1} is the inverse relation.

30. Show that the relation R on a set A is antisymmetric if and only if $R \cap R^{-1}$ is a subset of the diagonal relation $\Delta = \{(a, a) \mid a \in A\}$.

31. Show that the relation R on a set A is reflexive if and only if the inverse relation R^{-1} is reflexive.

32. Show that the relation R on a set A is reflexive if and only if the complementary relation $\overline{R}$ is irreflexive.

33. Let R be a relation that is reflexive and transitive. Prove that $R^n = R$ for all positive integers n.

34. Let R be the relation on the set $\{1, 2, 3, 4, 5\}$ containing the ordered pairs $(1, 1)$, $(1, 2)$, $(1, 3)$, $(2, 3)$, $(2, 4)$, $(3, 1)$, $(3, 4)$, $(3, 5)$, $(4, 2)$, $(4, 5)$, $(5, 1)$, $(5, 2)$, and $(5, 4)$. Find

a) R^2. **b)** R^3. **c)** R^4. **d)** R^5.

35. Let R be a reflexive relation on a set A. Show that R^n is reflexive for all positive integers n.

★36. Let R be a symmetric relation. Show that R^n is symmetric for all positive integers n.

37. Suppose that the relation R is irreflexive. Is R^2 necessarily irreflexive? Give a reason for your answer.

6.2 *n*-ary Relations and Their Applications

INTRODUCTION

Relationships among elements of more than two sets often arise. For instance, there is a relationship involving the name of a student, the student's major, and the student's grade point average. Similarly, there is a relationship involving the airline, flight number, starting point, destination, departure time, and arrival time of a flight. An example of such a relationship in mathematics involves three integers where the first integer is larger than the second integer, which is larger than the third. Another example is the betweenness relationship involving points on a line, such that three points are related when the second point is between the first and the third.

We will study relationships among elements from more than two sets in this section. These relationships are called ***n*-ary relations**. These relations are used to represent computer data bases. These representations help us answer queries about the information stored in data bases, such as: Which flights land at O'Hare Airport between 3 AM and 4 AM? Which students at your school are sophomores majoring in mathematics or computer science and have greater than a 3.0 average? Which employees of a company have worked for the company less than five years and make more than $50,000?

n-ARY RELATIONS

We begin with a definition.

> **DEFINITION 1.** Let $A_1, A_2, \ldots, A_n$ be sets. An *n-ary relation* on these sets is a subset of $A_1 \times A_2 \times \cdots \times A_n$. The sets $A_1, A_2, \ldots, A_n$ are called the *domains* of the relation, and n is called its *degree*.

EXAMPLE 1 Let R be the relation consisting of triples (a, b, c) where a, b, and c are integers with $a < b < c$. Then $(1, 2, 3) \in R$, but $(2, 4, 3) \notin R$. The degree of this relation is 3. Its domains are all equal to the set of integers. ■

EXAMPLE 2 Let R be the relation consisting of 5-tuples (A, N, S, D, T) representing airplane flights, where A is the airline, N is the flight number, S is the starting point, D is the destination, and T is the departure time. For instance, if Nadir Express Airlines has flight 963 from Newark to Bangor at 15:00, then (Nadir, 963, Newark, Bangor, 15:00) belongs to R. The degree of this relation is 5, and its domains are the set of all airlines, the set of flight numbers, the set of cities, the set of cities (again), and the set of times. ■

DATA BASES AND RELATIONS

The time required to manipulate information in a data base depends on how this information is stored. The operations of adding and deleting records, updating records, searching for records, and combining records from overlapping data bases are performed millions of times each day in a large data base. Because of the importance of these operations, various methods for representing data bases have been developed. We will discuss one of these methods, called the **relational data model**, based on the concept of a relation.

A data base consists of **records**, which are n-tuples, made up of **fields**. The fields are the entries of the n-tuples. For instance, a data base of student records may be made up of fields containing the name, student number, major, and grade point average of the student. The relational data model represents a data base of records as an n-ary relation. Thus, student records are represented as 4-tuples of the form

(*STUDENT NAME, ID NUMBER, MAJOR, GPA*). A sample data base of six such records is:

(Ackermann, 231455, Computer Science, 3.88)
(Adams, 888323, Physics, 3.45)
(Chou, 102147, Computer Science, 3.79)
(Goodfriend, 453876, Mathematics, 3.45)
(Rao, 678543, Mathematics, 3.90)
(Stevens, 786576, Psychology, 2.99)

Relations used to represent data bases are also called **tables**, since these relations are often displayed as tables. For instance, the same data base of students is displayed in Table 1.

A domain of an n-ary relation is called a **primary key** when the value of the n-tuple from this domain determines the n-tuple. That is, a domain is a primary key when no two n-tuples in the relation have the same value from this domain.

Records are often added to or deleted from data bases. Because of this, the property that a domain is a primary key is time-dependent. Consequently, a primary key should be chosen that remains one whenever the data base is changed. This can be done by using a primary key of the **intension** of the data base, which contains all the n-tuples that can ever be included in an n-ary relation representing this data base.

EXAMPLE 3 Which domains are primary keys for the n-ary relation displayed in Table 1, assuming that no n-tuples will be added in the future?

Solution: Since there is only one 4-tuple in this table for each student name, the domain of student names is a primary key. Similarly, the ID numbers in this table are unique, so that the domain of ID numbers is also a primary key. However, the domain of major fields of study is not a primary key, since more than one 4-tuple contains the same major field of study. The domain of grade point averages is also not a primary key, since there are two 4-tuples containing the same GPA (which ones?). ■

TABLE 1

Student Name	*ID Number*	*Major*	*GPA*
Ackermann	231455	Computer Science	3.88
Adams	888323	Physics	3.45
Chou	102147	Computer Science	3.79
Goodfriend	453876	Mathematics	3.45
Rao	678543	Mathematics	3.90
Stevens	786576	Psychology	2.99

Combinations of domains can also uniquely identify n-tuples in an n-ary relation. When the values of a set of domains determines an n-tuple in a relation, the Cartesian product of these domains is called a **composite key**.

EXAMPLE 4 Is the Cartesian product of the domain of major fields of study and the domain of GPAs a composite key for the n-ary relation from Table 1, assuming that no n-tuples are ever added?

Solution: Since no two 4-tuples from this table have both the same major and the same GPA, this Cartesian product is a composite key. ■

Since primary and composite keys are used to identify records uniquely in a data base, it is important that keys remain valid when new records are added to the data base. Hence, checks should be made to ensure that every new record has values that are different in the appropriate field, or fields, from all other records in this table. For instance, it makes sense to use student identification number as a key for student records if no two students ever have the same student identification number. A university should not use the name field as a key, since two students may have the same name (such as John Smith).

There are a variety of operations on n-ary relations that can be used to form new n-ary relations. Two of these operations will be discussed here, namely, the projection and join operations. Projections are used to form new n-ary relations by deleting the same fields in every record of the relation.

DEFINITION 2. The *projection* $P_{i_1 i_2, \ldots, i_m}$ maps the n-tuple $(a_1, a_2, \ldots, a_n)$ to the m-tuple $(a_{i_1}, a_{i_2}, \ldots, a_{i_m})$, where $m \leq n$.

In other words, the projection $P_{i_1, i_2, \ldots, i_m}$ deletes $n - m$ of the components of an n-tuple, leaving the i_1th, i_2th, $\ldots$, and i_mth components.

EXAMPLE 5 What results when the projection $P_{1,3}$ is applied to the 4-tuples $(2, 3, 0, 4)$, (Jane Doe, 234111001, Geography, 3.14), and (a_1, a_2, a_3, a_4)?

Solution: The projection $P_{1,3}$ sends these 4-tuples to $(2, 0)$, (Jane Doe, Geography), and (a_1, a_3), respectively. ■

The following example illustrates how new relations are produced using projections.

TABLE 2

Student Name	*GPA*
Ackermann	3.88
Adams	3.45
Chou	3.79
Goodfriend	3.45
Rao	3.90
Stevens	2.99

EXAMPLE 6 What relation results when the projection $P_{1,4}$ is applied to the relation in Table 1?

Solution: When the projection $P_{1,4}$ is used, the second and third columns of the table are deleted, and pairs representing student names and grade point averages are obtained. Table 2 displays the results of this projection. ■

Fewer rows may result when a projection is applied to the table for a relation. This happens when some of the n-tuples in the relation have identical values in each of the m components of the projection, and only disagree in components deleted by the projection. For instance, consider the following example.

EXAMPLE 7 What is the table obtained when the projection $P_{1,2}$ is applied to the relation in Table 3?

Solution: Table 4 displays the relation obtained when $P_{1,2}$ is applied to Table 3. Note that there are fewer rows after this projection is applied. ■

TABLE 3

Student	*Major*	*Course*
Glauser	Biology	BI 290
Glauser	Biology	MS 475
Glauser	Biology	PY 410
Marcus	Mathematics	MS 511
Marcus	Mathematics	MS 603
Marcus	Mathematics	CS 322
Miller	Computer Science	MS 575
Miller	Computer Science	CS 455

TABLE 4

Student	*Major*
Glauser	Biology
Marcus	Mathematics
Miller	Computer Science

The **join** operation is used to combine two tables into one when these tables share some identical fields. For instance, a table containing fields for airline, flight number, and gate, and another table containing fields for flight number, gate, and departure time can be combined into a table containing fields for airline, flight number, gate, and departure time.

DEFINITION 3. Let R be a relation of degree m and S a relation of degree n. The *join* $J_p(R, S)$, where $p \leq m$ and $p \leq n$, is a relation of degree $m+n-p$ that consists of all $(m+n-p)$-tuples $(a_1, a_2, \ldots, a_{m-p}, c_1, c_2, \ldots, c_p, b_1, b_2, \ldots, b_{n-p})$, where the m-tuple $(a_1, a_2, \ldots, a_{m-p}, c_1, c_2, \ldots, c_p)$ belongs to R and the n-tuple $(c_1, c_2, \ldots, c_p, b_1, b_2, \ldots, b_{n-p})$ belongs to S.

In other words, the join operator J_p produces a new relation from two relations by combining all m-tuples of the first relation with all n-tuples of the second relation, where the last p components of the m-tuples agree with the first p components of the n-tuples.

EXAMPLE 8 What relation results when the join operator J_2 is used to combine the relation displayed in Tables 5 and 6?

Solution: The join J_2 produces the relation shown in Table 7. ■

There are other operators besides projections and joins that produce new relations from existing relations. A description of these operations may be found in books on data base theory.

TABLE 5

Professor	*Department*	*Course Number*
Cruz	Zoology	335
Cruz	Zoology	412
Farber	Psychology	501
Farber	Psychology	617
Grammer	Physics	544
Grammer	Physics	551
Rosen	Computer Science	518
Rosen	Mathematics	575

TABLE 6

Department	*Course Number*	*Room*	*Time*
Computer Science	518	N521	2:00 P.M.
Mathematics	575	N502	3:00 P.M.
Mathematics	611	N521	4:00 P.M.
Physics	544	B505	4:00 P.M.
Psychology	501	A100	3:00 P.M.
Psychology	617	A110	11:00 A.M.
Zoology	335	A100	9:00 A.M.
Zoology	412	A100	8:00 A.M.

TABLE 7				
Professor	***Department***	***Course Number***	***Room***	***Time***
Cruz	Zoology	335	A100	9:00 A.M.
Cruz	Zoology	412	A100	8:00 A.M.
Farber	Psychology	501	A100	3:00 P.M.
Farber	Psychology	617	A110	11:00 A.M.
Grammer	Physics	544	B505	4:00 P.M.
Rosen	Computer Science	518	N521	2:00 P.M.
Rosen	Mathematics	575	N502	3:00 P.M.

Exercises

1. List the triples in the relation $\{(a, b, c) \mid a, b,$ and c are integers with $0 < a < b < c < 5\}$.

2. Which 4-tuples are in the relation $\{(a, b, c, d) \mid a, b,$ c, and d are positive integers with $abcd = 6\}$?

3. List the 5-tuples in the relation in Table 8.

4. Assuming that no new n-tuples are added, find all the primary keys for the relations displayed in
a) Table 3. **b)** Table 5.
c) Table 6. **d)** Table 8.

5. Assuming that no new n-tuples are added, find a composite key with two fields containing the AIRLINE field for the data base in Table 8.

6. What do you obtain when you apply the projection $P_{2,3,5}$ to the 5-tuple (a, b, c, d, e)?

7. Which projection mapping is used to delete the first, second, and fourth components of a 6-tuple?

8. Display the table produced by applying the projection $P_{1,2,4}$ to Table 8.

TABLE 8				
Airline	***Flight Number***	***Gate***	***Destination***	***Departure Time***
Nadir	122	34	Detroit	08:10
Acme	221	22	Denver	08:17
Acme	122	33	Anchorage	08:22
Acme	323	34	Honolulu	08:30
Nadir	199	13	Detroit	08:47
Acme	222	22	Denver	09:10
Nadir	322	34	Detroit	09:44

TABLE 9

Supplier	*Part Number*	*Project*
23	1092	1
23	1101	3
23	9048	4
31	4975	3
31	3477	2
32	6984	4
32	9191	2
33	1001	1

TABLE 10

Part Number	*Project*	*Quantity*	*Color Code*
1001	1	14	8
1092	1	2	2
1101	3	1	1
3477	2	25	2
4975	3	6	2
6984	4	10	1
9048	4	12	2
9191	2	80	4

9. Display the table produced by applying the projection $P_{1,4}$ to Table 8.

10. How many components are there in the n-tuples in the table obtained by applying the join operator J_3 to two tables with 5-tuples and 8-tuples, respectively?

11. Construct the table obtained by applying the join operator J_2 to the relations in Tables 9 and 10.

6.3 Representing Relations

INTRODUCTION

There are many ways to represent a relation between finite sets. As we have seen, one way is to list its ordered pairs. In this section we will discuss two alternative methods for representing relations. One method uses zero-one matrices. The other method uses directed graphs.

REPRESENTING RELATIONS USING MATRICES

A relation between finite sets can be represented using a zero-one matrix. Suppose that R is a relation from $A = \{a_1, a_2, \ldots, a_m\}$ to $B = \{b_1, b_2, \ldots, b_n\}$. (Here the elements of the sets A and B have been listed in a particular, but arbitrary, order. Furthermore, when $A = B$ we use the same ordering for A and B.) The relation R can be represented by the matrix $\mathbf{M}_R = [m_{ij}]$, where

$$m_{ij} = \begin{cases} 1 & \text{if } (a_i, b_j) \in R \\ 0 & \text{if } (a_i, b_j) \notin R. \end{cases}$$

In other words, the zero-one matrix representing R has a 1 as its (i, j) entry when a_i is related to b_j, and a 0 in this position if a_i is not related to b_j. (Such a representation depends on the orderings used for A and B.)

The use of matrices to represent relations is illustrated in the following examples.

EXAMPLE 1 Suppose that $A = \{1, 2, 3\}$ and $B = \{1, 2\}$. Let R be the relation from A to B containing (a, b) if $a \in A$, $b \in B$, and $a > b$. What is the matrix representing R if $a_1 = 1$, $a_2 = 2$, and $a_3 = 3$, and $b_1 = 1$ and $b_2 = 2$?

Solution: Since $R = \{(2, 1), (3, 1), (3, 2)\}$, the matrix for R is

$$\mathbf{M}_R = \begin{bmatrix} 0 & 0 \\ 1 & 0 \\ 1 & 1 \end{bmatrix}.$$

The 1s in $\mathbf{M}_R$ show that the pairs (2, 1), (3, 1), and (3, 2) belong to R. The 0s show that no other pairs belong to R. ■

EXAMPLE 2 Let $A = \{a_1, a_2, a_3\}$ and $B = \{b_1, b_2, b_3, b_4, b_5\}$. Which ordered pairs are in the relation R represented by the matrix

$$\mathbf{M}_R = \begin{bmatrix} 0 & 1 & 0 & 0 & 0 \\ 1 & 0 & 1 & 1 & 0 \\ 1 & 0 & 1 & 0 & 1 \end{bmatrix}?$$

Solution: Since R consists of those ordered pairs (a_i, b_j) with $m_{ij} = 1$, it follows that

$$R = \{(a_1, b_2), (a_2, b_1), (a_2, b_3), (a_2, b_4), (a_3, b_1), (a_3, b_3), (a_3, b_5)\}.$$ ■

The matrix of a relation on a set, which is a square matrix, can be used to determine whether the relation has certain properties. Recall that a relation R on A is reflexive if $(a, a) \in R$ whenever $a \in A$. Thus, R is reflexive if and only if $(a_i, a_i) \in R$ for $i = 1, 2, \ldots, n$. Hence, R is reflexive if and only if $m_{ii} = 1$, for $i = 1, 2, \ldots, n$. In other words, R is reflexive if all the elements on the main diagonal of $\mathbf{M}_R$ are equal to 1, as shown in Figure 1.

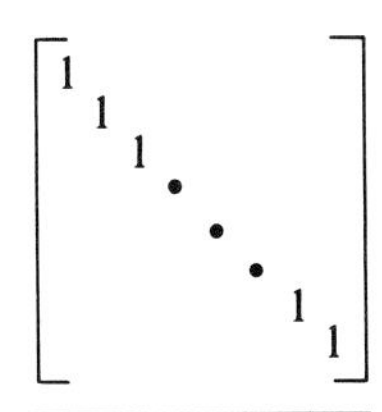

FIGURE 1 The Zero-One Matrix for a Reflexive Relation.

The relation R is symmetric if $(a, b) \in R$ implies that $(b, a) \in R$. Consequently, the relation R on the set $A = \{a_1, a_2, \ldots, a_n\}$ is symmetric if and only if $(a_j, a_i) \in R$ whenever $(a_i, a_j) \in R$. In terms of the entries of $\mathbf{M}_R$, R is symmetric if and only if $m_{ji} = 1$ whenever $m_{ij} = 1$. This also means $m_{ji} = 0$ whenever $m_{ij} = 0$. Consequently, R is symmetric if and only if $m_{ij} = m_{ji}$, for all pairs of integers i and j with $i = 1, 2, \ldots, n$ and $j = 1, 2, \ldots, n$. Recalling the definition of the transpose of a matrix from Section 2.6, we see that R is symmetric if and only if

$$\mathbf{M}_R = (\mathbf{M}_R)^t,$$

that is, if $\mathbf{M}_R$ is a symmetric matrix. The form of the matrix for a symmetric relation is illustrated in Figure 2(a).

The relation R is antisymmetric if and only if $(a, b) \in R$ and $(b, a) \in R$ imply that $a = b$. Consequently, the matrix of an antisymmetric relation has the property that if $m_{ij} = 1$ with $i \neq j$, then $m_{ji} = 0$. Or, in other words, either $m_{ij} = 0$ or $m_{ji} = 0$ when $i \neq j$. The form of the matrix for an antisymmetric relation is illustrated in Figure 2(b).

EXAMPLE 3 Suppose that the relation R on a set is represented by the matrix

$$\mathbf{M}_R = \begin{bmatrix} 1 & 1 & 0 \\ 1 & 1 & 1 \\ 0 & 1 & 1 \end{bmatrix}.$$

Is R reflexive, symmetric, and/or antisymmetric?

Solution: Since all the diagonal elements of this matrix are equal to 1, R is reflexive. Moreover, since $\mathbf{M}_R$ is symmetric, it follows that R is symmetric. It is also easy to see that R is not antisymmetric. ■

(a) Symmetric

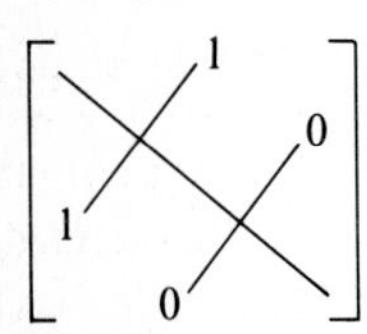

(b) Antisymmetric

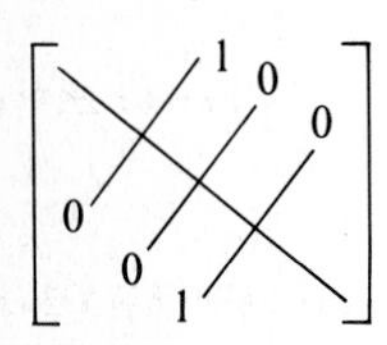

FIGURE 2 The Zero-One Matrices for Symmetric and Antisymmetric Relations.

The Boolean operations join and meet (discussed in Section 2.6) can be used to find the matrices representing the union and the intersection of two relations. Suppose that R_1 and R_2 are relations on a set A represented by the matrices $\mathbf{M}_{R_1}$ and $\mathbf{M}_{R_2}$, respectively. The matrix representing the union of these relations has a 1 in the positions where either $\mathbf{M}_{R_1}$ or $\mathbf{M}_{R_2}$ has a 1. The matrix representing the intersection of these relations has a 1 in the positions where both $\mathbf{M}_{R_1}$ and $\mathbf{M}_{R_2}$ have a 1. Thus, the matrices representing the union and intersection of these relations are

$$\mathbf{M}_{R_1 \cup R_2} = \mathbf{M}_{R_1} \vee \mathbf{M}_{R_2}$$

and

$$\mathbf{M}_{R_1 \cap R_2} = \mathbf{M}_{R_1} \wedge \mathbf{M}_{R_2}.$$

EXAMPLE 4 Suppose that the relations R_1 and R_2 on a set A are represented by the matrices

$$\mathbf{M}_{R_1} = \begin{bmatrix} 1 & 0 & 1 \\ 1 & 0 & 0 \\ 0 & 1 & 0 \end{bmatrix} \quad \text{and} \quad \mathbf{M}_{R_2} = \begin{bmatrix} 1 & 0 & 1 \\ 0 & 1 & 1 \\ 1 & 0 & 0 \end{bmatrix}.$$

What are the matrices representing $R_1 \cup R_2$ and $R_1 \cap R_2$?

Solution: The matrices of these relations are

$$\mathbf{M}_{R_1 \cup R_2} = \mathbf{M}_{R_1} \vee \mathbf{M}_{R_2} = \begin{bmatrix} 1 & 0 & 1 \\ 1 & 1 & 1 \\ 1 & 1 & 0 \end{bmatrix},$$

$$\mathbf{M}_{R_1 \cap R_2} = \mathbf{M}_{R_1} \wedge \mathbf{M}_{R_2} = \begin{bmatrix} 1 & 0 & 1 \\ 0 & 0 & 0 \\ 0 & 0 & 0 \end{bmatrix}.$$

■

We now turn our attention to determining the matrix for the composite of relations. This matrix can be found using the Boolean product of the matrices (discussed in Section 2.6) for these relations. In particular, suppose that R is a relation from A to B and S is a relation from B to C. Suppose that A, B, and C have m, p, and n elements, respectively. Let the zero-one matrices for $S \circ R$, R, and S be $\mathbf{M}_{S \circ R} = [t_{ij}]$, $\mathbf{M}_R = [r_{ij}]$, and $\mathbf{M}_S = [s_{ij}]$, respectively (these matrices have sizes $m \times p$, $m \times n$, and $n \times p$, respectively). The ordered pair (a_i, c_j) belongs to $S \circ R$ if and only if there is an element b_k such that (a_i, b_k) belongs to R and (b_k, c_j) belongs to S. It follows that $t_{ij} = 1$ if and only if $r_{ik} = s_{kj} = 1$ for some k. From the definition of the Boolean product, this means that

$$\mathbf{M}_{S \circ R} = \mathbf{M}_R \odot \mathbf{M}_S.$$

EXAMPLE 5 Find the matrix representing the relations $S \circ R$ where the matrices representing R and S are

$$\mathbf{M}_R = \begin{bmatrix} 1 & 0 & 1 \\ 1 & 1 & 0 \\ 0 & 0 & 0 \end{bmatrix} \quad \text{and} \quad \mathbf{M}_S = \begin{bmatrix} 0 & 1 & 0 \\ 0 & 0 & 1 \\ 1 & 0 & 1 \end{bmatrix}.$$

Solution: The matrix for $S \circ R$ is

$$\mathbf{M}_{S \circ R} = \mathbf{M}_R \odot \mathbf{M}_S = \begin{bmatrix} 1 & 1 & 1 \\ 0 & 1 & 1 \\ 0 & 0 & 0 \end{bmatrix}.$$

■

The matrix representing the composite of two relations can be used to find the matrix for $\mathbf{M}_{R^n}$. In particular,

$$\mathbf{M}_{R^n} = \mathbf{M}_R^{[n]},$$

from the definition of Boolean powers. Exercise 18 at the end of this section asks for a proof of this formula.

EXAMPLE 6 Find the matrix representing the relation R^2 where the matrix representing R is

$$\mathbf{M}_R = \begin{bmatrix} 0 & 1 & 0 \\ 0 & 1 & 1 \\ 1 & 0 & 0 \end{bmatrix}.$$

Solution: The matrix for R^2 is

$$\mathbf{M}_{R^2} = \mathbf{M}_R^{[2]} = \begin{bmatrix} 0 & 1 & 1 \\ 1 & 1 & 1 \\ 0 & 1 & 0 \end{bmatrix}.$$

■

REPRESENTING RELATIONS USING DIGRAPHS

We have shown that a relation can be represented by listing all of its ordered pairs or by using a zero-one matrix. There is another important way of representing a relation using a pictorial representation. Each element of the set is represented by a point, and each ordered pair is represented using an arc with its direction indicated by an arrow. We use such pictorial representations when we think of relations on a finite set as **directed graphs** or **digraphs.**

DEFINITION 1. A *directed graph*, or *digraph*, consists of a set V of *vertices* (or *nodes*) together with a set E of ordered pairs of elements of V called *edges* (or *arcs*). The vertex a is called the *initial vertex* of the edge (a, b), and the vertex b is called the *terminal vertex* of this edge.

An edge of the form (a, a) is represented using an arc from the vertex a back to itself. Such an edge is called a **loop**.

EXAMPLE 7 The directed graph with vertices a, b, c, and d, and edges (a, b), (a, d), (b, b), (b, d), (c, a), (c, b), and (d, b) is displayed in Figure 3. ■

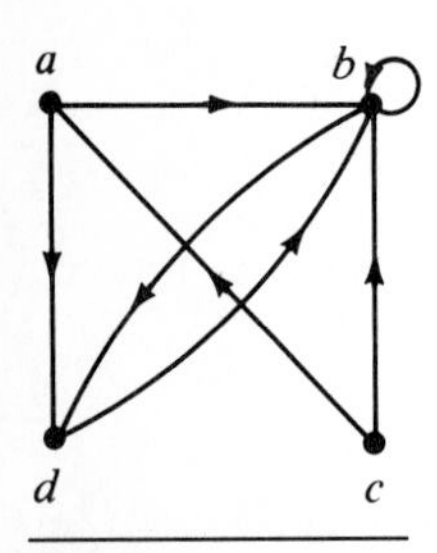

FIGURE 3 A Directed Graph.

The relation R on a set A is represented by the directed graph that has the elements of A as its vertices and the ordered pairs (a, b), where $(a, b) \in R$, as edges. This assignment sets up a one-to-one correspondence between the relations on a set A and the directed graphs with A as their set of vertices. Thus, every statement about relations corresponds to a statement about directed graphs, and vice versa. Directed graphs give a visual display of information about relations. As such, they are often used to study relations and their properties. Note that relations from a set A to a set B cannot be represented by directed graphs unless $A = B$. The use of directed graphs to represent relations is illustrated in the following examples.

EXAMPLE 8 The directed graph of the relation

$$R = \{(1, 1), (1, 3), (2, 1), (2, 3), (2, 4), (3, 1), (3, 2), (4, 1)\}$$

on the set $\{1, 2, 3, 4\}$ is shown in Figure 4. ■

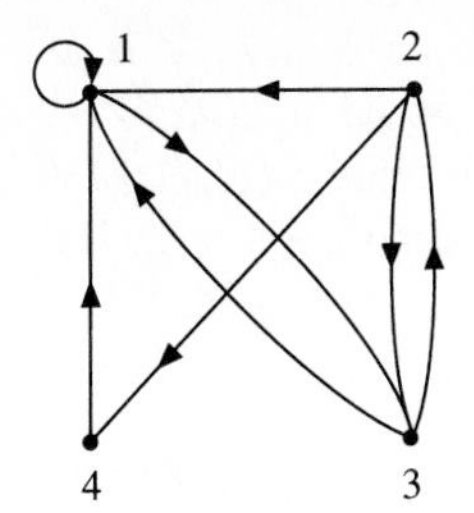

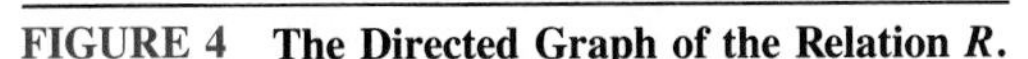

FIGURE 4 **The Directed Graph of the Relation R.**

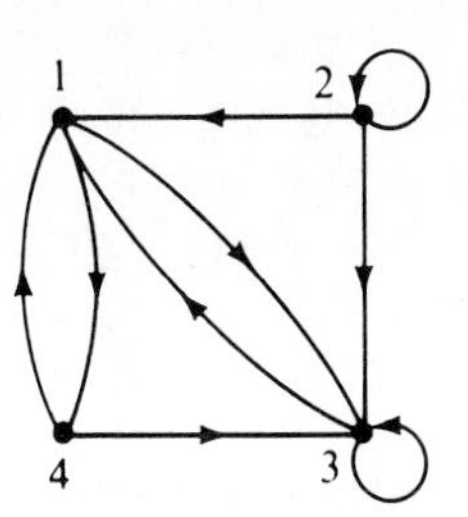

FIGURE 5 **The Directed Graph of the Relation R.**

EXAMPLE 9 What are the ordered pairs in the relation R represented by the directed graph shown in Figure 5?

Solution: The ordered pairs (x, y) in the relation are

$$R = \{(1, 3), (1, 4), (2, 1), (2, 2), (2, 3), (3, 1), (3, 3), (4, 1), (4, 3)\}.$$

Each of these pairs corresponds to an edge of the directed graph, with (2, 2) and (3, 3) corresponding to loops. ■

The directed graph representing a relation can be used to determine whether the relation has various properties. For instance, a relation is reflexive if and only if there is a loop at every vertex of the directed graph, so that every ordered pair of the form (x, x) occurs in the relation. A relation is symmetric if and only if for every edge between distinct vertices in its digraph there is an edge in the opposite direction, so that (y, x) is in the relation whenever (x, y) is in the relation. Similarly, a relation is antisymmetric if and only if there are never two edges in opposite directions between distinct vertices. Finally, a relation is transitive if and only if whenever there is an edge from a vertex x to a vertex y and an edge from a vertex y to a vertex z, there is an edge from x to z (completing a triangle where each side is a directed edge with the correct direction).

EXAMPLE 10 Determine whether the relations for the directed graphs shown in Figure 6 are reflexive, symmetric, antisymmetric, and/or transitive.

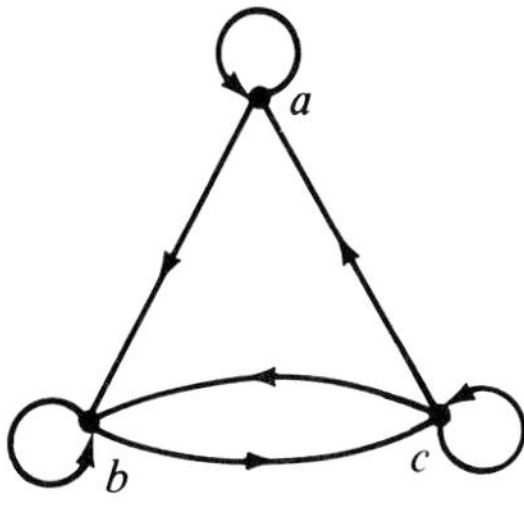

(a) Directed graph of R

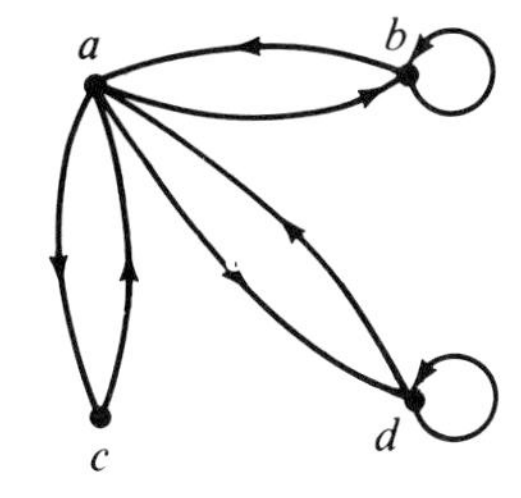

(b) Directed graph of S

FIGURE 6 **The Directed Graphs of the Relations R and S.**

Solution: Since there are loops at every vertex of the directed graph of R, it is reflexive. R is neither symmetric nor antisymmetric since there is an edge from a to b but not one from b to a, but there are edges in both directions connecting b and c. Finally, R is not transitive since there is an edge from a to b and an edge from b to c, but no edge from a to c.

Since loops are not present at all the vertices of the directed graph of S, this relation is not reflexive. It is symmetric and not antisymmetric, since every edge between distinct vertices is accompanied by an edge in the opposite direction. It is also not hard to see from the directed graph that S is not transitive, since (c, a) and (a, b) belong to S, but (c, b) does not belong to S. ■

Exercises

1. Represent each of the following relations on $\{1, 2, 3\}$ with a matrix (with the elements of this set listed in increasing order).
a) $\{(1, 1), (1, 2), (1, 3)\}$
b) $\{(1, 2), (2, 1), (2, 2), (3, 3)\}$
c) $\{(1, 1), (1, 2), (1, 3), (2, 2), (2, 3), (3, 3)\}$
d) $\{(1, 3), (3, 1)\}$

2. List the ordered pairs in the relations on $\{1, 2, 3\}$ corresponding to the following matrices (where the rows and columns correspond to the integers listed in increasing order).

a) $\begin{bmatrix} 1 & 0 & 1 \\ 0 & 1 & 0 \\ 1 & 0 & 1 \end{bmatrix}$ **b)** $\begin{bmatrix} 0 & 1 & 0 \\ 0 & 1 & 0 \\ 0 & 1 & 0 \end{bmatrix}$

c) $\begin{bmatrix} 1 & 1 & 1 \\ 1 & 0 & 1 \\ 1 & 1 & 1 \end{bmatrix}$

3. How can the matrix for a relation be used to determine whether the relation is irreflexive?

4. Determine whether the relations represented by the matrices in Exercise 2 are reflexive, irreflexive, symmetric, antisymmetric, and/or transitive.

5. How can the matrix for $\overline{R}$, the complement of the relation R, be found from the matrix representing R, when R is a relation on a finite set A?

6. How can the matrix for R^{-1}, the inverse of the relation R, be found from the matrix representing R, when R is a relation on a finite set A?

7. Let R be the relation represented by the matrix

$$\mathbf{M}_R = \begin{bmatrix} 0 & 1 & 1 \\ 1 & 1 & 0 \\ 1 & 0 & 1 \end{bmatrix}.$$

Find the matrix representing
a) R^{-1}. **b)** $\overline{R}$. **c)** R^2.

8. Let R_1 and R_2 be relations on a set A represented by the matrices

$$\mathbf{M}_{R_1} = \begin{bmatrix} 0 & 1 & 0 \\ 1 & 1 & 1 \\ 1 & 0 & 0 \end{bmatrix} \quad \text{and} \quad \mathbf{M}_{R_2} = \begin{bmatrix} 0 & 1 & 0 \\ 0 & 1 & 1 \\ 1 & 1 & 1 \end{bmatrix}.$$

Find the matrices that represent
a) $R_1 \cup R_2$. **b)** $R_1 \cap R_2$.
c) $R_2 \circ R_1$. **d)** $R_1 \circ R_1$.
e) $R_1 \oplus R_2$.

9. Let R be the relation represented by the matrix

$$\mathbf{M}_R = \begin{bmatrix} 0 & 1 & 0 \\ 0 & 0 & 1 \\ 1 & 1 & 0 \end{bmatrix}.$$

Find the matrices that represent
a) R^2. **b)** R^3. **c)** R^4.

10. Draw the directed graphs representing each of the relations from Exercise 1.

11. Draw the directed graphs representing each of the relations from Exercise 2.

12. Draw the directed graph that represents the relation $\{(a, a), (a, b), (b, c), (c, b), (c, d), (d, a), (d, b)\}$.

In Exercises 13–15 list the ordered pairs in the relations represented by the directed graphs.

13.

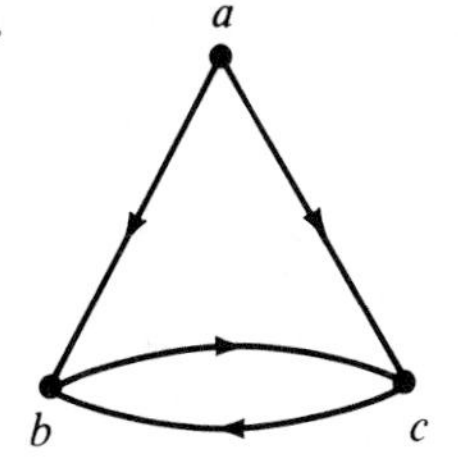

14.

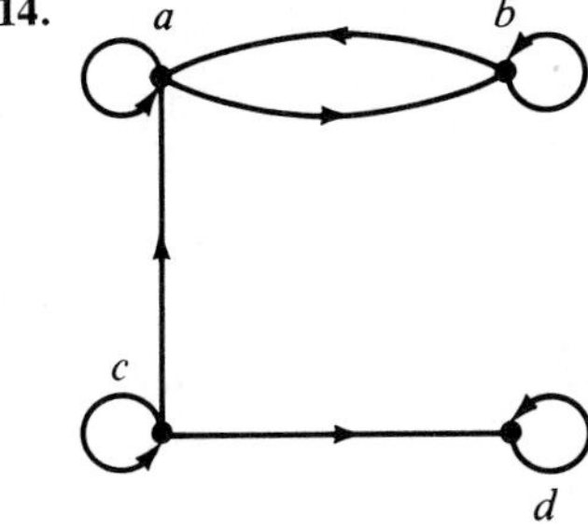

15.

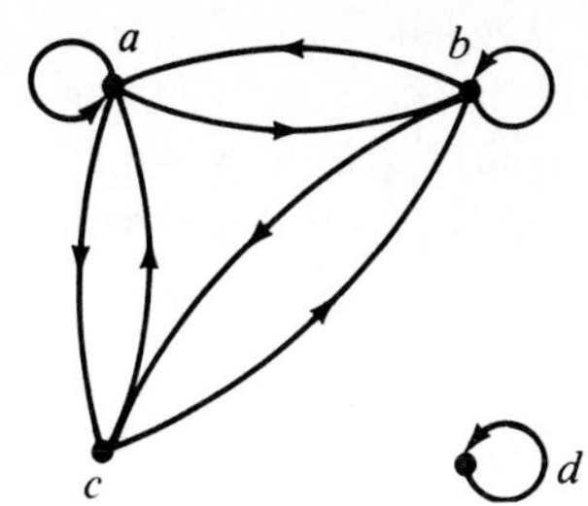

16. How can the directed graph of a relation R on a finite set A be used to determine if a relation is irreflexive?

17. Determine whether the relations represented by the directed graphs shown in Exercises 13–15 are reflexive, irreflexive, symmetric, antisymmetric, and/or transitive.

18. Given the directed graphs representing two relations, how can the directed graph of the union, intersection, symmetric difference, difference, and composition of these relations be found?

19. Show that if $\mathbf{M}_R$ is the matrix representing the relation R, then $\mathbf{M}_R^{[n]}$ is the matrix representing the relation R^n.

6.4 Closures of Relations

INTRODUCTION

A computer network has data centers in Boston, Chicago, Denver, Detroit, New York, and San Diego. There are direct, one-way telephone lines from Boston to Chicago, from Boston to Detroit, from Chicago to Detroit, from Detroit to Denver, and from New York to San Diego. Let R be the relation containing (a, b) if there is a telephone line from the data center in a to that in b. How can we determine if there is some (possibly indirect) link composed of one or more telephone lines from one center to another? Since not all links are direct, such as the link from Boston to Denver that goes through Detroit, R cannot be used directly to answer this. In the language of relations, R is not transitive, so it does not contain all the pairs that can be linked. As we will show in this section, we can find all pairs of data centers that have a link by constructing the smallest transitive relation that contains R. This relation is called the **transitive closure** of R.

In general, let R be a relation on a set A. R may or may not have some property **P**, such as reflexivity, symmetry, or transitivity. If there is a relation S with property

P containing R such that S is a subset of every relation with property **P** containing R, then S is called the **closure** of R with respect to **P**. (Note that the closure of a relation with respect to a property may not exist; see Exercises 15 and 35 at the end of this section.) We will show how reflexive, symmetric, and transitive closures of relations can be found.

CLOSURES

The relation $R = \{(1, 1), (1, 2), (2, 1), (3, 2)\}$ on the set $A = \{1, 2, 3\}$ is not reflexive. How can we produce a reflexive relation containing R that is as small as possible? This can be done by adding $(2, 2)$ and $(3, 3)$ to R, since these are the only pairs of the form (a, a) that are not in R. Clearly, this new relation contains R. Furthermore, *any* reflexive relation that contains R must also contain $(2, 2)$ and $(3, 3)$. Because this relation contains R, is reflexive, and is contained within every reflexive relation that contains R, it is called the **reflexive closure** of R.

As this example illustrates, given a relation R on a set A, the reflexive closure of R can be formed by adding to R all pairs of the form (a, a) with $a \in A$, not already in R. The addition of these pairs produces a new relation that is reflexive, contains R, and is contained within any reflexive relation containing R. We see that the reflexive closure of R equals $R \cup \Delta$, where $\Delta = \{(a, a) \mid a \in A\}$ is the **diagonal relation** on A. (The reader should verify this.)

EXAMPLE 1 What is the reflexive closure of the relation $R = \{(a, b) \mid a < b\}$ on the set of integers?

Solution: The reflexive closure of R is

$$R \cup \Delta = \{(a, b) \mid a < b\} \cup \{(a, a) \mid a \in \mathbf{Z}\} = \{(a, b) \mid a \leq b\}.$$

■

The relation $\{(1, 1), (1, 2), (2, 2), (2, 3), (3, 1), (3, 2)\}$ on $\{1, 2, 3\}$ is not symmetric. How can we produce a symmetric relation that is as small as possible and contains R? To do this, we need only add $(2, 1)$ and $(1, 3)$, since these are the only pairs of the form (b, a) with $(a, b) \in R$ that are not in R. This new relation is symmetric and contains R. Furthermore, *any* symmetric relation that contains R must contain this new relation, since a symmetric relation that contains R must contain $(2, 1)$ and $(1, 3)$. Consequently, this new relation is called the **symmetric closure** of R.

As this example illustrates, the symmetric closure of a relation R can be constructed by adding all ordered pairs of the form (b, a) where (a, b) is in the relation, that are not already present in R. Adding these pairs produces a relation that is symmetric, that contains R, and that is contained in any symmetric relation that contains R. The symmetric closure of a relation can be constructed by taking the union of a relation with its inverse; that is, $R \cup R^{-1}$ is the symmetric closure of R, where $R^{-1} = \{(b, a) \mid (a, b) \in R\}$. The reader should verify this statement.

EXAMPLE 2 What is the symmetric closure of the relation $R = \{(a, b) \mid a > b\}$ on the set of positive integers?

Solution: The symmetric closure of R is the relation

$$R \cup R^{-1} = \{(a, b) \mid a > b\} \cup \{(b, a) \mid a > b\} = \{(a, b) \mid a \neq b\}.$$ ■

Suppose that a relation R is not transitive. How can we produce a transitive relation that contains R such that this new relation is contained within any transitive relation that contains R? Can the transitive closure of a relation R be produced by adding all the pairs of the form (a, c) where (a, b) and (b, c) are already in the relation? Consider the relation $R = \{(1, 3), (1, 4), (2, 1), (3, 2)\}$ on the set $\{1, 2, 3, 4\}$. This relation is not transitive since it does not contain all pairs of the form (a, c) where (a, b) and (b, c) are in R. The pairs of this form not in R are $(1, 2)$, $(2, 3)$, $(2, 4)$, and $(3, 1)$. Adding these pairs does *not* produce a transitive relation, since the resulting relation contains $(3, 1)$ and $(1, 4)$ but does not contain $(3, 4)$. This shows that constructing the transitive closure of a relation is more complicated than constructing either the reflexive or symmetric closure. The rest of this section develops algorithms for constructing transitive closures. As will be shown, the transitive closure of a relation can be found by adding new ordered pairs that must be present and then repeating this process until no new ordered pairs are needed.

PATHS IN DIRECTED GRAPHS

We will see that representing relations by directed graphs helps in the construction of transitive closures. We now introduce some terminology that we will use for this purpose.

A path in a directed graph is obtained by traversing along edges (in the same direction as indicated by the arrow on the edge).

DEFINITION 1. A *path* from a to b in the directed graph G is a sequence of one or more edges $(x_0, x_1), (x_1, x_2), (x_2, x_3), \ldots, (x_{n-1}, x_n)$ in G where $x_0 = a$ and $x_n = b$, that is, a sequence of edges where the terminal vertex of an edge is the same as the initial vertex in the next edge in the path. This path is denoted by $x_0, x_1, x_2, \ldots, x_{n-1}, x_n$ and has *length n*. A path that begins and ends at the same vertex is called a *circuit* or *cycle*.

A path in a directed graph can pass through a vertex more than once. Moreover, an edge in a directed graph can occur more than once in a path. The reader should note that some authors allow paths of length zero, that is, paths consisting of no edges. In this book all paths must have length of at least one.

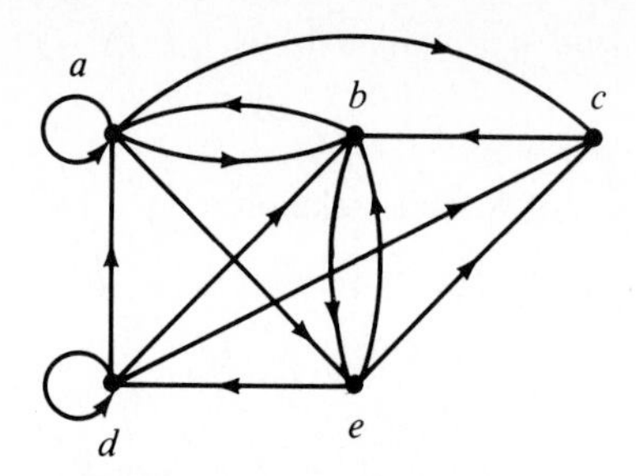

FIGURE 1 **A Directed Graph.**

EXAMPLE 3 Which of the following are paths in the directed graph shown in Figure 1: a, b, e, d; a, e, c, d, b; b, a, c, b, a, a, b; d, c; c, b, a; e, b, a, b, a, b, e? What are the lengths of those that are paths? Which of the paths in this list are circuits?

Solution: Since each of (a, b), (b, e), and (e, d) is an edge, a, b, e, d is a path of length three. Since (c, d) is not an edge, a, e, c, d, b is not a path. Also, b, a, c, b, a, a, b is a path of length six since (b, a), (a, c), (c, b), (b, a), (a, a), and (a, b) are all edges. We see that d, c is a path of length one, since (d, c) is an edge. Also c, b, a is a path of length two, since (c, b) and (b, a) are edges. All of (e, b), (b, a), (a, b), (b, a), (a, b), and (b, e) are edges, so that e, b, a, b, a, b, e is a path of length six.

The two paths b, a, c, b, a, a, b and e, b, a, b, a, b, e are circuits since they begin and end at the same vertex. The paths a, b, e, d; c, b, a; and d, c are not circuits. ∎

The term *path* also applies to relations. Carrying over the definition from directed graphs to relations, there is a **path** from a to b in R if there is a sequence of elements $a, x_1, x_2, \ldots, x_{n-1}, b$ with $(a, x_1) \in R$, $(x_1, x_2) \in R, \ldots,$ and $(x_{n-1}, b) \in R$. The following theorem can be obtained from the definition of a path in a relation.

THEOREM 1 Let R be a relation on a set A. There is a path of length n from a to b if and only if $(a, b) \in R^n$.

Proof: We will use mathematical induction. By definition, there is a path from a to b of length one if and only if $(a, b) \in R$, so the theorem is true when $n = 1$.

Assume that the theorem is true for the positive integer n. This is the inductive hypothesis. There is a path of length $n + 1$ from a to b if and only if there is an element $c \in A$ so that there is a path of length one from a to c, so that $(a, c) \in R$, and a path of length n from c to b, that is, $(c, b) \in R^n$. Consequently, by the induction hypothesis, there is a path of length $n + 1$ from a to b if and only if there is an element c with $(a, c) \in R$ and $(c, b) \in R^n$. But there is such an element if and only if $(a, b) \in R^{n+1}$. Therefore, there is a path of length $n + 1$ from a to b if and only if $(a, b) \in R^{n+1}$. This completes the proof. ∎

TRANSITIVE CLOSURES

We now show that finding the transitive closure of a relation is equivalent to determining which pairs of vertices in the associated directed graph are connected by a path. With this in mind, we define a new relation.

DEFINITION 2. Let R be a relation on a set A. The *connectivity relation* R^* consists of the pairs (a, b) such that there is a path between a and b in R.

Since R^n consists of the pairs (a, b) such that there is a path of length n from a to b, it follows that R^* is the union of all the sets R^n. In other words,

$$R^* = \bigcup_{n=1}^{\infty} R^n.$$

The connectivity relation is useful in many models.

EXAMPLE 4 Let R be the relation on the set of all people in the world that contains (a, b) if a has met b. What is R^n, where n is a positive integer greater than 2? What is R^*?

Solution: The relation R^2 contains (a, b) if there is a person c such that $(a, c) \in R$ and $(c, b) \in R$, that is, if there is a person c such that a has met c and c has met b. Similarly, R^n consists of those pairs (a, b) such that there are people $x_1, x_2, \ldots, x_{n-1}$ such that a has met x_1, x_1 has met $x_2, \ldots,$ and x_{n-1} has met b.

The relation R^* contains (a, b) if there is a sequence of people, starting with a and ending with b, such that each person in the sequence has met the next person in the sequence. (There are many interesting conjectures about R^*. Do you think that this connectivity relation includes the pair with you as the first element and the president of the USSR as the second element?) ■

EXAMPLE 5 Let R be the relation on the set of all subway stops in New York City that contains (a, b) if it is possible to travel from stop a to stop b without changing trains. What is R^n when n is a positive integer? What is R^*?

Solution: The relation R^n contains (a, b) if it is possible to travel from stop a to stop b by making exactly $n - 1$ changes of trains. The relation R^* consists of the ordered pairs (a, b) where it is possible to travel from stop a to stop b making as many changes of trains as necessary. (The reader should verify these statements.) ■

EXAMPLE 6 Let R be the relation on the set of all states in the United States that contains (a, b) if state a and state b have a common border. What is R^n where n is a positive integer? What is R^*?

Solution: The relation R^n consists of the pairs (a, b) where it is possible to go from state a to state b by crossing exactly n state borders. R^* consists of the ordered pairs (a, b) where it is possible to go from state a to state b crossing as many borders as necessary. (The reader should verify these statements.) The only ordered pairs not in R^* are those containing states that are not connected to the continental United States (i.e., those pairs containing Alaska or Hawaii). ■

The following theorem shows that the transitive closure of a relation and the associated connectivity relation are the same.

THEOREM 2 The transitive closure of a relation R equals the connectivity relation R^*.

Proof: Note that R^* contains R. To show that R^* is the transitive closure of R we must also show that R^* is transitive and that $R^* \subseteq S$ whenever S is a transitive relation that contains R.

First, we show that R^* is transitive. If $(a, b) \in R^*$ and $(b, c) \in R^*$, then there are paths from a to b and from b to c in R. We obtain a path from a to c by starting with the path from a to b and following it with the path from b to c. Hence, $(a, c) \in R^*$. It follows that R^* is transitive.

Now suppose that S is a transitive relation containing R. Since S is transitive, S^n also is transitive (the reader should verify this) and $S^n \subseteq S$ (by Theorem 1 of Section 6.1). Furthermore, since

$$S^* = \bigcup_{k=1}^{\infty} S^k,$$

and $S^k \subseteq S$, it follows that $S^* \subseteq S$. Now note that if $R \subseteq S$, then $R^* \subseteq S^*$, because any path in R is also a path in S. Consequently, $R^* \subseteq S^* \subseteq S$. Thus, any transitive relation that contains R must also contain R^*. Therefore, R^* is the transitive closure of R. ■

Now that we know that the transitive closure equals the connectivity relation, we turn our attention to the problem of computing this relation. We do not need to examine arbitrarily long paths to determine whether there is a path between two vertices in a finite directed graph. As the following lemma shows, it is sufficient to examine paths containing no more than n edges, where n is the number of elements in the set.

LEMMA 1 Let A be a set with n elements, and let R be a relation on A. If there is a path in R from a to b, then there is such a path with length not exceeding n. Moreover, when $a \neq b$, if there is a path in R from a to b, then there is such a path with length not exceeding $n - 1$.

Proof: Suppose there is a path from a to b in R. Let m be the length of the shortest such path. Suppose that $x_0, x_1, x_2, \ldots, x_{m-1}, x_m$, where $x_0 = a$ and $x_m = b$, is such a path.

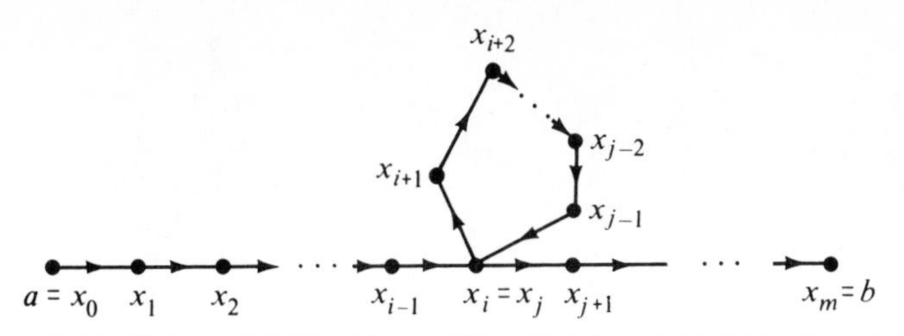

FIGURE 2 Producing a Path with Length Not Exceeding *n*.

Suppose that $a = b$ and that $m > n$, so that $m \geq n + 1$. By the pigeonhole principle, since there are n vertices in A, among the m vertices $x_0, x_1, \ldots, x_{m-1}$, at least two are equal (see Figure 2).

Suppose that $x_i = x_j$ with $0 \leq i < j \leq m - 1$. Then the path contains a circuit from x_i to itself. This circuit can be deleted from the path from a to b, leaving a path, namely, $x_0, x_1, \ldots, x_i, x_{j+1}, \ldots, x_{m-1}, x_m$, from a to b of shorter length. Hence, the path of shortest length must have length less than or equal to n.

The case where $a \neq b$ is left as an exercise for the reader. ■

From Lemma 1, we see that the transitive closure of R is the union of $R, R^2, R^3, \ldots,$ and R^n. This follows since there is a path in R^* between two vertices if and only if there is a path between these vertices in R^i, for some positive integer i with $i \leq n$. Since

$$R^* = R \cup R^2 \cup R^3 \cup \cdots \cup R^n,$$

and the zero-one matrix representing a union of relations is the join of the zero-one matrices of these relations, the zero-one matrix for the transitive closure is the join of the zero-one matrices of the first n powers of the zero-one matrix of R.

THEOREM 3 Let $\mathbf{M}_R$ be the zero-one matrix of the relation R. Then the zero-one matrix of the transitive closure R^* is

$$\mathbf{M}_{R^*} = \mathbf{M}_R \vee \mathbf{M}_R^{[2]} \vee \mathbf{M}_R^{[3]} \vee \cdots \vee \mathbf{M}_R^{[n]}.$$

EXAMPLE 7 Find the zero-one matrix of the transitive closure of the relation R where

$$\mathbf{M}_R = \begin{bmatrix} 1 & 0 & 1 \\ 0 & 1 & 0 \\ 1 & 1 & 0 \end{bmatrix}.$$

Solution: From Theorem 3, it follows that the zero-one matrix of R^* is

$$\mathbf{M}_{R*} = \mathbf{M}_R \vee \mathbf{M}_R^{[2]} \vee \mathbf{M}_R^{[3]}.$$

Since

$$\mathbf{M}_R^{[2]} = \begin{bmatrix} 1 & 1 & 1 \\ 0 & 1 & 0 \\ 1 & 1 & 1 \end{bmatrix} \quad \text{and} \quad \mathbf{M}_R^{[3]} = \begin{bmatrix} 1 & 1 & 1 \\ 0 & 1 & 0 \\ 1 & 1 & 1 \end{bmatrix},$$

it follows that

$$\mathbf{M}_{R^*} = \begin{bmatrix} 1 & 0 & 1 \\ 0 & 1 & 0 \\ 1 & 1 & 0 \end{bmatrix} \vee \begin{bmatrix} 1 & 1 & 1 \\ 0 & 1 & 0 \\ 1 & 1 & 1 \end{bmatrix} \vee \begin{bmatrix} 1 & 1 & 1 \\ 0 & 1 & 0 \\ 1 & 1 & 1 \end{bmatrix} = \begin{bmatrix} 1 & 1 & 1 \\ 0 & 1 & 0 \\ 1 & 1 & 1 \end{bmatrix}.$$

■

Theorem 3 can be used as a basis for an algorithm for computing the matrix of the relation R^*. To find this matrix, the successive Boolean powers of $\mathbf{M}_R$, up to the nth power, are computed. As each power is calculated, its join with the join of all smaller powers is formed. When this is done with the nth power, the matrix for R^* has been found. This procedure is displayed as Algorithm 1.

ALGORITHM 1 A Procedure for Computing the Transitive Closure.

procedure *transitive closure* ($\mathbf{M}_R$: zero-one $n \times n$ matrix)
$\mathbf{A} := \mathbf{M}_R$
$\mathbf{B} := \mathbf{A}$
for $i := 2$ **to** n
begin
  $\mathbf{A} := \mathbf{A} \odot \mathbf{M}_R$
  $\mathbf{B} := \mathbf{B} \vee \mathbf{A}$
end {B is the zero-one matrix for R^*}

We can easily find the number of bit operations used by Algorithm 1 to determine the transitive closure of a relation. Computing the Boolean powers $\mathbf{M}_R, \mathbf{M}_R^{[2]}, \ldots, \mathbf{M}_R^{[n]}$ requires that $n - 1$ Boolean products of $n \times n$ zero-one matrices be found. Each of these Boolean products can be found using n^3 bit operations. Hence, these products can be computed using $(n - 1)n^3$ bit operations.

To find $\mathbf{M}_{R*}$ from the n Boolean powers of $\mathbf{M}_R$, $n - 1$ joins of zero-one matrices need to be found. Computing each of these joins uses n^2 bit operations. Hence, $(n - 1)n^2$ bit operations are used in this part of the computation. Therefore, when Algorithm 1 is used, the matrix of the transitive closure of a relation on a set with n elements can be found using $(n-1)n^3+(n-1)n^2 = n^4-n^2 = O(n^4)$ bit operations. The remainder of this section describes a more efficient algorithm for finding transitive closures.

WARSHALL'S ALGORITHM

Warshall's algorithm, developed by Stephen Warshall in 1960, is an efficient method for computing the transitive closure of a relation. Algorithm 1 can find the transitive closure of a relation on a set with n elements using $n^4 - n^2$ bit operations. However, the transitive closure can be found by Warshall's algorithm using only $2n^3$ bit operations.

Suppose that R is a relation on a set with n elements. Let $v_1, v_2, \ldots, v_n$ be an arbitrary listing of these n elements. The concept of the **interior vertices** of a path is used in Warshall's algorithm. If $a, x_1, x_2, \ldots, x_{m-1}, b$ is a path, its interior vertices are $x_1, x_2, \ldots, x_{m-1}$, that is, all the vertices of the path that occur somewhere other than as the first and last vertices in the path. For instance, the interior vertices of a path a, c, d, f, g, h, b, j in a directed graph are c, d, f, g, h, and b. The interior vertices of a, c, d, a, f, b are c, d, a, and f. (Note that the first vertex in the path is not an interior vertex unless it is visited again by the path, except as the last vertex. Similarly, the last vertex in the path is not an interior vertex unless it was visited previously by the path, except as the first vertex.)

Warshall's algorithm is based on the construction of a sequence of zero-one matrices. These matrices are $\mathbf{W}_0, \mathbf{W}_1, \ldots, \mathbf{W}_n$, where $\mathbf{W}_0 = \mathbf{M}_R$ is the zero-one matrix of this relation, and $\mathbf{W}_k = [w_{ij}^{(k)}]$, where $w_{ij}^{(k)} = 1$ if there is a path from v_i to v_j such that all the interior vertices of this path are in the set $\{v_1, v_2, \ldots, v_k\}$ (the first k vertices in the list) and is 0 otherwise. (The first and last vertices in the path may be outside the set of the first k vertices in the list.) Note that $\mathbf{W}_n = \mathbf{M}_{R*}$, since the (i, j)th entry of $\mathbf{M}_{R*}$ is 1 if and only if there is a path from v_i to v_j, with all interior vertices in the set $\{v_1, v_2, \ldots, v_n\}$ (but these are the only vertices in the directed graph). The following example illustrates what the matrix $\mathbf{W}_k$ represents.

Stephen Warshall (born 1935). Stephen Warshall, born in New York City, went to public school in Brooklyn. He attended Harvard University, receiving his degree in mathematics in 1956. He never received an advanced degree, since at that time no programs were available in his areas of interest. However, he took graduate courses at several different universities and contributed to the development of computer science and software engineering.

After graduating from Harvard, Warshall worked at ORO (Operation Research Office), which was set up by Johns Hopkins to do research and development for the U.S. Army. In 1958 he left ORO to take a position at a company called Technical Operations, where he helped build a research and development laboratory for military software projects. In 1961 he left Technical Operations to found Massachusetts Computer Associates. Later, this company became part of Applied Data Research (ADR). After the merger, Warshall sat on the board of directors of ADR and managed a variety of projects and organizations. He retired from ADR in 1982.

During his career Warshall has carried out research and development in operating systems, compiler design, language design, and operations research. In the 1971–1972 academic year he presented lectures on software engineering at French universities. There is an interesting anecdote about his proof that the transitive closure algorithm, now known as Warshall's algorithm, is correct. He and a colleague at Technical Operations bet a bottle of rum on who first could determine whether this algorithm always works. Warshall came up with his proof overnight, winning the bet and the rum, which he shared with the loser of the bet. Because Warshall did not like sitting at a desk, he did much of his creative work in unconventional places, such as on a sailboat in the Indian Ocean or in a Greek lemon orchard.

EXAMPLE 8 Let R be the relation with directed graph shown in Figure 3. Let a, b, c, d be a listing of the elements of the set. Find the matrices $\mathbf{W}_0$, $\mathbf{W}_1$, $\mathbf{W}_2$, $\mathbf{W}_3$, and $\mathbf{W}_4$. The matrix $\mathbf{W}_4$ is the transitive closure of R.

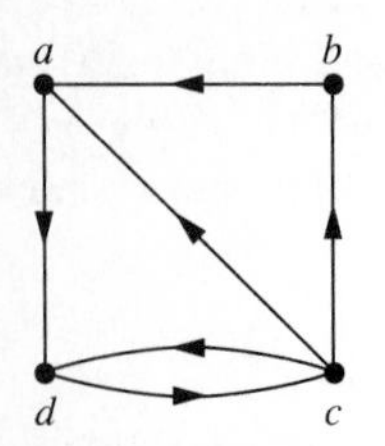

FIGURE 3 The Directed Graph of the Relation *R*.

Solution: Let $v_1 = a$, $v_2 = b$, $v_3 = c$, and $v_4 = d$. $\mathbf{W}_0$ is the matrix of the relation. Hence,

$$\mathbf{W}_0 = \begin{bmatrix} 0 & 0 & 0 & 1 \\ 1 & 0 & 1 & 0 \\ 1 & 0 & 0 & 1 \\ 0 & 0 & 1 & 0 \end{bmatrix}.$$

W_1 has 1 as its (i, j)th entry if there is a path from v_i to v_j that has only $v_1 = a$ as an interior vertex. Note that all paths of length one can still be used since they have no interior vertices. Also, there is now an allowable path from b to d, namely, b, a, d. Hence,

$$\mathbf{W}_1 = \begin{bmatrix} 0 & 0 & 0 & 1 \\ 1 & 0 & 1 & 1 \\ 1 & 0 & 0 & 1 \\ 0 & 0 & 1 & 0 \end{bmatrix}.$$

$\mathbf{W}_2$ has 1 as its (i, j)th entry if there is a path from v_i to v_j that has only $v_1 = a$ and/or $v_2 = b$ as its interior vertices, if any. Since there are no edges that have b as a terminal vertex, no new paths are obtained when we permit b to be an interior vertex. Hence, $\mathbf{W}_2 = W_1$.

$\mathbf{W}_3$ has 1 as its (i, j)th entry if there is a path from v_i to v_j that has only $v_1 = a$, $v_2 = b$, and/or $v_3 = c$ as its interior vertices, if any. We now have paths from d to a, namely, d, c, a, and from d to d, namely, d, c, d. Hence,

$$\mathbf{W}_3 = \begin{bmatrix} 0 & 0 & 0 & 1 \\ 1 & 0 & 1 & 1 \\ 1 & 0 & 0 & 1 \\ 1 & 0 & 1 & 1 \end{bmatrix}.$$

Finally, $\mathbf{W}_4$ has 1 as its (i, j)th entry if there is a path from v_i to v_j that has $v_1 = a$, $v_2 = b$, $v_3 = c$, and/or $v_4 = d$ as interior vertices, if any. Since these are all the vertices of the graph, this entry is 1 if and only if there is a path from v_i to v_j. Hence,

$$\mathbf{W}_4 = \begin{bmatrix} 1 & 0 & 1 & 1 \\ 1 & 0 & 1 & 1 \\ 1 & 0 & 1 & 1 \\ 1 & 0 & 1 & 1 \end{bmatrix}.$$

This last matrix, $\mathbf{W}_4$, is the matrix of the transitive closure. ■

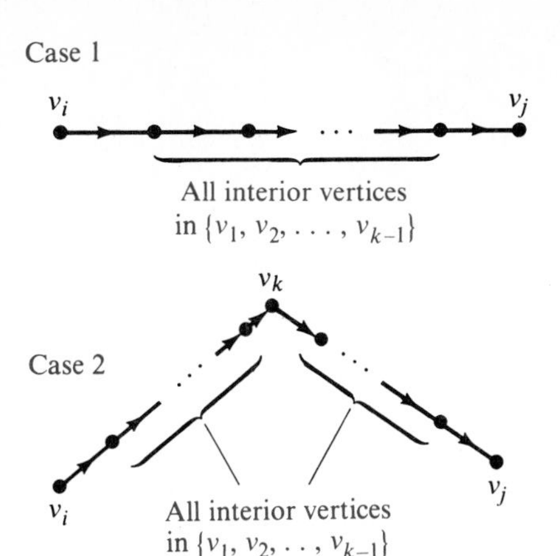

FIGURE 4 Adding v_k to the Set of Allowable Interior Vertices.

Warshall's algorithm computes $\mathbf{M}_{R^*}$ by efficiently computing $\mathbf{W}_0 = \mathbf{M}_R, \mathbf{W}_1, \mathbf{W}_2, \ldots, \mathbf{W}_n = \mathbf{M}_{R^*}$. The following observation shows that we can compute $\mathbf{W}_k$ directly from $\mathbf{W}_{k-1}$: There is a path from v_i to v_j with no vertices other than $v_1, v_2, \ldots, v_k$ as interior vertices if and only if either there is a path from v_i to v_j with its interior vertices among the first $k - 1$ vertices in the list, or there are paths from v_i to v_k and from v_k to v_j that have interior vertices only among the first $k - 1$ vertices in the list. That is, either a path from v_i to v_j already existed before v_k was permitted as an interior vertex, or allowing v_k as an interior vertex produces a path that goes from v_i to v_k and then from v_k to v_j. These two cases are shown in Figure 4.

The first type of path exists if and only if $w_{ij}^{[k-1]} = 1$, and the second type of path exists if and only if both $w_{ik}^{[k-1]}$ and $w_{kj}^{[k-1]}$ are 1. Hence, $w_{ij}^{[k]}$ is 1 if and only if either $w_{ij}^{[k-1]}$ is 1 or both $w_{ik}^{[k-1]}$ and $w_{kj}^{[k-1]}$ are 1. This gives us the following lemma.

LEMMA 2 Let $\mathbf{W}_k = [w_{ij}^{[k]}]$ be the zero-one matrix that has a 1 in its (i, j)th position if and only if there is a path from v_i to v_j with interior vertices from the set $\{v_1, v_2, \ldots, v_k\}$. Then

$$w_{ij}^{[k]} = w_{ij}^{[k-1]} \vee (w_{ik}^{[k-1]} \wedge w_{kj}^{[k-1]}),$$

whenever i, j, and k are positive integers not exceeding n.

Lemma 2 gives us the means efficiently to compute the matrices $\mathbf{W}_k$, $k = 1, 2, \ldots, n$. We display the pseudocode for Warshall's algorithm, using Lemma 2, as Algorithm 2.

ALGORITHM 2 Warshall Algorithm.

procedure *Warshall* ($\mathbf{M}_R$: $n \times n$ zero-one matrix)
$\mathbf{W} := \mathbf{M}_R$
for $k := 1$ **to** n
begin
  for $i := 1$ **to** n
  begin
    for $j := 1$ **to** n
    $w_{ij} := w_{ij} \vee (w_{ik} \wedge w_{kj})$
  end
end $\{\mathbf{W} = [w_{ij}]$ is $\mathbf{M}_{R^*}\}$

The computational complexity of Warshall's algorithm can easily be computed in terms of bit operations. To find the entry $w_{ij}^{[k]}$ from the entries $w_{ij}^{[k-1]}$, $w_{ik}^{[k-1]}$, and $w_{kj}^{[k-1]}$ using Lemma 2 requires two bit operations. To find all n^2 entries of $\mathbf{W}_k$ from those of $\mathbf{W}_{k-1}$ requires $2n^2$ bit operations. Since Warshall's algorithm begins with $\mathbf{W}_0 = \mathbf{M}_R$ and computes the sequence of n zero-one matrices $\mathbf{W}_1, \mathbf{W}_2, \ldots, \mathbf{W}_n = \mathbf{M}_{R*}$, the total number of bit operations used is $n \cdot 2n^2 = 2n^3$.

Exercises

1. Let R be the relation on the set $\{0, 1, 2, 3\}$ containing the ordered pairs $(0, 1)$, $(1, 1)$, $(1, 2)$, $(2, 0)$, $(2, 2)$, and $(3, 0)$. Find the
   a) reflexive closure of R.
   b) symmetric closure of R.
2. Let R be the relation $\{(a, b) \mid a \neq b\}$ on the set of integers. What is the reflexive closure of R?
3. Let R be the relation $\{(a, b) \mid a \text{ divides } b\}$ on the set of integers. What is the symmetric closure of R?
4. How can the directed graph representing the reflexive closure of a relation on a finite set be constructed from the directed graph of the relation?

In Exercises 5–7 draw the directed graph of the reflexive closure of the relations with the directed graph shown.

5.
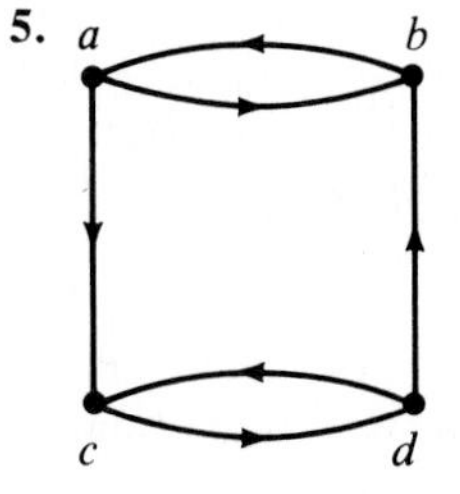

6.
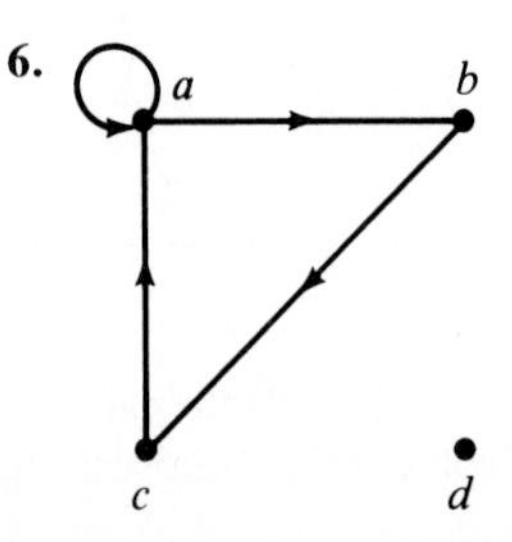

7.

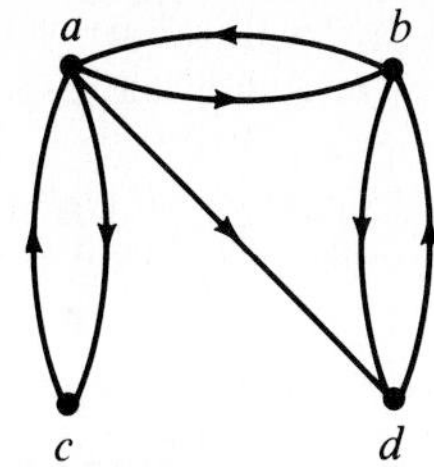

8. How can the directed graph representing the symmetric closure of a relation on a finite set be constructed from the directed graph for this relation?

9. Find the directed graphs of the symmetric closures of the relations with directed graphs shown in Exercises 5–7.

10. Find the smallest relation containing the relation in Example 2 that is both reflexive and symmetric.

11. Find the directed graph of the smallest relation that is both reflexive and symmetric for each of the relations with directed graphs shown in Exercises 5–7.

12. Suppose that the relation R on the finite set A is represented by the matrix $\mathbf{M}_R$. Show that the matrix that represents the reflexive closure of R is $\mathbf{M}_R \vee \mathbf{I}_n$.

13. Suppose that the relation R on the finite set A is represented by the matrix $\mathbf{M}_R$. Show that the matrix that represents the symmetric closure of R is $\mathbf{M}_R \vee \mathbf{M}_R^t$.

14. Show that the closure of a relation R with respect to a property **P**, if it exists, is the intersection of all the relations with property **P** that contain R.

15. When is it possible to define the "irreflexive closure" of a relation R, that is, a relation that contains R, is irreflexive, and is contained in every irreflexive relation that contains R?

16. Determine whether the following sequences of vertices are paths in the following directed graph.

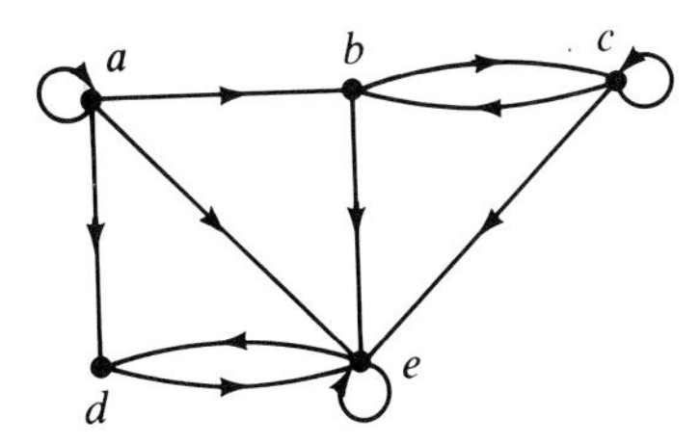

a) a, b, c, e **b)** b, e, c, b, e
c) a, a, b, e, d, e **d)** b, c, e, d, a, a, b
e) b, c, c, b, e, d, e, d **f)** $a, a, b, b, c, c, b, e, d$

17. Find all circuits of length three in the directed graph in Exercise 16.

18. Determine whether there is a path in the directed graph in Exercise 16 beginning at the first vertex given and ending at the second vertex given.
a) a, b **b)** b, a **c)** b, b
d) a, e **e)** b, d **f)** c, d
g) d, d **h)** e, a **i)** e, c

19. Let R be the relation on the set $\{1, 2, 3, 4, 5\}$ containing the ordered pairs $(1, 3)$, $(2, 4)$, $(3, 1)$, $(3, 5)$, $(4, 3)$, $(5, 1)$, $(5, 2)$, and $(5, 4)$. Find
a) R^2. **b)** R^3.
c) R^4. **d)** R^5.
e) R^6. **f)** R^*.

20. Let R be the relation that contains the pair (a, b) if a and b are cities such that there is a direct non-stop airline flight from a to b. When is (a,b) in
a) R^2? **b)** R^3? **c)** R^*?

21. Let R be the relation on the set of all students containing the ordered pair (a, b) if a and b are in at least one common class and $a \neq b$. When is (a, b) in
a) R^2? **b)** R^3? **c)** R^*?

22. Suppose that the relation R is reflexive. Show that R^* is reflexive.

23. Suppose that the relation R is symmetric. Show that R^* is symmetric.

24. Suppose that the relation R is irreflexive. Is the relation R^2 necessarily irreflexive?

25. Use Algorithm 1 to find the transitive closures of the following relations on $\{1, 2, 3, 4\}$.
a) $\{(1, 2), (2, 1), (2, 3), (3, 4), (4, 1)\}$
b) $\{(2, 1), (2, 3), (3, 1), (3, 4), (4, 1), (4, 3)\}$
c) $\{(1, 2), (1, 3), (1, 4), (2, 3), (2, 4), (3, 4)\}$
d) $\{(1, 1), (1, 4), (2, 1), (2, 3), (3, 1), (3, 2), (3, 4), (4, 2)\}$

26. Use Algorithm 1 to find the transitive closures of the following relations on $\{a, b, c, d, e\}$.
a) $\{(a, c), (b, d), (c, a), (d, b), (e, d)\}$
b) $\{(b, c), (b, e), (c, e), (d, a), (e, b), (e, c)\}$
c) $\{(a, b), (a, c), (a, e), (b, a), (b, c), (c, a), (c, b), (d, a), (e, d)\}$
d) $\{(a, e), (b, a), (b, d), (c, d), (d, a), (d, c), (e, a), (e, b), (e, c), (e, e)\}$

27. Use Warshall's algorithm to find the transitive closures of the relations in Exercise 25.

28. Use Warshall's algorithm to find the transitive closures of the relations in Exercise 26.

29. Find the smallest relation containing the relation $\{(1, 2), (1, 4), (3, 3), (4, 1)\}$ that is
a) reflexive and transitive.
b) symmetric and transitive.
c) reflexive, symmetric, and transitive.

30. Finish the proof of the case when $a \neq b$ in Lemma 1.

31. Algorithms have been devised that use $O(n^{2.8})$ bit operations to compute the Boolean product of two $n \times n$ zero-one matrices. Assuming that these algorithms can be used, give big-O estimates for the number of bit operations using Algorithm 1 and using Warshall's algorithm to find the transitive closure of a relation on a set with n elements.

★**32.** Devise an algorithm using the concept of interior vertices in a path to find the length of the shortest path between two vertices in a directed graph, if such a path exists.

33. Adapt Algorithm 1 to find the reflexive closure of the transitive closure of a relation on a set with n elements.

34. Adapt Warshall's algorithm to find the reflexive closure of the transitive closure of a relation on a set with n elements.

35. Show that the closure with respect to the property **P** of the relation $R = \{(0, 0), (0, 1), (1, 1), (2, 2)\}$ on the set $\{0, 1, 2\}$ does not exist if **P** is the property
a) is not reflexive.
b) has an odd number of elements.

6.5 Equivalence Relations

INTRODUCTION

Students at a college register for classes the day before the start of a semester. Those with last names beginning with a letter from A to G, from H to N, and from O to Z may register at any time in the periods 8 AM to 11 AM, 11 AM to 2 PM, and 2 PM to 5 PM, respectively. Let R be the relation containing (x, y) if and only if x and y are students with last names beginning with letters in the same block. Consequently, x and y can register at the same time if and only if (x, y) belongs to R. It is easy to see that R is reflexive, symmetric, and transitive. Furthermore, R divides the set of students into three classes, depending on the first letters of their last names. To know when a student can register we are concerned only with which of the three classes the student is in, and we do not care about the identity of the student.

The integers a and b are related by the "congruence modulo 4" relation when 4 divides $a - b$. We will show later that this relation is reflexive, symmetric, and transitive. It is not hard to see that a is related to b if and only if a and b have the same remainder when divided by 4. It follows that this relation splits the set of integers into four different classes. When we care only what remainder an integer leaves when it is divided by 4, we need only to know which class it is in, not its particular value.

These two relations, R and congruence modulo 4, are examples of equivalence relations, namely, relations that are reflexive, symmetric, and transitive. In this section we will show that such relations split sets into disjoint classes of equivalent elements. Equivalence relations arise whenever we care only whether an element of a set is in a certain class of elements, instead of caring about its particular identity.

EQUIVALENCE RELATIONS

In this section we will study relations that have a particularly useful combination of properties, used to relate objects that are similar in some way.

DEFINITION 1. A relation on a set A is called an *equivalence relation* if it is reflexive, symmetric, and transitive.

Two elements that are related by an equivalence relation are called **equivalent**. (This definition makes sense since an equivalence relation is symmetric.) Since an equivalence relation is reflexive, in an equivalence relation every element is equivalent to itself. Furthermore, since an equivalence relation is transitive, if a and b are equivalent and b and c are equivalent, it follows that a and c are equivalent.

The following examples illustrate the notion of an equivalence relation.

EXAMPLE 1 Suppose that R is the relation on the set of strings of English letters such that $a\,R\,b$ if and only if $l(a) = l(b)$, where $l(x)$ is the length of the string x. Is R an equivalence relation?

Solution: Since $l(a) = l(a)$, it follows that $a\ R\ a$ whenever a is a string, so that R is reflexive. Next, suppose that $a\,R\,b$, so that $l(a) = l(b)$. Then $b\,R\,a$, since $l(b) = l(a)$. Hence, R is symmetric. Finally, suppose that $a\,R\,b$ and $b\,R\,c$. Then $l(a) = l(b)$ and $l(b) = l(c)$. Hence, $l(a) = l(c)$, so that $a\,R\,c$. Consequently, R is transitive. Since R is reflexive, symmetric, and transitive, it is an equivalence relation. ■

EXAMPLE 2 Let R be the relation on the set of integers such that $a\,R\,b$ if and only if $a = b$ or $a = -b$. In Section 6.1 we showed that R is reflexive, symmetric, and transitive. It follows that R is an equivalence relation. ■

EXAMPLE 3 Let R be the relation on the set of real numbers such that aRb if and only if $a - b$ is an integer. Is R an equivalence relation?

Solution: Since $a - a = 0$ is an integer for all real numbers a, $a\,R\,a$ for all real numbers a. Hence, R is reflexive. Now suppose that $a\,R\,b$. Then $a - b$ is an integer, so that $b - a$ is also an integer. Hence, $b\,R\,a$. It follows that R is symmetric. If $a\,R\,b$ and $b\,R\,c$, then $a - b$ and $b - c$ are integers. Therefore, $a - c = (a - b) + (b - c)$ is also an integer. Hence, $a\,R\,c$. Thus, R is transitive. Consequently, R is an equivalence relation. ■

One of the most widely used equivalence relations is congruence modulo m, where m is a positive integer greater than 1.

EXAMPLE 4 Congruence Modulo m Let m be a positive integer greater than 1. Show that the relation

$$R = \{(a, b) \mid a \equiv b \pmod{m}\}$$

is an equivalence relation on the set of integers.

Solution: Recall from Section 2.3 that $a \equiv b \pmod{m}$ if and only if m divides $a - b$. Note that $a - a = 0$ is divisible by m, since $0 = 0 \cdot m$. Hence, $a \equiv a \pmod{m}$, so that congruence modulo m is reflexive. Now suppose that $a \equiv b \pmod{m}$. Then $a - b$ is divisible by m, so that $a - b = km$, where k is an integer. It follows that $b - a = (-k)m$, so that $b \equiv a \pmod{m}$. Hence, congruence modulo m is symmetric. Next, suppose that $a \equiv b \pmod{m}$ and $b \equiv c \pmod{m}$. Then m divides both $a - b$ and $b - c$. Therefore, there are integers k and l with $a - b = km$ and $b - c = lm$. Adding these two equations shows that $a - c = (a - b) + (b - c) = km + lm = (k + l)m$. Thus, $a \equiv c \pmod{m}$. Therefore, congruence modulo m is transitive. It follows that congruence modulo m is an equivalence relation. ■

EQUIVALENCE CLASSES

Let A be the set of all students in your school who graduated from high school. Consider the relation R on A that consists of all pairs (x, y) where x and y graduated from the same high school. Given a student x, we can form the set of all students equivalent to x with respect to R. This set consists of all students who graduated from the same high school as x did. This subset of A is called an equivalence class of the relation.

DEFINITION 2. Let R be an equivalence relation on a set A. The set of all elements that are related to an element a of A is called the *equivalence class* of a. The equivalence class of a with respect to R is denoted by $[a]_R$. When only one relation is under consideration, we will delete the subscript R and write $[a]$ for this equivalence class.

In other words, if R is an equivalence relation on a set A, the equivalence class of the element a is

$$[a]_R = \{s \mid (a, s) \in R\}.$$

If $b \in [a]_R$, b is called a **representative** of this equivalence class.

EXAMPLE 5 What is the equivalence class of an integer for the equivalence relation of Example 2?

Solution: Since an integer is equivalent to itself and its negative in this equivalence relation, it follows that $[a] = \{-a, a\}$. This set contains two distinct integers unless $a = 0$. For instance, $[7] = \{-7, 7\}$, $[-5] = \{-5, 5\}$, and $[0] = \{0\}$. ■

EXAMPLE 6 What are the equivalence classes of 0 and 1 for congruence modulo 4?

Solution: The equivalence class of 0 contains all integers a such that $a \equiv 0 \pmod 4$. The integers in this class are those divisible by 4. Hence, the equivalence class of 0 for this relation is

$$[0] = \{\ldots, -8, -4, 0, 4, 8, \ldots\}.$$

The equivalence class of 1 contains all the integers a such that $a \equiv 1 \pmod 4$. The integers in this class are those that have a remainder of 1 when divided by 4. Hence, the equivalence class of 1 for this relation is

$$[1] = \{\ldots, -7, -3, 1, 5, 9, \ldots\}.$$

■

In Example 6 the equivalence classes of 0 and 1 with respect to congruence modulo 4 were found. Example 4 can easily be generalized, replacing 4 with any positive integer m. The equivalence classes of the relation congruence modulo m are called the **congruence classes modulo** m. The congruence class of an integer a modulo m is denoted by $[a]_m$. For instance, from Example 6 it follows that $[0]_4 = \{\ldots, -8, -4, 0, 4, 8, \ldots\}$ and $[1]_4 = \{-7, -3, 1, 5, 9, \ldots\}$.

EQUIVALENCE CLASSES AND PARTITIONS

Let A be the set of students at your school who are majoring in exactly one subject, and let R be the relation on A consisting of pairs (x, y) where x and y are students with the same major. Then R is an equivalence relation, as the reader should verify. We can see that R splits all students in A into a collection of disjoint subsets, where each subset contains students with a specified major. For instance, one subset contains all students majoring (just) in computer science, and a second subset contains all students majoring in history. Furthermore, these subsets are equivalence classes of R. This example illustrates how the equivalence classes of an equivalence relation partition a set into disjoint, nonempty subsets. We will make these notions more precise in the following discussion.

Let R be a relation on the set A. The following theorem shows that the equivalence classes of two elements of A are either identical or disjoint.

THEOREM 1 Let R be an equivalence relation on a set A. The following statements are equivalent:

(i) $a\ R\ b$
(ii) $[a] = [b]$
(iii) $[a] \cap [b] \neq \emptyset$

Proof: We first show that (i) implies (ii). Assume that $a\ R\ b$. We will prove that $[a] = [b]$ by showing $[a] \subseteq [b]$ and $[b] \subseteq [a]$. Suppose $c \in [a]$. Then $a\ R\ c$. Since $a\ R\ b$ and R is symmetric, we know that $b\ R\ a$. Furthermore, since R is transitive and $b\ R\ a$ and $a\ R\ c$, it follows that $b\ R\ c$. Hence, $c \in [b]$. This shows that $[a] \subseteq [b]$. The proof that $[b] \subseteq [a]$ is similar; it is left as an exercise for the reader.

Second, we will show that (ii) implies (iii). Assume that $[a] = [b]$. It follows that $[a] \cap [b] \neq \emptyset$ since $[a]$ is nonempty (since $a \in [a]$ because R is reflexive).

Next, we will show that (iii) implies (i). Suppose that $[a] \cap [b] \neq \emptyset$. Then there is an element c with $c \in [a]$ and $c \in [b]$. In other words, $a\ R\ c$ and $b\ R\ c$. By the symmetric property, $c\ R\ b$. Then by transitivity, since $a\ R\ c$ and $c\ R\ b$, we have $a\ R\ b$.

Since (i) implies (ii), (ii) implies (iii), and (iii) implies (i), the three statements (i), (ii), and (iii) are equivalent. ■

We are now in a position to show how an equivalence relation *partitions* a set. Let R be an equivalence relation on a set A. The union of the equivalence classes of R is all of A, since an element a of A is in its own equivalence class, namely, $[a]_R$. In other words,

$$\bigcup_{a \in A} [a]_R = A.$$

In addition, from Theorem 1, it follows that these equivalence classes are either equal or disjoint, so that

$$[a]_R \cap [b]_R = \emptyset$$

when $[a]_R \neq [b]_R$.

These two observations show that the equivalence classes form a partition of A, since they split A into disjoint subsets. More precisely, a **partition** of a set S is a collection of disjoint nonempty subsets of S that have S as their union. In other words, the collection of subsets A_i, $i \in I$ (where I is an index set) forms a partition of S if and only if

$$A_i \neq \emptyset \text{ for } i \in I,$$

$$A_i \cap A_j = \emptyset, \text{ when } i \neq j,$$

and

$$\bigcup_{i \in I} A_i = S.$$

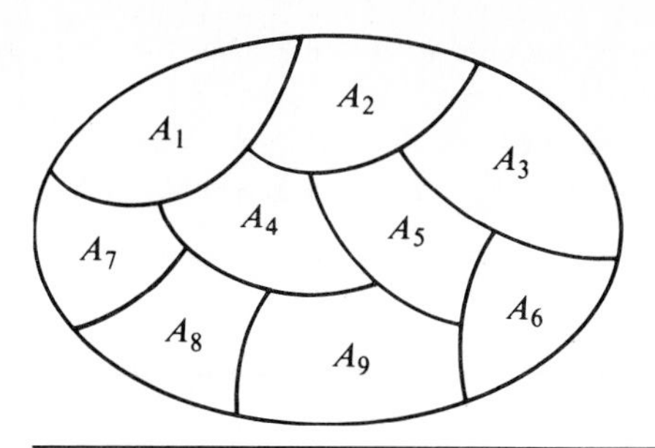

FIGURE 1 A Partition of a Set.

(Here the notation $\bigcup_{i \in I} A_i$ represents the union of the sets A_i for all $i \in I$.) Figure 1 illustrates the concept of a partition of a set.

EXAMPLE 7 Suppose that $S = \{1, 2, 3, 4, 5, 6\}$. The collection of sets $A_1 = \{1, 2, 3\}$, $A_2 = \{4, 5\}$, and $A_3 = \{6\}$ forms a partition of S, since these sets are disjoint and their union is S. ■

We have seen that the equivalence classes of an equivalence relation on a set form a partition of the set. The subsets in this partition are the equivalence classes. Conversely, every partition of a set can be used to form an equivalence relation. Two elements are equivalent with respect to this relation if and only if they are in the same subset of the partition.

To see this, assume that $\{A_i \mid i \in I\}$ is a partition on S. Let R be the relation on S consisting of the pair (x, y) where x and y belong to the same subset A_i in the partition. To show that R is an equivalence relation we must show that R Is reflexive, symmetric, and transitive.

We see that $(a, a) \in R$ for every $a \in S$, since a is in the same subset as itself. Hence, R is reflexive. If $(a, b) \in R$, then b and a are in the same subset of the partition, so that $(b, a) \in R$ as well. Hence, R is symmetric. If $(a, b) \in R$ and $(b, c) \in R$, then a and b are in the same subset in the partition, X, and b and c are in the same subset of the partition, Y. Since the subsets of the partition are disjoint, and b belongs to X and Y, it follows that $X = Y$. Consequently, a and c belong to the same subset of the partition, so that $(a, c) \in R$. Thus, R is transitive.

It follows that R is an equivalence relation. The equivalence classes of R consist of subsets of S containing related elements, and by the definition of R, these are the subsets of the partition. Theorem 2 summarizes the connections we have established between equivalence relations and partitions.

THEOREM 2 Let R be an equivalence relation on a set S. Then the equivalence classes of R form a partition of S. Conversely, given a partition $\{A_i \mid i \in I\}$ of the set S, there is an equivalence relation R that has the sets A_i, $i \in I$, as its equivalence classes.

The congruence classes modulo m provide a useful illustration of Theorem 2. There are m different congruence classes modulo m, corresponding to the m different remainders possible when an integer is divided by m. These m congruence classes are denoted by $[0]_m, [1]_m, \ldots, [m-1]_m$. They form a partition of the set of integers.

EXAMPLE 8 What are the sets in the partition of the integers arising from congruence modulo 4?

Solution: There are four congruence classes, corresponding to $[0]_4$, $[1]_4$, $[2]_4$, and $[3]_4$. They are the sets

$$\begin{aligned}[0]_4 &= \{\ldots, -8, -4, 0, 4, 8, \ldots\},\\ [1]_4 &= \{\ldots, -7, -3, 1, 5, 9, \ldots\},\\ [2]_4 &= \{\ldots, -6, -2, 2, 6, 10, \ldots\},\\ [3]_4 &= \{\ldots, -5, -1, 3, 7, 11, \ldots\}.\end{aligned}$$

These congruence classes are disjoint, and every integer is in exactly one of them. In other words, as Theorem 2 says, these congruence classes form a partition. ■

Exercises

1. Which of the following relations on $\{0, 1, 2, 3\}$ are equivalence relations? Determine the properties of an equivalence relation that the others lack.
  a) $\{(0,0), (1,1), (2,2), (3,3)\}$
  b) $\{(0,0), (0,2), (2,0), (2,2), (2,3), (3,2), (3,3)\}$
  c) $\{(0,0), (1,1), (1,2), (2,1), (2,2), (3,3)\}$
  d) $\{(0,0), (1,1), (1,3), (2,2), (2,3), (3,1), (3,2), (3,3)\}$
  e) $\{(0,0), (0,1), (0,2), (1,0), (1,1), (1,2), (2,0), (2,2), (3,3)\}$

2. Which of the following relations on the set of all people are equivalence relations? Determine the properties of an equivalence relation that the others lack.
  a) $\{(a, b) \mid a$ and b are the same age$\}$
  b) $\{(a, b) \mid a$ and b have the same parents$\}$
  c) $\{(a, b) \mid a$ and b share a common parent$\}$
  d) $\{(a, b) \mid a$ and b have met$\}$
  e) $\{(a, b) \mid a$ and b speak a common language$\}$

3. Which of the following relations on the set of all functions from $\mathbf{Z}$ to $\mathbf{Z}$ are equivalence relations? Determine the properties of an equivalence relation that the others lack.
  a) $\{(f, g) \mid f(1) = g(1)\}$
  b) $\{(f, g) \mid f(0) = g(0)$ or $f(1) = g(1)\}$
  c) $\{(f, g) \mid f(x) - g(x) = 1$ for all $x \in \mathbf{Z}\}$
  d) $\{(f, g) \mid f(x) - g(x) = C$ for some $C \in \mathbf{Z}$ for all $x \in \mathbf{Z}\}$
  e) $\{(f, g) \mid f(0) = g(1)$ and $f(1) = g(0)\}$

4. Define three equivalence relations on the set of students in your discrete mathematics class different from the relations discussed in the text. Determine the equivalence classes for these equivalence relations.

5. Suppose that A is a nonempty set, and f is a function that has A as its domain. Let R be the relation on A consisting of all ordered pairs (x, y) where $f(x) = f(y)$.
  a) Show that R is an equivalence relation on A.
  b) What are the equivalence classes of R?

6. Suppose that A is a nonempty set and R is an equivalence relation on A. Show that there is a function f with A as its domain such that $(x, y) \in R$ if and only if $f(x) = f(y)$.

7. Show that the relation R, consisting of all pairs (x, y) where x and y are bit strings of length 3 or more that agree in their first 3 bits, is an equivalence relation on the set of all bit strings of length three or more.

8. Show that the relation R, consisting of all pairs (x,y) where x and y are bit strings of length 3 or more that agree except perhaps in their first three bits, is an equivalence relation on the set of all bit strings.

9. Show that propositional equivalence is an equivalence relation on the set of all compound propositions.

10. Let R be the relation on the set of ordered pairs of positive integers such that $((a, b), (c, d)) \in R$ if and only if $ad = bc$. Show that R is an equivalence relation.

In Exercises 11–13 determine whether the relation with the directed graphs shown is an equivalence relation.

11.

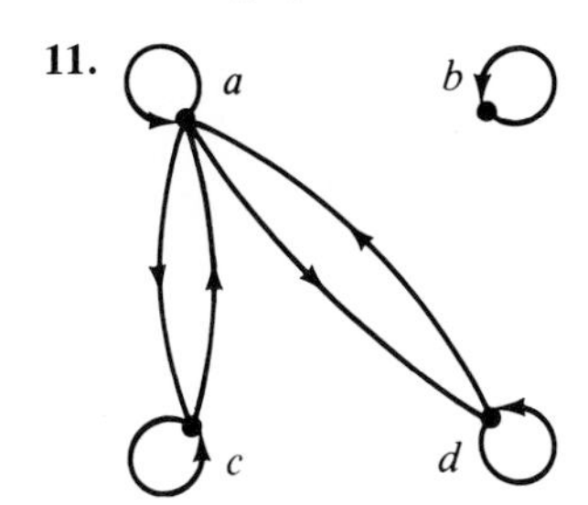

12.

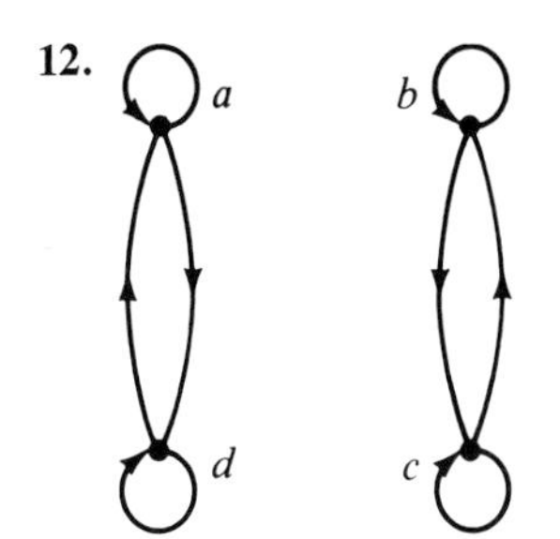

13.

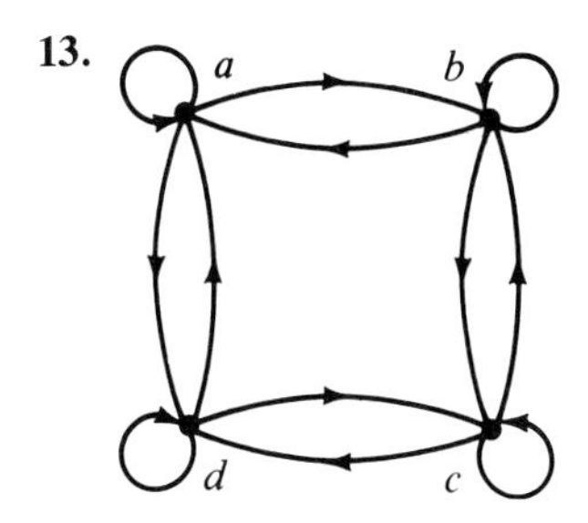

14. Determine whether the relations represented by the following zero-one matrices are equivalence relations.

a) $\begin{bmatrix} 1 & 1 & 1 \\ 0 & 1 & 1 \\ 1 & 1 & 1 \end{bmatrix}$ **b)** $\begin{bmatrix} 1 & 0 & 1 & 0 \\ 0 & 1 & 0 & 1 \\ 1 & 0 & 1 & 0 \\ 0 & 1 & 0 & 1 \end{bmatrix}$

c) $\begin{bmatrix} 1 & 1 & 1 & 0 \\ 1 & 1 & 1 & 0 \\ 1 & 1 & 1 & 0 \\ 0 & 0 & 0 & 1 \end{bmatrix}$

15. Show that the relation R on the set of all bit strings such that $s\ R\ t$ if and only if s and t contain the same number of 1s is an equivalence relation.

16. What are the equivalence classes of the equivalence relations in Exercise 1?

17. What are the equivalence classes of the equivalence relations in Exercise 2?

18. What are the equivalence classes of the equivalence relations in Exercise 3?

19. What is the equivalence class of the bit string 011 for the equivalence relation in Exercise 15?

20. What are the equivalence classes of the following bit strings for the equivalence relation in Exercise 7?

a) 010 **b)** 1011
c) 11111 **d)** 01010101

21. Describe the equivalence classes of the bit strings in Exercise 20 for the equivalence relation from Exercise 8.

22. What is the congruence class $[4]_m$ when m is

a) 2? **b)** 3? **c)** 6? **d)** 8?

23. Give a description of each of the congruence classes modulo 6.

24. **a)** What is the equivalence class of $(1, 2)$ with respect to the equivalence relation in Exercise 10?

b) Give an interpretation of the equivalence classes for the equivalence relation R in Exercise 10.

25. Which of the following collections of subsets are partitions of $\{1, 2, 3, 4, 5, 6\}$?

a) $\{1, 2\}$, $\{2, 3, 4\}$, $\{4, 5, 6\}$
b) $\{1\}$, $\{2, 3, 6\}$, $\{4\}$, $\{5\}$
c) $\{2, 4, 6\}$, $\{1, 3, 5\}$
c) $\{1, 4, 5\}$, $\{2, 6\}$

26. Which of the following collections of subsets are partitions of the set of integers?

a) the set of even integers and the set of odd integers

b) the set of positive integers and the set of negative integers

c) the set of integers divisible by 3, the set of integers leaving a remainder of 1 when divided by

3, and the set of integers leaving a remainder of 2 when divided by 3

d) the set of integers less than -100, the set of integers with absolute value not exceeding 100, and the set of integers greater than 100

e) the set of integers not divisible by 3, the set of even integers, and the set of integers that leave a remainder of 3 when divided by 6

A partition P_1 is called a **refinement** of the partition P_2 if every set in P_1 is a subset of one of the sets in P_2.

27. Show that the partition formed from the congruence classes modulo 6 is a refinement of the partition formed from the congruence classes modulo 3.

28. Suppose that R_1 and R_2 are equivalence relations on a set A. Let P_1 and P_2 be the partitions that correspond to R_1 and R_2, respectively. Show that $R_1 \subseteq R_2$ if and only if P_1 is a refinement of P_2.

29. Find the smallest equivalence relation on the set $\{a, b, c, d, e\}$ containing the relation $\{(a, b), (a, c), (d, e)\}$.

30. Suppose that R_1 and R_2 are equivalence relations on the set S. Determine whether each of the following combinations of R_1 and R_2 must be an equivalence relation.

a) $R_1 \cup R_2$ **b)** $R_1 \cap R_2$ **c)** $R_1 \oplus R_2$

31. Consider the equivalence relation from Example 3, namely, $R = \{(x, y) \mid x - y \text{ is an integer}\}$.

a) What is the equivalence class of 1 for this equivalence relation?

b) What is the equivalence class of 1/2 for this equivalence relation?

★32. How many different equivalence relations are there on a set with four elements?

★33. Do we necessarily get an equivalence relation when we form the transitive closure of the symmetric closure of the reflexive closure of a relation?

★34. Do we necessarily get an equivalence relation when we form the symmetric closure of the reflexive closure of the transitive closure of a relation?

35. Suppose we use Theorem 2 to form a partition P from an equivalence relation R. What is the equivalence relation R' that results if we use Theorem 2 again to form an equivalence relation from P?

36. Suppose we use Theorem 2 to form an equivalence relation R from a partition P. What is the partition P' that results if we use Theorem 2 again to form a partition from R?

37. Devise an algorithm to find the smallest equivalence relation containing a given relation.

6.6 Partial Orderings

INTRODUCTION

We often use relations to order some or all of the elements of sets. For instance, we order words using the relation containing pairs of words (x, y) where x comes before y in the dictionary. We schedule projects using the relation consisting of pairs (x, y) where x and y are tasks in a project such that x must be completed before y begins. We order the set of integers using the relation containing the pairs (x, y) where x is less than y. When we add all of the pairs of the form (x, x) to these relations, we obtain a relation that is reflexive, antisymmetric, and transitive. These are properties that characterize relations used to order the elements of sets using their relative size.

DEFINITION 1. A relation R on a set S is called a *partial ordering* or *partial order* if it is reflexive, antisymmetric, and transitive. A set S together with a partial ordering R is called a *partially ordered set*, or *poset*, and is denoted by (S, R).

EXAMPLE 1 Show that the "greater than or equal" relation ($\geq$) is a partial ordering on the set of integers.

Solution: Since $a \geq a$ for every integer a, $\geq$ is reflexive. If $a \geq b$ and $b \geq a$, then $a = b$. Hence, $\geq$ is antisymmetric. Finally, $\geq$ is transitive since $a \geq b$ and $b \geq c$ imply that $a \geq c$. It follows that $\geq$ is a partial ordering on the set of integers and $(\mathbf{Z}, \geq)$ is a poset. ■

EXAMPLE 2 The divisibility relation $|$ is a partial ordering on the set of positive integers, since it is reflexive, antisymmetric, and transitive, as was shown in Section 6.1. We see that $(\mathbf{Z}^+, |)$ is a poset. ($\mathbf{Z}^+$ denotes the set of positive integers.) ■

EXAMPLE 3 Show that the inclusion relation $\subseteq$ is a partial ordering on the power set of a set S.

Solution: Since $A \subseteq A$ whenever A is a subset of S, $\subseteq$ is reflexive. It is antisymmetric since $A \subseteq B$ and $B \subseteq A$ imply that $A = B$. Finally, $\subseteq$ is transitive, since $A \subseteq B$ and $B \subseteq C$ imply that $A \subseteq C$. Hence, $\subseteq$ is a partial ordering on $P(S)$, and $(P(S), \subseteq)$ is a poset. ■

In a poset the notation $a \preccurlyeq b$ denotes that $(a, b) \in R$. This notation is used because the "less than or equal to" relation is a paradigm for a partial ordering. (Note that the symbol $\preccurlyeq$ is used to denote the relation in *any* poset, not just the "less than or equals" relation.) The notation $a \prec b$ denotes that $a \preccurlyeq b$, but $a \neq b$. Also, we say "a is less than b" or "b is greater than a" if $a \prec b$.

When a and b are elements of the poset $(S, \preccurlyeq)$, it is not necessary that either $a \preccurlyeq b$ or $b \preccurlyeq a$. For instance, in $(P(\mathbf{Z}), \subseteq)$, $\{1, 2\}$ is not related to $\{1, 3\}$, and vice versa, since neither set is contained within the other. Similarly, in $(\mathbf{Z}, |)$, 2 is not related to 3 and 3 is not related to 2, since $2 \not| \, 3$ and $3 \not| \, 2$. This leads to the following definition.

DEFINITION 2. The elements a and b of a poset $(S, \prec)$ are called *comparable* if either $a \preccurlyeq b$ or $b \preccurlyeq a$. When a and b are elements of S such that neither $a \preccurlyeq b$ nor $b \preccurlyeq a$, a and b are called *incomparable*.

EXAMPLE 4 In the poset $(\mathbf{Z}^+, |)$, are the integers 3 and 9 comparable? Are 5 and 7 comparable?

Solution: The integers 3 and 9 are comparable, since $3 \mid 9$. The integers 5 and 7 are incomparable, because $5 \not| \, 7$ and $7 \not| \, 5$. ■

The adjective "partial" is used to describe partial orderings since pairs of elements may be incomparable. When every two elements in the set are comparable, the relation is called a **total ordering**.

DEFINITION 3. If $(S, \preccurlyeq)$ is a poset and every two elements of S are comparable, S is called a *totally ordered* or *linearly ordered set*, and $\preccurlyeq$ is called a *total order* or a *linear order*. A totally ordered set is also called a *chain*.

EXAMPLE 5 The poset $(\mathbf{Z}, \leq)$ is totally ordered, since $a \leq b$ or $b \leq a$ whenever a and b are integers. ■

EXAMPLE 6 The poset $(\mathbf{Z}^+, |)$ is not totally ordered since it contains elements that are incomparable, such as 5 and 7. ■

LEXICOGRAPHIC ORDER

The words in a dictionary are listed in alphabetic, or lexicographic, order, which is based on the ordering of the letters in the alphabet. This is a special case of an ordering of strings on a set constructed from a partial ordering on the set. We will show how this construction works in any poset.

First, we will show how to construct a partial ordering on the Cartesian product of two posets, $(A_1, \preccurlyeq_1)$ and $(A_2, \preccurlyeq_2)$. The **lexicographic ordering** $\preccurlyeq$ on $A_1 \times A_2$ is defined by specifying that one pair is less than a second pair if the first entry of the first pair is less than (in A_1) the first entry of the second pair, or if the first entries are equal, but the second entry of this pair is less than (in A_2) the second entry of the second pair. In other words, (a_1, a_2) is less than (b_1, b_2), that is

$$(a_1, a_2) \prec (b_1, b_2),$$

either if $a_1 \prec_1 b_1$ or if both $a_1 = b_1$ and $a_2 \prec_2 b_2$.

We obtain a partial ordering $\preccurlyeq$ by adding equality to the ordering $\prec$ on $A \times B$. The verification of this is left as an exercise.

EXAMPLE 7 Determine whether $(3, 5) \prec (4, 8)$, whether $(3, 8) \prec (4, 5)$, and whether $(4, 9) \prec (4, 11)$ in the poset $(\mathbf{Z} \times \mathbf{Z}, \preccurlyeq)$, where $\preccurlyeq$ is the lexicographic ordering constructed from the usual $\leq$ relation on $\mathbf{Z}$.

Solution: Since $3 < 4$, it follows that $(3, 5) \prec (4, 8)$ and that $(3, 8) \prec (4, 5)$. We have $(4, 9) \prec (4, 11)$, since the first entries of $(4, 9)$ and $(4, 11)$ are the same, but $9 < 11$. ■

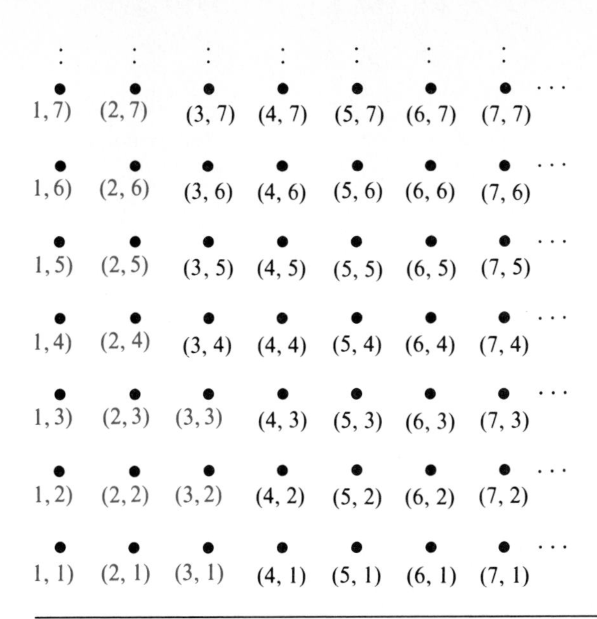

FIGURE 1 The Ordered Pairs Less than (3, 4) in Lexicographic Order.

In Figure 1 the set of ordered pairs in $\mathbf{Z}^+ \times \mathbf{Z}^+$ that are less than (3,4) are highlighted.

A lexicogaphic ordering can be defined on the Cartesian product of n posets $(A_1, \preccurlyeq_1), (A_2, \preccurlyeq_2), \ldots, (A_n, \preccurlyeq_n)$. Define the partial ordering $\preccurlyeq$ on $A_1 \times A_2 \times \cdots \times A_n$ by

$$(a_1, a_2, \ldots, a_n) \prec (b_1, b_2, \ldots, b_n)$$

if $a_1 \prec_1 b_1$, or if there is an integer $i > 0$ such that $a_1 = b_1, \ldots, a_i = b_i$, and $a_{i+1} \prec_{i+1} b_{i+1}$. In other words, one n-tuple is less than a second n-tuple if the entry of the first n-tuple in the first position where the two n-tuples disagree is less than the entry in that position in the second n-tuple.

EXAMPLE 8 Note that $(1, 2, 3, 5) \prec (1, 2, 4, 3)$, since the entries in the first two positions of these 4-tuples agree, but in the third position the entry in the first 4-tuple, 3, is less than that in the second 4-tuple, 4. (Here the ordering on 4-tuples is the lexicographic ordering that comes from the usual "less than or equals" relation on the set of integers.) ■

We can now define lexicographic ordering of strings. Consider the strings $a_1 a_2 \cdots a_m$ and $b_1 b_2 \cdots b_n$ on a partially ordered set S. Suppose these strings are not equal. Let t be the minimum of m and n. The definition of lexicographic ordering is that the string $a_1 a_2 \cdots a_m$ is less than $b_1 b_2 \cdots b_n$ if and only if

$$(a_1, a_2, \ldots, a_t) < (b_1, b_2, \ldots, b_t), \text{ or}$$
$$(a_1, a_2, \ldots, a_t) = (b_1, b_2, \ldots, b_t) \text{ and } m < n,$$

where $<$ in this inequality represents the lexicographic ordering of S^t. In other words, to determine the ordering of two different strings, the longer string is truncated to the length of the shorter string, namely, to $t = \min(m,n)$ terms. Then the t-tuples made up of the first t terms of each string are compared using the lexicographic ordering on S^t. One string is less than another string if the t-tuple corresponding to the first string is less than the t-tuple of the second string, or if these two t-tuples are the same, but the second string is longer. The verification that this is a partial ordering is left as an exercise for the reader.

EXAMPLE 9 Consider the set of strings of lowercase English letters. Using the ordering of letters in the alphabet, a lexicographic ordering on the set of strings can be constructed. A string is less than a second string if the letter in the first string in the first position where the strings differ comes before the letter in the second string in this position, or if the first string and the second string agree in all positions, but the second string has more letters. This ordering is the same as that used in dictionaries. For example,

$$\textit{discreet} < \textit{discrete},$$

since these strings differ first in the seventh position, and $e < t$. Also,

$$\textit{discreet} < \textit{discreetness},$$

since the first eight letters agree, but the second string is longer. Furthermore,

$$\textit{discrete} < \textit{discretion},$$

since

$$\textit{discrete} < \textit{discreti}.$$

HASSE DIAGRAMS

Many edges in the directed graph for a finite poset do not have to be shown since they must be present. For instance, consider the directed graph for the partial ordering $\{(a, b) \mid a \leq b\}$ on the set $\{1, 2, 3, 4\}$, shown in Figure 2(a). Since this relation is a partial ordering, it is reflexive, and its directed graph has loops at all vertices. Consequently, we do not have to show these loops since they must be present; in Figure 2(b) loops are not shown. Because a partial ordering is transitive, we do not have to show those edges that must be present because of transitivity. For example, in Figure 2(c) the edges $(1, 3)$, $(1, 4)$, and $(2, 4)$ are not shown since they must be present. If we assume that all edges are pointed "upward" (as they are drawn in the figure), we do not have to show the directions of the edges; Figure 2(c) does not show directions.

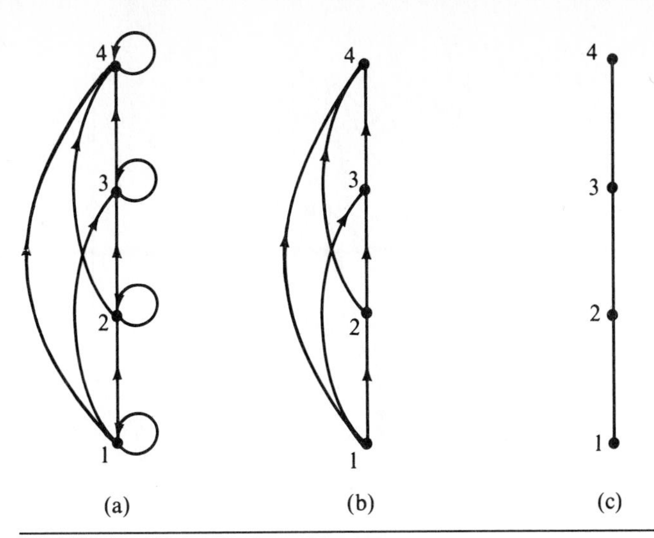

FIGURE 2 **Constructing the Hasse Diagram for** $(\{1, 2, 3, 4\}, \leq)$.

In general, we can represent a partial ordering on a set using the following procedure. Start with the directed graph for this relation. Because a partial ordering is reflexive, a loop is present at every vertex. Remove these loops. Remove all edges that must be present because of the transitivity, since they must be present since a partial ordering is transitive. For instance, if (a, b) and (b, c) are in the partial ordering, remove the edge (a, c), since it must be present also. Furthermore, if (c, d) is also in the partial ordering, remove the edge (a, d), since is must be present also. Finally, arrange each edge so that its initial edge is below its terminal edge (as it is drawn on paper). Remove all the arrows on the directed edges, since all edges point "upward" toward their terminal vertex.

These steps are well-defined, and only a finite number of steps need to be carried out for a finite poset. When all the steps have been taken, the resulting diagram contains sufficient information to find the partial ordering. This diagram is called a **Hasse diagram**, named after the 20th century German mathematician Helmut Hasse.

Helmut Hasse (1898–1979) Helmut Hasse began his university studies at Göttingen University in 1918, moving in 1920 to Marburg University to study under the number theorist Kurt Hensel. During this time, Hasse made fundamental contributions to algebraic number theory. He became Hensel's successor at Marburg, moving to Göttingen University in 1934, and later to Hamburg University in 1950. Hasse served for 50 years as an editor of *Crelle's Journal*, a famous German mathematics periodical, taking over the job of chief editor in 1936 when the Nazis forced Hensel to resign. During World War II Hasse worked on applied mathematics research for the German navy. He was noted for the clarity and personal style of his lectures and was devoted both to number theory and to his students. (Hasse has been controversial for connections with the Nazi party. Investigations have shown he was a strong German nationalist but not an ardent Nazi.)

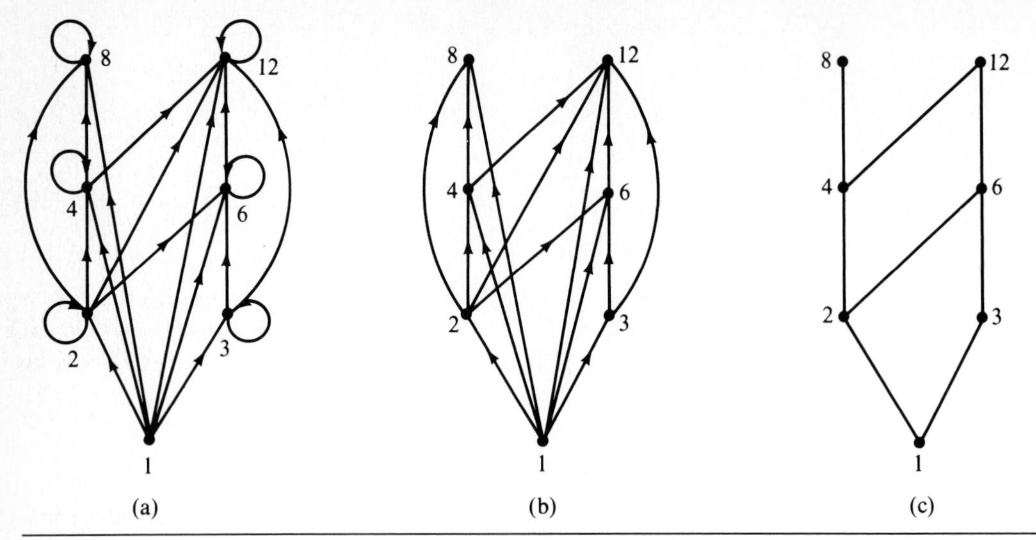

FIGURE 3 Constructing the Hasse Diagram of ({1, 2, 3, 4, 6, 8, 12}, |).

EXAMPLE 10 Draw the Hasse diagram representing the partial ordering $\{(a, b) \mid a \text{ divides } b\}$ on $\{1, 2, 3, 4, 6, 8, 12\}$.

Solution: Begin with the digraph for this partial order, as shown in Figure 3(a). Remove all loops, as shown in Figure 3(b). Then delete all the edges implied by the transitive property. These are $(1, 4)$, $(1, 6)$, $(1, 8)$, $(1, 12)$, $(2, 8)$, $(2, 12)$, and $(3, 12)$. Arrange all edges to point upward, and delete all arrows to obtain the Hasse diagram. The resulting Hasse diagram is shown in Figure 3(c). ■

EXAMPLE 11 Draw the Hasse diagram for the partial ordering $\{(A, B) \mid A \subseteq B\}$ on the power set $P(S)$ where $S = \{a, b, c\}$.

Solution: The Hasse diagram for this partial ordering is obtained from the associated digraph by deleting all the loops and all the edges that occur from transitivity, namely, $(\emptyset, \{a, b\})$, $(\emptyset, \{a, c\})$, $(\emptyset, \{b, c\})$, $(\emptyset, \{a, b, c\})$, $(\{a\}, \{a, b, c\})$, $(\{b\}, \{a, b, c\})$, and $(\{c\}, \{a, b, c\})$. Finally all edges point upward, and arrows are deleted. The resulting Hasse diagram is illustrated in Figure 4. ■

MAXIMAL AND MINIMAL ELEMENTS

Elements of posets that have certain extremal properties are important for many applications. An element of a poset is called maximal if it is not less than any element of the poset. That is, a is **maximal** in the poset $(S, \preccurlyeq)$ if there is no $b \in S$ such that

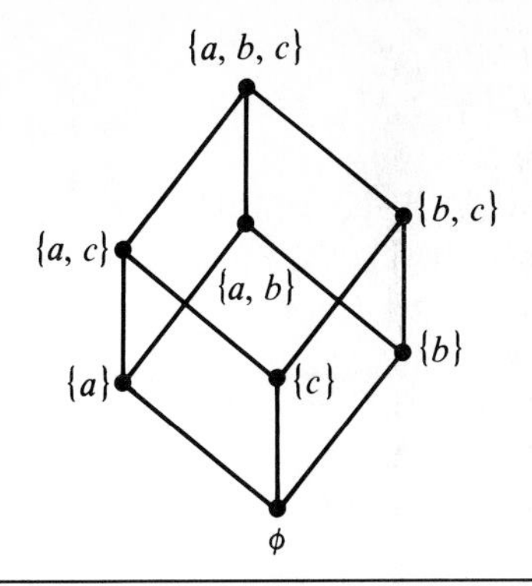

FIGURE 4 The Hasse Diagram of (P({a, b, c}), ⊆).

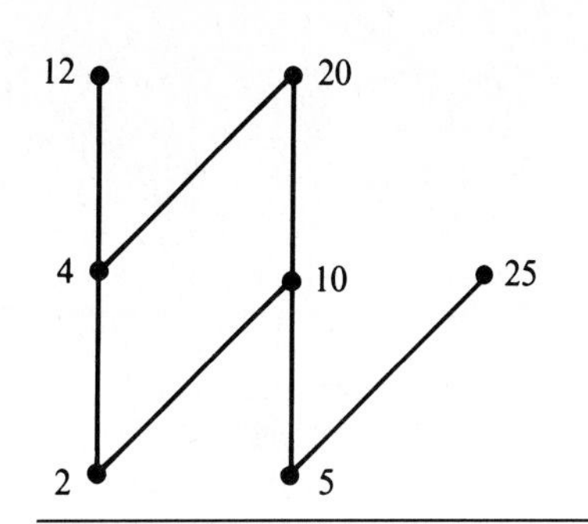

FIGURE 5 The Hasse Diagram of a Poset.

$a \prec b$. Similarly, an element of a poset is called minimal if it is not greater than any element of the poset. That is, a is **minimal** if there is no element $b \in S$ such that $b \prec a$. Maximal and minimal elements are easy to spot using a Hasse diagram. They are the "top" and "bottom" elements in the diagram.

EXAMPLE 12 Which elements of the poset $(\{2, 4, 5, 10, 12, 20, 25\}, |)$ are maximal, and which are minimal?

Solution: The Hasse diagram in Figure 5 for this poset shows that the maximal elements are 12, 20, and 25, and the minimal elements are 2 and 5. As this example shows, a poset can have more than one maximal element and more than one minimal element. ■

Sometimes there is an element in a poset that is greater than every other element. Such an element is called the greatest element. That is, a is the **greatest element** of the poset $(S, \preccurlyeq)$ if $b \preccurlyeq a$ for all $b \in S$. The greatest element is unique when it exists (see Exercise 26(a) at the end of this section). Likewise, an element is called the least element if it is less than all the other elements in the poset. That is, a is the **least element** of $(S, \preccurlyeq)$ if $a \preccurlyeq b$ for all $b \in S$. The least element is unique when it exists (see Exercise 26(b) at the end of the section).

EXAMPLE 13 Determine whether the posets represented by each of the Hasse diagrams in Figure 6 have a greatest element and a least element.

Solution: The least element of the poset with Hasse diagram (a) is a. This poset has no greatest element. The poset with Hasse diagram (b) has neither a least nor a greatest element. The poset with Hasse diagram (c) has no least element. Its greatest element is d. The poset with Hasse diagram (d) has least element a and greatest element d. ■

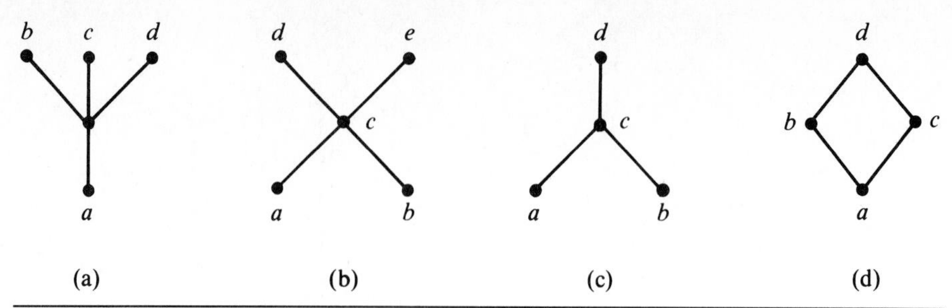

FIGURE 6 **Hasse Diagrams of Four Posets.**

EXAMPLE 14 Let S be a set. Determine whether there is a greatest element and a least element in the poset $(P(S), \subseteq)$.

Solution: The least element is the empty set, since $\emptyset \subseteq T$ for any subset T of S. The set S is the greatest element in this poset, since $T \subseteq S$ whenever T is a subset of S. ■

EXAMPLE 15 Is there a greatest element and a least element in the poset $(\mathbf{Z}^+, |)$?

Solution: The integer 1 is the least element since $1|n$ whenever n is a positive integer. Since there is no integer that is divisible by all positive integers, there is no greatest element. ■

Sometimes it is possible to find an element that is greater than all the elements in a subset A of a poset $(S, \preccurlyeq)$. If u is an element of S such that $a \preccurlyeq u$ for all elements $a \in A$, then u is called an **upper bound** of A. Likewise, there may be an element less than all the elements in A. If l is an element of S such that $l \preccurlyeq a$ for all elements $a \in A$, then l is called a **lower bound** of A.

EXAMPLE 16 Find the lower and upper bounds of the subsets $\{a, b, c\}$, $\{j, h\}$, and $\{a, c, d, f\}$ in the poset with the Hasse diagram shown in Figure 7.

Solution: The upper bounds of $\{a, b, c\}$ are e, f, j, and h, and its only lower bound is a. There are no upper bounds of $\{j, h\}$, and its lower bounds are a, b, c, d, e, and f. The upper bounds of $\{a, c, d, f\}$ are f, h, and j, and its lower bound is a. ■

The element x is called the **least upper bound** of the subset A if x is an upper bound that is less than every other upper bound of A. Since there is only one such

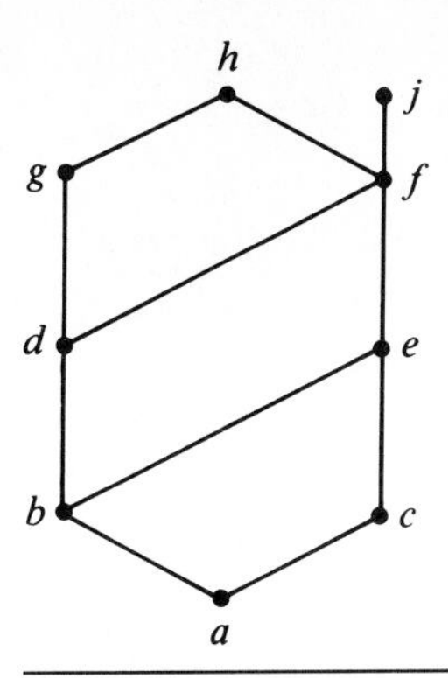

FIGURE 7 **The Hasse Diagram of a Poset.**

element, if it exists, it makes sense to call this element *the* least upper bound (see Exercise 28(a) at the end of this section). That is, x is the least upper bound of A if $a \preccurlyeq x$ whenever $a \in A$, and $x \preccurlyeq z$ whenever z is an upper bound of A. Similarly, the element y is called the **greatest lower bound** of A if y is a lower bound of A and $z \preccurlyeq y$ whenever z is a lower bound of A. The greatest lower bound of A is unique if it exists (see Exercise 28(b) at the end of this section). The greatest lower bound and least upper bound of a subset A are denoted by glb(A) and lub(A), respectively.

EXAMPLE 17 Find the greatest lower bound and the least upper bound of $\{b, d, g\}$, if they exist, in the poset shown in Figure 7.

Solution: The upper bounds of $\{b, d, g\}$ are g and h. Since $g \prec h$, g is the least upper bound. The lower bounds of $\{b, d, g\}$ are a and b. Since $a \prec b$, b is the greatest lower bound. ∎

EXAMPLE 18 Find the greatest lower bound and the least upper bound of the sets $\{3, 9, 12\}$ and $\{1, 2, 4, 5, 10\}$ if they exist, in the poset $(\mathbf{Z}+, |)$.

Solution: An integer is a lower bound of $\{3, 9, 12\}$ if 3, 9, and 12 are divisible by this integer. The only such integers are 1 and 3. Since $1 \mid 3$, 3 is the greatest lower bound of $\{3, 9, 12\}$. The only lower bound for the set $\{1, 2, 4, 5, 10\}$ with respect to $|$ is the element 1. Hence, 1 is the greatest lower bound for $\{1, 2, 4, 5, 10\}$.

An integer is an upper bound for $\{3, 9, 12\}$ if and only if it is divisible by 3, 9, and 12. The integers with this property are those divisible by the least common multiple of 3, 9, and 12, which is 36. Hence, 36 is the least upper bound of $\{3, 9, 12\}$. A positive integer is an upper bound for the set $\{1, 2, 4, 5, 10\}$ if and only if it is divisible by 1, 2, 4, 5, and 10. The integers with this property are those integers divisible by the least common multiple of these integers, which is 20. Hence, 20 is the least upper bound of $\{1, 2, 4, 5, 10\}$. ∎

In Chapter 3 we noted that $(\mathbf{Z}^+, \leq)$ is well-ordered, where $\leq$ is the usual "less than or equals" relation. We now define well-ordered sets.

DEFINITION 4. $(S, \preccurlyeq)$ is a *well-ordered set* if it is a poset such that $\preccurlyeq$ is a total ordering and such that every nonempty subset of S has a least element.

EXAMPLE 19 The set of ordered pairs of positive integers, $\mathbf{Z}^+ \times \mathbf{Z}^+$, with the lexicographic ordering, is a well-ordered set. The verification of this is left as an exercise at the end of this section. The set $\mathbf{Z}$, with the usual $\leq$ ordering, is not totally ordered since the set of negative integers, which is a subset of $\mathbf{Z}$, has no least element. ■

TOPOLOGICAL SORTING

Suppose that a project is made up of 20 different tasks. Some tasks can be completed only after others have been finished. How can an order be found for these tasks? To model this problem we set up a partial order on the set of tasks, so that $a \prec b$ if and only if a and b are tasks where b cannot be started until a has been completed. To produce a schedule for the project, we need to produce an order for all 20 tasks that is compatible with this partial order. We will show how this can be done.

We begin with a definition. A total ordering $\preccurlyeq$ is said to be **compatible** with the partial ordering R if $a \preccurlyeq b$ whenever $a\ R\ b$. Constructing a compatible total ordering from a partial ordering is called **topological sorting**. We will need to use the following lemma.

LEMMA 1 Every finite nonempty poset $(S, \preccurlyeq)$ has a minimal element.

Proof: Choose an element a_0 of S. If a_0 is not minimal, then there is an element a_1 with $a_1 \prec a_0$. If a_1 is not minimal, there is an element a_2 with $a_2 \prec a_1$. Continue this process, so that if a_n is not minimal, there is an element a_{n+1} with $a_{n+1} \prec a_n$. Since there are only a finite number of elements in the poset, this process must end with a minimal element a_n. ■

The topological sorting algorithm we will describe works for any finite nonempty poset. To define a total ordering on the poset $(A, \preccurlyeq)$, first choose a minimal element a_1; such an element exists by Lemma 1. Next, note that $(A - \{a_1\}, \preccurlyeq)$ is also a poset, as the reader should verify. If it is nonempty, choose a minimal element a_2 of this poset. Then remove a_2 as well, and if there are additional elements left, choose a minimal element a_3 in $A - \{a_1, a_2\}$. Continue this process by choosing a_{k+1} to be a minimal element in $A - \{a_1, a_2, \ldots, a_k\}$, as long as elements remain.

Since A is a finite set, this process must terminate. The end product is a sequence of elements $a_1, a_2, \ldots, a_n$. The desired total ordering is defined by

$$a_1 \preccurlyeq a_2 \preccurlyeq \cdots \preccurlyeq a_n.$$

This total ordering is compatible with the original partial ordering. To see this, note that if $b < c$ in the original partial ordering, c is chosen as the minimal element at a phase of the algorithm where b has already been removed, for otherwise c would not be a minimal element. Pseudocode for this topological sorting algorithm is shown in Algorithm 1.

ALGORITHM 1 Topological Sorting.

```
procedure topological sort (S: finite poset)
k := 1
while S ≠ ∅
begin
    a_k := a minimal element of S {such an element exists by Lemma 1}
    S := S − {a_k}
    k := k + 1
end {a_1, a_2, ..., a_n is a compatible total ordering of S}
```

EXAMPLE 20 Find a compatible total ordering for the poset $(\{1, 2, 4, 5, 12, 20\}, |)$.

Solution: The first step is to choose a minimal element. This must be 1, since it is the only minimal element. Next, select a minimal element of $(\{2, 4, 5, 12, 20\}, |)$. There are two minimal elements in this poset, namely, 2 and 5. We select 5. The remaining elements are $\{2, 4, 12, 20\}$. The only minimal element at this stage is 2. Next, 4 is chosen since it is the only minimal element of $(\{4, 12, 20\}, |)$. Since both 12 and 20 are minimal elements of $(\{12, 20\}, |)$, either can be chosen next. We select 20, which leaves 12 as the last element left. This produces the total ordering

$$1 < 5 < 2 < 4 < 20 < 12.$$

The steps used by this sorting algorithm are displayed in Figure 8. ∎

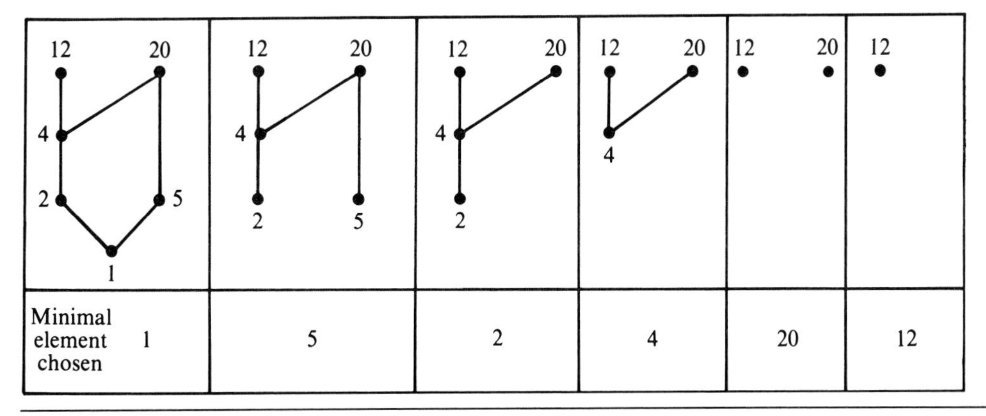

FIGURE 8 A Topological Sort of $(\{1, 2, 4, 5, 12, 20\}, |)$**.**

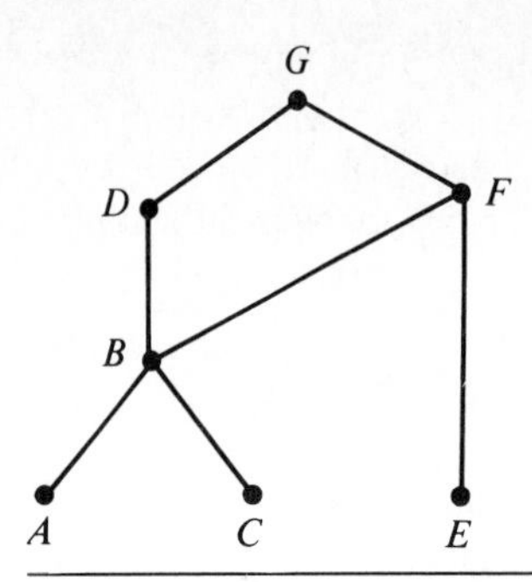

FIGURE 9 The Hasse Diagram for Seven Tasks.

Topological sorting has an application to the scheduling of projects.

EXAMPLE 21 A development project at a computer company requires the completion of seven tasks. Some of these tasks can be started only after other tasks are finished. A partial ordering on tasks is set up by considering task $X \prec$ task Y if task Y cannot be started until task X has been completed. The Hasse diagram for the seven tasks, with respect to this partial ordering, is shown in Figure 9. Find an order in which these tasks can be carried out to complete the project.

Solution: An ordering of the seven tasks can be obtained by performing a topological sort. The steps of a sort are illustrated in Figure 10. The result of this sort, $A \prec C \prec B \prec E \prec F \prec D \prec G$, gives one possible order for the tasks. ■

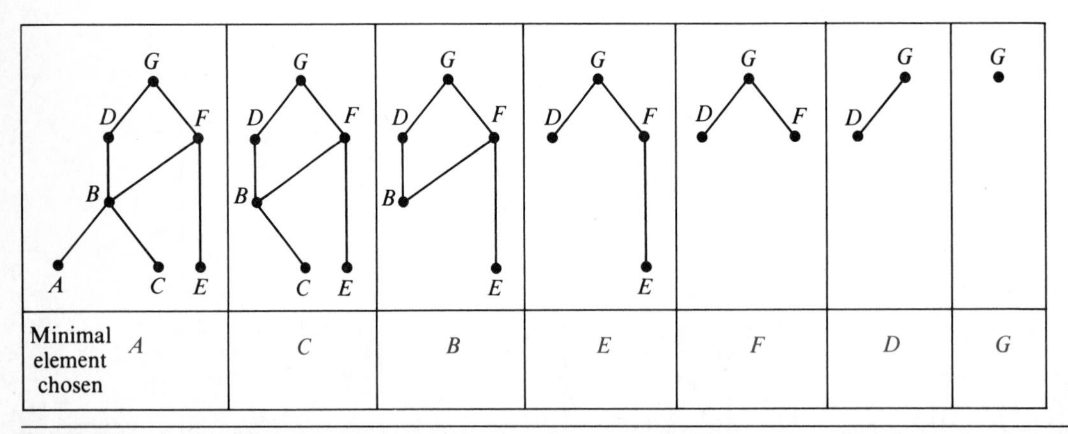

FIGURE 10 A Topological Sort of the Tasks.

Exercises

1. Which of the following are posets?
   a) $(\mathbf{Z}, =)$ **b)** $(\mathbf{Z}, \neq)$
   c) $(\mathbf{Z}, \geq)$ **d)** $(\mathbf{Z}, \nmid)$
2. Determine whether the relations represented by the following zero-one matrices are partial orders.

   a) $\begin{bmatrix} 1 & 0 & 1 \\ 1 & 1 & 0 \\ 0 & 0 & 1 \end{bmatrix}$ **b)** $\begin{bmatrix} 1 & 0 & 0 \\ 0 & 1 & 0 \\ 1 & 0 & 1 \end{bmatrix}$

   c) $\begin{bmatrix} 1 & 0 & 1 & 0 \\ 0 & 1 & 1 & 0 \\ 0 & 0 & 1 & 1 \\ 1 & 1 & 0 & 1 \end{bmatrix}$

In Exercises 3–5 determine whether the relation with the directed graph shown is a partial order.

3.
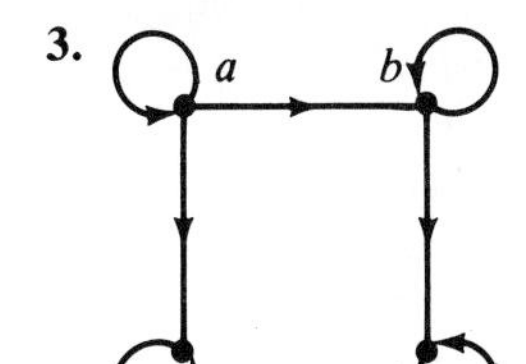

4.
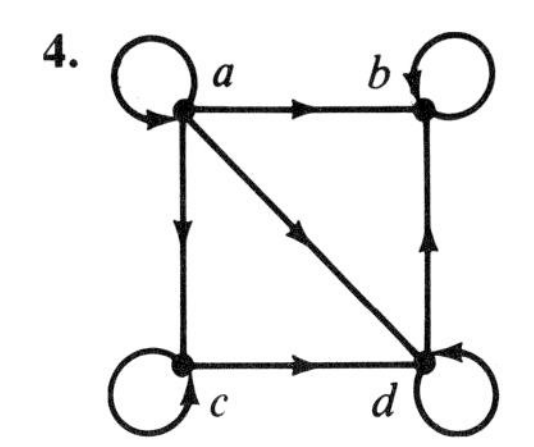

5.
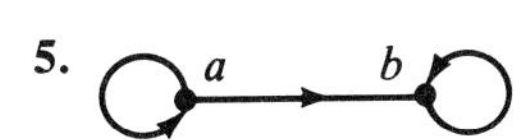

6. Let (S, R) be a poset. Show that (S, R^{-1}) is also a poset, where R^{-1} is the inverse of R. The poset (S, R^{-1}) is called the **dual** of (S, R).
7. Find the duals of the following posets.
   a) $(\{0, 1, 2\}, \leq)$ **b)** $(\mathbf{Z}, \geq)$
   c) $(P(\mathbf{Z}), \supseteq)$ **d)** $(\mathbf{Z}^+, |)$
8. Which of the following pairs of elements are comparable in the poset $(\mathbf{Z}^+, |)$?
   a) 5, 15 **b)** 6, 9
   c) 8, 16 **d)** 7, 7
9. Find two incomparable elements in the following posets.
   a) $(P(\{0, 1, 2\}), \subseteq)$ **b)** $(\{1, 2, 4, 6, 8\}, |)$
10. Let $S = \{1, 2, 3, 4\}$. With respect to the lexicographic order based on the usual "less than" relation,
    a) find all pairs in $S \times S$ less than (2, 3).
    b) find all pairs in $S \times S$ greater than (3, 1).
    c) draw the Hasse diagram of the poset $(S \times S, \preccurlyeq)$.
11. Find the lexicographic ordering of the following n-tuples.
    a) (1, 1, 2), (1, 2, 1)
    b) (0, 1, 2, 3), (0, 1, 3, 2)
    c) (1, 0, 1, 0, 1), (0, 1, 1, 1, 0)
12. Find the lexicographic ordering of the following strings of lowercase English letters:
    a) *quack, quick, quicksilver, quicksand, quacking*
    b) *open, opener, opera, operand, opened*
    c) *zoo, zero, zoom, zoology, zoological*
13. Find the lexicographic ordering of the bit strings 0, 01, 11, 001, 010, 011, 0001, and 0101 based on the ordering $0 < 1$.
14. Draw the Hasse diagram for the "greater than or equals" relation on $\{0, 1, 2, 3, 4, 5\}$.
15. Draw the Hasse diagram for divisibility on the set
    a) $\{1, 2, 3, 4, 5, 6, 7, 8\}$.
    b) $\{1, 2, 3, 5, 7, 11, 13\}$.
    c) $\{1, 2, 3, 6, 12, 24, 36, 48\}$.
    d) $\{1, 2, 4, 8, 16, 32, 64\}$.
16. Draw the Hasse diagram for inclusion on the set $P(S)$ where $S = \{a, b, c, d\}$.

In Exercises 17–19 list all ordered pairs in the partial ordering with the accompanying Hasse diagram.

17.
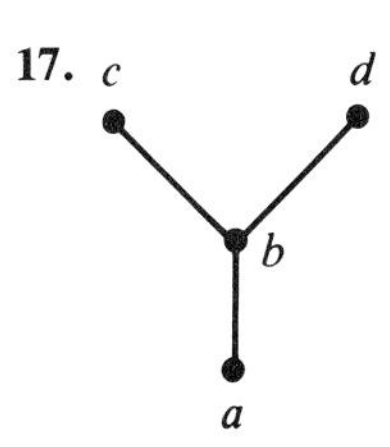

18.

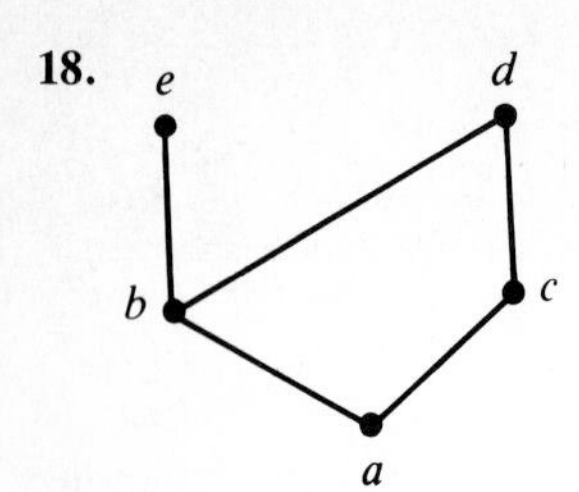

19.

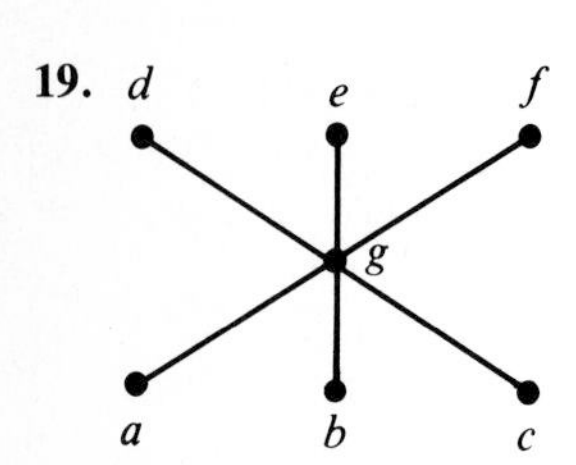

20. Answer the following questions for the partial order represented by the following Hasse diagram.

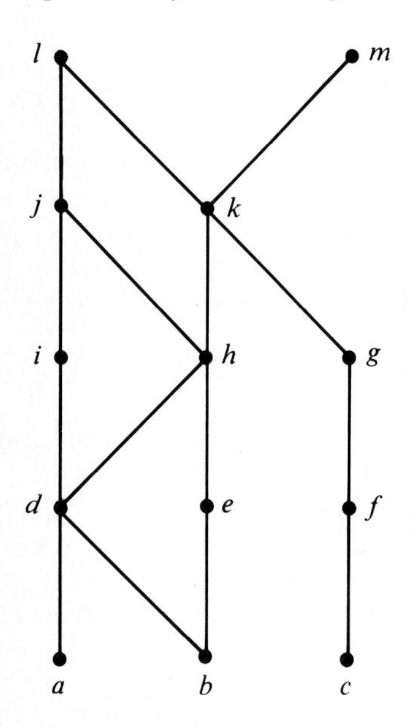

a) Find the maximal elements.
b) Find the minimal elements.
c) Is there a greatest element?
d) Is there a least element?
e) Find all upper bounds of $\{a, b, c\}$.
f) Find the least upper bound of $\{a, b, c\}$, if it exists.
g) Find all lower bounds of $\{f, g, h\}$.
h) Find the greatest lower bound of $\{f, g, h\}$, if it exists.

21. Answer the following questions concerning the poset $(\{3, 5, 9, 15, 24, 45\}, |)$.
a) Find the maximal elements.
b) Find the minimal elements.
c) Is there a greatest element?
d) Is there a least element?
e) Find all upper bounds of $\{3, 5\}$.
f) Find the least upper bound of $\{3, 5\}$, if it exists.
g) Find all lower bounds of $\{15, 45\}$.
h) Find the greatest lower bound of $\{15, 45\}$, if it exists.

22. Give a poset that has
a) a minimal element but no maximal element.
b) a maximal element but no minimal element.
c) neither a maximal nor a minimal element.

23. Show that lexicographic order is a partial ordering on the Cartesian product of two posets.

24. Show that lexicographic order is a partial ordering on the set of strings from a poset.

25. Suppose that $(S, \preccurlyeq_1)$ and $(T, \preccurlyeq_2)$ are posets. Show that $(S \times T, \preccurlyeq)$ is a poset where $(x, t) \preccurlyeq (u, v)$ if and only if $s \preccurlyeq_1 u$ and $t \preccurlyeq_2 v$.

26. **a)** Show that there is exactly one greatest element of a poset, if such an element exists.
b) Show that there is exactly one least element of a poset, if such an element exists.

27. **a)** Show that there is exactly one maximal element in a poset with a greatest element.
b) Show that there is exactly one minimal element in a poset with a least element.

28. **a)** Show that the least upper bound of a set in a poset is unique if it exists.
b) Show that the greatest lower bound of a set in a poset is unique if it exists.

29. Verify that $(\mathbf{Z}^+ \times \mathbf{Z}^+, \preccurlyeq)$ is a well-ordered set, where $\preccurlyeq$ is lexicographic order, as claimed in Example 19.

30. Show that a finite nonempty poset has a maximal element.

31. Find a compatible total order for the poset with Hasse diagram in Exercise 20.

32. Find a compatible total order for the divisibility relation on the set $\{1, 2, 3, 6, 8, 12, 24, 36\}$.

33. Find an order different from that constructed in Example 21 for completing the tasks in the development project.

34. Schedule the tasks needed to build a house, by specifying their order, if the Hasse diagram representing these tasks is as shown in the following figure.

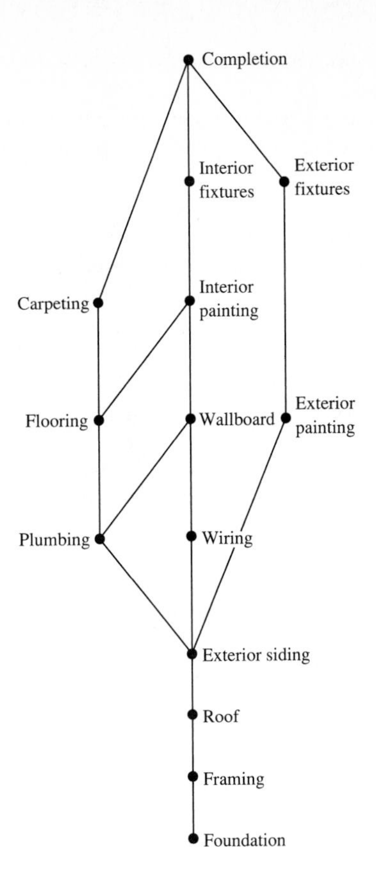

Key Terms and Results

TERMS

binary relation from A to B: a subset of $A \times B$
relation on A: a binary relation from A to itself (i.e., a subset of $A \times A$)
$S \circ R$: composite of R and S
R^{-1}: inverse relation of R
R^n: nth power of R
reflexive: a relation R on A is reflexive if $(a, a) \in R$ for all $a \in A$
symmetric: a relation R on A is symmetric if $(b, a) \in R$ whenever $(a, b) \in R$
antisymmetric: a relation R on A is antisymmetric if $a = b$ whenever $(a, b) \in R$ and $(b, a) \in R$
transitive: a relation R on A is transitive if $(a, b) \in R$ and $(b, c) \in R$ implies that $(a, c) \in R$
n-ary relation on $A_1, A_2, \ldots, A_n$: a subset of $A_1 \times A_2 \times \cdots \times A_n$
relational data model: a model for representing data bases using n-ary relations
primary key: a domain of an n-ary relation such that an n-tuple is uniquely determined by its value for this domain
composite key: the Cartesian product of domains of an n-ary relation such that an n-tuple is uniquely determined by its values in these domains

projection: a function that produces relations of smaller degree from an n-ary relation by deleting fields
join: a function that combines n-ary relations that agree on certain fields
directed graph or digraph: a set of elements called vertices and ordered pairs of these elements, called edges
loop: an edge of the form (a, a)
closure of a relation *R* with respect to a property P: the relation S (if it exists) that contains R, has property **P**, and is contained within any relation that contains R and has property **P**
path in a digraph: a sequence of edges $(a, x_1), (x_1, x_2), \ldots, (x_{n-2}, x_{n-1}), (x_{n-1}, b)$ such that the terminal vertex of each edge is the initial vertex of the succeeding edge in the sequence
circuit (or cycle) in a digraph: a path that begins and ends at the same vertex
R^* (connectivity relation): the relation consisting of those ordered pairs (a,b) such that there is a path from a to b
equivalence relation: a reflexive, symmetric, and transitive relation
equivalent: if R is an equivalence relation, a is equivalent to b if $a \, R \, b$
$[a]_R$ (equivalence class of a with respect to R): the set of all elements of A that are equivalent to a
$[a]_m$ (congruence class modulo *m*:) the set of integers congruent to a modulo m
partition of a set *S*: a collection of pairwise disjoint nonempty subsets that have S as their union
partial ordering: a relation that is reflexive, antisymmetric, and transitive
poset (S, R): a set S and a partial ordering R on this set
comparable: the elements a and b in the poset $(A, \preccurlyeq)$ are comparable if $a \preccurlyeq b$ or $b \preccurlyeq a$
incomparable: elements in a poset that are not comparable
total (or linear) ordering: a partial ordering for which every pair of elements are comparable
totally (or linearly) ordered set: a poset with a total (or linear) ordering
lexicographic order: a partial ordering of Cartesian products or strings (see page 388)
Hasse diagram: a graphical representation of a poset where loops and all edges resulting from the transitive property are not shown, and the direction of the edges is indicated by the position of the vertices
maximal element: an element of a poset that is not less than any other element of the poset
minimal element: an element of a poset that is not greater than any other element of the poset
least element: an element of a poset less than or equal to all other elements in this set
greatest element: an element of a poset greater than or equal to all other elements in this set
upper bound of a set: an element in a poset greater than all other elements in the set
lower bound of a set: an element in a poset less than all other elements in the set
least upper bound of a set: an upper bound of the set that is less than all other upper bounds
greatest lower bound of a set: a lower bound of the set that is greater than all other lower bounds
well-ordered set: a poset $(S, \preccurlyeq)$ where $\preccurlyeq$ is a total order and every non-empty subset of S has a least element
compatible total ordering for a partial ordering: a total ordering that contains the given partial ordering
topological sort: the construction of a total ordering compatible with a given partial ordering

RESULTS

The reflexive closure of a relation R on the set A equals $R \cup \Delta$, where $\Delta = (\{a, a) \mid a \in A\}$.
The symmetric closure of a relation R on the set A equals $R \cup R^{-1}$, where $R^{-1} = \{(b, a) \mid (a, b) \in R\}$.

The transitive closure of a relation equals the connectivity relation formed from this relation.

Warshall's algorithm for finding the transitive closure of a relation (see pages 373–376).

Let R be an equivalence relation. Then the following three statements are equivalent: (1) $a\ R\ b$; (2) $[a]_R \cap [b]_R \neq \emptyset$; (3) $[a]_R = [b]_R$.

The equivalence classes of an equivalence relation on a set A form a partition of A. Conversely, an equivalence relation can be constructed from any partition so that the equivalence classes are the subsets in the partition.

The topological sorting algorithm (see page 397-398).

Review Questions

1. a) What is a relation on a set?
b) How many relations are there on a set with n elements?

2. a) What is a reflexive relation?
b) What is a symmetric relation?
c) What is an antisymmetric relation?
d) What is a transitive relation?

3. Give an example of a relation on the set $\{1, 2, 3, 4\}$ that is
a) reflexive, symmetric, and not transitive.
b) not reflexive, symmetric, and transitive.
c) reflexive, antisymmetric, and not transitive.
d) reflexive, symmetric, and transitive.
e) reflexive, antisymmetric, and transitive.

4. a) How many reflexive relations are there on a set with n elements?
b) How many symmetric relations are there on a set with n elements?
c) How many antisymmetric relations are there on a set with n elements?

5. a) Explain how an n-ary relation can be used to represent information about students at a university.
b) How can the 5-ary relation containing names of students, their addresses, telephone numbers, majors, and grade point averages be used to form a 3-ary relation containing the names of students, their majors, and their grade point averages?
c) How can the 4-ary relation containing names of students, their addresses, telephone numbers, and majors and the 4-ary relation containing names of students, their student numbers, majors, and numbers of credit hours, be combined into a single n-ary relation?

6. a) Explain how to use a zero-one matrix to represent a relation on a finite set.
b) Explain how to use the zero-one matrix representing a relation to determine whether the relation is reflexive, symmetric, and/or antisymmetric.

7. a) Explain how to use a directed graph to represent a relation on a finite set.
b) Explain how to use the directed graph representing a relation to determine whether a relation is reflexive, symmetric, and/or antisymmetric.

8. a) Define the reflexive closure and the symmetric closure of a relation.
b) How can you construct the reflexive closure of a relation?
c) How can you construct the symmetric closure of a relation?
d) Find the reflexive closure and the symmetric closure of the relation $\{(1, 2), (2, 3), (2, 4), (3, 1)\}$ on the set $\{1, 2, 3, 4\}$.

9. a) Define the transitive closure of a relation.
b) Can the transitive closure of a relation be obtained by including all pairs (a, c) such that (a, b) and (b, c) belong to the relation?
c) Describe two algorithms for finding the transitive closure of a relation.
d) Find the transitive closure of the relation $\{(1, 1), (1,3), (2,1), (2,3), (2,4), (3,2), (3,4), (4, 1)\}$.

10. a) Define an equivalence relation.
b) Which relations on the set $\{a, b, c, d\}$ are equivalence relations and contain (a,b) and (b,d)?

11. a) Show that congruence modulo m is an equivalence relation whenever m is a positive integer.
b) Show that the relation $\{(a, b) \mid a \equiv \pm b \pmod 7\}$ is an equivalence relation on the set of integers.

12. a) What are the equivalence classes of an equivalence relation?
b) What are the equivalence classes of the congruent modulo 5 relation?
c) What are the equivalence classes of the equivalence relation in Question 11(b)?

13. Explain the relationship between equivalence relations on a set and partitions of this set.

14. **a)** Define a partial ordering.
b) Show that the divisibility relation on the set of positive integers is a partial order.

15. Explain how partial orderings on the sets A_1 and A_2 can be used to define a partial ordering on the set $A_1 \times A_2$.

16. **a)** Explain how to construct the Hasse diagram of a partial order on a finite set.
b) Draw the Hasse diagram of the divisibility relation on the set $\{2, 3, 5, 9, 12, 15, 18\}$.

17. **a)** Define a maximal element of a poset and the greatest element of a poset.
b) Give an example of a poset that has three maximal elements.
c) Give an example of a poset with a greatest element.

18. **a)** Define a well-ordered set.
b) Describe an algorithm for producing a well-ordered set from a partially ordered set.
c) Explain how the algorithm from (b) can be used to order the tasks in a project if each task can be done only after one or more of the other tasks have been completed.

Supplementary Exercises

1. Let S be the set of all strings of English letters. Determine whether the following relations are reflexive, irreflexive, symmetric, antisymmetric, and/or transitive.
a) $R_1 = \{(a, b) \mid a \text{ and } b \text{ have no letters in common}\}$
b) $R_2 = \{(a, b) \mid a \text{ and } b \text{ are not the same length}\}$
c) $R_3 = \{(a, b) \mid a \text{ is longer than } b\}$

2. Construct a relation on the set $\{a, b, c, d\}$ that is
a) reflexive, symmetric, but not transitive.
b) irreflexive, symmetric, and transitive.
c) irreflexive, antisymmetric, and not transitive.
d) reflexive, neither symmetric nor antisymmetric, transitive.
e) neither reflexive, irreflexive, symmetric, antisymmetric, nor transitive.

3. Show that the relation R on $\mathbf{Z} \times \mathbf{Z}$ defined by $(a, b)\, R\, (c, d)$ if and only if $a+d = b+c$ is an equivalence relation.

4. Show that a subset of an antisymmetric relation is also antisymmetric.

5. Let R be a reflexive relation on a set A. Show that $R \subseteq R^2$.

6. Suppose that R_1 and R_2 are reflexive relations on a set A. Show that $R_1 \oplus R_2$ is irreflexive.

7. Suppose that R_1 and R_2 are reflexive relations on a set A. Is $R_1 \cap R_2$ also reflexive? Is $R_1 \cup R_2$ also reflexive?

8. Suppose that R is a symmetric relation on a set A. Is $\overline{R}$ also symmetric?

9. Let R_1 and R_2 be symmetric relations. Is $R_1 \cap R_2$ also symmetric? Is $R_1 \cup R_2$ also symmetric?

10. A relation R is called **circular** if $a\, R\, b$ and $b\, R\, c$ imply that $c\, R\, a$. Show that R is reflexive and circular if and only if it is an equivalence relation.

11. Show that a primary key in an n-ary relation is a primary key in any projection of this relation that contains this key as one of its fields.

12. Is the primary key in an n-ary relation also a primary key in a larger relation obtained by taking the join of this relation with a second relation?

13. Show that the reflexive closure of the symmetric closure of a relation is the same as the symmetric closure of its reflexive closure.

14. Let R be the relation on the set of all mathematicians that contains the ordered pair (a, b) if and only if a and b have written a paper together.
a) Describe the relation R^2. **b)** Describe the relation R^*.
c) The **Erdös number** of a mathematician is 1 if this mathematician has written a paper

with the prolific Hungarian mathematician Paul Erdös, it is 2 if this mathematician has not written a joint paper with Erdös but has written a joint paper with someone who has written a joint paper with Erdös, and so on (except that the Erdös number of Erdös himself is 0). Give a definition of the Erdös number in terms of paths in R.

★15. a) Give an example to show that the transitive closure of the symmetric closure of a relation is not necessarily the same as the symmetric closure of the transitive closure of this relation.

b) Show, however, that the transitive closure of the symmetric closure of a relation must contain the symmetric closure of the transitive closure of this relation.

16. a) Let S be the set of subroutines of a computer program. Define the relation R by **P** R **Q** if subroutine **P** calls subroutine **Q** during its execution. Describe the transitive closure of R.

b) For which subroutines **P** does (**P**, **P**) belong to the transitive closure of R?

c) Describe the reflexive closure of the transitive closure of R.

17. Suppose that R and S are relations on a set A with $R \subseteq S$ such that the closures of R and S with respect to a property **P** both exist. Show that the closure of R with respect to **P** is a subset of the closure of S with respect to **P**.

18. Show that the symmetric closure of the union of two relations is the union of their symmetric closures.

★19. Devise an algorithm, based on the concept of interior vertices, that finds the length of the longest path between two vertices in a directed graph, or determines that there are arbitrarily long paths between these vertices.

20. Which of the following are equivalence relations on the set of all people?

Paul Erdös (born 1913). Paul Erdös, born in Budapest, Hungary, was the son of two high school mathematics teachers. He was a child prodigy; at 3 he could multiply three-digit numbers in his head, and at 4 he discovered negative numbers on his own. Because his mother did not want to expose him to contagious diseases, he did not attend school until high school. At 17 Erdös entered the University of Budapest, graduating four years later with a Ph.D. in mathematics. After graduating he spent four years at Manchester, England, on a postdoctoral fellowship. In 1938, because of the difficult political situation in Hungary, especially for Jews, he went to the United States. In the past half-century he has spent much of his time in the United States of America, except for 1954 to 1962, when he was banned as part of the paranoia of the McCarthy era. He has also spent considerable time in Israel.

Erdös has made many significant contributions to combinatorics and to number theory. One of the discoveries of which he is most proud is his elementary proof (in the sense that it does not use any complex analysis) of the prime number theorem, which provides an estimate for the number of primes not exceeding a fixed positive integer. He also participated in the modern development of Ramsey theory.

Erdös travels extensively throughout the world to work with other mathematicians, visiting conferences, universities, and research laboratories. He almost entirely devotes himself to mathematics, traveling from one mathematician to the next, proclaiming "My brain is open." Erdös is the author or coauthor of more than 1000 papers and has had more than 250 coauthors. Since he has no permanent home, copies of these articles are kept by Ron Graham, a famous discrete mathematician at AT&T Bell Laboratories, with whom he has collaborated extensively and who takes care of many of his worldly needs.

Erdös offers rewards, ranging from \$10 to \$10,000, for the solution of problems that he finds particularly interesting, with the size of the reward depending on the difficulty of the problem. So far, he has paid out close to \$4000. Erdös has his own special language, using such terms as "epsilon" (child), "boss" (woman), "slave" (man), "captured" (married), "liberated" (divorced), "Supreme Fascist" (God), "Sam" (United States), and "Joe" (Soviet Union). Although he is curious about many things, he concentrates almost all his energy on mathematical research. He has no hobbies and no full-time job. He has never married and has apparently remained celibate. Erdös is extremely generous, donating much of the money he has collected from prizes, awards, and stipends for scholarships and to worthwhile causes. He travels extremely lightly and does not like having many material possessions.

a) $\{(x, y) \mid x$ and y have the same sign of the zodiac$\}$
b) $\{(x, y) \mid x$ and y were born in the same year$\}$
c) $\{(x, y) \mid x$ and y have been in the same city$\}$

★**21.** How many different equivalence relations are there on a set with five elements?

22. Show that $\{(x, y) \mid x - y \in \mathbf{Q}\}$ is an equivalence relation on the set of real numbers where $\mathbf{Q}$ denotes the set of rational numbers. What are $[1]$, $[\frac{1}{2}]$, and $[\pi]$?

23. Suppose that $P_1 = \{A_1, A_2, \ldots, A_m\}$ and $P_2 = \{B_1, B_2, \ldots, B_n\}$ are both partitions of the set S. Show that the collection of nonempty subsets of the form $A_i \cap B_j$ is a partition of S that is a refinement of both P_1 and P_2 (see the preamble to Exercise 27 of Section 6.5).

★**24.** Show that the transitive closure of the symmetric closure of the reflexive closure of a relation R is the smallest equivalence relation that contains R.

25. Let $\mathbf{R}(S)$ be the set of all relations on a set S. Define the relation $\preccurlyeq$ on $\mathbf{R}(S)$ by $R_1 \preccurlyeq R_2$ if $R_1 \subseteq R_2$, where R_1 and R_2 are relations on S. Show that $(\mathbf{R}(S), \preccurlyeq)$ is a poset.

26. Let $\mathbf{P}(S)$ be the set of all partitions of the set S. Define the relation $\preccurlyeq$ on $\mathbf{P}(S)$ by $P_1 \preccurlyeq P_2$ if P_1 is a refinement of P_2 (see Exercise 27 of Section 6.5). Show that $(\mathbf{P}(S), \preccurlyeq)$ is a poset.

27. Find an ordering of the tasks of a software project if the Hasse diagram for the tasks of this project is as follows.

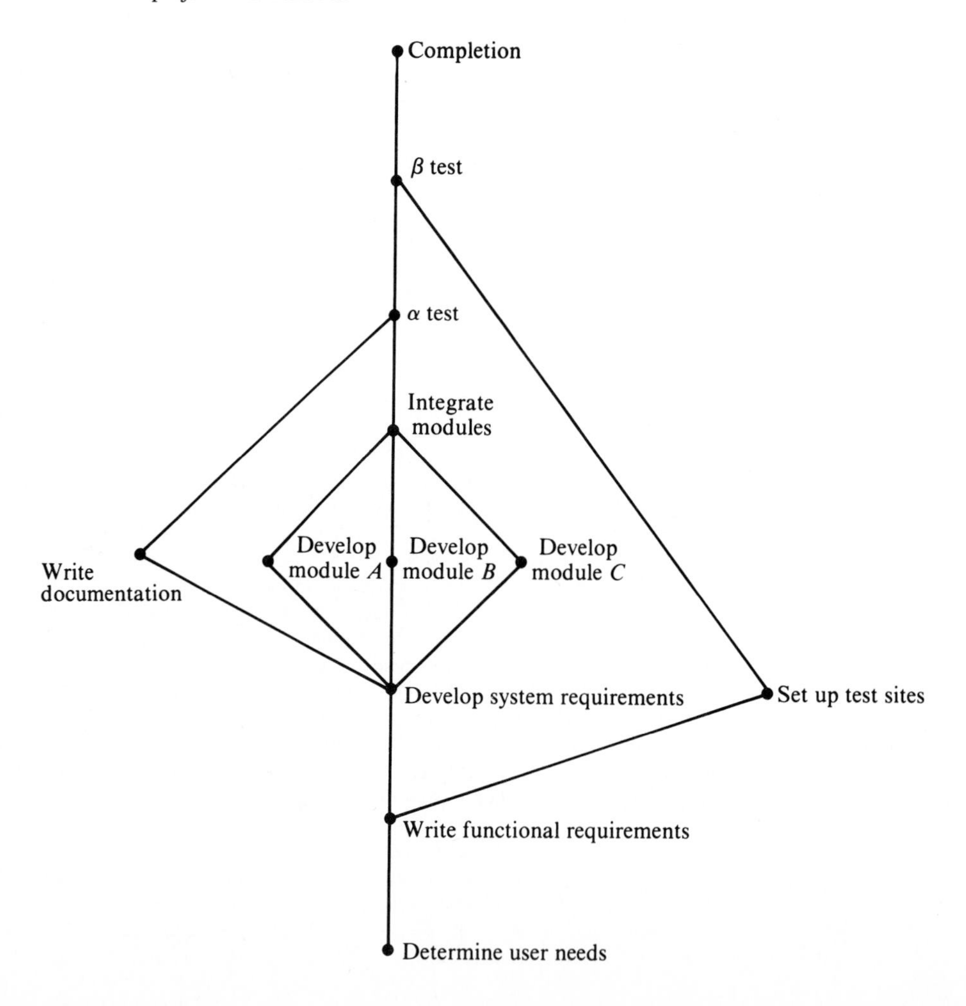

A subset of a poset such that every two elements of this subset are comparable is called a **chain**. A subset of a poset is called an **antichain** if every two elements of this subset are incomparable.

28. Find all chains in the posets with the Hasse diagrams shown in Exercises 17–19 in Section 6.6.

29. Find all antichains in the posets with the Hasse diagrams shown in Exercises 17–19 in Section 6.6.

30. Find an antichain with the greatest number of elements in the poset with the Hasse diagram of Exercise 20 in Section 6.6.

31. Show that every maximal chain in a finite poset $(S, \preccurlyeq)$ contains a minimal element of S. (A maximal chain is a chain that is not a subset of a larger chain.)

★★32. Show that a poset can be partitioned into k chains, where k is the largest number of elements in an antichain in this poset.

★33. Show that in any group of $mn + 1$ people there is either a list of $m + 1$ people where a person in the list (except for the first person listed) is a descendant of the previous person on the list, or there are $n + 1$ people such that none of these people is a descendant of any of the other n people. (*Hint:* Use Exercise 32.)

★34. Establish the **generalized induction principle**: $P(x)$ is true for every element x in a well-ordered set S if $P(x_0)$ is true, where x_0 is the least element of S (the basis case), and if $P(x)$ is true for all $x < y$, then $P(y)$ is true (the inductive step).

35. Use the generalized induction principle on the well-ordered set $(\mathbf{Z}^+ \cup \{0\} \times \mathbf{Z}^+ \cup \{0\})$ (with lexicographic ordering) to show that $a_{m,n} = [n(n+1)/2] + m$ where $a_{0,0} = 0$ and

$$a_{m,n} = \begin{cases} a_{m-1,n} + 1 & \text{if } n = 0 \\ a_{m,n-1} + n & \text{if } n \neq 0. \end{cases}$$

Computer Projects

WRITE PROGRAMS WITH THE FOLLOWING INPUT AND OUTPUT.

1. Given the matrix representing a relation on a finite set, determine whether the relation is reflexive and/or irreflexive.
2. Given the matrix representing a relation on a finite set, determine whether the relation is symmetric and/or antisymmetric.
3. Given the matrix representing a relation on a finite set, determine whether the relation is transitive.
4. Given an n-ary relation, find the projection of this relation when specified fields are deleted.
5. Give an m-ary relation and an n-ary relation, and a set of common fields, find the join of these relations with respect to these common fields.
6. Given the matrix representing a relation on a finite set, find the matrix representing the reflexive closure of this relation.
7. Given the matrix representing a relation on a finite set, find the matrix representing the symmetric closure of this relation.
8. Given the matrix representing a relation on a finite set, find the matrix representing the transitive closure of this relation by computing the join of the powers of the matrix representing the relation.
9. Given the matrix representing a relation on a finite set, find the matrix representing the transitive closure of this relation using Warshall's algorithm.
10. Given the matrix representing a relation on a finite set, find the matrix representing the smallest equivalence relation containing this relation.
11. Given a partial ordering on a finite set, find a total ordering compatible with it using topological sorting.

7

Graphs

Graph theory is an old subject with many modern applications. Its basic ideas were introduced in the 18th century by the great Swiss mathematician Leonhard Euler. He used graphs to solve the famous Königsberg bridge problem, which we will discuss in this chapter.

Graphs are used to solve problems in many fields. For instance, graphs can be used to determine whether a circuit can be implemented on a planar circuit board. We can distinguish between two chemical compounds with the same molecular formula but different structures using graphs. We can determine whether two computers are connected by a communications link using graph models of computer networks. Graphs with weights assigned to their edges can be used to solve problems such as finding the shortest path between two cities in a transportation network. We can also use graphs to schedule exams and assign channels to television stations.

7.1 Introduction to Graphs

Graphs are discrete structures consisting of vertices and edges that connect these vertices. There are several different types of graphs that differ with respect to the kind and number of edges that can connect a pair of vertices. Problems in almost every conceivable discipline can be solved using graph models. We will give examples to show how graphs are used as models in a variety of areas. For instance, we will show how graphs are used to represent the competition of different species in an ecological niche, how graphs are used to represent who influences whom in an organization, and how graphs are used to represent the outcome of tournaments. Later we will show how graphs can be used to solve many types of problems, such as computing the number of different combinations of flights between two cities in an airline network, determining whether it is possible to walk down all the streets in a city without going down a street twice, and finding the number of colors needed to color the regions of a map.

TYPES OF GRAPHS

We will introduce the different types of graphs by showing how each can be used to model a computer network. Suppose that a network is made up of computers and telephone lines between computers. We can represent the location of each computer by a point and each telephone line by an arc, as shown in Figure 1.

We make the following observations about the network in Figure 1. There is at most one telephone line between two computers in this network, each line operates in both directions, and no computer has a telephone line to itself. Consequently this network can be modeled using a **simple graph**, consisting of vertices which represent the computers, and undirected edges, which represent telephone lines, where each edge connects two distinct vertices and no two edges connect the same pair of vertices.

DEFINITION 1. A *simple graph* $G = (V, E)$ consists of V, a nonempty set of *vertices*, and E, a set of unordered pairs of distinct elements of V called *edges*.

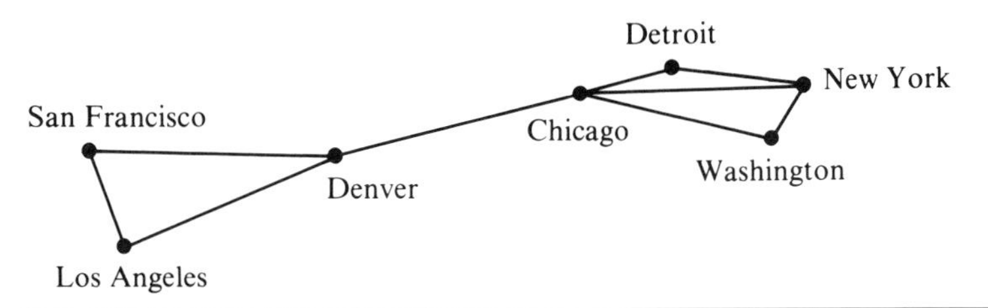

FIGURE 1 **A Computer Network.**

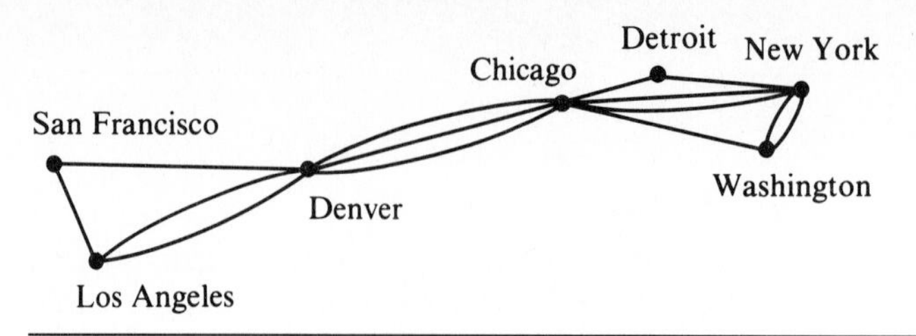

FIGURE 2 **A Computer Network with Multiple Lines.**

Sometimes there are multiple telephone lines between computers in a network. This is the case when there is heavy traffic between computers. A network with multiple lines is displayed in Figure 2. Simple graphs are not sufficient to model such networks. Instead, **multigraphs** are used, which consist of vertices and undirected edges between these vertices, with multiple edges between pairs of vertices allowed. Every simple graph is also a multigraph. However, not all multigraphs are simple graphs, since in a multigraph two or more edges may connect the same pair of vertices.

We cannot use a pair of vertices to specify an edge of a graph when multiple edges are present. This makes the formal definition of multigraphs somewhat complicated.

DEFINITION 2. A *multigraph* $G = (V, E)$ consists of a set V of vertices, a set E of edges, and a function f from E to $\{\{u, v\} \mid u, v \in V, u \neq v\}$. The edges e_1 and e_2 are called *multiple* or *parallel edges* if $f(e_1) = f(e_2)$.

A computer network may contain a telephone line from a computer to itself (perhaps for diagnostic purposes). Such a network is shown in Figure 3. We cannot use multigraphs to model such networks, since **loops**, which are edges from a vertex to itself, are not allowed in multigraphs. Instead, **pseudographs** are used. Pseudographs are more general than multigraphs, since an edge in a pseudograph may connect a vertex with itself.

To formally define pseudograph we must be able to associate edges to sets containing just one vertex.

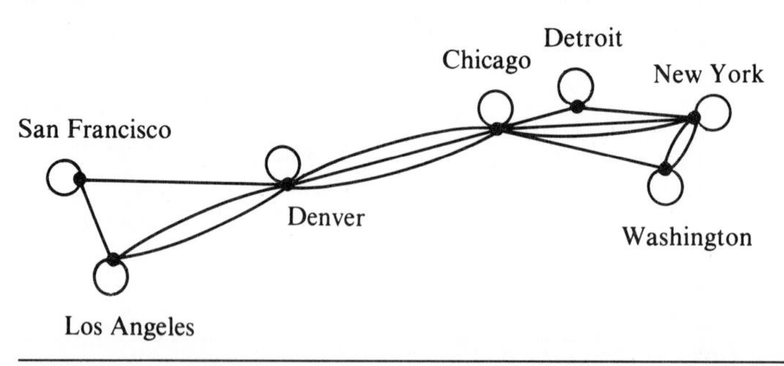

FIGURE 3 **A Computer Network with Diagnostic Lines.**

DEFINITION 3. A *pseudograph* $G = (V, E)$ consists of a set V vertices, a set E of edges, and a function f from E to $\{\{u, v\} \mid u, v \in V\}$. An edge is a *loop* if $f(e) = \{u\}$ for some $u \in V$.

The reader should note that multiple edges in a pseudograph are associated to the same pair of vertices. However, we will say that $\{u, v\}$ is an edge of a graph $G = (V, E)$ if there is at least one edge e with $f(e) = \{u, v\}$. We will not distinguish between the edge e and the set $\{u, v\}$ associated to it unless the identity of individual multiple edges is important.

To summarize, pseudographs are the most general type of undirected graphs since they may contain loops and multiple edges. Multigraphs are undirected graphs that may contain multiple edges but may not have loops. Finally, simple graphs are undirected graphs with no multiple edges or loops.

The telephone lines in a computer network may not operate in both directions. For instance, in Figure 4 the host computer in New York can only receive data from other computers and cannot send out data. The other telephone lines operate in both directions and are represented by pairs of edges in opposite directions.

We use directed graphs (which were studied in Chapter 6) to model such networks. The edges of a directed graph are ordered pairs. Loops, ordered pairs of the same element, are allowed, but multiple edges in the same direction between two vertices are not. Recall the following definition.

DEFINITION 4. A *directed graph* (V, E) consists of a set of vertices V and a set of edges E that are ordered pairs of elements of V.

Finally, multiple lines may be present in the computer network, so that there may be several one-way lines to the host in New York from each location, and perhaps more than one line back to each remote computer from the host. Such a network is shown in Figure 5. Directed graphs are not sufficient for modeling such a network, since multiple edges are not allowed in these graphs. Instead, **directed multigraphs**, which may have multiple directed edges from a vertex to a second (possibly the same) vertex, are needed. The formal definition of a directed multigraph follows.

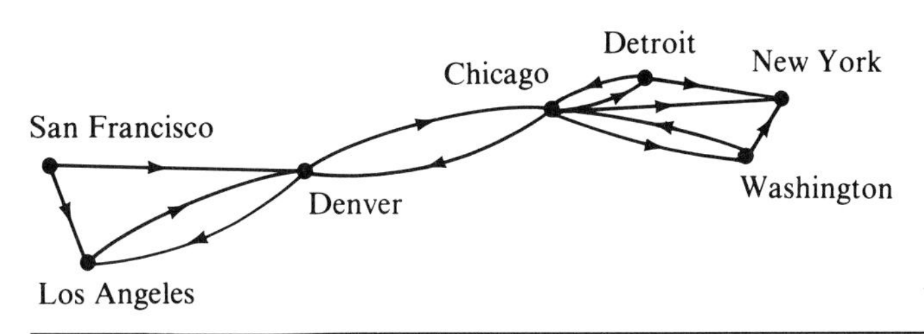

FIGURE 4 **A Communications Network with One-Way Telephone Lines.**

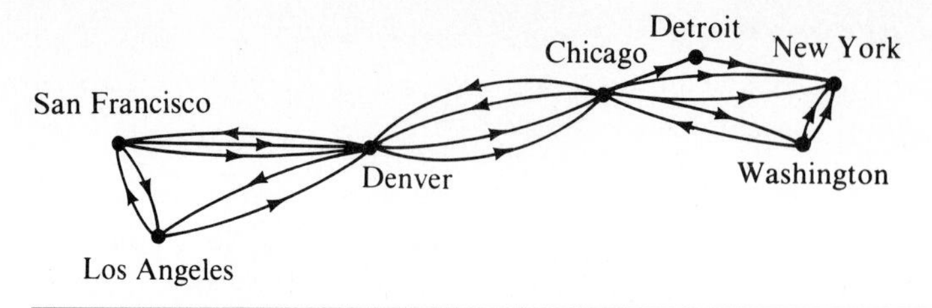

FIGURE 5 **A Computer Network with Multiple One-Way Lines.**

DEFINITION 5. A directed multigraph $G = (V, E)$ consists of a set V of vertices, a set E of edges, and a function f from E to $\{(u, v) \mid u, v \in V\}$. The edges e_1 and e_2 are *multiple edges* if $f(e_1) = f(e_2)$.

The reader should note that multiple directed edges are associated to the same pair of vertices. However, we will say that (u, v) is an edge of $G = (V, E)$ as long as there is at least one edge e with $f(e) = (u, v)$. We will not make the distinction between the edge e and the ordered pair (u, v) associated to it unless the identity of individual multiple edges is important.

This terminology for the various types of graphs makes clear whether the edges of a graph are associated to ordered or unordered pairs, whether multiple edges are allowed, and whether loops are allowed. We will use **graph** to describe graphs with directed or undirected edges, with or without loops and multiple edges. We will use the terms **undirected graph** or **pseudograph** for an undirected graph that may have multiple edges and loops. We will always use the adjective **directed** when referring to graphs that have ordered pairs associated to their edges. The definitions of the various types of graphs are summarized in Table 1. Because of the relatively modern interest in graph theory, and because it has applications to a wide variety of disciplines, many different terminologies of graph theory are commonly used. The reader should determine how such terms are being used whenever they are encountered. Perhaps this terminology will become standardized someday.

TABLE 1 **Graph Terminology**

Type	*Edges*	*Multiple Edges Allowed?*	*Loops Allowed?*
Simple graph	Undirected	No	No
Multigraph	Undirected	Yes	No
Pseudograph	Undirected	Yes	Yes
Directed graph	Directed	No	Yes
Directed multigraph	Directed	Yes	Yes

GRAPH MODELS

Graphs are used in a wide variety of models. We will present a few graph models from diverse fields here. Others will be introduced in subsequent sections of this and the following chapters.

EXAMPLE 1 Niche Overlap Graphs in Ecology Graphs are used in many models involving the interaction of different species of animals. For instance, the competition between species in an ecosystem can be modeled using a **niche overlap graph**. Each species is represented by a vertex. An undirected edge connects two vertices if the two species represented by these vertices compete (that is, some of the food resources they use are the same). The graph in Figure 6 models the ecosystem of a forest. We see from this graph that squirrels and raccoons compete but that crows and shrews do not. ■

EXAMPLE 2 Influence Graphs In studies of group behavior it is observed that certain people can influence the thinking of others. A directed graph called an **influence graph** can be used to model this behavior. Each person of the group is represented by a vertex. There is a directed edge from vertex a to vertex b when the person represented by vertex a influences the person represented by vertex b. An example of an influence graph for members of a group is shown in Figure 7. In the group modeled by this influence graph, Deborah can influence Brian, Fred, and Linda, but no one can influence her. Also, Yvonne and Brian can influence each other. ■

EXAMPLE 3 Round-Robin Tournaments A tournament where each team plays each other team exactly once is called a **round-robin tournament**. Such tournaments can be modeled

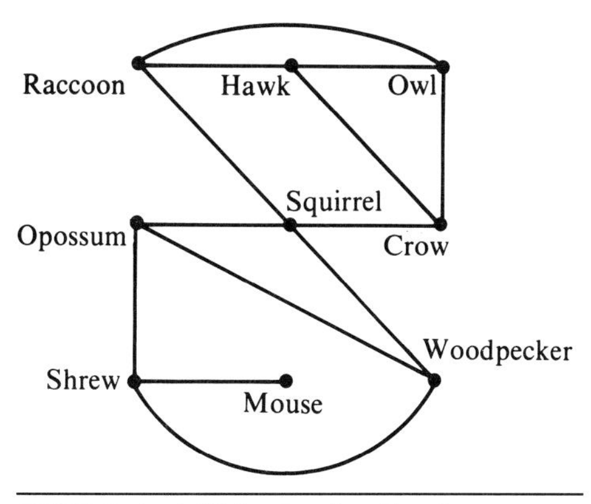

FIGURE 6 **A Niche Overlap Graph.**

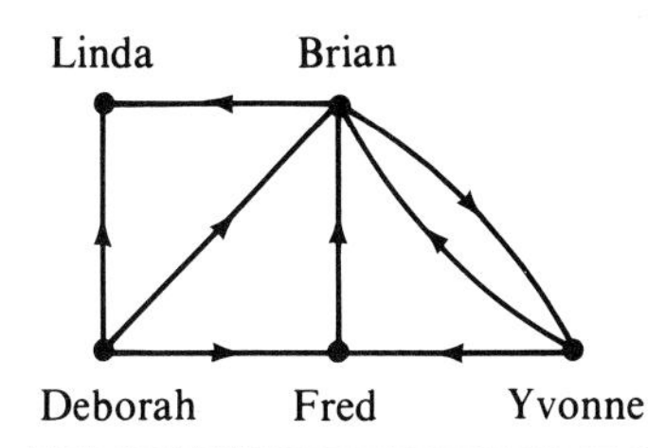

FIGURE 7 **An Influence Graph.**

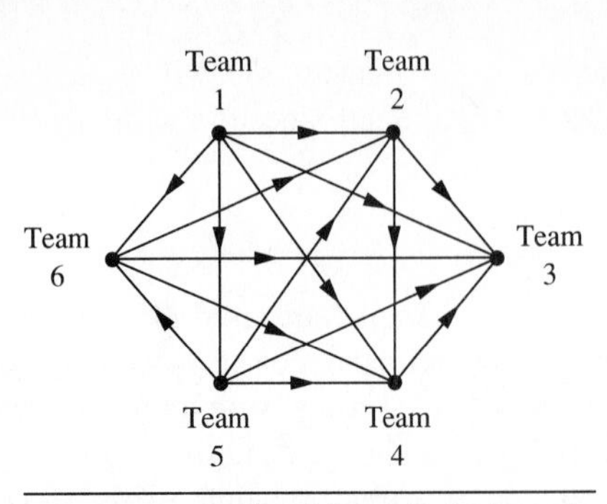

FIGURE 8 **A Graph Model of a Round-Robin Tournament.**

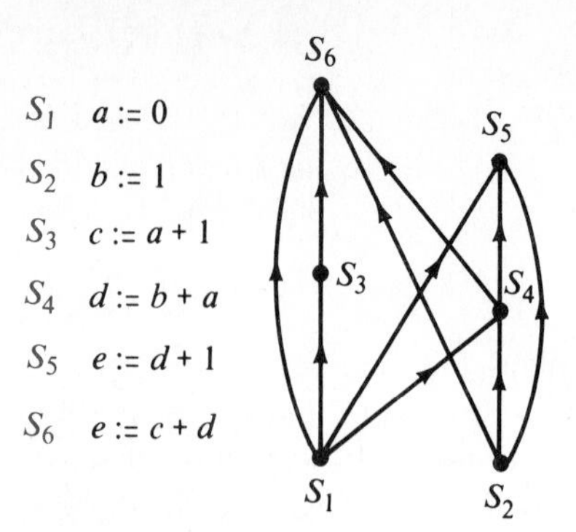

FIGURE 9 **A Precedence Graph.**

using directed graphs where each team is represented by a vertex. Note that (a, b) is an edge if team a beats team b. Such a directed graph model is presented in Figure 8. Note that Team 1 is undefeated in this tournament, and Team 3 is winless. ■

EXAMPLE 4 Precedence Graphs and Concurrent Processing Computer programs can be executed more rapidly by executing certain statements concurrently. It is important not to execute a statement that requires results of statements not yet executed. The dependence of statements on previous statements can be represented by a directed graph. Each statement is represented by a vertex, and there is an edge from one vertex to a second vertex if the statement represented by the second vertex cannot be executed before the statement represented by the first vertex has been executed. This graph is called a **precedence graph**. A computer program and its graph are displayed in Figure 9. For instance, the graph shows that statement S_5 cannot be executed before statements S_1, S_2, and S_4 are executed. ■

Exercises

1. Draw graph models, stating the type of graph used, to represent airline routes where every day there are four flights from Boston to Newark, two flights from Newark to Boston, three flights from Newark to Miami, two flights from Miami to Newark, one flight from Newark to Detroit, two flights from Detroit to Newark, three flights from Newark to Washington, two flights from Washington to Newark, and one flight from Washington to Miami, with

a) an edge between vertices representing cities that have a flight between them (in either direction).

b) an edge between vertices representing cities for each flight that operates between them (in either direction).

c) an edge between vertices representing cities for each flight that operates between them (in either direction), plus a loop for a special sightseeing trip that takes off and lands in Miami.

d) an edge from a vertex representing a city where a flight starts to the vertex representing the city where it ends.

e) an edge for each flight from a vertex representing a city where the flight begins to the vertex representing the city where the flight ends.

2. What kind of graph can be used to model a highway system between major cities where
   a) there is an edge between the vertices representing cities if there is an interstate highway between them?
   b) there is an edge between the vertices representing cities for each interstate highway between them?
   c) there is an edge between the vertices representing cities for each interstate highway between them, and there is a loop at the vertex representing a city if there is an interstate highway that circles this city?

For Exercises 3–9, determine whether the graph shown is a simple graph, a multigraph (and not a simple graph), a pseudograph (and not a multigraph), a directed graph, or a directed multigraph (and not a directed graph).

3.
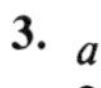
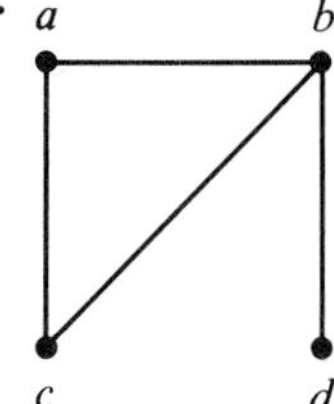

4.
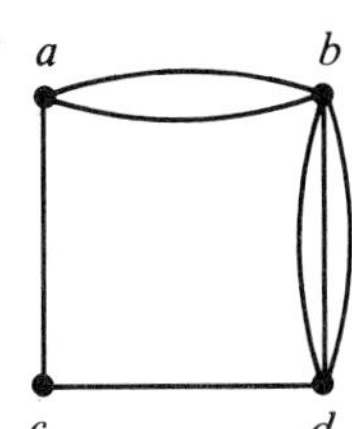

5.

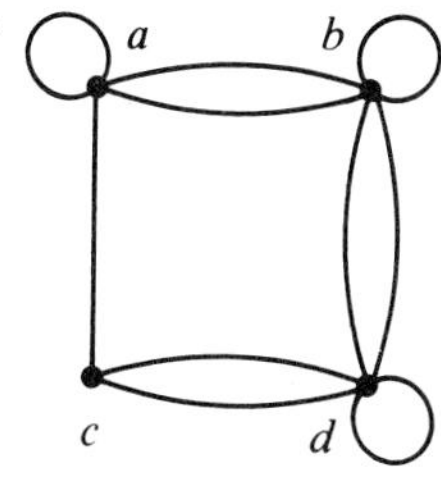

6.
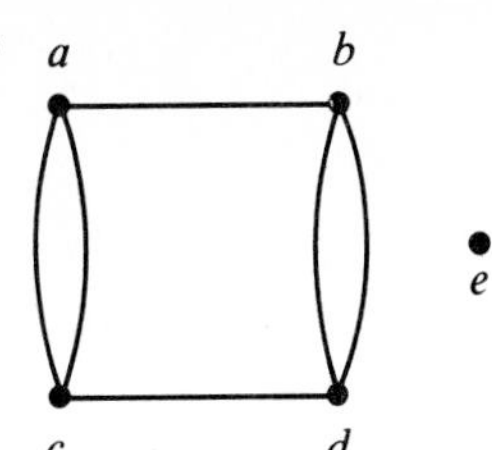

7.
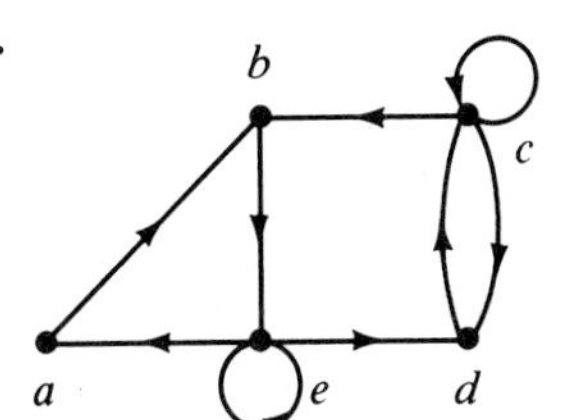

8.
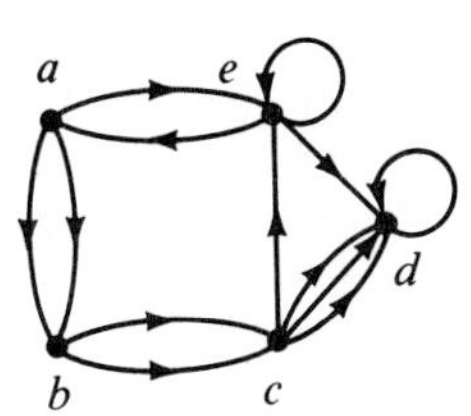

9.
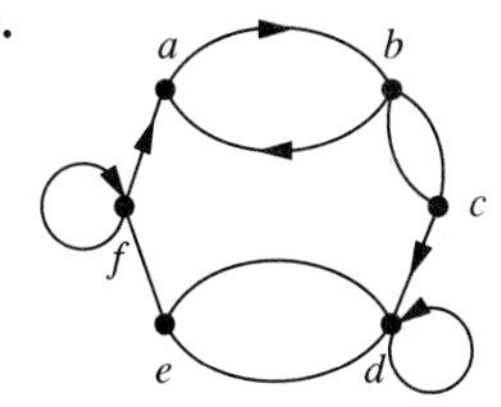

10. For each undirected graph in Exercises 3–9 that is not simple, find a set of edges to remove to make it simple.

11. The **intersection graph** of a collection of sets $A_1, A_2, \ldots, A_n$ is the graph that has a vertex for each of these sets and has an edge connecting the vertices representing two sets if these sets have a nonempty intersection. Construct the intersection graph of the following collections of sets.
   a) $A_1 = \{0, 2, 4, 6, 8\}$, $A_2 = \{0, 1, 2, 3, 4\}$, $A_3 = \{1, 3, 5, 7, 9\}$, $A_4 = \{5, 6, 7, 8, 9\}$, $A_5 = \{0, 1, 8, 9\}$
   b) $A_1 = \{\ldots, -4, -3, -2, -1, 0\}$, $A_2 = \{\ldots, -2, -1, 0, 1, 2, \ldots\}$, $A_3 = \{\ldots, -6, -4, -2, 0, 2, 4, 6, \ldots\}$,

$A_4 = \{\ldots, -5, -3, -1, 1, 3, 5, \ldots\}$,
$A_5 = \{\ldots, -6. -3, 0, 3, 6, \ldots\}$

c) $A_1 = \{x \mid x < 0\}$,
$A_2 = \{x \mid -1 < x < 0\}$,
$A_3 = \{x \mid 0 < x < 1\}$,
$A_4 = \{x \mid -1 < x < 1\}$,
$A_5 = \{x \mid x > -1\}$,
$A_6 = \mathbf{R}$

12. Use the niche overlap graph in Figure 6 to determine the species that compete with hawks.

13. Construct a niche overlap graph for six species of birds where the hermit thrush competes with the robin and with the blue jay, the robin also competes with the mockingbird, the mockingbird also competes with the blue jay, and the nuthatch competes with the hairy woodpecker.

14. Who can influence Fred and whom can Fred influence in the influence graph in Example 2?

15. Construct an influence graph for the board members of a company if the President can influence the Director of Research and Development, the Director of Marketing, and the Director of Operations; the Director of Research and Development can influence the Director of Operations; the Director of Marketing can influence the Director of Operations; and no one can influence, or be influenced by, the Chief Financial Officer.

16. Which other teams did Team four beat and which teams beat Team four in the round-robin tournament represented by the graph in Figure 8?

17. In a round-robin tournament the Tigers beat the Blue Jays, the Tigers beat the Cardinals, the Tigers beat the Orioles, the Blue Jays beat the Cardinals, the Blue Jays beat the Orioles, and the Cardinals beat the Orioles. Model this outcome with a directed graph.

18. Which statements must be executed before S_6 is executed in the program in Example 4? (Use the precedence graph in Figure 9.)

19. Construct a precedence graph for the following program:

$S_1 : x := 0$
$S_2 : x := x + 1$
$S_3 : y := 2$
$S_4 : z := y$
$S_5 : x := x + 2$
$S_6 : y := x + z$
$S_7 : z := 4$

20. Describe a discrete structure based on a graph that can be used to model airline routes and their flight times. (*Hint:* Add structure to a directed graph.)

21. Describe a discrete structure based on a graph that can be used to model relationships between pairs of individuals in a group, where each individual may either like, dislike, or be neutral about another individual, and the reverse relationship may be different. (*Hint:* Add structure to a directed graph. Treat separately the edges in opposite directions between vertices representing two individuals.)

7.2 Graph Terminology

INTRODUCTION

We introduce some of the basic vocabulary of graph theory in this section. We will use this vocabulary when we solve many different types of problems. One such problem involves determining whether a graph can be drawn in the plane so that no two of its edges cross. Another example is deciding whether there is a one-to-one correspondence between the vertices of two graphs that produces a one-to-one correspondence between the edges of the graphs. We will also introduce several important families of graphs often used as examples and in models.

BASIC TERMINOLOGY

First, we give some terminology that describes the vertices and edges of undirected graphs.

DEFINITION 1. Two vertices u and v in an undirected graph G are called *adjacent* (or *neighbors*) in G if $\{u, v\}$ is an edge of G. If $e = \{u, v\}$, the edge e is called *incident with* the vertices u and v. The edge e is also said to *connect* u and v. The vertices u and v are called *endpoints* of the edge $\{u, v\}$.

To keep track of how many edges are incident to a vertex, we make the following definition.

DEFINITION 2. The *degree* of a vertex in an undirected graph is the number of edges incident with it, except that a loop at a vertex contributes twice to the degree of that vertex. The degree of the vertex v is denoted by $\deg(v)$.

EXAMPLE 1 What are the degrees of the vertices in the graphs G and H displayed in Figure 1?

Solution: In G, $\deg(a) = 2$, $\deg(b) = \deg(c) = \deg(f) = 4$, $\deg(d) = 1$, $\deg(e) = 3$, and $\deg(g) = 0$. In H, $\deg(a) = 4$, $\deg(b) = \deg(e) = 6$, $\deg(c) = 1$, and $\deg(d) = 5$. ■

A vertex of degree 0 is called **isolated**. It follows that an isolated vertex is not adjacent to any vertex. Vertex g in graph G in Example 1 is isolated. A vertex is **pendant** if and only if it has degree 1. Consequently, a pendant vertex is adjacent to exactly one other vertex. Vertex d in graph G in Example 1 is pendant.

What do we get when we add the degrees of all the vertices of a graph $G = (V, E)$? Each edge contributes 2 to the sum of the degrees of the vertices since an edge is incident with exactly two (possibly equal) vertices. This means that the sum of the degrees of the vertices is twice the number of edges. We have the following result, which is sometimes called the Handshaking Theorem, because of the analogy between an edge having two endpoints and a handshake involving two hands.

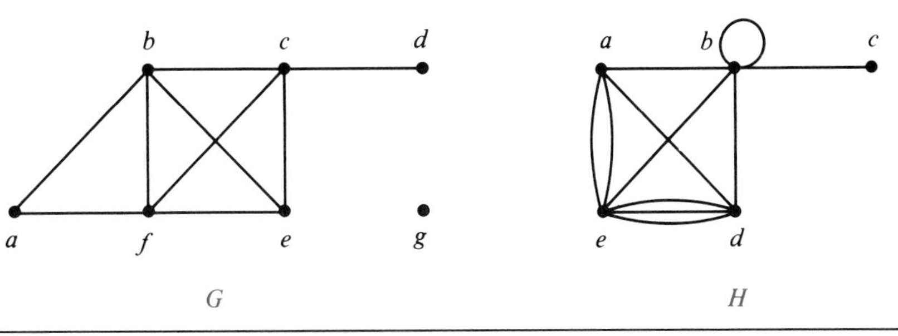

FIGURE 1 **The Undirected Graphs G and H.**

THEOREM 1 **THE HANDSHAKING THEOREM** Let $G = (V, E)$ be an undirected graph with e edges. Then

$$2e = \sum_{v \in V} \deg(v).$$

(Note that this applies even if multiple edges and loops are present.)

EXAMPLE 2 How many edges are there in a graph with 10 vertices each of degree 6?

Solution: Since the sum of the degrees of the vertices is $6 \cdot 10 = 60$, it follows that $2e = 60$. Therefore, $e = 30$. ■

Theorem 1 shows that the sum of the degrees of the vertices of an undirected graph is even. This simple fact has many consequences, one of which is given as Theorem 2.

THEOREM 2 An undirected graph has an even number of vertices of odd degree.

Proof: Let V_1 and V_2 be the set of vertices of even degree and the set of vertices of odd degree, respectively, in an undirected graph $G = (V, E)$. Then

$$2e = \sum_{v \in V} \deg(v) = \sum_{v \in V_1} \deg(v) + \sum_{v \in V_2} \deg(v).$$

Since $\deg(v)$ is even for $v \in V_1$, the first term in the right-hand side of the last equality is even. Furthermore, the sum of the two terms on the right-hand side of the last equality is even, since this sum is $2e$. Hence, the second term in the sum is also even. Since all the terms in this sum are odd, there must be an even number of such terms. Hence, there are an even number of vertices of odd degree. ■

There is also some useful terminology for graphs with directed edges.

DEFINITION 3. When (u, v) is an edge of the graph G with directed edges, u is said to be *adjacent to* v and v is said to be *adjacent from* u. The vertex u is called the *initial vertex* of (u, v), and v is called the *terminal* or *end vertex* of (u, v). The initial vertex and terminal vertex of a loop are the same.

Since the edges in graphs with directed edges are ordered pairs, the definition of the degree of a vertex can be refined to reflect the number of edges with this vertex as the initial vertex and as the terminal vertex.

DEFINITION 4. In a graph with directed edges the *in-degree* of a vertex v, denoted by $\deg^-(v)$, is the number of edges with v as their terminal vertex. The *out-degree* of v, denoted by $\deg^+(v)$, is the number of edges with v as their initial vertex. (Note that a loop at a vertex contributes 1 to both the in-degree and the out-degree of this vertex.)

EXAMPLE 3 Find the in-degree and out-degree of each vertex in the graph G with directed edges shown in Figure 2.

Solution: The in-degrees in G are: $\deg^-(a) = 2$, $\deg^-(b) = 2$, $\deg^-(c) = 3$, $\deg^-(d) = 2$, $\deg^-(e) = 3$, and $\deg^-(f) = 0$. The out-degrees are: $\deg^+(a) = 4$, $\deg^+(b) = 1$, $\deg^+(c) = 2$, $\deg^+(d) = 2$, $\deg^+(e) = 3$, and $\deg^-(f) = 0$. ■

Since each edge has an initial vertex and a terminal vertex, the sum of the in-degrees and the sum of the out-degrees of all vertices in a graph with directed edges are the same. Both of these sums are the number of edges in the graph. This result is stated as the following theorem.

THEOREM 3 Let $G = (V, E)$ be a graph with directed edges. Then

$$\sum_{v \in V} \deg^-(v) = \sum_{v \in V} \deg^+(v) = |E|.$$

There are many properties of a graph with directed edges that do not depend on the direction of its edges. Consequently, it is often useful to ignore these directions. The undirected graph that results from ignoring directions of edges is called the **underlying undirected graph**. A graph with directed edges and its underlying undirected graph have the same number of edges.

SOME SPECIAL SIMPLE GRAPHS

We will now introduce several classes of simple graphs. These graphs are often used as examples and arise in many applications.

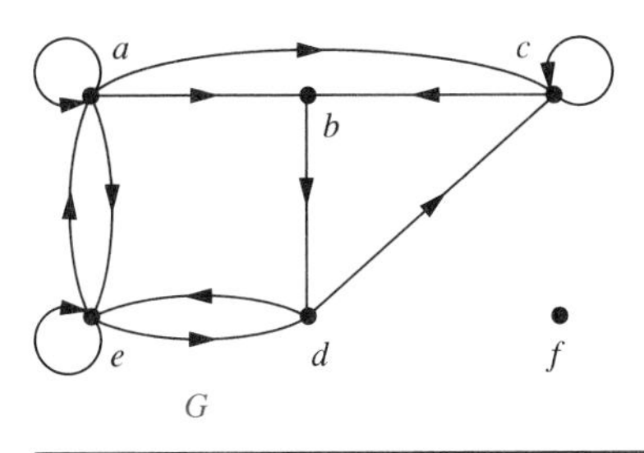

FIGURE 2 **The Directed Graph G.**

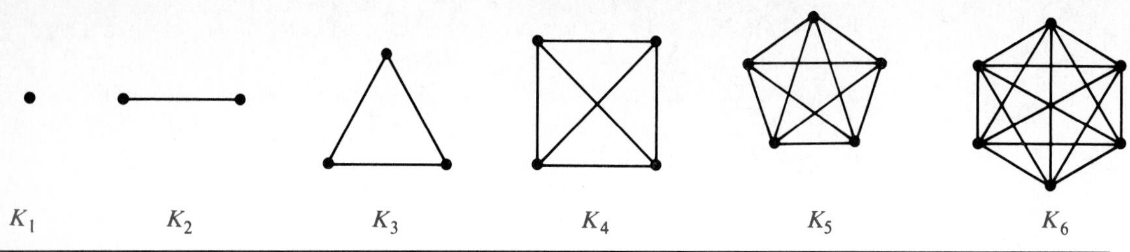

FIGURE 3 The Graphs K_n, $1 \leq n \leq 6$.

EXAMPLE 4 Complete Graphs The *complete graph* on n vertices, denoted by K_n, is the simple graph that contains exactly one edge between each pair of distinct vertices. The graphs K_n, for $n = 1, 2, 3, 4, 5, 6$, are displayed in Figure 3. ■

EXAMPLE 5 Cycles The *cycle* C_n, $n \geq 3$, consists of n vertices $v_1, v_2, \ldots, v_n$ and edges $\{v_1, v_2\}, \{v_2, v_3\}, \ldots, \{v_{n-1}, v_n\}$, and $\{v_n, v_1\}$. The cycles C_3, C_4, C_5, and C_6 are displayed in Figure 4. ■

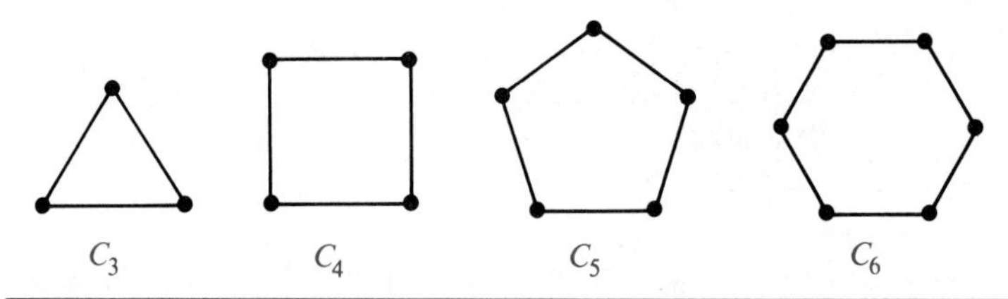

FIGURE 4 The Cycles C_3, C_4, C_5, and C_6.

EXAMPLE 6 Wheels We obtain the *wheel* W_n when we add an additional vertex to the cycle C_n, for $n \geq 3$, and connect this new vertex to each of the n vertices in C_n, by new edges. The wheels W_3, W_4, W_5, and W_6 are displayed in Figure 5. ■

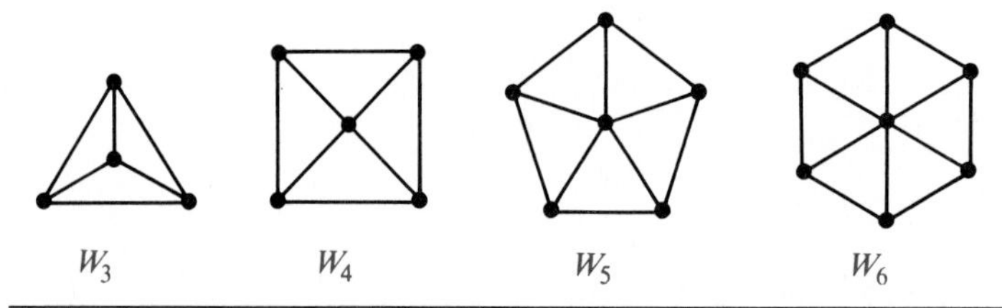

FIGURE 5 The Wheels, W_3, W_4, W_5, and W_6.

EXAMPLE 7 n-Cubes The *n-cube*, denoted by Q_n, is the graph that has vertices representing the 2^n bit strings of length n. Two vertices are adjacent if and only if the bit strings

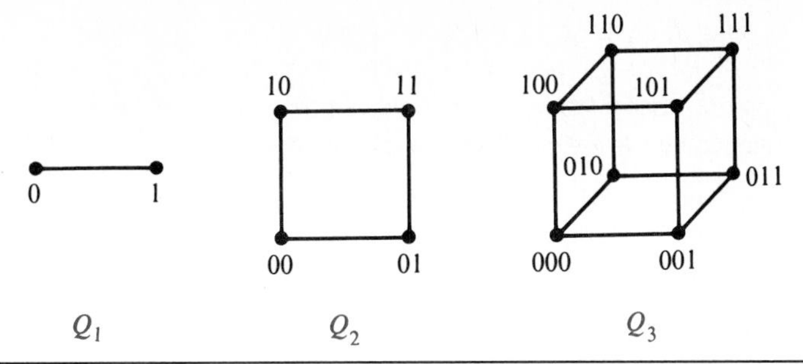

FIGURE 6 **The n-cube Q_n for $n = 1, 2,$ and 3.**

that they represent differ in exactly one bit. The graphs Q_1, Q_2, and Q_3 are displayed in Figure 6. ■

BIPARTITE GRAPHS

Sometimes a graph has the property that its vertex set can be divided into two disjoint subsets such that each edge connects a vertex in one of these subsets to a vertex in the other subset. For example, consider the graph representing marriages between people in a village, where each person is represented by a vertex and a marriage is represented by an edge. In this graph, each edge connects a vertex in the subset of vertices representing males and a vertex in the subset of vertices representing females. This leads us to the following definition.

DEFINITION 5. A simple graph G is called *bipartite* if its vertex set V can be partitioned into two disjoint nonempty sets V_1 and V_2 such that every edge in the graph connects a vertex in V_1 and a vertex in V_2 (so that no edge in G connects either two vertices in V_1 or two vertices in V_2).

In Example 8 we will show that C_6 is bipartite, and in Example 9 we will show that K_3 is not bipartite.

EXAMPLE 8 C_6 is bipartite, as shown in Figure 7, since its vertex set can be partitioned into the two sets $V_1 = \{v_1, v_3, v_5\}$ and $V_2 = \{v_2, v_4, v_6\}$, and every edge of C_6 connects a vertex in V_1 and a vertex in V_2. ■

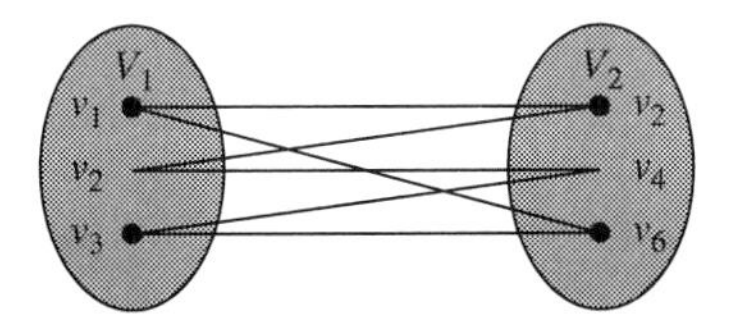

FIGURE 7 **Showing that C_6 Is Bipartite.**

EXAMPLE 9 K_3 is not bipartite. To see this, note that if we divide the vertex set of K_3 into two disjoint sets, one of the two sets must contain two vertices. If the graph were bipartite, these two vertices could not be connected by an edge, but in K_3 each vertex is connected to every other vertex by an edge. ■

EXAMPLE 10 Are the graphs G and H displayed in Figure 8 bipartite?

Solution: Graph G is bipartite, since its vertex set is the union of two disjoint sets, $\{a, b, d\}$ and $\{c, e, f, g\}$, and each edge connects a vertex in one of these subsets to a vertex in the other subset. (Note that for G to be bipartite it is not necessary that every vertex in $\{a, b, d\}$ be adjacent to every vertex in $\{c, e, f, g\}$. For instance, b and g are not adjacent.)

Graph H is not bipartite since its vertex set cannot be partitioned into two subsets so that edges do not connect two vertices from the same subset. (The reader should verify this by considering the vertices a, b, and f.) ■

EXAMPLE 11 Complete Bipartite Graphs The *complete bipartite graph* $K_{m,n}$ is the graph that has its vertex set partitioned into two subsets of m and n vertices, respectively. There is an edge between two vertices if and only if one vertex is in the first subset and the other vertex is in the second subset. The complete bipartite graphs $K_{2,3}$, $K_{3,3}$, $K_{3,5}$, and $K_{2,6}$ are displayed in Figure 9. ■

SOME APPLICATIONS OF SPECIAL TYPES OF GRAPHS

We will show how special types of graphs are used in models for data communications and parallel processing.

EXAMPLE 12 Local Area Networks The various computers, such as minicomputers and personal computers, as well as peripheral devices such as printers and plotters, in a building can be connected using a *local area network*. Some of these networks are based on a *star topology*, where all devices are connected to a central control device. A local area

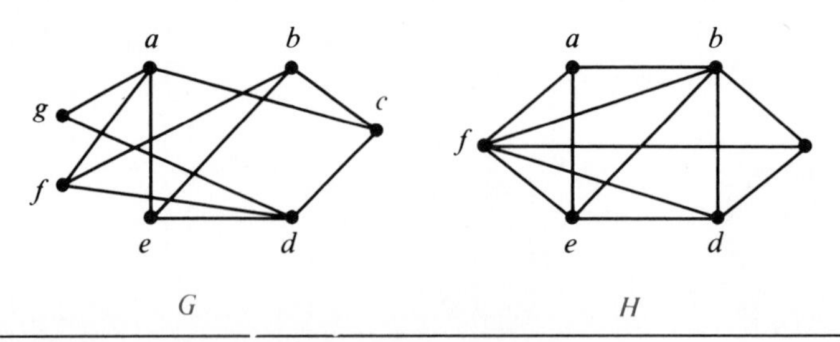

FIGURE 8 The Undirected Graphs G and H.

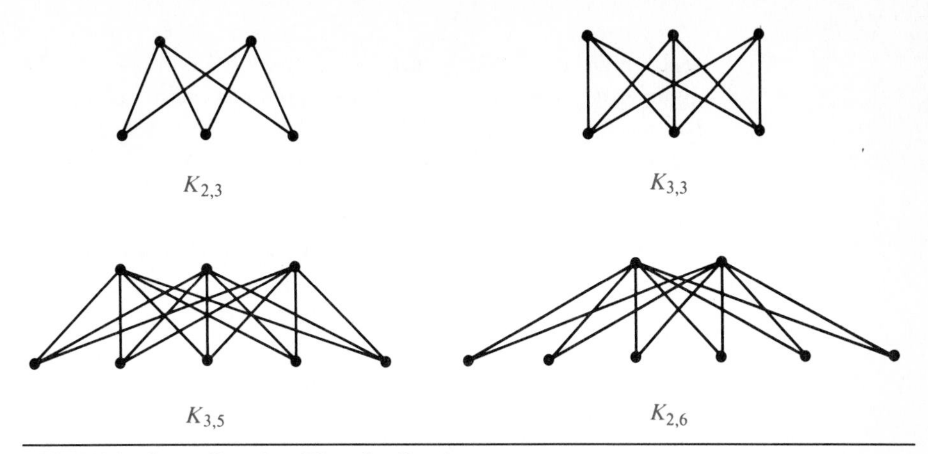

FIGURE 9 Some Complete Bipartite Graphs.

network can be represented using a complete bipartite graph $K_{1,n}$, as shown in Figure 10(a). Messages are sent from device to device through the central control device.

Other local area networks are based on a *ring topology*, where each device is connected to exactly two others. Local area networks with a ring topology are modeled using n-cycles, C_n, as shown in Figure 10(b). Messages are sent from device to device around the cycle until the intended recipient of a message is reached.

Finally, some local area networks use a hybrid of these two topologies. Messages may be sent around the ring, or through a central device. This redundancy makes the network more reliable. Local area networks with this redundancy can be modeled using wheels W_n, as shown in Figure 10. ■

EXAMPLE 13 Interconnection Networks for Parallel Computation Until recently, computers executed programs one operation at a time. Consequently, the algorithms written to solve problems were designed to perform one step at a time; such algorithms are called **serial**. (Almost all algorithms described in this book are serial.) However, many computationally intense problems, such as weather simulations, medical imaging, and cryptanalysis, cannot be solved in a reasonable amount of time using serial operations, even on a supercomputer. Furthermore, there is a physical limit to how fast a computer can carry out basic operations, so that there will always be problems that cannot be solved in a reasonable length of time using serial operations.

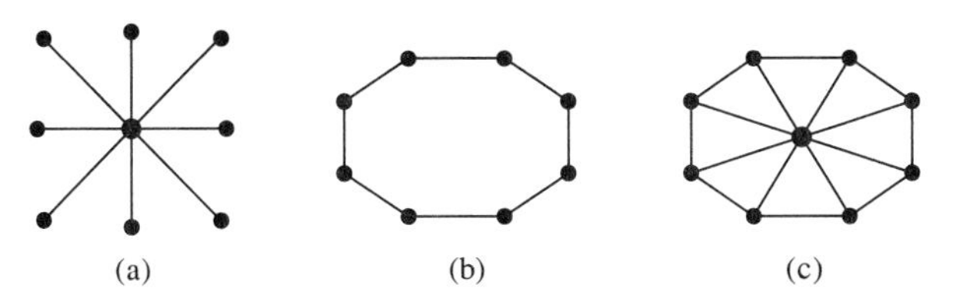

FIGURE 10 Star, Ring, and Hybrid Topologies for Local Area Networks.

Parallel processing, which uses computers made up of many separate processors, each with its own memory, helps overcome the limitations of serial computers. **Parallel algorithms**, which break a problem into a number of subproblems that can be solved concurrently, can then be devised to rapidly solve problems using a computer with multiple processors. In a parallel algorithm, a single instruction stream controls the execution of the algorithm, sending subproblems to different processors, and directs the input and output of these subproblems to the appropriate processors.

When parallel processing is used, one processor may need output generated by another processor. Consequently, these processors need to be interconnected. We can use the appropriate type of graph to represent the interconnection network of the processors in a computer with multiple processors. In the following discussion, we will describe the most commonly used types of interconnection networks for parallel processors. The type of interconnection network used to implement a particular parallel algorithm depends on the requirements for exchange of data between processors, the desired speed, and of course, the available hardware.

The simplest, but most expensive, network interconnecting processors includes a two-way link between each pair of processors. This network can be represented by K_n, the complete graph on n vertices, when there are n processors. However, there are serious problems with this type of interconnection network because the required number of connections is so large. In reality, the number of direct connections to a processor is limited, so that when there are a large number of processors, a processor cannot be linked directly to all others. For example, when there are 64 processors, $C(64, 2) = 2016$ connections would be required, and each processor would have to be directly connected to 63 others.

On the other hand, perhaps the simplest way to interconnect n processors is to use an arrangement known as a **linear array**. Each processor P_i, other than P_1 and P_n, is connected to its neighbors P_{i-1} and P_{i+1} via a two-way link. P_1 is connected only to P_2, and P_n is connected only to P_{n-1}. The linear array for six processors is shown in Figure 11. The advantage of a linear array is that each processor has at most two direct connections to other processors. The disadvantage is that it is sometimes necessary to use a large number of intermediate links, called **hops**, for processors to share information.

The **mesh network** (or **two-dimensional array**) is a commonly used interconnection network. In such a network, the number of processors is a perfect square, say $n = m^2$. The n processors are labeled $P(i, j)$, $0 \leq i \leq m-1$, $0 \leq j \leq m-1$. Two-way links connect processor $P(i, j)$ with its four neighbors, processors $P(i \pm 1, j)$ and $P(i, j \pm 1)$, as long as these are processors in the mesh. (Note that four processors, on the corners of the mesh, have only two adjacent processors, and other processors on the boundaries have only three neighbors. Sometimes a variant of a mesh network in which every processor has exactly four connections is used; see Exercise 40 at the end of this section.) The mesh network limits the number of links for each processor.

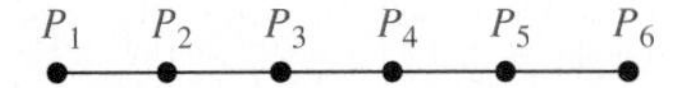

FIGURE 11 **A Linear Array for Six Processors.**

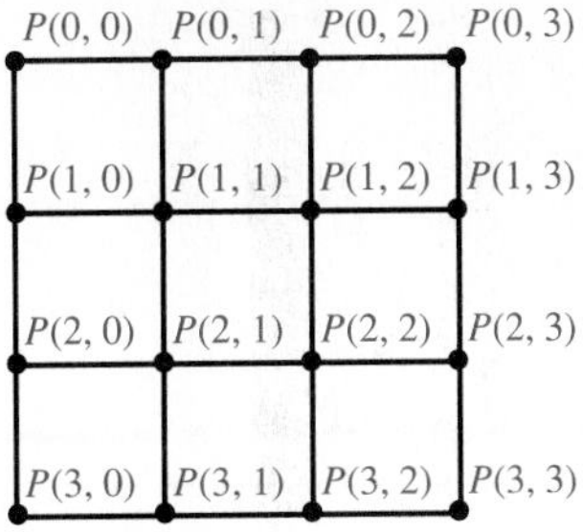

$P(0, 0)$ $P(0, 1)$ $P(0, 2)$ $P(0, 3)$
$P(1, 0)$ $P(1, 1)$ $P(1, 2)$ $P(1, 3)$
$P(2, 0)$ $P(2, 1)$ $P(2, 2)$ $P(2, 3)$
$P(3, 0)$ $P(3, 1)$ $P(3, 2)$ $P(3, 3)$

FIGURE 12 **A Mesh Network for 16 Processors.**

Communication between some pairs of processors requires $O(\sqrt{n}) = O(m)$ intermediate links. (See Exercise 41 at the end of this section.) The graph representing the mesh network for 16 processors is shown in Figure 12.

Perhaps the most important interconnection network is the hypercube. For such a network, the number of processors is a power of 2, $n = 2^m$. The n processors are labeled $P_0, P_1, \ldots, P_{n-1}$. Each processor has two-way connections to m other processors. Processor P_i is linked to the processors with indices whose binary representations differ from the binary representation of i in exactly one bit. The hypercube network balances the number of direct connections for each processor and the number of intermediate connections required so that processors can communicate. Many computers have been built using a hypercube network, and many parallel algorithms have been devised that use a hypercube network. The graph Q_n, the n-cube, represents the hypercube network with n processors. Figure 13 displays the hypercube network for eight processors. (Figure 13 displays a different way to draw Q_3 than was shown in Figure 6.) ■

NEW GRAPHS FROM OLD

Sometimes we need only part of a graph to solve a problem. For instance, we may care only about the part of a large computer network that involves the computer centers in New York, Denver, Detroit, and Atlanta. Then we can ignore the other computer centers and all telephone lines not linking two of these specific four computer centers. In the graph model for the large network, we can remove the vertices corresponding to the computer centers other than the four of interest, and we can remove all edges incident with a vertex that was removed. When edges and vertices are removed from a graph, without removing endpoints of any remaining edges, a smaller graph is obtained. Such a graph is called a **subgraph** of the original graph.

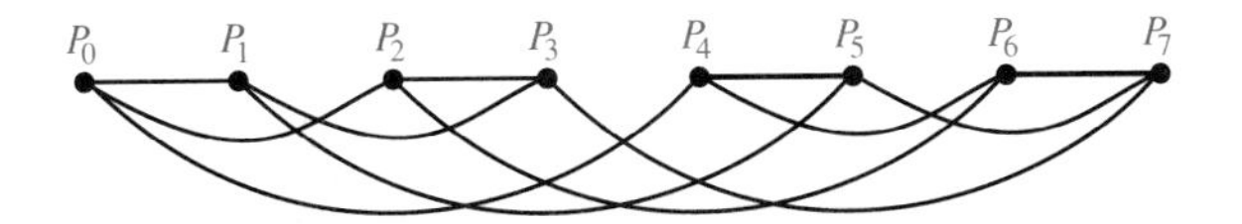

FIGURE 13 **A Hypercube Network for Eight Processors.**

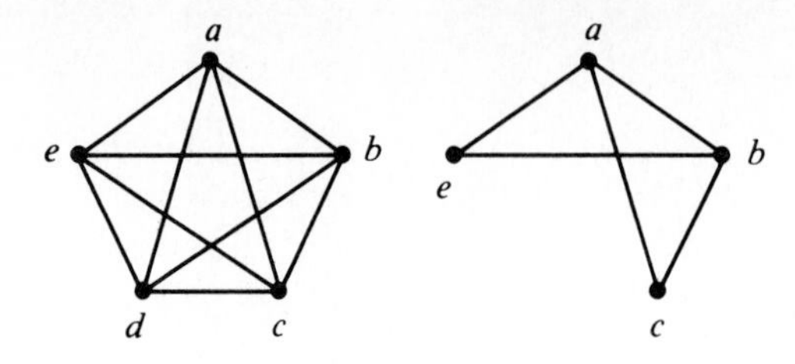

FIGURE 14 **A Subgraph of K_5.**

DEFINITION 6. A *subgraph* of a graph $G = (V, E)$ is a graph $H = (W, F)$ where $W \subseteq V$ and $F \subseteq E$.

EXAMPLE 14 The graph G shown in Figure 14 is a subgraph of K_5. ■

Two or more graphs can be combined in various ways. The new graph that contains all the vertices and edges of these graphs is called the **union** of the graphs. We will give a more formal definition for the union of two simple graphs.

DEFINITION 7. The *union* of two simple graphs $G_1 = (V_1, E_1)$ and $G_2 = (V_2, E_2)$ is the simple graph with vertex set $V_1 \cup V_2$ and edge set $E_1 \cup E_2$. The union of G_1 and G_2 is denoted by $G_1 \cup G_2$.

EXAMPLE 15 Find the union of the graphs G_1 and G_2 shown in Figure 15(a).

Solution: The vertex set of the union $G_1 \cup G_2$ is the union of the two vertex sets, namely, $\{a, b, c, d, e, f\}$. The edge set of the union is the union of the two edge sets. The union is displayed in Figure 15(b). ■

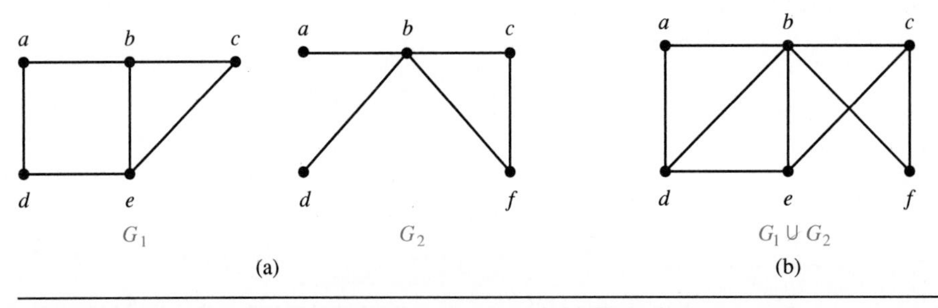

FIGURE 15 **(a) The Simple Graphs G_1 and G_2, and (b) Their Union $G_1 \cup G_2$.**

Exercises

In Exercises 1–3 find the number of vertices, the number of edges, and the degree of each vertex in the given undirected graph. Identify all isolated and pendant vertices.

1.

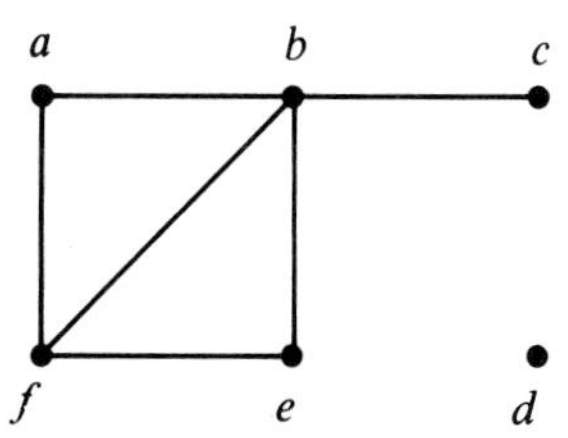

2.

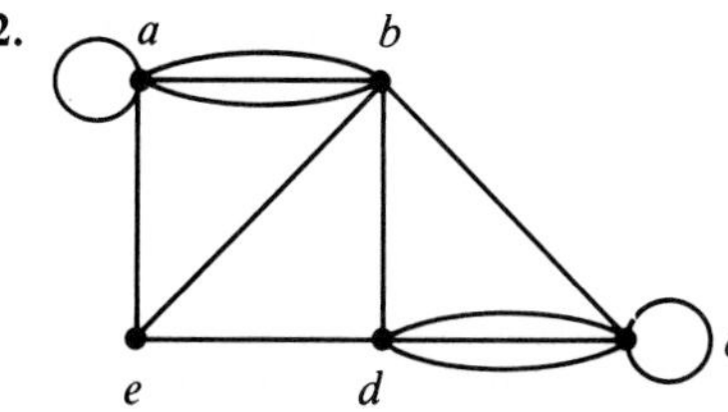

3.

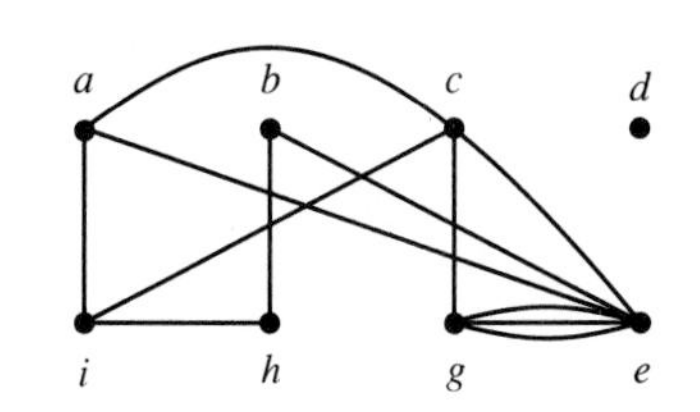

4. Find the sum of the degrees of the vertices of each graph in Exercises 1–3 and verify that it equals twice the number of edges in the graph.

5. Can a simple graph exist with 15 vertices each of degree 5?

6. Show that the sum, over the set of people at a party, of the number of people a person has shaken hands with, is even. Assume that no one shakes his or her own hand.

In Exercises 7–9 determine the number of vertices and edges and find the in-degree and out-degree of each vertex for the given directed multigraph.

7.

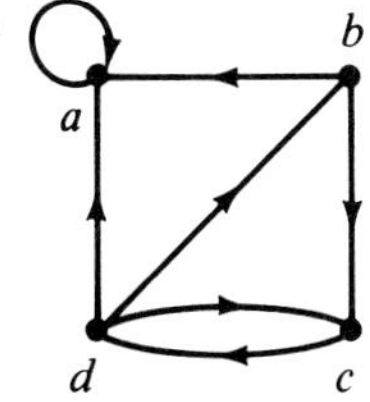

8.

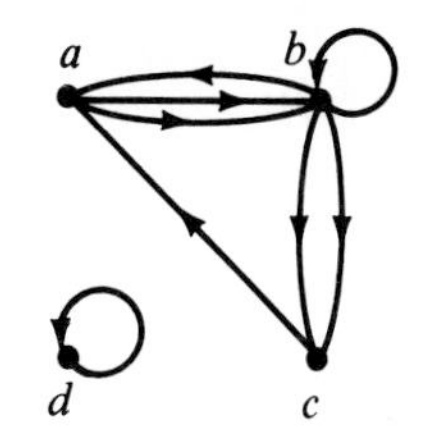

9.

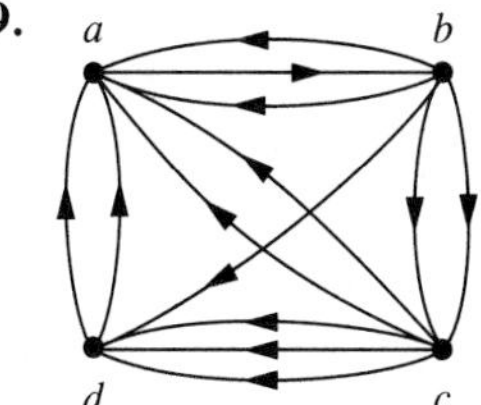

10. For each of the graphs in Exercises 7–9 determine the sum of the in-degrees of the vertices and the sum of the out-degrees of the vertices directly. Show that they are both equal to the number of edges in the graph.

11. Construct the underlying undirected graph for the graph with directed edges in Figure 2.

12. Draw the following graphs.

a) K_7 **b)** $K_{1,8}$ **c)** $K_{4,4}$
d) C_7 **e)** W_7 **f)** Q_4

In Exercises 13–17 determine whether the graph is bipartite.

13.

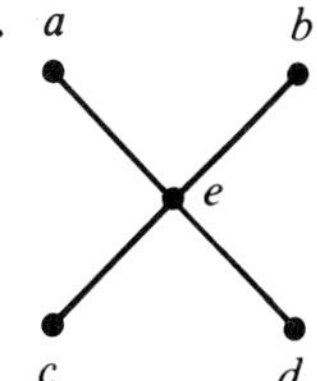

14.

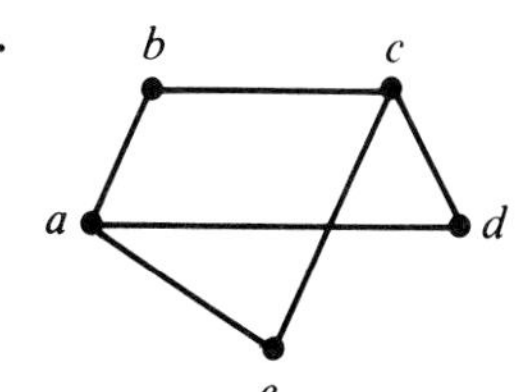

15.

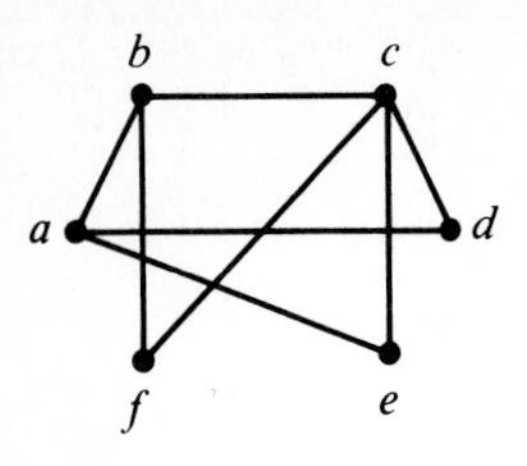

16.

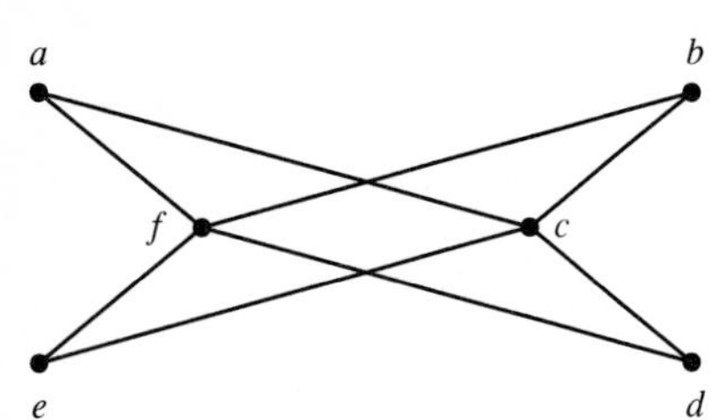

17.

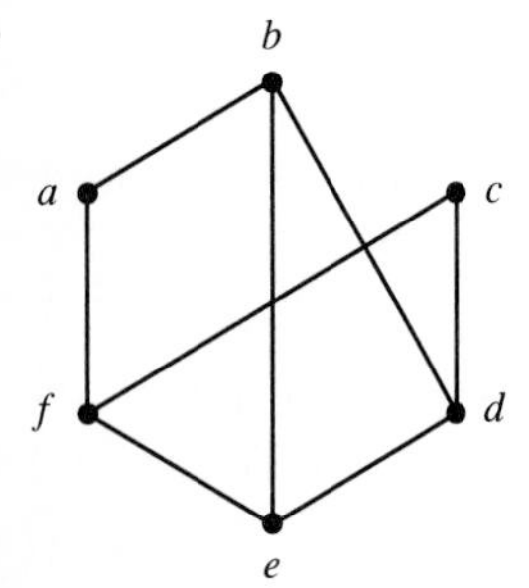

18. For which values of n are the following graphs bipartite?
a) K_n **b)** C_n **c)** W_n **d)** Q_n

19. How many vertices and how many edges do the following graphs have?
a) K_n **b)** C_n **c)** W_n
d) $K_{m,n}$ **e)** Q_n

20. How many edges does a graph have if it has vertices of degree 4, 3, 3, 2, 2? Draw such a graph.

21. Does there exist a simple graph with five vertices of the following degrees? If so, draw such a graph.
a) 3, 3, 3, 3, 2 **b)** 1, 2, 3, 4, 5
c) 1, 2, 3, 4, 4 **d)** 3, 4, 3, 4, 3
e) 0, 1, 2, 2, 3 **f)** 1, 1, 1, 1, 1

22. How many subgraphs with at least one vertex does K_2 have?

23. How many subgraphs with at least one vertex does K_3 have?

24. How many subgraphs with at least one vertex does W_3 have?

25. Draw all subgraphs of the following graph.

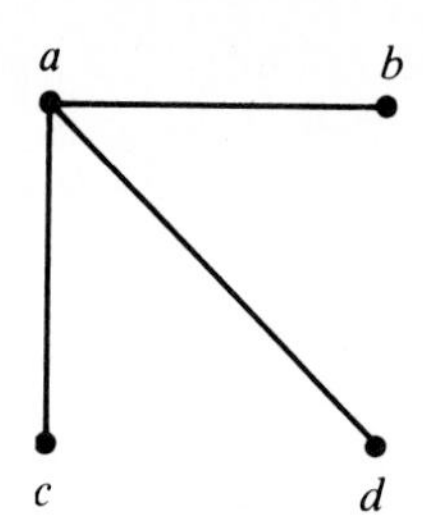

26. Let G be a graph with v vertices and e edges. Let M be the maximum degree of the vertices of G, and let m be the minimum degree of the vertices of G. Show that
a) $2e/v \geq m$. **b)** $2e/v \leq M$.

A simple graph is called **regular** if every vertex of this graph has the same degree. A regular graph is called n-**regular** if every vertex in this graph has degree n.

27. For which values of n are the following graphs regular?
a) K_n **b)** C_n **c)** W_n **d)** Q_n

28. For which values of m and n is $K_{m,n}$ regular?

29. How many vertices does a regular graph of degree 4 with 10 edges have?

In Exercises 30–32 find the union of the given pair of simple graphs. (Assume edges with the same endpoints are the same.)

30.

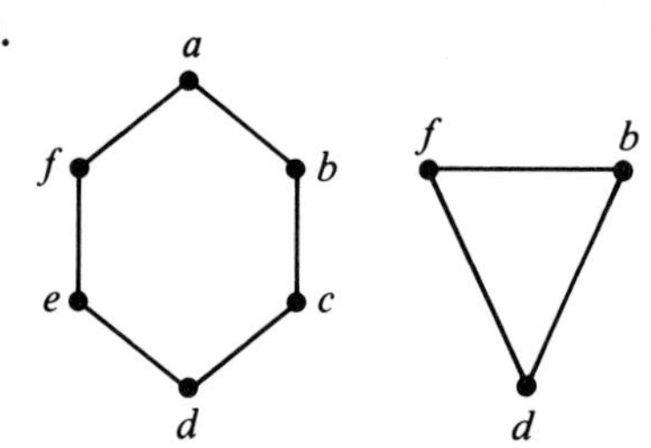

31.

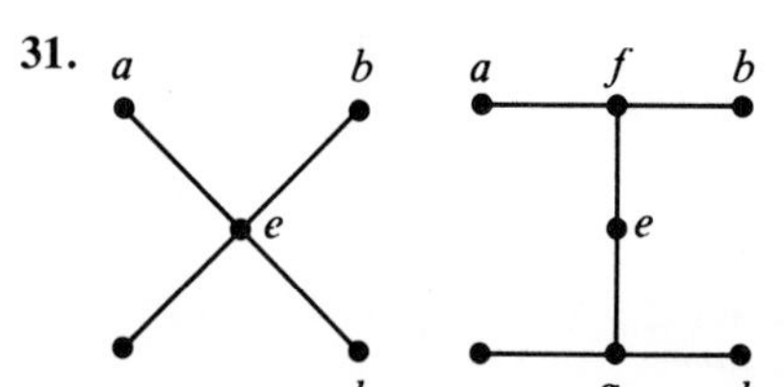

32.

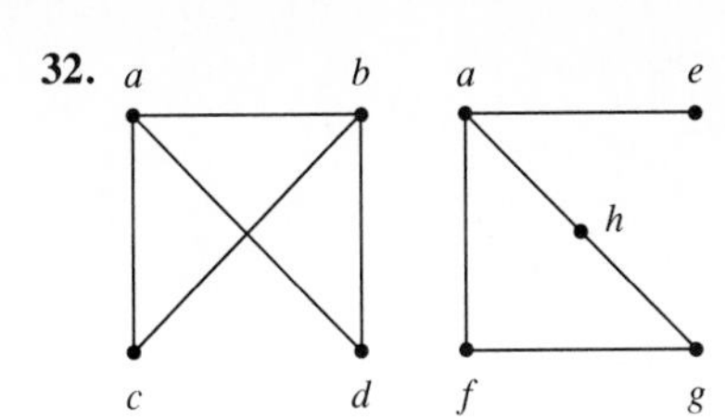

33. The **complementary graph** $\overline{G}$ of a simple graph G has the same vertices as G. Two vertices are adjacent in $\overline{G}$ if and only if they are not adjacent in G. Find the following.

a) $\overline{K_n}$ **b)** $\overline{K_{m,n}}$ **c)** $\overline{C_n}$ **d)** $\overline{Q_n}$

34. If G is a simple graph with 15 edges and $\overline{G}$ has 13 edges, how many vertices does G have?

35. If the simple graph G has v vertices and e edges, how many edges does $\overline{G}$ have?

★**36.** Show that if G is a bipartite simple graph with v vertices and e edges, then $e \leq v^2/4$.

37. Show that if G is a simple graph with n vertices, then the union of G and $\overline{G}$ is K_n.

★**38.** Describe an algorithm to decide whether a graph is bipartite.

39. Draw the mesh network for interconnecting nine parallel processors.

40. In a variant of a mesh network for interconnecting $n = m^2$ processors, processor $P(i, j)$ is connected to the four processors $P((i \pm 1) \textbf{ mod } m, j)$, $P(i, (j \pm 1) \textbf{ mod } m)$, so that connections wrap around the edges of the mesh. Draw this variant of the mesh network for 16 processors.

41. Show that every pair of processors in a mesh network of $n = m^2$ processors can communicate using $O(\sqrt{n}) = O(m)$ hops between directly connected processors.

7.3 Representing Graphs and Graph Isomorphism

INTRODUCTION

There are many useful ways to represent graphs. As we will see throughout this chapter, in working with a graph, it is helpful to be able to choose its most convenient representation. In this section we will show how to represent graphs in several different ways.

Sometimes, two graphs have exactly the same form, in the sense that there is a one-to-one correspondence between their vertex sets that preserves edges. In such a case, we say that the two graphs are **isomorphic**. Determining whether two graphs are isomorphic is an important problem of graph theory that we will study in this section.

REPRESENTING GRAPHS

One way to represent a graph without multiple edges is to list all the edges of this graph. Another way to represent a graph with no multiple edges is to use **adjacency lists**, which specify the vertices that are adjacent to each vertex of the graph.

EXAMPLE 1 Use adjacency lists to describe the simple graph given in Figure 1.

Solution. Table 1 lists those vertices adjacent to each of the vertices of the graph. ■

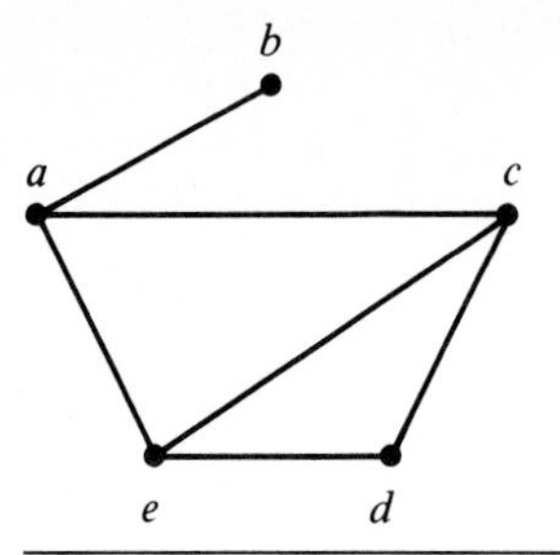

FIGURE 1 A Simple Graph.

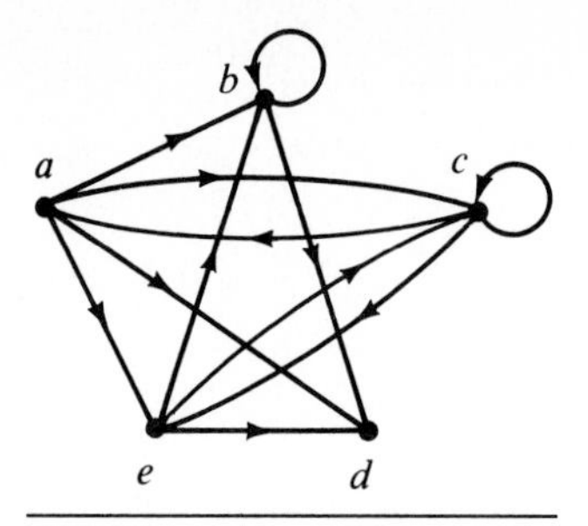

FIGURE 2 A Directed Graph.

EXAMPLE 2 Represent the directed graph shown in Figure 2 by listing all the vertices that are the terminal vertices of edges starting at each vertex of the graph.

Solution: Table 2 represents the directed graph shown in Figure 2. ■

ADJACENCY MATRICES

Carrying out graph algorithms using the representation of graphs by lists of edges, or by adjacency lists, can be cumbersome if there are many edges in the graph. To simplify computation, graphs can be represented using matrices. Two types of matrices commonly used to represent graphs will be presented here. One is based on the adjacency of vertices, and the other is based on incidence of vertices and edges.

Suppose that $G = (V, E)$ is a simple graph where $|V| = n$. Suppose that the vertices of G are listed arbitrarily as $v_1, v_2, \ldots, v_n$. The **adjacency matrix A** (or $\mathbf{A}_G$) of G, with respect to this listing of the vertices, is the $n \times n$ zero-one matrix with 1 as its (i, j)th entry when v_i and v_j are adjacent, and 0 as its (i, j)th entry when they

TABLE 1 **An Edge List for a Simple Graph.**

Vertex	*Adjacent Vertices*
a	b, c, e
b	a
c	a, d, e
d	c, e
e	a, c, d

TABLE 2 **An Edge List for a Directed Graph.**

Initial Vertex	*Terminal Vertices*
a	b, c, d, e
b	b, d
c	a, c, e
d	
e	b, c, d

are not adjacent. In other words, if its adjacency matrix is $\mathbf{A} = [a_{ij}]$, then

$$a_{ij} = \begin{cases} 1 & \text{if } \{v_i, v_j\} \text{ is an edge of } G \\ 0 & \text{otherwise.} \end{cases}$$

Note that an adjacency matrix of a graph is based on the ordering chosen for the vertices. Hence, there are as many as $n!$ different adjacency matrices for a graph with n vertices, since there are $n!$ different orderings of n vertices.

The adjacency matrix of a simple graph is symmetric, that is, $a_{ij} = a_{ji}$, since both of these entries are 1 when v_i and v_j are adjacent, and both are 0 otherwise. Furthermore, since a simple graph has no loops, each entry a_{ii}, $i = 1, 2, 3, \ldots, n$, is 0.

EXAMPLE 3 Use an adjacency matrix to represent the graph shown in Figure 3.

Solution: We order the vertices as a, b, c, d. The matrix representing this graph is

$$\begin{bmatrix} 0 & 1 & 1 & 1 \\ 1 & 0 & 1 & 0 \\ 1 & 1 & 0 & 0 \\ 1 & 0 & 0 & 0 \end{bmatrix}.$$

■

EXAMPLE 4 Draw a graph with the adjacency matrix

$$\begin{bmatrix} 0 & 1 & 1 & 0 \\ 1 & 0 & 0 & 1 \\ 1 & 0 & 0 & 1 \\ 0 & 1 & 1 & 0 \end{bmatrix}.$$

with respect to the ordering of vertices a, b, c, d.

Solution. A graph with this adjacency matrix is shown in Figure 4. ■

Adjacency matrices can also be used to represent undirected graphs with loops and with multiple edges. A loop at the vertex a_i is represented by a 1 at the (i, i)th

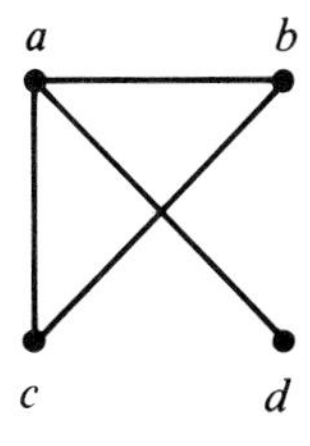

FIGURE 3 **A Simple Graph.**

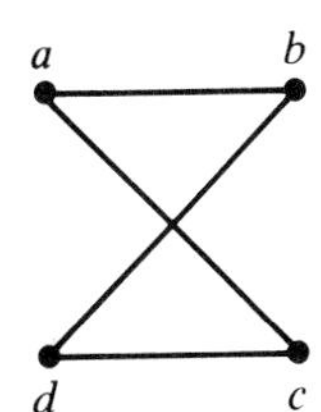

FIGURE 4 **A Graph with the Given Adjacency Matrix.**

position of the adjacency matrix. When multiple edges are present, the adjacency matrix is no longer a zero-one matrix, since the (i, j)th entry of this matrix equals the number of edges that are associated to $\{a_i, a_j\}$. All undirected graphs, including multigraphs and pseudographs, have symmetric adjacency matrices.

EXAMPLE 5 Use an adjacency matrix to represent the pseudograph shown in Figure 5.

Solution: The adjacency matrix using the ordering of vertices a, b, c, d is

$$\begin{bmatrix} 0 & 3 & 0 & 2 \\ 3 & 0 & 1 & 1 \\ 0 & 1 & 1 & 2 \\ 2 & 1 & 2 & 0 \end{bmatrix}.$$

■

We used zero-one matrices in Chapter 6 to represent directed graphs. The matrix for a directed graph $G = (V, E)$ has a 1 in its (i, j)th position if there is an edge from v_i to v_j, where $v_1, v_2, \ldots, v_n$ is an arbitrary listing of the vertices of the directed graph. In other words, if $\mathbf{A} = [a_{ij}]$ is the adjacency matrix for the directed graph with respect to this listing of the vertices, then

$$a_{ij} = \begin{cases} 1 & \text{if } (v_i, v_j) \text{ is an edge of } G \\ 0 & \text{otherwise.} \end{cases}$$

The adjacency matrix for a directed graph does not have to be symmetric, since there may not be an edge from a_j to a_i when there is an edge from a_i to a_j.

Adjacency matrices can also be used to represent directed multigraphs. Again, such matrices are not zero-one matrices when there are multiple edges in the same direction connecting two vertices. In the adjacency matrix for a directed multigraph, a_{ij} equals the number of edges that are associated to (v_i, v_j).

INCIDENCE MATRICES

Another common way to represent graphs is to use **incidence matrices**. Let $G = (V, E)$ be an undirected graph. Suppose that $v_1, v_2, \ldots, v_n$ are the vertices and

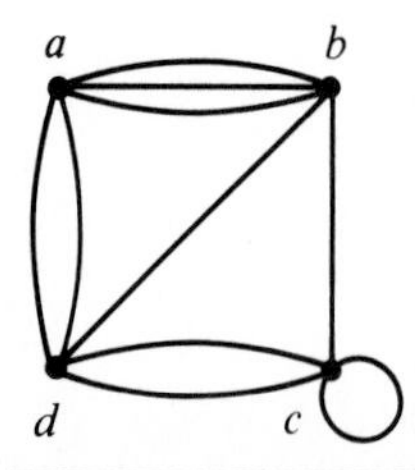

FIGURE 5 **A Pseudograph.**

$e_1, e_2, \ldots, e_m$ are the edges of G. Then the incidence matrix with respect to this ordering of V and E is the $n \times m$ matrix $\mathbf{M} = [m_{ij}]$ where

$$m_{ij} = \begin{cases} 1 & \text{when edge } e_j \text{ is incident with } v_i \\ 0 & \text{otherwise.} \end{cases}$$

EXAMPLE 6 Represent the graph shown in Figure 6 with an incidence matrix.

Solution. The incidence matrix is

$$\begin{array}{c} \\ v_1 \\ v_2 \\ v_3 \\ v_4 \\ v_5 \end{array} \begin{array}{c} \begin{array}{cccccc} e_1 & e_2 & e_3 & e_4 & e_5 & e_6 \end{array} \\ \left[\begin{array}{cccccc} 1 & 1 & 0 & 0 & 0 & 0 \\ 0 & 0 & 1 & 1 & 0 & 1 \\ 0 & 0 & 0 & 0 & 1 & 1 \\ 1 & 0 & 1 & 0 & 0 & 0 \\ 0 & 1 & 0 & 1 & 1 & 0 \end{array}\right] \end{array}.$$

■

Incidence matrices can also be used to represent multiple edges and loops. Multiple edges are represented in the incidence matrix using columns with identical entries, since these edges are incident with the same pair of vertices. Loops are represented using a column with exactly one entry equal to 1, corresponding to the vertex that is incident with this loop.

EXAMPLE 7 Represent the pseudograph shown in Figure 7 using an incidence matrix.

Solution: The incidence matrix for this graph is

$$\begin{array}{c} \\ v_1 \\ v_2 \\ v_3 \\ v_4 \\ v_5 \end{array} \begin{array}{c} \begin{array}{cccccccc} e_1 & e_2 & e_3 & e_4 & e_5 & e_6 & e_7 & e_8 \end{array} \\ \left[\begin{array}{cccccccc} 1 & 1 & 1 & 0 & 0 & 0 & 0 & 0 \\ 0 & 1 & 1 & 1 & 0 & 1 & 1 & 0 \\ 0 & 0 & 0 & 1 & 1 & 0 & 0 & 0 \\ 0 & 0 & 0 & 0 & 0 & 0 & 1 & 1 \\ 0 & 0 & 0 & 0 & 1 & 1 & 0 & 0 \end{array}\right] \end{array}.$$

■

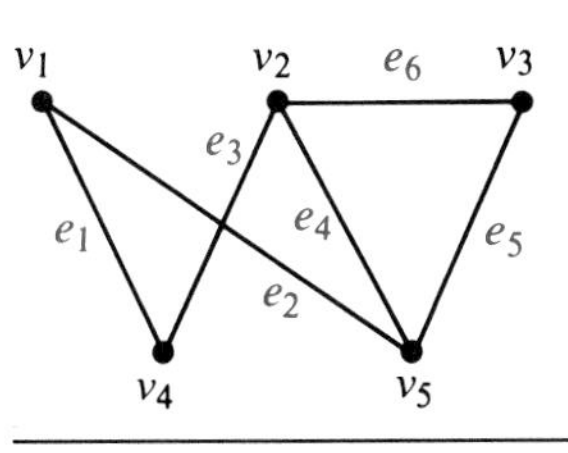

FIGURE 6 **An Undirected Graph.**

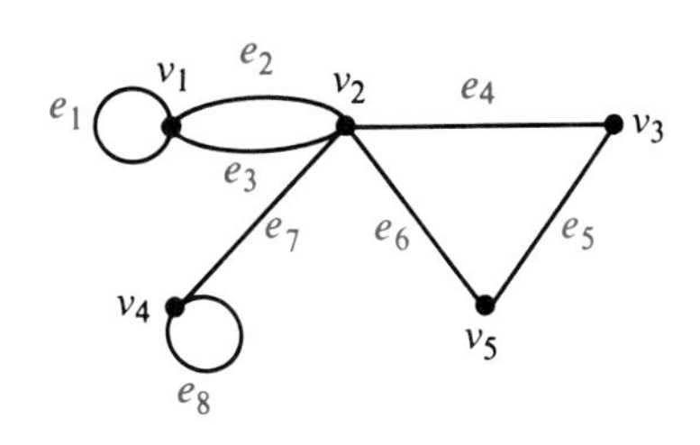

FIGURE 7 **A Pseudograph.**

ISOMORPHISM OF GRAPHS

We often need to know whether it is possible to draw two graphs in the same way. For instance, in chemistry, graphs are used to model compounds. Different compounds can have the same molecular formula but can differ in structure. Such compounds will be represented by graphs that cannot be drawn in the same way. The graphs representing previously known compounds can be used to determine whether a supposedly new compound has been studied before.

There is a useful terminology for graphs with the same structure.

DEFINITION 1. The simple graphs $G_1 = (V_1, E_1)$ and $G_2 = (V_2, E_2)$ are *isomorphic* if there is a one-to-one and onto function f from V_1 to V_2 with the property that a and b are adjacent in G_1 if and only if $f(a)$ and $f(b)$ are adjacent in G_2, for all a and b in V_1. Such a function f is called an *isomorphism*.*

In other words, when two simple graphs are isomorphic, there is a one-to-one correspondence between vertices of the two graphs that preserves the adjacency relationship.

EXAMPLE 8 Show that the graphs $G = (V, E)$ and $H = (W, F)$, displayed in Figure 8, are isomorphic.

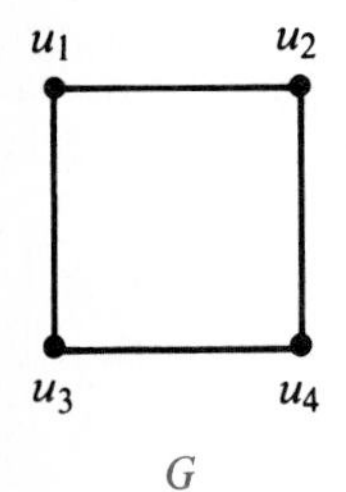

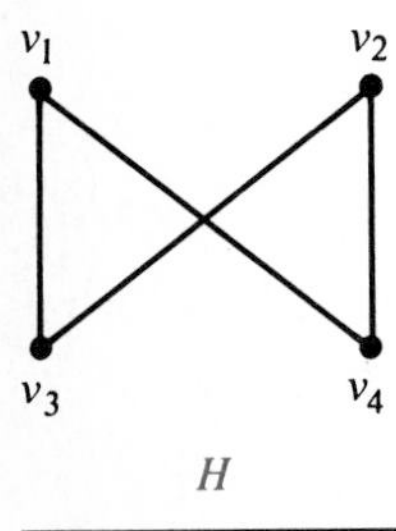

FIGURE 8 The Graphs G and H.

Solution: The function f with $f(u_1) = v_1$, $f(u_2) = v_4$, $f(u_3) = v_3$, and $f(u_4) = v_2$ is a one-to-one correspondence between V and W. To see that this correspondence preserves adjacency, note that adjacent vertices in G are u_1 and u_2, u_1 and u_3, u_2 and u_4, and u_3 and u_4, and each of the pairs $f(u_1) = v_1$ and $f(u_2) = v_4$, $f(u_1) = v_1$ and $f(u_3) = v_3$, $f(u_2) = v_4$ and $f(u_4) = v_2$, and $f(u_3) = v_3$ and $f(u_4) = v_2$ are adjacent in H. ■

It is often difficult to determine whether two simple graphs are isomorphic. There are $n!$ possible one-to-one correspondences between the vertex sets of two simple graphs with n vertices. Testing each such correspondence to see whether it preserves adjacency and nonadjacency is impractical if n is at all large.

However, we can often show that two simple graphs are not isomorphic by showing that they do not share a property that isomorphic simple graphs must both have. Such a property is called an **invariant** with respect to isomorphism of simple graphs. For instance, isomorphic simple graphs must have the same number of vertices, since there is a one-to-one correspondence between the sets of vertices of the graphs. Furthermore, isomorphic simple graphs must have the same number of edges, because the

* The word *isomorphism* comes from the Greek roots *isos* for "equal" and *morphe* for "form."

one-to-one correspondence between vertices establishes a one-to-one correspondence between edges. In addition, the degrees of the vertices in isomorphic simple graphs must be the same. That is, a vertex v of degree d in G must correspond to a vertex $f(v)$ of degree d in H, since a vertex w in G is adjacent to v if and only if $f(v)$ and $f(w)$ are adjacent in H.

EXAMPLE 9 Show that the graphs displayed in Figure 9 are not isomorphic.

Solution: Both G and H have five vertices and six edges. However, H has a vertex of degree 1, namely e, whereas G has no vertices of degree 1. It follows that G and H are not isomorphic. ■

The number of vertices, the number of edges, and the degrees of the vertices are all invariants under isomorphism. If any of these quantities differ in two simple graphs, these graphs cannot be isomorphic. However, when these invariants are the same, it does not necessarily mean that the two graphs are isomorphic. There are no useful sets of invariants currently known that can be used to determine whether simple graphs are isomorphic.

EXAMPLE 10 Determine whether the graphs shown in Figure 10 are isomorphic.

Solution: The graphs G and H both have eight vertices and 10 edges. They also both have four vertices of degree 2 and four of degree 3. Since these invariants all agree, it is still conceivable that these graphs are isomorphic.

However G and H are not isomorphic. To see this note that since $\deg(a) = 2$ in G, a must correspond to either t, u, x, or y in H, since these are the vertices of degree 2 in H. However, each of these four vertices in H is adjacent to another vertex of degree 2 in H, which is not true for a in G.

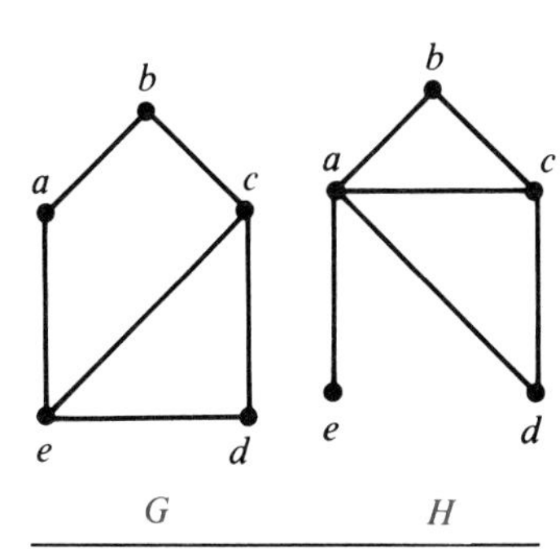

FIGURE 9 **The Graphs G and H.**

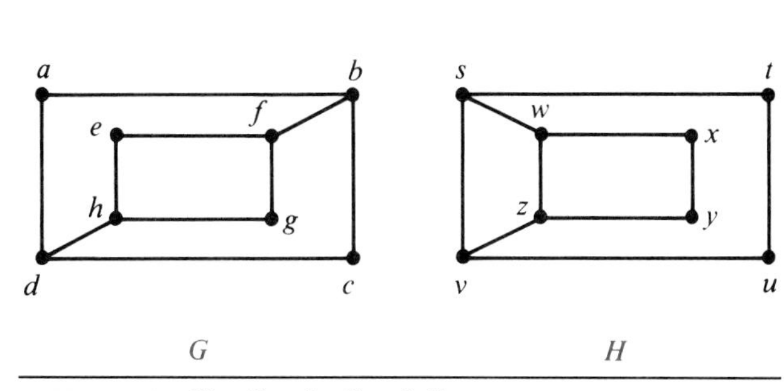

FIGURE 10 **The Graphs G and H.**

Another way to see that G and H are not isomorphic is to note that the subgraphs of G and H made up of vertices of degree 3 and the edges connecting them must be isomorphic if these two graphs are isomorphic (the reader should verify this). However, these subgraphs, shown in Figure 11, are not isomorphic. ■

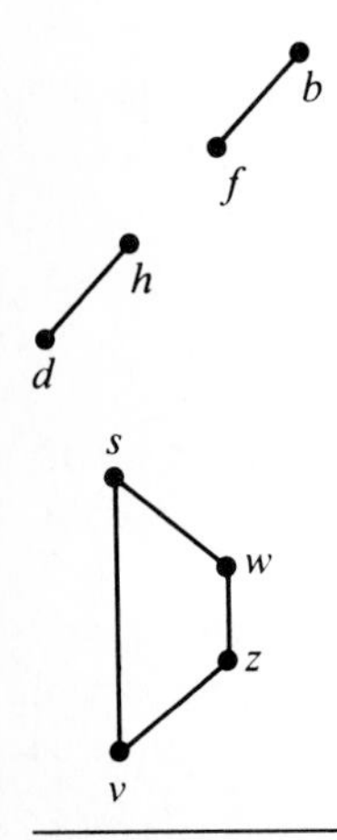

FIGURE 11 The Subgraphs of G and H Made up of Vertices of Degree Three and the Edges Connecting Them.

To show that a function f from the vertex set of a graph G to the vertex set of a graph H is an isomorphism, we need to show that f preserves edges. One helpful way to do this is to use adjacency matrices. In particular, to show that f is an isomorphism, we can show that the adjacency matrix of G is the same as the adjacency matrix of H, when rows and columns are labeled to correspond to the images under f of the vertices in G that are the labels of these rows and columns in the adjacency matrix of G. We illustrate how this is done in the following example.

EXAMPLE 11 Determine whether the graphs G and H displayed in Figure 12 are isomorphic.

Solution: Both G and H have six vertices and seven edges. Both have four vertices of degree 2 and two vertices of degree 3. It is also easy to see that the subgraphs of G and H consisting of all vertices of degree 2 and the edges connecting them are isomorphic (as the reader should verify). Since G and H agree with respect to these invariants, it is reasonable to try to find an isomorphism f.

We now will define a function f and then determine whether it is an isomorphism. Since $\deg(u_1) = 2$ and since u_1 is not adjacent to any other vertex of degree 2, the image of u_1 must be either v_4 or v_6, the only vertices of degree 2 in H not adjacent to a vertex of degree 2. We arbitrarily set $f(u_1) = v_6$. (If we found that this choice did not lead to isomorphism, we would then try $f(u_1) = v_4$.) Since u_2 is adjacent to u_1, the possible images of u_2 are v_3 and v_5. We arbitrarily set $f(u_2) = v_3$. Continuing in this way, using adjacency of vertices and degrees as a guide, we set $f(u_3) = v_4$, $f(u_4) = v_5$, $f(u_5) = v_1$, and $f(u_6) = v_2$. We now have a one-to-one correspondence between the vertex set of G and the vertex set of H, namely: $f(u_1) = v_6$, $f(u_2) = v_3$, $f(u_3) = v_4$, $f(u_4) = v_5$, $f(u_5) = v_1$, $f(u_6) = v_2$.

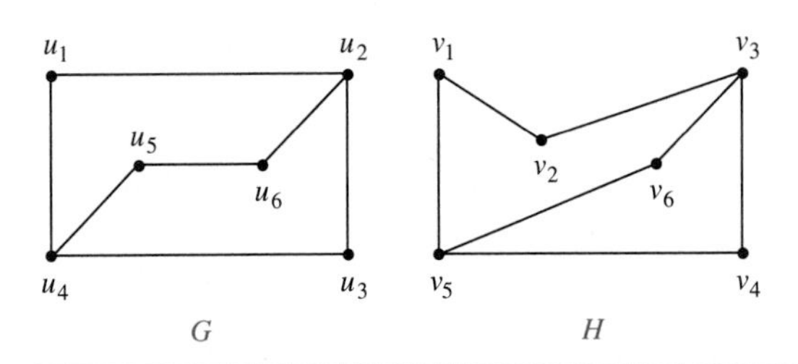

FIGURE 12 Graphs G and H.

To see whether f preserves edges, we examine the adjacency matrix of G,

$$\mathbf{A}_G = \begin{array}{c} \\ u_1 \\ u_2 \\ u_3 \\ u_4 \\ u_5 \\ u_6 \end{array} \begin{array}{c} \begin{array}{cccccc} u_1 & u_2 & u_3 & u_4 & u_5 & u_6 \end{array} \\ \left[\begin{array}{cccccc} 0 & 1 & 0 & 1 & 0 & 0 \\ 1 & 0 & 1 & 0 & 0 & 1 \\ 0 & 1 & 0 & 1 & 0 & 0 \\ 1 & 0 & 1 & 0 & 1 & 0 \\ 0 & 0 & 0 & 1 & 0 & 1 \\ 0 & 1 & 0 & 0 & 1 & 0 \end{array}\right] \end{array},$$

and the adjacency matrix of H with the rows and columns labeled by the images of the corresponding vertices in G,

$$\mathbf{A}_H = \begin{array}{c} \\ v_6 \\ v_3 \\ v_4 \\ v_5 \\ v_1 \\ v_2 \end{array} \begin{array}{c} \begin{array}{cccccc} v_6 & v_3 & v_4 & v_5 & v_1 & v_2 \end{array} \\ \left[\begin{array}{cccccc} 0 & 1 & 0 & 1 & 0 & 0 \\ 1 & 0 & 1 & 0 & 0 & 1 \\ 0 & 1 & 0 & 1 & 0 & 0 \\ 1 & 0 & 1 & 0 & 1 & 0 \\ 0 & 0 & 0 & 1 & 0 & 1 \\ 0 & 1 & 0 & 0 & 1 & 0 \end{array}\right] \end{array}.$$

Since $\mathbf{A}_G = \mathbf{A}_H$, it follows that f preserves edges. We conclude that f is an isomorphism, so that G and H are isomorphic. Note that if f turned out not to be an isomorphism, we would *not* have established that G and H are not isomorphic, since another correspondence of the vertices in G and H may be an isomorphism. ■

Exercises

In Exercises 1–4 use an adjacency list to represent the given graph.

1.

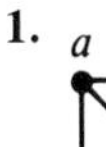
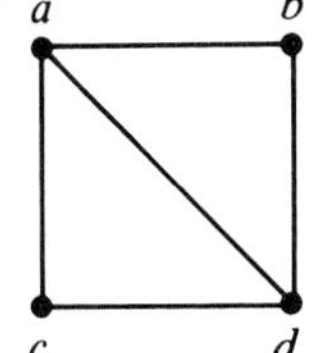

2.

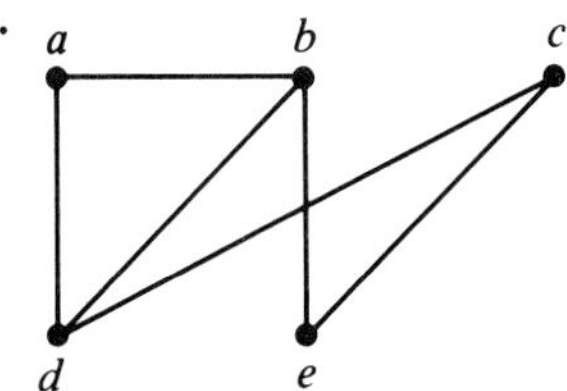

3.

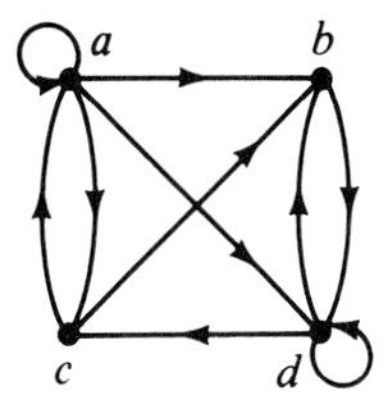

4.

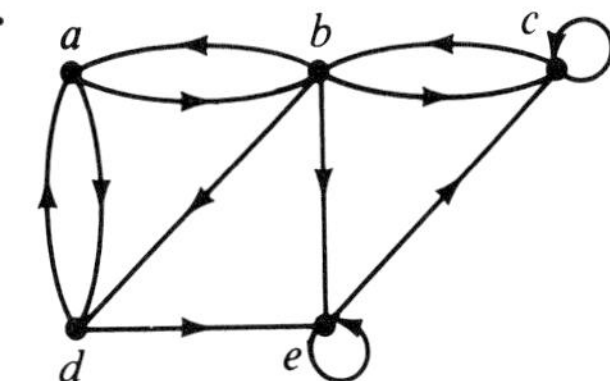

5. Represent the graph in Exercise 1 with an adjacency matrix.

6. Represent the graph in Exercise 2 with an adjacency matrix.
7. Represent the graph in Exercise 3 with an adjacency matrix.
8. Represent the graph in Exercise 4 with an adjacency matrix.
9. Represent each of the following graphs with an adjacency matrix.
   a) K_4 b) $K_{1,4}$ c) $K_{2,3}$
   d) C_4 e) W_4 f) Q_3

In Exercises 10–12 draw a graph with the given adjacency matrix.

10. $\begin{bmatrix} 0 & 1 & 0 \\ 1 & 0 & 1 \\ 0 & 1 & 0 \end{bmatrix}$

11. $\begin{bmatrix} 0 & 0 & 1 & 1 \\ 0 & 0 & 1 & 0 \\ 1 & 1 & 0 & 1 \\ 1 & 1 & 1 & 0 \end{bmatrix}$

12. $\begin{bmatrix} 1 & 1 & 1 & 0 \\ 0 & 0 & 1 & 0 \\ 1 & 0 & 1 & 0 \\ 1 & 1 & 1 & 0 \end{bmatrix}$

In Exercises 13–15 represent the given graph using an adjacency matrix.

13.

a b

c d

14.

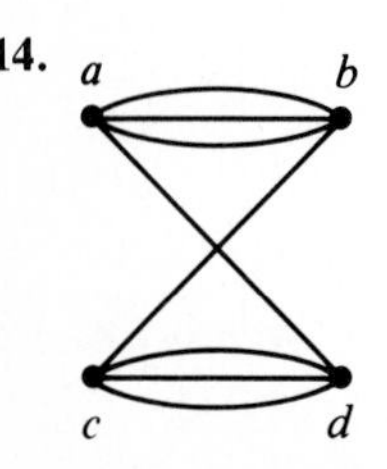

15.

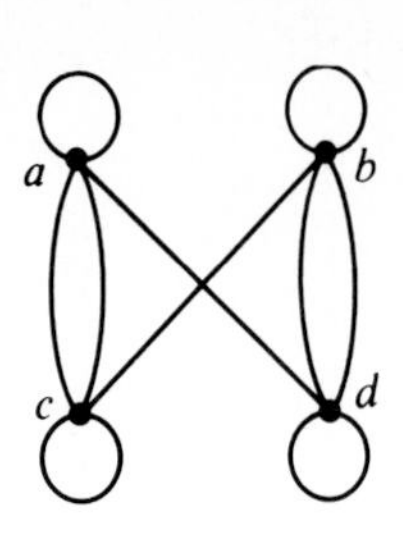

In Exercises 16–18 draw an undirected graph represented by the given adjacency matrix.

16. $\begin{bmatrix} 1 & 3 & 2 \\ 3 & 0 & 4 \\ 2 & 4 & 0 \end{bmatrix}$

17. $\begin{bmatrix} 1 & 2 & 0 & 1 \\ 2 & 0 & 3 & 0 \\ 0 & 3 & 1 & 1 \\ 1 & 0 & 1 & 0 \end{bmatrix}$

18. $\begin{bmatrix} 0 & 1 & 3 & 0 & 4 \\ 1 & 2 & 1 & 3 & 0 \\ 3 & 1 & 1 & 0 & 1 \\ 0 & 3 & 0 & 0 & 2 \\ 4 & 0 & 1 & 2 & 3 \end{bmatrix}$

In Exercises 19–21 find the adjacency matrix of the given directed multigraph.

19.

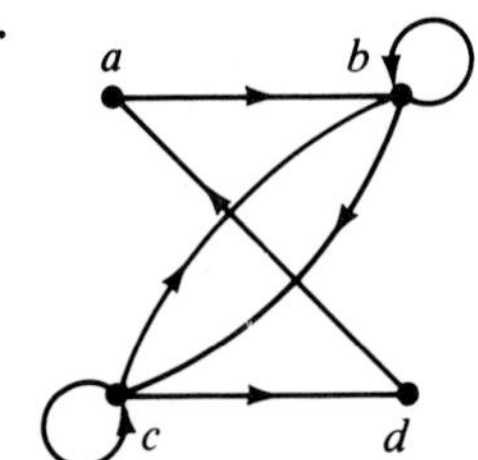

20.

a b

c d

21.

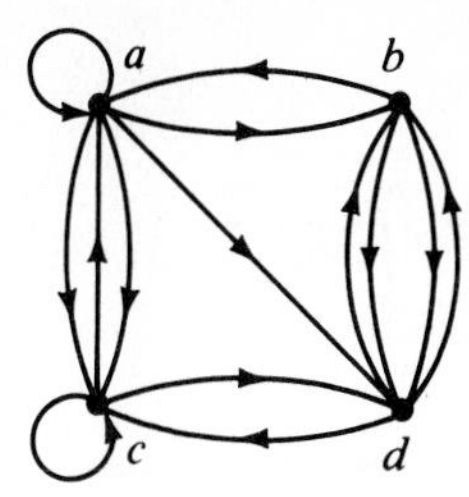

In Exercises 22–24 draw the graph represented by the given adjacency matrix.

22. $\begin{bmatrix} 1 & 0 & 1 \\ 0 & 0 & 1 \\ 1 & 1 & 1 \end{bmatrix}$

23. $\begin{bmatrix} 1 & 2 & 1 \\ 2 & 0 & 0 \\ 0 & 2 & 2 \end{bmatrix}$

24. $\begin{bmatrix} 0 & 2 & 3 & 0 \\ 1 & 2 & 2 & 1 \\ 2 & 1 & 1 & 0 \\ 1 & 0 & 0 & 2 \end{bmatrix}$

25. Is every zero-one square matrix that is symmetric and has zeros on the diagonal the adjacency matrix of a simple graph?
26. Use an incidence matrix to represent the graphs in Exercises 1 and 2.
27. Use an incidence matrix to represent the graphs in Exercises 13–15.

★28. What is the sum of the entries in a row of the adjacency matrix for an undirected graph? For a directed graph?

★29. What is the sum of the entries in a column of the adjacency matrix for an undirected graph? For a directed graph?

30. What is the sum of the entries in a row of the incidence matrix for an undirected graph?
31. What is the sum of the entries in a column of the incidence matrix for an undirected graph?

★32. Find an adjacency matrix for each of the following.
  a) K_n b) C_n c) W_n
  d) $K_{m,n}$ e) Q_n

★33. Find incidence matrices for the graphs in parts (a)–(d) of Exercise 32.

In Exercises 34–44 determine whether the given pair of graphs is isomorphic.

34.

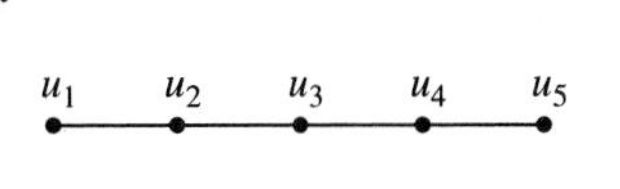

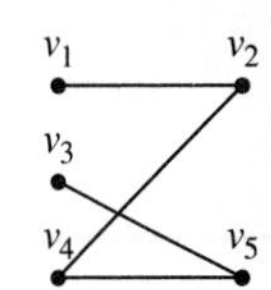

35.

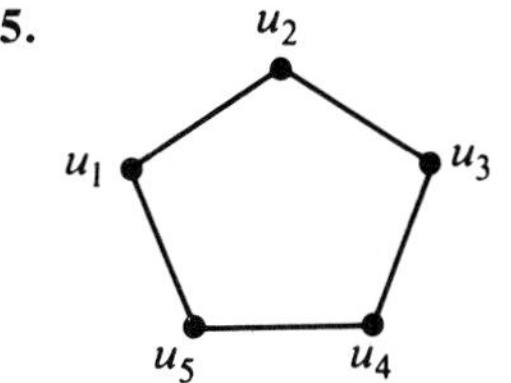

36.

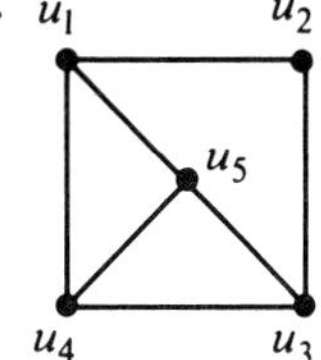

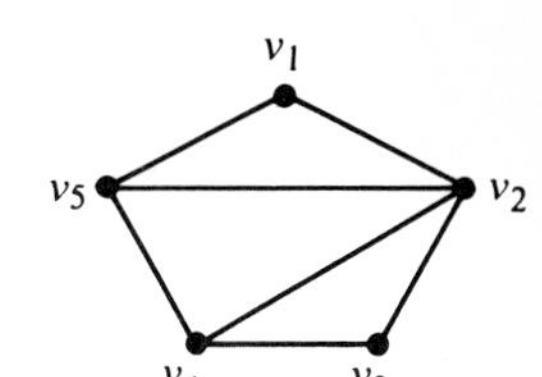

37.

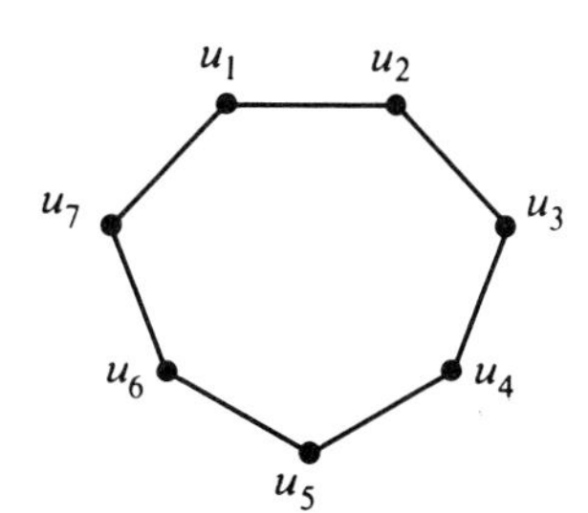

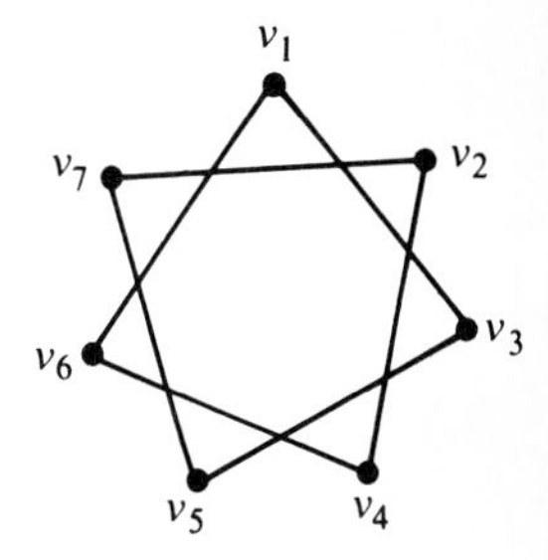

38.

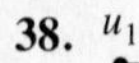

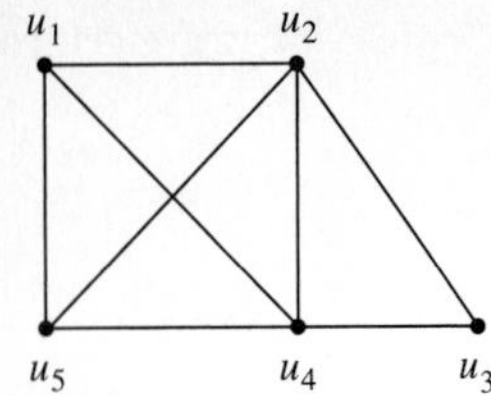

39.

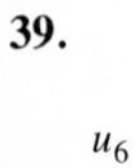

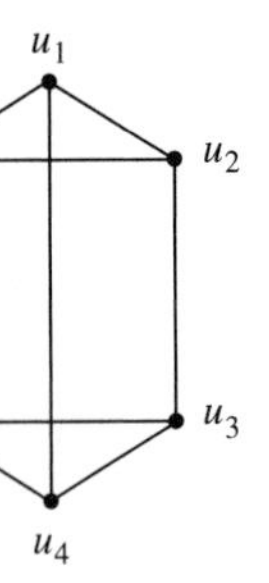

40.

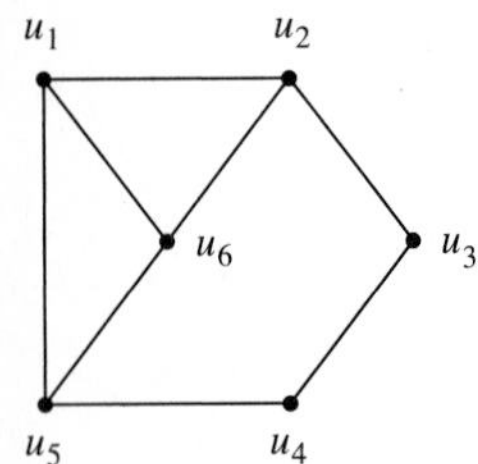

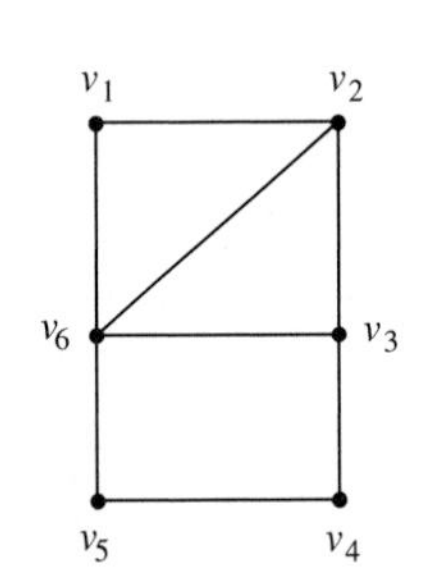

41.

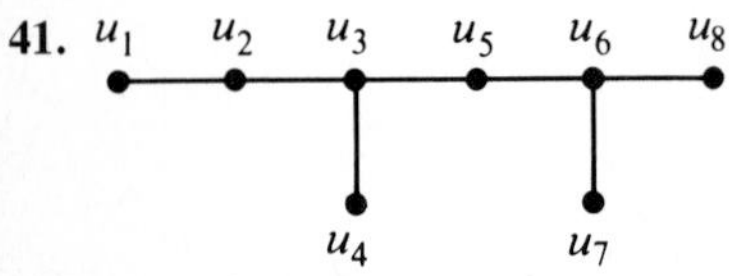

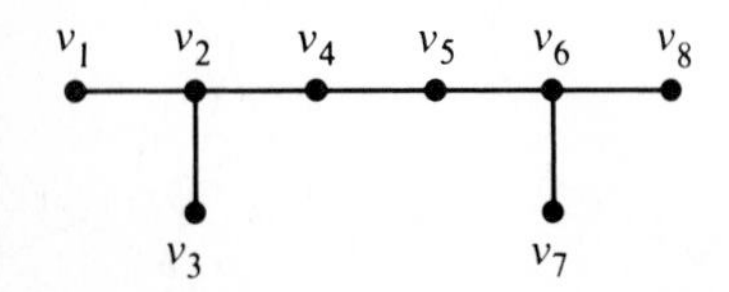

42.

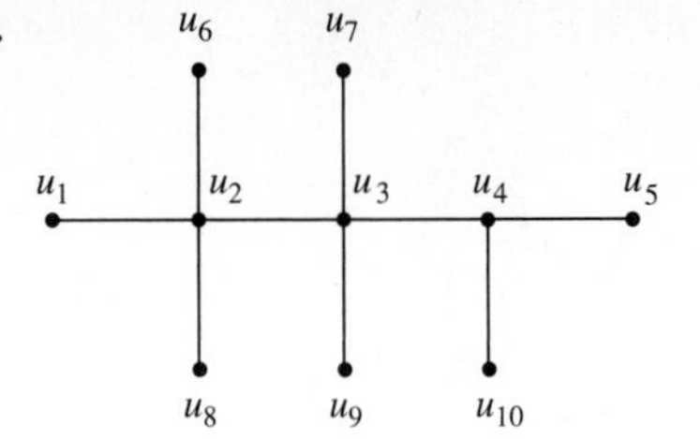

43.

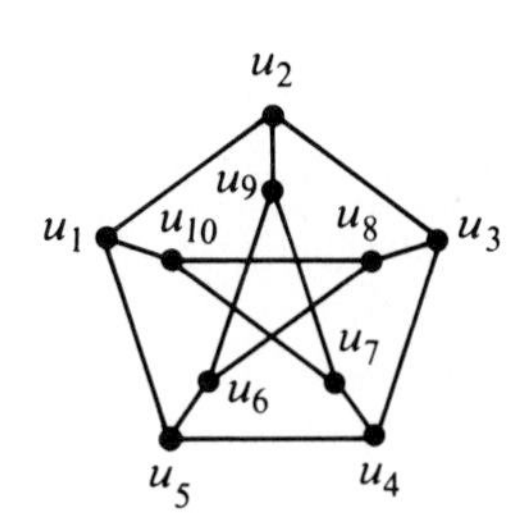

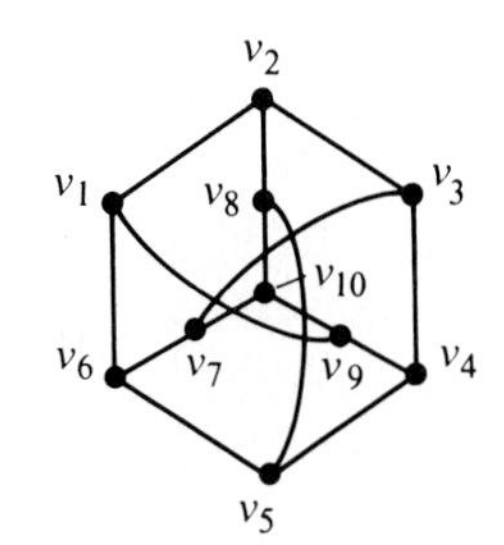

44.

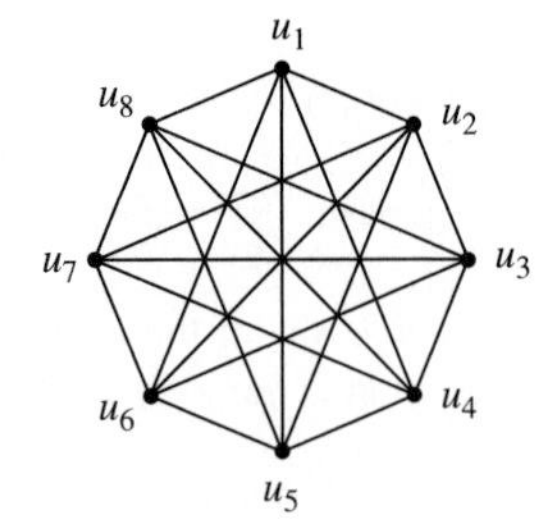

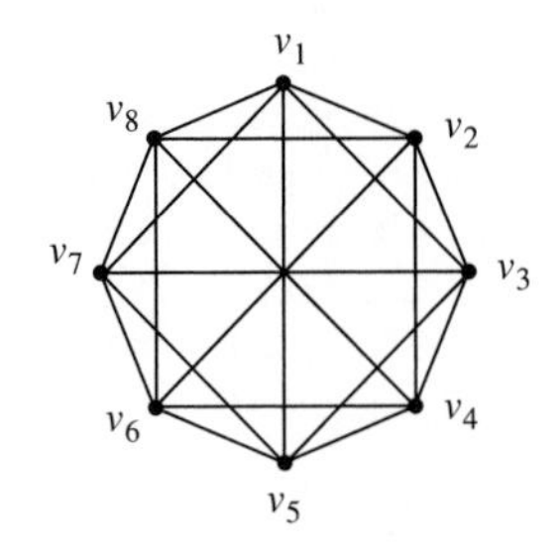

45. Show that isomorphism of simple graphs is an equivalence relation.

46. Suppose that G and H are isomorphic simple graphs. Show that their complementary graphs $\overline{G}$ and $\overline{H}$ are also isomorphic.

47. Describe the row and column of an adjacency matrix of a graph corresponding to an isolated vertex.

48. Describe the row of an incidence matrix of a graph corresponding to an isolated vertex.

49. Show that the vertices of a bipartite graph with two or more vertices can be ordered so that its adjacency matrix has the form

$$\begin{bmatrix} \mathbf{0} & \mathbf{A} \\ \mathbf{B} & \mathbf{0} \end{bmatrix}$$

where the four entries shown are rectangular blocks.

A simple graph G is called **self-complementary** if G and G are isomorphic.

50. Show that the following graph is self-complementary.

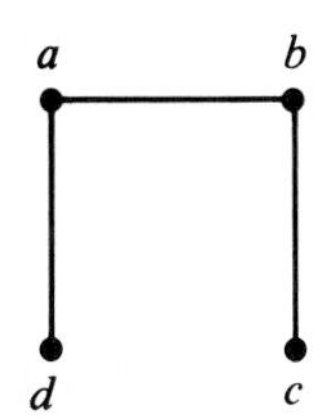

51. Find a self-complementary simple graph with five vertices.

★52. Show that if G is a self-complementary simple graph with v vertices, then $v \equiv 0$ or $1 \pmod 4$.

53. For which integers n is C_n self-complementary?

54. How many nonisomorphic simple graphs are there with n vertices, when n is

a) 2? **b)** 3? **c)** 4?

55. How many nonisomorphic simple graphs are there with five vertices and three edges?

56. How many nonisomorphic simple graphs are there with six vertices and four edges?

57. Are the simple graphs with the following adjacency matrices isomorphic?

a) $\begin{bmatrix} 0 & 0 & 1 \\ 0 & 0 & 1 \\ 1 & 1 & 0 \end{bmatrix}, \begin{bmatrix} 0 & 1 & 1 \\ 1 & 0 & 0 \\ 1 & 0 & 0 \end{bmatrix}$

b) $\begin{bmatrix} 0 & 1 & 0 & 1 \\ 1 & 0 & 0 & 1 \\ 0 & 0 & 0 & 1 \\ 1 & 1 & 1 & 0 \end{bmatrix}, \begin{bmatrix} 0 & 1 & 1 & 1 \\ 1 & 0 & 0 & 1 \\ 1 & 0 & 0 & 1 \\ 1 & 1 & 1 & 0 \end{bmatrix}$

c) $\begin{bmatrix} 0 & 1 & 1 & 0 \\ 1 & 0 & 0 & 1 \\ 1 & 0 & 0 & 1 \\ 0 & 1 & 1 & 0 \end{bmatrix}, \begin{bmatrix} 0 & 1 & 0 & 1 \\ 1 & 0 & 0 & 0 \\ 0 & 0 & 0 & 1 \\ 1 & 0 & 1 & 0 \end{bmatrix}$

58. Determine whether the graphs without loops with the following incidence matrices are isomorphic.

a) $\begin{bmatrix} 1 & 0 & 1 \\ 0 & 1 & 1 \\ 1 & 1 & 0 \end{bmatrix}, \begin{bmatrix} 1 & 1 & 0 \\ 1 & 0 & 1 \\ 0 & 1 & 1 \end{bmatrix}$

b) $\begin{bmatrix} 1 & 1 & 0 & 0 & 0 \\ 1 & 0 & 1 & 0 & 1 \\ 0 & 0 & 0 & 1 & 1 \\ 0 & 1 & 1 & 1 & 0 \end{bmatrix}, \begin{bmatrix} 0 & 1 & 0 & 0 & 1 \\ 0 & 1 & 1 & 1 & 0 \\ 1 & 0 & 0 & 1 & 0 \\ 1 & 0 & 1 & 0 & 1 \end{bmatrix}$

59. Extend the definition of isomorphism of simple graphs to undirected graphs containing loops and multiple edges.

60. Define isomorphism of directed graphs.

In Exercises 61–63 determine whether the given pair of directed graphs is isomorphic.

61.

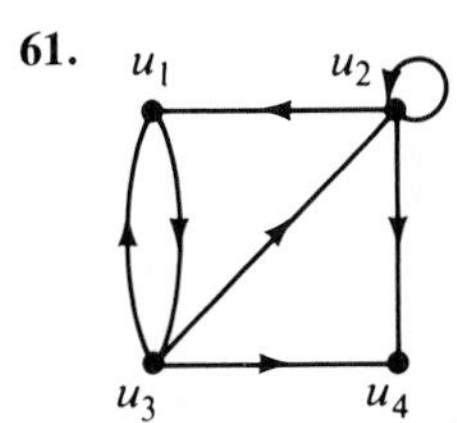

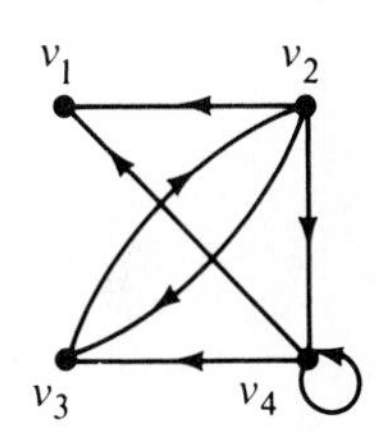

62.

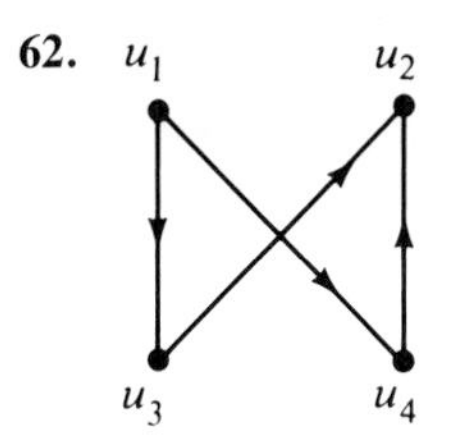

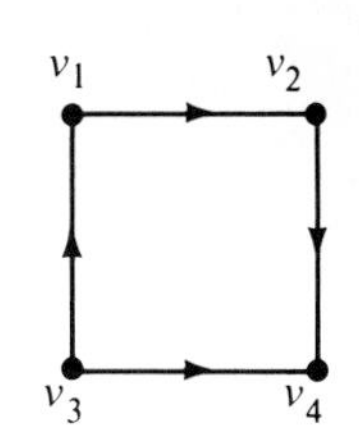

63.

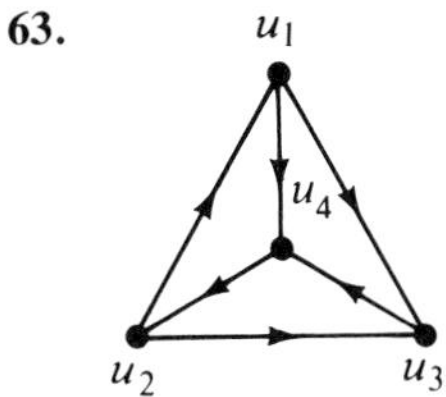

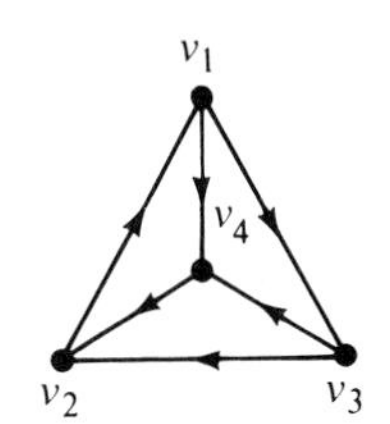

★64. How many nonisomorphic directed graphs are there with n vertices, when n is

a) 2? **b)** 3? **c)** 4?

★65. What is the product of the incidence matrix and its transpose for an undirected graph?

★66. How much storage is needed to represent a simple graph with v vertices and e edges using

a) adjacency lists? **b)** an adjacency matrix?

c) an incidence matrix?

7.4 Connectivity

INTRODUCTION

Many problems can be modeled with paths formed by traveling along the edges of graphs. For instance, the problem of determining whether a message can be sent between two computers using intermediate links can be studied with a graph model. Problems of efficiently planning routes for mail delivery, garbage pickup, diagnostics in computer networks, and so on, can be solved using models that involve paths in graphs.

PATHS

We begin by defining the basic terminology of graph theory that deals with paths.

DEFINITION 1. A *path* of *length* n from u to v, where n is a positive integer, in an undirected graph is a sequence of edges $e_1, \ldots, e_n$ of the graph such that $f(e_1) = \{x_0, x_1\}, f(e_2) = \{x_1, x_2\}, \ldots, f(e_n) = \{x_{n-1}, x_n\}$, where $x_0 = u$ and $x_n = v$. When the graph is simple, we denote this path by its vertex sequence $x_0, x_1, \ldots, x_n$ (since listing these vertices uniquely determines the path). The path is a *circuit* if it begins and ends at the same vertex, that is, if $u = v$. The path or circuit is said to *pass through* or *traverse* the vertices $x_1, x_2, \ldots, x_{n-1}$. A path or circuit is *simple* if it does not contain the same edge more than once.

When it is not necessary to distinguish between multiple edges, we will denote a path $e_1, e_2, \ldots, e_n$ where $f(e_i) = \{x_{i-1}, x_i\}$ for $i = 1, 2, \ldots, n$ by its vertex sequence $x_0, x_1, \ldots, x_n$. This notation identifies a path only up to the vertices it passes through. There may be more than one path that passes through this sequence of vertices.

EXAMPLE 1 In the simple graph shown in Figure 1, a, d, c, f, e is a simple path of length 4, since $\{a, d\}$, $\{d, c\}$, $\{c, f\}$, and $\{f, e\}$ are all edges. However, d, e, c, a is not a path, since $\{e, c\}$ is not an edge. Note that b, c, f, e, b is a circuit of length 4 since $\{b, c\}$, $\{c, f\}$, $\{f, e\}$, and $\{e, b\}$ are edges, and this path begins and ends at b. The path a, b, e, d, a, b, which is of length 5, is not simple since it contains the edge $\{a, b\}$ twice. ■

Paths and circuits in directed graphs were introduced in Chapter 6. We now define such paths for directed multigraphs.

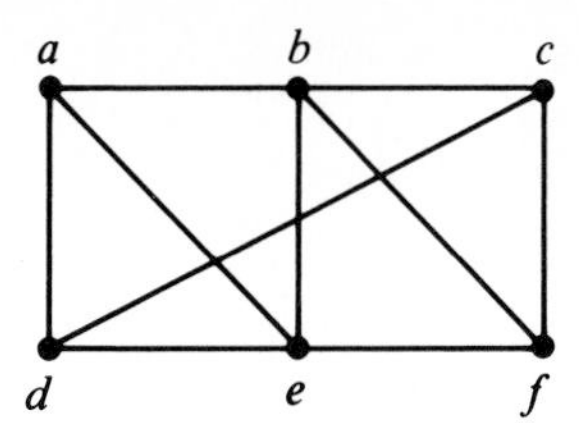

FIGURE 1 **A Simple Graph.**

DEFINITION 2. A *path* of length n, where n is a positive integer, from u to v in a directed multigraph is a sequence of edges $e_1, e_2, \ldots, e_n$ of the graph such that $f(e_1) = (x_0, x_1), f(e_2) = (x_1, x_2), \ldots, f(e_n) = (x_{n-1}, x_n)$ where $x_0 = u$ and $x_n = v$. When there are no multiple edges in the graph, this path is denoted by its vertex sequence $x_0, x_1, x_2, \ldots, x_n$. A path that begins and ends at the same vertex is called a *circuit* or *cycle*. A path or circuit is called *simple* if it does not contain the same edge more than once.

When it is not necessary to distinguish between multiple edges, we will denote a path $e_1, e_2, \ldots, e_n$ where $f(e_i) = (x_{i-1}, x_i)$ for $i = 1, 2, \ldots, n$ by its vertex sequence $x_0, x_1, \ldots, x_n$. The notation identifies a path only up to the vertices it passes through. There may be more than one path that passes through this sequence of vertices.

CONNECTEDNESS IN UNDIRECTED GRAPHS

When does a computer network have the property that every pair of computers can share information, if messages can be sent through one or more intermediate computers? When a graph is used to represent this computer network, where vertices represent the computers and edges represent the communications links, this question becomes: When is there always a path between two vertices in the graph?

DEFINITION 3. An undirected graph is called *connected* if there is a path between every pair of distinct vertices of the graph.

Thus, any two computers in the network can communicate if and only if the graph of this network is connected.

EXAMPLE 2 The graph G in Figure 2 is connected, since for every pair of distinct vertices there is a path between them (the reader should verify this). However, the graph H in Figure 2 is not connected. For instance, there is no path in H between vertices a and d. ■

We will need the following theorem in Chapter 8.

THEOREM 1 There is a simple path between every pair of distinct vertices of a connected undirected graph.

Proof: Let u and v be two distinct vertices of the connected undirected graph $G = (V, E)$. Since G is connected, there is at least one path between u and v. Let $x_0, x_1, \ldots, x_n$, where $x_0 = u$ and $x_n = v$, be the vertex sequence of a path of least length. This path of least length is simple. To see this, suppose it is not simple. Then $x_i = x_j$ for some i and j with $0 \leq i < j$. This means that there is a path from u to v of shorter length with vertex sequence $x_0, x_1, \ldots, x_{i-1}, x_j, \ldots, x_n$ obtained by deleting the edges corresponding to the vertex sequence $x_i, \ldots, x_{j-1}$. ■

A graph that is not connected is the union of two or more connected subgraphs, each pair of which has no vertex in common. These disjoint connected subgraphs are called the **connected components** of the graph.

EXAMPLE 3 What are the connected components of the graph G shown in Figure 3?

Solution: The graph G is the union of three disjoint connected subgraphs G_1, G_2, and G_3, shown in Figure 3. These three subgraphs are the connected components of G. ■

Sometimes the removal of a vertex and all edges incident with it produces a subgraph with more connected components than in the original graph. Such vertices are called **cut vertices** (or **articulation points**). The removal of a cut vertex from a connected graph produces a subgraph that is not connected. Analogously, an edge whose removal produces a graph with more connected components than in the original graph is called a **cut edge** or **bridge**.

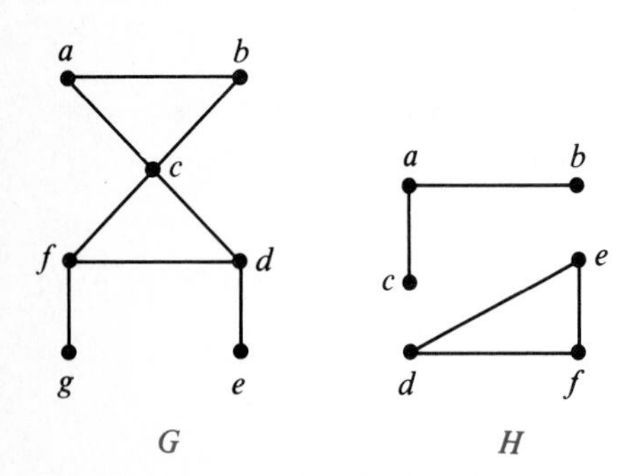

FIGURE 2 **The Graphs G and H.**

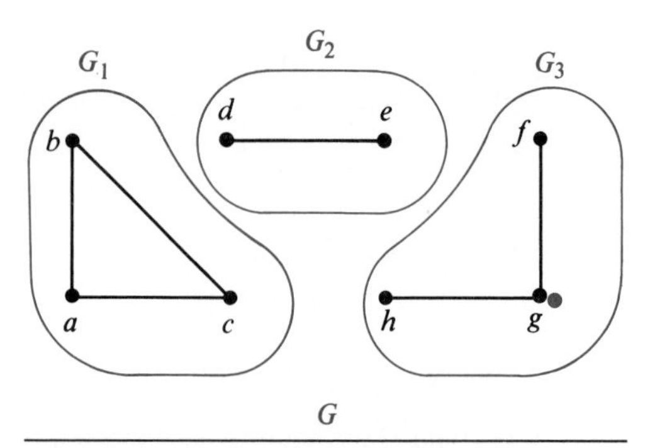

FIGURE 3 **The Graph G and Its Connected Components $G_1, G_2,$ and G_3.**

EXAMPLE 4 Find the cut vertices and cut edges in the graph G shown in Figure 4.

Solution: The cut vertices of G are b, c, and e. The removal of one of these vertices (and its adjacent edges) disconnects the graph. The cut edges are $\{a, b\}$ and $\{c, e\}$. Removing either one of these edges disconnects G. ■

CONNECTEDNESS IN DIRECTED GRAPHS

There are two notions of connectedness in directed graphs, depending on whether the directions of the edges are considered.

DEFINITION 4. A directed graph is *strongly connected* if there is a path from a to b and from b to a whenever a and b are vertices in the graph.

For a directed graph to be strongly connected there must be a sequence of directed edges from any vertex in the graph to any other vertex. A directed graph can fail to be strongly connected but still be in "one piece." To make this precise, the following definition is given.

DEFINITION 5. A directed graph is *weakly connected* if there is a path between any two vertices in the underlying undirected graph.

That is, a directed graph is weakly connected if and only if there is always a path between two vertices when the directions of the edges are disregarded. Clearly, any strongly connected directed graph is also weakly connected.

EXAMPLE 5 Are the directed graphs G and H shown in Figure 5 strongly connected? Are they weakly connected?

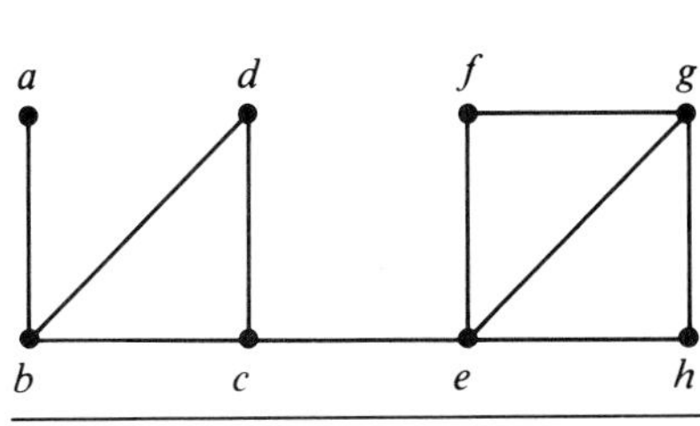

FIGURE 4 **The Graph G.**

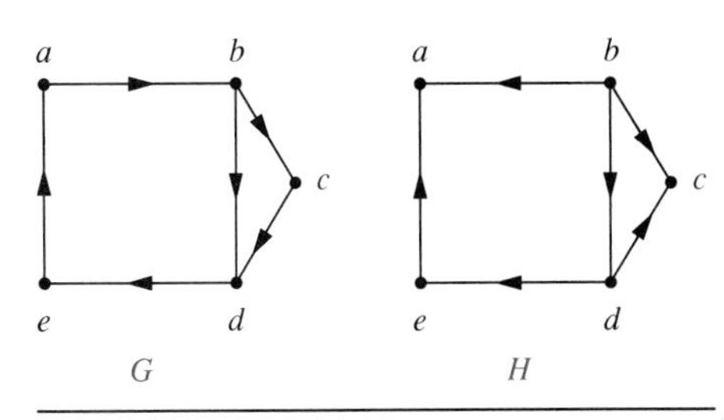

FIGURE 5 **The Directed Graphs G and H.**

Solution: G is strongly connected because there is a path between any two vertices in this directed graph (the reader should verify this). Hence, G is also weakly connected. The graph H is not strongly connected. There is no directed path from a to b in this graph. However, H is weakly connected, since there is a path between any two vertices in the underlying undirected graph of H (the reader should verify this). ■

PATHS AND ISOMORPHISM

There are several ways that paths and circuits can help determine whether two graphs are isomorphic. For example, the existence of a simple circuit of a particular length is a useful invariant that can be used to show that two graphs are not isomorphic. In addition, paths can be used to construct mappings that may be isomorphisms.

As we mentioned, a useful isomorphic invariant for simple graphs is the existence of a simple circuit of length k, where k is a positive integer greater than 2. (The proof that this is an invariant is left as Exercise 36 at the end of this section.) Example 6 illustrates how this invariant can be used to show that two graphs are not isomorphic.

EXAMPLE 6 Determine whether the graphs G and H shown in Figure 6 are isomorphic.

Solution: Both G and H have six vertices and eight edges. Each has four vertices of degree 3, and two vertices of degree 2. So, the three invariants—number of vertices, number of edges, and degrees of vertices—all agree for the two graphs. However, H has a simple circuit of length 3, namely, v_1, v_2, v_6, v_1 whereas G has no simple circuit of length 3, as can be determined by inspection (all simple circuits in G have length at least 4). Since the existence of a simple circuit of length 3 is an isomorphic invariant, G and H are not isomorphic. ■

We have shown how the existence of a type of path, namely, a simple circuit of a particular length, can be used to show that two graphs are not isomorphic. We can also use paths to find mappings that are potential isomorphisms.

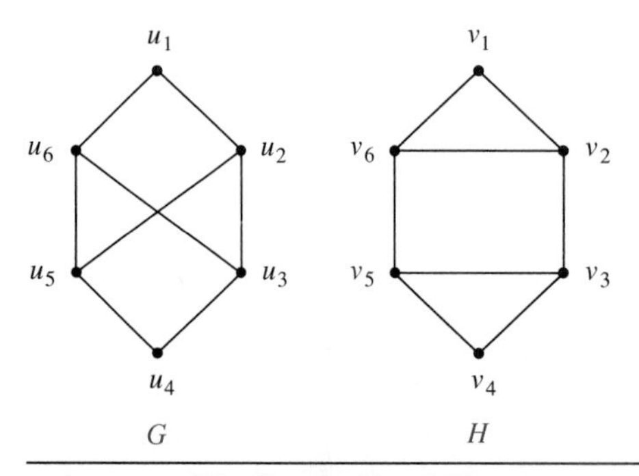

FIGURE 6 The Graphs *G* and *H*.

EXAMPLE 7 Determine whether the graphs G and H shown in Figure 7 are isomorphic.

Solution: Both G and H have five vertices and seven edges, both have three vertices of degree 3 and one vertex of degree 2, and both have a simple circuit of length 3, a simple circuit of length 4, and a simple circuit of length 5. Since all these isomorphic invariants agree, G and H may be isomorphic. To find a possible isomorphism, we can follow paths that go through all vertices so that the corresponding vertices in the two graphs have the same degree. For example, the paths u_1, u_4, u_3, u_2, u_5 in G and v_3, v_1, v_2, v_5, v_4 in H both go through every vertex in the graph and end at the vertex of degree 2. By following these paths through the graphs, we define the mapping f with $f(u_1) = v_3$, $f(u_4) = v_1$, $f(u_3) = v_2$, $f(u_2) = v_5$, and $f(u_5) = v_4$. The reader can show that f is an isomorphism, so that G and H are isomorphic, either by showing that f preserves edges, or by showing that with the appropriate orderings of vertices, the adjacency matrices of G and H are the same. ■

COUNTING PATHS BETWEEN VERTICES

The number of paths between two vertices in a graph can be determined using its adjacency matrix.

THEOREM 2 Let G be a graph with adjacency matrix $\mathbf{A}$ with respect to the ordering $v_1, v_2, \ldots, v_n$ (with directed or undirected edges, with multiple edges and loops allowed). The number of different paths of length from v_i to v_j, where r is a positive integer, equals the (i, j)th entry of $\mathbf{A}^r$.

Proof: The theorem will be proved using mathematical induction. Let G be a graph with adjacency matrix $\mathbf{A}$ (assuming an ordering $v_1, v_2, \ldots, v_n$ of the vertices of G). The number of paths from v_i to v_j of length 1 is the (i, j)th entry of $\mathbf{A}$, since this entry is the number of edges from v_i to v_j.

Assume that the (i, j)th entry of $\mathbf{A}^r$ is the number of different paths of length r from v_i to v_j. This is the induction hypothesis. Since $\mathbf{A}^{r+1} = \mathbf{A}^r\mathbf{A}$, the (i, j)th entry of $\mathbf{A}^{r+1}$ equals

$$\mathbf{A}^{r+1} b_{i1}a_{1j} + b_{i2}a_{2j} + \cdots + b_{in}a_{nj}$$

where b_{ik} is the (i, k)th entry of $\mathbf{A}^r$. By the induction hypothesis, b_{ik} is the number of paths of length r from v_i to v_k.

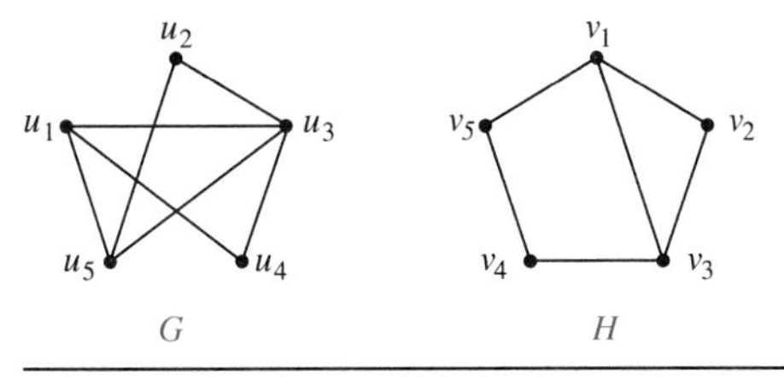

FIGURE 7 **The Graphs G and H.**

A path of length $r + 1$ from v_i to v_j is made up of a path of length r from v_i to some intermediate vertex v_k, and an edge from v_k to v_j. By the product rule for counting, the number of such paths is the product of the number of paths of length r from v_i to v_k, namely, b_{ik}, and the number of edges from v_k to v_j, namely, a_{kj}. When these products are added for all possible intermediate vertices v_k, the desired result follows by the sum rule for counting. ■

EXAMPLE 8 How many paths of length 4 are there from a to d in the simple graph G shown in Figure 8?

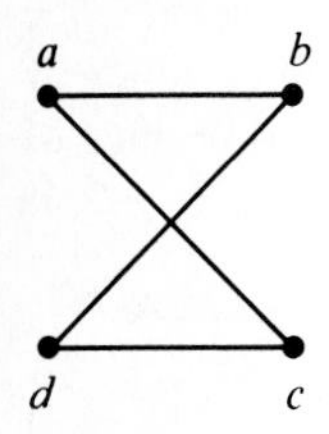

FIGURE 8 The Graph *G*.

Solution: The adjacency matrix of G (ordering the vertices as a, b, c, d) is

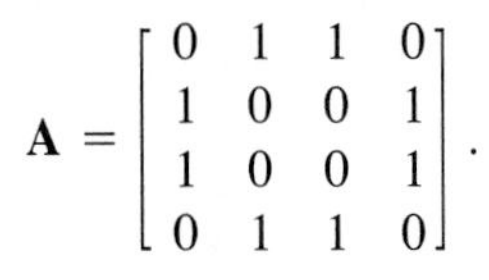

$$\mathbf{A} = \begin{bmatrix} 0 & 1 & 1 & 0 \\ 1 & 0 & 0 & 1 \\ 1 & 0 & 0 & 1 \\ 0 & 1 & 1 & 0 \end{bmatrix}.$$

Hence, the number of paths of length 4 from a to d is the $(1, 4)$th entry of $\mathbf{A}^4$. Since

$$\mathbf{A}^4 = \begin{bmatrix} 8 & 0 & 0 & 8 \\ 0 & 8 & 8 & 0 \\ 0 & 8 & 8 & 0 \\ 8 & 0 & 0 & 8 \end{bmatrix},$$

there are exactly eight paths of length 4 from a to d. By inspection of the graph, we see that a, b, a, b, d; a, b, a, c, d; a, b, d, b, d; a, b, d, c, d; a, c, a, b, d; a, c, a, c, d; a, c, d, b, d; and a, c, d, c, d are the eight paths from a to d. ■

Theorem 2 can be used to find the length of the shortest path between two vertices of a graph (see Exercise 32), and it can also be used to determine whether a graph is connected (see Exercises 37 and 38).

Exercises

1. Does each of the following lists of vertices form a path in the following graph? Which paths are simple? Which are circuits? What are the lengths of those that are paths?

a) a, e, b, c, b **b)** a, e, a, d, b, c, a
c) e, b, a, d, b, e **d)** c, b, d, a, e, c

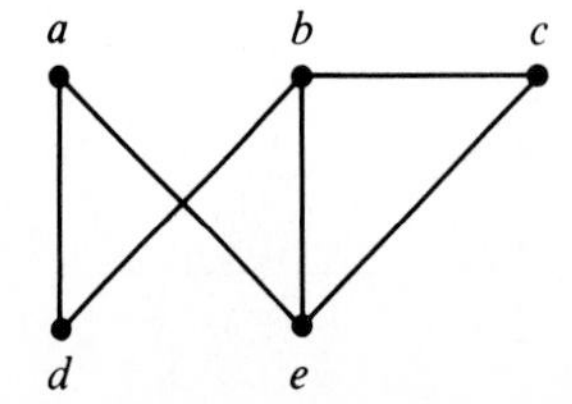

2. Does each of the following lists of vertices form a path in the following graph? Which paths are simple? Which are circuits? What are the lengths of those that are paths?

a) a, b, e, c, b **b)** a, d, a, d, a
c) a, d, b, e, a **d)** a, b, e, c, b, d, a

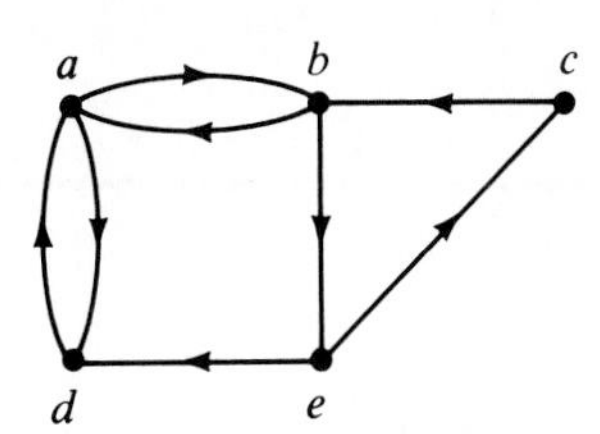

In Exercises 3–5 determine whether the given graph is connected.

3.

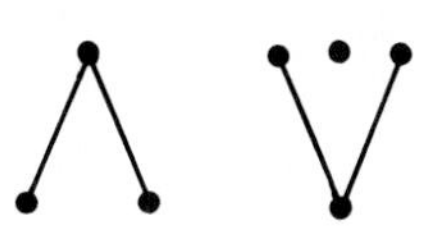

4.

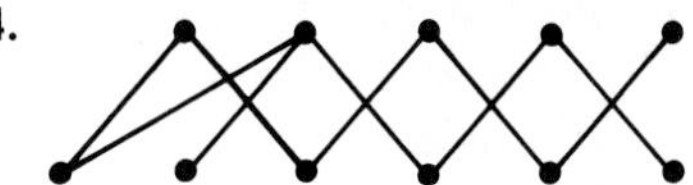

5.

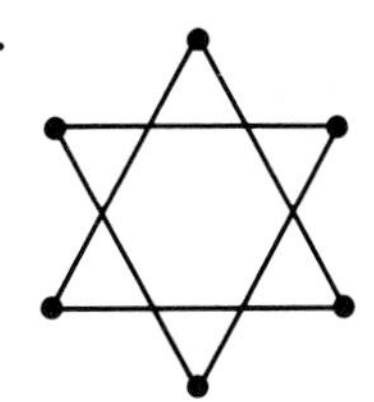

6. How many connected components does each of the graphs in Exercises 3–5 have? For each graph find each of its connected components.

★7. Find the number of paths of length n between two different vertices in K_4 if n is
a) 2. **b)** 3. **c)** 4. **d)** 5.

★8. Find the number of paths of length n between any two adjacent vertices in $K_{3,3}$ for the values of n in Exercise 7.

★9. Find the number of paths of length n between any two nonadjacent vertices in $K_{3,3}$ for the values of n in Exercise 7.

10. Find the number of paths between c and d in the graph in Figure 1 of length
a) 2. **b)** 3. **c)** 4. **d)** 5. **e)** 6. **f)** 7.

11. Find the number of paths from a to e in the directed graph in Exercise 2 of length
a) 2. **b)** 3. **c)** 4. **d)** 5. **e)** 6. **f)** 7.

★12. Show that a connected graph with n vertices has at least $n - 1$ edges.

13. Let $G = (V, E)$ be a simple graph. Let R be the relation on V consisting of pairs of vertices (u, v) such that there is a path from u to v or such that $u = v$. Show that R is an equivalence relation.

★14. Show that in any simple graph there is a path from any vertex of odd degree to some other vertex of odd degree.

In Exercises 15–17 find all the cut vertices of the given graph.

15.

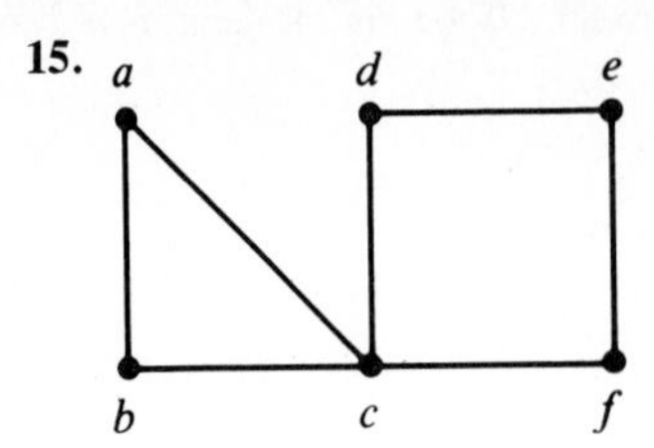

16.

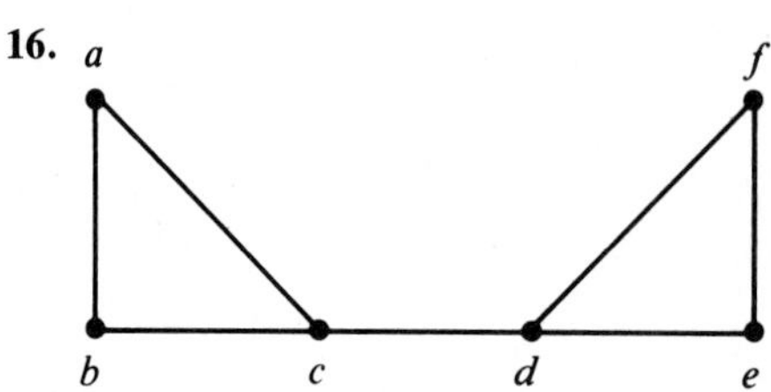

17.

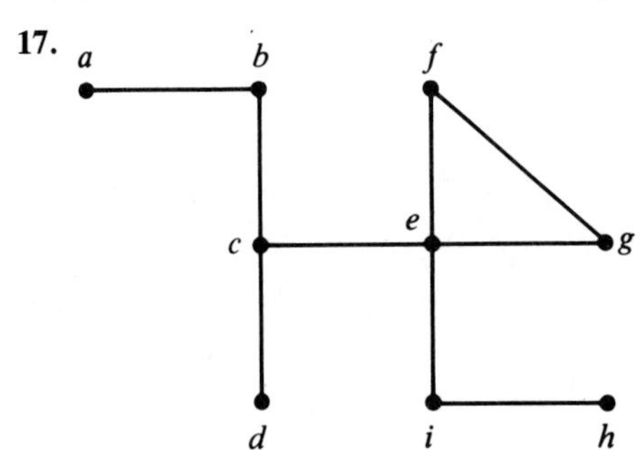

18. Find all the cut edges in the graphs in Exercises 15–17.

★19. Suppose that v is an endpoint of a cut edge. Prove that v is a cut vertex if and only if this vertex is not pendant.

★20. Show that a vertex c in the connected simple graph G is a cut vertex if and only if there are vertices u and v, both different from c, such that every path between u and v passes through c.

★21. Show that a simple graph with at least two vertices has at least two vertices that are not cut vertices.

★22. Show that an edge in a simple graph is a cut edge if and only if this edge is not part of any simple circuit in the graph.

23. A communications link in a network should be provided with a backup link if its failure makes it impossible for some message to be sent. For each of the following communications networks, determine those links that should be backed up.

a)

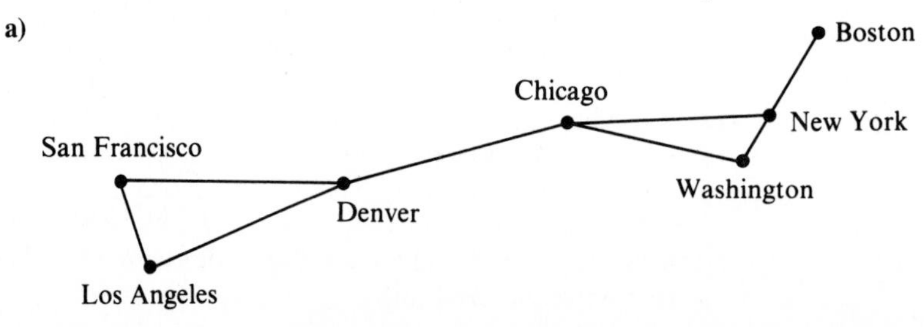

b)

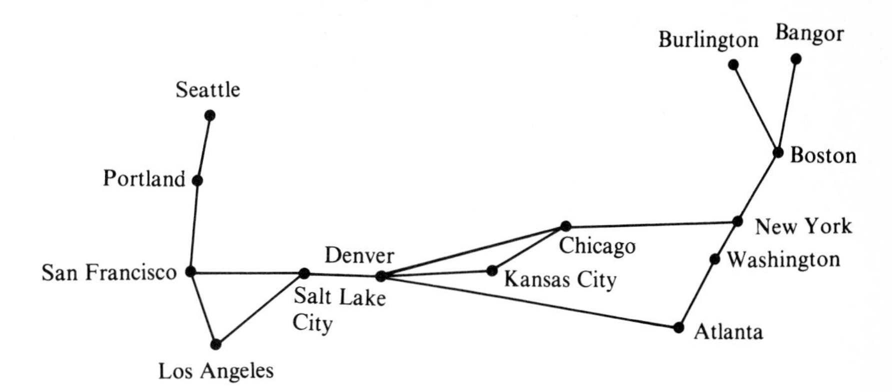

A **vertex basis** in a directed graph is a set of vertices such that there is a path to every vertex in the directed graph not in the set from some vertex in this set.

24. Find a vertex basis for each of the directed graphs in Exercises 7–9 of Section 7.2.

25. What is the significance of a vertex basis in an influence graph (described in Example 2 of Section 7.1)? Find a vertex basis in the influence graph in this example.

☞ **26.** Show that if a connected simple graph G is the union of the graphs G_1 and G_2, then G_1 and G_2 have at least one common vertex.

★**27.** Show that if a simple graph G has k connected components and these components have $n_1, n_2, \ldots, n_k$ vertices, respectively, then the number of edges of G does not exceed

$$\sum_{i=1}^{k} C(n_i, 2).$$

★**28.** Use Exercise 27 to show that a simple graph with n vertices and k connected components has at most $(n-k)(n-k+1)/2$ edges. (*Hint:* First show that

$$\sum_{i=1}^{k} n_i^2 \leq n^2 - (k-1)(2n-k)$$

where n_i is the number of vertices in the ith connected component.)

★**29.** Show that a simple graph G with n vertices is connected if it has more than $(n-1)(n-2)/2$ edges.

30. Describe the adjacency matrix of a graph with n connected components when the vertices of the graph are listed so that vertices in each connected component are listed successively.

31. How many nonisomorphic connected simple graphs are there with n vertices when n is
a) 2? **b)** 3? **c)** 4? **d)** 5?

32. Explain how Theorem 2 can be used to find the length of the shortest path from a vertex v to a vertex w in a graph.

33. Use Theorem 2 to find the length of the shortest path between a and f in the multigraph in Figure 1.

34. Use Theorem 2 to find the length of the shortest path from a and c in the directed graph in Exercise 2.

☞**35.** Let P_1 and P_2 be two simple simple paths between the vertices u and v in the simple graph G that do not contain the same set of edges. Show that there is a simple circuit in G.

36. Show that the existence of a simple circuit of length k, where k is a positive integer greater than 2, is an isomorphic invariant.

37. Explain how Theorem 2 can be used to determine whether a graph is connected.

38. Use Exercise 37 to show that the graph G in Figure 2 is connected whereas the graph H in that figure is not connected.

7.5 Euler and Hamilton Paths

INTRODUCTION

The town of Königsberg, Prussia (now called Kaliningrad and part of the USSR), was divided into four sections by the branches of the Pregel River. These four sections included the two regions on the banks of the Pregel, Kneiphof Island, and the region between the two branches of the Pregel. In the 18th century seven bridges connected these regions. Figure 1 depicts these regions and bridges.

The townspeople took long walks through town on Sundays. They wondered whether it was possible to start at some location in the town and travel across all the bridges without crossing any bridge twice, and return to the starting point.

The Swiss mathematician Leonhard Euler solved this problem. His solution, published in 1736, may be the first use of graph theory. Euler studied this problem using the multigraph obtained when the four regions are represented by vertices and the bridges by edges. This multigraph is shown in Figure 2.

The problem of traveling across every bridge without crossing any bridge more than once can be rephrased in terms of this model. The question becomes: Is there a simple circuit in this multigraph that contains every edge?

Leonhard Euler (1707–1783). Leonhard Euler was the son of a Calvinist minister from the vicinity of Basel, Switzerland. At 13, he entered the University of Basel, pursuing a career in theology, as his father wished. At the university Euler was tutored by Johann Bernoulli of the famous Bernoulli family of mathematicians. His interest and skills led him to abandon his theological studies and take up mathematics. Euler obtained his master's degree in philosophy at the age of 16. In 1727 Peter the Great invited him to join the Academy at St. Petersburg. In 1741 he moved to the Berlin Academy, where he stayed until 1766. He then returned to St. Petersburg, where he remained for the rest of his life.

Euler was incredibly prolific, contributing to many areas of mathematics, including number theory, combinatorics, analysis, as well as its applications to such areas as music and naval architecture. He wrote over 700 books and papers and left so much unpublished work that it took 47 years after he died for all his work to be published. During his life his papers accumulated so quickly that he kept a large pile of articles awaiting publication. The Berlin Academy published the papers on top of this pile, so that later results were often published before results they depended on or superceded. Euler had 13 children and was able to continue his work while a child or two bounced on his knees. He was blind for the last 17 years of his life, but because of his fantastic memory this did not diminish his mathematical output. The project of publishing his collected works, undertaken by the Swiss Society of Natural Science, is still going on and will require more than 75 volumes.

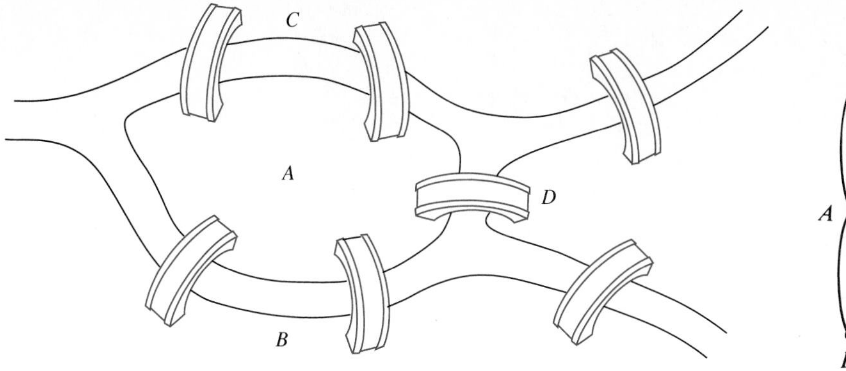

FIGURE 1 The 18th-Century Town of Königsberg.

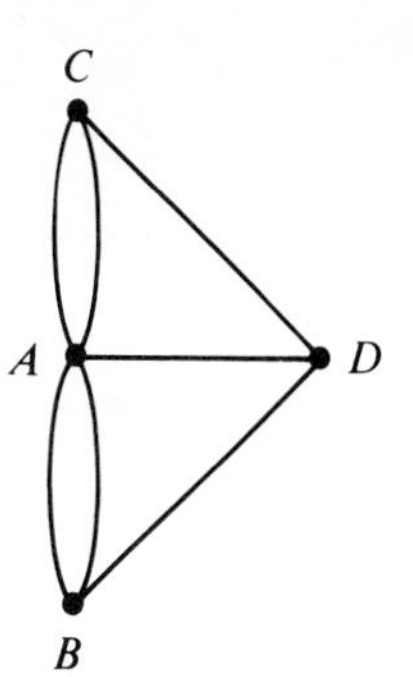

FIGURE 2 A Multigraph Model of the Town of Königsberg.

DEFINITION 1. An *Euler circuit* in a graph G is a simple circuit containing every edge of G. An *Euler path* in G is a simple path containing every edge of G.

The following examples illustrate the concept of Euler circuits and paths.

EXAMPLE 1 Which of the undirected graphs in Figure 3 have an Euler circuit? Of those that do not, which have an Euler path?

Solution: The graph G_1 has an Euler circuit, for example a, e, c, d, e, b, a. Neither of the graphs G_2 or G_3 has an Euler circuit (the reader should verify this). However, G_3 has an Euler path, namely, a, c, d, e, b, d, a, b. G_2 does not have an Euler path (as the reader should verify). ■

EXAMPLE 2 Which of the directed graphs in Figure 4 have an Euler circuit? Of those that do not, which have an Euler path?

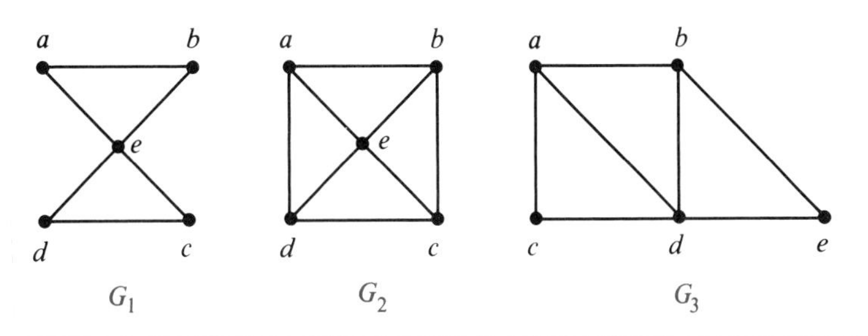

FIGURE 3 The Undirected Graphs G_1, G_2, and G_3.

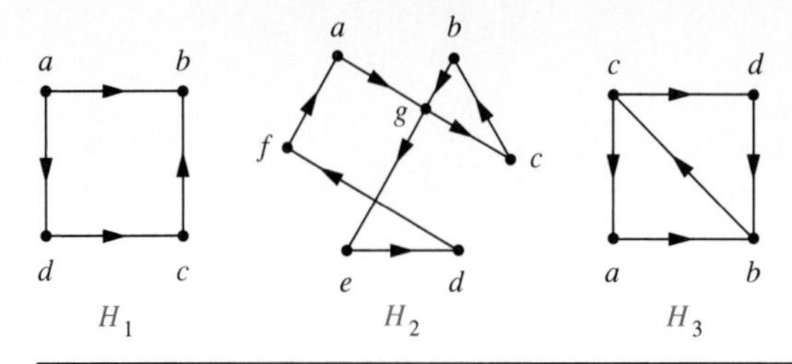

FIGURE 4 **The Directed Graphs H_1, H_2, and H_3.**

Solution: The graph H_2 has an Euler circuit, for example, $a, g, c, b, g, e, d, f, a$. Neither H_1 nor H_3 has an Euler circuit (as the reader should verify). H_3 has an Euler path, namely c, a, b, c, d, b but H_1 does not (as the reader should verify). ■

NECESSARY AND SUFFICIENT CONDITIONS FOR EULER CIRCUITS AND PATHS

There are simple criteria for determining whether a multigraph has an Euler circuit or an Euler path. Euler discovered them when he solved the famous Königsberg bridge problem. We will assume that all graphs discussed in this section have a finite number of vertices and edges.

What we can say if a connected multigraph has an Euler circuit? What we can show is that every vertex must have even degree. To do this, first note that an Euler circuit begins with a vertex a and continues with an edge incident to a, say $\{a, b\}$. The edge $\{a, b\}$ contributes 1 to deg(a). Each time the circuit passes through a vertex it contributes 2 to its degree, since the circuit enters via an edge incident with this vertex and leaves via another such edge. Finally, the circuit terminates where it started, contributing 1 to deg(a). Therefore, deg(a) must be even, because the circuit contributes 1 when it begins, 1 when it ends, and 2 every time it passes through a (if it ever does). A vertex other than a has even degree because the circuit contributes 2 to its degree each time it passes through the vertex. We conclude that if a connected graph has an Euler circuit, then every vertex must have even degree.

Is this necessary condition for the existence of an Euler circuit also sufficient? That is, must an Euler circuit exist in a connected multigraph if all vertices have even degree? This question can be settled affirmatively with a construction.

Suppose that G is a connected multigraph and the degree of every vertex of G is even. We will form a simple circuit that begins at an arbitrary vertex a of G. Let $x_0 = a$. First, we arbitrarily choose an edge $\{x_0, x_1\}$ incident with a. We continue by building a simple path $\{x_0, x_1\}, \{x_1, x_2\}, \ldots, \{x_{n-1}, x_n\}$ that is as long as possible. For instance, in the graph G in Figure 5 we begin at a and choose in succession the edges $\{a, f\}$, $\{f, c\}$, $\{c, b\}$, and $\{b, a\}$.

The path terminates since the graph has a finite number of edges. It begins at a with an edge of the form $\{a, x\}$, and it terminates at a with an edge of the form $\{y, a\}$. This follows since each time the path goes through a vertex with even degree,

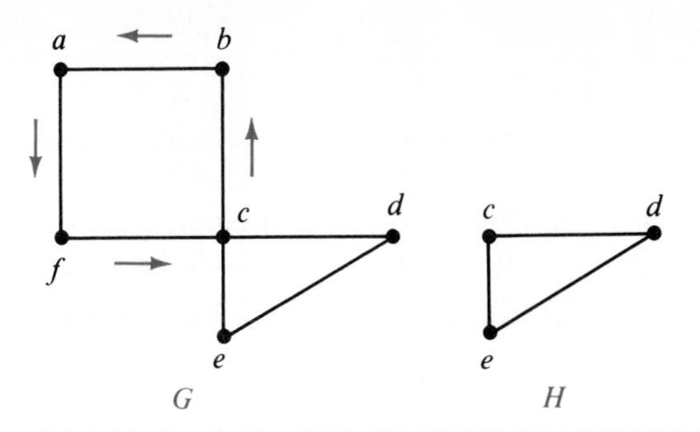

FIGURE 5 **Constructing an Euler Circuit in G.**

it uses only one edge to enter this vertex, so that at least one edge remains for the path to leave the vertex. This path may use all the edges, or it may not.

An Euler circuit has been constructed if all the edges have been used. Otherwise, consider the subgraph H obtained from G by deleting the edges already used and vertices that are not incident with any remaining edges. When we delete the circuit a, f, c, b, a from the graph in Figure 5, we obtain the subgraph labeled as H.

Since G is connected, H has at least one vertex in common with the circuit that has been deleted. Let w be such a vertex. (In our example, c is the vertex.)

Every vertex in H has even degree (because in G all vertices had even degree, and for each vertex, pairs of edges incident with this vertex have been deleted to form H.) Note that H may not be connected. Beginning at w, construct a simple path in H by choosing edges as long as possible, as was done in G. This path must terminate at w. For instance, in Figure 5, c, d, e, c is a path in H. Next, form a circuit in G by splicing the circuit in H with the original circuit in G (this can be done since w is one of the vertices in this circuit). When this is done in the graph in Figure 5, we obtain the circuit a, f, c, d, e, c, b, a.

Continue this process until all edges have been used. (The process must terminate since there are only a finite number of edges in the graph.) This produces an Euler circuit. The construction shows that if the vertices of a connected multigraph all have even degree, then the graph has an Euler circuit.

We summarize these results in Theorem 1.

THEOREM 1 A connected multigraph has an Euler circuit if and only if each of its vertices has even degree.

We can now solve the Königsberg bridge problem. Since the multigraph representing these bridges, shown in Figure 2, has four vertices of odd degree, it does not have an Euler circuit. There is no way to start at a given point, cross each bridge exactly once, and return to the starting point.

Algorithm 1 gives the constructive procedure for finding Euler circuits given in the discussion preceding Theorem 1. (Since the circuits in the procedure are chosen arbitrarily, there is some ambiguity. We will not bother to remove this ambiguity by specifying the steps of the procedure more precisely.)

ALGORITHM 1 Constructing Euler Circuits.

procedure *Euler*(G: connected multigraph with all vertices of even degree)
circuit := a circuit in G beginning at an arbitrarily chosen vertex
  with edges successively added to form a path that returns to this vertex
H := G with the edges of this circuit removed
while H has edges
begin
  subcircuit := a circuit in H beginning at a vertex in H that also is an
    endpoint of an edge of *circuit*
  H := H with edges of *subcircuit* and all isolated vertices removed
  circuit := *circuit* with *subcircuit* inserted at the appropriate vertex
end {*circuit* is an Euler circuit}

The next example shows how Euler paths and circuits can be used to solve a type of puzzle.

EXAMPLE 3 Many puzzles ask you to draw a picture in a continuous motion without lifting a pencil so that no part of the picture is retraced. We can solve such puzzles using Euler circuits and paths. For example, can **Mohammed's scimitars**, shown in Figure 6, be drawn in this way, where the drawing begins and ends at the same point?

Solution: We can solve this problem since the graph G shown in Figure 6 has an Euler circuit. It has such a circuit since all its vertices have even degree. We will use Algorithm 1 to construct an Euler circuit. First, we form the circuit $a, b, d, c, b, e, i, f, e, a$. We obtain the subgraph H by deleting the edges in this circuit and all vertices that become isolated when these edges are removed. Then we form the circuit $d, g, h, j, i, h, k, g, f, d$ in H. After forming this circuit we have used all edges in G. Splicing this new circuit into the first circuit at the appropriate place produces the Euler circuit $a, b, d, g, h, j, i, h, k, g, f, d, c, b, e, i, f, e, a$. This circuit gives a way to draw the scimitars without lifting the pencil or retracing part of the picture ■

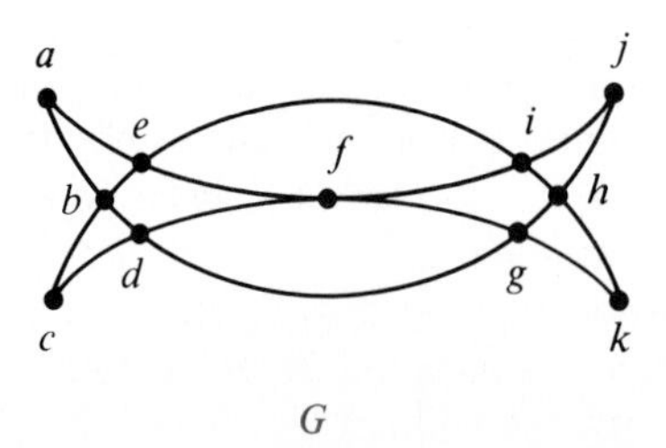

FIGURE 6 **Mohammed's Scimitars.**

Another algorithm for constructing Euler circuits, called Fleury's algorithm, is described in the exercises at the end of this section.

We will now show that a connected multigraph has an Euler path (and not an Euler circuit) if and only if it has exactly two vertices of odd degree. First, suppose that a connected multigraph does have an Euler path from a to b, but not an Euler circuit. The first edge of the path contributes 1 to the degree of a. A contribution of 2 to the degree of a is made every time the path passes through a. The last edge in the path contributes 1 to the degree of b. Every time the path goes through b there is a contribution of 2 to its degree. Consequently, both a and b have odd degree. Every other vertex has even degree, since the path contributes 2 to the degree of a vertex whenever it passes through it.

Now consider the converse. Suppose that a graph has exactly two vertices of odd degree, say a and b. Consider the larger graph made up of the original graph with the addition of an edge $\{a, b\}$. Every vertex of this larger graph has even degree, so that there is an Euler circuit. The removal of the new edge produces an Euler path in the original graph. The following theorem summarizes these results.

THEOREM 2 A connected multigraph has an Euler path but not an Euler circuit if and only if it has exactly two vertices of odd degree. ■

EXAMPLE 4 Which graphs shown in Figure 7 have an Euler path?

Solution: G_1 contains exactly two vertices of odd degree, namely, b and d. Hence, it has an Euler path that must have b and d as its endpoints. One such Euler path is d, a, b, c, d, b. Similarly, G_2 has exactly two vertices of odd degree, namely, b and f. So it has an Euler path that must have b and f as endpoints. One such Euler path is $b, a, g, f, e, d, c, g, b, c, f$. G_3 has no Euler path since it has six vertices of odd degree. ■

Returning to 18th century Königsberg, is it possible to start at some point in the town, travel across all the bridges, and end up at some other point in town? This question can be answered by determining whether there is an Euler path in the multigraph representing the bridges in Königsberg. Since there are four vertices of odd degree in this multigraph, there is no Euler path, so that such a trip is impossible.

Necessary and sufficient conditions for Euler paths and circuits in directed graphs are discussed in the exercises at the end of this section.

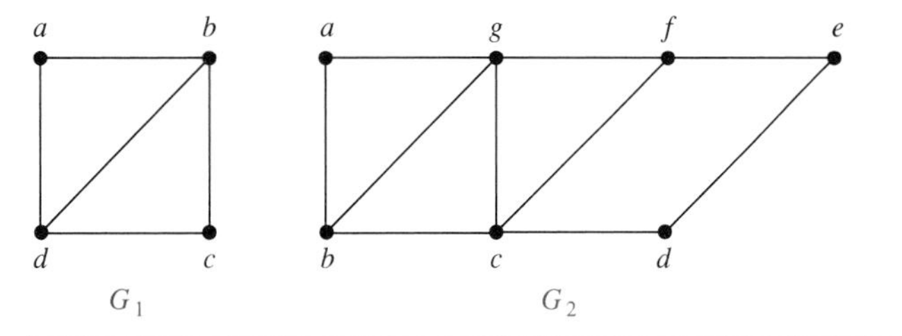

FIGURE 7 Three Undirected Graphs.

HAMILTON PATHS AND CIRCUITS

We have developed necessary and sufficient conditions for the existence of paths and circuits that contain every edge of a multigraph exactly once. Can we do the same for simple paths and circuits that contain every vertex of the graph exactly once?

DEFINITION 2. A path $x_0, x_1, \ldots, x_{n-1}, x_n$ in the graph $G = (V, E)$ is called a *Hamilton path* if $V = \{x_0, x_1, \ldots, x_{n-1}, x_n\}$ and $x_i \neq x_j$ for $0 \leq i < j \leq n$. A circuit $x_0, x_1, \ldots, x_{n-1}, x_n, x_0$ (with $n > 1$) in a graph $G = (V, E)$ is called a *Hamilton circuit* if $x_0, x_1, \ldots, x_{n-1}, x_n$ is a Hamilton path.

This terminology comes from a puzzle invented in 1857 by the Irish mathematician Sir William Rowan Hamilton. Hamilton's puzzle consisted of a wooden dodecahedron (a polyhedron with 12 regular pentagons as faces, as shown in Figure 8(a)), with a peg at each vertex of the dodecahedron, and string. The 20 vertices of the dodecahedron were labeled with different cities in the world. The object of the puzzle was to start at a city and travel along the edges of the dodecahedron, visiting each of the other 19 cities exactly once, and end back at the first city. The circuit traveled was marked off using the strings and pegs.

Since the author cannot supply each reader with a wooden solid with pegs and string, we will consider the equivalent question: Is there a circuit in the graph shown

---

William Rowan Hamilton (1805–1865). William Rowan Hamilton, the most famous Irish scientist ever to have lived, was born in 1805 in Dublin. His father was a successful lawyer, his mother came from a family noted for their intelligence, and he was a child prodigy. By the age of 3, he was an excellent reader and had mastered advanced arithmetic. Because of his brilliance, he was sent off to live with his uncle James, a noted linguist. By 8 Hamilton had learned Latin, Greek, and Hebrew; by 10 he had also learned Italian and French and he began his study of oriental languages, including Arabic, Sanskrit, and Persian. During this period he took pride in knowing as many languages as his age. At 17, no longer devoted to learning new languages and having mastered calculus and much mathematical astronomy, he began original work in optics, and he also found an important mistake in Laplace's work on celestial mechanics. Before entering Trinity College, Dublin, at 18, Hamilton had not attended school; rather, he received private tutoring. At Trinity, he was a superior student in both the sciences and the classics. Prior to receiving his degree, because of his brilliance he was appointed the Astronomer Royal of Ireland, beating out several famous astronomers for the post. He held this position until his death, living and working at Dunsink Observatory outside of Dublin. Hamilton made important contributions to optics, abstract algebra, and dynamics. Hamilton invented algebraic objects called quaternions as an example of a noncommutative system. He discovered the appropriate way to multiply quaternions while walking along a canal in Dublin. In his excitement, he carved the formula in the stone of a bridge crossing the canal, a spot marked today by a plaque. Later, Hamilton remained obsessed with quaternions, working to apply them to other areas of mathematics, instead of moving to new areas of research.

In 1857 Hamilton invented "The Icosian Game" based on his work in noncommutative algebra. He sold the idea for 25 pounds to a dealer in games and puzzles. (Since the game never sold well, this turned out to be a bad investment for the dealer.) The "Traveler's Dodecahedron," also called "A Voyage Round the World," the puzzle described in this section, is a variant of that game.

Hamilton married his third love in 1833, but his marriage worked out poorly, since his wife, a semi-invalid, was unable to cope with his household affairs. He suffered from alcoholism and lived reclusively for the last two decades of his life. He died from gout in 1865, leaving masses of papers containing unpublished research. Mixed in with these papers were a large number of dinner plates, many containing the remains of desiccated, uneaten chops.

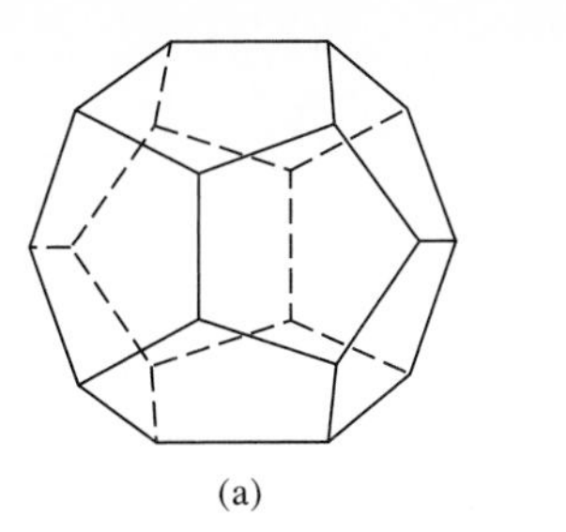

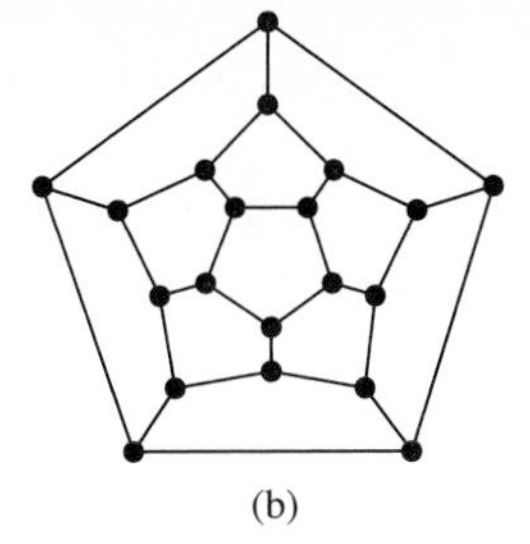

FIGURE 8 **Hamilton's "Round the World" Puzzle.**

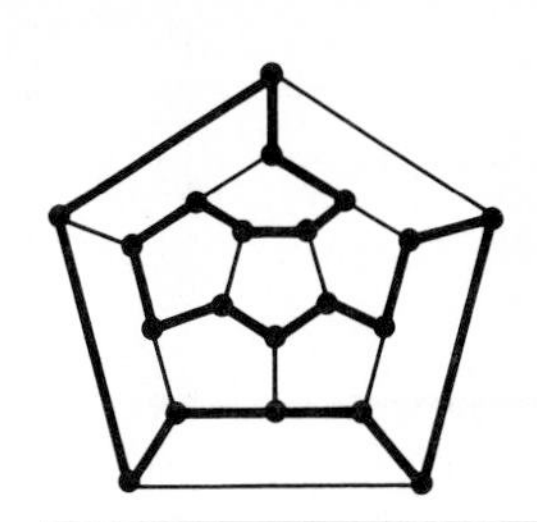

FIGURE 9 **A Solution to the "Round the World" Puzzle.**

in Figure 8(b) that passes through each vertex exactly once? This solves the puzzle since this graph is isomorphic to the graph consisting of the vertices and edges of the dodecahedron. A solution of Hamilton's puzzle is shown in Figure 9.

EXAMPLE 5 Which of the simple graphs in Figure 10 have a Hamilton circuit or, if not, a Hamilton path?

Solution: G_1 has a Hamilton circuit: a, b, c, d, e, a. There is no Hamilton circuit in G_2 (this can be seen by noting that any circuit containing every vertex must contain the edge $\{a, b\}$ twice), but G_2 does have a Hamilton path, namely a, b, c, d. G_3 has neither a Hamilton circuit nor a Hamilton path, since any path containing all vertices must contain one of the edges $\{a, b\}$, $\{e, f\}$, and $\{c, d\}$ more than once. ■

Is there a simple way to determine whether a graph has a Hamilton circuit or path? At first, it might seem that there should be an easy way to determine this, since there is a simple way to answer the similar question of whether a graph has an Euler circuit. Surprisingly, there are no known simple necessary and sufficient criteria for the existence of Hamilton circuits. However, many theorems are known that give sufficient conditions for the existence of Hamilton circuits. Also, certain properties can be used to show that a graph has no Hamilton circuit. For instance, a graph with a vertex of degree 1 cannot have a Hamilton circuit, since in a Hamilton circuit each vertex is incident with two edges in the circuit. Moreover, if a vertex in the graph

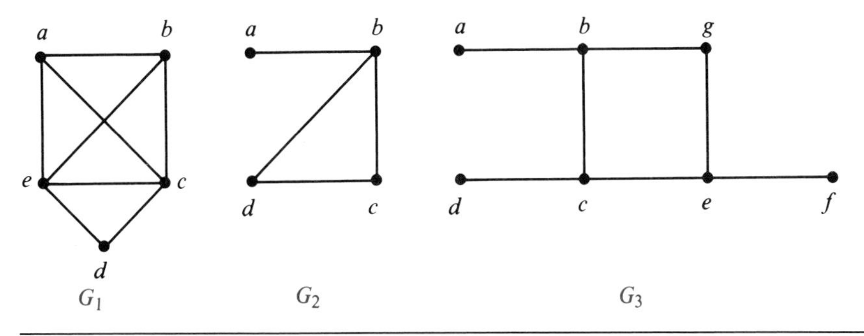

FIGURE 10 **Three Simple Graphs.**

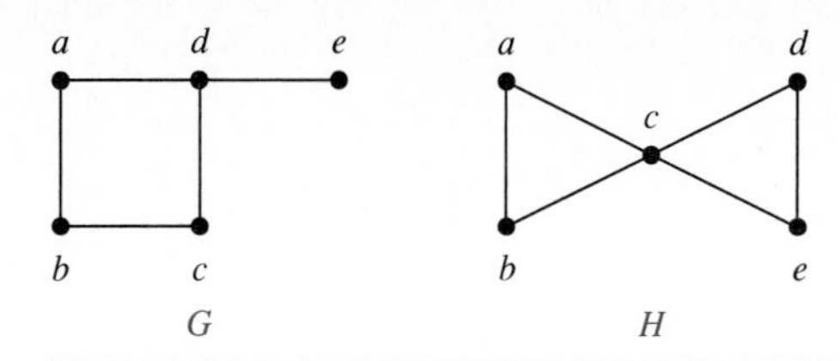

FIGURE 11 **Two Graphs That Do Not Have a Hamilton Circuit.**

has degree 2, then both edges that are incident with this vertex must be part of any Hamilton circuit. Also, note that when a Hamilton circuit is being constructed and this circuit has passed through a vertex, then all remaining edges incident with this vertex, other than the two used in the circuit, can be removed from consideration. Furthermore, a Hamilton circuit cannot contain a smaller circuit within it.

EXAMPLE 6 Show that neither graph displayed in Figure 11 has a Hamilton circuit.

Solution: There is no Hamilton circuit in G since G has a vertex of degree 1, namely, e.

Now consider H. Since the degrees of the vertices a, b, d, and e are all 2, every edge incident with these vertices must be part of any Hamilton circuit. It is now easy to see that no Hamilton circuit can exist in H, for any Hamilton circuit would have to contain four edges incident with c, which is impossible. ■

EXAMPLE 7 Show that K_n has a Hamilton circuit whenever $n \geq 3$.

Solution: We can form a Hamilton circuit in K_n beginning at any vertex. Such a circuit can be built by visiting vertices in any order we choose, as long as the path begins and ends at the same vertex and visits each other vertex exactly once. This is possible since there are edges in K_n between any two vertices. ■

We now state a theorem that gives sufficient conditions for the existence of Hamilton circuits. This is just one of many such theorems known.

THEOREM 3 If G is a connected simple graph with n vertices where $n \geq 3$, then G has a Hamilton circuit if the degree of each vertex is at least $n/2$.

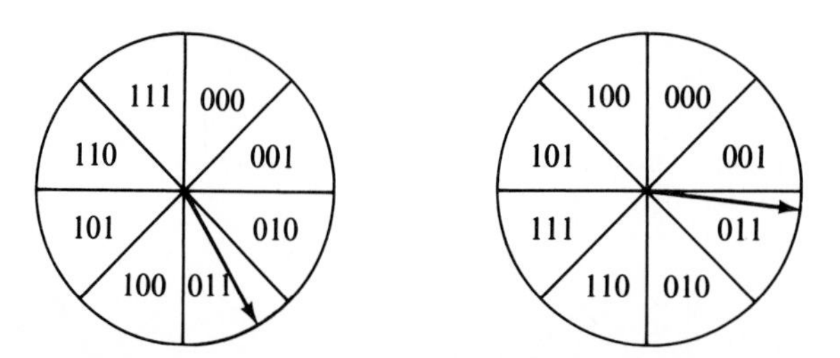

FIGURE 12 **Converting the Position of a Pointer into Digital Form.**

We will now give an application of Hamilton circuits to coding.

EXAMPLE 8 Gray Codes The position of a rotating pointer can be represented in digital form. One way to do this is to split the circle into 2^n arcs of equal length and to assign a bit string of length n to each arc. Two ways to do this using bit strings of length three are shown in Figure 12.

The digital representation of the position of the pointer can be determined using a set of n contacts. Each contact is used to read one bit in the digital representation of the position. This is illustrated in Figure 13 for the two assignments from Figure 12.

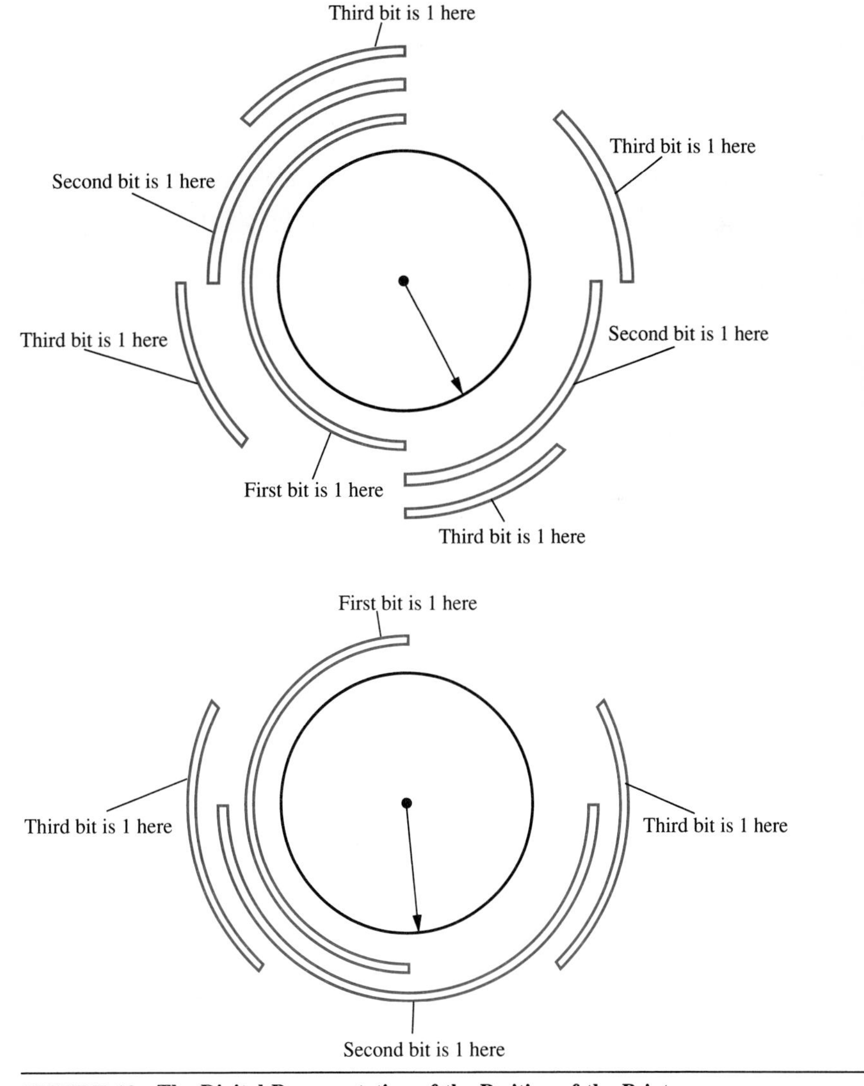

FIGURE 13 The Digital Representation of the Position of the Pointer.

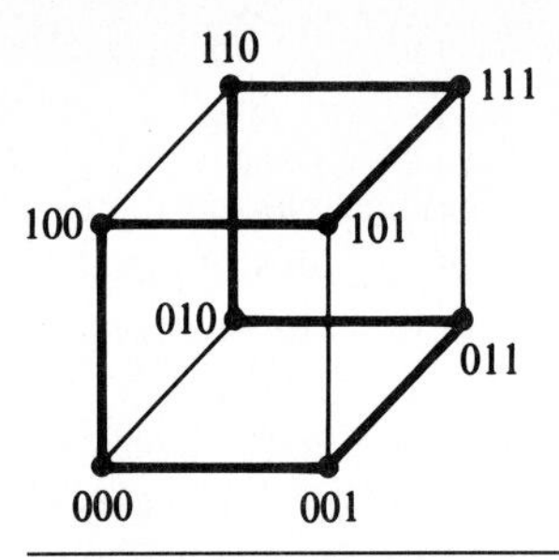

FIGURE 14 **A Hamilton Circuit for Q_3.**

When the pointer is near the boundary of two arcs, a mistake may be made in reading its position. This may result in a major error in the bit string read. For instance, in the coding scheme in Figure 12(a), if a small error is made in determining the position of the pointer, the bit string 100 is read instead of 011. All three bits are incorrect! To minimize the effect of an error in determining the position of the pointer, the assignment of the bit strings to the 2^n arcs should be made so that only one bit is different in the bit strings represented by adjacent arcs. This is exactly the situation in the coding scheme in Figure 12(b). An error in determining the position of the pointer gives the bit string 010 instead of 011. Only one bit is wrong.

A **Gray code** is a labeling of the arcs of the circle so that adjacent arcs are labeled with bit strings that differ in exactly one bit. The assignment in Figure 12(b) is a Gray code. We can find a Gray code by listing all bit strings of length n in such a way that each string differs in exactly one position from the preceding bit string, and the last string differs from the first in exactly one position. We can model this problem using the n-cube Q_n. What is needed to solve this problem is a Hamilton circuit in Q_n. Such Hamilton circuits are easily found. For instance, a Hamilton circuit for Q_3 is displayed in Figure 14. The sequence of bit strings differing in exactly one bit produced by this Hamilton circuit is 000, 001, 011, 010, 110, 111, 101, 100. ■

Exercises

In Exercises 1–7 determine whether each graph has an Euler circuit. Construct such a circuit when one exists.

1.

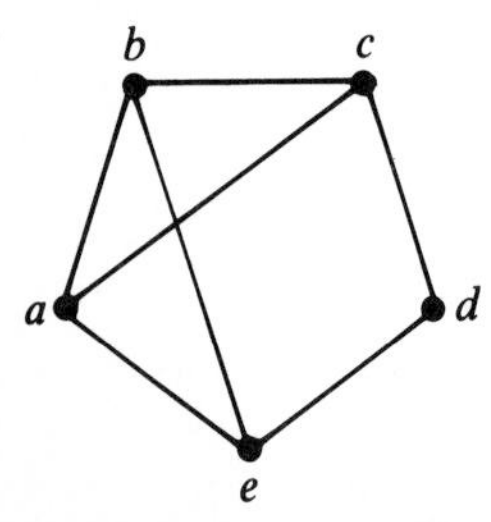

2.

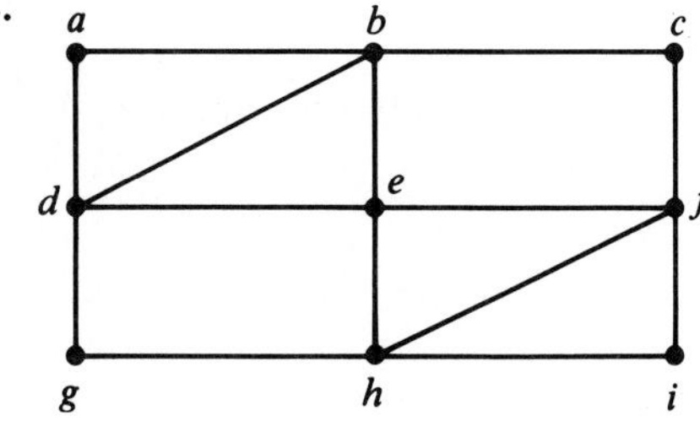

3.

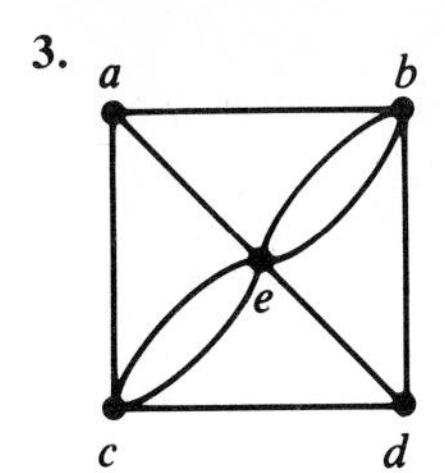

4.

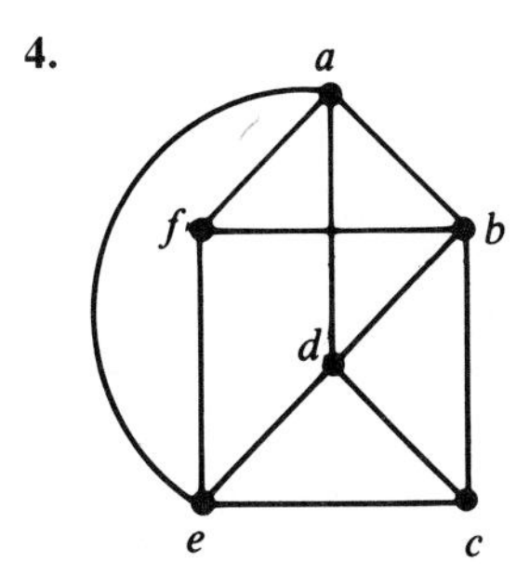

5.

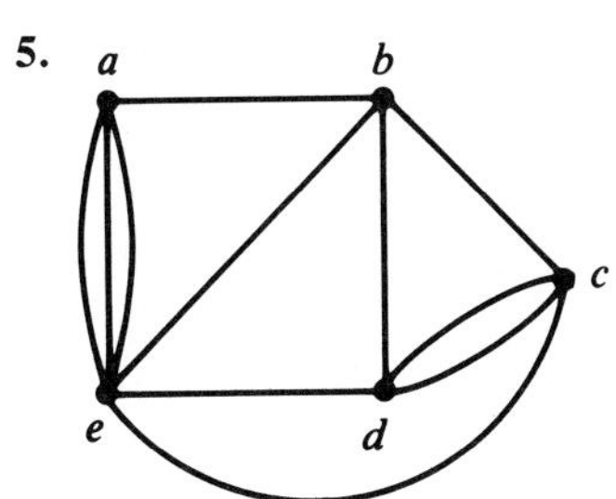

6.

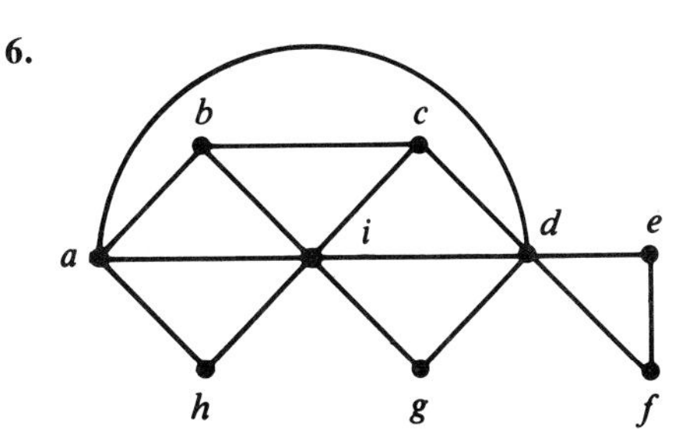

7.

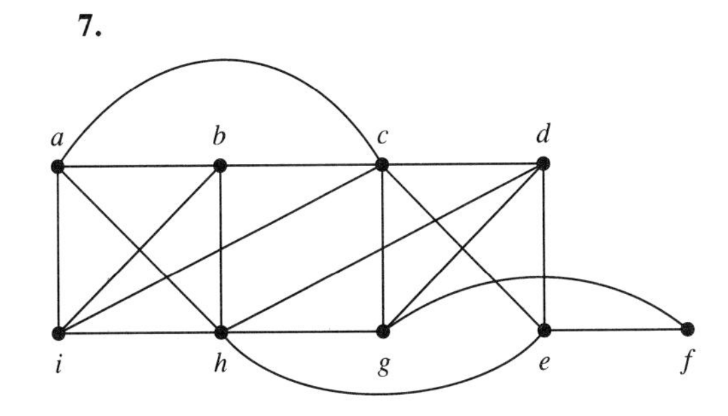

8. Determine whether the graph in Exercise 1 has an Euler path. Construct such a path if it exists.

9. Determine whether the graph in Exercise 2 has an Euler path. Construct such a path if it exists.

10. Determine whether the graph in Exercise 3 has an Euler path. Construct such a path if it exists.

11. Determine whether the graph in Exercise 4 has an Euler path. Construct such a path if it exists.

12. Determine whether the graph in Exercise 5 has an Euler path. Construct such a path if it exists.

13. Determine whether the graph in Exercise 6 has an Euler path. Construct such a path if it exists.

14. Determine whether the graph in Exercise 7 has an Euler path. Construct such a path if it exists.

15. In Kaliningrad (the Russian name for Königsberg) there are two additional bridges, besides the seven that were present in the 18th century. These new bridges connect regions B and C and regions B and D, respectively. Can someone cross all nine bridges in Kaliningrad exactly once and return to the starting point?

16. Can someone cross all the bridges shown in the following map exactly once and return to the starting point?

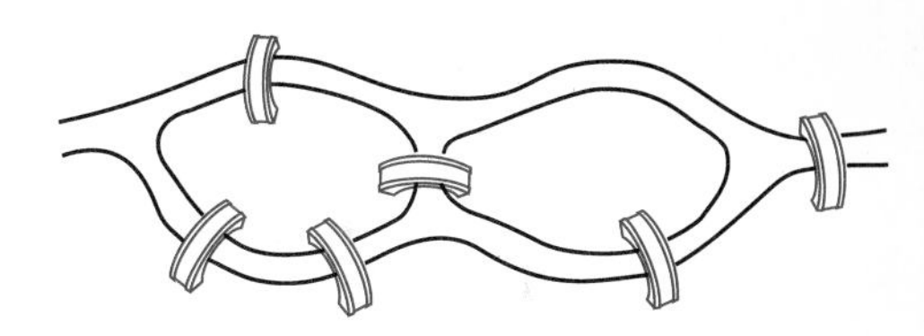

17. When can the center lines of the streets in a city be painted without traveling a street more than once? (Assume that all the streets are two-way streets.)

18. Devise a procedure, similar to Algorithm 1, for constructing Euler paths in multigraphs.

In Exercises 19–21 determine whether the picture shown can be drawn with a pencil in a continuous motion without lifting the pencil or retracing part of the picture.

19.

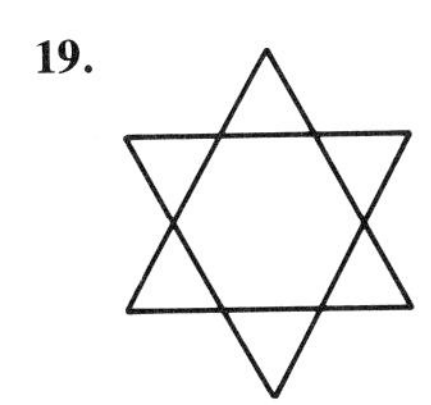

20.

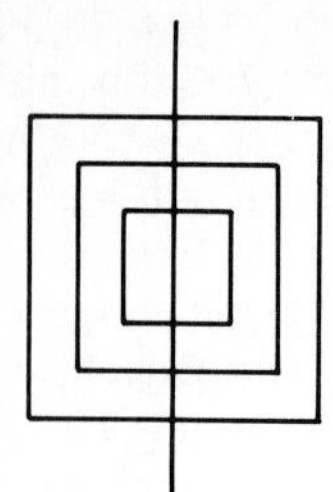

21.

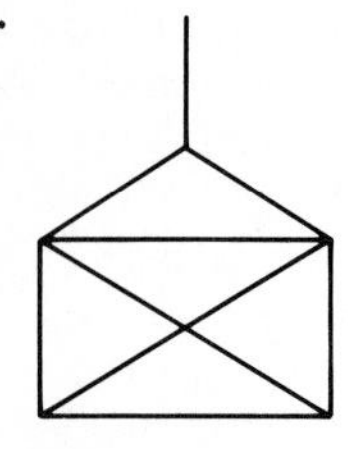

★**22.** Show that a directed multigraph having no isolated vertices has an Euler circuit if and only if the graph is weakly connected and the in-degree and out-degree of each vertex are equal.

★**23.** Show that a directed multigraph having no isolated vertices has an Euler path but not an Euler circuit if and only if the graph is weakly connected and the in-degree and out-degree of each vertex are equal for all but two vertices, one that has in-degree 1 larger than its out-degree and the other that has out-degree 1 larger than its in-degree.

In Exercises 24–28 determine whether the directed graph shown has an Euler circuit. Construct an Euler circuit if it exists.

24.

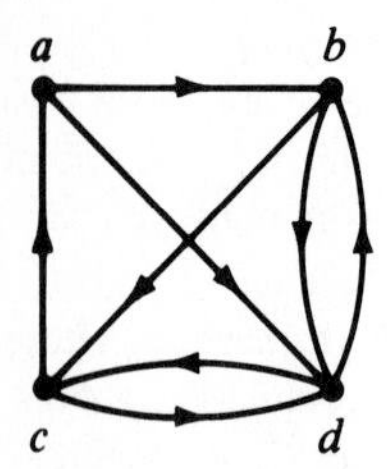

25.

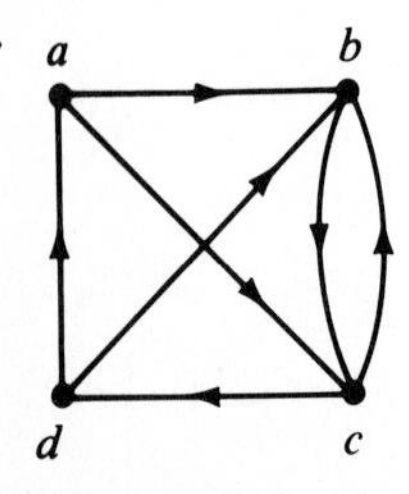

26.

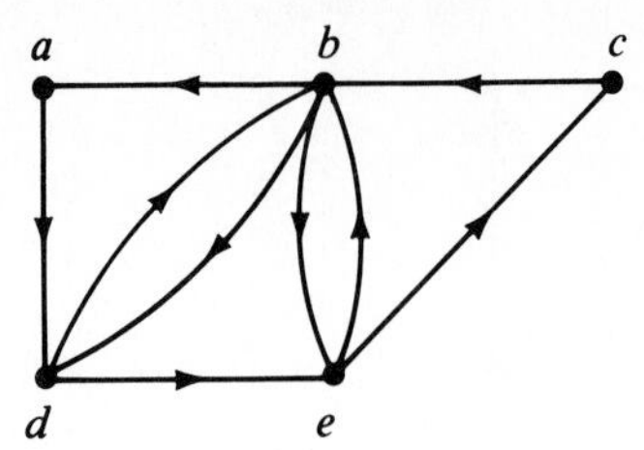

27.

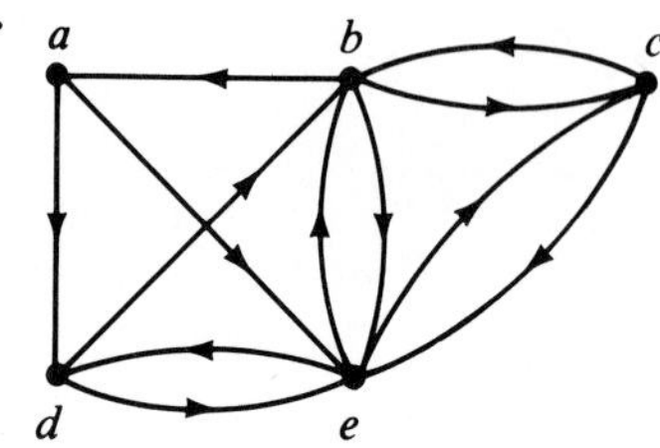

28.

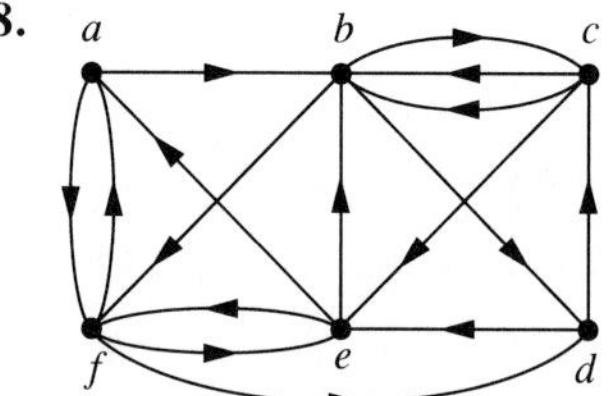

29. Determine whether the directed graph in Exercise 24 has an Euler path. Construct an Euler path if one exists.

30. Determine whether the directed graph in Exercise 25 has an Euler path. Construct an Euler path if one exists.

31. Determine whether the directed graph in Exercise 26 has an Euler path. Construct an Euler path if one exists.

32. Determine whether the directed graph in Exercise 27 has an Euler path. Construct an Euler path if one exists.

33. Determine whether the directed graph in Exercise 28 has an Euler path. Construct an Euler path if one exists.

★**34.** Devise an algorithm for constructing Euler circuits in directed graphs.

35. Devise an algorithm for constructing Euler paths in directed graphs.

36. For which values of n do the following graphs have an Euler circuit?

a) K_n **b)** C_n
c) W_n **d)** Q_n

37. For which values of n do the graphs in Exercise 36 have an Euler path but no Euler circuit?

38. For which values of m and n does the complete bipartite graph $K_{m,n}$ have an

a) Euler circuit?

b) Euler path?

39. Find the least number of times it is necessary to lift a pencil from the paper when drawing each of the graphs in Exercises 1–7 without retracing any part of the graph.

In Exercises 40–44 determine whether the given graph has a Hamilton circuit. If it does, find such a circuit. If it does not, give an argument to show why no such circuit exists.

40.

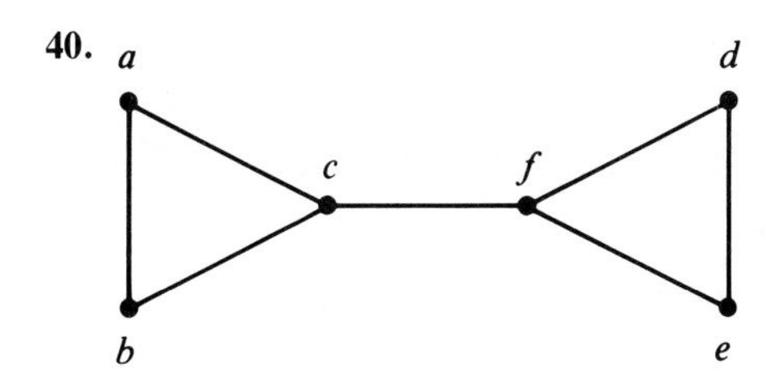

41.

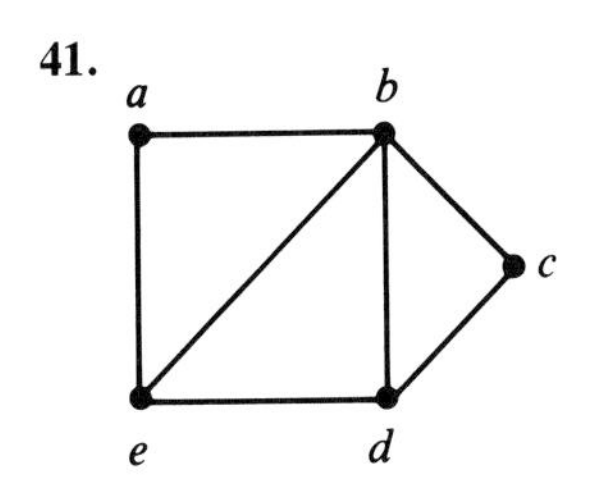

42.

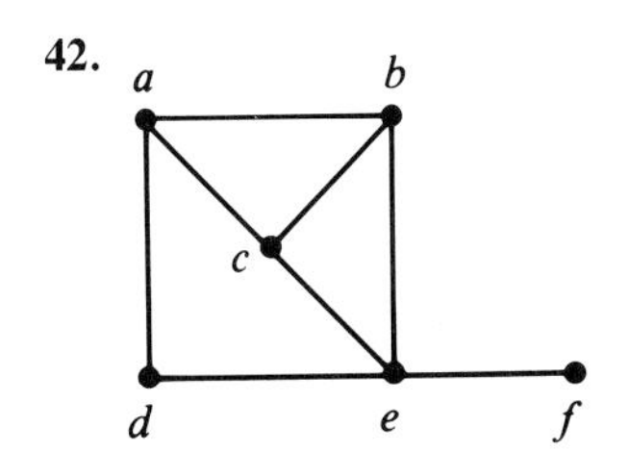

43.

44.

45. Does the graph in Exercise 40 have a Hamilton path? If it does, find such a path. If it does not, give an argument to show why no such path exists.

46. Does the graph in Exercise 41 have a Hamilton path? If it does, find such a path. If it does not, give an argument to show why no such path exists.

47. Does the graph in Exercise 42 have a Hamilton path? If it does, find such a path. If it does not, give an argument to show why no such path exists.

48. Does the graph in Exercise 43 have a Hamilton path? If it does, find such a path. If it does not, give an argument to show why no such path exists.

★49. Does the graph in Exercise 44 have a Hamilton path? If it does, find such a path. If it does not, give an argument to show why no such path exists.

50. For which values of n do the graphs in Exercise 36 have a Hamilton circuit?

51. For which values of m and n does the complete bipartite graph $K_{m,n}$ have a Hamilton circuit?

★52. Show that the Petersen graph shown in the following diagram, does not have a Hamilton circuit, but

Julius Peter Christian Petersen (1839–1910). Julius Petersen, born in the Danish town of Soro, obtained his Ph.D. in 1871. He first taught at a polytechnic and military academy. Later he moved to a professorship at the University of Copenhagen. He was known in Denmark as the author of a series of textbooks for high schools and universities. Petersen worked in a wide range of areas, including algebra, analysis, geometry, mechanics, and number theory. His contributions to graph theory, including results on regular graphs, are his best-known work.

that the subgraph obtained by deleting a vertex v, and all edges incident with v, does have a Hamilton circuit.

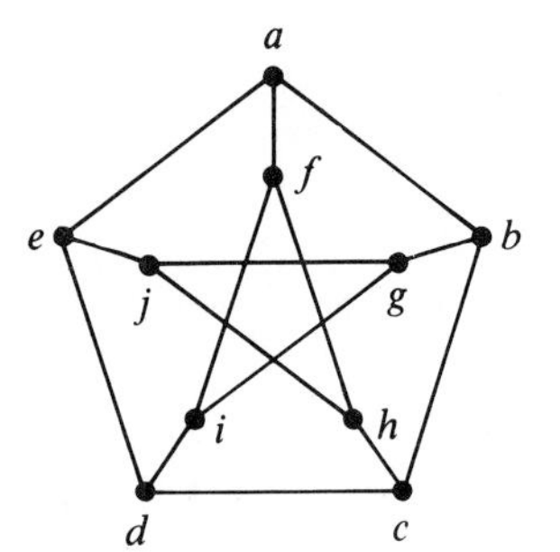

The Petersen Graph.

★**53.** Show that there is a Gray code of order n whenever n is a positive integer, or equivalently, show that the n-cube Q_n, $n > 1$, always has a Hamilton circuit. (*Hint:* Use mathematical induction. Show how to produce a Gray code of order n from one of order $n - 1$.)

Fleury's algorithm for constructing Euler circuits begins with an arbitrary vertex of a connected multigraph and forms a circuit by choosing edges successively. Once an edge is chosen, it is removed. Edges are chosen successively so that each edge begins where the last edge ends, and so that this edge is not a cut edge unless there is no alternative.

54. Use Fleury's algorithm to find an Euler circuit in the graph G in Example 5.

★**55.** Express Fleury's algorithm in pseudocode.

★★**56.** Prove that Fleury's algorithm always produces an Euler circuit.

★**57.** Give a variant of Fleury's algorithm to produce Euler paths.

58. A diagnostic message can be sent out over a computer network to perform tests over all links and in all devices. What sort of paths should be used to test all links? To test all devices?

59. Show that a bipartite graph with an odd number of vertices does not have a Hamilton circuit.

A **knight** is a chess piece that can move either two spaces horizontally and one space vertically, or one space horizontally and two spaces vertically. That is, a knight on square (x,y) can move to any of the eight squares $(x \pm 2, y \pm 1)$, $(x \pm 1, y \pm 2)$, if these squares are on the chessboard, as illustrated here.

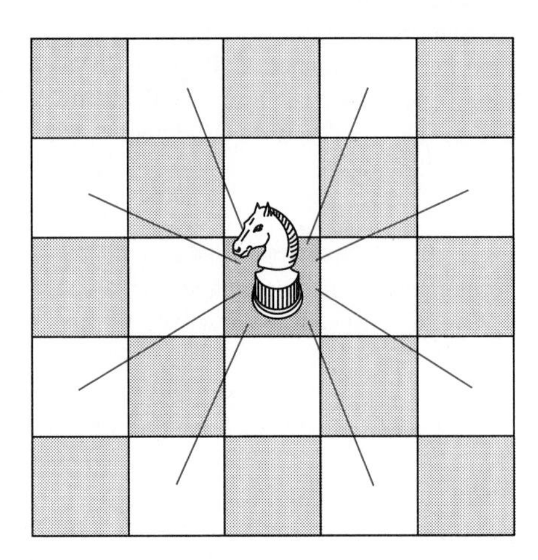

A **knight's tour** is a sequence of legal moves by a knight starting at some square and visiting each square exactly once. A knight's tour is called **reentrant** if there is a legal move that takes the knight from the last square of the tour back to where the tour began. We can model knight's tours using the graph that has a vertex for each square on the board, with an edge connecting two vertices if a knight can legally move between the squares represented by these vertices.

60. Draw the graph that represents the legal moves of a knight on a 3×3 chessboard.

61. Draw the graph that represents the legal moves of a knight on a 3×4 chessboard.

62. **a)** Show that finding a knight's tour on an $m \times n$ chessboard is equivalent to finding a Hamilton path on the graph representing the legal moves of a knight on that board.

b) Show that the finding a reentrant knight's tour on an $m \times n$ chessboard is equivalent to finding a Hamilton circuit on the corresponding graph.

★**63.** Show that there is a knight's tour on a 3×4 chessboard.

★**64.** Show that there is a no knight's tour on a 3×3 chessboard.

*65. Show that there is a no knight's tour on a 4×4 chessboard.

66. Show that the graph representing the legal moves of a knight on an $m \times n$ chessboard, whenever m and n are positive integers, is bipartite.

67. Show that there is no reentrant knight's tour on an $m \times n$ chessboard when m and n are both odd. (*Hint;* Use Exercises 59 and 66.)

*68. Show that there is a knight's tour on an 8×8 chessboard. (*Hint:* You can construct a knight's tour using the following method invented by Warnsdorff. Start in any square, and then always move to a square connected to the fewest number of unused squares. Although this method may not always produce a knight's tour, it often does.)

7.6 Shortest Path Problems

INTRODUCTION

Many problems can be modeled using graphs with weights assigned to their edges. As an illustration, consider how an airline system can be modeled. We set up the basic graph model by representing cities by vertices and flights by edges. Problems involving distances can be modeled by assigning distances between cities to the edges. Problems involving flight time can be modeled by assigning flight times to edges. Problems involving fares can be modeled by assigning fares to the edges. Figure 1 displays three different assignments of weights to the edges of a graph representing distances, flight times, and fares respectively.

Graphs that have a number assigned to each edge are called **weighted graphs.** Weighted graphs are used to model computer networks. Communications costs (such as the monthly cost of leasing a telephone line), the response times of the computers over these lines, or the distance between computers, can all be studied using weighted graphs. Figure 2 displays weighted graphs that represent three ways to assign weights to the edges of a graph of a computer network, corresponding to costs, response times over the lines, and distance.

Several types of problems involving weighted graphs arise frequently. Determining the path of least length between two vertices in a network is one such problem. To be more specific, let the **length** of a path in a weighted graph be the sum of the weights of the edges of this path. (The reader should note that this use of the term *length* is different from the use of *length* to denote the number of edges in a path in a graph without weights.) The question is: What is the shortest path, that is, the path of least length, between two given vertices? For instance, in the airline system represented by the weighted graph shown in Figure 1, what is the shortest path in air distance between Boston and Los Angeles? What combinations of flights has the smallest total flight time (that is, total time in the air, not including time between flights) between Boston and Los Angeles? What is the cheapest fare between these two cities? In the computer network shown in Figure 2, what is the least expensive set of telephone lines needed to connect the computers in San Francisco with those in New York? Which set of telephone lines gives the fastest response time for commu-

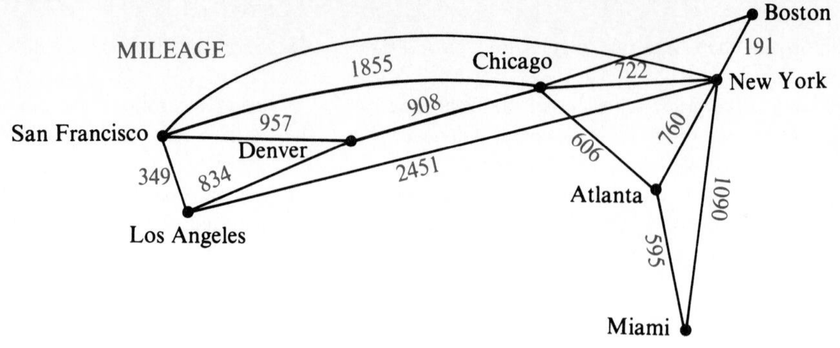

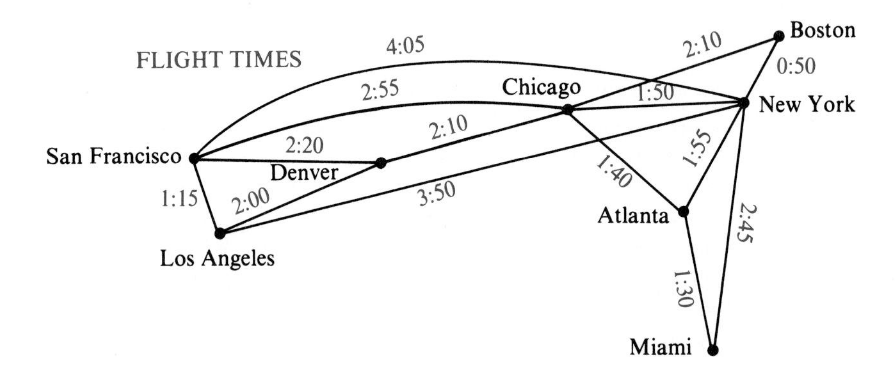

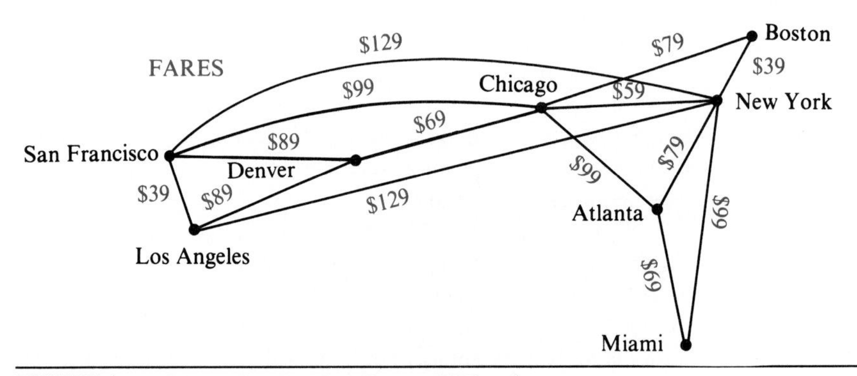

FIGURE 1 **Weighted Graphs Modeling an Airline System.**

nications between San Francisco and New York? Which set of lines has the shortest overall distance?

A SHORTEST PATH ALGORITHM

There are several different algorithms that find the shortest path between two vertices in a weighted graph. We will present an algorithm discovered by the Dutch mathematician E. Dijkstra in 1959. The version we will describe solves this problem in undirected

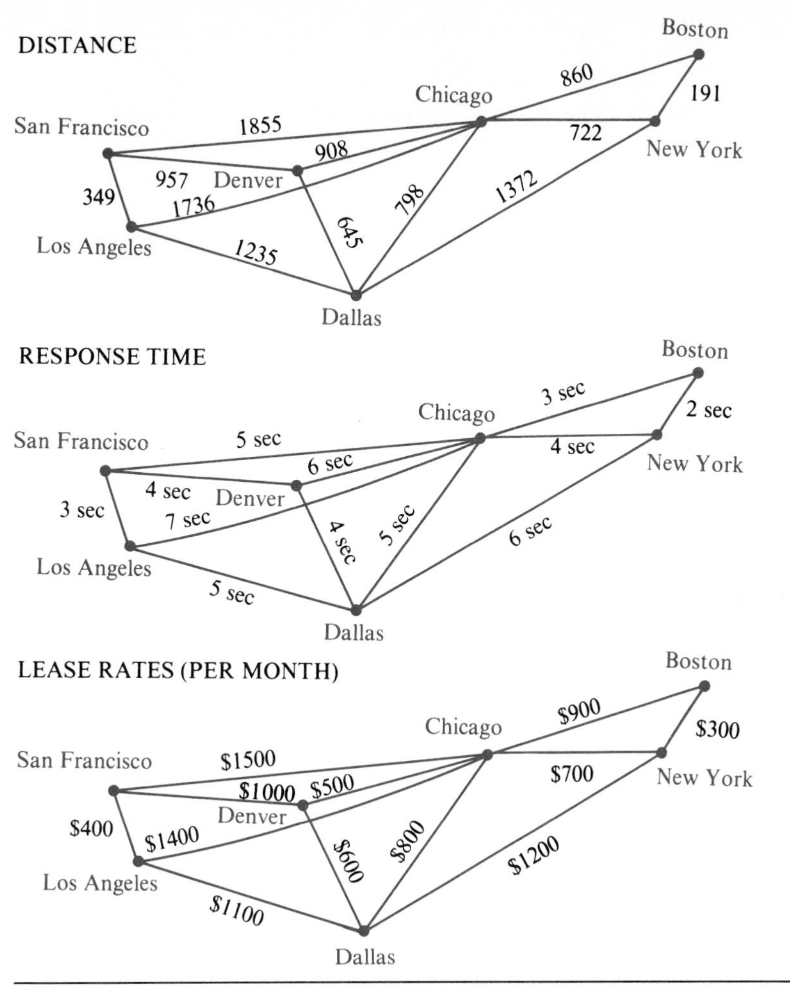

FIGURE 2 Weighted Graphs Modeling a Computer Network.

weighted graphs where all the weights are positive. It is easy to adapt it to solve shortest-path problems in directed graphs.

Before giving a formal presentation of the algorithm, we will give a motivating example.

Edsger Wybe Dijkstra (born 1930). Edsger Dijkstra, born in the Netherlands, began programming computers in the early 1950s while studying theoretical physics at the University of Leiden. In 1952, realizing that he was more interested in programming than in physics, he quickly completed the requirements for his physics degree and began his career as a programmer, even though programming was not a recognized profession. (In 1957, the authorities in Amsterdam refused to accept "programming" as his profession on his marriage license. However, they did accept "theoretical physicist" when he changed his entry to this.)

Dijkstra has been one of the most forceful proponents of programming as a scientific discipline. He has made fundamental contributions to the areas of operating systems, including deadlock avoidance; programming languages, including the notion of structured programming; and algorithms. In 1972 Dijkstra received the Turing Award from the Association for Computing Machinery, one of the most prestigious awards in computer science. Dijkstra became a Burroughs Research Fellow in 1973, and in 1984 he was appointed to a chair in Computer Science at the University of Texas, Austin.

EXAMPLE 1 What is the length of the shortest path between a and z in the weighted graph shown in Figure 3?

Solution: Although the shortest path is easily found by inspection, we will develop some ideas useful in understanding Dijkstra's algorithm. We will solve this problem by finding the length of the shortest path from a to successive vertices, until z is reached.

The only paths starting at a that contain no vertex other than a (until the terminal vertex is reached), are a, b and a, d. Since the lengths of a, b and a, d are 4 and 2, respectively, it follows that d is the closest vertex to a.

We can find the next closest vertex by looking at all paths that go through only a and d (until the terminal vertex is reached). The shortest such path to b is still a, b, with length 4, and the shortest such path to e is a, d, e, with length 5. Consequently, the next closest vertex to a is b.

To find the third closest vertex to a, we need to examine only paths that go through only a, d, and b (until the terminal vertex is reached). There is a path of length 7 to c, namely, a, b, c, and a path of length 6 to z, namely, a, d, e, z. Consequently, z is the next closest vertex to a, and the length of the shortest path to z is 6. ■

Example 1 illustrates the general principles used in Dijkstra's algorithm. Note that the shortest path from a to z could have been found by inspection. However, inspection is impractical for both humans and computers for graphs with large numbers of edges.

We will now consider the general problem of finding the length of the shortest path between a and z in an undirected connected simple weighted graph. Dijkstra's algorithm proceeds by finding the length of the shortest path from a to a first vertex, the length of the shortest path from a to a second vertex, and so on, until the length of the shortest path from a to z is found.

The algorithm relies on a series of iterations. A distinguished set of vertices is constructed by adding one vertex at each iteration. A labeling procedure is carried out at each iteration. In this labeling procedure, a vertex w is labeled with the length of the shortest path from a to w that contains only vertices already in the distinguished set. The vertex added to the distinguished set is one with a minimal label among those vertices not already in the set.

We now give the details of Dijkstra's algorithm. It begins by labeling a with 0 and the other vertices with ∞. We use the notation $L_0(a) = 0$ and $L_0(v) = \infty$

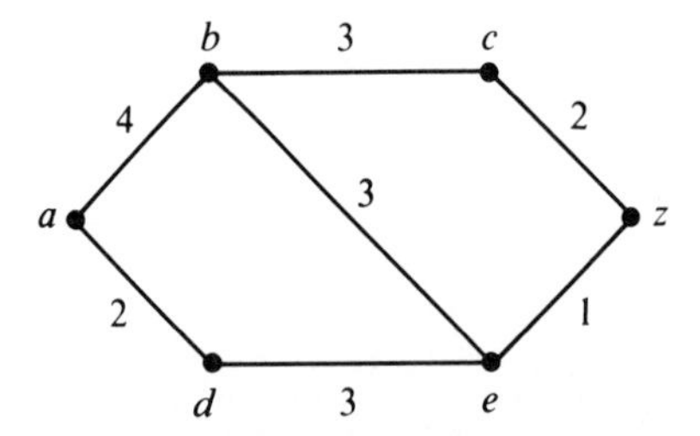

FIGURE 3 A Weighted Simple Graph.

for these labels before any iterations have taken place (the subscript 0 stands for the "0th" iteration). These labels are the lengths of the shortest paths from a to the vertices, where the paths contain only the vertex a. (Since no path from a to a vertex different from a exists, ∞ is the length of the shortest path between a and this vertex.)

Dijkstra's algorithm proceeds by forming a distinguished set of vertices. Let S_k denote this set after k iterations of the labeling procedure. We begin with $S_0 = \emptyset$. The set S_k is formed from S_{k-1} by adding a vertex u not in S_{k-1} with the smallest label. Once u is added to S_k, we update the labels of all vertices not in S_k, so that $L_k(v)$, the label of the vertex v at the kth stage, is the length of the shortest path from a to v that contains vertices only in S_k (that is, vertices that were already in the distinguished set together with u).

Let v be a vertex not in S_k. To update the label of v, note that $L_k(v)$ is the length of the shortest path from a to v containing only vertices in S_k. The updating can be carried out efficiently when the following observation is used: The shortest path from a to v containing only elements of S_k is either the shortest path from a to v that contains only elements of S_{k-1} (that is, the distinguished vertices not including u), or it is the shortest path from a to u at the $(k-1)$st stage with the edge (u, v) added. In other words,

$$L_k(a, v) = \min\{L_{k-1}(a, v), L_{k-1}(a, u) + w(u, v)\}.$$

This procedure is iterated by successively adding vertices to the distinguished set until z is added. When z is added to the distinguished set, its label is the length of the shortest path from a to z. Dijkstra's algorithm is given in Algorithm 1. Later we will give a proof that this algorithm is correct.

ALGORITHM 1 Dijkstra's Algorithm.

procedure *Dijkstra*(G: weighted connected simple graph, with all weights positive)
{G has vertices $a = v_0, v_1, \ldots, v_n = z$ and weights $w(v_i, v_j)$ where $w(v_i, v_j) = \infty$ if $\{v_i, v_j\}$ is not an edge in G}
for $i := 1$ **to** n
  $L(v_i) := \infty$
$L(a) := 0$
$S := \emptyset$
{the labels are now initialized so that the label of a is zero and all other labels are ∞, and S is the empty set}
while $z \notin S$
begin
  $u :=$ a vertex not in S with $L(u)$ minimal
  $S := S \cup \{u\}$
  for all vertices v not in S
    if $L(u) + w(u, v) < L(v)$ **then** $L(v) := L(u) + w(u, v)$
  {this adds a vertex to S with minimal label and updates the labels of vertices not in S}
end {$L(z)$ = length of shortest path from a to z}

The following example illustrates how Dijkstra's algorithm works. Afterward, we will show that this algorithm always produces the length of the shortest path between two vertices in a weighted graph.

EXAMPLE 2 Use Dijkstra's algorithm to find the length of the shortest path between the vertices a and z in the weighted graph displayed in Figure 4(a).

Solution: The steps used by Dijkstra's algorithm to find the shortest path between a and z are shown in Figure 4. At each iteration of the algorithm the vertices of the set S_k are circled. The shortest path from a to each vertex containing only vertices in S_k is indicated for each iteration. The algorithm terminates when z is circled. We find that the shortest path from a to z is a, c, b, d, e, z, with length 13. ■

Remark: In performing Dijkstra's algorithm it is sometimes more convenient to keep track of labels of vertices in each step using a table instead of redrawing the graph for each step.

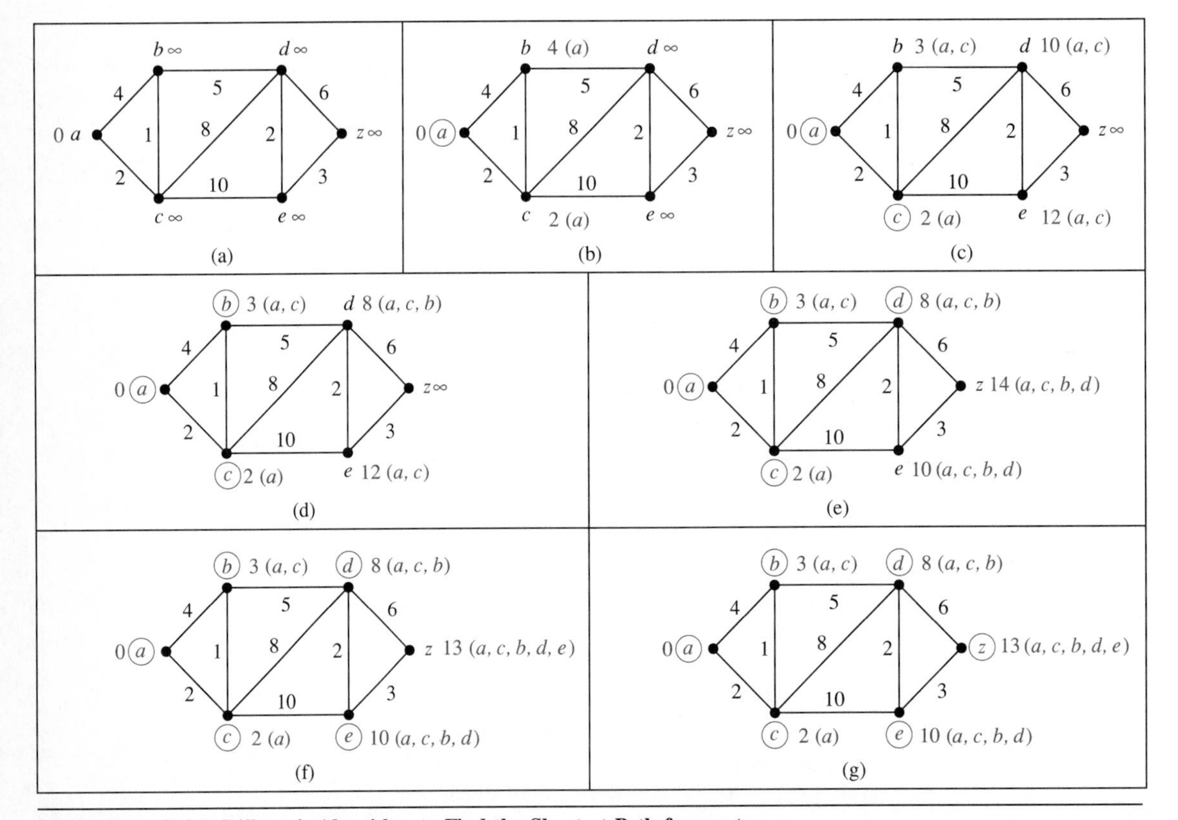

FIGURE 4 Using Dijkstra's Algorithm to Find the Shortest Path from a to z.

Next, we use an inductive argument to show that Dijkstra's algorithm produces the length of the shortest path between two vertices a and z in an undirected connected weighted graph. Take as the induction hypothesis the following assertion: At the kth iteration

(i) the label of a vertex v, $v \neq 0$, in S is the length of the shortest path from a to this vertex, and

(ii) the label of a vertex not in S is the length of the shortest path from a to this vertex that contains only (besides the vertex itself) vertices in S.

When $k = 0$, before any iterations are carried out, $S = \{a\}$, so the length of the shortest path from a to a vertex other than a is ∞, and the length of the shortest path from a to itself is 0 (here we are allowing a path to have no edges in it). Hence, the basis case is true.

Assume that the inductive hypothesis holds for the kth iteration. Let v be the vertex added to S at the $(k + 1)$st iteration, so that v is a vertex not in S at the end of the kth iteration with the smallest label (in the case of ties, any vertex with smallest label may be used).

From the inductive hypothesis we see that the vertices in S before the $(k + 1)$st iteration are labeled with the length of the shortest path from a. Also, v must be labeled with the length of the shortest path to it from a. If this were not the case, at the end of the kth iteration there would be a path of length less than $L_k(v)$ containing a vertex not in S (because $L_k(v)$ is the length of the shortest path from a to v containing only vertices in S after the kth iteration). Let u be the first vertex not in S in such a path. There is a path with length less than $L_k(v)$ from a to u containing only vertices of S. This contradicts the choice of v. Hence, *(i)* holds at the end of the $(k + 1)$st iteration.

Let u be a vertex not in S after $k + 1$ iterations. A shortest path from a to u containing only elements of S either contains v or it does not. If it does not contain v, then by the inductive hypothesis its length is $L_k(u)$. If it does contain v, then it must be made up of a path from a to v of shortest possible length containing elements of S other than v, followed by the edge from v to u. In this case its length would be $L_k(v) + w(v, u)$. This shows that *(ii)* is true, since $L_{k+1}(u) = \min\{L_k(u), L_k(v) + w(v, u)\}$.

The following theorem has been proved.

THEOREM 1 Dijkstra's algorithm finds the length of a shortest path between two vertices in a connected simple undirected weighted graph.

We can now estimate the computational complexity of Dijkstra's algorithm (in terms of additions and comparisons). The algorithm uses no more than $n - 1$ iterations, since one vertex is added to the distinguished set at each iteration. We are done if we can estimate the number of operations used for each iteration. We can identify the vertex not in S_k with the smallest label using no more than $n - 1$ comparisons. Then we use an addition and a comparison to update the label of each vertex not in S_k. It follows that no more than $2(n - 1)$ operations are used at each iteration, since there are no more than $n - 1$ labels to update at each iteration. Since we use no more than $n - 1$ iterations, each using no more than $2(n - 1)$ operations, we have the following theorem.

THEOREM 2 Dijkstra's algorithm uses $O(n^2)$ operations (additions and comparisons) to find the length of the shortest path between two vertices in a connected simple undirected weighted graph.

Exercises

1. For each of the following problems about a subway system, describe a weighted graph model that can be used to solve it.
   a) What is the least amount of time required to travel between two stops?
   b) What is the minimum distance that can be traveled to reach a stop from another stop?
   c) What is the least fare required to travel between two stops if fares between stops are added to give the total fare?

In Exercises 2–4 find the length of the shortest path between a and z in the given weighted graph.

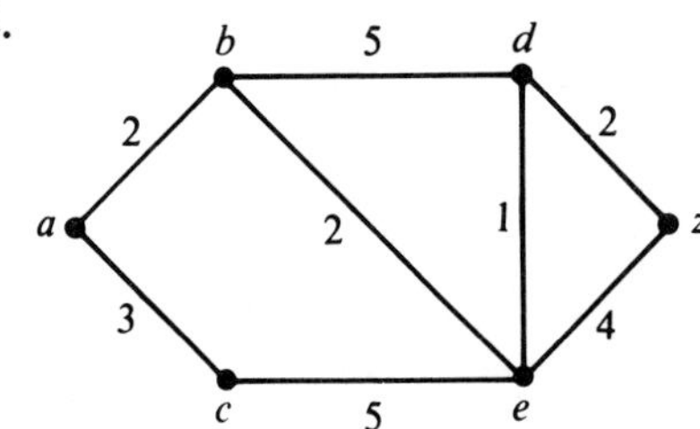

3.

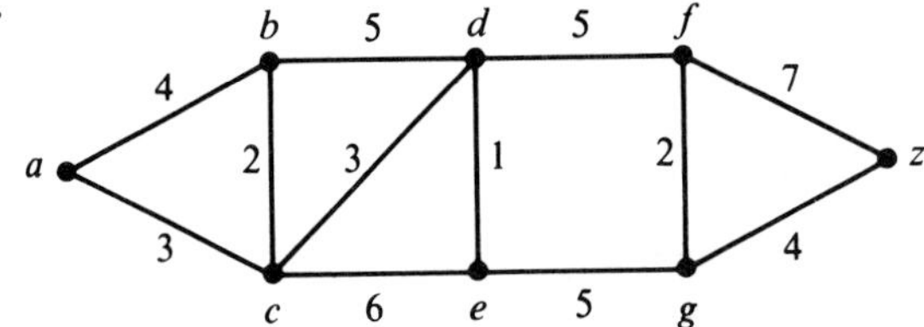

4.

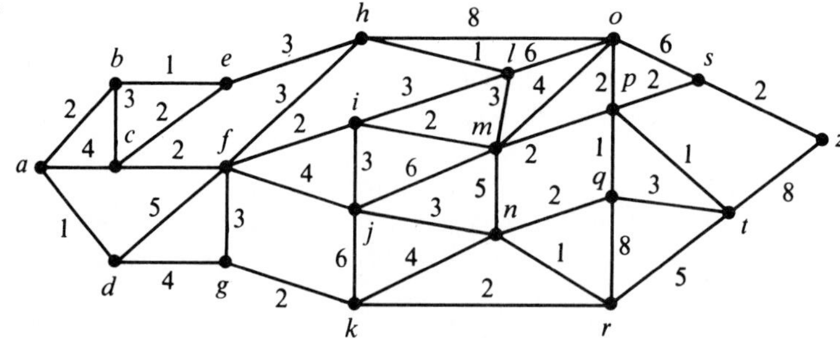

5. What is the shortest path between a and z in each of the weighted graphs in Exercises 2–4?
6. Find the length of the shortest path between the following pairs of vertices in the weighted graph in Exercise 3.
   a) a and d **b)** a and f **c)** c and f **d)** b and z

7. Find the shortest paths in the weighted graph in Exercise 3 between the pairs of vertices in Exercise 6.
8. Find the shortest path (in mileage) between each of the following pairs of cities in the airline system shown in Figure 1.
   a) New York and Los Angeles b) Boston and San Francisco
   c) Miami and Denver d) Miami and Los Angeles
9. Find the combination of flights with the least total air time between the pairs of cities in Exercise 8, using the flight times shown in Figure 1.
10. Find the least expensive combination of flights connecting the pairs of cities in Exercise 8, using the fares shown in Figure 1.
11. Find the shortest route (in distance) between computer centers in each of the following pairs of cities in the communications network shown in Figure 2.
   a) Boston and Los Angeles b) New York and San Francisco
   c) Dallas and San Francisco d) Denver and New York
12. Find the route with the shortest response time between the pairs of computer centers in Exercise 11 using the response times given in Figure 2.
13. Find the least expensive route, in monthly lease charges, between the pairs of computer centers in Exercise 11 using the lease charges given in Figure 2.
14. Explain how to find the path with the least number of edges between two vertices in an undirected graph by considering it as a shortest-path problem in a weighted graph.
15. Extend Dijkstra's algorithm for finding the length of the shortest path between two vertices in a weighted simple connected graph so that the length of the shortest path between the vertex a and every other vertex of the graph is found.
16. Extend Dijkstra's algorithm for finding the length of the shortest path between two vertices in a weighted simple connected graph so that the shortest path between these vertices is constructed.
17. The weighted graphs in the accompanying figure show some major roads in New Jersey. In part (a) the distances between cities on these roads are shown. Part (b) shows the tolls of the roads.
   a) Find the shortest route in distance between Newark and Camden, and between Newark and Cape May, using these roads.
   b) Find the least expensive route in terms of total tolls using the roads in the graph between the pairs of cities in part (a) of this exercise.

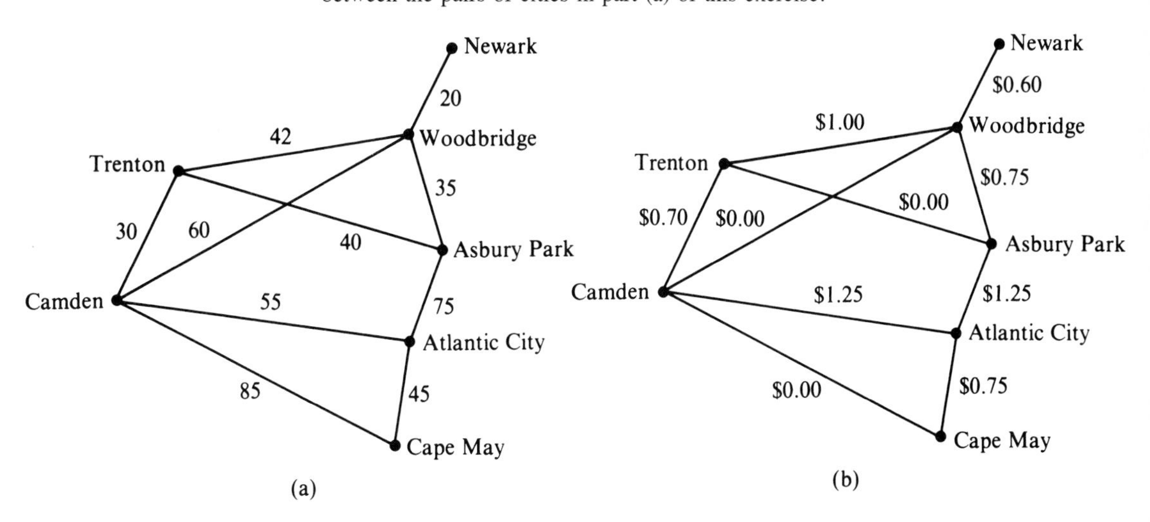

18. Is the shortest path between two vertices in a weighted graph unique if the weights of edges are distinct?

19. What are some applications where it is necessary to find the length of the longest simple path between two vertices in a weighted graph?

20. What is the length of the longest simple path in the weighted graph in Figure 4 between a and z? Between c and z?

Floyd's algorithm can be used to find the length of the shortest path between all pairs of vertices in a weighted connected simple graph. However, this algorithm cannot be used to construct shortest paths. (In the following, assign an infinite weight to any pair of vertices not connected by an edge in the graph.)

21. Use Floyd's algorithm to find the distance between all pairs of vertices in the weighted graph in Figure 4.

ALGORITHM 2 Floyd's Algorithm.

procedure *Floyd*(G: weighted simple graph)
{G has vertices $v_1, v_2, \ldots, v_n$ and weights $w(v_i, v_j)$ with $w(v_i, v_j) = \infty$
  if (v_i, v_j) is not an edge}
for $i := 1$ **to** n
  for $j := 1$ **to** n
    $d(v_i, v_j) := w(v_i, v_j)$
for $i := 1$ **to** n
  for $j := 1$ **to** n
    for $k := 1$ **to** n
      if $d(v_j, v_i) + d(v_i, v_k) < d(v_j, v_k)$ **then** $d(v_j, v_k) :=$
          $d(v_j, v_i) + d(v_i, v_k)$
{$d(v_i, v_j)$ is the length of the shortest path between v_i and v_j}

★22. Prove that Floyd's algorithm determines the shortest distance between all pairs of vertices in a weighted simple graph.

★23. Give a big-O estimate of the number of operations (comparisons and additions) used by Floyd's algorithm to determine the shortest distance between every pair of vertices in a weighted simple graph with n vertices.

★24. Show that Dijkstra's algorithm may not work if edges can have negative weights.

7.7 Planar Graphs

INTRODUCTION

Consider the problem of joining three houses to each of three separate utilities, as shown in Figure 1. Is it possible to join these houses and utilities so that none of the connections cross? This problem can be modeled using the complete bipartite graph

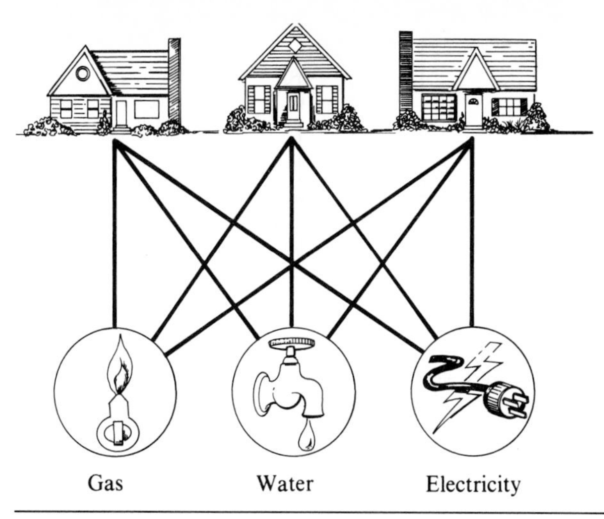

FIGURE 1 **Three Houses and Three Utilities.**

$K_{3,3}$. The original question can be rephrased as: Can $K_{3,3}$ be drawn in the plane so that no two of its edges cross?

In this section we will study the question of whether a graph can be drawn in the plane without edges crossing in this section. In particular, we will answer the houses-and-utilities problem.

There are always many ways to represent a graph. When is it possible to find at least one way to represent this graph in a plane without any edges crossing?

DEFINITION 1. A graph is called *planar* if it can be drawn in the plane without any edges crossing (where a crossing of edges is the intersection of the lines or arcs representing them at a point other than their common endpoint). Such a drawing is called a *planar representation* of the graph.

A graph may be planar even if it is usually drawn with crossings, since it may be possible to draw it in a different way without crossings.

EXAMPLE 1 Is K_4 (shown in Figure 2 with two edges crossing) planar?

Solution: K_4 is planar because it can be drawn without crossings, as shown in Figure 3. ■

EXAMPLE 2 Is Q_3, shown in Figure 4, planar?

Solution: Q_3 is planar, because it can be drawn without any edges crossing, as shown in Figure 5. ■

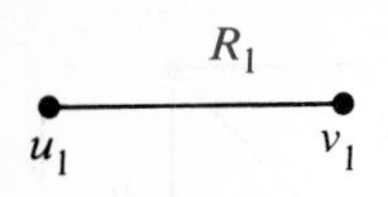

FIGURE 9 The Basis Case of the Proof of Euler's Formula.

The proof will now proceed by induction. The relationship $r_1 = e_1 - v_1 + 2$ is true for G_1, since $e_1 = 1$, $v_1 = 2$, and $r_1 = 1$. This is shown in Figure 9.

Now assume that $r_n = e_n - v_n + 2$. Let $\{a_{n+1}, b_{n+1}\}$ be the edge that is added to G_n to obtain G_{n+1}. There are two possibilities to consider. In the first case, both a_{n+1} and b_{n+1} are already in G_n. These two vertices must be on the boundary of a common region R, or else it would be impossible to add the edge $\{a_{n+1}, b_{n+1}\}$ to G_n without two edges crossing (and G_{n+1} is planar). The addition of this new edge splits R into two regions. Consequently, in this case, $r_{n+1} = r_n + 1$, $e_{n+1} = e_n + 1$, and $v_{n+1} = v_n$. Thus, each side of the formula relating the number of regions, edges, and vertices increases by exactly one, so that this formula is still true. In other words, $r_{n+1} = e_{n+1} - v_{n+1} + 2$. This case is illustrated in Figure 10(a).

In the second case, one of the two vertices of the new edge is not already in G_n. Suppose that a_{n+1} is in G_n, but that b_{n+1} is not. Adding this new edge does not produce any new regions, since b_{n+1} must be in a region that has a_{n+1} on its boundary. Consequently, $r_{n+1} = r_n$. Moreover, $e_{n+1} = e_n + 1$ and $v_{n+1} = v_n + 1$. Each side of the formula relating the number of regions, edges, and vertices, remains the same, so that the formula is still true. In other words, $r_{n+1} = e_{n+1} - v_{n+1} + 2$. This case is illustrated in Figure 10(b).

We have completed the induction argument. Hence $r_n = e_n - v_n + 2$ for all n. Since the original graph is the graph G_e, obtained after e edges have been added, the theorem is true. ■

Euler's formula is illustrated in the following example.

EXAMPLE 4 Suppose that a connected planar simple graph has 20 vertices, each of degree 3. Into how many regions does a representation of this planar graph split the plane?

Solution: This graph has 20 vertices, each of degree 3, so that $v = 20$. Since the sum of the degrees of the vertices, $3v = 3 \cdot 20 = 60$, is equal to twice the number

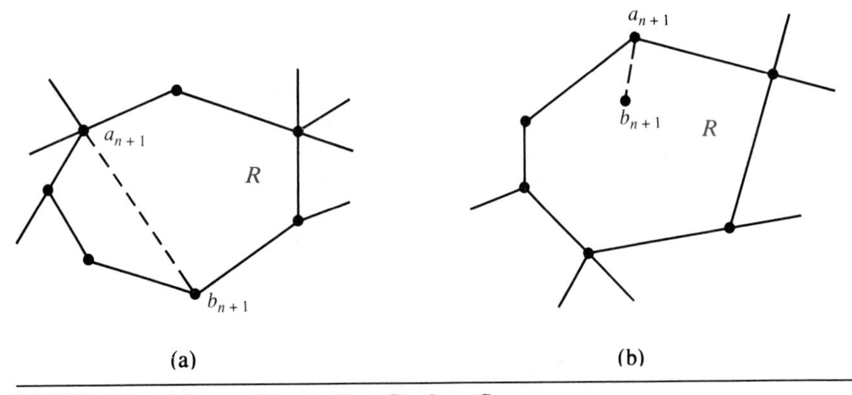

FIGURE 10 Adding an Edge to G_n to Produce G_{n+1}.

of edges, $2e$, we have $2e = 60$, or $e = 30$. Consequently, from Euler's formula, the number of regions is

$$r = e - v + 2 = 30 - 20 + 2 = 12.$$

■

Euler's formula can be used to establish some inequalities that must be satisfied by planar graphs. One such inequality is given in the following corollary.

COROLLARY 1 If G is a connected planar simple graph with e edges and v vertices where $v \geq 3$, then $e \leq 3v - 6$.

The proof of Corollary 1 is based on the concept of the **degree** of a region, which is defined to be the number of edges on the boundary of this region. When an edge occurs twice on the boundary (so that it is traced out twice when the boundary is traced out), it contributes 2 to the degree. The degrees of the regions of the graph shown in Figure 11 are displayed in the figure.

The proof of Corollary 1 can now be given.

Proof: A connected planar simple graph drawn in the plane divides the plane into regions, say r of them. The degree of each region is at least 3. (Since the graphs discussed here are simple graphs, no multiple edges that could produce regions of degree 2, or loops that could produce regions of degree 1, are permitted.) In particular, note that the degree of the unbounded region is at least 3 since there are at least three vertices in the graph.

Note that the sum of the degrees of the regions is exactly twice the number of edges in the graph, because each edge occurs on the boundary of a region exactly twice (either in two different regions, or twice in the same region). Since each region has degree greater than or equal to 3, it follows that

$$2e = \sum_{\text{all regions } R} \deg(R) \geq 3r.$$

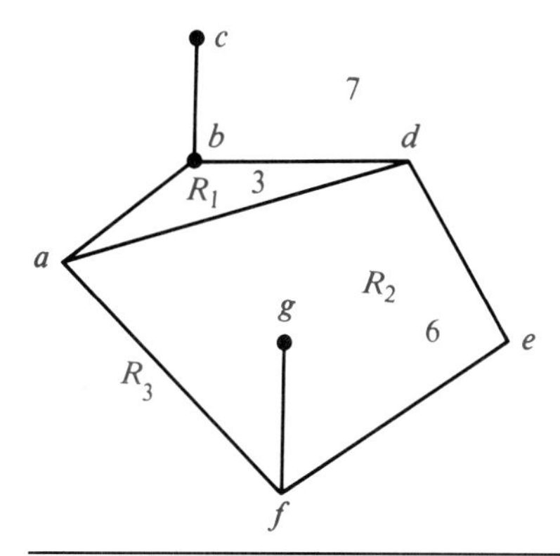

FIGURE 11 The Degrees of Regions.

Hence,

$$(2/3)e \geq r.$$

Using $r = e - v + 2$ (Euler's formula), we obtain

$$e - v + 2 \leq (2/3)e.$$

It follows that $e/3 \leq v - 2$. This shows that $e \leq 3v - 6$. ■

This corollary can be used to demonstrate that K_5 is nonplanar.

EXAMPLE 5 Show that K_5 is nonplanar using Corollary 1.

Solution: The graph K_5 has five vertices and 10 edges. However the inequality $e \leq 3v - 6$ is not satisfied for this graph since $e = 10$ and $3v - 6 = 9$. Therefore, K_5 is not planar. ■

It was previously shown that $K_{3,3}$ is not planar. Note, however, that this graph has six vertices and nine edges. This means that the inequality $e = 9 \leq 12 = 3 \cdot 6 - 6$ is satisfied. Consequently, the fact that the inequality $e \leq 3v - 6$ is satisfied does *not* imply that a graph is planar. However, the following corollary of Theorem 1 can be used to show that $K_{3,3}$ is nonplanar.

COROLLARY 2 If a connected planar simple graph has e edges and v vertices with $v \geq 3$ and no circuits of length 3, then $e \leq 2v - 4$.

The proof of Corollary 2 is similar to that of Corollary 1, except that in this case the fact that there are no circuits of length 3 implies that the degree of a region must be at least 4. The details of this proof are left for the reader (see Exercise 13 at the end of this section).

EXAMPLE 6 Use Corollary 2 to show that $K_{3,3}$ is nonplanar.

Solution: Since $K_{3,3}$ has no circuits of length 3 (this is easy to see since it is bipartite), Corollary 2 can be used. $K_{3,3}$ has six vertices and nine edges. Since $e = 9$ and $2v - 4 = 8$, Corollary 2 shows that $K_{3,3}$ is nonplanar. ■

KURATOWSKI'S THEOREM

We have seen that $K_{3,3}$ and K_5 are not planar. Clearly, a graph is not planar if it contains either of these two graphs as a subgraph. Furthermore, all nonplanar graphs must contain a subgraph that can be obtained from $K_{3,3}$ or K_5 using certain permitted operations.

If a graph is planar, so will be any graph obtained by removing an edge $\{u, v\}$ and adding a new vertex w together with edges $\{u, w\}$ and $\{w, v\}$. Such an operation is

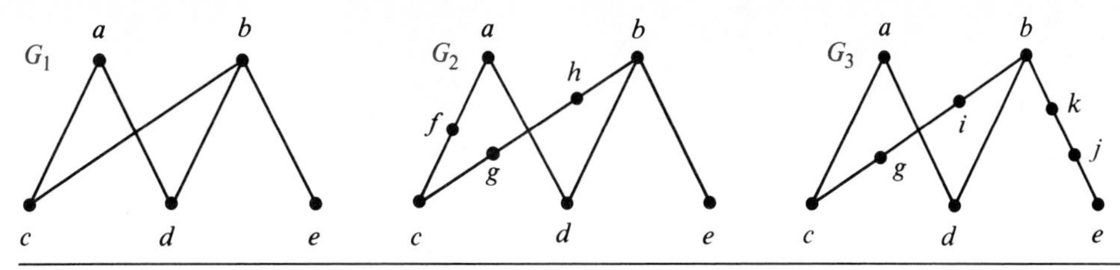

FIGURE 12 Homeomorphic Graphs.

called an **elementary subdivision.** The graphs $G_1 = (V_1, E_1)$ and $G_2 = (V_2, E_2)$ are called **homeomorphic** if they can be obtained from the same graph by a sequence of elementary subdivisions. The three graphs displayed in Figure 12 are homeomorphic, since all can be obtained from the first graph by elementary subdivisions. (The reader should determine the sequences of elementary subdivisions needed to obtain G_2 and G_3 from G_1.)

The Polish mathematician Kuratowski established the following theorem in 1930, which characterizes planar graphs using the concept of graph homeomorphism.

THEOREM 2 A graph is nonplanar if and only if it contains a subgraph homeomorphic to $K_{3,3}$ or K_5.

It is clear that a graph containing a subgraph homeomorphic to $K_{3,3}$ or K_5 is nonplanar. However, the proof of the converse, namely that every nonplanar graph contains a subgraph homeomorphic to $K_{3,3}$ or K_5, is complicated and will not be given here. The following examples illustrate how Kuratowski's theorem is used.

EXAMPLE 7 Determine whether the graph G shown in Figure 13 is planar.

Solution: G has a subgraph H homeomorphic to K_5. H is obtained by deleting h, j, and k and all edges incident with these vertices. H is homeomorphic to K_5 since it can be obtained from K_5 (with vertices a, b, c, g, and i) by a sequence of elementary subdivisions, adding the vertices d, e, and f. (The reader should construct such a sequence of elementary subdivisions.) Hence, G is nonplanar. ■

Kazimierz Kuratowski (1896–1980). Kazimierz Kuratowski, the son of a famous Warsaw lawyer, attended secondary school in Warsaw. He studied in Glasgow, Scotland, from 1913 to 1914 but could not return there after the outbreak of World War I. In 1915 he entered Warsaw University, where he was active in the Polish patriotic student movement. He published his first paper in 1919 and received his Ph.D. in 1921. He was an active member of the group known as the Warsaw School of Mathematics, working in the areas of the foundations of set theory and topology. He was appointed associate professor at the Lwow Polytechnical University, where he stayed for seven years, collaborating with the important Polish mathematicians Banach and Ulam. In 1930, while at Lwow, Kuratowski completed his work characterizing planar graphs.

In 1934 he returned to Warsaw University as a full professor. Until the start of World War II, he was active in research and teaching. During the war, because of the persecution of educated Poles, he went into hiding under an assumed name and taught at the clandestine Warsaw University. After the war he helped revive Polish mathematics, serving as director of the Polish National Mathematics Institute. He wrote over 180 papers and three widely used textbooks.

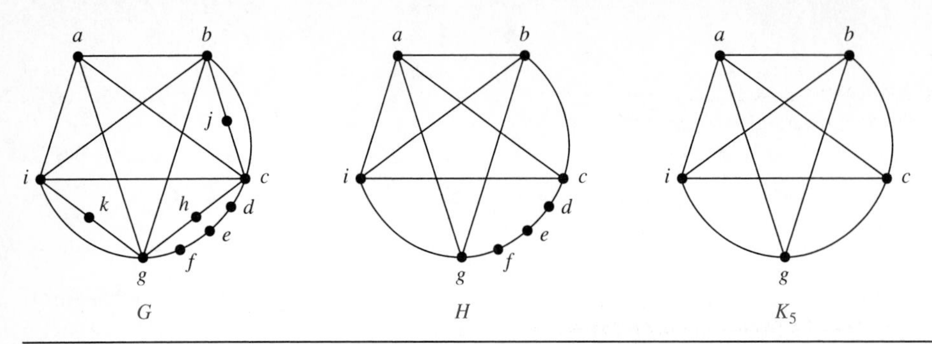

FIGURE 13 The Undirected Graph G, a subgraph H homeomorphic to K_5, and K_5.

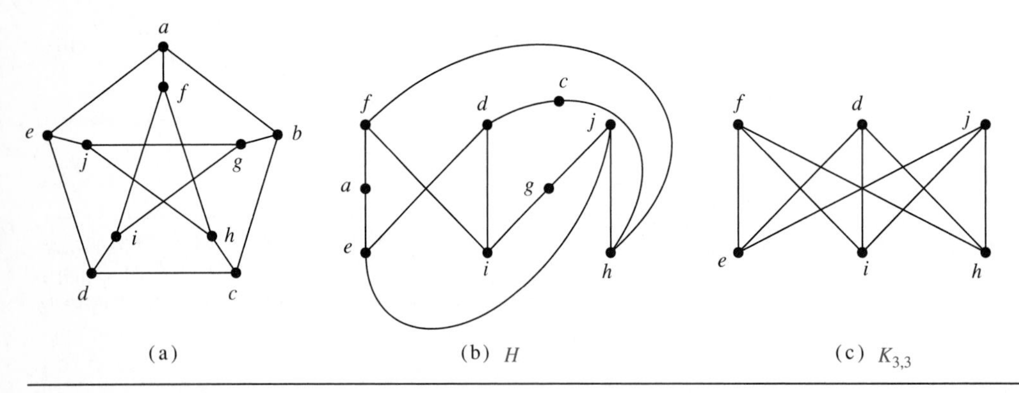

FIGURE 14 (a) The Peterson Graph, (b) a Subgraph H Homeomorphic to $K_{3,3}$, and (c) $K_{3,3}$.

EXAMPLE 8 Is the Petersen graph, shown in Figure 14(a), planar? (The Danish mathematician Julius Petersen introduced this graph in 1891; it is often used to illustrate various theoretical properties of graphs.)

Solution: The subgraph H of the Petersen graph obtained by deleting b and the three edges that have b as an endpoint, shown in Figure 14(b), is homeomorphic to $K_{3,3}$, with vertex sets $\{f, d, j\}$ and $\{e, i, h\}$, since it can be obtained by a sequence of elementary subdivisions, deleting $\{d, h\}$ and adding $\{c, h\}$ and $\{c, d\}$, deleting $\{e, f\}$ and adding $\{a, e\}$ and $\{a, f\}$, and deleting $\{i, j\}$ and adding $\{g, i\}$ and $\{g, j\}$. Hence, the Petersen graph is not planar. ■

Exercises

1. Can five houses be connected to two utilities without connections crossing?

In Exercises 2–4 draw the given planar graph without any crossings.

2. 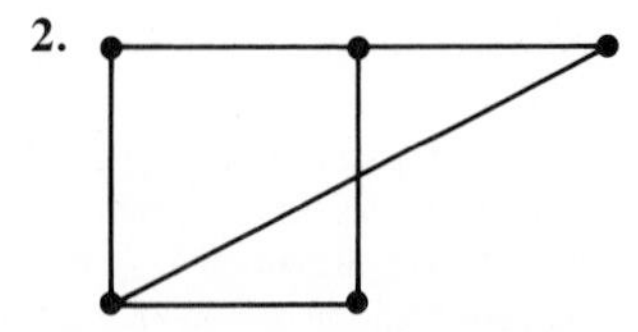

3.

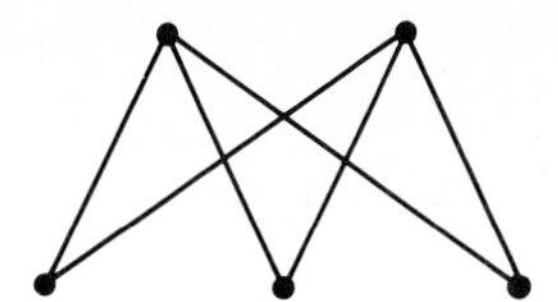

4.

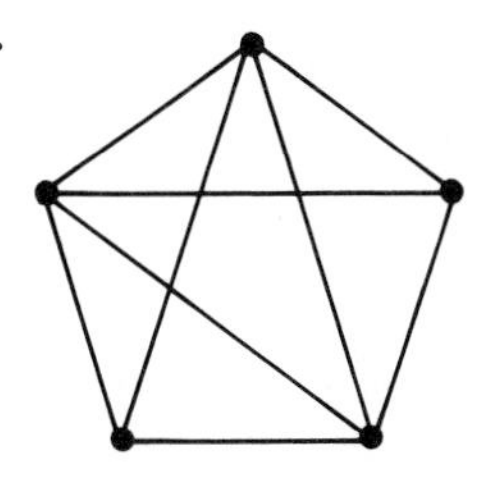

In Exercises 5–7 determine whether the given graph is planar. If so, draw it so that no edges cross.

5.

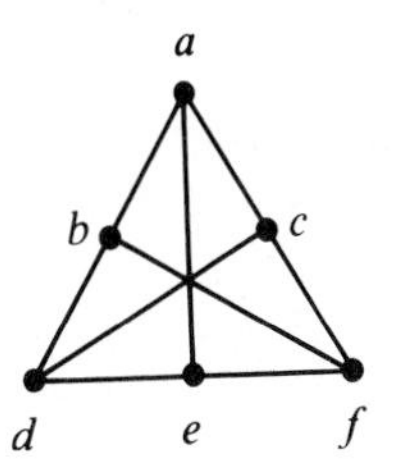

6.

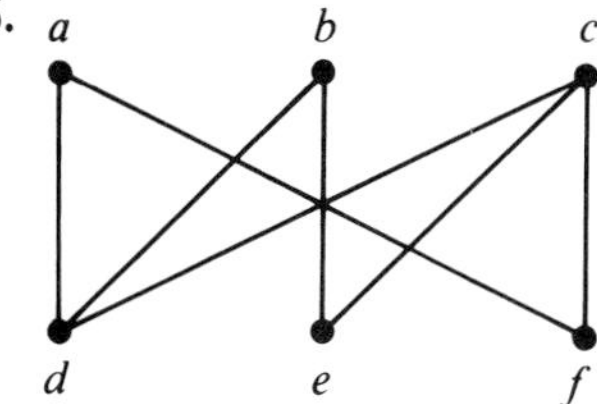

7.

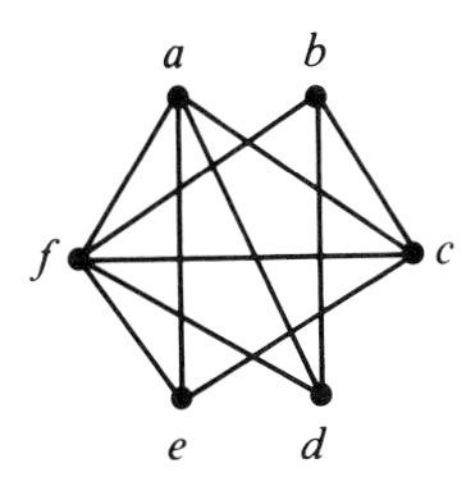

8. Complete the argument in Example 3.

9. Show that K_5 is nonplanar using an argument similar to that given in Example 3.

10. Suppose that a connected planar graph has eight vertices, each of degree 3. Into how many regions is the plane divided by a planar representation of this graph?

11. Suppose that a connected planar graph has six vertices, each of degree 4. Into how many regions is the plane divided by a planar representation of this graph?

12. Suppose that a connected planar graph has 30 edges. If a planar representation of this graph divides the plane into 20 regions, how many vertices does this graph have?

13. Prove Corollary 2.

14. Suppose that a connected bipartite planar simple graph has e edges and v vertices. Show that $e \leq 2v - 4$ if $v \geq 3$.

★15. Suppose that a connected planar simple graph with e edges and v vertices contains no simple circuits of length 4 or less. Show that $e \leq (5/3)v - (10/3)$ if $v \geq 4$.

16. Suppose that a planar graph has k connected components, e edges, and v vertices. Also suppose that the plane is divided into r regions by a planar representation of the graph. Find a formula for r in terms of e, v, and k.

17. Which of the following nonplanar graphs have the property that the removal of any vertex and all edges incident with that vertex produces a planar graph?

a) K_5 **b)** K_6 **c)** $K_{3,3}$ **d)** $K_{3,4}$

In Exercises 18–20 determine whether the given graph is homeomorphic to $K_{3,3}$.

18.

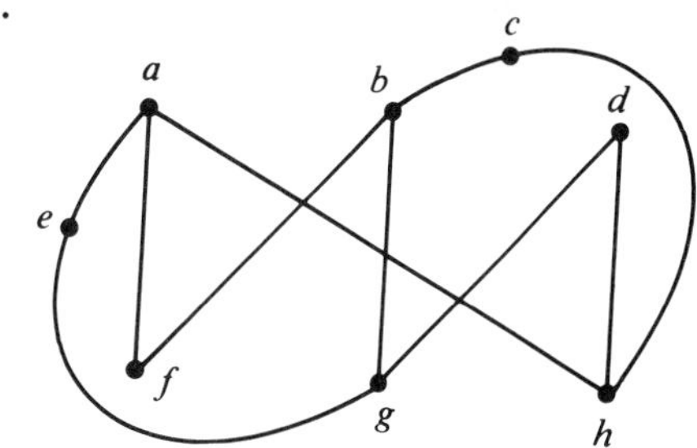

19.

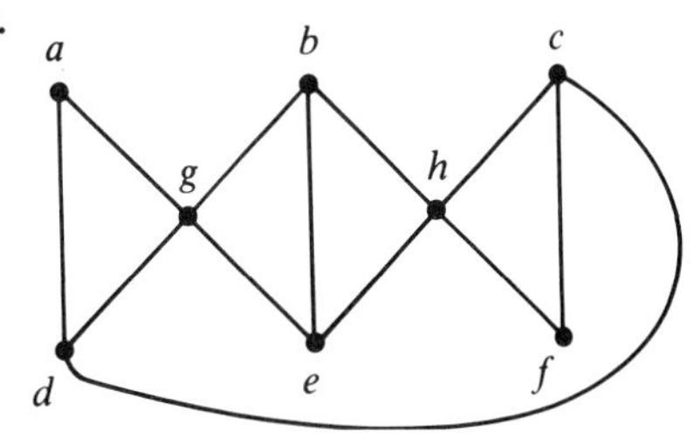

20.

In Exercises 21–23 use Kuratowski's theorem to determine whether the given graph is planar.

21.

22.

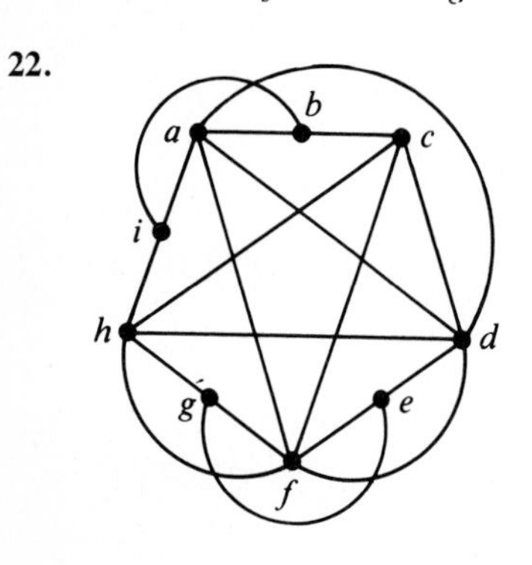

23.

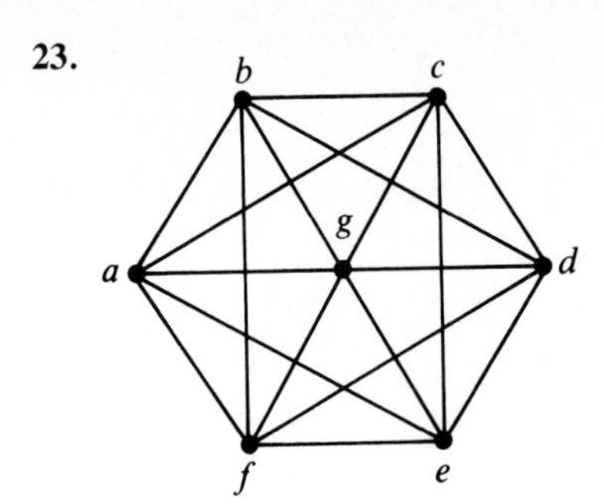

The **crossing number** of a simple graph is the minimum number of crossings that can occur in a planar representation of this graph, where no three arcs representing edges can cross at the same point.

24. Show that $K_{3,3}$ has 1 as its crossing number.

★★25. Find the crossing numbers of each of the following nonplanar graphs.

a) K_5 **b)** K_6 **c)** K_7
d) $K_{3,4}$ **e)** $K_{4,4}$ **f)** $K_{5,5}$

The **thickness** of a simple graph G is the smallest number of planar subgraphs of G that have G as their union.

26. Show that $K_{3,3}$ has 2 as its thickness.

★27. Find the thickness of the graphs in Exercise 25.

★28. Draw K_5 on the surface of a torus (a doughnut-shaped solid) so no edges cross.

★29. Draw $K_{3,3}$ on the surface of a torus so no edges cross.

7.8 Graph Coloring

INTRODUCTION

Problems related to the coloring of maps of regions, such as maps of parts of the world, have generated many results in graph theory. When a map* is colored, two regions with a common border are customarily assigned different colors. One way to ensure that two adjacent regions never have the same color is to use a different

*We will assume that all regions in a map are connected. This eliminates any problems presented by such geographical entities as Michigan.

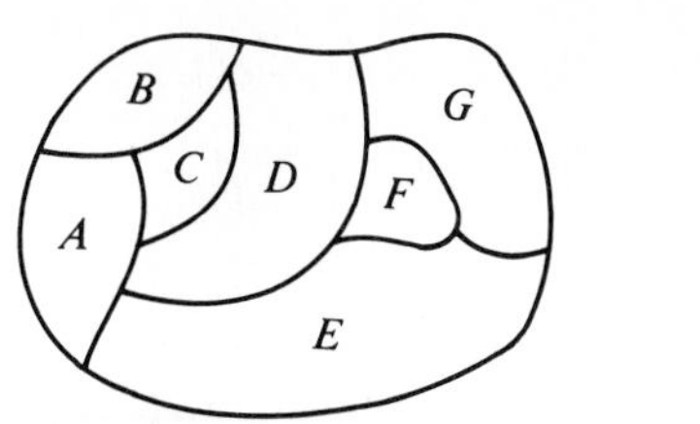

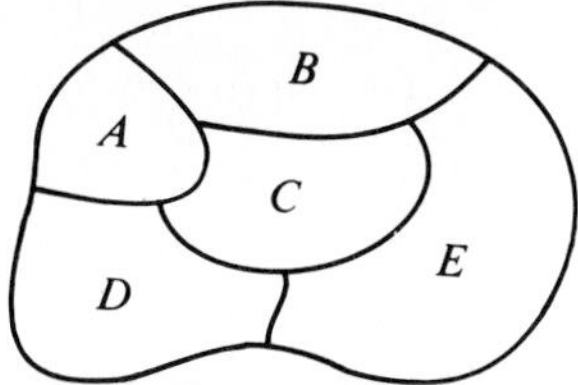

FIGURE 1 **Two Maps.**

color for each region. However, this is inefficient, and on maps with many regions it would be hard to distinguish similar colors. Instead, a small number of colors should be used whenever possible. Consider the problem of determining the least number of colors that can be used to color a map so that adjacent regions never have the same color. For instance, for the map shown on the left in Figure 1, four colors suffice, but three colors are not enough. (The reader should check this.) In the map on the right in Figure 1, three colors are sufficient (but two are not).

Each map in the plane can be represented by a graph. To set up this correspondence, each region of the map is represented by a vertex. Edges connect two vertices if the regions represented by these vertices have a common border. Two regions that touch at only one point are not considered adjacent. The resulting graph is called the **dual graph** of the map. By the way in which dual graphs of maps are constructed, it is clear that any map in the plane has a planar dual graph. Figure 2 displays the dual graphs that correspond to the maps shown in Figure 1.

The problem of coloring the regions of a map is equivalent to the problem of coloring the vertices of the dual graph so that no two adjacent vertices in this graph have the same color. We give the following definition.

DEFINITION 1. A *coloring* of a simple graph is the assignment of a color to each vertex of the graph so that no two adjacent vertices are assigned the same color.

A graph can be colored by assigning a different color to each of its vertices. However, for most graphs a coloring can be found that uses fewer colors than the number of vertices in the graph. What is the least number of colors necessary?

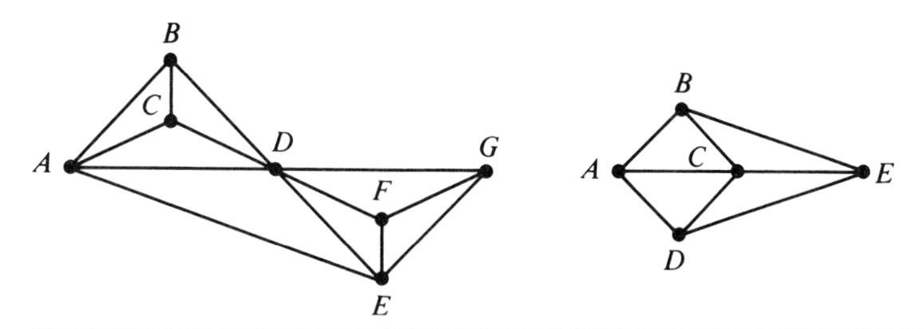

FIGURE 2 **Dual Graphs of the Maps in Figure 1.**

DEFINITION 2. The *chromatic number* of a graph is the least number of colors needed for a coloring of this graph.

The four-color theorem is one of the most famous theorems in mathematics. It asserts that the chromatic number of a planar graph is no larger than 4. That is, it states that every planar graph can be colored using four or fewer colors. This theorem was originally posed as a conjecture in the 1850s. It was finally proved by the American mathematicians Kenneth Appel and Wolfgang Haken in 1976. Prior to 1976, many incorrect proofs were published, often with hard-to-find errors. In addition, many futile attempts were made to construct counterexamples by drawing maps that require more than four colors.

Perhaps the most notorious fallacious proof in all of mathematics is an incorrect proof of the four-color theorem published in 1879 by a London barrister and amateur mathematician, Alfred Kempe. Mathematicians accepted his proof as correct until 1890, when Percy Heawood found an error that made Kempe's argument incomplete. However, Kempe's line of reasoning turned out to be the basis of the successful proof given by Appel and Haken. Their proof relies on a careful case-by-case analysis carried out by computer. They showed that if the four-color theorem were false, there would have to be a counterexample of one of approximately 2000 different types, and they then showed that none of these types could lead to a counterexample. They used over 1000 hours of computer time in their proof. This proof generated a large amount of controversy, since computers played such an important role in it. For example, could there be an error in a computer program that led to incorrect results? Was their argument really a proof if it depended on what could be unreliable computer output?

The four-color theorem applies only to planar graphs. Nonplanar graphs can have arbitrarily large chromatic numbers, as will be shown in Example 2.

Two things are required to show that the chromatic number of a graph is n. First, we must show that the graph can be colored with n colors. This can be done by constructing such a coloring. Second, we must show that the graph cannot be colored using fewer than n colors. The following examples illustrate how chromatic numbers can be found.

Alfred Bray Kempe (1849–1922). Kempe was a barrister and a leading authority on ecclesiastical law. However, having studied mathematics at Cambridge University, he retained his interest in it, and later in life he devoted considerable time to mathematical research. Kempe made contributions to kinematics, the branch of mathematics dealing with motion, and to mathematical logic. However, Kempe is best remembered for his fallacious proof of the four-color theorem.

Historical Note: In 1852, an ex-student of Augustus De Morgan, Francis Guthrie, noticed that the counties in England could be colored using four colors so that no adjacent counties were assigned the same color. On this evidence, he conjectured that the four-color theorem was true. Francis told his brother Frederick, at that time a student of De Morgan, about this problem. Frederick in turn asked his teacher De Morgan about his brother's conjecture. De Morgan was extremely interested in this problem and publicized it throughout the mathematical community. In fact, the first written reference to the conjecture can be found in a letter from De Morgan to Sir William Rowan Hamilton. Although De Morgan thought Hamilton would be interested in this problem, Hamilton apparently was not interested in it, since it had nothing to do with quaternions.

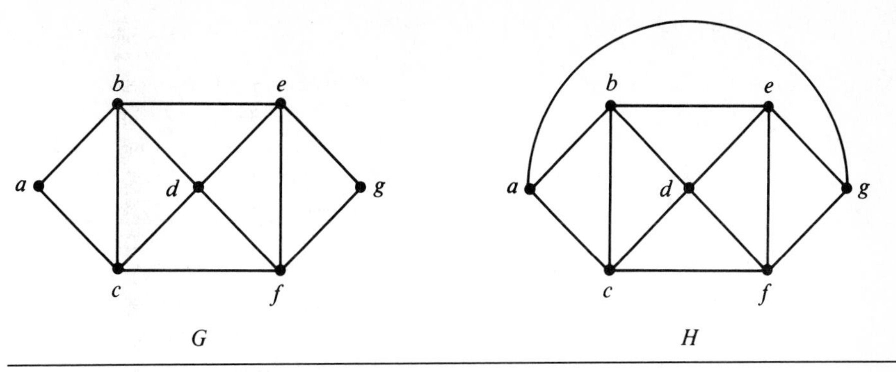

FIGURE 3 **The Simple Graphs G and H.**

EXAMPLE 1 What are the chromatic numbers of the graphs G and H shown in Figure 3?

Solution: The chromatic number of G is at least 3, since the vertices a, b, and c must be assigned different colors. To see if G can be colored with three colors, assign red to a, blue to b, and green to c. Then, d can (and must) be colored red since it is adjacent to b and c. Furthermore, e can (and must) be colored green since it is adjacent only to vertices colored red and blue, and f can (and must) be colored blue since it is adjacent only to vertices colored red and green. Finally, g can (and must) be colored red, since it is adjacent only to vertices colored blue and green. This produces a coloring of G using exactly three colors. Figure 4 displays such a coloring.

The graph H is made up of the graph G with an edge connecting a and g. Any attempt to color H using three colors must follow the same reasoning as that used to color G, except at the last stage, when all vertices other than g have been colored. Then, since g is adjacent (in H) to vertices colored red, blue, and green, a fourth color, say brown, needs to be used. Hence, H has chromatic number equal to 4. A coloring of H is shown in Figure 3. ■

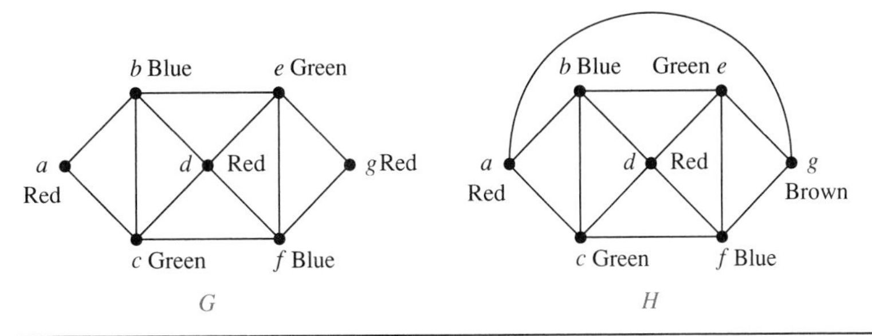

FIGURE 4 **Colorings of the Graphs G and H.**

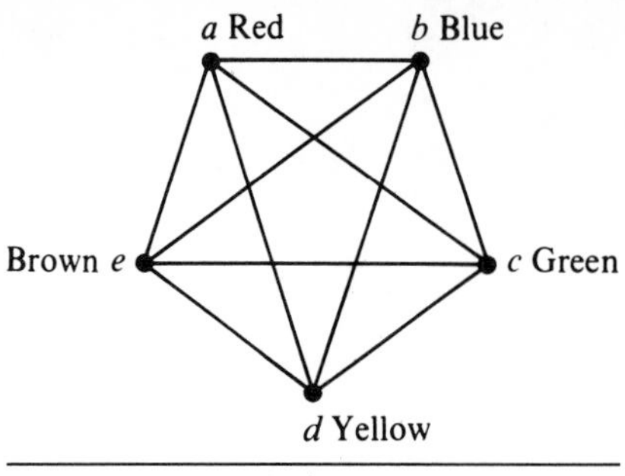

FIGURE 5 **A Coloring of K_5.**

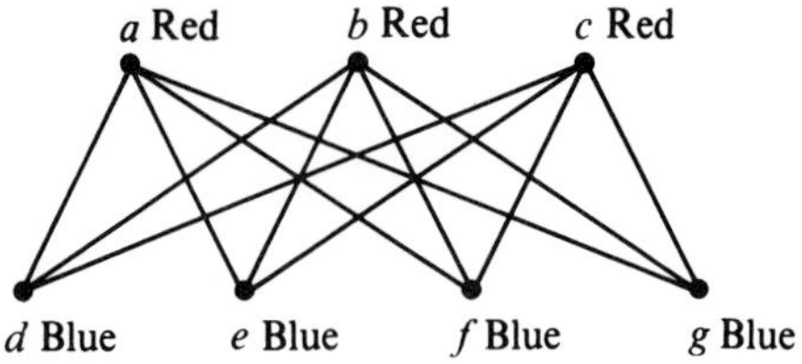

a Red b Red c Red

d Blue e Blue f Blue g Blue

FIGURE 6 **A Coloring of $K_{3,4}$.**

EXAMPLE 2 What is the chromatic number of K_n?

Solution: A coloring of K_n can be constructed using n colors by assigning a different color to each vertex. Is there a coloring using fewer colors? The answer is no. No two vertices can be assigned the same color, since every two vertices of this graph are adjacent. Hence, the chromatic number of $K_n = n$. (Recall that K_n is not planar when $n \geq 5$, so that this result does not contradict the four-color theorem). A coloring of K_5 using five colors is shown in Figure 5. ■

EXAMPLE 3 What is the chromatic number of the complete bipartite graph $K_{m,n}$, where m and n are positive integers?

Solution: The number of colors needed may seem to depend on m and n. However, only two colors are needed. Color the set of m vertices with one color and the set of n vertices with a second color. Since edges connect only a vertex from the set of m vertices and a vertex from the set of n vertices, no two adjacent vertices have the same color. A coloring of $K_{3,4}$ with two colors is displayed in Figure 6. ■

Every connected bipartite simple graph has a chromatic number of 2, or 1, since the reasoning used in Example 3 applies to any such graph. Conversely, every graph with a chromatic number of 2 is bipartite. (See Exercises 23 and 24 at the end of this section.)

EXAMPLE 4 What is the chromatic number of the graph C_n? (Recall that C_n is the cycle with n vertices.)

Solution: We will first consider some individual cases. To begin, let $n = 6$. Pick a vertex and color it red. Proceed clockwise in the planar depiction of C_6 shown in Figure 7. It is necessary to assign a second color, say blue, to the next vertex reached. Continue in the clockwise direction; the third vertex can be colored red, the fourth vertex blue, and the fifth vertex red. Finally, the sixth vertex, which is adjacent to the

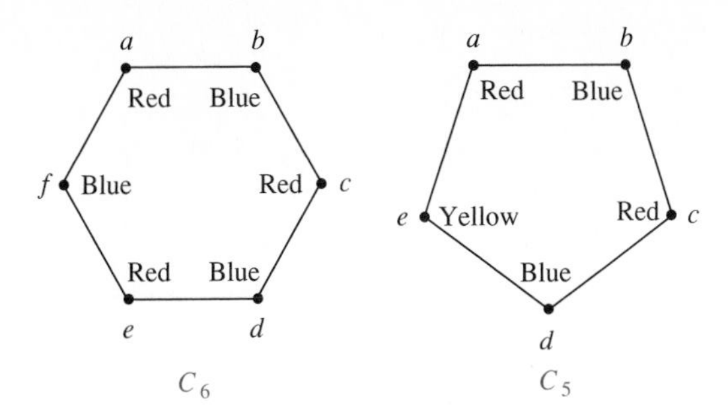

FIGURE 7 Colorings of C_5 and C_6.

first, can be colored blue. Hence, the chromatic number of C_6 is 2. Figure 7 displays the coloring constructed here.

Next, let $n = 5$ and consider C_5. Pick a vertex and color it red. Proceeding clockwise, it is necessary to assign a second color, say blue, to the next vertex reached. Continuing in the clockwise direction, the third vertex can be colored red, and the fourth vertex can be colored blue. The fifth vertex cannot be colored either red or blue, since it is adjacent to the fourth vertex and the first vertex. Consequently, a third color is required for this vertex. Note that we would have also needed three colors if we had colored vertices in the counterclockwise direction. Thus, the chromatic number of C_5 is 3. A coloring of C_5 using three colors is displayed in Figure 7.

In general, two colors are needed to color C_n when n is even. To construct such a coloring, simply pick a vertex and color it red. Then proceed around the graph in a clockwise direction (using a planar representation of the graph) coloring the second vertex blue, the third vertex red, and so on. The nth vertex can be colored blue, since the two vertices adjacent to it, namely the $(n - 1)$st and the first vertices, are both colored red.

When n is odd and $n > 1$, the chromatic number of C_n is 3. To see this, pick an initial vertex. To use only two colors, it is necessary to alternate colors as the graph is traversed in a clockwise direction. However, the nth vertex reached is adjacent to two vertices of different colors, namely, the first and $(n - 1)$st. Hence, a third color must be used. ■

APPLICATIONS OF GRAPH COLORINGS

Graph coloring has a variety of applications to problems involving scheduling and assignments. Examples of such applications will be given here. The first application deals with the scheduling of final exams.

EXAMPLE 5 Scheduling Final Exams How can the final exams at a university be scheduled so that no student has two exams at the same time?

Solution: This scheduling problem can be solved using a graph model, with vertices representing courses and with an edge between two vertices if there is a common

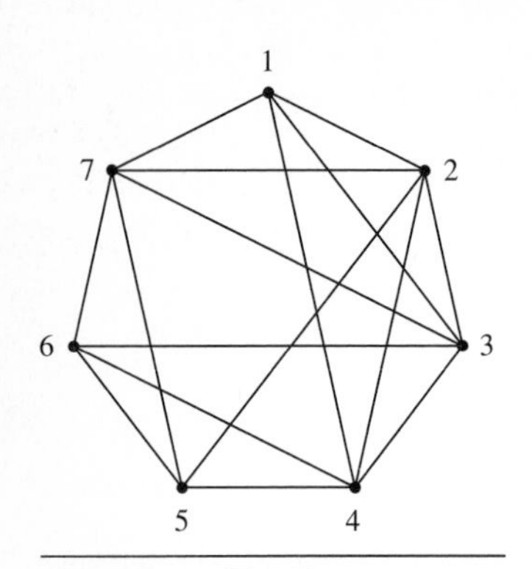

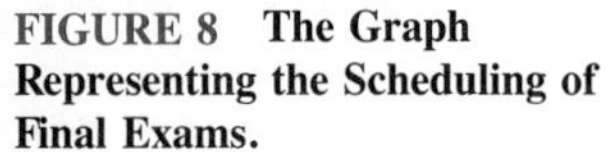
FIGURE 8 The Graph Representing the Scheduling of Final Exams.

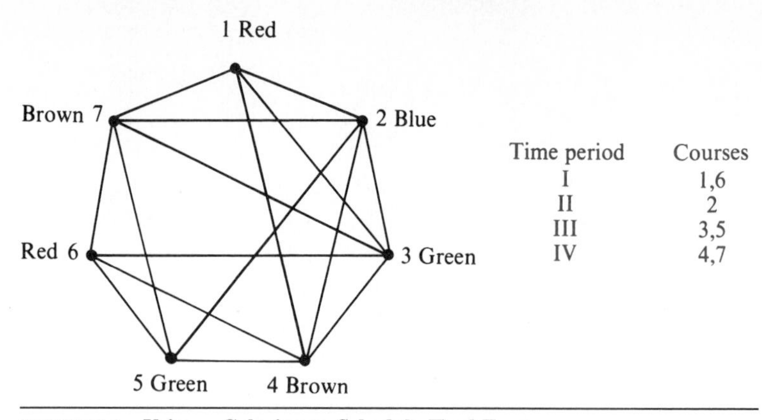

Time period	Courses
I	1,6
II	2
III	3,5
IV	4,7

FIGURE 9 Using a Coloring to Schedule Final Exams.

student in the courses they represent. Each time slot for a final exam is represented by a different color. A scheduling of the exams corresponds to a coloring of the associated graph.

For instance, suppose there are seven finals to be scheduled. Suppose the courses are numbered 1 through 7. Suppose that the following pairs of courses have common students: 1 and 2, 1 and 3, 1 and 4, 1 and 7, 2 and 3, 2 and 4, 2 and 5, 2 and 7, 3 and 4, 3 and 6, 3 and 7, 4 and 5, 4 and 6, 5 and 6, 5 and 7, and 6 and 7. In Figure 8 the graph associated with this set of classes is shown. A scheduling consists of a coloring of this graph.

Since the chromatic number of this graph is 4 (the reader should verify this), four time slots are needed. A coloring of the graph using four colors and the associated schedule are shown in Figure 9. ■

Now consider an application to the assignment of television channels.

EXAMPLE 6 Frequency Assignments Television channels 2 through 13 are assigned to stations in North America so that no two stations within 150 miles can operate on the same channel. How can the assignment of channels be modeled by graph coloring?

Solution: Construct a graph by assigning a vertex to each station. Two vertices are connected by an edge if they are located within 150 miles of each other. An assignment of channels corresponds to a coloring of the graph, where each color represents a different channel. ■

An application of graph coloring to compilers follows.

EXAMPLE 7 Index Registers In efficient compilers the execution of loops is speeded up when frequently used variables are stored temporarily in index registers in the central processing unit, instead of in regular memory. For a given loop, how many index registers are needed? This problem can be addressed using a graph coloring model. To set up the model, let each vertex of a graph represent a variable in the loop. There is an edge between two vertices if the variables they represent must be stored in index registers at the same time during the execution of the loop. Thus, the chromatic number of the graph gives the number of index registers needed, since different registers must be assigned to variables when the vertices representing these variables are adjacent in the graph. ■

Exercises

1. Construct the dual graphs for each of the following maps.

a)

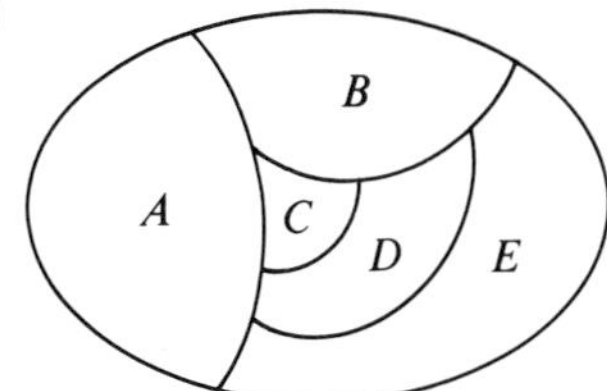

b)

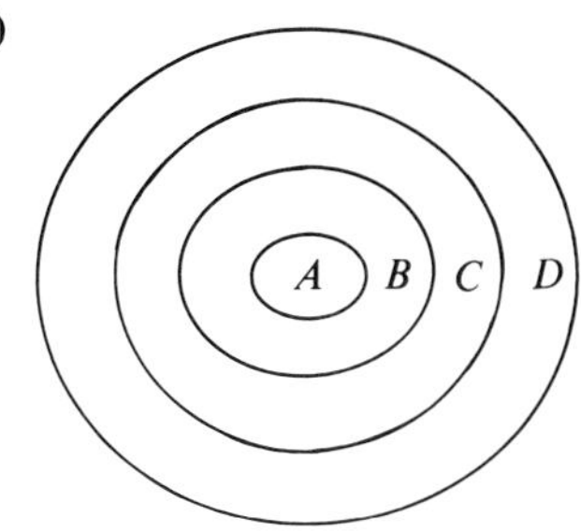

c)

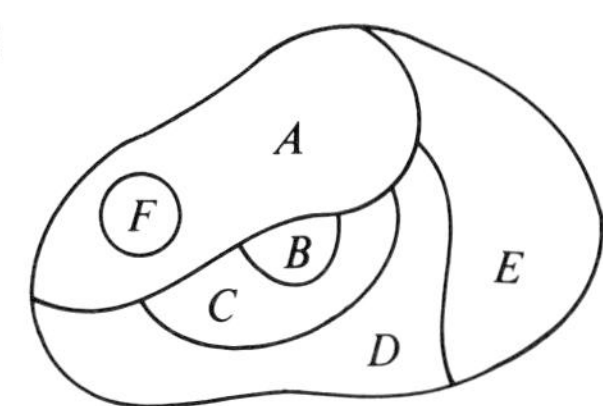

2. Find the number of colors needed to color the maps in Exercise 1 so that no two adjacent regions have the same color.

In Exercises 3–9 find the chromatic number of the given graph.

3.

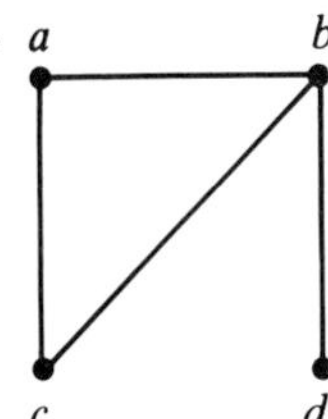

4.

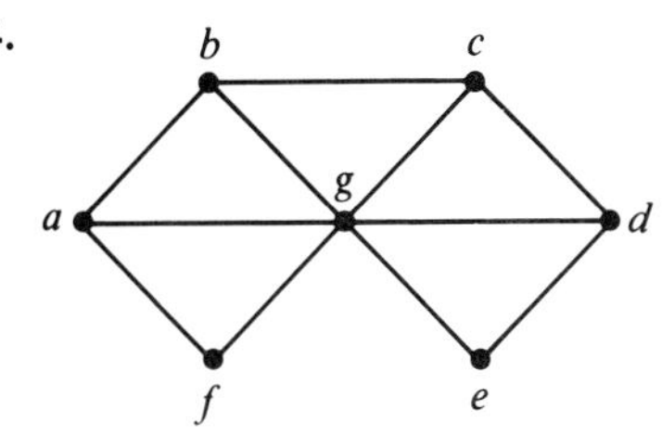

5.

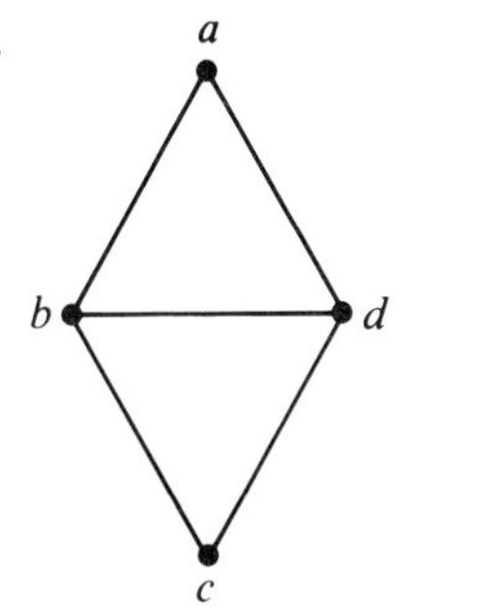

6.

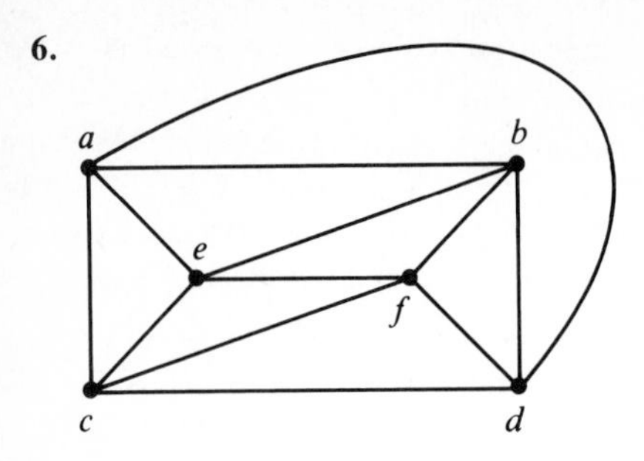

7.

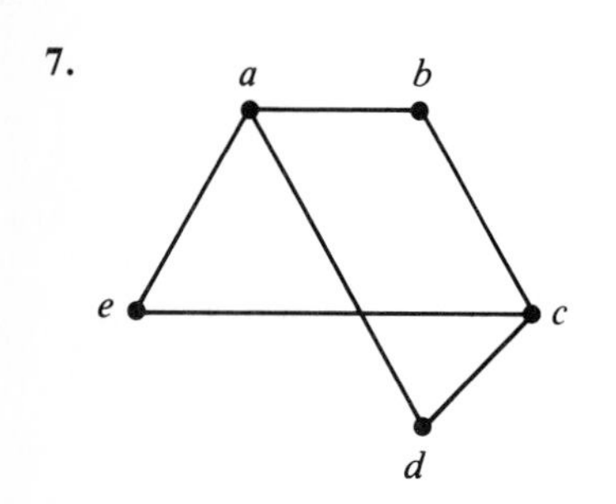

8.

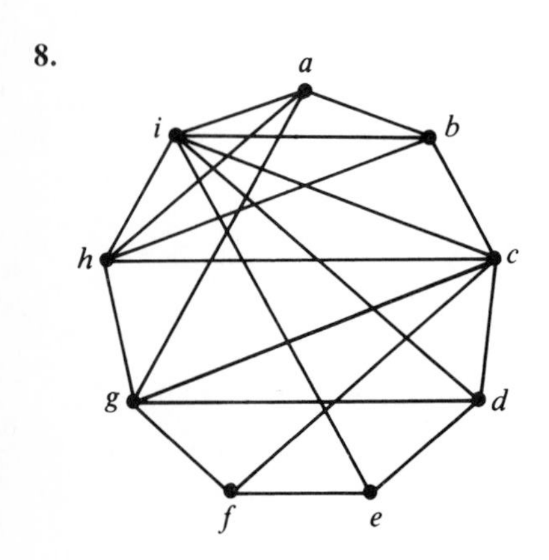

9.

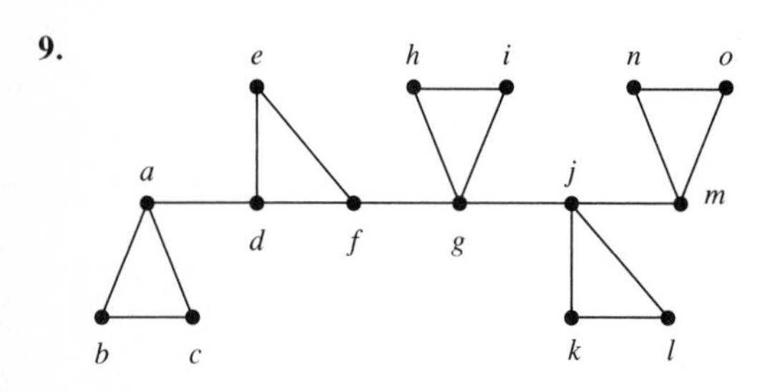

10. For the graphs in Exercises 3–9, decide whether it is possible to decrease the chromatic number by removing a single vertex and all edges incident with it.

11. Which graphs have a chromatic number of 1?

12. What is the least number of colors needed to color a map of the United States? Do not consider adjacent states that meet only at a corner. Suppose that Michigan is one region. Consider the vertices representing Alaska and Hawaii as isolated vertices.

13. What is the chromatic number of W_n?

14. Show that a simple graph that has a circuit with an odd number of vertices in it cannot be colored using two colors.

15. Schedule the final exams for Math 115, Math 116, Math 185, Math 195, CS 101, CS 102, CS 273, and CS 473, using the fewest number of different time slots, if there are no students taking both Math 115 and CS 473, both Math 116 and CS 473, both Math 195 and CS 101, both Math 195 and CS 102, both Math 115 and Math 116, both Math 115 and Math 185, and both Math 185 and Math 195, but there are students in every other combination of courses.

16. How many different channels are needed for six stations located at the distances shown in the table, if two stations cannot use the same channel when they are within 150 miles of each other?

	1	*2*	*3*	*4*	*5*	*6*
1	—	85	175	200	50	100
2	85	—	125	175	100	160
3	175	125	—	100	200	250
4	200	175	100	—	210	220
5	50	100	200	210	—	100
6	100	160	250	220	100	—

17. The mathematics department has six committees that meet once a month. How many different meeting times must be used to ensure that no one is scheduled to be at two meetings at the same time if the committees are C_1 = {*Arlinghaus, Brand, Zaslavsky*}, C_2 = {*Brand, Lee, Rosen*}, C_3 = {*Arlinghaus, Rosen, Zaslavsky*}, C_4 = {*Lee, Rosen,*

Zaslavsky}, C_5 = {*Arlinghaus, Brand*}, and C_6 = {*Brand, Rosen, Zaslavsky*}.

18. A zoo wants to set up natural habitats in which to exhibit its animals. Unfortunately, some animals will eat some of the others when given the opportunity. How can a graph model and a coloring be used to determine the number of different habitats needed and the placement of the animals in these habitats?

An **edge coloring** of a graph is an assignment of colors to edges so that edges incident with a common vertex are assigned different colors. The **edge chromatic number** of a graph is the smallest number of colors that can be used in an edge coloring of the graph.

19. Find the edge chromatic numbers of each of the graphs in Exercises 3–9.

★20. Find the edge chromatic numbers of
a) K_n. **b)** $K_{m,n}$. **c)** C_n. **d)** W_n.

21. Seven variables occur in a loop of a computer program. The variables and the steps during which they must be stored are: *t*: steps 1 through 6, *u*: step 2, *v*: steps 2 through 4, *w*: steps 1, 3, and 5, *x*: steps 1 and 6, *y*: steps 3 through 6, and *z*: steps 4 and 5. How many different index registers are needed to store these variables during execution?

22. What can be said about the chromatic number of a graph that has K_n as a subgraph?

23. Show that a simple graph with a chromatic number of 2 is bipartite.

24. Show that a connected bipartite graph has a chromatic number of 2.

The following algorithm can be used to color a simple graph. First, list the vertices $v_1, v_2, v_3, \ldots, v_n$ in order of decreasing degree, so that $\deg(v_1) \geq \deg(v_2) \geq \cdots \geq \deg(v_n)$. Assign color 1 to v_1 and to the next vertex in the list not adjacent to v_1 (if one exists), and successively to each vertex in the list not adjacent to a vertex already assigned color 1. Then assign color 2 to the first vertex in the list not already colored. Successively assign color 2 to vertices in the list that have not been already colored and are not adjacent to vertices assigned color 2. If uncolored vertices remain, assign color 3 to the first vertex in the list not yet colored, and use color 3 to successively color those vertices not already colored and not adjacent to vertices assigned color 3. Continue this process until all vertices are colored.

25. Construct a coloring of the graph illustrated using this algorithm.

★26. Use pseudocode to describe this coloring algorithm.

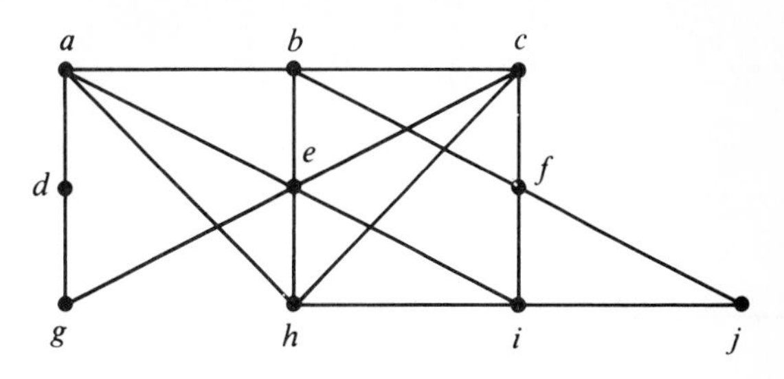

★27. Show that the coloring produced in this algorithm may use more colors than are necessary to color a graph.

A **k-tuple coloring** of a graph *G* is an assignment of a set of *k* different colors to the vertices of *G* so that no two adjacent vertices are assigned a common color. We denote by $\chi_k(G)$ the smallest positive integer *n* such that *G* has a *k*-tuple coloring using *n* colors. For example, $\chi_2(C_4) = 4$. To see this, note that using only four colors we can assign 2 colors to each vertex of C_4, as illustrated, so that no two adjacent vertices are assigned the same color. Furthermore, no fewer than four colors suffice because the vertices v_1 and v_2 each must be assigned two colors, and a common color cannot be assigned to both v_1 and v_2.

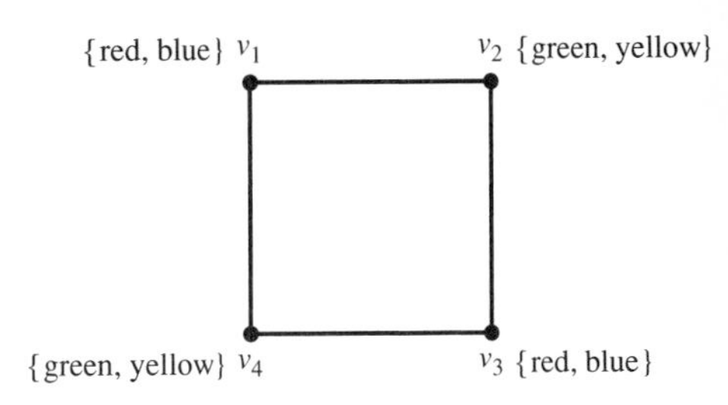

28. Find the following values:
a) $\chi_2(K_3)$ **b)** $\chi_2(K_4)$ **c)** $\chi_2(W_4)$
d) $\chi_2(C_5)$ **e)** $\chi_2(K_{3,4})$ **f)** $\chi_3(K_5)$
★g) $\chi_3(C_5)$ **h)** $\chi_3(K_{4,5})$

★29. Let *G* and *H* be the graphs displayed in Figure 3. Find
a) $\chi_2(G)$ **b)** $\chi_2(H)$ **c)** $\chi_3(G)$ **d)** $\chi_3(H)$

30. What is $\chi_k(G)$ if *G* is a bipartite graph and *k* is a positive integer?

31. Frequencies for mobile radio (or cellular) telephones are assigned by zones. Each zone is assigned a set of frequencies to be used by vehicles in that zone. The same frequency cannot be used in zones where interference is a problem. Explain how a *k*-tuple coloring can be used to assign *k* frequencies to each mobile radio zone in a region.

Key Terms and Results

TERMS

undirected edge: an edge associated to a set $\{u, v\}$ where u and v are vertices

directed edge: an edge associated to an ordered pair (u, v) where u and v are vertices

multiple edges: distinct edges connecting the same vertices

loop: an edge connecting a vertex with itself

undirected graph: a set of vertices together with a set of undirected edges that connect these vertices

simple graph: an undirected graph with no multiple edges or loops

multigraph: an undirected graph that may contain multiple edges but no loops

pseudograph: an undirected graph that may contain multiple edges and loops

directed graph: a set of vertices together with a set of directed edges that connect these vertices

directed multigraph: a graph with directed edges that may contain multiple directed edges

adjacent: two vertices are adjacent if there is an edge between them

incident: an edge is incident with a vertex if the vertex is an endpoint of that edge

deg(v) (degree of the vertex v in an undirected graph): the number of edges incident with v

$\deg^-(v)$ (the in-degree of the vertex v in a graph with directed edges): the number of edges with v as their terminal vertex

$\deg^+(v)$ (the out-degree of the vertex v in a graph with directed edges): the number of edges with v as their initial vertex

underlying undirected graph of a graph with directed edges: the undirected graph obtained by ignoring the directions of the edges

K_n (complete graph on n vertices): the undirected graph with n vertices where each pair of vertices is connected by an edge

bipartite graph: a graph with vertex set partitioned into subsets V_1 and V_2 such that each edge connects a vertex in V_1 and a vertex in V_2

$K_{m,n}$ (complete bipartite graph): the graph with vertex set partitioned into a subset of m elements and a subset of n elements such that two vertices are connected by an edge if and only if one is in the first subset and the other is in the second subset

C_n (cycle of size n), $n \geq 3$: the graph with n vertices $v_1, v_2, \ldots, v_n$ and edges $\{v_1, v_2\}$, $\{v_2, v_3\}, \ldots, \{v_{n-1}, v_n\}, \{v_n, v_1\}$

W_n (wheel of size n), $n \geq 3$: the graph obtained from C_n by adding a vertex and edges from this vertex to the original vertices in C_n

Q_n (n-cube), $n \geq 1$: the graph that has the 2^n bit strings of length n as its vertices and edges connecting every pair of bit strings that differ by exactly one bit

isolated vertex: a vertex of degree 0

pendant vertex: a vertex of degree 1

regular graph: a graph where all vertices have the same degree

subgraph of a graph $G = (V, E)$: a graph (W, F) where W is a subset of V and F is a subset of E

$G_1 \cup G_2$ (union of G_1 and G_2): the graph $(V_1 \cup V_2, E_1 \cup E_2)$ where $G_1 = (V_1, E_1)$ and $G_2 = (V_2, E_2)$

adjacency matrix: a matrix representing a graph using the adjacency of vertices

incidence matrix: a matrix representing a graph using the incidence of edges and vertices

isomorphic simple graphs: the simple graphs $G_1 = (V_1, E_1)$ and $G_2 = (V_2, E_2)$ are isomorphic if there is a one-to-one correspondence f from V_1 to V_2 such that $\{f(v_1), f(v_2)\} \in E_2$ if and only if $\{v_1, v_2\} \in E_1$ for all v_1 and v_2 in V_1

invariant: a property that isomorphic graphs either both have or both do not have

path from u to v in an undirected graph: a sequence of one or more edges $e_1, e_2, \ldots, e_n$ where e_i is associated to $\{x_i, x_{i+1}\}$ for $i = 0, 1, \ldots, n$ where $x_0 = u$ and $x_1 = v$

path from u to v in a graph with directed edges: a sequence of one or more edges $e_1, e_2, \ldots, e_n$ where e_i is associated to (x_i, x_{i+1}) for $i = 0, 1, \ldots, n$ where $x_0 = u$ and $x_{n+1} = v$

simple path: a path that does not contain an edge more than once

circuit: a path that begins and ends at the same vertex

connected graph: an undirected graph with the property that there is a path between every pair of vertices in the graph

connected components: the set of connected subgraphs of a graph such that no two of these subgraphs have a vertex in common

Euler circuit: a circuit that contains every edge of a graph exactly once

Euler path: a path that contains every edge of a graph exactly once

Hamilton path: a path $x_0, x_1, \ldots, x_n$ in a simple graph $G = (V, E)$ such that $\{x_0, x_1, \ldots, x_n\} = V$ and $x_i \neq x_j$ for $0 \leq i < j \leq n$

Hamilton circuit: a circuit $x_0, x_1, \ldots, x_n, x_0$ in a simple graph such that $x_0, x_1, \ldots, x_n$ is a Hamilton path

weighted graph: a graph with numbers assigned to its edges

shortest-path problem: the problem of determining the path in a weighted graph such that the sum of the weights of the edges in this path is a minimum over all paths between specified vertices

planar graph: a graph that can be drawn in the plane with no crossings

regions of a representation of a planar graph: the regions the plane is divided into by the planar representation of the graph

elementary subdivision: the removal of an edge $\{u, v\}$ of an undirected graph and the addition of a new vertex w together with edges $\{u, w\}$ and $\{w, v\}$

homeomorphic: two undirected graphs are homeomorphic if they can be obtained from the same graph by a sequence of elementary subdivisions

graph coloring: an assignment of colors to the vertices of a graph so that no two adjacent vertices have the same color

chromatic number: the minimum number of colors needed in a coloring of a graph

RESULTS

There is an Euler circuit in a connected multigraph if and only if every vertex has even degree.

There is an Euler path in a connected multigraph if and only if at most two vertices have odd degree.

Dijkstra's algorithm: a procedure for finding the shortest path between two vertices in a weighted graph (see page 471).

Euler's formula: $r = e - v + 2$ where r, e, and v are the number of regions of a planar representation, the number of edges, and the number of vertices, respectively, of a planar graph.

Kuratowski's theorem: A graph is nonplanar if and only if it contains a subgraph homeomorphic to $K_{3,3}$ or K_5. (Proof beyond scope of this book.)

The four-color theorem: Every planar graph can be colored using no more than four colors. (Proof far beyond scope of this book!)

Review Questions

1. a) Define a simple graph, a multigraph, a pseudograph, a directed graph, and a directed multigraph.

   b) Use an example to show how each of the types of graph in (a) can be used in modeling. For example, explain how to model different aspects of a computer network or airline routes.

2. Give at least four examples of how graphs are used in modeling.

3. What is the relationship between the sum of the de grees of the vertices in an undirected graph and the number of edges in this graph? Explain why this relationship holds.

4. Why must there be an even number of vertices of odd degree in an undirected graph?

5. What is the relationship between the sum of the in-degrees and the sum of the out-degrees of the vertices in a directed graph? Explain why this relationship holds.

6. Describe the following families of graphs.
   a) K_n, the complete graph on n vertices
   b) $K_{m,n}$, the complete bipartite graph on m and n vertices
   c) C_n, the cycle with n vertices
   d) W_n, the wheel of size n
   e) Q_n, the n-cube

7. How many vertices and how many edges are there in each of the graphs in the families in Question 6?

8. **a)** What is a bipartite graph?
   b) Which of the graphs K_n, C_n, and W_n are bipartite?
   c) How can you determine whether an undirected graph is bipartite?

9. **a)** Describe three different methods that can be used to represent a graph.
   b) Draw a simple graph with at least five vertices and eight edges. Illustrate how it can be represented using the methods you described in (a).

10. **a)** What does it mean for two simple graphs to be isomorphic?
   b) What is meant by an invariant with respect to isomorphism for simple graphs? Give at least five examples of such invariants.
   c) Give an example of two graphs that have the same numbers of vertices, edges, and degrees of vertices, but that are not isomorphic.
   d) Is a set of invariants known that can be used to efficiently determine whether two simple graphs are isomorphic?

11. **a)** What does it mean for a graph to be connected?
   b) What are the connected components of a graph?

12. **a)** Explain how an adjacency matrix can be used to represent a graph.
   b) How can adjacency matrices be used to determine whether a function from the vertex set of a graph G to the vertex set of a graph H is an isomorphism?
   c) How can the adjacency matrix of a graph be used to determine the number of paths of length r, where r is a positive integer, between two vertices of a graph?

13. **a)** Define an Euler circuit and an Euler path in an undirected graph.
   b) Describe the famous Königsberg bridge problem and explain how to rephrase it in terms of an Euler circuit.
   c) How can it be determined whether an undirected graph has an Euler path?
   d) How can it be determined whether an undirected graph has an Euler circuit?

14. **a)** Define a Hamilton circuit in a simple graph.
   b) Give some properties of a simple graph that imply that it does not have a Hamilton circuit.

15. Give examples of at least two problems that can be solved by the finding the shortest path in a weighted graph.

16. **a)** Describe Dijkstra's algorithm for finding the shortest path in a weighted graph between two vertices.
   b) Draw a weighted graph with at least 10 vertices and 20 edges. Use Dijkstra's algorithm to find the shortest path between 2 vertices of your choice in the graph.

17. **a)** What does it mean for a graph to be planar?
   b) Give an example of a nonplanar graph.

18. **a)** What is Euler's formula for planar graphs?
   b) How can Euler's formula for planar graphs be used to show that a simple graph is nonplanar?

19. State Kuratowski's theorem on the planarity of graphs and explain how it characterizes which graphs are planar.

20. **a)** Define the chromatic number of a graph.
   b) What is the chromatic number of the graph K_n when n is a positive integer?
   c) What is the chromatic number of the graph C_n when n is a positive integer greater than 2?
   d) What is the chromatic number of the graph $K_{m,n}$ when m and n are positive integers?

21. What is the four-color theorem? Are there graphs that cannot be colored with four colors?

22. Explain how graph coloring can be used in modeling. Use at least two different examples.

Supplementary Exercises

1. How many edges does a 50-regular graph with 100 vertices have?
2. How many nonisomorphic subgraphs does K_3 have?

In Exercises 3–5 determine whether the given pair of graphs is isomorphic.

3.

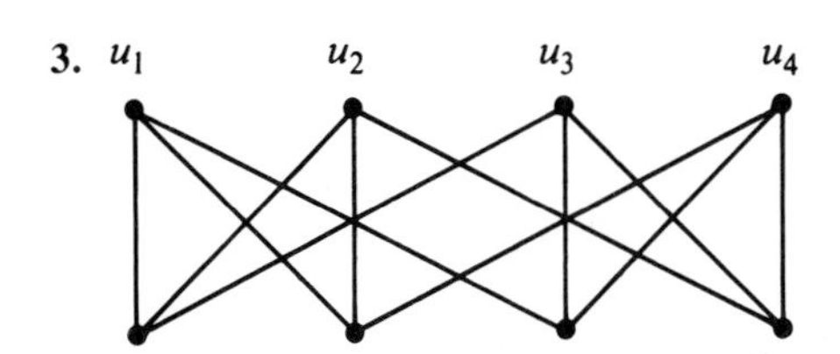

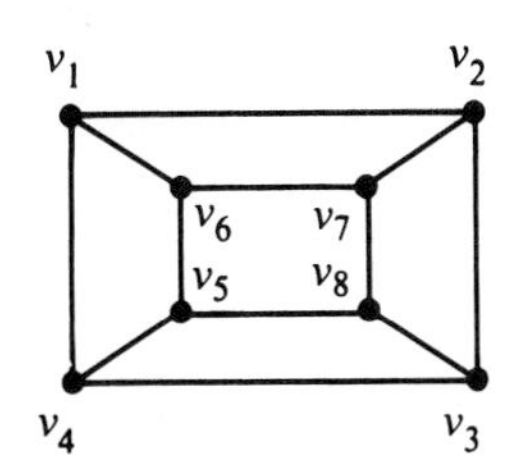

4.

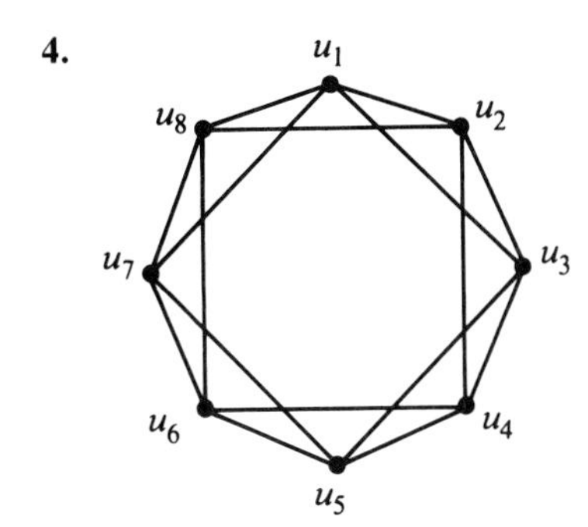

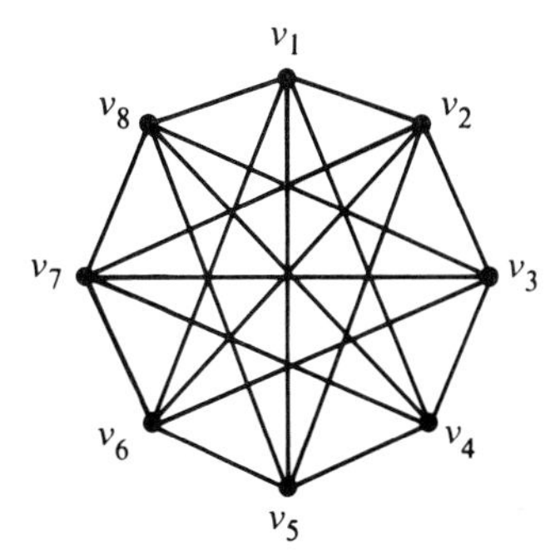

★5.

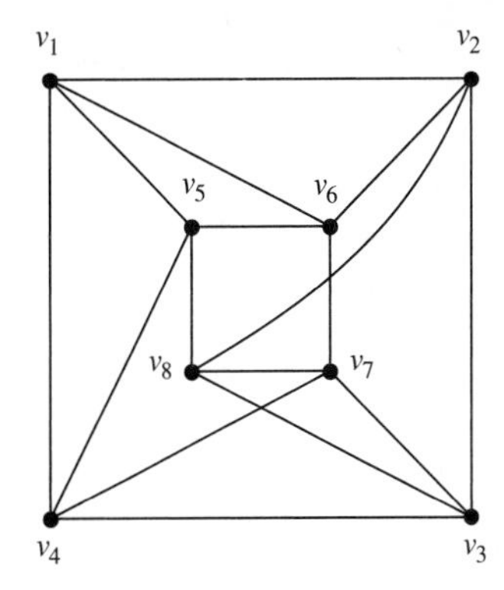

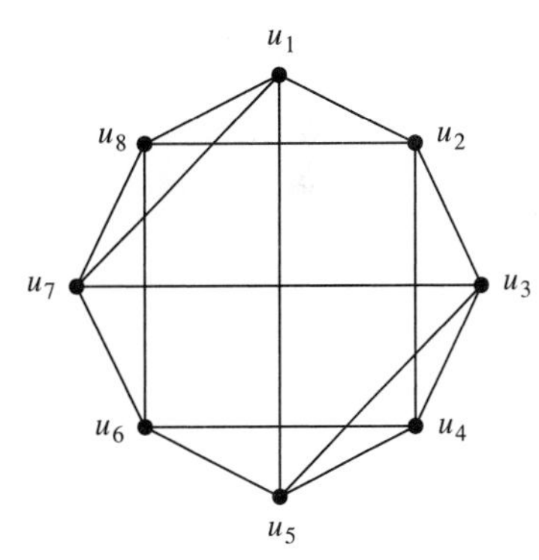

The **complete *m*-partite graph** $K_{n_1,n_2,\ldots,n_m}$ has vertices partitioned into m subsets of $n_1, n_2, \ldots, n_m$ elements each, and vertices are adjacent if and only if they are in different subsets in the partition.

6. Draw the following graphs.
   a) $K_{1,2,3}$ b) $K_{2,2,2}$ c) $K_{1,2,2,3}$

★7. How many vertices and how many edges does the complete m-partite graph $K_{n_1,n_2,\ldots,n_m}$ have?

★8. a) Prove or disprove that there are always two vertices with the same degree in a finite simple graph having at least two vertices.
   b) Do the same as in part (a) for finite multigraphs.

Let $G = (V, E)$ be a simple graph. The **subgraph induced** by a subset W of the vertex set V is the graph (W, F), where the edge set F contains an edge in E if and only if both endpoints of this edge are in W.

9. Consider the graph shown in Figure 3 of Section 7.4. Find the subgraphs induced by
a) $\{a, b, c\}$. **b)** $\{a, e, g\}$.
c) $\{b, c, f, g, h\}$.

10. Let n be a positive integer. Show that a subgraph induced by a nonempty subset of the vertex set of K_n is a complete graph.

A **clique** in a simple undirected graph is a complete subgraph that is not contained in any larger complete subgraph. In Exercises 11–13 find all cliques in the given graph.

11.

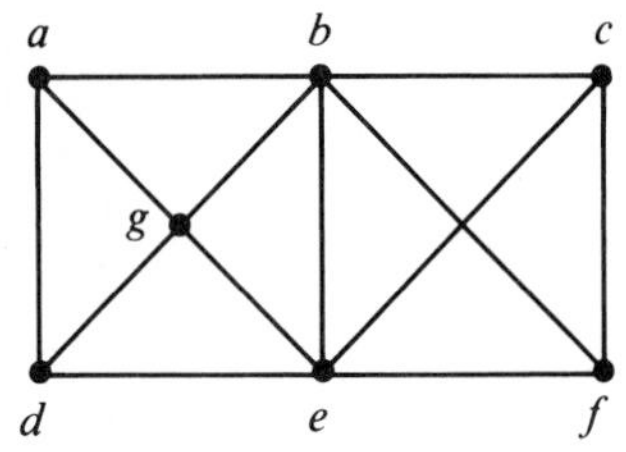

12.

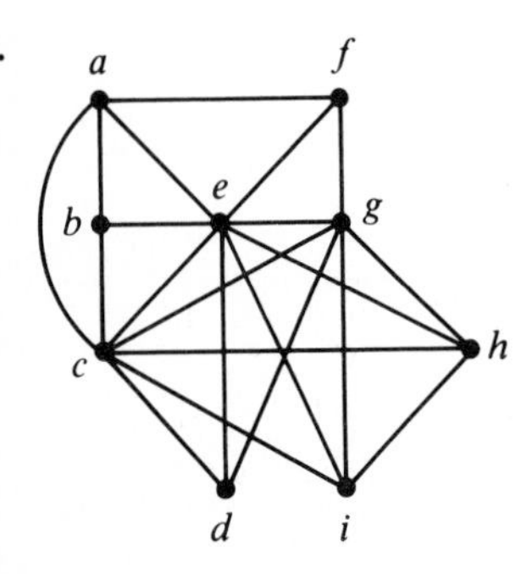

13.

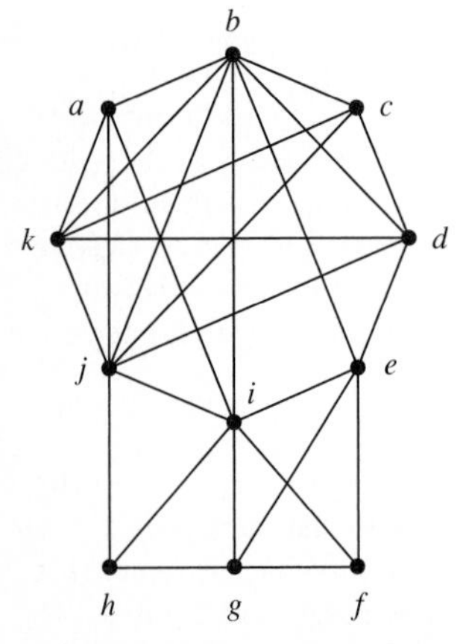

A **dominating set** of vertices in a simple graph is a set of vertices such that every other vertex is adjacent to at least one vertex of this set. A dominating set with the least number of vertices is called a **minimum dominating set**. In Exercises 14–16 find a minimum dominating set for the given graph.

14.

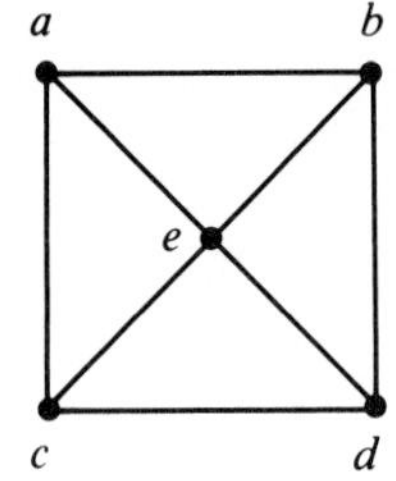

15.

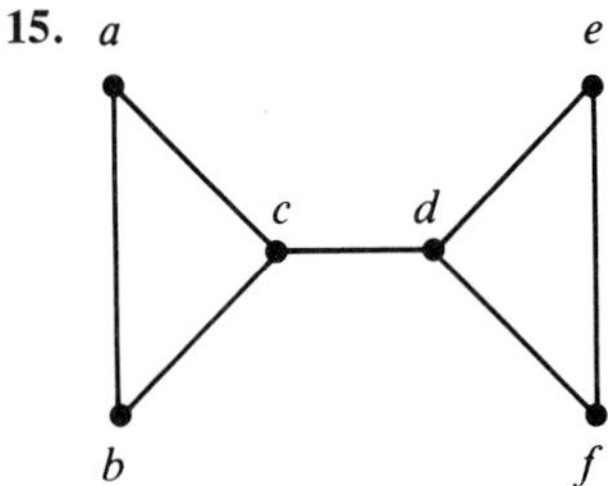

16.

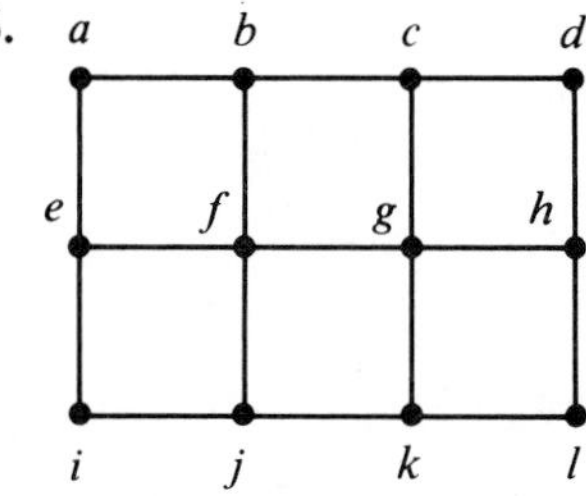

A simple graph can be used to determine the minimum number of queens on a chessboard that control the entire chessboard. An $n \times n$ chessboard has n^2 squares in an $n \times n$ configuration. A queen in a given position controls all squares in the same row, the same column, and on the two diagonals containing this square, as illustrated. The appropriate simple graph has n^2 vertices, one for each square, and two vertices are adjacent if a queen in the square represented by one of the vertices controls the square represented by the other vertex.

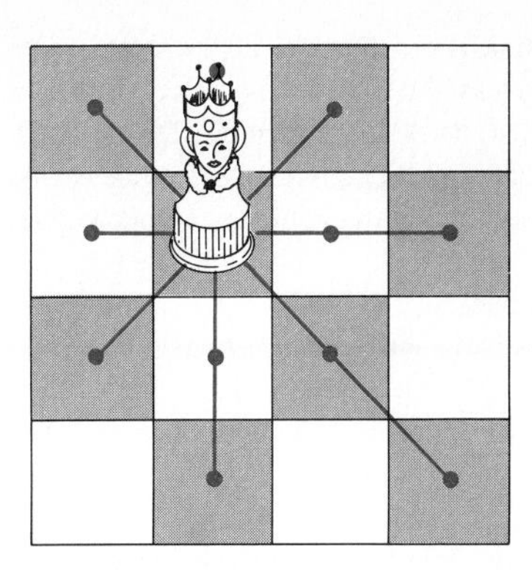

The Squares Controlled by a Queen

17. Construct the simple graph representing the $n \times n$ chessboard with edges representing the control of squares by queens for
a) $n = 3$. **b)** $n = 4$.

18. Explain how the concept of a minimum dominating set applies to the problem of determining the minimum number of queens controlling an $n \times n$ chessboard.

⋆⋆19. Find the minimum number of queens controlling an $n \times n$ chessboard for
a) $n = 3$. **b)** $n = 4$. **c)** $n = 5$.

20. Suppose that G_1 and H_1 are isomorphic and that G_2 and H_2 are isomorphic. Prove or disprove that $G_1 \cup G_2$ and $H_1 \cup H_2$ are isomorphic.

21. Show that each of the following properties is an invariant that isomorphic simple graphs either both have or both do not have:
a) connectedness
b) the existence of a Hamilton circuit
c) the existence of an Euler circuit
d) having crossing number C
e) having n isolated vertices
f) being bipartite

22. How can the adjacency matrix of $\overline{G}$ be found from the adjacency matrix of G, where G is a simple graph?

23. How many nonisomorphic connected bipartite simple graphs are there with four vertices?

⋆24. How many nonisomorphic simple connected graphs with five vertices are there
a) with no vertex of degree more than 2?
b) with chromatic number equal to 4?
c) that are nonplanar?

An **orientation** of an undirected simple graph is an assignment of directions to its edges so that the resulting directed graph is strongly connected. When an orientation of an undirected graph exists, this graph is called **orientable**.

In Exercises 25–27 determine whether the given simple graph is orientable.

25.

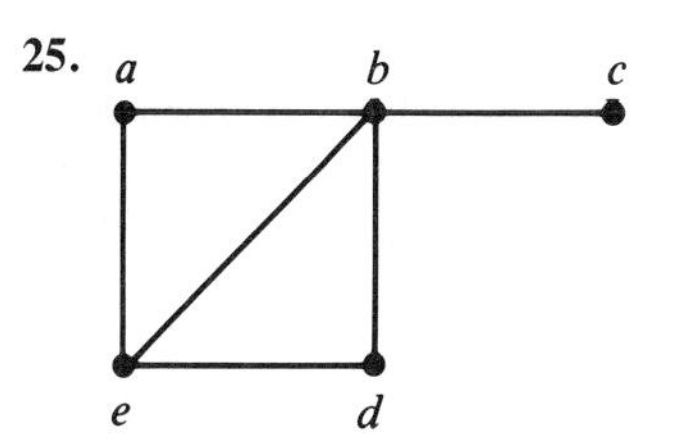

26.

a f
b c d e

27.

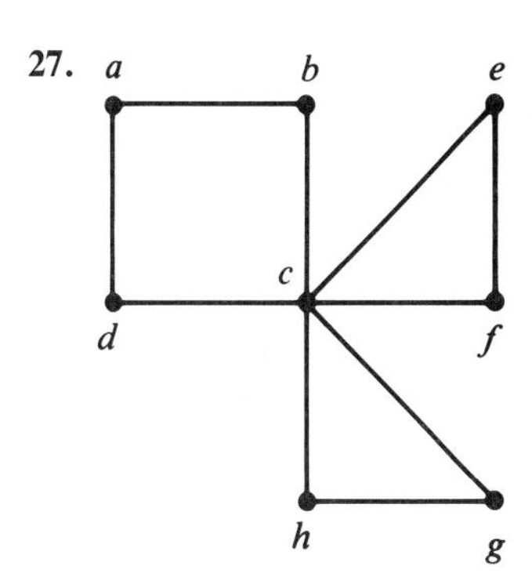

28. Because traffic is growing heavy in the central part of a city, traffic engineers are planning to change all the streets, which are currently two-way, into one-way streets. Explain how to model this problem.

⋆29. Show that a graph is not orientable if it has a cut edge.

A **tournament** is a directed graph such that if u and v are vertices in the graph, exactly one of (u, v) and (v, u) is an edge of the graph.

30. How many different tournaments are there with n vertices?

31. What is the sum of the in-degree and out-degree of a vertex in a tournament?

★32. Show that every tournament has a Hamilton path.

33. Given two chickens in a flock, one of them is dominant. This defines the **pecking order** of the flock. How can a tournament be used to model pecking order?

34. Suppose that G is a connected multigraph with $2k$ vertices of odd degree. Show that there exist k subgraphs that have G as their union, where each of these subgraphs has an Euler path and where no two of these subgraphs have an edge in common. (*Hint:* Add k edges to the graph connecting pairs of vertices of odd degree and use an Euler circuit in this larger graph.)

★35. Let G be a simple graph with n vertices. The **bandwidth** of G, denoted by $B(G)$, is the minimum, over all permutations, $a_1, a_2, \ldots, a_n$ of the vertices of G, of $\max\{|i - j| \mid a_i$ and a_j are adjacent$\}$. That is, the bandwidth is the minimum over all listings of the vertices of the maximum difference in the indices assigned to adjacent vertices. Find the bandwidths of the following graphs.

a) K_5 **b)** $K_{1,3}$ **c)** $K_{2,3}$
d) $K_{3,3}$ **e)** Q_3 **f)** C_5

★36. The **distance** between two distinct vertices v_1 and v_2 of a connected simple graph is the length (number of edges) of the shortest path between v_1 and v_2. The **radius** of a graph is the minimum over all vertices v of the maximum distance from v to another vertex. The **diameter** of a graph is the maximum distance between two distinct vertices. Find the radius and diameter of

a) K_6. **b)** $K_{4,5}$. **c)** Q_3. **d)** C_6.

★37. **a)** Show that if the diameter of the simple graph G is at least 4, then the diameter of its complement $\overline{G}$ is no more than 2.

b) Show that if the diameter of the simple graph G is at least 3, then the diameter of its complement $\overline{G}$ is no more than 3.

★38. Suppose that a multigraph has $2m$ vertices of odd degree. Show that any circuit that contains every edge of the graph must contain at least m edges more than once.

39. Find the second shortest path between the vertices a and z in Figure 3 of Section 7.6.

40. Devise an algorithm for finding the second shortest path between two vertices in a simple connected weighted graph.

41. Find the shortest path between the vertices a and z that passes through the vertex e in the weighted graph in Figure 4 in Section 7.6.

42. Devise an algorithm for finding the shortest path between two vertices in a simple connected weighted graph that passes through a specified third vertex.

★43. Show that if G is a simple graph with at least 11 vertices, then either G or $\overline{G}$, the complement of G, is nonplanar.

A set of vertices in a graph is called **independent** if no two vertices in the set are adjacent. The **independence number** of a graph is the maximum number of vertices in an independent set of vertices for the graph.

★44. What is the independence number of

a) K_n? **b)** C_n? **c)** Q_n? **d)** $K_{m,n}$?

45. Show that the number of vertices in a simple graph is less than or equal to the product of the independence number and the chromatic number of the graph.

46. Show that the chromatic number of a graph is less than or equal to $v - i + 1$, where v is the number of vertices in the graph and i is the independence number of this graph.

Computer Projects

WRITE PROGRAMS WITH THE FOLLOWING INPUT AND OUTPUT.

1. Given the vertex pairs associated to the edges of an undirected graph, determine the degree of each vertex.

2. Given the ordered pairs of vertices associated to the edges of a directed graph, determine the in-degree and out-degree of each vertex.

3. Given the list of edges of a simple graph, determine whether the graph is bipartite.
4. Given the vertex pairs associated to the edges of a graph, construct an adjacency matrix for the graph. (Produce a version that works when loops, multiple edges, or directed edges are present.)
5. Given an adjacency matrix of a graph, list the edges of this graph and give the number of times each edge appears.
6. Given the vertex pairs associated to the edges of an undirected graph and the number of times each edge appears, construct an incidence matrix for the graph.
7. Given an incidence matrix of an undirected graph, list its edges and give the number of times each edge appears.
8. Given the lists of edges of two simple graphs with no more than six vertices, determine whether the graphs are isomorphic.
9. Given an adjacency matrix of a graph and a positive integer n, find the number of paths of length n between two vertices. (Produce a version that works for directed and undirected graphs.)

★10. Given the list of edges of a simple graph, determine whether it is connected and find the number of connected components if it is not connected.

11. Given the vertex pairs associated to the edges of a multigraph, determine whether it has an Euler circuit and, if not, whether it has an Euler path. Construct an Euler path or circuit if it exists.

★12. Given the ordered pairs of vertices associated to the edges of a directed multigraph, construct an Euler path or Euler circuit, if such a path or circuit exists.

★★13. Given the list of edges of a simple graph, produce a Hamilton circuit, or determine that the graph does not have such a circuit.

★★14. Given the list of edges of a simple graph, produce a Hamilton path, or determine that the graph does not have such a path.

15. Given the list of edges and weights of these edges of a weighted connected simple graph and two vertices in this graph, find the length of the shortest path between them using Dijkstra's algorithm. Also, find this path.
16. Given the list of edges of an undirected graph, find a coloring of this graph using the algorithm given in the exercise set of Section 7.8.
17. Given a list of students and the courses that they are enrolled in, construct a schedule of final exams.
18. Given the distances between pairs of television stations, assign frequencies to these stations.

8

Trees

A connected graph that contains no simple circuits is called a tree. Trees were used as long ago as 1857, when the English mathematician Arthur Cayley used them to count certain types of chemical compounds. Since that time, trees have been employed to solve problems in a wide variety of disciplines, as the examples in this chapter will show.

Trees are particularly useful in computer science. For instance, trees are employed to construct efficient algorithms for locating items in a list. They are used to construct networks with the least expensive set of telephone lines linking distributed computers. Trees can be used to construct efficient codes for storing and transmitting data. Trees can model procedures that are carried out using a sequence of decisions. This makes trees valuable in the study of sorting algorithms.

8.1 Introduction to Trees

A genealogical chart of the Bernoullis, a famous family of Swiss mathematicians, is shown in Figure 1. Such a chart is also called a family tree. A family tree is a graph where the vertices represent family members and the edges represent parent-child relationships. The undirected graph that represents a genealogical chart is an example of a special type of graph called a **tree.**

DEFINITION 1. A *tree* is a connected undirected graph with no simple circuits.

Since a tree cannot have a simple circuit, a tree cannot contain multiple edges or loops. Therefore any tree must be a simple graph.

EXAMPLE 1 Which of the graphs shown in Figure 2 are trees?

Solution: G_1 and G_2 are trees, since both are connected graphs with no simple circuits. G_3 is not a tree because e, b, a, d, e is a simple circuit in this graph. Finally, G_4 is not a tree since it is not connected. ■

Any connected graph that contains no simple circuits is a tree. What about graphs containing no simple circuits that are not necessarily connected? These graphs are

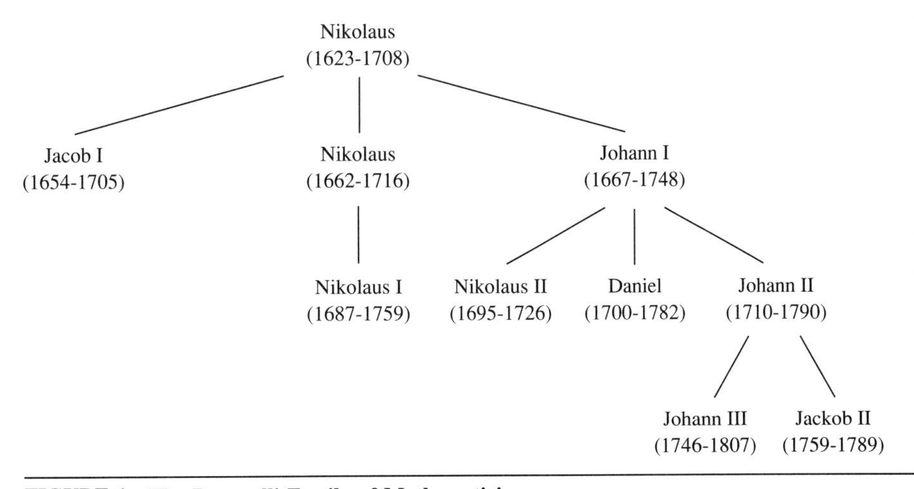

FIGURE 1 **The Bernoulli Family of Mathematicians.**

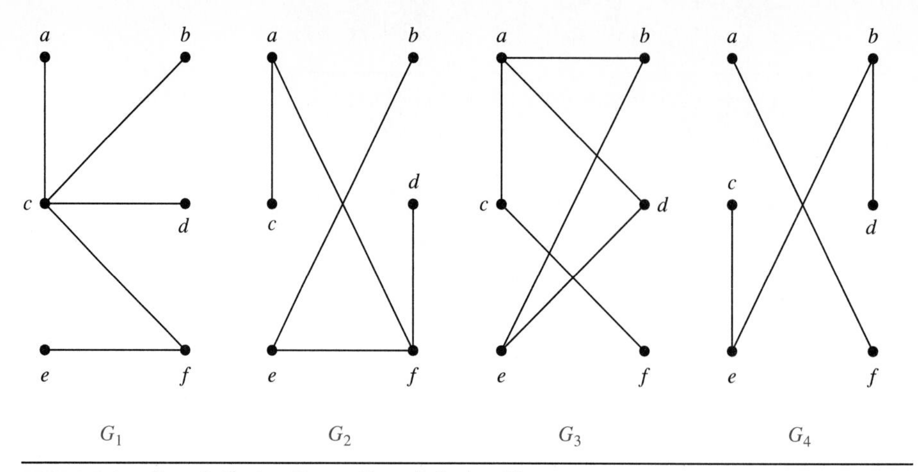

FIGURE 2 **G_1 and G_2 Are Trees; G_3 and G_4 Are Not.**

called **forests** and have the property that each of their connected components is a tree. Figure 3 displays some forests.

Trees are often defined as undirected graphs with the property that there is a unique simple path between every pair of vertices. The following theorem shows that this alternative definition is equivalent to our definition.

THEOREM 1 An undirected graph is a tree if and only if there is a unique simple path between any two of its vertices.

Proof: First assume that T is a tree. Then T is a connected graph with no simple circuits. Let x and y be two vertices of T. Since T is connected, by Theorem 1 of Section 7.4 there is a simple path between x and y. Moreover, this path must be

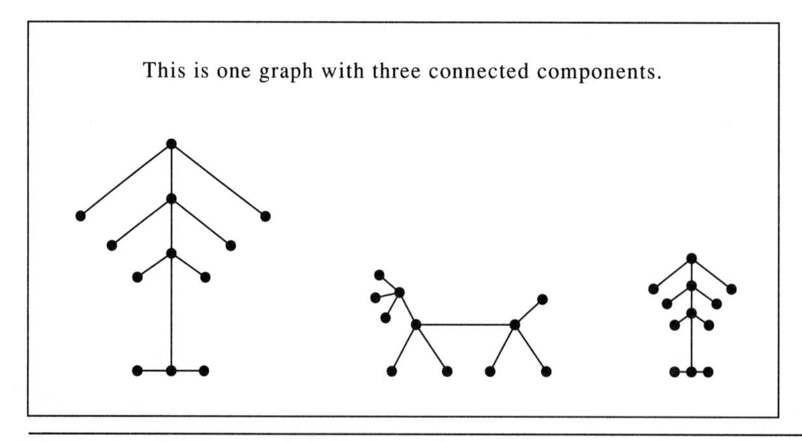

FIGURE 3 **Example of a Forest.**

unique, for if there were a second such path, the path formed by combining the first path from x to y followed by the path from y to x obtained by reversing the order of the second path from x to y would form a circuit. This implies, using Exercise 35 of Section 7.4, that there is a simple circuit in T. Hence, there is a unique simple path between any two vertices of a tree.

Now assume that there is a unique simple path between any two vertices of a graph T. Then T is connected, since there is a path between any two of its vertices. Furthermore, T can have no simple circuits. To see that this is true, suppose T had a simple circuit that contained the vertices x and y. Then there would be two simple paths between x and y, since the simple circuit is made up of a simple path from x to y and a second simple path from y to x. Hence, a graph with a unique simple path between any two vertices is a tree. ■

In many applications of trees a particular vertex of a tree is designated as the **root.** Once we specify a root, we can assign a direction to each edge as follows. Since there is a unique path from the root to each vertex of the graph (from Theorem 1), we direct each edge away from the root. Thus, a tree together with its root produces a directed graph called a **rooted tree.** We can change an unrooted tree into a rooted tree by choosing any vertex as the root. Note that different choices of the root produce different rooted trees. For instance, Figure 4 displays the rooted trees formed by designating a to be the root and c to be the root, respectively, in the tree T. We usually draw a rooted tree with its root at the top of the graph. The arrows indicating the directions of the edges in a rooted tree can be omitted, since the choice of root determines the directions of the edges.

The terminology for trees has botanical and genealogical origins. Suppose that T is a rooted tree. If v is a vertex in T other than the root, the **parent** of v is the unique vertex u such that there is a directed edge from u to v (the reader should show that such a vertex is unique). When u is the parent of v, v is called a **child** of u. Vertices with the same parent are called **siblings.** The **ancestors** of a vertex other than the root are the vertices in the path from the root to this vertex, excluding the vertex itself and including the root (that is, its parent, its parent's parent, and so on, until the root is reached). The **descendants** of a vertex v are those vertices that

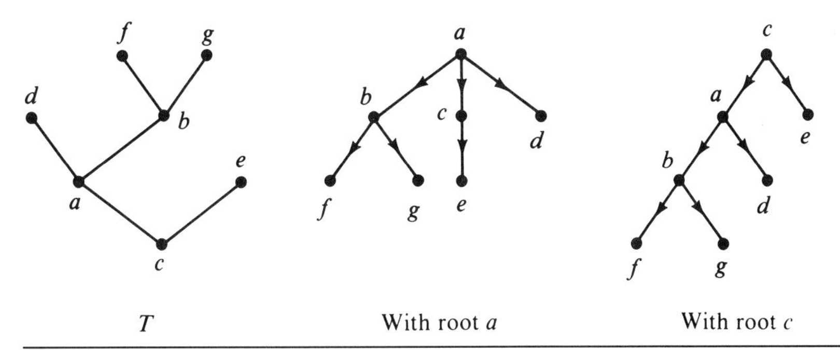

FIGURE 4 **A Tree and Rooted Trees Formed by Designating Two Roots.**

have v as an ancestor. A vertex of a tree is called a **leaf** if it has no children. Vertices that have children are called **internal vertices.** The root is an internal vertex unless it is the only vertex in the graph, in which case it is a leaf.

If a is a vertex in a tree, the **subtree** with a as its root is the subgraph of the tree consisting of a and its descendants and all edges incident to these descendants.

EXAMPLE 2 In the rooted tree T (with root a) shown in Figure 5, find the parent of c, the children of g, the siblings of h, all ancestors of e, all descendants of b, all internal vertices, and all leaves. What is the subtree rooted at g?

Solution: The parent of c is b. The children of g are h, i, and j. The siblings of h are i and j. The ancestors of e are c, b, and a. The descendants of b are c, d, and e. The internal vertices are a, b, c, g, h, and j. The leaves are d, e, f, i, k, l, and m. The subtree rooted at g is shown in Figure 6. ■

Rooted trees with the property that all of their internal vertices have the same number of children are used in many different applications. Later in this chapter we will use such trees to study problems involving searching, sorting, and coding.

DEFINITION 2. A rooted tree is called an *m-ary tree* if every internal vertex has no more than m children. The tree is called a *full m-ary tree* if every internal vertex has exactly m children. An m-ary tree with $m = 2$ is called a *binary tree.*

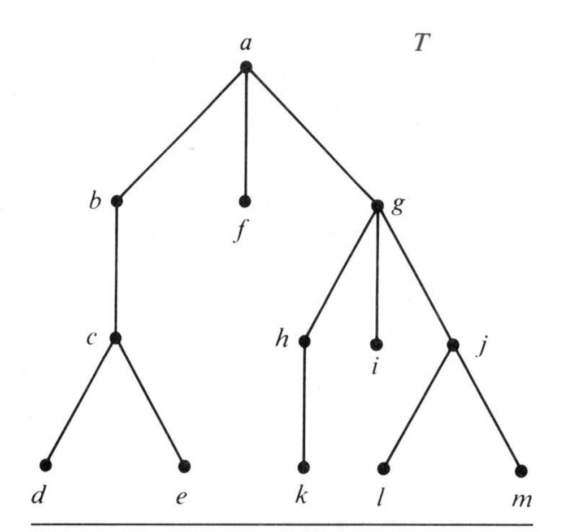

FIGURE 5 A Rooted Tree T.

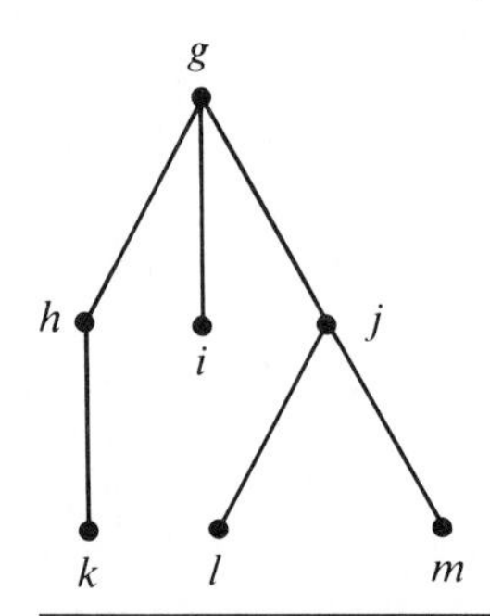

FIGURE 6 The Subtree Rooted at g.

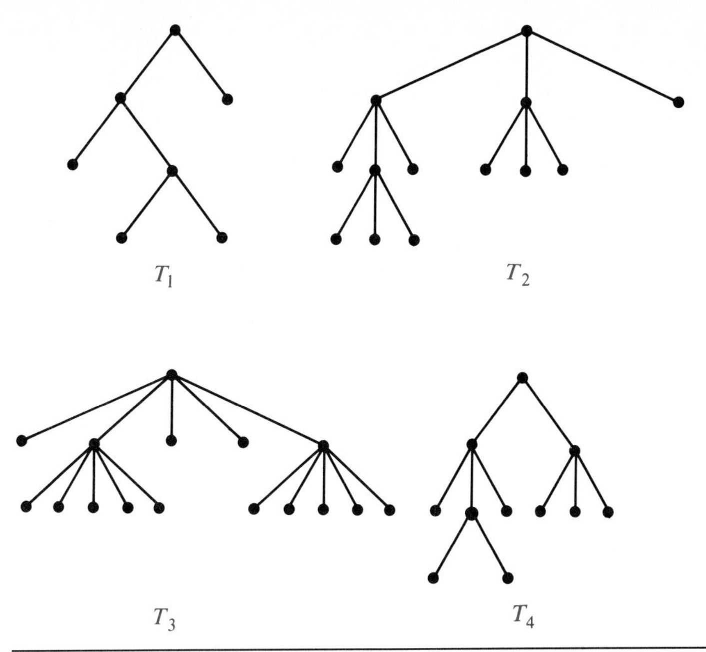

FIGURE 7 **Four Rooted Trees.**

EXAMPLE 3 Are the rooted trees in Figure 7 full m-ary trees for some positive integer m?

Solution: T_1 is a full binary tree since each of its internal vertices has two children. T_2 is a full 3-ary tree since each of its internal vertices has three children. In T_3 each internal vertex has five children, so T_3 is a full 5-ary tree. T_4 is not a full m-ary tree for any m since some of its internal vertices have two children and others have three children. ■

An **ordered rooted tree** is a rooted tree where the children of each internal vertex are ordered. Ordered rooted trees are drawn so that the children of each internal vertex are shown in order from left to right. Note that a representation of a rooted tree in the conventional way determines an ordering for its edges. We will use such orderings of edges in drawings without explicitly mentioning that we are considering a rooted tree to be ordered.

In an ordered binary tree, if an internal vertex has two children, the first child is called the **left child** and the second child is called the **right child.** The tree rooted at the left child of a vertex is called the **left subtree** of this vertex, and the tree rooted at the right child of a vertex is called the **right subtree** of the vertex. The reader should note that for some applications every vertex of a binary tree, other than the root, is designated as a right or a left child of its parent. This is done even when

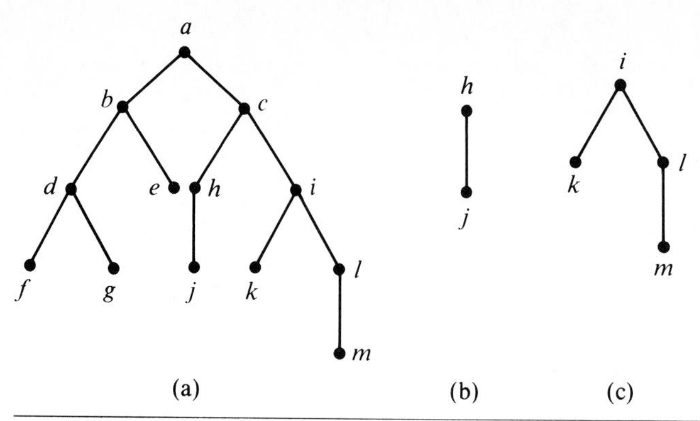

FIGURE 8 **A Binary Tree T and Left and Right Subtrees of the Vertex c.**

some vertices have only one child. We will make such designations whenever it is necessary, but not otherwise.

EXAMPLE 4 What are the left and right children of d in the binary tree T shown in Figure 8(a) (where the order is that implied by the drawing)? What are the left and right subtrees of c?

Solution: The left child of d if f and the right child is g. We show the left and right subtrees of c in Figures 8(b) and 8(c), respectively. ■

Just as in the case of graphs, there is no standard terminology used to describe trees, rooted trees, ordered rooted trees, and binary trees. This nonstandard terminology occurs since trees are used extensively throughout computer science, which is a relatively young field. The reader should carefully check meanings given to terms dealing with trees whenever they occur.

TREES AS MODELS

Trees are used as models in such diverse areas as computer science, chemistry, geology, botany, and psychology. We will describe a variety of such models based on trees.

EXAMPLE 5 Saturated Hydrocarbons and Trees Graphs can be used to represent molecules, where atoms are represented by vertices and bonds between them by edges. The English mathematician Arthur Cayley discovered trees in 1857 when he was trying to enumerate the isomers of compounds of the form C_nH_{2n+2}, which are called saturated hydrocarbons.

In graph models of saturated hydrocarbons, each carbon atom is represented by a vertex of degree 4, and each hydrogen atom is represented by a vertex of degree 1. There are $3n + 2$ vertices in a graph representing a compound of the form C_nH_{2n+2}. The number of edges in such a graph is half the sum of the degrees of the vertices. Hence, there are $(4n + 2n + 2)/2 = 3n + 1$ edges in this graph. Since the graph is connected and the number of edges is one less than the number of vertices, it must be a tree (see Exercise 9 at the end of this section).

The nonisomorphic trees with n vertices of degree 4 and $2n + 2$ of degree 1 represent the different isomers of C_nH_{2n+2}. For instance, when $n = 4$, there are exactly two nonisomorphic trees of this type (the reader should verify this). Hence, there are exactly two different isomers of C_4H_{10}. Their structures are displayed in Figure 9. These two isomers are called butane and isobutane. ■

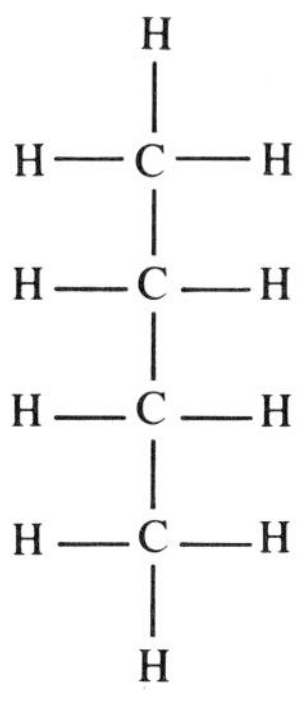

Butane

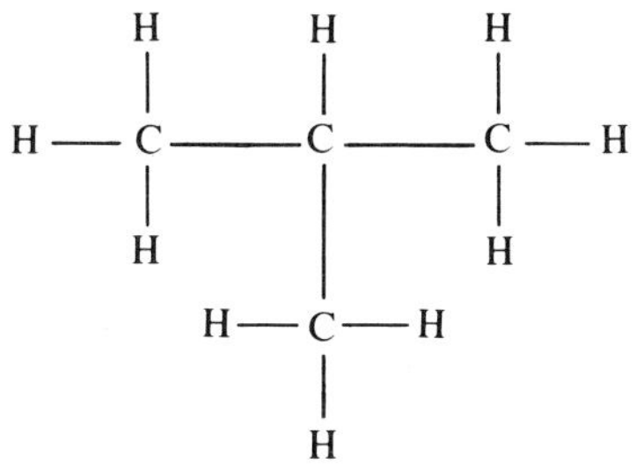

Isobutane

FIGURE 9 The Two Isomers of Butane.

Arthur Cayley (1821–1895) Arthur Cayley, the son of a merchant, displayed his mathematical talents at an early age with amazing skill in numerical calculations. Cayley entered Trinity College, Cambridge, when he was 17. While in college he developed a passion for reading novels. Cayley excelled at Cambridge and was elected to a three-year appointment as Fellow of Trinity and assistant tutor. During this time Cayley began his study of n-dimensional geometry and made a variety of contributions to geometry and to analysis. He also developed an interest in mountaineering, which he enjoyed during vacations in Switzerland. Since no position as a mathematician was available to him, Cayley left Cambridge, entering the legal profession and gaining admittance to the bar in 1849. Although Cayley limited his legal work to be able to continue his mathematics research, he developed a reputation as a legal specialist. During his legal career he was able to write more than 300 mathematical papers. In 1863 Cambridge University established a new post in mathematics and offered it to Cayley. He took this job, even though it paid less money than he made as a lawyer.

EXAMPLE 6 Representing Organizations The structure of a large organization can be modeled using a rooted tree. Each vertex in this tree represents a position in the organization. An edge from one vertex to another indicates that the person represented by the initial vertex is the (direct) boss of the person represented by the terminal vertex. The graph shown in Figure 10 displays such a tree. In the organization represented by this tree, the Director of Hardware Development works directly for the Vice President of R&D. ■

EXAMPLE 7 Computer File Systems Files in computer memory can be organized into directories. A directory can contain both files and subdirectories. The root directory contains the entire file system. Thus, a file system may be represented by a rooted tree, where the root represents the root directory, internal vertices represent subdirectories, and leaves represent ordinary files or empty directories. One such file system is shown in Figure 11. In this system, the file khr is in the directory rje. ■

EXAMPLE 8 Tree-Connected Parallel Processors In Example 13 of Section 7.2 we described several interconnection networks for parallel processing. A **tree-connected network** is another important way to interconnect processors. The graph representing such a network is a complete binary tree. Such a network interconnects $n = 2^k - 1$ processors, where k is a positive integer. A processor represented by the vertex v that is not a root or a leaf has three two-way connections—one to the processor represented by the parent of v and two to the processors represented by the two children of v. The processor represented by the root has two two-way connections to the processors represented by its two children. A processor represented by a leaf v has a single two-

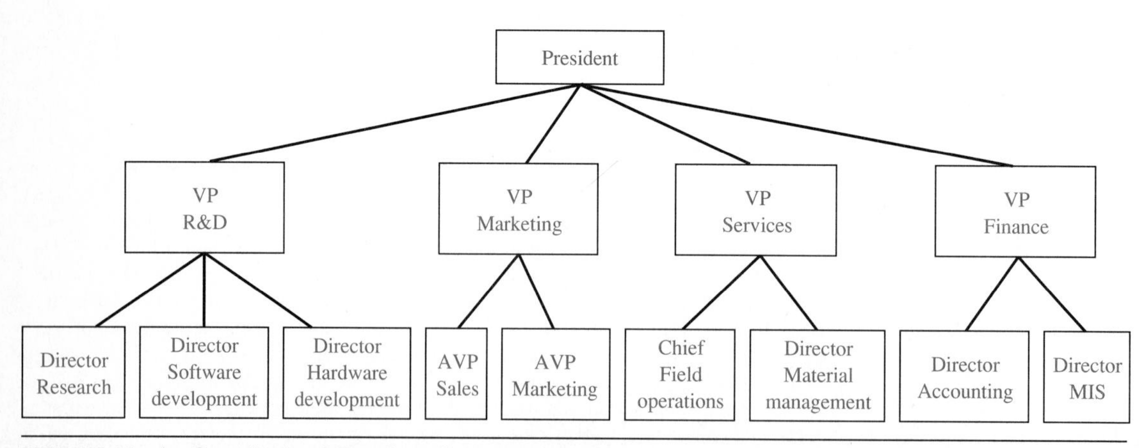

FIGURE 10 An Organizational Tree for a Computer Company.

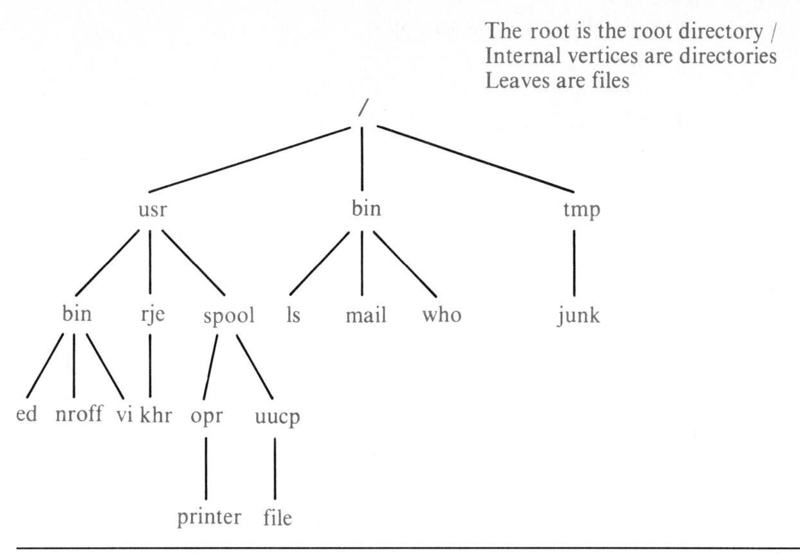

FIGURE 11 A Computer File System.

way connection to the parent of v. We display a tree-connected network with seven processors in Figure 12.

We will illustrate how a tree-connected network can be used for parallel computation. In particular, we will show how the processors in Figure 12 can be used to add eight numbers, using three steps. In the first step, we add x_1 and x_2 using P_4, x_3 and x_4 using P_5, x_5 and x_6 using P_6, and x_7 and x_8 using P_7. In the second step, we add $x_1 + x_2$ and $x_3 + x_4$ using P_2 and $x_5 + x_6$ and $x_7 + x_8$ using P_3. Finally, in the third step, we add $x_1 + x_2 + x_3 + x_4$ and $x_5 + x_6 + x_7 + x_8$ using P_1. The three steps used to add eight numbers compares favorably to the seven steps required to add eight numbers serially, where the steps are the addition of one number to the sum of the previous numbers in the list. ■

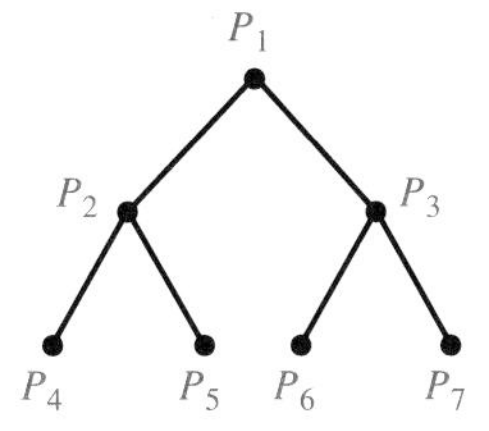

FIGURE 12 A Tree-Connected Network of Seven Processors.

PROPERTIES OF TREES

We will often need results relating the numbers of edges and vertices of various types in trees.

THEOREM 2 A tree with n vertices has $n - 1$ edges.

Proof: Choose the vertex r as the root of the tree. We set up a one-to-one correspondence between the edges and the vertices other than r by associating the terminal vertex of an edge to that edge. Since there are $n - 1$ vertices other than r, there are $n - 1$ edges in the tree. ■

The number of vertices in a full m-ary tree with a specified number of internal vertices is determined, as the following theorem shows. As in Theorem 2, we will use n to denote the number of vertices in a tree.

THEOREM 3 A full m-ary tree with i internal vertices contains $n = mi + 1$ vertices.

Proof: Every vertex, except the root, is the child of an internal vertex. Since each of the i internal vertices has m children, there are mi vertices in the tree other than the root. Therefore, the tree contains $n = mi + 1$ vertices. ■

Suppose that T is a full m-ary tree. Let i be the number of internal vertices and l the number of leaves in this tree. Once one of n, i, and l is known, the other two quantities are determined. How to find the other two quantities from the one that is known is given in the following theorem.

THEOREM 4 A full m-ary tree with

(i) n vertices has $i = (n - 1)/m$ internal vertices and $l = [(m - 1)n + 1]/m$ leaves,

(ii) i internal vertices has $n = mi + 1$ vertices and $l = (m - 1)i + 1$ leaves,

(iii) l leaves has $n = (ml - 1)/(m - 1)$ vertices and $i = (l - 1)/(m - 1)$ internal vertices.

Proof: Let n represent the number of vertices, i the number of internal vertices, and l the number of leaves. The three parts of the theorem can all be proved using the equality given in Theorem 3, that is, $n = mi + 1$, together with the equality $n = l + i$, which is true because each vertex is either a leaf or an internal vertex. We will prove (i) here. The proofs of (ii) and (iii) are left as exercises for the reader.

Solving for i in $n = mi + 1$ gives $i = (n - 1)/m$. Then inserting this expression for i into the equation $n = l + i$ shows that $l = n - i = n - (n - 1)/m = [(m - 1)n + 1]/m$. ■

The following example illustrates how Theorem 4 can be used.

EXAMPLE 9 Suppose that someone starts a chain letter. Each person who receives the letter is asked to send it on to four other people. Some people do this, but others do not send any letters. How many people have seen the letter, including the first person, if no one receives more than one letter and if the chain letter ends after there have been 100 people who read it but did not send it out? How many people sent out the letter?

Solution: The chain letter can be represented using a 4-ary tree. The internal vertices correspond to people who sent out the letter, and the leaves correspond to people who did not send it out. Since 100 people did not send out the letter, the number of leaves in this rooted tree is $l = 100$. Hence, part (iii) of Theorem 4 shows that the number of people who have seen the letter is $n = (4 \cdot 100 - 1)/(4 - 1) = 133$. Also, the number of internal vertices is $133 - 100 = 33$, so that 33 people sent out the letter. ■

It is often desirable to use rooted trees that are "balanced," so that the subtrees at each vertex contain paths of approximately the same length. To make this concept clear, some definitions are needed. The **level** of a vertex v in a rooted tree is the length of the unique path from the root to this vertex. The level of the root is defined to be zero. The **height** of a rooted tree is the maximum of the levels of vertices. In other words, the height of a rooted tree is the length of the longest path from the root to any vertex.

EXAMPLE 10 Find the level of each vertex in the rooted tree shown in Figure 13. What is the height of this tree?

Solution: The root a is at level 0. Vertices b, j, and k are at level 1. Vertices c, e, f, and l are at level 2. Vertices d, g, i, m, and n are at level 3. Finally, vertex h is at level 4. Since the largest level of any vertex is 4, this tree has height 4. ■

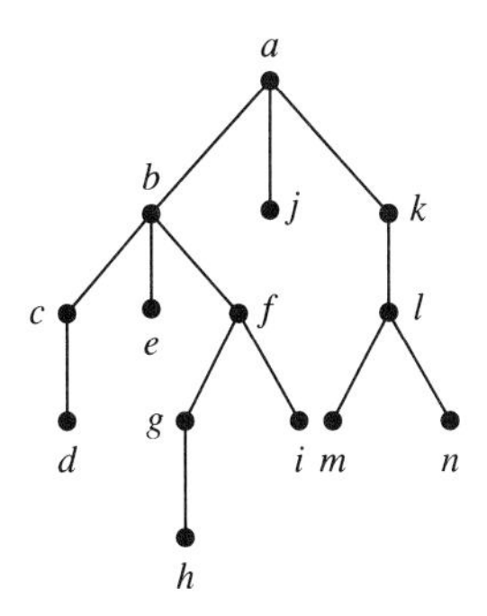

FIGURE 13 **A Rooted Tree.**

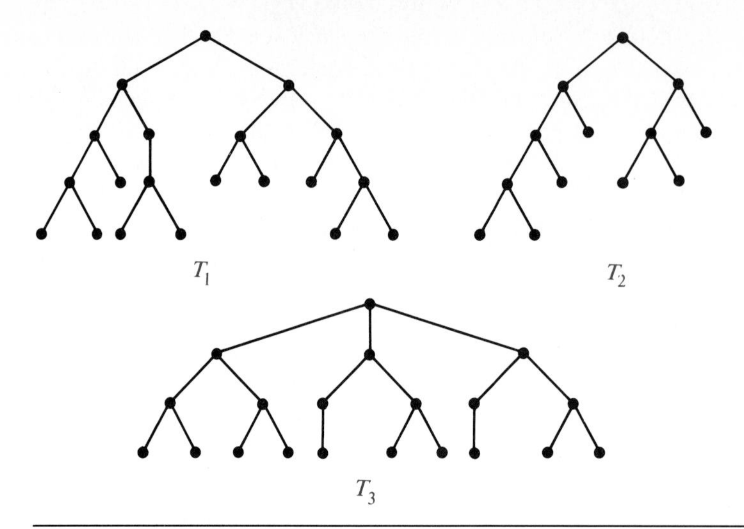

FIGURE 14 **Some Rooted Trees.**

A rooted m-ary tree of height h is called **balanced** if all leaves are at levels h or $h-1$.

EXAMPLE 11 Which of the rooted trees shown in Figure 14 are balanced?

Solution: T_1 is balanced, since all its leaves are at levels 3 and 4. However, T_2 is not balanced, since it has leaves at levels 2, 3, and 4. Finally, T_3 is balanced, since all its leaves are at level 3. ■

The following results relate the height and the number of leaves in m-ary trees.

THEOREM 5 There are at most m^h leaves in an m-ary tree of height h.

Proof: The proof uses mathematical induction on the height. First, consider m-ary trees of height 1. These trees consist of a root with no more than m children, each of which is a leaf. Hence there are no more than $m^1 = m$ leaves in an m-ary tree of height 1. This is the basis step of the inductive argument.

Now assume that the result is true for all m-ary trees of height less than h; this is the inductive hypothesis. Let T be an m-ary tree of height h. The leaves of T are the leaves of the subtrees of T obtained by deleting the edges from the root to each of the vertices at level 1, as shown in Figure 15.

Each of these subtrees has height less than or equal to $h-1$. So by the inductive hypothesis, each of these rooted trees has at most m^{h-1} leaves. Since there are at

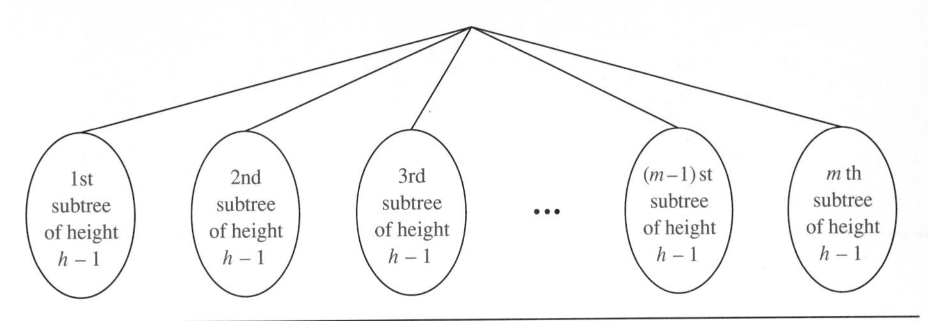

FIGURE 15 **The Inductive Step of the Proof.**

most m such subtrees, each with a maximum of m^{h-1} leaves, there are at most $m \cdot m^{h-1} = m^h$ leaves in the rooted tree. This finishes the inductive argument. ■

COROLLARY 1 If an m-ary tree of height h has l leaves, then $h \geq \lceil \log_m l \rceil$. If the m-ary tree is full and balanced, then $h = \lceil \log_m l \rceil$. (We are using the ceiling function here. Recall that $\lceil x \rceil$ is the smallest integer greater than or equal to x.)

Proof: We know that $l \leq m^h$ from Theorem 5. Taking logarithms to the base m shows that $\log_m l \leq h$. Since h is an integer, we have $h \geq \lceil \log_m l \rceil$. Now suppose that the tree is balanced. Then each leaf is at level h or $h - 1$, and since the height is h, there is at least one leaf at level h. It follows that there must be more than m^{h-1} leaves (see Exercise 22 at the end of this section). Since $l \leq m^h$, we have $m^{h-1} < l \leq m^h$. Taking logarithms to the base m in this inequality gives $h - 1 < \log_m l \leq h$. Hence, $h = \lceil \log_m l \rceil$. ■

Exercises

1. Which of the following graphs are trees?

a)

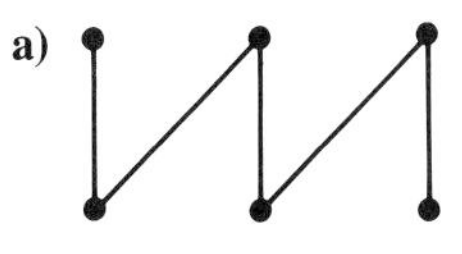

b)

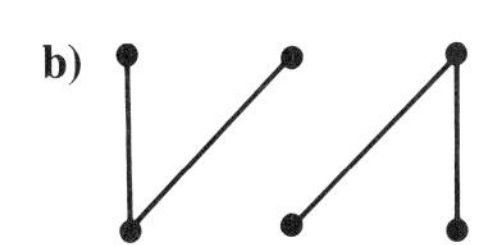

c)

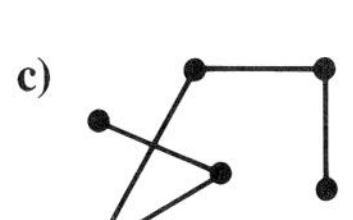

d)

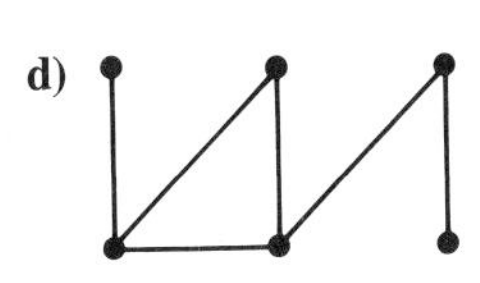

e)

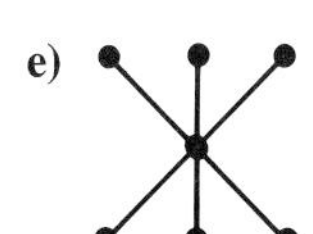

f)

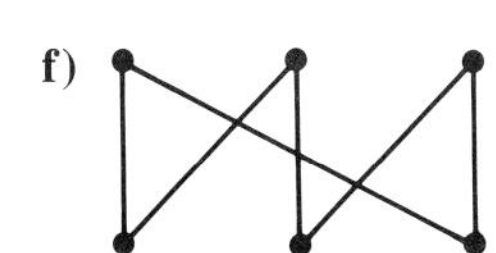

2. Answer the following questions about the rooted tree illustrated.

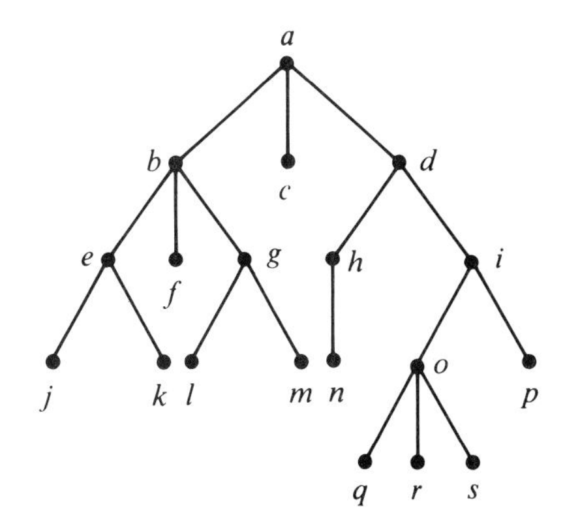

a) Which vertex is the root?
b) Which vertices are internal?
c) Which vertices are leaves?
d) Which vertices are children of i?
e) Which vertex is the parent of h?
f) Which vertices are siblings of o?
g) Which vertices are ancestors of m?
h) Which vertices are descendants of b?

3. Is the rooted tree in Exercise 2 a full m-ary tree for some positive integer m?

4. What is the level of each vertex of the tree in Exercise 2?

5. Draw the subtree of the tree in Exercise 2 that is rooted at
a) a. b) c. c) e.

★6. How many nonisomorphic unrooted trees are there with n vertices if
a) $n = 3$? b) $n = 4$? c) $n = 5$?

★7. Answer the same question as that given in Exercise 6 for rooted trees (using isomorphism for directed graphs).

★8. Show that a simple graph is a tree if and only if it is connected, but the deletion of any of its edges produces a graph that is not connected.

★9. Let G be a simple graph with n vertices. Show that G is a tree if and only if G is connected and has $n - 1$ edges.

10. Which complete bipartite graphs $K_{m,n}$, where m and n are positive integers, are trees?

11. How many edges does a tree with 10,000 vertices have?

12. How many vertices does a full 5-ary tree with 100 internal vertices have?

13. How many edges does a full binary tree with 1000 internal vertices have?

14. How many leaves does a full 3-ary tree with 100 vertices have?

15. Suppose 1000 people enter a chess tournament. Use a rooted tree model of the tournament to determine how many games must be played to determine a champion, if a player is eliminated after one loss and games are played until only one entrant has not lost. (Assume there are no ties.)

16. A chain letter starts when a person sends a letter to five others. Each person who receives the letter either sends it to five other people who have never received it or does not send it to anyone. Suppose that 10,000 people send out the letter before the chain ends and that no one receives more than one letter. How many people receive the letter, and how many do not send it out?

17. A chain letter starts with a person sending a letter out to 10 others. Each person is asked to send the letter out to 10 others, and each letter contains a list of the previous 6 people in the chain. Unless there are fewer than 6 names in the list, each person sends one dollar to the first person in this list, removes the name of this person from the list, moves up each of the other 5 names one position, and inserts his or her name at the end of this list. If no person breaks the chain and no one receives more than one letter, how much money will a person in the chain ultimately receive?

★18. A full m-ary tree T has 84 leaves and height 3.
a) Give the upper and lower bounds for m.
b) What is m if T is also balanced?

A **complete m-ary tree** is a full m-ary tree where every leaf is at the same level.

19. Construct a complete binary tree of height 4 and a complete 3-ary tree of height 3.

20. How many vertices and how many leaves does a complete m-ary tree of height h have?

21. Prove
a) part (ii) of Theorem 4.
b) part (iii) of Theorem 4.

☞22. Show that a full m-ary balanced tree of height h has more than m^{h-1} leaves.

23. How many edges are there in a forest of t trees containing a total of n vertices?

24. Explain how a tree can be used to represent the table of contents of a book organized into chapters, where each chapter is organized into sections, and each section is organized into subsections.

25. How many different isomers do the following saturated hydrocarbons have?
a) C_3H_8 b) C_5H_{12} c) C_6H_{14}

26. What does each of the following represent in an organizational tree?
a) the parent of a vertex
b) a child of a vertex
c) a sibling of a vertex
d) the ancestors of a vertex
e) the descendants of a vertex
f) the level of a vertex
g) the height of the tree

27. Answer the same questions as those given in Exercise 26 for a rooted tree representing a computer file system.

28. **a)** Draw the complete binary tree with 15 vertices that represents a tree-connected network of 15 processors.
b) Show how 16 numbers can be added using the 15 processors in part (a) using four steps.

29. Let n be a power of 2. Show that n numbers can be added in $\log n$ steps using a tree-connected network of $n - 1$ processors.

★30. A **labeled tree** is a tree where each vertex is assigned a label. Two labeled trees are considered isomorphic when there is an isomorphism between them that preserves the labels of vertices. How many nonisomorphic trees are there with three vertices labeled with different integers from the set $\{1, 2, 3\}$? How many nonisomorphic trees are there with four vertices labeled with different integers from the set $\{1, 2, 3, 4\}$?

The **eccentricity** of a vertex in an unrooted tree is the length of the longest simple path beginning at this vertex. A vertex is called a **center** if no vertex in the tree has smaller eccentricity than this vertex. In Exercises 31–33 find every vertex that is a center in the given tree.

31.

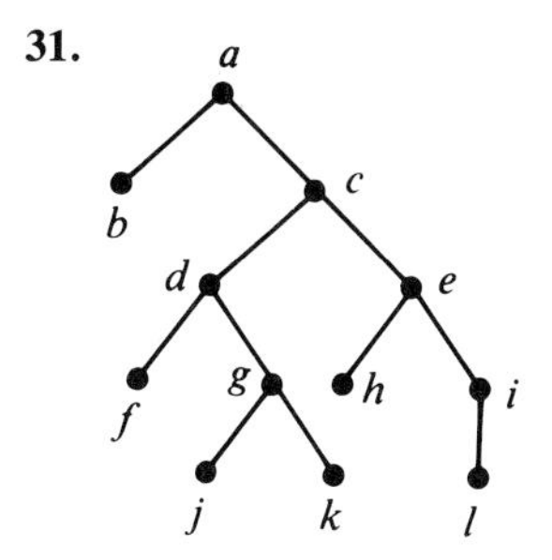

32.

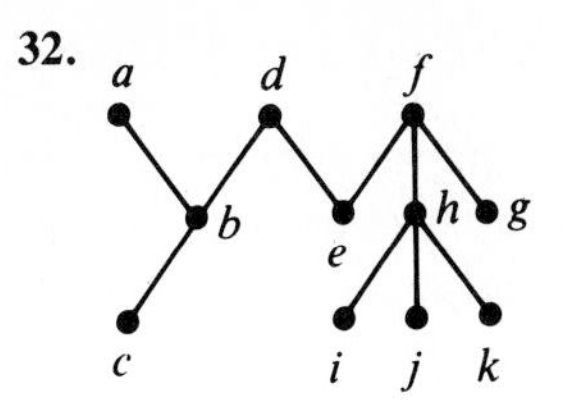

33.

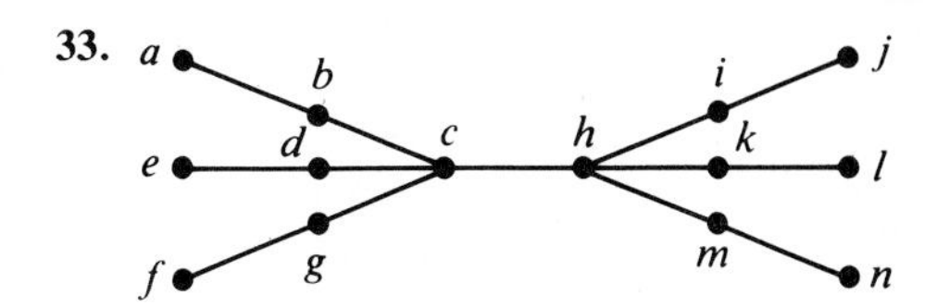

34. Show that a center should be chosen as the root to produce a rooted tree of minimal height from an unrooted tree.

★35. Show that a tree has either one center or two centers that are adjacent.

36. Show that every tree can be colored using two colors.

The **rooted Fibonacci trees** T_n are defined recursively in the following way. T_1 and T_2 are both the rooted tree consisting of a single vertex, and for $n = 3, 4, \ldots$, the rooted tree T_n is constructed from a root with T_{n-1} as its left subtree and T_{n-2} as its right subtree.

37. Draw the first seven rooted Fibonacci trees.

★38. How many vertices, leaves, and internal vertices does the rooted Fibonacci tree T_n have, where n is a positive integer? What is its height?

8.2 Applications of Trees

INTRODUCTION

We will discuss three problems that can be studied using trees. The first problem is: How should items in a list be stored so that an item can be easily located? The second problem is: What series of decisions should be made to find an object with a certain property in a collection of objects of a certain type? The third problem is: How should a set of characters be efficiently coded by bit strings?

BINARY SEARCH TREES

Searching for items in a list is one of the most important tasks that arises in computer science. Our primary goal is to implement a searching algorithm that finds items efficiently when the items are totally ordered. This can be accomplished through the use of a **binary search tree,** which is a binary tree in which each child of a vertex is designated as a right or left child, no vertex has more than one right child or left child, and each vertex is labeled with a key, which is one of the items. Furthermore, vertices are assigned keys so that the key of a vertex is both larger than the keys of all vertices in its left subtree and smaller than the keys of all vertices in its right subtree.

The following recursive procedure is used to form the binary search tree for a list of items. Start with a tree containing just one vertex, namely, the root. The first item in the list is assigned as the key of the root. To add a new item, first compare it with the keys of vertices already in the tree, starting at the root and moving to the left if the item is less than the key of the respective vertex if this vertex has a left child, or moving to the right if the item is greater than the key of the respective vertex if this vertex has a right child. When the item is less than the respective vertex and this vertex has no left child, then a new vertex with this item as its key is inserted as a new left child. Similarly, when the item is greater than the respective vertex and this vertex has no right child, then a new vertex with this item as its key is inserted as a new right child. We illustrate this procedure with the following example.

Form a binary search tree for the words *mathematics, physics, geography, zoology, meteorology, geology, psychology,* and *chemistry* (using alphabetical order).

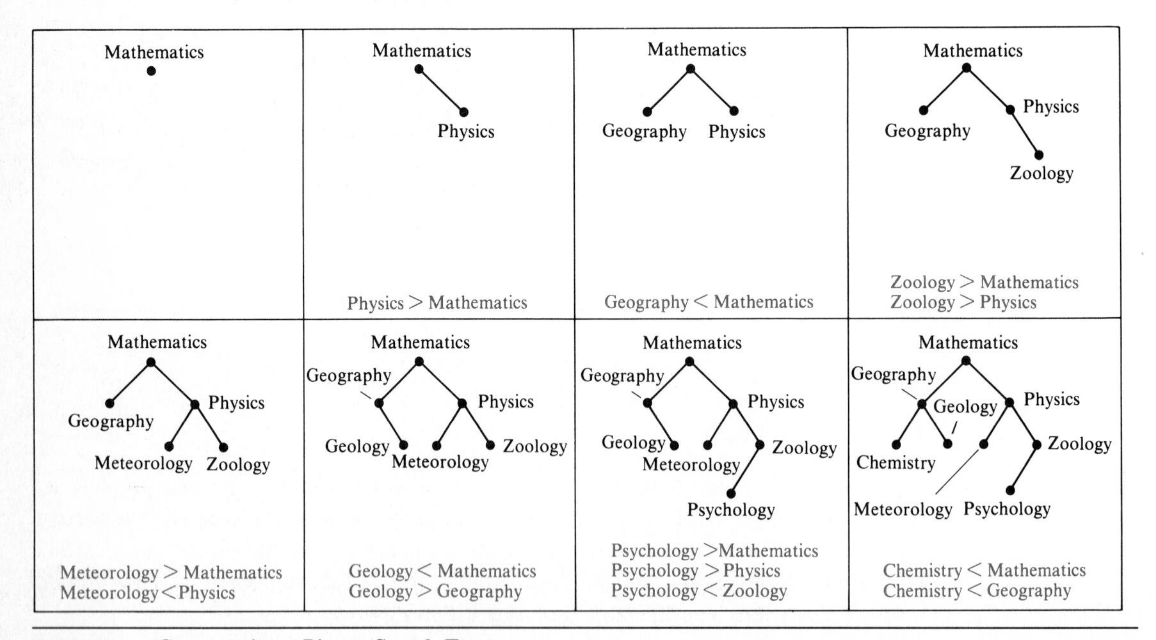

FIGURE 1 Constructing a Binary Search Tree.

Solution: Figure 1 displays the steps used to construct this binary search tree. The word *mathematics* is the key of the root. Since *physics* comes after *mathematics* (in alphabetical order), add a right child of the root with key *physics*. Since *geography* comes before *mathematics,* add a left child of the root with key *geography.* Next, add a right child of the vertex with key *physics,* and assign it the key *zoology,* since *zoology* comes after *mathematics* and after *physics*. Similarly, add a left child of the vertex with key *physics* and assign this new vertex the key *meteorology.* Add a right child of the vertex with key *geography* and assign this new vertex the key *geology.* Add a left child of the vertex with key *zoology* and assign it the key *psychology.* Add a left child of the vertex with key *geography* and assign it the key *chemistry.* (The reader should work through all the comparisons needed at each step.) ■

ALGORITHM 1 Binary Search Tree Algorithm.

procedure *insertion*(T: binary search tree, x: item)
$v :=$ root of T
{a vertex not present in T has the value *null*}
while $v \neq null$ and $label(v) \neq x$
begin
  if $x < label(v)$ **then**
  if left child of $v \neq null$ **then** $v :=$ left child of v
  else add *new vertex* as a left child of v and set $v := null$
  else
  if right child of $v \neq null$ **then** $v :=$ right child of v
  else add *new vertex* as a right child of v to T and set $v := null$
end
if root of $T = null$ **then** add a vertex r to the tree and label it with x
else if $label(v) \neq x$ **then** label *new vertex* with x
{$v =$ location of x}

To locate an item we try to add it to a binary search tree. We locate it if it is present. Algorithm 1 gives pseudocode for locating an item in a binary search tree and adding a new vertex with this item as its key if the item is not found. Algorithm 1 locates x if it is already the key of a vertex. When x is not a key, a new vertex with key x is added to the tree. In the pseudocode, v is the vertex that has x as its key, and $label(v)$ represents the key of vertex v.

We will now determine the computational complexity of this procedure. Suppose we have a binary search tree T for a list of n items. We can form a full binary tree U from T by adding unlabeled vertices whenever necessary so that every vertex with a key has two children. This is illustrated in Figure 2. Once we have done this, we can easily locate or add a new item as a key without adding a vertex.

The most comparisons needed to add a new item is the length of the longest path in U from the root to a leaf. The internal vertices of U are the vertices of T. It follows that U has n internal vertices. We can now use part (ii) of Theorem 4 in

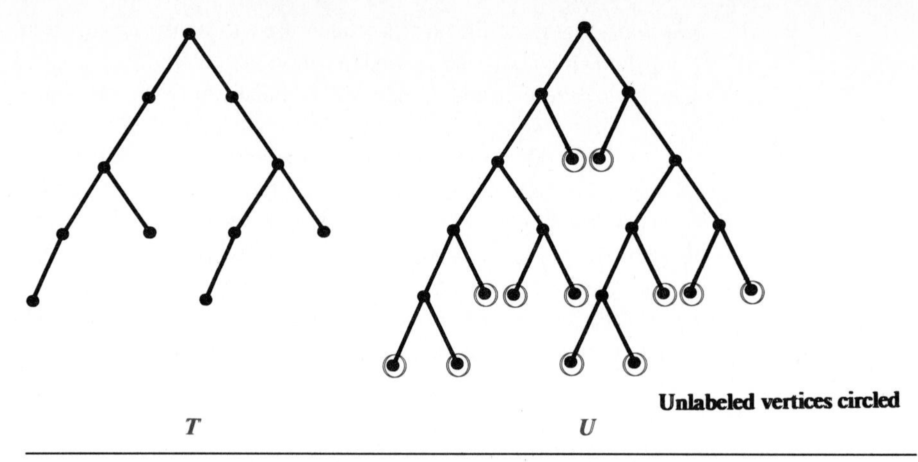

FIGURE 2 **Adding Unlabeled Vertices to Make a Binary Search Tree Full.**

Section 8.1 to conclude that U has $n + 1$ leaves. Using Corollary 1 of Section 8.1, we see that the height of U is greater than or equal to $h = \lceil \log(n + 1) \rceil$. Consequently, it is necessary to perform at least $\lceil \log(n + 1) \rceil$ comparisons to add some item. Note that if U is balanced, its height is $\lceil \log(n + 1) \rceil$ (from Corollary 1 of Section 8.1). Thus, if a binary search tree is balanced, locating or adding an item requires no more than $\lceil \log(n + 1) \rceil$ comparisons. A binary search tree can become unbalanced as items are added to it. Since balanced binary search trees give optimal worst-case complexity for binary searching, algorithms have been devised that rebalance binary search trees as items are added. The interested reader can consult references on data structures for the description of such algorithms.

DECISION TREES

Rooted trees can be used to model problems in which a series of decisions leads to a solution. For instance, a binary search tree can be used to locate items based on a series of comparisons, where each comparison tells us whether we have located the item, or whether we should go right or left in a subtree. A rooted tree in which each internal vertex corresponds to a decision, with a subtree at these vertices for each possible outcome of the decision, is called a **decision tree.** The possible solutions of the problem correspond to the paths to the leaves of this rooted tree. The next example illustrates an application of decision trees.

EXAMPLE 2 Suppose there are seven coins, all with the same weight, and a counterfeit coin that weighs less than the others. How many weighings are necessary using a balance scale to determine which of the eight coins is the counterfeit one? Give an algorithm for finding this counterfeit coin.

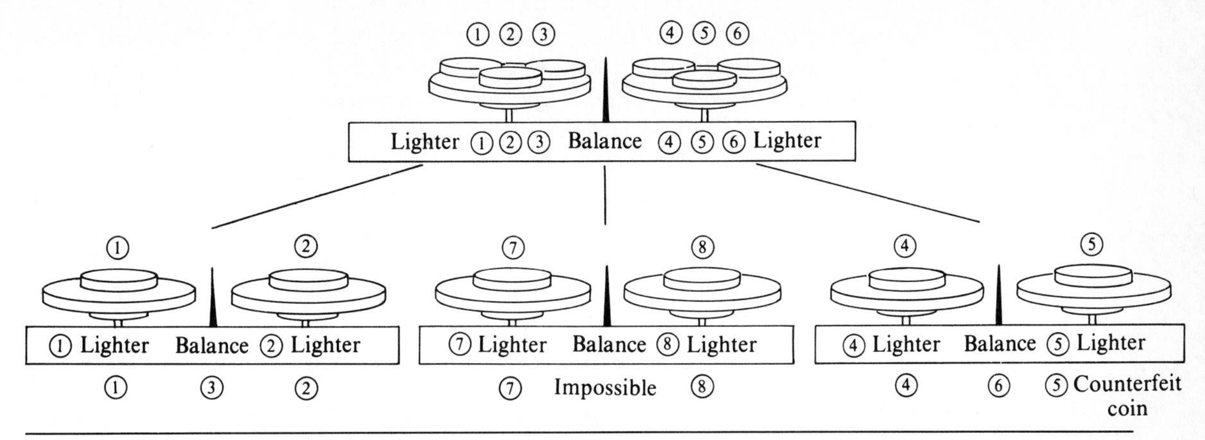

FIGURE 3 **A Decision Tree for Locating a Counterfeit Coin.**

Solution: There are three possibilities for each weighing on a balance scale. The two pans can have equal weight, the first pan can be heavier, or the second pan can be heavier. Consequently, the decision tree for the sequence of weighings is a 3-ary tree. There are at least eight leaves in the decision tree since there are eight possible outcomes (since each of the eight coins can be the counterfeit lighter coin), and each possible outcome must be represented by at least one leaf. The largest number of weighings needed to determine the counterfeit coin is the height of the decision tree. From Corollary 1 of Section 8.1 it follows that the height of the decision tree is at least $\lceil \log_3 8 \rceil = 2$. Hence, at least two weighings are needed.

It is possible to determine the counterfeit coin using two weighings. The decision tree that illustrates how this is done is shown in Figure 3. ■

In Section 4 of this chapter we will study sorting algorithms using decision trees.

PREFIX CODES

Consider the problem of using bit strings to encode the letters of the English alphabet (where no distinction is made between lowercase and uppercase letters). We can represent each letter with a bit string of length five, since there are only 26 letters and there are 32 bit strings of length five. The total number of bits used to encode data is five times the number of characters in the text when each character is encoded with five bits. Is it possible to find a coding scheme of these letters so that, when data are coded, fewer bits are used? We can save memory and reduce transmittal time if this can be done.

Consider using bit strings of different lengths to encode letters. Letters that occur more frequently should be encoded using short bit strings, and longer bit strings should be used to encode rarely occurring letters. When letters are encoded using varying numbers of bits, some method must be used to determine where the bits for each character start and end. For instance, if *e* were encoded with 0, *a* with 1, and *t* with 01, then the bit string 0101 could correspond to *eat, tea, eaea,* or *tt.*

To ensure that no bit string corresponds to more than one sequence of letters, the bit string for a letter must never occur as the first part of the bit string for another letter. Codes with this property are called **prefix codes.** For instance, the encoding of *e* as 0, *a* as 10, and *t* as 11 is a prefix code. A word can be recovered from the unique bit string that encodes its letters. For example the string 10110 is the encoding of *ate*. To see this, note that the initial 1 does not represent a character, but 10 does represent *a* (and could not be the first part of the bit string of another letter). Then, the next 1 does not represent a character, but 11 does represent *t*. The final bit, 0, represents *e*.

A prefix code can be represented using a binary tree, where the characters are the labels of the leaves in the tree. The edges of the tree are labeled so that an edge leading to a left child is assigned a 0 and an edge leading to a right child is assigned a 1. The bit string used to encode a character is the sequence of labels of the edges in the unique path from the root to the leaf that has this character as its label. For instance, the tree in Figure 4 represents the encoding of *e* by 0, *a* by 10, *t* by 110, *n* by 1110, and *s* by 1111.

The tree representing a code can be used to decode a bit string. For instance, consider the word encoded by 11111011100 using the code in Figure 4. This bit string can be decoded by starting at the root, using the sequence of bits to form a path that stops when a leaf is reached. Each 0 bit takes the path down the edge leading to the left child of the last vertex in the path, and each 1 bit corresponds to the right child of this vertex. Consequently, the initial 1111 corresponds to the path starting at the root, going right four times, leading to a leaf in the graph that has *s* as its label, since the string 1111 is the code for *s*. Continuing with the fifth bit, we reach a leaf next after going right then left, when the vertex labeled with *a,* which is encoded by 10, is visited. Starting with the seventh bit, we reach a leaf next after going right three times and then left, when the vertex labeled with *n,* which is encoded by 1110, is visited. Finally, the last bit, 0, leads to the leaf that is labeled with *e*. Therefore, the original word is *sane*.

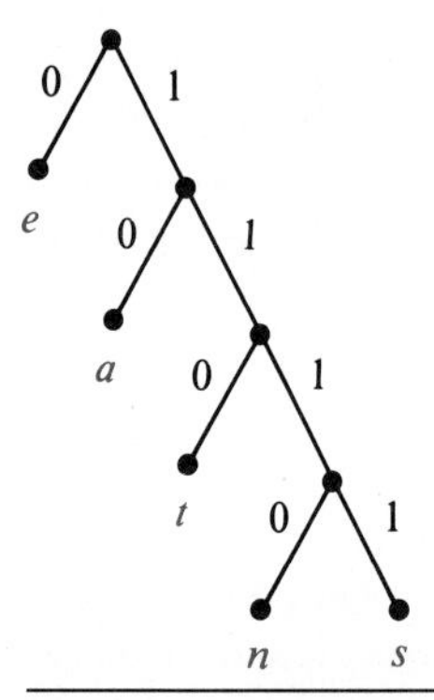

FIGURE 4 The Binary Tree with a Prefix Code.

We can construct a prefix code from any binary tree where the left edge at each internal vertex is labeled by 0 and the right edge by a 1 and where the leaves are labeled by characters. Characters are encoded with the bit string constructed using the labels of the edges in the unique path from the root to the leaves.

There are algorithms, such as Huffman coding, that can be used to produce efficient codes based on the frequencies of occurrences of characters. We will not present the details of such algorithms here. (The interested reader can find the details of such algorithms in the references given for this section at the end of the book.)

Exercises

1. Build a binary search tree for the words *banana, peach, apple, pear, coconut, mango,* and *papaya* using alphabetical order.

2. Build a binary search tree for the words *oenology, phrenology, campanology, ornithology, ichthyology, limnology, alchemy,* and *astrology* using alphabetical order.

3. How many comparisons are needed to locate or to add each of the following words in the search tree for Exercise 1, starting fresh each time?
a) *pear* **b)** *banana*
c) *kumquat* **d)** *orange*

4. How many comparisons are needed to locate or to add each of the following words in the search tree for Exercise 2, starting fresh each time?
a) *palmistry* **b)** *etymology*
c) *paleontology* **d)** *glaciology*

5. Using alphabetical order, construct a binary search tree for the words in the sentence, "*The quick brown fox jumps over the lazy dog.*"

6. How many weighings of a balance scale are needed to find a lighter counterfeit coin among four coins? Describe an algorithm to find the lighter coin using this number of weighings.

7. How many weighings of a balance scale are needed to find a counterfeit coin among four coins if the counterfeit coin may be either heavier or lighter than the others? Describe an algorithm to find the counterfeit coin using this number of weighings.

★8. How many weighings of a balance scale are needed to find a counterfeit coin among eight coins if the counterfeit coin is either heavier or lighter than the others? Describe an algorithm to find the counterfeit coin using this number of weighings.

★9. How many weighings of a balance scale are needed to find a counterfeit coin among 12 coins if the counterfeit coin is lighter than the others? Describe an algorithm to find the lighter coin using this number of weighings.

★10. One of four coins may be counterfeit. If it is counterfeit, it may be lighter or heavier than the others. How many weighings are needed, using a balance scale, to determine whether there is a counterfeit coin, and if there is, whether it is lighter or heavier than the others? Describe an algorithm to find the counterfeit coin and determine whether it is lighter or heavier using this number of weighings.

11. Which of the following codes are prefix codes?
a) a:11, e:00, t:10, s:01
b) a:0, e:1, t:01, s:001
c) a:101, e:11, t:001, s:011, n:010
d) a:010, e:11, t:011, s:1011, n:1001, i:10101

12. Construct the binary tree with prefix codes representing the following coding schemes.
a) a:11, e:0, t:101, s:100
b) a:1, e:01, t:001, s:0001, n:00001
c) a:1010, e:0, t:11, s:1011, n:1001, i:100001

13. What are the codes for a, e, i, k, o, p, and u if the coding scheme is represented by the following tree?

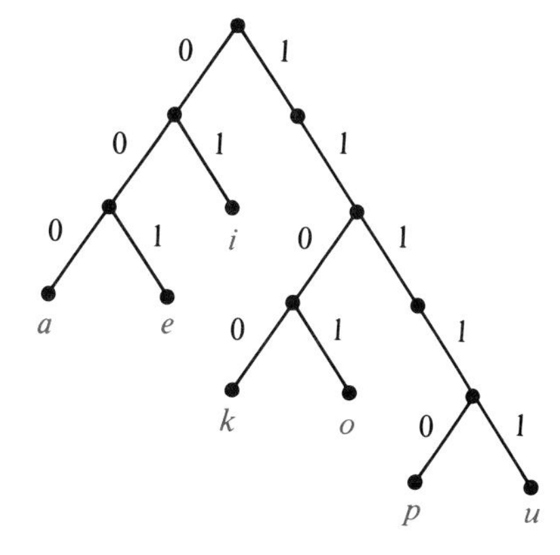

14. Given the coding scheme
a:001, b:0001, e:1, r:0000, s:0100,
t:011, x:01010,
find the word represented by

a) 01110100011.
b) 0001110000.
c) 0100101010.
d) 01100101010.

8.3 Tree Traversal

INTRODUCTION

Ordered rooted trees are often used to store information. We need procedures for visiting each vertex of an ordered rooted tree to access data. We will describe several important algorithms for visiting all the vertices of an ordered rooted tree. Ordered rooted trees can also be used to represent various types of expressions, such as arithmetic expressions involving numbers, variables, and operations. The different listings of the vertices of ordered rooted trees used to represent expressions are useful in the evaluation of these expressions.

UNIVERSAL ADDRESS SYSTEMS

Procedures for traversing all vertices of an ordered rooted tree rely on the orderings of children. In ordered rooted trees, the children of an internal vertex are shown from left to right in the drawings representing these directed graphs.

We will describe one way to order totally the vertices of an ordered rooted tree. To produce this ordering, we must first label all the vertices. We do this recursively as follows.

1. Label the root with the integer 0. Then label its k children (at level 1) from left to right with $1, 2, 3, \ldots, k$.
2. For each vertex v at level n with label A, label its k_v children, as they are drawn from left to right, with $A.1, A.2, \ldots, A.k_v$.

Following this procedure, a vertex v at level n, for $n \geq 1$, is labeled $x_1.x_2.\cdots.x_n$, where the unique path from the root to v goes through the x_1th vertex at level 1, the x_2th vertex at level 2, and so on. This labeling is called the **universal address system** of the ordered rooted tree.

We can totally order the vertices using the lexicographic ordering of their labels in the universal address system. The vertex labeled $x_1.x_2.\cdots.x_n$ is less than the vertex labeled $y_1.y_2.\cdots.y_m$ if there is an i, $0 \leq i \leq n$, with $x_1 = y_1$, $x_2 = y_2, \ldots$, $x_{i-1} = y_{i-1}$, and $x_i < y_i$; or if $n < m$ and $x_i = y_i$ for $i = 1, 2, \ldots, n$.

EXAMPLE 1 We display the labelings of the universal address system next to the vertices in the ordered rooted tree shown in Figure 1. The lexicographic ordering of the labelings is:

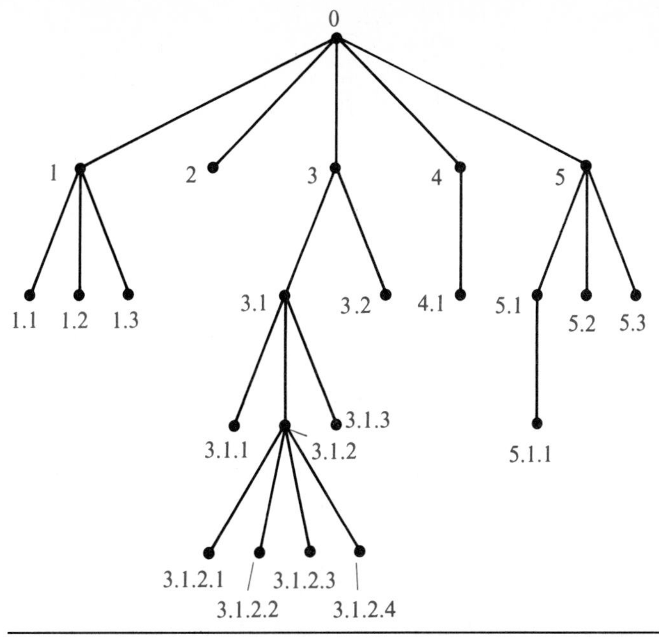

FIGURE 1 **The Universal Address System of an Ordered Rooted Tree.**

$$0 < 1 < 1.1 < 1.2 < 1.3 < 2 < 3 < 3.1 < 3.1.1 < 3.1.2 < 3.1.2.1 < 3.1.2.2$$
$$< 3.1.2.3 < 3.1.2.4 < 3.1.3 < 3.2 < 4 < 4.1 < 5 < 5.1 < 5.1.1 < 5.2 < 5.3$$ ■

TRAVERSAL ALGORITHMS

Procedures for systematically visiting every vertex of an ordered rooted tree are called **traversal algorithms.** We will describe three of the most commonly used such algorithms, **preorder traversal, inorder traversal,** and **postorder traversal.** Each of these algorithms can be defined recursively. We first define preorder traversal.

DEFINITION 1. Let T be an ordered rooted tree with root r. If T consists only of r, then r is the *preorder traversal of* T. Otherwise, suppose that $T_1, T_2, \ldots, T_n$ are the subtrees at r from left to right in T. The *preorder traversal* begins by visiting r. It continues by traversing T_1 in preorder, then T_2 in preorder, and so on, until T_n is traversed in preorder.

The reader should verify that the preorder traversal of an ordered rooted tree gives the same ordering of the vertices as the ordering obtained using a universal address system. Figure 2 indicates how a preorder traversal is carried out.

The following example illustrates preorder traversal.

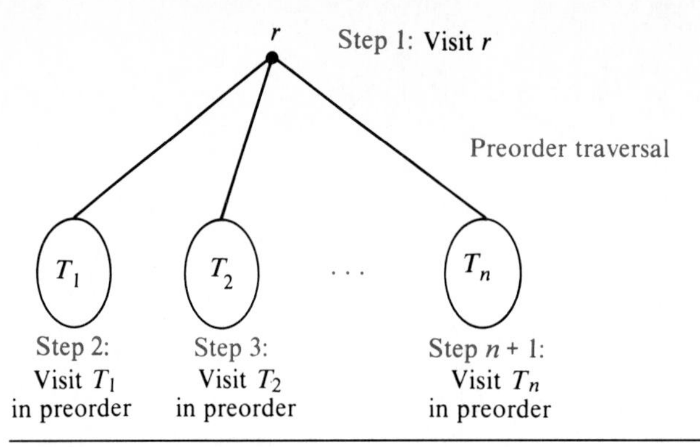

FIGURE 2 **Preorder Traversal.**

EXAMPLE 2 In which order does a preorder traversal visit the vertices in the ordered rooted tree T shown in Figure 3?

Solution: The steps of the preorder traversal of T are shown in Figure 4. We traverse T in preorder by first listing the root a, followed by the preorder list of the subtree with root b, the preorder list of the subtree with root c (which is just c) and the preorder list of the subtree with root d.

The preorder list of the subtree with root b begins by listing b, then the vertices of the subtree with root e in preorder, and then the subtree with root f in preorder (which is just f). The preorder list of the subtree with root d begins by listing d, followed by the preorder list of the subtree with root g, followed by the subtree with root h (which is just h), followed by the subtree with root i (which is just i).

The preorder list of the subtree with root e begins by listing e, followed by the preorder listing of the subtree with root j (which is just j), followed by the preorder listing of the subtree with root k. The preorder listing of the subtree with root g is g

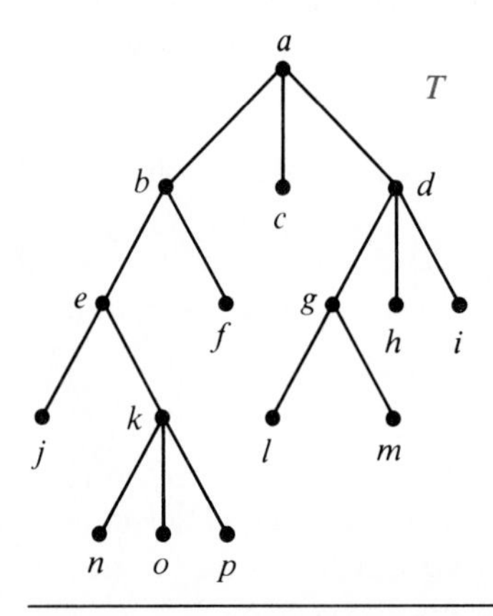

FIGURE 3 **The Ordered Rooted Tree T.**

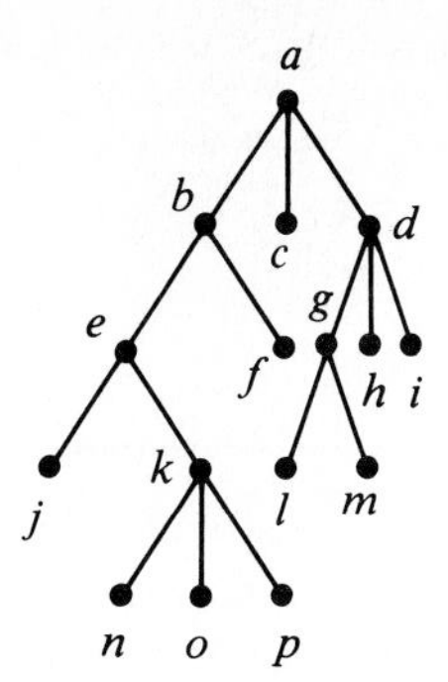

Preorder traversal: Visit root, visit subtrees left to right

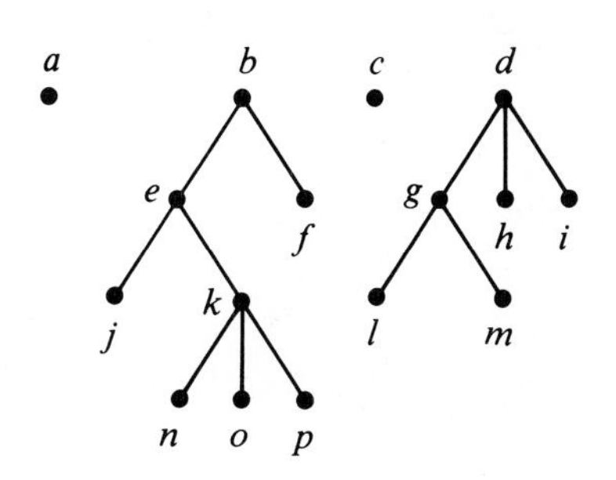

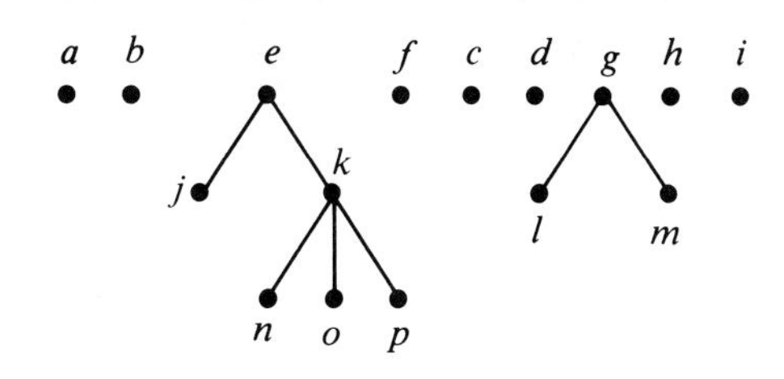

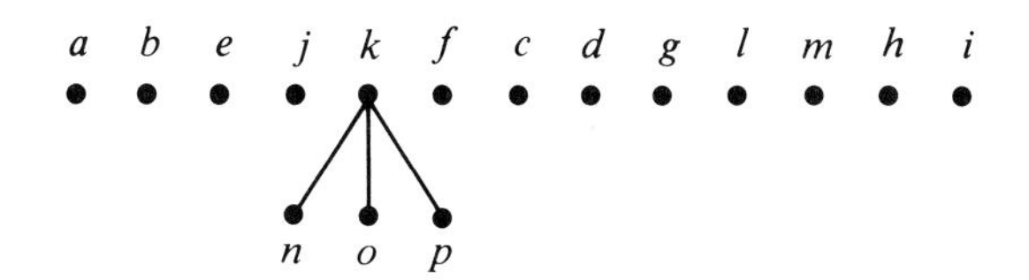

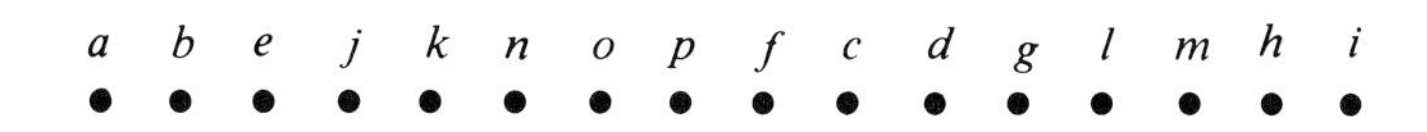

FIGURE 4 **The Preorder Traversal of T.**

followed by l, followed by m. The preorder listing of the subtree with root k is k, n, o, p. Consequently, the preorder traversal of T is a, b, e, j, k, n, o, p, f, c, d, g, l, m, h, i. ■

We will now define inorder traversal.

DEFINITION 2. Let T be an ordered rooted tree with root r. If T consists only of r, then r is the *inorder traversal* of T. Otherwise, suppose that $T_1, T_2, \ldots, T_n$ are the subtrees at r from left to right. The *inorder traversal* begins by traversing T_1 in inorder, then visiting r. It continues by traversing T_2 in inorder, then T_3 in inorder, . . . , and finally T_n in inorder.

Figure 5 indicates how inorder traversal is carried out.

The following example illustrates how inorder traversal is carried out.

EXAMPLE 3 In which order does an inorder traversal visit the vertices of the ordered rooted tree T in Figure 3?

Solution: The steps of the inorder traversal of the ordered rooted tree T are shown in Figure 6. The inorder traversal begins with an inorder traversal of the subtree with root b, the root a, the inorder listing of the subtree with root c, which is just c, and the inorder listing of the subtree with root d.

The inorder listing of the subtree with root b begins with the inorder listing of the subtree with root e, the root b, and f. The inorder listing of the subtree with root d begins with the inorder listing of the subtree with root g, followed by the root d, followed by h, followed by i.

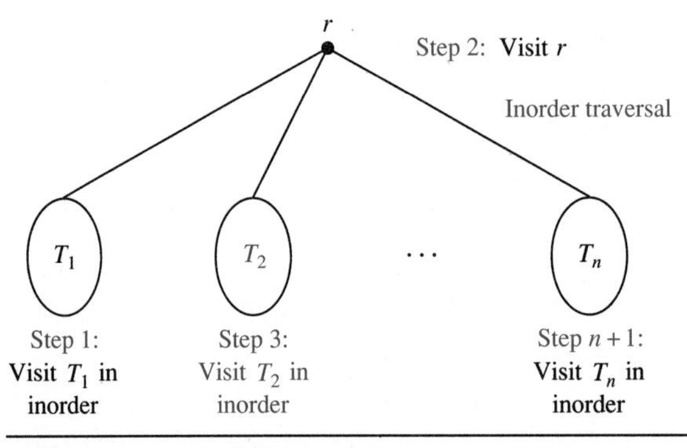

FIGURE 5 Inorder Traversal.

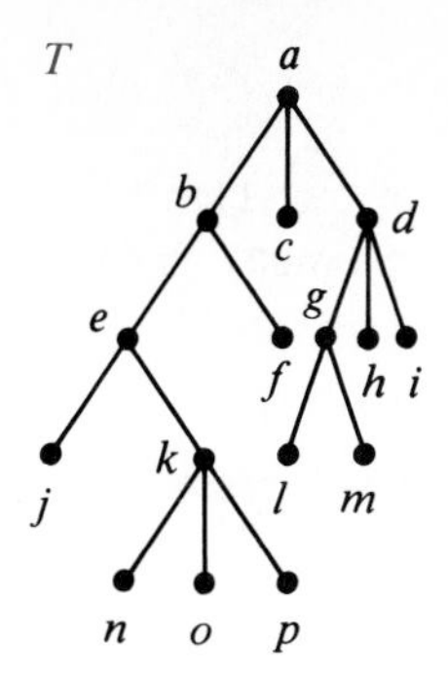

Inorder traversal: Visit leftmost subtree, visit root, visit other subtrees left to right

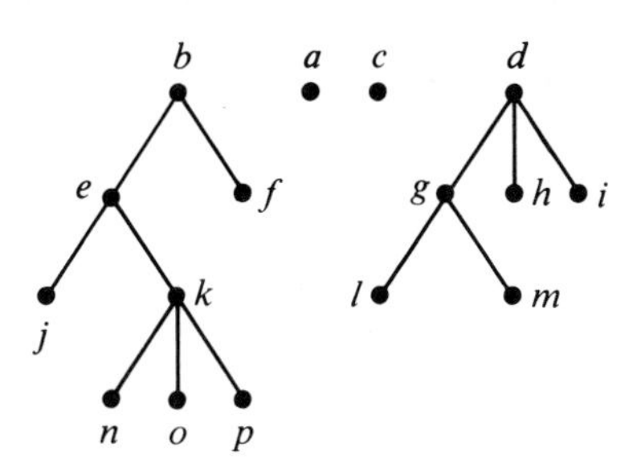

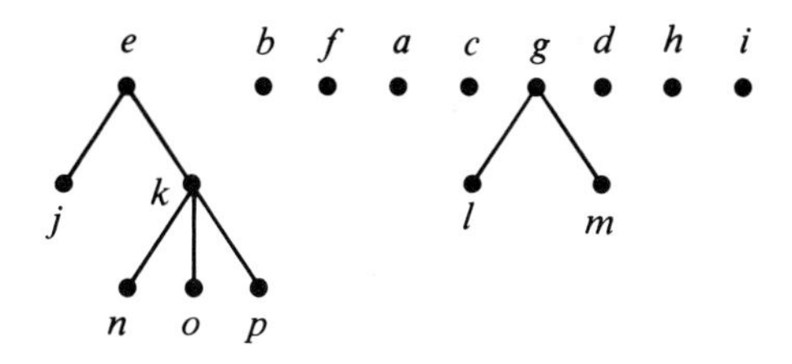

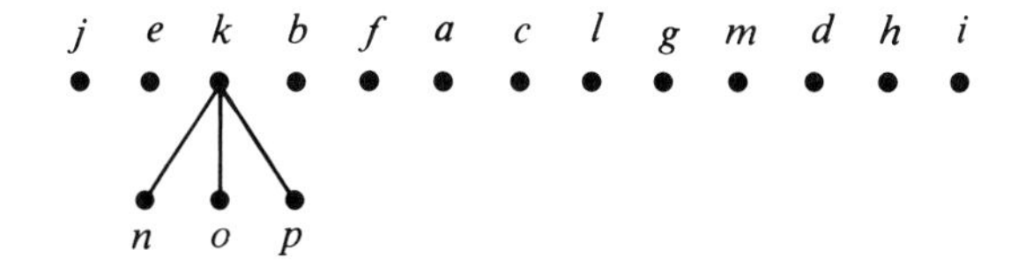

j e n k o p b f a c l g m d h i

FIGURE 6 **The Inorder Traversal of *T*.**

The inorder listing of the subtree with root e is j, followed by the root e, followed by the inorder listing of the subtree with root k. The inorder listing of the subtree with root g is l, g, m. The inorder listing of the subtree with root k is n, k, o, p. Consequently, the inorder listing of the ordered rooted tree is j, e, n, k, o, p, b, f, a, c, l, g, m, d, h, i. ■

The definition of postorder traversal follows.

DEFINITION 3. Let T be an ordered rooted tree with root r. If T consists only of r, then r is the *postorder traversal* of T. Otherwise, suppose that $T_1, T_2, \ldots, T_n$ are the subtrees at r from left to right. The *postorder traversal* begins by traversing T_1 in postorder, then T_2 in postorder, . . . , then T_n in postorder, and ends by visiting r.

Figure 7 illustrates how postorder traversal is done. The following example illustrates how postorder traversal works.

EXAMPLE 4 In which order does a postorder traversal visit the vertices of the ordered rooted tree T shown in Figure 3?

Solution: The steps of the postorder traversal of the ordered rooted tree T are shown in Figure 8. The postorder traversal begins with the postorder traversal of the subtree with root b, the postorder traversal of the subtree with root c, which is just c, the postorder traversal of the subtree with root d, followed by the root a.

The postorder traversal of the subtree with root b begins with the postorder traversal of the subtree with root e, followed by f, followed by the root b. The postorder traversal of the rooted tree with root d begins with the postorder traversal of the subtree with root g, followed by h, followed by i, followed by the root d.

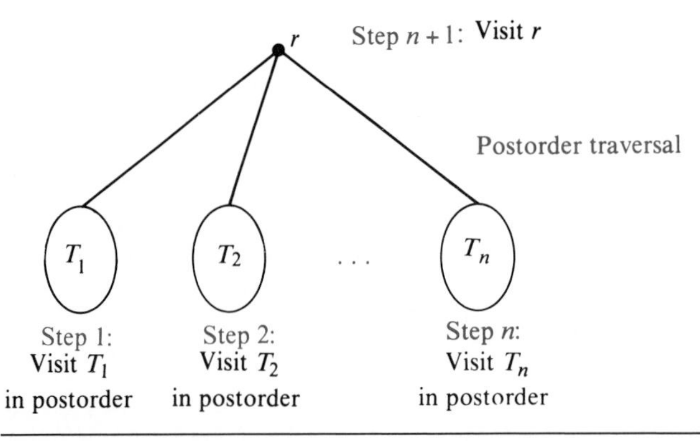

FIGURE 7 **Postorder Traversal.**

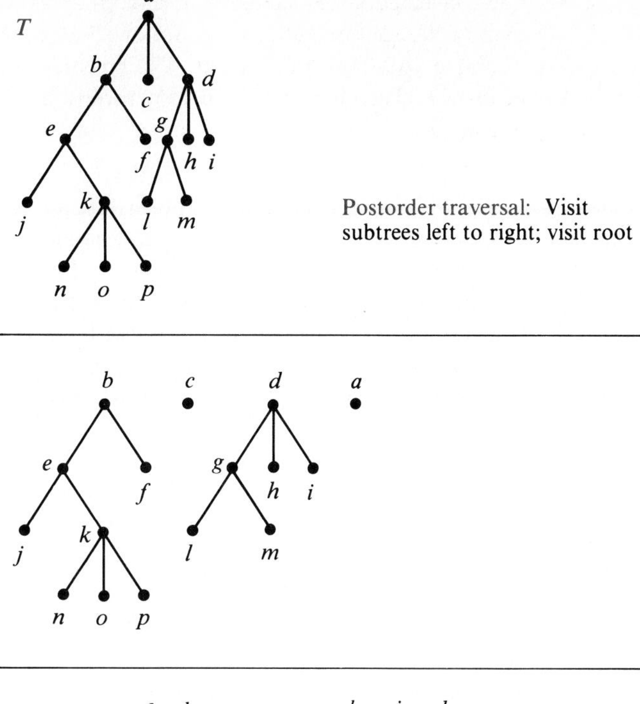

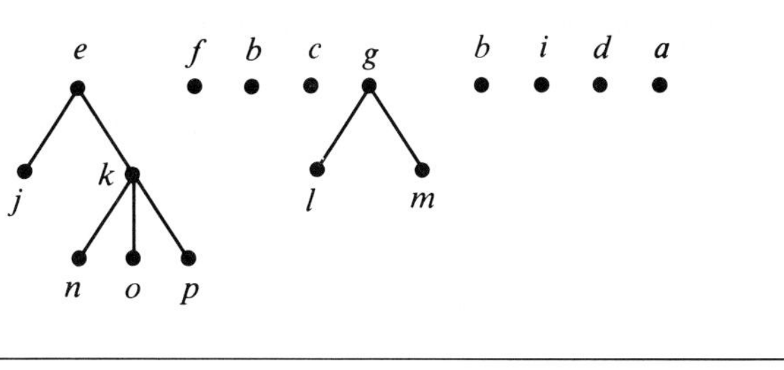

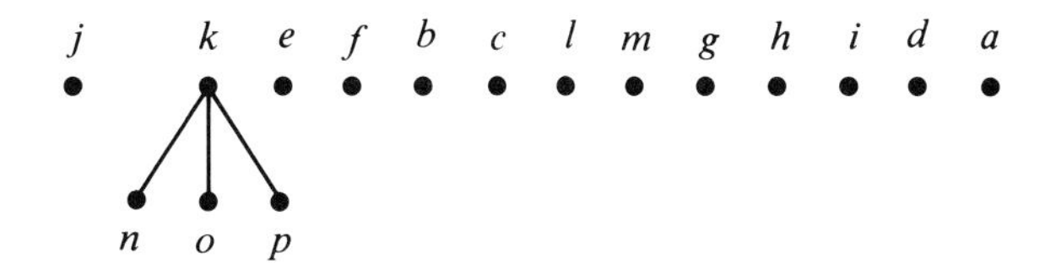

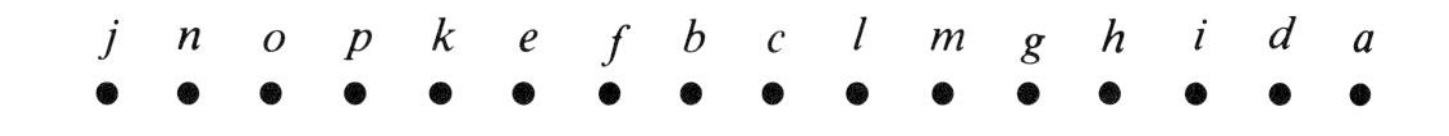

FIGURE 8 **The Postorder Traversal of *T*.**

The postorder traversal of the subtree with root e begins with j, followed by the postorder traversal of the subtree with root k, followed by the root e. The postorder traversal of the subtree with root g is l, m, g. The postorder traversal of the subtree with root k is n, o, p, k. Therefore, the postorder traversal of T is j, n, o, p, k, e, f, b, c, l, m, g, h, i, d, a. ■

There are easy ways to list the vertices of an ordered rooted tree in preorder, inorder, and postorder. To do this, first draw a curve around the ordered rooted tree starting at the root, moving along the edges as shown in the example in Figure 9. We can list the vertices in preorder by listing each vertex the first time this curve passes it. We can list the vertices in inorder by listing a leaf the first time the curve passes it and listing each internal vertex the second time the curve passes it. We can list the vertices in postorder by listing a vertex the last time it is passed on the way back up to its parent. When this is done in the rooted tree in Figure 9, it follows that the preorder traversal gives a, b, d, h, e, i, j, c, f, g, k, the inorder traversal gives h, d, b, i, e, j, a, f, c, k, g, and the postorder traversal gives h, d, i, j, e, b, f, k, g, c, a.

Algorithms for traversing ordered rooted trees in preorder, inorder, or postorder are most easily expressed recursively.

ALGORITHM 1 Preorder Traversal.

procedure *preorder*(T: ordered rooted tree)
r := root of T
list r
for each child c of r from left to right
begin
  $T(c)$:= subtree with c as its root
  preorder($T(c)$)
end

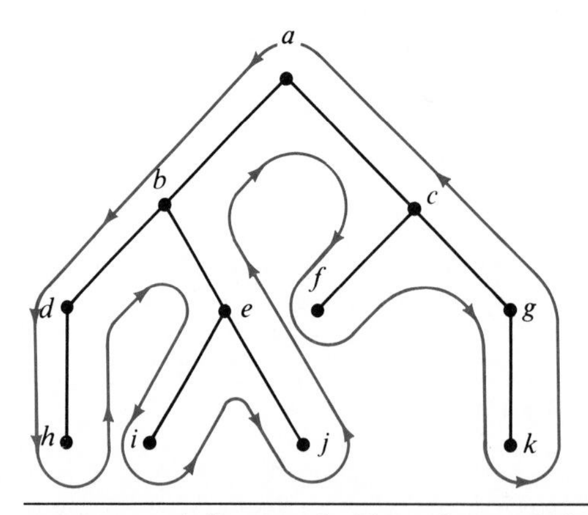

FIGURE 9 **A Shortcut for Traversing an Ordered Rooted Tree in Preorder, Inorder, and Postorder.**

ALGORITHM 2 Inorder Traversal.

```
procedure inorder(T: ordered rooted tree)
r := root of T
if r is a leaf then list r
else
begin
   l := first child of r from left to right
   T(l) := subtree with l as its root
   inorder(T(l))
   list r
   for each child c of r except for l from left to right
      T(c) := subtree with c as its root
      inorder(T(c))
end
```

ALGORITHM 3 Postorder Traversal.

```
procedure postorder(T: ordered rooted tree)
r := root of T
for each child c of r from left to right
begin
   T(c) := subtree with c as its root
   postorder(T(c))
end
list r
```

INFIX, PREFIX, AND POSTFIX NOTATION

We can represent complicated expressions, such as compound propositions, combinations of sets, and arithmetic expressions using ordered rooted trees. For instance, consider the representation of an arithmetic expression involving the operators $+$ (addition), $-$ (subtraction), $*$ (multiplication), $/$ (division), and $\uparrow$ (exponentiation). We will use parentheses to indicate the order of the operations. An ordered rooted tree can be used to represent such expressions, where the internal vertices represent operations, and the leaves represent the variables or numbers. Each operation operates on its left and right subtrees (in that order).

EXAMPLE 5 What is the ordered rooted tree that represents the expression $((x + y) \uparrow 2) + ((x - 4)/3)$?

Solution: The binary tree for this expression can be built from the bottom up. First, a subtree for the expression $x + y$ is constructed. Then this is incorporated as part of the larger subtree representing $(x + y) \uparrow 2$. Also, a subtree for $x - 4$ is constructed, and then this is incorporated into a subtree representing $(x - 4)/3$. Finally the subtrees representing $(x + y) \uparrow 2$ and $(x - 4)/3$ are combined to form the ordered rooted tree representing $((x + y) \uparrow 2) + ((x - 4)/3)$. These steps are shown in Figure 10. ■

An inorder traversal of the binary tree representing an expression produces the original expression with the elements and operations in the same order as they originally occurred, except for unary operations, which instead immediately follow their operands. For instance, infix traversals of the binary trees in Figure 11, which represent the expressions $(x + y)/(x + 3)$, $(x + (y/x)) + 3$, and $x + (y/(x + 3))$, all lead to the infix expression $x + y/x + 3$. To make such expressions unambiguous it is necessary to include parentheses in the inorder traversal whenever we encounter an operation. The fully parenthesized expression obtained in this way is said to be in **infix form.**

We obtain the **prefix form** of an expression when we traverse its rooted tree in preorder. Expressions written in prefix form are said to be in **Polish notation,**

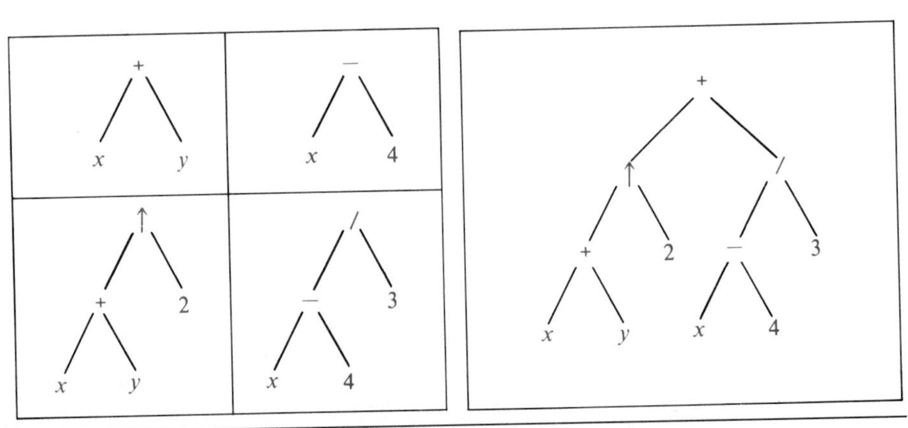

FIGURE 10 A Binary Tree Representing $((x + y) \uparrow 2) + ((x - 4)/3)$.

Jan Łukasiewicz (1878–1956). Jan Łukasiewicz, born in Lwów, studied and began his professional career at the University of Lwów. He later moved to a professorship in Warsaw. After World War II, he was appointed to a position at the Royal Irish Academy in Dublin. Łukasiewicz worked in the area of many-valued logic; his 1921 paper on a three-valued logic was an important contribution to this subject. Nevertheless, he is best known in the mathematical community for his introduction of parenthesis-free notation, now called Polish notation.

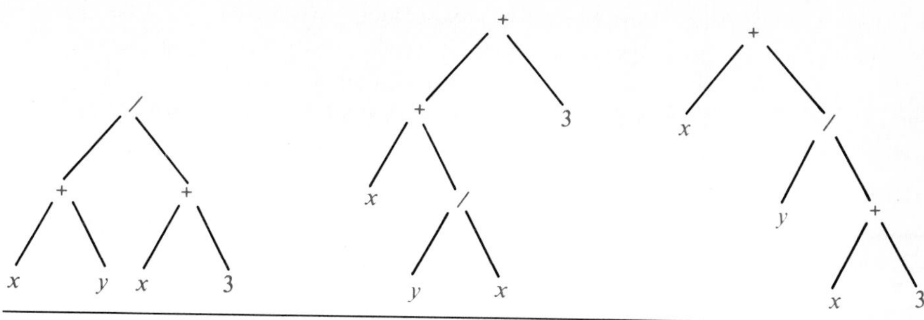

FIGURE 11 **Rooted Trees Representing** $(x + y)/(x + 3)$, $(x + (y/x)) + 3$, **and** $x + (y/(x + 3))$.

which is named after the logician Jan Łukasiewicz (who was actually Ukrainian and not Polish). An expression in prefix notation (where each operation has a specified number of operands), is unambiguous, so that no parentheses are needed in such an expression. The verification of this is left as an exercise for the reader.

EXAMPLE 6 What is the prefix form for $((x + y) \uparrow 2) + ((x - 4)/3)$?

Solution: We obtain the prefix form for this expression by traversing the binary tree that represents it, shown in Figure 10. This produces $+ \uparrow + x\ y\ 2\ /\ -\ x\ 4\ 3$. ■

In the prefix form of an expression, a binary operator, such as $+$, precedes its two operands. Hence, we can evaluate an expression in prefix form by working from right to left. When we encounter an operator, we perform the corresponding operation with the two operands immediately to the right of this operand. Also, whenever an operation is performed, we consider the result a new operand.

EXAMPLE 7 What is the value of the prefix expression $+ - * 2\ 3\ 5\ / \uparrow 2\ 3\ 4$?

Solution: The steps used to evaluate this expression by working right to left, and performing operations using the operands on the right, are shown in Figure 12. The value of this expression is 3. ■

We obtain the **postfix form** of an expression by traversing its binary tree in postorder. Expressions written in postfix form are said to be in **reverse Polish notation.** Expressions in reverse Polish notation are unambiguous, so that parentheses are not needed. The verification of this is left to the reader.

EXAMPLE 8 What is the postfix form of the expression $((x + y) \uparrow 2) + ((x - 4)/3)$?

Solution: The postfix form of the expression is obtained by carrying out a postorder traversal of the binary tree for this expression, shown in Figure 10. This produces the postfix expression: $x\ y\ +\ 2\ \uparrow\ x\ 4\ -\ 3\ /\ +$. ■

In the postfix form of an expression, a binary operator follows its two operands. So, to evaluate an expression from its postfix form, work from left to right, carrying out operations whenever an operator follows two operands. After an operation is carried out, the result of this operation becomes a new operand.

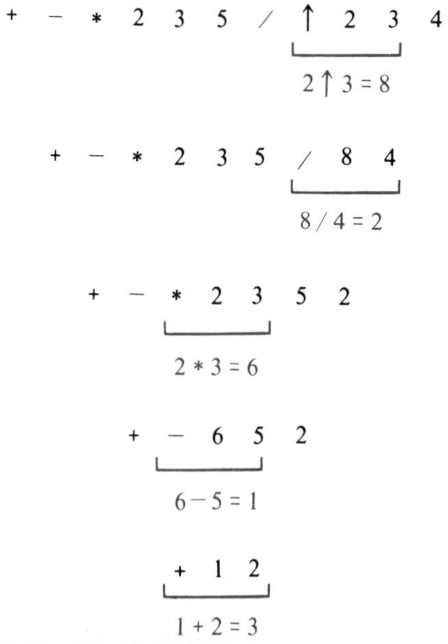

$+ \ - \ * \ 2 \ 3 \ 5 \ / \ \uparrow \ 2 \ 3 \ 4$

$2 \uparrow 3 = 8$

$+ \ - \ * \ 2 \ 3 \ 5 \ / \ 8 \ 4$

$8 / 4 = 2$

$+ \ - \ * \ 2 \ 3 \ 5 \ 2$

$2 * 3 = 6$

$+ \ - \ 6 \ 5 \ 2$

$6 - 5 = 1$

$+ \ 1 \ 2$

$1 + 2 = 3$

Value of expression: 3

FIGURE 12 **Evaluating a Prefix Expression.**

EXAMPLE 9 What is the value of the postfix expression $7\ 2\ 3 * -\ 4 \uparrow 9\ 3\ /\ +$?

Solution: The steps used to evaluate this expression by starting at the left and carrying out operations when two operands are followed by an operator are shown in Figure 13. The value of this expression is 4. ■

Rooted trees can be used to represent other types of expressions, such as those representing compound propositions and combinations of sets. In these examples unary operators, such as the negation of a proposition, occur. To represent such operators and their operands, a vertex representing the operator and a child of this vertex representing the operand are used.

EXAMPLE 10 Find the ordered rooted tree representing the compound proposition $(\neg(p \wedge q)) \leftrightarrow (\neg p \vee \neg q)$. Then use this rooted tree to find the prefix, postfix, and infix forms of this expression.

Solution: The rooted tree for this compound proposition is constructed from the bottom up. First, subtrees for $\neg p$ and $\neg q$ are formed (where $\neg$ is considered a unary

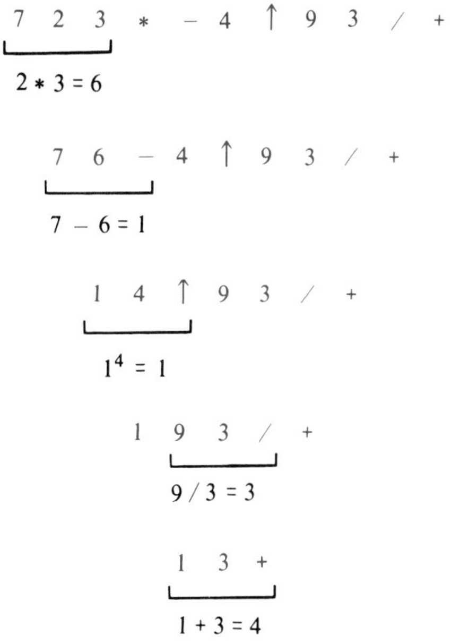

$7\ 2\ 3 * -\ 4 \uparrow 9\ 3\ /\ +$

$2 * 3 = 6$

$7\ 6 -\ 4 \uparrow 9\ 3\ /\ +$

$7 - 6 = 1$

$1\ 4 \uparrow 9\ 3\ /\ +$

$1^4 = 1$

$1\ 9\ 3\ /\ +$

$9 / 3 = 3$

$1\ 3\ +$

$1 + 3 = 4$

FIGURE 13 **Evaluating a Postfix Expression.**

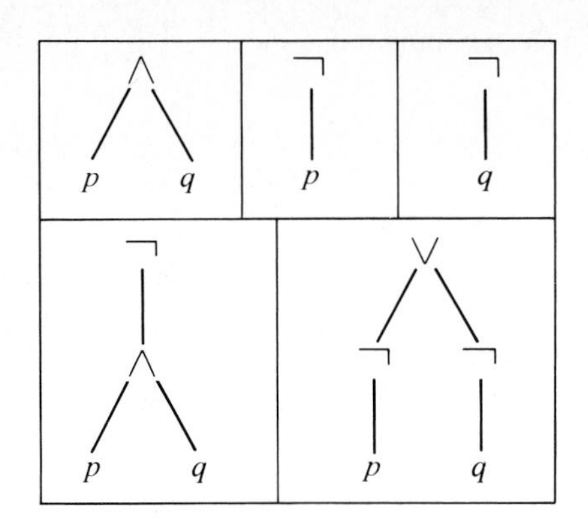

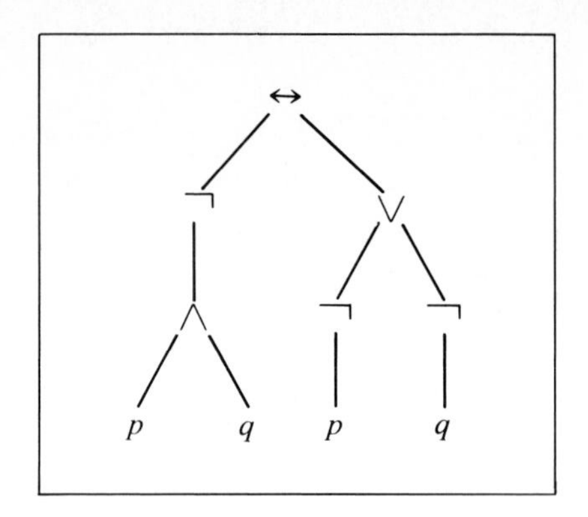

FIGURE 14 Constructing the Rooted Tree for a Compound Proposition.

The prefix, postfix, and infix forms of this expression are found by traversing this rooted tree in preorder, postorder, and inorder (including parentheses), respectively. These traversals give $\leftrightarrow \neg \wedge pq \vee \neg p \neg q$, $pq \wedge \neg p \neg q \neg \vee \leftrightarrow$, and $(\neg(p \wedge q)) \leftrightarrow ((\neg p) \vee (\neg q))$, respectively. ■

Because prefix and postfix expressions are unambiguous and because they can easily be evaluated without scanning back and forth, they are used extensively in computer science. Such expressions are especially useful in the construction of compilers.

Exercises

In Exercises 1–3 construct the universal address system for the given ordered rooted tree. Then use this to order its vertices using the lexicographic order of their labels.

1.

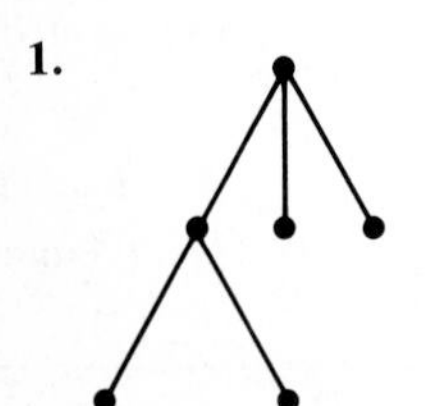

2.

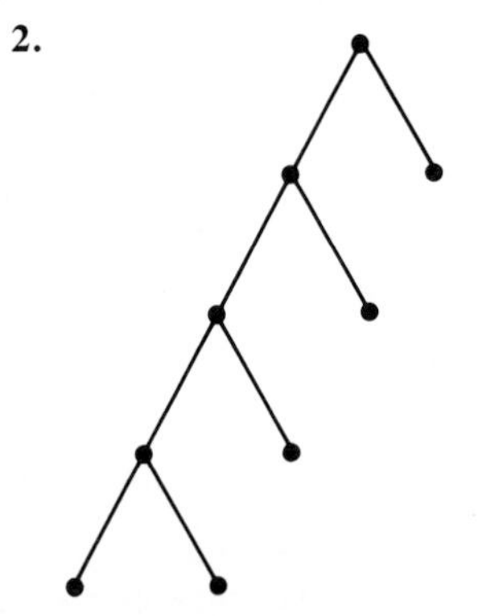

3.

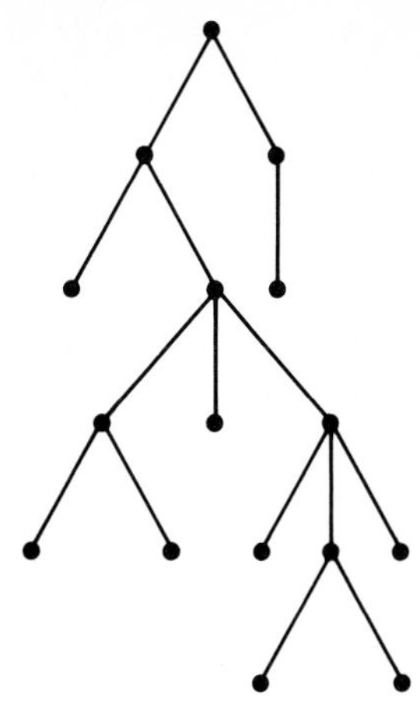

4. Suppose that the address of the vertex v in the ordered rooted tree T is 3.4.5.2.4.
   a) At what level is v?
   b) What is the address of the parent of v?
   c) What is the least number of siblings v can have?
   d) What is the smallest possible number of vertices in T if v has this address?
   e) Find the other addresses that must occur.
5. Suppose that the vertex with the largest address in an ordered rooted tree T has address 2.3.4.3.1. Is it possible to determine the number of vertices in T?
6. Can the leaves of an ordered rooted tree have the following list of universal addresses? If so, construct such an ordered rooted tree.
   a) 1.1.1, 1.1.2, 1.2, 2.1.1.1, 2.1.2, 2.1.3, 2.2, 3.1.1, 3.1.2.1, 3.1.2.2, 3.2
   b) 1.1, 1.2.1, 1.2.2, 1.2.3, 2.1, 2.2.1, 2.3.1, 2.3.2, 2.4.2.1, 2.4.2.2, 3.1, 3.2.1, 3.2.2
   c) 1.1, 1.2.1, 1.2.2, 1.2.2.1, 1.3, 1.4, 2, 3.1, 3.2, 4.1.1.1

In Exercises 7–9 determine the order in which a preorder traversal visits the vertices of the given ordered rooted tree.

7.

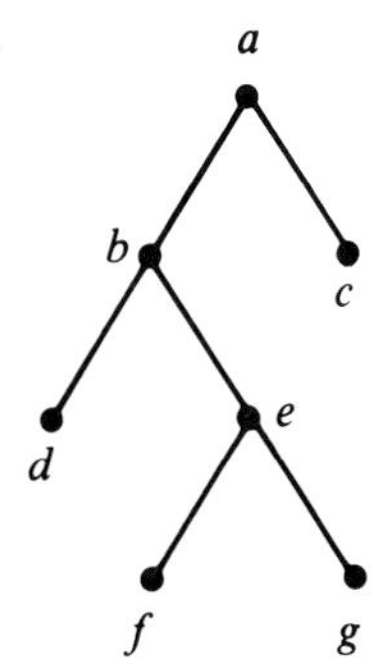

8.

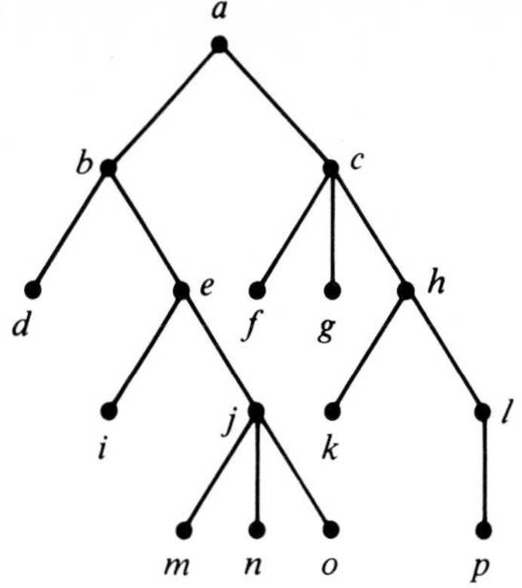

9.

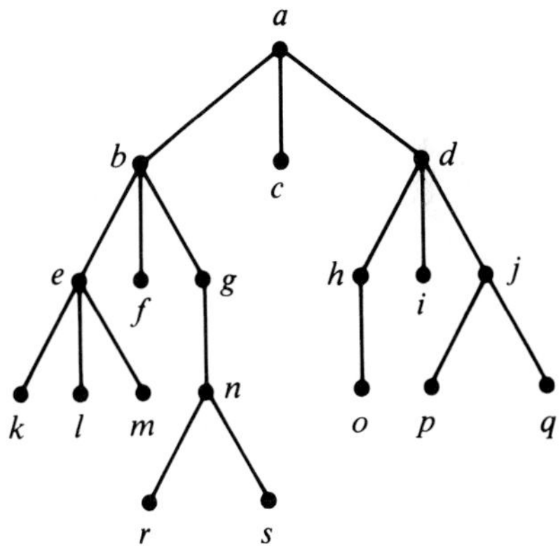

10. In which order are the vertices of the ordered rooted tree in Exercise 7 visited using an inorder traversal?
11. In which order are the vertices of the ordered rooted tree in Exercise 8 visited using an inorder traversal?
12. In which order are the vertices of the ordered rooted tree in Exercise 9 visited using an inorder traversal?
13. In which order are the vertices of the ordered rooted tree in Exercise 7 visited using a postorder traversal?
14. In which order are the vertices of the ordered rooted tree in Exercise 8 visited using a postorder traversal?
15. In which order are the vertices of the ordered rooted tree in Exercise 9 visited using a postorder traversal?
16. Represent the expression $((x+2) \uparrow 3)*(y-(3+x))-5$ using a binary tree.

17. Write the expression in Exercise 16 in
 a) prefix notation.
 b) postfix notation.
 c) infix notation.
18. Represent the expressions $(x + xy) + (x/y)$ and $x + ((xy + x)/y)$ using binary trees.
19. Write the expression in Exercise 18 in
 a) prefix notation.
 b) postfix notation.
 c) infix notation.
20. Represent the compound propositions $\neg(p \wedge q) \leftrightarrow (\neg p \vee \neg q)$ and $(\neg p \wedge (q \leftrightarrow \neg p)) \vee \neg q$ using ordered rooted trees.
21. Write the expressions in Exercise 20 in
 a) prefix notation.
 b) postfix notation.
 c) infix notation.
22. Represent $(A \cap B) - (A \cup (B - A))$ using an ordered rooted tree.
23. Write the expression in Exercise 22 in
 a) prefix notation.
 b) postfix notation.
 c) infix notation.

★24. How many ways can the string $\neg p \wedge q \leftrightarrow \neg p \vee \neg q$ be fully parenthesized to yield an infix expression?

★25. How many ways can the string $A \cap B - A \cap B - A$ be fully parenthesized to yield an infix expression?

26. Draw the ordered rooted tree corresponding to each of the following arithmetic expressions written in prefix notation. Then write each expression using infix notation.
 a) $+ * + - 5\ 3\ 2\ 1\ 4$
 b) $\uparrow + 2\ 3 - 5\ 1$
 c) $* / 9\ 3 + * 2\ 4 - 7\ 6$
27. What is the value of each of the following prefix expressions?
 a) $- * 2 / 8\ 4\ 3$
 b) $\uparrow - * 3\ 3 * 4\ 2\ 5$
 c) $+ - \uparrow 3\ 2 \uparrow 2\ 3 / 6 - 4\ 2$
 d) $* + 3 + 3 \uparrow 3 + 3\ 3\ 3$
28. What is the value of each of the following postfix expressions?
 a) $5\ 2\ 1 - - 3\ 1\ 4 + + *$
 b) $9\ 3 / 5 + 7\ 2 - *$
 c) $3\ 2 * 2 \uparrow 5\ 3 - 8\ 4 / * -$
29. Construct the ordered rooted tree whose preorder traversal is $a, b, f, c, g, h, i, d, e, j, k, l$, where a has four children, c has three children, j has two children, b and e have one child each, and all other vertices are leaves.

★30. Show that an ordered rooted tree is uniquely determined when a list of vertices generated by a preorder traversal of the tree and the number of children of each vertex are specified.

★31. Show that an ordered rooted tree is uniquely determined when a list of vertices generated by a postorder traversal of the tree and the number of children of each vertex are specified.

32. Show that preorder traversals of the two ordered rooted trees displayed below produce the same list of vertices. Note that this does not contradict the statement in Exercise 30, since the numbers of children of internal vertices in the two ordered rooted trees differ.

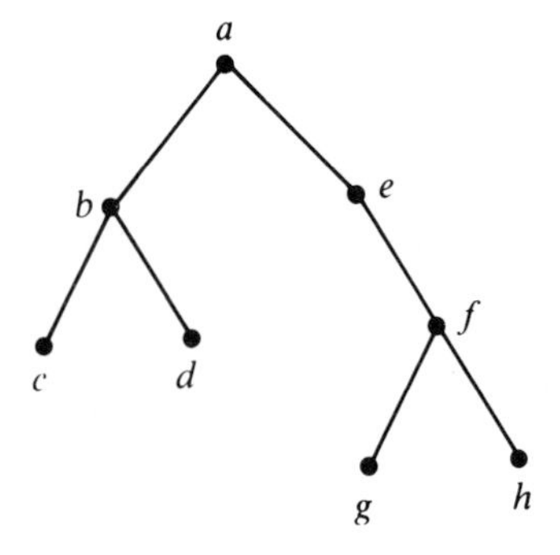

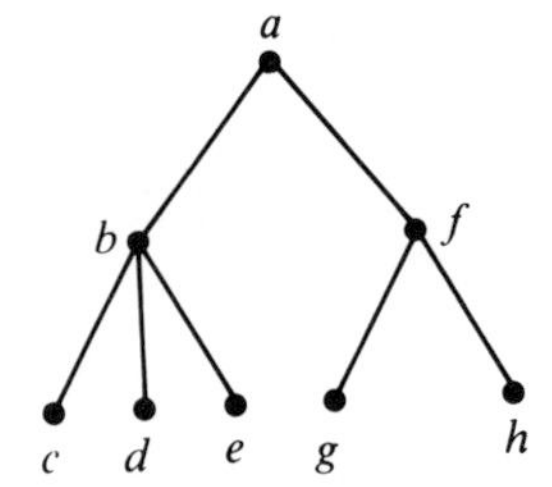

33. Show that postorder traversals of the two ordered rooted trees displayed on page 543 produce the same list of vertices. Note that this does not contradict the statement in Exercise 31, since the numbers of children of internal vertices in the two ordered rooted trees differ.

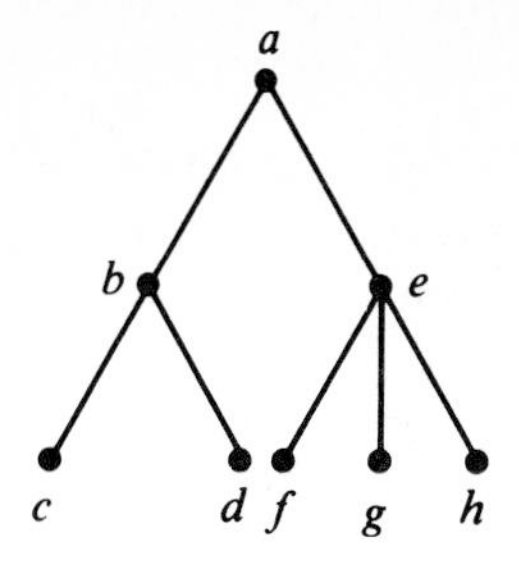

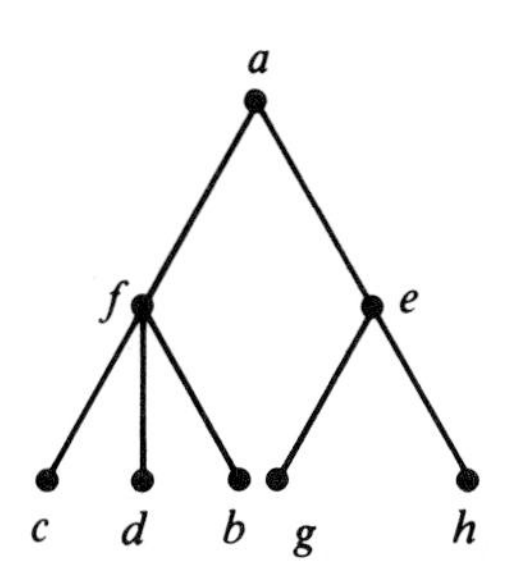

Well-formed formulae in prefix notation over a set of symbols and a set of binary operators are defined recursively by the following rules:

(i) if x is a symbol, then x is a well-formed formula in prefix notation;

(ii) if X and Y are well-formed formulae and $*$ is an operator, then $* XY$ is a well-formed formula.

34. Which of the following are well-formed formulae over the symbols $\{x, y, z\}$ and the set of binary operators $\{\times, +, \circ\}$?

a) $\times + + x\ y\ z$ **b)** $\circ\ x\ y \times x\ z$

c) $\times \circ x\ z \times \times x\ y$ **d)** $\times + \circ x\ x \circ x\ x\ x$

★35. Show that any well-formed formula in prefix notation over a set of symbols and a set of binary operators contains exactly one more symbol than the number of operators.

36. Give a definition of well-formed formulae in postfix notation over a set of symbols and a set of binary operators.

37. Give six examples of well-formed formulae with three or more operators in postfix notation over the set of symbols $\{x, y, z\}$ and the set of operators $\{+, \times, \circ\}$.

38. Extend the definition of well-formed formulae in prefix notation to sets of symbols and operators where the operators may not be binary.

8.4 Trees and Sorting

INTRODUCTION

The problem of ordering the elements in a set occurs in many contexts. For instance, to produce a printed telephone directory it is necessary to alphabetize the names of the subscribers.

Suppose that there is a total ordering of the elements of a set. Initially the elements in a set may be in any order. A **sorting** is a reordering of these elements into a list in which the elements are in increasing order. For instance, sorting the list 7, 2, 1, 4, 5, 9 produces the list 1, 2, 4, 5, 7, 9. Sorting the list d, h, c, a, f (using alphabetical order) produces the list a, c, d, f, h.

A large percentage of computer use is devoted to sorting one thing or another. Hence, much effort has been devoted to the development of efficient sorting algorithms. In this section several sorting algorithms and their computational complexity will be discussed. As will be seen in this section, trees are used to describe sorting algorithms and are used in the analysis of their complexity.

THE COMPLEXITY OF SORTING

Many different sorting algorithms have been developed. To decide whether a particular sorting algorithm is efficient, its complexity is determined. Using trees as models, a lower bound for the worst-case complexity of sorting algorithms can be found.

There are $n!$ possible orderings of n elements, since each of the $n!$ permutations of these elements can be the correct order. The sorting algorithms we will study are based on binary comparisons, that is, the comparison of two elements at a time. The result of each such comparison narrows down the set of possible orderings. Thus, a sorting algorithm based on binary comparisons can be represented by a binary decision tree in which each internal vertex represents a comparison of two elements. Each leaf represents one of the $n!$ permutations of n elements.

EXAMPLE 1 We display in Figure 1 a decision tree that orders the elements of the list a, b, c. ■

The complexity of a sort based on binary comparisons is measured in terms of the number of such comparisons used. The most binary comparisons ever needed to sort a list with n elements gives the worst-case performance of the algorithm. The most comparisons used equals the longest path length in the decision tree representing the sorting procedure. In other words, the most comparisons ever needed is equal to the height of the decision tree. Since the height of a binary tree with $n!$ leaves is at least $\lceil \log n! \rceil$ (using Corollary 1 in Section 8.1), at least $\lceil \log n! \rceil$ comparisons are needed, as stated in Theorem 1.

THEOREM 1 A sorting algorithm based on binary comparisons requires at least $\lceil \log n! \rceil$ comparisons.

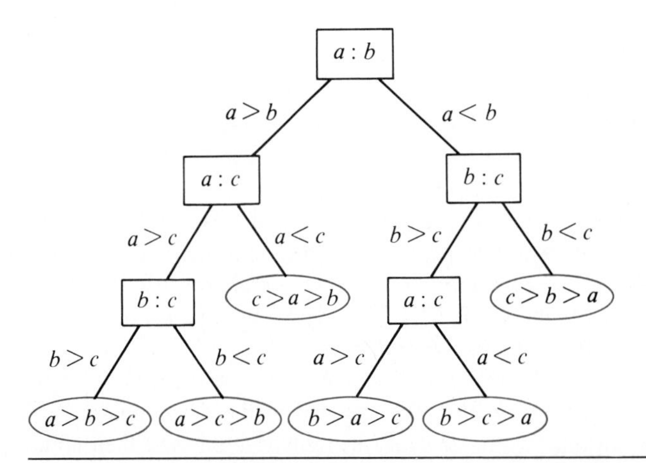

FIGURE 1 A Decision Tree for Sorting Three Distinct Elements.

From Example 6 of Section 1.9, it follows that $\lceil \log n! \rceil$ is $O(n \log n)$. In fact, since $\log n!$ is greater than $(n \log n)/4$ for $n > 4$ (see Exercise 18), it follows that no sorting algorithm that uses comparisons as the method of sorting can have worst-case time complexity that is better than $O(n \log n)$. Consequently, a sorting algorithm is as efficient as possible (in the sense of a big-O estimate of time complexity) if it has $O(n \log n)$ time complexity.

THE BUBBLE SORT

The **bubble sort** is one of the simplest sorting algorithms but not one of the most efficient. It puts a list into increasing order by successively comparing adjacent elements, interchanging them if they are in the wrong order. To carry out the bubble sort, we perform the basic operation, that is, interchanging a larger element with a smaller one following it, starting at the beginning of the list, for a full pass. We iterate this procedure until the sort is complete. We can imagine the elements in the list placed in a column. In the bubble sort, the smaller elements "bubble" to the top as they are interchanged with larger elements. The larger elements "sink" to the bottom. This is illustrated in the following example.

EXAMPLE 2 Use the bubble sort to put 3, 2, 4, 1, 5 into increasing order.

Solution: Begin by comparing the first two elements, 3 and 2. Since $3 > 2$, interchange 3 and 2, producing the list 2, 3, 4, 1, 5. Since $3 < 4$, continue by comparing 4 and 1. Since $4 > 1$, interchange 1 and 4, producing the list 2, 3, 1, 4, 5. Since $4 < 5$, the first pass is complete. The first pass guarantees that the largest element, 5, is in the correct position.

The second pass begins by comparing 2 and 3. Since these are in the correct order, 3 and 1 are compared. Since $3 > 1$, these numbers are interchanged, producing 2, 1, 3, 4, 5. Since $3 < 4$, these numbers are in the correct order. It is not necessary to do any more comparisons for this pass because 5 is already in the correct position. The second pass guarantees that the two largest elements, 4 and 5, are in their correct positions.

The third pass begins by comparing 2 and 1. These are interchanged since $2 > 1$, producing 1, 2, 3, 4, 5. Because $2 < 3$, these two elements are in the correct order. It is not necessary to do any more comparisons for this pass because 4 and 5 are already in the correct positions. The third pass guarantees that the three largest elements, 3, 4, and 5, are in their correct positions.

The fourth pass consists of one comparison, namely, the comparison of 1 and 2. Since $1 < 2$, these elements are in the correct order. This completes the bubble sort.

The steps of this algorithm are illustrated in Figure 2. ■

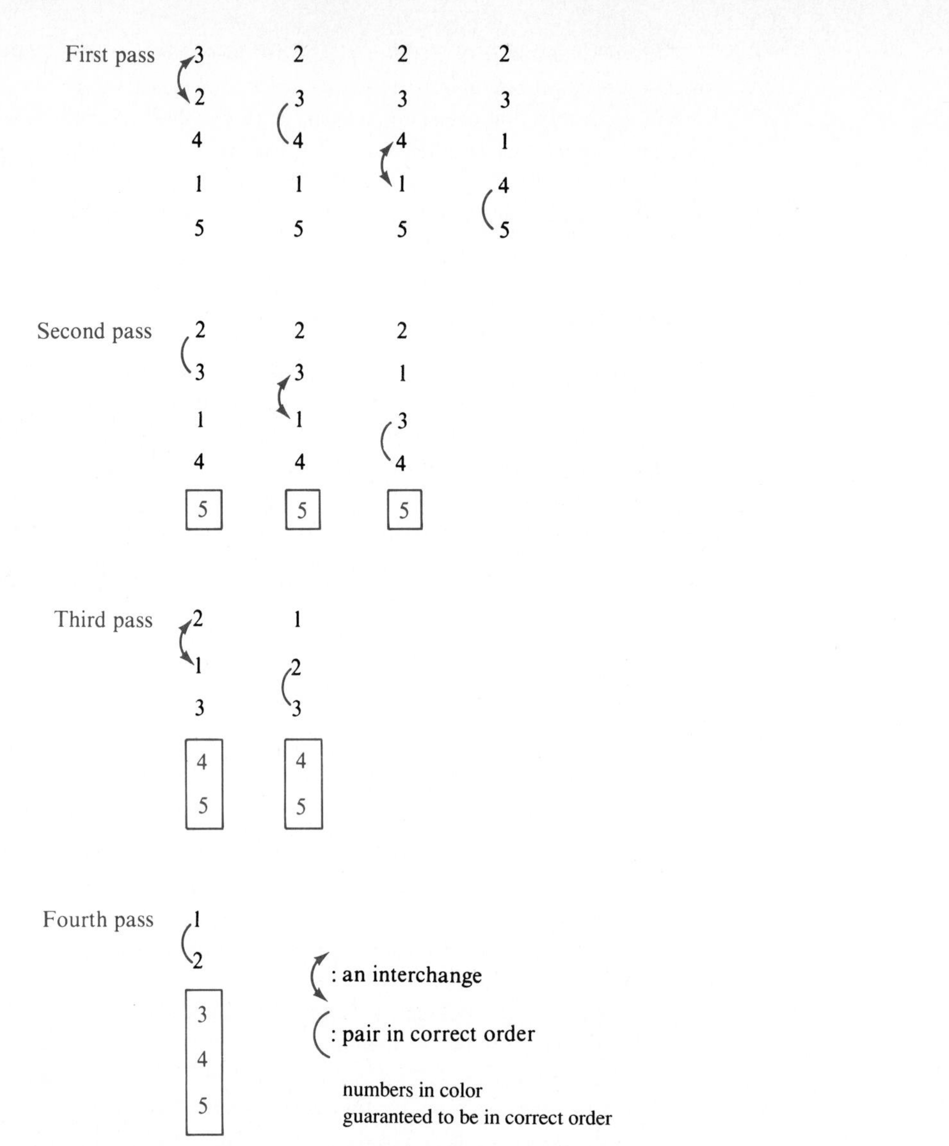

FIGURE 2 **The Steps of a Bubble Sort.**

A pseudocode description of the bubble sort is given in Algorithm 1.

How efficient is the bubble sort? Since $n - i$ comparisons are used during the ith pass, the total number of comparisons used in a bubble sort of a list of n elements is

$$(n - 1) + (n - 2) + \cdots + 2 + 1.$$

This is the sum of the $n - 1$ smallest integers. From Example 9 of Section 3.2, this equals $(n-1)n/2$. Consequently, the bubble sort uses $n(n-1)/2$ comparisons to order

ALGORITHM 1 The Bubble Sort.

procedure *bubblesort*$(a_1, \ldots, a_n)$
for $i := 1$ **to** $n - 1$
begin
  for $j := 1$ **to** $n - i$
    if $a_j > a_{j+1}$ **then** interchange a_j and a_{j+1}
end
$\{a_1, \ldots, a_n$ is in increasing order$\}$

a list of n elements. (Note that the bubble sort always uses this many comparisons, since it continues even if the list becomes completely sorted at some intermediate step.) Hence, the bubble sort algorithm has worst-case complexity $O(n^2)$. Since for every positive real number c, $n(n - 1)/2 > cn \log n$ for some sufficiently large positive integer n, it follows that the bubble sort does not have $O(n \log n)$ worst-case time complexity. We need to find another algorithm in order to achieve this optimal estimate of worst-case complexity.

THE MERGE SORT

Many different sorting algorithms achieve the best possible worst-case complexity for a sorting algorithm, namely, $O(n \log n)$ comparisons to sort n elements. We will describe one of these algorithms, called the **merge sort** algorithm, here. We will demonstrate how the merge sort algorithm works with an example before describing it in generality.

EXAMPLE 3 We will sort the list 8, 2, 4, 6, 9, 7, 10, 1, 5, 3 using the merge sort. A merge sort begins by splitting the list into individual elements by successively splitting lists in two. The progression of sublists for this example is represented with the balanced binary tree of height 4 shown in the upper half of Figure 3.

Sorting is done by successively merging pairs of lists. At the first stage, pairs of individual elements are merged into lists of length two in increasing order. Then successive merges of pairs of lists are performed until the entire list is put into increasing order. The succession of merged lists in increasing order is represented by the balanced binary tree of height 4 shown in the lower half of Figure 3 (note that this tree is displayed "upside down"). ■

In general, a merge sort proceeds by iteratively splitting lists into two sublists of equal length (or where one sublist has one more element than the other), until

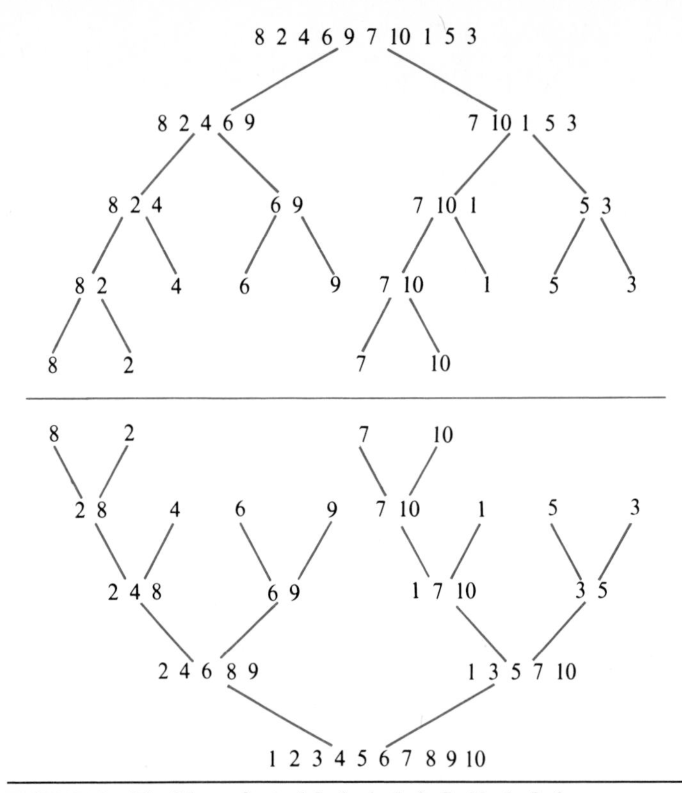

FIGURE 3 **The Merge Sort of 8, 2, 4, 6, 9, 7, 10, 1, 5, 3.**

each sublist contains one element. This succession of sublists can be represented by a balanced binary tree. The procedure continues by successively merging pairs of lists, where both lists are in increasing order, into a larger list with elements in increasing order, until the original list is put into increasing order. The succession of merged lists can be represented by a balanced binary tree.

We can also describe the merge sort recursively. To do a merge sort, we split a list into two sublists of equal, or approximately equal, size, sorting each sublist using the merge sort algorithm, and then merging the two lists. We leave it for the reader to give a complete specification of the recursive version of the merge sort.

An efficient algorithm for merging two ordered lists into a larger ordered list is needed to implement the merge sort. We will now describe such a procedure.

EXAMPLE 4 We will describe how to merge the two lists 2, 3, 5, 6 and 1, 4. Table 1 illustrates the steps we use.

TABLE 1 Merging the Lists 2, 3, 5, 6; and 1, 4.

First List	*Second List*	*Merged List*	*Comparison*
2 3 5 6	1 4		1 < 2
2 3 5 6	4	1	2 < 4
3 5 6	4	1 2	3 < 4
5 6	4	1 2 3	4 < 5
5 6		1 2 3 4	
		1 2 3 4 5 6	

First, compare the smallest elements in the two lists, 2 and 1, respectively. Since 1 is the smaller, put it at the beginning of the merged list and remove it from the second list. At this stage, the first list is 2, 3, 5, 6, the second is 4, and the combined list is 1.

Next, compare 2 and 4, the smallest elements of the two lists. Since 2 is the smaller, add it to the combined list and remove it from the first list. At this stage the first list is 3, 5, 6, the second is 4, and the combined list is 1, 2.

Continue by comparing 3 and 4, the smallest elements of their respective lists. Since 3 is the smaller of these two elements, add it to the combined list and remove it from the first list. At this stage the first list is 5, 6, and the second is 4. The combined list is 1, 2, 3.

Then compare 5 and 4, the smallest elements in the two lists. Since 4 is the smaller of these two elements, add it to the combined list and remove it from the second list. At this stage the first list is 5, 6, the second list is empty, and the combined list is 1, 2, 3, 4.

Finally, since the second list is empty, all elements of the first list can be appended to the end of the combined list in the order they occur in the first list. This produces the ordered list 1, 2, 3, 4, 5, 6. ■

We will now consider the general problem of merging two ordered lists L_1 and L_2, into an ordered list L. We can use the following procedure. Start with an empty list L. Compare the smallest elements of the two lists. Put the smaller of these two elements at the end of L, and remove it from the list it was in. Next, if one of L_1 and L_2 is empty, append the other (nonempty) list to L, which completes the merging. If neither L_1 nor L_2 is empty, repeat this process. Algorithm 2 gives a pseudocode description of this procedure.

ALGORITHM 2 Merging Two Lists.

procedure *merge*(L_1, L_2: lists)
$L :=$ empty list
while L_1 and L_2 are both nonempty
begin
  remove smaller of first element of L_1 and L_2 from the list it is in
    and put it at the end of L
  if removal of this element makes one list empty **then** remove all
    elements from the other list and append them to L
end {L is the merged list with elements in increasing order}

We will need estimates for the number of comparisons used to merge two ordered lists in the analysis of the merge sort. We can easily obtain such an estimate for Algorithm 2. Each time a comparison of an element from L_1 and an element from L_2 is made, an additional element is added to the merged list L. However, when either L_1 or L_2 is empty, no more comparisons are needed. Hence, Algorithm 2 is least efficient when $m + n - 2$ comparisons are carried out, leaving one element in each of L_1 and L_2. The next comparison will be the last one needed, because it will make one of these lists empty. Hence, Algorithm 2 uses no more than $m + n - 1$ comparisons. The following lemma summarizes this estimate.

LEMMA 1 Two sorted lists with m elements and n elements can be merged into a sorted list using no more than $m + n - 1$ comparisons.

Sometimes two sorted lists of length m and n can be merged using far fewer than $m + n - 1$ comparisons. For instance, when $m = 1$, a binary search procedure can be applied to put the one element in the first list into the second list. This requires only $\lceil \log n \rceil$ comparisons, which is much smaller than $m + n - 1 = n$, for $m = 1$. On the other hand, for some values of m and n, Lemma 1 gives the best possible bound. That is, there are lists with m and n elements that cannot be merged using fewer than $m + n - 1$ comparisons. (See Exercise 7 at the end of this section.)

We can now analyze the complexity of the merge sort. Instead of studying the general problem, we will assume that n, the number of elements in the list, is a power of 2, say 2^m. This will make the analysis less complicated, but when this is not the case, various modifications can be applied that will yield the same estimate.

At the first stage of the splitting procedure, the list is split into two sublists, of 2^{m-1} elements each, at level 1 of the tree generated by the splitting. This process continues, splitting the two sublists with 2^{m-1} elements into four sublists of 2^{m-2} elements each at level 2, and so on. In general, there are 2^{k-1} lists at level $k - 1$, each with 2^{m-k+1} elements. These lists at level $k - 1$ are split into 2^k lists at level k, each with 2^{m-k} elements. At the end of this process, we have 2^m lists each with one element at level m.

We start merging by combining pairs of the 2^m lists of one element into 2^{m-1} lists, at level $m-1$, each with two elements. To do this, 2^{m-1} pairs of lists with one element each are merged. The merger of each pair requires exactly one comparison.

The procedure continues, so that at level k ($k = m, m-1, m-2, \ldots, 3, 2, 1$), 2^k lists each with 2^{m-k} elements are merged into 2^{k-1} lists, each with 2^{m-k+1} elements, at level $k-1$. To do this a total of 2^{k-1} mergers of two lists, each with 2^{m-k} elements, are needed. But, by Lemma 1, each of these mergers can be carried out using at most $2^{m-k} + 2^{m-k} - 1 = 2^{m-k+1} - 1$ comparisons. Hence, going from level k to $k-1$ can be accomplished using at most $2^{k-1}(2^{m-k+1} - 1)$ comparisons. Summing all these estimates shows that the number of comparisons required for the merge sort is at most

$$\begin{aligned}\sum_{k=1}^{m} 2^{k-1}(2^{m-k+1} - 1) &= \sum_{k=1}^{m} 2^m - \sum_{k=1}^{m} 2^{k-1} \\ &= m2^m - (2^m - 1) \\ &= n \log n - n + 1,\end{aligned}$$

since $m = \log n$ and $n = 2^m$. (We evaluated $\sum_{k=1}^{m} 2^m$ by noting that it is the sum of m identical terms, each equal to 2^m. We evaluated $\sum_{k=1}^{m} 2^{k-1}$ using the formula for the sum of the terms of a geometric progression from Example 6 of Section 3.2.)

This analysis shows that the merge sort achieves the best possible big-O estimate for the number of comparisons needed by sorting algorithms, as stated in the following theorem.

THEOREM 2 The number of comparisons needed to merge-sort a list with n elements is $O(n \log n)$.

We describe another efficient algorithm, the quick sort, in the exercises.

Exercises

1. Use a bubble sort to sort 3, 1, 5, 7, 4, showing the lists obtained at each step.

2. Use a bubble sort to sort *d, f, k, m, a, b,* showing the lists obtained at each step.

★3. Adapt the bubble sort algorithm so that it stops when no interchanges are required. Express this more efficient version of the algorithm in pseudocode.

4. Use a merge sort to sort 4, 3, 2, 5, 1, 8, 7, 6. Show all the steps used by the algorithm.

5. Use a merge sort to sort *b, d, a, f, g, h, z, p, o, k.* Show all the steps used by the algorithm.

6. How many comparisons are required to merge the following pairs of lists using Algorithm 2?

a) 1, 3, 5, 7, 9; 2, 4, 6, 8, 10

b) 1, 2, 3, 4, 5; 6, 7, 8, 9, 10

c) 1, 5, 6, 7, 8; 2, 3, 4, 9, 10

7. Show that there are lists with m elements and n elements such that they cannot be merged into one sorted list using Algorithm 2 with fewer than $m + n - 1$ comparisons.

★8. What is the least number of comparisons needed to merge any two lists in increasing order into one list in increasing order when the number of elements in the two lists are

a) 1, 4? **b)** 2, 4? **c)** 3, 4? **d)** 4, 4?

The **selection sort** begins by finding the least element in the list. This element is moved to the front. Then the least element among the remaining elements is found and put into the second position. This procedure is repeated until the entire list has been sorted.

9. Sort the following lists using the selection sort.
a) 3, 5, 4, 1, 2
b) 5, 4, 3, 2, 1
c) 1, 2, 3, 4, 5

10. Write the selection sort algorithm in pseudocode.

11. How many comparisons are used to perform a selection sort of n items?

The **quick sort** is an efficient algorithm. To sort $a_1, a_2, \ldots, a_n$, this algorithm begins by taking the first element a_1 and forming two sublists, the first containing those elements that are less than a_1, in the order they arise, and the second containing those elements greater than a_1, in the order they arise. Then a_1 is put at the end of the first sublist. This procedure is repeated recursively for each sublist, until all sublists contain one item. The ordered list of n items is obtained by combining the sublists of one item in the order they occur.

12. Sort 3, 5, 7, 8, 1, 9, 2, 4, 6 using the quick sort.

13. Let $a_1, a_2, \ldots, a_n$ be a list of n distinct real numbers. How many comparisons are needed to form two sublists from this list, the first containing elements less than a_1 and the second containing elements greater than a_1?

14. Describe the quick sort algorithm using pseudocode.

15. What is the largest number of comparisons needed to order a list of four elements using the quick sort algorithm?

16. What is the least number of comparisons needed to order a list of four elements using the quick sort algorithm?

17. Determine the worst-case complexity of the quick sort algorithm in terms of the number of comparisons used.

★18. Show that $\log n!$ is greater than $(n \log n)/4$ for $n > 4$. (*Hint:* Begin with the inequality $n! > n(n-1)(n-2)\cdots\lceil n/2 \rceil$.)

★19. Write the merge sort algorithm in pseudocode.

8.5 Spanning Trees

INTRODUCTION

Consider the system of roads in Maine represented by the simple graph shown in Figure 1(a). The only way the roads can be kept open in the winter is by frequently plowing them. The highway department wants to plow the fewest roads so that there will always be cleared roads connecting any two towns. How can this be done?

At least five roads must be plowed to ensure that there is a path between any two towns. Figure 1(b) shows one such set of roads. Note that the subgraph representing these roads is a tree, since it is connected and contains six vertices and five edges.

This problem was solved with a connected subgraph with the minimum number of edges containing all vertices of the original simple graph. Such a graph must be a tree.

DEFINITION 1. Let G be a simple graph. A *spanning tree* of G is a subgraph of G that is a tree containing every vertex of G.

A simple graph with a spanning tree must be connected, since there is a path in the spanning tree between any two vertices. The converse is also true; that is, every connected simple graph has a spanning tree. We will give an example before proving this result.

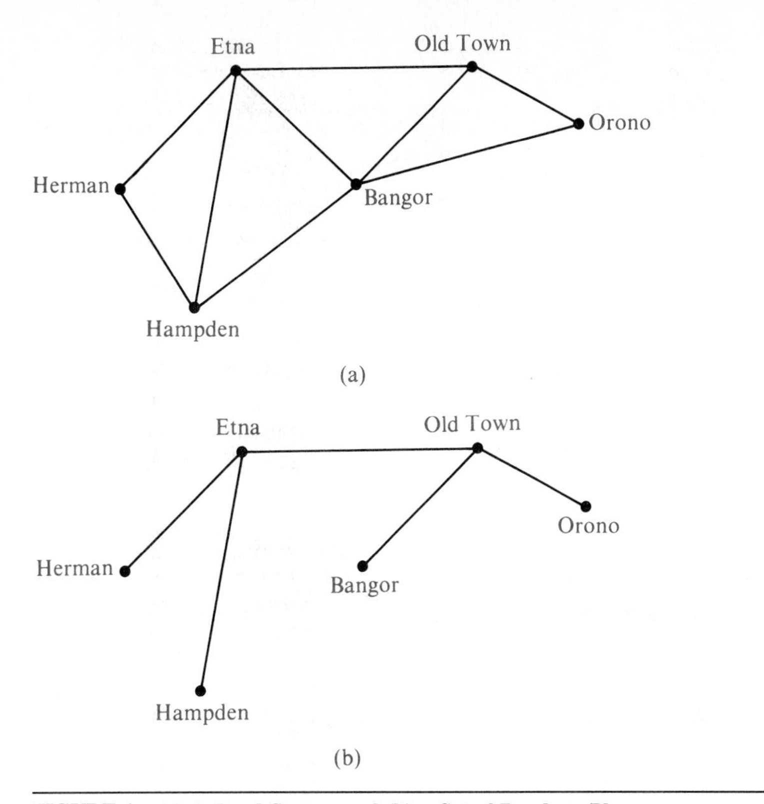

FIGURE 1 **(a) A Road System and (b) a Set of Roads to Plow.**

EXAMPLE 1 Find a spanning tree of the simple graph G shown in Figure 2.

Solution: The graph G is connected, but it is not a tree because it contains simple circuits. Remove the edge $\{a, e\}$. This eliminates one simple circuit, and the resulting subgraph is still connected and still contains every vertex of G. Next remove the edge $\{e, f\}$ to eliminate a second simple circuit. Finally, remove edge $\{c, g\}$ to produce a simple graph with no simple circuits. This subgraph is a spanning tree, since it is a tree that contains every vertex of G. The sequence of edge removals used to produce the spanning tree is illustrated in Figure 3.

The tree shown in Figure 3 is not the only spanning tree of G. For instance, each of the trees shown in Figure 4 is a spanning tree of G. ■

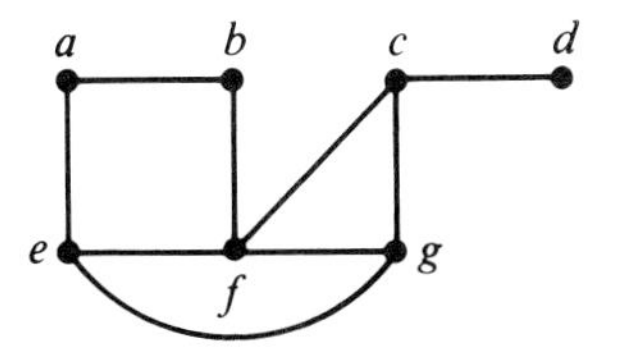

FIGURE 2 **The Simple Graph G.**

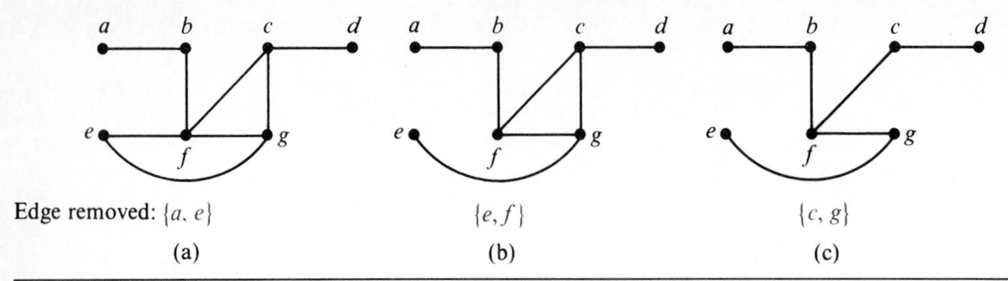

FIGURE 3 **Producing a Spanning Tree for G by Removing Edges that Form Simple Circuits.**

THEOREM 1 A simple graph is connected if and only if it has a spanning tree.

Proof: First, suppose that a simple graph G has a spanning tree T. T contains every vertex of G. Furthermore, there is a path in T between any two of its vertices. Since T is a subgraph of G, there is a path in G between any two of its vertices. Hence, G is connected.

Now suppose that G is connected. If G is not a tree, it must contain a simple circuit. Remove an edge from one of these simple circuits. The resulting subgraph has one fewer edge but still contains all the vertices of G and is connected. If this subgraph is not a tree, it has a simple circuit; so as before, remove an edge that is in a simple circuit. Repeat this process until no simple circuits remain. This is possible because there are only a finite number of edges in the graph. The process terminates when no simple circuits remain. A tree is produced since the graph stays connected as edges are removed. This tree is a spanning tree since it contains every vertex of G. ∎

ALGORITHMS FOR CONSTRUCTING SPANNING TREES

The proof of Theorem 1 gives an algorithm for finding spanning trees by removing edges from simple circuits. This algorithm is inefficient, since it requires that simple circuits be identified. Instead of constructing spanning trees by removing edges,

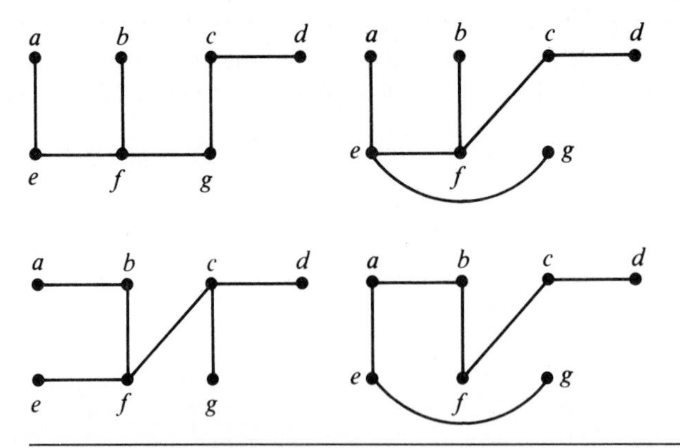

FIGURE 4 **Spanning Trees of G.**

spanning trees can be built up by successively adding edges. Two algorithms based on this principle will be presented here.

We can build a spanning tree for a connected simple graph using a **depth-first search.** We will form a rooted tree, and the spanning tree will be the underlying undirected graph of this rooted tree. Arbitrarily choose a vertex of the graph as the root. Form a path starting at this vertex by successively adding edges, where each new edge is incident with the last vertex in the path and a vertex not already in the path. Continue adding edges to this path as long as possible. If the path goes through all vertices of the graph, the tree consisting of this path is a spanning tree. However, if the path does not go through all vertices, more edges must be added. Move back to the next to last vertex in the path, and, if possible, form a new path starting at this vertex passing through vertices that were not already visited. If this cannot be done, move back another vertex in the path, that is, two vertices back in the path, and try again. Repeat this procedure, beginning at the last vertex visited, moving back up the path one vertex at a time, forming new paths that are as long as possible until no more edges can be added. Since the graph has a finite number of edges and is connected, this process ends with the production of a spanning tree. Each vertex that ends a path at a stage of the algorithm will be a leaf in the rooted tree, and each vertex where a path is constructed starting at this vertex will be an internal vertex. The reader should note the recursive nature of this procedure. Also, note that if the vertices in the graph are ordered, the choices of edges at each stage of the procedure are all determined. However, we will not always explicitly order the vertices of a graph.

Depth-first search is also called **backtracking,** since the algorithm returns to vertices previously visited to add paths. The following example illustrates backtracking.

EXAMPLE 2 Use a depth-first search to find a spanning tree for the graph G shown in Figure 5.

Solution: The steps used by a depth-first search to produce a spanning tree of G are shown in Figure 6. We arbitrarily start with the vertex f. A path is built by successively adding edges incident with vertices not already in the path, as long as this is possible. This produces a path f, g, h, k, j (note that other paths could have been built). Next, backtrack to k. There is no path beginning at k containing vertices not already visited. So we backtrack to h. Form the path h, i. Then backtrack to h, and then to f. From f build the path f, d, e, c, a. Then backtrack to c and form the path c, b. This produces the spanning tree. ■

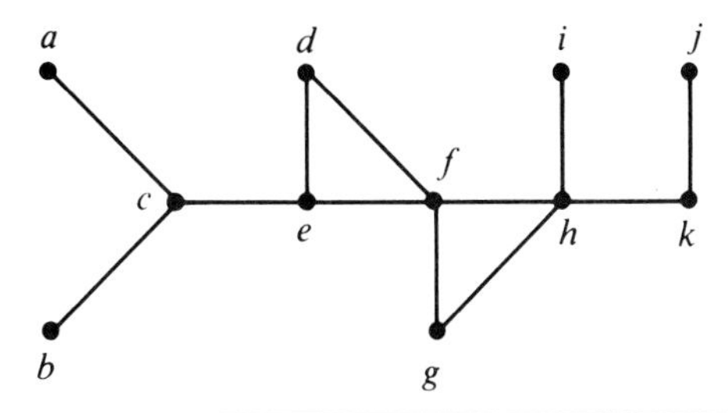

FIGURE 5 **The Graph G.**

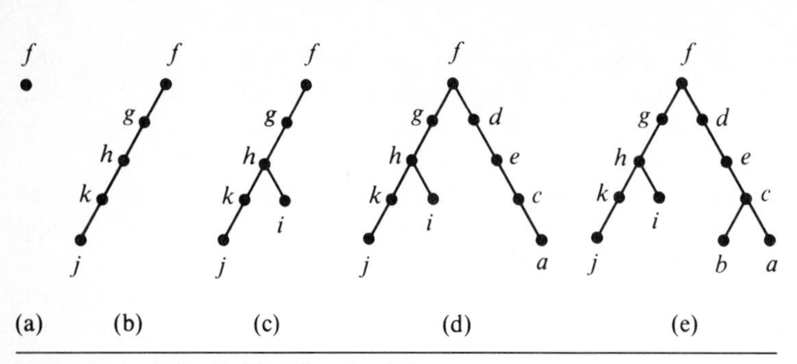

FIGURE 6 A Depth-First Search of G.

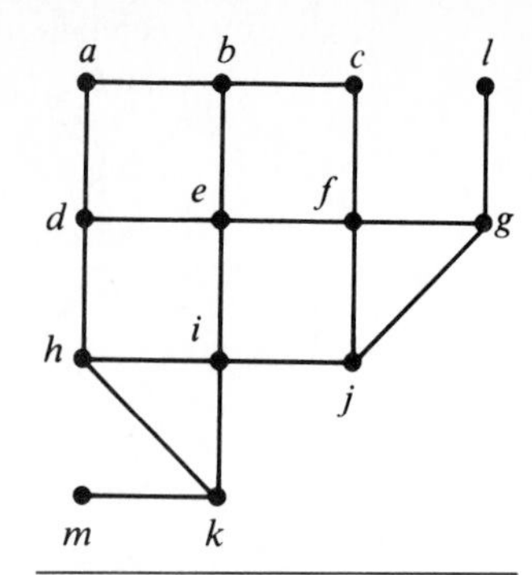

FIGURE 7 A Graph G.

We can also produce a spanning tree of a simple graph by the use of a **breadth-first search.** Again, a rooted tree will be constructed, and the underlying undirected graph of this rooted tree forms the spanning tree. Arbitrarily choose a root from the vertices of the graph. Then add all edges incident to this vertex. The new vertices added at this stage become the vertices at level 1 in the spanning tree. Arbitrarily order them. Next, for each vertex at level 1, visited in order, add each edge incident to this vertex to the tree as long as it does not produce a simple circuit. Arbitrarily order the children of each vertex at level 1. This produces the vertices at level 2 in the tree. Follow the same procedure until all the vertices in the tree have been added. The procedure ends since there are only a finite number of edges in the graph. A spanning tree is produced since we have produced a tree containing every vertex of the graph. An example of a breadth-first search follows.

EXAMPLE 3 Use a breadth-first search to find a spanning tree for the graph shown in Figure 7.

Solution: The steps of the breadth-first search procedure are shown in Figure 8. We choose the vertex e to be the root. Then we add edges incident with all vertices adjacent to e, so that edges from e to b, d, f, and i are added. These four vertices are at level 1 in the tree. Next, add the edges from these vertices at level 1 to adjacent vertices not already in the tree. Hence, the edges from b to a and c are added, as are

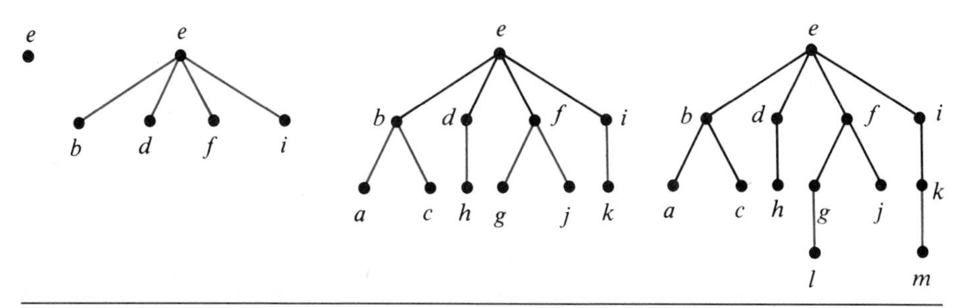

FIGURE 8 A Breadth-First Search of G.

edges from d to h, from f to j and g, and from i to k. The new vertices a, c, h, j, g, and k are at level 2. Next, add edges from these vertices to adjacent vertices not already in the graph. This adds edges from g to l and from k to m. ■

BACKTRACKING

There are problems that can be solved only by performing an exhaustive search of all possible solutions. One way to search systematically for a solution is to use a decision tree, where each internal vertex represents a decision and each leaf a possible solution. To find a solution via backtracking, first make a sequence of decisions in an attempt to reach a solution as long as this is possible. The sequence of decisions can be represented by a path in the decision tree. Once it is known that no solution can result from any further sequence of decisions, backtrack to the parent of the current vertex and work toward a solution with another series of decisions, if this is possible. The procedure continues until a solution is found, or it is established that no solution exists. The following examples illustrate the usefulness of backtracking.

EXAMPLE 4 Graph Colorings How can backtracking be used to decide whether a graph can be colored using n colors?

Solution: We can solve this problem using backtracking in the following way. First pick some vertex a and assign it color 1. Then pick a second vertex b, and if b is not adjacent to a, assign it color 1. Otherwise, assign color 2 to b. Then go on to a third vertex c. Use color 1, if possible, for c. Otherwise use color 2, if this is possible. Only if neither color 1 nor color 2 can be used should color 3 be used. Continue this process as long as it is possible to assign one of the n colors to each additional vertex, always using the first allowable color in the list. If a vertex is reached that cannot be colored by any of the n colors, backtrack to the last assignment made and change the coloring of the last vertex colored, if possible, using the next allowable color in the list. If it is not possible to change this coloring, backtrack further to previous assignments, one step back at a time, until is is possible to change a coloring of a vertex. Then continue assigning colors of additional vertices as long as possible. If a coloring using n colors exists, backtracking will produce it. (Unfortunately this procedure can be extremely inefficient.)

In particular, consider the problem of coloring the graph shown in Figure 9 with three colors. The tree shown in Figure 9 illustrates how backtracking can be used to construct a 3-coloring. In this procedure, red is used first, then blue, and finally green. This simple example can obviously be done without backtracking, but it is a good illustration of the technique.

In this tree, the initial path from the root, which represents the assignment of red to a, leads to a coloring with a red, b blue, c red, and d green. It is impossible to color e using any of the three colors when a, b, c, and d are colored in this way. So, backtrack to the parent of the vertex representing this coloring. Since no other color can be used for d, backtrack one more level. Then change the color of c to green. We obtain a coloring of the graph by then assigning red to d and green to e. ■

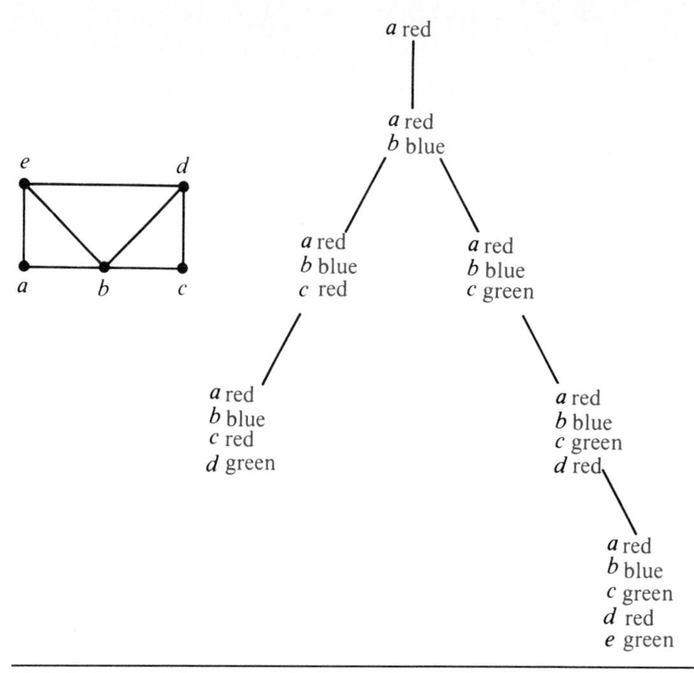

FIGURE 9 Coloring a Graph Using Backtracking.

EXAMPLE 5 The n-Queens Problem The n-queens problem asks how n queens can be placed on an $n \times n$ chessboard so that no two queens can attack one another. How can backtracking be used to solve the n-queens problem?

Solution: To solve this problem we must find n positions on an $n \times n$ chess board so that no two of these positions are in the same row, same column, or in the same diagonal (a diagonal consists of all positions (i, j) with $i + j = m$ for some m, or $i - j = m$ for some m). We will use backtracking to solve the n-queens problem. We start with an empty chessboard. At stage $k + 1$ we attempt putting an additional queen on the board in the $(k + 1)$st column, where there are already queens in the first k columns. We examine squares in the $(k + 1)$st column starting with the square in the first row, looking for a position to place this queen so that is not in the same row or on the same diagonal as a queen already on the board. (We already know it is not in the same column.) If it is impossible to find a position to place the queen in the $(k + 1)$st column, backtrack to the placement of the queen in the kth column, and place this queen in the next allowable row in this column, if such a row exists. If no such row exists, backtrack further.

In particular, Figure 10 displays a backtracking solution to the four-queens problem. In this solution, we place a queen in the first row and column. Then we put a queen in the third row of the second column. However, this makes it impossible to place a queen in the third column. So we backtrack and put a queen in the fourth row

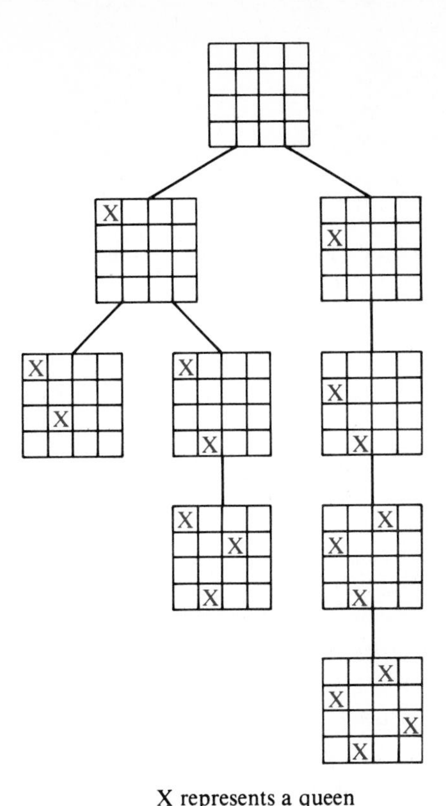

X represents a queen

FIGURE 10 A Backtracking Solution of the Four-Queens Problem.

of the second column. When we do this, we can place a queen in the second row of the third column. But there is no way to add a queen to the fourth column. This shows that no solution results when a queen is placed in the first row and column. We backtrack to the empty chessboard, and place a queen in the second row of the first column. This leads to a solution as shown in Figure 10. ■

EXAMPLE 6 Sums of Subsets Consider the following problem. Given a set of positive integers $x_1, x_2, \ldots, x_n$, find a subset of this set of integers that has M as its sum. How can backtracking be used to solve this problem?

Solution: We start with a sum with no terms. We build up the sum by successively adding terms. An integer in the sequence is included if the sum remains less than M when this integer is added to the sum. If a sum is reached so that the addition of any term is greater than M, backtrack by dropping the last term of the sum.

Figure 11 displays a backtracking solution to the problem of finding a subset of $\{31, 27, 15, 11, 7, 5\}$ with sum equal to 39. ■

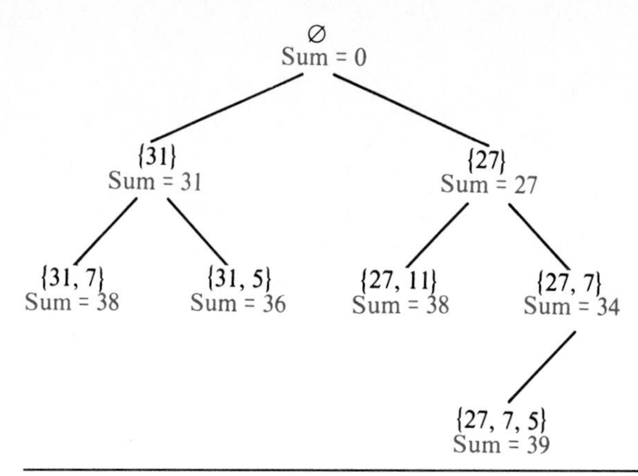

FIGURE 11 Find a Sum Equal to 39 Using Backtracking.

Exercises

1. How many edges must be removed from a connected graph with n vertices and m edges to produce a spanning tree?

In Exercises 2–4 find a spanning tree for the graph shown by removing edges in simple circuits.

2.

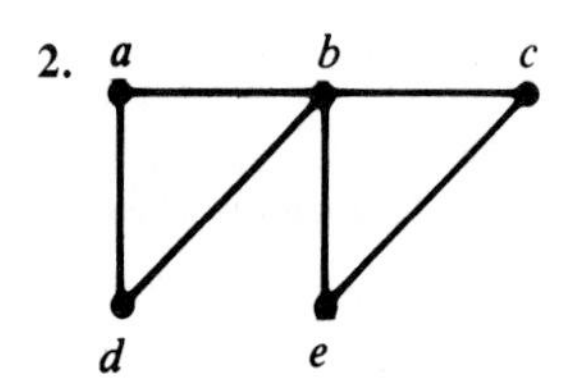

3.

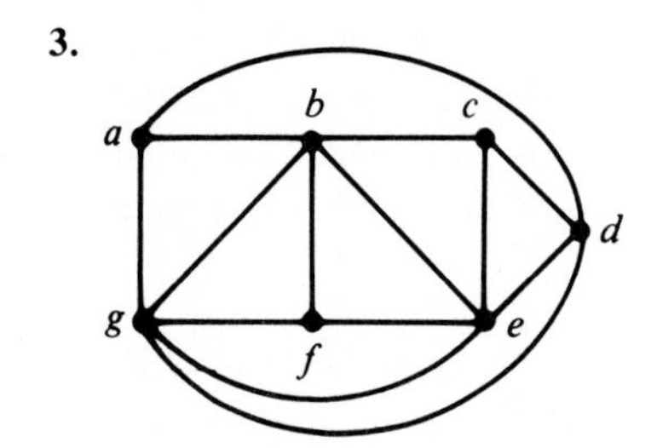

4.

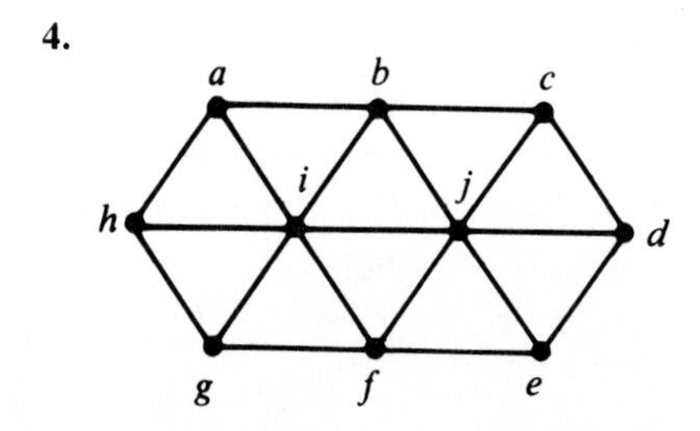

5. Find a spanning tree for each of the following graphs.
a) K_5 **b)** $K_{4,4}$ **c)** $K_{1,6}$ **d)** Q_3 **e)** C_5 **f)** W_5

In Exercises 6–8 draw all the spanning trees of the given simple graphs.

6.

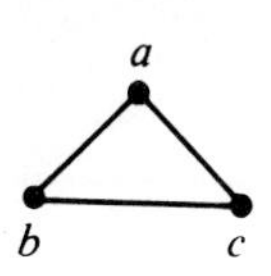

7.

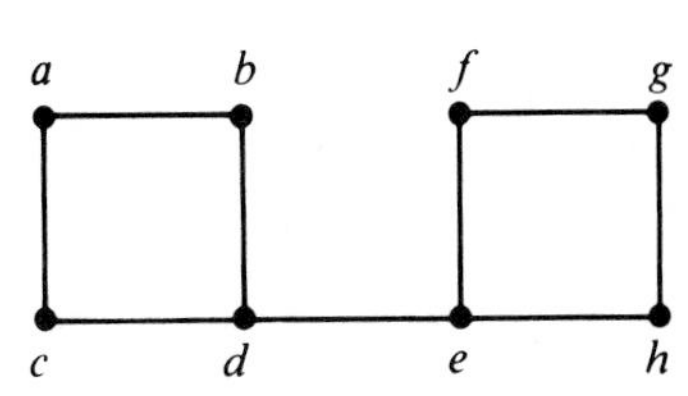

8.

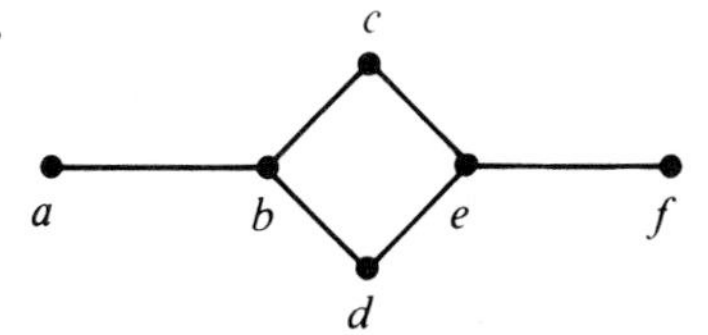

★9. How many different spanning trees does each of the following simple graphs have?
a) K_3 **b)** K_4 **c)** $K_{2,2}$ **d)** C_5

★10. How many nonisomorphic spanning trees does each of the following simple graphs have?
a) K_3 **b)** K_4 **c)** K_5

In Exercises 11–13 use a depth-first search to produce a spanning tree for the given simple graph. Choose a as the root of this spanning tree and assume that the vertices are ordered alphabetically.

11.

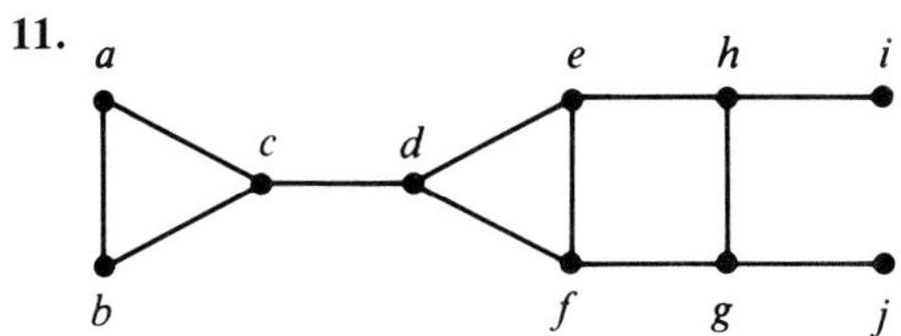

12.

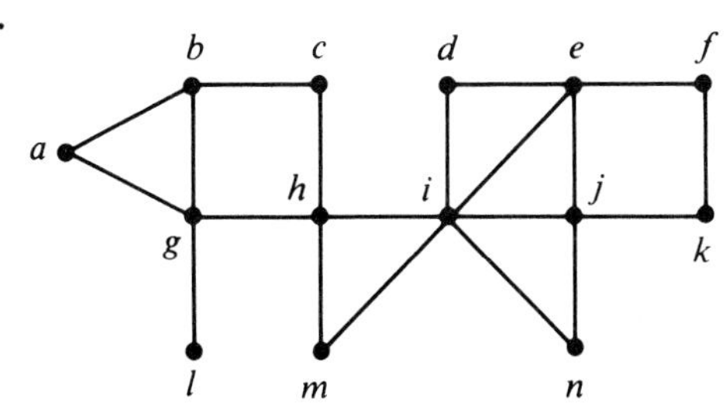

13.

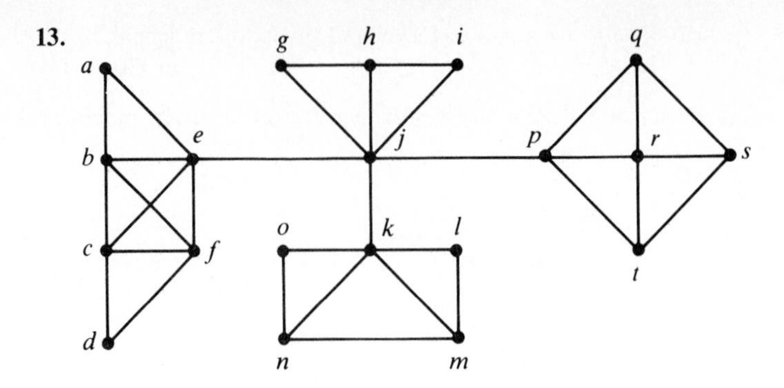

14. Use a breadth-first search to produce a spanning tree for each of the simple graphs in Exercises 11–13. Choose a as the root of each spanning tree.

15. Suppose that an airline must reduce its flight schedule to save money. If its original routes are as illustrated, which flights can be discontinued to retain service between all pairs of cities (where it may be necessary to combine flights to fly from one city to another)?

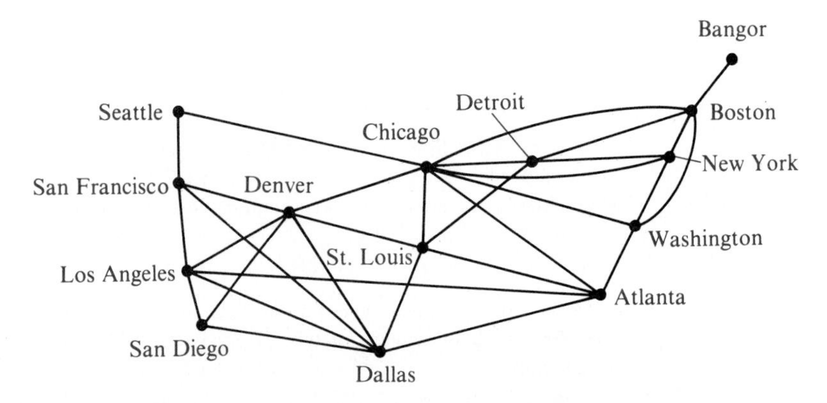

16. When must an edge of a connected simple graph be in every spanning tree for this graph?

17. Which connected simple graphs have exactly one spanning tree?

18. Explain how a breadth-first search or a depth-first search can be used to order the vertices of a connected graph.

★19. Write the depth-first search procedure in pseudocode.

★20. Write the breadth-first search procedure in pseudocode.

★21. Show that the length of the shortest path between vertices v and u in a connected simple graph equals the level number of u in the breadth-first spanning tree of G with root v.

22. Use backtracking to try to find a coloring of each of the graphs in Exercises 5–7 of Section 7.8 using three colors.

23. Use backtracking to solve the n-queens problem for the following values of n.

a) $n = 3$ **b)** $n = 5$ **c)** $n = 6$

24. Use backtracking to find a subset, if it exists, of the set $\{27, 24, 19, 14, 11, 8\}$ with sum

a) 20. **b)** 41. **c)** 60.

25. Explain how backtracking can be used to find a Hamilton path or circuit in a graph.

26. a) Explain how backtracking can be used to find the way out of a maze, given a starting position and the exit position. Consider the maze divided into positions, where at each position the set of available moves includes one to four possibilities (up, down, right, left).

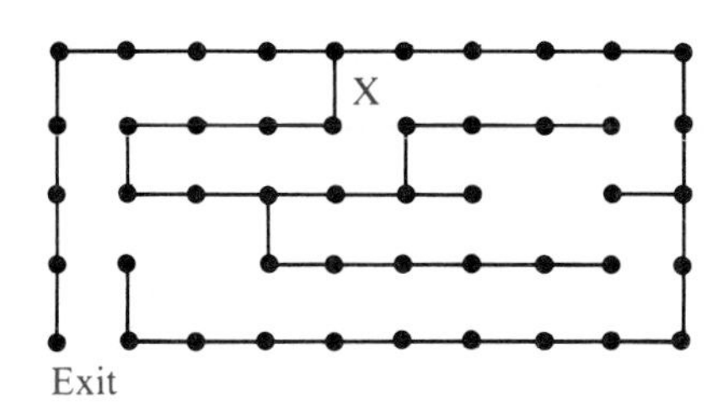

A **spanning forest** of graph G is a forest that contains every vertex of G such that two vertices are in the same tree of the forest when there is a path in G between these two vertices.

27. Show that every finite simple graph has a spanning forest.

28. How many trees are in the spanning forest of a graph?

29. How many edges must be removed to produce the spanning forest of a graph with n vertices, m edges, and c connected components?

30. Devise an algorithm for constructing the spanning forest of a graph based on deleting edges that form simple circuits.

31. Devise an algorithm for constructing the spanning forest of a graph based on depth-first searching.

32. Devise an algorithm for constructing the spanning forest of a graph based on breadth-first searching.

Let T_1 and T_2 be spanning trees of a graph. The **distance** between T_1 and T_2 is the number of edges that are in exactly one of the trees.

33. Find the distance between each pair of spanning trees shown in Figures 3 and 4 of the graph G shown in Figure 2.

★34. Suppose that T_1, T_2, and T_3 are spanning trees of the simple graph G. Show that the distance between T_1 and T_3 does not exceed the sum of the distance between T_1 and T_2 and the distance between T_2 and T_3.

★★35. Suppose that T_1 and T_2 are spanning trees of a simple graph G. Moreover, suppose that e_1 is an edge in T_1 that is not in T_2. Show that there is an edge e_2 in T_2 that is not in T_1 such that T_1 remains a spanning tree if e_1 is removed from it and e_2 is added to it, and T_2 remains a spanning tree if e_2 is removed from it and e_1 is added to it.

★36. Show that is possible to find a sequence of spanning trees leading from any spanning tree to any other by successively removing one edge and adding another.

A **rooted spanning tree** of a directed graph is a rooted tree containing edges of the graph such that every vertex of the graph is an endpoint of one of the edges in the tree.

37. For each of the directed graphs in Exercises 24–28 of Section 7.5 either find a rooted spanning tree of the graph or determine that no such tree exists.

★38. Show that a connected directed graph in which each vertex has the same in-degree and out-degree has a rooted spanning tree. (*Hint:* Use an Euler circuit.)

★39. Give an algorithm to build a rooted spanning tree for connected directed graphs in which each vertex has the same in-degree and out-degree.

8.6 Minimal Spanning Trees

INTRODUCTION

A company plans to build a communications network connecting its five computer centers. Any pair of these centers can be linked with a leased telephone line. Which links should be made to ensure that there is a path between any two computer centers so that the total cost of the network is minimized? We can model this problem using the weighted graph shown in Figure 1, where vertices represent computer centers, edges represent possible leased lines, and the weights on edges are the monthly lease rates of the lines represented by the edges. We can solve this problem by finding a spanning tree so that the sum of the weights of the edges of the tree is minimized. Such a spanning tree is called a **minimal spanning tree.**

ALGORITHMS FOR MINIMAL SPANNING TREES

A wide variety of problems are solved by finding a spanning tree in a weighted graph such that the sum of the weights of the edges in the tree is a minimum.

DEFINITION 1. A *minimal spanning tree* in a connected weighted graph is a spanning tree that has the smallest possible sum of weights of its edges.

We will present two algorithms for constructing minimal spanning trees. Both proceed by successively adding edges of smallest weight from those edges with a specified property that have not already been used. These algorithms are examples of **greedy algorithms.** A greedy algorithm is a procedure that makes an optimal choice at each of its steps. Optimizing at each stage of an algorithm does not guarantee that the optimal overall solution is produced. However, the two algorithms presented in this section for constructing minimal spanning trees are greedy algorithms that do produce optimal solutions.

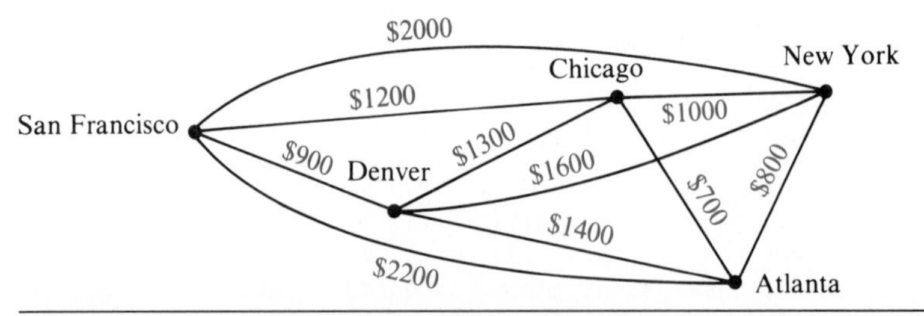

FIGURE 1 A Weighted Graph Showing Monthly Lease Costs for Lines in a Computer Network.

The first algorithm that we will discuss was given by Robert Prim in 1957. To carry out **Prim's algorithm,** begin by choosing any edge with smallest weight, putting it into the spanning tree. Successively add to the tree edges of minimum weight that are incident to a vertex already in the tree and not forming a simple circuit with those edges already in the tree. Stop when $n - 1$ edges have been added.

Later in this section, we will prove that this algorithm produces a minimal spanning tree for any connected weighted graph. Algorithm 1 gives a pseudocode description of Prim's algorithm.

ALGORITHM 1 Prim's Algorithm.

procedure *Prim*(G: weighted connected undirected graph with n vertices)
$T :=$ a minimum-weight edge
for $i := 1$ **to** $n - 2$
begin
  $e :=$ an edge of minimum weight incident to a vertex in T and not forming a simple circuit in T if added to T
  $T := T$ with e added
end {T is a minimal spanning tree of G}

Note that the choice of an edge to add at a stage of the algorithm is not determined when there is more than one edge with the same weight that satisfies the appropriate criteria. We need to order the edges to make the choices deterministic. We will not worry about this in the remainder of the section. Also note that there may be more than one minimal spanning tree for a given connected weighted simple graph. (See Exercise 9.) The following examples illustrate how Prim's algorithm is used.

EXAMPLE 1 Use Prim's algorithm to design a minimum-cost communications network connecting all the computers represented by the graph in Figure 1.

Solution: We solve this problem by finding a minimal spanning tree in the graph in Figure 1. Prim's algorithm is carried out by choosing an initial edge of minimum weight and successively adding edges of minimum weight that are incident to a vertex in the tree and that do not form simple circuits. The edges in color in Figure 2 show a minimal spanning tree produced by Prim's algorithm, with the choice made at each step displayed. ■

Robert Clay Prim (born 1921). Robert Prim, born at Sweetwater, Texas, received his B.S. in Electrical Engineering in 1941 and his Ph.D. in Mathematics from Princeton University in 1949. He was an engineer at the General Electric Company from 1941 until 1944, an engineer and mathematician at the United States Naval Ordnance Lab from 1944 until 1949, and a research associate at Princeton University from 1948 until 1949. Among the other positions he has held are: director of mathematics and mechanics research at Bell Telephone Laboratories from 1958 until 1961 and vice president of research at Sandia Corporation. He is currently retired.

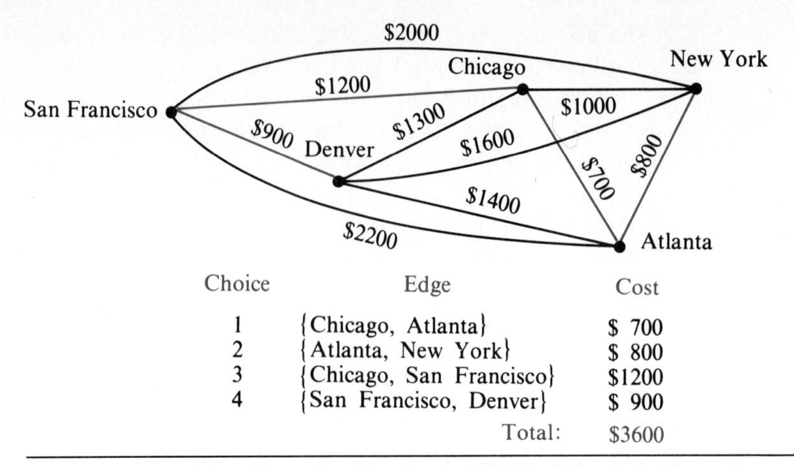

Choice	Edge	Cost
1	{Chicago, Atlanta}	$ 700
2	{Atlanta, New York}	$ 800
3	{Chicago, San Francisco}	$1200
4	{San Francisco, Denver}	$ 900
	Total:	$3600

FIGURE 2 A Minimal Spanning Tree for the Weighted Graph in Figure 1.

EXAMPLE 2 Use Prim's algorithm to find a minimal spanning tree in the graph shown in Figure 3.

Solution: A minimal spanning tree constructed using Prim's algorithm is shown in Figure 4. The successive edges chosen are displayed. ■

The second algorithm we will discuss was discovered by Joseph Kruskal in 1956. To carry out **Kruskal's algorithm,** choose an edge in the graph with minimum weight.

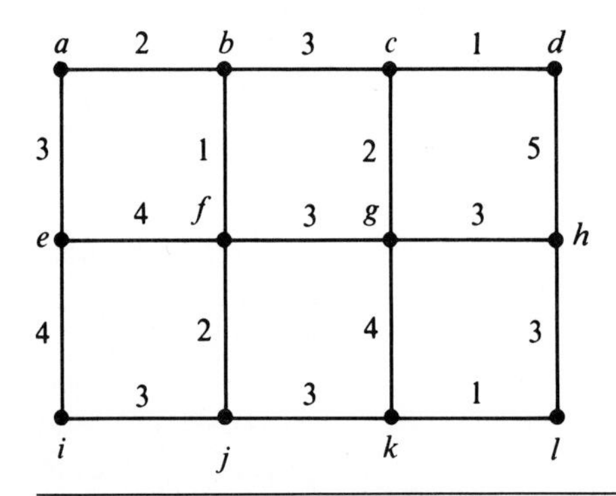

FIGURE 3 A Weighted Graph.

Joseph Bernard Kruskal (born 1928). Joseph Kruskal, born in New York City, attended the University of Chicago and received his Ph.D. from Princeton University in 1954. He was an instructor in mathematics at Princeton and at the University of Wisconsin, and later he was an assistant professor at the University of Michigan. In 1959 he became a member of the technical staff at Bell Laboratories, a position he continues to hold. His current research interests include statistical linguistics and psychometrics. Besides his work on minimal spanning trees, Kruskal is also known for contributions to multidimensional scaling. Kruskal discovered his algorithm for producing minimal spanning trees when he was a second-year graduate student. He was not sure his two-and-a-half-page paper on this subject was worthy of publication but was convinced by others to submit it.

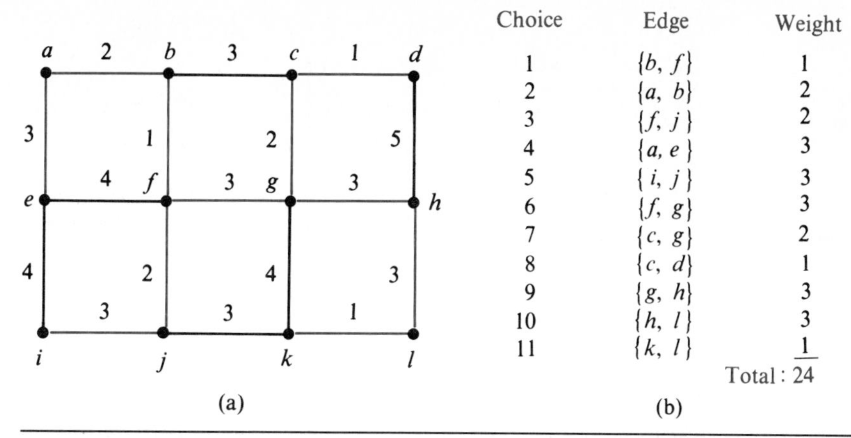

Choice	Edge	Weight
1	$\{b, f\}$	1
2	$\{a, b\}$	2
3	$\{f, j\}$	2
4	$\{a, e\}$	3
5	$\{i, j\}$	3
6	$\{f, g\}$	3
7	$\{c, g\}$	2
8	$\{c, d\}$	1
9	$\{g, h\}$	3
10	$\{h, l\}$	3
11	$\{k, l\}$	1
		Total: 24

(b)

FIGURE 4 A Minimal Spanning Tree Produced Using Prim's Algorithm.

Successively add edges with minimum weight that do not form a simple circuit with those edges already chosen. Stop after $n - 1$ edges have been selected.

The proof that Kruskal's algorithm produces a minimal spanning tree for every connected weighted graph is left as an exercise at the end of this section. Pseudocode for Kruskal's algorithm is given in Algorithm 2.

ALGORITHM 2 Kruskal's Algorithm.

procedure *Kruskal*(G: weighted connected undirected graph with n
 vertices)
$T :=$ empty graph
for $i := 1$ **to** $n - 1$
begin
 $e :=$ any edge in G with smallest weight that does not form a
 simple circuit when added to T
 $T := T$ with e added
end $\{T$ is a minimal spanning tree of $G\}$

The reader should note the difference between Prim's and Kruskal's algorithms. In Prim's algorithm edges of minimum weight that are incident to a vertex already in the tree, and not forming a circuit, are chosen; whereas, in Kruskal's algorithm edges of minimum weight that are not necessarily incident to a vertex already in the tree, and that do not form a circuit, are chosen. Note that as in Prim's algorithm, if the edges

Historical Note: Joseph Kruskal and Robert Prim developed their algorithms for constructing minimal spanning trees in the mid-1950s. However, they were not the first people to discover such algorithms. For example, the work of the anthropologist Jan Czekanowski, in 1909, contains many of the ideas required to find minimal spanning trees. In 1926, Otakar Boruvka described methods for constructing minimal spanning trees in work relating to the construction of electric power networks.

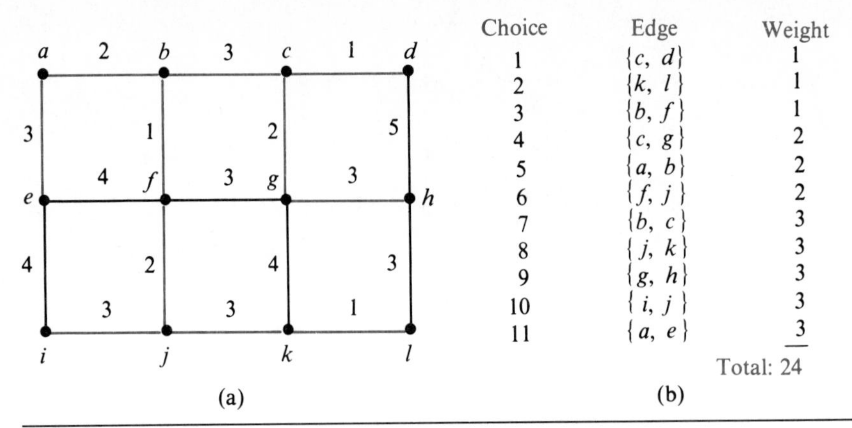

Choice	Edge	Weight
1	$\{c, d\}$	1
2	$\{k, l\}$	1
3	$\{b, f\}$	1
4	$\{c, g\}$	2
5	$\{a, b\}$	2
6	$\{f, j\}$	2
7	$\{b, c\}$	3
8	$\{j, k\}$	3
9	$\{g, h\}$	3
10	$\{i, j\}$	3
11	$\{a, e\}$	3
		Total: 24

FIGURE 5 A Minimal Spanning Tree Produced by Kruskal's Algorithm.

are not ordered, there may be more than one choice for the edge to add at a stage of this procedure. Consequently, the edges need to be ordered for the procedure to be deterministic. The following example illustrates how Kruskal's algorithm is used.

EXAMPLE 3 Use Kruskal's algorithm to find a minimal spanning tree in the weighted graph shown in Figure 3.

Solution: A minimal spanning tree and the choices of edges at each stage of Kruskal's algorithm are shown in Figure 5. ■

We will now prove that Prim's algorithm produces a minimal spanning tree of a connected weighted graph.

Proof: Let G be a connected weighted graph. Suppose that the successive edges chosen by Prim's algorithm are $e_1, e_2, \ldots, e_{n-1}$. Let S be the tree with $e_1, e_2, \ldots, e_{n-1}$ as its edges, and let S_k be the tree with $e_1, e_2, \ldots, e_k$ as its edges. Let T be a minimal spanning tree of G containing the edges $e_1, e_2, \ldots, e_k$, where k is the maximum integer with the property that a minimal spanning tree exists containing the first k edges chosen by Prim's algorithm. The theorem follows if we can show that $S = T$.

Suppose that $S \neq T$, so that $k < n - 1$. Consequently, T contains $e_1, e_2, \ldots, e_k$, but not e_{k+1}. Consider the graph made up of T together with e_{k+1}. Since this graph is connected and has n edges, too many edges to be a tree, it must contain a simple circuit. This simple circuit must contain e_{k+1} since there was no simple circuit in T. Furthermore, there must be an edge in the simple circuit that does not belong to S_{k+1} since S_{k+1} is a tree. By starting at an endpoint of e_{k+1} that is also an endpoint of one of the edges $e_1, \ldots, e_k$, and following the circuit until it reaches an edge not in S_{k+1}, we can find an edge e not in S_{k+1} that has an endpoint that is also an endpoint of one of the edges $e_1, e_2, \ldots, e_k$. By deleting e from T and adding e_{k+1}, we obtain a tree

T' with $n - 1$ edges (it is a tree since it has no simple circuits). Note that the tree T' contains $e_1, e_2, \ldots, e_k, e_{k+1}$. Furthermore, since e_{k+1} was chosen by Prim's algorithm at the $(k + 1)$st step, and e was also available at that step, the weight of e_{k+1} is less than or equal to the weight of e. From this observation it follows that T' is also a minimal spanning tree, since the sum of the weights of its edges does not exceed the sum of the weights of the edges of T. This contradicts the choice of k as the maximum integer so that a minimal spanning tree exists containing $e_1, \ldots, e_k$. Hence, $k = n - 1$, and $S = T$. It follows that Prim's algorithm produces a minimal spanning tree. ■

Exercises

1. The roads represented by the following graph are all unpaved. The lengths of the roads between pairs of towns are represented by edge weights. Which roads should be paved so that there is a path of paved roads between each pair of towns so that a minimum road length is paved? (*Note:* These towns are in Nevada.)

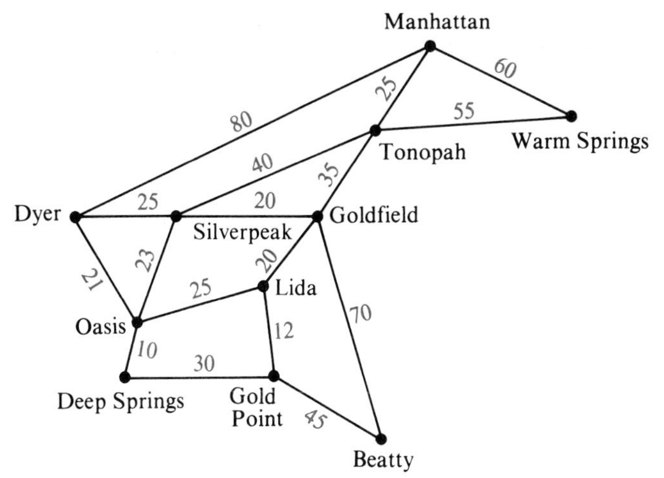

In Exercises 2–4 use Prim's algorithm to find a minimal spanning tree for the given weighted graph.

2.

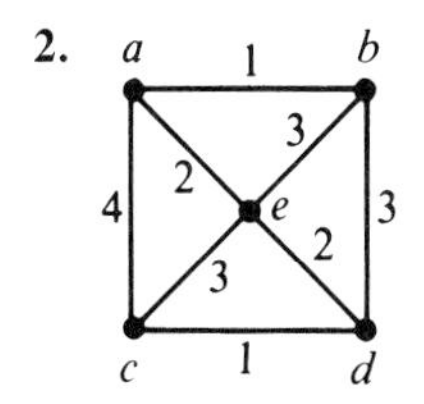

3.

4.

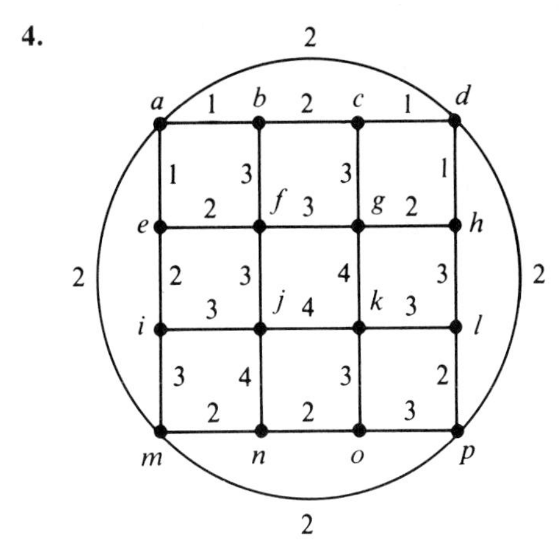

5. Use Kruskal's algorithm to design the communications network described at the beginning of the section.
6. Use Kruskal's algorithm to find a minimal spanning tree for the weighted graph in Exercise 2.
7. Use Kruskal's algorithm to find a minimal spanning tree for the weighted graph in Exercise 3.

8. Use Kruskal's algorithm to find a minimal spanning tree for the weighted graph in Exercise 4.

9. Find a connected weighted simple graph with the fewest edges possible that has more than one minimal spanning tree.

10. A **minimal spanning forest** in a weighted graph is a spanning forest with minimal weight. Explain how Prim's and Kruskal's algorithms can be adapted to construct minimal spanning forests.

A **maximal spanning tree** of a connected weighted undirected graph is a spanning tree with the largest possible weight.

11. Devise an algorithm similar to Prim's algorithm for constructing a maximal spanning tree of a connected weighted graph.

12. Devise an algorithm similar to Kruskal's algorithm for constructing a maximal spanning tree of a connected weighted graph.

13. Find a maximal spanning tree for the weighted graph in Exercise 2.

14. Find a maximal spanning tree for the weighted graph in Exercise 3.

15. Find a maximal spanning tree for the weighted graph in Exercise 4.

16. Find the second least expensive communications network connecting the five computer centers in the problem posed at the beginning of the section.

★17. Devise an algorithm for finding the second shortest spanning tree in a connected weighted graph.

★18. Show that an edge with smallest weight in a connected weighted graph must be part of any minimal spanning tree.

19. Show that there is a unique minimal spanning tree in a connected weighted graph if the weights of the edges are all different.

20. Suppose that the computer network connecting the cities in Figure 1 must contain a direct link between New York and Denver. What other links should be included so that there is a link between every two computer centers and the cost is minimized?

21. Find a spanning tree with minimal total weight containing the edges $\{e, i\}$ and $\{g, k\}$ in the weighted graph in Figure 3.

22. Describe an algorithm for finding a spanning tree with minimal weight containing a specified set of edges in a connected weighted undirected simple graph.

23. Express the algorithm devised in Exercise 22 in pseudocode.

Sollin's algorithm produces a minimal spanning tree from a connected weighted simple graph $G = (V, E)$ by successively adding groups of edges. Suppose that the vertices in V are ordered. This produces an ordering of the edges where $\{u_0, v_0\}$ precedes $\{u_1, v_1\}$ if u_0 precedes u_1 or if $u_0 = u_1$ and v_0 precedes v_1. The algorithm begins by simultaneously choosing the edge of least weight incident to each vertex. The first edge in the ordering is taken in the case of ties. This produces a graph with no simple circuits, that is, a forest of trees (Exercise 24 asks for a proof of this fact). Next, simultaneously choose for each tree in the forest the shortest edge between a vertex in this tree and a vertex in a different tree. Again the first edge in the ordering is chosen in the case of ties. (This produces a graph with no simple circuits containing fewer trees than were present before this step; see Exercise 24.) Continue the process of simultaneously adding edges connecting trees until $n - 1$ edges have been chosen. At this stage a minimal spanning tree has been constructed.

★24. Show that the addition of edges at each stage of Sollin's algorithm produces a forest.

25. Use Sollin's algorithm to produce a minimal spanning tree for the weighted graph shown in
a) Figure 1. **b)** Figure 3.

★26. Express Sollin's algorithm in pseudocode.

★★27. Prove that Sollin's algorithm produces a minimal spanning tree in a connected undirected weighted graph.

★28. Show that the first step of Sollin's algorithm produces a forest containing at least $\lceil n/2 \rceil$ edges.

★29. Show that if there are r trees in the forest at some intermediate step of Sollin's algorithm, then at least $\lceil r/2 \rceil$ edges are added by the next iteration of the algorithm.

★30. Show that no more than $\lfloor n/2^k \rfloor$ trees remain after the first step of Sollin's algorithm has been carried out and the second step of the algorithm has been carried out $k - 1$ times.

★31. Show that Sollin's algorithm requires at most $\log n$ iterations to produce a minimal spanning tree from a connected undirected weighted graph with n vertices.

32. Prove that Kruskal's algorithm produces minimal spanning trees.

Key Terms and Results

TERMS

tree: a connected undirected graph with no simple circuits
forest: an undirected graph with no simple circuits
rooted tree: a directed graph with a specified vertex, called the root, such that there is a unique path to any other vertex from this root
subtree: a subgraph of a tree that is also a tree
parent of v in a rooted tree: the vertex u such that (u,v) is an edge of the rooted tree
child of a vertex v in a rooted tree: any vertex with v as its parent
sibling of a vertex v in a rooted tree: a vertex with the same parent as v
ancestor of a vertex v in a rooted tree: any vertex on the path from the root to v
descendant of a vertex v in a rooted tree: any vertex that has v as an ancestor
internal vertex: a vertex that has children
leaf: a vertex with no children
level of a vertex: the length of the path from the root to this vertex
height of a tree: the largest level of the vertices of a tree
m-ary tree: a tree with the property that every internal vertex has no more than m children
full m-ary tree: a tree with the property that every internal vertex has exactly m children
binary tree: an m-ary tree with $m = 2$ (and where each child may be designated as a left or a right child of its parent)
ordered tree: a tree in which the children of each internal vertex are linearly ordered
balanced tree: a tree in which every vertex is at level h or $h - 1$, where h is the height of the tree
binary search tree: a binary tree in which the vertices are labeled with items so that a label of a vertex is greater than the labels of all vertices in the left subtree of this vertex and is less than the labels of all vertices in the right subtree of this vertex
decision tree: a rooted tree where each vertex represents a possible outcome of a decision and the leaves represent the possible solutions
prefix code: a code that has the property that the code of a character is never a prefix of the code of another character
tree traversal: a listing of the vertices of a tree
preorder traversal: a listing of the vertices of an ordered rooted tree defined recursively by specifying that the root is listed, followed by the first subtree, followed by the other subtrees in the order they occur from left to right
inorder traversal: a listing of the vertices of an ordered rooted tree defined recursively by specifying that the first subtree is listed, followed by the root, followed by the other subtrees in the order they occur from left to right
postorder traversal: a listing of the vertices of an ordered rooted tree defined recursively by specifying that the subtrees are listed in the order they occur from left to right, followed by the root
infix notation: the form of an expression (including a full set of parentheses) obtained from an inorder traversal of the binary tree representing this expression
prefix (or Polish) notation: the form of an expression obtained from a preorder traversal of the tree representing this expression
postfix (or reverse Polish) notation: the form of an expression obtained from a postorder traversal of the tree representing this expression

sorting problem: a problem in which a list of items is to be put into increasing order
spanning tree: a tree that contains every vertex of a given graph
minimal spanning tree: a spanning tree with smallest possible sum of weights of its edges
greedy algorithm: an algorithm that optimizes by making the optimal choice at each step

RESULTS

A graph is a tree if and only if there is a unique simple path between any of its vertices.

A tree with n vertices has $n - 1$ edges.

A full m-ary tree with i internal vertices has $mi + 1$ vertices.

The relationships between the numbers of vertices, leaves, and internal vertices in a full m-ary tree (see Theorem 4 in Section 8.1).

There are at most m^h leaves in an m-ary tree of height h.

If an m-ary tree has l leaves, its height h is at least $\lceil \log_m l \rceil$. If the tree is also full and balanced, then its height is $\lceil \log_m l \rceil$.

the bubble sort: a sorting procedure that is carried out using passes where successive items that are out of order are interchanged
the merge sort: a sorting procedure that is carried out by successively merging pairs of sublists of the original list
depth-first search, or backtracking: a procedure for constructing a spanning tree by adding edges that form a path until this is not possible, and then moving back up the path until a vertex is found where a new path can be formed
breadth-first search: a procedure for constructing a spanning tree that successively adds all edges incident to the last set of edges added, unless a simple circuit is formed
Prim's algorithm: a procedure for producing a minimal spanning tree in a weighted graph that successively adds edges with minimal weight among all edges incident to a vertex already in the tree such that no edge produces a simple circuit when it is added
Kruskal's algorithm: a procedure for producing a minimal spanning tree in a weighted graph that successively adds edges of least weight that are not already in the tree such that no edge produces a simple circuit when it is added

Review Questions

1. **a)** Define a tree. **b)** Define a forest.

2. Can there be two different simple paths between the vertices of a tree?

3. Give at least three examples of how trees are used in modeling.

4. **a)** Define a rooted tree and the root of such a tree.
b) Define the parent of a vertex and a child of a vertex in a rooted tree.
c) What are an internal vertex, a leaf, and a subtree in a rooted tree?
d) Draw a rooted tree with at least 10 vertices, where the degree of each vertex does not exceed 3. Identify the root, the parent of each vertex, the children of each vertex, the internal vertices, and the leaves.

5. **a)** How many edges does a tree with n vertices have?
b) What do you need to know to determine the number of edges in a forest with n vertices?

6. **a)** Define a full m-ary tree.
b) How many vertices does a full m-ary tree have if it has i internal vertices? How many leaves does the tree have?

7. a) What is the height of a rooted tree?
   b) What is a balanced tree?
   c) How many leaves can an m-ary tree of height h have?
8. a) What is a binary search tree?
   b) Describe an algorithm for constructing a binary search tree.
   c) Form a binary search tree for the words *vireo, warbler, egret, crossbeak, nuthatch,* and *kingfisher.*
9. a) What is a prefix code?
   b) How can a prefix code be represented by a binary tree?
10. a) Define preorder, inorder, and postorder tree traversal.
    b) Give an example of preorder, postorder, and inorder traversal of a binary tree of your choice with at least 12 vertices.
11. a) Explain how to use preorder, inorder, and postorder traversals to find the prefix, infix, and postfix forms of an arithmetic expression.
    b) Draw the ordered rooted tree that represents $((x - 3) + ((x/4) + (x - y) \uparrow 3))$.
    c) Find the prefix and postfix forms of the expression in part (b).
12. Show that the number of comparisons used by a sorting algorithm is at least $\lceil \log n! \rceil$.
13. a) Describe the bubble sort algorithm.
    b) Use the bubble sort algorithm to put the list 5, 2, 4, 1, 3 in increasing order.
    c) Give a big-O estimate for the number of comparisons used by the bubble sort.
14. a) Describe the merge sort algorithm.
    b) Use the merge sort algorithm to put the list 4, 10, 1, 5, 3, 8, 7, 2, 6, 9 in increasing order.
    c) Give a big-O estimate for the number of comparisons used by the merge sort.
15. a) What is a spanning tree of a simple graph?
    b) Which simple graphs have spanning trees?
    c) Describe at least two different applications that require that a spanning tree of a simple graph be found.
16. a) Describe two different algorithms for finding a spanning tree in a simple graph.
    b) Illustrate how the two algorithms you described in (a) can be used to find the spanning tree of a simple graph, using a graph of your choice with at least eight vertices and 15 edges.
17. a) Explain how backtracking can be used to determine whether a simple graph can be colored using n colors.
    b) Show, with an example, how backtracking can be used to show that a graph with chromatic number equal to 4 cannot be colored with three colors, but can be colored with four colors.
18. a) What is a minimal spanning tree of a simple graph?
    b) Describe at least two different applications that require that a minimal spanning tree of a simple graph be found.
19. a) Describe Kruskal's algorithm and Prim's algorithm for finding minimal spanning trees.
    b) Illustrate how Kruskal's algorithm and Prim's algorithm are used to find a minimal spanning tree, using a weighted graph with at least eight vertices and 15 edges.

Supplementary Exercises

★1. Show that a simple graph is a tree if and only if it contains no simple circuits and the addition of an edge connecting two nonadjacent vertices produces a new graph that has exactly one simple circuit (where circuits that contain the same edges are not considered different).

★2. How many nonisomorphic rooted trees are there with six vertices?

3. Show that every tree with at least one edge must have at least two pendant vertices.

4. Show that a tree with n vertices that has $n - 1$ pendant vertices must be isomorphic to $K_{1,n-1}$.

5. What is the sum of the degrees of the vertices of a tree with n vertices?

★6. Suppose that $d_1, d_2, \ldots, d_n$ are n positive integers with sum $2n - 2$. Show that there is a tree that has n vertices so that the degrees of these vertices are $d_1, d_2, \ldots, d_n$.

7. Show that every tree is a planar graph.

8. Show that every tree is bipartite.

9. Show that every forest can be colored using two colors.

A **B-tree of degree** k is a rooted tree such that all its leaves are at the same level, its root has at least two and at most k children unless it is a leaf, and every internal vertex other than the root has at least $\lceil k/2 \rceil$, but no more than k, children. Computer files can be accessed efficiently when B-trees are used to represent them.

10. Draw three different B-trees of degree 3 with height 4.

★11. Give an upper bound and a lower bound for the number of leaves in a B-tree of degree k with height h.

★12. Give an upper bound and a lower bound for the height of a B-tree of degree k with n leaves.

A rooted tree T is called an S_k**-tree** if it satisfies the following recursive definition. It is an S_0-tree if it has one vertex. For $k > 0$, T is an S_k-tree if it can be built from two S_{k-1}-trees by making the root of one the root of the S_k-tree and making the root of the other the child of the root of the first S_k-tree.

13. Draw an S_k-tree for $k = 0, 1, 2, 3, 4$.

14. Show that an S_k-tree has 2^k vertices and a unique vertex at level k. This vertex at level k is called the **handle.**

★15. Suppose that T is an S_k-tree with handle v. Show that T can be obtained from disjoint trees $T_0, T_1, \ldots, T_{k-1}$, where v is not in any of these trees, where T_i is an S_i-tree for $i = 0, 1, \ldots, k-1$, by connecting v to r_0 and r_i to r_{i+1} for $i = 0, 1, \ldots, k-2$.

The listing of the vertices of an ordered rooted tree in **level order** begins with the root, followed by the vertices at level 1 from left to right, followed by the vertices at level 2 from left to right, and so on.

16. List the vertices of the ordered rooted trees in Figures 3 and 9 of Section 8.3 in level order.

17. Devise an algorithm for listing the vertices of an ordered rooted tree in level order.

★18. Devise an algorithm for determining if a set of universal addresses can be the addresses of the leaves of a rooted tree.

19. Devise an algorithm for constructing a rooted tree from the universal addresses of its leaves.

The **insertion sort** operates by considering the elements in a list one at a time, beginning with the second element. Each element is compared to the previous elements in the list, which have been put in the correct order, and this element is put in the correct position among these, moving the element that was in this position, and all elements to the right of this, one position to the right.

20. Sort the list 3, 2, 4, 5, 1 using an insertion sort.

21. Write the insertion sort in pseudocode.

22. Determine the worst-case complexity of the insertion sort in terms of the number of comparisons used.

23. Suppose that e is an edge in a simple graph that is incident to a pendant vertex. Show that e must be in any spanning tree.

A **cut set** of a graph is a set of edges such that the removal of these edges produces a subgraph with more connected components than in the original graph, but no proper subset of this set of edges has this property.

24. Show that a cut set of a graph must have at least one edge in common with any spanning tree of this graph.

A **cactus** is a connected graph in which no edge is in more than one simple circuit not passing through any vertex other than its initial vertex more than once or its initial vertex other than at its terminal vertex (where two circuits that contain the same edges are not considered different).

25. Which of the following graphs are cacti?

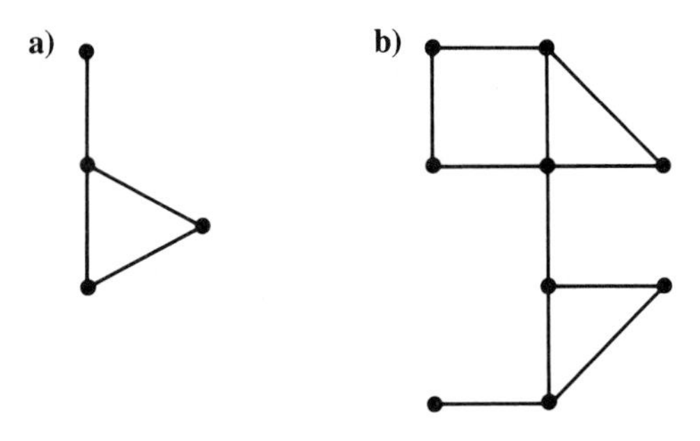

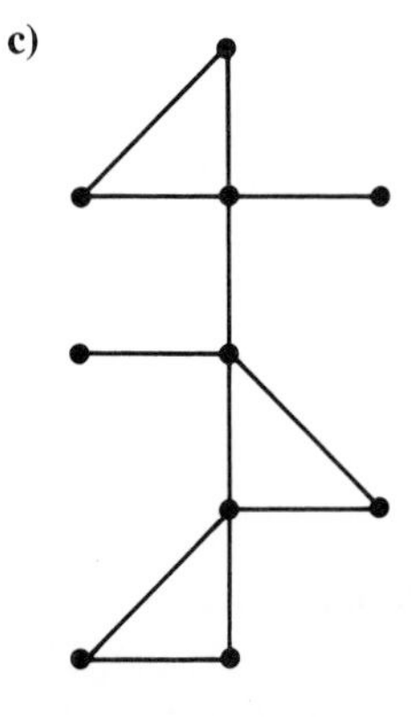

26. Is a tree necessarily a cactus?

27. Show that a cactus is formed if we add a circuit containing new edges beginning and ending at a vertex of a tree.

★28. Show that if every circuit not passing through any vertex other than its initial vertex more than once in a connected graph contains an odd number of edges, then this graph must be a cactus.

A **degree-constrained spanning tree** of a simple graph G is a spanning tree with the property that the degree of a vertex in this tree cannot exceed some specified bound. Degree-constrained spanning trees are useful in models of transportation systems where the number of roads at an intersection is limited, models of communications networks where the number of links entering a node is limited, and so on.

In Exercises 29–31 find a degree-constrained spanning tree of the given graph where each vertex has degree less than or equal to 3, or show that such a spanning tree does not exist.

29.

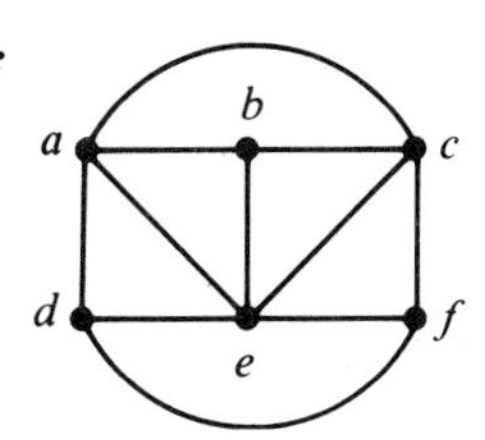

30.

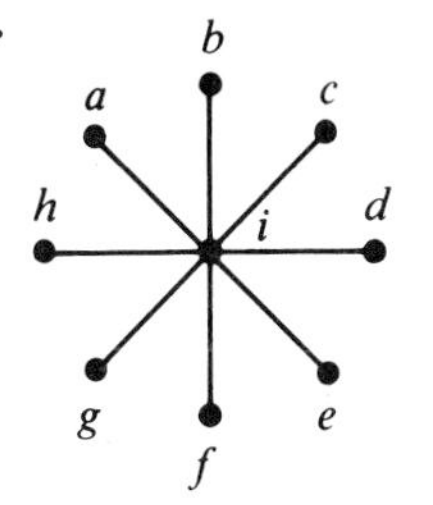

31.

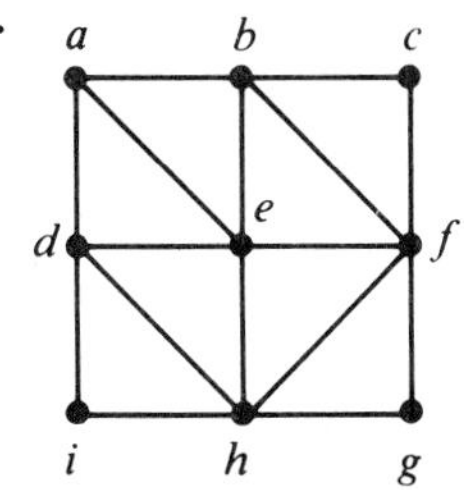

32. Show that a degree-constrained spanning tree of a simple graph in which each vertex has degree not exceeding 2 consists of a single Hamilton path in the graph.

33. A tree with n vertices is called **graceful** if its vertices can be labeled with the integers $1, 2, \ldots, n$ such that the absolute value of the difference of the labels of adjacent vertices are all different. Show that the following trees are graceful.

a)

b)

c)

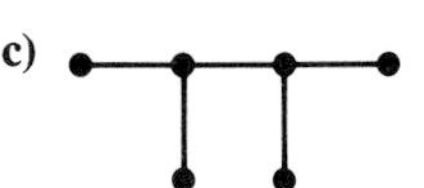

d)

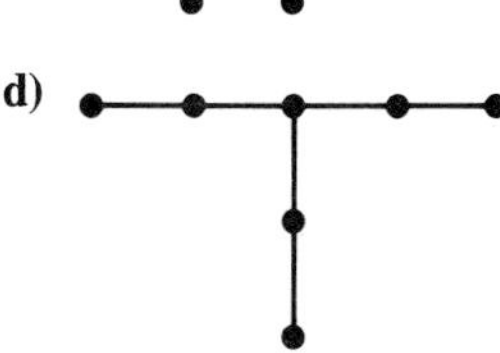

A **caterpillar** is a tree that contains a path such that every vertex not contained in this path is adjacent to a vertex in the path.

34. Which of the graphs in Exercise 33 are caterpillars?

35. How many nonisomorphic caterpillars are there with six vertices?

★36. **a)** Prove or disprove that all trees whose edges form a single path are graceful.

★★b) Prove or disprove that all caterpillars are graceful.

37. Suppose that the first four moves of a tic-tac-toe game are as shown. Explain how a tree can be used to show the possible successive moves of this game. If the player using X goes first, does this player have a strategy that will always win?

a)

X	O	X
	O	

b)

X	O	X
	O	

c)

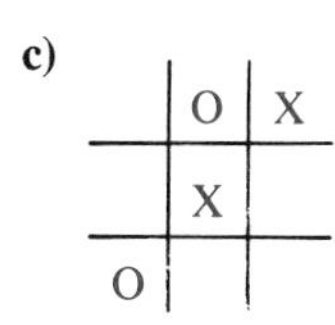

38. Three couples arrive at the bank of a river. Each of the wives is jealous and does not trust her husband when he is with one of the other wives (and perhaps with other people), but not with her. How can six people cross to the other side of the river using a boat that can hold no more than two people so that no husband is alone with a woman other than his wife? Use a graph theory model.

★39. Suppose that e is an edge in a weighted graph that is incident to a vertex v so that the weight of e does not exceed the weight of any other edge incident to v. Show that there exists a minimal spanning tree containing this edge.

★40. Show that if no two edges in a weighted graph have the same weight, then the edge with least weight incident to a vertex v is included in every minimal spanning tree.

41. Find a minimal spanning tree of each of the following graphs where the degree of each vertex in the spanning tree does not exceed 2.

a)

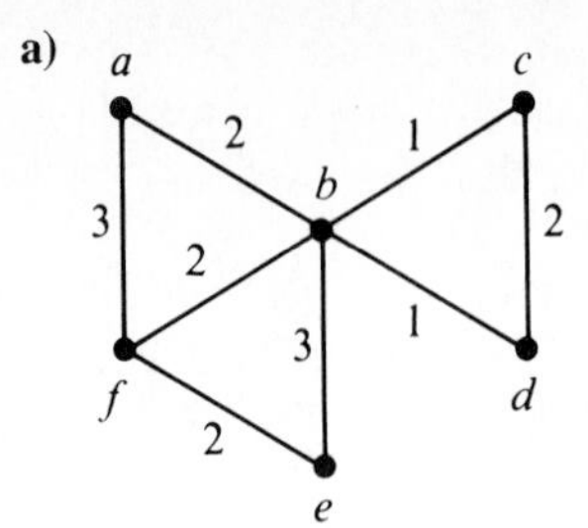

b)

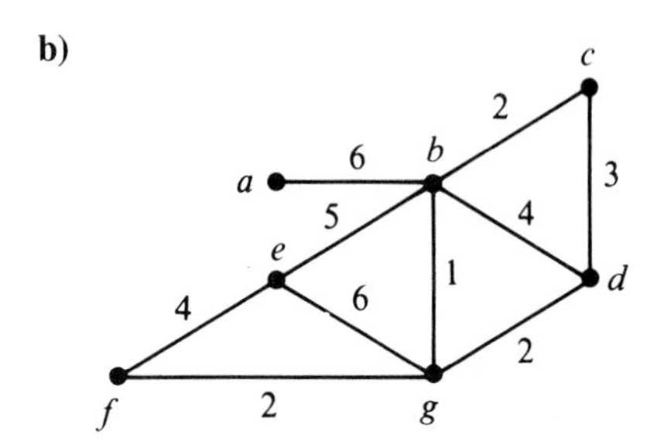

Computer Projects

WRITE PROGRAMS WITH THE FOLLOWING INPUT AND OUTPUT.

1. Given the adjacency matrix of an undirected simple graph, determine whether the graph is a tree.

2. Given the adjacency matrix of a rooted tree and a vertex in the tree, find the parent, children, ancestors, descendants, and level of this vertex.

3. Given the list of edges of a rooted tree and a vertex in the tree, find the parent, children, ancestors, descendants, and level of this vertex.

4. Given a list of items, construct a binary search tree containing these items.

5. Given a binary search tree and an item, locate or add this item to the binary search tree.

6. Given the ordered list of edges of an ordered rooted tree, find the universal addresses of its vertices.

7. Given the ordered list of edges of an ordered rooted tree, list its vertices in preorder, inorder, and postorder.

8. Given an arithmetic expression in prefix form, find its value.

9. Given an arithmetic expression in postfix form, find its value.

10. Given a set of n integers, sort them using a bubble sort.

11. Given two sorted lists of integers, merge them into one sorted list, keeping track of the number of comparisons used.

12. Given a set of n integers, sort them using a merge sort.

13. Given the adjacency matrix of a connected undirected simple graph, find a spanning tree for this graph using a depth-first search.

14. Given the adjacency matrix of a connected undirected simple graph, find a spanning tree for this graph using a breadth-first search.

15. Given a set of positive integers and a positive integer N, use backtracking to find a subset of these integers that have N as their sum.

★16. Given the adjacency matrix of an undirected simple graph, use backtracking to color the graph with three colors, if this is possible.

★17. Given a positive integer n, solve the n-queens problem using backtracking.

18. Given the list of edges and their weights of a weighted undirected connected graph, use Prim's algorithm to find a minimal spanning tree of this graph.

19. Given the list of edges and their weights of a weighted undirected connected graph, use Kruskal's algorithm to find a minimal spanning tree of this graph.

9

Boolean Algebra

The circuits in computers and other electronic devices have inputs, each of which is either a 0 or a 1, and produce outputs that are also 0s and 1s. Circuits can be constructed using any basic element that has two different states. Such elements include switches that can be in either the on or the off position and optical devices that can either be lit or unlit. In 1938 Claude Shannon showed how the basic rules of logic, first given by George Boole in 1854 in his *The Laws of Thought*, could be used to design circuits. These rules form the basis for Boolean algebra. In this chapter we develop the basic properties of Boolean algebra. The operation of a circuit is defined by a Boolean function that specifies the value of an output for each set of inputs. The first step in constructing a circuit is to represent its Boolean function by an expression built up using the basic operations of Boolean algebra. We will provide an algorithm for producing such expressions. The expression that we obtain may contain many more operations than are necessary to represent the function. Later in the chapter we will describe methods for finding an expression with the minimum number of sums and products that represents a Boolean function. The procedures that we will develop, Karnaugh maps and the Quine–McCluskey method, are important in the design of efficient circuits.

Claude Elwood Shannon (born 1916). Claude Shannon was born in Gaylord, Michigan, and attended the University of Michigan, graduating in 1936. He received his M.S. in 1936 and his Ph.D. in 1940, both from M.I.T. His thesis contains new results on switching circuits. He was the first person to apply Boolean algebra to the design of circuits. Shannon joined Bell Laboratories in 1940, where he worked on transmitting data efficiently. He was one of the first people to use bits to represent information. At Bell Laboratories he worked on determining the amount of traffic that telephone lines can carry. Shannon has made many fundamental contributions to information theory. In the early 1950s he was one of the founders of the study of artificial intelligence. He joined the M.I.T. faculty in 1956, where he continued his study of information theory.

9.1 Boolean Functions

INTRODUCTION

Boolean algebra provides the operations and the rules for working with the set {0, 1}. Electronic and optical switches can be studied using this set and the rules of Boolean algebra. The three operations in Boolean algebra that we will use most are complementation, the Boolean sum, and the Boolean product. The **complement** of an element, denoted with a bar, is defined by $\overline{0} = 1$ and $\overline{1} = 0$. The Boolean sum, denoted by $+$ or by *OR*, has the following values:

$$1 + 1 = 1 \qquad 1 + 0 = 1 \qquad 0 + 1 = 1 \qquad 0 + 0 = 0$$

The Boolean product, denoted by $\cdot$ or by *AND*, has the following values:

$$1 \cdot 1 = 1 \qquad 1 \cdot 0 = 0 \qquad 0 \cdot 1 = 0 \qquad 0 \cdot 0 = 0$$

When there is no danger of confusion, the symbol $\cdot$ can be deleted, just as in writing algebraic products. Unless parentheses are used, the rules of precedence for Boolean operators are: first, all complements are computed, followed by Boolean products, followed by all Boolean sums. This is illustrated in Example 1.

EXAMPLE 1 Find the value of $1 \cdot 0 + \overline{(0 + 1)}$.

Solution: Using the definitions of complementation, the Boolean sum, and the Boolean product, it follows that

$$\begin{aligned}(1 \cdot 0) + \overline{(0 + 1)} &= 0 + \overline{1}\\ &= 0 + 0\\ &= 0.\end{aligned}$$

■

The complement, Boolean sum, and Boolean product correspond to the logical operators, $\neg$, $\vee$, and $\wedge$, respectively, where 0 corresponds to F (false) and 1 corresponds to T (true). The results of Boolean algebra can be directly translated into results about propositions. Conversely, results about propositions can be translated into statements about Boolean algebra.

BOOLEAN EXPRESSIONS AND BOOLEAN FUNCTIONS

Let $B = \{0, 1\}$. The variable x is called a **Boolean variable** if it assumes values only from B. A function from B^n, the set $\{(x_1, x_2, \ldots, x_n) \mid x_i \in B, 1 \leq i \leq n\}$, to B is called a **Boolean function of degree** n. The values of a Boolean function are often displayed in tables. For instance, the Boolean function $F(x,y)$ with the value 1 when $x = 1$ and $y = 0$ and the value 0 for all other choices of x and y can be represented by Table 1.

TABLE 1

x	y	$F(x,y)$
1	1	0
1	0	1
0	1	0
0	0	0

Boolean functions can be represented using expressions made up from the variables and Boolean operations. The **Boolean expressions** in the variables $x_1, x_2, \ldots, x_n$ are defined recursively as follows:

> $0, 1, x_1, x_2, \ldots, x_n$ are Boolean expressions;
> if E_1 and E_2 are Boolean expressions, then $\overline{E}_1$, (E_1E_2), and (E_1+E_2) are Boolean expressions.

Each Boolean expression represents a Boolean function. The values of this function are obtained by substituting 0 and 1 for the variables in the expression. In Section 9.2 we will show that every Boolean function can be represented by a Boolean expression.

EXAMPLE 2 Find the values of the Boolean function represented by $F(x, y, z) = xy + \overline{z}$.

Solution: The values of this function are displayed in Table 2. ■

The Boolean functions F and G of n variables are equal if and only if $F(b_1, b_2, \ldots, b_n) = G(b_1, b_2, \ldots, b_n)$ whenever $b_1, b_2, \ldots, b_n$ belong to B. Two different Boolean expressions that represent the same function are called **equivalent**. For instance, all the Boolean expressions xy, $xy + 0$, and $xy \cdot 1$ are equivalent. The **complement** of the Boolean function F is the function $\overline{F}$, where $\overline{F}(x_1, \ldots, x_n) = \overline{F(x_1, \ldots, x_n)}$. Let F and G be Boolean functions of degree n. The **Boolean sum** $F + G$ and **Boolean product** FG are defined by

TABLE 2

x	y	z	xy	$\overline{z}$	$F(x, y, z) = xy + \overline{z}$
1	1	1	1	0	1
1	1	0	1	1	1
1	0	1	0	0	0
1	0	0	0	1	1
0	1	1	0	0	0
0	1	0	0	1	1
0	0	1	0	0	0
0	0	0	0	1	1

TABLE 3 **The Boolean Functions of Degree 2**

x	y	F_1	F_2	F_3	F_4	F_5	F_6	F_7	F_8	F_9	F_{10}	F_{11}	F_{12}	F_{13}	F_{14}	F_{15}	F_{16}
1	1	1	1	1	1	1	1	1	1	0	0	0	0	0	0	0	0
1	0	1	1	1	1	0	0	0	0	1	1	1	1	0	0	0	0
0	1	1	1	0	0	1	1	0	0	1	1	0	0	1	1	0	0
0	0	1	0	1	0	1	0	1	0	1	0	1	0	1	0	1	0

$$(F + G)(x_1, \ldots, x_n) = F(x_1, \ldots, x_n) + G(x_1, \ldots, x_n),$$
$$(FG)(x_1, \ldots, x_n) = F(x_1, \ldots, x_n)G(x_1, \ldots, x_n).$$

A Boolean function of degree 2 is a function from a set with four elements, namely, pairs of elements from $B = \{0, 1\}$, to B, a set with two elements. Hence, there are 16 different Boolean functions of degree 2. In Table 3 we display the values of the 16 different Boolean functions of degree 2, labeled $F_1, F_2, \ldots, F_{16}$.

EXAMPLE 3 How many different Boolean functions of degree n are there?

Solution: From the product rule for counting, it follows that there are 2^n different n-tuples of 0s and 1s. Since a Boolean function is an assignment of 0 or 1 to each of these 2^n different n-tuples, the product rule shows that there are 2^{2^n} different Boolean functions. ■

Table 4 displays the number of different Boolean functions of degrees 1 through 6. The number of such functions grows extremely rapidly.

TABLE 4 **The Number of Boolean Functions of Degree n**

Degree	*Number*
1	2
2	16
3	256
4	65,536
5	4,294,967,296
6	18,446,744,073,709,551,616

TABLE 5 Boolean Identities

Identity	*Name*
$\overline{\overline{x}} = x$	Law of the double complement
$x + x = x$ $x \cdot x = x$	Idempotent laws
$x + 0 = x$ $x \cdot 1 = x$	Identity laws
$x + 1 = 1$ $x \cdot 0 = 0$	Dominance laws
$x + y = y + x$ $xy = yx$	Commutative laws
$x + (y + z) = (x + y) + z$ $x(yz) = (xy)z$	Associative laws
$x + yz = (x + y)(x + z)$ $x(y + z) = xy + xz$	Distributive laws
$\overline{(xy)} = \overline{x} + \overline{y}$ $\overline{(x + y)} = \overline{x}\,\overline{y}$	De Morgan's laws

IDENTITIES OF BOOLEAN ALGEBRA

There are many identities in Boolean algebra. The most important of these are displayed in Table 5. These identities are particularly useful in simplifying the design of circuits. Each of the identities in Table 5 can be proved using a table. We will prove one of the distributive laws in this way in the following example. The proofs of the remaining properties are left as exercises for the reader.

EXAMPLE 4 Show that the distributive law $x(y + z) = xy + xz$ is valid.

Solution: The verification of this identity is shown in Table 6. The identity holds because the last two columns of the table agree. ■

TABLE 6

x	y	z	$y+z$	xy	xz	$x(y+z)$	$xy+xz$
1	1	1	1	1	1	1	1
1	1	0	1	1	0	1	1
1	0	1	1	0	1	1	1
1	0	0	0	0	0	0	0
0	1	1	1	0	0	0	0
0	1	0	1	0	0	0	0
0	0	1	1	0	0	0	0
0	0	0	0	0	0	0	0

The identities in Table 5 can be used to prove further identities. We demonstrate this in the following example.

EXAMPLE 5 Prove the **absorption law** $x(x + y) = x$ using the identities of Boolean algebra. (This is called an absorption law since absorbing $x + y$ into x leaves x unchanged.)

Solution: The steps used to derive this identity and the law used in each step follow:

$x(x + y) = (x + 0)(x + y)$ Identity law for the Boolean sum
$= x + 0 \cdot y$ Distributive law of the Boolean sum over the Boolean product
$= x + y \cdot 0$ Commutative law for the Boolean product
$= x + 0$ Dominance law for the Boolean product
$= x$ Identity law for the Boolean sum ■

DUALITY

The identities in Table 5 come in pairs (except for the law of the double complement). To explain the relationship between the two identities in each pair we use the concept of a dual. The **dual** of a Boolean expression is obtained by interchanging Boolean sums and Boolean products and interchanging 0s and 1s.

EXAMPLE 6 Find the duals of $x(y + 0)$ and $\overline{x} \cdot 1 + (\overline{y} + z)$.

Solution: Interchanging $\cdot$ signs and $+$ signs and interchanging 0s and 1s in these expressions produces their duals. The duals are $x + (y \cdot 1)$ and $(\overline{x} + 0)(\overline{y}z)$, respectively. ■

The dual of a Boolean function F represented by a Boolean expression is the function represented by the dual of this expression. This dual function, denoted by F^d, does not depend on the particular Boolean expression used to represent F. An identity between functions represented by Boolean expressions remains valid when the duals of both sides of the identity are taken. (See Exercise 22 for the reason this is true.) This result, called the **duality principle**, is useful for obtaining new identities.

EXAMPLE 7 Construct an identity from the absorption law $x(x + y) = x$ given in Example 5 by taking duals.

Solution: Taking the duals of both sides of this identity produces the identity $x + xy = x$, which is also called an absorption law. ∎

Exercises

1. Find the values of the following expressions.
a) $1 \cdot \overline{0}$ **b)** $1 + \overline{1}$
c) $\overline{0} \cdot 0$ **d)** $\overline{(1 + 0)}$

2. Find the values, if any, of the Boolean variable x that satisfy the following equations.
a) $x \cdot 1 = 0$ **b)** $x + x = 0$
c) $x \cdot 1 = x$ **d)** $x \cdot \overline{x} = 1$

3. What values of the Boolean variables x and y satisfy $xy = x + y$?

4. How many different Boolean functions are there of degree 7?

5. Prove the **absorption law** $x + xy = x$ using the laws in Table 5.

☞**6.** Show that $F(x, y, z) = xy + xz + yz$ has the value 1 if and only if at least two of the variables x, y, and z have the value 1.

7. Show that $x\overline{y} + y\overline{z} + \overline{x}z = \overline{x}y + \overline{y}z + x\overline{z}$.

8. Verify the law of the double complement.

9. Verify the idempotent laws.

10. Verify the identity laws.

11. Verify the dominance laws.

12. Verify the commutative laws.

13. Verify the associative laws.

14. Verify the first distributive law in Table 5.

15. Verify De Morgan's laws.

The Boolean operator $\oplus$, called the *XOR* operator, is defined by $1 \oplus 1 = 0$, $1 \oplus 0 = 1$, $0 \oplus 1 = 1$, and $0 \oplus 0 = 0$.

16. Simplify the following expressions.
a) $x \oplus 0$ **b)** $x \oplus 1$
c) $x \oplus x$ **d)** $x \oplus \overline{x}$

17. Show that the following identities hold.
a) $x \oplus y = (x + y)\overline{(xy)}$
b) $x \oplus y = (x\overline{y}) + (\overline{x}y)$

18. Show that $x \oplus y = y \oplus x$.

19. Prove or disprove the following equalities.
a) $x \oplus (y \oplus z) = (x \oplus y) \oplus z$
b) $x + (y \oplus z) = (x + y) \oplus (x + z)$
c) $x \oplus (y + z) = (x \oplus y) + (x \oplus z)$

20. Find the duals of the following Boolean expressions.
a) $x + y$ **b)** $\overline{x}\,\overline{y}$
c) $x y z + \overline{x}\,\overline{y}\,\overline{z}$ **d)** $x\overline{z} + x \cdot 0 + \overline{x} \cdot 1$

★**21.** Suppose that F is a Boolean function represented by a Boolean expression in the variables $x_1, \ldots, x_n$. Show that $F^d(x_1, \ldots, x_n) = \overline{F(\overline{x}_1, \ldots, \overline{x}_n)}$.

★**22.** Show that if F and G are Boolean functions represented by Boolean expressions in n variables and $F = G$, then $F^d = G^d$, where F^d and G^d are the Boolean functions represented by the duals of the Boolean expressions representing F and G, respectively. (*Hint:* Use the result of Exercise 21.)

★**23.** How many different Boolean functions $F(x, y, z)$ are there so that $F(\overline{x}, \overline{y}, \overline{z}) = F(x, y, z)$ for all values of the Boolean variables x, y, and z?

★**24.** How many different Boolean functions $F(x, y, z)$ are there so that $F(\overline{x}, y, z) = F(x, \overline{y}, z) = F(x, y, \overline{z})$ for all values of the Boolean variables x, y, z?

9.2 Representing Boolean Functions

Two important problems of Boolean algebra will be studied in this section. The first problem is: Given the values of a Boolean function, how can a Boolean expression that represents this function be found? This problem will be solved by showing that any Boolean function may be represented by a Boolean sum of Boolean products of the variables and their complements. The solution of this problem shows that every Boolean function can be represented using the three Boolean operators $\cdot$, $+$, and $^-$. The second problem is: Is there a smaller set of operators that can be used to represent all Boolean functions? We will answer this problem by showing that all Boolean functions can be represented using only one operator. Both of these problems have practical importance in circuit design.

SUM-OF-PRODUCTS EXPANSIONS

We will use examples to illustrate one important way to find a Boolean expression that represents a Boolean function.

EXAMPLE 1 Find Boolean expressions that represent the functions $F(x, y, z)$ and $G(x, y, z)$, which are given in Table 1.

Solution: An expression that has the value 1 when $x = z = 1$ and $y = 0$, and the value 0 otherwise, is needed to represent F. Such an expression can be formed by taking the Boolean product of x, $\overline{y}$, and z. This product, $x\overline{y}z$, has the value 1 if and only if $x = \overline{y} = z = 1$, which holds if and only if $x = z = 1$ and $y = 0$.

TABLE 1

x	y	z	F	G
1	1	1	0	0
1	1	0	0	1
1	0	1	1	0
1	0	0	0	0
0	1	1	0	0
0	1	0	0	1
0	0	1	0	0
0	0	0	0	0

To represent G, we need an expression that equals 1 when $x = y = 1$ and $z = 0$, or when $x = z = 0$ and $y = 1$. We can form an expression with these values by taking the Boolean sum of two different Boolean products. The Boolean product $xy\overline{z}$ has the value 1 if and only if $x = y = 1$ and $z = 0$. Similarly, the product $\overline{x}y\overline{z}$ has the value 1 if and only if $x = z = 0$ and $y = 1$. The Boolean sum of these two products, $xy\overline{z} + \overline{x}y\overline{z}$, represents G, since it has the value 1 if and only if $x = y = 1$ and $z = 0$ or $x = z = 0$ and $y = 1$. ■

Example 1 illustrates a procedure for constructing a Boolean expression representing a function with given values. Each combination of values of the variables for which the function has the value 1 leads to a Boolean product of the variables or their complements.

DEFINITION 1. A *minterm* of the Boolean variables $x_1, x_2, \ldots, x_n$ is a Boolean product $y_1 y_2 \cdots y_n$ where $y_i = x_i$ or $y_i = \overline{x}_i$. A *literal* is a Boolean variable or its complement. Hence, a minterm is a product of n literals, with one literal for each variable.

A minterm has the value 1 for one and only one combination of values of its variables. More precisely, the minterm $y_1 y_2 \cdots y_n$ is 1 if and only if each y_i is 1, and this occurs if and only if $x_i = 1$ when $y_i = x_i$ and $x_i = 0$ when $y_i = \overline{x}_i$.

EXAMPLE 2 Find a minterm that equals 1 if $x_1 = x_3 = 0$ and $x_2 = x_4 = x_5 = 1$, and equals 0 otherwise.

Solution: The minterm $\overline{x}_1 x_2 \overline{x}_3 x_4 x_5$ has the correct set of values. ■

By taking Boolean sums of distinct minterms we can build up a Boolean expression with a specified set of values. In particular, a Boolean sum of minterms has the value 1 when exactly one of the minterms in the sum has the value 1. It has the value 0 for all other combinations of values of the variables. Consequently, given a Boolean function, a Boolean sum of minterms can be formed that has the value 1 when this Boolean function has the value 1, and has the value 0 when the function has the value 0. The minterms in this Boolean sum correspond to those combinations of values for which the function has the value 1. The sum of minterms that represents the function is called the **sum-of-products expansion** or the **disjunctive normal form** of the Boolean function.

EXAMPLE 3 Find the sum-of-products expansion for the function $F(x, y, z) = (x + y)\overline{z}$.

TABLE 2

x	y	z	$x + y$	$\overline{z}$	$(x + y)\overline{z}$
1	1	1	1	0	0
1	1	0	1	1	1
1	0	1	1	0	0
1	0	0	1	1	1
0	1	1	1	0	0
0	1	0	1	1	1
0	0	1	0	0	0
0	0	0	0	1	0

Solution: The first step is to find the values of F. These are found in Table 2. The sum-of-products expansion of F is the Boolean sum of three minterms corresponding to the three rows of this table that give the value 1 for the function. This gives

$$F(x, y, z) = xy\overline{z} + x\overline{y}\,\overline{z} + \overline{x}y\overline{z}.$$

■

It is also possible to find a Boolean expression that represents a Boolean function by taking a Boolean product of Boolean sums. The resulting expansion is called the **conjunctive normal form** or **product-of-sums expansion** of the function. These expansions can be found from sum-of-products expansions by taking duals. How to find such expansions directly is described in Exercise 10 at the end of this section.

FUNCTIONAL COMPLETENESS

Every Boolean function can be expressed as a Boolean sum of minterms. Each minterm is the Boolean product of Boolean variables or their complements. This shows that every Boolean function can be represented using the Boolean operators $\cdot$, $+$, and $^{-}$. Since every Boolean function can be represented using these operators we say that the set $\{\cdot, +, ^{-}\}$ is **functionally complete**. Can we find a smaller set of functionally complete operators? We can do so if one of the three operators of this set can be expressed in terms of the other two. This can be done using one of De Morgan's laws. We can eliminate all Boolean sums using the identity

$$x + y = \overline{\overline{x}\,\overline{y}},$$

which is obtained by taking complements of both sides in the second De Morgan's law, given in Table 5 in Section 9.1, and then applying the double complementation law. This means that the set $\{\cdot, ^{-}\}$ is functionally complete. Similarly, we could eliminate all Boolean products using the identity

$$xy = \overline{\overline{x} + \overline{y}},$$

which is obtained by taking complements of both sides in the first De Morgan's law, given in Table 5 in Section 9.1, and then applying the double complementation law. Consequently $\{+, ^-\}$ is functionally complete. Note that the set $\{+, \cdot\}$ is not functionally complete, since it is impossible to express the Boolean function $F(x) = \overline{x}$ using these operators (see Exercise 19).

We have found sets containing two operators that are functionally complete. Can we find a smaller set of functionally complete operators, namely, a set containing just one operator? Such sets exist. Define two operators, the $|$ or *NAND* operator, defined by $1 \mid 1 = 0$ and $1 \mid 0 = 0 \mid 1 = 0 \mid 0 = 1$; and the $\downarrow$ or *NOR* operator, defined by $1 \downarrow 1 = 1 \downarrow 0 = 0 \downarrow 1 = 0$ and $0 \downarrow 0 = 1$. Both of the sets $\{|\}$ and $\{\downarrow\}$ are functionally complete. To see that $\{|\}$ is functionally complete, since $\{\cdot, ^-\}$ is functionally complete, all that we have to do is show that both of the operators $\cdot$ and $^-$ can be expressed using just the $|$ operator. This can be done as follows:

$$\overline{x} = x \mid x$$
$$xy = (x \mid y) \mid (x \mid y).$$

The reader should verify these identities (see Exercise 14). We leave the demonstration that $\{\downarrow\}$ is functionally complete for the reader (see Exercises 15 and 16).

Exercises

1. Find a Boolean product of the Boolean variables x, y, and z, or their complements, that has the value 1 if and only if
a) $x = y = 0, z = 1$
b) $x = 0, y = 1, z = 0$
c) $x = 0, y = z = 1$
d) $x = y = z = 0$

2. Find the sum-of-products expansions of the following Boolean functions.
a) $F(x, y) = \overline{x} + y$ **b)** $F(x, y) = x\,\overline{y}$
c) $F(x, y) = 1$ **d)** $F(x, y) = \overline{y}$

3. Find the sum-of-products expansions of the following Boolean functions.
a) $F(x, y, z) = x + y + z$
b) $F(x, y, z) = (x + z)y$
c) $F(x, y, z) = x$
d) $F(x, y, z) = x\,\overline{y}$

4. Find the sum-of-products expansions of the Boolean function $F(x, y, z)$ that equals 1 if and only if
a) $x = 0$. **b)** $x\,y = 0$.
c) $x + y = 0$. **d)** $x\,y\,z = 0$.

5. Find the sum-of-products expansion of the Boolean function $F(w, x, y, z)$ that has the value 1 if and only if an odd number of w, x, y, and z have the value 1.

6. Find the sum-of-products expansion of the Boolean function $F(x_1, x_2, x_3, x_4, x_5)$ that has the value 1 if and only if three or more of the variables x_1, x_2, x_3, x_4, and x_5 have the value 1.

Another way to find a Boolean expression that represents a Boolean function is to form a Boolean product of Boolean sums of literals. Exercises 7 to 11 are concerned with representations of this kind.

7. Find a Boolean sum containing either x or $\overline{x}$, either y or $\overline{y}$, and either z or $\overline{z}$ that has the value 0 if and only if
a) $x = y = 1, z = 0$.
b) $x = y = z = 0$.
c) $x = z = 0, y = 1$.

8. Find a Boolean product of Boolean sums of literals that has the value 0 if and only if either $x = y = 1$ and $z = 0$, $x = z = 0$ and $y = 1$, or $x = y = z = 0$. (*Hint:* Take the Boolean product of the Boolean sums found in parts (a), (b), and (c) in Exercise 7.)

9. Show that the Boolean sum $y_1 + y_2 + \cdots + y_n$, where $y_i = x_i$ or $y_i = \overline{x}_i$, has the value 0 for exactly one combination of the values of the variables, namely when $x_i = 0$ if $y_i = x_i$ and $x_i = 1$ if $y_i = \overline{x}_i$. This Boolean sum is called a **maxterm**.

10. Show that a Boolean function can be represented as a Boolean product of maxterms. This representation is called the **product-of-sums expansion or conjunctive normal form** of the function. (*Hint:* Include one maxterm in this product for each combination of the variables where the function has the value 0.)

11. Find the product-of-sums expansion of each of the Boolean functions in Exercise 3.

12. Express each of the following Boolean functions using the operators $\cdot$ and $^{-}$.

a) $x + y + z$ **b)** $x + \overline{y}(\overline{x} + z)$.
c) $\overline{(x + \overline{y})}$ **d)** $\overline{x}(x + \overline{y} + \overline{z})$

13. Express each of the Boolean functions in Exercise 12 using the operators $+$ and $^{-}$.

14. Show that

a) $\overline{x} = x \mid x$.
b) $xy = (x \mid y) \mid (x \mid y)$.
c) $x + y = (x \mid x) \mid (y \mid y)$.

15. Show that

a) $\overline{x} = x \downarrow x$.
b) $xy = (x \downarrow x) \downarrow (y \downarrow y)$.
c) $x + y = (x \downarrow y) \downarrow (x \downarrow y)$.

16. Show that $\{\downarrow\}$ is functionally complete using Exercise 15.

17. Express each of the Boolean functions in Exercise 3 using the operator $\mid$.

18. Express each of the Boolean functions in Exercise 3 using the operator $\downarrow$.

19. Show that the set of operators $\{+, \cdot\}$ is not functionally complete.

20. Are the following sets of operators functionally complete?

a) $\{+, \oplus\}$ **b)** $\{^{-}, \oplus\}$ **c)** $\{\cdot, \oplus\}$

9.3 Logic Gates

INTRODUCTION

Boolean algebra is used to model the circuitry of electronic devices. Each input and each output of such a device can be thought of as a member of the set $\{0, 1\}$. A computer, or other electronic device, is made up of a number of circuits. Each circuit can be designed using the rules of Boolean algebra that were studied in Sections 9.1 and 9.2. The basic elements of circuits are called **gates**. Each type of gate implements a Boolean operation. In this section we define several types of gates. Using these gates, we will apply the rules of Boolean algebra to design circuits that perform a variety of tasks. The circuits that we will study in this chapter give output that depends only on the input, and not on the current state of the circuit. In other words, these circuits have no memory capabilities. Such circuits are called **combinatorial circuits**.

We will construct combinatorial circuits using three types of elements. The first is an **inverter**, which accepts the value of one Boolean variable as input and produces the complement of this value as its output. The symbol used for an inverter is shown in Figure 1(a). The input to the inverter is shown on the left side entering the element, and the output is shown on the right side leaving the element.

The next type of element we will use is the *OR* gate. The inputs to this gate are the values of two or more Boolean variables. The output is the Boolean sum of their values. The symbol used for an *OR* gate is shown in Figure 1(b). The inputs to the *OR* gate are shown on the left side entering the element, and the output is shown on the right side leaving the element.

The third type of element we will use is the *AND* gate. The inputs to this gate are the values of two or more Boolean variables. The output is the Boolean product of their values. The symbol used for an *AND* gate is shown in Figure 1(c). The inputs

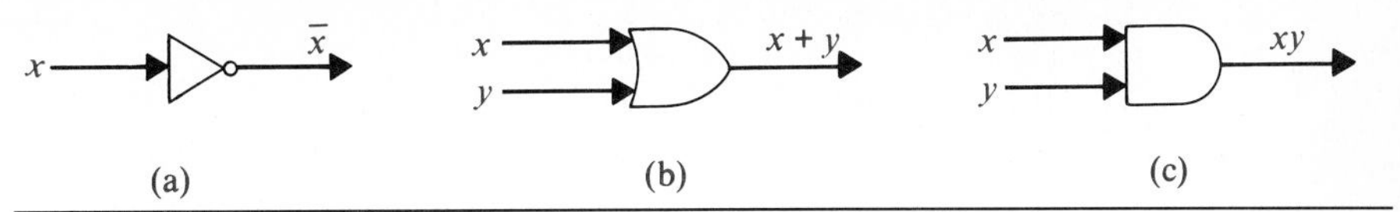

FIGURE 1 Basic Types of Gates.

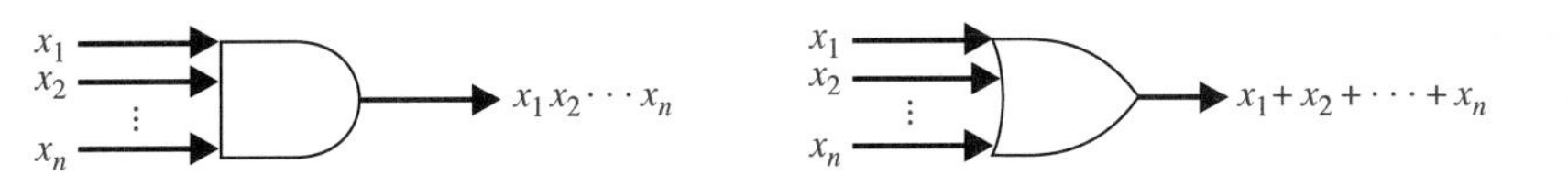

FIGURE 2 Gates with *n* Inputs.

to the *AND* gate are shown on the left side entering the element, and the output is shown on the right side leaving the element.

We will permit multiple inputs to *AND* and *OR* gates. The inputs to each of these gates are shown on the left side entering the element, and the output is shown on the right side. Examples of *AND* and *OR* gates with n inputs are shown in Figure 2.

COMBINATIONS OF GATES

Combinatorial circuits can be constructed using a combination of inverters, *OR* gates, and *AND* gates. When combinations of circuits are formed, some gates may share inputs. This is shown in one of two ways in depictions of circuits. One method is to use branchings that indicate all the gates that use a given input. The other method is to indicate this input separately for each gate. Figure 3 illustrates the two ways of showing gates with the same input values. Note also that output from a gate may be

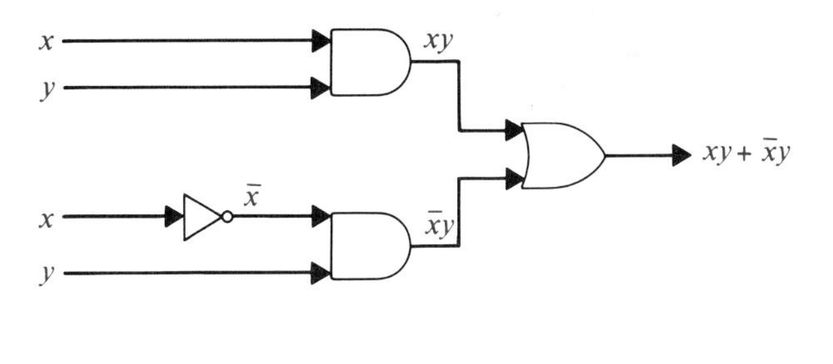

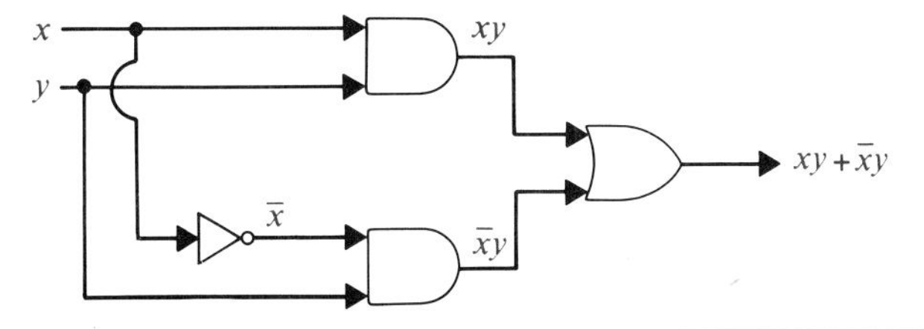

FIGURE 3 Two Ways to Draw the Same Circuit.

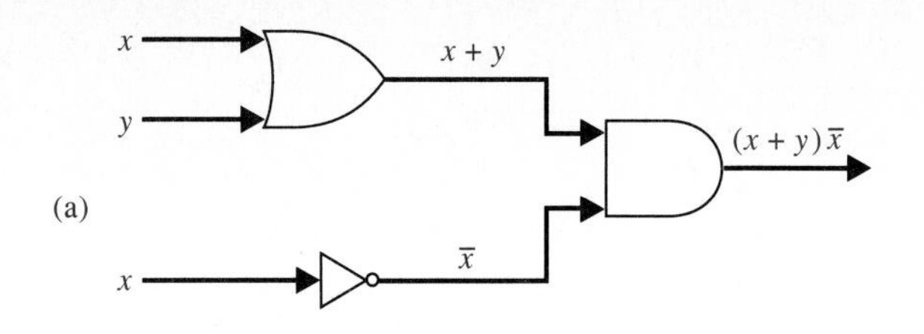

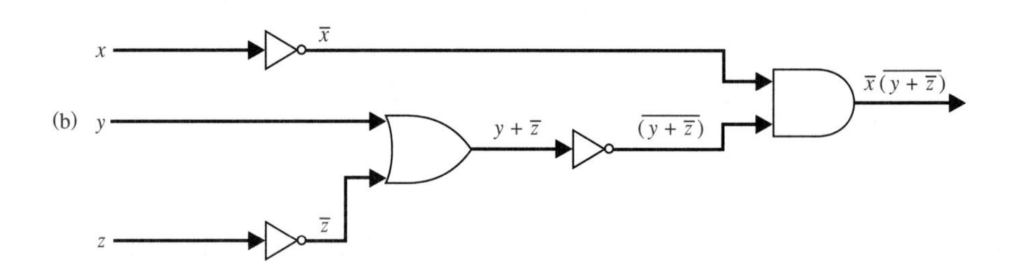

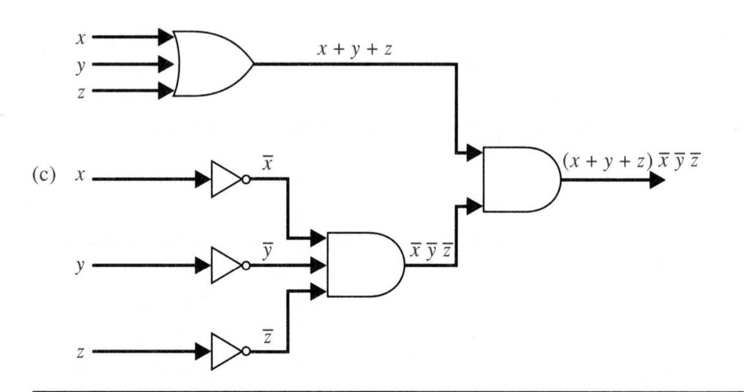

FIGURE 4 Circuits that Produce the Outputs Specified in Example 1.

used as input by one or more other elements, as shown in Figure 3. Both drawings in Figure 3 depict the circuit that produces the output $xy + \overline{x}y$.

EXAMPLE 1 Construct circuits that produce the following outputs: (a) $(x + y)\overline{x}$, (b) $\overline{x}\,\overline{(y + \overline{z})}$, and (c) $(x + y + z)(\overline{x}\,\overline{y}\,\overline{z})$.

Solution: Circuits that produce these outputs are shown in Figure 4. ■

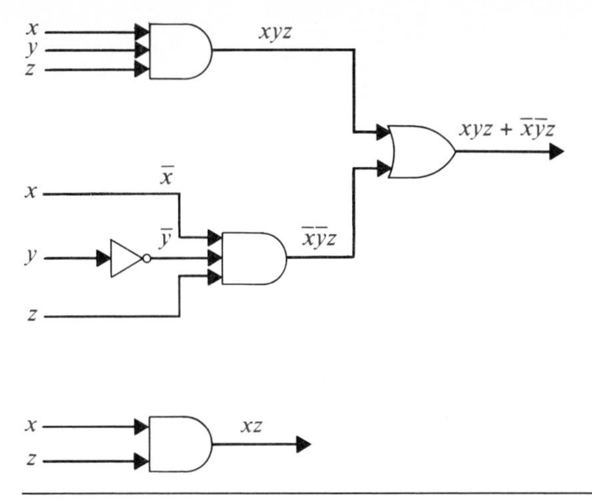

FIGURE 1 Two Circuits with the Same Output.

$$\begin{aligned} xyz + x\overline{y}z &= (y + \overline{y})(xz) \\ &= 1 \cdot (xz) \\ &= xz \end{aligned}$$

Hence, xz is a Boolean expression with fewer operators that represents the circuit. We show two different implementations of this circuit in Figure 1. The second circuit uses only one gate, whereas the first circuit uses three gates and an inverter.

This example shows that combining terms in the sum-of-products expansion of a circuit leads to a simpler expression for the circuit. We will describe two procedures that simplify sum-of-products expansions. The goal of both of these procedures is to produce Boolean sums of Boolean products that contain the least number of products of literals such that these products contain the least number of literals possible among all sums of products that represent a Boolean function.

KARNAUGH MAPS

To reduce the number of terms in a Boolean expression representing a circuit, it is necessary to find terms to combine. There is a graphical method, called a **Karnaugh map**, for finding terms to combine for Boolean functions involving a relatively small number of variables. The method we will describe was introduced by M. Karnaugh in

Maurice Karnaugh (born 1924). Maurice Karnaugh, born in New York City, received his B.S. from the City College of New York and his M.S. and Ph.D. from Yale University. He was a member of the technical staff at Bell Laboratories from 1952 until 1966 and Manager of Research and Development at the Federal Systems Division of AT&T from 1966 to 1970. In 1970 he joined IBM as a member of the research staff. Karnaugh has made fundamental contributions to the application of digital techniques in both computing and telecommunications. His current interests include knowledge-based systems in computers and heuristic search methods.

	y	$\overline{y}$
x	xy	$x\overline{y}$
$\overline{x}$	$\overline{x}y$	$\overline{x}\,\overline{y}$

FIGURE 2 Karnaugh Maps in Two Variables.

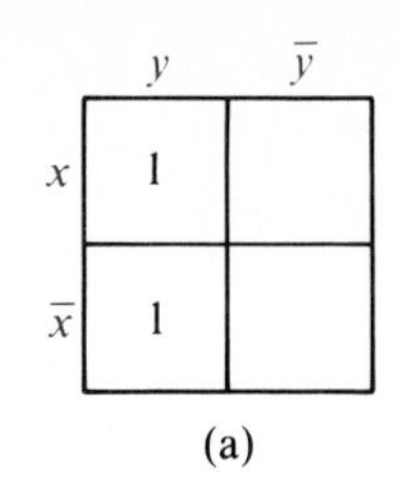

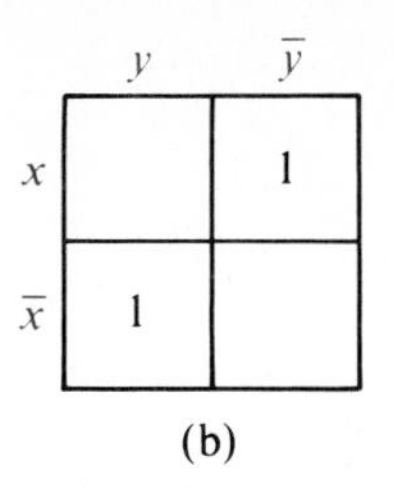

	y	$\overline{y}$
x		1
$\overline{x}$	1	1

(c)

FIGURE 3 Karnaugh Maps for the Sum-of-Products Expansions in Example 1.

1953. His method is based on earlier work by E.W. Veitch. (This method is usually applied only when the function involves six or fewer variables.) Karnaugh maps give us a visual method for simplifying sum-of-products expansions; they are not suited for mechanizing this process. We will first illustrate how Karnaugh maps are used to simplify expansions of Boolean functions in two variables.

There are four possible minterms in the sum-of-products expansion of a Boolean function in the two variables x and y. A Karnaugh map for a Boolean function in these two variables consists of four squares, where a 1 is placed in the square representing a minterm if this minterm is present in the expansion. Squares are said to be **adjacent** if the minterms that they represent differ in exactly one literal. For instance the square representing $\overline{x}y$ is adjacent to the squares representing xy and $\overline{x}\,\overline{y}$. The four squares and the terms that they represent are shown in Figure 2.

EXAMPLE 1 Find the Karnaugh maps for (a) $xy + \overline{x}y$, (b) $x\overline{y} + \overline{x}y$, and (c) $x\overline{y} + \overline{x}y + \overline{x}\,\overline{y}$.

Solution: We include a 1 in a square when the minterm represented by this square is present in the sum-of-products expansion. The three Karnaugh maps are shown in Figure 3. ∎

We can identify minterms that can be combined from the Karnaugh map. Whenever there are 1s in two adjacent squares in the Karnaugh map, the minterms represented by these squares can be combined into a product involving just one of the variables. For instance, $x\overline{y}$ and $\overline{x}\,\overline{y}$ are represented by adjacent squares and can be combined into $\overline{y}$, since $x\overline{y} + \overline{x}\,\overline{y} = (x + \overline{x})\overline{y} = \overline{y}$. Moreover, if 1s are in all four squares, the four minterms can be combined into one term, namely, the Boolean expression 1 that involves none of the variables. We circle blocks of squares in the Karnaugh map that represent minterms that can be combined and then find the corresponding sum of products. The goal is to identify the largest possible blocks, and to cover all the 1s with the fewest blocks using the largest blocks first and always using the largest possible blocks.

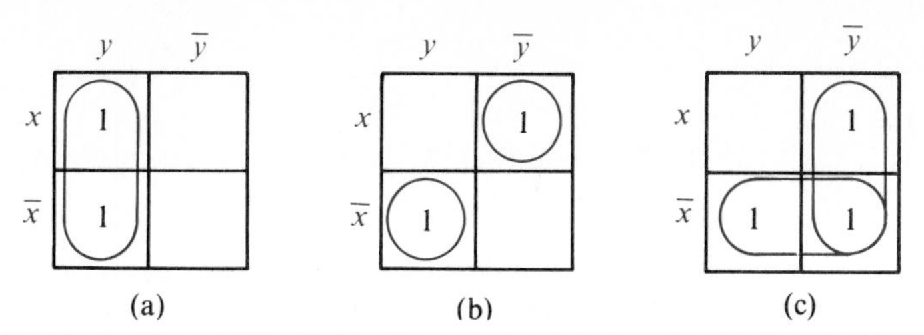

FIGURE 4 **Simplifying the Sum-of-Products Expansion from Example 1.**

EXAMPLE 2 Simplify the sum-of-products expansions given in Example 1.

Solution: The grouping of minterms is shown in Figure 4 using the Karnaugh maps for these expansions. Minimal expansions for these sums-of-products are (a) y, (b) $x\overline{y} + \overline{x}y$, and (c) $\overline{x} + \overline{y}$. ■

A Karnaugh map in three variables is a rectangle divided into eight squares. The squares represent the eight possible minterms in three variables. Two squares are said to be adjacent if the minterms that they represent differ in exactly one literal. One of the ways to form a Karnaugh map in three variables is shown in Figure 5(a). This Karnaugh map can be thought of as lying on a cylinder, as shown in Figure 5(b). On the cylinder two squares have a common border if and only if they are adjacent.

To simplify a sum-of-products expansion in three variables, we use the Karnaugh map to identify blocks of minterms that can be combined. Blocks of two adjacent squares represent pairs of minterms that can be combined into a product of two literals; 2×2 and 4×1 blocks of squares represent minterms that can be combined into a single literal; and the block of all eight squares represents a product of no literals, namely, the function 1. In Figure 6, 1×2, 2×1, 2×2, 4×1, and 4×2 blocks and the products they represent are shown.

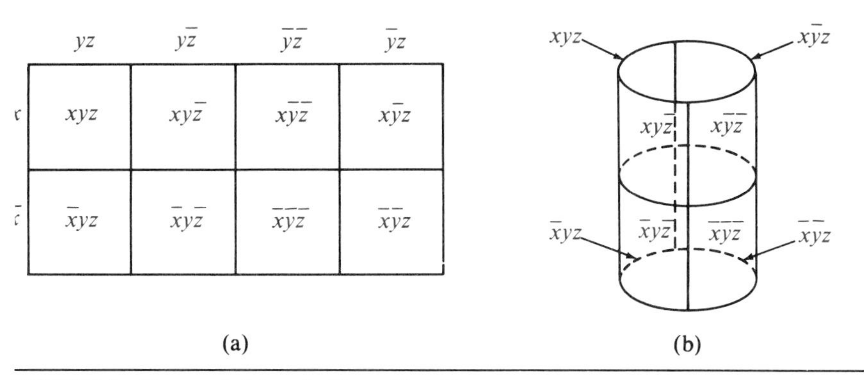

FIGURE 5 **Karnaugh Maps in Three Variables.**

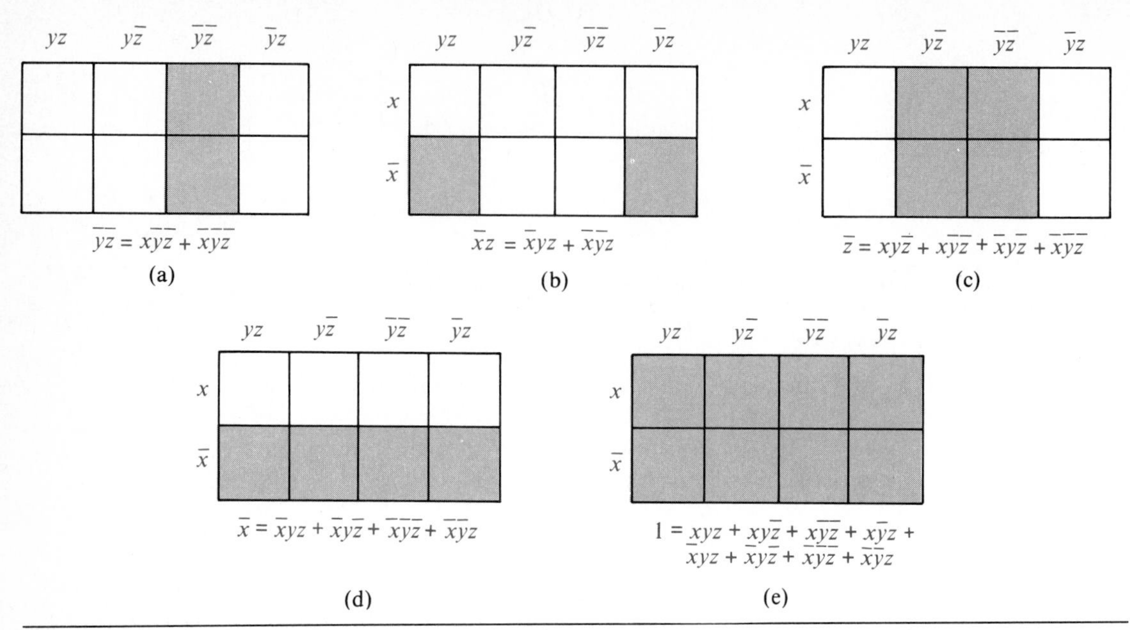

yz $y\bar{z}$ $\bar{y}\bar{z}$ $\bar{y}z$
x
$\bar{x}$
$\bar{y}\bar{z} = x\bar{y}\bar{z} + \bar{x}\bar{y}\bar{z}$
(a)
$\bar{x}z = \bar{x}yz + \bar{x}\bar{y}z$
(b)
$\bar{z} = xy\bar{z} + x\bar{y}\bar{z} + \bar{x}y\bar{z} + \bar{x}\bar{y}\bar{z}$
(c)
$\bar{x} = \bar{x}yz + \bar{x}y\bar{z} + \bar{x}\bar{y}\bar{z} + \bar{x}\bar{y}z$
(d)
$1 = xyz + xy\bar{z} + x\bar{y}\bar{z} + x\bar{y}z +$
$\bar{x}yz + \bar{x}y\bar{z} + \bar{x}\bar{y}\bar{z} + \bar{x}\bar{y}z$
(e)

FIGURE 6 Blocks in Karnaugh Maps in Three Variables.

The goal is to identify the largest possible blocks in the map and cover all the 1s in the map with the least number of blocks, using the largest blocks first. The largest possible blocks are always chosen. Note that there may be more than one way to do this. The following example illustrates how Karnaugh maps in three variables are used.

EXAMPLE 3 Use Karnaugh maps to simplify the sum-of-products expansions (a) $xy\bar{z} + x\bar{y}\bar{z} + \bar{x}yz + \bar{x}\bar{y}\bar{z}$, (b) $x\bar{y}z + x\bar{y}\bar{z} + \bar{x}yz + \bar{x}\bar{y}z + \bar{x}\bar{y}\bar{z}$, and (c) $xyz + xy\bar{z} + x\bar{y}z + x\bar{y}\bar{z} + \bar{x}yz + \bar{x}\bar{y}z + \bar{x}\bar{y}\bar{z}$.

Solution: The Karnaugh maps for these sum-of-products expansions are shown in Figure 7. The grouping of blocks shows that minimal expansions into Boolean sums of Boolean products are (a) $x\bar{z} + \bar{y}\bar{z} + \bar{x}yz$, (b) $\bar{y} + \bar{x}z$, and (c) $x + \bar{y} + z$. ■

A Karnaugh map in four variables is a square that is divided into 16 squares. The squares represent the 16 possible minterms in four variables. One of the ways to form a Karnaugh map in four variables is shown in Figure 8.

Two squares are adjacent if and only if the minterms they represent differ in one literal. Consequently, each square is adjacent to four other squares. The Karnaugh map of a sum-of-products expansion in four variables can be thought of as lying

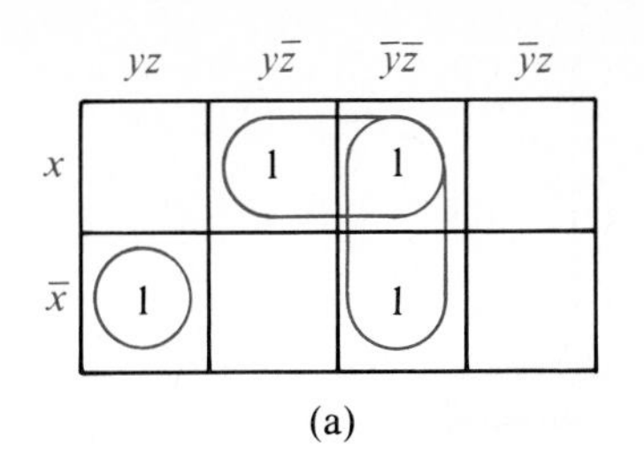

(a)

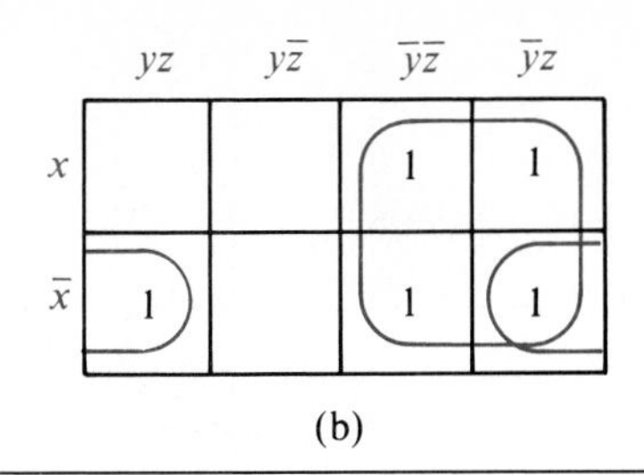

(b)

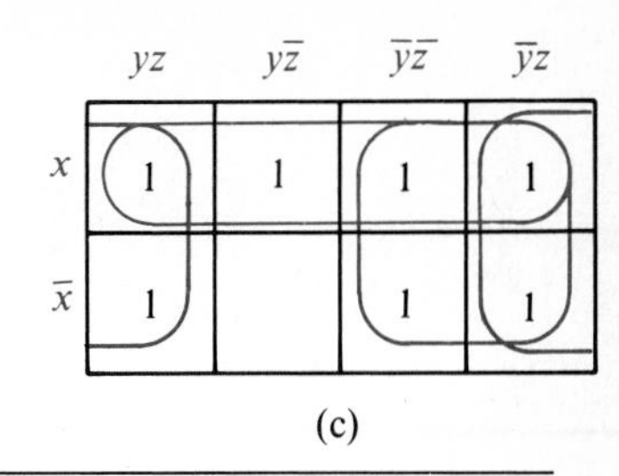

(c)

FIGURE 7 Using Karnaugh Maps in Three Variables.

on a torus, so that adjacent squares have a common boundary (see Exercise 20). The simplification of a sum-of-products expansion in four variables is carried out by identifying those blocks of 2, 4, 8, or 16 squares that represent minterms that can be combined. Each square representing a minterm must either be used to form a product using fewer literals, or be included in the expansion. In Figure 9 some examples of blocks that represent products of three literals, products of two literals, and a single literal are illustrated.

As is the case in Karnaugh maps in two and three variables, the goal is to identify the largest blocks of 1s in the map and to cover all the 1s using the fewest blocks needed, using the largest blocks first. The largest possible blocks are always used. The following example illustrates how Karnaugh maps in four variables are used.

EXAMPLE 4 Use Karnaugh maps to simplify the sum-of-products expansions (a) $wxyz + wxy\overline{z} + wx\overline{y}\,\overline{z} + w\overline{x}yz + w\overline{x}\,\overline{y}z + w\overline{x}\,\overline{y}\,\overline{z} + \overline{w}x\overline{y}z + \overline{w}\,\overline{x}yz + \overline{w}\,\overline{x}y\overline{z}$, (b) $wx\overline{y}\,\overline{z} + w\overline{x}yz + w\overline{x}y\overline{z} + w\overline{x}\,\overline{y}\,\overline{z} + \overline{w}x\overline{y}\,\overline{z} + \overline{w}\,\overline{x}y\overline{z} + \overline{w}\,\overline{x}\,\overline{y}\,\overline{z}$, and (c) $wxy\overline{z} + wx\overline{y}\,\overline{z} + w\overline{x}yz + w\overline{x}y\overline{z} + w\overline{x}\,\overline{y}\,\overline{z} + \overline{w}xyz + \overline{w}xy\overline{z} + \overline{w}x\overline{y}\,\overline{z} + \overline{w}x\overline{y}z + \overline{w}\,\overline{x}y\overline{z} + \overline{w}\,\overline{x}\,\overline{y}\,\overline{z}$.

Solution: The Karnaugh maps for these expansions are shown in Figure 10. Using the blocks shown leads to the sum of products (a) $wyz + wx\overline{z} + w\overline{x}\,\overline{y} + \overline{w}\,\overline{x}y + \overline{w}x\overline{y}z$,

	yz	$y\overline{z}$	$\overline{y}\,\overline{z}$	$\overline{y}z$
wx	$wxyz$	$wxy\overline{z}$	$wx\overline{y}\,\overline{z}$	$wx\overline{y}z$
$w\overline{x}$	$w\overline{x}yz$	$w\overline{x}y\overline{z}$	$w\overline{x}\,\overline{y}\,\overline{z}$	$w\overline{x}\,\overline{y}z$
$\overline{w}\,\overline{x}$	$\overline{w}\,\overline{x}yz$	$\overline{w}\,\overline{x}y\overline{z}$	$\overline{w}\,\overline{x}\,\overline{y}\,\overline{z}$	$\overline{w}\,\overline{x}\,\overline{y}z$
$\overline{w}x$	$\overline{w}xyz$	$\overline{w}xy\overline{z}$	$\overline{w}x\overline{y}\,\overline{z}$	$\overline{w}x\overline{y}z$

FIGURE 8 Karnaugh Map in Four Variables.

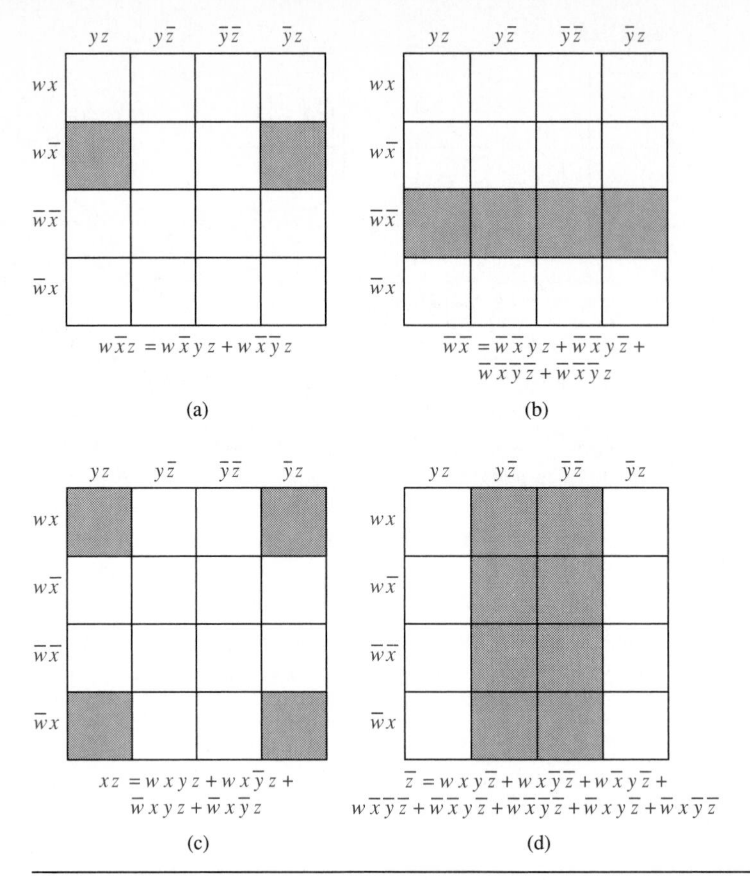

FIGURE 9 Blocks in Karnaugh Maps in Four Variables.

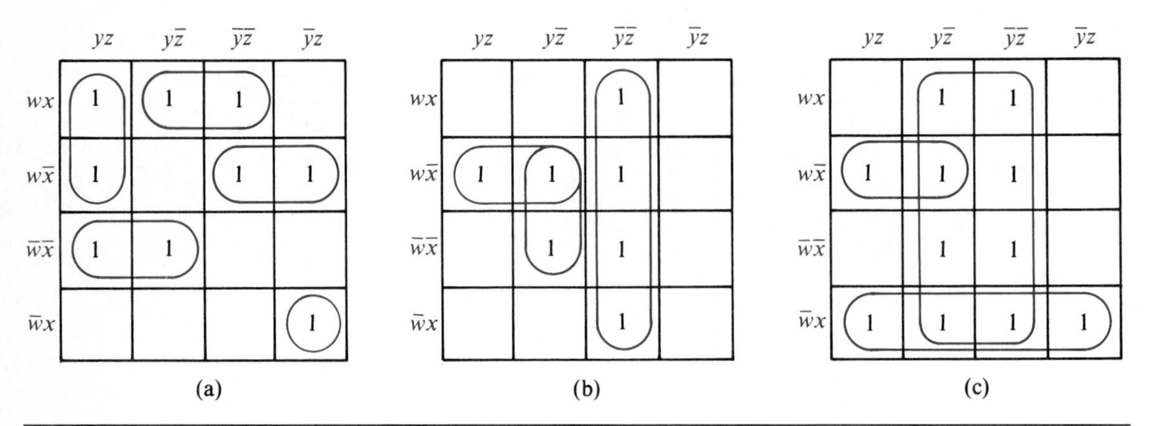

FIGURE 10 Using Karnaugh Maps in Four Variables.

(b) $\overline{y}\,\overline{z} + w\overline{x}y + \overline{x}y\overline{z}$, and (c) $\overline{z} + \overline{w}x + w\overline{x}y$. The reader should determine whether there are other choices of blocks in each part that lead to different sums of products representing these Boolean functions. ■

DON'T CARE CONDITIONS

In some circuits we care only about the output for some combinations of input values, since the other combinations of input values never arise. This gives us freedom in producing a simple circuit with the desired output since the output values for all those combinations that never occur can be arbitrarily chosen. The values of the function for these combinations are called ***don't care* conditions**. A d is used in a Karnaugh map to mark those combinations of values of the variables for which the function can be arbitrarily assigned. In the simplification process we can assign 1s as values to those combinations of the input values that lead to the largest blocks in the Karnaugh map. This is illustrated in the following example.

EXAMPLE 5 One way to code decimal expansions using bits is to use the four bits of the binary expansion of each digit in the decimal expansion. For instance, 873 is encoded as 100001110011. This encoding of a decimal expansion is called a **binary coded decimal expansion**. Since there are 16 blocks of four bits and only 10 decimal digits, there are six combinations of four bits that are not used to encode digits. Suppose that a circuit is to be built that produces an output of 1 if the decimal digit is 5 or greater and an output of 0 if the decimal digit is less than 5. How can this circuit be simply built using *OR* gates, *AND* gates, and inverters?

Solution: Let $F(w, x, y, z)$ denote the output of the circuit, where $wxyz$ is a binary expansion of a decimal digit. The values of F are shown in Table 1. The Karnaugh map for F, with ds in the *don't care* positions, is shown in Figure 11(a). We can either

TABLE 1

Digit	w	x	y	z	F
0	0	0	0	0	0
1	0	0	0	1	0
2	0	0	1	0	0
3	0	0	1	1	0
4	0	1	0	0	0
5	0	1	0	1	1
6	0	1	1	0	1
7	0	1	1	1	1
8	1	0	0	0	1
9	1	0	0	1	1

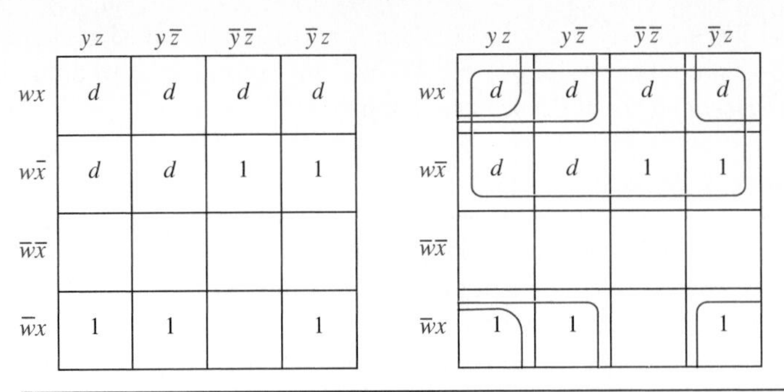

FIGURE 11 The Karnaugh Map for F Showing Its *Don't Care* Positions.

include or exclude squares with ds from blocks. This gives us many possible choices for the blocks. The blocks shown in Figure 11(b) produce the simplest expansion possible, namely, $F(x, y, z) = w + xy + xz$. ■

THE QUINE–McCLUSKEY METHOD

We have seen that Karnaugh maps can be used to produce minimal expansions of Boolean functions as Boolean sums of Boolean products. However, Karnaugh maps are awkward to use when there are more than four variables. Furthermore, the use of Karnaugh maps relies on visual inspection to identify terms to group. For these reasons there is a need for a procedure for simplifying sum-of-products expansions that can be mechanized. The Quine–McCluskey method is such a procedure. It can be used for Boolean functions in any number of variables. It was developed in the 1950s by W. V. Quine and E. J. McClusky, Jr. Basically, the Quine–McCluskey method consists of two parts. The first part finds those terms that are candidates for inclusion in a minimal expansion as a Boolean sum of Boolean products. The second part determines which of these terms to actually use. We will show how this procedure works using an example.

EXAMPLE 6 We will show how the Quine–McCluskey method can be used to find a minimal expansion equivalent to

$$xyz + x\overline{y}z + \overline{x}yz + \overline{x}\,\overline{y}z + \overline{x}\,\overline{y}\,\overline{z}.$$

Edward J. McCluskey (born 1929). Edward McCluskey attended Bowdoin College and M.I.T., where he received his doctorate in Electrical Engineering in 1956. He joined Bell Telephone Laboratories in 1955, remaining there until 1959. McClusky was professor of Electrical Engineering at Princeton University from 1959 until 1966, also serving as Director of the Computer Center at Princeton from 1961 to 1966. In 1967 he took a position as professor of Computer Science and Electrical Engineering at Stanford University, where he also served as director of the Digital Systems Laboratory from 1969 to 1978.

Willard Van Orman Quine (born 1908). Willard Quine, born in Akron, Ohio, attended Oberlin College and later Harvard University, where he received his Ph.D. in Philosophy in 1932. He joined the Department of Philosophy at Harvard in 1936, where he remained until 1978. Quine has made many fundamental contributions to logic and philosophy.

We will represent the minterms in this expansion by bit strings. The first bit will be 1 if x occurs and 0 if $\overline{x}$ occurs. The second bit will be 1 if y occurs and 0 if $\overline{y}$ occurs. The third bit will be 1 if z occurs and 0 if $\overline{z}$ occurs. We then group these terms according to the number of 1s in the corresponding bit strings. This information is shown in Table 2.

Minterms that can be combined are those that differ in exactly one literal. Hence, two terms that can be combined differ by exactly one in the number of 1s in the bit strings that represent them. When two minterms are combined into a product, this product contains two literals. A product in two literals is represented using a dash to denote the variable that does not occur. For instance, the minterms $x\overline{y}z$ and $\overline{x}\,\overline{y}z$, represented by bit strings 101 and 001, can be combined into $\overline{y}z$, represented by the string – 01. All pairs of minterms that can be combined and the product formed from these combinations are shown in Table 3.

Next, all pairs of products of two literals that can be combined are combined into one literal. Two such products can be combined if they contain literals for the same two variables, and literals for only one of the two variables differ. In terms of the strings representing the products, these strings must have a dash in the same position and must differ in exactly one of the other two slots. We can combine the products yz and $\overline{y}z$, represented by strings – 11 and – 01, into z, represented by the string – –1. We show all the combinations of terms that can be formed in this way in Table 3.

In Table 3 we also indicate which terms have been used to form products with fewer literals; these terms will not be needed in a minimal expansion. The next step is to identify a minimal set of products needed to represent the Boolean function. We begin with all those products that were not used to construct products with fewer literals. Next, we form Table 4, which has a row for each candidate product formed by combining original terms, and a column for each original term; and we put an X in a position if the original term in the sum-of-products expansion was used to form this candidate product. In this case, we say that the candidate product **covers** the original minterm. We need to include at least one product that covers each of the original minterms. Consequently, whenever there is only one X in a column in the table, the

TABLE 2

Minterm	*Bit String*	*Number of 1s*
xyz	111	3
$x\overline{y}z$	101	2
$\overline{x}yz$	011	2
$\overline{x}\,\overline{y}z$	001	1
$\overline{x}\,\overline{y}\,\overline{z}$	000	0

TABLE 3

	Term	*Bit String*	*Step 1*			*Step 2*		
				Term	*String*		*Term*	*String*
1	xyz	111	(1,2)	xz	1–1	(1,2,3,4)	z	– –1
2	$x\overline{y}z$	101	(1,3)	yz	–11			
3	$\overline{x}yz$	011	(2,4)	$\overline{y}z$	–01			
4	$\overline{x}\,\overline{y}z$	001	(3,4)	$\overline{x}z$	0–1			
5	$\overline{x}\,\overline{y}\,\overline{z}$	000	(4,5)	$\overline{x}\,\overline{y}$	00–			

TABLE 4

	xyz	$x\overline{y}z$	$\overline{x}yz$	$\overline{x}\overline{y}z$	$\overline{x}\overline{y}\overline{z}$
z	X	X	X	X	
$\overline{x}\overline{y}$				X	X

product corresponding to the row this X is in must be used. From Table 4 we see that both z and $\overline{x}\,\overline{y}$ are needed. Hence, the final answer is $z + \overline{x}\,\overline{y}$. ■

As was illustrated in Example 6, the Quine–McCluskey method uses the following sequence of steps to simplify a sum-of-products expression.

1. Express each minterm in n variables by a bit string of length n with a 1 in the ith position if x_i occurs and a 0 in this position if $\overline{x}_i$ occurs.
2. Group the bit strings according to the number of 1s in them.
3. Determine all products in $n - 1$ variables that can be formed by taking the Boolean sum of minterms in the expansion. Minterms that can be combined are represented by bit strings that differ in exactly one position. Represent these products in $n - 1$ variables with strings that have a 1 in the ith position if x_i occurs in the product, a 0 in this position if $\overline{x}_i$ occurs, and a dash in this position if there is no literal involving x_i in the product.
4. Determine all products in $n - 2$ variables that can be formed by taking the Boolean sum of the products in $n - 1$ variables found in the previous step. Products in $n - 1$ variables that can be combined are represented by bit strings that have a dash in the same position, and differ in exactly one position.
5. Continue combining Boolean products into products in fewer variables as long as possible.
6. Find all the Boolean products that arose that were not used to form a Boolean product in one fewer literal.
7. Find the smallest set of these Boolean products so that the sum of these products represents the Boolean function. This is done by forming a table showing which minterms are covered by which products. Every minterm must be covered by at least one product. (This is the most difficult part of the procedure. It can be mechanized using a backtracking procedure.)

A final example will illustrate how this procedure is used to simplify a sum-of-products expansion in four variables.

EXAMPLE 7 Use the Quine–McCluskey method to simplify the sum-of-products expansion $wxy\overline{z} + w\overline{x}yz + w\overline{x}y\overline{z} + \overline{w}xyz + \overline{w}x\overline{y}z + \overline{w}\,\overline{x}yz + \overline{w}\,\overline{x}\,\overline{y}z$.

Solution: We first represent the minterms by bit strings, and then group these terms together according to the number of 1s in the bit strings. This is shown in Table 5. All

TABLE 5

Term	*Bit String*	*Number of 1s*
$wxy\overline{z}$	1110	3
$w\overline{x}yz$	1011	3
$\overline{w}xyz$	0111	3
$w\overline{x}y\overline{z}$	1010	2
$\overline{w}x\overline{y}z$	0101	2
$\overline{w}\overline{x}yz$	0011	2
$\overline{w}\overline{x}\overline{y}z$	0001	1

TABLE 6

		Step 1		*Step 2*	
Term	*Bit String*	*Term*	*String*	*Term*	*String*
1 $wxy\overline{z}$	1110	(1,4) $wy\overline{z}$	1–10	(3,5,6,7) $\overline{w}z$	0––1
2 $w\overline{x}yz$	1011	(2,4) $w\overline{x}y$	101–		
3 $\overline{w}xyz$	0111	(2,6) $\overline{x}yz$	–011		
4 $w\overline{x}y\overline{z}$	1010	(3,5) $\overline{w}xz$	01–1		
5 $\overline{w}x\overline{y}z$	0101	(3,6) $\overline{w}yz$	0–11		
6 $\overline{w}\overline{x}yz$	0011	(5,7) $\overline{w}\overline{y}z$	0–01		
7 $\overline{w}\overline{x}\overline{y}z$	0001	(6,7) $\overline{w}\overline{x}z$	00–1		

the Boolean products that can be formed by taking Boolean sums of these products are shown in Table 6.

The only products that were not used to form products in fewer variables are $\overline{w}z$, $wy\overline{z}$, $w\overline{x}y$, and $\overline{x}yz$. In Table 7 we show the minterms covered by each of these products. To cover these minterms we must include $\overline{w}z$ and $wy\overline{z}$, since these products are the only products that cover $\overline{w}xyz$ and $wxy\overline{z}$, respectively. Once these two products are included, we see that only one of the two products left is needed. Consequently, we can take either $\overline{w}z + wy\overline{z} + w\overline{x}y$ or $\overline{w}z + wy\overline{z} + \overline{x}yz$ as the final answer. ■

TABLE 7

	$wxy\overline{z}$	$w\overline{x}yz$	$\overline{w}xyz$	$w\overline{x}y\overline{z}$	$\overline{w}x\overline{y}z$	$\overline{w}\overline{x}yz$	$\overline{w}\overline{x}\overline{y}z$
$\overline{w}z$			X		X	X	X
$wy\overline{z}$	X			X			
$w\overline{x}y$		X		X			
$\overline{x}yz$		X				X	

Exercises

1. **a)** Draw a Karnaugh map for a function in two variables and put a 1 in the square representing $\overline{x}y$.

   b) What are the minterms represented by squares adjacent to this square?

2. Find the sum-of-products expansions represented by each of the following Karnaugh maps.

a)

	y	$\overline{y}$
x	1	
$\overline{x}$	1	1

b)

	y	$\overline{y}$
x	1	1
$\overline{x}$		

c)

	y	$\overline{y}$
x	1	1
$\overline{x}$	1	1

3. Draw the Karnaugh maps of the following sum-of-products expansions in two variables.
a) $x\overline{y}$ **b)** $xy + \overline{x}\,\overline{y}$ **c)** $xy + x\overline{y} + \overline{x}y + \overline{x}\,\overline{y}$

4. Use a Karnaugh map to find a minimal expansion as a Boolean sum of Boolean products of each of the following functions of the Boolean variables x and y.
a) $\overline{x}y + \overline{x}\,\overline{y}$ **b)** $xy + x\overline{y}$ **c)** $xy + x\overline{y} + \overline{x}y + \overline{x}\,\overline{y}$

5. **a)** Draw a Karnaugh map for a function in three variables. Put a 1 in the square that represents $\overline{x}y\overline{z}$.
b) Which minterms are represented by squares adjacent to this square?

6. Use Karnaugh maps to find simpler circuits with the same output as each of the following circuits.

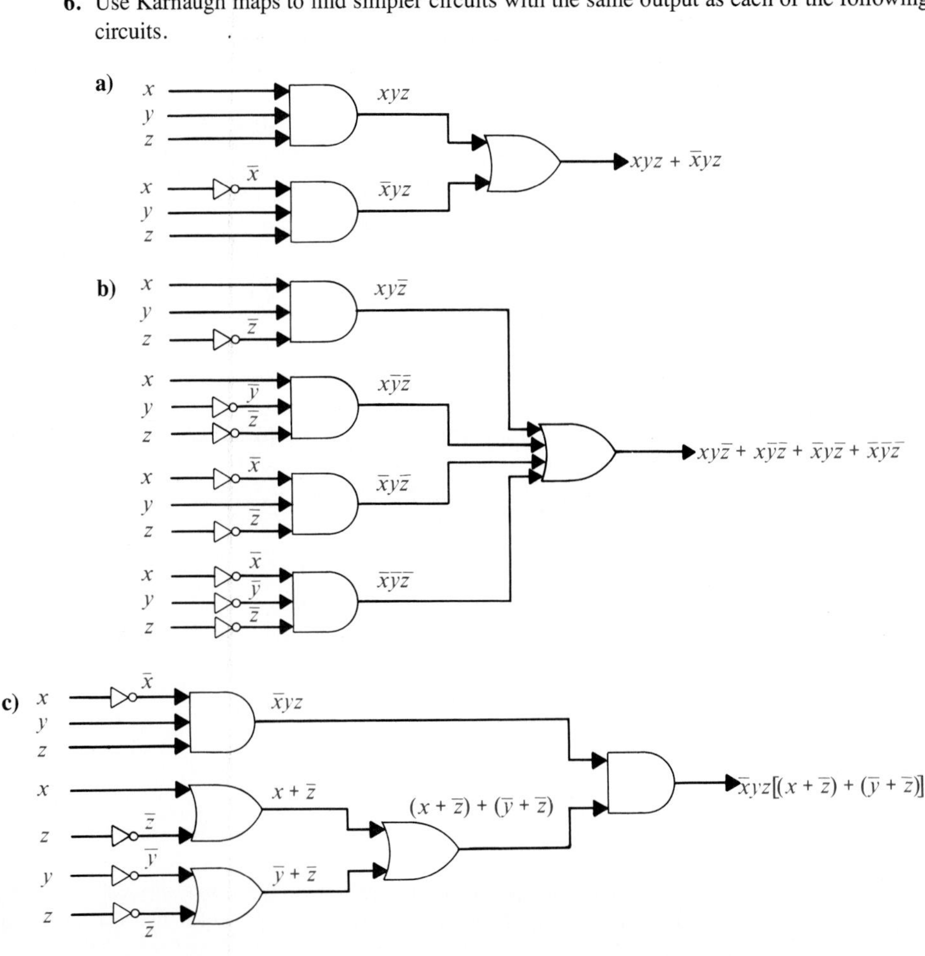

7. Draw the Karnaugh maps of the following sum-of-products expansions in three variables.
   a) $x\overline{y}\,\overline{z}$ b) $\overline{x}yz + \overline{x}\,\overline{y}\,\overline{z}$ c) $xyz + xy\overline{z} + \overline{x}y\overline{z} + \overline{x}\,\overline{y}z$
8. Use a Karnaugh map to find a minimal expansion as a Boolean sum of Boolean products of each of the following functions in the variables x, y, and z.
   a) $\overline{x}yz + \overline{x}\,\overline{y}z$ b) $xyz + xy\overline{z} + \overline{x}yz + \overline{x}y\overline{z}$
   c) $xy\overline{z} + x\overline{y}z + x\overline{y}\,\overline{z} + \overline{x}yz + \overline{x}\,\overline{y}z$ d) $xyz + x\overline{y}z + x\overline{y}\,\overline{z} + \overline{x}yz + \overline{x}y\overline{z} + \overline{x}\,\overline{y}\,\overline{z}$
9. a) Draw a Karnaugh map for a function in four variables. Put a 1 in the square that represents $\overline{w}xy\overline{z}$.
   b) Which minterms are represented by squares adjacent to this square?
10. Use a Karnaugh map to find a minimal expansion as a Boolean sum of Boolean products of each of the following functions in the variables w, x, y, and z.
    a) $wxyz + wx\overline{y}z + wx\overline{y}\,\overline{z} + w\overline{x}y\overline{z} + w\overline{x}\,\overline{y}z$
    b) $wxy\overline{z} + wx\overline{y}z + w\overline{x}yz + \overline{w}x\overline{y}z + \overline{w}\,\overline{x}y\overline{z} + \overline{w}\,\overline{x}\,\overline{y}z$
    c) $wxyz + wxy\overline{z} + wx\overline{y}z + w\overline{x}\,\overline{y}z + w\overline{x}\,\overline{y}\,\overline{z} + \overline{w}x\overline{y}z + \overline{w}\,\overline{x}y\overline{z} + \overline{w}\,\overline{x}\,\overline{y}z$
    d) $wxyz + wxy\overline{z} + wx\overline{y}z + w\overline{x}yz + w\overline{x}y\overline{z} + \overline{w}xyz + \overline{w}\,\overline{x}yz + \overline{w}\,\overline{x}y\overline{z} + \overline{w}\,\overline{x}\,\overline{y}z$
11. a) How many squares does a Karnaugh map in five variables have?
    b) How many squares are adjacent to a given square in a Karnaugh map in five variables?

★12. Use Karnaugh maps to find a minimal expansion as a Boolean sum of Boolean products of Boolean functions that have as input the binary code for each decimal digit and produce as output a 1 if and only if the digit corresponding to the input is
   a) odd. b) not divisible by 3.
   c) not 4, 5, or 6.

★13. Suppose that there are five members on a committee, but that Smith and Jones always vote the opposite of Marcus. Design a circuit that implements majority voting of the committee using this relationship between votes.

14. Use the Quine–McCluskey method to simplify the sum-of-products expansions in Example 3.
15. Use the Quine–McCluskey method to simplify the sum-of-products expansions in Exercise 8.
16. Use the Quine–McCluskey method to simplify the sum-of-products expansions in Example 4.
17. Use the Quine–McCluskey method to simplify the sum-of-products expansions in Exercise 10.

★18. Explain how Karnaugh maps can be used to simplify product-of-sums expansions in three variables. (*Hint:* Mark with a 0 all the maxterms in an expansion and combine blocks of maxterms.)

19. Use the method from Exercise 18 to simplify the product-of-sums expansion $(x + y + z) \cdot (x + y + \overline{z})(x + \overline{y} + z)(x + \overline{y} + z)(\overline{x} + y + z)$.

★20. Draw a Karnaugh map for the 16 minterms in four Boolean variables on the surface of a torus.

Key Terms and Results

TERMS

Boolean variable: a variable that assumes only the values 0 and 1

$\overline{x}$ (complement of x): an expression with the value 1 when x has the value 0 and the value 0 when x has the value 1

$x \cdot y$ (or xy) (Boolean product or conjunction of x and y): an expression with the value 1 when both x and y have the value 1 and the value 0 otherwise

$x + y$ (Boolean sum or disjunction of x and y): an expression with the value 1 when either x or y, or both, has the value 1, and 0 otherwise

Boolean expressions: the expressions obtained recursively by specifying that $0, 1, x_1, \ldots, x_n$ are Boolean expressions and $\overline{E}_1$, $(E_1 + E_2)$, and (E_1E_2) are Boolean expressions if E_1 and E_2 are

dual of a Boolean expression: the expression obtained by interchanging + signs and $\cdot$ signs and interchanging 0s and 1s

Boolean function of degree n: a function from B^n to B where $B = \{0, 1\}$

literal of the Boolean variable x: either x or $\overline{x}$

minterm of $x_1, x_2, \ldots, x_n$: a Boolean product $y_1y_2 \cdots y_n$ where each y_i is either x_i or $\overline{x}_i$

sum-of-products expansion (or disjunctive normal form): the representation of a Boolean function as a disjunction of minterms

functionally complete: a set of Boolean operators is called functionally complete if every Boolean function can be represented using these operators

$x \mid y$ (or x *NAND* y): the expression that has the value 0 when both x and y have the value 1 and the value 1 otherwise

$x \downarrow y$ (or x *NOR* y): the expression that has the value 0 when either x or y or both have the value 1 and the value 0 otherwise

inverter: a device that accepts the value of a Boolean variable as input and produces the complement of the input

***OR* gate:** a device that accepts the values of two or more Boolean variables as input and produces their Boolean sum as output

***AND* gate:** a device that accepts the values of two or more Boolean variables as input and produces their Boolean product as output.

half adder: a circuit that adds two bits, producing a sum bit and a carry bit

full adder: a circuit that adds two bits and a carry, producing a sum bit and a carry bit

Karnaugh map for n variables: a rectangle divided into 2^n squares where each square represents a minterm in the variables

RESULTS

The identities for Boolean algebra (see Table 5 in Section 9.1).

An identity between Boolean functions represented by Boolean expressions remains valid when the duals of both sides of the identity are taken.

Every Boolean function can be represented by a sum-of-products expansion.

Each of the sets $\{+, ^-\}$ and $\{\cdot, ^-\}$ is functionally complete.

Each of the sets $\{\downarrow\}$ and $\{\mid\}$ is functionally complete.

The use of Karnaugh maps to minimize Boolean expressions.

The Quine–McCluskey method for minimizing Boolean expressions.

Review Questions

1. Define a Boolean function of degree n.
2. How many Boolean functions of degree 2 are there?
3. Give a recursive definition of the set of Boolean expressions.
4. a) What is the dual of a Boolean expression?
   b) What is the duality principle? How can it be used to find new identities involving Boolean expressions?

5. Explain how to construct the sum-of-products expansion of a Boolean function.
6. a) What does it mean for a set of operators to be functionally complete?
   b) Is the set $\{+, \cdot\}$ functionally complete?
   c) Are there sets of a single operator that are functionally complete?
7. Explain how to build a circuit for a light controlled by two switches using *OR* gates, *AND* gates, and inverters.
8. Construct a half adder using *OR* gates, *AND* gates, and inverters.
9. Is there a single type of logic gate that can be used to build all circuits that can be built using *OR* gates, *AND* gates, and inverters?
10. a) Explain how Karnaugh maps can be used to simplify sum-of-products expansions in three Boolean variables.
    b) Use a Karnaugh map to simplify the sum-of-products expansion $xyz + x\overline{y}z + x\overline{y}\,\overline{z} + \overline{x}yz + \overline{x}\,\overline{y}\,\overline{z}$.
11. a) Explain how Karnaugh maps can be used to simplify sum-of-products expansions in four Boolean variables.
    b) Use a Karnaugh map to simplify the sum-of-products expansion $wxyz + wxy\overline{z} + wx\overline{y}z + wx\overline{y}\,\overline{z} + w\overline{x}yz + w\overline{x}\,\overline{y}z + \overline{w}xyz + \overline{w}\,\overline{x}yz + \overline{w}\,\overline{x}y\overline{z}$.
12. a) What is a *don't care* condition?
    b) Explain how *don't care* conditions can be used to build a circuit using *OR* gates, *AND* gates, and inverters that produces an output of 1 if a decimal digit is 6 or greater, and an output of 0 if this digit is less than 6.
13. a) Explain how to use the Quine–McCluskey method to simplify sum-of-products expansions.
    b) Use this method to simplify $xy\overline{z} + x\overline{y}\,\overline{z} + \overline{x}y\overline{z} + \overline{x}\,\overline{y}z + \overline{x}\,\overline{y}\,\overline{z}$.

Supplementary Exercises

1. For which values of the Boolean variables x, y, and z does
   a) $x + y + z = xyz$? b) $x(y + z) = x + yz$? c) $\overline{x}\,\overline{y}\,\overline{z} = x + y + z$?
2. Let x and y belong to $\{0, 1\}$. Does it necessarily follow that $x = y$ if there exists a value z in $\{0, 1\}$ such that
   a) $xz = yz$? b) $x + z = y + z$? c) $x \oplus z = y \oplus z$?
   d) $x \downarrow z = y \downarrow z$? e) $x \mid z = y \mid z$?

A Boolean function F is called **self-dual** if and only if $F(x_1, \ldots, x_n) = \overline{F(\overline{x}_1, \ldots, \overline{x}_n)}$.

3. Which of the following functions are self-dual?
   a) $F(x, y) = x$ b) $F(x, y) = xy + \overline{x}\,\overline{y}$
   c) $F(x, y) = x + y$ d) $F(x, y) = xy + \overline{x}y$.
4. Give an example of a self-dual Boolean function of three variables.
★5. How many Boolean functions of degree n are self-dual?

We define the relation $\leq$ on the set of Boolean functions of degree n so that $F \leq G$ where F and G are Boolean functions if and only $G(x_1, x_2, \ldots, x_n) = 1$ whenever $F(x_1, x_2, \ldots, x_n) = 1$.

6. Determine whether $F \leq G$ or $G \leq F$ for the following pairs of functions.
   a) $F(x, y) = x, G(x, y) = x + y$ b) $F(x, y) = x + y, G(x, y) = xy$
   c) $F(x, y) = \overline{x}, G(x, y) = x + y$.
7. Show that if F and G are Boolean functions of degree n, then
   a) $F \leq F + G$. b) $FG \leq F$.
8. Show that if F, G, and H are Boolean functions of degree n, then $F + G \leq H$ if and only if $F \leq H$ and $G \leq H$.
★9. Show that the relation $\leq$ is a partial ordering on the set of Boolean functions of degree n.
★10. Draw the Hasse diagram for the poset consisting of the set of the 16 Boolean functions of degree 2 (shown in Table 3 of section 9.1) with the partial ordering $\leq$.

★11. For each of the following equalities either prove it is an identity or find a set of values of the variables for which it does not hold.

a) $x \mid (y \mid z) = (x \mid y) \mid z$ **b)** $x \downarrow (y \downarrow z) = (x \downarrow y) \downarrow (x \downarrow z)$

c) $x \downarrow (y \mid z) = (x \downarrow y) \mid (x \downarrow z)$

Define the Boolean operator $\odot$ as follows: $1 \odot 1 = 1$, $1 \odot 0 = 0$, $0 \odot 1 = 0$, and $0 \odot 0 = 1$.

12. Show that $x \odot y = xy + \overline{x}\,\overline{y}$.

13. Show that $x \odot y = \overline{(x \oplus y)}$.

14. Show that each of the following identities holds.

a) $x \odot x = 1$ **b)** $x \odot \overline{x} = 0$ **c)** $x \odot y = y \odot x$

15. Is it always true that $(x \odot y) \odot z = x \odot (y \odot z)$?

★16. Determine whether the set $\{\odot\}$ is functionally complete.

★17. How many of the 16 Boolean functions in two variables x and y can be represented using only the following set of operators, the variables x and y, and the values 0 and 1?

a) $\{\overline{\ }\}$ **b)** $\{\cdot\}$ **c)** $\{+\}$ **d)** $\{\cdot, +\}$

The notation for an *XOR* gate, which produces the output $x \oplus y$ from x and y, is as follows:

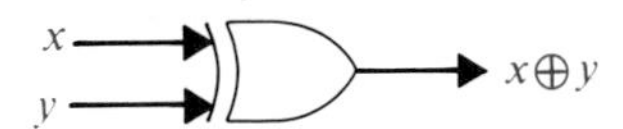

18. Determine the output of each of the following circuits.

a)

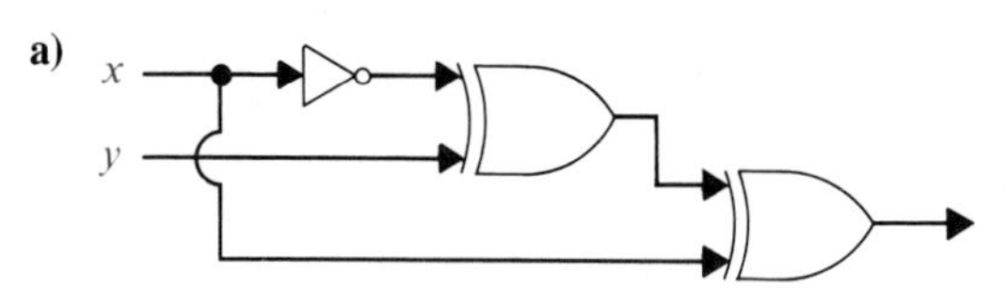

b)

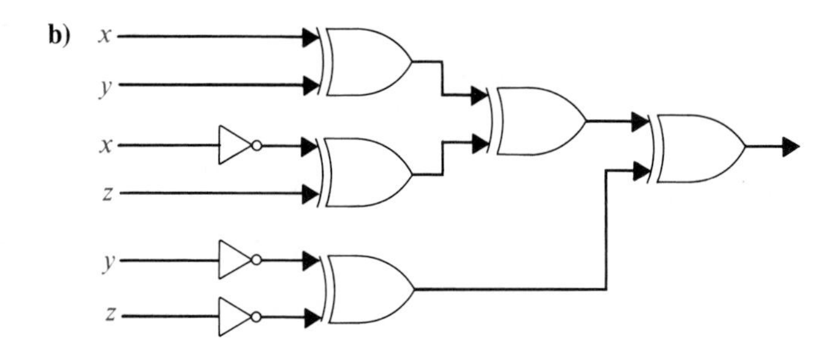

19. Show how a half adder can be constructed using fewer gates than are used in Figure 8 of Section 9.3 when *XOR* gates can be used in addition to *OR* gates, *AND* gates, and inverters.

20. Design a circuit that determines whether three or more of four individuals on a committee vote yes on an issue, where each individual uses a switch for the voting.

A **threshold gate** produces an output y that is either 0 or 1 given a set of input values for the Boolean variables $x_1, x_2, \ldots, x_n$. A threshold gate has a **threshold value** T, which is a real number, and **weights** $w_1, w_2, \ldots, w_n$, each of which is a real number. The output y of the threshold gate is 1 if and only if $w_1x_1 + w_2x_2 + \cdots + w_nx_n \geq T$. The threshold gate with threshold value T and weights $w_1, w_2, \ldots, w_n$ is represented by the following diagram. Threshold gates are useful in modeling in neurophysiology and in artificial intelligence.

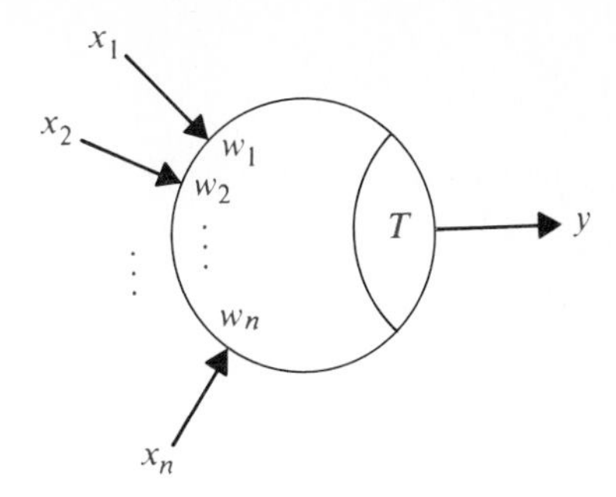

21. A threshold gate represents a Boolean function. Find a Boolean expression for the Boolean function represented by the following threshold gate.

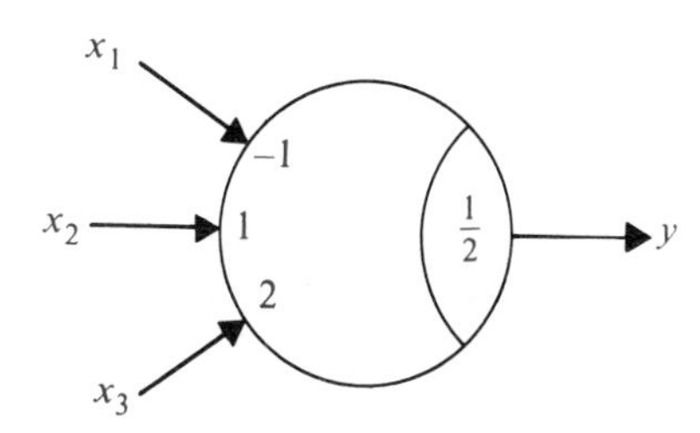

22. A Boolean function that can be represented by a threshold gate is called a **threshold function**. Show that each of the following functions is a threshold function.

a) $F(x) = \overline{x}$ **b)** $F(x, y) = x + y$ **c)** $F(x, y) = xy$
d) $F(x, y) = x \mid y$ **e)** $F(x, y) = x \downarrow y$ **f)** $F(x, y, z) = x + yz$
g) $F(w, x, y, z) = w + xy + z$ **h)** $F(w, x, y, z) = wxz + x\overline{y}z$

★23. Show that $F(x, y) = x \oplus y$ is not a threshold function.

★24. Show that $F(w, x, y, z) = wx + yz$ is not a threshold function.

Computer Projects

WRITE PROGRAMS WITH THE FOLLOWING INPUT AND OUTPUT.

1. Given the values of two Boolean variables x and y, find the values of $x + y$, xy, $x \oplus y$, $x \mid y$, and $x \downarrow y$.

2. Construct a table listing the set of values of all 256 Boolean functions of degree 3.

3. Given the values of a Boolean function in n variables, where n is a positive integer, construct the sum-of-products expansion of this function.

4. Given the table of values of a Boolean function, express this function using only the operators $\cdot$ and $\overline{}$.

5. Given the table of values of a Boolean function, express this function using only the operators $+$ and $\overline{}$.

★6. Given the table of values of a Boolean function, express this function using only the operator $\mid$.

★7. Given the table of values of a Boolean function, express this function using only the operator $\downarrow$.

8. Given the table of values of a Boolean function of degree 3, construct its Karnaugh map.

9. Given the table of values of a Boolean function of degree 4, construct its Karnaugh map.

★★10. Given the table of values of a Boolean function, use the Quine–McCluskey method to find a minimal sum-of-products representation of this function.

11. Given a threshold value and a set of weights for a threshold gate and the values of the n Boolean variables in the input, determine the output of this gate.

10

Modeling Computation

Computers can perform many tasks. Given a task, two questions arise. The first is: Can it be carried out using a computer? Once we know that this first question has an affirmative answer, we can ask the second question: How can the task be carried out? Models of computation are used to help answer these questions.

We will study two types of structures used in models of computation, namely, grammars and finite-state machines. Grammars are used to generate the words of a language and to determine whether a word is in a language. Formal languages, which are generated by grammars, provide models for natural languages, such as English, and for programming languages, such as Pascal, Fortran, Prolog, and C. In particular, grammars are extremely important in the construction and theory of compilers. The grammars that we will discuss were first used by the American linguist Chomsky in the 1950s.

Finite-state machines provide us with a general model for computing. Various types of finite-state machines are used in modeling. All finite-state machines have a set of states, including a starting state, an input alphabet, and a transition function that assigns a next state to every pair of a state and an input. The states of a finite-state machine give it limited memory capabilities. Some finite-state machines produce an output symbol for each transition; these machines can be used to model many kinds of machines, including vending machines, delay machines, binary adders, and language recognizers. We will also study finite-state machines that have no output but do have final states. Such machines are extensively used in language recognition. The strings that are recognized are those that take the starting state to a final state.

Finally, we will tie together the concept of grammars and finite-state machines. We will characterize those sets that are recognized by a finite-state machine and show that these are precisely the sets that are generated by a certain type of grammar.

10.1 Languages and Grammars

INTRODUCTION

Words in the English language can be combined in various ways. The grammar of English tells us whether a combination of words is a valid sentence. For instance, *the frog writes neatly* is a valid sentence, since it is formed from a noun phrase, *the frog*, made up of the article *the* and the noun *frog*, followed by a verb phrase, *writes neatly*, made up of the verb *writes* and the adverb *neatly*. We do not care that this is a nonsensical statement, because we are concerned only with the **syntax**, or form, of the sentence, and not its **semantics**, or meaning. We also note that the combination of words *swims quickly mathematics* is not a valid sentence, because it does not follow the rules of English grammar.

The syntax of a **natural language**, that is, a spoken language, such as English, French, German, or Spanish, is extremely complicated. In fact, it does not seem possible to specify all the rules of syntax for a natural language. Research in the automatic translation of one language to another has led to the concept of a **formal language**, which, unlike a natural language, is specified by a well-defined set of rules of syntax. Rules of syntax are important not only in linguistics, the study of natural languages, but also in the study of programming languages.

We will describe the sentences of a formal language using a grammar. The use of grammars helps when we consider the two classes of problems that arise most frequently in applications to programming languages: (1) How can we determine whether a combination of words is a valid sentence in a formal language? (2) How can we generate the valid sentences of a formal language?

Before giving a technical definition of a grammar, we will describe an example of a grammar that generates a subset of English. This subset of English is defined using a list of rules that describe how a valid sentence can be produced. We specify that

1. a **sentence** is made up of a **noun phrase** followed by a **verb phrase**;
2. a **noun phrase** is made up of an **article** followed by an **adjective** followed by a **noun**, or
3. a **noun phrase** is made up of an **article** followed by a **noun**;
4. a **verb phrase** is made up of a **verb** followed by an **adverb**, or
5. a **verb phrase** is made up of a **verb**;
6. an article is *a*, or
7. an article is *the*;
8. an adjective is *large*, or
9. an adjective is *hungry*;
10. a noun is *rabbit*, or

11. a noun is *mathematician*;

12. a verb is *eats*, or

13. a verb is *hops*;

14. an adverb is *quickly*, or

15. an adverb is *wildly*.

From these rules we can form valid sentences using a series of replacements until no more rules can be used. For instance, we can follow the sequence of replacements:

sentence
noun phrase **verb phrase**
article **adjective** **noun** **verb phrase**
article **adjective** **noun** **verb** **adverb**
the **adjective** **noun** **verb** **adverb**
the large **noun** **verb** **adverb**
the large rabbit **verb** **adverb**
the large rabbit hops **adverb**
the large rabbit hops quickly

to obtain a valid sentence. It is also easy to see that some other valid sentences are: *a hungry mathematician eats wildly, a large mathematician hops, the rabbit eats quickly*, and so on. Also, we can see that *the quickly eats mathematician* is not a valid sentence.

PHRASE-STRUCTURE GRAMMARS

Before we give a formal definition of a grammar, we introduce a little terminology.

DEFINITION 1. A *vocabulary* (or *alphabet*) V is a finite, nonempty set of elements called *symbols*. A *word* (or *sentence*) over V is a string of finite length of elements of V. The *empty string* or *null string*, denoted by λ, is the string containing no symbols. The set of all words over V is denoted by V^*. A *language over* V is a subset of V^*.

Note that λ, the empty string, is the string containing no symbols. It is different from $\emptyset$, the empty set. It follows that $\{\lambda\}$ is the set containing exactly one string, namely, the empty string.

Languages can be specified in various ways. Once way is to list all the words in the language. Another is to give some criteria that a word must satisfy to be in the language. In this section we describe another important way to specify a language, namely, through the use of a grammar, such as the set of rules we gave in the introduction to this section. A grammar provides a set of symbols of various types and a set of rules for producing words. More precisely, a grammar has a **vocabulary** V, which is a set of symbols used to derive members of the language. Some of the elements of the vocabulary cannot be replaced by other symbols. These are called **terminals**, and the other members of the vocabulary, which can be replaced by other symbols, are called **nonterminals**. The sets of terminals and nonterminals are usually denoted by T and N, respectively. In the example given in the introduction of the section, the set of terminals is {*a, the, rabbit, mathematician, hops, eats, quickly, wildly*}, and the set of nonterminals is {**sentence, noun phrase, verb phrase, adjective, article, noun, verb, adverb**}. There is a special member of the vocabulary called the **start symbol**, denoted by S, which is the element of the vocabulary that we always begin with. In the example in the introduction, the start symbol is **sentence**. The rules that specify when we can replace a string from V^*, the set of all strings of elements in the vocabulary, with another string are called the **productions** of the grammar. We denote by $w_0 \rightarrow w_1$ the production that specifies that w_0 can be replaced by w_1. The productions in the grammar given in the introduction of this section were listed. The first production, written using this notation, is **sentence** $\rightarrow$ **noun phrase verb phrase**. We summarize with the following definition.

DEFINITION 2. A *phrase-structure grammar* $G = (V, T, S, P)$ consists of a vocabulary V, a subset T of V consisting of terminal elements, a start symbol S from V, and a set of productions P. The set $V - T$ is denoted by N. Elements of N are called *nonterminal symbols*. Every production in P must contain at least one nonterminal on its left side.

EXAMPLE 1 Let $G = \{V, T, S, P\}$ where $V = \{a, b, A, B, S\}$, $T = \{a, b\}$, S is the start symbol, and $P = \{S \rightarrow ABa, A \rightarrow BB, B \rightarrow ab, AB \rightarrow b\}$. G is an example of a phrase-structure grammar. ■

We will be interested in the words that can be generated by the productions of a phrase-structure grammar.

DEFINITION 3. Let $G = (V, T, S, P)$ be a phrase-structure grammar. Let $w_0 = lz_0r$ (that is, the concatenation of l, z_0, and r) and $w_1 = lz_1r$ be strings over V. If $z_0 \rightarrow z_1$ is a production of G, we say that w_1 is *directly derivable* from w_0 and we write $w_0 \Rightarrow w_1$. If $w_0, w_1, \ldots, w_n$, $n \geq 0$, are strings over V such that $w_0 \Rightarrow w_1, w_1 \Rightarrow w_2, \ldots, w_{n-1} \Rightarrow w_n$, then we say that w_n is *derivable from* w_0, and we write $w_0 \overset{*}{\Rightarrow} w_n$. The sequence of steps used to obtain w_n from w_0 is called a *derivation*.

EXAMPLE 2 The string $Aaba$ is directly derivable from ABa in the grammar in Example 1 since $B \rightarrow ab$ is a production in the grammar. The string $abababa$ is derivable from ABa

since $ABa \Rightarrow Aaba \Rightarrow BBaba \Rightarrow Bababa \Rightarrow abababa$, using the productions $B \to ab$, $A \to BB$, $B \to ab$, and $B \to ab$ in succession. ■

DEFINITION 4. Let $G = \{V, T, S, P\}$ be a phrase-structure grammar. The *language generated by* G (or the *language of* G), denoted by *L(G)*, is the set of all strings of terminals that are derivable from the starting state *S*. In other words,

$$L(G) = \{w \in T^* \mid S \overset{*}{\Rightarrow} w\}.$$

In the following two examples we find the language generated by a phrase-structure grammar.

EXAMPLE 3 Let *G* be the grammar with vocabulary $V = \{S, A, a, b\}$, set of terminals $T = \{a, b\}$, starting symbol *S*, and productions $P = \{S \to aA, S \to b, A \to aa\}$. What is *L(G)*, the language of this grammar?

Solution: From the start state *S* we can derive aA using the production $S \to aA$. We can also use the production $S \to b$ to derive *b*. From *aA* the production $A \to aa$ can be used to derive aaa. No additional words can be derived. Hence $L(G) = \{b, aaa\}$. ■

EXAMPLE 4 Let *G* be the grammar with vocabulary $V = \{S, 0, 1\}$, set of terminals $T = \{0, 1\}$, starting symbol *S*, and productions $P = \{S \to 11S, S \to 0\}$. What is *L(G)*, the language of this grammar?

Solution: From *S* we can derive 0 using $S \to 0$, or $11S$ using $S \to 11S$. From $11S$ we can derive either 110 or $1111S$. From $1111S$ we can derive 11110 and $111111S$. At any stage of a derivation we can either add two 1s at the end of the string or terminate the derivation by adding a 0 at the end of the string. We surmise that $L(G) = \{0, 110, 11110, 1111110, \ldots\}$, the set of all strings that begin with an even number of 1s and end with a 0. This can be proved using an inductive argument that shows that after n productions have been used, the only strings of terminals generated are those consisting of $n - 1$ or fewer concatenations of 11 followed by 0. (This is left as an exercise for the reader.) ■

The problem of constructing a grammar that generates a given language often arises. The next three examples describe problems of this kind.

EXAMPLE 5 Give a phrase-structure grammar that generates the set $\{0^n 1^n \mid n = 0, 1, 2, \ldots\}$.

Solution: Two productions can be used to generate all strings consisting of a string of 0s followed by a string of the same number of 1s, including the null string. The first builds up successively longer strings in the language by adding a 0 at the start of the string and a 1 at the end. The second production replaces S with the empty string. The solution is the grammar $G = (V, T, S, P)$, where $V = \{0, 1, S\}$, $T = \{0, 1\}$, S is the starting symbol, and the productions are

$$S \rightarrow 0S1$$
$$S \rightarrow \lambda.$$

The verification that this grammar generates the correct set is left as an exercise for the reader. ■

The last example involved the set of strings made up of 0s followed by 1s, where the number of 0s and 1s are the same. The next example considers the set of strings consisting of 0s followed by 1s, where the number of 0s and 1s may differ.

EXAMPLE 6 Find a phrase-structure grammar to generate the set $\{0^m 1^n \mid m \text{ and } n \text{ are nonnegative integers}\}$.

Solution: We will give two grammars G_1 and G_2 that generate this set. This will illustrate that two grammars can generate the same language.

The grammar G_1 has alphabet $V = \{S, 0, 1\}$, terminals $T = \{0, 1\}$, and productions $S \rightarrow 0S$, $S \rightarrow S1$, and $S \rightarrow \lambda$. G_1 generates the correct set, since using the first production m times puts m 0s at the beginning of the string, and using the second production n times puts n 1s at the end of the string. The details of this verification are left to the reader.

The grammar G_2 has alphabet $V = \{S, A, 0, 1\}$, terminals $T = \{0, 1\}$, and productions $S \rightarrow 0S$, $S \rightarrow 1A$, $S \rightarrow 1$, $A \rightarrow 1A$, $A \rightarrow 1$, $S \rightarrow \lambda$. The details that this grammar generates the correct set are left as an exercise for the reader. ■

It is sometimes the case that a set that is easy to describe can be generated only by a complicated grammar. The next example illustrates this.

EXAMPLE 7 One grammar that generates the set $\{0^n 1^n 2^n \mid n = 0, 1, 2, 3, \ldots\}$ is $G = (V, T, S, P)$ with $V = \{0, 1, 2, S, A, B\}$, $T = \{0, 1, 2\}$, starting state S, and productions $S \rightarrow 0SAB$, $S \rightarrow \lambda$, $BA \rightarrow AB$, $0A \rightarrow 01$, $1A \rightarrow 11$, $1B \rightarrow 12$, $2B \rightarrow 22$. We leave it as an exercise for the reader to show that this statement is correct. The grammar given is the simplest type of grammar that generates this set, in a sense that will be made clear later in this section. The reader may wonder where this grammar came from, since it seems difficult to come up with this grammar from scratch. It may be comforting to know that this grammar can be systematically constructed using techniques from the theory of computation that are beyond the scope of this book. ■

TYPES OF PHRASE-STRUCTURE GRAMMARS

Phrase-structure grammars can be classified according to the types of productions that are allowed. We will describe the classification scheme introduced by Noam Chomsky. In Section 10.4 we will see that the different types of languages defined in this scheme correspond to the classes of languages that can be recognized using different models of computing machines.

A **type 0** grammar has no restrictions on its productions. A **type 1** grammar can have productions only of the form $w_1 \to w_2$, where the length of w_2 is greater than or equal to the length of w_1, or of the form $w_1 \to \lambda$. A **type 2** grammar can have productions only of the form $w_1 \to w_2$, where w_1 is a single symbol that is not a terminal symbol. A **type 3** grammar can have productions only of the form $w_1 \to w_2$ with $w_1 = A$ and either $w_2 = aB$ or $w_2 = a$, where A and B are nonterminal symbols and a is a terminal symbol, or with $w_1 = S$ and $w_2 = \lambda$.

From these definitions we see that every type 3 grammar is a type 2 grammar, every type 2 grammar is a type 1 grammar, and every type 1 grammar is a type 0 grammar. Type 2 grammars are called **context-free grammars** since a nonterminal symbol that is the left side of a production can be replaced in a string whenever it occurs, no matter what else is in the string. A language generated by a type 2 grammar is called a **context-free language**. When there is a production of the form $lw_1r \to lw_2r$ (but not of the form $w_1 \to w_2$), the grammar is called type 1 or **context-sensitive** since w_1 can be replaced by w_2 only when it is surrounded by the strings l and r. Type 3 grammars are also called **regular grammars**. A language generated by a regular grammar is called **regular**. Section 10.4 deals with the relationship between regular languages and finite-state machines. The Venn diagram in Figure 1 shows the relationship among different types of grammars.

EXAMPLE 8 From Example 6 we know that $\{0^m 1^n \mid m, n = 0, 1, 2, \ldots\}$ is a regular language, since it can be generated by a regular grammar, namely, the grammar G_2 in Example 6. ■

EXAMPLE 9 It follows from Example 5 that $\{0^n 1^n \mid n = 0, 1, 2, \ldots\}$ is a context-free language,

---

Avram Noam Chomsky (born 1928). Noam Chomsky, born in Philadelphia, is the son of a Hebrew scholar. He received his B.A., M.A, and Ph.D. in linguistics, all from the University of Pennsylvania. He was on the staff of the University of Pennsylvania from 1950 until 1951. In 1955 he joined the faculty at M.I.T, beginning his M.I.T. career teaching engineers French and German. Chomsky is currently the Ferrari P. Ward Professor of foreign languages and linguistics at M.I.T. He is known for his many fundamental contributions to linguistics, including the study of grammars. Chomsky is also widely known for his outspoken political activism.

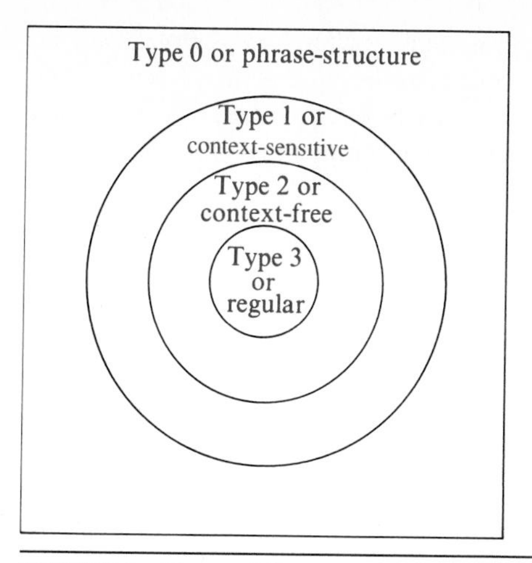

FIGURE 1 Types of Grammars.

since the productions in this grammar are $S \to 0S1$ and $S \to \lambda$. However, it is not a regular language. This will be shown in Section 10.4. ■

EXAMPLE 10 The set $\{0^n 1^n 2^n \mid n = 0, 1, 2, \ldots\}$ is a context-sensitive language, since it can be generated by a type 1 language, as Example 7 shows, but not by any type 2 language. (This is shown in Exercise 28 in the supplementary exercises at the end of the chapter.) ■

Table 1 summarizes the terminology used to classify phrase-structure grammars.

DERIVATION TREES

A derivation in the language generated by a context-free grammar can be represented graphically using an ordered rooted tree, called a **derivation**, or **parse, tree**. The root

TABLE 1 Types of Grammars.

Type	*Restrictions on Productions* $w_1 \to w_2$
0	No restrictions
1	$l(w_1) \leq l(w_2)$, or $w_2 = \lambda$
2	$w_1 = A$ where A is nonterminal symbol
3	$w_1 = A$ and $w_2 = aB$ or $w_2 = a$, where $A \in N$, $B \in N$, and $a \in T$, or $S \to \lambda$

of this tree represents the starting symbol. The internal vertices of the tree represent the nonterminal symbols that arise in the derivation. The leaves of the tree represent the terminal symbols that arise. If the production $A \rightarrow w$ arises in the derivation, where w is a word, the vertex that represents A has as children vertices that represent each symbol in w, in order from left to right.

EXAMPLE 11 Construct a derivation tree for the derivation of *the hungry rabbit eats quickly*, given in the introduction of this section.

Solution: The derivation tree is shown in Figure 2. ■

The problem of determining whether a string is in the language generated by a context-free grammar arises in many applications, such as in the construction of compilers. Two approaches to this problem are indicated in the following example.

EXAMPLE 12 Determine whether the word *cbab* belongs to the language generated by the grammar $G = (V, T, S, P)$ where $V = \{a, b, c, A, B, C, S\}$, $T = \{a, b, c\}$, S is the starting symbol, and the productions are

$$S \rightarrow AB$$
$$A \rightarrow Ca$$
$$B \rightarrow Ba$$
$$B \rightarrow Cb$$
$$B \rightarrow b$$
$$C \rightarrow cb$$
$$C \rightarrow b.$$

Solution: One way to approach this problem is to begin with S and attempt to derive *cbab* using a series of productions. Since there is only one production with S on its

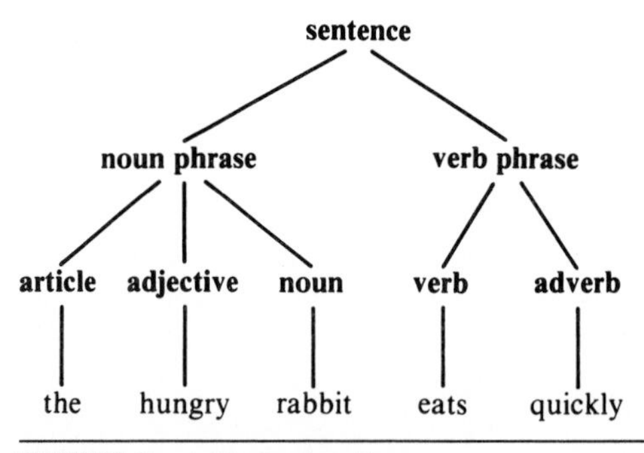

FIGURE 2 **A Derivation Tree.**

left-hand side, we must start with $S \Rightarrow AB$. Next we use the only production that has A on its left-hand side, namely $A \rightarrow Ca$, to obtain $S \Rightarrow AB \Rightarrow CaB$. Since *cbab* begins with the symbols *cb*, we use the production $C \rightarrow cb$. This gives us $S \Rightarrow Ab \Rightarrow CaB \Rightarrow cbaB$. We finish by using the production $B \rightarrow b$, to obtain $S \Rightarrow AB \Rightarrow CaB \Rightarrow cbaB \Rightarrow cbab$. The approach that we have used is called **top-down parsing**, since it begins with the starting symbol and proceeds by successively applying productions.

There is another approach to this problem, called **bottom-up parsing**. In this approach, we work backwards. Since *cbab* is the string to be derived, we can use the production $C \rightarrow cb$, so that $Cab \Rightarrow cbab$. Then, we can use the production $A \rightarrow Ca$, so that $Ab \Rightarrow Cab \Rightarrow cbab$. Using the production $B \rightarrow b$, we have $AB \Rightarrow Ab \Rightarrow Cab \Rightarrow cbab$. Finally, using $S \rightarrow AB$ shows that a complete derivation for *cbab* is $S \Rightarrow AB \Rightarrow Ab \Rightarrow Cab \Rightarrow cbab$. ■

BACKUS–NAUR FORM

There is another notation that is sometimes used to specify a type 2 grammar, called the **Backus–Naur form**, after John Backus, who invented it, and Peter Naur, who refined it for use in the specification of the programming language ALGOL. The productions in a type 2 grammar have a single nonterminal symbol as their left-hand side. Instead of listing all the productions separately, we can combine all those with the same nonterminal symbol on the left-hand side into one statement. Instead of using the symbol $\rightarrow$ in a production, we use the symbol $::=$. We enclose all nonterminal symbols in brackets, $\langle\,\rangle$, and we list all the right-hand sides of productions in the same statement, separating them by bars. For instance, the productions $A \rightarrow Aa$, $A \rightarrow a$, and $A \rightarrow AB$ can be combined into $\langle A\rangle ::= \langle A\rangle a \mid a \mid \langle A\rangle\langle B\rangle$.

John Backus (born 1924). John Backus was born in Philadelphia. He received his B.S. and master's degree in mathematics from Columbia University. Backus joined IBM as a programmer in 1950. He participated in the design and development of two of IBM's early computers. From 1954 to 1958 he led the IBM group that developed FORTRAN. Backus became a staff member at the IBM Watson Research Center in 1958. He was part of the committees that designed the programming language ALGOL, employing what is now called Backus–Naur form for the description of the syntax of this language. Later, Backus worked on the mathematics of families of sets and on a functional style of programming. Backus became an IBM Fellow in 1963, and he received the National Medal of Science in 1974 and the prestigious Turing Award from the Association of Computing Machinery in 1977.

Peter Naur (born in 1928). Peter Naur was born in Frederiksberg, near Copenhagen, Denmark. As a boy he became interested in astronomy. Not only did he observe heavenly bodies, but he also computed the orbits of comets and asteroids. Naur attended Copenhagen University, receiving his degree in 1949. He spent 1950 and 1951 in Cambridge, where he used an early computer to calculate the motions of comets and planets. After returning to Denmark he continued working in astronomy but kept his ties to computing. In 1955 he served as a consultant to the building of the first Danish computer. In 1959 Naur made the switch from astronomy to computing as a full-time activity. His first job as a full-time computer scientist was participating in the development of the programming language ALGOL. From 1960 to 1967 he worked on the development of compilers for ALGOL and COBOL. In 1969 he became professor of computer science at Copenhagen University, where he has worked in the area of programming methodology.

EXAMPLE 13 What is the Backus–Naur form of the grammar for a subset of English described in the introduction to this section?

Solution: The Backus–Naur form of this grammar is:

⟨*sentence*⟩ ::= ⟨*noun phrase*⟩⟨*verb phrase*⟩
⟨*noun phrase*⟩ ::= ⟨*article*⟩⟨*adjective*⟩⟨*noun*⟩ | ⟨*article*⟩⟨*noun*⟩
⟨*verb phrase*⟩ ::= ⟨*verb*⟩⟨*adverb*⟩ | ⟨*verb*⟩
⟨*article*⟩ ::= *a* | *the*
⟨*adjective*⟩ ::= *large* | *hungry*
⟨*noun*⟩ ::= *rabbit* | *mathematician*
⟨*verb*⟩ ::= *eats* | *hops*
⟨*adverb*⟩ ::= *quickly* | *wildly* ■

EXAMPLE 14 Give the Backus–Naur form for the production of signed integers in decimal notation. (A **signed integer** is a nonnegative integer preceded by a plus sign or a minus sign.)

Solution: The Backus–Naur form for a grammar that produces signed integers follows:

⟨*signed integer*⟩ ::= ⟨*sign*⟩⟨*integer*⟩
⟨*sign*⟩ ::= + | −
⟨*integer*⟩ ::= ⟨*digit*⟩ | ⟨*digit*⟩⟨*integer*⟩
⟨*digit*⟩ ::= 0 | 1 | 2 | 3 | 4 | 5 | 6 | 7 | 8 | 9 ■

Exercises

Exercises 1–3 refer to the grammar with start symbol **sentence**, set of terminals T = {*the, sleepy, happy, tortoise, hare, passes, runs, quickly, slowly*}, set of nonterminals N = {**noun phrase, transitive verb phrase, intransitive verb phrase, article, adjective, noun, verb, adverb**}, and productions

sentence → **noun phrase transitive verb phrase noun phrase**
sentence → **noun phrase intransitive verb phrase**
noun phrase → **article adjective noun**
noun phrase → **article noun**
transitive verb phrase → **transitive verb**
intransitive verb phrase → **intransitive verb adverb**
intransitive verb phrase → **intransitive verb**
article → *the*
adjective → *sleepy*
adjective → *happy*
noun → *tortoise*
noun → *hare*
transitive verb → *passes*
intransitive verb → *runs*
adverb → *quickly*
adverb → *slowly*

1. Use the set of products to show that each of the following is a valid sentence.
  a) *the happy hare runs*
  b) *the sleepy tortoise runs quickly*
  c) *the tortoise passes the hare*
  d) *the sleepy hare passes the happy tortoise*

2. Find five other valid sentences, besides those given in Exercise 1.

3. Show that *the hare runs the sleepy tortoise* is not a valid sentence.

★4. Let $V = \{S, A, B, a, b\}$ and $T = \{a, b\}$. Find the language generated by the grammar $\{V, T, S, P\}$ when the set P of products consists of
  a) $S \to AB$, $A \to ab$, $B \to bb$.
  b) $S \to AB$, $S \to aA$, $A \to a$, $B \to ba$.
  c) $S \to AB$, $S \to AA$, $A \to aB$, $A \to ab$, $B \to b$.
  d) $S \to AA$, $S \to B$, $A \to aaA$, $A \to aa$, $B \to bB$, $B \to b$.
  e) $S \to AB$, $A \to aAb$, $B \to bBa$, $A \to \lambda$, $B \to \lambda$.

5. Construct a derivation of $0^3 1^3$ using the grammar given in Example 5.

6. Show that the grammar given in Example 5 generates the set $\{0^n1^n \mid n = 0, 1, 2, \ldots\}$.

7. a) Construct a derivation of 0^21^4 using the grammar G_1 in Example 6.

b) Construct a derivation of 0^21^4 using the grammar G_2 in Example 6.

8. a) Show that the grammar G_1 given in Example 6 generates the set $\{0^m1^n \mid m, n = 0, 1, 2, \ldots\}$.

b) Show that the grammar G_2 in Example 6 generates the same set.

9. Construct a derivation of $0^21^22^2$ in the grammar given in Example 7.

★10. Show that the grammar given in Example 7 generates the set $\{0^n1^n2^n \mid n = 0, 1, 2, \ldots\}$.

★11. Find a phrase-structure grammar for each of the following languages.

a) the set of all bit strings containing an even number of 0s and no 1s

b) the set of all bit strings made up of a 1 followed by an odd number of 0s

c) the set of all bit strings containing an even number of 0s and an even number of 1s

d) the set of all strings containing 10 or more 0s and no 1s

e) the set of all strings containing more 0s than 1s

f) the set of all strings containing an equal number of 0s and 1s

g) the set of all strings containing an unequal number of 0s and 1s

12. Construct phrase-structure grammars to generate each of the following sets.

a) $\{01^{2n} \mid n \geq 0\}$

b) $\{0^n1^{2n} \mid n \geq 0\}$

c) $\{0^n1^m0^n \mid m \geq 0, n \geq 0\}$

13. Let $V = \{S, A, B, a, b\}$ and $T = \{a, b\}$. Determine whether $G = (V, T, S, P)$ is a type 0 grammar but not a type 1 grammar, a type 1 grammar but not a type 2 grammar, or a type 2 grammar but not a type 3 grammar if P, the set of productions, is

a) $S \to aAB$, $A \to Bb$, $B \to \lambda$.

b) $S \to aA$, $A \to a$, $A \to b$.

c) $S \to ABa$, $AB \to a$.

d) $S \to ABA$, $A \to aB$, $B \to ab$.

e) $S \to bA$, $A \to B$, $B \to a$.

f) $S \to aA$, $aA \to B$, $B \to aA$, $A \to b$.

g) $S \to bA$, $A \to b$, $S \to \lambda$.

h) $S \to AB$, $B \to aAb$, $aAb \to b$.

i) $S \to aA$, $A \to bB$, $B \to b$, $B \to \lambda$.

j) $S \to A$, $A \to B$, $B \to \lambda$.

14. A **palindrome** is a string that reads the same backward as it does forward, that is, a string w where $w = w^R$, where w^R is the reversal of the string w. Find a context-free grammar that generates the set of all palindromes over the alphabet $\{0, 1\}$.

★15. Let G_1 and G_2 be context-free grammars, generating the languages $L(G_1)$ and $L(G_2)$, respectively. Show that there is a context-free grammar generating each of the following sets.

a) $L(G_1) \cup L(G_2)$ **b)** $L(G_1)L(G_2)$

c) $L(G_1)^*$

16. Find the strings constructed using the following derivation trees.

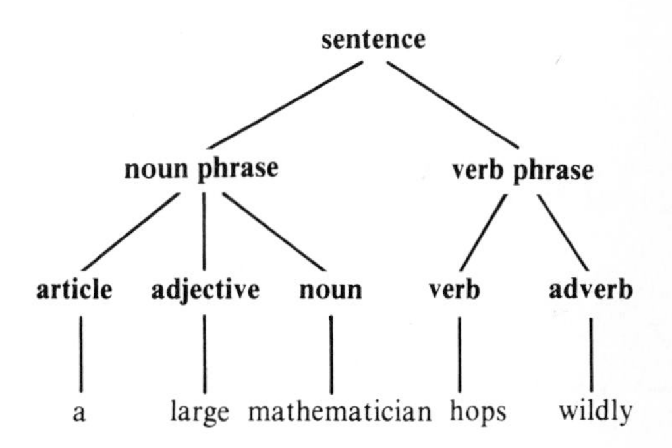

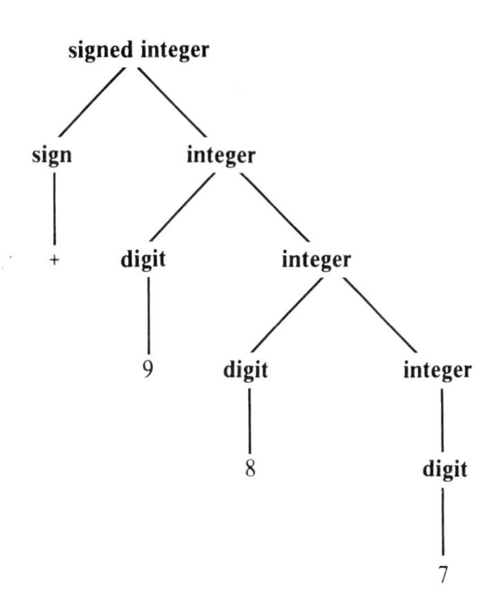

17. Construct derivation trees for the sentences in Exercise 1.

18. Let G be the grammar with $V = \{a, b, c, S\}$, $T = \{a, b, c\}$, starting symbol S, and productions $S \rightarrow abS$, $S \rightarrow bcS$, $S \rightarrow bbS$, $S \rightarrow a$, $S \rightarrow cb$. Construct derivation trees for

a) $bcbba$ **b)** $bbbcbba$

c) $bcabbbbbcb$

★19. Use top-down parsing to determine whether each of the following strings belongs to the language generated by the grammar in Example 12.

a) $baba$ **b)** $abab$

c) $cbaba$ **d)** $bbbcba$

★20. Use bottom-up parsing to determine whether the strings in Exercise 19 belong to the language generated by the grammar in Example 12.

21. Construct a derivation tree for -109 using the grammar given in Example 14.

22. a) What are the productions in a grammar if the Backus–Naur form for productions is as follows?

$\langle expression \rangle ::= (\langle expression \rangle) \mid$
$\langle expression \rangle + \langle expression \rangle \mid$
$\langle expression \rangle * \langle expression \rangle \mid$
$\langle variable \rangle$
$\langle variable \rangle ::= x \mid y$

b) Find a derivation tree for $(x * y) + x$ in this grammar.

23. a) Construct a phrase-structure grammar that generates all signed decimal numbers, consisting of a sign, either + or −; a nonnegative integer; and a decimal fraction that is either the empty string or a decimal point followed by a positive integer, where initial zeros in an integer are allowed.

b) Give the Backus–Naur form of this grammar.

c) Construct a derivation tree for -31.4 in this grammar.

24. a) Construct a phrase-structure grammar for the set of all fractions of the form a/b where a is a signed integer in decimal notation and b is a positive integer.

b) What is the Backus–Naur form for this grammar?

c) Construct a derivation tree for $+311/17$ in this grammar.

25. Let G be a grammar and let R be the relation containing the ordered pair (w_0, w_1) if and only if w_1 is directly derivable from w_0 in G. What is the reflexive transitive closure of R?

10.2 Finite-State Machines with Output

INTRODUCTION

Many kinds of machines, including components in computers, can be modeled using a structure called a finite-state machine. Several types of finite-state machines are commonly used in models. All these versions of finite-state machines include a finite set of states, with a designated starting state, an input alphabet, and a transition function that assigns a next state to every state and input pair. In this section we will study those finite-state machines that produce output. We will show how finite-state machines can be used to model a vending machine, a machine that delays input, a machine that adds integers, and a machine that determines whether a bit string contains a specified pattern.

Before giving formal definitions, we will show how a vending machine can be modeled. A vending machine accepts nickels (5 cents), dimes (10 cents), and quarters (25 cents). When a total of 30 cents or more has been deposited, the machine immediately returns the amount in excess of 30 cents. When 30 cents has been deposited and any excess refunded, the customer can push an orange button and receive an orange juice or push a red button and receive an apple juice. We can describe how the machine works by specifying its states, how it changes states when input is received, and the output that is produced for every combination of input and current state.

The machine can be in any of seven different states $s_i, i = 0, 1, 2, \ldots, 6$, where s_i is the state where the machine has collected $5i$ cents. The machine starts in state s_0, with 0 cents received. The possible inputs are 5 cents, 10 cents, 25 cents, the orange button, and the red button. The possible outputs are nothing (n) , 5 cents, 10 cents, 15 cents, 20 cents, 25 cents, an orange juice, and an apple juice.

We illustrate how this model of the machine works with the following example. Suppose that a student puts in a dime followed by a quarter, receives 5 cents back, and then pushes the orange button for an orange juice. The machine starts in state s_0. The first input is 10 cents, which changes the state of the machine to s_2 and gives no output. The second input is 25 cents. This changes the state from s_2 to s_6, and gives 5 cents as output. The next input is the orange button, which changes the state from s_6 back to s_0 (since the machine returns to the start state) and gives an orange juice as its output.

We can display all the state changes and output of this machine in a table. To do this we need to specify for each combination of state and input the next state and the output obtained. Table 1 shows the transitions and outputs for each pair of a state and an input.

TABLE 1 State Table for a Vending Machine

	Next State					***Output***				
	Input					***Input***				
State	5	10	25	O	R	5	10	25	O	R
s_0	s_1	s_2	s_5	s_0	s_0	n	n	n	n	n
s_1	s_2	s_3	s_6	s_1	s_1	n	n	n	n	n
s_2	s_3	s_4	s_6	s_2	s_2	n	n	5	n	n
s_3	s_4	s_5	s_6	s_3	s_3	n	n	10	n	n
s_4	s_5	s_6	s_6	s_4	s_4	n	n	15	n	n
s_5	s_6	s_6	s_6	s_5	s_5	n	5	20	n	n
s_6	s_6	s_6	s_6	s_0	s_0	5	10	25	OJ	AJ

Another way to show the actions of a machine is to use a directed graph with labeled edges, where each state is represented by a circle, edges represent the transitions, and edges are labeled with the input and the output for that transition. Figure 1 shows such a directed graph for the vending machine.

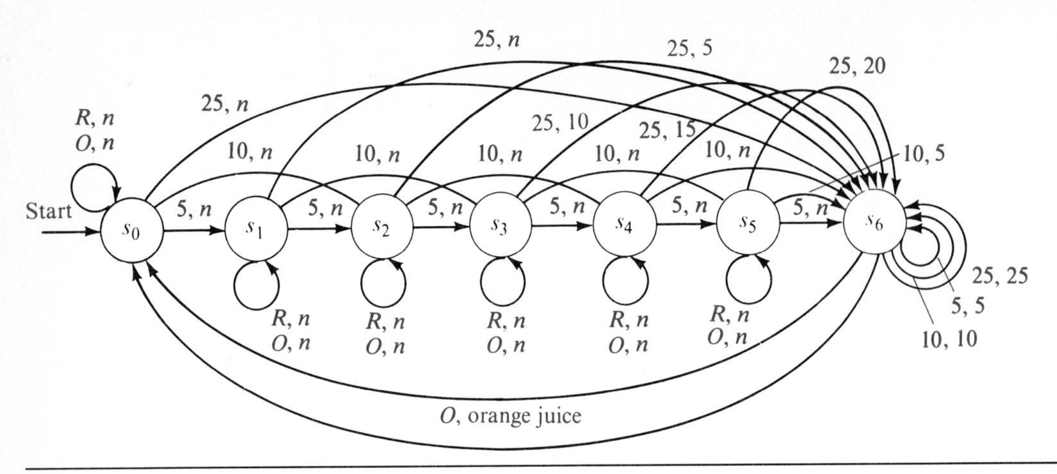

FIGURE 1 **A Vending Machine.**

FINITE-STATE MACHINES WITH OUTPUTS

We will now give the formal definition of a finite-state machine with output.

DEFINITION 1. A *finite-state machine* $M = (S, I, O, f, g, s_0)$ consists of a finite set S of *states*, a finite *input alphabet* I, a finite *output alphabet* O, a *transition function* f that assigns to each state and input pair a new state, an *output function* g that assigns to each state and input pair an output, and an initial state s_0.

Let $M = (S, I, O, f, g, s_0)$ be a finite-state machine. We can use a **state table** to represent the values of the transition function f and the output function g for all pairs of states and input. We previously constructed a state table for the vending machine discussed in the introduction to this section.

EXAMPLE 1 The state table shown in Table 2 describes a finite-state machine with $S = \{s_0, s_1, s_2, s_3\}$, $I = \{0, 1\}$, and $O = \{0, 1\}$. The values of the transition function f are displayed in the first two columns, and the values of the output function g are displayed in the last two columns. ■

Another way to represent a finite-state machine is to use a **state diagram**, which is a directed graph with labeled edges. In this diagram, each state is represented by a circle. Arrows labeled with the input and output pair are shown for each transition.

TABLE 2				
	f		*g*	
	Input		*Input*	
State	0	1	0	1
s_0	s_1	s_0	1	0
s_1	s_3	s_0	1	1
s_2	s_1	s_2	0	1
s_3	s_2	s_1	0	0

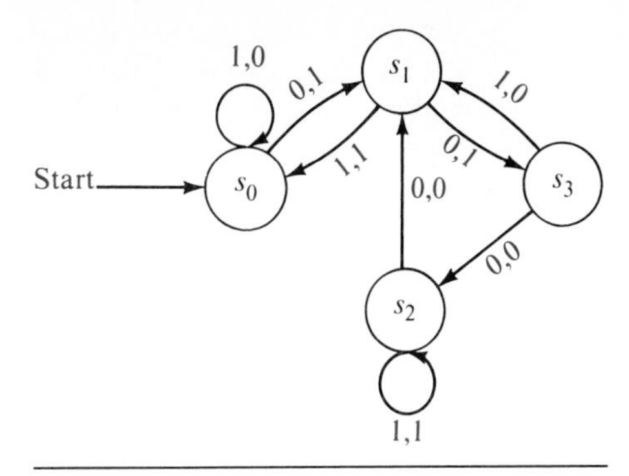

FIGURE 2 The State Diagram for the Finite-State Machine Shown in Table 2.

EXAMPLE 2 Construct the state diagram for the finite-state machine with the state table shown in Table 2.

Solution: The state diagram for this machine is shown in Figure 2. ■

EXAMPLE 3 Construct the state table for the finite-state machine with the state diagram shown in Figure 3.

Solution: The state table for this machine is shown in Table 3. ■

An input string takes the starting state through a sequence of states, as determined by the transition function. As we read the input string symbol by symbol (from left to right), each input symbol takes the machine from one state to another. Because each transition produces an output, an input string also produces an output string.

Suppose that the input string is $x = x_1x_2 \cdots x_k$. Then, reading this input takes the machine from state s_0 to state s_1, where $s_1 = f(s_0, x_1)$, then to state s_2,

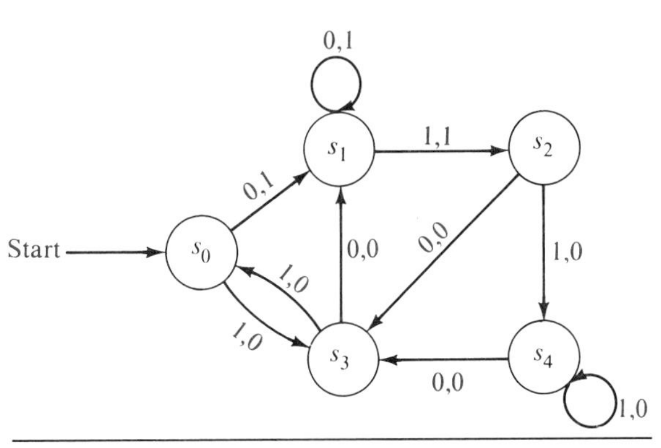

FIGURE 3 A Finite-State Machine.

TABLE 3				
	f		*g*	
	Input		*Input*	
State	0	1	0	1
s_0	s_1	s_3	1	0
s_1	s_1	s_2	1	1
s_2	s_3	s_4	0	0
s_3	s_1	s_0	0	0
s_4	s_3	s_4	0	0

where $s_2 = f(s_1, x_2)$, and so on, ending at state $s_k = f(s_{k-1}, x_k)$. This sequence of transitions produces an output string $y_1 y_2 \cdots y_k$, where $y_1 = g(s_0, x_1)$ is the output corresponding to the transition from s_0 to s_1, $y_2 = g(s_1, x_2)$ is the output corresponding to the transition from s_1 to s_2, and so on. In general $y_j = g(s_{j-1}, x_j)$ for $j = 1, 2, \ldots, k$. Hence, we can extend the definition of the output function g to input strings so that $g(x) = y$, where y is the output corresponding to the input string x. This notation is useful in many applications.

EXAMPLE 4 Find the output string generated by the finite-state machine in Figure 3 if the input string is 101011.

Solution: The output obtained is 001000. The successive states and outputs are shown in Table 4. ■

We can now give some examples of useful finite-state machines. These examples illustrate that the states of a finite-state machine give it limited memory capabilities. The states can be used to remember the properties of the symbols that have been read by the machine. However, since there are only finitely many different states, finite-state machines cannot be used for some important purposes. This will be illustrated in Section 10.4.

EXAMPLE 5 An important element in many electronic devices is a *unit-delay machine*, which produces as output the input string delayed by a specified amount of time. How can a finite-state machine be constructed that delays an input string by one unit of time, that is, produces as output the bit string $0x_1 x_2 \cdots x_{k-1}$ given the input bit string $x_1 x_2 \cdots x_k$?

Solution: A delay machine can be constructed that has two possible inputs, namely, 0 and 1. The machine must have a start state s_0. Since the machine has to remember whether the previous input was a 0 or a 1, two other states s_1 and s_2 are needed, where the machine is in state s_1 if the previous input was 1 and in state s_2 if the previous input was 0. An output of 0 is produced for the initial transition from s_0. Each transition from s_1 gives an output of 1, and each transition from s_2 gives an output of 0. The output corresponding to the input of a string $x_1 \cdots x_k$ is the string that begins with 0, followed by x_1, followed by $x_2, \ldots,$ ending with x_{k-1}. The state diagram for this machine is shown in Figure 4. ■

TABLE 4

Input	1	0	1	0	1	1	–
State	s_0	s_3	s_1	s_2	s_3	s_0	s_3
Output	0	0	1	0	0	0	–

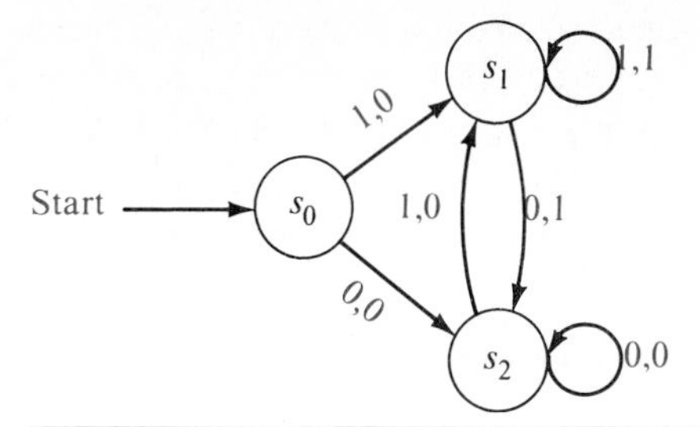

FIGURE 4 **A Unit-Delay Machine.**

EXAMPLE 6 Produce a finite-state machine that adds two integers using their binary expansions.

Solution: When $(x_n \cdots x_1 x_0)_2$ and $(y_n \cdots y_1 y_0)_2$ are added, the following procedure (as described in Section 2.4) is followed. First, the bits x_0 and y_0 are added, producing a sum bit z_0 and a carry bit c_0. This carry bit is either 0 or 1. Then, the bits x_1 and y_1 are added, together with the carry c_0. This gives a sum bit z_1 and a carry bit c_1. This procedure is continued until the nth stage, where x_n, y_n, and the previous carry c_{n-1} are added to produce the sum bit z_n and the carry bit c_n, which is equal to the sum bit z_{n-1}.

A finite-state machine to carry out this addition can be constructed using just two states. For simplicity we assume that both the initial bits x_n and y_n are 0 (otherwise we have to make special arrangements concerning the sum bit z_{n+1}). The start state s_0 is used to remember that the previous carry is 0 (or for the addition of the rightmost bits). The other state, s_1, is used to remember that the previous carry is 1. Since the inputs to the machine are pairs of bits, there are four possible inputs. We represent these possibilities by 00 (when both bits are 0), 01 (when the first bit is 0 and the second is 1), 10 (when the first bit is 1 and the second is 0), and 11 (when both bits are 1). The transitions and the outputs are constructed from the sum of the two bits represented by the input and the carry represented by the state. For instance, when the machine is in state s_1 and receives 01 as input, the next state is s_1 and the output is 0, since the sum that arises is $0 + 1 + 1 = (10)_2$. The state diagram for this machine is shown in Figure 5. ■

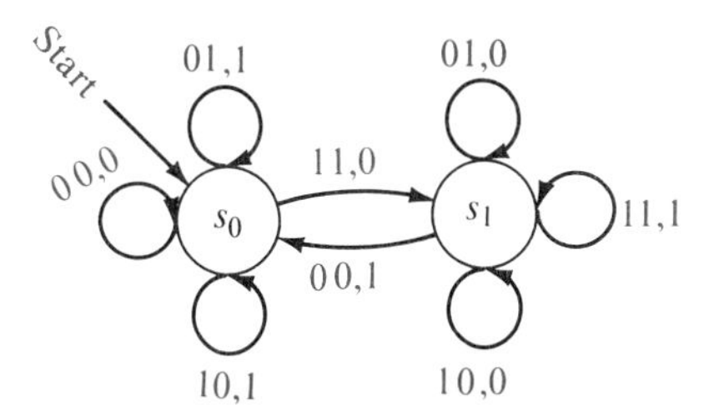

FIGURE 5 **A Finite-State Machine for Addition.**

EXAMPLE 7 In a certain coding scheme, when three consecutive 1s appear in a message, the receiver of the message knows that there has been a transmission error. Construct a finite-state machine that gives a 1 as its output bit if and only if the last three bits received are all 1s.

Solution: Three states are needed in this machine. The start state s_0 remembers that the previous input value, if it exists, was not a 1. The state s_1 remembers that the previous input was a 1, but the input before the previous input, if it exists, was not a 1. The state s_2 remembers that the previous two inputs were 1s. An input of 1 takes s_0 to s_1, since now a 1, and not two consecutive 1s, have been read; it takes s_1 to s_2, since now two consecutive 1s have been read; and it takes s_2 to itself, since at least two consecutive 1s have been read. An input of 0 takes every state to s_0, since this breaks up any string of consecutive 1s. The output for the transition from s_2 to itself when a 1 is read is 1, since this combination of input and state show that three consecutive 1s have been read. All other outputs are O. The state diagram of this machine is shown in Figure 6. ■

The machine in Figure 6 is an example of a **language recognizer**, because it produces an output of 1 if and only if the input string read so far has a specified property. Language recognition is an important application of finite-state machines.

Types of Finite-State Machines Many different kinds of finite-state machines have been developed to model computing machines. In this section we have given a definition of one type of finite-state machine. In the type of machine introduced in this section outputs correspond to transitions between states. Machines of this type are known as **Mealy machines**, since they were first studied by G. H. Mealy in 1955. There is another important type of finite-state machine with output, where the output is determined only by the state. This type of finite-state machine is known as a **Moore machine**, since E. F. Moore introduced this type of machine in 1956. Moore machines are considered in a sequence of exercises at the end of this section.

In Example 7 we showed how a Mealy machine can be used for language recognition. However, another type of finite-state machine, giving no output, is usually used for this purpose. Finite-state machines with no output, also known as finite-state automata, have a set of final states and recognize a string if and only if it takes the start state to a final state. We will study this type of finite-state machine in Section 10.3.

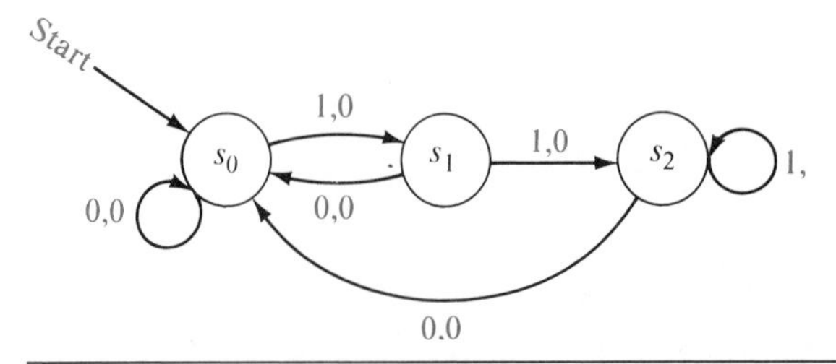

FIGURE 6 A Finite-State Machine that Gives an Output of 1 if and Only if the Input String Read So Far Ends with 111.

Exercises

1. Draw the state diagrams for the finite-state machines with the following state tables.

a)

State	f Input 0	f Input 1	g Input 0	g Input 1
s_0	s_1	s_0	0	1
s_1	s_0	s_2	0	1
s_2	s_1	s_1	0	0

b)

State	f Input 0	f Input 1	g Input 0	g Input 1
s_0	s_1	s_0	0	0
s_1	s_2	s_0	1	1
s_2	s_0	s_3	0	1
s_3	s_1	s_2	1	0

c)

State	f Input 0	f Input 1	g Input 0	g Input 1
s_0	s_0	s_4	1	1
s_1	s_0	s_3	0	1
s_2	s_0	s_2	0	0
s_3	s_1	s_1	1	1
s_4	s_1	s_0	1	0

2. Give the state tables for the finite-state machines with the following state diagrams.

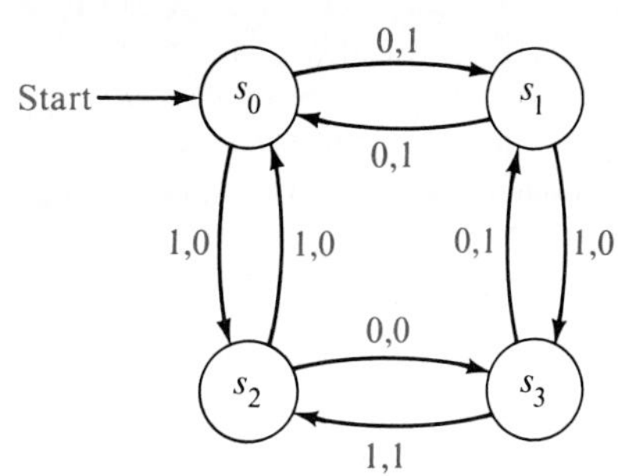

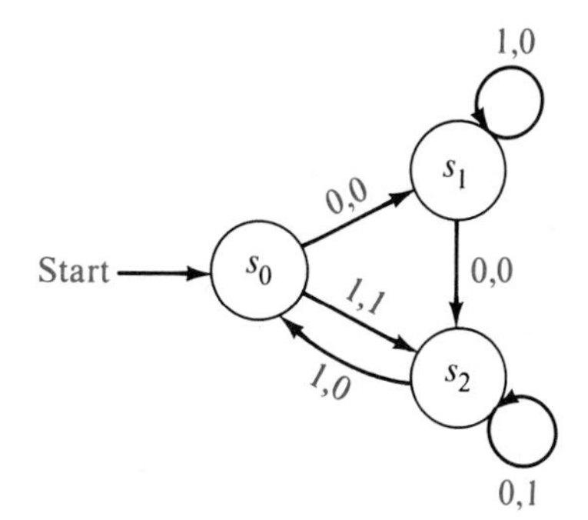

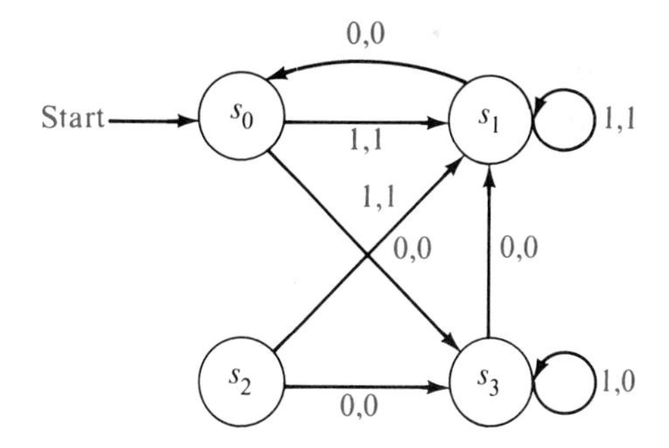

3. Given the finite-state machine shown in Example 2, determine the output for each of the following input strings.
a) 0111 **b)** 11011011 **c)** 01010101010

4. Given the finite-state machine shown in Example 3, determine the output for each of the following input strings.
a) 0000 **b)** 101010 **c)** 11011100010

5. Construct a finite-state machine that models a soda machine that accepts nickels, dimes, and quarters. The soda machine accepts change until 35 cents has been put in. It gives change back for any amount greater than 35 cents. Then the customer can push buttons to receive either a cola, a root beer, or a ginger ale.

6. Construct a finite-state machine that models a newspaper vending machine that has a door that can be opened only after either three dimes (and any number of other coins) or a quarter and a nickel (and any number of other coins) have been inserted. Once the door can be opened, the customer opens it and takes a paper, closing the door. No change is ever returned no matter how much extra money has been inserted. The next customer starts with no credit.

7. Construct a finite-state machine that delays an input string two bits, giving 00 as the first two bits of output.

8. Construct a finite-state machine that changes every other bit, starting with the second bit, of an input string, and leaves the other bits unchanged.

9. Construct a finite-state machine for the log on procedure for a computer, where the user logs in by entering a user identification number, which is considered to be a single input, and then a password, which is considered to be a single input. If the password is incorrect, the user is asked for the user identification number again.

10. Construct a finite-state machine for a combination lock that contains numbers 1 through 40 and that opens only when the correct combination, 10 right, 8 second left, 37 right, is entered. Each input is a triple consisting of a number, the direction of the turn, and the number of times the lock is turned in that direction.

11. Construct a finite-state machine for a toll machine that opens a gate after 25 cents, in nickels, dimes, and quarters, has been deposited. No change is given for overpayment, and no credit is given to the next driver when more than 25 cents has been deposited.

12. Construct a finite-state machine that gives an output of 1 if the number of input symbols read so far is divisible by 3 and an output of 0 otherwise.

13. Construct a finite-state machine that determines whether the input string has a 1 in the last position and a 0 in the third to the last position read so far.

14. Construct a finite-state machine that determines whether the input string read so far ends in at least five consecutive 1s.

15. Construct a finite-state machine that determines whether the word *computer* has been read as the last eight characters in the input read so far, where the input can be any string of English letters.

A **Moore machine** $M = (S, I, O, f, g, s_0)$ consists of a finite set of states, an input alphabet I, an output alphabet O, a transition function f that assigns a next state to every pair of a state and an input, an output function g that assigns an output to every state, and a starting state s_0. A Moore machine can be represented either by a table listing the transitions for each pair of state and input and the outputs for each state, or by a state diagram that displays the states, the transitions between states, and the output for each state. In the diagram, transitions are indicated with arrows labeled with the input, and the outputs are shown next to the states.

16. Construct the state diagram for the Moore machine with the following state table.

	f		
	Input		
State	0	1	*g*
s_0	s_0	s_2	0
s_1	s_3	s_0	1
s_2	s_2	s_1	1
s_3	s_2	s_0	1

17. Construct the state table for the Moore machine with the following state diagram. Each input string to a Moore machine M produces an output string. In particular, the output corresponding to the input string $a_1 a_2 \cdots a_k$ is the string $g(s_0)g(s_1), \ldots g(s_k)$ where $s_1 = f(s_{i-1}, a_i)$ for $i = 1, 2, \ldots, k$.

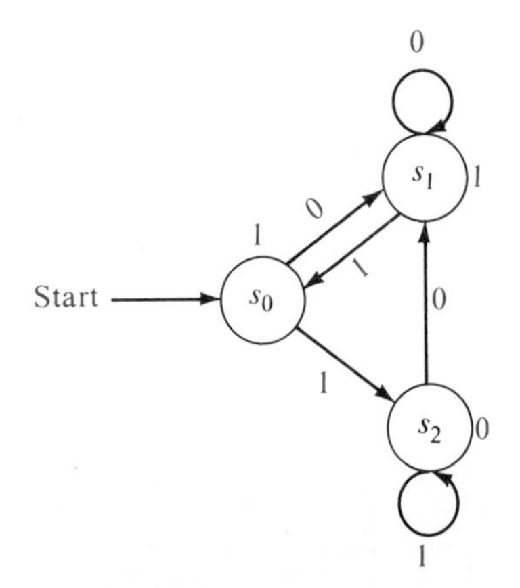

18. Find the output string generated by the Moore machine in Exercise 16 with each of the following input strings.
 a) 0101 **b)** 111111
 c) 11101110111

19. Find the output string generated by the Moore machine in Exercise 17 with each of the input strings in Exercise 18.

20. Construct a Moore machine that gives an output of 1 whenever the number of symbols in the input string read so far is divisible by 4.

21. Construct a Moore machine that determines whether an input string contains an even or odd number of 1s. The machine should give 1 as output if an even number of 1s are in the string and 0 as output if an odd number of 1s are in the string.

10.3 Finite-State Machines with No Output

INTRODUCTION

One of the most important applications of finite-state machines is in language recognition. This application plays a fundamental role in the design and construction of compilers for programming languages. In Section 10.2 we showed that a finite-state machine with output can be used to recognize a language, by giving an output of 1 when a string from the language has been read and a 0 otherwise. However, there are other types of finite-state machines that are specially designed for recognizing languages. Instead of producing output, these machines have final states. A string is recognized if and only if it takes the starting state to one of these final states.

SET OF STRINGS

Before discussing finite-state machines with no output, we will introduce some important background material on sets of strings. The operations that will be defined here will be used extensively in our discussion of language recognition by finite-state machines.

DEFINITION 1. Suppose that A and B are subsets of V^*, where V is a vocabulary. The *concatenation* of A and B, denoted by AB, is the set of all strings of the form xy where x is a string in A and y is a string in B.

EXAMPLE 1 Let $A = \{0, 11\}$ and $B = \{1, 10, 110\}$. Find AB and BA.

Solution: The set AB contains every concatenation of a string in A and a string in B. Hence, $AB = \{01, 010, 0110, 111, 1110, 11110\}$. The set BA contains every concatenation of a string in B and a string in A. Hence, $BA = \{10, 111, 100, 1011, 1100, 11011\}$. ■

Note that it is not necessarily the case that $AB = BA$ when A and B are subsets of V^*, where V is an alphabet, as Example 1 illustrates.

From the definition of the concatenation of two sets of strings, we can define A^n, for $n = 0, 1, 2, \ldots$. This is done recursively by specifying that

$$A^0 = \{\lambda\}$$
$$A^{n+1} = A^n A \text{ for } n = 0, 1, 2, \ldots.$$

EXAMPLE 2 Let $A = \{1,00\}$. Find A^n for $n = 0, 1, 2,$ and 3.

Solution: We have $A^0 = \{\lambda\}$ and $A^1 = A^0A = \{\lambda\}A = \{1, 00\}$. To find A^2 we take concatenations of pairs of elements of A. This gives $A^2 = \{11, 100, 001, 0000\}$. To find A^3 we take concatenations of elements in A^2 and A; this gives $A^3 = \{111, 1100, 1001, 10000, 0011, 00100, 00001, 000000\}$. ■

DEFINITION 2. Suppose that A is a subset of V^*. Then the *Kleene closure* of A, denoted by A^*, is the set consisting of concatenations of arbitrarily many strings from A. That is, $A^* = \bigcup_{k=0}^{\infty} A^k$.

EXAMPLE 3 What are the Kleene closures of the sets $A = \{0\}$, $B = \{0, 1\}$, and $C = \{11\}$?

Solution: The Kleene closure of A is the concatenation of the string 0 with itself an arbitrary finite number of times. Hence $A^* = \{0^n \mid n = 0, 1, 2, \ldots\}$. The Kleene closure of B is the concatenation of an arbitrary number of strings where each string is either 0 or 1. This is the set of all strings over the alphabet $V = \{0, 1\}$. That is, $B^* = V^*$. Finally, the Kleene closure of C is the concatenation of the string 11 with itself an arbitrary number of times. Hence, C^* is the set of string consisting of an even number of 1s. That is, $C^* = \{1^{2n} \mid n = 0, 1, 2, \ldots\}$. ■

Stephen Cole Kleene (born 1909). Stephen Kleene was born in Hartford, Connecticut. His mother, Alice Lena Cole, was a poet, and his father, Gustav Adolph Kleene, was an economics professor. Kleene attended Amherst College and received his Ph.D. from Princeton in 1934, where he studied under the famous logician Alonzo Church. Kleene joined the faculty of the University of Wisconsin in 1935, where he has remained except for several one-year leaves, including stays at the Institute for Advanced Study in Princeton. Kleene has made significant contributions to the theory of recursive functions and to automata theory. He has served as the Acting Director of the Mathematics Research Center and as Dean of the College of Letters and Sciences at the University of Wisconsin.

FINITE-STATE AUTOMATA

We will now give a definition of a finite-state machine with no output. Such machines are also called **finite-state automata**, and that is the terminology we will use for them here. (*Note:* The singular of *automata* is *automaton*.) These machines differ from the finite-state machines studied in Section 10.2 in that they do not produce output, but they do have a set of final states. As we will see, they recognize strings that take the starting state to a final state.

DEFINITION 3. A *finite-state automaton* $M = (S, I, f, s_0, F)$ consists of a finite set S of *states*, a finite *input alphabet* I, a transition function f that assigns a next state to every pair of state and input, an initial state s_0, and a subset F of S consisting of *final states*.

We can represent finite-state automata using either state tables or state diagrams. Final states are indicated in state diagrams by using double circles.

EXAMPLE 4 Construct the state diagram for the finite-state automaton $M = (S, I, f, s_0, F)$ where $S = \{s_0, s_1, s_2, s_3\}$, $I = \{0, 1\}$, $F = \{s_0, s_3\}$, and the transition function f is given in Table 1.

Solution: The state diagram is shown in Figure 1. Note that since both the inputs 0 and 1 take s_2 to s_0, we write 0,1 over the edge from s_2 to s_0. ■

The transition function f can be extended so that it is defined for all pairs of states and strings. Let $x = x_1x_2 \cdots x_k$ be a string in I^*. Then $f(s_1, x)$ is the state obtained by using each successive symbol of x, from left to right, as input, starting with state s_1. From s_1 we go on to state $s_2 = f(s_1, x_1)$, then to state $s_2 = f(s_2, x_2)$, and so on, with $f(s_1, x) = f(s_k, x_k)$.

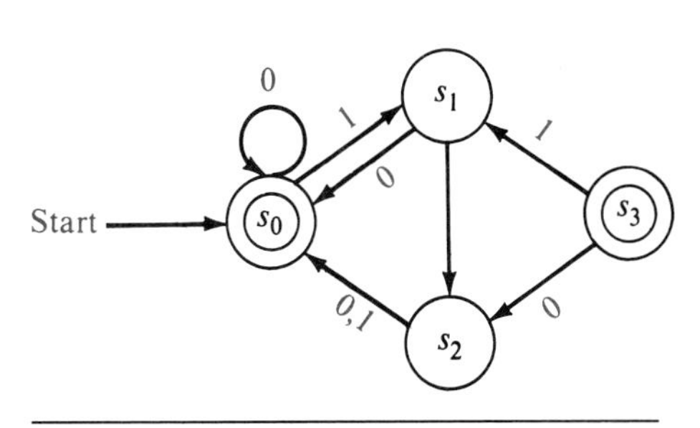

FIGURE 1 The State Diagram for a Finite-State Automaton.

TABLE 1

	f	
	Input	
State	0	1
s_0	s_0	s_1
s_1	s_0	s_2
s_2	s_0	s_0
s_3	s_2	s_1

A string x is said to be **recognized** or **accepted** by the machine $M = (S, I, f, s_0, F)$ if it takes the initial state s_0 to a final state, that is, $f(s_0, x)$ is a state in F. The **language recognized** or **accepted** by the machine M, denoted by $L(M)$, is the set of all strings that are recognized by M. Two finite-state automata are called **equivalent** if they recognize the same language.

EXAMPLE 5 Determine the language recognized by the finite-state automata M_1, M_2, and M_3 in Figure 2.

Solution: The only final state of M_1 is s_0. The strings that take s_0 to itself are those consisting of zero or more consecutive 1s. Hence, $L(M_1) = \{1^n \mid n = 0, 1, 2, \ldots\}$.

The only final state of M_2 is s_2. The only strings that take s_0 to s_2 are 1 and 01. Hence, $L(M_2) = \{1, 01\}$.

The final states of M_3 are s_0 and s_3. The only strings that take s_0 to itself are $\lambda, 0, 00, 000 \ldots$, that is, any string of zero or more consecutive 0s. The only strings

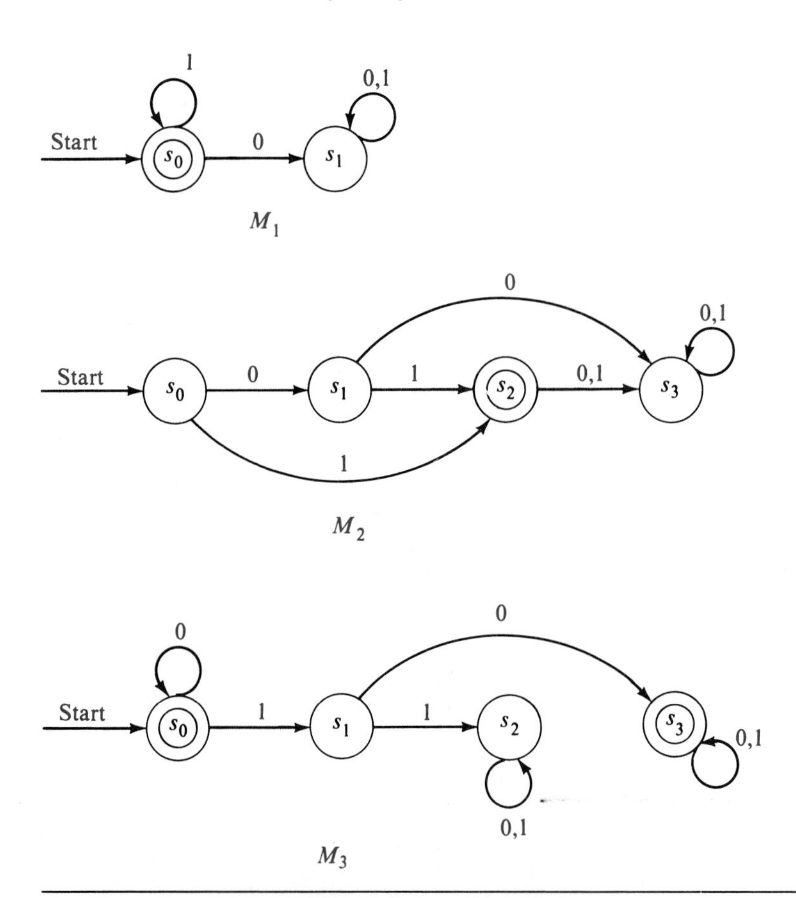

FIGURE 2 **Some Finite-State Automata.**

that take s_0 to s_3 are a string of zero or more consecutive 0s, followed by 10, followed by any string. Hence, $L(M_3) = \{0^n, 0^n 10x \mid n = 0, 1, 2, \ldots,$ and x is any string$\}$. ■

The finite-state automata discussed so far are **deterministic**, since for each pair of state and input value there is a unique next state given by the transition function. There is another important type of finite-state automaton in which there may be several possible next states for each pair of input value and state. Such machines are called **nondeterministic**. Nondeterministic finite-state automata are important in determining which languages can be recognized by a finite-state automaton.

DEFINITION 4. A *nondeterministic finite-state automaton* $M = (S, I, f, s_0, F)$ consists of a set S of states, an input alphabet I, a transition function f that assigns a set of states to each pair of state and input, a starting state s_0, and a subset F of S consisting of the final states.

We can represent nondeterministic finite-state automata using state tables or state diagrams. When we use a state table, for each pair of state and input value we give a list of possible next states. In the state diagram we include an edge from each state to all possible next states, labeling edges with the input or inputs that lead to this transition.

EXAMPLE 6 Find the state diagram for the nondeterministic finite-state automaton with state table shown in Table 2. The final states are s_2 and s_3.

Solution: The state diagram for this automaton is shown in Figure 3. ■

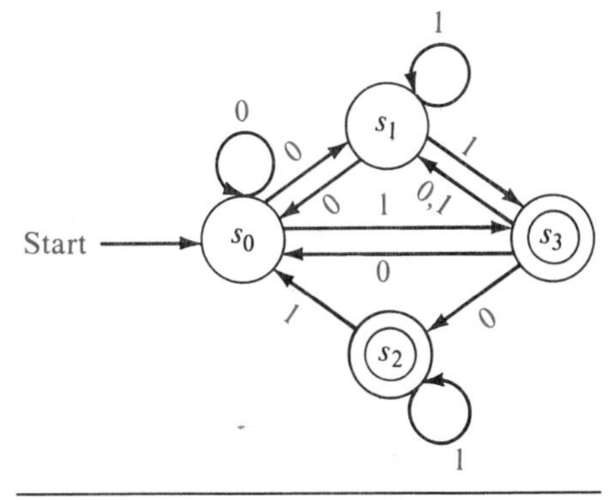

FIGURE 3 The Nondeterministic Finite-State Automaton with State Table Given in Table 2.

TABLE 2

	f	
	Input	
State	0	1
s_0	s_0, s_1	s_3
s_1	s_0	s_1, s_3
s_2		s_0, s_2
s_3	s_0, s_1, s_2	s_1

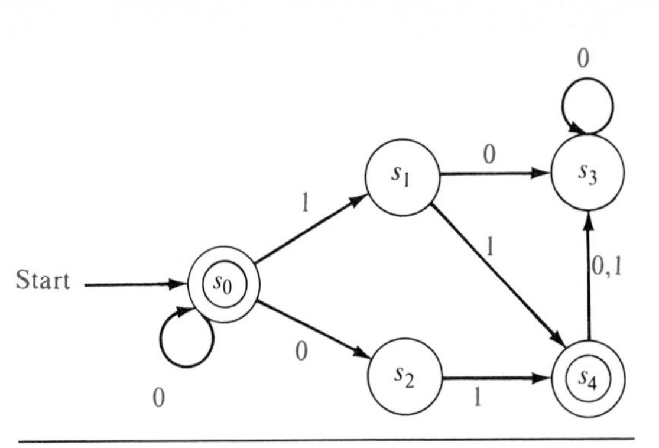

FIGURE 4 A Nondeterministic Finite-State Automaton.

TABLE 3

State	*f* Input 0	*f* Input 1
s_0	s_0, s_2	s_1
s_1	s_3	s_4
s_2		s_4
s_3	s_3	
s_4	s_3	s_3

EXAMPLE 7 Find the state table for the nondeterministic finite-state automaton with state diagram shown in Figure 4.

Solution: The state table is given as Table 3. ■

What does it mean for a nondeterministic finite-state automaton to recognize a string $x = x_1x_2\cdots x_k$? The first input symbol x_1 takes the starting state s_0 to a set S_1 of states. The next input symbol x_2 takes each of the states in S_1 to a set of states. Let S_2 be the union of these sets. We continue this process, including at a stage all states obtained using a state obtained at the previous stage and the current input symbol. We **recognize**, or **accept**, the string x, if there is a final state in the set of all states that can be obtained from s_0 using x. The **language recognized** by a nondeterministic finite-state automaton is the set of all strings recognized by this automaton.

EXAMPLE 8 Find the language recognized by the nondeterministic finite-state automaton shown in Figure 4.

Solution: Since s_0 is a final state, and there is a transition from s_0 to itself when 0 is the input, the machine recognizes all strings consisting of zero or more consecutive 0s. Furthermore, since s_4 is a final state, any string that has s_4 in the set of states that can be reached from s_0 with this input string is recognized. The only such strings are strings consisting of zero or more consecutive 0s followed by 01 or 11. Since s_0 and s_4 are the only final states, the language recognized by the machine is $\{0^n, 0^n01, 0^n11 \mid n \geq 0\}$. ■

One important fact is that a language recognized by a nondeterministic finite-state automaton is also recognized by a deterministic finite-state automaton. We will take advantage of this fact in the next section when we will determine which languages are recognized by finite-state automata.

THEOREM 1 If the language L is recognized by a nondeterministic finite-state automaton M_0, then L is also recognized by a deterministic finite-state automaton M_1.

Proof: We will describe how to construct the deterministic finite-state automaton M_1 that recognizes L from M_0, the nondeterministic finite-state automaton that recognizes this language. Each state in M_1 will be made up of a set of states in M_0. The start symbol of M_1 is $\{s_0\}$, which is the set containing the start state of M_0. The input set of M_1 is the same as the input set of M_0. Given a state $\{s_{i_1}, s_{i_2}, \ldots, s_{i_k}\}$ of M_1, the input symbol x takes this state to the union of the sets of next states for the elements of this set, that is, the union of the sets $f(s_{i_1}), f(s_{i_2}), \ldots, f(s_{i_k})$. The states of M_1 are all the subsets of S, the set of states of M_0, that are obtained in this way starting with s_0. (There are as many as 2^n states in the deterministic machine, where n is the number of states in the nondeterministic machine, since all subsets may occur as states, including the empty set, although usually far fewer states occur.) The final states of M_1 are those sets that contain a final state of M.

Suppose that an input string is recognized by M_0. Then one of the states that can be reached from s_0 using this input string is a final state (the reader should provide an inductive proof of this). This means that in M_1, this input string leads from $\{s_0\}$ to a set of states of M_0 that contains a final state. This subset is a final state of M_1, so this string is also recognized by M_1. Also, an input string not recognized by M_0 does not lead to any final states in M_0. (The reader should provide the details that prove this statement.) Consequently, this input string does not lead from $\{s_0\}$ to a final state in M_1. ■

EXAMPLE 9 Find a deterministic finite-state automaton that recognizes the same language as the nondeterministic finite-state automaton in Example 7.

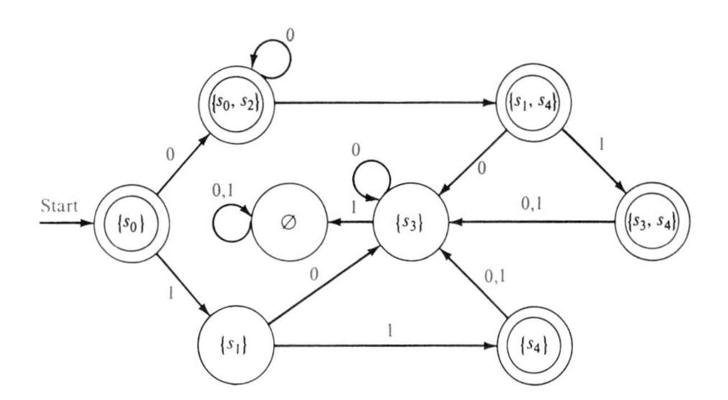

FIGURE 5 A Deterministic Automaton Equivalent to the Nondeterministic Automaton in Example 7.

Solution: The deterministic automaton shown in Figure 5 is constructed from the nondeterministic automaton in Example 7. The states of this deterministic automaton are subsets of the set of all states of the nondeterministic machine. The next state of a subset under an input symbol is the subset containing the next states in the nondeterministic machine of all elements in this subset. For instance, on input of 0, $\{s_0\}$ goes to $\{s_0, s_2\}$ since s_0 has transitions to itself and to s_2 in the nondeterministic machine; the set $\{s_0, s_2\}$ goes to $\{s_1, s_4\}$ on input of 1, since s_0 goes just to s_1 and s_2 goes just to s_4 on input of 1 in the nondeterministic machine; and the set $\{s_1, s_4\}$ goes to $\{s_3\}$ on input of 0, since s_1 and s_4 both go to just s_3 on input of 0 in the deterministic machine. All subsets that are obtained in this way are included in the deterministic finite-state machine. Note that the empty set is one of the states of this machine, since it is the subset containing all the next states of $\{s_3\}$ on input of 1. The start state is $\{s_0\}$ and the set of final states are all those that include s_0 or s_4. ■

Exercises

1. Let $A = \{0, 11\}$ and $B = \{00, 01\}$. Find each of the following sets.
a) AB **b)** BA **c)** A^2 **d)** B^3

2. Show that if A is a set of strings, then $A\emptyset = \emptyset A = \emptyset$.

3. Find all pairs of sets of strings A and B for which $AB = \{10, 111, 1010, 1000, 10111, 101000\}$.

4. Show that the following equalities hold.
a) $\{\lambda\}^* = \{\lambda\}$ **b)** $(A^*)^* = A^*$ for every set of strings A

5. Describe the elements of the set A^* for the following values of A.
a) $\{10\}$ **b)** $\{111\}$ **c)** $\{0, 01\}$ **d)** $\{1, 101\}$

6. Let V be an alphabet, and let A and B be subsets of V^*. Show that $|AB| \leq |A||B|$.

7. Let V be an alphabet, and let A and B be subsets of V^* with $A \subseteq B$. Show that $A^* \subseteq B^*$.

8. Suppose that A is a subset of V^* where V is an alphabet. Prove or disprove each of the following statements.
a) $A \subseteq A^2$ **b)** if $A = A^2$, then $\lambda \in A$ **c)** $A\{\lambda\} = A$
d) $(A^*)^* = A^*$ **e)** $A^*A = A^*$ **f)** $|A^n| = |A|^n$

9. Determine whether the string 11101 is in each of the following sets.
a) $\{0, 1\}^*$ **b)** $\{1\}^*\{0\}^*\{1\}^*$ **c)** $\{11\}\{1\}^*\{01\}$
d) $\{11\}^*\{01\}^*$ **e)** $\{111\}^*\{0\}^*\{1\}$ **f)** $\{111, 000\}\{00, 01\}$

10. Determine whether each of the following strings is recognized by the deterministic finite-state automaton in Figure 1.
a) 010 **b)** 1101 **c)** 1111110 **d)** 010101010

11. Determine whether all the strings in each of the following sets are recognized by the deterministic finite-state automaton in Figure 1.
a) $\{0\}^*$ **b)** $\{0\}\{0\}^*$ **c)** $\{1\}\{0\}^*$ **d)** $\{01\}^*$ **e)** $\{0\}^*\{1\}^*$ **f)** $\{1\}\{0, 1\}^*$

In Exercises 12–16 find the language recognized by the given deterministic finite-state automaton.

12.

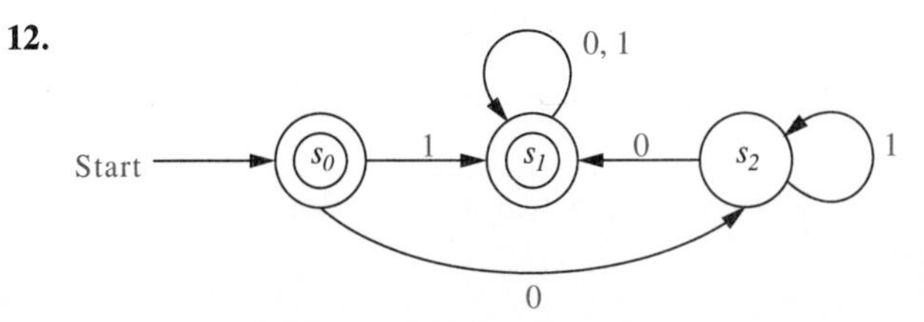

13.

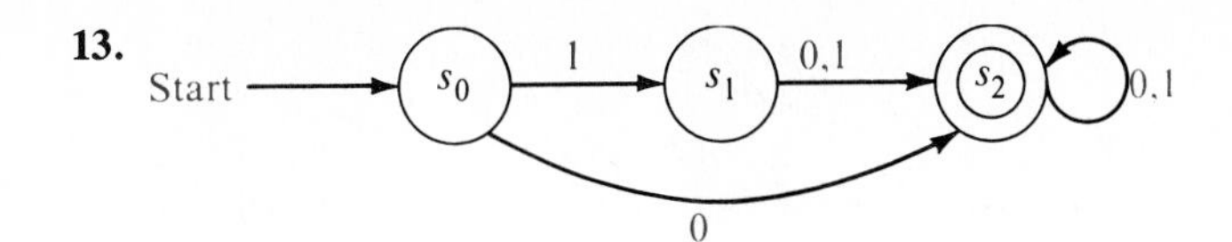

14.

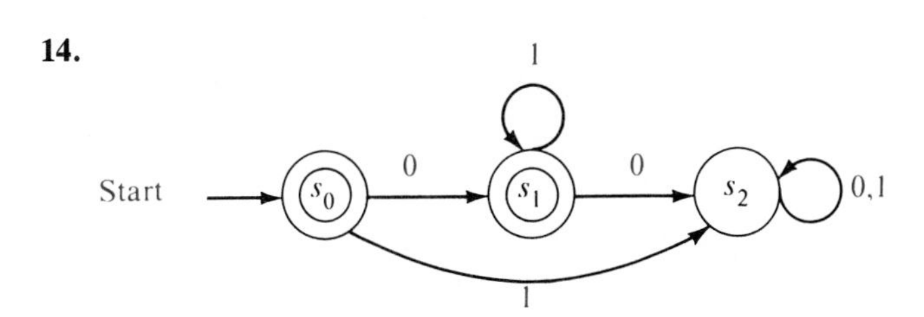

15.

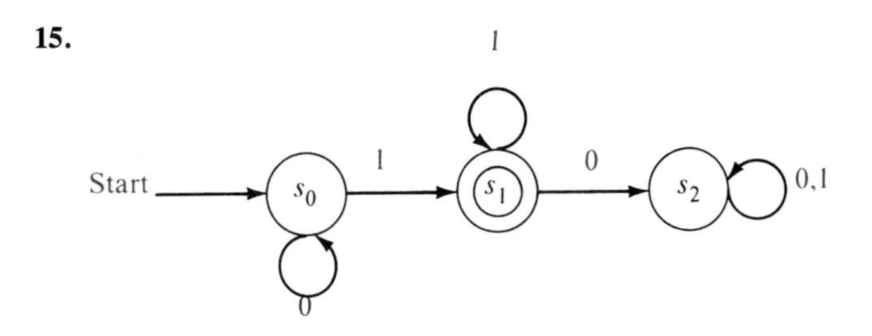

16.

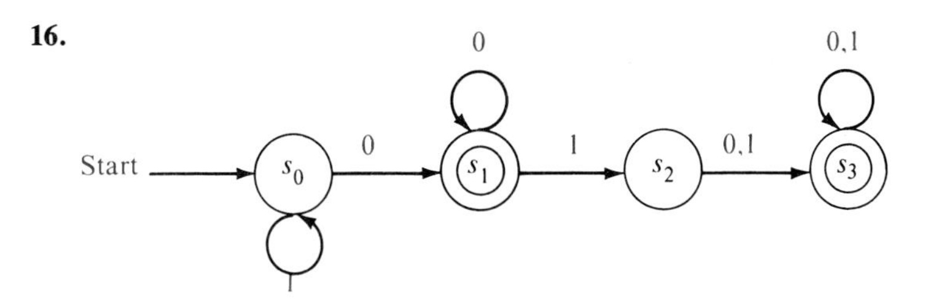

In Exercises 17–21 find the language recognized by the given nondeterministic finite-state automaton.

17.

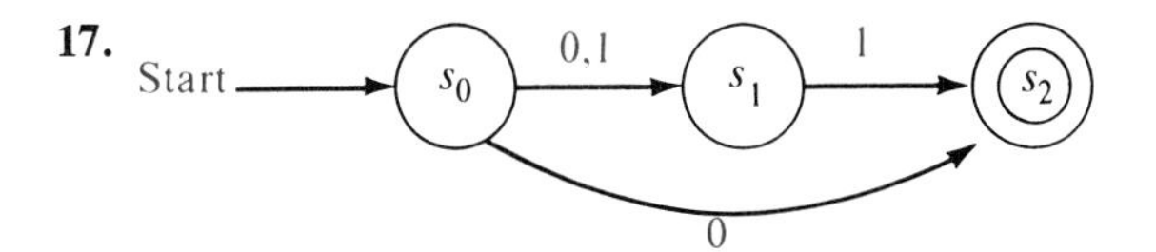

18.

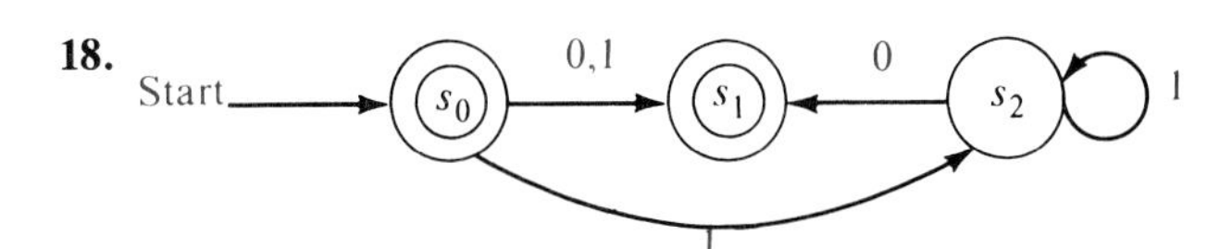

19.

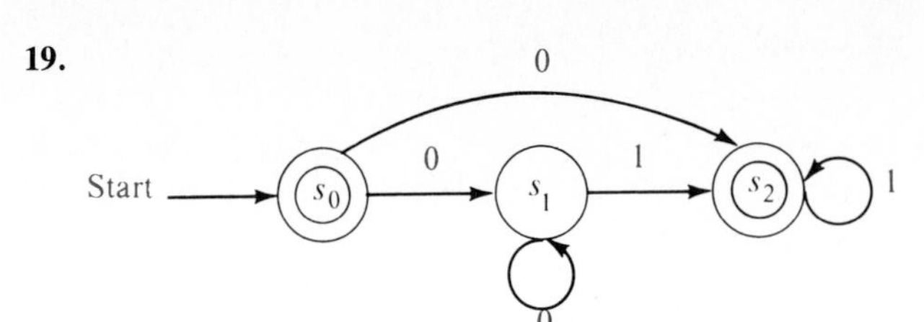

20.

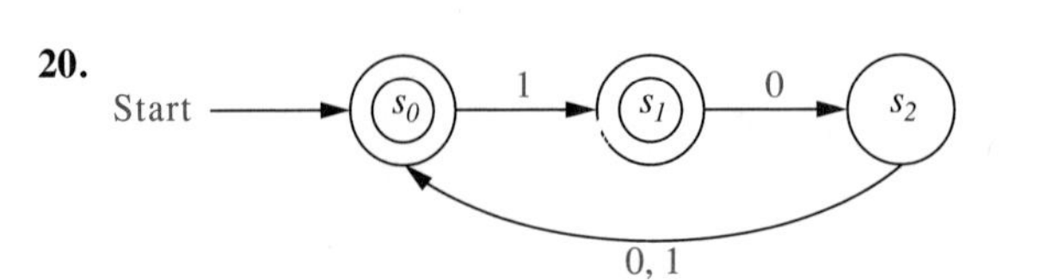

21.

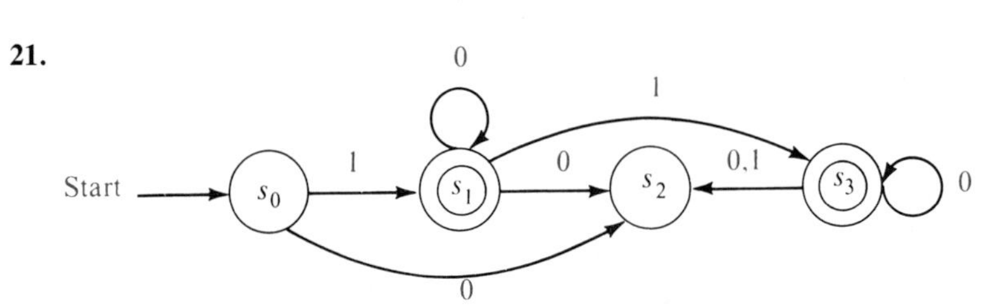

22. Find a deterministic finite-state automaton that recognizes the same language as the nondeterministic finite-state automaton in Exercise 17.

23. Find a deterministic finite-state automaton that recognizes the same language as the nondeterministic finite-state automaton in Exercise 18.

24. Find a deterministic finite-state automaton that recognizes the same language as the nondeterministic finite-state automaton in Exercise 19.

25. Find a deterministic finite-state automaton that recognizes the same language as the nondeterministic finite-state automaton in Exercise 20.

26. Find a deterministic finite-state automaton that recognizes the same language as the nondeterministic finite-state automaton in Exercise 21.

27. Find a deterministic finite-state automaton that recognizes each of the following sets.
a) $\{0\}$ **b)** $\{1, 00\}$ **c)** $\{1^n \mid n = 2, 3, 4, \ldots\}$

28. Find a nondeterministic finite-state automaton that recognizes each of the languages in Exercise 27, and has fewer states, if possible, than the deterministic automaton you found in that exercise.

10.4 Language Recognition

INTRODUCTION

We have seen that finite-state automata can be used as language recognizers. What sets can be recognized by these machines? Although this seems like an extremely difficult problem, there is a simple characterization of the sets that can be recognized by finite-state automata. This problem was first solved in 1956 by the American mathematician Stephen Kleene. He showed that there is a finite-state automaton that recognizes a set if and only if this set can be built up from the null set, the empty string, and singleton strings by taking concatenations, unions, and Kleene closures, in arbitrary order. Sets that can be built up in this way are called **regular sets**.

Regular grammars were defined in Section 10.1. Because of the terminology used, it is not surprising that there is a connection between regular sets, which are the sets recognized by finite-state automata, and regular grammars. In particular, a set is regular if and only if it is generated by a regular grammar.

Finally, there are sets that cannot be recognized by any finite-state automata. We will give an example of such a set. We will briefly discuss more powerful models of computation, such as pushdown automata and Turing machines, at the end of this section.

REGULAR SETS

The regular sets are those that can be formed using the operations of concatenation, union, and Kleene closure in arbitrary order, starting with the empty set, the empty string, and singleton sets. We will see that the regular sets are those that can be recognized using a finite-state automaton. To define regular sets we first need to define regular expressions.

DEFINITION 1. The *regular expressions* over a set I are defined recursively by:

the symbol $\emptyset$ is a regular expression;
the symbol λ is a regular expression;
the symbol x is a regular expression whenever $x \in I$;
the symbols **(AB)**,**(A ∪ B)**, and **A*** are regular expressions whenever **A** and **B** are regular expressions.

Each regular expression represents a set specified by the following rules:

$\emptyset$ represents the empty set, that is, the set with no strings;
$\boldsymbol{\lambda}$ represents the set $\{\lambda\}$, which is the set containing the empty string;
$\boldsymbol{x}$ represents the set $\{x\}$ containing the string with one symbol x;
(AB) represents the concatenation of the sets represented by **A** and by **B**;
(A ∪ B) represents the union of the sets represented by **A** and by **B**;
A* represents the Kleene closure of the set represented by **A**.

Sets represented by regular expressions are called **regular sets**. Henceforth regular expressions will be used to describe regular sets, so that when we refer to the regular set **A**, we will mean the regular set represented by the regular expression **A**. The following example shows how regular expressions are used to specify regular sets.

EXAMPLE 1 What are the strings in the regular sets specified by the regular expressions **10***, **(10)***, **0 ∪ 01, 0(0 ∪ 1)***, and **(0*1)***?

Solution: The regular sets represented by these expressions are as follows, as the reader should verify. ■

Expression	*Strings*
10*	A 1 followed by any number of 0s (including no zeros)
(10)*	Any number of copies of 10 (including the null string)
0 ∪ 01	The string 0 or the string 01
0(0 ∪ 1)*	Any string beginning with 0
(0*1)*	Any string not ending with 0

KLEENE'S THEOREM

In 1956 Kleene proved that the regular sets are the sets that are recognized by a finite-state automaton. Consequently, this important result is called Kleene's theorem.

THEOREM 1 **KLEENE'S THEOREM** A set is regular if and only if it is recognized by a finite-state automaton.

Kleene's theorem is one of the central results in automata theory. We will prove the *only if* part of this theorem, namely, that every regular set is recognized by a finite-state automaton. The proof of the *if* part, that a set recognized by a finite-state automaton is regular, is left as an exercise for the reader.

Proof: Recall that a regular set is defined in terms of regular expressions, which are defined recursively. We can prove that every regular set is recognized by a finite-state automaton if we can do the following things.

1. Show that $\emptyset$ is recognized by a finite-state automaton.
2. Show that $\{\lambda\}$ is recognized by a finite-state automaton.
3. Show that $\{a\}$ is recognized by a finite-state automaton whenever a is a symbol in I.
4. Show that AB is recognized by a finite-state automaton whenever both A and B are.
5. Show that $A \cup B$ is recognized by a finite-state automaton whenever both A and B are.
6. Show that A^* is recognized by a finite-state automaton whenever A is.

We now consider each of these tasks. First, we show that $\emptyset$ is recognized by a nondeterministic finite-state automaton. To do this, all we need is an automaton with no final states. Such an automaton is shown in Figure 1(a).

Second, we show that $\{\lambda\}$ is recognized by a finite-state automaton. To do this, all we need is an automaton that recognizes λ, the null string, but not any other string. This can be done by making the start state s_0 a final state and having no transitions, so that no other string takes s_0 to a final state. The nondeterministic automaton in Figure 1(b) shows such a machine.

Third, we show that $\{a\}$ is recognized by a nondeterministic finite-state automaton. To do this, we can use a machine with a starting state s_0 and a final state s_1. We have a transition from s_0 to s_1 when the input is a, and no other transitions. The only string recognized by this machine is a. This machine is shown in Figure 1(c).

Next, we show that AB and $A \cup B$ can be recognized by finite-state automata if A and B are languages recognized by finite-state automata. Suppose that A is recognized by $M_A = (S_A, I, f_A, s_A, F_A)$ and B is recognized by $M_B = (S_B, I, f_B, s_B, F_B)$.

We begin by constructing a finite-state machine $M_{AB} = (S_{AB}, I, f_{AB}, s_{AB}, F_{AB})$ that recognizes AB, the concatenation of A and B. We build such a machine by combining the machines for A and B in series, so that a string in A takes the combined machine from s_A, the start state of M_A, to s_B, the start state of M_B. A string in B should take the combined machine from s_B to a final state of the combined machine. Consequently, we make the following construction. Let S_{AB} be $S_A \cup S_B$. The starting state s_{AB} is the same as s_A. The set of final states, F_{AB}, is the set of final states of M_B with s_{AB} included if and only if $\lambda \in A \cap B$. The transitions in M_{AB} include all transitions in M_A and in M_B, as well as some new transitions. For every transition in M_A that leads to a final state, we form a transition in M_{AB} from the same state to s_B, on the same input. In this way, a string in A takes M_{AB} from s_{AB} to s_B, and then a string in B takes s_B to a final state of M_{AB}. Moreover, for every transition from s_B we form a transition in M_{AB} from s_{AB} to the same state. Figure 2(a) contains an illustration of this construction.

We now construct a machine $M_{A \cup B} = \{S_{A \cup B}, I, f_{A \cup B}, s_{A \cup B}, F_{A \cup B}\}$ that recognizes $A \cup B$. This automaton can be constructed by combining M_A and M_B in parallel, using a new start state that has the transitions that both s_A and s_B have. Let $S_{A \cup B} = S_A \cup S_B \cup \{s_{A \cup B}\}$, where $s_{A \cup B}$ is a new state that is the start state of $M_{A \cup B}$. Let the set of final states $F_{A \cup B}$ be $F_A \cup F_B \cup \{s_{A \cup B}\}$ if $\lambda \in A \cup B$, and $F_A \cup F_B$ otherwise. The transitions in $M_{A \cup B}$ include all those in M_A and in M_B. Also, for each transition from s_A to a state s on input i we include a transition from $s_{A \cup B}$ to s on input i, and for each transition from s_B to a state s on input i we

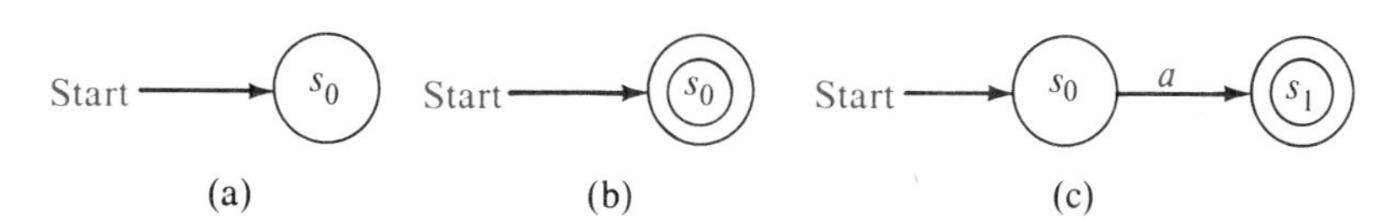

FIGURE 1 **Nondeterministic Finite-State Automata that Recognize Some Basic Sets.**

(a)

Transition to final state in M_A produces a transition to s_B.

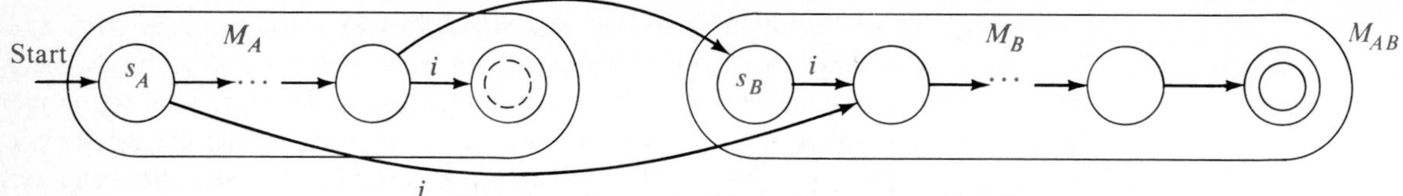

Transition from s_B in M_B produces a transition from $s_{AB} = s_A$.

Start state is $s_{AB} = s_A$, which is final if s_A and s_B are final. Final states include all finite states of M_B.

(b)

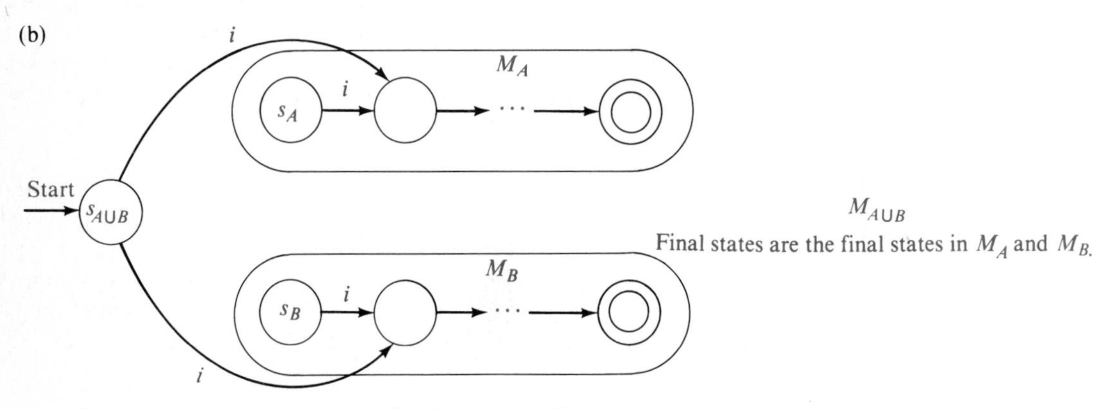

$M_{A \cup B}$

Final states are the final states in M_A and M_B.

$s_{A \cup B}$ is the new start state, which is a final if s_A or s_B is final.

(c)

Transitions from s_A produce A transitions from s_{A*} and all final states of M_A.

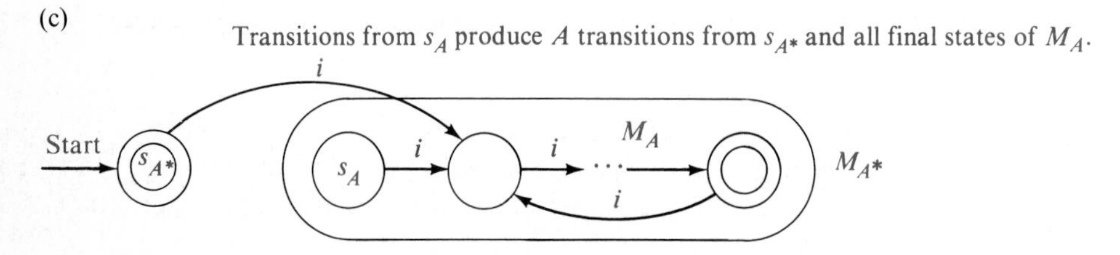

s_{A*} is the new start state, which is a final state. Final states include all final states in M_A.

FIGURE 2 Building Automata to Recognize Concatenations, Unions, and Kleene Closures.

include a transition from $s_{A \cup B}$ to s on input i. In this way, a string in A leads from $s_{A \cup B}$ to a final state in the new machine, and a string in B leads from $s_{A \cup B}$ to a final state in the new machine. Figure 2(b) illustrates the construction of $M_{A \cup B}$.

Finally, we construct $M_{A*} = (S_{A*}, I, f_{A*}, s_{A*}, F_{A*})$, a machine that recognizes A^*, the Kleene closure of A. Let S_{A*} include all states in S_A and one additional state s_{A*}, which is the starting state for the new machine. The set of final states F_{A*} includes all states in F_A as well as the start state s_{A*}, since λ must be recognized. To recognize concatenations of arbitrarily many strings from A, we include all the transitions in

M_A, as well as transitions from s_{A^*} that match the transitions from s_A, and transitions from each final state that match the transitions from s_A. With this set of transitions, a string made up of concatenations of strings from A will take s_{A^*} to a final state when the first string in A has been read, returning to a final state when the second string in A has been read, and so on. Figure 2(c) illustrates the construction we used. ■

A nondeterministic finite-state automaton can be constructed for any regular set using the procedure described in this proof. We illustrate how this is done with the following example.

EXAMPLE 2 Construct a nondeterministic finite-state automaton that recognizes the regular set $\mathbf{1}^* \cup \mathbf{01}$.

Solution: We begin by building a machine that recognizes $\mathbf{1}^*$. This is done using the machine that recognizes $\mathbf{1}$ and then using the construction for M_{A^*} described in the proof. Next, we build a machine that recognizes $\mathbf{01}$, using machines that recognize $\mathbf{0}$ and $\mathbf{1}$ and the construction in the proof for M_{AB}. Finally, using the construction in the proof for $M_{A \cup B}$, we construct the machine for $\mathbf{1}^* \cup \mathbf{01}$. The finite-state automata used in this construction are shown in Figure 3. The states in the successive machines have been labeled using different subscripts, even when a state is formed from one previously used in another machine. Note that the construction given here does not produce the simplest machine that recognizes $\mathbf{1}^* \cup \mathbf{01}$. A much simpler machine that recognizes this set is shown in Figure 3(b). ■

REGULAR SETS AND REGULAR GRAMMARS

In Section 10.1 we introduced phrase-structure grammars and defined different types of grammars. In particular we defined regular, or type 3, grammars, which are grammars of the form $G = (V, T, S, P)$ where each production is of the form $S \rightarrow \lambda$, $A \rightarrow a$, or $A \rightarrow aB$, where a is a terminal symbol, and A and B are nonterminal symbols. As the terminology suggests, there is a close connection between regular grammars and regular sets.

THEOREM 2 A set is generated by a regular grammar if and only if it is a regular set.

Proof: First we show that a set generated by a regular grammar is a regular set. Suppose that $G = (V, T, S, P)$ is a regular grammar generating the set $L(G)$. To show that $L(G)$ is regular we will build a nondeterministic finite-state machine $M = (S, I, f, s_0, F)$ that recognizes $L(G)$. Let S, the set of states, contain a state s_A for each nonterminal symbol A of G and an additional state s_F, which is a final state. The start state s_0 is the state formed from the start symbol S. The transitions of M are formed from the productions of G in the following way. A transition from s_A to s_F on input of a is included if $A \rightarrow a$ is a production, and a transition from s_A to s_B on input of a is included if $A \rightarrow aB$ is a production. The set of final states

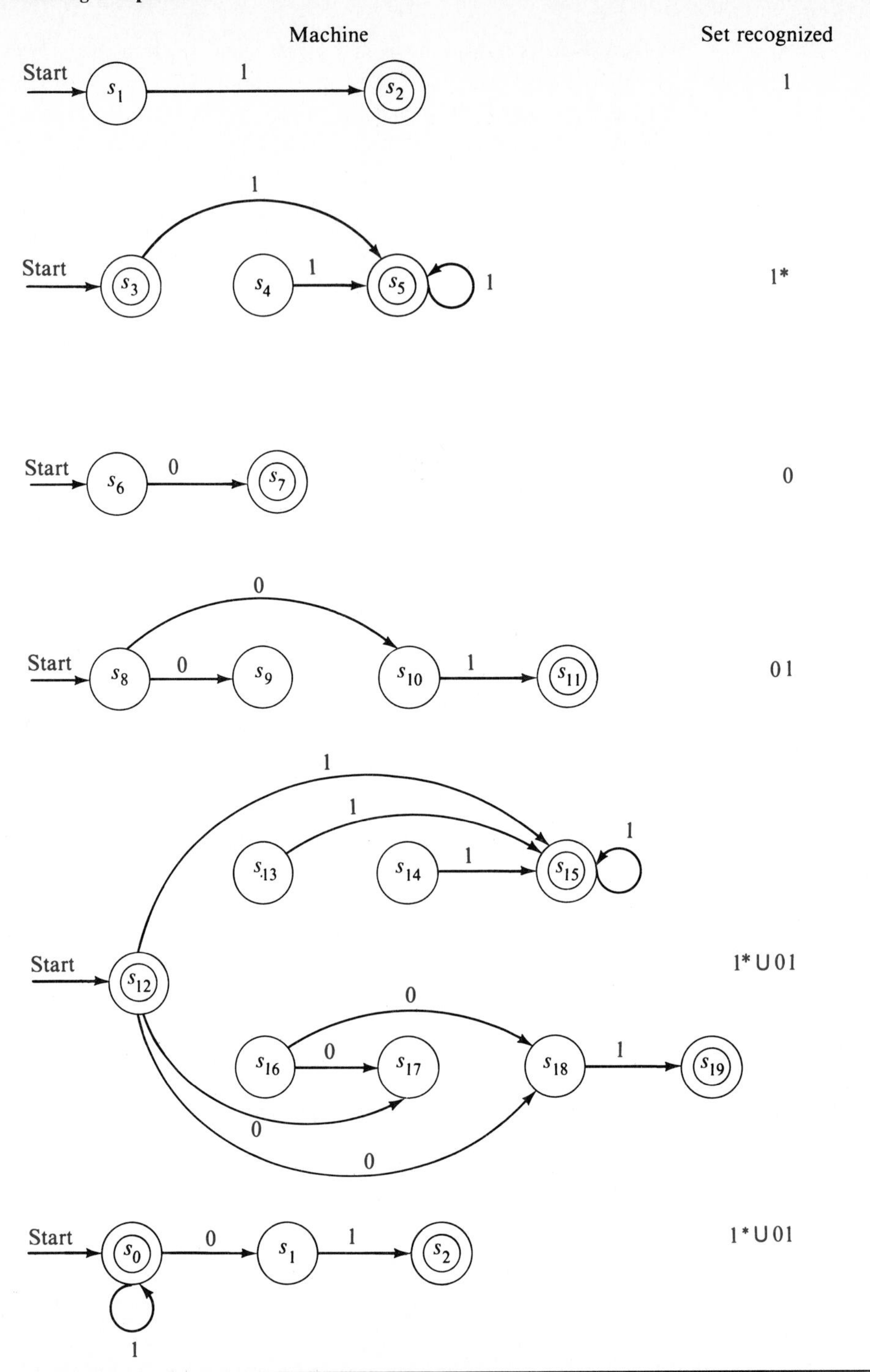

FIGURE 3 Nondeterministic Finite-State Automata Recognizing $1^* \cup 01$.

includes s_F and also includes s_0 if $S \rightarrow \lambda$ is a production in G. It is not hard to show that the language recognized by M equals the language generated by the grammar G, that is, the $L(M) = L(G)$. This can be done by determining the words that lead to a final state. The details are left as an exercise for the reader. ■

Before giving the proof of the converse, we illustrate how a nondeterministic machine is constructed that recognizes the same set as a regular grammar.

EXAMPLE 3 Construct a nondeterministic finite-state automaton that recognizes the language generated by the regular grammar $G = (V, T, S, P)$ where $V = \{0, 1, A, S\}$, $T = \{0, 1\}$, and the productions in P are $S \rightarrow 1A$, $S \rightarrow 0$, $S \rightarrow \lambda$, $A \rightarrow 0A$, $A \rightarrow 1A$, and $A \rightarrow 1$.

Solution: The state diagram for a nondeterministic finite-state automaton that recognizes $L(G)$ is shown in Figure 4. This automaton is constructed following the procedure described in the proof. In this automaton s_0 is the state corresponding to S, s_1 is the state corresponding to A, and s_2 is the final state. ■

We now complete the proof of Theorem 2.

Proof: We now show that if a set is regular, then there is a regular grammar that generates it. Suppose that M is a finite-state machine that recognizes this set with the property that s_0, the starting state of M, is never the next state for a transition. (We can find such a machine by Exercise 14.) The language $G = (V, T, S, P)$ is defined as follows. The set V of symbols of G is formed by assigning a symbol to each state of S and each input symbol in I. The set T of terminal symbols of G is the symbols of G formed from the input symbols in I. The start symbol S is the symbol formed from the start state s_0. The set P of productions in G is formed from the transitions in M. In particular, if the state s goes to a final state under input a, then the production $A_s \rightarrow a$ is included in P, where A_s is the nonterminal symbol formed from

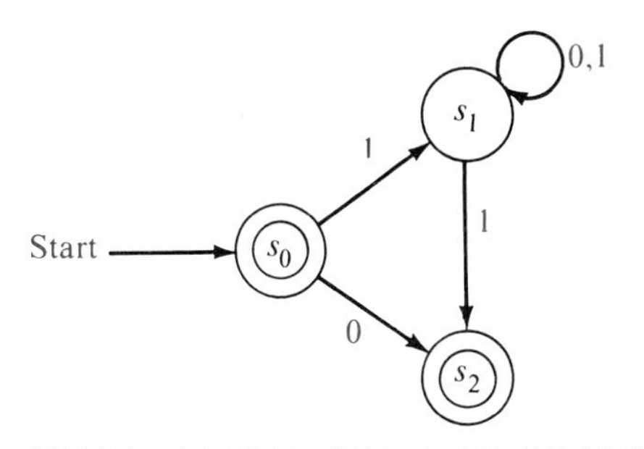

FIGURE 4 **A Nondeterministic Finite-State Automaton Recognizing $L(G)$.**

the state s. If the state s goes to the state t on input a, then the production $A_s \rightarrow aA_t$ is included in P. The production $S \rightarrow \lambda$ is included in P if and only if $\lambda \in L(M)$. Since the productions of G correspond to the transitions of M and the productions leading to terminals correspond to transitions to final states, it is not hard to show that $L(G) = L(M)$. We leave the details as an exercise for the reader. ■

The following example illustrates the construction used to produce a grammar from an automaton that generates the language recognized by this automaton.

EXAMPLE 4 Find a regular grammar that generates the regular set recognized by the finite-state automaton shown in Figure 5.

Solution: The grammar $G = (V, T, S, P)$ generates the set recognized by this automaton where $G = \{S, A, B, 0, 1\}$; where the symbols S, A, and B correspond to the states s_0, s_1, and s_2, respectively; $T = \{0, 1\}$; S is the start symbol; and the productions are $S \rightarrow 0A$, $S \rightarrow 1B$, $S \rightarrow 1$, $S \rightarrow \lambda$, $A \rightarrow 0A$, $A \rightarrow 1B$, $A \rightarrow 1$, $B \rightarrow 0A$, $B \rightarrow 1B$, and $B \rightarrow 1$. ■

A SET NOT RECOGNIZED BY A FINITE-STATE AUTOMATON

We have seen that a set is recognized by a finite-state automaton if and only if it is regular. We will now show that there are sets that are not regular by describing one such set. The technique used to show that this set is not regular illustrates an important method for showing that certain sets are not regular.

EXAMPLE 5 Show that the set $\{0^n 1^n | n = 0, 1, 2, \ldots\}$, made up of all strings consisting of a block of 0s followed by a block of an equal number of 1s, is not regular.

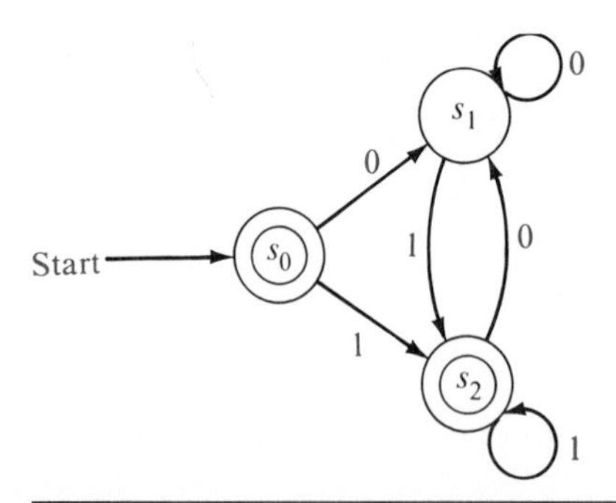

FIGURE 5 A Finite-State Automaton.

Solution: Suppose that this set were regular. Then there would be a deterministic finite-state automaton $M = (S, I, f, s_0, F)$ recognizing it. Let N be the number of states in this machine, that is, $N = |S|$. Since M recognizes all strings made up of a number of 0s followed by an equal number of 1s, M must recognize $0^N 1^N$. Let $s_0, s_1, s_2, \ldots, s_{2N}$ be the sequence of states that is obtained starting at s_0 and using the symbols of $0^N 1^N$ as input, so that $s_1 = f(s_0, 0)$, $s_2 = f(s_1, 0), \ldots, s_N = f(s_{N-1}, 0)$, $s_{N+1} = f(s_N, 1), \ldots, s_{2N} = f(s_{2N-1}, 1)$. Note that s_{2N} is a final state.

Since there are only N states, the pigeonhole principle shows that at least two of the first $N + 1$ of the states, which are $s_0, \ldots, s_N$, must be the same. Say that s_i and s_j are two such identical states, with $0 \leq i < j \leq N$. This means that $f(s_i, 0^t) = s_j$ where $t = i - j$. It follows that there is a loop leading from s_i back to itself, obtained using the input 0 a total of t times, in the state diagram shown in Figure 6.

Now consider the input string $0^N 0^t 1^N = 0^{N+t} 1^N$. There are t more consecutive 0s at the start of this block than there are consecutive 1s that follow it. Since this string is not of the form $0^n 1^n$ (since it has more 0s that 1s), it is not recognized by M. Consequently, $f(s_0, 0^{N+t} 1^N)$ cannot be a final state. However, when we use the string $0^{N+t} 1^N$ as input, we end up in the same state as before, namely s_{2N}. The reason for this is that the extra t 0s in this string take us around the loop from s_i back to itself an extra time, as shown in Figure 6. Then the rest of the string leads us to exactly the same state as before. This contradiction shows that $\{0^n 1^n \mid n = 1, 2, \ldots\}$ is not regular. ■

MORE POWERFUL TYPES OF MACHINES

Finite-state automata are unable to carry out many computations. The main limitation of these machines is their finite amount of memory. This prevents them from recognizing languages that are not regular, such as $\{0^n 1^n \mid n = 0, 1, 2, \ldots\}$. Since a set is regular if and only if it is the language generated by a regular grammar, Example 5 shows that there is no regular grammar that generates the set $\{0^n 1^n \mid n = 0, 1, 2, \ldots\}$.

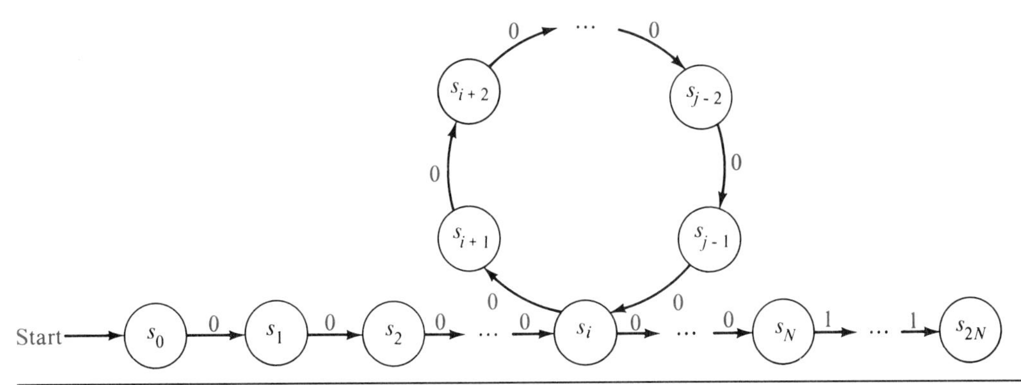

FIGURE 6 The Path Produced by $0^N 1^N$.

However, there is a context-free grammar that recognizes this set. Such a grammar was given in Example 5 in Section 10.1.

Because of the limitations of finite-state machines, it is necessary to use other, more powerful, models of computation. One such model is the **pushdown automaton**. A pushdown automaton includes everything in a finite-state automaton, as well as a stack, which provides unlimited memory. Symbols can be placed on the top or taken off the top of the stack. A set is recognized in one of two ways by a pushdown automaton. First, a set is recognized if the set consists of all the strings that produce an empty stack when they are used as input. Second, a set is recognized if it consists of all the strings that lead to a final state when used as input. It can be shown that a set is recognized by a pushdown automaton if and only if it is the language generated by a context-free grammar.

However, there are sets that cannot be expressed as the language generated by a context-free grammar. One such set is $\{0^n1^n2^n \mid n = 0, 1, 2, \ldots\}$. We will indicate why this set cannot be recognized by a pushdown automaton, but we will not give a proof, since we have not developed the machinery needed. (However, one method of proof is given in Exercise 28 of the supplementary exercises at the end of this chapter.) The stack can be used to show that a string begins with a sequence of 0s followed by an equal number of 1s by placing a symbol on the stack for each 0 (as long as only 0s are read), and removing one of these symbols for each 1 (as long as only 1s following the 0s are read). But once this is done, the stack is empty, and there is no way to determine that there are the same number of 2s in the string as 0s.

There are other machines called **linear bounded automata**, more powerful than pushdown automata, that can recognize sets such as $\{0^n1^n2^n \mid n = 0, 1, 2, \ldots\}$. In particular, linear bounded automata can recognize context-sensitive languages. However, these machines cannot recognize all the languages generated by phrase-structure grammars. To avoid the limitations of special types of machines, the model known as a **Turing machine**, named after the British mathematician Alan Turing, is used. A Turing machine is made up of everything included in a finite-state machine together with a tape, which is infinite in both directions. A Turing machine has read and write capabilities on the tape, and it can move back and forth along this tape. Turing machines can recognize all languages generated by phrase-structure grammars. In addition, Turing machines can model all the computations that can be performed on a computing machine. Because of their power, Turing machines are extensively studied in theoretical computer science.

Exercises

1. Describe in words the strings in each of the following regular sets.

a) **1*0** **b)** **1*00***
c) **111 ∪ 001** **d)** **(1 ∪ 00)***
e) **(00*1)*** **f)** **(0 ∪ 1)(0 ∪ 1)*00**

2. Determine whether 1011 belongs to each of the following regular sets.

a) **10*1*** **b)** **0*(10 ∪ 11)***
c) **1(01)*1*** **d)** **1*01(0 ∪ 1)**
e) **(10)*(11)*** **f)** **1(00)*(11)***
g) **(10)*1011** **h)** **(1 ∪ 00)(01 ∪ 0)1***

3. Express each of the following sets using a regular expression.

a) the set of strings of one or more 0s followed by a 1

b) the set of strings of two or more symbols followed by three or more 0s

c) the set of strings with either no 1 preceding a 0 or no 0 preceding a 1

d) the set of strings containing a string of 1s so that the number of 1s equals 2 modulo 3, followed by an even number of 0s

4. Construct deterministic finite-state automata that recognize the following sets from I^*, where I is an alphabet.

a) $\emptyset$ **b)** $\{\lambda\}$ **c)** $\{a\}$, where $a \in I$

★5. Show that if A is a regular set, then A^R, the set of all reversals of strings in A, is also regular.

6. Find a finite-state automaton that recognizes

a) $\{\lambda, 0\}$. **b)** $\{0, 11\}$. **c)** $\{0, 11, 000\}$.

7. Using the constructions described in the proof of Kleene's theorem, find nondeterministic finite-state automata that recognize each of the following sets.

a) $\mathbf{0^*1^*}$ **b)** $\mathbf{(0 \cup 11)^*}$ **c)** $\mathbf{01^* \cup 00^*1}$

8. Construct a nondeterministic finite-state automaton that recognizes the language generated by the regular grammar $G = (V, T, S, P)$ where $V = \{0, 1, S, A, B\}$, $T = \{0, 1\}$, S is the start symbol, and the set of productions is

a) $S \rightarrow 0A$, $S \rightarrow 1B$, $A \rightarrow 0$, $B \rightarrow 0$.

b) $S \rightarrow 1A$, $S \rightarrow 0$, $S \rightarrow \lambda$, $A \rightarrow 0B$, $B \rightarrow 1B$, $B \rightarrow 1$.

c) $S \rightarrow 1B$, $S \rightarrow 0$, $A \rightarrow 1A$, $A \rightarrow 0B$, $A \rightarrow 1$, $A \rightarrow 0$, $B \rightarrow 1$.

In Exercises 9–11 construct a regular grammar $G = (V, T, S, P)$ that generates the language recognized by the given finite-state machine.

9.

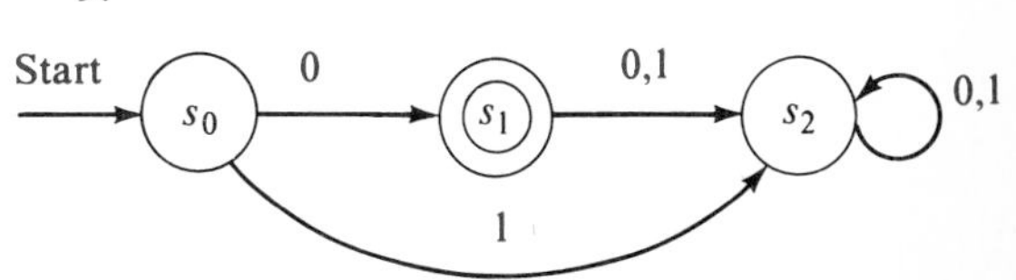

10.

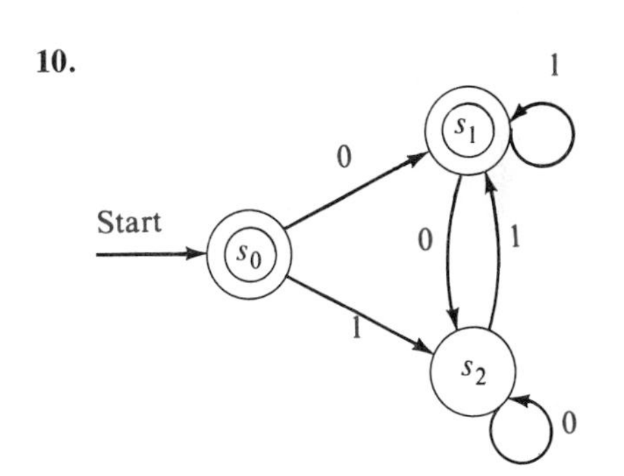

Alan Mathison Turing (1912–1954). Alan Turing was born in London, although he was conceived in India, where his father was employed in the Indian Civil Service. As a boy, he was fascinated by chemistry, performing a wide variety of experiments, and by machinery. Turing attended Sherborne, an English boarding school. In 1931 he won a scholarship to King's College, Cambridge. After completing his dissertation, which included a rediscovery of the central limit theorem, a famous theorem in statistics, he was elected a fellow of his college. In 1935 Turing became fascinated with the decision problem, a problem posed by the great German mathematician Hilbert, which asked whether there is a general method that can be applied to any assertion to determine whether the assertion is true. Turing enjoyed running (later in life running as a serious amateur in competitions), and one day, while resting after a run, he discovered the key ideas needed to solve the decision problem. In his solution, he invented what is now called a **Turing machine** as the most general model of a computing machine. Using these machines, he found a problem, involving what he called computable numbers, that could not be decided using a general method.

From 1936 to 1938 Turing visited Princeton University to work with Alonzo Church, who had also solved Hilbert's decision problem. In 1939 Turing returned to King's College. However, at the outbreak of World War II, he joined the Foreign Office, performing cryptanalysis of German ciphers. His contribution to the breaking the code of the Enigma, a mechanical German cipher machine, played an important role in winning the war.

After the war, Turing worked on the development of early computers. He was interested in the ability of machines to think, proposing that if a computer could not be distinguished from a person based on written replies to questions, it should be considered to be "thinking." He was also interested in biology, having written on morphogenesis, the development of form in organisms. In 1954 Turing committed suicide by taking cyanide, without leaving a clear explanation. Legal troubles related to a homosexual relationship and hormonal treatments mandated by the court to lessen his sex drive may have been factors in his decision to end his life.

11.

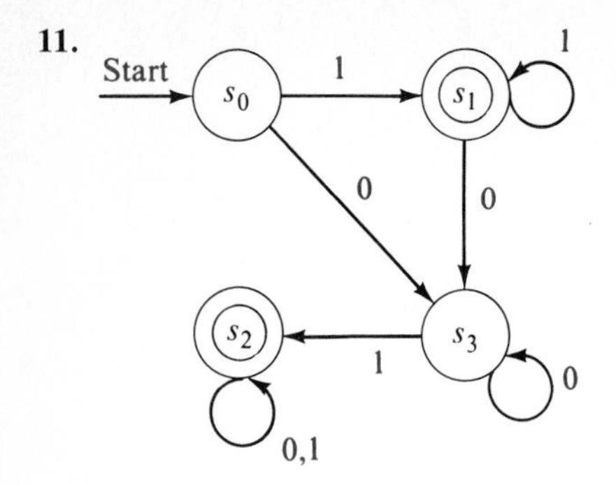

12. Show that the finite-state automaton constructed from a regular grammar in the proof of Theorem 2 recognizes the set generated by this grammar.
13. Show that the regular grammar constructed from a finite-state automaton in the proof of Theorem 2 generates the set recognized by this automaton.
14. Show that every nondeterministic finite-state automaton is equivalent to another such automaton that has the property that its starting state is never revisited.
★15. Let $M = (S, I, f, s_0, F)$ be a deterministic finite-state automaton. Show that the language recognized by M, $L(M)$, is infinite if and only if there is a word x recognized by M with $l(x) \geq |S|$.
★16. One important technique used to prove that certain sets are not regular is the **pumping lemma**. The pumping lemma states that if $M = (S, I, f, s_0, F)$ is a deterministic finite-state automaton and if x is a string in $L(M)$, the language recognized by M, with $l(x) \geq |S|$, then there are strings u, v, and w in I^* such that $x = uvw$, $l(uv) \leq |S|$ and $l(v) \geq 1$, and $uv^iw \in L(M)$ for $i = 0, 1, 2, \ldots$. Prove the pumping lemma. (*Hint:* Use the same idea as was used in Example 5.)
★17. Show that the set $\{0^{2n}1^n\}$ is not regular. You may use the pumping lemma given in Exercise 16.
★18. Show that the set $\{1^{n^2} \mid n = 0, 1, 2, \ldots\}$ is not regular. You may use the pumping lemma given in Exercise 16.
★19. Show that the set of palindromes over $\{0, 1\}$ is not regular. You may use the pumping lemma given in Exercise 16. (*Hint:* Consider strings of the form 0^N10^N.)
★★20. Show that a set recognized by a finite-state automaton is regular. (This is the *if* part of Kleene's theorem.)

Key Terms and Results

TERMS

alphabet (or vocabulary): a set that contains elements used to form strings

language: a subset of the set of all strings over an alphabet

phrase-structure grammar (V,T,S,P): a description of a language containing an alphabet V, a set of terminal symbols T, a start symbol S, and a set of productions P

the production $w \rightarrow w_1$: w can be replaced by w_1 whenever it occurs in a string in the language

$w_1 \Rightarrow w_2$ (w_2 is directly derivable from w_1): w_2 can be obtained from w_1 using a production to replace a string in w_1 with another string

$w_1 \overset{*}{\Rightarrow} w_2$ (w_2 is derivable from w_1): w_2 can be obtained from w_1 using a sequence of productions to replace strings by other strings

type 0 grammar: any phrase-structure grammar

type 1 grammar: a phrase-structure grammar in which every production is of the form $w_1 \rightarrow w_2$ where $l(w_1) \leq l(w_2)$ or $w_2 = \lambda$

type 2, or context-free, grammar: a phrase-structure grammar in which every production is of the form $A \rightarrow w_1$ where A is a nonterminal symbol

type 3, or regular, grammar: a phrase-structure grammar where every production is of the form $A \rightarrow aB$, $A \rightarrow a$, or $S \rightarrow \lambda$, where A and B are nonterminal symbols, S is the start symbol, and a is a terminal symbol

derivation (or parse) tree: an ordered rooted tree where the root represents the starting symbol of a type 2 grammar, internal vertices represent nonterminals, leaves represent terminals, and the children of a vertex are the symbols on a right side of a production, in order from left to right, where the symbol represented by the parent is on the left-hand side

Backus–Naur form: a description of a context-free grammar in which all productions having the same nonterminal as their left-hand side are combined with the different right-hand sides of these productions, each separated by a bar, with nonterminal symbols enclosed in angular brackets and the symbol $\rightarrow$ replaced by ::=

finite-state machine (S, I, O, f, g, s_0) (or a Mealy machine): a six-tuple containing a set S of states, an input alphabet I, an output alphabet O, a transition function f that assigns a next state to every pair of a state and an input, an output function g that assigns an output to every pair of a state and an input, and a starting state s_0

AB (concatenation of A and B): the set of all strings formed by concatenating a string in A and a string in B in that order

A^* (Kleene closure of A): the set of all strings made up by concatenating arbitrarily many strings from A

deterministic finite-state automaton (S, I, f, s_0, F): a five-tuple containing a set S of states, an input alphabet I, a transition function f that assigns a next state to every pair of a state and an input, a starting state s_0, and a set of final states F

nondeterministic finite-state automaton (S, I, f, s_0, F): a five-tuple containing a set S of states, an input alphabet I, a transition function f that assigns a set of possible next states to every pair of a state and an input, a starting state s_0, and a set of final states F

language recognized by an automaton: the set of input strings that take the start state to a final state of the automaton

regular expression: an expression defined recursively by specifying that $\emptyset$, $\boldsymbol{\lambda}$, and $\boldsymbol{x}$, for all x in the input alphabet, are regular expressions, and that (AB), (A $\cup$ B), and (A)* are regular expressions when A and B are regular expressions

regular set: a set defined by a regular expression (see page 645)

RESULTS

For any nondeterministic finite-state automaton there is a deterministic finite-state automaton that recognizes the same set.

Kleene's theorem: A set is regular if and only if there is a finite-state automaton that recognizes it.

A set is regular if and only if it is generated by a regular grammar.

Review Questions

1. a) Define a phrase-structure grammar.
   b) What does it mean for a string to be derivable from a string w by a phrase-structure grammar G?

2. a) What is the language generated by a phrase-structure grammar G?
   b) What is the language generated by the grammar G with vocabulary $\{S, 0, 1\}$, set of terminals T =

$\{0, 1\}$, starting symbol S, and productions $S \rightarrow 000S$, $S \rightarrow 1$?

c) Give a phrase-structure grammar that generates the set $\{01^n \mid n = 0, 1, 2, \ldots\}$.

3. a) Define a type 1 grammar.
   b) Give an example of a grammar that is not a type 1 grammar.
   c) Define a type 2 grammar.
   d) Give an example of a grammar that is not a type 2 grammar, but is a type 1 grammar.
   e) Define a type 3, or context-free, grammar.
   f) Give an example of a grammar that is not a type 3 grammar, but is a type 2 grammar.
4. a) Define a regular grammar.
   b) Define a regular language.
   c) Give an example of a grammar that is not regular but is a type 3 grammar.
   d) Show that the set $\{0^m 1^n \mid m, n = 0, 1, 2, \ldots\}$ is a regular language.
5. a) What is Backus–Naur form?
   b) Give an example of the Backus–Naur form of the grammar for a subset of English of your choice.
6. a) What is a finite-state machine?
   b) Show how a vending machine that accepts only quarters and dispenses a soft drink after 75 cents has been deposited can be modeled using a finite-state machine.
7. a) What is the Kleene closure of a set of strings?
   b) Find the Kleene closure of the set $\{11, 0\}$.
8. a) Define a finite-state automaton.
   b) What does it mean for a string to be recognized by a finite-state automaton?
9. a) Define a nondeterministic finite-state automaton.
   b) Show that given a nondeterministic finite-state automaton, there is a deterministic finite-state automaton that recognizes the same language.
10. a) Define the set of regular expressions over a set I.
    b) Explain how regular expressions are used to represent regular sets.
11. State Kleene's Theorem.
12. Show that a set is generated by a regular grammar if and only if it is a regular set.
13. Give an example of a set not recognized by a finite-state automaton. Show that no finite-state automaton recognizes it.

Supplementary Exercises

★1. Find a phrase-structure grammar that generates each of the following languages.
   a) the set of bit strings of the form $0^{2n}1^{3n}$ where n is a nonnegative integer
   b) the set of bit strings with twice as many 0s as 1s
   c) the set of bit strings of the form w^2 where w is a bit string

★2. Find a phrase-structure grammar that generates the set $\{0^{2^n} \mid n \geq 0\}$.

For Exercises 3 and 4, let $G = (V, T, S, P)$ be the context-free grammar with $V = \{(,), S, A, B\}$, $T = \{(,)\}$, starting symbol S, and productions $S \rightarrow A$, $A \rightarrow AB$, $A \rightarrow B$, $B \rightarrow (A)$, and $B \rightarrow ()$, $S \rightarrow \lambda$.

3. Construct the derivation trees of the following.
   a) (( ))   b) ( )(( ))   c) ((( )( )))

★4. Show that $L(G)$ is the set of all well-formed strings of parentheses, defined in Chapter 3.

A context-free grammar is **ambiguous** if there is a word in $L(G)$ with two derivations that produce different derivation trees, considered as ordered, rooted trees.

5. Show that the grammar $G = (V, T, S, P)$ with $V = \{0,S\}$, $T = \{0\}$, starting state S, and productions $S \rightarrow 0S$, $S \rightarrow S0$, and $S \rightarrow 0$ is ambiguous by constructing two different derivation trees for 0^3.
6. Show that the grammar $G = (V, T, S, P)$ with $V = \{0,S\}$, $T = \{0\}$, starting state S, and productions $S \rightarrow 0S$ and $S \rightarrow 0$ is unambiguous.
7. Suppose that A and B are finite subsets of V^*, where V is an alphabet. Is it necessarily true that $|AB| = |BA|$?

8. Prove or disprove each of the following statements for subsets A, B, and C of V^*, where V is an alphabet.
a) $A(B \cup C) = AB \cup AC$ **b)** $A(B \cap C) = AB \cap AC$
c) $(AB)C = A(BC)$ **d)** $(A \cup B)^* = A^* \cup B^*$

9. Suppose that A and B are subsets of V^*, where V is an alphabet. Does it follow that $A \subseteq B$ if $A^* \subseteq B^*$?

10. What set of strings with symbols in the set $\{0, 1, 2\}$ is represented by the regular expression $(\mathbf{2}^*)(\mathbf{0} \cup (\mathbf{12}^*))^*$?

The **star height** $h(E)$ of a regular expression is defined recursively by

$h(\emptyset) = 0$
$h(\mathbf{x}) = 0$ if $\mathbf{x} \in I$
$h((\mathbf{E}_1 \cup \mathbf{E}_2)) = h((\mathbf{E}_1\mathbf{E}_2)) = \max(h(\mathbf{E}_1), h(\mathbf{E}_2))$ if $\mathbf{E}_1$ and $\mathbf{E}_2$ are regular expressions
$h(\mathbf{E}^*) = h(\mathbf{E}) + 1$ if $\mathbf{E}$ is a regular expression.

11. Find the star height of each of the following regular expressions.
a) $\mathbf{0^*1}$ **b)** $\mathbf{0^*1^*}$ **c)** $\mathbf{(0^*01)^*}$
d) $\mathbf{((0^*1)^*)^*}$ **e)** $\mathbf{(010^*)(1^*01^*)^*((01)^*(10)^*)^*}$ **f)** $\mathbf{(((((0^*)1)^*0)^*)1)^*}$

★12. For each of the following regular expressions find a regular expression that represents the same language with minimum star height.
a) $\mathbf{(0^*1^*)^*}$ **b)** $\mathbf{(0(01^*0)^*)^*}$ **c)** $\mathbf{(0^* \cup (01)^* \cup 1^*)^*}$

13. Construct a finite-state machine with output that produces an output of 1 if the bit string read so far as input contains four or more 1s. Then construct a deterministic finite-state automaton that recognizes this set.

14. Construct a finite-state machine with output that produces an output of 1 if the bit string read so far as input contains four or more consecutive 1s. Then construct a deterministic finite-state automaton that recognizes this set.

15. Construct a finite-state machine with output that produces an output of 1 if the bit string read so far as input ends with four or more consecutive 1s. Then construct a deterministic finite-state automaton that recognizes this set.

16. A state s in a finite-state machine is said to be **reachable** from state s' if there is an input string x such that $f(s, x) = s'$. A state s is called **transient** if there is no nonempty input string x with $f(s, x) = s$. A state s is called a **sink** if $f(s, x) = s$ for all input strings x. Answer the following questions about the finite-state machine with the state diagram illustrated.

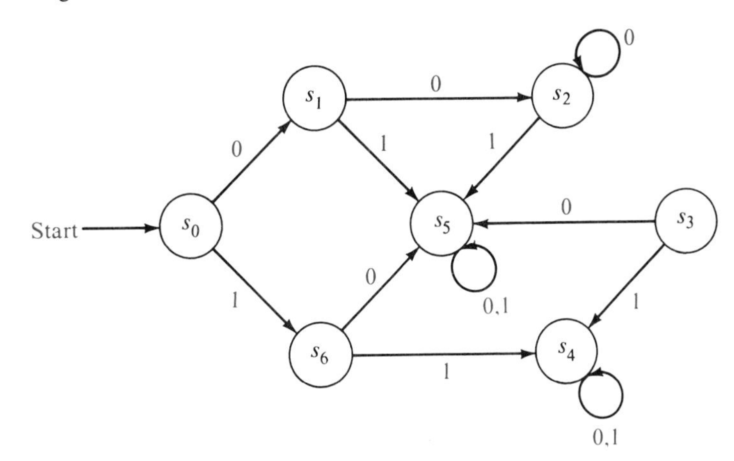

a) Which states are reachable from s_0?
b) Which states are reachable from s_2?
c) Which states are transient?
d) Which states are sinks?

★17. Suppose that S, I, and O are finite sets such that $|S| = n$, $|I| = k$, and $|O| = m$.
a) How many different finite-state machines (Mealy machines) $M = (S, I, O, f, g, s_0)$ can be constructed, where the starting state s_0 can be arbitrarily chosen?
b) How many different Moore machines $M = (S, I, O, f, g, s_0)$ can be constructed, where the starting state s_0 can be arbitrarily chosen?

★18. Suppose that S and I are finite sets such that $|S| = n$ and $|I| = k$. How many different finite-state automata $M = (S, I, f, s_0, F)$ are there where the starting state s_0 and the subset F of S consisting of final states can be chosen arbitrarily
a) if the automata are deterministic? **b)** if the automata may be nondeterministic? (*Note:* This includes deterministic automata.)

19. Construct a deterministic finite-state automaton that is equivalent to the nondeterministic automaton with the state diagram illustrated.

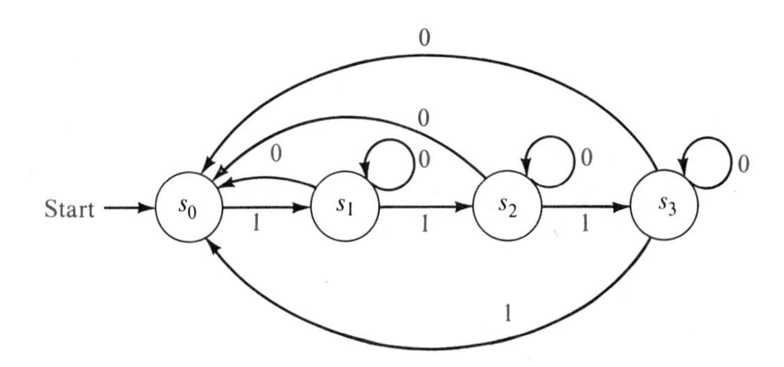

20. What is the language recognized by the automaton in Exercise 19?

21. Construct finite-state automata that recognize the following sets.
a) **0*(10)*** **b)** **(01 ∪ 111)*10*(0 ∪ 1)** **c)** **(001 ∪ (11)*)***

★22. Find regular expressions that represent the set of all string of 0s and 1s
a) made up of blocks of even numbers of 1s interspersed with odd numbers of 0s.
b) with at least two consecutive 0s or three consecutive 1s.
c) with no three consecutive 0s or two consecutive 1s.

★23. Show that if A is a regular set, then so is $\overline{A}$.

★24. Show that if A and B are regular sets, then so is $A \cap B$.

★25. Find finite-state automata that recognize the following sets of strings of 0s and 1s.
a) the set of all strings that start with no more than three consecutive 0s and contain at least two consecutive 1s
b) the set of all strings with an even number of symbols that do not contain the pattern 101
c) the set of all strings with at least three blocks of two or more 1s and at least two 0s

★26. Show that $\{0^{2^n} \mid n \in \mathbf{Z}\}$ is not regular. You may use the pumping lemma given in Exercise 16 of Section 10.4.)

★27. Show that $\{1^p \mid p \text{ is prime}\}$ is not regular. You may use the pumping lemma given in Exercise 16 of Section 10.4.

★28. There is a result for context-free languages analogous to the pumping lemma for regular sets. Suppose that $L(G)$ is the language recognized by a context-free language G. This result states that there is a constant N such that if z is a word in $L(G)$ with $l(w) \geq N$, then z can be written as $uvwxy$ where $l(vwx) \leq N$, $l(vx) \geq 1$, and uv^iwx^iy belongs to $L(G)$ for $i = 0, 1, 2, 3, \ldots$. Use this result to show that there is no context-free grammar G with $L(G) = \{0^n1^n2^n \mid n = 0, 1, 2, \ldots\}$.

Computer Projects

WRITE PROGRAMS WITH THE FOLLOWING INPUT AND OUTPUT

1. Given the productions in a phrase-structure grammar, determine which type this grammar is in the Chomsky classification scheme.

★2. Given the productions of a context-free grammar and a string, produce a derivation tree for this string if it is in the language generated by this grammar.

3. Given the state table of a Moore machine and an input string, produce the output string generated by the machine.

4. Given the state table of a Mealy machine and an input string, produce the output string generated by the machine.

5. Given the state table of a deterministic finite-state automaton and a string, decide whether this string is recognized by the automaton.

6. Given the state table of a nondeterministic finite-state automaton and a string, decide whether this string is recognized by the automaton.

★7. Given the state table of a nondeterministic finite-state automaton, construct the state table of a deterministic finite-state automaton that recognizes the same language.

★★8. Given a regular expression, construct a nondeterministic finite-state automaton that recognizes the set that this expression represents.

9. Given a regular grammar, construct a finite-state automaton that recognizes the language generated by this grammar.

10. Given a finite-state automaton, construct a regular grammar that generates the language recognized by this automaton.

Appendix 1

Exponential and Logarithmic Functions

In this appendix we review some of the basic properties of exponential functions and logarithms. These properties are used throughout the text. Students requiring further review of this material consult precalculus or calculus books, such as those mentioned in the Suggested Readings.

EXPONENTIAL FUNCTIONS

Let n be a positive integer, and let b be a fixed positive real number. The function $f_b(n) = b^n$ is defined by

$$f_b(n) = b^n = b \cdot b \cdot b \cdots b$$

where there are n factors of b multiplied together on the right-hand side of the equation.

We can define the function $f_b(x) = b^x$ for all real numbers x using techniques from calculus. The function $f_b(x) = b^x$ is called the **exponential function to the base b**. We will not discuss how to find the values of exponential functions to the base b when x is not an integer.

Two of the important properties satisfied by exponential functions are given in Theorem 1. Proofs of these and other related properties can be found in calculus texts.

THEOREM 1 Let b be a real number. Then

1. $b^{x+y} = b^x b^y$, and

2. $(b^x)^y = b^{xy}$.

We display the graphs of some exponential functions in Figure 1.

LOGARITHMIC FUNCTIONS

Suppose that b is a real number with $b > 1$. Then the exponential function b^x is strictly increasing (a fact shown in calculus). It is a one-to-one correspondence from

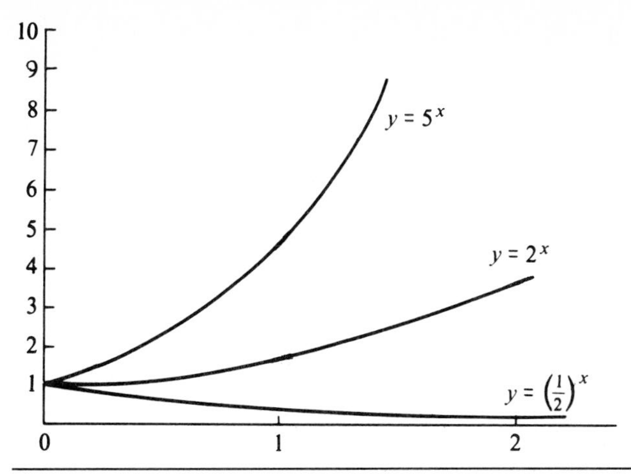

FIGURE 1 Graphs of the Exponential Functions to the Bases 1/2, 2, and 5.

the set of real numbers to the set of nonnegative real numbers. Hence, this function has an inverse, called **the logarithmic function to the base b**. In other words, if b is a real number greater than 1 and x is a positive real number, then

$$b^{\log_b x} = x.$$

The value of this function at x is called the **logarithm of x to the base b**.

From the definition, it follows that

$$\log_b b^x = x.$$

We give several important properties of logarithms in Theorem 2.

THEOREM 2 Let b be a real number greater than 1. Then

1. $\log_b(xy) = \log_b x + \log_b y$ whenever x and y are positive real numbers, and
2. $\log_b(x^y) = y \log_b x$ whenever x is a positive real number.

Proof: Since $\log_b(xy)$ is the unique real number with $b^{\log_b(xy)} = xy$, to prove part 1 it suffices to show that $b^{\log_b x + \log_b y} = xy$. By part 1 of Theorem 1, we have

$$\begin{aligned} b^{\log_b x + \log_b y} &= b^{\log_b x} b^{\log_b y} \\ &= xy. \end{aligned}$$

To prove part 2, it suffices to show that $b^{y \log_b x} = x^y$. By part 2 of Theorem 1, we have

$$\begin{aligned} b^{y \log_b x} &= (b^{\log_b x})^y \\ &= x^y. \end{aligned}$$

■

The following theorem relates logarithms to two different bases.

THEOREM 3 Let a and b be real numbers greater than 1, and let x be a positive real number. Then

$$\log_a x = \log_b x / \log_b a.$$

Proof: To prove this result, it suffices to show that

$$b^{\log_a x \cdot \log_b a} = x.$$

By part 2 of Theorem 1, we have

$$\begin{aligned} b^{\log_a x \cdot \log_b a} &= (b^{\log_b a})^{\log_a x} \\ &= a^{\log_a x} \\ &= x. \end{aligned}$$

This completes the proof. ■

Since the base used most often for logarithms in this text is $b = 2$, the notation $\log x$ is used throughout the test to denote $\log_2 x$.

The graph of the function $f(x) = \log x$ is displayed in Figure 2. From Theorem 3, when a base b other than 2 is used, a function that is a constant multiple of the function $\log x$, namely, $(1/\log b)\log x$, is obtained.

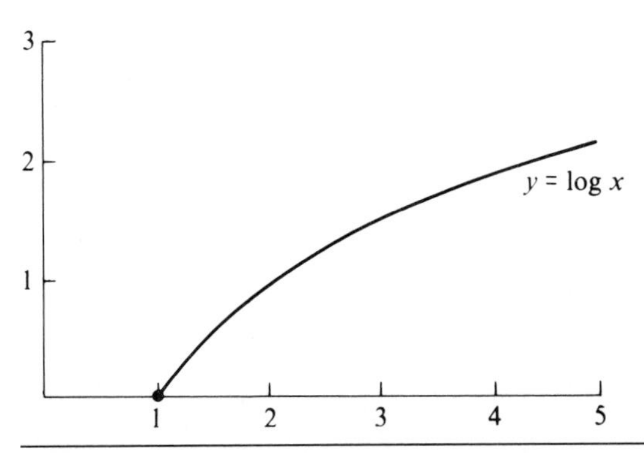

FIGURE 2 The Graph of $f(x) = \log x$.

Exercises

1. Express each of the following quantities as powers of 2.
   a) $2 \cdot 2^2$ b) $(2^2)^3$ c) $2^{(2^2)}$
2. Find each of the following quantities.
   a) $\log_2 1024$ b) $\log_2 1/4$ c) $\log_4 8$
3. Suppose that $\log_4 x = y$. Find each of the following quantities.
   a) $\log_2 x$ b) $\log_8 x$ c) $\log_{16} x$
4. ☞ Let a, b, and c be positive real numbers. Show that $a^{\log_b c} = c^{\log_b a}$.
5. Draw the graph of $f(x) = b^x$ if b is
   a) 3. b) 1/3. c) 1.
6. Draw the graph of $f(x) = \log_b x$ if b is
   a) 4. b) 100. c) 1000.

Appendix 2

Pseudocode

The algorithms in this text are described both in English and in pseudocode. Pseudocode is an intermediate step between an English language description of the steps of a procedure and a specification of this procedure using an actual programming language. The advantages of using pseudocode include the simplicity with which it can be written and understood and the ease of producing actual computer code (in a variety of programming languages) from the pseudocode.

This appendix describes the format and syntax of the pseudocode used in the text. This pseudocode is designed so that its basic structure resembles that of Pascal, which is currently the most commonly taught programming language. However, the pseudocode we use will be a lot looser than a formal programming language since a lot of English language descriptions of steps will be allowed.

This appendix is not meant for formal study. Rather, it should serve as a reference guide for students when they study the descriptions of algorithms given in the text and when they write pseudocode solutions to exercises.

PROCEDURE STATEMENTS

The pseudocode for an algorithm begins with a **procedure** statement that gives the name of an algorithm, lists the input variables, and describes what kind of variable each input is. For instance, the statement

procedure *maximum*(*L*: list of integers)

is the first statement in the pseudocode description of the algorithm, which we have named *maximum*, that finds the maximum of a list L of integers.

ASSIGNMENTS AND OTHER TYPES OF STATEMENTS

An assignment statement is used to assign values to variables. In an assignment statement the left-hand side is the name of the value and the right-hand side is an expression that involves constants, variables that have been assigned values, or functions defined by procedures. The right-hand side may contain any of the usual arithmetic operations. However, in the pseudocode in this book it may include any well-defined operation, even if this operation can be carried out only by using a large number of statements in an actual programming language.

The symbol := is used for assignments. Thus, an assignment statement has the form

```
variable := expression
```

For example, the statement

```
max := a
```

assigns to the variable *max* the value of a. A statement such as

```
x := largest integer in the list L
```

can also be used. This sets x equal to the largest integer in the list L. To translate this statement into an actual programming language would require more than one statement. Also, the instruction

```
interchange a and b
```

can be used to interchange a and b. We could also express this one statement with several assignment statements (see Exercise 2 at the end of this appendix), but for simplicity, we will often prefer this abbreviated form of pseudocode.

BLOCKS OF STATEMENTS

Statements can be grouped into blocks to carry out complicated procedures. These blocks are set off using **begin** and **end** statements, and the statements in the block are all indented the same amount.

```
begin
   statement 1
   statement 2
   statement 3
   .
   .
   .
   statement n
end
```

The statements in the block are executed sequentially.

COMMENTS

In the pseudocode in this book, statements enclosed in curly braces are not executed. Such statements serve as comments or reminders that help explain how the procedure works. For instance, the statement

```
{x is the largest element in L}
```

can be used to remind the reader that at that point in the procedure the variable x equals the largest element in the list L.

CONDITIONAL CONSTRUCTIONS

The simplest form of the conditional construction that we will use is

```
if condition then statement
```

or

```
if condition then
begin
   block of statements
end
```

Here, condition 1 is checked, and if it is true, then the statement given is carried out. For example, in Algorithm 1 in Section 2.1, which finds the maximum of a set of integers, we use a conditional statement to check whether $max < a_i$ for each variable; if it is, we assign the value of a_i to max.

Often, we require the use of a more general type of construction. This is used when we wish to do one thing when the indicated condition is true, but another when it is false. We use the construction

```
if condition then statement 1
else statement 2
```

Note that either one or both of statement 1 and statement 2 can be replaced with a block of statements. Sometimes, we require the use of an even more general form of a conditional. The general form of the conditional construction that we will use is

```
if condition 1 then statement 1
else if condition 2 then statement 2
else if condition 3 then statement 3
  .
  .
  .
else if condition n then statement n
else statement n + 1
```

When this construction is used, if condition 1 is true, then statement 1 is carried out, and the program exits this construction. In addition, if condition 1 is false, the program checks whether condition 2 is true; if it is, statement 2 is carried out, and so on. Thus, if none of the first $n - 1$ conditions hold, but condition n does, statement n is carried out. Finally, if none of condition 1, condition 2, condition 3, . . . , condition n is true, then statement $n + 1$ is executed. Note that any of the $n + 1$ statements can be replaced by a block of statements.

LOOP CONSTRUCTIONS

There are two types of loop construction in the pseudocode in this book. The first is the "for" construction, which has the form

```
for variable := initial value to final value
  statement
```

or

for *variable* := *initial value* **to** *final value*
begin
  block of statements
end

Here, at the start of the loop, *variable* is assigned *initial value* if *initial value* is less than or equal to *final value,* and the statements at the end of this construction are carried out with this value of *variable*. Then *variable* is increased by one, and the statement, or the statements in the block, are carried out with this new value of *variable*. This is repeated until *variable* reaches *final value*. After the instructions are carried out with *variable* equal to *final value,* the algorithm proceeds to the next statement. When *initial value* exceeds *final value,* none of the statements in the loop is executed.

We can use the "for" loop construction to find the sum of the positive integers from 1 to n with the following pseudocode.

sum := 0
for i := 1 **to** n
  sum := *sum* + i

Also, the more general "for" statement, of the form

for all elements with a certain property

is used in this text. This means that the statement or block of statements that follow are carried out successively for the elements with the given property.

The second type of loop construction that we will use is the "while" construction. This has the form

while condition
  statement

or

while condition
begin
  block of statements
end

When this construction is used, the condition given is checked, and if it is true, the statements that follow are carried out, which may change the values of the variables that are part of the condition. If the condition is still true after these instructions have been carried out, the instructions are carried out again. This is repeated until the condition becomes false. As an example, we can find the sum of the integers from 1 to n using the following block of pseudocode including a "while" construction.

```
sum := 0
while n > 0
begin
  sum := sum + n
  n := n − 1
end
```

Note that any "for" construction can be turned into a "while" construction (see Exercise 3 at the end of this appendix). However, it is often easier to understand the "for" construction. So, when it makes sense, we will use the "for" construction in preference to the corresponding "while" construction.

USING PROCEDURES IN OTHER PROCEDURES

We can use a procedure from within another procedure (or within itself in a recursive program) simply by writing the name of this procedure followed by the inputs to this procedure. For instance,

```
max(L)
```

will carry out the procedure *max* with the input list L. After all the steps of this procedure have been carried out, execution carries on with the next statement in the procedure.

Exercises

1. What is the difference between the following blocks of two assignments statements?

   $a := b$
   $b := c$

   and

   $b := c$
   $a := b$

2. Give a procedure using assignment statements to interchange the values of the variables x and y. What is the minimum number of assignment statements needed to do this?

3. Show how a loop of the form

   for $i :=$ *initial value* **to** *final value*
   statement

   can be written using the "while" construction.

Appendix 3

Generating Functions

Generating functions are important tools for solving counting problems. We will discuss two ways that generating functions are used to solve such problems. First, we will show how generating functions model counting problems. Then we will show how recurrence relations can be solved using generating functions.

The **generating function** for the sequence $a_0, a_1, a_2, \ldots, a_k, \ldots$ of real numbers is the infinite series

$$G(x) = a_0 + a_1 x + a_2 x^2 + \cdots + a_k x^k + \cdots = \sum_{k=0}^{\infty} a_k x^k.$$

We can define generating functions for finite sequences of real numbers by extending a finite sequence $a_0, a_1, \ldots, a_n$ into an infinite sequence by setting $a_{n+1} = 0$, $a_{n+2} = 0, \ldots$. The generating function $G(x)$ of this infinite sequence $\{a_n\}$ is a polynomial of degree n since no terms of the form $a_j x^j$ with $j > n$ occur, that is

$$G(x) = a_0 + a_1 x + \cdots + a_n x^n.$$

EXAMPLE 1 What is the generating function for the sequence 1, 1, 1, 1, 1, 1?

Solution: The generating function of 1, 1, 1, 1, 1, 1 is

$$1 + x + x^2 + x^3 + x^4 + x^5.$$

From Example 6 of Section 2.3 we have

$$(x^6 - 1)/(x - 1) = 1 + x + x^2 + x^3 + x^4 + x^5.$$

Consequently, $f(x) = (x^6 - 1)/(x - 1)$ is the generating function of the sequence 1, 1, 1, 1, 1, 1. ■

EXAMPLE 2 Let m be a positive integer. Let $a_k = C(m, k)$, for $k = 0, 1, 2, \ldots, m$. What is the generating function for the sequence $a_0, a_1, \ldots, a_m$?

Solution: The generating function for this sequence is

$$G(x) = C(m,0) + C(m,1)x + C(m,2)x^2 + \cdots + C(m,m)x^m.$$

The binomial theorem shows that $G(x) = (1+x)^m$. ■

SOME FACTS ABOUT SERIES

We will state some important facts about infinite series used when working with generating functions. A discussion of these and related results can be found in calculus texts.

EXAMPLE 3 The function $f(x) = 1/(1-x)$ is the generating function of the sequence $1, 1, 1, 1, \ldots,$ since

$$1/(1-x) = 1 + x + x^2 + \cdots$$

for $|x| < 1$. ■

EXAMPLE 4 The function $f(x) = 1/(1-ax)$ is the generating function of the sequence $1, a, a^2, a^3, \ldots,$ since

$$1/(1-ax) = 1 + ax + a^2x^2 + \cdots$$

when $|ax| < 1$, or equivalently, for $|x| < 1/|a|$ for $a \neq 0$. ■

We also will need some results on how to add and how to multiply two generating functions. Proofs of these results can be found in calculus texts.

THEOREM 1 Let $f(x) = \sum_{k=0}^{\infty} a_k x^k$ and $g(x) = \sum_{k=0}^{\infty} b_k x^k$. Then

$$f(x) + g(x) = \sum_{k=0}^{\infty} (a_k + b_k)x^k$$

and

$$f(x)g(x) = \sum_{k=0}^{\infty} \left(\sum_{j=0}^{k} a_j b_{k-j} \right) x^k.$$

We will illustrate how Theorem 1 can be used with the following example.

EXAMPLE 5 Let $f(x) = 1/(1-x)^2$. Use Example 3 to find the coefficients $a_0, a_1, a_2, \ldots$ in the expansion $f(x) = \sum_{k=0}^{\infty} a_k x^k$.

Solution: From Example 3 we see that

$$1/(1-x) = 1 + x + x^2 + x^3 + \cdots.$$

Hence, from Theorem 1, we have

$$1/(1-x)^2 = \sum_{k=0}^{\infty}\left(\sum_{j=0}^{k} 1\right)x^k = \sum_{k=0}^{\infty}(k+1)x^k. \qquad \blacksquare$$

Remark: This result also can be derived from Example 3 by differentiation. Taking derivatives is a useful technique for producing new identities from existing identities for generating functions.

COUNTING PROBLEMS AND GENERATING FUNCTIONS

In Chapters 4 and 5 we developed techniques to count the r-combinations from a set with n elements when repetition is allowed and additional constraints may exist. Such problems are equivalent to counting the solutions to equations of the form

$$e_1 + e_2 + \cdots + e_n = C,$$

where C is a constant and each e_i is an integer that may be subject to a specified constraint. Generating functions can also be used to solve counting problems of this type, as the following examples show.

EXAMPLE 6 Find the number of solutions of

$$e_1 + e_2 + e_3 = 17,$$

where e_1, e_2, and e_3 are nonnegative integers with $2 \leq e_1 \leq 5$, $3 \leq e_2 \leq 6$, and $4 \leq e_3 \leq 7$.

Solution: The number of solutions with the indicated constraints is the coefficient of x^{17} in the expansion of

$$(x^2 + x^3 + x^4 + x^5)(x^3 + x^4 + x^5 + x^6)(x^4 + x^5 + x^6 + x^7).$$

This follows since we obtain a term equal to x^{17} in the product by picking a term in the first sum x^{e_1}, a term in the second sum x^{e_2}, and a term in the third sum x^{e_3}, where the exponents e_1, e_2, and e_3 satisfy the equation $e_1 + e_2 + e_3 = 17$ and the given constraints.

It is not hard to see that the coefficient of x^{17} in this product is 3. Hence, there are three solutions. (Note that the computation of this coefficient involves about as much work as enumerating all the solutions of the equation with the given constraints. However, the method that this illustrates often can be used to solve wide classes of counting problems with special formulae, as we will see.) ■

EXAMPLE 7 In how many different ways can eight identical cookies be distributed among three distinct children if each child receives at least two cookies and no more than four cookies?

Solution: Since each child receives at least two but no more than four cookies, for each child there is a factor equal to

$$(x^2 + x^3 + x^4)$$

in the generating function for the sequence $\{c_n\}$, where c_n is the number of ways to distribute n cookies. Since there are three children, this generating function is

$$(x^2 + x^3 + x^4)^3.$$

We need the coefficient of x^8 in this product. The reason is that the x^8 terms in the expansion correspond to the ways that three terms can be selected, with one from each factor, that have exponents adding up to 8. Furthermore, the exponents of the term from the first, second, and third factors are the numbers of cookies the first, second, and third children receive, respectively. Computation shows that this coefficient equals 6. Hence, there are six ways to distribute the cookies so that each child receives at least two, but no more than four, cookies. ■

EXAMPLE 8 Use generating functions to find the number of k-combinations of a set with n elements. Assume that the binomial theorem has already been established.

Solution: Each of the n elements in the set contributes the term $(1 + x)$ to the generating function $f(x) = \sum_{k=0}^{n} a_k x^k$. Here $f(x)$ is the generating function for $\{a_k\}$, where a_k represents the number of k-combinations of a set with n elements. Hence,

$$f(x) = (1 + x)^n.$$

But by the binomial theorem, we have

$$f(x) = \sum_{k=0}^{n} \binom{n}{k} x^k,$$

where

$$\binom{n}{k} = \frac{n!}{k!(n-k)!}.$$

Hence, $C(n, k)$, the number of k-combinations of a set with n elements, is

$$\frac{n!}{k!(n-k)!}.$$

■

Remark: We proved the binomial theorem in Section 4.3 using the formula for the number of r-combinations of a set with n elements. This example shows that the binomial theorem, which can be proved by mathematical induction, can be used to derive the formula for the number of r-combinations of a set with n elements.

USING GENERATING FUNCTIONS TO SOLVE RECURRENCE RELATIONS

We can find the solution to a recurrence relation and its initial conditions by finding an explicit formula for the associated generating function. This is illustrated in the following examples.

EXAMPLE 9 Solve the recurrence relation $a_k = 3a_{k-1}$ for $k = 1, 2, 3, \ldots$ and initial condition $a_0 = 2$.

Solution: Let $G(x)$ be the generating function for the sequence $\{a_k\}$, that is, $G(x) = \sum_{k=0}^{\infty} a_k x^k$. First note that

$$xG(x) = \sum_{k=0}^{\infty} a_k x^{k+1} = \sum_{k=1}^{\infty} a_{k-1} x^k.$$

Using the recurrence relation, we see that

$$\begin{aligned} G(x) - 3xG(x) &= \sum_{k=0}^{\infty} a_k x^k - 3\sum_{k=1}^{\infty} a_{k-1} x^k \\ &= a_0 + \sum_{k=1}^{\infty} (a_k - 3a_{k-1}) x^k. \\ &= 2, \end{aligned}$$

since $a_0 = 2$ and $a_k = 3a_{k-1}$. Thus,

$$G(x) - 3xG(x) = (1 - 3x)G(x) = 2.$$

Solving for $G(x)$ shows that $G(x) = 2/(1 - 3x)$. Using the identity $1/(1 - ax) = \sum_{k=0}^{\infty} a^k x^k$, we have

$$G(x) = 2\sum_{k=0}^{\infty} 3^k x^k = \sum_{k=0}^{\infty} 2 \cdot 3^k x^k.$$

Consequently, $a_k = 2 \cdot 3^k$. ■

EXAMPLE 10 Suppose that a valid code word is an n-digit number in decimal notation containing an even number of 0s. Let a_n denote the number of valid code words of length n. In Example 7 of Section 5.1 we showed that the sequence $\{a_n\}$ satisfies the recurrence relation

$$a_n = 8a_{n-1} + 10^{n-1}$$

and the initial condition $a_1 = 9$. Use generating functions to find an explicit formula for a_n.

Solution: To make our work with generating functions simpler, we extend this sequence by setting $a_0 = 1$; when we assign this value to a_0 and use the recurrence relation, we have $a_1 = 8a_0 + 10^0 = 8 + 1 = 9$, which is consistent with our original initial condition.

We multiply both sides of the recurrence relation by x^n to obtain

$$a_n x^n = 8a_{n-1}x^n + 10^{n-1}x^n.$$

Let $G(x) = \sum_{n=0}^{\infty} a_n x^n$ be the generating function of the sequence $a_0, a_1, a_2, \ldots$. We sum both sides of the last equation starting with $n = 1$, to find that

$$\begin{aligned}
G(x) - 1 = \sum_{n=1}^{\infty} a_n x^n &= \sum_{n=1}^{\infty} (8a_{n-1}x^n + 10^{n-1}x^n) \\
&= 8\sum_{n=1}^{\infty} a_{n-1}x^n + \sum_{n=1}^{\infty} 10^{n-1}x^n \\
&= 8x\sum_{n=1}^{\infty} a_{n-1}x^{n-1} + x\sum_{n=1}^{\infty} 10^{n-1}x^{n-1} \\
&= 8x\sum_{n=0}^{\infty} a_n x^n + x\sum_{n=0}^{\infty} 10^n x^n \\
&= 8xG(x) + x/(1 - 10x),
\end{aligned}$$

where we have used Example 4 to evaluate the second summation. Therefore, we have

$$G(x) - 1 = 8xG(x) + x/(1 - 10x).$$

Solving for $G(x)$ shows that

$$G(x) = \frac{1 - 9x}{(1 - 8x)(1 - 10x)}.$$

Expanding the right-hand side of this equation into partial fractions (as is done in the integration of rational functions studied in calculus) gives

$$G(x) = \frac{1}{2}\left(\frac{1}{1 - 8x} + \frac{1}{1 - 10x}\right).$$

Using Example 4 twice (once with $a = 8$ and once with $a = 10$) gives

$$\begin{aligned}
G(x) &= \frac{1}{2}\left(\sum_{n=0}^{\infty} 8^n x^n + \sum_{n=0}^{\infty} 10^n x^n\right) \\
&= \sum_{n=0}^{\infty} \frac{1}{2}(8^n + 10^n)x^n.
\end{aligned}$$

Consequently, we have shown that

$$a_n = \frac{1}{2}(8^n + 10^n).$$

■

Exercises

1. Find the generating function of the sequence 2, 2, 2, 2, 2, 2.
2. Find the generating function of the sequence 1, 4, 16, 64, 256.
3. In how many different ways can 10 balloons be given to four children if each child receives at least 2 balloons?
4. Use generating functions to find the number of ways to choose a dozen bagels from three varieties—egg, salty, and plain—if there are at least two bagels of each kind, but no more than three salty bagels.
5. What is the generating function for c_k where c_k represents the number of ways to make change for k dollars using \$1 bills, \$2 bills, \$5 bills, and \$10 bills?
6. Give a combinatorial interpretation of the coefficient a_6 of x^6 in the expansion of $(1 + x + x^2 + x^3 + \cdots)^n$. Use this interpretation to find a_6.
7. What is the generating function for a_k where a_k is the number of solutions of $x_1 + x_2 + x_3 = k$ where x_1, x_2, and x_3 are integers with $x_1 \geq 2$, $0 \leq x_2 \leq 3$, and $2 \leq x_3 \leq 5$?
8. Determine the series with the following functions as their generating functions.
   a) $(1 + x)^4$ **b)** $(3 + x)^5$
   c) $(x^2 + 1)^3$ **d)** $1/(1 - 2x)$
   e) $5/(1 - x^2)$ **f)** $1/(1 - 2x)^2$
9. Find a simple expression for the generating function of $\{a_k\}$, $k = 0, 1, 2, \ldots$, if
   a) $a_k = 3$. **b)** $a_k = 5^k$. **c)** $a_k = k + 1$.

★10. Use generating functions to prove Pascal's identity, $C(n, r) = C(n-1, r) + C(n-1, r-1)$. (*Hint:* Use the identity $(1 + x)^n = (1 + x)^{n-1} + x(1 + x)^{n-1}$.)

11. Use a generating function to solve the recurrence relation $a_k = 7a_{k-1}$ and the initial condition $a_0 = 5$.
12. Use a generating function to solve the recurrence relation $a_k = 3a_{k-1} + 2$ and the initial condition $a_0 = 1$.
13. Use a generating function to solve the recurrence relation $a_k = 3a_{k-1} + 4^{k-1}$ and the initial condition $a_0 = 1$.
14. Use a generating function to solve the recurrence relation $a_k = 5a_{k-1} - 6a_{k-2}$ and the initial conditions $a_0 = 6$ and $a_1 = 30$.
15. Use a generating function to find an explicit formula for the Fibonacci numbers.

Suggested Readings

CHAPTER 1

An entertaining way to study logic is to read Lewis Carroll's book [23]. General references for logic include Mendelson [76], Stoll [112], and Suppes [114]. Lin and Lin [67] is an easily read text on sets and their applications. Axiomatic developments of set theory can be found in Halmos [46], Monk [77], and Stoll [112]. Brualdi [21] and Reingold, Nievergelt, and Deo [90] contain introductions to multisets. Fuzzy sets and their application to expert systems and artificial intelligence are treated in Negoita [79]. Calculus books, such as Apostol [6], Spivak [109], and Thomas and Finney [116], contain discussions of functions. Stanat and McAllister [110] has a thorough section on countability. Extensive material on big-O estimates of functions can be found in Knuth [58]. Discussions of the mathematical foundations needed for computer science can be found in Arbib, Kfoury, and Moll [8], Bobrow and Arbib [17], Beckman [12], and Tremblay and Manohar [117]. Biographical information on many of the mathematicians and computer scientists mentioned in this book can be found in Gillispie [38].

CHAPTER 2

The articles by Knuth [57] and Wirth [129] are accessible introductions to the subject of algorithms. General references for algorithms and their complexity include Aho, Hopcroft, and Ullman [1], Baase [9], Gonnet [40], Goodman and Hedetniemi [41], Horowitz and Sahni [53], the famous series of books by Knuth on the art of computer programming [58], [59], and [60], Kronsjö [62], Pohl and Shaw [81], Purdom and Brown [87], Sedgewick [105], Wilf [123], and Wirth [128]. References for number theory include Hardy and Wright [49], LeVeque [65], Rosen [99], and Stark [111]. Applications of number theory to cryptography are covered in Denning [29], Rosen

introduced in Rivest, Shamir, and Adleman [93]. Algorithms for computer arithmetic are discussed in Knuth [59] and Pohl and Shaw [81]. Matrices and their operations are covered in any linear algebra books, such as Curtis [27] and Strang [113].

CHAPTER 3

The science and art of constructing proofs is discussed in a delightful way in three books by Pólya: [82], [83], and [84]. An accessible introduction to mathematical induction can be found in the English translation (from the Russian original) of Sominskii [108]. Problems involving the covering of some or all of the squares of a chessboard using L-shaped pieces or other types of pieces are treated by Golomb [39]. Books that contain thorough treatments of mathematical induction and recursive definitions include Liu [68], Sahni [103], Stanat and McAllister [117], and Tremblay and Manohar [117]. The Ackermann function, introduced in 1928 by W. Ackermann, arises in the theory of recursive function (see Beckman [12] and McNaughton [74], for instance) and in the analysis of the complexity of certain set theoretic algorithms (see Tarjan [115]). Recursion is studied in Roberts [94], Rohl [98], and Wand [121]. Discussions of program correctness and the logical machinery used to prove programs are correct can be found in Alagic and Arbib [3], Anderson [5], Backhouse [10], Sahni [103], and Stanat and McAllister [110].

CHAPTER 4

General references for counting techniques and their applications include Anderson [4], Berman and Fryer [15], Bogart [18], Bose and Manvel [20], Brualdi [21], Cohen [26], Grimaldi [45], Liu [69], Pólya, Tarjan, and Woods [85], Riordan [91], Roberts [95], Tucker [118], and Williamson [124]. Vilenkin [120] contains a selection of combinatorial problems and their solutions. A selection of more difficult combinatorial problems can be found in Lovász [70]. Applications of the pigeonhole principle can be found in Brualdi [21], Liu [68], and Roberts [95]. References for discrete probability include Feller [36] and Ross [100]. A wide selection of combinatorial identities can be found in Riordan [92]. Combinatorial algorithms, including algorithms for generating permutations and combinations, are described by Even [34], Lehmer [64], and Reingold, Nievergelt, and Deo [90].

CHAPTER 5

Many different models using recurrence relations can be found in Roberts [95] and Tucker [118]. Exhaustive treatments of linear homogeneous recurrence relations with constant coefficients, and related inhomogeneous recurrence relations, can be found in Brualdi [21] and Liu [69]. Divide-and-conquer algorithms and their complexity are covered in Roberts [95] and Stanat and McAllister [110]. Descriptions of fast multiplication of integers and matrices can be found in Aho, Hopcroft, and Ullman [1] and Knuth [59]. Additional applications of the principle of inclusion-exclusion can be found in Liu [68] and [69], Roberts [95], and Ryser [101].

CHAPTER 6

General references for relations, including treatments of equivalence relations and partial orders, include Bobrow and Arbib [17], Grimaldi [45], Sanhi [103], and Tremblay and Manohar [117]. Discussions of relational models for data bases are given in Date [28]. Warshall's original paper for finding transitive closures can be found in [122]. Directed graphs are studied in Behzad, Chartrand, and Lesniak-Foster [14], Grimaldi [45], Robinson and Foulds [97], Roberts [95], and Tucker [118].

CHAPTER 7

General references for graph theory include Behzad and Chartrand [13], Behzad, Chartrand, and Lesniak-Foster [14], Bondy and Murty [19], Graver and Watkins [44], Grimaldi [45], Harary [48], Ore [80], Roberts [95], Tucker [118], Wilson [125], and Wilson and Watkins [127]. A wide variety of applications of graph theory can be found in Chartrand [25], Deo [31], Roberts [95] and [96], and Wilson and Beineke [126], and in McHugh [73]. A comprehensive description of algorithms in graph theory can be found in Gibbons [37]. Other references for algorithms in graph theory include Buckley and Harary [22], Chachra, Ghare, and Moore [24], Even [34] and [35], Hu [54], and Reingold, Nievergelt, and Deo [90]. A translation of Euler's original paper on the Königsberg bridge problem can be found in [33]. Dijkstra's algorithm is studied in Gibbons [37], Liu [68], and Reingold, Nievergelt, and Deo [90]. Dijkstra's original paper can be found in [32]. A proof of Kuratowski's theorem can be found in Harary [48] and in Liu [69]. Crossing numbers and thicknesses of graphs are studied in Behzad, Chartrand, and Lesniak-Foster [14]. References for graph coloring and the four-color theorem are included in Barnette [11] and Saaty and Kainen [102]. The original conquest of the four-color theorem is reported in Appel and Haken [7]. Applications of graph coloring are described by Roberts [95]. The History of graph theory is covered in Biggs, Lloyd, and Wilson [16]. Interconnection networks for parallel processing are discussed in Akl [2] and Siegel and Hsu [106].

CHAPTER 8

Trees are studied in Deo [31], Grimaldi [45], Knuth [58], Roberts [95], and Tucker [118]. The use of trees in computer science is described by Gotlieb and Gotlieb [42], Horowitz and Sahni [53], and Knuth [58] and [60]. Roberts [95] covers applications of trees to many different areas. Prefix codes and Huffman coding are covered in Hamming [47]. Sorting and searching algorithms and their complexity are studied in detail in Knuth [60]. Backtracking is an old technique; its use to solve maze puzzles can be found in the 1891 book by Lucas [71]. An extensive discussion of how to solve problems using backtracking can be found in Reingold, Nievergelt, and Deo [90]. Gibbons [37] and Reingold, Nievergelt, and Deo [90] contain discussions of algorithms for constructing spanning trees and minimal spanning trees. The background and history of algorithms for finding minimal spanning trees in covered in Graham and Hell [43]. Prim and Kruskal described their algorithms for finding minimal spanning trees in [86] and [63], respectively. Sollin's algorithm is an example of an algorithm well suited for parallel processing; although Sollin never published a description of it, his algorithm has been described by Even [34] and Goodman and Hedetniemi [41].

CHAPTER 9

Boolean algebra is studied in Grimaldi [45], Hohn [50], Kohavi [61], and Tremblay and Manohar [117]. Applications of Boolean algebra to logic circuits and switching circuits are described by Hohn [50] and Kohavi [61]. The original papers dealing with the minimization of sum-of-products expansions using maps are Karnaugh [55] and Veitch [119]. The Quine-McCluskey method was introduced in McCluskey [72], Quine [88], and Quine [89]. Threshold functions are covered in Kohavi [61].

CHAPTER 10

General references for languages and automata theory include Denning, Dennis, and Qualitz [30], Hopcroft and Ullman [51], Hopkin and Moss [52], Lewis and Papadimitriou [66], and McNaughton [74]. Mealy machines and Moore machines were originally introduced in Mealy [75] and Moore [78]. The original proof of Kleene's theorem can be found in [56]. Powerful models of computation, including pushdown automata and Turing machines, are discussed in Hopcroft and Ullman [51] and in Hopkin and Moss [52].

APPENDIXES

Detailed treatments of exponential and logarithmic functions can be found in calculus books such as Apostol [6], Spivak [109], and Thomas [116].

The pseudocode described in Appendix 2 resembles Pascal. Pohl and Shaw [81] uses a similar type of language to describe algorithms. Wirth [128] describes how to use algorithms and data structures to build programs using Pascal. Also, Rohl [98] shows how to implement recursive programs using Pascal.

An excellent introduction to generating functions can be found in Pólya, Tarjan, and Woods [85]. Generating functions are studied in detail in Brualdi [21], Cohen [26], Grimaldi [45], and Roberts [95].

REFERENCES

1. A.V. Aho, J.E. Hopcroft, and J.D. Ullman, *The Design and Analysis of Computer Algorithms*, Addison-Wesley, Reading, Mass., 1974.
2. S.G. Akl, *The Design and Analysis of Parallel Algorithms*, Prentice-Hall, Englewood Cliffs, N.J., 1989.
3. S. Alagic and M.A. Arbib, *The Design of Well-Structured and Correct Programs*, Springer–Verlag, New York, 1978.
4. I. Anderson, *A First Course in Combinatorial Mathematics*, Clarendon, Oxford, England, 1974.
5. R.B. Anderson, *Proving Programs Correct*, Wiley, New York, 1979.
6. T.M. Apostol, *Calculus*, vol. I, 2d ed., Wiley, New York, 1967.
7. K. Appel and W. Haken, "Every Planar Map Is 4-colorable," *Bulletin of the AMS*, 82 (1976), 711–712.
8. M.A. Arbib, A.J. Kfoury, and R.N. Moll, *A Basis for Theoretical Computer Science*, Springer–Verlag, New York, 1980.
9. S. Baase, *Computer Algorithms*, 2nd ed., Addison-Wesley, Reading, Mass., 1988.
10. R.C. Backhouse, *Program Construction and Verification*, Prentice-Hall International, Englewood Cliffs, N.J., 1986.
11. D. Barnette, *Map Coloring, Polyhedra, and the Four-Color Problem*, Mathematical Association of America, Washington, D.C., 1983.
12. F.S. Beckman, *Mathematical Foundations of Programming*, Addison-Wesley, Reading, Mass., 1980.
13. M. Behzad and G. Chartrand, *Introduction to the Theory of Graphs*, Allyn & Bacon, Boston, 1971.
14. M. Behzad and G. Chartrand, and L. Lesniak-Foster, *Graphs and Digraphs*, Wadsworth, Belmont, Cal., 1979.
15. G. Berman and K.D. Fryer, *Introduction to Combinatorics*, Academic Press, New York, 1972.
16. N.L. Biggs, E.K. Lloyd, and R.J. Wilson, *Graph Theory 1736–1936*, Clarendon, Oxford, England, 1986.
17. L.S. Bobrow and M.A. Arbib, *Discrete Mathematics*, Saunders, Philadelphia, 1974.
18. K.P. Bogart, *Introductory Combinatorics*, Wiley, New York, 1986.
19. J.A. Bondy and U.S.R. Murty, *Graph Theory with Applications*, North-Holland, New York, 1976.
20. R.C. Bose and B. Manvel, *Introduction to Combinatorial Theory*, Wiley, New York, 1986.

21. R.A. Brualdi, *Introductory Combinatorics*, North-Holland, New York, 1977.

22. F. Buckley and F. Harary, *Distance in Graphs*, Addison-Wesley, Redwood City, Cal., 1990.

23. L. Carroll, *Symbolic Logic*, Crown, New York, 1978.

24. V. Chachra, P.M. Ghare, J.M. Moore, *Applications of Graph Theory Algorithms*, North Holland, New York, 1979.

25. G. Chartrand, *Graphs as Mathematical Models*, Prindle, Weber & Schmidt, Boston, 1977.

26. D.I.A. Cohen, *Basic Techniques of Combinatorial Theory*, Wiley, New York, 1978.

27. C.W. Curtis, *Linear Algebra*, Springer–Verlag, New York, 1984.

28. C.J. Date, *An Introduction to Database Systems*, 3d ed., Addison-Wesley, Reading, Mass., 1982.

29. D.E.R. Denning, *Cryptography and Data Security*, Addison-Wesley, Reading, Mass., 1982.

30. P.J. Denning, J.B. Dennis, and J.E. Qualitz, *Machines, Languages, and Computation*, Prentice-Hall, Englewood Cliffs, N.J., 1981.

31. N. Deo, *Graph Theory with Applications to Engineering and Computer Science*, Prentice-Hall, Englewood Cliffs, N.J., 1974.

32. E. Dijkstra, "Two Problems in Connexion with Graphs," *Numerische Mathematik*, 1 (1959), 269–271.

33. L. Euler, "The Koenigsberg Bridges," *Scientific American*, 189, no. 1 (July 1953), 66–70.

34. S. Even, *Algorithmic Combinatorics*, Macmillan, New York, 1973.

35. S. Even, *Graph Algorithms*, Computer Science Press, Rockville, Md., 1979.

36. W. Feller, *An Introduction to Probability Theory and Its Applications*, vol. 1, 3d ed., Wiley, New York, 1968.

37. A. Gibbons, *Algorithmic Graph Theory*, Cambridge University Press, Cambridge, England, 1985.

38. C.C. Gillispie, ed. *Dictionary of Scientific Biography*, Scribner's, New York, 1970.

39. S.W. Golomb, *Polyominoes*, Scribner's, New York, 1965.

40. G.H. Gonnet, *Handbook of Algorithms and Data Structures*, Addison-Wesley, London, 1984.

41. S.E. Goodman and S.T. Hedetniemi, *Introduction to the Design and Analysis of Algorithms*, McGraw-Hill, New York, 1977.

42. C.C. Gotlieb and L.R. Gotlieb, *Data Types and Structures*, Prentice-Hall, Englewood Cliffs, N.J., 1978.

43. R.L. Graham and P. Hell, "On the History of the Minimum Spanning Tree Problem," *Annals of the History of Computing*, 7 (1985), 43–57.

44. J.E. Graver and Mark E. Watkins, *Combinatorics with Emphasis on the Theory of Graphs*, Springer-Verlag, New York, 1977.

45. R.P. Gimaldi, *Discrete and Combinatorial Mathematics*, 2d ed., Addison-Wesley, Reading, Mass., 1989.

46. P.R. Halmos, *Naive Set Theory*, D. Van Nostrand, New York, 1960.

47. R.W. Hamming, *Coding and Information Theory*, Prentice-Hall, Englewood Cliffs, N.J., 1980.

48. F. Harary, *Graph Theory*, Addison-Wesley, Reading, Mass., 1969.

49. G.H. Hardy and E.M. Wright, *An Introduction to the Theory of Numbers*, 5th ed., Oxford University Press, Oxford, England, 1979.

50. F.E. Hohn, *Applied Boolean Algebra*, 2d ed., Macmillan, New York, 1966.

51. J.E. Hopcroft and J.D. Ullman, *Introduction to Automata Theory, Languages, and Computation*, Addison-Wesley, Reading, Mass., 1979.

52. D. Hopkin and B. Moss, *Automata*, Elsevier, North-Holland, New York, 1976.

53. E. Horowitz and S. Sahni, *Fundamentals of Computer Algorithms*, Computer Science Press, Rockville, Md., 1982.

54. T.C. Hu, *Combinatorial Algorithms*, Addison-Wesley, Reading, Mass., 1982.

55. M. Karnaugh, "The Map Method for Synthesis of Combinatorial Logic Circuits," *Transactions of the AIEE*, part I, 72 (1953), 593–599.

56. S.C. Kleene, "Representation of Events by Nerve Nets," in *Automata Studies*, 3–42, Princeton University Press, Princeton, New Jersey, 1956.

57. D.E. Knuth, "Algorithms," *Scientific American*, 236, no. 4 (April 1977), 63–80.

58. D.E. Knuth, *The Art of Computer Programming, Vol. I: Fundamental Algorithms*, 2d ed., Addison-Wesley, Reading, Mass., 1973.

59. D.E. Knuth, *The Art of Computer Programming, Vol. II: Seminumerical Algorithms*, 2d ed., Addison-Wesley, Reading, Mass., 1981.

60. D.E. Knuth, *The Art of Computer Programming, Vol. III: Sorting and Searching*, Addison-Wesley, Reading, Mass., 1973.

61. Z. Kohavi, *Switching and Finite Automata Theory*, 2d ed., McGraw-Hill, New York, 1978.

62. L. Kronsjö, *Algorithms: Their Complexity and Efficiency*, 2d ed., Wiley, New York, 1987.

63. J.B. Kruskal, "On the Shortest Spanning Subtree of a Graph and the Traveling Salesman Problem," *Proceedings of the AMS*, 1 (1956), 48–50.

64. D.H. Lehmer, "The Machine Tools of Combinatorics," in *Applied Combinatorial Mathematics*, edited by E.F. Beckenbach, Wiley, New York, 1964.

65. W.J. LeVeque, *Fundamentals of Number Theory*, Addison-Wesley, Reading, Mass., 1977.

66. H.R. Lewis and C.H. Papadimitriou, *Elements of the Theory of Computation*, Prentice-Hall, Englewood Cliffs, N.J., 1981.

67. Y. Lin and S.Y.T. Lin, *Set Theory with Applications*, 2d ed., Mariner, Tampa, Fla., 1981.

68. C.L. Liu, *Elements of Discrete Mathematics*, 2d ed., McGraw-Hill, New York, 1985.

69. C.L. Liu, *Introduction to Combinatorial Mathematics*, MacGraw-Hill, New York, 1968.

70. L. Lovász, *Combinatorial Problems and Exercises*, North-Holland, Amsterdam, 1979.

71. E. Lucas, *Récreations Mathématiques*, Gauthier-Villars, Paris, 1891.

72. E.J. McCluskey, Jr., "Minimization of Boolean Functions," *Bell System Technical Journal*, 35 (1956), 1417–1444.

73. J.A. McHugh, *Algorithmic Graph Theory*, Prentice-Hall, Englewood Cliffs, N.J., 1990.

74. R. McNaughton, *Elementary Computability, Formal Languages, and Automata*, Prentice-Hall, Englewood Cliffs, N.J., 1982.

75. G.H. Mealy, "A Method for Synthesizing Sequential Circuits," *Bell System Technical Journal*, 34 (1955), 1045–1079.

76. E. Mendelson, *Introduction to Mathematical Logic*, Van Nostrand Reinhold, New York, 1964.

77. J.R. Monk, *Introduction to Set Theory*, McGraw-Hill, New York, 1969.

78. E.F. Moore, "Gedanken-Experiments on Sequential Machines," in *Automata Studies*, 129–153, Princeton University Press, Princeton, N.J., 1956.

79. C.V. Negoita, *Expert Systems and Fuzzy Systems*, Benjamin Cummings, Menlo Park, Cal., 1985.

80. O. Ore, *Graphs and Their Uses*, Mathematical Association of America, Washington, D.C., 1963.

81. I. Pohl and A. Shaw, *The Nature of Computation: An Introduction to Computer Science*, Computer Science Press, Rockville, Md., 1981.

82. G. Pólya, *How to Solve It*, Doubleday, Garden City, N.Y. 1957.

83. G. Pólya, *Mathematical Discovery*, Wiley, New York, 1962.

84. G. Pólya, *Mathematics and Plausible Reasoning*, Princeton University Press, Princeton, N.J., 1954.

85. G. Pólya, R.E. Tarjan, and D.R. Woods, *Notes on Introductory Combinatorics*, Birkhäuser, Boston, 1983.

86. R.C. Prim, "Shortest Connection Networks and Some Generalizations," *Bell System Technical Journal*, 36 (1957), 1389–1401.

87. P.W. Purdom, Jr. and C.A. Brown, *The Analysis of Algorithms*, Holt, Rinehart & Winston, New York, 1985.

88. W.V. Quine, "The Problem of Simplifying Truth Functions," *American Mathematical Monthly*, 59 (1952), 521–531.

89. W.V. Quine, "A Way to Simplify Truth Functions," *American Mathematical Monthly*, 62 (1955), 627–631.

90. E.M. Reingold, J. Nievergelt, and N. Deo, *Combinatorial Algorithms: Theory and Practice*, Prentice-Hall, Englewood Cliffs, N.J., 1977.

91. J. Riordan, *An Introduction to Combinatorial Analysis*, Wiley, New York, 1958.

92. J. Riordan, *Combinatorial Identities*, Wiley, New York, 1968.

93. R. Rivest, A. Shamir, and L. Adleman,"A Method for Obtaining Digital Signatures and Public-Key Cryptosystems," *Communications of the Association for Computing Machinery*, 31, no. 2 (1978), 120–128.

94. E.S. Roberts, *Thinking Recursively*, Wiley, New York, 1986.

95. F.S. Roberts, *Applied Combinatorics*, Prentice-Hall, Englewood Cliffs, N.J., 1984.

96. F.S. Roberts, *Discrete Mathematics Models*, Prentice-Hall, Englewood Cliffs, N.J., 1976.

97. D.F. Robinson and L.R. Foulds, *Digraphs: Theory and Techniques*, Gordon and Breach, New York, 1980.

98. J.S. Rohl, *Recursion via Pascal*, Cambridge University Press, Cambridge, England, 1984.

99. K.H. Rosen, *Elementary Number Theory and its Applications,* 2d ed., Addison-Wesley, Reading, Mass., 1988.

100. S. Ross, *A First Course in Probability*, 2d ed., Macmillan, New York, 1984.

101. H. Ryser, *Combinatorial Mathematics*, Mathematical Association of America, Washington, D.C., 1963.

102. T.L. Saaty and P.C. Kainen, *The Four-Color Problem: Assaults and Conquest*, Dover, New York, 1986.

103. S. Sahni, *Concepts in Discrete Mathematics*, Camelot, Minneapolis, 1985.

104. J. Seberry and J. Pieprzyk, *Cryptography: An Introduction to Computer Security*, Prentice-Hall, Englewood Cliffs, N.J., 1989.

105. R. Sedgewick, *Algorithms*, 2d ed., Addison-Wesley, Reading, Mass., 1988.

106. H.J. Siegel and W.T. Hsu, "Interconnection Networks," in *Computer Architectures*, edited by V.M. Milutinovic, North-Holland, New York, 1988, pp. 225–264.

107. A. Sinkov, *Elementary Cryptanalysis*, Mathematical Association of America, Washington, D.C., 1966.

108. I.S. Sominskii, *Method of Mathematical Induction*, Blaisdell, New York, 1961.

109. M. Spivak, *Calculus*, 2d ed., Publish or Perish, Wilmington, Del., 1980.

110. D. Stanat and D.F. McAllister, *Discrete Mathematics in Computer Science*, Prentice-Hall, Englewood Cliffs, N.J., 1977.

111. H.M. Stark, *An Introduction to Number Theory*, MIT Press, Cambridge, Mass., 1978.

112. R.R. Stoll, *Sets, Logic, and Axiomatic Theories*, 2d ed., W.H. Freeman, San Francisco, 1974.

113. G.W. Strang, *Linear Algebra and its Applications*, 3d ed., Harcourt Brace Jovanovich, San Diego, 1988.

114. P. Suppes, *Introduction to Logic*, D. Van Nostrand, Princeton, N.J., 1987.

115. R.E. Tarjan, *Data Structures and Network Algorithms*, Society for Industrial and Applied Mathematics, Philadelphia, 1983.

116. G.B. Thomas and R.L. Finney, *Calculus and Analytic Geometry*, 6th ed., Addison-Wesley, Reading, Mass., 1984.

117. J.P. Tremblay and R.P. Manohar, *Discrete Mathematical Structures with Applications to Computer Science*, McGraw-Hill, New York, 1975.

118. A. Tucker, *Applied Combinatorics*, 2d ed., Wiley, New York, 1985.

119. E.W. Veitch, "A Chart Method for Simplifying Truth Functions," *Proceedings of the ACM* 1952, 127–133.

120. N.Y. Vilenkin, *Combinatorics*, Academic Press, New York, 1971.

121. M. Wand, *Induction, Recursion, and Programming*, North-Holland, New York, 1980.

122. S. Warshall, "A Theorem on Boolean Matrices," *Journal of the ACM*, 9 (1962), 11–12.

123. Herbert S. Wilf, *Algorithms and Complexity*, Prentice-Hall, Englewood Cliffs, N.J., 1986.

124. S.G. Williamson, *Combinatorics for Computer Science*, Computer Science Press, Rockville, Md., 1985.

125. R.J. Wilson, *Introduction to Graph Theory*, 3d ed., Longman, Essex, England, 1985.

126. R.J. Wilson and L.W. Beineke, *Applications of Graph Theory*, Academic Press, London, 1979.

127. R.J. Wilson and J.J. Watkins, *Graphs, An Introductory Approach*, Wiley, New York, 1990.

128. N. Wirth, *Algorithms + Data Structures = Programs*, Prentice-Hall, Englewood Cliffs, N.J., 1976.

129. N. Wirth, "Data Structures and Algorithms," *Scientific American*, 251 (September 1984), 60–69.

Solutions to Odd-Numbered Exercises

CHAPTER 1

Section 1.1

1. a) Yes, T b) Yes, F c) Yes, T d) Yes, F
   e) No f) No g) Yes, T
3. a) Today is not Thursday. b) There is pollution in New Jersey.
   c) $2 + 1 \neq 3$. d) The summer in Maine is not hot or it is not sunny.
5. a) $p \wedge q$ b) $p \wedge \neg q$ c) $\neg p \wedge \neg q$ d) $p \vee q$
   e) $p \to q$ f) $(p \vee q) \wedge (p \to \neg q)$ g) $q \leftrightarrow p$
7. a) $\neg p$ b) $p \wedge \neg q$ c) $p \to q$ d) $\neg p \to \neg q$
   e) $p \to q$ f) $q \wedge \neg p$ g) $q \to p$
9. a) Inclusive or: It is allowable to take discrete mathematics if you have had calculus or computer science, or both. Exclusive or: It is allowable to take discrete mathematics if you have had calculus or computer science, but not if you have had both. Most likely the inclusive or is intended.
   b) Inclusive or: You can take the rebate, or you can get a low-interest loan, or you can get both the rebate and a low-interest loan. Exclusive or: You can take the rebate, or you can get a low-interest loan, but you cannot get both the rebate and a low-interest loan. Most likely the exclusive or is intended.
   c) Inclusive or: You can order two items from column A and none from column B, or three items from column B and none from column A, or five items including two from column A and three from column B. Exclusive or: You can order two items from column A or three items from column B, but not both. Almost certainly the exclusive or is intended.
   d) Inclusive or: More than two feet of snow or wind chill below -100, or both, will close school. Exclusive or: More than two feet of snow or wind chill below -100, but not both, will close school. Certainly the inclusive or is intended.
11. a) If the wind blows from the northeast, then it snows.
    b) If it stays warm for a week, then the apple trees will bloom.
    c) If the Pistons win the championship, then they beat the Lakers.
    d) If you get to the top of Longs Peak, then you must have walked eight miles.
    e) If you are world-famous, then you will get tenure as a professor.
    f) If you drive more than 400 miles, then you will need to buy gasoline.
    g) If your guarantee is good, then you must have bought your CD player less than 90 days ago.

13. a) Converse: "I will ski tomorrow only if it snows today." Contrapositive: "If I do not ski tomorrow, then it will not have snowed today."

b) Converse: "If I come to class, then there will be a quiz." Contrapositive: "If I do not come to class, then there will not be a quiz."

c) Converse: "A positive integer is a prime if it has no divisors other than 1 and itself." Contrapositive: "If a positive integer has a divisor other than 1 and itself, then it is not prime."

15. a)

p	$\neg p$	$p \wedge \neg p$
T	F	F
F	T	F

b)

p	$\neg p$	$p \vee \neg p$
T	F	T
F	T	T

c)

p	q	$\neg q$	$p \vee \neg q$	$(p \vee \neg q) \rightarrow q$
T	T	F	T	T
T	F	T	T	F
F	T	F	F	T
F	F	T	T	F

d)

p	q	$p \vee q$	$p \wedge q$	$(p \vee q) \rightarrow (p \wedge q)$
T	T	T	T	T
T	F	T	F	F
F	T	T	F	F
F	F	F	F	T

e)

p	q	$p \rightarrow q$	$\neg q$	$\neg p$	$\neg q \rightarrow \neg p$	$(p \rightarrow q) \leftrightarrow (\neg q \rightarrow \neg p)$
T	T	T	F	F	T	T
T	F	F	T	F	F	T
F	T	T	F	T	T	T
F	F	T	T	T	T	T

f)

p	q	$p \rightarrow q$	$q \rightarrow p$	$(p \rightarrow q) \rightarrow (q \rightarrow p)$
T	T	T	T	T
T	F	F	T	T
F	T	T	F	F
F	F	T	T	T

17. **a)**

p	q	$p \to \neg q$
T	T	F
T	F	T
F	T	T
F	F	T

b)

p	q	$\neg p \leftrightarrow q$
T	T	F
T	F	T
F	T	T
F	F	F

c)

p	q	$(p \to q) \vee (\neg p \to q)$
T	T	T
T	F	T
F	T	T
F	F	T

d)

p	q	$(p \to q) \wedge (\neg p \to q)$
T	T	T
T	F	F
F	T	T
F	F	F

e)

p	q	$(p \leftrightarrow q) \vee (\neg p \leftrightarrow q)$
T	T	T
T	F	T
F	T	T
F	F	T

f)

p	q	$(\neg p \to \neg q) \leftrightarrow (p \leftrightarrow q)$
T	T	T
T	F	T
F	T	T
F	F	T

19. **a)** Bitwise *OR* is 11 11111; bitwise *AND* is 00 00000; bitwise *XOR* is 11 11111.
b) Bitwise *OR* is 111 11011; bitwise *AND* is 101 00000; bitwise *XOR* is 010 11010.
c) Bitwise *OR* is 10011 11001; bitwise *AND* is 00010 00000; bitwise *XOR* is 10001 11001.
d) Bitwise *OR* is 11111 11111; bitwise *AND* is 00000 00000; bitwise *XOR* is 11111 11111.

21. 0.2, 0.6

23. 0.8, 0.6

Section 1.2

1. The equivalences follow by showing that the appropriate pairs of columns of the following table agree.

p	$p \wedge \mathbf{T}$	$p \vee \mathbf{F}$	$p \wedge \mathbf{F}$	$p \vee \mathbf{T}$	$p \vee p$	$p \wedge p$
T	T	T	F	T	T	T
F	F	F	F	T	F	F

3. a)

p	q	$p \vee q$	$q \vee p$
T	T	T	T
T	F	T	T
F	T	T	T
F	F	F	F

b)

p	q	$p \wedge q$	$q \wedge p$
T	T	T	T
T	F	F	F
F	T	F	F
F	F	F	F

5.

p	q	r	$q \vee r$	$p \wedge (q \vee r)$	$p \wedge q$	$p \wedge r$	$(p \wedge q) \vee (p \wedge r)$
T	T	T	T	T	T	T	T
T	T	F	T	T	T	F	T
T	F	T	T	T	F	T	T
T	F	F	F	F	F	F	F
F	T	T	T	F	F	F	F
F	T	F	T	F	F	F	F
F	F	T	T	F	F	F	F
F	F	F	F	F	F	F	F

7. a)

p	q	$p \wedge q$	$(p \wedge q) \rightarrow p$
T	T	T	T
T	F	F	T
F	T	F	T
F	F	F	T

b)

p	q	$p \vee q$	$p \rightarrow (p \vee q)$
T	T	T	T
T	F	T	T
F	T	T	T
F	F	F	T

c)

p	q	$\neg p$	$p \rightarrow q$	$\neg p \rightarrow (p \rightarrow q)$
T	T	F	T	T
T	F	F	F	T
F	T	T	T	T
F	F	T	T	T

d)

p	q	$p \wedge q$	$p \rightarrow q$	$(p \wedge q) \rightarrow (p \rightarrow q)$
T	T	T	T	T
T	F	F	F	T
F	T	F	T	T
F	F	F	T	T

e)

p	q	$p \to q$	$\neg(p \to q)$	$\neg(p \to q) \to p$
T	T	T	F	T
T	F	F	T	T
F	T	T	F	T
F	F	T	F	T

f)

p	q	$p \to q$	$\neg(p \to q)$	$\neg q$	$\neg(p \to q) \to \neg q$
T	T	T	F	F	T
T	F	F	T	T	T
F	T	T	F	F	T
F	F	T	F	T	T

9. In each case we will show that if the hypothesis is true, then the conclusion is also.

a) If the hypothesis $p \wedge q$ is true, then by the definition of conjunction, the conclusion p must also be true.

b) If the hypothesis p is true, by the definition of disjunction, the conclusion $p \vee q$ is also true.

c) If the hypothesis $\neg p$ is true, that is, if p is false, then the conclusion $p \to q$ is true.

d) If the hypothesis $p \wedge q$ is true, then both p and q are true so that the conclusion $p \to q$ is also true.

e) If the hypothesis $\neg(p \to q)$ is true, then $p \to q$ is false, so that the conclusion p is true (and q is false).

f) If the hypothesis $\neg(p \to q)$ is true, then $p \to q$ is false, so that p is true and q is false. Hence, the conclusion $\neg q$ is true.

11. a) If p is true, then $p \vee (p \wedge q)$ is true since the first proposition in the disjunction is true. On the other hand, if p is false, then $p \wedge q$ is also false, so that $p \vee (p \wedge q)$ is false. Since p and $p \vee (p \wedge q)$ always have the same truth value, they are equivalent.

b) If p is false, then $p \wedge (p \vee q)$ is false since the first part of the conjunction is false. On the other hand, if p is true, then both parts of the conjunction are true since $p \vee q$ is also true. Since p and $p \wedge (p \vee q)$ always have the same truth value they are equivalent.

13. The only way this implication can be false is when $\neg q \wedge (p \to q)$ is true and $\neg p$ is false. For $\neg p$ to be false, p must be true. For $\neg q \wedge (p \to q)$ to be true, $\neg q$ must be true, so that q is false. Since p is true, this makes $p \to q$ false, which is impossible.

15. These are not logically equivalent since when p, q, and r are all false, $(p \to q) \to r$ is false, but $p \to (q \to r)$ is true.

17. If we take duals twice, every $\vee$ changes to an $\wedge$ and then back to an $\vee$, every $\wedge$ changes to an $\vee$ and then back to an $\wedge$, every **T** changes to an **F** and then back to a **T**, every **F** changes to a **T** and then back to an **F**. Hence, $(s^*)^* = s$.

19. Let p and q be equivalent compound propositions involving only the operators $\wedge$, $\vee$, and $\neg$, and **T** and **F**. Note that $\neg p$ and $\neg q$ are also equivalent. Use De Morgan's laws as many times as necessary to push negations in as far as possible within these compound propositions, changing $\vee$s to $\wedge$s and vice versa, and changing **T**s to **F**s and vice versa. This shows that $\neg p$ and $\neg q$ are the same as p^* and q^* except that each atomic proposition p_i within them is replaced by its negation. From this we can conclude that p^* and q^* are equivalent since $\neg p$ and $\neg q$ are.

21. $(p \wedge q \wedge \neg r) \vee (p \wedge \neg q \wedge r) \vee (\neg p \wedge q \wedge r)$

23. Given a compound proposition p, form its truth table and then write down a proposition q in disjunctive normal form that is logically equivalent to p. Since q involves only $\neg$, $\wedge$, and $\vee$, this shows that these three operators form a functionally complete set.

25. By Exercise 23, given a compound proposition p, we can write down a proposition q that is logically equivalent to p and involves only $\neg$, $\wedge$, and $\vee$. By De Morgan's law we can eliminate all the $\wedge$s by replacing each occurrence of $p_1 \wedge p_2 \wedge \cdots \wedge p_n$ with $\neg(\neg p_1 \vee \neg p_2 \vee \cdots \vee \neg p_n)$.

27. $\neg(p \wedge q)$ is true when either p or q, or both, are false, and is false when both p and q are true. Since this was the definition of $p \mid q$, the two compound propositions are logically equivalent.

29. $\neg(p \vee q)$ is true when both p and q are false, and is false otherwise. Since this was the definition of $p \downarrow q$, the two are logically equivalent.

31. $((p \downarrow p) \downarrow q) \downarrow ((p \downarrow p) \downarrow q)$

33. This follows immediately from the truth table or definition of $p \mid q$.

35. 16

37. $p \rightarrow q$ is false if and only if p is true and q is false. Similarly, $\neg q \rightarrow \neg p$ is false if and only if $\neg q$ is true and $\neg p$ is false, that is, if p is true and q is false. Hence, $p \rightarrow q$ and $\neg q \rightarrow \neg p$ are logically equivalent.

Section 1.3

1. a) T **b)** T **c)** F

3. a) T **b)** F **c)** F **d)** F

5. a) There is a student who spends more than five hours every weekday in class.
b) Every student spends more than five hours every weekday in class.
c) There is a student who does not spend more than five hours every weekday in class.
d) No student spends more than five hours every weekday in class.

7. a) $\forall x P(x)$, where $P(x)$ is "x needs a course in discrete mathematics" and the universe of discourse is the set of all computer science students
b) $\exists x P(x)$, where $P(x)$ is "x owns a personal computer" and the universe is the set of students in this class
c) $\forall x \exists y P(x, y)$, where $P(x, y)$ is "x has taken y," the universe of discourse for x is the set of students in this class, and the universe of discourse for y is the set of computer science classes
d) $\exists x \exists y P(x, y)$, where $P(x, y)$ and universes of discourse are the same as in (c)
e) $\forall x \forall y P(x, y)$, where $P(x, y)$ is "x has been in y," the universe of discourse for x is the set of students in this class, and the universe of discourse for y is the set of buildings on campus
f) $\exists x \exists y \forall z (P(z, y) \rightarrow Q(x, z))$, where $P(z, y)$ is "z is in y" and $Q(x, z)$ is "x has been in z"; the universe of discourse for x is the set of students in the class, the universe of discourse for y is the set of buildings on campus, and the universe of discourse of z is the set of rooms
g) $\forall x \forall y \exists z (P(z, y) \wedge Q(x, z))$, with same environment as in (f)

9. a) T **b)** T **c)** F **d)** F **e)** T **f)** F

11. a) $P(1, 3) \vee P(2, 3) \vee P(3, 3)$
b) $P(1, 1) \wedge P(1, 2) \wedge P(1, 3)$
c) $P(1, 1) \wedge P(1, 2) \wedge P(1, 3) \wedge P(2, 1) \wedge P(2, 2) \wedge P(2, 3) \wedge P(3, 1) \wedge P(3, 2) \wedge P(3, 3)$
d) $P(1, 1) \vee P(1, 2) \vee P(1, 3) \vee P(2, 1) \vee P(2, 2) \vee P(2, 3) \vee P(3, 1) \vee P(3, 2) \vee P(3, 3)$
e) $(P(1, 1) \wedge P(1, 2) \wedge P(1, 3)) \vee (P(2, 1) \wedge P(2, 2) \wedge P(2, 3)) \vee (P(3, 1) \wedge P(3, 2) \wedge P(3, 3))$
f) $(P(1, 1) \vee P(2, 1) \vee P(3, 1)) \wedge (P(1, 2) \vee P(2, 2) \vee P(3, 2)) \wedge (P(1, 3) \vee P(2, 3) \vee P(3, 3))$

13. a) $\forall x(P(x) \rightarrow \neg Q(x))$
b) $\forall x(Q(x) \rightarrow R(x))$
c) $\forall x(P(x) \rightarrow \neg R(x))$
d) The conclusion does not follow. There may be vain professors, since the premises do not rule out the possibility that there are other vain people besides ignorant ones.

15. a) $\forall x(P(x) \rightarrow \neg Q(x))$
b) $\forall x(R(x) \rightarrow \neg S(x))$
c) $\forall x(\neg Q(x) \rightarrow S(x))$
d) $\forall x(P(x) \rightarrow \neg R(x))$

e) The conclusion follows. Suppose x is a baby. Then by the first premise, x is illogical, so that by the third premise, x is despised. The second premise says that if x could manage a crocodile, then x would not be despised. Therefore, x cannot manage a crocodile.

17. $\neg(\exists x \forall y P(x, y)) \leftrightarrow \forall x(\neg \forall y P(x, y)) \leftrightarrow \forall x \exists y \neg P(x, y)$

19. Both statements are true precisely when at least one of $P(x)$ and $Q(x)$ is true for at least one value of x in the universe of discourse.

21. a) If A is true, then both sides are logically equivalent to $\forall x P(x)$. If A is false, the left-hand side is clearly false. Furthermore, for every x, $P(x) \wedge A$ is false, so that the right-hand side is false. Hence, the two sides are logically equivalent.

b) If A is true, then both sides are logically equivalent to $\exists x P(x)$. If A is false, the left-hand side is clearly false. Furthermore, for every x, $P(x) \wedge A$ is false, so that $\exists x(P(x) \wedge A)$ is false. Hence, the two sides are logically equivalent.

23. To show these are not logically equivalent, let $P(x)$ be the statement "x is positive," and let $Q(x)$ be the statement "x is negative" with universe of discourse the set of integers. Then $\exists x P(x) \wedge \exists x Q(x)$ is true, but $\exists x(P(x) \wedge Q(x))$ is false.

25. a) Suppose that $\forall x P(x) \wedge \exists x Q(x)$ is true. Then $P(x)$ is true for all x and there is an element y for which $Q(y)$ is true. Since $P(x) \wedge Q(y)$ is true for all x and there is a y for which $Q(y)$ is true, $\forall x \exists y(P(x) \wedge Q(y))$ is true. Conversely, suppose that the second proposition is true. Let x be an element in the universe of discourse. There is a y such that $Q(y)$ is true, so that $\exists x Q(x)$ is true. Since $\forall x P(x)$ is also true, it follows that the first proposition is true.

b) Suppose that $\forall x P(x) \vee \exists x Q(x)$ is true. Then either $P(x)$ is true for all x, or there exists a y for which $Q(y)$ is true. In the former case, $P(x) \vee Q(y)$ is true for all x, so that $\forall x \exists y(P(x) \vee Q(y))$ is true. In the latter case, $Q(y)$ is true for a particular y, so that $P(x) \vee Q(y)$ is true for all x and consequently $\forall x \exists y(P(x) \vee Q(y))$ is true. Conversely, suppose that the second proposition is true. If $P(x)$ is true for all x, then the first proposition is true. If not, $P(x)$ is false for some x, and for this x there must be a y such that $P(x) \vee Q(y)$ is true. Hence, $Q(y)$ must be true, so that $\exists y Q(y)$ is true. It follows that the first proposition must hold.

27. a) True

b) False, unless the universe of discourse consists of just one element

c) True

29. $\exists x P(x) \wedge \forall x \forall y((P(x) \wedge P(y)) \rightarrow x = y)$

31. We will show how an expression can be put into prenex normal form (PNF) if subexpressions in it can be put into PNF. Then, working from the inside out, any expression can be put in PNF. (To formalize the argument, it is necessary to use the method of mathematical induction for sets that will be discussed in Section 3.3.) By Exercise 25 of Section 1.2, we can assume that the proposition uses only $\vee$ and $\neg$ as logical connectives. Now note that any proposition with no quantifiers is already in PNF. (This is the basis case of the argument.) Now suppose that the proposition is of the form $QxP(x)$, where Q is a quantifier. Since $P(x)$ is a shorter expression than the original proposition, we can put it into PNF. Then Qx followed by this PNF is again in PNF and is equivalent to the original proposition. Next, suppose that the proposition is of the form $\neg P$. If P is already in PNF, we slide the negation sign past all the quantifiers using the equivalences in Table 3. Finally, assume that proposition is of the form $P \vee Q$, where each of P and Q is in PNF. If only one of P and Q has quantifiers, then we can use Exercise 20 to bring the quantifier in front of both. If both P and Q have quantifiers, we can use Exercise 19, Exercise 24, or Exercise 25b to rewrite $P \vee Q$ with two quantifiers preceding the disjunction of a proposition of the form $R \vee S$, and then put $R \vee S$ into PNF.

Section 1.4

1. a) $\{-1, 1\}$

b) $\{1, 2, 3, 4, 5, 6, 7, 8, 9, 10, 11\}$

c) $\{0, 1, 4, 9, 16, 25, 36, 49, 64, 81\}$

d) $\emptyset$

3. a) Yes **b)** No **c)** No

5. a) True **b)** True **c)** False **d)** True **e)** True **f)** False

7. Suppose that $x \in A$. Since $A \subseteq B$, this implies that $x \in B$. Since $B \subseteq C$, we see that $x \in C$. Since $x \in A$ implies that $x \in C$, it follows that $A \subseteq C$.

9. **a)** 1 **b)** 1 **c)** 2 **d)** 3

11. **a)** $\{\emptyset, \{a\}\}$ **b)** $\{\emptyset, \{a\}, \{b\}, \{a, b\}\}$ **c)** $\{\emptyset, \{\emptyset\}, \{\{\emptyset\}\}, \{\emptyset, \{\emptyset\}\}\}$

13. **a)** 8 **b)** 16 **c)** 2

15. **a)** $\{(a, y), (b, y), (c, y), (d, y), (a, z), (b, z), (c, z), (d, z)\}$ **b)** $\{(y, a), (y, b), (y, c), (y, d), (z, a), (z, b), (z, c), (z, d)\}$

17. The set of triples (a, b, c), where a is an airline and b and c are cities.

19. $\emptyset \times A = \{(x, y) \mid x \in \emptyset \text{ and } y \in A\} = \emptyset = \{(x, y) \mid x \in A \text{ and } y \in \emptyset\} = A \times \emptyset$

21. mn.

23. We must show that $\{\{a\}, \{a, b\}\} = \{\{c\}, \{c, d\}\}$ if and only if $a = c$ and $b = d$. The "if" part is immediate. So assume these two sets are equal. First, consider the case when $a \neq b$. Then $\{\{a\}, \{a, b\}\}$ contains exactly two elements, one of which contains one element. Thus, $\{\{c\}, \{c, d\}\}$ must have the same property, so that $c \neq d$ and $\{c\}$ is the element containing exactly one element. Hence, $\{a\} = \{c\}$, which implies that $a = c$. Also, the two-element sets $\{a, b\}$ and $\{c, d\}$ must be equal. Since $a = c$ and $a \neq b$, it follows that $b = d$. Second, suppose that $a = b$. Then $\{\{a\}, \{a, b\}\} = \{\{a\}\}$, a set with one element. Hence $\{\{c\}, \{c, d\}\}$ has only one element, which can happen only when $c = d$, and the set is $\{\{c\}\}$. It then follows that $a = c$ and $b = d$.

25. Let $S = \{a_1, a_2, \ldots, a_n\}$. Represent each subset of S with a bit string of length n, where the ith bit is 1 if and only if $a_i \in S$. To generate all subsets of S, list all 2^n bit strings of length n (for instance, in increasing order), and write down the corresponding subsets.

Section 1.5

1. **a)** the set of students who live within one mile of school and who walk to classes; **b)** the set of students who live within one mile of school or who walk to classes (or who do both); **c)** the set of students who live within one mile of school but do not walk to classes; **d)** the set of students who walk to classes but live more than one mile away from school.

3. **a)** $\{0, 1, 2, 3, 4, 5, 6\}$ **b)** $\{3\}$ **c)** $\{1, 2, 4, 5\}$ **d)** $\{0, 6\}$

5. $\overline{\overline{A}} = \{x \mid \neg(x \in \overline{A})\} = \{x \mid \neg(\neg x \in A)\} = \{x \mid x \in A\} = A$.

7. **a)** $A \cup B = \{x \mid x \in A \vee x \in B\} = \{x \mid x \in B \vee x \in A\} = B \cup A$; **b)** $A \cap B = \{x \mid x \in A \wedge x \in B\} = \{x \mid x \in B \wedge x \in A\} = B \cap A$.

9. **a)** $x \in \overline{(A \cup B)} \Leftrightarrow x \notin (A \cup B) \Leftrightarrow \neg(x \in A \vee x \in B) \Leftrightarrow \neg(x \in A) \wedge \neg(x \in B) \Leftrightarrow x \notin A \wedge x \notin B \Leftrightarrow x \in \overline{A} \wedge x \in \overline{B} \Leftrightarrow x \in \overline{A} \cap \overline{B}$.

b)

A	B	$A \cup B$	$\overline{(A \cup B)}$	$\overline{A}$	$\overline{B}$	$\overline{A} \cap \overline{B}$
1	1	1	0	0	0	0
1	0	1	0	0	1	0
0	1	1	0	1	0	0
0	0	0	1	1	1	1

11. Both sides equal $\{x \mid x \in A \wedge x \notin B\}$.

13. **a)** $x \in A \cup (B \cup C) \Leftrightarrow (x \in A) \vee (x \in (B \cup C)) \Leftrightarrow (x \in A) \vee (x \in B \vee x \in C) \Leftrightarrow (x \in A \vee x \in B) \vee (x \in C) \Leftrightarrow x \in (A \cup B) \cup C$ **b)** same as (a) with $\cup$ replaced by $\cap$ and $\vee$ replaced by $\wedge$ **c)** $x \in A \cup (B \cap C) \Leftrightarrow (x \in A) \vee (x \in (B \cap C)) \Leftrightarrow (x \in A) \vee (x \in B \wedge x \in C) \Leftrightarrow (x \in A \vee x \in B) \wedge (x \in A \vee x \in C) \Leftrightarrow x \in (A \cup B) \cap (A \cup C)$.

15. **a)** $\{4, 6\}$ **b)** $\{0, 1, 2, 3, 4, 5, 6, 7, 8, 9, 10\}$ **c)** $\{4, 5, 6, 8, 10\}$ **d)** $\{0, 2, 4, 5, 6, 7, 8, 9, 10\}$

17. **a)** $B \subseteq A$ **b)** $A \subseteq B$ **c)** $A \cap B = \emptyset$ **d)** nothing, since this is always true **e)** $A = B$

19. $A \subseteq B \Leftrightarrow \forall x(x \in A \rightarrow x \in B) \Leftrightarrow \forall x(x \notin B \rightarrow x \notin A) \Leftrightarrow \forall x(x \in \overline{B} \rightarrow x \in \overline{A}) \Leftrightarrow \overline{B} \subseteq \overline{A}$

21. the set of students who are computer science majors but not mathematics majors or who are mathematics majors but not computer science majors.

23. $A \oplus B = \{x \mid (x \in A) \oplus (x \in B)\} = \{x \mid (x \in A \wedge \neg(x \in B)) \vee (\neg(x \in A) \wedge x \in B)\} = \{x \mid (x \in A \wedge x \notin B) \vee (x \notin A \wedge x \in B)\} = \{x \mid x \in A \wedge x \notin B\} \cup \{x \mid x \notin A \wedge x \in B\} = (A - B) \cup (B - A)$

25. a) $A \oplus A = (A - A) \cup (A - A) = \emptyset \cup \emptyset = \emptyset$
b) $A \oplus \emptyset = (A - \emptyset) \cup (\emptyset - A) = A \cup \emptyset = A$
c) $A \oplus U = (A - U) \cup (U - A) = \emptyset \cup \overline{A} = \overline{A}$
d) $A \oplus \overline{A} = (A - \overline{A}) \cup (\overline{A} - A) = A \cup \overline{A} = U$

27. $B = \emptyset$

29. Yes. Suppose that $x \in A$ but $x \notin B$. If $x \in C$, then $x \notin A \oplus C$ but $x \in B \oplus C$, a contradiction. If $x \notin C$ then $x \in A \oplus C$ but $x \notin B \oplus C$, a contradiction. Hence, $A \subseteq B$. Similarly, $B \subseteq A$, so that $A = B$.

31. a) $\{1, 2, 3, \ldots, n\}$ **b)** $\{1\}$

33. a) A_n
b) $\{\lambda, 0, 1\}$, where λ is the empty string

35. a) $\{1, 2, 3, 4, 7, 8, 9, 10\}$
b) $\{2, 4, 5, 6, 7\}$
c) $\{1, 10\}$

37. The bit in the ith position of the bit string of the difference of two sets is 1 if the ith bit of the first string is 1 and the ith bit of the second string is 0, and is 0 otherwise.

39. a) 1 11110 00000 00000 00000 00000 $\vee$ 0 11100 10000 00001 00010 10000 $=$ 1 11110 10000 00001 00010 10000, representing $\{a, b, c, d, e, g, p, t, v\}$
b) 1 11110 00000 00000 00000 00000 $\wedge$ 0 11100 10000 00001 00010 10000 $=$ 0 11100 00000 00000 00000 00000, representing $\{b, c, d\}$
c) (1 11110 00000 00000 00000 00000 $\vee$ 0 00110 01100 00110 00011 00110) $\wedge$ (0 11100 10000 00001 00010 10000 $\vee$ 0 01010 00100 00010 00001 00111) $=$ 1 11110 01100 00110 00011 00110 $\wedge$ 0 11110 10100 00011 00011 10111 $=$ 0 11110 00100 00010 00011 00110, representing $\{b, c, d, e, i, o, t, u, x, y\}$
d) 1 11110 00000 00000 00000 00000 $\vee$ 0 11100 10000 00001 00010 10000 $\vee$ 0 01010 00100 00010 00001 00111 $\vee$ 0 00110 01100 00110 00011 00110 $=$ 1 11110 11100 00111 00011 10111, representing $\{a, b, c, d, e, g, h, i, n, o, p, t, u, v, x, y, z\}$

41. a) $\{1, 2, 3, \{1, 2, 3\}\}$
b) $\{\emptyset\}$
c) $\{\emptyset, \{\emptyset\}\}$
d) $\{\emptyset, \{\emptyset\}, \{\emptyset, \{\emptyset\}\}\}$

43. a) $\{3 \cdot a,\ 3 \cdot b,\ 1 \cdot c,\ 4 \cdot d\}$
b) $\{2 \cdot a,\ 2 \cdot b\}$
c) $\{1 \cdot a,\ 1 \cdot c\}$
d) $\{1 \cdot b,\ 4 \cdot d\}$
e) $\{5 \cdot a,\ 5 \cdot b,\ 1 \cdot c,\ 4 \cdot d\}$

45. $\overline{F} =$ {0.4 Alice, 0.1 Brian, 0.6 Fred, 0.9 Oscar, 0.5 Rita}, $\overline{R} =$ {0.6 Alice, 0.2 Brian, 0.8 Fred, 0.1 Oscar, 0.3 Rita}

47. $F \cap R =$ {0.4 Alice, 0.8 Brian, 0.2 Fred, 0.1 Oscar, 0.5 Rita}

Section 1.6

1. a) $f(0)$ is not defined.
b) $f(x)$ is not defined for $x < 0$.
c) $f(x)$ is not well defined since there are two distinct values assigned to each x.

3. a) not a function
b) a function
c) not a function

5. a) 1 **b)** 0 **c)** 0
d) -1 **e)** 3 **f)** -1

7. only the function in part (a)

9. only the functions in parts (a) and (d)

11. a) Yes **b)** No **c)** Yes **d)** No

13. a) Let x and y be distinct elements of A. Since g is one-to-one, $g(x)$ and $g(y)$ are distinct elements of B. Since f is one-to-one, $f(g(x)) = (f \circ g)(x)$ and $f(g(y)) = (f \circ g)(y)$ are distinct elements of C. Hence, $f \circ g$ is one-to-one.
b) Let $y \in C$. Since f is onto, $y = f(b)$ for some $b \in B$. Now since g is onto, $b = g(x)$ for some $x \in A$. Hence, $y = f(b) = f(g(x)) = (f \circ g)(x)$. It follows that $f \circ g$ is onto.

15. No. For example, suppose that $A = \{a\}$, $B = \{b, c\}$, and $C = \{d\}$. Let $g(a) = b$, $f(b) = d$, and $f(c) = d$. f and $f \circ g$ are onto, but g is not.

17. $(f+g)(x) = x^2+x+3$, $(fg)(x) = x^3+2x^2+x+2$.

19. f is one-to-one since $f(x_1) = f(x_2) \Longleftrightarrow ax_1 + b = ax_2 + b \Longleftrightarrow ax_1 = ax_2 \Longleftrightarrow x_1 = x_2$. f is onto since $f((y-b)/a) = y$. $f^{-1}(y) = (y-b)/a$.

21. Let $f(1) = a$, $f(2) = a$. Let $S = \{1\}$ and $T = \{2\}$. Then $f(S \cap T) = f(\emptyset) = \emptyset$, but $f(S) \cap f(T) = \{a\} \cap \{a\} = \{a\}$.

23. a) $\{x \mid 0 \le x < 1\}$
b) $\{x \mid -1 \le x < 2\}$
c) $\emptyset$

25. $f^{-1}(\overline{S}) = \{x \in A \mid f(x) \notin S\} = \overline{\{x \in A \mid f(x) \in S\}} = \overline{f^{-1}(S)}$

27. Suppose that $N \le x < N+1$. If $N + \frac{1}{2} \le x$, then $\lfloor 2x \rfloor = 2N+1$, $\lfloor x \rfloor = N$, and $\lfloor x + \frac{1}{2} \rfloor = N+1$, so that $\lfloor 2x \rfloor = \lfloor x \rfloor + \lfloor x + \frac{1}{2} \rfloor$. If $x < N + \frac{1}{2}$, then $\lfloor 2x \rfloor = 2N$ and $\lfloor x \rfloor = \lfloor x + \frac{1}{2} \rfloor = N$, and again the identity follows.

29.

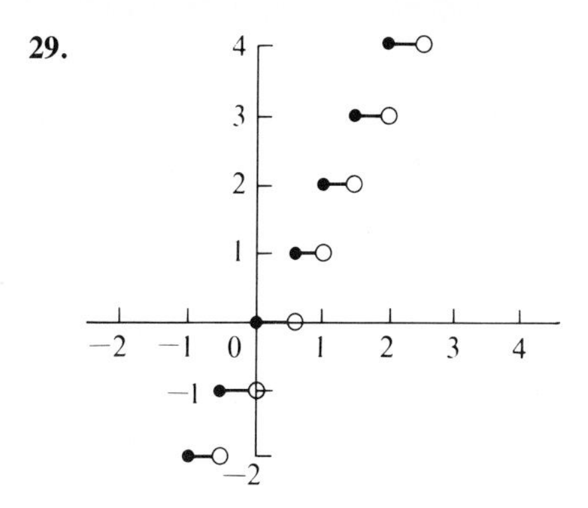

31. $f^{-1}(y) = (y-1)^{1/3}$

33. **a)** $f_{A\cap B}(x) = 1 \Leftrightarrow x \in A \cap B \Leftrightarrow x \in A$ and $x \in B \Leftrightarrow f_A(x) = 1$ and $f_B(x) = 1 \Leftrightarrow f_A(x)f_B(x) = 1$

b) $f_{A\cup B}(x) = 1 \Leftrightarrow x \in A \cup B \Leftrightarrow x \in A$ or $x \in B \Leftrightarrow f_A(x) = 1$ or $f_B(x) = 1 \Leftrightarrow f_A(x) + f_B(x) - f_A(x)f_B(x) = 1$

c) $f_{\overline{A}}(x) = 1 \Leftrightarrow x \in \overline{A} \Leftrightarrow x \notin A \Leftrightarrow f_A(x) = 0 \Leftrightarrow 1 - f_A(x) = 1$

d) $f_{A\oplus B}(x) = 1 \Leftrightarrow x \in A \oplus B \Leftrightarrow (x \in A$ and $x \notin B)$ or $(x \notin A$ and $x \in B) \Leftrightarrow f_A(x) + f_B(x) - 2f_A(x)f_B(x) = 1$

35. **a)** Domain is **Z**; codomain is **R**; domain of definition is the set of nonzero integers; the set of values for which f is undefined is $\{0\}$; not a total function.

b) Domain is **Z**; codomain is **Z**; domain of definition is **Z**; set of values for which f is undefined is $\emptyset$; total function.

c) Domain is $\mathbf{Z} \times \mathbf{Z}$; codomain is **Q**; domain of definition is $\mathbf{Z} \times (\mathbf{Z} - \{0\})$; set of values for which f is undefined is $\mathbf{Z} \times \{0\}$; not a total function.

d) Domain is $\mathbf{Z} \times \mathbf{Z}$; codomain is **Z**; domain of definition is $\mathbf{Z} \times \mathbf{Z}$; set of values for which f is undefined is $\emptyset$; total function;

e) Domain is $\mathbf{Z} \times \mathbf{Z}$; codomain is **Z**; domain of definitions is $\{(m, n) \mid m > n\}$; set of values for which f is undefined is $\{(m, n) \mid m \leq n\}$; not a total function.

Section 1.7

1. **a)** 3 **b)** -1 **c)** 787 **d)** 2639

3. **a)** $a_0 = 2,\ a_1 = 3,\ a_2 = 5,\ a_3 = 9$ **b)** $a_0 = 1,\ a_1 = 4,\ a_2 = 27,\ a_3 = 256$ **c)** $a_0 = 0,\ a_1 = 0,\ a_2 = 1,\ a_3 = 1$ **d)** $a_0 = 0,\ a_1 = 1,\ a_2 = 2,\ a_3 = 3$

5. **a)** 20 **b)** 11 **c)** 30 **d)** 511

7. **a)** 1533 **b)** 510 **c)** 4923 **d)** 9842

9. **a)** 21 **b)** 78 **c)** 18 **d)** 18

11. $\sum_{j=1}^{n}(a_j - a_{j-1}) = a_n - a_0$

13. **a)** n^2 **b)** $n(n+1)/2$

15. **a)** 0 **b)** 1680 **c)** 1 **d)** 1024

17. 34

19. **a)** countable, $-1, -2, -3, -4, \ldots$ **b)** countable, $0, 2, -2, 4, -4, \ldots$ **c)** uncountable **d)** countable, $0, 7, -7, 14, -14, \ldots$

21. Assume that B is countable. Then the elements of B can be listed as $b_1, b_2, b_3, \ldots$. Since A is a subset of B, taking the subsequence of $\{b_n\}$ that contains the terms that are in A gives a listing of the elements of A. Since A is uncountable, this is impossible.

23. Suppose that $A_1, A_2, A_3, \ldots$ are countable sets. Since A_i is countable, we can list its elements in a sequence as $a_{i1}, a_{i2}, a_{i3}, \ldots$. The elements of the set $\bigcup_{i=1}^{n} A_i$ can be listed by listing all terms a_{ij} with $i + j = 2$, then all terms a_{ij} with $i + j = 3$, then all terms a_{ij} with $i + j = 4$, and so on.

25. There are a finite number, namely 2^m, bit strings of length m. The set of all bit strings is the union of the bit strings of lengths m over $m = 0, 1, 2, \ldots$. Since the union of a countable number of countable sets is countable, there are a countable number of bit strings.

27. For any finite alphabet there are a finite number of strings of length n, whenever n is a positive integer. It follows by the result of Exercise 23 that there are only a finite number of strings from any given finite alphabet. Since the set of all computer programs in a particular language is a subset of the set of all strings of a finite alphabet, which is a countable set, by the result of Exercise 20, it is itself a countable set.

29. Exercise 27 shows that there are only a countable number of computer programs. Consequently, there are only a countable number of computable functions. Since, as Exercise 28 shows, there are an uncountable number of functions, not all functions are computable.

Section 1.8

1. a) Yes **b)** Yes **c)** No **d)** Yes **e)** Yes **f)** Yes

3. $x^4 + 9x^3 + 4x + 7 \leq 4x^4$ for all $x > 9$, so $x^4 + 9x^3 + 4x + 7 = O(x^4)$.

5. $(x^2 + 1)/(x + 1) = x - 1 + 2/(x + 1) < x$ for all $x > 1$. Hence $(x^2 + 1)/(x + 1) = O(x)$.

7. $x^2 + 4x + 17 \leq 3x^3$ for all $x > 17$, so that $x^2 + 4x + 17 = O(x^3)$. However, if x^3 were $O(x^2 + 4x + 17)$, then $x^3 < C(x^2 + 4x + 17) < 3Cx^2$ for some C, for all sufficiently large x, which implies that $x < C$ for sufficiently large x, which is impossible. Hence, x^3 is not $O(x^2 + 4x + 17)$.

9. $3x^4 + 1 \leq 4x^4 = 8(x^4/2)$ for all $x > 1$, so $3x^4 + 1 = O(x^4/2)$. Also, $x^4/2 \leq 3x^4 + 1$ for all $x > 0$, so $x^4/2 = O(3x^4 + 1)$.

11. Since $2^n < 3^n$ for all $n > 0$, it follows that 2^n is $O(3^n)$. However, if 3^n were $O(2^n)$, then for some C, $3^n \leq C \cdot 2^n$ for all sufficiently large n. This says that $C \geq (3/2)^n$ for all sufficiently large n, which is impossible. Hence, 3^n is not $O(2^n)$.

13. All functions for which there exist real numbers k and C with $|f(x)| \leq C$ for $x > k$. These are the functions $f(x)$ that are bounded for all sufficiently large x.

15. There are constants C_1, C_2, k_1, and k_2 such that $|f(x)| \leq C_1|g(x)|$ for all $x > k_1$ and $|g(x)| \leq C_2|h(x)|$ for all $x > k_2$. Hence, for $x > \max(k_1, k_2)$ it follows that $|f(x)| \leq C_1|g(x)| \leq C_1C_2|h(x)|$. This shows that $f(x) = O(h(x))$.

17. a) $O(n^3)$ **b)** $O(n^5)$ **c)** $O(n^3 \cdot n!)$

19. a) $O(n^2 \log n)$ **b)** $O(n^2(\log n)^2)$ **c)** $O(n^{2^n})$

21. If $f(x) = \Theta(g(x))$, then there exist constants C_1 and C_2 with $C_1|g(x)| \leq |f(x)| \leq C_2|g(x)|$. It follows that $|f(x)| \leq C_2|g(x)|$ and $|g(x)| \leq (1/C_1)|f(x)|$ for $x > k$. Thus, $f(x) = O(g(x))$ and $g(x) = O(f(x))$. Conversely, suppose that $f(x) = O(g(x))$ and $g(x) = O(f(x))$. Then there are constants C_1, C_2, k_1, and k_2 such that $|f(x)| \leq C_1|g(x)|$ for $x > k_1$ and $|g(x)| \leq C_2|f(x)|$ for $x > k_2$. We can assume that $C_2 > 0$ (we can always make C_2 larger). Then we have $(1/C_2)|g(x)| \leq |f(x)| \leq C_1|g(x)|$ for $x > \max(k_1, k_2)$. Hence, $f(x) = \Theta(g(x))$.

23.

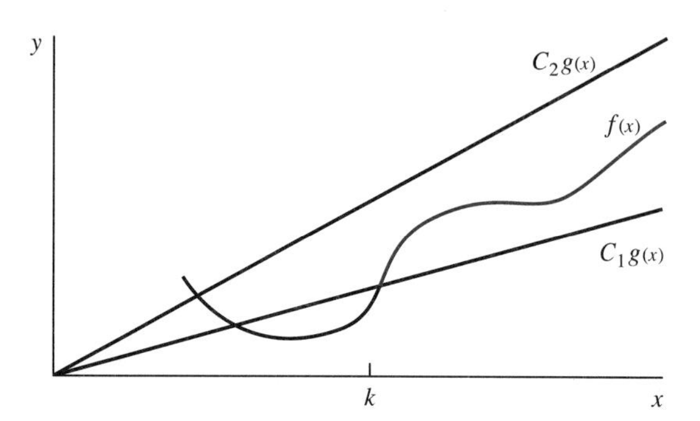

25. Since $f(x) = O(g(x))$ there are constants C and l such that $|f(x)| \leq C|g(x)|$ for $x > l$. Hence $|f^k(x)| \leq C^k|g^k(x)|$ for $x > l$, so $f^k(x) = O(g^k(x))$ by taking the constant to be C^k.

27. Since $f(x)$ and $g(x)$ are increasing and unbounded, we can assume $f(x) \geq 1$ and $g(x) \geq 1$ for sufficiently large x. There are constants C and k with $f(x) \leq Cg(x)$ for $x > k$. This implies that $\log f(x) \leq \log C + \log g(x) < 2 \log g(x)$ for sufficiently large x. Hence $\log f(x) = O(\log g(x))$.

29. a) $\lim_{x\to\infty} x^2/x^3 = \lim_{x\to\infty} 1/x = 0$

b) $\displaystyle\lim_{x\to\infty} \frac{x \log x}{x^2} = \lim_{x\to\infty} \frac{\log x}{x} = \lim_{x\to\infty} \frac{1}{x \ln 2} = 0$ (using L'Hôpital's rule)

c) $\displaystyle\lim_{x\to\infty} \frac{x^2}{2^x} = \lim_{x\to\infty} \frac{2x}{2^x \cdot \ln 2} = \lim_{x\to\infty} \frac{2}{2^x \cdot (\ln 2)^2} = 0$ (using L'Hôpital's rule)

d) $\displaystyle\lim_{x\to\infty} \frac{x^2 + x + 1}{x^2} = \lim_{x\to\infty}\left(1 + \frac{1}{x} + \frac{1}{x^2}\right) = 1 \neq 0$

31.

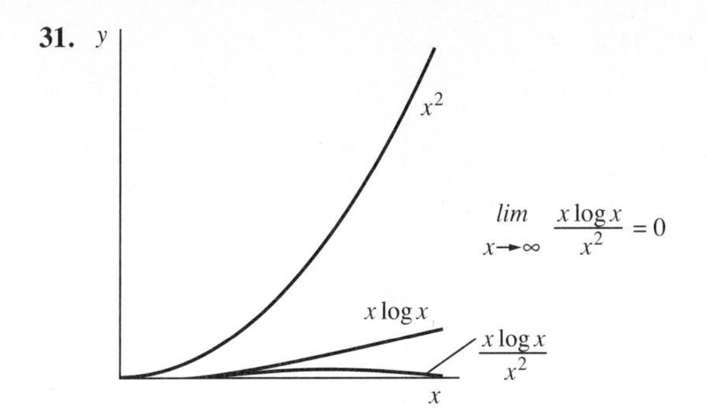

33. No. Take $f(x) = 1/x^2$ and $g(x) = 1/x$.

35. a) Since $\lim_{x\to\infty} f(x)/g(x) = 0$, $|f(x)|/|g(x)| < 1$ for sufficiently large x. Hence, $|f(x)| < |g(x)|$ for $x > k$ for some constant k. Therefore, $f(x) = O(g(x))$.
b) Let $f(x) = g(x) = x$. Then $f(x) = O(g(x))$, but $f(x) \neq o(g(x))$ since $f(x)/g(x) = 1$.

37. Since $f_2(x) = o(g(x))$, from Exercise 35(a) it follows that $f_2(x) = O(g(x))$. By Corollary 1, we have $f_1(x) + f_2(x) = O(g(x))$.

39. We can easily show that $(n - i)(i + 1) \geq n$ for $i = 0, 1, \ldots, n - 1$. Hence, $(n!)^2 = (n \cdot 1)((n - 1) \cdot 2) \cdot ((n - 2) \cdot 3) \cdots (2 \cdot (n - 1)) \cdot (1 \cdot n) \geq n^n$. Therefore, $2 \log n! \geq n \log n$.

Supplementary Exercises

1. a) $q \to p$ **b)** $q \wedge p$ **c)** $\neg q \vee \neg p$ **d)** $q \leftrightarrow p$

3. a) The proposition cannot be false unless $\neg p$ is false, so that p is true. If p is true and q is true then $\neg q \wedge (p \to q)$ is false and the implication is true. If p is true and q is false then $p \to q$ is false so that $\neg q \wedge (p \to q)$ is false and the implication is true.
b) The proposition cannot be false unless q is false. If q is false and p is true then $(p \vee q) \wedge \neg p$ is false, and the implication is true. If q is false and p is false then $(p \vee q) \wedge \neg p$ is false and the implication is true.

5. $(p \wedge q \wedge r \wedge \neg s) \vee (p \wedge q \wedge \neg r \wedge s) \vee (p \wedge \neg q \wedge r \wedge s) \vee (\neg p \wedge q \wedge r \wedge s)$

7. a) F **b)** T **c)** F **d)** T **e)** F **f)** T

9. Suppose that $\exists x(P(x) \to Q(x))$ is true. Then either $Q(x_0)$ is true for some x_0, in which case $\forall x P(x) \to \exists x Q(x)$ is true; or $P(x_0)$ is false for some x_0, in which case $\forall x P(x) \to \exists x Q(x)$ is true. Conversely, suppose that $\exists x(P(x) \to Q(x))$ is false. That means that $\forall x(P(x) \wedge \neg Q(x))$ is true, which implies $\forall x P(x)$ and $\forall x(\neg Q(x))$. This latter proposition is equivalent to $\neg \exists x Q(x)$. Thus, $\forall x P(x) \to \exists x Q(x)$ is false.

11. No

13. $\forall x \forall z \exists y T(x, y, z)$, where $T(x, y, z)$ is the statement that student x has taken class y in department z, where the universes of discourse are the set of students in the class, the set of courses at this university, and the set of departments in the school of mathematical sciences.

15. a) $\overline{A}$ **b)** $A \cap B$ **c)** $A - B$ **d)** $\overline{A} \cap \overline{B}$ **e)** $A \oplus B$

17. Yes

19. a) $A \cap \overline{A} = \{x | x \in A \wedge x \notin A\} = \emptyset$
b) $A \cup \overline{A} = \{x | x \in A \vee x \notin A\} = U$

21. $A - (A - B) = A - (A \cap \overline{B}) = A \cap \overline{(A \cap \overline{B})} = A \cap (\overline{A} \cup B) = (A \cap \overline{A}) \cup (A \cap B) = \emptyset \cup (A \cap B) = A \cap B$

23. Let $A = \{1\}$, $B = \emptyset$, $C = \{1\}$. Then $(A - B) - C = \emptyset$ but $A - (B - C) = \{1\}$.

25. No. For example, let $A = B = \{a, b\}$, $C = \emptyset$, and $D = \{a\}$. Then $(A - B) - (C - D) = \emptyset - \emptyset = \emptyset$, but $(A - C) - (B - D) = \{a, b\} - \{b\} = \{a\}$.

27. a) $|\emptyset| \leq |A \cap B| \leq |A| \leq |A \cup B| \leq |U|$
b) $|\emptyset| \leq |A - B| \leq |A \oplus B| \leq |A \cup B| \leq |A| + |B|$

29. a) yes, no
b) yes, no
c) f has inverse with $f^{-1}(a) = 3$, $f^{-1}(b) = 4$, $f^{-1}(c) = 2$, $f^{-1}(d) = 1$; g has no inverse

31. Let $f(a) = f(b) = 1$, $f(c) = f(d) = 2$, $S = \{a, c\}$, $T = \{b, d\}$. Then $f(S \cap T) = f(\emptyset) = \emptyset$, but $f(S) \cap f(T) = \{1, 2\} \cap \{1, 2\} = \{1, 2\}$.

33. a) 60 **b)** 6144 **c)** 20 **d)** 0

35. Let P_n be the set of polynomials of degree at most n with integer coefficients with absolute value at most

n. Then P_n is finite for all n. Since any polynomial of degree n has at most n distinct real roots, there are only a finite number of algebraic numbers that are roots of polynomials in P_n. Since the set of algebraic numbers is the union of the sets of roots of polynomials in P_n for $n = 1, 2, 3, \ldots$, by Exercise 23 of Section 1.7, it is a countable set.

37. $O(x^2 2^x)$

39. Note that

$$\frac{n!}{2^n} = \frac{n}{2} \cdot \frac{n-1}{2} \cdots \frac{3}{2} \cdot \frac{2}{2} \cdot \frac{1}{2} > \frac{n}{2} \cdot 1 \cdot 1 \cdots 1 \cdot \frac{1}{2} = \frac{n}{4}.$$

Since $n!/2^n$ increases without bound as n increases, $n!$ cannot be bounded by a constant times 2^n for sufficiently large n. Hence, $n!$ is not $O(2^n)$.

CHAPTER 2

Section 2.1

1. $max := 1$, $i := 2$, $max := 8$, $i := 3$, $max := 12$, $i := 4$, $i := 5$, $i := 6$, $i := 7$, $max := 14$, $i := 8$, $i := 9$, $i := 10$, $i := 11$

3.
```
procedure sum(a_1, ..., a_n : integers)
sum := a_1
for i := 2 to n
   sum := sum + a_i
{sum has desired value}
```

5.
```
procedure interchange(x, y : real numbers)
z := x
x := y
y := z
```
The minimum number of assignments needed is three.

7. Linear search: $i := 1$, $i := 2$, $i := 3$, $i := 4$, $i := 5$, $i := 6$, $i := 7$, $location := 7$; binary search: $i := 1$, $j := 8$, $m := 4$, $i := 5$, $m := 6$, $i := 7$, $m := 7$, $j := 7$, $location := 7$

9.
```
procedure insert(x, a_1, a_2, ..., a_n : integers)
{the list is in order : a_1 ≤ a_2 ≤ ··· ≤ a_n}
a_{n+1} := x + 1
i := 1
while x > a_i
   i := i + 1
for j := 0 to n − i
   a_{n−j+1} := a_{n−j}
a_i := x
{x has been inserted into correct position}
```

11.
```
procedure first largest(a_1, ..., a_n : integers)
max := a_1
location := 1
for i := 2 to n
begin
   if max < a_i then
   begin
      max := a_i
      location := i
   end
end
```

13.
```
procedure mean-median-max-min(a, b, c : integers)
mean := (a + b + c)/3
{the six different orderings of a, b, c with respect to
      ≥ will be handled separately}
if a > b then
begin
   if b > c then
   median := b; max := a; min := c
end
   . . .
```
(The rest of the algorithm is similar.)

15.
```
procedure first-three (a_1, a_2, ..., a_n : integers)
if a_1 > a_2 then interchange a_1 and a_2
   if a_2 > a_3 then interchange a_2 and a_3
   if a_1 > a_2 then interchange a_1 and a_2
```

17.
```
procedure onto(f: function from A to B where A =
      {a_1, ..., a_n}, B = {b_1, ..., b_m}, a_1, ..., a_n,
      b_1, ..., b_m are integers)
for i := 1 to m
   hit(b_i) := 0
count := 0
for j := 1 to n
   if hit(f(a_j)) = 0 then
   begin
      hit(f(a_j)) := 1
      count := count + 1
   end
if count = m then onto := true
else onto := false
```

19.
```
procedure ones(a : bit string, a = a_1 a_2 ··· a_n)
ones := 0
for i := 1 to n
begin
  if a_i := 1 then
    ones := ones + 1
end {ones is the number of ones in the bit strings a}
```

21.
```
procedure ternary search(s : integer, a_1, a_2, ..., a_n:
    increasing integers)
i := 1
j := n
while i < j - 1
begin
  l = ⌊(i + j)/3⌋
  u = ⌊2(i + j)/3⌋
  if x > a_u then i := u + 1
  else if x > a_t then
  begin
    i := l + 1
    j := u
  end
  else j := l
end
if x = a_i then location := i
else if x = a_j then location := j
else location := 0
{location is the subscript of the term equal to x (0 if
    not found)}
```

23.
```
procedure find a mode(a_1, a_2, ..., a_n: nondecreas-
    ing integers)
modecount := 0
i := 1
while i ≤ n
begin
  value := a_i
  count := 1
  while i ≤ n and a_i = value
  begin
    count := count + 1
    i := i + 1
  end
  if count > modecount then
  begin
    modecount := count
    mode := value
  end
end
{mode is the first value occurring most often }
```

Section 2.2

1. $2n - 1$
3. Linear
5. $O(n)$
7. **a)** $power := 1$, $y := 1$; $i := 1$, $power := 2$, $y := 3$; $i := 2$, $power := 4$, $y := 15$ **b)** $2n$ multiplications and n additions
9. **a)** $2^{10^9} \sim 10^{3\times10^8}$ **b)** 10^9 **c)** 3.96×10^7 **d)** 3.16×10^4 **e)** 29 **f)** 12
11. **a)** 36 years **b)** 13 days **c)** 19 minutes
13. The average number of comparisons is $(3n + 4)/2$.
15. $O(\log n)$
17. $O(n)$

Section 2.3

1. **a)** Yes **b)** No **c)** Yes **d)** No
3. Suppose that $a \mid b$. Then there exists an integer k such that $ka = b$. Since $a(ck) = bc$ it follows that $a \mid bc$.
5. If $a \mid b$ and $b \mid a$, there are integers c and d such that $b = ac$ and $a = bd$. Hence $a = acd$. Since $a \neq 0$ it follows that $cd = 1$. Thus either $c = d = 1$ or $c = d = -1$. Hence either $a = b$ or $a = -b$.
7. Since $ac \mid bc$ there is an integer k such that $ack = bc$. Hence $ak = b$, so $a \mid b$.
9. **a)** 2, 5 **b)** −11, 10 **c)** 34, 7 **d)** 77, 0 **e)** 0, 0 **f)** 0, 3 **g)** −1, 2 **h)** 4, 0
11. $2^8 \cdot 3^4 \cdot 5^2 \cdot 7$
13. Suppose that $\log_2 3 = a/b$ where $a, b \in \mathbf{Z}^+$ and $b \neq 0$. Then $2^{a/b} = 3$, so that $2^a = 3^b$. This violates the fundamental theorem of arithmetic. Hence $\log_2 3$ is irrational.
15. **a)** Yes **b)** No **c)** Yes **d)** Yes
17. If $a \textbf{ mod } m = b \textbf{ mod } m$, then a and b have the same remainder when divided by m. Hence $a = q_1 m + r$ and $b = q_2 m + r$, where $0 \leq r < m$. It follows that $a - b = (q_1 - q_2)m$ so that $m \mid (a - b)$. It follows that $a \equiv b \pmod{m}$.
19. Suppose that n is not prime, so that $n = ab$, where a and b are integers greater than 1. Since $a > 1$, by

the identity in the hint, $2^a - 1$ is a factor of $2^n - 1$ that is greater than 1, and the second factor in this identity is also greater than 1. Hence $2^n - 1$ is not prime.

21. **a)** 2 **b)** 4 **c)** 12

23. $\phi(p^k) = p^k - p^{k-1}$

25. There is some b with $(b-1)k < n \le bk$. Hence $(b-1)k \le n-1 < bk$. Divide by k to obtain $b-1 < n/k \le b$ and $b-1 \le (n-1)/k < b$. Hence $\lceil n/k \rceil = b$ and $\lfloor (n-1)/k \rfloor = b-1$.

27. **a)** 1 **b)** 2 **c)** 3 **d)** 9

29. **a)** No **b)** No **c)** Yes **d)** No

31. Since $\min(x, y) + \max(x, y) = x + y$, the exponent of p_i in the prime factorization of $gcd(a, b) \cdot lcm(a, b)$ is the sum of the exponents of p_i in the prime factorizations of a and b.

33. Let $m = tn$. Since $a \equiv b \pmod{m}$ there exists an integer s such that $a = b + sm$. Hence $a = b + (st)n$, so that $a \equiv b \pmod{n}$.

35. Let $m = c = 2$, $a = 0$, and $b = 1$. Then $0 = ac \equiv bc = 2 \pmod{2}$, but $0 = a \not\equiv b = 1 \pmod{2}$.

37. Since $a \equiv b \pmod{m}$, there exists an integer s such that $a = b + sm$, so that $a - b = sm$. Then $a^k - b^k = (a-b)(a^{k-1} + a^{k-2}b + \cdots + ab^{k-2} + b^{k-1})$, $k \ge 2$, is also a multiple of m. It follows that $a^k \equiv b^k \pmod{m}$.

39. **a)** 7, 19, 7, 7, 18, 0
b) Take the next available space **mod** 31.

41. 2, 6, 7, 10, 8, 2, 6, 7, 10, 8, . . .

43. **a)** GR QRW SDVV JR **b)** QB ABG CNFF TB
c) QX UXM AHJJ ZX

Section 2.4

1. **a)** 6 **b)** 3 **c)** 11 **d)** 3

3. 8

5. **a)** 111 00111
b) 100 01101 10100
c) 10 11111 01011 01100

7. **a)** 31 **b)** 513 **c)** 341 **d)** 26,896

9. Convert each hex digit to a block of four bits.

11. **a)** 10 00000 01110
b) 10 01101 01101 01011
c) 1 01010 11101 11010
d) 110 11110 11111 01011 00111 01101

13. The binary expansion of the integer is the unique such sum.

15. Let $a = (a_{n-1}a_{n-2}\cdots a_1a_0)_{10}$. Then $a = 10^{n-1}a_{n-1} + 10^{n-2}a_{n-2} + \cdots + 10a_1 + a_0 \equiv a_{n-1} + a_{n-2} + \cdots + a_1 + a_0 \pmod{3}$, since $10^j \equiv 1 \pmod{3}$ for all nonnegative integers j. It follows that $3 \mid a$ if and only if 3 divides the sum of the decimal digits of a.

17. **a)** −6 **b)** 13 **c)** −14 **d)** 0

19. The one's complement of the sum is found by adding the one's complements of the two integers except that a carry in the leading bit is used as a carry to the last bit of the sum.

21. $4n$

23.
```
procedure Cantor(x : positive integer)
  n := 1; f := 1
  while (n + 1) * f ≤ x
  begin
    n := n + 1
    f := f * n
  end
  y := x
  while n > 0
  begin
    a_n := ⌊y/f⌋
    y := y − a_n * f
    f := f/n
    n := n − 1
  end {x = a_n n! + a_{n−1}(n − 1)! + ··· + a_1 1!}
```

25. First step: $c = 0$, $d = 0$, $s_0 = 1$; second step: $c = 0$, $d = 1$, $s_1 = 0$; third step: $c = 1$, $d = 1$, $s_2 = 0$; fourth step: $c = 1$, $d = 1$, $s_3 = 0$; fifth step: $c = 1$, $d = 1$, $s_4 = 1$; sixth step: $c = 1$, $s_5 = 1$

27.
```
procedure subtract(a, b : positive integers, a > b,
    a = (a_{n−1}a_{n−2}···a_1a_0)_2,
    b = (b_{n−1}b_{n−2}···b_1b_0)_2)
B := 0 {B is the borrow}
for j := 0 to n − 1
begin
  if a_j ≥ b_j + B then
  begin
    s_j := a_j − b_j − B
    B := 0
  end
  else
  begin
    s_j := a_j + 2 − b_j − B
    B := 1
  end
end {(s_{n−1}s_{n−2}···s_1s_0)_2 is the difference }
```

29. **procedure** *compare*$(a, b$: positive integers and
$a = (a_n a_{n-1} \cdots a_1 a_0)_2$,
$b = (b_n b_{n-1} \cdots b_1 b_0)_2)$
$k := n$
while $a_k = b_k$ and $k > 0$
$k := k - 1$
if $a_k = b_k$ **then** print "a equals b"
if $a_k > b_k$ **print** "a is greater than b"
if $a_k < b_k$ **then** print "a is less than b"

31. $O(\log n)$

Section 2.5

1. **a)** $1 = (-1) \cdot 10 + 1 \cdot 11$
b) $1 = 21 \cdot 21 + (-10) \cdot 44$
c) $12 = (-1) \cdot 36 + 48$
d) $1 = 13 \cdot 55 + (-21) \cdot 34$
e) $3 = 11 \cdot 213 + (-20) \cdot 117$
f) $223 = 1 \cdot 0 + 1 \cdot 223$
g) $1 = 37 \cdot 2347 + (-706) \cdot 123$
h) $2 = 1128 \cdot 3454 + (-835) \cdot 4666$
i) $1 = 2468 \cdot 9999 + (-2221) \cdot 11111$

3. $15 \cdot 7 = 105 \equiv 1 \pmod{26}$

5. 7

7. 52

9. Suppose that b and c are both inverses of a modulo m. Then $ba \equiv 1 \pmod{m}$ and $ca \equiv 1 \pmod{m}$. Hence $ba \equiv ca \pmod{m}$. Since $gcd(a, m) = 1$ it follows by Theorem 2 that $b \equiv c \pmod{m}$.

11. $x \equiv 8 \pmod{9}$

13. Let $m' = m/gcd(c, m)$. Since all the common factors of m and c are divided out of m to obtain m', it follows that m' and c are relatively prime. Since m divides $(ac - bc) = (a - b)c$, it follows that m' divides $(a - b)c$. By Lemma 1, we see that m' divides $a - b$, so that $a \equiv b \pmod{m'}$.

15. Suppose that $x^2 \equiv 1 \pmod{p}$. Then p divides $x^2 - 1 = (x + 1)(x - 1)$. By Lemma 2 it follows that $p \mid (x + 1)$ or $p \mid (x - 1)$, so that $x \equiv -1 \pmod{p}$ or $x \equiv 1 \pmod{p}$.

17. **a)** Suppose that $ia \equiv ja \pmod{p}$ where $1 \le i < j < p$. Then p divides $ja - ia = a(j - i)$. By Theorem 1, since a is not divisible by p, p divides $j - i$, which is impossible since $j - i$ is a positive integer less than p.
b) By (a), since no two of $a, 2a, \ldots, (p-1)a$ are congruent modulo p, each must be congruent to a different number from 1 to $p - 1$. It follows that $a \cdot 2a \cdot 3a \cdot \cdots \cdot (p-1) \cdot a \equiv 1 \cdot 2 \cdot 3 \cdot \cdots \cdot (p-1) \pmod{p}$. It follows that $(p-1)! \cdot a^{p-1} \equiv (p-1)! \pmod{p}$.
c) By Wilson's theorem and (c), if p does not divide a, it follows that $(-1) \cdot a^{p-1} \equiv -1 \pmod{p}$. Hence $a^{p-1} \equiv 1 \pmod{p}$.
d) If $p \mid a$, then $p \mid a^p$. Hence $a^p \equiv a \equiv 0 \pmod{p}$. If p does not divide a, then $a^{p-1} \equiv a \pmod{p}$, by (c). Multiplying both sides of this congruence by a gives $a^p \equiv a \pmod{p}$.

19. Suppose that p is a prime appearing in the prime factorization of $m_1 m_2 \cdots m_n$. Since the m_is are relatively prime, p is a factor of exactly one of the m_is, say m_j. Since m_j divides $a - b$, it follows that $a - b$ has the factor p in its prime factorization to a power at least as large as the power to which it appears in the prime factorization of m_j. It follows that $m_1 m_2 \cdots m_n$ divides $a - b$, so that $a \equiv b \pmod{m_1 m_2 \cdots m_n}$.

21. $x \equiv 1 \pmod{6}$

23. **a)** By Fermat's Little Theorem, we have $2^{10} \equiv 1 \pmod{11}$. Hence $2^{340} = (2^{10})^{34} \equiv 1^{34} = 1 \pmod{11}$.
b) Since $32 \equiv 1 \pmod{31}$, it follows that $2^{340} = (2^5)^{68} = 32^{68} \equiv 1^{68} = 1 \pmod{31}$.
c) Since 11 and 31 are relatively prime, and $11 \cdot 31 = 341$, it follows by (a) and (b) and Exercise 19 that $2^{340} \equiv 1 \pmod{341}$.

25. $0 = (0, 0)$, $1 = (1, 1)$, $2 = (2, 2)$, $3 = (0, 3)$, $4 = (1, 4)$, $5 = (2, 0)$, $6 = (0, 1)$, $7 = (1, 2)$, $8 = (2, 3)$, $9 = (0, 4)$, $10 = (1, 0)$, $11 = (2, 1)$, $12 = (0, 2)$, $13 = (1, 3)$, $14 = (2, 4)$

27. We have $m_1 = 99$, $m_2 = 98$, $m_3 = 97$, and $m_4 = 95$, so that $m = 99 \cdot 98 \cdot 97 \cdot 95 = 89{,}403{,}930$. We find that $M_1 = m/m_1 = 903{,}070$, $M_2 = m/m_2 = 912{,}285$, $M_3 = m/m_3 = 921{,}690$, and $M_4 = m/m_4 = 941{,}094$. Using the Euclidean algorithm, we compute that $y_1 = 37$, $y_2 = 33$, $y_3 = 24$, and $y_4 = 4$ are inverses of M_k modulo m_k for $k = 1, 2, 3, 4$, respectively. It follows that the solution is $65 \cdot 903{,}070 \cdot 37 + 2 \cdot 912{,}285 \cdot 33 + 51 \cdot 921{,}690 \cdot 24 + 10 \cdot 941{,}094 \cdot 4 = 3{,}397{,}886{,}480 \equiv 537{,}140 \pmod{89{,}403{,}930}$.

29. By Exercise 28 it follows that $gcd(2^b - 1, (2^a - 1) \bmod (2^b - 1)) = gcd(2^b - 1, 2^{a \bmod b} - 1)$. Since the exponents involved in the calculation are b and a **mod** b, the same as the quantities involved in computing $gcd(a, b)$, the steps used by the Euclidean algorithm to compute $gcd(2^a - 1, 2^b - 1)$ run in parallel to those used to compute $gcd(a, b)$ and show that $gcd(2^a - 1, 2^b - 1) = 2^{gcd(a,b)} - 1$.

31. Suppose we know both $n = pq$ and $(p-1)(q-1)$. To find p and q, first note that $(p-1)(q-1) = pq - p - q + 1 = n - (p+q) + 1$. From this we can find $s = (p+q)$. Since $q = s - p$, we have $n = p(s-p)$. Hence $p^2 - ps + n = 0$. We now can use the quadratic formula to find p. Once we have found p we can find q since $q = n/p$.

33. SILVER

Section 2.6

1. a) 3×4 **b)** $\begin{bmatrix} 1 \\ 4 \\ 3 \end{bmatrix}$ **c)** $[2\ \ 0\ \ 4\ \ 6]$ **d)** 1 **e)** $\begin{bmatrix} 1 & 2 & 1 \\ 1 & 0 & 1 \\ 1 & 4 & 3 \\ 3 & 6 & 7 \end{bmatrix}$

3. a) $\begin{bmatrix} 1 & 11 \\ 2 & 18 \end{bmatrix}$ **b)** $\begin{bmatrix} 2 & -2 & -3 \\ 1 & 0 & 2 \\ 9 & -4 & 4 \end{bmatrix}$

5. $\begin{bmatrix} 9/5 & -6/5 \\ -1/5 & 4/5 \end{bmatrix}$

7. $\mathbf{0} + \mathbf{A} = [0 + a_{ij}] = [a_{ij} + 0] = \mathbf{0} + \mathbf{A}$

9. $\mathbf{A} + (\mathbf{B} + \mathbf{C}) = [a_{ij} + (b_{ij} + c_{ij})] = [(a_{ij} + b_{ij}) + c_{ij}] = (\mathbf{A} + \mathbf{B}) + \mathbf{C}$

11. The number of rows of **A** equals the number of columns of **B**, and the number of columns of **A** equals the number of rows of **B**.

13.
$$\mathbf{A}(\mathbf{BC}) = \left[\sum_q a_{iq}\left(\sum_r b_{qr}c_{rl}\right)\right] = \left[\sum_q \sum_r a_{iq}b_{qr}c_{rl}\right] = \left[\sum_r \sum_q a_{iq}b_{qr}c_{rl}\right] = \left[\sum_r \left(\sum_q a_{iq}b_{qr}\right) c_{rl}\right] = (\mathbf{AB})\mathbf{C}$$

15. $\mathbf{A}^n = \begin{bmatrix} 1 & n \\ 0 & 1 \end{bmatrix}$

17. a) Let $\mathbf{A} = [a_{ij}]$ and $\mathbf{B} = [b_{ij}]$. Then $(\mathbf{A} + \mathbf{B}) = [a_{ij} + b_{ij}]$. We have $(\mathbf{A} + \mathbf{B})^t = [a_{ji} + b_{ji}] = [a_{ji}] + [b_{ji}] = \mathbf{A}^t + \mathbf{B}^t$. **b)** Using the same notation as in (a), we have

$$\mathbf{B}^t\mathbf{A}^t = \left[\sum_q b_{qi}a_{jq}\right] = \left[\sum_q a_{jq}b_{qi}\right] = (\mathbf{AB})^t,$$

since the (i, j)th entry is the (j, i)th entry of **AB**.

19. The result follows since

$$\begin{bmatrix} a & b \\ c & d \end{bmatrix}\begin{bmatrix} d & -b \\ -c & a \end{bmatrix} = \begin{bmatrix} ad - bc & 0 \\ 0 & ad - bc \end{bmatrix} = (ad - bc)\mathbf{I}_2 = \begin{bmatrix} d & -b \\ -c & a \end{bmatrix}\begin{bmatrix} a & b \\ c & d \end{bmatrix}.$$

21. $\mathbf{A}^n(\mathbf{A}^{-1})^n = \mathbf{A}(\mathbf{A} \cdots (\mathbf{A}(\mathbf{A}\mathbf{A}^{-1})\mathbf{A}^{-1}) \cdots \mathbf{A}^{-1})\mathbf{A}^{-1}$ by the associative law. Since $\mathbf{A}\mathbf{A}^{-1} = \mathbf{I}$, working from the inside shows that $\mathbf{A}^n(\mathbf{A}^{-1})^n = \mathbf{I}$. Similarly $(\mathbf{A}^{-1})^n\mathbf{A}^n = \mathbf{I}$. Therefore $(\mathbf{A}^n)^{-1} = (\mathbf{A}^{-1})^n$.

23. There are m_2 multiplications used to find each of the m_1m_3 entries of the product. Hence $m_1m_2m_3$ multiplications are used.

25. $\mathbf{A}_1((\mathbf{A}_2\mathbf{A}_3)\mathbf{A}_4)$

27. $x_1 = 1,\ x_2 = -1,\ x_3 = -2$

29. a) $\begin{bmatrix} 1 & 1 & 1 \\ 1 & 1 & 1 \\ 1 & 0 & 1 \end{bmatrix}$ **b)** $\begin{bmatrix} 0 & 0 & 1 \\ 1 & 0 & 0 \\ 0 & 0 & 1 \end{bmatrix}$ **c)** $\begin{bmatrix} 1 & 1 & 1 \\ 1 & 1 & 1 \\ 1 & 0 & 1 \end{bmatrix}$

31. a) $\begin{bmatrix} 1 & 0 & 0 \\ 1 & 1 & 0 \\ 1 & 0 & 1 \end{bmatrix}$ **b)** $\begin{bmatrix} 1 & 0 & 0 \\ 1 & 0 & 1 \\ 1 & 1 & 0 \end{bmatrix}$ **c)** $\begin{bmatrix} 1 & 0 & 0 \\ 1 & 1 & 1 \\ 1 & 1 & 1 \end{bmatrix}$

33. a) $\mathbf{A} \vee \mathbf{B} = [a_{ij} \vee b_{ij}] = [b_{ij} \vee a_{ij}] = \mathbf{B} \vee \mathbf{A}$ **b)** $\mathbf{A} \wedge \mathbf{B} = [a_{ij} \wedge b_{ij}] = [b_{ij} \wedge a_{ij}] = \mathbf{B} \wedge \mathbf{A}$

35. **a)** $\mathbf{A} \vee (\mathbf{B} \wedge \mathbf{C}) = [a_{ij}] \vee [b_{ij} \wedge c_{ij}]$
$= [a_{ij} \vee (b_{ij} \wedge c_{ij})]$
$= [(a_{ij} \vee b_{ij}) \wedge (a_{ij} \vee c_{ij})]$
$= [a_{ij} \vee b_{ij}] \wedge [a_{ij} \vee c_{ij}]$
$= (\mathbf{A} \vee \mathbf{B}) \wedge (\mathbf{A} \vee \mathbf{C})$

b) $\mathbf{A} \wedge (\mathbf{B} \vee \mathbf{C}) = [a_{ij}] \wedge [b_{ij} \vee c_{ij}]$
$= [a_{ij} \wedge (b_{ij} \vee c_{ij})]$
$= [(a_{ij} \wedge b_{ij}) \vee (a_{ij} \wedge c_{ij})]$
$= [a_{ij} \wedge b_{ij}] \vee [a_{ij} \wedge c_{ij}]$
$= (\mathbf{A} \wedge \mathbf{B}) \vee (\mathbf{A} \wedge \mathbf{C})$

37. $\mathbf{A} \odot (\mathbf{B} \odot \mathbf{C}) = \left[\bigvee_q a_{iq} \wedge \left(\bigvee_r (b_{qr} \wedge c_{rl})\right)\right]$
$= \left[\bigvee_q \bigvee_r (a_{iq} \wedge b_{qr} \wedge c_{rl})\right]$
$= \left[\bigvee_r \bigvee_q (a_{iq} \wedge b_{qr} \wedge c_{rl})\right]$
$= \left[\bigvee_r \left(\bigvee_q (a_{iq} \wedge b_{qr})\right) \wedge c_{rl}\right]$
$= (\mathbf{A} \odot \mathbf{B}) \odot \mathbf{C}$

Supplementary Exercises

1. **a)** **procedure** *last max*($a_1, \ldots, a_n$: integers)
max := a_1
last := 1
i := 2
while $i \leq n$
begin
if $a_i \geq max$ **then**
begin
max := a_i
last := i
end
$i := i + 1$
end {*last* is the location of final occurrence of largest integer in list}

b) $2n - 1 = O(n)$ comparisons

3. **a)** **procedure** *pair zeros*($b_1 b_2 \cdots b_n$: bit string, $n \geq 2$)
$x := b_1$
$y := b_2$
$k := 2$
while ($k < n$ and ($x \neq 0$ or $y \neq 0$))
begin
$k := k + 1$
$x := y$
$y := b_k$
end
if ($x = 0$ and $y = 0$) **then** print "YES"
else print "NO"

b) $O(n)$ comparisons

5. 5, 22, −12, −29

7. Since $ac \equiv bc \pmod{m}$ there is an integer k such that $ac = bc + km$. Hence $a - b = km/c$. Since $a - b$ is an integer, $c \mid km$. Letting $d = \gcd(m, c)$, write $c = de$. Since no factor of e divides m/d, it follows that $d \mid m$ and $e \mid k$. Thus $a - b = (k/e)(m/d)$, where $k/e \in \mathbf{Z}$ and $m/d \in \mathbf{Z}$. Therefore $a \equiv b \pmod{m/d}$.

9. 1

11. 1

13. $(a_n a_{n-1} \cdots a_1 a_0)_{10} = \sum_{k=0}^{n} 10^k a_k \equiv \sum_{k=0}^{n} a_k \pmod{9}$ since $10^k \equiv 1 \pmod{9}$ for every nonnegative integer k.

15. **a)** Not mutually relatively prime **b)** Mutually relatively prime **c)** Mutually relatively prime **d)** Mutually relatively prime

17. **a)** The decryption function is $g(q) = \bar{a}(q - b)$ **mod** 26, where $\bar{a}$ is an inverse of a modulo 26. **b)** PLEASE SEND MONEY

19. $x \equiv 28 \pmod{30}$

21. $\mathbf{A}^{4n} = \begin{bmatrix} 1 & 0 \\ 0 & 1 \end{bmatrix}$, $\mathbf{A}^{4n+1} = \begin{bmatrix} 0 & 1 \\ -1 & 0 \end{bmatrix}$,
$\mathbf{A}^{4n+2} = \begin{bmatrix} -1 & 0 \\ 0 & -1 \end{bmatrix}$, $\mathbf{A}^{4n+3} = \begin{bmatrix} 0 & -1 \\ 1 & 0 \end{bmatrix}$,
for $n \geq 0$

23. Suppose that
$$\mathbf{A} = \begin{bmatrix} a & b \\ c & d \end{bmatrix}.$$
Let
$$\mathbf{B} = \begin{bmatrix} 0 & 1 \\ 0 & 0 \end{bmatrix}.$$
Since $\mathbf{AB} = \mathbf{BA}$, it follows that $c = 0$ and $a = d$. Let
$$\mathbf{B} = \begin{bmatrix} 0 & 0 \\ 1 & 0 \end{bmatrix}.$$
Since $\mathbf{AB} = \mathbf{BA}$, it follows that $b = 0$. Hence
$$\mathbf{A} = \begin{bmatrix} a & 0 \\ 0 & a \end{bmatrix} = a\mathbf{I}.$$

25. **procedure** *triangular matrix multiplication*(**A**,**B**: upper triangular $n \times n$ matrices, $\mathbf{A} = [a_{ij}]$, $\mathbf{B} = [b_{ij}]$)

```
for i := 1 to n
begin
  for j := i to n
  begin
    c_ij := 0
    for k := i to j
      c_ij := c_ij + a_ik b_kj
  end
end
```

27. $(\mathbf{AB})(\mathbf{B}^{-1}\mathbf{A}^{-1}) = \mathbf{A}(\mathbf{BB}^{-1})\mathbf{A}^{-1} = \mathbf{AIA}^{-1} = \mathbf{AA}^{-1} = \mathbf{I}$. Similarly, $(\mathbf{B}^{-1}\mathbf{A}^{-1})(\mathbf{AB}) = \mathbf{I}$. Hence $(\mathbf{AB})^{-1} = \mathbf{B}^{-1}\mathbf{A}^{-1}$.

29. **a)** Let $\mathbf{A} \odot \mathbf{0} = [b_{ij}]$. Then $b_{ij} = (a_{i1} \wedge 0) \vee \cdots \vee (a_{ip} \wedge 0) = 0$. Hence $\mathbf{A} \odot \mathbf{0} = \mathbf{0}$. Similarly $\mathbf{0} \odot \mathbf{A} = 0$.
b) $\mathbf{A} \vee \mathbf{0} = [a_{ij} \vee 0] = [a_{ij}] = \mathbf{A}$. Hence $\mathbf{A} \vee \mathbf{0} = \mathbf{A}$. Similarly $\mathbf{0} \vee \mathbf{A} = \mathbf{A}$.
c) $\mathbf{A} \wedge \mathbf{0} = [a_{ij} \wedge 0] = [0] = \mathbf{0}$. Hence $\mathbf{A} \wedge \mathbf{0} = \mathbf{0}$. Similarly $\mathbf{0} \wedge \mathbf{A} = \mathbf{0}$.

CHAPTER 3

Section 3.1

1. **a)** Addition **b)** Simplification **c)** Modus ponens **d)** Modus tollens **e)** Hypothetical syllogism

3. **a)** Fallacy of affirming the conclusion **b)** Fallacy of begging the question **c)** Valid argument using modus tollens **d)** Valid argument using disjunctive syllogism **e)** Fallacy of denying the hypothesis

5. The proposition is vacuously true since 0 is not a positive integer. Vacuous proof.

7. $P(1)$ is true since $(a+b)^1 = a + b \geq a^1 + b^1 = a + b$. Direct proof.

9. Let $n = 2k + 1$ and $m = 2l + 1$ be odd integers. Then $n + m = 2(k + l + 1)$ is even.

11. Suppose that r is rational and i is irrational and $s = r + i$ is rational. Then by Exercise 10, $s + (-r) = i$ is rational, which is a contradiction.

13. Since $\sqrt{2} \cdot \sqrt{2} = 2$ is rational and $\sqrt{2}$ is irrational, the product of two irrational numbers is not necessarily irrational.

15. $41^2 - 41 + 41 = 41^2$ is composite, so that $n^2 - n + 41$ is not necessarily prime.

17. Suppose that $3^{1/3} = a/b$ where $a, b \in \mathbf{Z}$, $b \neq 0$, and $gcd(a, b) = 1$. Then $3 = a^3/b^3$, so that $3b^3 = a^3$. Hence $3 \mid a^3$, which can happen only if $3 \mid a$. Let $a = 3m$. Then $3b^3 = 27m^3$, or $b^3 = 9m^3$. Thus $3 \mid b^3$, which shows that $3 \mid b$. This is a contradiction of the assumption that $gcd(a, b) = 1$.

19. If $x \leq y$, then $\max(x, y) + \min(x, y) = y + x = x + y$. If $x \geq y$, then $\max(x, y) + \min(x, y) = x + y$. Since these are the only two cases, the equality always holds.

21. There are four cases. Case 1: $x \geq 0$ and $y \geq 0$. Then $|x| + |y| = x + y = |x + y|$. Case 2: $x < 0$ and $y < 0$. Then $|x| + |y| = -x + (-y) = -(x + y) = |x + y|$ since $x + y < 0$. Case 3: $x \geq 0$ and $y < 0$. Then $|x| + |y| = x + (-y)$. If $x \geq -y$, then $|x + y| = x + y$. But since $y < 0$, $-y > y$, so that $|x| + |y| = x + (-y) > x + y = |x + y|$. If $x < -y$, then $|x + y| = -(x + y) = -x + (-y)$. But since $x \geq 0$, $x \geq -x$, so that $|x| + |y| = x + (-y) \geq -x + (-y) = |x + y|$. Case 4: $x < 0$ and $y \geq 0$. Identical to Case 3 with the roles of x and y reversed.

23. First, assume that n is odd, so that $n = 2k + 1$ for some integer k. Then $5n + 6 = 5(2k + 1) + 6 = 10k + 11 = 2(5k + 5) + 1$. Hence $5n + 6$ is odd. To prove the converse, suppose that n is even, so that $n = 2k$ for some integer k. Then $5n + 6 = 10k + 6 = 2(5k + 3)$, so that $5n + 6$ is even. Hence n is odd if and only if $5n + 6$ is odd.

25. $a^2 \equiv b^2 \pmod{p}$ if and only if $p \mid (a^2 - b^2) = (a + b)(a - b)$. By the uniqueness of prime factorization, this is equivalent to $p \mid (a - b)$ or $p \mid (a + b)$, which is the same as $a \equiv b \pmod{p}$ or $a \equiv -b \pmod{p}$.

27. This proposition is true. Suppose that m is neither 1 nor -1. Then mn has a factor m larger than 1. On the other hand, $mn = 1$, and 1 has no such factor. Hence $m = 1$ or $m = -1$. In the first case $n = 1$, and in the second case $n = -1$, since $n = 1/m$.

29. The positive integer 3 is not the sum of two squares of integers, so that the proposition is false.

31. We will give a proof by contradiction. Suppose that $a_1, a_2, \ldots, a_n$ are all less than A, where A is the average of these numbers. Then $a_1 + a_2 + \cdots + a_n < nA$. Dividing both sides by n shows that $A = (a_1+a_2+\cdots+a_n)/n < A$, which is a contradiction.

33. We will show that the four statements are equivalent by showing that (i) implies (ii), (ii) implies (iii), (iii) implies (iv), and (iv) implies (i). First, assume that n is even. Then $n = 2k$ for some integer k. Then $n+1 = 2k+1$, so that $n+1$ is odd. This shows that (i) implies (ii). Next, suppose that $n+1$ is odd, so that $n+1 = 2k+1$ for some integer k. Then $3n+1 = 2n+(n+1) = 2(n+k)+1$, which shows that $3n+1$ is odd, showing that (ii) implies (iii). Next, suppose that $3n+1$ is odd, so that $3n+1 = 2k+1$ for some integer k. Then $3n = (2k+1)-1 = 2k$, so that $3n$ is even. This shows that (iii) implies (iv). Finally, suppose that n is not even. Then n is odd, so $n = 2k+1$ for some integer k. Then $3n = 3(2k+1) = 6k+3 = 2(3k+1)+1$, so that $3n$ is odd. This completes an indirect proof that (iv) implies (i).

35. The integers 3, 5, and 7 are three primes of the desired form.

37. By the second premise, there is some lion that does not drink coffee. Let Leo be such a creature. By simplification we know that Leo is a lion. By modus ponens we know from the first premise that Leo is fierce. Hence Leo is fierce and does not drink coffee. By the definition of the existential quantifier, there exist fierce creatures that do not drink coffee, that is, some fierce creatures do not drink coffee.

39. Assume we have the first $n+1$ prime numbers $p_1, p_2, \ldots, p_{n+1}$. Then $p_1 p_2 \cdots p_{n+1}$ is divisible by more than n primes.

41. Suppose that $p_1, p_2, \ldots, p_n$ are all the primes congruent to 3 module 4, except for 3. Let $q = 4p_1p_2\cdots p_n + 3$. Then $q \equiv 3 \pmod 4$, and q is not divisible by p_i, $i = 1, 2, \ldots, n$, or by 3. Since q must have at least one prime factor that is congruent to 3 module 4, there must be a prime of this type not in our list. This is a nonconstructive existence proof.

43. Suppose that $p_1 \to p_4 \to p_2 \to p_5 \to p_3 \to p_1$. To prove that one of these propositions implies any of the others, just use hypothetical syllogism repeatedly.

Section 3.2

1. $n(n+1)$

3. Let $P(n)$ be "$\sum_{j=0}^{n} 3\cdot 5^j = 3(5^{n+1}-1)/4$." Basis step: $P(0)$ is true since $\sum_{j=0}^{0} 3\cdot 5^j = 3 = 3(5^1-1)/4$. Inductive step: Assume that $\sum_{j=0}^{n} 3\cdot 5^j = 3(5^{n+1}-1)/4$. Then $\sum_{j=0}^{n+1} 3\cdot 5^j = (\sum_{j=0}^{n} 3\cdot 5^j) + 3\cdot 5^{n+1} = 3(5^{n+1}-1)/4 + 3\cdot 5^{n+1} = 3(5^{n+1}+4\cdot 5^{n+1}-1)/4 = 3(5^{n+2}-1)/4$.

5. Let $P(n)$ be "$\sum_{j=1}^{n} j^2 = n(n+1)(2n+1)/6$." Basis step: $P(1)$ is true since $\sum_1^1 j^2 = 1 = 1(1+1)\cdot(2\cdot 1+1)/6$. Inductive step: Assume that $\sum_{j=1}^{n} j^2 = n(n+1)(2n+1)/6$. Then $\sum_{j=1}^{n+1} j^2 = (\sum_{j=1}^{n} j^2) + (n+1)^2 = n(n+1)(2n+1)/6 + (n+1)^2 = (n+1)[2n^2+n+6n+6]/6 = (n+1)(n+2)\cdot(2n+3)/6 = (n+1)((n+1)+1)(2(n+1)+1)/6$.

7. Let $P(n)$ be "$1^2+3^2+\cdots+(2n+1)^2 = (n+1)(2n+1)(2n+3)/3$." Basis step: $P(0)$ is true since $1^2 = 1 = (0+1)(2\cdot 0+1)(2\cdot 0+3)/3$. Inductive step: Assume that $P(n)$ is true. Then $1^2+3^2+\cdots+(2n+1)^2+(2(n+1)+1)^2 = (n+1)(2n+1)(2n+3)/3+(2n+3)^2 = (2n+3)[(n+1)(2n+1)/3+(2n+3)] = (2n+3)(2n^2+9n+10)/3 = (2n+3)(2n+5)(n+2)/3 = ((n+1)+1)(2(n+1)+1)\cdot(2(n+1)+3)/3$.

9. Let $P(n)$ be "$1+nh \le (1+h)^n, h > -1$." Basis step: $P(0)$ is true since $1+0\cdot h = 1 \le 1 = (1+h)^0$. Inductive step: Assume $1+nh \le (1+h)^n$. Then since $(1+h) > 0$, $(1+h)^{n+1} = (1+h)(1+h)^n \ge (1+h)(1+nh) = 1+(n+1)h+nh^2 \ge 1+(n+1)h$.

11. Let $P(n)$ be "$2^n > n^2$." Basis step: $P(5)$ is true since $2^5 = 32 > 25 = 5^2$. Inductive step: Assume that $P(n)$ is true, that is, $2^n > n^2$. Then $2^{n+1} = 2\cdot 2^n > n^2+n^2 > n^2+4n \ge n^2+2n+1 = (n+1)^2$ since $n > 4$.

13. Let $P(n)$ be "$1\cdot 2+2\cdot 3+\cdots+n(n+1) = n(n+1)(n+2)/3$." Basis step: $P(1)$ is true since $1\cdot 2 = 2 = 1(1+1)(1+2)/3$. Inductive step: Assume that $P(n)$ is true. Then $1\cdot 2+2\cdot 3+\cdots+n(n+1)+(n+1)(n+2) = [n(n+1)(n+2)/3]+(n+1)(n+2) = (n+1)(n+2)[(n/3)+1] = (n+1)(n+2)(n+3)/3$.

15. Let $P(n)$ be "$1^2-2^2+3^2-\cdots+(-1)^{n-1}n^2 = (-1)^{n-1}n(n+1)/2$." Basis step: $P(1)$ is true since $1^2 = 1 = (-1)^0 1^2$. Inductive step: Assume that $P(n)$ is true. Then $1^2-2^2+3^2-\cdots+(-1)^{n-1}n^2+(-1)^n(n+1)^2 = (-1)^{n-1}n(n+1)/2+(-1)^n$

$(n+1)^2 = (-1)^n(n+1)[-n/2+(n+1)] = (-1)^n (n+1)[(n/2)+1] = (-1)^n(n+1)(n+2)/2$.

17. Let $P(n)$ be "a postage of n cents can be formed using 3-cent and 5-cent stamps." Basis step: $P(8)$ is true since 8 cents postage can be formed with one 3-cent and one 5-cent stamp. Inductive step: Assume that $P(n)$ is true, that is, postage of n cents can be formed. We will show how to form postage of $n+1$ cents. By the inductive hypothesis postage of n cents can be formed. If this included a 5-cent stamp, replace this with two 3-cent stamps to obtain $n+1$ cents postage. Otherwise, only 3-cent stamps were used and $n \geq 9$. Remove three of these 3-cent stamps and replace them with two 5-cent stamps to obtain $n+1$ cents postage.

19. Let $P(n)$ be "$n^5 - n$ is divisible by 5." Basis step: $P(0)$ is true since $0^5 - 0 = 0$ is divisible by 5. Inductive step: Assume that $P(n)$ is true, that is, $n^5 - 5$ is divisible by 5. Then $(n+1)^5 - (n+1) = (n^5 + 5n^4 + 10n^3 + 10n^2 + 5n + 1) - (n+1) = (n^5 - n) + 5(n^4 + 2n^3 + 2n^2 + n)$ is also divisible by 5, since both terms in this sum are divisible by 5.

21. Let $P(n)$ be the proposition that $(2n-1)^2 - 1$ is divisible by 8. The basis case $P(1)$ is true since $8 \mid 0$. Now assume that $P(n)$ is true. Since $((2(n+1)-1)^2 - 1) = ((2n-1)^2 - 1) + 8n$, $P(n+1)$ is true since both terms on the right-hand side are divisible by 8. This shows that $P(n)$ is true for all positive integers so that $m^2 - 1$ is divisible by 8 whenever m is an odd positive integer.

23. Let $P(n)$ be the statement that a set with n elements has $n(n-1)/2$ two-element subsets. $P(2)$, the basis case, is true, since a set with two elements has one subset with two elements—namely, itself—and $2(2-1)/2 = 1$. Now assume that $P(n)$ is true. Let S be a set with $n+1$ elements. Choose an element a in S and let $T = S - \{a\}$. A two-element subset of S either contains a or it does not. Those subsets not containing a are the subsets of 2 elements of T; by the inductive hypothesis there are $n(n-1)/2$ of these. There are n subsets of two elements of S that contain a, since such a subset contains a and one of the n elements in T. Hence there are $n(n-1)/2 + n = (n+1)n/2$ two-element subsets of S. This completes the inductive proof.

25. Let $P(n)$ be the statement that $1^4 + 2^4 + 3^4 + \cdots + n^4 = n(n+1)(2n+1)(3n^2+3n-1)/30$. $P(1)$ is true since $1 \cdot 2 \cdot 3 \cdot 5/30 = 1$. Assume that $P(n)$ is true. Then $(1^4 + 2^4 + 3^4 + \cdots + n^4) + (n+1)^4 = n(n+1)(2n+1)(3n^2+3n-1)/30 + (n+1)^4 = ((n+1)/30)(n(2n+1)(3n^2+3n-1)+30(n+1)^3) = ((n+1)/30)(6n^4 + 39n^3 + 91n^2 + 89n + 30) = ((n+1)/30)(n+2)(2n+3)(3(n+1)^2+3(n+1)-1)$. This demonstrates that $P(n+1)$ is true.

27. By inspection we find that the inequality $2n+3 \leq 2^n$ does not hold for $n = 0, 1, 2, 3$. Let $P(n)$ be the proposition that this inequality holds for the positive integer n. $P(4)$, the basis case, is true since $2 \cdot 4 + 3 = 11 \leq 16 = 2^4$. For the inductive step assume that $P(n)$ is true. Then, by the inductive hypothesis, $2(n+1) + 3 = (2n+3) + 2 < 2^n + 2$. But since $n \geq 1$, $2^n + 2 \leq 2^n + 2^n = 2^{n+1}$. This shows that $P(n+1)$ is true.

29. **a)** The postages that can be formed using 5-cent and 6-cent stamps are 5 cents, 6 cents, 10 cents, 11 cents, 12 cents, 15 cents, 16 cents, 17 cents, 18 cents, and all postages of 20 cents or more. **b)** We will prove that all postages of 20 cents or more can be formed using 5-cent and 6-cent stamps. Let $P(n)$ be the statement that postage of n cents can be formed. $P(20)$ is true since postage of 20 cents can be formed using four 5-cent stamps. Now assume that $P(n)$ is true. If a 5-cent stamp was used to form postage of n cents, then replace it by a 6-cent stamp to form postage of $n+1$ cents. Otherwise, if only 6-cent stamps were used, since $n \geq 20$ at least four 6-cent stamps were used. Replace four 6-cent stamps by five 5-cent stamps to get postage of $n+1$ cents. Hence $P(n+1)$ is true. This completes the proof by mathematical induction. **c)** Let $P(n)$ be as in (b). The basis cases are $P(20)$, $P(21)$, $P(22)$, $P(23)$, and $P(24)$. These are true since postage of 20 cents, 21 cents, 22 cents, 23 cents, and 24 cents can be formed using four 5-cent stamps, three 5-cent stamps and one 6-cent stamp, two 5-cent stamps and two 6-cent stamps, one 5-cent and three 6-cent stamps, and four 6-cent stamps, respectively. Now assume that $P(k)$ is true for $20 \leq k \leq n$ where $n \geq 24$. Since $n+1 \geq 25$, it follows that $n-4 \geq 20$, so that by the inductive hypothesis postage of $n-4$ can be formed. Add a 5-cent stamp to obtain postage of $n+1$ cents, showing that $P(n+1)$ is true. This completes the proof by the second principle of mathematical induction.

31. All multiples of \$10 greater than or equal to \$40 can be formed as well as \$20. Let $P(n)$ be the statement that $10n$ dollars can be formed. $P(4)$ is true since \$40 can be formed using two \$20s. Now assume that $P(n)$ is true with $n \geq 4$. If a \$50 bill is used to form $10n$ dollars, replace it by three \$20 bills to obtain $10(n+1)$ dollars. Otherwise, at least two \$20 bills were used since $10n$ is at least \$40. Replace two \$20 bills with a \$50 bill to obtain $\$10(n+1)$. This shows that $P(n+1)$ is true.

33. Let $P(n)$ be the statement that $\mathbf{AB}^n = \mathbf{B}^n\mathbf{A}$. $P(1)$ is true since $\mathbf{AB} = \mathbf{BA}$. Now assume that $P(n)$ is true. Then $\mathbf{AB}^{n+1} = \mathbf{AB}^n\mathbf{B} = \mathbf{B}^n\mathbf{AB} = \mathbf{B}^n\mathbf{BA} = \mathbf{B}^{n+1}\mathbf{A}$. It follows that $P(n+1)$ is true.

35. Let $P(n)$ be "$(A_1 \cup A_2 \cup \cdots \cup A_n) \cap B = (A_1 \cap B) \cup (A_2 \cap B) \cup \cdots \cup (A_n \cap B)$." Basis step: $P(1)$ is trivially true. Inductive step: Assume that $P(n)$ is true. Then $(A_1 \cup A_2 \cup \cdots \cup A_n \cup A_{n+1}) \cap B = [(A_1 \cup A_2 \cup \cdots \cup A_n) \cup A_{n+1}] \cap B = [(A_1 \cup A_2 \cup \cdots \cup A_n) \cap B] \cup (A_{n+1} \cap B) = [(A_1 \cap B) \cup (A_2 \cap B) \cup \cdots \cup (A_n \cap B)] \cup (A_{n+1} \cap B) = (A_1 \cap B) \cup (A_2 \cap B) \cup \cdots \cup (A_n \cap B) \cup (A_{n+1} \cap B)$.

37. Let $P(n)$ be

$$\text{“}\overline{\bigcup_{k=1}^{n} A_k} = \bigcap_{k=1}^{n} \overline{A_k}.\text{”}$$

Basis step: $P(1)$ is trivially true. Inductive step: Assume that $P(n)$ is true. Then

$$\overline{\bigcup_{k=1}^{n+1} A_k} = \overline{\left(\bigcup_{k=1}^{n} A_k\right) \cup A_{n+1}} = \overline{\left(\bigcup_{k=1}^{n} A_k\right)} \cap \overline{A_{n+1}} = \left(\bigcap_{k=1}^{n} \overline{A_k}\right) \cap \overline{A_{n+1}} = \bigcap_{k=1}^{n+1} \overline{A_k}.$$

39. Let $P(n)$ be "$[(p_1 \to p_2) \wedge (p_2 \to p_3) \wedge \cdots \wedge (p_{n-1} \to p_n)] \to [(p_1 \wedge \cdots \wedge p_{n-1}) \to p_n]$." Basis step: $P(2)$ is true since $(p_1 \to p_2) \to (p_1 \to p_2)$ is a tautology. Inductive step: Assume $P(n)$ is true. To show $[(p_1 \to p_2) \wedge \cdots \wedge (p_{n-1} \to p_n) \wedge (p_n \to p_{n+1})] \to [(p_1 \wedge \cdots \wedge p_{n-1} \wedge p_n) \to p_{n+1}]$ is a tautology, assume the hypothesis of this implication is true. Since both the hypothesis and $P(n)$ are true, it follows that $(p_1 \wedge \cdots \wedge p_{n-1}) \to p_n$ is true. Since this is true, and since $p_n \to p_{n+1}$ is true (it is part of the assumption) it follows by hypothetical syllogism that $(p_1 \wedge \cdots \wedge p_{n-1}) \to p_{n+1}$ is true. The weaker statement $(p_1 \wedge \cdots \wedge p_{n-1} \wedge p_n) \to p_{n+1}$ follows from this.

41. The two sets do not overlap if $n+1 = 2$. In fact, the implication $P(1) \to P(2)$ is false.

43. Assume that the well-ordering property holds. Suppose that $P(1)$ is true and that the implication $(P(1) \wedge P(2) \wedge \cdots \wedge P(n)) \to P(n+1)$ is true for every positive integer $n \geq 1$. Let S be the set of integers n for which $P(n)$ is false. We will show $S = \emptyset$. Assume that $S \neq \emptyset$. Then by the well-ordering property there is a least integer m in S. We know that m cannot be 1 because $P(1)$ is true. Since $n = m$ is the least integer such that $P(n)$ is false, $P(1), P(2), \ldots, P(m-1)$ are true, and $m - 1 \geq 1$. Since $(P(1) \wedge P(2) \wedge \cdots \wedge P(m-1)) \to P(m)$ is true, it follows that $P(m)$ must also be true, which is a contradiction. Hence $S = \emptyset$.

45. Let $P(n)$ be "$H_{2^n} \leq 1 + n$." Basis step: $P(0)$ is true since $H_{2^0} = H_1 = 1 \leq 1 + 0$. Inductive step: Assume that $H_{2^n} \leq 1 + n$. Then

$$H_{2^{n+1}} = H_{2^n} + \sum_{j=2^n+1}^{2^{n+1}} 1/j \leq 1 + n + 2^n(1/2^{n+1}) < 1 + n + 1 = 1 + (n+1).$$

47. Let $P(n)$ be "$1/\sqrt{1} + 1/\sqrt{2} + 1/\sqrt{3} + \cdots + 1/\sqrt{n} > 2(\sqrt{n+1} - 1)$." Basis step: $P(1)$ is true since $1 > 2(\sqrt{2} - 1)$. Inductive step: Assume that $P(n)$ is true. Then $1 + 1/\sqrt{2} + \cdots + 1/\sqrt{n} + 1/\sqrt{n+1} > 2(\sqrt{n+1} - 1) + 1/\sqrt{n+1}$. If we show that $2(\sqrt{n+1} - 1) + 1/\sqrt{n+1} > 2(\sqrt{n+2} - 1)$, it follows that $P(n+1)$ is true. This inequality is equivalent to $2(\sqrt{n+2} - \sqrt{n+1}) < 1/\sqrt{n+1}$, which is equivalent to $2(\sqrt{n+2} - \sqrt{n+1})(\sqrt{n+2} + \sqrt{n+1}) < \sqrt{n+1}/\sqrt{n+1} + \sqrt{n+2}/\sqrt{n+1}$. This is equivalent to $2 < 1 + \sqrt{n+2}/\sqrt{n+1}$, which is clearly true.

49. We will first prove the result when n is a power of 2, that is, if $n = 2^k, k = 1, 2, \cdots$. Let $P(k)$ be the statement $A \geq G$ where A and G are the arithmetic and geometric means of a set of $n = 2^k$ positive real numbers. Basis step: $k = 1$ and $n = 2^1 = 2$. Note that $(\sqrt{a_1} - \sqrt{a_2})^2 \geq 0$. Expanding this shows that $a_1 - 2\sqrt{a_1a_2} + a_2 \geq 0$, that is, $(a_1 + a_2)/2 \geq (a_1a_2)^{1/2}$. Inductive step: Assume that $P(k)$ is true, with $n = 2^k$. We will show that $P(k+1)$ is true. We have $2^{k+1} = 2n$. Now $(a_1 + a_2 + \cdots + a_{2n})/(2n) = ((a_1 + a_2 + \cdots + a_n)/n + (a_{n+1} + a_{n+2} + \cdots + a_{2n})/n)/2$ and similarly $(a_1a_2 \cdots a_{2n})^{1/(2n)} = [(a_1 \cdots a_n)^{1/n}(a_{n+1} \cdots a_{2n})^{1/n}]^{1/2}$. To simplify the notation, let $A(x, y, \ldots)$ and $G(x, y, \ldots)$ denote the arithmetic mean and geometric mean of $x, y, \ldots$, respectively. Also, if $x \leq x'$, $y \leq y'$, and so on, then $A(x, y, \ldots) \leq A(x', y', \ldots)$ and $G(x, y, \ldots) \leq G(x', y', \ldots)$. Hence $A(a_1, \ldots, a_{2n}) = A(A(a_1, \ldots, a_n), \quad A(a_{n+1}, \ldots, a_{2n})) \geq$

$A(G(a_1, \ldots, a_n),\ G(a_{n+1}, \ldots, a_{2n})) \geq G(G(a_1, \ldots, a_n), G(a_{n+1}, \ldots, a_{2n})) = G(a_1, \ldots, a_{2n})$. This finishes the proof for powers of 2. Now if n is not a power of 2, let m be the next higher power of 2, and let $a_{n+1}, \ldots, a_m$ all equal $A(a_1, \ldots, a_n) = \overline{a}$. Then we have $((a_1 a_2 \cdots a_n)\overline{a}^{m-n})^{1/m} \leq A(a_1, \ldots, a_m)$, since m is a power of 2. Since $A(a_1, \ldots, a_m) = \overline{a}$, it follows that $(a_1 \cdots a_n)^{1/m}\overline{a}^{1-n/m} \leq \overline{a}^{n/m}$. Raising both sides to the (m/n)th power gives $G(a_1, \ldots, a_n) \leq A(a_1, \ldots, a_n)$.

51. There is nothing to prove for the basis case when $n = 1$. Now assume the inductive hypothesis. Suppose that $p \mid a_1 a_2 \cdots a_n a_{n+1}$. Note that $gcd(p, a_1, a_2, \ldots, a_n) = 1$ or p. If it is 1, by Lemma 1 in Section 2.5, $p \mid a_{n+1}$. If it is p, then $p \mid a_1 a_2 \cdots a_n$, so that by the inductive hypothesis, $p \mid a_i$ for some $i \leq n$. This completes the proof.

53. Let $P(n)$ be the statement that if $x_1, x_2, \ldots, x_n$ are n distinct real numbers, then $n - 1$ multiplications are used to find the product of these numbers no matter how parentheses are inserted in the product. We will prove that $P(n)$ is true using the second principle of mathematical induction. The basis case $P(1)$ is true since $1 - 1 = 0$ multiplications are required to find the product of x_1, a product with only one factor. Suppose that $P(k)$ is true for $1 \leq k \leq n$. The last multiplication used to find the product of the $n + 1$ distinct real numbers $x_1, x_2, \ldots, x_n, x_{n+1}$ is a multiplication of the product of the first k of these numbers for some k and the product of the last $n+1-k$ of them. By the inductive hypothesis, $k - 1$ multiplications are used to find the product of k of the numbers, no matter how parentheses were inserted in the product of these numbers, and $n - k$ multiplications are used to find the product of the other $n + 1 - k$ of them, no matter how parentheses were inserted in the product of these numbers. Since one more multiplication is required to find the product of all $n + 1$ numbers, the total number of multiplications used equals $(k - 1) + (n - k) + 1 = n$. Hence $P(n + 1)$ is true. The proof is complete.

55.

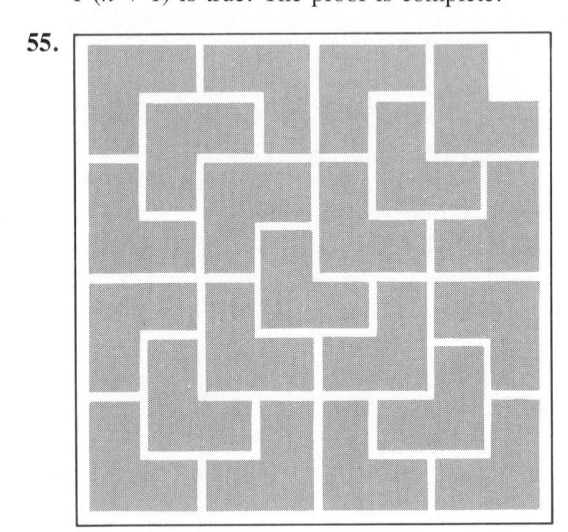

57. Assume that $a = dq + r = dq' + r'$ such that $0 \leq r < d$ and $0 \leq r' < d$. Then $d(q - q') = r' - r$. It follows that d divides $r' - r$. Since $-d < r' - r < d$, we have $r' - r = 0$. Hence $r' = r$. It follows that $q = q'$.

Section 3.3

1. a) $f(1) = 3, f(2) = 5, f(3) = 7, f(4) = 9$
b) $f(1) = 3, f(2) = 9, f(3) = 27, f(4) = 81$
c) $f(1) = 2, f(2) = 4, f(3) = 16, f(4) = 65{,}536$
d) $f(1) = 3, f(2) = 13, f(3) = 183, f(4) = 33{,}673$

3. There are many possible correct answers. We will supply relatively simple ones.
a) $a_{n+1} = a_n + 6$ for $n \geq 1$ and $a_1 = 6$
b) $a_{n+1} = a_n + 2$ for $n \geq 1$ and $a_1 = 3$
c) $a_{n+1} = 10a_n$ for $n \geq 1$ and $a_1 = 10$
d) $a_{n+1} = a_n$ for $n \geq 1$ and $a_1 = 5$

5. $F(0) = 0, F(n) = F(n - 1) + n$ for $n \geq 1$

7. $P_m(0) = 0, P_m(n + 1) = P_m(n) + m$

9. Let $P(n)$ be "$f_1 + f_3 + \cdots + f_{2n-1} = f_{2n}$." Basis step: $P(1)$ is true since $f_1 = 1 = f_2$. Inductive step: Assume that $P(n)$ is true. Then $f_1 + f_3 + \cdots + f_{2n-1} + f_{2n+1} = f_{2n} + f_{2n+1} = f_{2n+2} + f_{2(n+1)}$.

11. The number of divisions used by the Euclidean algorithm to find $gcd(f_{n+1}, f_n)$ is 0 for $n = 0$, 1 for $n = 1$, and $n - 1$ for $n \geq 2$. To prove this result for $n \geq 2$ we use mathematical induction. For $n = 2$, one division shows that $gcd(f_3, f_2) = gcd(2, 1) = gcd(1, 0) = 1$. Now assume that $n - 1$ divisions are used to find $gcd(f_{n+1}, f_n)$. To find $gcd(f_{n+2}, f_{n+1})$, first divide f_{n+2} by f_{n+1} to obtain $f_{n+2} = 1 \cdot f_{n+1} + f_n$. After one division we have $gcd(f_{n+2}, f_{n+1}) = gcd(f_{n+1}, f_n)$. By the inductive hypothesis it follows that exactly $n-1$ more divisions are required. This shows that n divisions are required to find $gcd(f_{n+2}, f_{n+1})$, finishing the inductive proof.

13. $|A| = -1$. Hence $|A^n| = (-1)^n$. It follows that $f_{n+1}f_{n-1} - f_n^2 = (-1)^n$.

15. a) Proof by induction. Basis step: For $n = 1$, $\max(-a_1) = -a_1 = -\min(a_1)$. For $n = 2$, there are two cases. If $a_2 \geq a_1$ then $-a_1 \geq -a_2$, so $\max(-a_1, -a_2) = -a_1 = -\min(a_1, a_2)$. If $a_2 < a_1$, then $-a_1 < -a_2$, so $\max(-a_1, -a_2) = -a_2 = -\min(a_1, a_2)$. Inductive step: Assume true for n with $n \geq 2$. Then $\max(-a_1, -a_2, \ldots, -a_n, -a_{n+1}) = \max(\max(-a_1, \ldots, -a_n), -a_{n+1}) = \max(-\min(a_1, \ldots, a_n), -a_{n+1}) = -\min(\min(a_1, \ldots, a_n), a_{n+1}) = -\min(a_1, \ldots, a_{n+1})$.

b) Proof by mathematical induction. For $n = 1$, the result is the identity $a_1 + b_1 = a_1 + b_1$. For $n = 2$, first consider the case in which $a_1 + b_1 \geq a_2 + b_2$. Then $\max(a_1 + b_1, a_2 + b_2) = a_1 + b_1$. Also note that $a_1 \leq \max(a_1, a_2)$ and $b_1 \leq \max(b_1, b_2)$, so that $a_1 + b_1 \leq \max(a_1, a_2) + \max(b_1, b_2)$. Therefore $\max(a_1 + b_1, a_2 + b_2) = a_1 + b_1 \leq \max(a_1, a_2) + \max(b_1, b_2)$. The case with $a_1 + b_1 < a_2 + b_2$ is similar. For the inductive step, assume that the result is true for n. Then $\max(a_1 + b_1, a_2 + b_2, \ldots, a_n + b_n, a_{n+1} + b_{n+1}) = \max(\max(a_1 + b_1, a_2 + b_2, \ldots, a_n + b_n), a_{n+1} + b_{n+1}) \leq \max(\max(a_1, a_2, \ldots, a_n) + \max(b_1, b_2, \ldots, b_n), a_{n+1} + b_{n+1}) \leq \max(\max(a_1, a_2, \ldots, a_n), a_{n+1}) + \max(\max(b_1, b_2, \ldots, b_n), b_{n+1}) = \max(a_1, a_2, \ldots, a_n, a_{n+1}) + \max(b_1, b_2, \ldots, b_n, b_{n+1})$.

c) Same as (b), but replace every occurrence of "max" by "min" and invert each inequality.

17. $5 \in S$, and $x + y \in S$ if $x, y \in S$.

19. a) $0 \in S$, and if $x \in S$, then $x + 2 \in S$ and $x - 2 \in S$.

b) $2 \in S$, and if $x \in S$, then $x + 3 \in S$.

c) $1 \in S, 2 \in S, 3 \in S, 4 \in S$, and if $x \in S$, then $x + 5 \in S$.

21. If x is a set or a variable representing a set, then x is a well-formed formula. If x and y are well-formed formulae, then so are $\overline{x}$, $(x \cup y)(x \cap y)$, and $(x - y)$.

23. $\lambda^R = \lambda$ and $(ux)^R = xu^R$ for $x \in \Sigma$, $u \in \Sigma^*$.

25. $w^0 = \lambda$ and $w^{n+1} = ww^n$.

27. When the string consists of n 0s followed by n 1s for some nonnegative integer n

29. Let $P(i)$ be "$l(w^i) = i \cdot l(w)$." $P(0)$ is true since $l(w^0) = 0 = 0 \cdot l(w)$. Assume $P(i)$ is true. Then $l(w^{i+1}) = l(ww^i) = l(w) + l(w^i) = l(w) + i \cdot l(w) = (i+1) \cdot l(w)$.

31. a) $P_{m,m} = P_m$ since a number exceeding m cannot be used in a partition of m.

b) Since there is only one way to partition 1, namely, $1 = 1$, it follows that $P_{1,n} = 1$. Since there is only one way to partition m into 1s, $P_{m,1} = 1$. When $n > m$ it follows that $P_{m,n} = P_{m,m}$ since a number exceeding m cannot be used. $P_{m,m} = 1 + P_{m,m-1}$ since one extra partition, namely, $m = m$, arises when m is allowed in the partition. $P_{m,n} = P_{m,n-1} + P_{m-n,n}$ if $m > n$ since a partition of m into integers not exceeding n either does not use any ns and hence is counted in $P_{m,n-1}$ or else uses an n and a partition of $m - n$, and hence is counted in $P_{m-n,n}$.

c) $P_5 = 7, P_6 = 11$

33. Let $P(n)$ be "$A(n, 2) = 4$". Basis step: $P(1)$ is true since $A(1, 2) = A(0, A(1, 1)) = A(0, 2) = 2 \cdot 2 = 4$. Inductive step: Assume that $P(n)$ is true, that is, $A(n, 2) = 4$. Then $A(n+1, 2) = A(n, A(n+1, 1)) = A(n, 2) = 4$.

35. a) 16

b) 65,536

37. Use a double induction argument to prove the stronger statement: $A(m, k) > A(m, l)$ when $k > l$. Basis step: When $m = 0$ the statement is true since $k > l$ implies that $A(0, k) = 2k > 2l = A(0, l)$. Inductive step: Assume that $A(m, x) > A(m, y)$ for all nonnegative integers x and y with $x > y$. We will show that this implies that $A(m+1, k) > A(m+1, l)$ if $k > l$. Basis steps: When $l = 0$ and $k > 0$, $A(m+1, l) = 0$ and either $A(m+1, k) = 2$ or $A(m+1, k) = A(m, A(m+1, k-1))$. If $m = 0$, this is $2A(1, k-1) = 2^k$. If $m > 0$ this is greater than 0 by the inductive hypothesis. In all cases $A(m+1, k) > 0$, and in fact, $A(m+1, k) \geq 2$. If $l = 1$ and $k > 1$ then $A(m+1, l) = 2$ and $A(m+1, k) = A(m, A(m+1, k-1))$, with $A(m+1, k-1) \geq 2$. Hence by the inductive hypothesis, $A(m, A(m+1, k-1)) \geq A(m, 2) > A(m, 1) = 2$. Inductive step: Assume that $A(m+1, r) > A(m+1, s)$ for all $r > s$, $s = 0, 1, \ldots, l$. Then if $k+1 > l+1$ it follows that $A(m+1, k+1) = A(m, A(m+1, k)) > A(m, A(m+1, k)) = A(m+1, l+1)$.

39. From Exercise 38 it follows that $A(i, j) \geq A(i-1, j) \geq \cdots \geq A(0, j) = 2j \geq j$.

41. Let $P(n)$ be "$F(n)$ is well-defined." Then $P(0)$ is true since $F(0)$ is specified. Assume that $P(k)$ is true for all $k < n$. Then $F(n)$ is well-defined at n since $F(n)$ is given in terms of $F(0), F(1), \ldots, F(n-1)$. So $P(n)$ is true for all integers n.

Section 3.4

1. **procedure** *mult*(n: positive integer, x: integer)
if $n = 1$ **then** $mult(n, x) := x$
else $mult(n, x) := x + mult(n-1, x)$

3. **procedure** *sum of odds* (n: positive integer)
if $n = 1$ **then** *sum of odds* $(n) := 1$
else *sum of odds*$(n) :=$ *sum of odds* $(n-1) + 2n - 1$

5. **procedure** $smallest(a_1, \ldots, a_n$: integers)
if $n = 1$ **then** $smallest(a_1, \ldots, a_n) = a_1$
else $smallest(a_1, \ldots, a_n) :=$
$\min(smallest(a_1, \ldots, a_{n-1}), a_n)$

7. **procedure** *modfactorial*(n, m: positive integers)
if $n = 1$ **then** $modfactorial(n, m) := 1$
else $modfactorial(n, m) :=$
$(n{*}modfactorial(n-1, m))$ **mod** m

9. **procedure** *gcd*(a,b: nonnegative integers)
{$a < b$ assumed to hold}
if $a = 0$ **then** $gcd(a, b) := b$
else if $a = b - a$ **then** $gcd(a, b) := a$
else if $a < b - a$ **then** $gcd(a, b) := gcd(a, b - a)$
else $gcd(a, b) := gcd(b - a, a)$

11. n multiplications versus 2^n

13. $O(\log n)$ versus n

15. **procedure** a(n: nonnegative integer)
if $n = 0$ **then** $a(n) := 1$
else if $n = 1$ **then** $a(n) := 2$
else $a(n) := a(n-1) * a(n-2)$

17. Iterative

19. **procedure** *reverse* (w: bit string)
$n :=$ length(w)
if $n \le 1$ **then** $reverse(w) := w$
else $reverse(w) :=$
$substr(w, n, n)reverse(substr(w, 1, n-1))$
{$substr(w, a, b)$ is the substring of w consisting of the symbols in the ath through bth positions}

21. **procedure** A(m,n: nonnegative integers)
if $m = 0$ **then** $A(m, n) := 2n$
else if $n = 0$ **then** $A(m, n) := 0$
else if $n = 1$ **then** $A(m, n) := 2$
else $A(m, n) := A(m-1, A(m, n-1))$

Section 3.5

1. Suppose that $x = 0$. The program segment first assigns the value 1 to y and then assigns the value $x + y = 0 + 1 = 1$ to z.

3. Suppose that $y = 3$. The program segment assigns the value 2 to x and then assigns the value $x + y = 2 + 3 = 5$ to z. Since $y = 3 > 0$ it then assigns the value $z + 1 = 5 + 1 = 6$ to z.

5.
$$(p \wedge condition1)\{S_1\}q$$
$$(p \wedge \neg condition1 \wedge condition2)\{S_2\}q$$
$$\vdots$$
$$(p \wedge \neg condition1 \wedge \neg condition2 \cdots \wedge \neg condition(n-1)\{S_n\}q$$

$\therefore\ p$\{**if** *condition*1 **then** S_1;
else if *condition*2 **then** $S_2; \ldots;$ **else** $S_n\}q$

7. We will show that p : "$power = x^{i-1}$ and $i \le n+1$" is a loop invariant. Note that p is true initially, since before the loop starts, $i = 1$ and $power = 1 = x^0 = x^{1-1}$. Next, we must show that if p is true and $i \le n$ after an execution of the loop, then p remains true after one more execution. The loop increments i by 1. Hence since $i \le n$ before this pass, $i \le n + 1$ after this pass. Also the loop assigns $power \cdot x$ to $power$. By the inductive hypothesis we see that $power$ is assigned the value $x^{i-1} \cdot x = x^i$. Hence p remains true. Furthermore, the loop terminates after n traversals of the loop with $i = n + 1$ since i is assigned the value 1 prior to entering the loop, is incremented by 1 on each pass, and the loop terminates when $i > n$. Consequently, at termination $power = x^n$, as desired.

9. Suppose that p is "m and n are integers." Then if the condition $n < 0$ is true, $a = -n = |n|$ after S_1 is executed. If the condition $n < 0$ is false, then $a = n = |n|$ after S_1 is executed. Hence $p\{S_1\}q$ is true where q is $p \wedge (a = |n|)$. Since S_2 assigns the value 0 to both k and x it is clear that $q\{S_2\}r$ is true where r is $q \wedge (k = 0) \wedge (x = 0)$. Suppose that r is true. Let $P(k)$ be "$x = mk$ and $k \le a$." We can show that $P(k)$ is a loop invariant for the loop in S_3. $P(0)$ is true, since before the loop is entered $x = 0 = m \cdot 0$ and $0 \le a$. Now assume $P(k)$ is true and $k < a$. Then $P(k+1)$ is true since x is assigned the value $x + m = mk + m = m(k+1)$. The loop terminates when $k = a$, and at that point $x = ma$. Hence $r\{S_3\}s$ is true where s is "$a = |n|$ and $x = ma$." Now assume that s is true. Then if $n < 0$ it follows that $a = -n$, so that $x = -mn$. In this case S_4 assigns $-x = mn$ to *product*. If $n > 0$ then $x = ma = mn$, so that S_4 assigns mn to *product*. Hence $s\{S_4\}t$ is true.

11. Suppose that the initial assertion p is true. Then since $p\{S\}q_0$ is true, q_0 is true after the segment S is executed. Since $q_0 \to q_1$ is true, it also follows that q_1 is true after S is executed. Hence $p\{S\}q_1$ is true.

13. We will use the proposition p, "$gcd(a, b) = gcd(x, y)$ and $y \ge 0$," as the loop invariant. Note that p is true before the loop is entered, since at that point $x = a$, $y = b$, and y is a positive integer, using the initial assertion. Now assume that p is true and $y > 0$; then the loop will be executed again. Inside the loop, x and y are replaced by y and x **mod** y, respectively. By Lemma 1 of Section 2.4, $gcd(x, y) = gcd(y, x \textbf{ mod } y)$. Therefore, after execution of the loop, the value of $gcd(x, y)$ is the same as it was before. Moreover, since y is the remainder, it is at least 0. Hence p remains true, so that it is a loop invariant. Furthermore, if the loop terminates, then $y = 0$. In this case, we have $gcd(x, y) = x$, the final assertion. Therefore the program, which gives x as its output, has correctly computed $gcd(a, b)$. Finally, we can prove the loop must terminate, since each iteration causes the value of y to decrease by at least 1. Therefore, the loop can be iterated at most b times.

Supplementary Exercises

1. Let $a = 2n + 1$ and $b = 2m + 1$. Then $ab = (2n+1)(2m+1) = 2(2nm + m + n) + 1$, which is odd.

3. False. $\sqrt{2} + (-\sqrt{2}) = 0$ is a counterexample.

5. This is an example of the fallacy of affirming the conclusion.

7. Proof by cases. Case 1: $x \ge 0$ and $y \ge 0$. Then $|xy| = xy = |x||y|$. Case 2: $x \ge 0$ and $y < 0$. Then $|xy| = -xy = x(-y) = |x||y|$. Case 3: $x < 0$ and $y \ge 0$. Then $|xy| = -xy = (-x)y = |x||y|$. Case 4: $x < 0$ and $y < 0$. Then $|xy| = xy = (-x)(-y) = |x||y|$.

9. Assume that the x_js are distinct. Let $P(x)$ be as in the hint. Then $P(x)$ is a polynomial (of degree $n-1$, in fact); and if $x = x_m$, then $\prod_{i \ne j}(x - x_j)/(x_i - x_j) = 0$ unless $i = m$. Thus $P(x_m) = \prod_{j \ne m} y_m(x_m - x_j)/(x_m - x_j) = 1 \cdot y_m = y_m$.

11. Let $P(n)$ be "$1 \cdot 1 + 2 \cdot 2 + \cdots + n \cdot 2^{n-1} = (n-1)2^n + 1$." Basis step: $P(1)$ is true since $1 \cdot 1 = 1 = (1-1)2^1 + 1$. Inductive step: Assume that $P(n)$ is true. Then $1 \cdot 1 + 2 \cdot 2 + \cdots + n \cdot 2^{n-1} + (n+1) \cdot 2^n = (n-1)2^n + 1 + (n+1)2^n = 2n \cdot 2^n + 1 = ((n+1) - 1)2^{n+1} + 1$.

13. Let $P(n)$ be "$1/1 \cdot 4 + \cdots + 1/[(3n-2)(3n+1)] = n/(3n+1)$." Basis step: $P(1)$ is true since $1/1 \cdot 4 = 1/4$. Inductive step: Assume $P(n)$ is true. Then $1/1 \cdot 4 + \cdots + 1/[(3n-2)(3n+1)] + 1/[(3n+1)(3n+4)] = n/(3n+1) + 1/[(3n+1)(3n+4)] = [n(3n+4)+1]/[(3n+1)(3n+4)] = [(3n+1)(n+1)]/[(3n+1)(3n+4)] = (n+1)/(3n+4)$.

15. Let $P(n)$ be "$2^n > n^3$." Basis step: $P(10)$ is true since $1024 > 1000$. Inductive step: Assume $P(n)$ is true. Then $(n+1)^3 = n^3 + 3n^2 + 3n + 1 \le n^3 + 9n^2 \le n^3 + n^3 = 2n^3 < 2 \cdot 2^n = 2^{n+1}$.

17. Let $P(n)$ be "$a - b$ is a factor of $a^n - b^n$." Basis step: $P(1)$ is trivially true. Assume $P(n)$ is true. Then $a^{n+1} - b^{n+1} = a^{n+1} - ab^n + ab^n - b^{n+1} = a(a^n - b^n) + b^n(a - b)$. Then since $a - b$ is a factor of $a^n - b^n$ and $a - b$ is a factor of $a - b$, it follows that $a - b$ is a factor of $a^{n+1} - b^{n+1}$.

19. Let $P(n)$ be "$a + (a+d) + \cdots + (a+nd) = (n+1)(2a+nd)/2$." Basis step: $P(1)$ is true since $a + (a+d) = 2a + d = 2(2a+d)/2$. Inductive step: Assume that $P(n)$ is true. Then $a + (a+d) + \cdots + (a+nd) + (a+(n+1)d) = (n+1)(2a+nd)/2 + a + (n+1)d = \frac{1}{2}[2an + 2a + n^2d + nd + 2a + 2nd + 2d] = \frac{1}{2}[2an + 4a + n^2d + 3nd + 2d] = \frac{1}{2}(n+2)(2a + (n+1)d)$.

21. We will use the second principle of mathematical induction to show that f_n is even if $n \equiv 0 \pmod 3$ and is odd otherwise. The basis step follows since $f_0 = 0$ is even and $f_1 = 1$ is odd. Now assume that if $k \le n$, then f_k is even if $k \equiv 0 \pmod 3$ and is odd otherwise. Now suppose $n + 1 \equiv 0 \pmod 3$. Then $f_{n+1} = f_n + f_{n-1}$ is even since f_n and f_{n-1} are both odd. If $n + 1 \equiv 1 \pmod 3$, then $f_{n+1} = f_n + f_{n-1}$ is odd since f_n is even and f_{n-1} is odd. Finally, if $n + 1 \equiv 2 \pmod 3$, then $f_{n+1} = f_n + f_{n-1}$ is odd, since f_n is odd and f_{n-1} is even. This completes the inductive proof.

23. Let $P(n)$ be the statement that $f_k f_n + f_{k+1} f_{n+1} = f_{n+k+1}$ for every nonnegative integer k. The basis steps consist of showing that $P(0)$ and $P(1)$ both hold. $P(0)$ is true since $f_k f_0 + f_{k+1} f_1 = f_{k+1} \cdot 0 + f_{k+1} \cdot 1 = f_1$. Since $f_k f_1 + f_{k+1} f_2 = f_k + f_{k+1} = f_{k+2}$, it follows that $P(1)$ is true. Now assume that $P(n)$ holds. Then, by the inductive hypothesis and the recursive definition of the Fibonacci numbers, it follows that $f_{k+1} f_{n+1} + f_{k+2} f_{n+2} = f_k(f_{n-1} + f_n) + f_{k+1}(f_n + f_{n+1}) = (f_k f_{n-1} + f_{k+1} f_n) + (f_k f_n + f_{k+1} f_{n+1}) = f_{n-1+k+1} + f_{n+k+1} = f_{n+k+2}$. This shows that $P(n+1)$ is true and completes the proof.

25. Let $P(n)$ be the statement $l_0^2 + l_1^2 + \cdots + l_n^2 = l_n l_{n+1} + 2$. The basis cases $P(0)$ and $P(1)$ both hold since $l_0^2 = 2^2 = 2 \cdot 1 + 2 = l_0 l_1 + 2$ and $l_0^2 + l_1^2 = 2^2 + 1^2 = 1 \cdot 3 + 2 = l_1 l_3 + 2$. Now assume that $P(n)$ holds. Then by the inductive hypothesis $l_0^2 + l_1^2 + \cdots + l_n^2 + l_{n+1}^2 = l_n l_{n+1} + 2 + l_{n+1}^2 = l_{n+1}(l_n + l_{n+1}) + 2 = l_{n+1} l_{n+2} + 2$. This shows that $P(n+1)$ holds and completes the proof.

27. Let $P(n)$ be the statement that the identity holds for the integer n. The basis case $P(1)$ is obviously true. Assume that $P(n)$ is true. Then $\cos(n+1)x + i\sin(n+1)x = \cos(nx + x) + i\sin(nx + x) = \cos nx \cos x - \sin nx \sin x + i(\sin nx \cos x + \cos nx \sin x) = \cos x(\cos nx + i \sin nx)(\cos x + i \sin x) = (\cos x + i \sin x)^n(\cos x + i \sin x) = (\cos x + i \sin x)^{n+1}$. It follows that $P(n+1)$ is true, completing the proof.

29. **a)** 92 **b)** 91 **c)** 91 **d)** 91 **e)** 91 **f)** 91

31. The basis step is incorrect since $n \neq 1$ for the sum shown.

33. Let $P(n)$ be "the plane is divided into $n^2 - n + 2$ regions by n circles if every two of these circles have two common points but no three have a common point." Basis step: $P(1)$ is true since a circle divides the plane into $2 = 1^2 - 1 + 2$ regions. Assume that $P(n)$ is true, that is, n circles with the specified properties divide the plane into $n^2 - n + 2$ regions. Suppose that an $(n+1)$th circle is added. This circle intersects each of the other n circles in two points, so that these points of intersection form $2n$ new arcs, each of which splits an old region. Hence there are $2n$ regions split, which shows that there are $2n$ more regions than there were previously. Hence $n+1$ circles satisfying the specified properties divide the plane into $n^2 - n + 2 + 2n = (n^2 + 2n + 1) - (n+1) + 2 = (n+1)^2 - (n+1) + 2$ regions.

35. Suppose $\sqrt{2}$ were rational. Then $\sqrt{2} = a/b$ where a and b are positive integers. It follows that the set $S = \{n\sqrt{2} \mid n \in \mathbf{N}\} \cap \mathbf{N}$ is a nonempty set of positive integers, since $b\sqrt{2} = a$ belongs to S. Let t be the least element of S, which exists by the well-ordering property. Then $t = s\sqrt{2}$ for some integer s. We have $t - s = s\sqrt{2} - s = s(\sqrt{2} - 1)$, so that $t - s$ is a positive integer since $\sqrt{2} > 1$. Hence $t - s$ belongs to S. This is a contradiction since $t - s = s\sqrt{2} - s < s$. Hence $\sqrt{2}$ is irrational.

37. Suppose that the well-ordering property were false. Let S be a nonempty set of nonnegative integers that has no least element. Let $P(n)$ be the statement "$i \notin S, i = 0, 1, \ldots, n$." $P(0)$ is true because if $0 \in S$ then S has a least element, namely, 0. Now suppose that $P(n)$ is true. Thus $0 \notin S, 1 \notin S, \ldots, n \notin S$. Clearly $n+1$ cannot be in S, for it were, it would be its least element. Thus $P(n+1)$ is true. So by the principle of mathematical induction, $n \notin S$ for all nonnegative integers n. Thus $S = \emptyset$, a contradiction.

39. **a)** Let $d = gcd(a_1, a_2, \ldots, a_n)$. Then d is a divisor of each a_i and so must be a divisor of $gcd(a_{n-1}, a_n)$. Hence d is a common divisor of $a_1, a_2, \ldots, a_{n-2}$, and $gcd(a_{n-1}, a_n)$. To show it is the greatest common divisor of these numbers, suppose that c is a common divisor of them. Then c is a divisor of a_i for $i = 1, 2, \ldots, n-2$ and a divisor of $gcd(a_{n-1}, a_n)$, so that it is a divisor of a_{n-1} and a_n. Hence c is a common divisor of $a_1, a_2, \ldots, a_{n-1}, a_n$. Hence it is a divisor of d, the greatest common divisor of $a_1, a_2, \ldots, a_n$. It follows that d is the greatest common divisor, as claimed.
b) If $n = 2$, apply the Euclidean algorithm. Otherwise, apply the Euclidean algorithm to a_{n-1} and a_n, obtaining $d = gcd(a_{n-1}, a_n)$, and then apply the algorithm recursively to $a_1, a_2, \ldots, a_{n-2}, d$.

41. $f(n) = n^2$. Let $P(n)$ be "$f(n) = n^2$." Basis step: $P(1)$ is true since $f(1) = 1 = 1^2$, which follows from the definition of f. Inductive step: Assume $f(n) = n^2$. Then $f(n+1) = f((n+1)-1) + 2(n+1) - 1 = f(n) + 2n + 1 = n^2 + 2n + 1 = (n+1)^2$.

43. **a)** λ, 0, 1, 00, 01, 11, 000, 001, 011, 111, 0000, 0001, 0011, 0111, 1111, 00000, 00001, 00011, 00111, 01111, 11111
b) $S = \{\alpha\beta \mid \alpha$ is a string of m 0s and β is a string of n 1s, $m \geq 0, n \geq 0\}$

45. λ, ( ), (( )), ( )( )

47. **a)** 0 **b)** -2 **c)** 2 **d)** 0

49. **procedure** *generate*(n: nonnegative integer)
if n is odd **then**
begin
$S := S(n-1); T := T(n-1)$
end
else if $n = 0$ **then**
begin
$S := \emptyset; T := \{\lambda\}$
end
else
begin
$T_1 := T(n-2); S_1 := S(n-2)$
$T := T_1 \cup \{(x) \mid x \in T_1 \cup S_1 \text{ and } l(x) = n-2\}$
$S := S_1 \cup \{xy \mid x \in T_1 \text{ and } y \in T_1 \cup S_1$ and $l(xy) = n\}$
end $\{T \cup S$ is the set of balanced strings of length at most $n\}$

51. If $x \le y$ initially, $x := y$ is not executed, so $x \le y$ is a true final assertion. If $x > y$ initially, then $x := y$ is executed, so $x \le y$ is again a true final assertion.

CHAPTER 4

Section 4.1

1. a) 5850 b) 343
3. a) 4^{10} b) 5^{10}
5. 42
7. 26^3
9. 676
11. 2^8
13. $n + 1$ (counting the empty string)
15. 475,255 (counting the empty string)
17. a) 128 b) 450 c) 9 d) 675 e) 450 f) 450 g) 225 h) 75
19. 3^{50}
21. 52,457,600
23. 20,077,200
25. a) 0 b) 120 c) 720 d) 2520
27. a) 2 if $n = 1$, 2 if $n = 2$, 0 if $n \ge 3$
b) 2^{n-2} for $n > 1$; 1 if $n = 1$
c) $2(n-1)$
29. If n is even $2^{n/2}$; if n is odd $2^{(n+1)/2}$
31. a) 240 b) 480 c) 360
33. 352
35. 147
37. 33
39. 7,104,000,000,000
41. 18
43. 17
45. Let $P(m)$ be the sum rule for m tasks. For the basis case take $m = 2$. This is just the sum rule for two tasks. Now assume that $P(m)$ is true. Consider $m+1$ tasks, $T_1, T_2, \ldots, T_m, T_{m+1}$, which can be done in $n_1, n_2, \ldots, n_m, n_{m+1}$ ways respectively, such that no two of these tasks can be done at the same time. To do one of these tasks, we can either do one of the first m of these, or do task T_{m+1}. By the sum rule for two tasks, the number of ways to do this is the sum of the number of ways to do one of the first m tasks, plus n_{m+1}. By the inductive hypothesis this is $n_1 + n_2 + \cdots + n_m + n_{m+1}$, as desired.

Section 4.2

1. Since there are six classes, but only five weekdays, the pigeonhole principle shows that at least two classes must be held on the same day.
3. 3
5. Let $a, a+1, \ldots, a+n-1$ be the integers in the sequence. The integers $(a+i)$ **mod** n, $i = 0, 1, 2, \ldots, n-1$, are distinct, since $0 < (a+j)-(a+k) < n$ whenever $0 \le k < j \le n-1$. Since there are n possible values for $(a+i)$ **mod** n, and there are n different integers in the set, each of these values is taken on exactly once. It follows that there is exactly one integer in the sequence that is divisible by n.
7. 4951
9. The midpoint of the segment joining the points (a, b, c) and (d, e, f) is $((a+d)/2, (b+e)/2, (c+f)/2)$. It has integer coefficients if and only if a and d have the same parity, b and e have the same parity, and c and f have the same parity. Since there are eight possible triples of parity (such as (*even*, *odd*, *even*)), by the pigeonhole principle at least two of the

nine points have the same triple of parities. The midpoint of the segment joining two such points has integer coefficients.

11. **a)** Group the first eight positive integers into four subsets of two integers each so that the integers of each subset add up to 9: $\{1, 8\}$, $\{2, 7\}$, $\{3, 6\}$, and $\{4, 5\}$. If five integers are selected from the first eight positive integers, by the pigeonhole principle at least two of them come from the same subset. Two such integers have a sum of 9, as desired.

b) No. Take $\{1, 2, 3, 4\}$, for example.

13. 21,251

15. Let d_j be $jx - N(jx)$ where $N(jx)$ is the integer closest to jx for $1 \leq j \leq n$. Each d_j is an irrational number between $-1/2$ and $1/2$. We will assume that n is even; the case where n is odd is messier. Consider the n intervals $\{x \mid j/n < x < (j+1)/n\}$, $\{x \mid -(j+1)/n < x < -j/n\}$ for $j = 0, 1, \ldots, (n/2)-1$. If d_j belongs to the interval $\{x \mid 0 < x < 1/n\}$ or to the interval $\{x \mid -1/n < x < 0\}$ for some j, we are done. If not, since there are $n - 2$ intervals and n numbers d_j, the pigeonhole principle tells us that there is an interval $\{x \mid (k-1)/n < x < k/n\}$ containing d_r and d_s with $r < s$. The proof can be finished by showing that $(s - r)x$ is within $1/n$ of its nearest integer.

17. 4, 3, 2, 1, 8, 7, 6, 5, 12, 11, 10, 9, 16, 15, 14, 13

19.
```
procedure long(a_1, ..., a_n: positive integers)
{first find longest increasing subsequence}
max := 0; set := 00...00 {n bits}
for i := 1 to 2^n
begin
  last := 0; count := 0, OK := true
  for j := 1 to n
  begin
    if set(j) = 1 then
    begin
      if a_j > last then last := a_j
      count := count + 1
    end
    else OK := false
  end
  if count > max then
  begin
    max := count
    best := set
  end
  set := set + 1 (binary addition)
end {max is length and best indicates the sequence}
{repeat for decreasing subsequence with only changes
    being a_j < last instead of a_j > last and
    last := ∞ instead of last := 0}
```

21. By symmetry we need prove only the first statement. Let A be one of the people. Either A has at least four friends, or A has at least six enemies among the other nine people (since $3 + 5 < 9$). Suppose, in the first case, that B, C, D, and E are all A's friends. If any two of these are friends with each other, then we have found three mutual friends. Otherwise $\{B, C, D, E\}$ is a set of four mutual enemies. In the second case, let $\{B, C, D, E, F, G\}$ be a set of enemies of A. By Example 11, among B, C, D, E, F, and G there are either three mutual friends or three mutual enemies, who form, with A, a set of four mutual enemies.

23. There are 6,432,816 possibilities for the three initials and a birthday. So, by the generalized pigeonhole principle, there are at least $\lceil 25{,}000{,}000/6{,}432{,}816 \rceil = 4$ people who share the same initials and birthday.

25. 18

27. Since there are six computers, the number of other computers a computer is connected to is an integer between 0 and 5, inclusive. However, 0 and 5 cannot both occur. To see this, note that if some computer is connected to no others, then no computer is connected to all five others, and if some computer is connected to all five others, then no computer is connected to no others. Hence, by the pigeonhole principle, since there are at most five possibilities for the number of computers a computer is connected to, there are at least two computers in the set of six connected to the same number of others.

29. Let a_i be the number of matches completed by hour i. Then $1 \leq a_1 < a_2 < \cdots < a_{75} \leq 125$. Also $25 \leq a_1 + 24 < a_2 + 24 < \cdots < a_{75} + 24 \leq 149$. There are 150 numbers $a_1, \ldots, a_{75}, a_1 + 24, \ldots, a_{75} + 24$. By the pigeonhole principle, at least two are equal. Since all the a_is are distinct and all the $(a_i + 24)$s are distinct, it follows that $a_i = a_j + 24$ for some $i > j$. Thus, in the period from the $(j + 1)$st to the ith hour, there are exactly 24 matches.

31. Use the generalized pigeonhole principle, placing the $|S|$ objects $f(s)$ for $s \in S$ in $|T|$ boxes, one for each element of T.

33. **a)** If there were fewer than 9 freshmen, fewer than 9 sophomores, and fewer than 9 juniors in the class, there would be no more than 8 of each of these three class standings, for a total of at most 24 students, contradicting the fact that there are 25 students in the class.
b) If there were fewer than 3 freshmen, fewer than 19 sophomores, and fewer than 5 juniors, then there would be at most 2 freshmen, at most 18 sophomores, and at most 4 juniors, for a total of at most 24 students. This contradicts the fact that there are 25 students in the class.

35. **a)** Assume that $i_k \leq n$ for all k. Then by the generalized pigeonhole principle, at least $\lceil (n^2+1)/n \rceil = n+1$ of the numbers $i_1, i_2, \ldots, i_{n^2+1}$ are equal.
b) If $a_{k_j} < a_{k_{j+1}}$, then the subsequence consisting of a_{k_j}, followed by the increasing subsequence of length $i_{k_{j+1}}$ starting at $a_{k_{j+1}}$ contradicts the fact that $i_{k_j} = i_{k_{j+1}}$. Hence $a_{k_j} > a_{k_{j+1}}$.
c) If there is no increasing subsequence of length greater than n, then parts (a) and (b) apply. Therefore we have $a_{k_{n+1}} > a_{k_n} > \cdots > a_{k_2} > a_{k_1}$, a decreasing sequence of length $n+1$.

Section 4.3

1. *abc*, *acb*, *bac*, *bca*, *cab*, *cba*
3. 720
5. **a)** 120 **b)** 720 **c)** 8 **d)** 6720 **e)** 40,320 **f)** 3,628,800
7. 15,120
9. 1320
11. $2(n!)^2$
13. 65,780
15. $2^{100} - 5051$
17. **a)** 94,109,400 **b)** 941,094 **c)** 3,764,376 **d)** 90,345,024 **e)** 114,072 **f)** 2328 **g)** 24 **h)** 79,727,040 **i)** 3,764,376 **j)** 109,440
19. **a)** 12,650 **b)** 303,600
21. 18,915
23. **a)** 122,523,030 **b)** 72,930,375 **c)** 223,149,655 **d)** 100,626,625
25. 54,600
27. 45
29. 912
31. 11,232,000

33.
$$C(n+1,k) = \frac{(n+1)!}{k!(n+1-k)!} = \frac{(n+1)}{k}\,\frac{n!}{(k-1)!(n-(k-1))!} = (n+1)C(n,k-1)/k$$

This identity together with $C(n,0)=1$ gives a recursive definition.

35. $x^5 + 5x^4y + 10x^3y^2 + 10x^2y^3 + 5xy^4 + y^5$
37. 101
39. $-2^{10}C(19,9) = -94{,}595{,}072$
41. $-2^{101}3^{99}C(200,99)$
43. $(-1)^{(200-k)/3}C(100,(200-k)/3)$ if $k \equiv 2 \pmod 3$ and $-100 \leq k \leq 200$; 0 otherwise
45. 1 9 36 84 126 126 84 36 9 1
47.
$$C(n,k-1) + C(n,k) = \frac{n!}{(k-1)!(n-k+1)!} + \frac{n!}{k!(n-k)!} = \frac{n!}{k!(n-k+1)!}\cdot[k+(n-k+1)] = \frac{(n+1)!}{k!(n+1-k)!} = C(n+1,k)$$

49. **a)** $C(n+r+1,r)$ counts the number of ways to choose a sequence of r 0s and $n+1$ 1s by choosing the positions of the 0s. Alternately, suppose that the $(j+1)$st term is the last term equal to 1, so that $n \leq j \leq n+r$. Once we have determined where the last 1 is, we decide where the 0s are to be placed in the j spaces before the last 1. There are n 1s and $j-n$ 0s in this range. By the sum rule it follows that there are $\sum_{j=n}^{n+r} C(j, j-n) = \sum_{k=0}^{r} C(n+k,k)$ ways to do this.
b) Let $P(r)$ be the statement to be proved. The basis step is the equation $C(n,0) = C(n+1,0)$, which is just $1 = 1$. Assume that $P(r)$ is true. Then $\sum_{k=0}^{r+1} C(n+k,k) = \sum_{k=0}^{r} C(n+k,k) + C(n+r+1,\ r+1) = C(n+r+1,r) + C(n+r+1,\ r+1) = C(n+r+2,\ r+1)$, using the inductive hypothesis and Pascal's identity.

51. We can choose the leader first in n different ways. We can then choose the rest of the committee in 2^{n-1} ways. Hence there are $n2^{n-1}$ ways to choose the committee and its leader. Meanwhile, the number of ways to select a committee with k people is $C(n,k)$. Once we have chosen a committee with k people, there are k ways to choose its leader. Hence there are $\sum_{k=1}^{n} kC(n,k)$ ways to choose the committee and its leader. Hence $\sum_{k=1}^{n} kC(n,k) = n2^{n-1}$.

53. Let the set have n elements. From Theorem 7 we have $C(n,0) - C(n,1) + C(n,2) - \cdots + (-1)^n C(n,n) = 0$. It follows that $C(n,0) + C(n,2) + C(n,4) + \cdots = C(n,1) + C(n,3) + C(n,5) + \cdots$. The left-hand side gives the number of subsets with an even number of elements, and the right-hand side gives the number of subsets with an odd number of elements.

55. **a)** A path of the desired type consists of m moves to the right and n moves up. Each such path can be represented by a bit string of length $m+n$ with m 0s and n 1s, where a 0 represents a move to the right and a 1 a move up.
b) The number of bit strings of length $m+n$ containing exactly n 1s equals $C(m+n,n) = C(m+n,m)$ since such a string is determined by specifying the positions of the n 1s or by specifying the positions of the m 0s.

57. By Exercise 55 the number of paths of length n of the type described in that exercise equals 2^n, the number of bit strings of length n. On the other hand, a path of length n of the type described in Exercise 55 must end at a point that has n as the sum of its coordinates, say $(n-k,k)$ for some k between 0 and n, inclusive. By Exercise 55, the number of such paths ending at $(n-k,k)$ equals $C(n-k+k,k) = C(n,k)$. Hence $\sum_{k=0}^{n} C(n,k) = 2^n$.

59. By Exercise 55 the number of paths from $(0,0)$ to $(n+1,r)$ of the type described in that exercise equals $C(n+r+1,r)$. But such a path starts by going j steps vertically for some j with $0 \le j \le r$. The number of these paths beginning with j vertical steps equals the number of paths of the type described in Exercise 55 that go from $(1,j)$ to $(n+1,r)$. This is the same as the number of such paths that go from $(0,0)$ to $(n,r-j)$, which by Exercise 55 equals $C(n+r-j,r-j)$. Since $\sum_{j=0}^{r} C(n+r-j,r-j) = \sum_{k=0}^{r} C(n+k,k)$, it follows that $\sum_{k=1}^{r} C(n+k,k) = C(n+r+1,r)$.

Section 4.4

1. 1/13
3. 1/2
5. 1/2
7. 1/64
9. 47/52
11. $1/C(52,5)$
13. $1-(C(48,5)/C(52,5))$
15. $C(13,2)C(4,2)C(4,2)C(44,1)/C(52,5)$
17. $10{,}240/C(52,5)$
19. $1{,}302{,}540/C(52,5)$
21. 1/64
23. 8/25
25. **a)** $1/C(50,6) = 1/15{,}890{,}700$
b) $1/C(52,6) = 1/20{,}358{,}520$
c) $1/C(56,6) = 1/32{,}468{,}436$
d) $1/C(60,6) = 1/50{,}063{,}860$
27. $1/C(100,8)$
29. **a)** 9/19 **b)** 81/361 **c)** 1/19 **d)** 1,889,568/2,476,099 **e)** 48/361

Section 4.5

1. $p(T) = 1/4$, $p(H) = 3/4$

3. $p(1) = p(3) = p(5) = p(6) = 1/16$; $p(2) = p(4) = 3/8$

5. 9/49

7. Since $p(E \cup F) = p(E) + p(F) - p(E \cap F)$ and $p(E \cup F) \le 1$, it follows that $1 \ge p(E) + p(F) - p(E \cap F)$. From this inequality we conclude that $p(E) + p(F) \le 1 + p(E \cap F)$.

9. We will use mathematical induction to prove that the inequality holds for $n \ge 2$. Let $P(n)$ be the statement that $p(\bigcup_{j=1}^{n} E_j) \le \sum_{j=1}^{n} p(E_j)$. $P(2)$ is true since $p(E_1 \cup E_2) = p(E_1) + p(E_2) - p(E_1 \cap E_2) \le p(E_1) + p(E_2)$. Now assume that $P(n)$ is true. Using the basis case and the inductive hypothesis, it follows that $p(\bigcup_{j=1}^{n+1} E_j) \le p(\bigcup_{j=1}^{n} E_j) + p(E_{n+1}) \le \sum_{j=1}^{n+1} p(E_j)$. This shows that $P(n+1)$ is true, completing the proof by mathematical induction.

11. **a)** $1 - 365/366 \cdot 364/366 \cdots (367-n)/366$
b) 23

13. 1/4

15. 3/8

17. **a)** Not independent **b)** Not independent **c)** Not independent

19. 3/16

21. **a)** $1/32 = 0.03125$ **b)** $0.49^5 \approx 0.02825$ **c)** 0.03795012

23. **a)** 5/8 **b)** 0.627649 **c)** 0.6431

25. **a)** p^n **b)** $1 - p^n$ **c)** $p^n + n \cdot p^{n-1} \cdot (1 - p)$ **d)** $1 - [p^n + n \cdot p^{n-1} \cdot (1 - p)]$

27. 5/3

29. 336/49

31. 170

33. $(4n + 6)/3$

35. $p \cdot q^{n-1}$

37. $1 - (1 - p)^n$

Section 4.6

1. 243

3. 26^6

5. 125

7. 35

9. **a)** 1716 **b)** 50,388 **c)** 2,629,575 **d)** 330 **e)** 9,724

11. 9

13. 4,504,501

15. **a)** 10,626 **b)** 1,365 **c)** 11,649 **d)** 106

17. 2,520

19. 302,702,400

21. 30,492

23. $C(59, 50)$

25. 35

27. 83,160

29. 63

31. 19,635

33. 210

35. 27,720

37. $52!/(7!^5 17!)$

39. $24 \cdot 13^4/(52 \cdot 51 \cdot 50 \cdot 49)$

41. **a)** $C(k + n - 1, n)$ **b)** $(k + n - 1)!/(k - 1)!$

43. There are $C(n, n_1)$ ways to choose n_1 objects for the first box. Once these objects are chosen, there are $C(n - n_1, n_2)$ ways to choose objects for the second box. Similarly, there are $C(n - n_1 - n_2, n_3)$ ways to choose objects for the third box. Continue in this way until there is $C(n - n_1 - n_2 - \cdots - n_{k-1}, n_k) = C(n_k, n_k) = 1$ way to choose the objects for the last box (since $n_1 + n_2 + \cdots + n_k = n$). By the product rule, the number of ways to make the entire assignment is $C(n, n_1)C(n - n_1, n_2)C(n - n_1 - n_2, n_3) \cdots C(n - n_1 - n_2 - \cdots - n_{k-1}, n_k)$, which equals $n!/(n_1!n_2! \cdots n_k!)$, as straightforward simplification shows.

45. **a)** Since $x_1 \le x_2 \le \cdots \le x_r$, it follows that $x_1 + 0 < x_2 + 1 < \cdots < x_r + r - 1$. The inequalities are strict since $x_j + j - 1 < x_{j+1} + j$ as long as $x_j \le x_{j+1}$. Since $1 \le x_j \le n + r - 1$, this sequence is made up of r distinct elements from T. **b)** Suppose that $1 \le x_1 < x_2 < \cdots < x_r \le n + r - 1$. Let $y_k = x_k - (k - 1)$. Then it is not hard to see that $y_k \le y_{k+1}$ for $k = 1, 2, \ldots, r-1$ and that $1 \le y_k \le n$ for $k = 1, 2, \ldots r$. It follows that $\{y_1, y_2, \ldots, y_r\}$ is an r-combination with repetition allowed of S. **c)** From (a) and (b) it follows that there is a one-to-one correspondence of r-combinations with repetition allowed of S and r-combinations of T, a set with $n + r - 1$ elements. We conclude that there are $C(n + r - 1, r)$ r-combinations with repetitions allowed of S.

47. 5

49. The terms in the expansion are of the form $x_1^{n_1} x_2^{n_2} \cdots x_m^{n_m}$, where $n_1 + n_2 + \cdots + n_m = n$. Such a term arises from choosing the x_1 in n_1 factors, the x_2 in n_2 factors, . . . , and the x_m in n_m factors. This can be done in $C(n; n_1, n_2, \ldots, n_m)$ ways, since a choice is a permutation of n_1 labels "1," n_2 labels "2," . . . and n_m labels "m."

51. 2520

Section 4.7

1. **a)** 2134 **b)** 54132 **c)** 12534 **d)** 45312 **e)** 6714253 **f)** 31542678

3. 1234, 1243, 1324, 1342, 1423, 1432, 2134, 2143, 2314, 2341, 2413, 2431, 3124, 3142, 3214, 3241, 3412, 3421, 4123, 4132, 4213, 4231, 4312, 4321

5. $\{1,2,3\}$, $\{1,2,4\}$, $\{1,2,5\}$, $\{1,3,4\}$, $\{1,3,5\}$, $\{1,4,5\}$, $\{2,3,4\}$, $\{2,3,5\}$, $\{2,4,5\}$, $\{3,4,5\}$

7. The bit string representing the next largest r-combination must differ from the bit string representing the original one in position i since positions $i+1, \ldots, r$ are occupied by the largest possible numbers. Also $a_i + 1$ is the smallest possible number we can put in position i if we want a combination greater than the original one. Then $a_i+2, \ldots, a_i+r-i+1$ are the smallest allowable numbers for positions $i+1$ to r. Thus we have produced the next r-combination.

9. 123, 132, 213, 231, 312, 321, 124, 142, 214, 241, 412, 421, 125, 152, 215, 251, 512, 521, 134, 143, 314, 341, 413, 431, 135, 153, 315, 351, 513, 531, 145, 154, 415, 451, 514, 541, 234, 243, 324, 342, 423, 432, 235, 253, 325, 352, 523, 532, 245, 254, 425, 452, 524, 542, 345, 354, 435, 453, 534, 543

11. We will show it is a bijection by showing it has an inverse. Given a positive integer less than $n!$, let $a_1, a_2, \ldots, a_{n-1}$ be its Cantor digits. Put n in position $n - a_{n-1}$, so clearly a_{n-1} is the number of integers less than n that follow n in the permutation. Then put $n-1$ in free position $(n-1)-a_{n-2}$, where we have numbered the free positions $1, 2, \ldots, n-1$ (excluding the position that n is already in). Continue until 1 is placed in the only free position left. Since we have constructed an inverse, the correspondence is a bijection.

13.
```
procedure Cantor permutation(n, i: integers with
    n ≥ 1 and 0 ≤ i < n!)
x := n
for j := 1 to n
    p_j := 0
for k := 1 to n − 1
begin
    c := ⌊x/(n − k)!⌋; x := x − c(n − k)!; h := n
    while p_h ≠ 0
        h := h − 1
    for j := 1 to c
    begin
        h := h − 1
        while p_h ≠ 0
            h := h − 1
    end
    p_h := n − k + 1
end
h := 1
while p_h ≠ 0
    h := h + 1
p_h := 1
{p_1 p_2 ··· p_n is the permutation corresponding to i}
```

Supplementary Exercises

1. **a)** 151,200 **b)** 1,000,000 **c)** 210 **d)** 5005

3. 3^{100}

5. 24,600

7. **a)** 4,060 **b)** 2688 **c)** 25,009,600

9. **a)** 192 **b)** 301 **c)** 300 **d)** 300

11. 639

13. The maximum possible sum is 240, and the minimum possible sum is 15. So the number of possible sums is 226. Since there are 252 subsets with five elements of a set with 10 elements, by the pigeonhole principle it follows that at least two have the same sum.

15. **a)** 50 **b)** 50 **c)** 14 **d)** 5

17. Let $a_1, a_2, \ldots, a_m$ be the integers, and let $d_i = \sum_{j=1}^{i} a_j$. If $d_i \equiv 0 \pmod{m}$ for some i, we are done. Otherwise $d_1 \textbf{ mod } m, d_2 \textbf{ mod } m, \ldots, d_m \textbf{ mod } m$ are m integers with values in $\{1, 2, \ldots, m-1\}$. By the pigeonhole principle $d_k = d_l$ for some $1 \le k < l \le m$. Then $\sum_{j=k+1}^{l} a_j = d_l - d_k \equiv 0 \pmod{m}$.

19. The decimal expansion of the rational number a/b can be obtained by division of b into a where a is written with a decimal point and an arbitrarily long string of 0s following it. The basic step is finding the next digit of the quotient, namely, $\lfloor r/b \rfloor$, where r is the remainder with the next digit of the dividend brought down. The current remainder is obtained from the previous remainder by subtracting b times the previous digit of the quotient. Eventually the dividend has nothing but 0s to bring down. Furthermore, there are only b possible remainders. Thus, at some point, by the pigeonhole principle, we will have the same situation as had previously arisen. From that point onward, the calculation must follow the same pattern. In particular, the quotient will repeat.

21. **a)** 125,970 **b)** 20 **c)** 141,120,525 **d)** 141,120,505 **e)** 177,100 **f)** 141,078,021

23. $4^{13}/C(52, 13)$

25. **a)** 10 **b)** 8 **c)** 7

27. There are the same number of ways to decide which r elements of an n-element set to select as there are ways to decide which $n-r$ elements to not select.

29. $C(n+2, r+1) = C(n+1, r+1) + C(n+1, r) = 2C(n+1, r+1) - C(n+1, r+1) + C(n+1, r) = 2C(n+1, r+1) - (C(n, r+1) + C(n, r)) + (C(n, r) + C(n, r-1)) = 2C(n+1, r+1) - C(n, r+1) + C(n, r-1)$

31. By the binomial theorem, $3^n = (2+1)^n = \sum_{k=0}^{n} C(n,k)1^{n-k}2^k = \sum_{k=0}^{n} C(n,k)2^k$.

33. $C(n+1, 5)$

35. **a)** $1/C(52, 13)$
b) $4/C(52, 13)$
c) $2{,}944{,}656/C(52, 13)$
d) $35{,}335{,}872/C(52, 13)$
e) $1{,}244{,}117{,}160/C(52, 13)$
f) $29{,}858{,}811{,}840/C(52, 13)$

37. $\dfrac{(m-1)(n-1) + gcd(m,n) - 1}{mn - 1}$

39. **a)** $p(E_1 \cap E_2) = p(E_1)p(E_2)$, $p(E_1 \cap E_3) = p(E_1)p(E_3)$, $p(E_2 \cap E_3) = p(E_2)p(E_3)$, $p(E_1 \cap E_2 \cap E_3) = p(E_1)p(E_2)p(E_3)$
b) Yes
c) Yes
d) $2^n - n - 1$

41. The events $E \cap F_i$, $i = 1, 2, \ldots, n$, are mutually exclusive and cover all the conditions under which E occurs. Therefore, $p(E) = \sum_{i=1}^{n} p(E \cap F_i)$. By the definition of conditional probability, we know that $p(E \cap F_i) = p(E \mid F_i)p(F_i)$ and $p(F_i \mid E) = p(E \cap F_i)/p(E)$. Hence

$$p(F_j \mid E) = \frac{p(E \cap F_j)}{p(E)} = \frac{p(E \mid F_j)p(F_j)}{\sum_{i=1}^{n} p(E \cap F_i)} = \frac{p(E \mid F_j)p(F_j)}{\sum_{i=1}^{n} p(E \mid F_i)p(F_i)}.$$

43. 3,491,888,400

45. 5^{24}

47. **a)** 45 **b)** 57 **c)** 12

49. **a)** 386 **b)** 56 **c)** 512

51. 0 if $n < m$; $C(n-1, n-m)$ if $n \geq m$

53.
```
procedure next permutation(n: positive integer,
    a1, a2, ..., ar: positive integers not exceeding n
    with a1a2···ar ≠ nn···n)
i := r
while ai = n
begin
    ai := 1
    i := i − 1
end
ai := ai + 1
{a1a2···ar is the next permutation in lexicographic
    order}
```

CHAPTER 5

Section 5.1

1. **a)** 2, 12, 72, 432, 2592
b) 2, 4, 16, 256, 65,536
c) 1, 2, 5, 11, 26
d) 1, 1, 6, 27, 204
e) 1, 2, 0, 1, 3

3. **a)** Yes **b)** No **c)** No **d)** Yes
e) Yes **f)** Yes **g)** No **h)** No

5. **a)** $a_n = 2 \cdot 3^n$ **b)** $a_n = 2n + 3$
c) $a_n = 1 + n(n+1)/2$ **d)** $a_n = n^2 + 4n + 4$
e) $a_n = 1$ **f)** $a_n = (3^{n+1} - 1)/2$
g) $a_n = 5n!$ **h)** $a_n = 2^n n!$

7. **a)** $a_n = 3a_{n-1}$ **b)** 5,904,900

9. **a)** $a_n = n + a_{n-1}$, $a_0 = 0$
b) $a_{12} = 78$
c) $a_n = n(n+1)/2$

11. Let $P(n)$ be "$H_n = 2^n - 1$." Basis step: $P(1)$ is true since $H_1 = 1$. Inductive step: Assume that $H_n = 2^n - 1$. Then since $H_{n+1} = 2H_n + 1$, it follows that $H_{n+1} = 2(2^n - 1) + 1 = 2^{n+1} - 1$.

13. **a)** $a_n = 2a_{n-1} + a_{n-5}$ for $n \geq 5$
b) $a_0 = 1$, $a_1 = 2$, $a_3 = 8$, $a_4 = 16$
c) 1217

15. **a)** $a_n = a_{n-1} + a_{n-2} + 2^{n-2}$ for $n \geq 2$
b) $a_0 = 0$, $a_1 = 0$
c) 94

17. **a)** $a_n = a_{n-1} + a_{n-2} + a_{n-3}$ for $n \geq 3$
b) $a_0 = 1$, $a_1 = 2$, $a_2 = 4$
c) 81

19. **a)** $a_n = a_{n-1} + a_{n-2}$ for $n \geq 2$
b) $a_0 = 1$, $a_1 = 1$
c) 34

21. **a)** $a_n = 2a_{n-1} + 2a_{n-2}$ for $n \geq 2$
b) $a_0 = 1$, $a_1 = 3$
c) 448

23. a) $a_n = 2a_{n-1} + a_{n-2}$ for $n \geq 2$
b) $a_0 = 1$, $a_1 = 3$
c) 239

25. a) $a_n = 2a_{n-1}$ for $n \geq 2$
b) $a_1 = 3$
c) 96

27. a) $a_n = a_{n-1} + a_{n-2}$ for $n \geq 2$
b) $a_0 = 1$, $a_1 = 1$
c) 89

29. a) $R_n = n + R_{n-1}$, $R_0 = 1$
b) $R_n = n(n+1)/2 + 1$

31. 64

33. Clearly $S(m, 1) = 1$ for $m \geq 1$. If $m \geq n$, then a function that is not onto from the set with m elements to the set with n elements can be specified by picking the size of the range, which is an integer between 1 and $n - 1$ inclusive, picking the elements of the range, which can be done in $C(n, k)$ ways, and picking an onto function onto the range, which can be done in $S(m, k)$ ways. Hence there are $\sum_{k=1}^{n-1} C(n, k)S(m, k)$ functions that are not onto. But there are n^m functions altogether, so that $S(m, n) = n^m - \sum_{k=1}^{n-1} C(n, k)S(m, k)$.

35. a) 0 **b)** 0 **c)** 2 **d)** $2^{n-1} - 2^{n-2}$

37. $a_n - 2\nabla a_n + \nabla^2 a_n$
$= a_n - 2(a_n - a_{n-1}) + (\nabla a_n - \nabla a_{n-1})$
$= -a_n + 2a_{n-1} + ((a_n - a_{n-1}) - (a_{n-1} - a_{n-2}))$
$= -a_n + 2a_{n-1} + (a_n - 2a_{n-1} + a_{n-2})$
$= a_{n-2}$

39. $a_n = a_{n-1} + a_{n-2}$
$= (a_n - \nabla a_n) + (a_n - 2\nabla a_n + \nabla^2 a_n)$
$= 2a_n - 3\nabla a_n + \nabla^2 a_n$,
or $a_n = 3\nabla a_n - \nabla^2 a_n$

Section 5.2

1. a) Degree 3 **b)** No **c)** Degree 4 **d)** No **e)** No **f)** Degree 2

3. a) $a_n = 3 \cdot 2^n$
b) $a_n = 2$
c) $a_n = 3 \cdot 2^n - 2 \cdot 3^n$
d) $a_n = 6 \cdot 2^n - 2 \cdot n2^n$
e) $a_n = n(-2)^{n-1}$
f) $a_n = 2^n - (-2)^n$
g) $a_n = (1/2)^{n+1} - (-1/2)^{n+1}$

5. $a_n = \dfrac{1}{\sqrt{5}}\left(\dfrac{1+\sqrt{5}}{2}\right)^{n+1} - \dfrac{1}{\sqrt{5}}\left(\dfrac{1-\sqrt{5}}{2}\right)^{n+1}$

7. $(2^{n+1} + (-1)^n)/3$

9. a) $P_n = 1.2P_{n-1} + 0.45P_{n-2}$, $P_0 = 100{,}000$, $P_1 = 120{,}000$
b) $P_n = (250{,}000/3)(3/2)^n + (50{,}000/3)(-3/10)^n$

11. a) Basis step: For $n = 1$ we have $1 = 0 + 1$, and for $n = 2$ we have $3 = 1 + 2$. Inductive step: Assume true for $k \leq n$. Then $L_{n+1} = L_n + L_{n-1} = f_{n-1} + f_{n+1} + f_{n-2} + f_n = (f_{n-1} + f_{n-2}) + (f_{n+1} + f_n) = f_n + f_{n+2}$.

b) $L_n = \left(\dfrac{1+\sqrt{5}}{2}\right)^n + \left(\dfrac{1-\sqrt{5}}{2}\right)^n$

13. $a_n = 8(-1)^n - 3(-2)^n + 4 \cdot 3^n$

15. $a_n = 5 + 3(-2)^n - 3^n$

17. Let $a_n = C(n, 0) + C(n-1, 1) + \cdots + C(n-k, k)$ where $k = \lfloor n/2 \rfloor$. First, assume that n is even, so that $k = n/2$, and the last term is $C(k, k)$. By Pascal's identity we have $a_n = 1 + C(n-2, 0) + C(n-2, 1) + C(n-3, 1) + C(n-3, 2) + \cdots + C(n-k, k-2) + C(n-k, k-1) + 1 = 1 + C(n-2, 1) + C(n-3, 2) + \cdots + C(n-k, k-1) + C(n-2, 0) + C(n-3, 1) + \cdots + C(n-k, k-2) + 1 = a_{n-1} + a_{n-2}$ since $\lfloor (n-1)/2 \rfloor = k - 1 = \lfloor (n-2)/2 \rfloor$. A similar calculation works when n is odd. Hence $\{a_n\}$ satisfies the recurrence relation $a_n = a_{n-1} + a_{n-2}$ for all positive integers n, $n \geq 2$. Also, $a_1 = C(1, 0) = 1$ and $a_2 = C(2, 0) + C(1, 1) = 2$, which are f_2 and f_3. It follows that $a_n = f_{n+1}$ for all positive integers n.

19. a) $3a_{n-1} + 2^n = 3(-2)^n + 2^n = 2^n(-3 + 1) = -2^{n+1} = a_n$
b) $a_n = \alpha 3^n - 2^{n+1}$
c) $a_n = 3^{n+1} - 2^{n+1}$

21. a) $A = -1$, $B = -7$
b) $a_n = \alpha 2^n - n - 7$
c) $a_n = 11 \cdot 2^n - n - 7$

23. a) $1, -1, i, -i$
b) $a_n = \frac{1}{4} - \frac{1}{4}(-1)^n + \frac{2+i}{4} i^n + \frac{2-i}{4}(-i)^n$

Section 5.3

1. 14

3. The first step is $(1110)_2(1010)_2 = (2^4 + 2^2)(11)_2(10)_2 + 2^2((11)_2 - (10)_2)((10)_2 - (10)_2) + (2^2 + 1)(10)_2 \cdot (10)_2$. The product is $(10001100)_2$.

5. $C = 50,665C + 729 = 33,979$
7. a) 2 **b)** 4 **c)** 7
9. a) 79 **b)** 48,829 **c)** 30,517,579
11. $O(\log n)$
13. $O(n^{\log_3 2})$
15. 5
17. With $k = \log_b n$, it follows that $f(n) = a^k f(1) + \sum_{j=0}^{k-1} a^j c(n/b^j)^d = a^k f(1) + \sum_{j=0}^{k-1} cn^d = a^k f(1) + kcn^d = a^{\log_b n} f(1) + c(\log_b n)\, n^d = n^{\log_b a} f(1) + cn^d \log_b n = n^d f(1) + cn^d \log_b n$.
19. Let $k = \log_b n$ where n is a power of b. Basis step: if $n = 1$ and $k = 0$, then $c_1 n^d + c_2 n^{\log_b a} = c_1 + c_2 = b^d c/(b^d - a) + f(1) + b^d c/(a - b^d) = f(1)$. Inductive step: Assume true for k, where $n = b^k$. Then for $n = b^{k+1}$, $f(n) = af(n/b) + cn^d = a\{[b^d c/(b^d - a)](n/b)^d + [f(1) + b^d c/(a - b^d)] \cdot (n/b)^{\log_b a}\} + cn^d = b^d c/(b^d - a) n^d a/b^d + [f(1) + (b^d c/(a - b^d)] n^{\log_b a} + cn^d = n^d[ac/(b^d - a) + c(b^d - a)/(b^d - a)] + [f(1) + b^d c/(a - b^d c)] n^{\log_b a} = [(b^d c)/(b^d - a)] n^d + [f(1) + b^d c/(a - b^d)] n^{\log_b a}$.
21. If $a > b^d$, then $\log_b a > d$, so the second term dominates, giving $O(n^{\log_b a})$.
23. $O(n^{\log_4 5})$
25. $O(n^3)$

Section 5.4

1. a) 30 **b)** 29 **c)** 24 **d)** 18
3. 1%
5. a) 300 **b)** 150 **c)** 175 **d)** 100
7. 492
9. 974
11. 55
13. 248
15. 50,138
17. 234
19. $|A_1 \cup A_2 \cup A_3 \cup A_4 \cup A_5| = |A_1| + |A_2| + |A_3| + |A_4| + |A_5| - |A_1 \cap A_2| - |A_1 \cap A_3| - |A_1 \cap A_4| - |A_1 \cap A_5| - |A_2 \cap A_3| - |A_2 \cap A_4| - |A_2 \cap A_5| - |A_3 \cap A_4| - |A_3 \cap A_5| - |A_4 \cap A_5| + |A_1 \cap A_2 \cap A_3| + |A_1 \cap A_2 \cap A_4| + |A_1 \cap A_2 \cap A_5| + |A_1 \cap A_2 \cap A_5| + |A_1 \cap A_3 \cap A_5| + |A_1 \cap A_4 \cap A_5| + |A_2 \cap A_3 \cap A_4| + |A_2 \cap A_3 \cap A_5| + |A_2 \cap A_4 \cap A_5| + |A_3 \cap A_4 \cap A_5| - |A_1 \cap A_2 \cap A_3 \cap A_4| - |A_1 \cap A_2 \cap A_3 \cap A_5| - |A_1 \cap A_2 \cap A_4 \cap A_5| - |A_1 \cap A_3 \cap A_4 \cap A_5| - |A_2 \cap A_3 \cap A_4 \cap A_5| + |A_1 \cap A_2 \cap A_3 \cap A_4 \cap A_5|$
21. $|A_1 \cup A_2 \cup A_3 \cup A_4 \cup A_5 \cup A_6| = |A_1| + |A_2| + |A_3| + |A_4| + |A_5| + |A_6| - |A_1 \cap A_2| - |A_1 \cap A_3| - |A_1 \cap A_4| - |A_1 \cap A_5| - |A_1 \cap A_6| - |A_2 \cap A_3| - |A_2 \cap A_4| - |A_2 \cap A_5| - |A_2 \cap A_6| - |A_3 \cap A_4| - |A_3 \cap A_5| - |A_3 \cap A_6| - |A_4 \cap A_5| - |A_4 \cap A_6| - |A_5 \cap A_6|$
23. $p(E_1 \cup E_2 \cup E_3) = p(E_1) + p(E_2) + p(E_3) - p(E_1 \cap E_2) - p(E_1 \cap E_3) - p(E_2 \cap E_3) + p(E_1 \cap E_2 \cap E_3)$
25. 4972/71,295
27. $p(E_1 \cup E_2 \cup E_3 \cup E_4 \cup E_5) = p(E_1) + p(E_2) + p(E_3) + p(E_4) + p(E_5) - p(E_1 \cap E_2) - p(E_1 \cap E_3) - p(E_1 \cap E_4) - p(E_1 \cap E_5) - p(E_2 \cap E_3) - p(E_2 \cap E_4) - p(E_2 \cap E_5) - p(E_3 \cap E_4) - p(E_3 \cap E_5) - p(E_4 \cap E_5) + p(E_1 \cap E_2 \cap E_3) + p(E_1 \cap E_2 \cap E_4) + p(E_1 \cap E_2 \cap E_5) + p(E_1 \cap E_3 \cap E_4) + p(E_1 \cap E_3 \cap E_5) + p(E_1 \cap E_4 \cap E_5) + p(E_2 \cap E_3 \cap E_4) + P(E_2 \cap E_3 \cap E_5) + p(E_2 \cap E_4 \cap E_5) + p(E_3 \cap E_4 \cap E_5)$
29. $p\left(\bigcup_{i=1}^{n} E_i\right) = \sum_{1 \le i \le n} p(E_i) - \sum_{1 \le i < j \le n} p(E_i \cap E_j) + \sum_{1 \le i < j < k \le n} p(E_i \cap E_j \cap E_k) - \cdots + (-1)^{n+1} p\left(\bigcap_{i=1}^{n} E_i\right)$

Section 5.5

1. 75
3. 6
5. 46
7. 9875
9. 540
11. 2100
13. 1854
15. a) $D_{100}/100!$ **b)** $100D_{99}/100!$ **c)** $C(100,2)/100!$ **d)** 0 **e)** $1/100!$
17. 2,170,680
19. By Exercise 18 we have $D_n - nD_{n-1} = -(D_{n-1} - (n-1)D_{n-2})$. Iterating, we have $D_n - nD_{n-1} = -(D_{n-1} - (n-1)D_{n-2}) = -(-(D_{n-2} - (n-2) \cdot D_{n-3})) = D_{n-2} - (n-2)D_{n-3} = \cdots = (-1)^n \cdot (D_2 - 2D_1) = (-1)^n$ since $D_2 = 1$ and $D_1 = 0$.
21. When n is odd

23. $\phi(n) = n - \sum_{i=1}^{m} \frac{n}{p_i} + \sum_{1 \le i < j \le m} \frac{n}{p_i p_j} - \cdots \pm \frac{n}{p_1 p_2 \cdots p_m} = n \prod_{i=1}^{m} \left(a - \frac{1}{p_i} \right)$

25. 4

27. There are n^m functions from a set with m elements to a set with n elements, $C(n, 1)(n-1)^m$ functions from a set with m elements to a set with n elements that miss exactly one element, $C(n, 2)(n-2)^m$ functions from a set with m elements to a set with n elements that miss exactly two elements, and so on, with $C(n, n-1)1^m$ functions from a set with m elements to a set with n elements that miss exactly $n-1$ elements. Hence by the principle of inclusion-exclusion there are $n^m - C(n, 1)(n-1)^m + C(n, 2)(n-2)^m - \cdots + (-1)^{n-1}C(n, n-1) \cdot 1^m$ onto functions.

Supplementary Problems

1. **a)** $A_n = 4A_{n-1}$ **b)** $A_1 = 40$
c) $A_n = 10 \cdot 4^n$

3. **a)** $M_n = M_{n-1} + 160{,}000$
b) $M_1 = 186{,}000$
c) $M_n = 160{,}000n + 26{,}000$
d) $T_n = T_{n-1} + 160{,}000n + 26{,}000$
e) $T_n = 80{,}000n^2 + 106{,}000n$

5. **a)** $a_n = a_{n-2} + a_{n-3}$
b) $a_1 = 0,\ a_2 = 1,\ a_3 = 1$
c) $a_{12} = 12$

7. **a)** 2 **b)** 5 **c)** 8 **d)** 16

9. $a_n = 2^n$

11. Suppose the characteristic roots are $r_1, r_2, \ldots, r_t$ with multiplicities $m_1, m_2, \ldots, m_t$, respectively. Then all solutions are of the form $a_n = \sum_{k=1}^{t} \sum_{j=1}^{m_t} \alpha_{kj} n^{j-1} r_k^n$ where the α_{kj}s are constants.

13. $a_n = a_{n-2} + a_{n-3}$

15. $O(n^4)$

17. $O(n)$

19. **a)** $18n + 18$ **b)** 18 **c)** 0

21. $\Delta(a_n b_n) = a_{n+1}b_{n+1} - a_n b_n = a_{n+1}(b_{n+1} - b_n) + b_n(a_{n+1} - a_n) = a_{n+1}\Delta b_n + b_n \Delta a_n$

23. 7

25. 110

27. 0

29. **a)** 19 **b)** 65 **c)** 122 **d)** 167 **e)** 168

31. $D_{n-1}/(n-1)!$

33. 11/32

CHAPTER 6

Section 6.1

1. **a)** $\{(0,0), (1,1), (2,2), (3,3)\}$
b) $\{(1,3), (2,2), (3,1), (4,0)\}$
c) $\{(1,0), (2,0), (2,1), (3,0), (3,1), (3,2), (4,0), (4,1), (4,2), (4,3)\}$
d) $\{(1,0), (1,1), (1,2), (1,3), (2,0), (2,2), (3,0), (3,3), (4,0)\}$ (assuming that 0 does not divide 0)
e) $\{(0,1), (1,0), (1,1), (1,2), (1,3), (2,1), (2,3), (3,1), (3,2), (4,1), (4,3)\}$
f) $\{(1,2), (2,1), (2,2)\}$

3. **a)** Transitive
b) Reflexive, symmetric, transitive
c) Symmetric
d) Antisymmetric
e) Reflexive, symmetric, antisymmetric, transitive
f) None of these properties

5. **a)** Symmetric
b) Symmetric, transitive
c) Symmetric
d) Reflexive, symmetric, transitive
e) Reflexive, transitive
f) Reflexive, symmetric, transitive
g) Antisymmetric
h) Antisymmetric, transitive

7. (c), (d), (f)

9. Yes, for instance $\{(1,1)\}$ on $\{1,2\}$

11. (a)

13. 2^{mn}

15. **a)** $\{(a,b) \mid b \text{ divides } a\}$
b) $\{(a,b) \mid a \text{ does not divide } b\}$

17. The graph of f^{-1}

19. **a)** $\{(a, b) \mid a$ is required to read or has read $b\}$
b) $\{(a, b) \mid a$ is required to read and has read $b\}$
c) $\{(a, b) \mid$ either a is required to read b but has not read it or a has read b but is not required to$\}$
d) $\{(a, b) \mid a$ is required to read b but has not read it$\}$
e) $\{(a, b) \mid a$ has read b but is not required to$\}$

21. $S \circ R = \{(a, b) \mid a$ is a parent of b and b has a sibling$\}$, $R \circ S = \{(a, b) \mid a$ is an aunt or uncle of $b\}$

23. 8

25. **a)** $2^{n(n+1)/2}$ **b)** $2^n 3^{n(n-1)/2}$
c) $3^{n(n-1)/2}$ **d)** $2^{n(n-1)}$
e) $2^{n(n-1)/2}$ **f)** $2^{n^2} - 2 \cdot 2^{n(n-1)}$

27. There may be no such b.

29. If R is symmetric and $(a, b) \in R$, then $(b, a) \in R$, so that $(a, b) \in R^{-1}$. Hence $R \subseteq R^{-1}$. Similarly $R^{-1} \subseteq R$. So $R = R^{-1}$. Conversely, if $R = R^{-1}$ and $(a, b) \in R$, then $(a, b) \in R^{-1}$, so that $(b, a) \in R$. Thus R is symmetric.

31. R is reflexive if and only if $(a, a) \in R$ for all $a \in A$ if and only if $(a, a) \in R^{-1}$ (since $(a, a) \in R$ if and only if $(a, a) \in R^{-1}$) if and only if R^{-1} is reflexive.

33. Use mathematical induction. The result is trivial for $n = 1$. Assume R^n is reflexive and transitive. By Theorem 1, $R^{n+1} \subseteq R$. To see that $R \subseteq R^{n+1} = R^n \circ R$, let $(a, b) \in R$. By the inductive hypothesis, $R^n = R$ and hence is reflexive. Thus $(b, b) \in R^n$. Therefore $(a, b) \in R^{n+1}$.

35. Use mathematical induction. The result is trivial for $n = 1$. Assume R^n is reflexive. Then $(a, a) \in R^n$ for all $a \in A$ and $(a, a) \in R$. Thus $(a, a) \in R^n \circ R = R^{n+1}$ for all $a \in A$.

37. No, for instance take $R = \{(1, 2), (2, 1)\}$.

Section 6.2

1. $\{(1, 2, 3), (1, 2, 4), (1, 3, 4), (2, 3, 4)\}$

3. (Nadir, 122, 34, Detroit, 08:10), (Acme, 221, 22, Denver, 08:17), (Acme, 122, 33, Anchorage, 08:22), (Acme, 323, 34, Honolulu 08:30), (Nadir, 199, 13, Detroit, 08:47), (Acme, 222, 22, Denver, 09:10), (Nadir, 322, 34, Detroit, 09:44)

5. Airline and flight number, airline and departure time

7. $P_{3,5,6}$

9.

Airline	*Destination*
Nadir	Detroit
Acme	Denver
Acme	Anchorage
Acme	Honolulu

11.

Supplier	*Part Number*	*Project*	*Quantity*	*Color Code*
23	1092	1	2	2
23	1101	3	1	1
23	9048	4	12	2
31	4975	3	6	2
31	3477	2	25	2
32	6984	4	10	1
32	9191	2	80	4
33	1001	1	14	8

Section 6.3

1. a) $\begin{bmatrix} 1 & 1 & 1 \\ 0 & 0 & 0 \\ 0 & 0 & 0 \end{bmatrix}$

b) $\begin{bmatrix} 0 & 1 & 0 \\ 1 & 1 & 0 \\ 0 & 0 & 1 \end{bmatrix}$

c) $\begin{bmatrix} 1 & 1 & 1 \\ 0 & 1 & 1 \\ 0 & 0 & 1 \end{bmatrix}$

d) $\begin{bmatrix} 0 & 0 & 1 \\ 0 & 0 & 0 \\ 1 & 0 & 0 \end{bmatrix}$

3. R is irreflexive if and only if each diagonal entry is 0.

5. Change each 0 to a 1 and each 1 to a 0.

7. a) $\begin{bmatrix} 0 & 1 & 1 \\ 1 & 1 & 0 \\ 1 & 0 & 1 \end{bmatrix}$

b) $\begin{bmatrix} 1 & 0 & 0 \\ 0 & 0 & 1 \\ 0 & 1 & 0 \end{bmatrix}$

c) $\begin{bmatrix} 1 & 1 & 1 \\ 1 & 1 & 1 \\ 1 & 1 & 1 \end{bmatrix}$

9. a) $\begin{bmatrix} 0 & 0 & 1 \\ 1 & 1 & 0 \\ 0 & 1 & 1 \end{bmatrix}$

b) $\begin{bmatrix} 1 & 1 & 0 \\ 0 & 1 & 1 \\ 1 & 1 & 1 \end{bmatrix}$

c) $\begin{bmatrix} 0 & 1 & 1 \\ 1 & 1 & 1 \\ 1 & 1 & 1 \end{bmatrix}$

11. a)

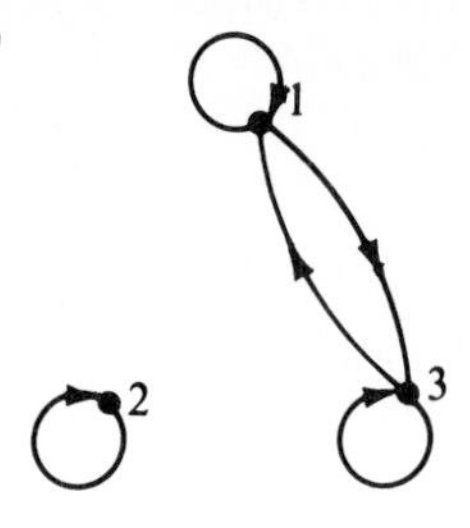

b)

c)

13. $\{(a, b), (a, c), (b, c), (c, b)\}$

15. $\{(a, a), (a, b), (a, c), (b, a), (b, b), (b, c), (c, a), (c, b), (d, d)\}$

17. a) Irreflexive only
b) Reflexive only
c) Symmetric only

19. Proof by mathematical induction. Trivial for $n = 1$. Assume true for n. Since $R^{n+1} = R^n \circ R$, its matrix is $\mathbf{M}_R \odot \mathbf{M}_{R^n}$. By the inductive hypothesis this is $\mathbf{M}_R \odot \mathbf{M}_R^{[n]} = \mathbf{M}_R^{[n+1]}$.

Section 6.4

1. a) $\{(0, 0), (0, 1), (1, 1), (1, 2), (2, 0), (2, 2), (3, 0), (3, 3)\}$

b) $\{(0, 1), (0, 2), (0, 3), (1, 0), (1, 1), (1, 2), (2, 0), (2, 1), (2, 2), (3, 0)\}$

3. $\{(a, b) \mid a \text{ divides } b \text{ or } b \text{ divides } a\}$

5.

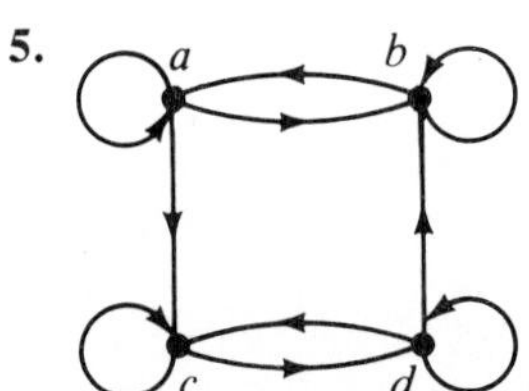

7.

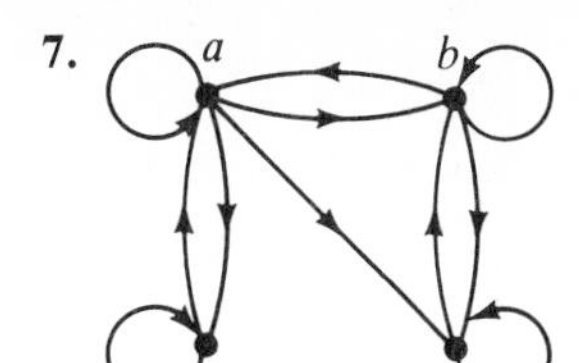

9. a)

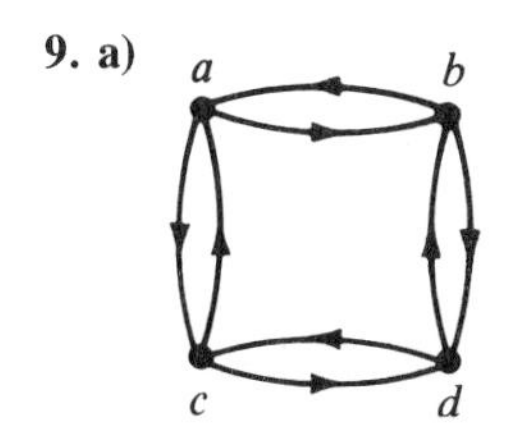

b)

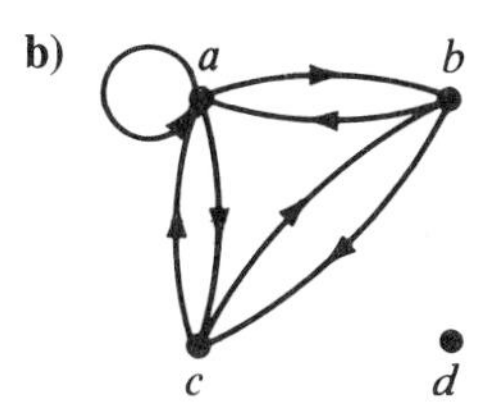

c)

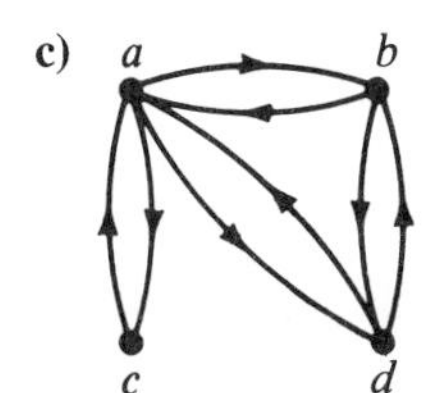

11. a)

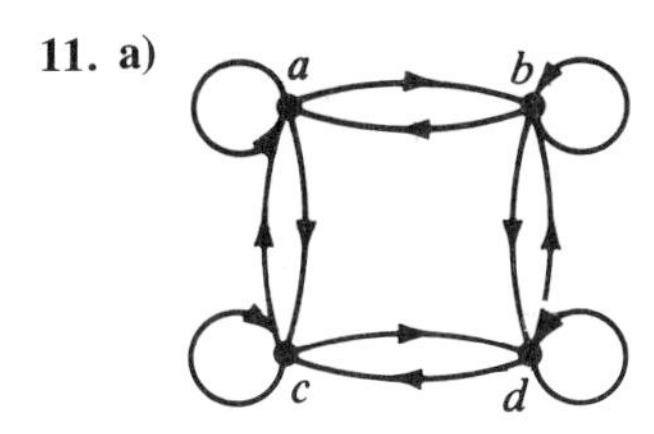

b)

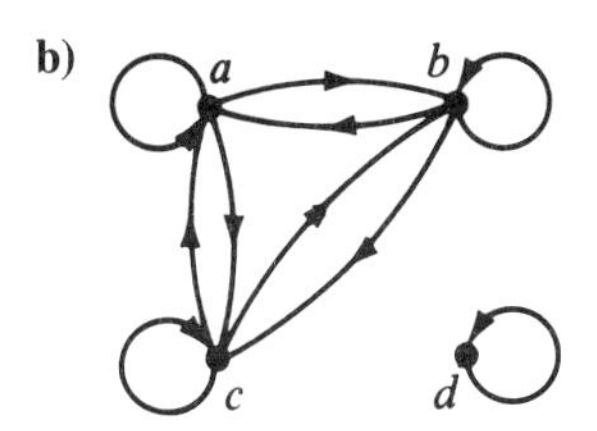

c)

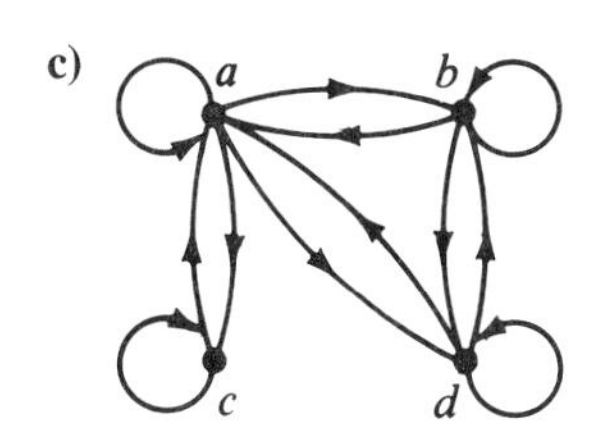

13. The symmetric closure of R is $R \cup R^{-1}$. $\mathbf{M}_{R \cup R^{-1}} = \mathbf{M}_R \vee \mathbf{M}_{R^{-1}} = \mathbf{M}_R \vee \mathbf{M}_R^t$.

15. Only when R is irreflexive, in which case it is its own closure.

17. $a, a, a, a; b, c, c, b; c, b, c, c; c, c, b, c; c, c, c, c; d, e, e, d; e, d, e, e; e, e, d, e; e, e, e, e$

19. a) $\{(1, 1), (1, 5), (2, 3), (3, 1), (3, 2), (3, 3), (3, 4), (4, 1), (4, 5), (5, 3), (5, 4)\}$

b) $\{(1, 1), (1, 2), (1, 3), (1, 4), (2, 1), (2, 5), (3, 1), (3, 3), (3, 4), (3, 5), (4, 1), (4, 2), (4, 3), (4, 4), (5, 1), (5, 3), (5, 5)\}$

c) $\{(1, 1), (1, 3), (1, 4), (1, 5), (2, 1), (2, 2), (2, 3), (2, 4), (3, 1), (3, 2), (3, 3), (3, 4), (3, 5), (4, 1), (4, 3), (4, 4), (4, 5), (5, 1), (5, 2), (5, 3), (5, 4), (5, 5)\}$

d) $\{(1, 1), (1, 2), (1, 3), (1, 4), (1, 5), (2, 1), (2, 3), (2, 4), (2, 5), (3, 1), (3, 2), (3, 3), (3, 4), (3, 5), (4, 1), (4, 2), (4, 3), (4, 4), (4, 5), (5, 1), (5, 2), (5, 3), (5, 4), (5, 5)\}$

e) $\{(1, 1), (1, 2), (1, 3), (1, 4), (1, 5), (2, 1), (2, 2), (2, 3), (2, 4), (2, 5), (3, 1), (3, 2), (3, 3), (3, 4), (3, 5), (4, 1), (4, 2), (4, 3), (4, 4), (4, 5), (5, 1), (5, 2), (5, 3), (5, 4), (5, 5)\}$

f) $\{(1, 1), (1, 2), (1, 3), (1, 4), (1, 5), (2, 1), (2, 2), (2, 3), (2, 4), (2, 5), (3, 1), (3, 2), (3, 3), (3, 4), (3, 5), (4, 1), (4, 2), (4, 3), (4, 4), (4, 5), (5, 1), (5, 2), (5, 3), (5, 4), (5, 5)\}$

21. a) If there is a student c who shares a class with a and a class with b

b) If there are two students c and d such that a and c share a class, c and d share a class, and d and b share a class

c) If there is a sequence $s_0, \ldots, s_n$ of students with $n \geq 1$ such that $s_0 = a$, $s_n = b$, and for each $i = 1, 2, \ldots, n$, s_i and s_{i-1} share a class

23. The result follows from $(R^*)^{-1} = (\bigcup_{n=1}^{\infty} R^n)^{-1} = \bigcup_{n=1}^{\infty} (R^n)^{-1} = \bigcup_{n=1}^{\infty} R^n = R^*$.

25. a) $\begin{bmatrix} 1 & 1 & 1 & 1 \\ 1 & 1 & 1 & 1 \\ 1 & 1 & 1 & 1 \\ 1 & 1 & 1 & 1 \end{bmatrix}$ **b)** $\begin{bmatrix} 0 & 0 & 0 & 0 \\ 1 & 0 & 1 & 1 \\ 1 & 0 & 1 & 1 \\ 1 & 0 & 1 & 1 \end{bmatrix}$

c) $\begin{bmatrix} 0 & 1 & 1 & 1 \\ 0 & 0 & 1 & 1 \\ 0 & 0 & 0 & 1 \\ 0 & 0 & 0 & 0 \end{bmatrix}$ **d)** $\begin{bmatrix} 1 & 1 & 1 & 1 \\ 1 & 1 & 1 & 1 \\ 1 & 1 & 1 & 1 \\ 1 & 1 & 1 & 1 \end{bmatrix}$

27. Answers same as for Exercise 25.

29. a) $\{(1, 1), (1, 2), (1, 4), (2, 2), (3, 3), (4, 1), (4, 2), (4, 4)\}$

b) $\{(1, 1), (1, 2), (1, 4), (2, 1), (2, 2), (2, 4), (3, 3), (4, 1), (4, 2), (4, 4)\}$

c) $\{(1, 1), (1, 2), (1, 4), (2, 1), (2, 2), (2, 4), (3, 3), (4, 1), (4, 2), (4, 4)\}$

31. Algorithm 1: $O(n^{3.8})$; Algorithm 2: $O(n^3)$
33. Initialize with **A:** $= \mathbf{M}_R \vee \mathbf{I}_n$ and loop only for $i :=$ 2 to $n - 1$.
35. a) Since R is reflexive, every relation containing it must also be reflexive.
b) Both $\{(0,0), (0,1), (0,2), (1,1), (2,2)\}$ and $\{(0,0), (0,1), (1,0), (1,1), (2,2)\}$ contain R and have an odd number of elements, but neither is a subset of the other.

Section 6.5

1. a) Equivalence relation
b) Not reflexive, not transitive
c) Equivalence relation
d) Not transitive
e) Not symmetric, not transitive
3. a) Equivalence relation
b) Not transitive
c) Not reflexive, not symmetric, not transitive
d) Equivalence relation
e) Not reflexive, not transitive
5. a) $(x, x) \in R$ since $f(x) = f(x)$. Hence R is reflexive. $(x, y) \in R$ if and only if $f(x) = f(y)$, which holds if and only if $f(y) = f(x)$ if and only if $(y, x) \in R$. Hence R is symmetric. If $(x, y) \in R$ and $(y, z) \in R$, then $f(x) = f(y)$ and $f(y) = f(z)$. Hence $f(x) = f(z)$. Thus $(x, z) \in R$. It follows that R is transitive.
b) The sets $f^{-1}(b)$ for b in the range of f.
7. Let x be a string of length 3 or more. Because x agrees with itself in the first three bits, $(x, x) \in R$. Hence R is reflexive. Suppose that $(x, y) \in R$. Then x and y agree in the first three bits. Hence y and x agree in the first three bits. Thus $(y, x) \in R$. If (x, y) and (y, z) are in R then x and y agree in the first three bits, as do y and z. Hence x and z agree in the first three bits. Hence $(x, z) \in R$. It follows that R is transitive.
9. The statement p is equivalent to q means that p and q have the same entries in their truth tables. R is reflexive, since p has the same truth table as p. R is symmetric, for if p and q have the same truth table, then q and p have the truth table. If p and q have the same entries in their truth tables and q and r have the same entries in their truth tables, then p and r also do, so R is transitive.
11. No
13. No
15. R is reflexive since a bit string s has the same number of 1s as itself. R is symmetric since s and t having the same number of 1s implies that t and s do. R is transitive since s and t having the same number of 1s, and t and u having the same number of 1s implies that s and u have the same number of 1s.
17. a) The sets of people of the same age
b) The sets of people with the same two parents
19. The set of all bit strings with exactly two 1s.
21. a) $\{s \mid s$ is a bit string of length 3$\}$
b) $\{s1 \mid s$ is a bit string of length 3$\}$
c) $\{s11 \mid s$ is a bit string of length 3$\}$
d) $\{s10101 \mid s$ is a bit string of length 3$\}$
23. $\{6n + k \mid n \in \mathbf{Z}\}$ for $k \in \{0, 1, 2, 3, 4, 5\}$
25. a) No **b)** Yes
c) Yes **d)** No
27. $[0]_6 \subseteq [0]_3$, $[1]_6 \subseteq [1]_3$, $[2]_6 \subseteq [2]_3$, $[3]_6 \subseteq [0]_3$, $[4]_6 \subseteq [1]_3$, $[5]_6 \subseteq [2]_3$
29. $\{(a, a), (a, b), (a, c), (b, a), (b, b), (b, c), (c, a), (c, b), (c, c), (d, d), (d, e), (e, d), (e, e)\}$
31. a) $\mathbf{Z}$ **b)** $\{n + \frac{1}{2} \mid n \in \mathbf{Z}\}$
33. Yes
35. R
37. First form the reflexive closure of R, then form the symmetric closure of the reflexive closure, and finally form the transitive closure of the symmetric closure of the reflexive closure.

Section 6.6

1. a) Yes **b)** No
c) Yes **d)** No
3. No
5. Yes
7. a) $\{(0,0), (1,0), (1,1), (2,0), (2,1), (2,2)\}$
b) $(\mathbf{Z}, \leq)$
c) $(P(\mathbf{Z}), \subseteq)$
d) $(\mathbf{Z}^+$, "is a multiple of")
9. a) $\{0\}$ and $\{1\}$, for instance
b) 4 and 6, for instance

11. a) $(1, 1, 2) < (1, 2, 1)$
b) $(0, 1, 2, 3) < (0, 1, 3, 2)$
c) $(0, 1, 1, 1, 0) < (1, 0, 1, 0, 1)$

13. $0 < 0001 < 001 < 01 < 010 < 0101 < 011 < 11$

15. a)

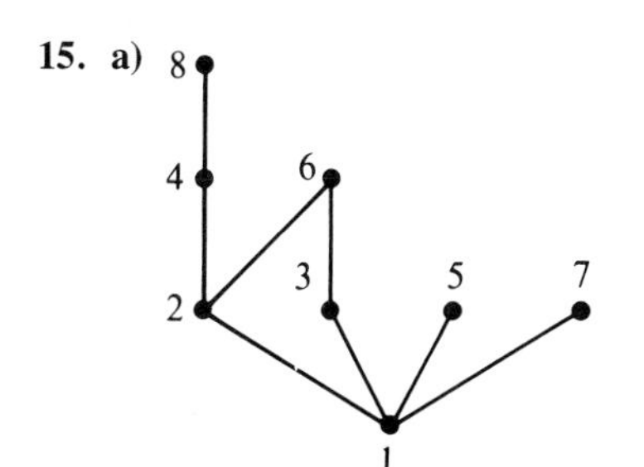

b)

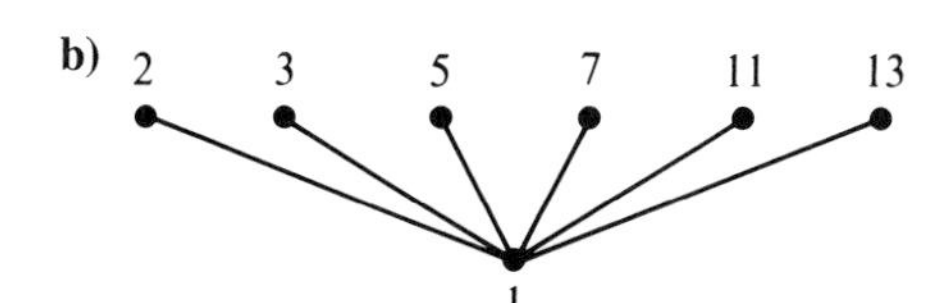

c) **d)**

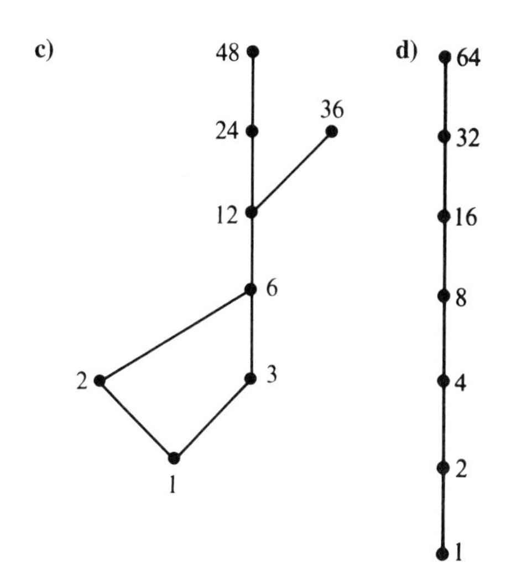

17. (a, b), (a, c), (a, d), (b, c), (b, d), (a, a), (b, b), (c, c), (d, d)

19. (a, a), (a, g), (a, d), (a, e), (a, f), (b, b), (b, g), (b, d), (b, e), (b, f), (c, c), (c, g), (c, d), (c, e), (c, f), (g, d), (g, e), (g, f), (g, g), (d, d), (e, e), (f, f)

21. **(a)** 24, 45 **(b)** 3, 5 **(c)** No **(d)** No **(e)** 15, 45 **(f)** 15 **(g)** 15, 5, 3 **(h)** 15

23. Since $(a, b) \preccurlyeq (a, b)$, $\preccurlyeq$ is reflexive. If $(a_1, a_2) \preccurlyeq (b_1, b_2)$ and $(a_1, a_2) \neq (b_1, b_2)$, either $a_1 \prec b_1$, or $a_1 = b_1$ and $a_2 \prec b_2$. In either case, (b_1, b_2) is not less than or equal to (a_1, a_2). Hence $\preccurlyeq$ is antisymmetric. Suppose that $(a_1, a_2) \prec (b_1, b_2) \prec (c_1, c_2)$. Then if $a_1 \prec b_1$ or $b_1 \prec c_1$, we have $a_1 \prec c_1$, so $(a_1, a_2) \prec (c_1, c_2)$, but if $a_1 = b_1 = c_1$, then $a_2 \prec b_2 \prec c_2$, which implies that $(a_1, a_2) \prec (c_1, c_2)$. Hence $\preccurlyeq$ is transitive.

25. Since $(s, t) \preccurlyeq (s, t)$, $\preccurlyeq$ is reflexive. If $(s, t) \preccurlyeq (u, v)$ and $(u, v) \preccurlyeq (s, t)$, then $s \preccurlyeq u \preccurlyeq s$ and $t \preccurlyeq v \preccurlyeq t$; hence $s = u$ and $t = v$. Hence $\preccurlyeq$ is antisymmetric. Suppose that $(s, t) \preccurlyeq (u, v) \preccurlyeq (w, x)$. Then $s \preccurlyeq u$, $t \preccurlyeq v$, $u \preccurlyeq w$, and $v \preccurlyeq x$. It follows that $s \preccurlyeq w$ and $t \preccurlyeq x$. Hence $(s, t) \preccurlyeq (w, x)$. Hence $\preccurlyeq$ is transitive.

27. a) Suppose that x is maximal and that y is the largest element. Then $x \preccurlyeq y$. Since x is not less than y, it follows that $x = y$. By Exercise 26(a) y is unique. Hence x is unique.
b) Suppose that x is minimal and that y is the smallest element. Then $x \succcurlyeq y$. Since x is not greater than y, it follows that $x = y$. By Exercise 26(b) y is unique. Hence x is unique.

29. The least element of a subset of $\mathbf{Z}^+ \times \mathbf{Z}^+$ is that pair which has the smallest possible first coordinate, and, if there is more than one such pair, that pair among those that has smallest second coordinate.

31. $a \prec b \prec c \prec d \prec e \prec f \prec g \prec h \prec i \prec j \prec k \prec l \prec m$

33. $C \prec A \prec B \prec D \prec E \prec F \prec G$

Supplementary Exercises

1. a) Irreflexive (we do not include the empty string), symmetric
b) Irreflexive, symmetric
c) Irreflexive, antisymmetric, transitive

3. $((a, b), (a, b)) \in R$ since $a + b = a + b$. Hence R is reflexive. If $((a, b), (c, d)) \in R$ then $a + d = b + c$, so that $c + b = d + a$. It follows that $((c, d), (a, b)) \in R$. Hence R is symmetric. Suppose that $((a, b), (c, d))$ and $((c, d), (e, f))$ belong to R. Then $a + d = b + c$ and $c + f = d + e$. Adding these two equations and subtracting $c + d$ from both sides gives $a + f = b + e$. Hence $((a, b), (e, f))$ belongs to R. Hence R is transitive.

5. Suppose that $(a, b) \in R$. Since $(b, b) \in R$ it follows that $(a, b) \in R^2$.

7. Yes, yes

9. Yes, yes

11. Two records with identical keys in the projection would have identical keys in the original.

13. $(\Delta \cup R)^{-1} = \Delta^{-1} \cup R^{-1} = \Delta \cup R^{-1}$

15. **a)** $R = \{(a, b), (a, c)\}$. The transitive closure of the symmetric closure of R is $\{(a, a), (a, b), (a, c), (b, a), (b, b), (b, c), (c, a), (c, b), (c, c)\}$ and is different from the symmetric closure of the transitive closure of R, which is $\{(a, b), (a, c), (b, a), (c, a)\}$.

b) Suppose (a, b) is in the symmetric closure of the transitive closure of R. We must show that (a, b) is in the transitive closure of the symmetric closure of R. We know that at least one of (a, b) and (b, a) is in the transitive closure of R. Hence, there is either a path from a to b in R or a path from b to a in R (or both). In the former case, there is a path from a to b in the symmetric closure of R. In the latter case, we can form a path from a to b in the symmetric closure of R by reversing the directions of all the edges in a path from b to a, going backwards. Hence (a, b) is in the transitive closure of the symmetric closure of R.

17. The closure of S with respect to property **P** is a relation with property **P** that contains R since $R \subseteq S$. Hence the closure of S with respect to property **P** contains the closure of R with respect to property **P**.

19. Use the basic idea of Warshall's algorithm, except let $w_{ij}^{[k]}$ equal the length of the longest path from v_i to v_j using interior vertices with subscripts not exceeding k, and equal to -1 if there is no such path. To find $w_{ij}^{[k]}$ from the entries of $\mathbf{W}_{k-1}$, determine for each pair (i, j) whether there are paths from v_i to v_k and from v_k to v_j using no vertices labeled greater than k. If either $w_{ik}^{[k-1]}$ or $w_{kj}^{[k-1]}$ is -1, then such a pair of paths does not exist, so set $w_{ij}^{[k]} = w_{ij}^{[k-1]}$. If such a pair of paths exists, then there are two possibilities. If $w_{kk}^{[k-1]} > 0$, there are paths of arbitrary long length from v_i to v_j, so set $w_{ij}^{[k]} = \infty$. If $w_{kk}^{[k-1]} = 0$, set $w_{ij}^{[k-1]} = \max(w_{ij}^{[k-1]}, w_{ik}^{[k-1]} + w_{kj}^{[k-1]})$. (Initially take $\mathbf{W}_0 = \mathbf{M}_R$.)

21. 52

23. Since $A_i \cap B_j$ is a subset of A_i and of B_j, the collection of subsets is a refinement of each of the given partitions. We must show that it is a partition. By construction, each of these sets is nonempty. To see that their union is S, suppose that $s \in S$. Since P_1 and P_2 are partitions of S, there are sets A_i and B_j such that $s \in A_i$ and $s \in B_j$. Therefore $s \in A_i \cap B_j$. Hence the union of these sets is S. To see that they are pairwise disjoint, note that unless $i = i'$ and $j = j'$, $(A_i \cap B_j) \cap (A_{i'} \cap B_{j'}) = (A_i \cap A_{i'}) \cap (B_j \cap B_{j'}) = \emptyset$.

25. The subset relation is a partial ordering on any collection of sets, since it is reflexive, antisymmetric, and transitive. Here the collection of sets is $\mathbf{R}(S)$.

27. Determine user needs $\prec$ write functional requirements $\prec$ set up test sites $\prec$ develop system requirements $\prec$ write documentation $\prec$ develop module A $\prec$ develop module B $\prec$ develop module C $\prec$ integrate modules $\prec$ alpha test $\prec$ beta test $\prec$ completion

29. **a)** The only antichain with more than one element is $\{c, d\}$.

b) The only antichains with more than one element are $\{b, c\}$, $\{c, e\}$, and $\{d, e\}$.

c) The only antichains with more than one element are $\{a, b\}, \{a, c\}, \{b, c\}, \{a, b, c\}, \{d, e\}, \{d, f\}, \{e, f\}, \{d, e, f\}$.

31. Let $(S, \preccurlyeq)$ be a finite poset, and let A be a maximal chain. Since $(A, \preccurlyeq)$ is also a poset it must have a minimal element m. Suppose that m is not minimal in S. Then there would be an element a of S with $a \prec m$. However, this would make the set $A \cup \{a\}$ a larger chain than A. To show this, we must show that a is comparable with every element of A. Since m is comparable with every element of A and m is minimal, it follows that $m \prec x$ when x is in A and $x \neq m$. Since $a \prec m$ and $m \prec x$, the transitive law shows that $a \prec x$ for every element of A.

33. Let $a\ R\ b$ denote that a is a descendant of b. By Exercise 32, if there is not a set of $n + 1$ people none of whom is a descendant of any other (an antichain), then $k \leq n$, so the set can be partitioned into $k \leq n$ chains. By the pigeonhole principle, at least one of these chains has at least $m + 1$ people in it.

35. Basis step: $a_{0,0} = [0(0 + 1)/2] + 0 = 0$. Inductive step: Assume true for all pairs less than (m, n). If $n = 0$ then $a_{m,n} = a_{m-1,n} + 1 = [n(n + 1)/2] + m - 1 + 1 = [n(n + 1)/2] + m$ as desired, since $(m-1, n) \prec (m, n)$. If $n \neq 0$, then $a_{m,n} = a_{m,n-1} + n = [n(n - 1)/2] + m + n = [n(n + 1)/2] + m$ since $(m, n - 1) \prec (m, n)$.

CHAPTER 7

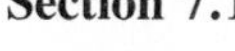

Section 7.1

1. a)

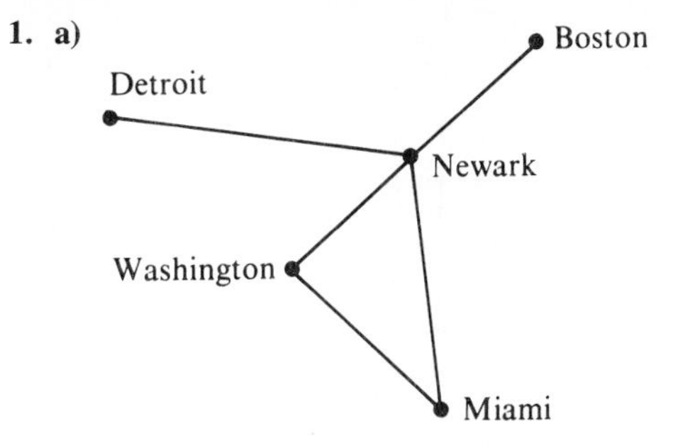

b)

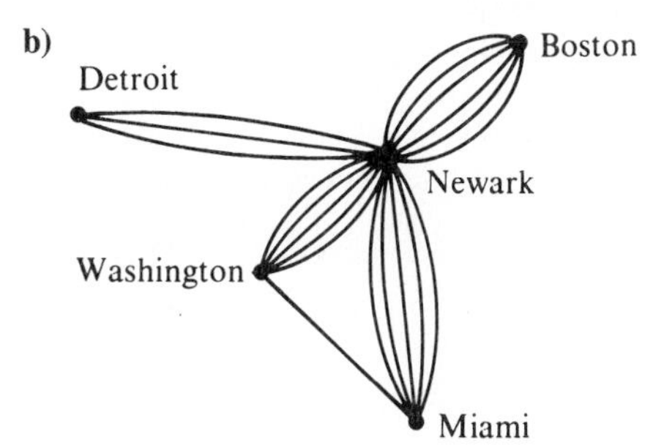

c)

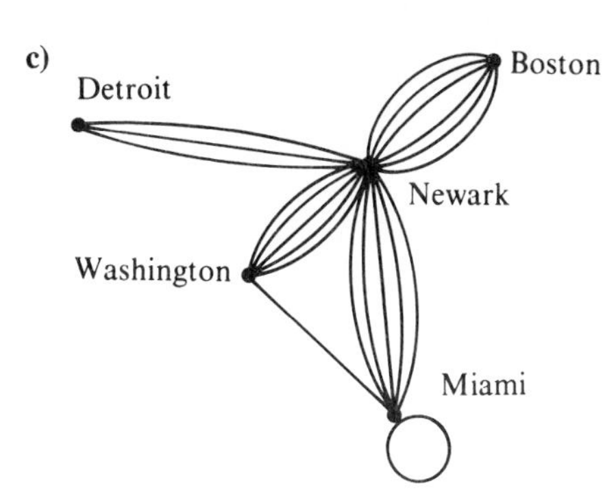

d)

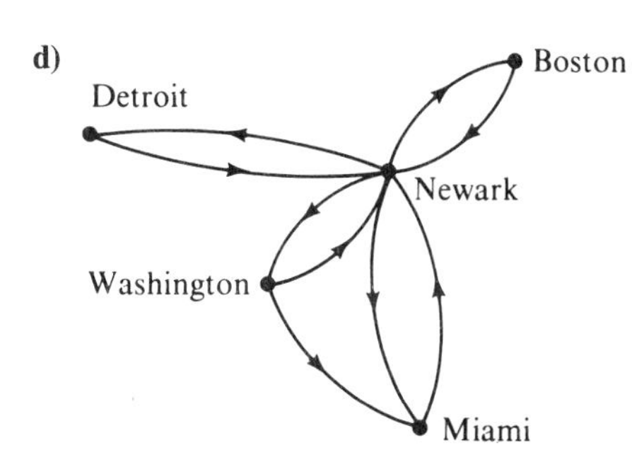

e)

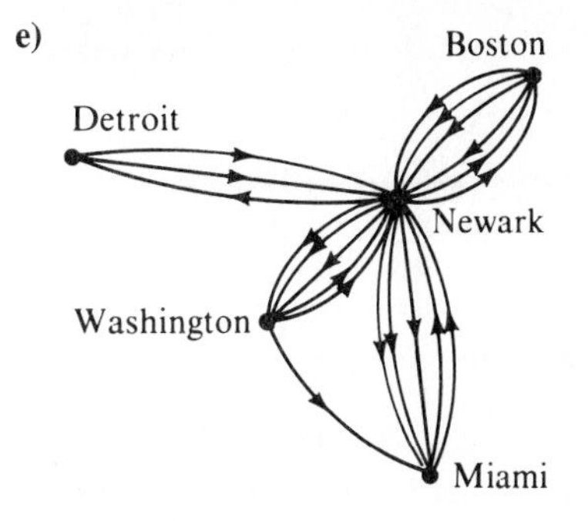

3. Simple graph
5. Pseudograph
7. Directed graph
9. Directed multigraph
11. a)

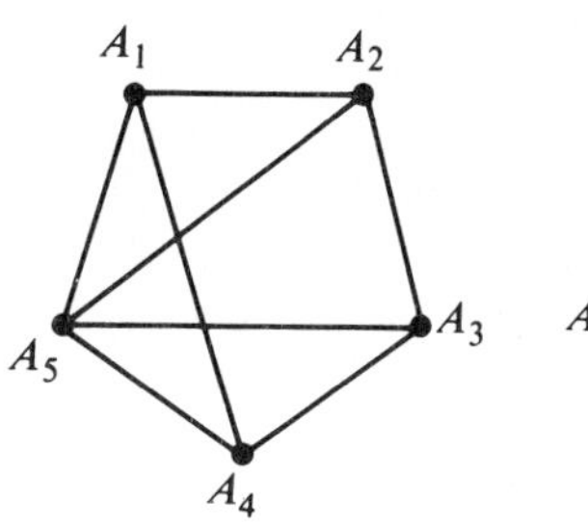

b)

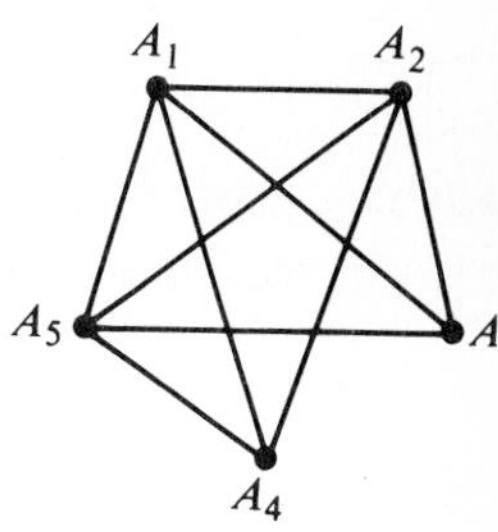

c)

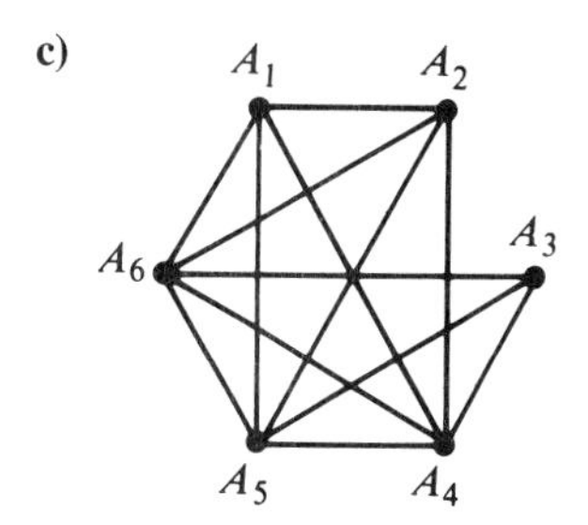

13.

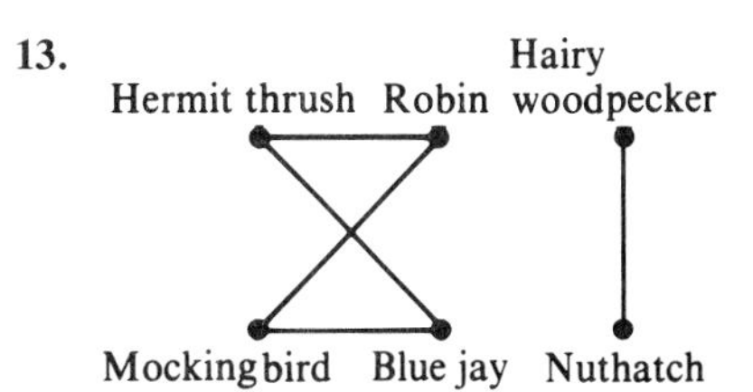

15.

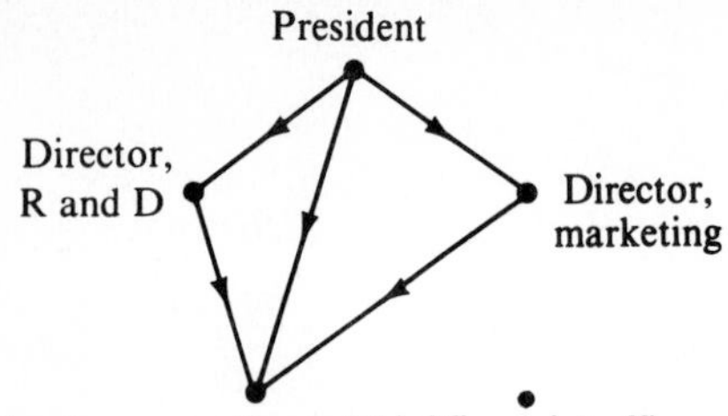

17.

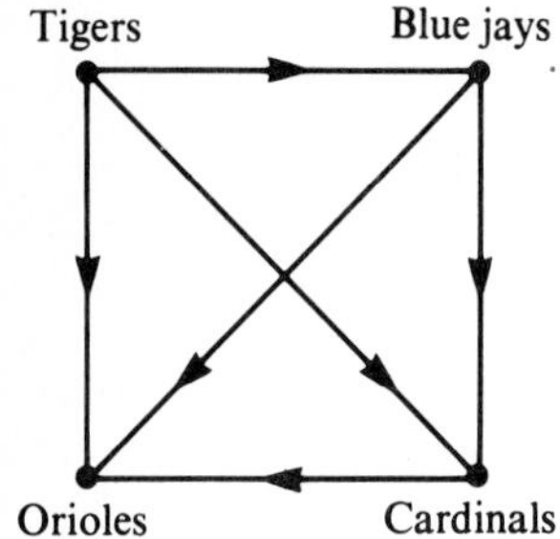

19.

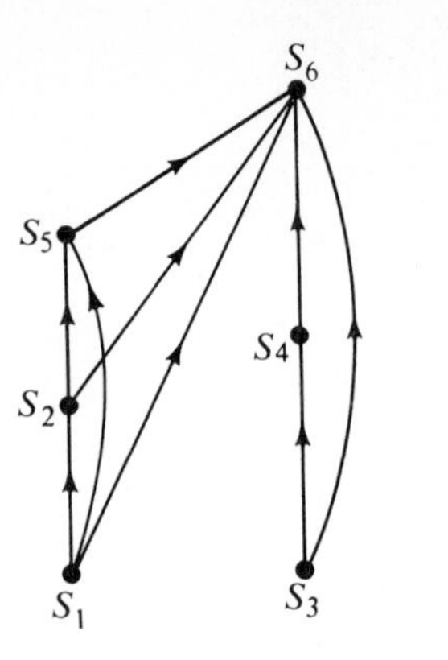

21. Represent people in the group by vertices. Put a directed edge into the graph for every pair of vertices. Label the edge from the vertex representing A to the vertex representing B with a + (plus) if A likes B, a − (minus) if A dislikes B, and a 0 if A is neutral about B.

Section 7.2

1. $v = 6$; $e = 6$; $\deg(a) = 2$, $\deg(b) = 4$, $\deg(c) = 1$, $\deg(d) = 0$, $\deg(e) = 2$, $\deg(f) = 3$; c is pendant; d is isolated.

3. $v = 9$; $e = 12$; $\deg(a) = 3$, $\deg(b) = 2$, $\deg(c) = 4$, $\deg(d) = 0$, $\deg(e) = 6$, $\deg(f) = 0$; $\deg(g) = 4$; $\deg(h) = 2$; $\deg(i) = 3$; d and f are isolated.

5. No, since the sum of the degrees of the vertices cannot be odd.

7. $v = 4$; $e = 7$; $\deg^-(a) = 3$, $\deg^-(b) = 1$, $\deg^-(c) = 2$, $\deg^-(d) = 1$, $\deg^+(a) = 1$, $\deg^+(b) = 2$, $\deg^+(c) = 1$, $\deg^+(d) = 3$.

9. 5 vertices, 13 edges; $\deg^-(a) = 6$, $\deg^+(a) = 1$, $\deg^-(b) = 1$, $\deg^+(b) = 5$, $\deg^-(c) = 2$, $\deg^+(c) = 5$, $\deg^-(d) = 4$, $\deg^+(d) = 2$, $\deg^-(e) = 0$, $\deg^+(e) = 0$.

11.

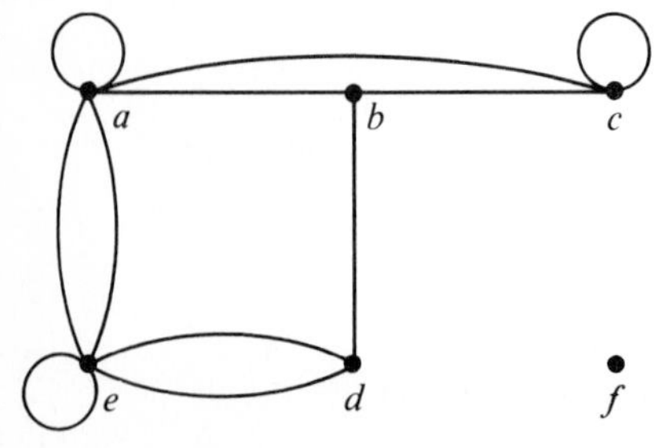

13. Bipartite

15. Not bipartite

17. Not bipartite

19. **a)** n vertices, $n(n-1)/2$ edges
b) n vertices, n edges
c) $n+1$ vertices, $2n$ edges
d) $m+n$ vertices, mn edges
e) 2^n vertices, $n2^{n-1}$ edges

21. **a)** Yes

b) No, sum of degrees is odd.
c) No
d) No, sum of degrees is odd.
e) Yes

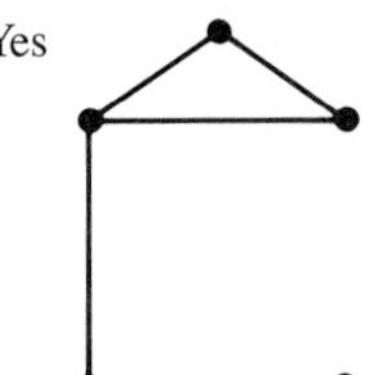

f) No, sum of degrees is odd.

23. 17

25.

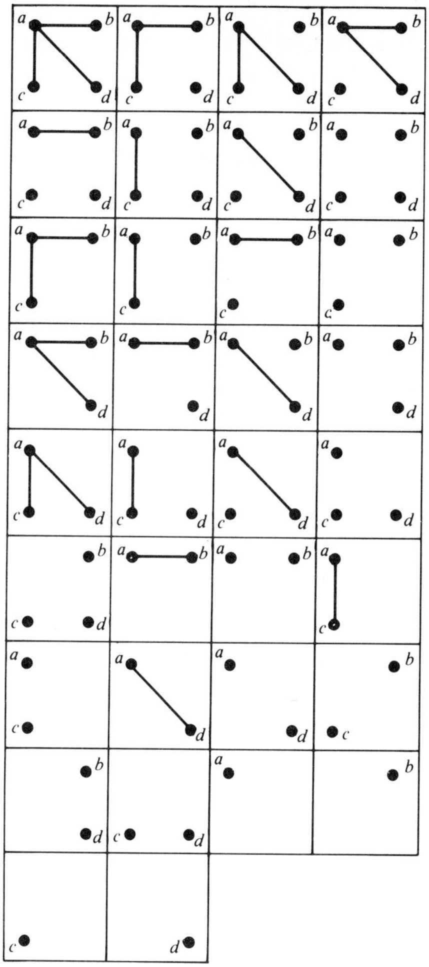

27. **a)** For all $n \geq 1$ **b)** For all $n \geq 3$
c) For $n = 3$ **d)** For all $n \geq 0$

29. 5

31.

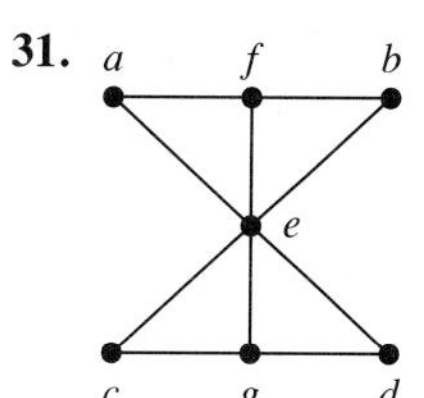

33. **a)** The graph with n vertices and no edges
b) The disjoint union of K_m and K_n
c) The graph with vertices $\{v_1, \cdots, v_n\}$ with an edge between v_i and v_j unless $i \equiv j \pm 1 \pmod{n}$
d) The graph whose vertices are represented by bit strings of length n with an edge between two vertices if the associated bit strings differ in more than one bit

35. $v(v-1)/2 - e$

37. The union of G and $\bar{G}$ contains an edge between each pair of the n vertices. Hence this union is K_n.

39.

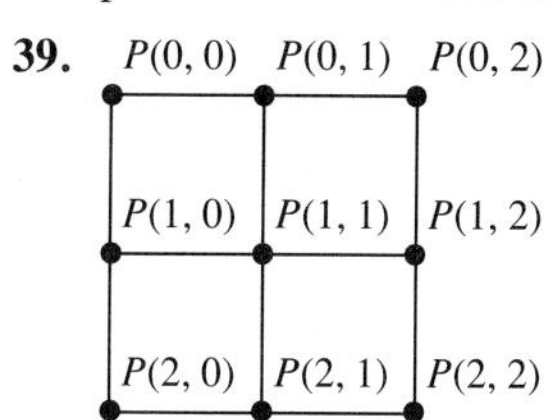

41. We can connect $P(i, j)$ and $P(k, l)$ by using $|i - k|$ hops to connect $P(i, j)$ and $P(k, j)$ and $|j - l|$ hops to connect $P(k, j)$ and $P(k, l)$. Hence the total number of hops required to connect $P(i, j)$ and $P(k, l)$ does not exceed $|i-k| + |j-l|$. This is less than or equal to $m + m = 2m$, which is $O(m)$.

Section 7.3

1.

Vertex	*Adjacent Vertices*
a	b, c, d
b	a, d
c	a, d
d	a, b, c

3.

Vertex	*Terminal Vertices*
a	a, b, c, d
b	d
c	a, b
d	b, c, d

5. $\begin{bmatrix} 0 & 1 & 1 & 1 \\ 1 & 0 & 0 & 1 \\ 1 & 0 & 0 & 1 \\ 1 & 1 & 1 & 0 \end{bmatrix}$, where vertices are listed in alphabetical order

7. $\begin{bmatrix} 1 & 1 & 1 & 1 \\ 0 & 0 & 0 & 1 \\ 1 & 1 & 0 & 0 \\ 0 & 1 & 1 & 1 \end{bmatrix}$

9. a) $\begin{bmatrix} 0 & 1 & 1 & 1 \\ 1 & 0 & 1 & 1 \\ 1 & 1 & 0 & 1 \\ 1 & 1 & 1 & 0 \end{bmatrix}$

b) $\begin{bmatrix} 0 & 1 & 1 & 1 & 1 \\ 1 & 0 & 0 & 0 & 0 \\ 1 & 0 & 0 & 0 & 0 \\ 1 & 0 & 0 & 0 & 0 \\ 1 & 0 & 0 & 0 & 0 \end{bmatrix}$

c) $\begin{bmatrix} 0 & 0 & 1 & 1 & 1 \\ 0 & 0 & 1 & 1 & 1 \\ 1 & 1 & 0 & 0 & 0 \\ 1 & 1 & 0 & 0 & 0 \\ 1 & 1 & 0 & 0 & 0 \end{bmatrix}$

d) $\begin{bmatrix} 0 & 1 & 0 & 1 \\ 1 & 0 & 1 & 0 \\ 0 & 1 & 0 & 1 \\ 1 & 0 & 1 & 0 \end{bmatrix}$

e) $\begin{bmatrix} 0 & 1 & 0 & 1 & 1 \\ 1 & 0 & 1 & 0 & 1 \\ 0 & 1 & 0 & 1 & 1 \\ 1 & 0 & 1 & 0 & 1 \\ 1 & 1 & 1 & 1 & 0 \end{bmatrix}$

f) $\begin{bmatrix} 0 & 1 & 1 & 0 & 1 & 0 & 0 & 0 \\ 1 & 0 & 0 & 1 & 0 & 1 & 0 & 0 \\ 1 & 0 & 0 & 1 & 0 & 0 & 1 & 0 \\ 0 & 1 & 1 & 0 & 0 & 0 & 0 & 1 \\ 1 & 0 & 0 & 0 & 0 & 1 & 1 & 0 \\ 0 & 1 & 0 & 0 & 1 & 0 & 0 & 1 \\ 0 & 0 & 1 & 0 & 1 & 0 & 0 & 1 \\ 0 & 0 & 0 & 1 & 0 & 1 & 1 & 0 \end{bmatrix}$

11.

13. $\begin{bmatrix} 0 & 0 & 1 & 0 \\ 0 & 0 & 1 & 2 \\ 1 & 1 & 0 & 1 \\ 0 & 2 & 1 & 0 \end{bmatrix}$

15. $\begin{bmatrix} 1 & 0 & 2 & 1 \\ 0 & 1 & 1 & 2 \\ 2 & 1 & 1 & 0 \\ 1 & 2 & 0 & 1 \end{bmatrix}$

17.

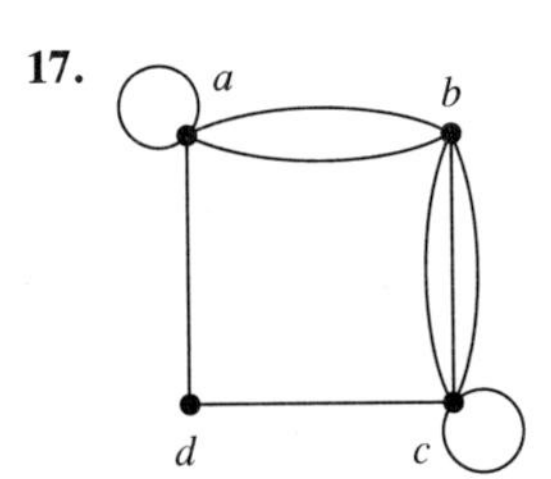

19. $\begin{bmatrix} 0 & 1 & 0 & 0 \\ 0 & 1 & 1 & 0 \\ 0 & 1 & 1 & 1 \\ 1 & 0 & 0 & 0 \end{bmatrix}$

21. $\begin{bmatrix} 1 & 1 & 2 & 1 \\ 1 & 0 & 0 & 2 \\ 1 & 0 & 1 & 1 \\ 0 & 2 & 1 & 0 \end{bmatrix}$

23.

25. Yes

27. Exercise 13: $\begin{bmatrix} 1 & 0 & 0 & 0 & 0 \\ 0 & 1 & 1 & 1 & 0 \\ 1 & 1 & 0 & 0 & 1 \\ 0 & 0 & 1 & 1 & 1 \end{bmatrix}$

Exercise 14: $\begin{bmatrix} 1 & 1 & 1 & 1 & 0 & 0 & 0 & 0 \\ 1 & 1 & 1 & 0 & 1 & 0 & 0 & 0 \\ 0 & 0 & 0 & 0 & 1 & 1 & 1 & 1 \\ 0 & 0 & 0 & 1 & 0 & 1 & 1 & 1 \end{bmatrix}$

Exercise 15: $\begin{bmatrix} 1 & 1 & 1 & 1 & 0 & 0 & 0 & 0 & 0 & 0 \\ 0 & 0 & 0 & 0 & 1 & 1 & 1 & 1 & 0 & 0 \\ 0 & 1 & 1 & 0 & 0 & 1 & 0 & 0 & 1 & 0 \\ 0 & 0 & 0 & 1 & 0 & 0 & 1 & 1 & 0 & 1 \end{bmatrix}$

29. $\deg(v)-$ number of loops at v; $\deg^-(v)$.

31. 2 if e is not a loop, 1 if e is a loop.

33. **a)**
$$\begin{bmatrix} 1 & 1 & \cdots & 1 & 0 & \cdots & 0 \\ 1 & 0 & \cdots & 0 & 1 & \cdots & 0 \\ 0 & 1 & \cdots & 0 & 1 & \cdots & 0 \\ \vdots & \vdots & & \vdots & \vdots & & \vdots \\ 0 & 0 & \cdots & 0 & 0 & \cdots & 1 \\ 0 & 0 & \cdots & 1 & 0 & \cdots & 1 \end{bmatrix}$$

b)
$$\begin{bmatrix} 1 & 0 & \cdots & 0 & 1 \\ 1 & 1 & \cdots & 0 & 0 \\ 0 & 1 & \cdots & 0 & 0 \\ \vdots & \vdots & & \vdots & \vdots \\ 0 & 0 & \cdots & 1 & 0 \\ 0 & 0 & \cdots & 1 & 1 \end{bmatrix}$$

c)
$$\begin{bmatrix} 0 \;\; 0 \;\; \cdots \;\; 0 & 1 & 1 & \cdots & 1 \\ & 1 & 0 & \cdots & 0 \\ \mathbf{B} & 0 & 1 & \cdots & 0 \\ & \vdots & \vdots & & \vdots \\ & 0 & 0 & \cdots & 1 \end{bmatrix}$$

where **B** is the answer to (b)

d)
$$\begin{bmatrix} 1 & 1 & \cdots & 1 & 0 & \cdots & 0 \\ 0 & 0 & \cdots & 0 & 1 & \cdots & 0 \\ \vdots & \vdots & & \vdots & \vdots & & \vdots \\ 0 & 0 & \cdots & 0 & 0 & \cdots & 1 \\ 1 & 0 & & 0 & 1 & \cdots & 0 \\ 0 & 1 & & 0 & 0 & & \\ \vdots & \vdots & & \vdots & \vdots & & \\ 0 & 0 & \cdots & 1 & 0 & \cdots & 1 \end{bmatrix}$$

35. Isomorphic

37. Isomorphic

39. Isomorphic

41. Not isomorphic

43. Isomorphic

45. G is isomorphic to itself by the identity function, so isomorphism is reflexive. Suppose that G is isomorphic to H. Then there exists a one-to-one correspondence f from G to H that preserves adjacency and nonadjacency. It follows that f^{-1} is a one-to-one correspondence from H to G that preserves adjacency and nonadjacency. Hence isomorphism is symmetric. If G is isomorphic to H and H is isomorphic to K, then there are one-to-one correspondences f and g from G to H and from H to K that preserve adjacency and nonadjacency. It follows that $g \circ f$ is a one-to-one correspondence from G to K that preserves adjacency and nonadjacency. Hence isomorphism is transitive.

47. All zeros

49. Label the vertices in order so that all of the vertices in the first set of the partition of the vertex set come first. Since no edges join vertices in the same set of the partition, the matrix has the desired form.

51. C_5

53. $n = 5$ only

55. 4

57. **a)** Yes
b) No
c) No

59. $G = (V_1, E_1)$ is isomorphic to $H = (V_2, E_2)$ if and only if there exist functions f from V_1 to V_2 and g from E_1 to E_2 such that each is a one-to-one correspondence and for every edge e in E_1 the endpoints of $g(e)$ are $f(v)$ and $f(w)$ where v and w are the endpoints of e.

61. Yes

63. Yes

65. The product is $[a_{ij}]$ where a_{ij} is the number of edges from v_i to v_j when $i \neq j$ and a_{ii} is the number of edges incident to v_i.

Section 7.4

1. **a)** Path of length 4; not a circuit; not simple
b) Not a path
c) Not a path
d) Simple circuit of length 5.

3. No

5. No

7. **a)** 2 **b)** 7 **c)** 20 **d)** 61

9. **a)** 3 **b)** 0 **c)** 27 **d)** 0

11. **a)** 1 **b)** 0 **c)** 2 **d)** 1 **e)** 5 **f)** 3

13. R is reflexive by definition. Assume that $(u, v) \in R$; then there is a path from u to v. Then $(v, u) \in R$ since there is a path from v to u, namely, the path from u

to v traversed backwards. Assume that $(u, v) \in R$ and $(v, w) \in R$; then there are paths from u to v and from v to w. Putting these two paths together gives a path from u to w. Hence $(u, w) \in R$. It follows that R is transitive.

15. c

17. b, c, e, i.

19. If a vertex is pendant it is clearly not a cut vertex. So an endpoint of a cut edge that is a cut vertex is not pendant. Removal of a cut edge produces a graph with more connected components than in the original graph. If an endpoint of a cut edge is not pendant, the connected component it is in after the removal of the cut edge contains more than just this vertex. Consequently, removal of that vertex, and all edges incident to it, including the original cut edge, produces a graph with more connected components than were in the original graph. Hence an endpoint of a cut edge that is not pendant is a cut vertex.

21. Assume there exists a connected graph G with at most one vertex that is not a cut vertex. Define the distance between the vertices u and v, denoted by $d(u, v)$, to be the length of the shortest path between u and v in G. Let s and t be vertices in G such that $d(s, t)$ is a maximum. Either s or t (or both) is a cut vertex, so without loss of generality suppose that s is a cut vertex. Let w belong to the connected component that does not contain t of the graph obtained by deleting s and all edges incident to s from G. Since every path from w to t contains s, $d(w, t) > d(s, t)$, which is a contradiction.

23. a) Denver-Chicago, Boston-New York **b)** Seattle-Portland, Portland-San Francisco, Salt Lake City-Denver, New York-Boston, Boston-Burlington, Boston-Bangor.

25. A set of people who collectively influence everyone (directly or indirectly); {Deborah, Yvonne}

27. An edge cannot connect two vertices in different connected components. Since there are at most $C(n_i, 2)$ edges in the connected component with n_i vertices, it follows that there are at most $\sum_{i=1}^{k} C(n_i, 2)$ edges in the graph.

29. Suppose that G is not connected. Then it has a component of k vertices for some k, $1 \le k \le n-1$. The most edges G could have is $C(k, 2) + C(n-k, 2) = (k(k-1) + (n-k)(n-k-1))/2 = k^2 - nk + (n^2 - n)/2$. This quadratic function of f is minimized at $k = n/2$ and maximized at $k = 1$ or $k = n-1$. Hence if G is not connected, the number of edges does not exceed the value of this function at 1 and at $n-1$, namely, $(n-1)(n-2)/2$.

31. a) 1 **b)** 2 **c)** 6 **d)** 21

33. 2

35. Let the paths P_1 and P_2 be $u = x_0, x_1, \ldots, x_n = v$ and $u = y_0, y_1, \ldots, y_m = v$, respectively. Since P_1 and P_2 do not contain the same set of edges, they must eventually diverge. If this happens only after one of them has ended, the rest of the other path is a simple circuit from v to v. Otherwise, we can suppose that $x_0 = y_0, x_1 = y_1 \ldots, x_t = y_t$, but $x_{i+1} \neq y_{i+1}$. Follow the path y_1, y_{t+1}, y_{t+2}, and so on, until it once again encounters a vertex on P_1. Once we are back on P_1, follow it, forwards or backwards as necessary, back to x_t. Since $x_i = y_i$, this forms a circuit that must be simple since no edge among the x_ks can be repeated and no edge among the x_ks can equal one of the Y_is that we used.

37. The graph G is connected if and only if every off-diagonal entry of $\mathbf{A} + \mathbf{A}^2 + \mathbf{A}^3 + \cdots + \mathbf{A}^{n-1}$ is positive where $\mathbf{A}$ is the adjacency matrix of G

Section 7.5

1. No

3. No

5. $a, b, c, d, c, e, d, b, e, a, e, a$

7. $a, i, h, g, d, e, f, g, c, e, h, d, c, a, b, i, c, b, h, a$

9. An Euler path exists: $f, a, b, c, d, e, f, b, d$ is one such Euler path.

11. An Euler path exists: $b, c, d, e, f, d, g, i, d, a, h, i, a, b, i, c$ is one such Euler path.

13. An Euler path exists: $b, c, d, e, f, d, g, i, d, a, h, i, a, b, i, c$

15. No, A still has odd degree.

17. When the graph in which vertices represent intersection and edges streets has an Euler path

19. Yes

21. No

23. If there is an Euler path, as we follow it each vertex except the starting and ending vertices must have equal in-degree and out-degree, since whenever we come to a vertex along an edge, we leave it along another edge. The starting vertex must have out-degree 1 larger than its in-degree, since we use one edge leading out of this vertex and whenever we visit it again we use one edge leading into it and one leaving it. Similarly, the ending vertex must have in-degree 1 greater than its out-degree. Since the Euler path with directions erased produces a path between any two vertices, in the underlying undirected graph, the graph is weakly connected. Conversely, suppose the graph meets the degree conditions stated. If we add one more edge from the vertex of deficient out-degree to the vertex of deficient in-degree, then the graph has every vertex with equal in-degree and out-degree. Since the graph is still weakly connected, by Exercise 22 this new graph has an Euler circuit. Now delete the added edge to obtain the Euler path.

25. No

27. No

29. $a, b, d, b, c, d, c, a, d$

31. $a, d, b, d, e, b, e, c, b, a$

33. $a, b, c, e, b, d, c, b, f, d, e, f, e, a, f, a$

35. Follow the same procedure as Algorithm 1, taking care to follow the directions of edges.

37. **a)** $n = 2$ **b)** None **c)** None **d)** $n = 1$

39. Exercise 1: 1 time;
Exercise 2–7: 0 times

41. a, b, c, d, e, a is a Hamilton circuit.

43. No Hamilton circuit exists, because once a purported circuit has reached e it would have nowhere to go.

45. a, b, c, f, d, e is a Hamilton path.

47. f, e, d, a, b, c is a Hamilton path.

49. No Hamilton path exists. There are eight vertices of degree 2, and only two of them can be end vertices of a path. For each of the other six, their two incident edges must be in the path. It is not hard to see that if there is to be a Hamilton path, exactly one of the inside corner vertices must be an end, and that this is impossible.

51. $m = n \geq 2$

53. The result is trivial for $n = 1$: code is 0, 1. Assume we have a Gray code of order n. Let $c_1, \ldots, c_k, k = 2^n$ be such a code. Then $0c_1, \ldots, 0c_k, 1c_k, \ldots, 1c_l$ is a Gray code of order $n + 1$.

55. **procedure** *Fleury* ($G = (V, E)$: connected multigraph with the degree of all vertices even, $V = \{v_l, \ldots, v_n\}$)
$v := v_1$
circuit $:= v$
$H := G$
while H has edges
begin
$e :=$ first edge with endpoint V in H (with respect to listing of V) such that e is not a cut edge of H, if one exists, and simply the first edge in H with endpoint v otherwise
$w :=$ other endpoint of e
circuit $:=$ *circuit* with edge e, w added
$v := w$
$H := H - e$
end {*circuit* is an Euler circuit}

57. If G has an Euler circuit, it is also an Euler path. If not, add an edge between the two vertices of odd degree and apply the algorithm to get an Euler circuit. Then delete the new edge.

59. Suppose $G = (V, E)$ is a bipartite graph with $V = V_1 \cup V_2$, where no edge connects a vertex in V_1 and a vertex in V_2. Suppose that V has a Hamilton circuit. Such a circuit must be of the form $a_1, b_1, a_2, b_2, \ldots, a_k, b_k, a_1$, where $a_i \in V_1$ and $b_i \in V_2$ for $i = 1, 2, \ldots, k$. Since the Hamilton circuit visits each vertex exactly once, except for v_1, where it begins and ends, the number of vertices in the graph equals $2k$, an even number. Hence, a bipartite graph with an odd number of vertices cannot have a Hamilton circuit.

61.

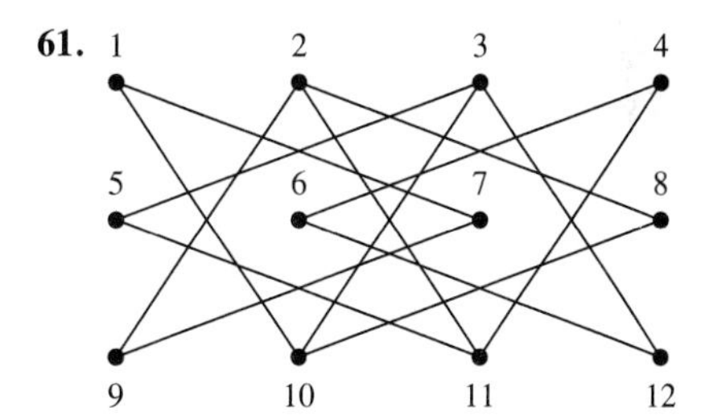

63. We represent the squares of a 3×4 chessboard as follows:

1	2	3	4
5	6	7	8
9	10	11	12

A knight's tour can be made by following the moves 8, 10, 1, 7, 9, 2, 11, 5, 3, 12, 6, 4

65. We represent the squares of a 4×4 chessboard as follows:

1	2	3	4
5	6	7	8
9	10	11	12
13	14	15	16

There are only two moves from each of the four corner squares. If we include all the edges 1–10, 1–7, 16–10, and 16–7, a circuit is completed too soon, so at least one of these edges must be missing. Without loss of generality, assume the path starts 1–10, 10–16, and 16–7. Now the only moves from square 3 are to squares 5, 10, and 12 and square 10 already has two incident edges. Therefore, 3–5 and 3–12 must be in the Hamilton circuit. Similarly, edges 8–2 and 8–15 must be in the circuit. Now the only moves from square 9 are to squares 2, 7, and 15. If there were edges from square 9 to both squares 2 and 15, a circuit would be completed too soon. Therefore the edge 9-7 must be in the circuit giving square 7 its full complement of edges. But now square 14 is forced to be joined to squares 5 and 12, completing a circuit too soon (5–14–12–3–5). This contradiction shows that there is no knight's tour on the 4×4 board.

67. Since there are mn squares on an $m \times n$ board, if both m and n are odd, there are an odd number of squares. Since by Exercise 66 the corresponding graph is bipartite, by Exercise 59 it has no Hamilton circuit. Hence there is no re-entrant knight's tour.

Section 7.6

1. a) Vertices are the stops, edges join adjacent stops, weights are the times required to travel between adjacent stops.
b) Same as (a), except weights are distances between adjacent stops.
c) Same as (a), except weights are fares between stops.

3. 16

5. Exercise 2: a, b, e, d, z
Exercise 3: a, c, d, e, g, z
Exercise 4: $a, b, e, h, l, m, p, s, z$

7. a) a, c, d **b)** a, c, d, f **c)** c, d, f **e)** b, d, e, g, z

9. a) Direct
b) Via New York
c) Via Atlanta and Chicago
d) Via New York

11. a) Via Chicago **b)** Via Chicago **c)** Via Los Angeles **d)** Via Chicago

13. a) Via Chicago **b)** Via Chicago **c)** Via Los Angeles **d)** Via Chicago

15. Do not stop the algorithm when z is added to the set S.

17. a) Via Woodbridge, via Woodbridge and Camden
b) Via Woodbridge, via Woodbridge and Camden

19. For instance, sightseeing tours, street cleaning.

21.

	a	b	c	d	e	z
a	4	3	2	8	10	13
b	3	2	1	5	7	10
c	2	1	2	6	8	11
d	8	5	6	4	2	5
e	10	7	8	2	4	3
z	13	10	11	5	3	6

23. $O(n^3)$

Section 7.7

1. Yes

3.

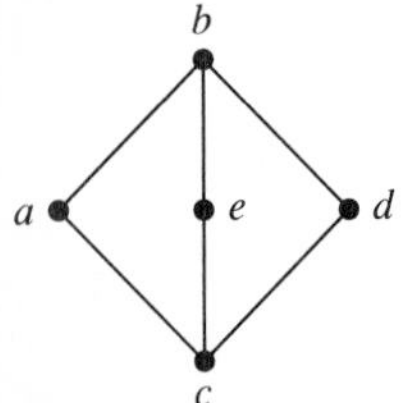

5. No

7. Yes

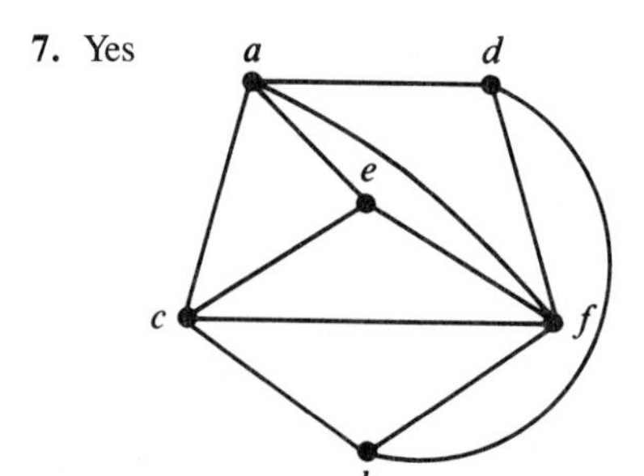

9. A triangle is formed by the planar representation of the subgraph of K_5 consisting of the edges connecting v_1, v_2, and v_3. The vertex v_4 must be placed either within the triangle or outside of it. We will consider only the case when v_4 is inside the triangle; the other case is similar. Drawing the three edges from v_1, v_2, and v_3 to v_4 forms four regions. No matter which of these four regions v_5 is in, it is possible to join it to only three, and not all four, of the other vertices.

11. 8

13. Since there are no loops or multiple edges and no simple circuits of length 3, and the degree of the unbounded region is at least 4, each region has degree at least 4. Thus $2e \geq 4r$, or $r \leq e/2$. But $r = e - v + 2$, so we have $e - v + 2 \leq e/2$, which implies that $e \leq 2v - 4$.

15. As in Corollary 2, we have $2e \geq 5r$ and $r = e - v + 2$. Thus $e - v + 2 \leq 2e/5$, which implies that $e \leq (5/3)v - (10/3)$.

17. Only (a) and (c)

19. Not homeomorphic to $K_{3,3}$

21. Planar

23. Nonplanar

25. **a)** 1 **b)** 3 **c)** 9 **d)** 2 **e)** 4 **f)** 16

27. **a)** 2 **b)** 2 **c)** 2 **d)** 2 **e)** 2 **f)** 2

29.

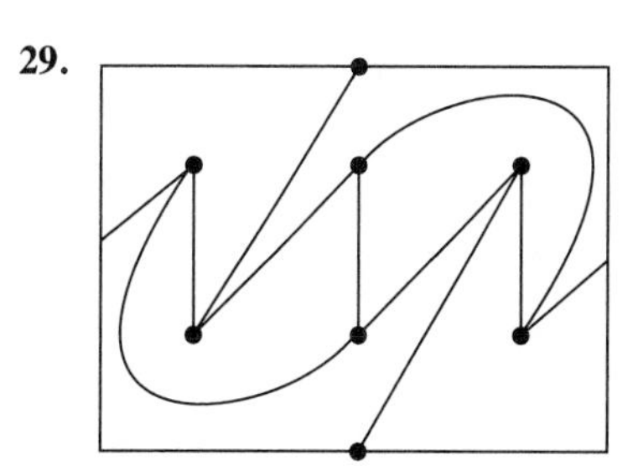

Section 7.8

1. **a)**

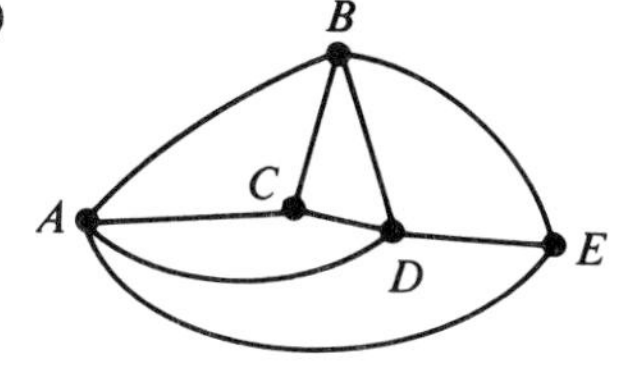

b)

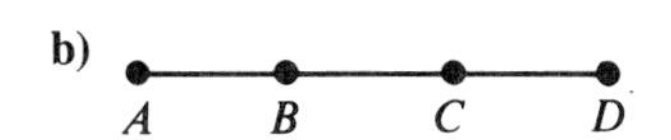

c)

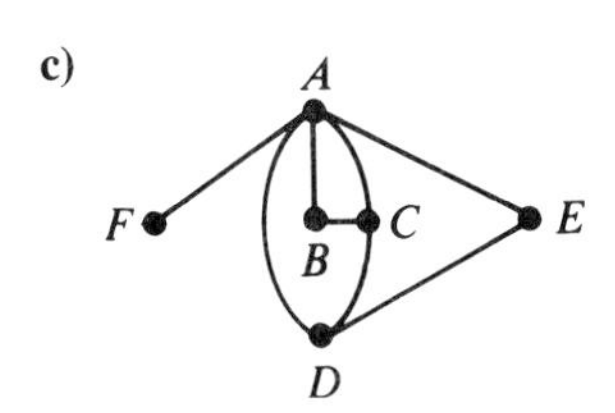

3. 3

5. 3

7. 2

9. 3

11. Graphs with no edges

13. 3 if n is even, 4 if n is odd

15. Period 1: Math 115, Math 185; period 2: Math 116, CS 473; period 3: Math 195, CS 101; period 4: CS 102; period 5: CS 273

17. 5

19. Exercise 3: 3
Exercise 4: 6
Exercise 5: 3
Exercise 6: 4
Exercise 7: 3
Exercise 8: 6
Exercise 9: 4

21. 5

23. The set of vertices with one of the colors is one of the parts, and the set of vertices with the other color is the other part. Since no edge can join vertices of the same color, there are no edges between vertices in the same part.

25. Color 1: *e, f, d*; color 2: *c, a, i, g*; color 3: *h, b, j*

27. Color C_6

29. **a)** 6 **b)** 7 **c)** 9 **d)** 11

31. Represent frequencies by colors and zones by vertices. Join two vertices with an edge if the zones these vertices represent interfere with one another. Then a k-tuple coloring is precisely an assignment of frequencies that avoids interference.

Supplementary Exercises

1. 2500

3. Yes

5. Yes

7. $\sum_{i=1}^{m} n_i$ vertices, $\sum_{i<j} n_i n_j$ edges

9. a)

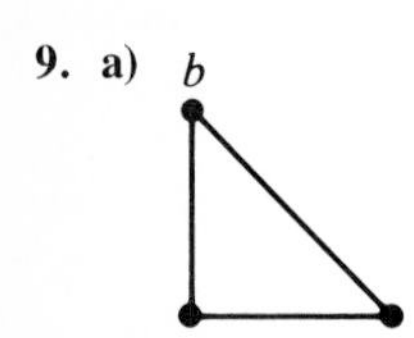

b)

c)

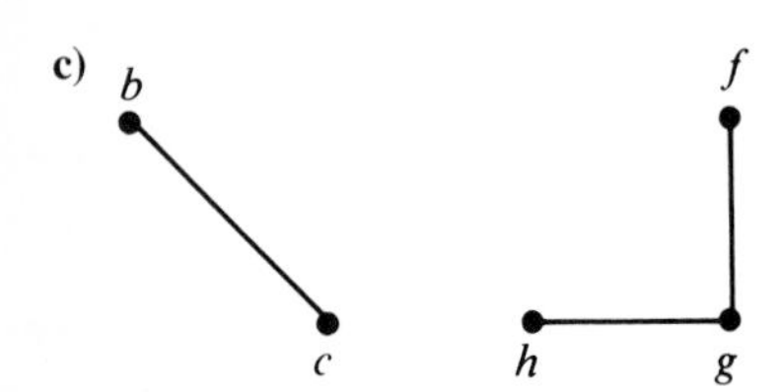

11. Complete subgraphs containing the following sets of vertices: $\{b, c, e, f\}$, $\{a, b, g\}$, $\{a, d, g\}$, $\{d, e, g\}$, $\{b, e, g\}$

13. Complete subgraphs containing the following sets of vertices: $\{b, c, d, j, k\}$, $\{a, b, j, k\}$, $\{e, f, g, i\}$, $\{a, b, i\}$, $\{a, i, j\}$, $\{b, d, e\}$, $\{b, e, i\}$, $\{b, i, j\}$, $\{g, h, i\}$, $\{h, i, j\}$

15. $\{c, d\}$ is a minimum dominating set.

17. a)

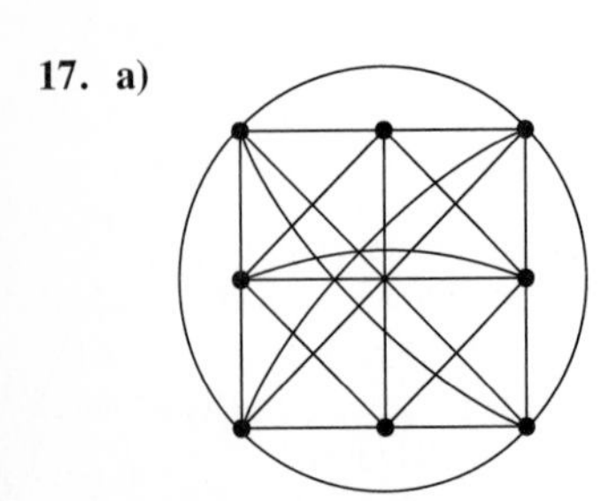

b)

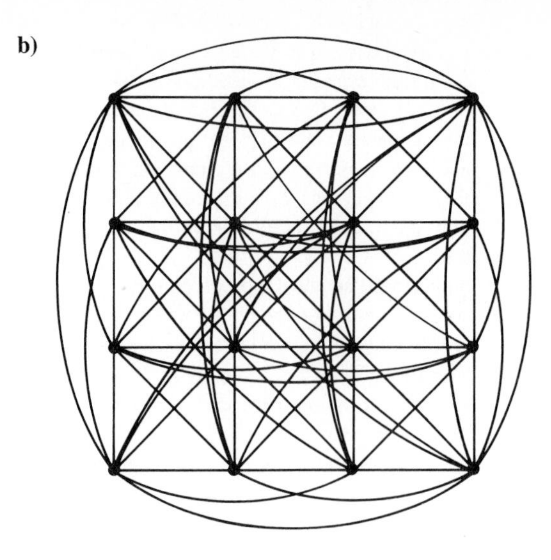

19. a) 1 **b)** 2 **c)** 3

21. a) A path from u to v in a graph G induces a path from $f(u)$ to $f(v)$ in an isomorphic graph H.

b) Suppose f is an isomorphism from G to H. If $v_0, v_1, \ldots, v_n, v_0$ is a Hamilton circuit in G, then $f(v_0), f(v_1), \ldots, f(v_n), f(v_0)$ must be a Hamilton circuit in H since it is still a circuit and $f(v_i) \neq f(v_j)$ for $0 \leq i < j \leq n$.

c) Suppose f is an isomorphism from G to H. Then if $v_0, v_1, \ldots, v_n, v_0$ is an Euler circuit in G, then $f(v_0), f(v_1), \ldots, f(v_n), f(v_0)$ must be an Euler circuit in H since it is a circuit that contains each edge exactly once.

d) Two isomorphic graphs must have the same crossing number since they can be drawn exactly the same way in the plane.

e) Suppose f is an isomorphism from G to H. Then v is isolated in G if and only if $f(v)$ is isolated in H. Hence the graphs must have the same number of isolated vertices.

f) Suppose f is an isomorphism from G to H. If G is bipartite, then the vertex set of G can be partitioned into V_1 and V_2 with no edge connecting a vertex of V_1 with one in V_2. Then the vertex set of H can be partitioned into $f(V_1)$ and $f(V_2)$ with no edge connecting a vertex in $f(V_1)$ and one in $f(V_2)$.

23. 3

25. No

27. Yes

29. If e is a cut edge with endpoints u and v, then if we direct e from u to v, there will be no path in the directed graph from v to u, or else e would not have been a cut edge. Similar reasoning works if we direct e from v to u.

31. $n - 1$

33. Let the vertices represent the chickens. We include the edge (u, v) in the graph if and only if chicken u dominates chicken v.

35. **a)** 4 **b)** 2 **c)** 3 **d)** 4 **e)** 4 **f)** 2

37. **a)** Suppose that $G = (V, E)$. Let $a, b \in V$. We must show that the distance between a and b in $\overline{G}$ is at most 2. If $\{a, b\} \notin E$ this distance is 1, so assume $\{a, b\} \in E$. Since the diameter of G is greater than 3, there are vertices u and v such that the distance in G between u and v is greater than 3. Either u or v, or both, is not in the set $\{a, b\}$. Assume that u is different from both a and b. Either $\{a, u\}$ or $\{b, u\}$ belongs to E; otherwise a, u, b would be a path in $\overline{G}$ of length 2. So, without loss of generality, assume $\{a, u\} \in E$. Thus v cannot be a or b, and by the same reasoning either $\{a, v\} \in E$ or $\{b, v\} \in E$. In either case, this gives a path of length less than or equal to 3 from u to v in G, a contradiction.

b) Suppose $G = (V, E)$. Let $a, b \in V$. We must show that the distance between a and b in $\overline{G}$ does not exceed 3. If $\{a, b\} \notin E$, the result follows, so assume that $\{a, b\} \in E$. Since the diameter of G is greater than or equal to 3, there exist vertices u and v such that the distance in G between u and v is greater than or equal to 3. Either u or v, or both, is not in the set $\{a, b\}$. Assume u is different from both a and b. Either $\{a, u\} \in E$ or $\{b, u\} \in E$; otherwise a, u, b is a path of length 2 in $\overline{G}$. So, without loss of generality, assume $\{a, u\} \in E$. Thus v is different from a and from b. If $\{a, v\} \in E$, then u, a, v is a path of length 2 in G, so $\{a, v\} \notin E$ and thus $\{b, v\} \in E$ (or else there would be a path a, v, b of length 2 in $\overline{G}$). Hence $\{u, b\} \notin E$; otherwise u, b, v is a path of length 2 in G. Thus, a, v, u, b is a path of length 3 in $\overline{G}$, as desired.

39. a, b, e, z

41. a, c, b, d, e, z

43. If G is planar, then since $e \leq 3v - 6$, G has at most 27 edges. Similarly, $\overline{G}$ has at most 27 edges. But the union of G and $\overline{G}$, is K_{11} which has 55 edges, and $55 > 27 + 27$.

45. Suppose that G is colored with k colors and has independence number i. Since each color class must be an independent set, each color class has no more than i elements. Thus there are at most ki vertices.

CHAPTER 8

Section 8.1

1. (a), (c), (e)

3. No

5. **a)**

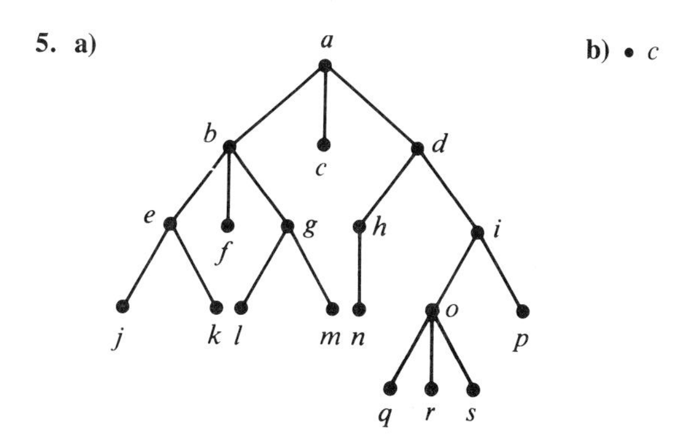

b) • c

c)

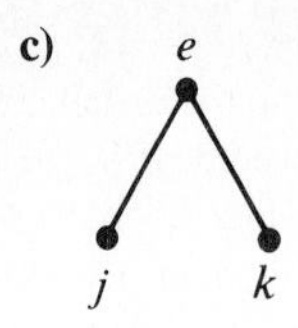

7. a) 2 **b)** 4 **c)** 9

9. The "only if" part is Theorem 2 and the definition of a tree. Suppose G is a connected simple graph with n vertices and $n - 1$ edges. If G is not a tree, it contains, by Exercise 8, an edge whose removal produces a graph G', which is still connected. If G' is not a tree, remove an edge to produce a connected graph G''. Repeat this procedure until the result is a tree. This requires at most $n - 1$ steps since there are only $n - 1$ edges. By Theorem 2, the resulting graph has $n - 1$ edges since it has n vertices. It follows that no edges were deleted, so that G was already a tree.

11. 9,999

13. 2,000

15. 999

17. 1,000,000 dollars

19. Complete binary tree of height 4:

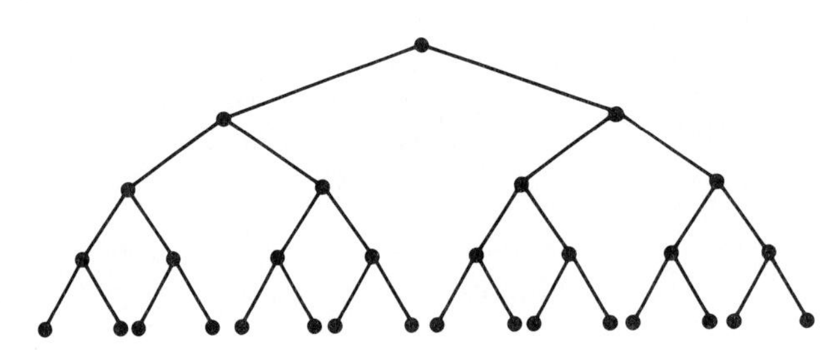

Complement 3-ary tree of height 3:

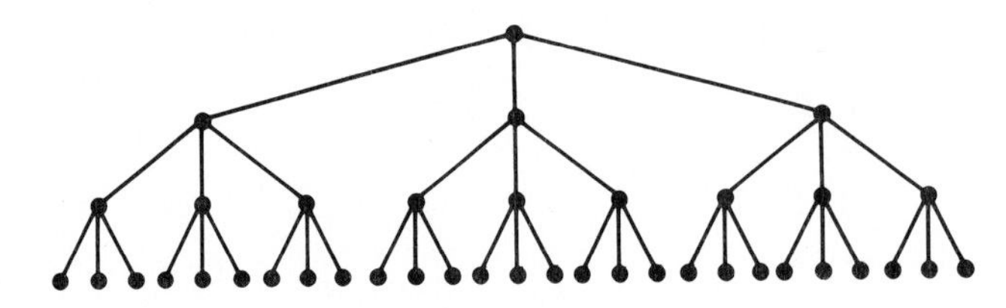

21. a) By Theorem 3 it follows that $n = mi + 1$. Since $i + l = n$, we have $l = n - i$, so that $l = (mi + 1) - i = (m - 1)i + 1$.

b) We have $n = mi + 1$ and $i + l = n$. Hence $i = n - l$. It follows that $n = m(n - l) + 1$. Solving for n gives $n = (ml - 1)/(m - 1)$. From $i = n - l$ we obtain $i = [(ml - 1)/(m - 1)] - l = (l - 1)/(m - 1)$.

23. $n - t$

25. a) 1 **b)** 3 **c)** 5

27. a) The parent directory

b) A subdirectory or contained file

c) A subdirectory or contained file in the same parent directory

d) All directories in the path name

e) All subdirectories and files continued in the directory or a subdirectory of this directory, and so on

f) The length of the path to this directory or file

g) The depth of the system, i.e., the length of the longest path

29. Let $n = 2^k$, where k is a positive integer. If $k = 1$, there is nothing to prove since we can add two numbers with $n - 1 = 1$ processor in $\log 2 = 1$ step. Assume we can add $n = 2^k$ numbers in $\log n$ steps using a tree-connected network of $n - 1$ processors. Let $x_1, x_2, \ldots, x_{2n}$ be $2n = 2^{k+1}$ numbers that we wish to add. The tree-connected network of $2n - 1$ processors consists of the tree-connected network of $n - 1$ processors together with two new processors as children of each leaf. In one step we can use the leaves of the larger network to find $x_1 + x_2, x_3 + x_4, \ldots, x_{2n-1} + x_{2n}$, giving us n numbers, which, by the inductive hypothesis, we can add in $\log n$ steps using the rest of the network. Since we have used $\log n + 1$ steps and $\log(2n) = \log 2 + \log n = 1 + \log n$, this completes the proof.

31. c only

33. c and h

35. Suppose a tree T has at least two centers. Let u and v be distinct centers, both with eccentricity e, with u and v not adjacent. Since T is connected, there is a simple path P from u to v. Let c be any other vertex on this path. Since the eccentricity of c is at least e, there is a vertex w such that the unique simple path from c to w has length at least e. Clearly this path cannot contain both u and v or else there would be a simple circuit. In fact, this path from c to w leaves P and does not return to P once it, possibly, follows part of P toward either u or v. Without loss of generality, assume this path does not follow P toward u. Then the path from u to c to w is simple, and of length more than e, a contradiction. Hence u and v are adjacent. Now since any two centers are adjacent, if there were more than two centers, T would contain K_3, a simple circuit, as a subgraph, which is a contradiction.

37.

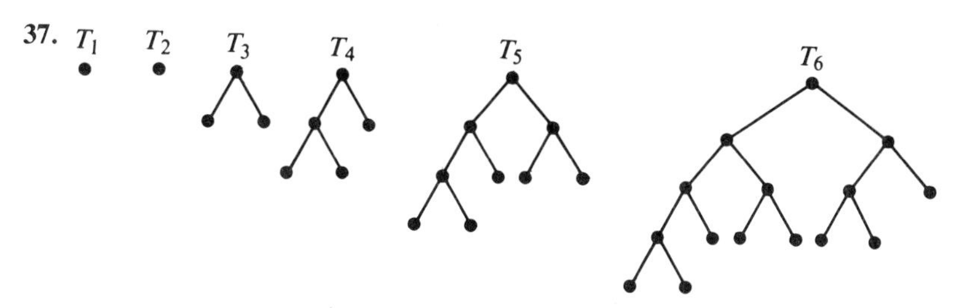

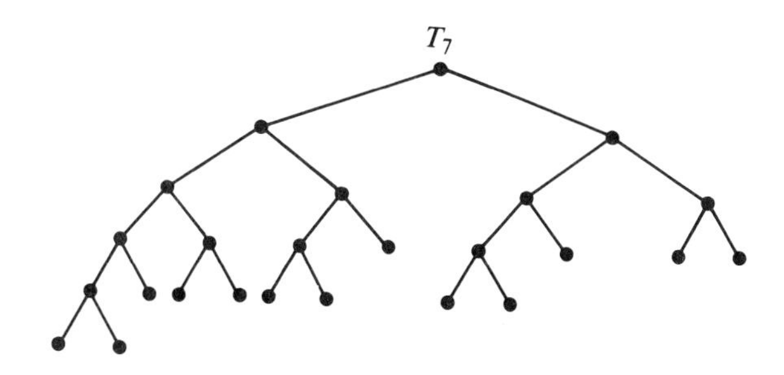

Section 8.2

1.

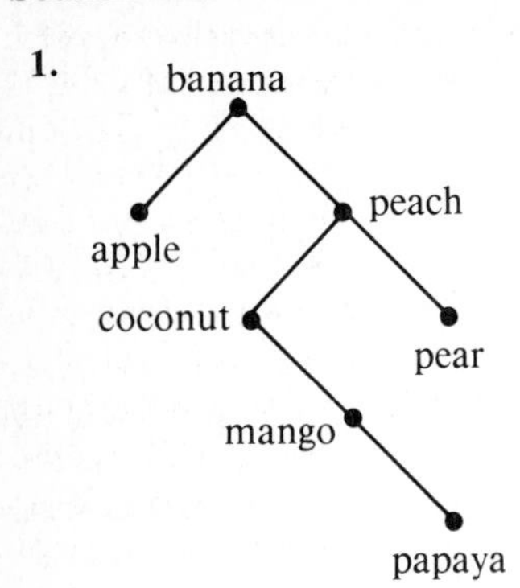

3. **a)** 3 **b)** 1 **c)** 4 **d)** 5

5.

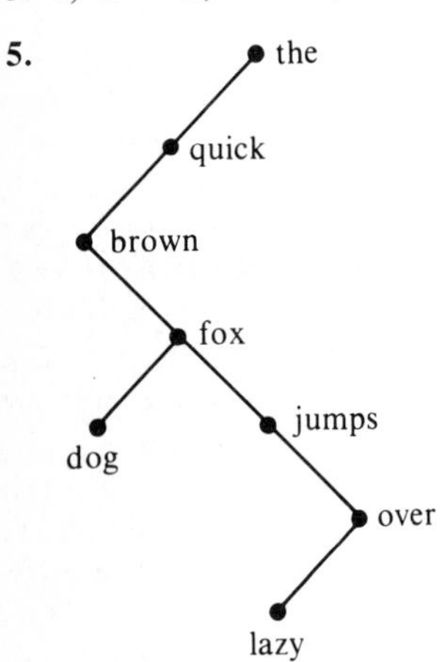

7. At least $\lceil \log_3 4 \rceil = 2$ weighings are needed, since there are only four outcomes (since it is not required to determine whether the coin is lighter or heavier). In fact, two weighings suffice. Begin by weighing coin 1 against coin 2. If they balance, weigh coin 1 against coin 3. If coin 1 and coin 3 are the same weight, coin 4 is the counterfeit coin, and if they are not the same weight, then coin 3 is the counterfeit coin. If coin 1 and coin 2 are not the same weight, again weigh coin 1 against coin 3. If they balance, coin 2 is the counterfeit coin; if they do not balance, coin 1 is the counterfeit coin.

9. At least $\lceil \log_3 13 \rceil = 3$ weighings are needed. In fact, three weighings suffice. Start by putting coins 1, 2, and 3 on the left-hand side of the balance and coins 4, 5, and 6 on the right-hand side. If equal, apply Example 2 to coins 1, 2, 7, 8, 9, 10, 11, and 12. If unequal, apply Example 2 to 1, 2, 3, 4, 5, 6, 7, and 8.

11. **a)** Yes
b) No
c) Yes
d) Yes

13. a: 000, e: 001, i: 01, k: 1100, o: 1101, p: 11110, u: 11111

Section 8.3

1.

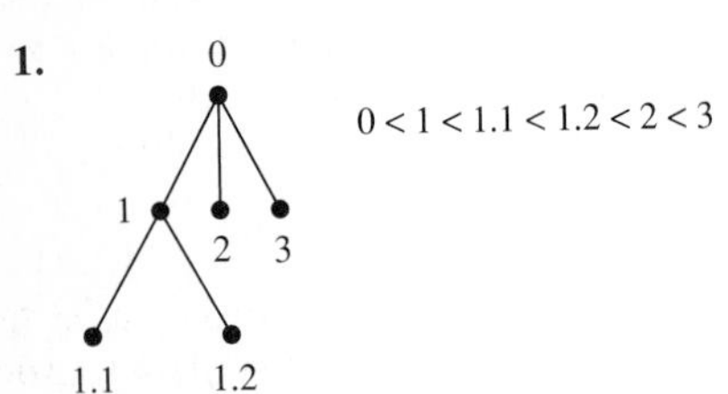

3.

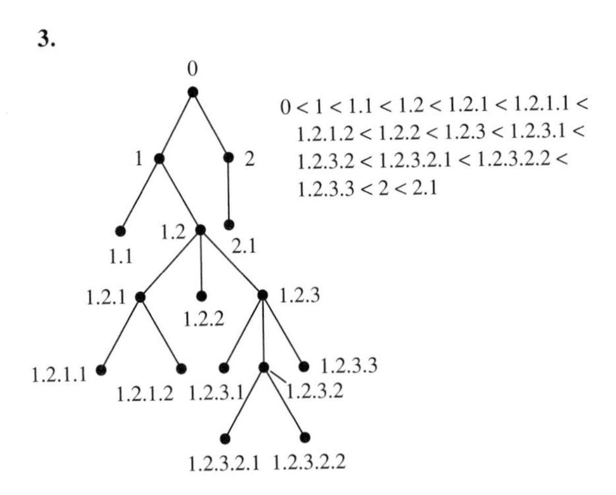

5. No

7. a, b, d, e, f, g, c

9. $a, b, e, k, l, m, f, g, n, r, s, c, d, h, o, i, j, p, q$

11. $d, b, i, e, m, j, n, o, a, f, c, g, k, h, p, l$

13. d, f, g, e, b, c, a

15. $k, l, m, e, f, r, s, n, g, b, c, o, h, i, p, q, j, d, a$

17. **a)** $- * \uparrow + x\, 2\, 3 - y + 3x5$
b) $x2 + 3 \uparrow y3x + - * 5-$
c) $((((x + 2) \uparrow 3) * (y - (3 + x))) - 5)$

19. **a)** $+ + x * x\, y / x\, y,\ + x / + * x\, y\, x\, y$
b) $x\, x\, y * + x\, y / +,\ x\, x\, y * x + y / +$
c) $((x + (x * y)) + (x/y)),\ (x + (((x * y) + x)/y))$

21. **a)** $\leftrightarrow \neg \wedge p\, q \vee \neg p \neg q,\ \vee \wedge \neg p \leftrightarrow q \neg p \neg q$
b) $p\, q \wedge \neg p \neg q \neg \vee \leftrightarrow,\ p \neg q\, p \neg \leftrightarrow \wedge q \neg \vee$
c) $(((p \wedge q) \neg) \leftrightarrow ((p \neg) \vee (q \neg))),\quad (((p \neg) \wedge (q \leftrightarrow (p \neg))) \vee (q \neg))$ (with unary operators following their operands)

23. **a)** $- \cap AB \cup A - BA$ **b)** $A\, B \cap A\, B\, A - \cup -$
c) $((A \cap B) - (A \cup (B - A)))$

25. 14

27. **a)** 1 **b)** 1 **c)** 4 **d)** 2205

29.

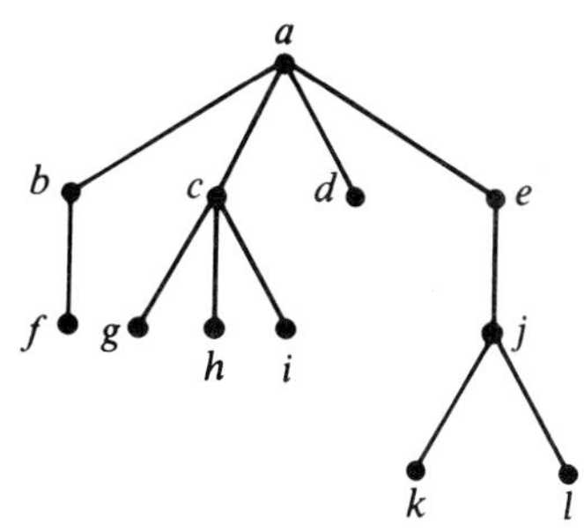

31. Use mathematical induction. The result is trivial for a list with one element. Assume the result is true for a list with n elements. For the inductive step, start at the end. Find the sequence of vertices at the end of the list starting with the last leaf, ending with the root, each vertex being the last child of the one following it. Remove this leaf and apply the inductive hypothesis.

33. c, d, b, f, g, h, e, a in each case.

35. Proof by mathematical induction. Let $S(X)$ and $O(X)$ represent the number of symbols and number of operators in the well-formed formula X, respectively. The statement is true for well-formed formulae of length 1, since they have 1 symbol and 0 operators. Assume the statement is true for all well-formed formulae of length less than n. A well-formed formula of length n must be of the form $*XY$ where $*$ is an operator and X and Y are well-formed formulae of length less than n. Then by the inductive hypothesis $S(*XY) = S(X) + S(Y) = (O(X)+1)+(O(Y)+1) = O(X)+O(Y)+2$. Since $O(*XY) = 1 + O(X) + O(Y)$, it follows that $S(*XY) = O(*XY) + 1$.

37. $xy + zx \circ + x\circ,\ xyz + + yx + +,\ xyxy \circ\circ xy \circ\circ z \circ +,\ xz \times zz + \circ,\ yyyy \circ \circ\circ,\ zx + yz + \circ$, for instance

Section 8.4

1. At the end of the first pass: 1, 3, 5, 4, 7; at the end of the second pass: 1, 3, 4, 5, 7; at the end of the third pass: 1, 3, 4, 5, 7; at the end of the fourth pass: 1, 3, 4, 5, 7

3.

```
procedure better bubblesort(a_1, ..., a_n: integers)
i := 1; done := false
while (i < n and done = false)
begin
   done := true
   for j := 1 to n - i
      if a_j > a_{j+1} then
      begin
         interchange a_j and a_{j+1}
         done := false
      end
   i := i + 1
end {a_1, ..., a_n is in increasing order}
```

5.

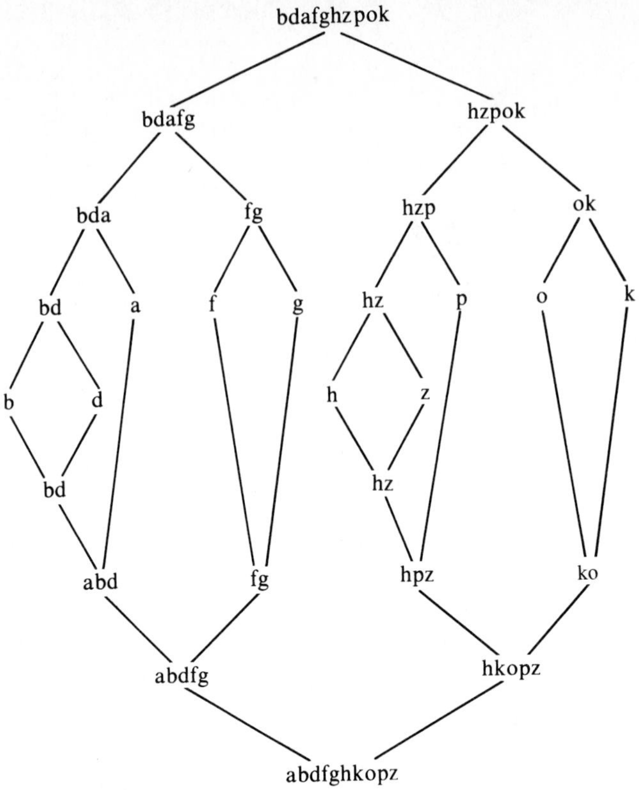

bdafghzpok

bdafg hzpok

bda fg hzp ok

bd a f g hz p o k

b d h z

bd hz

abd fg hpz ko

abdfg hkopz

abdfghkopz

7. Let the two lists be $1, 2, \ldots, m-1, m+n-1$ and $m, m+1, \ldots, m+n-2, m+n$, respectively.

9. a) 1, 5, 4, 3, 2; 1, 2, 4, 3, 5; 1, 2, 3, 4, 5; 1, 2, 3, 4, 5
b) 1, 4, 3, 2, 5; 1, 2, 3, 4, 5; 1, 2, 3, 4, 5; 1, 2, 3, 4, 5
c) 1, 2, 3, 4, 5; 1, 2, 3, 4, 5; 1, 2, 3, 4, 5; 1, 2, 3, 4, 5

11. $O(n^2)$

13. $n-1$

15. 6

17. $O(n^2)$ in worst case

19. **procedure** *mergesort* ($a_1, \ldots, a_n$: integers)
$m := \lceil n/2 \rceil$
if $n > 1$ **then**
begin
  $L_1 := (a_1, \ldots, a_m)$
  $L_2 := (a_{m+1}, \ldots, a_n)$
  $L_1 := mergesort(L_1)$; $L_2 = mergesort(L_2)$
  $L := merge(L_1, L_2)$
end
else $L := (a_1)$ {the list with one element is already sorted}
{L is sorted}

Section 8.5

1. $m - n + 1$

3.

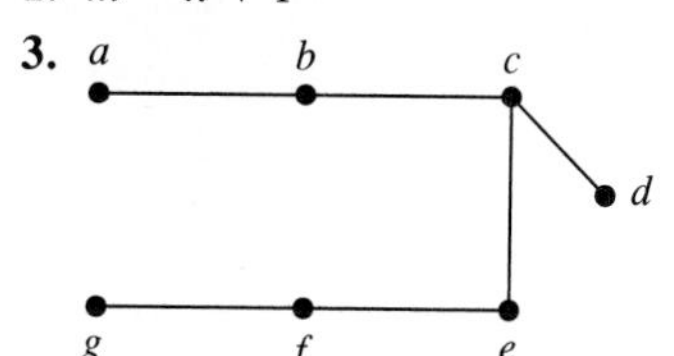

5. a)

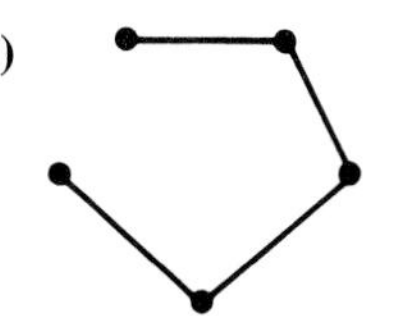

b)

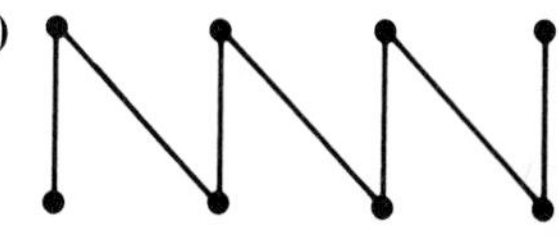

c)

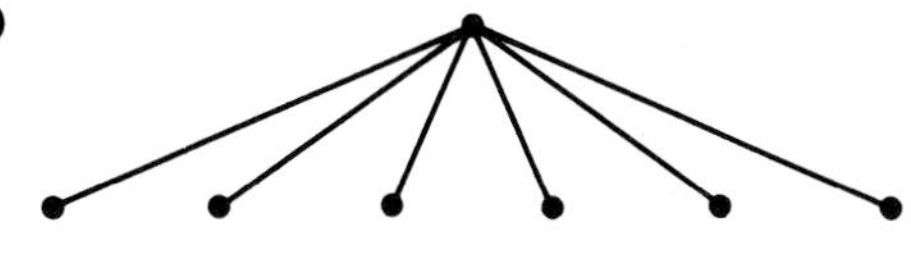

d)

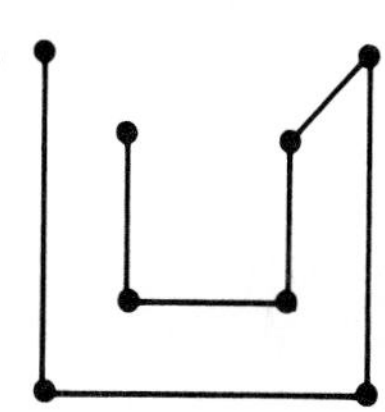

e)

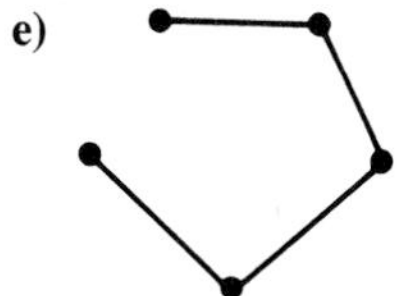

f)

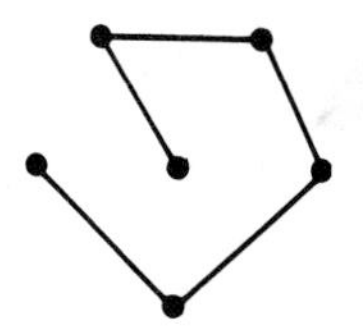

7.

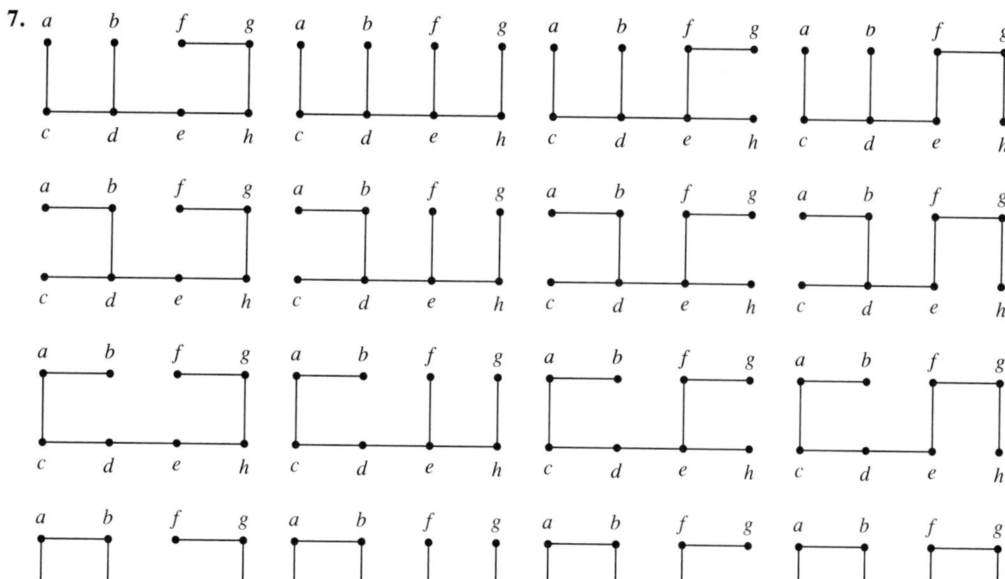

9. **a)** 3 **b)** 16 **c)** 4 **d)** 5

11.

13.

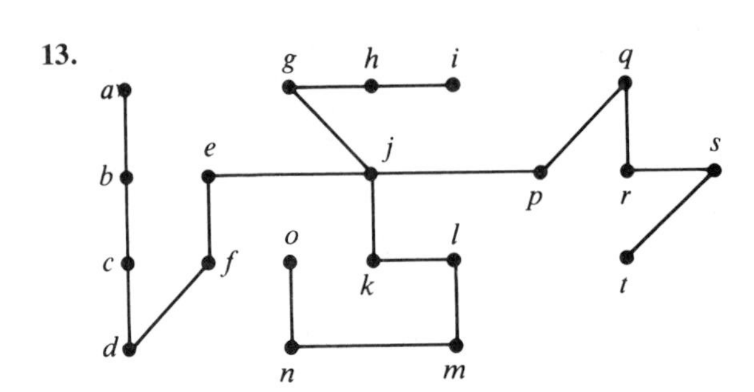

15. A possible set of flights to discontinue are: Boston–New York, Detroit–Boston, Boston–Washington, New York–Washington, New York–Chicago, Atlanta–Washington, Atlanta–Dallas, Atlanta–Los Angeles, Atlanta–St. Louis, St. Louis–Dallas, St. Louis–Detroit, St. Louis–Denver, Dallas–San Diego, Dallas–Los Angeles, Dallas–San Francisco, San Diego–Los Angeles, Los Angeles–San Francisco, San Francisco–Seattle.

17. Trees

19.
procedure *depth first search*(G: simple graph with ordered vertices $v_1, \ldots, v_n$)
$T :=$ rooted tree with v_1 as root and no other vertices
visit(v_1)
{T is the desired tree}
procedure *visit*(v)
for each neighbor w of v
begin
  if w is not in T **then**
  begin
    put the vertex w and the edge $\{v, w\}$ into T
    visit(w)
  end
end

21. Proof by induction on the length of the path: If the path has length 0, then the result is trivial. If the length is 1, then u is adjacent to v, so u is at level 1 in the breadth-first spanning tree. Assume that the result is true for paths of length l. If the length of a path is $l + 1$, let u' be the next-to-last vertex in a shortest path from v to u. By the inductive hypothesis, u' is at level l in the breadth-first spanning tree. If u were at a level not exceeding l, then clearly the length of the shortest path from v to u would also not exceed l. So u has not been added to the breadth-first spanning tree yet after the vertices of level l have been added. Since u is adjacent to u', it will be added at level $l + 1$ (although the edge connecting u' and u is not necessarily added).

23. **a)** No solution

b)

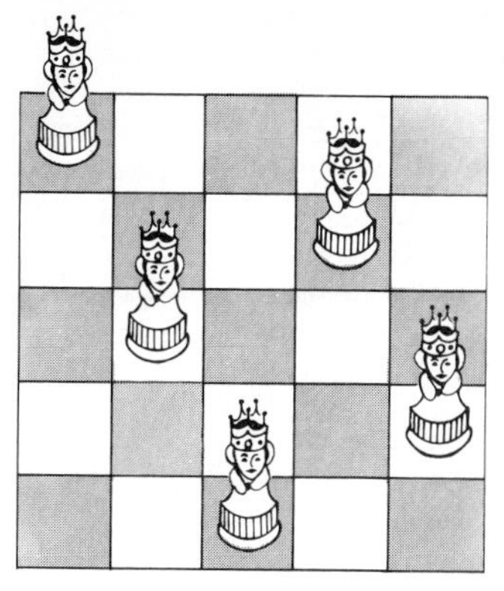

c)

25. Start at a vertex and proceed along a path without repeating vertices as long as possible, allowing the return to the start after all vertices have been visited. When it is impossible to continue along a path, backtrack and try another extension of the current path.

27. Take the union of the spanning trees of the connected components of G. They are disjoint so the result is a forest.

29. $m - n + c$

31. Use depth-first search on each component.

33. Let T be the spanning tree constructed in Figure 3 and T_1, T_2, T_3, and T_4 the spanning trees in Figure 4. Denote by $d(T', T'')$ the distance between T' and T''. Then $d(T, T_1) = 6$, $d(T, T_2) = 4$, $d(T, T_3) = 4$, $d(T, T_4) = 2$, $d(T_1, T_2) = 4$, $d(T_1, T_3) = 4$, $d(T_1, T_4) = 6$, $d(T_2, T_3) = 4$, $d(T_2, T_4) = 2$, and $d(T_3, T_4) = 4$.

35. Suppose $e_1 = \{u, v\}$ is as specified. Then $T_2 \cup \{e_1\}$ contains a simple circuit C containing e_1. The graph $T_1 - \{e_1\}$ has two connected components; the endpoints of e_1 are in different components. Travel C from u in the direction opposite to e_1 until you come to the first vertex in the same component as v. The edge just crossed is e_2. Clearly $T_2 \cup \{e_1\} - \{e_2\}$ is a tree, since e_2 was on C. Also $T_1 - \{e_1\} \cup \{e_2\}$ is a tree, since e_2 reunited the two components.

37. Exercise 24:

a b

c d

Exercise 25:

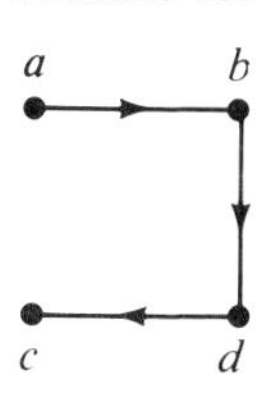

Exercise 26:

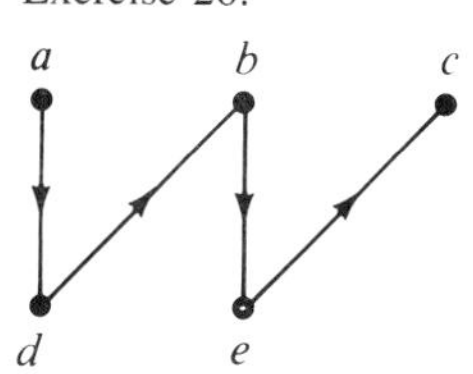

Exercise 27:

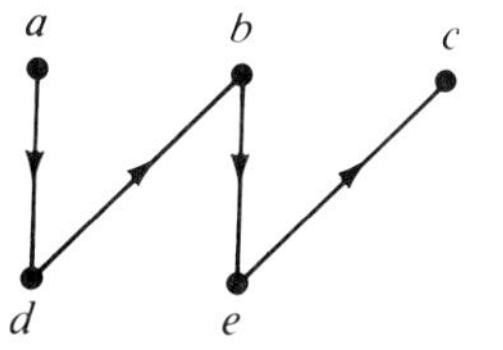

Exercise 28:

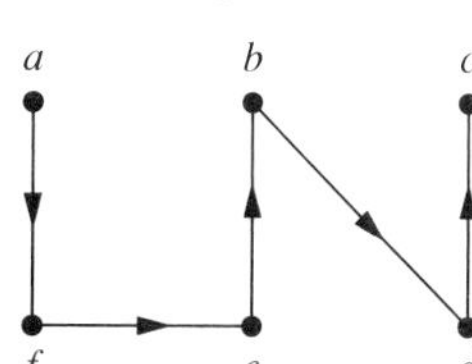

39. First construct an Euler circuit in the directed graph. Then delete from this circuit every edge that goes to a vertex previously visited.

Section 8.6

1. Deep Springs–Oasis, Oasis–Dyer, Oasis–Silverspeak, Silverspeak–Goldfield, Lida–Gold Point, Gold Point–Beatty, Lida–Goldfield, Goldfield–Tonopah, Tonopah–Manhattan, Tonopah–Warm Springs

3. $\{e, f\}$, $\{c, f\}$, $\{e, h\}$, $\{h, i\}$, $\{b, c\}$, $\{b, d\}$, $\{a, d\}$, $\{g, h\}$

5.

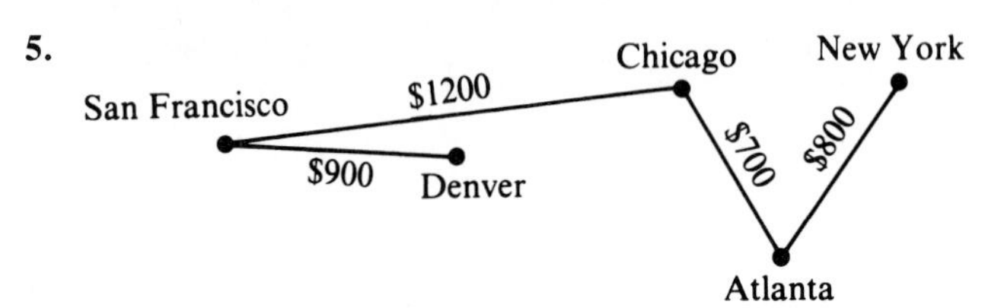

7. $\{e, f\}$, $\{a, d\}$, $\{h, i\}$, $\{b, d\}$, $\{c, f\}$, $\{e, h\}$, $\{b, c\}$, $\{g, h\}$

9.

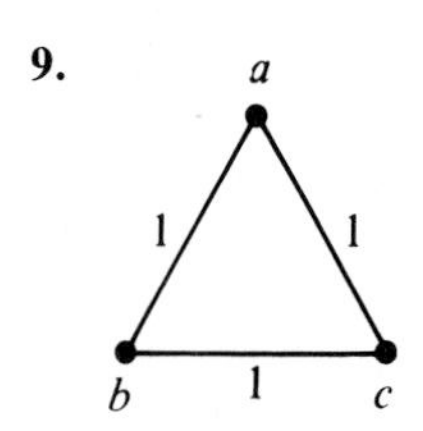

11. Instead of choosing minimum-weight edges at each stage, choose maximum-weight edges at each stage with the same properties.

13.

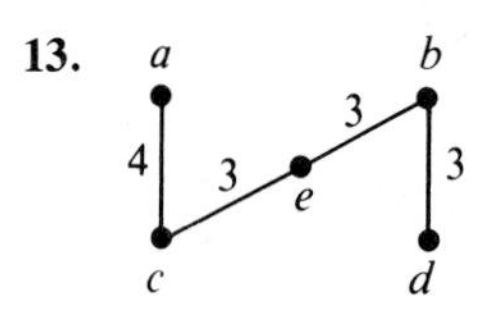

15.

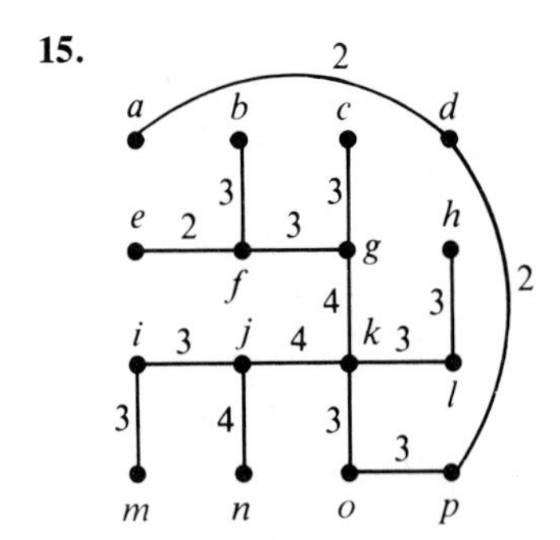

17. First find a minimal spanning tree T of the graph G with n edges. Then for $i = 1$ to $n - 1$, only delete the ith edge of T from G and find a minimal spanning tree of the remaining graph. Pick the one of these $n - 1$ trees with the shortest length.

19. If all edges have different weights a contradiction is obtained in the proof that Prim's algorithm works when an edge e_{k+1} is added to T and an edge e is deleted, instead of possibly producing another spanning tree.

21.

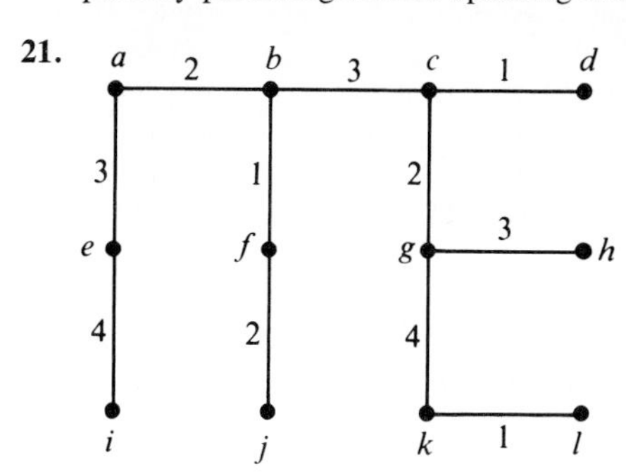

23. Same as Kruskal's algorithm, except start with $T :=$ this set of edges and iterate from $i = 1$ to $n - 1 - s$ where s is the number of edges you start with.

25. a)

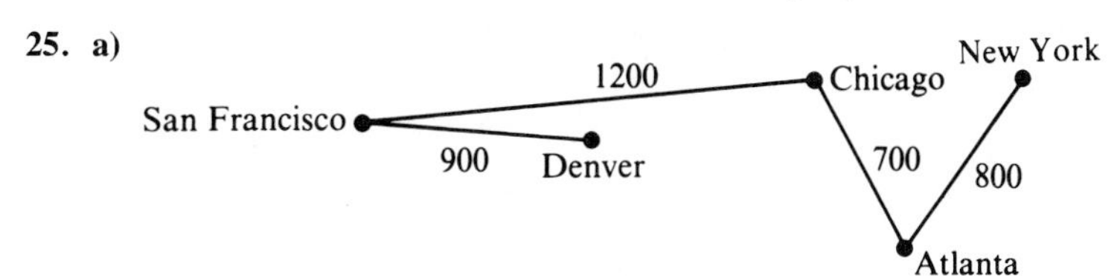

b)

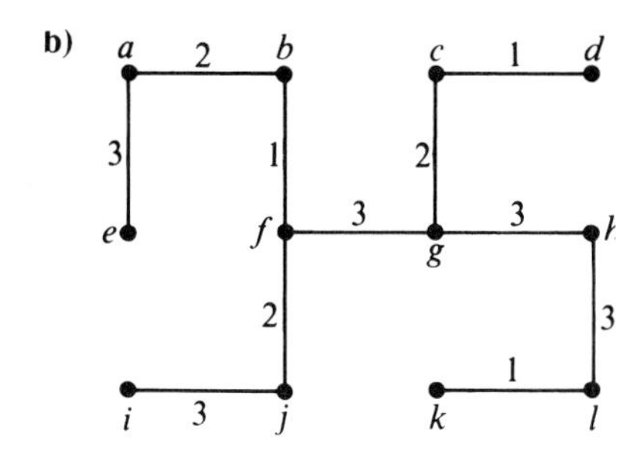

27. By Exercise 24, at each stage of Sollin's algorithm a forest results. Hence after $n - 1$ edges are chosen, a tree results. It remains to show that this tree is a minimal spanning tree. Let T be a minimal spanning tree with as many edges in common with Sollin's tree S as possible. If $T \neq S$, then there is an edge $e \in S - T$ added at some stage in the algorithm, where prior to that stage all edges in S are also in T. $T \cup \{e\}$ contains a unique simple circuit. Find an edge $e' \in S - T$ and an edge $e'' \in T - S$ on this circuit and "adjacent" when viewing the trees of this stage as "supervertices." Then by the algorithm, $w(e') \leq w(e'')$. So replace T by $T - \{e''\} \cup \{e'\}$ to produce a minimal spanning tree closer to S than T was.

29. Each of the r trees is joined to at least one other tree by a new edge. Hence there are at most $r/2$ trees in the result (each new tree contains two or more old trees). To accomplish this, we need to add $r - (r/2) = r/2$ edges. Since the number of edges added is integral, it is at least $\lceil r/2 \rceil$.

31. If $k \geq \log n$, then $n/2^k \leq 1$, so $\lceil n/2^k \rceil = 1$, so by Exercise 30 the algorithm is finished after at most $\log n$ iterations.

Supplementary Exercises

1. Suppose T is a tree. Then clearly T has no simple circuits. If we add an edge e connecting two nonadjacent vertices u and v, then obviously a simple circuit is formed, since when e is added to T the resulting graph has too many edges to be a tree. The only simple circuit formed is made up of the edge e together with the unique path P in T from v to u. Suppose T satisfies the given conditions. All that is needed is to show that T is connected, since there are no simple circuits in the graph. Assume that T is not connected. Then let u and v be in separate connected components. Adding $e = \{u, v\}$ does not satisfy the conditions.

3. Suppose that a tree T has n vertices of degrees $d_1, d_2, \ldots, d_n$, respectively. Since $2e = \sum_{i=1}^{n} d_i$ and $e = n - 1$, we have $2(n-1) = \sum_{i=1}^{n} d_i$. Since each $d_i \geq 1$, it follows that $2(n-1) = n + \sum_{i=1}^{n}(d_i - 1)$, or that $n - 2 = \sum_{i=1}^{n}(d_i - 1)$. Hence at most $n - 2$ of the terms of this sum can be 1 or more. Hence, at least two of them are 0. It follows that $d_i = 1$ for at least two values of i.

5. $2n - 2$

7. T has no circuits, so it cannot have a subgraph homeomorphic to $K_{3,3}$ or K_5.

9. Color each connected component separately. For each of these connected components, first root the tree, then color all vertices at even levels red and all vertices at odd levels blue.

11. Upper bound: k^h; lower bound: $2\lceil k/2 \rceil^{h-1}$

13.

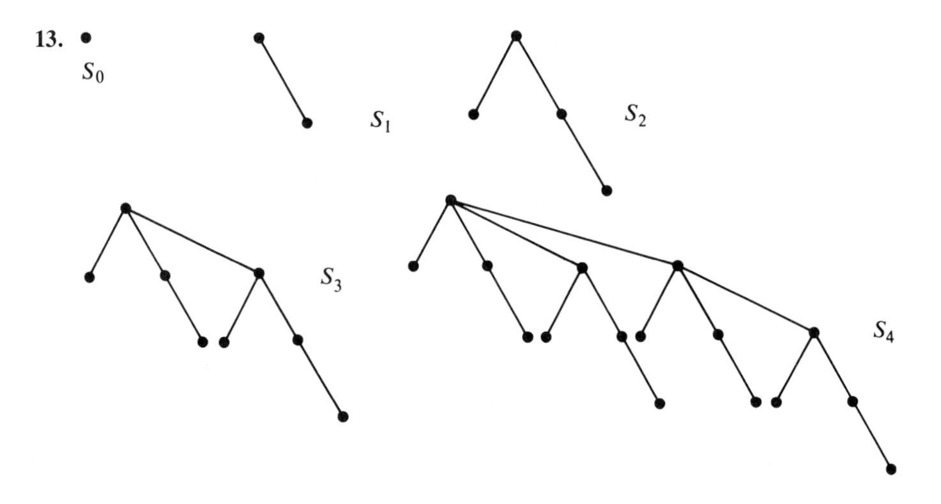

15. Use mathematical induction. The result is trivial for $k = 0$. Suppose it is true for $k - 1$. T_{k-1} is the parent tree for T. By induction, the child tree for T can be obtained from $T_0, \ldots, T_{k-2}$ in the manner stated. The final connection of r_{k-2} to r_{k-1} is as stated in the definition of S_k-tree.

17.

```
procedure level(T: ordered rooted tree with root r)
queue := sequence consisting of just the root r
while queue contains at least one term
begin
   v := first vertex in queue
   list v
   remove v from queue and put children of v onto
     the end of queue
end
```

19. Build the tree by inserting a root for the address 0, and then inserting a subtree for each vertex labeled i, for i a positive integer, built up from subtrees for each vertex labeled $i.j$ for j a positive integer, and so on.

21. **procedure** *insertion*($a_1, \ldots, a_n$: real numbers)
for $j := 2$ **to** n
begin
  $i := 1$
  while $a_j > a_i$
    $i := i + 1$
  $m := a_j$
  for $k := 0$ **to** $j - i - 1$
    $a_{j-k} := a_{j-k-1}$
  $a_i := m$
end $\{a_1, \ldots, a_n$ are sorted$\}$

23. If u is pendant and $e = \{u, v\}$ is an edge of the graph incident to u, then there are no other edges of the graph incident to u. Consequently e must be part of any spanning tree, for if it were not, the spanning tree would not contain an edge incident to u.

25. **a)** Yes **b)** No **c)** Yes

27. The resulting graph has no edge that is in more than one simple circuit of the type described. Hence it is a cactus.

29.

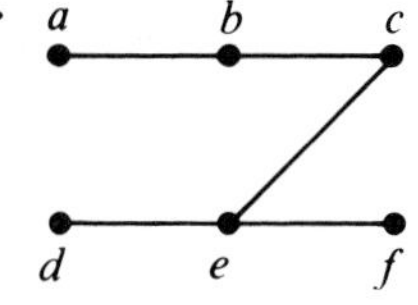

31.

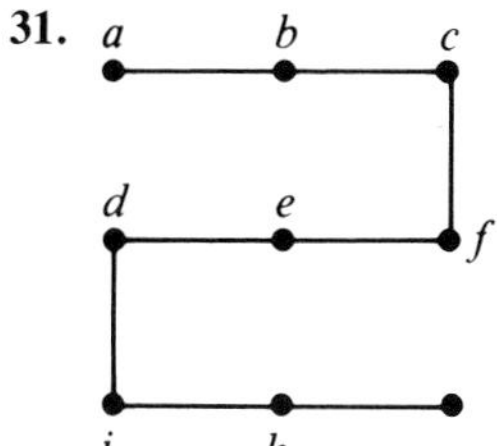

33. **a)**

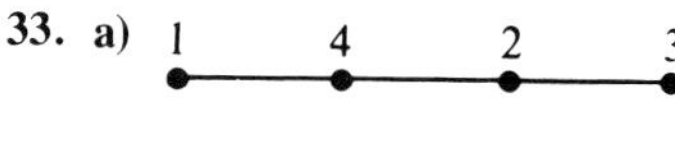

b)

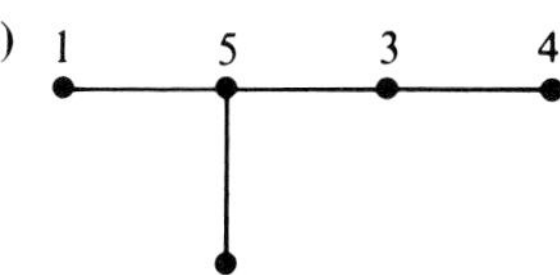

c)

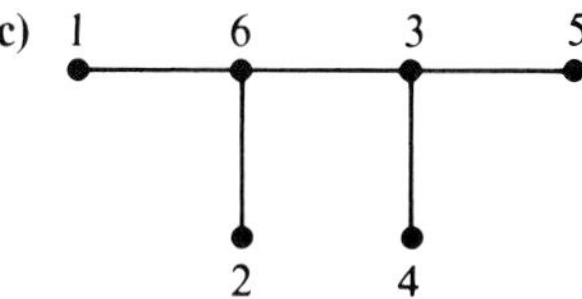

d)

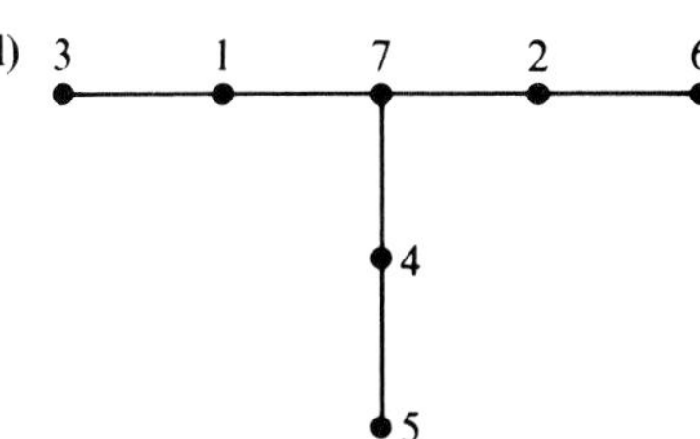

35. 6

37. **a)** Yes **b)** No **c)** Yes

39. Let G' be the graph obtained by deleting from G the vertex v and all edges incident to v. A minimal spanning tree of G can be obtained by taking an edge of minimal weight incident to v together with a minimal spanning tree of G'.

41. 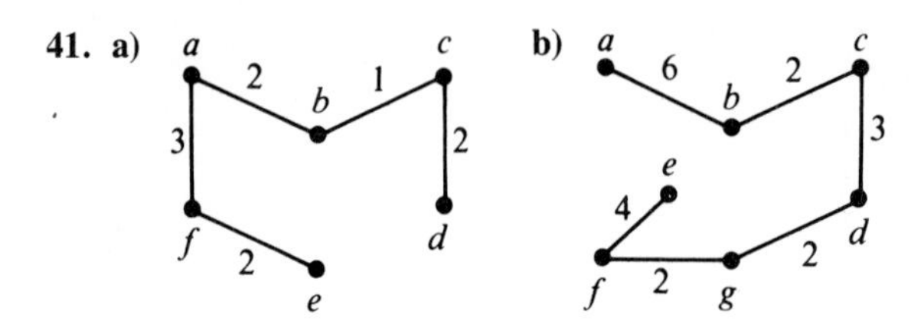

CHAPTER 9

Section 9.1

1. **a)** 1 **b)** 1 **c)** 0 **d)** 0

3. (0,0) and (1,1)

5. $x + xy = x \cdot 1 + xy = x(1 + y) = x(y + 1) = x \cdot 1 = x$

7.

x	y	z	$x\bar{y}$	$y\bar{z}$	$\bar{x}z$	$x\bar{y} + y\bar{z} + \bar{x}z$	$\bar{x}y$	$\bar{y}z$	$x\bar{z}$	$\bar{x}y + \bar{y}z + x\bar{z}$
1	1	1	0	0	0	0	0	0	0	0
1	1	0	0	1	0	1	0	0	1	1
1	0	1	1	0	0	1	0	1	0	1
1	0	0	1	0	0	1	0	0	1	1
0	1	1	0	0	1	1	1	0	0	1
0	1	0	0	1	0	1	1	0	0	1
0	0	1	0	0	1	1	0	1	0	1
0	0	0	0	0	0	0	0	0	0	0

9.

x	$x + x$	$x \cdot x$
0	0	0
1	1	1

11.

x	$x + 1$	$x \cdot 0$
0	1	0
1	1	0

13.

x	y	z	$y+z$	$x+(y+z)$	$x+y$	$(x+y)+z$	yz	$x(yz)$	xy	$(xy)z$
1	1	1	1	1	1	1	1	1	1	1
1	1	0	1	1	1	1	0	0	1	0
1	0	1	1	1	1	1	0	0	0	0
1	0	0	0	1	1	1	0	0	0	0
0	1	1	1	1	1	1	1	0	0	0
0	1	0	1	1	1	1	0	0	0	0
0	0	1	1	1	0	1	0	0	0	0
0	0	0	0	0	0	0	0	0	0	0

15.

x	y	xy	$\overline{(xy)}$	$\overline{x}$	$\overline{y}$	$\overline{x}+\overline{y}$	$x+y$	$\overline{(x+y)}$	$\overline{x}\,\overline{y}$
1	1	1	0	0	0	0	1	0	0
1	0	0	1	0	1	1	1	0	0
0	1	0	1	1	0	1	1	0	0
0	0	0	1	1	1	1	0	1	1

17.

x	y	$x \oplus y$	$x+y$	xy	$\overline{(xy)}$	$(x+y)\overline{(xy)}$	$x\overline{y}$	$\overline{x}y$	$x\overline{y}+\overline{x}y$
1	1	0	1	1	0	0	0	0	0
1	0	1	1	0	1	1	1	0	1
0	1	1	1	0	1	1	0	1	1
0	0	0	0	0	1	0	0	0	0

19. **a)** True, as a table of values can show
b) False; take $x = 1, y = 1, z = 1$, for instance
c) False; take $x = 1, y = 1, z = 0$, for instance

21. By De Morgan's laws, the complement of an expression is like the dual except that the complement of each variable has been taken.

23. 16

Section 9.2

1. **a)** $\overline{x}\,\overline{y}z$ **b)** $\overline{x}y\overline{z}$ **c)** $\overline{x}yz$ **d)** $\overline{x}\,\overline{y}\,\overline{z}$

3. **a)** $xyz + xy\overline{z} + x\overline{y}z + x\overline{y}\,\overline{z} + \overline{x}yz + \overline{x}y\overline{z} + \overline{x}\,\overline{y}z$
b) $xyz + xy\overline{z} + \overline{x}yz$
c) $xyz + xy\overline{z} + x\overline{y}z + x\overline{y}\,\overline{z}$
d) $x\overline{y}z + x\overline{y}\,\overline{z}$

5. $wxy\overline{z} + wx\overline{y}z + w\overline{x}yz + \overline{w}xyz + \overline{w}x\overline{y}\,\overline{z} + \overline{w}\,\overline{x}\,\overline{y}z + \overline{w}\,\overline{x}y\overline{z} + w\overline{x}\,\overline{y}\,\overline{z}$

7. **a)** $\overline{x} + \overline{y} + z$ **b)** $x + y + z$ **c)** $x + \overline{y} + z$

9. $y_1 + y_2 + \cdots + y_n = 0$ if and only if $y_i = 0$ for $i = 1, 2, \ldots, n$. This holds if and only if $x_i = 0$ when $y_i = x_i$ and $x_i = 1$ when $y_i = \overline{x_i}$.

11. **a)** $x + y + z$
b) $(x+y+z)(x+y+\overline{z})(x+\overline{y}+z)(\overline{x}+y+z)(\overline{x}+y+\overline{z})$
c) $(x+y+z)(x+y+\overline{z})(x+\overline{y}+z)(x+\overline{y}+\overline{z})$
d) $(x+y+z)(x+y+\overline{z})(x+\overline{y}+z)(x+\overline{y}+\overline{z})(\overline{x}+\overline{y}+z)(\overline{x}+\overline{y}+\overline{z})$

13. **a)** $x + y + z$
b) $x + \overline{(y + \overline{(\overline{x} + z)})}$
c) $\overline{(x + \overline{y})}$
d) $\overline{(x + \overline{(x + \overline{y} + \overline{z})})}$

15. a)

x	$\overline{x}$	$x \downarrow x$
1	0	0
0	1	1

b)

x	y	xy	$x \downarrow x$	$y \downarrow y$	$(x \downarrow x) \downarrow (y \downarrow y)$
1	1	1	0	0	1
1	0	0	0	1	0
0	1	0	1	0	0
0	0	0	1	1	0

c)

x	y	$x + y$	$(x \downarrow y)$	$(x \downarrow y) \downarrow (x \downarrow y)$
1	1	1	0	1
1	0	1	0	1
0	1	1	0	1
0	0	0	1	0

17. a) $(((x \mid x) \mid (y \mid y)) \mid ((x \mid x) \mid (y \mid y))) \mid (z \mid z)$
b) $(((x \mid x) \mid (z \mid z)) \mid y) \mid (((x \mid x) \mid (z \mid z)) \mid y)$
c) x
d) $(x \mid (y \mid y)) \mid (x \mid (y \mid y))$

19. It is impossible to represent $\overline{x}$ using $+$ and $\cdot$ since there is no way to get the value 0 if the input is 1.

Section 9.3

1. $\overline{(x + y)\overline{y}}$
3. $\overline{(xy)} + (\overline{z} + x)$
5. $(x + y + z) + (\overline{x} + y + z) + (\overline{x} + \overline{y} + \overline{z})$
7. (See answer on facing page)

9.

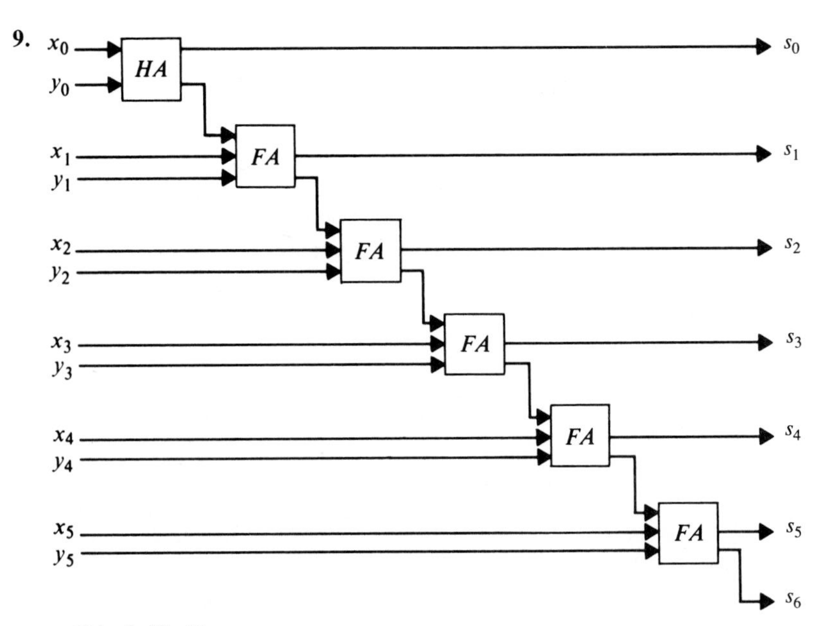

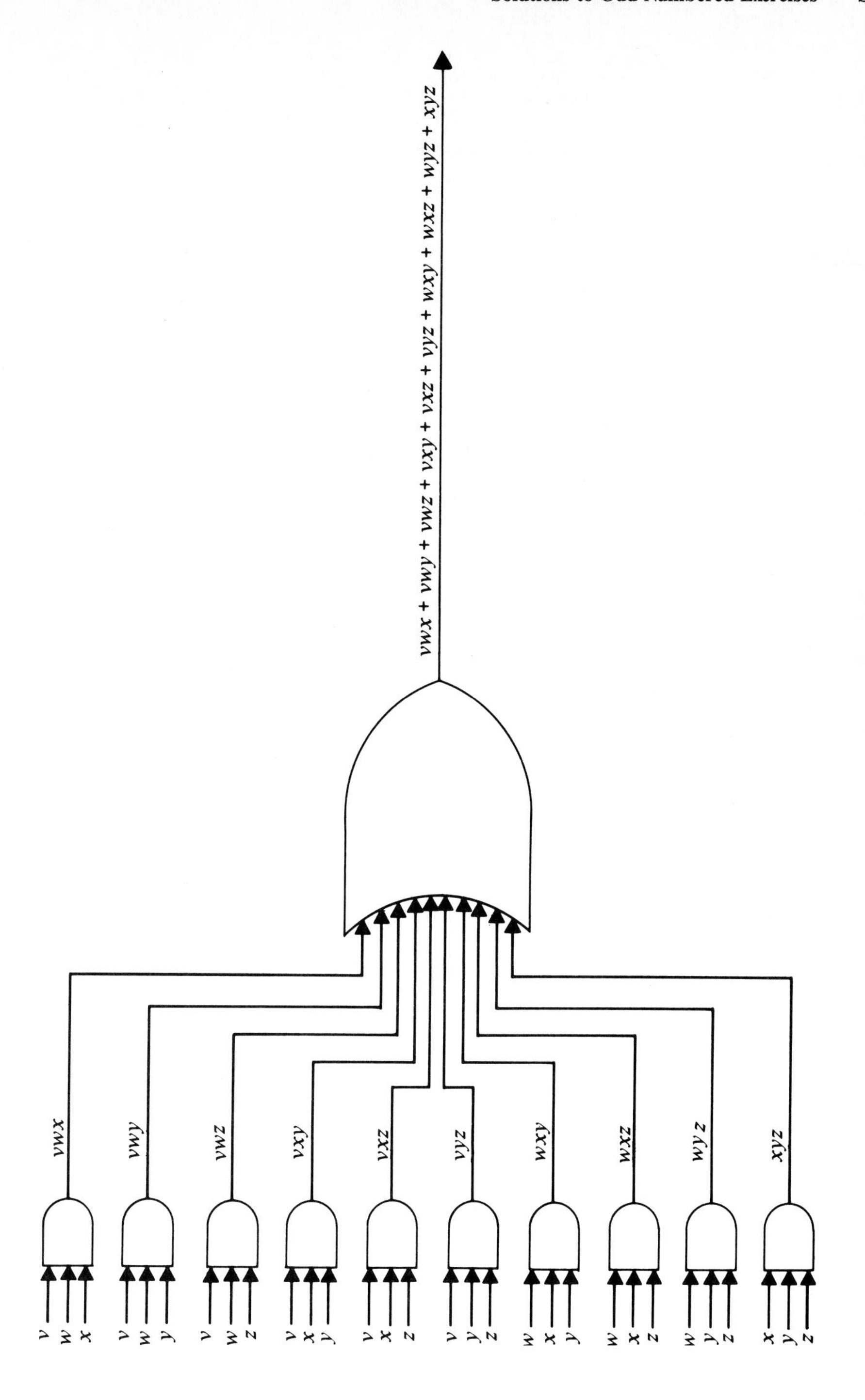
vwx + vwy + vwz + vxy + vxz + vyz + wxy + wxz + wyz + xyz
vwx
vwy
vwz
vxy
vxz
vyz
wxy
wxz
wyz
xyz
v w x
v w y
v w z
v x y
v x z
v y z
w x y
w x z
w y z
x y z

11.

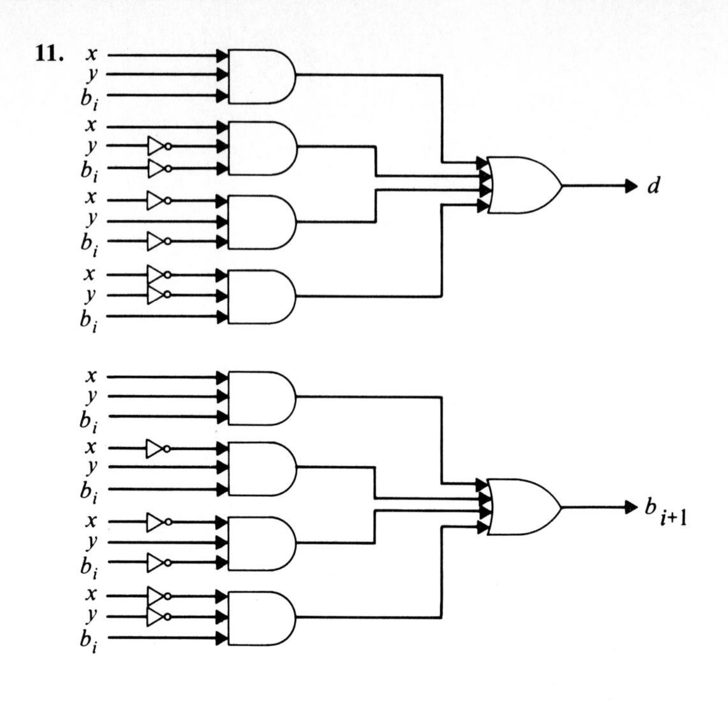
x
y
b_i
d
b_{i+1}

13.

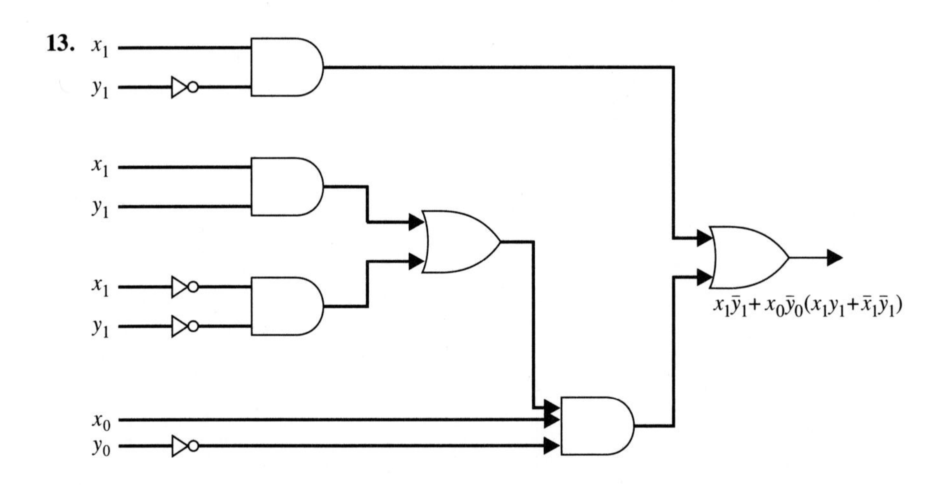
x_1
y_1
x_0
y_0
$x_1\bar{y}_1 + x_0\bar{y}_0(x_1y_1 + \bar{x}_1\bar{y}_1)$

15. **a)** **b)**

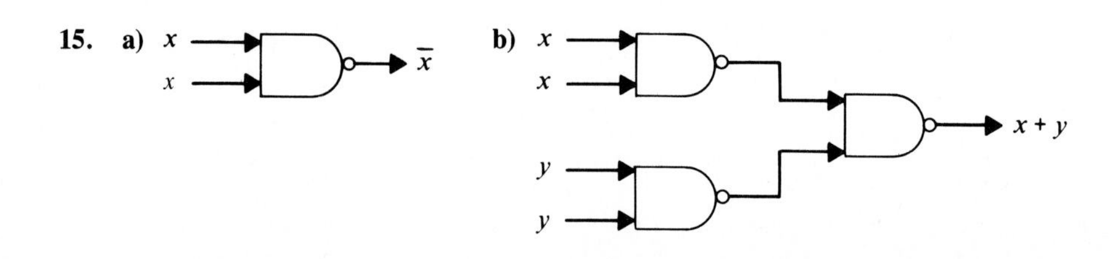
x
$\bar{x}$
y
x + y

c)

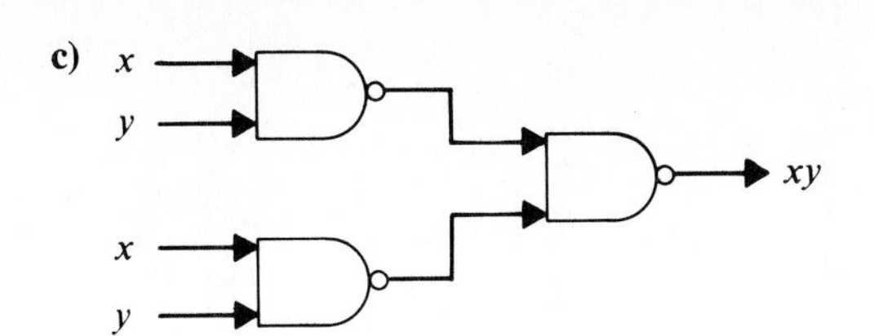

d)

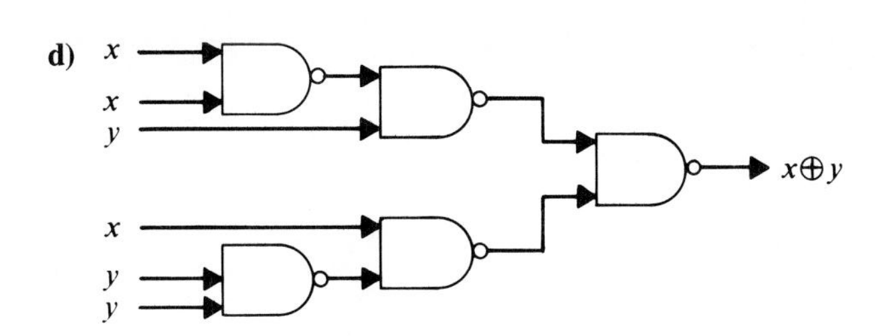

17.

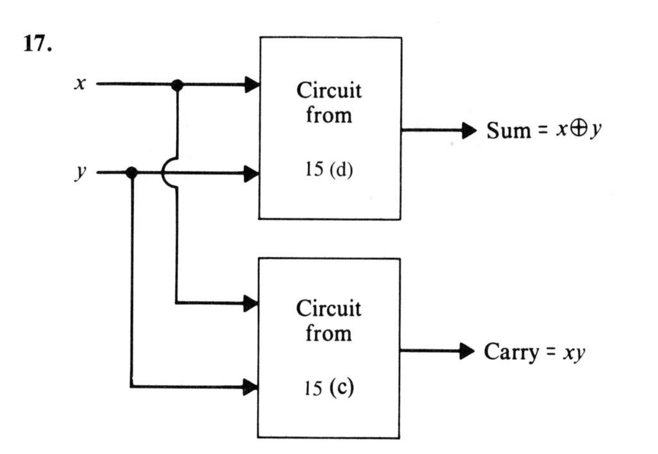

Section 9.4

1. a)

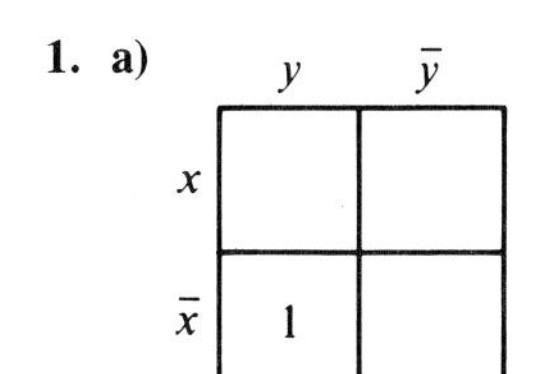

c)

	y	$\overline{y}$
x	1	1
$\overline{x}$	1	1

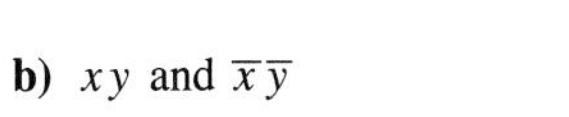

b) xy and $\overline{x}\,\overline{y}$

3. a) **b)**

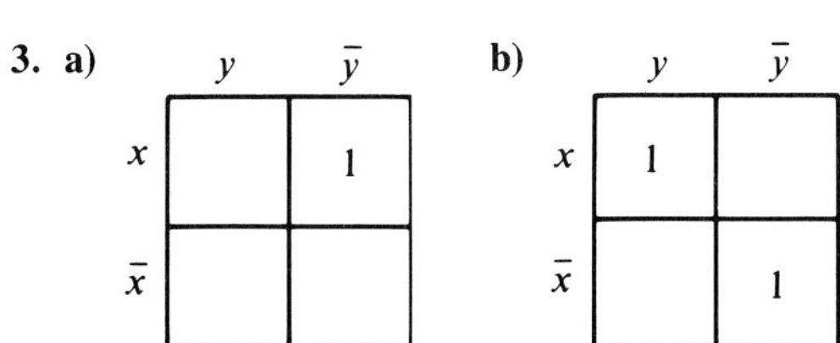

5. a)

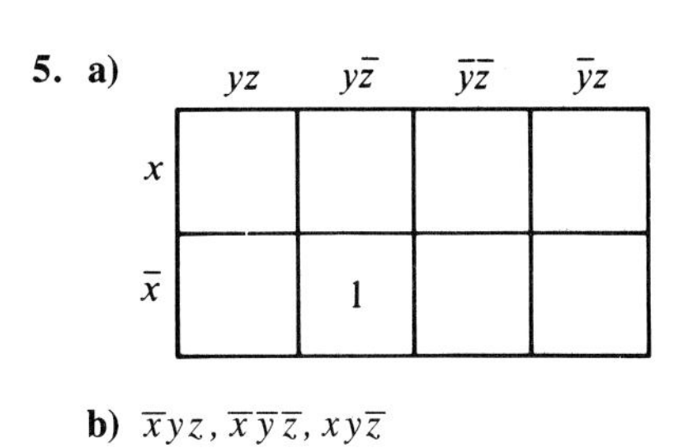

b) $\overline{x}yz,\ \overline{x}\,\overline{y}\,\overline{z},\ xy\overline{z}$

7. a)

	yz	$y\overline{z}$	$\overline{y}\overline{z}$	$\overline{y}z$
x			1	
$\overline{x}$				

b)

	yz	$y\overline{z}$	$\overline{y}\overline{z}$	$\overline{y}z$
x				
$\overline{x}$	1		1	

c)

	yz	$y\overline{z}$	$\overline{y}\overline{z}$	$\overline{y}z$
x	1	1		
$\overline{x}$		1		1

9. a)

	yz	$y\overline{z}$	$\overline{y}\overline{z}$	$\overline{y}z$
wx				
$w\overline{x}$				
$\overline{w}\overline{x}$				
$\overline{w}x$		1		

b) $wxy\overline{z}, \overline{w}xyz, \overline{w}x\overline{y}\overline{z}, \overline{w}\overline{x}y\overline{z}$

11. a) 32
b) 5

13.

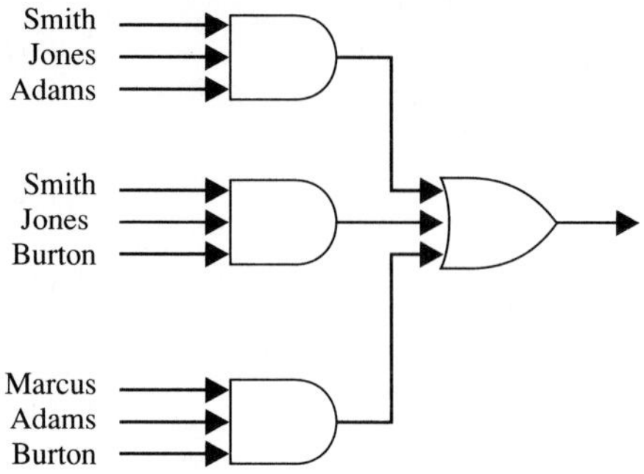

15. a) $\overline{x}z$
b) y
c) $x\overline{z} + \overline{x}z + \overline{y}z$
d) $xz + \overline{x}y + \overline{y}\overline{z}$

17. a) $wxz + wx\overline{y} + w\overline{y}z + w\overline{x}y\overline{z}$
b) $x\overline{y}z + \overline{w}\overline{y}z + wxy\overline{z} + w\overline{x}yz + \overline{w}\overline{x}y\overline{z}$
c) $\overline{y}z + wxz + w\overline{x}\overline{y} + \overline{w}\overline{x}y\overline{z}$
d) $wy + yz + \overline{x}y + wxz + \overline{w}\overline{x}z$

19. $x(y + z)$

Supplementary Exercises

1. a) $x = 0, y = 0, z = 0; x = 1, y = 1, z = 1$
b) $x = 0, y = 0, z = 0; x = 0, y = 0, z = 1; x = 0, y = 1, z = 0; x = 1, y = 0, z = 1; x = 1, y = 1, z = 0; x = 1, y = 1, z = 1$
c) No values

3. a) Yes **b)** No **c)** No **d)** Yes

5. $2^{2^{n-1}}$

7. a) If $F(x_1, \ldots, x_n) = 1$, then $(F + G)(x_1, \ldots, x_n) = F(x_1, \ldots, x_n) + G(x_1, \ldots, x_n) = 1$ by the dominance law. Hence $F \leq F + G$.
b) If $(FG)(x_1, \ldots, x_n) = 1$, then $F(x_1, \ldots, x_n) \cdot G(x_1, \ldots, x_n) = 1$. Hence $F(x_1, \ldots, x_n) = 1$. It follows that $FG \leq F$.

9. Since $F(x_1, \ldots, x_n) = 1$ implies that $F(x_1, \ldots, x_n) = 1$, $\leq$ is reflexive. Suppose that $F \leq G$ and $G \leq F$. Then $F(x_1, \ldots, x_n) = 1$ if and only if $G(x_1, \ldots, x_n) = 1$. This implies that $F = G$. Hence $\leq$ is antisymmetric. Suppose that $F \leq G \leq H$. Then if $F(x_1, \ldots, x_n) = 1$, it follows that $G(x_1, \ldots, x_n) = 1$, which implies that $H(x_1, \ldots, x_n) = 1$. Hence $F \leq H$, so that $\leq$ is transitive.

11. a) $x = 1, y = 0, z = 0$
b) $x = 1, y = 0, z = 0$
c) $x = 1, y = 0, z = 0$

13.

x	y	$x \odot y$	$x \oplus y$	$\overline{(x \oplus y)}$
1	1	1	0	1
1	0	0	1	0
0	1	0	1	0
0	0	1	0	1

15. Yes, as a truth table shows.

17. **a)** 6
b) 5
c) 5
d) 6

19.

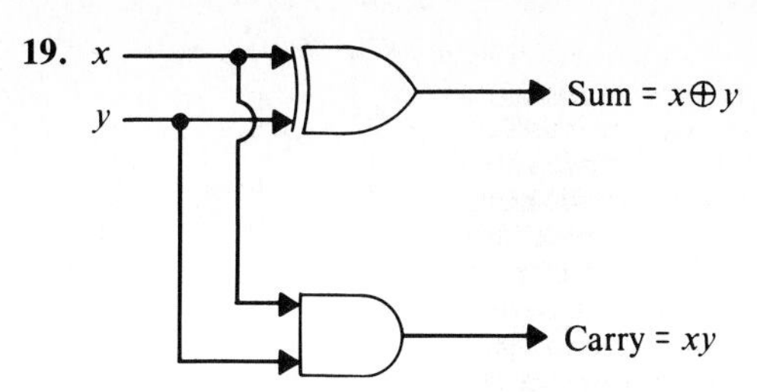

21. $x_3 + x_2\overline{x_1}$

23. Suppose it were with weights a and b. Then there would be a real number T such that $xa + yb \geq T$ for (1,0) and (0,1), but with $xa + yb < T$ for (0,0) and (1,1). Hence $a \geq T, b \geq T, 0 < T$, and $a + b < T$. Thus a and b are positive, which implies that $a + b > a \geq T$, a contradiction.

CHAPTER 10

Section 10.1

1. **a)** **sentence** ⇨ **noun phrase intransitive verb phrase** ⇨ **article adjective noun intransitive verb phrase** ⇨ **article adjective noun intransitive verb** ⇨ . . .(after 3 steps). . . ⇨ *the happy hare runs.*

b) **sentence** ⇨ **noun phrase intransitive verb phrase** ⇨ **article adjective noun intransitive verb phrase** ⇨ **article adjective noun intransitive verb adverb**. . . (after 5 steps). . . ⇨ *the sleepy tortoise runs quickly*

c) **sentence** ⇨ **noun phrase transitive verb phrase noun phrase** ⇨ **article noun transitive verb phrase noun phrase** ⇨ **article noun transitive verb noun phrase** ⇨ **article noun transitive verb article noun** ⇨ . . .(after 5 steps). . . ⇨ *the tortoise passes the hare*

d) **sentence** ⇨ **noun phrase transitive verb phrase noun phrase** ⇨ **article adjective noun transitive verb phrase noun phrase** ⇨ **article adjective noun transitive verb noun phrase** ⇨ **article adjective noun transitive verb article adjective noun** ⇨ . . .(after 6 steps). . . ⇨ *the sleepy hare passes the happy tortoise*

3. The only way to get a noun, such as tortoise, at the end is to have a noun phrase at the end, which can be achieved only via the production **sentence** → **noun phrase transitive verb phrase noun phrase**. However, **transitive verb phrase** → **transitive verb** → *passes*, and this sentence does not contain *passes*.

5. $S \Rightarrow 0S1 \Rightarrow 00S11 \Rightarrow 000S111 \Rightarrow 000111$

7. **a)** $S \Rightarrow 0S \Rightarrow 00S \Rightarrow 00S1 \Rightarrow 00S11 \Rightarrow 00S111 \Rightarrow 00S1111 \Rightarrow 001111$
b) $S \Rightarrow 0S \Rightarrow 00S \Rightarrow 001A \Rightarrow 0011A \Rightarrow 00111A \Rightarrow 001111$

9. $S \Rightarrow 0SAB \Rightarrow 00SABAB \Rightarrow 00ABAB \Rightarrow 00AABB \Rightarrow 001ABB \Rightarrow 0011BB \Rightarrow 00112B \Rightarrow 001122$

11. **a)** $S \to 00S, S \to \lambda$
b) $S \to 10A, A \to 00A, A \to \lambda$
c) $S \to AAS, S \to BBS, AB \to BA, BA \to AB, S \to \lambda, A \to 0, B \to 1$
d) $S \to 0000000000A, A \to 0A, A \to \lambda$

e) $S \to AS, S \to ABS, S \to A, AB \to BA, BA \to AB, A \to 0, B \to 1$
f) $S \to ABS, S \to \lambda, AB \to BA, BA \to AB, A \to 0, B \to 1$
g) $S \to ABS, S \to T, S \to U, T \to AT, T \to A, U \to BU, U \to B, AB \to BA, BA \to AB, A \to 0, B \to 1$

13. **a)** Type 2, not type 3 **b)** Type 3, not type 2 **c)** Type 0, not type 1
d) Type 2, not type 3 **e)** Type 2 **f)** Type 0, not type 1
g) Type 3 **h)** Type 0, not type 1 **i)** Type 2, not type 3
j) Type 2, not type 3

15. Set S_1 and S_2 be the start symbols of G_1 and G_2, respectively. Let S be a new start symbol.
a) Add S and productions $S \to S_1$ and $S \to S_2$.
b) Add S and production $S \to S_1S_2$.
c) Add S and production $S \to \lambda$ and $S \to S_1S$.

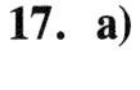

17. **a)**

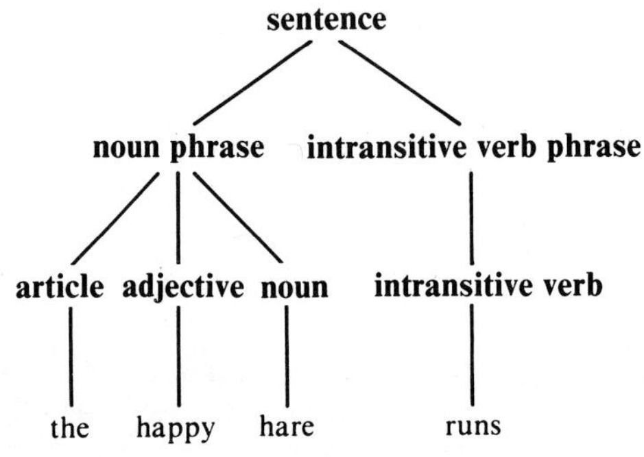

b)

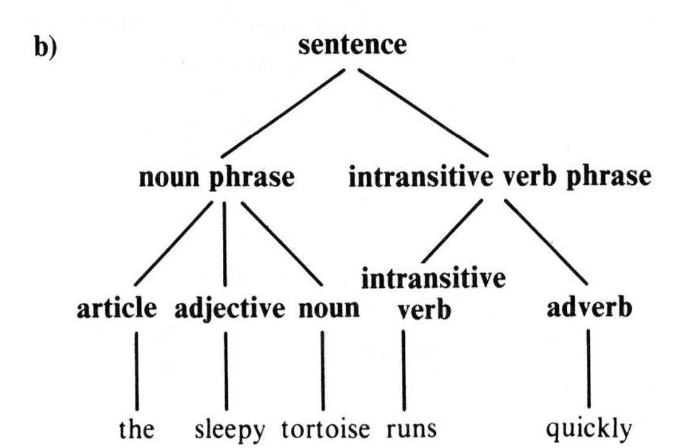

c)

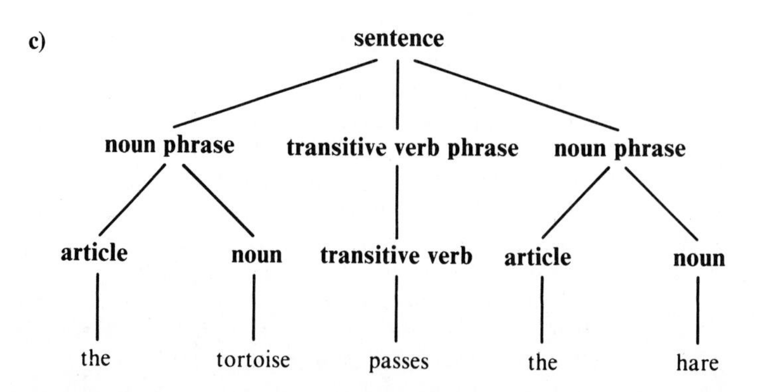

d)

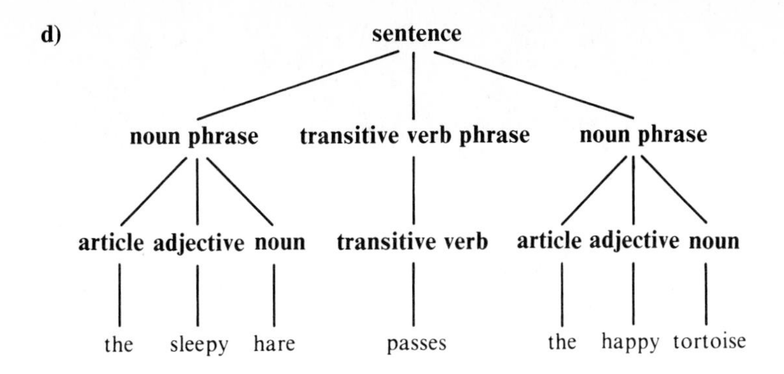

19. **a)** Yes **b)** No **c)** Yes **d)** No

21. signed integer

sign integer

— digit integer

1 digit integer

0 digit

9

23. **a)** $S \rightarrow$ ⟨sign⟩⟨integer⟩
$S \rightarrow$ ⟨sign⟩⟨integer⟩ . ⟨positive integer⟩
⟨sign⟩ $\rightarrow +$
⟨sign⟩ $\rightarrow -$
⟨integer⟩ $\rightarrow$ ⟨digit⟩
⟨integer⟩ $\rightarrow$ ⟨integer⟩⟨digit⟩
⟨digit⟩ $\rightarrow i, i = 1, 2, 3, 4, 5, 6, 7, 8, 9, 0$
⟨positive integer⟩ $\rightarrow$ ⟨integer⟩⟨nonzero digit⟩
⟨positive integer⟩ $\rightarrow$ ⟨nonzero digit⟩⟨integer⟩
⟨positive integer⟩ $\rightarrow$ ⟨integer⟩⟨nonzero digit⟩⟨integer⟩
⟨positive integer⟩ $\rightarrow$ ⟨nonzero digit⟩
⟨nonzero digit⟩ $\rightarrow i, i = 1, 2, 3, 4, 5, 6, 7, 8, 9$

b) ⟨signed decimal number⟩ ::= ⟨sign⟩⟨integer⟩ | ⟨sign⟩⟨integer⟩ . ⟨positive integer⟩
⟨sign⟩ ::= $+ | -$
⟨integer⟩ ::= ⟨digit⟩ | ⟨integer⟩⟨digit⟩

⟨digit⟩ ::= 0 | 1 | 2 | 3 | 4 | 5 | 6 | 7 | 8 | 9
⟨nonzero digit⟩ ::= 1 | 2 | 3 | 4 | 5 | 6 | 7 | 8 | 9
⟨positive integer⟩ ::= ⟨integer⟩⟨nonzero digit⟩ | ⟨nonzero digit⟩⟨integer⟩
| ⟨integer⟩⟨nonzero integer⟩⟨integer⟩ | ⟨nonzero digit⟩

c)

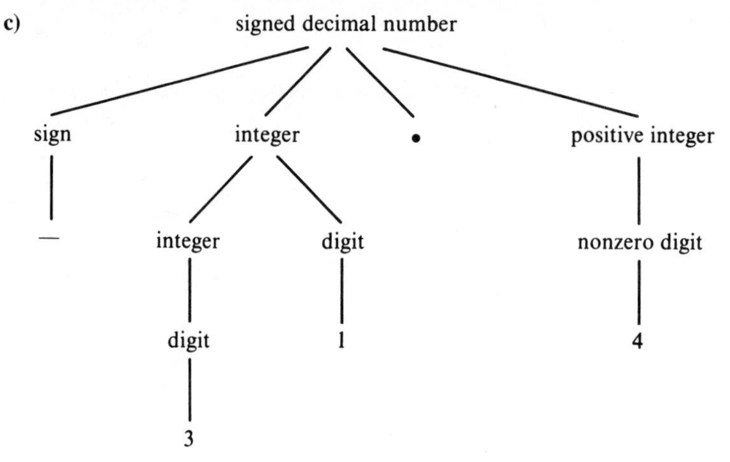

25. $\{(u, v) \mid v \text{ is derivable from } u\}$

Section 10.2

1. **a)** **b)**

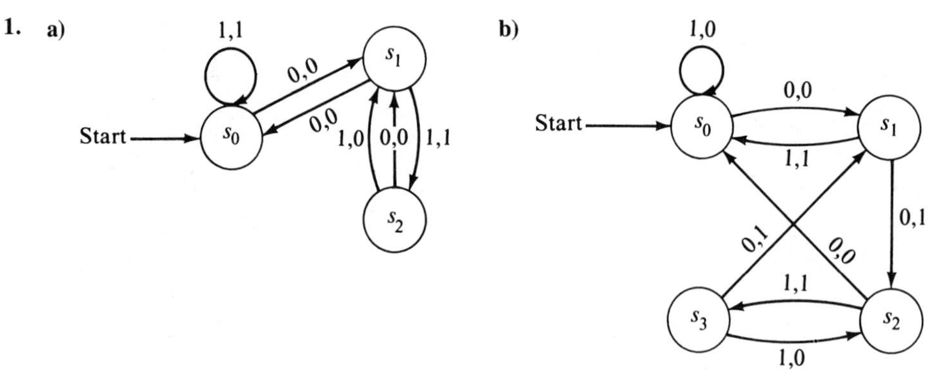

c)

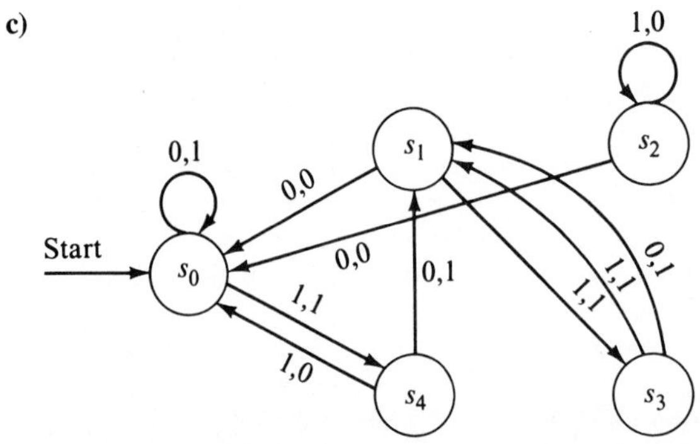

3. **a)** 1100 **b)** 00110110 **c)** 11111111111

5.

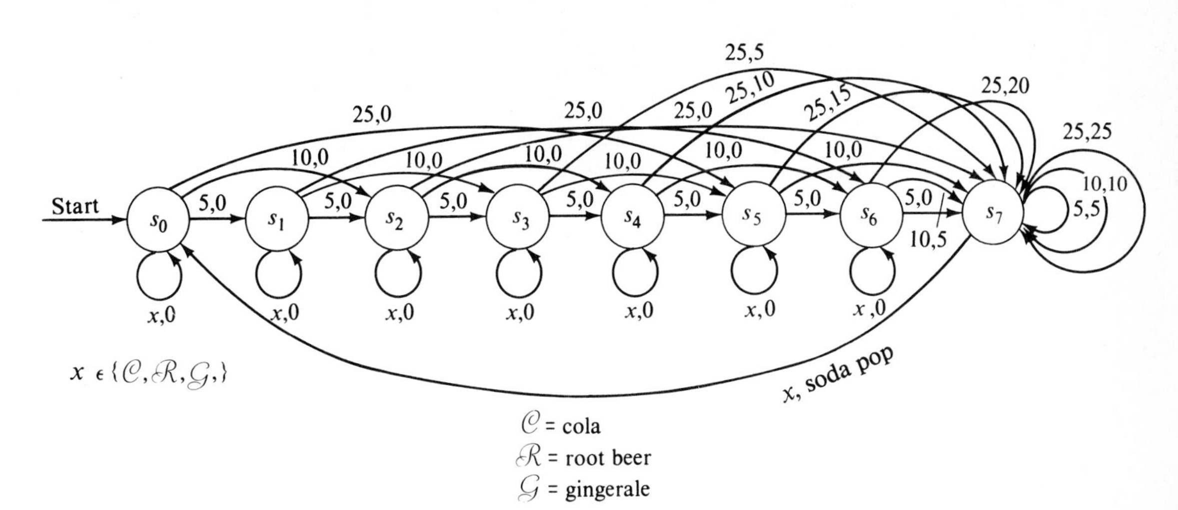

7.

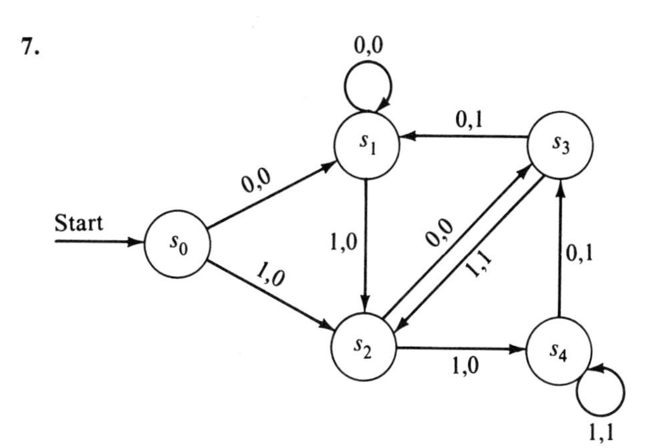

9.

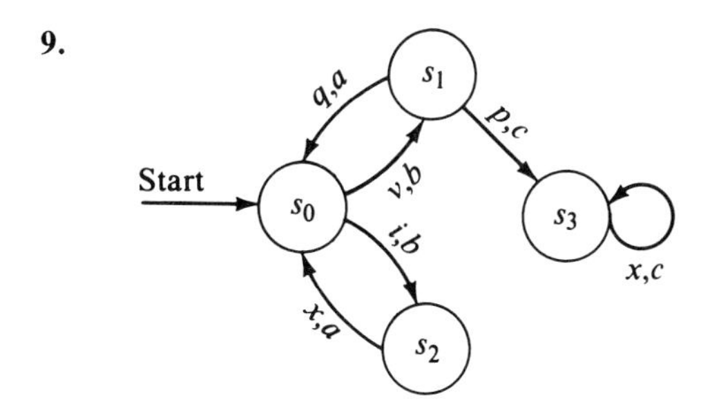

v = Valid ID
i = Invalid ID
p = Valid password
q = Invalid password

a = "Enter user ID
b = "Enter password"
c = Prompt
x = Any input

11.

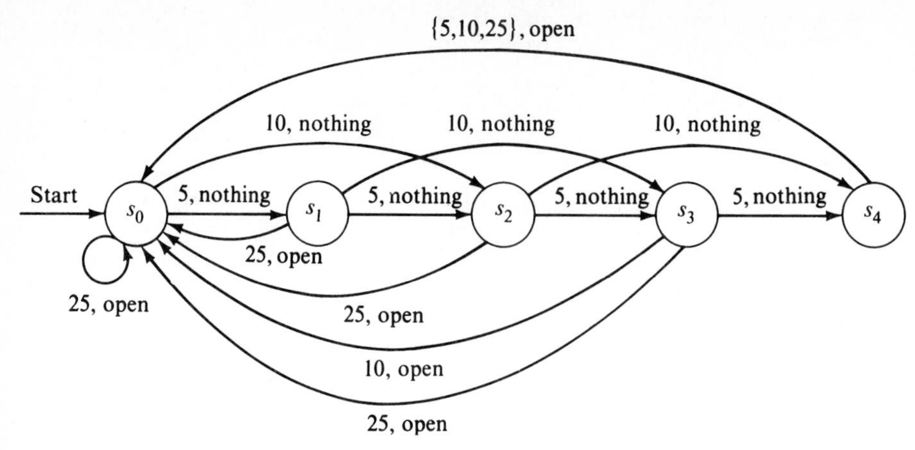
{5,10,25}, open
10, nothing
10, nothing
10, nothing
Start
s_0
5, nothing
s_1
5, nothing
s_2
5, nothing
s_3
5, nothing
s_4
25, open
25, open
25, open
10, open
25, open

13.

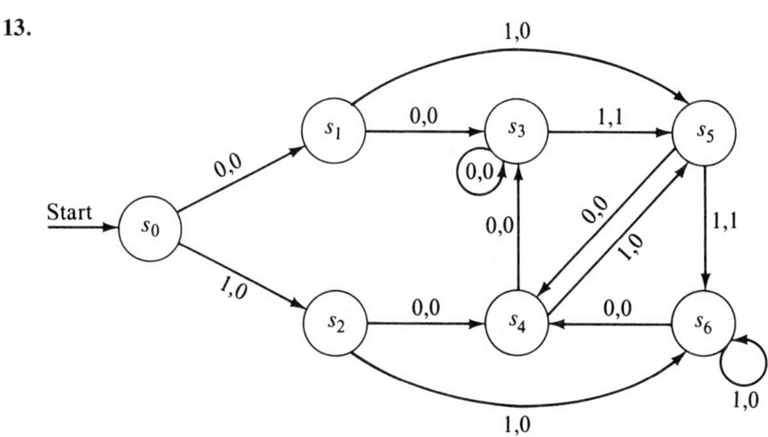
1,0
s_1
0,0
s_3
1,1
s_5
0,0
0,0
Start
s_0
0,0
0,0
1,0
1,1
1,0
s_2
0,0
s_4
0,0
s_6
1,0
1,0

15.

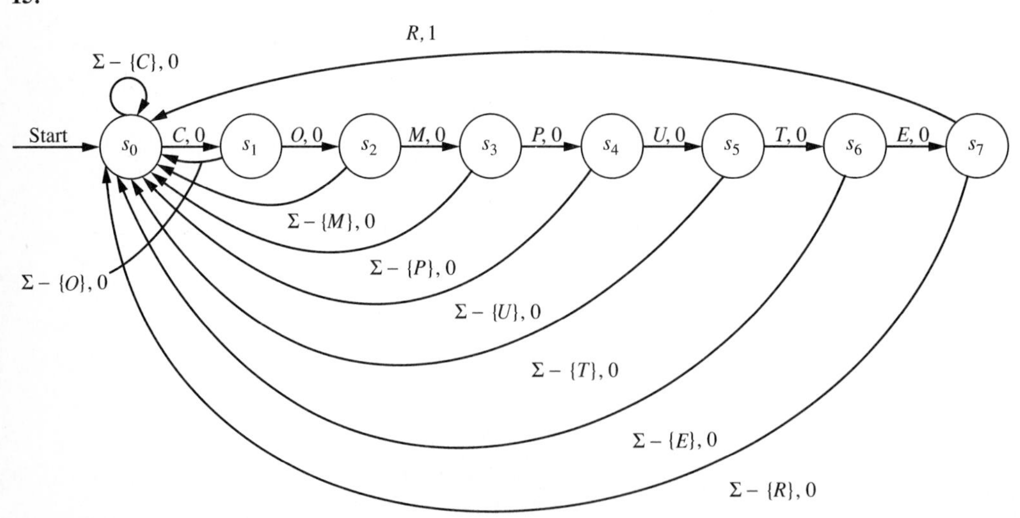
R, 1
Σ – {C}, 0
Start
s_0
C, 0
s_1
O, 0
s_2
M, 0
s_3
P, 0
s_4
U, 0
s_5
T, 0
s_6
E, 0
s_7
Σ – {M}, 0
Σ – {O}, 0
Σ – {P}, 0
Σ – {U}, 0
Σ – {T}, 0
Σ – {E}, 0
Σ – {R}, 0

17.

State	*f* Input 0	1	*g*
s_0	s_1	s_2	1
s_1	s_1	s_0	1
s_2	s_1	s_2	0

19. a) 11111
b) 1000000
c) 100011001100

21.

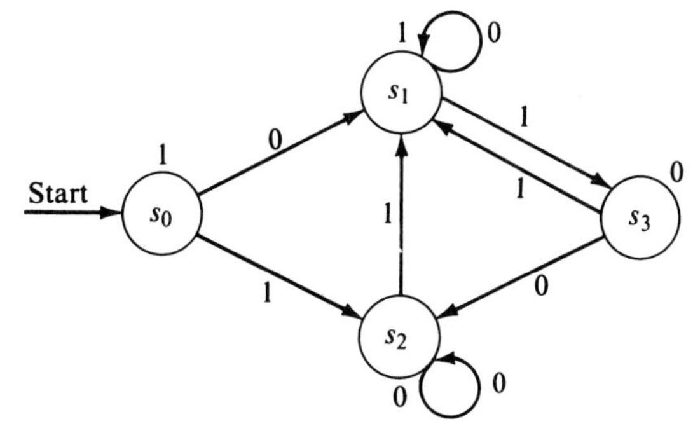

Section 10.3

1. a) {000, 001, 1100, 1101}
b) {000, 0011, 010, 0111}
c) {00, 011, 110, 1111}
d) {000000, 000001, 000100, 000101, 010000, 010001, 010100, 010101}

3. $A = \{1, 101\}, B = \{0, 11, 000\}; A = \{10, 111, 1010, 1000, 10111, 101000\}, B = \{\lambda\}; A = \{\lambda, 10\}, B = \{10, 111, 1000\}$ or $A = \{\lambda\}, B = \{10, 111, 1010, 1000, 10111, 101000\}$

5. a) The set of all strings consisting of zero or more consecutive bit pairs 10
b) The set of all strings consisting of all 1s such that the number of 1s is divisible by 3, including the null string
c) The set of all strings that begin and end with a 1 and that have at least two 1s between every pair of 0s
d) The set of all strings that begin and end with a 1 and have at least two 1s between every pair of 0s

7. A string is in A^* if and only if it is a concatenation of an arbitrary number of strings in A. Since each string in A is also in B, it follows that a string in A^* is also a concatenation of strings in B. Hence $A^* \subseteq B^*$.

9. a) Yes **b)** Yes
c) Yes **d)** No
e) Yes **f)** Yes

11. a) Yes **b)** Yes
c) No **d)** No
e) No **f)** No

13. $\{0, 10, 11\}\ \{0, 1\}^*$

15. $\{0^m 1^n | m \geq 0 \text{ and } n \geq 1\}$

17. $\{0, 01, 11\}$

19. $\{\lambda, 0\} \cup \{0^m 1^n | m \geq 1, n \geq 1\}$

21. $\{10^n | n \geq 0\} \cup \{10^n 10^m | n, m \geq 0\}$

23.

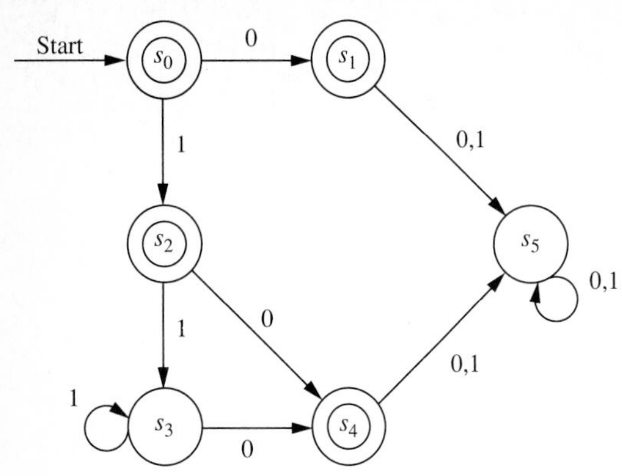

25. Add a non-final state s_3 sith transitions to s_3 from s_0 on input 0, from s_1 on input 1, and from s_3 on input 0 or 1.

27. a)

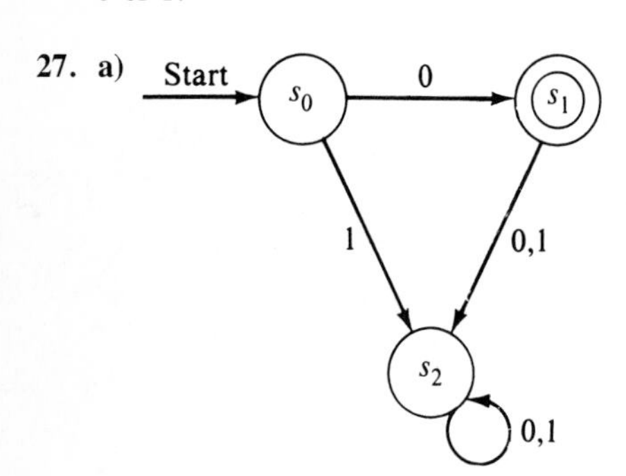

b)

c)

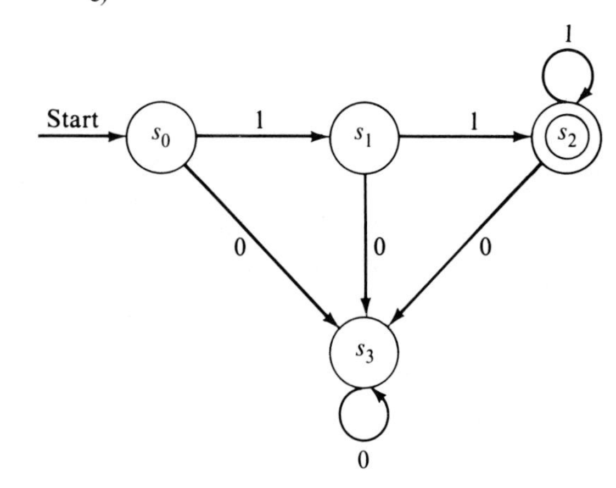

Section 10.4

1. a) Any number of 1s followed by a 0
b) Any number of 1s followed by one or more 0s
c) 111 or 001
d) A string of any number of 1s or 00s or some of each in a row
e) λ or a string that ends with a 1 and has one or more 0s before each 1
f) A string of length at least 3 that ends with 00

3. a) $\mathbf{00^*1}$ **b)** $\mathbf{(0 \cup 1)(0 \cup 1)(0 \cup 1)^*0000^*}$
c) $\mathbf{0^*1^* \cup 1^*0^*}$ **d)** $\mathbf{11(111)^*(00)^*}$

5. Use an inductive proof. If the regular expression for A is $\emptyset$, λ, or $\boldsymbol{x}$, the result is trivial. Otherwise, suppose the regular expression for A is $\mathbf{BC}$. Then $A = BC$ where B is the set generated by $\mathbf{B}$ and C is the set generated by $\mathbf{C}$. By the inductive hypothesis there are regular expressions $\mathbf{B}'$ and $\mathbf{C}'$ that generate B^R and C^R, respectively. Since $A^R = (BC)^R = C^RB^R$, $\mathbf{C'B'}$ is a regular expression for A^R. If the regular expression for A is $\mathbf{B} \cup \mathbf{C}$, then the regular expression for A is $\mathbf{B}' \cup \mathbf{C}'$ since $(B \cup C)^R = (B^R) \cup (C^R)$. Finally, if the regular expression for A is $\mathbf{B}^*$, then it is easy to see that $(\mathbf{B}')^*$ is a regular expression for A^R.

7. a)

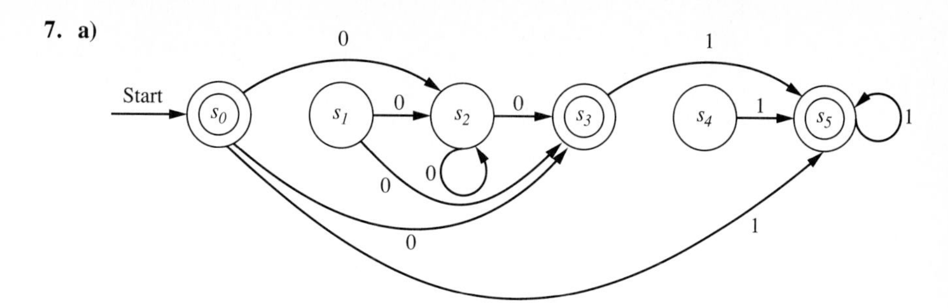
Start
s_0
s_1
s_2
s_3
s_4
s_5

b)

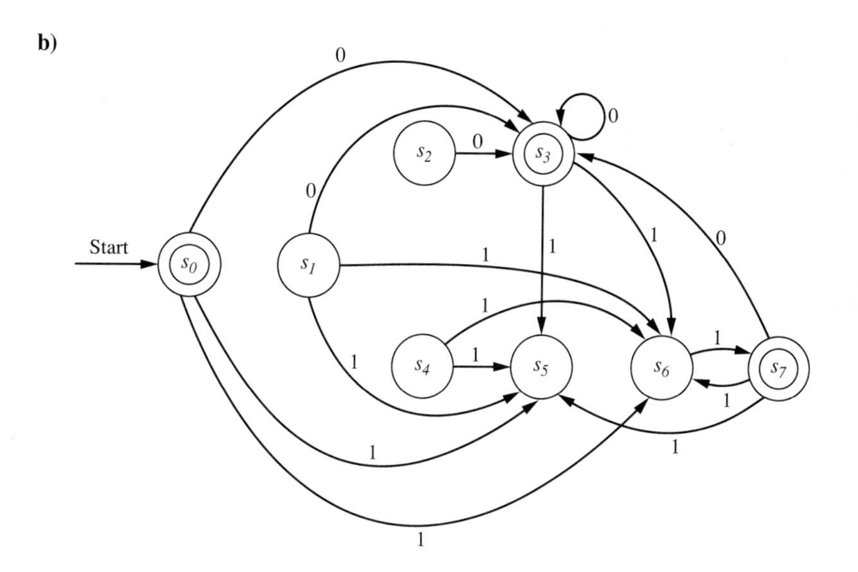
Start
s_0
s_1
s_2
s_3
s_4
s_5
s_6
s_7

c)

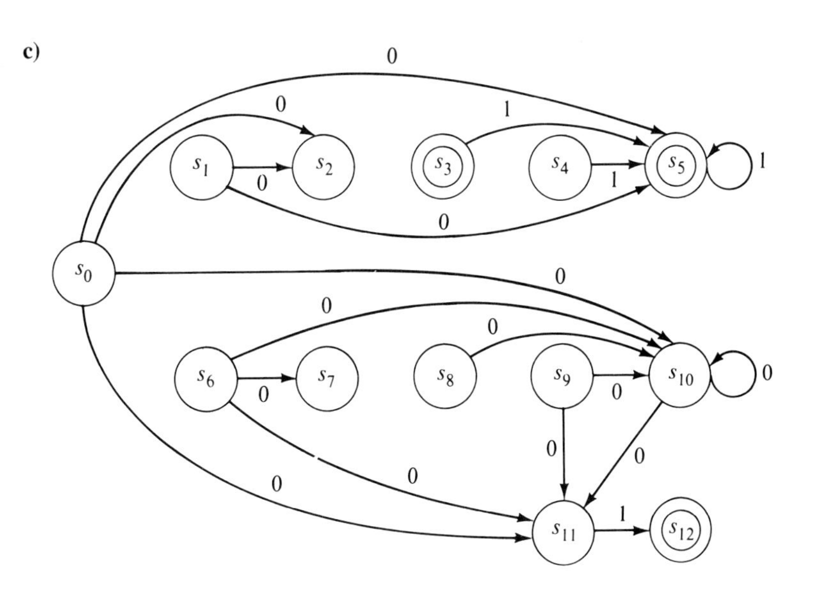
s_0
s_1
s_2
s_3
s_4
s_5
s_6
s_7
s_8
s_9
s_{10}
s_{11}
s_{12}

9. $S \to 0A$, $S \to 1B$, $S \to 0$, $A \to 0B$, $A \to 1B$, $B \to 0B$, $B \to 1B$

11. $S \to 0C$, $S \to 1A$, $S \to 1$, $A \to 1A$, $A \to 0C$, $A \to 1$, $B \to 0B$, $B \to 1B$, $B \to 0$, $B \to 1$, $C \to 0C$, $C \to 1B$, $C \to 1$.

13. This follows since input that leads to a final state in the automaton corresponds uniquely to a derivation in the grammar.

15. The "only if" part is clear since I is finite. For the "if" part let the states be $s_{i_0}, s_{i_1}, s_{i_2}, \ldots, s_{i_n}$, where $n = l(x)$. Since $n \geq |S|$, some state is repeated by the pigeonhole principle. Let y be the part of x that causes the loop, so that $x = uyv$ and y sends s_j to s_j, for some j. Then $uy^kv \in L(M)$ for all k. Hence $L(M)$ is infinite.

17. Suppose that $L = \{0^{2n}1^n\}$ were regular. Let S be the set of states of a finite-state machine recognizing this set. Let $z = 0^{2n}1^n$ where $3n \geq |S|$. Then by the pumping lemma, $z = 0^{2n}1^n = uvw$, $l(v) \geq 1$, and $uv^iw \in \{0^{2n}1^n | n \geq 0\}$. Obviously v cannot contain both 0 and 1, since v^2 would then contain 10. So v is all 0s or all 1s, and hence uv^2w contains too many 0s or too many 1s, so it is not in L. This contradiction shows that L is not regular.

19. Suppose that the set of palindromes over $\{0, 1\}$ were regular. Let S be the set of states of a finite-state machine recognizing this set. Let $z = 0^n10^n$, where $n > |S|$. Apply the pumping lemma to get $uv^iw \in L$ for all nonnegative integers i where $l(v) \geq 1$, and $l(uv) \leq |S|$, and $z = 0^n10^n = uvw$. Then v must be a string of 0s (since $|n| > |S|$), so uv^2w is not a palindrome. Hence the set of palindromes is not regular.

Supplementary Exercises

1. a) $S \to 00S111$, $S \to \lambda$

b) $S \to AABS$, $AB \to BA$, $BA \to AB$, $A \to 0$, $B \to 1$, $S \to \lambda$

c) $S \to ET$, $T \to 0TA$, $T \to 1TB$, $T \to \lambda$, $0A \to A0$, $1A \to A1$, $0B \to B0$, $1B \to B1$, $EA \to E0$, $EB \to E1$, $E \to \lambda$

3. a) **b)** **c)**

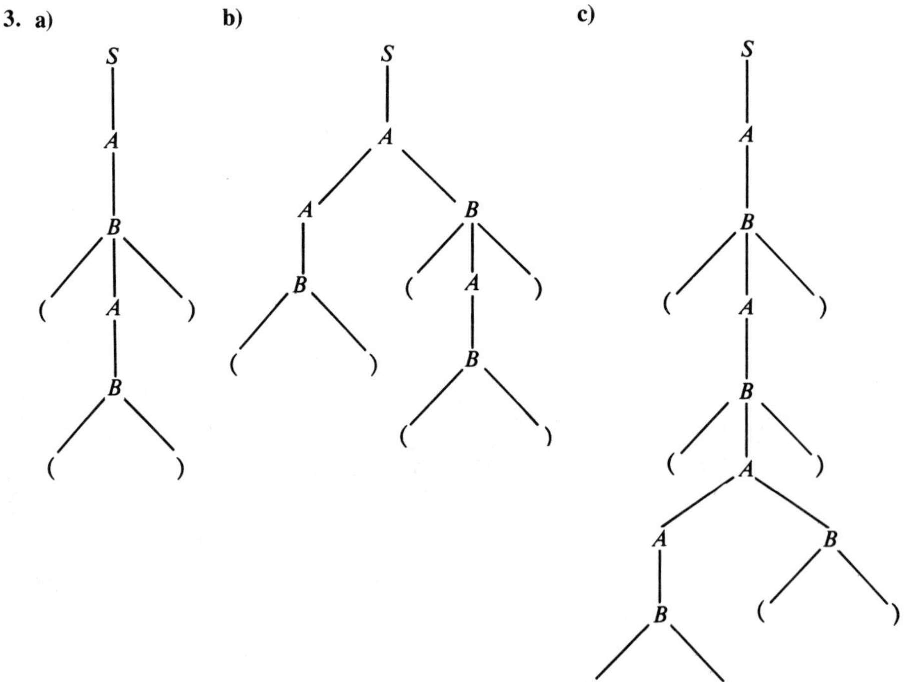

5.

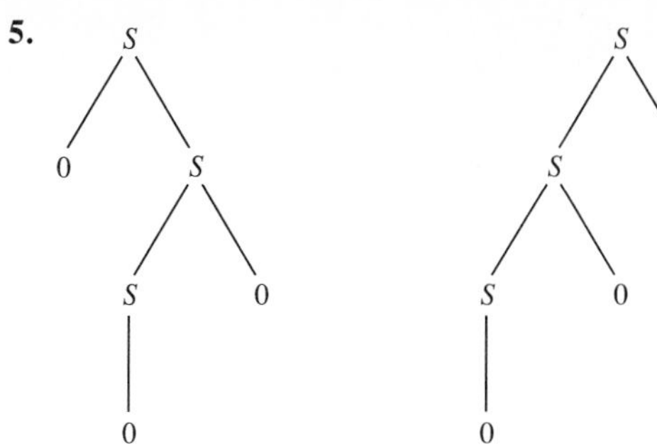

7. No, take $A = \{1, 10\}$ and $B = \{0, 00\}$

9. No, take $A = \{00, 000, 00000\}$ and $B = \{00, 000\}$.

11. **a)** 1 **b)** 1 **c)** 2 **d)** 3 **e)** 2 **f)** 4

13

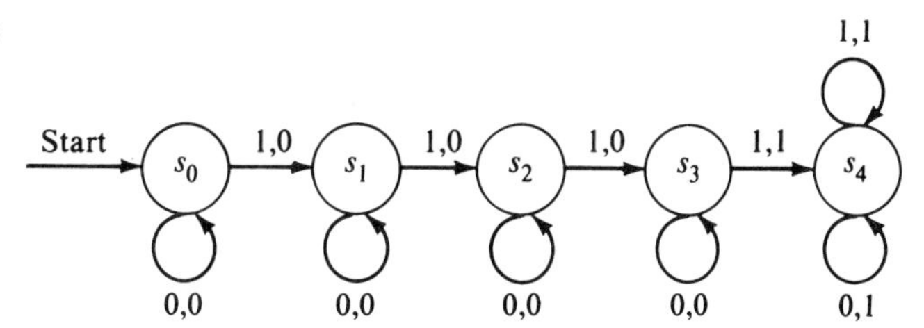

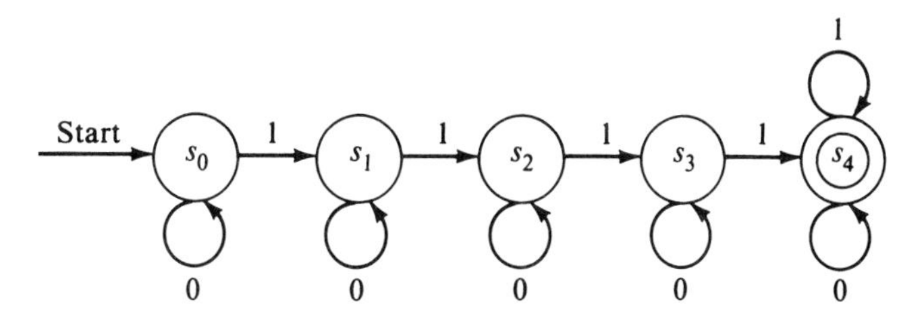

15.

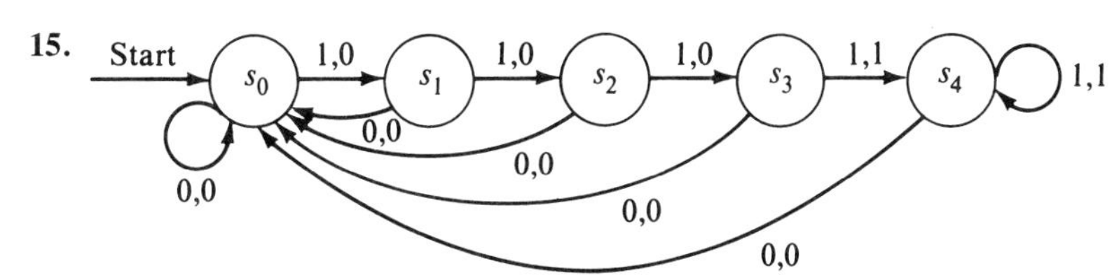

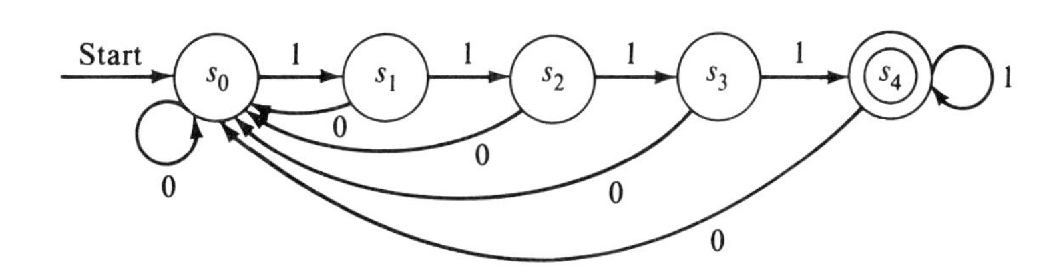

17. **a)** $n^{nk+1}m^{nk}$ **b)** $n^{nk+1}m^{n}$

19.

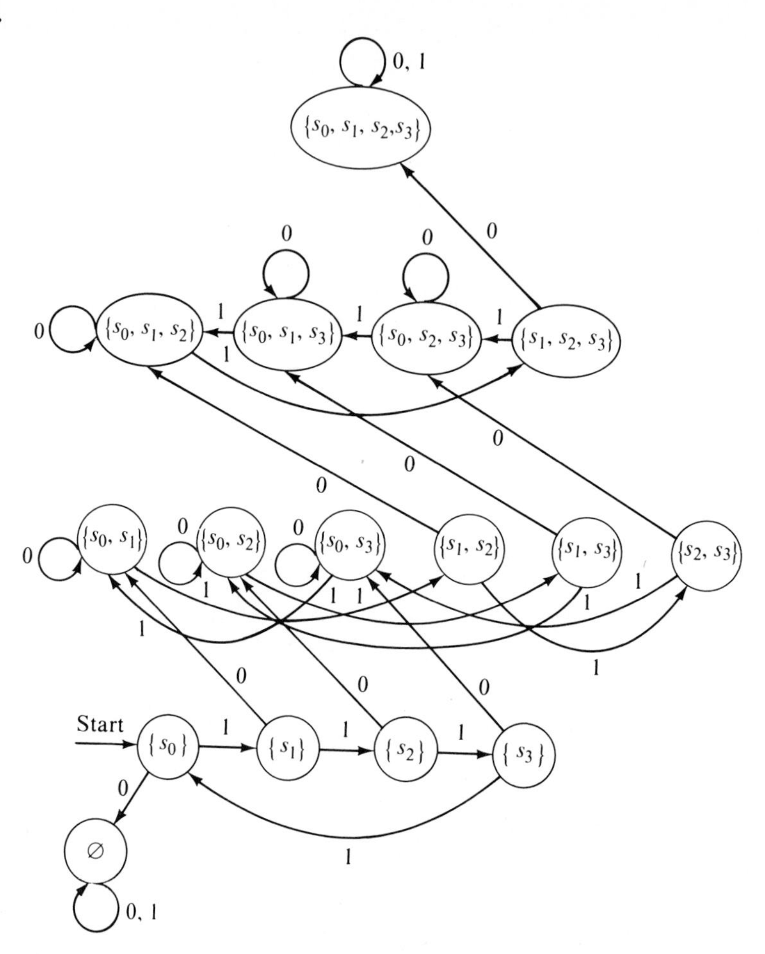

21. **a)**

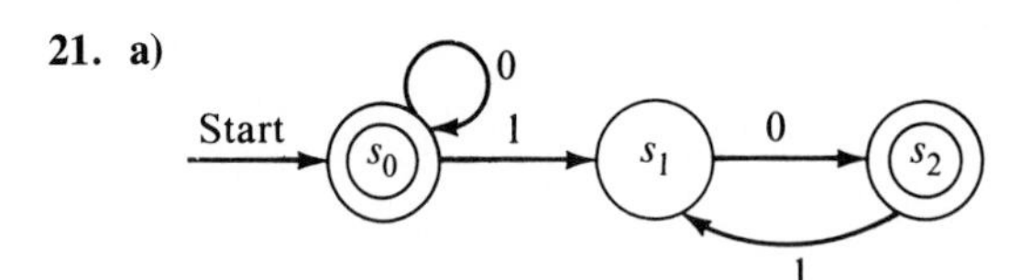

b)

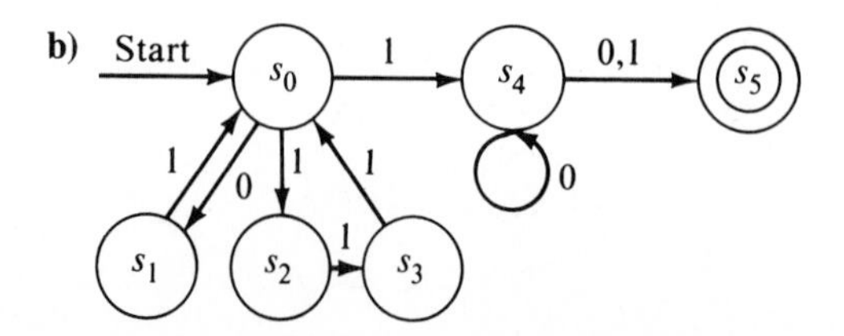

c)

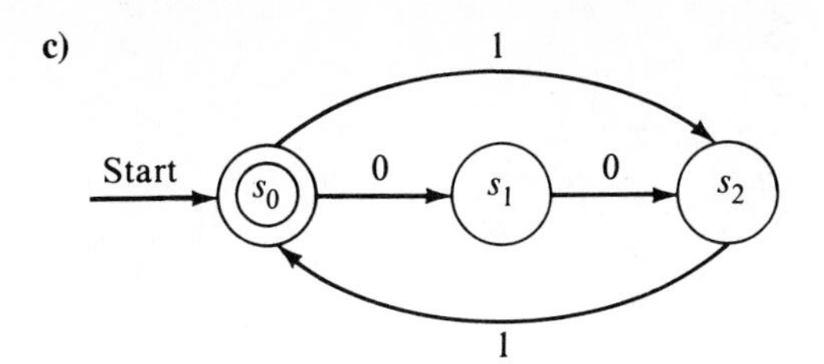

23. Construct the deterministic finite automaton for A with states S and final states F. For $\overline{A}$ use the same automaton but with final states $S - F$.

25. a)

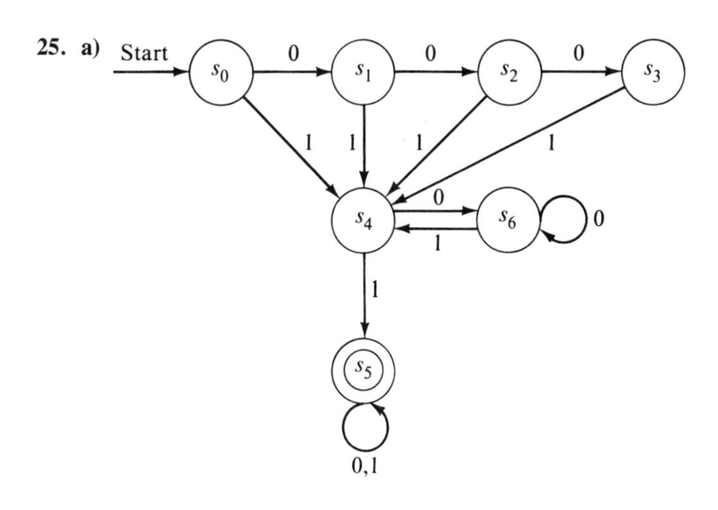

b)

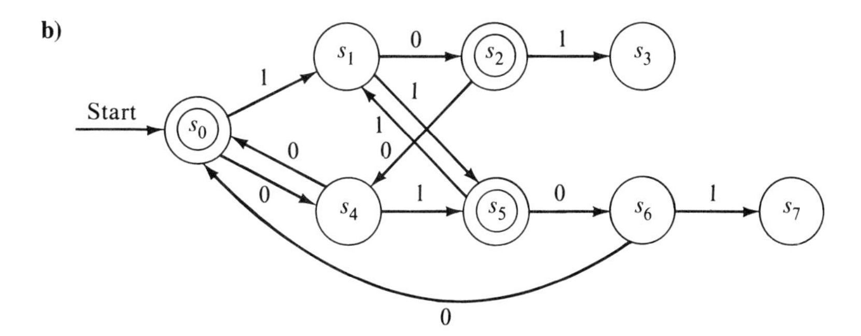

c)

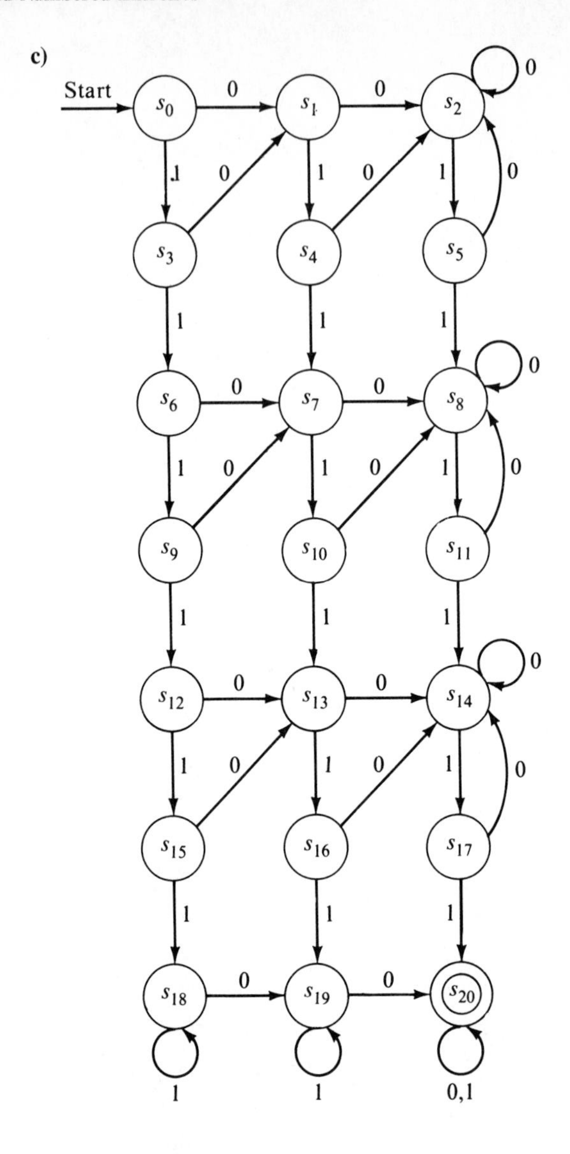

27. Suppose that $L = \{1^p \mid p \text{ is prime}\}$ is regular, and let S be the set of states in a finite-state machine recognizing L. Let $z = 1^p$ where p is a prime with $p > |S|$ (such a prime exists since there are infinitely many primes). By the pumping lemma it must be possible to write $z = uvw$ with $l(uv) \le |S|$, $l(v) \ge 1$ and for all nonnegative integers i, $uv^iw \in L$. Since z is a string of all 1s, $u = 1^a$, $v = 1^b$, and $w = 1^c$, where $a + b + c = p$, $a + b \le n$, and $b \ge 1$. This means that $uv^iw = 1^a1^{bi}1^c = 1^{(a+b+c)+b(i-1)} = 1^{p+b(i-1)}$. Now take $i = p + 1$. Then $uv^iw = 1^{p(1+b)}$. Since $p(1 + b)$ is not prime, $uv^iw \notin L$, which is a contradiction.

APPENDICES

Appendix 1

1. a) 2^3 **b)** 2^6 **c)** 2^4

3. a) $2y$ **b)** $2y/3$ **c)** $y/2$

5. a)

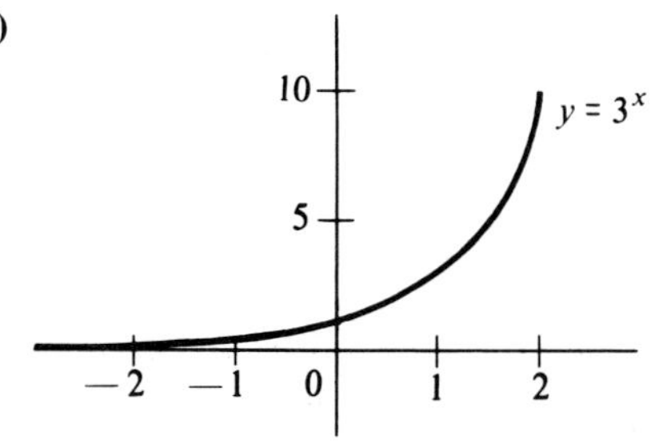

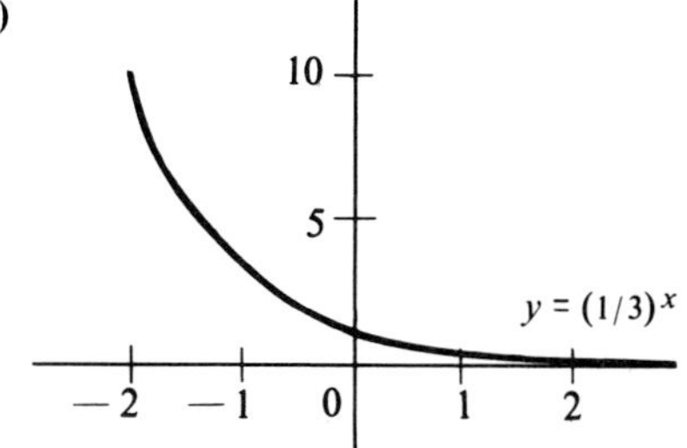

c)

Appendix 2

1. After the first block is executed, a has been assigned the original value of b and b has been assigned the original value of c, whereas after the second block is executed, b is assigned the original value of c and a the original value of c as well.

3. The following **while** construction does the same thing.

i := *initial value*
while $i \leq$ *final value*
begin
  statement
  $i := i + 1$
end

Appendix 3

1. $f(x) = 2(x^6 - 1)/(x - 1)$

3. 10

5. $f(x) = (1 + x + x^2 + x^3 + \cdots)(1 + x^2 + x^4 + x^6 + \cdots)(1 + x^5 + x^{10} + x^{15} + \cdots)(1 + x^{10} + x^{20} + x^{30} + \cdots)$

7. $f(x) = (x^2 + x^3 + \cdots)(1 + x + x^2 + x^3)(x^2 + x^3 + x^4 + x^5)$

9. a) $f(x) = 3/(1 - x)$ **b)** $f(x) = 1/(1 - 5x)$ **c)** $f(x) = 1/(1 - x)^2$

11. $a_k = 5 \cdot 7^k$

13. $a_k = 4^k$

15. Let $G(x) = \sum_{k=0}^{\infty} f_k x^k$. After shifting indices of summation and adding series, we see that $G(x) - xG(x) - x^2G(x) = f_0 + (f_1 - f_0)x + \sum_{k=2}^{\infty}(f_k - f_{k-1} - f_{k-2})x^k = 0 + x + \sum_{k=2}^{\infty} 0x^k$. Hence $G(x) - xG(x) - x^2G(x) = x$. Solving for $G(x)$ gives $G(x) = x/(1 - x - x^2)$. By the method of partial fractions, it can be shown that $x/(1 - x - x^2) = (1/\sqrt{5})[1/(1 - \alpha x) - 1/(1 - \beta x)]$ where $\alpha = (1 + \sqrt{5})/2$ and $\beta = (1 - \sqrt{5})/2$. Using the fact that $1/(1 - ax) = \sum_{k=0}^{\infty} a^k x^k$, it follows that $G(x) = (1/\sqrt{5}) \cdot \sum_{k=0}^{\infty}(\alpha^k - \beta^k)x^k$. Hence $f_k = (1/\sqrt{5}) \cdot (\alpha^k - \beta^k)$.

Index of Biographies

Augusta, Ada (Countess of Lovelace), 16
Bachmann, Paul, 76
Backus, John, 623
Bernoulli, James, 263
Boole, George, 3
Cantor, Georg, 33
Carroll, Lewis (Charles Dodgson), 24
Cayley, Arthur, 511
Chomsky, Noam, 620
De Morgan, Augustus, 13
Descartes, René, 39
Dijkstra, Edsger, 469
Dirichlet, G. Lejeune, 234
Dodgson, Charles (Lewis Carroll), 24
Eratosthenes, 329
Erdös, Paul, 405
Euclid, 120
Euler, Leonhard, 452
Fermat, Pierre de, 138
Fibonacci, 195
Gauss, Karl Friedrich, 113
Goldbach, Christian, 173
Hamilton, Sir Williman Rowan, 458
Hasse, Helmut, 391
Karnaugh, Maurice, 597
Kempe, Alfred, 488
al-Khowarizmi, Abu Ja'far Mohammed ibn Musa, 92
Kleene, Stephen, 636
Kruskal, Joseph, 566
Kuratowski, Kazimierz, 483
Lamé, Gabriel, 197
Landau, Edmund, 76
Laplace, Pierre Simon, 251
Łukasiewicz, Jan, 536
McCluskey, Edward, 604
Naur, Peter, 623
Pascal, Blaise, 244
Petersen, Julius, 465
Prim, Robert Clay, 565
Quine, William, 604
Ramsay, Frank Plumpton, 238
Russell, Betrand, 33
Shannon, Claude, 577
Turing, Alan, 655
Vandermonde, Alexandre-Théophile, 245
Venn, John, 35
Warshall, Stephen, 373

Index

Absorption laws:
  for Boolean variables, 582, 583
  for propositions, 17
Accepted string, 638, 640
Ackermann, W., 202, B-1
Ackermann's function, 202, B-1
Ada, 16
Adder, 593–595, 610
  full, 594–595, 610
  half, 593
Addition rule of inference, 163
Address system, universal, 526
Adjacency list, 429–430
Adjacency matrix, 430–432, 496
Adjacent square in a Karnaugh map, 598
Adjacent vertex, 417, 496
  from, 418
  to, 418
Algebra, Boolean, 577–609
Algorithm, 92–98, 156, B-1, B-2
  for base *b* expansions, 124
  binary search, 95–97, 100–101, 313, 316
  for binary search tree, 521
  for Boolean product of zero-one matrices, 152
  bubble sort, 547
  for coloring a graph, 495
  for computing $x^n (\bmod m)$, 159
  definition of, 92
  Dijkstra's, 471–474, B-2
  divide-and-conquer, 312–314
  division, 109
  Euclidean, 120–122, 196–198, 203–204
  for Euler circuits, 455
  for Euler paths, 455–456
  for evaluating a polynomial, 104
  fast integer multiplication, 313–314, B-1–B-2
  fast matrix multiplication, 314, B-1–B-2
  for finding maximum element, 93–95, 99–100
  Fleury's, 466
  Floyd's, 476
  for generating combinations, 285, B-1
  for generating permutations, 284, B-1
  in graph theory, B-2
  greedy, 564
  history of the word, 92
  inorder traversal, 535
  integer addition, 125–126
  integer multiplication, 126–127
  for integer operations, 125–128
  Kruskal's, 566–569, B-2
  linear search, 95
  matrix multiplication, 148
  merge sort, 547–551
  optimal, 105
  postorder traversal, 535
  preorder traversal, 534
  Prim's, 565, B-3
  properties of, 94
  recursive, 203
  Sollin's, 570, B-3
  for spanning tree, 555–557
  topological sorting, 397
  transitive closure, 372
  Warshall's, 373–376, B-2
Alphabet, 616, 626, 656
Alphanumeric character, 228
Ambiguous grammar, 658
Ancestor of an internal vertex in a tree, 507, 571
AND, 9
AND gate, 588–589, 610
Antecedent, 6
Antichain, 407
Antisymmetric property, 345, 360, 401
Arc, 362
Argument, valid, 164
Arithmetic:
  computer, with large integers, 136–138
  modular, 112–114
Arithmetic mean, 192
Arithmetic progression, 218

Array:
linear, 424
two-dimensional, 424
Articulation point, 444
Artificial intelligence, B-0
Assertion:
final, 209, 216
initial, 209, 216
Assignment statement, A-3–A-4
Associative laws:
for Boolean variables, 581
for propositions, 15
for sets, 46
Asymmetric property, 350
Augusta, Ada (Countess of Lovelace), 13, 16
Automata:
deterministic finite-state, 639
equivalent, 641
finite-state, 637
linear bounded, 654
nondeterministic finite-state, 639
pushdown, 654
theory, 626–654, B-3
Average-case complexity, 101, 268–269
of linear search, 101–102, 269
Axiom, 33, 85, 162

Babbage, Charles, 16
Bachmann, Paul, 76
Backtracking, 555, 557–560, B-3
for graph coloring, 558–559
solving the *n*-queens problem, 559
Backus, John, 623
Backus-Naur form, 623, 657
Backward difference, 304
Bacteria, 303
Bagel, 279
Balanced string of parentheses, 220
Balanced ternary expansion, 129
Balanced tree, 516, 571
Bandwidth of a graph, 502
Baseball, 230–231
Base *b* expansion, 123
BASIC, counting variable names in, 228
Basis step, 177, 188, 216
Basis, vertex, 450
Bayes' formula, 292
Begging the question, 166, 215
Begin statement, A-5
Bernoulli, James, 263
Bernoulli family tree, 505
Bernoulli's inequality, 190
Bernoulli trial, 263, 287
probability of *k* successes in *n* trials, 264
Biconditional, 8, 85
Big-*O* notation, 74–82
Bijection, 59, 86
Binary coded decimal expansion, 129, 603
Binary digit, 8
Binary expansion, 123
Binary relation, 341, 401
Binary representation, 123
Binary search algorithm, 95–97, 100–101, 156, 204–205, 313, 316
Binary search tree, 520–522, 571
Binary search tree algorithm, 521
Binary tree, 508, 571
Binomial coefficient, 243–244, 287, A-12
Binomial expression, 246
Binomial theorem, 243, 246
Bipartite graph, 421–422, 496
Birthday problem, 269
Bit, 8, 85
operation, 8–9, 85
string, 9, 50, 85
Bitwise operation, 9, 85
AND, 9
OR, 9
XOR, 9
Blocks of statements, A-4–A-5
Bonferroni's inequality, 269
Boole, George, 3, 577
Boolean algebra, 577–609, B-3
Boolean expression, 579, 587, 610
Boolean function, 578–580, 610
complement of, 579
counting, 580
dual of, 582, 610
equality of, 579
Boolean identities, 581
Boolean power, 153
Boolean product, 151–152, 157, 578, 579–580, 610
of zero-one matrices, 151–152
Boolean sum, 578, 579, 610
Boolean variable, 8, 578, 609
Boole's inequality, 269
Bottom-up parsing, 623
Bound:
greatest lower, 395
least upper, 394
lower, 394
upper, 394
Bound variable, 25
Box, distributing objects into, 278
Breadth-first search, 556
Bridge:
in a graph, 444
Königsberg, problem, 451–452
B-tree of degree *k*, 574
Bubble sort, 545
complexity, 546–547
Bubble sort algorithm, 547
Byte, 123

Cactus, 574
Caesar, Julius, 117
Caesar cipher, 117
Calculus, 29, A-10, B-0
Cantor digits, 286
Cantor expansion, 129

Cantor, Georg, 33
Cardinality, 37, 71–72, 85
Cards, deck of, 253
Carroll, Lewis, 24, B-0
Carry, 125
Cartesian product, 38–40
Cases, proof by, 169, 216
Caterpillar, 575
Cayley, Arthur, 504, 511
Ceiling function, 63–64, 86
Center of a tree, 519
Chain, 388, 407
Chain letter, 515
Characteristic equation, 306
Characteristic function, 66
Characteristic roots, 306, 335
Chessboard, 186–187, 465–466, 501, 559, B-1
Chicken, 502
Child, 507, 509, 571
  left, 509
  right, 509
  of a vertex in a tree, 507, 571
Chinese remainder theorem, 134–136, 157
Chomsky, Noam, 620
Chomsky classification of grammars, 620
Chromatic number, 488–490, 497, 558
  edge, 495
Cipher:
  affine, 117–118
  Caesar, 117
  RSA, 140–142
  shift, 117
Circuit, 497
  combinational, 588
  in a directed graph, 367, 402, 443
  Euler, 452–457
  in a graph, 442
  Hamilton, 457–462
  multiple output, 593
  simple, 442
Circular reasoning, 166, 215
Circular relation, 404
Class:
  congruence, 381
  equivalence, 380
Clique, 500
Closure:
  Kleene, 636, 657
  reflexive, 366
  of a relation, 365–376, 402
  symmetric, 366
  transitive, 369
Code, Gray, 460
Code, prefix, 524
Codeword enumeration, 301–302
Coding, Huffman, 525, B-2
Codomain, 55, 86
Coefficient:
  binomial, 243
  multinomial, 281
Collision, 115
Coloring:
  of a graph, 487–493, 558–559
  k-tuple, 495
  of a map, 486–487
Column of a matrix, 145
Combination, 242–243
  generating, 284–286
  linear, 130
  with repetition, 272–276
Combinatorial circuit, 588
Combinatorial identity, 243–245, B-1
Combinatorial proof, 243, 245, 246
Combinatorics, 222, 287
Comment in pseudocode, A-5
Commutative laws:
  for Boolean variables, 581
  for propositions, 15
  for sets, 46
Comparable elements in a poset, 387, 402
Compatible total ordering, 396, 402
Complement:
  of a Boolean function, 579
  of a Boolean variable, 578, 609
  of a fuzzy set, 54
  of a set, 44–45, 86
Complementary event, 254
Complementary graph, 429
Complementary relation, 350
Complementation law, 46
Complete bipartite graph, 422, 496
Complete graph, 420, 496
Complete *m*-ary tree, 518
Complete *m*-partite graph, 499
Complexity of algorithms, 98–103
  of algorithm for maximum element, 99–100
  average-case, 101, 156, 268–269
  of binary search algorithm, 100–101
  of Boolean product of zero-one matrices, 153–154
  of bubble sort, 546–547
  computational, 99
  of Dijkstra's algorithm, 473
  of Euclidean algorithm, 120–122
  exponential, 102
  factorial, 102
  of integer addition, 126
  of integer multiplication, 128
  linear, 102
  of linear search algorithm, 100, 101–102, 268–269
  logarithmic, 102
  of matrix multiplication, 148
  of merge sort, 551
  $n \log n$, 102
  polynomial, 102
  of a sorting algorithm, 544
  space, 99, 156
  time, 99, 156
  worst-case, 100, 156

Components, connected, 444
Composite integer, 107, 156
  consecutive, 171–172
Composite key, 354, 401
Composite of relations, 348, 401
Composition of functions, 61, 86
Composition rule for program correctness, 210
Compound interest, 296–297
Compound proposition, 3, 84
Computational complexity of an algorithm, 98–103
Computer arithmetic, B-1
  with large integers, 136–138
Computer file system, 512
Computer representation of sets, 49–51
Concatenation, 199, 635, 657
  of sets of strings, 635
Conclusion of an implication, 6
Concurrent processing, 414
Condition, don't care, 603–604
Conditional constructions in pseudocode, A-5–A-6
Conditional probability, 260–261, 287
Conditional statements, partial correctness of, 210–211
Congruence, 113
Congruence, linear, 133, 156
Congruence class, 381, 402
Congruence modulo m, 113, 156, 380
Conjecture, Goldbach's, 173
Conjunction, 4, 85, 610
Conjunctive normal form, 586, 588
Connected component of a graph, 444, 497
Connected graph, 443, 497
  strongly, 445
  weakly, 445
Connectives, logical, 4
Connectivity relation, 369, 402
Connect vertices, 417
Consequence, 6
Constant complexity, 102
Constructive existence proof, 171
Contain, 34
Context-free grammar, 620, 656
Context-free language, 620
Context-sensitive grammar, 620
Contingency, 12, 85
Contradiction, 12, 85
  proof by, 168, 216
Contrapositive, 7, 85
Converse, 7, 85
Cookie, 274
Corollary, 162, 215
Correspondence, one-to-one, 59, 86
Countability, 71
Countable set, 71, 86
Counterexample, 172, 216
Counterfeit coins, 522–523
Counting, 222–329, A-10–A-14, B-1
  antisymmetric relations, 351
  basic rules of, 223–231
  Boolean functions, 580
  functions, 226
  one-to-one functions, 226
  onto functions, 329–331
  paths between vertices, 447–448
  relations, 347
  reflexive relations, 347
  symmetric relations, 351
  subsets, 227
  using tree diagrams, 230–231
Counting problems, using generating functions to solve, A-13–A-14
Covers, 605
C Programming language, 233
Crossing number of a graph, 486, B-2
Cryptography, 116–118, B-0–B-1
  private key, 139
  public key, 130
Cut edge, 444
Cut set, 574
Cut vertex, 444
Cycle in a directed graph, 367, 402
Cycle with n vertices, 420, 496
Cylinder, 599

Data base, 352–357
  intension of, 353
  key of, 353
  records, 352
  relational model of, 352–357
Decision tree, 522–523, 571
Deck of cards, 253
Decryption, 117, 156
Definition:
  inductive, 194
  recursive, 194
Degree-constrained spanning tree, 575
Degree of a linear homogeneous recurrence relation, 305
Degree of membership in a fuzzy set, 53
Degree of an n-ary relation, 352
Degree of a region, 480
Degree of a vertex, 417, 496
De Morgan, Augustus, 13
De Morgan's laws:
  for Boolean variables, 581
  for propositions, 13, 15
  for sets, 46
Depth-first search, 555
Derangement, 331–334, 335, 336
Derivable from, 617, 656
Derivation, 617, 657
Derivation tree, 621
Descartes, René, 39
Descendant of a vertex in a tree, 507–508, 571
Description of a set, 32–34
Detachment, law of, 162
Deterministic finite-state automaton, 639, 657
Diagonal matrix, 154
Diagonal relation, 366
Diagram:
  Hasse, 390–392, 402
  state, for a finite-state machine, 628
  tree, 230–231
  Venn, 35, 85

Diameter of a graph, 502
Dice, 251, 252
Die, 251
Difference:
  backward, 304
  equation, 305
  first forward, 338
  forward, 304, 338
  of sets, 44, 86
  symmetric, 52
Digraph, 362
Digit:
  binary, 123
  Cantor, 286
Dijkstra, Edsger, 469
Dijkstra's algorithm, 469–473
Directed edge, 362, 496
Directed graph, 362, 402, 411, 496
  circuit in, 367
  cycle in, 367
  path in, 367
  representing relation, 362–365
Directed multigraph, 411, 496
Directly derivable, 617, 656
Directly derivable from, 617
Direct proof, 167, 215
Dirichlet, G. Lejeune, 234
Dirichlet drawer principle, 234
Discrete mathematics, definition of, *xxi*
Discrete probability, 250–269
Discrete structure, definition of, *xii*
Disjoint sets, 43
Disjunction, 4, 84, 610
Disjunctive normal form:
  for Boolean variables, 585
  for propositions, 18
Disjunctive syllogism, 163
Distance:
  between spanning trees, 563
  between vertices in a graph, 502
Distributing objects into boxes, 278
Distributive laws:
  for Boolean variables, 581
  for propositions, 14, 15
  for sets, 46
Divide-and-conquer algorithm, 312, 335, B-1–B-2
Divide-and-conquer recurrence relation, 312
Dividend, 109
Divides, 106, 156
Divisibility relation, 346
Division algorithm, 109, 157
Division of integers, 106
Divisor:
  greatest common, 110
Dodecahedron, 457–459
Dodgson, Charles, 24
Domain of a function, 55, 86
Domains of an n-ary relation, 352
Dominating set, 500
  minimum, 500
Domination laws:
  for Boolean variables, 581
  for propositions, 15
  for sets, 46
Dominos, 178
Don't care condition, 603–604
Double negation law, 15
Dual:
  of a Boolean function, 582, 610
  graph of a map, 487
  of a proposition, 18
Duality principle:
  for Boolean identities, 583
  for propositional identities, 18

Eccentricity of a vertex of a tree, 519
Ecology, 413
Edge:
  chromatic number, 495
  coloring, 495
  cut, 444
  of a directed graph, 362
  list, 430
  multiple edges, 410, 412
  parallel, 410
  of a simple graph, 409
Element, 34, 85
  greatest of a poset, 393
  least of a poset, 393
  of a matrix, 145
  maximal of a poset, 392
  minimal of a poset, 393
  of a set, 34
Elementary subdivision, 483, 497
Ellipsis, 34
Empty set, 35, 85
Empty string, 68, 616
Encryption, 117, 156
Endpoint of an edge, 417
End statement, A-4
End vertex, 418
Entry of a matrix, 145
Enumeration, 222, 287
Equality of sets, 34, 85
Equivalence, logical, 13, 85
Equivalence class,
  definition of, 380, 402
  representative of, 380
Equivalence relation, 378–384, 402, B-2
  definition of, 379
Equivalent Boolean functions, 579
Equivalent deterministic automaton for a nondeterministic automaton, 641
Equivalent elements, 379, 402
Equivalent finite-state automata, 638
Equivalent partition, 383
Equivalent propositions, 13
Eratosthenes, 329
  sieve of, 328–329
Erdös, Paul, 405
Erdös number, 404
Euclid, 120, 172

Euclidean algorithm, 120–122, 157, 196–198, 203–204
Euler, Leonhard, 451, B-2
Euler circuit, 452–457, 497
Euler path, 452, 497
Euler phi-function, 119
Euler's formula, 479, 497
Event, 251
Exams, scheduling, 491–492
Exclusive or, 5, 85
Existence proof, 171
  constructive, 171
  nonconstructive, 171
Existential quantification, 22, 85, 171
  negation of, 29, 30
Expansion:
  balanced ternary, 129
  base b, 123
  binary, 123
  binary coded decimal, 129
  Cantor, 129
  hexadecimal, 123
  one's complement, 129
Expected value, 265–268, 287
Experiment, 251
Expert system, B-0
Exponential complexity, 102
Exponential function, A-0, B-3
Expression:
  binomial, 246
  Boolean, 579
  infix, 540
  postfix, 540
  prefix, 540
  regular, 645
  tree representing, 536

Factorial complexity, 102
Factorial function, 73, 78, 194, 205–206, 212–213
  big-O estimate for, 78
  inductive definition of, 194
Factorization into primes, 107–108
Factor of an integer, 106
Fallacy, 162, 165–166, 215
  of affirming the conclusion, 165
  of begging the question, 166
  of circular reasoning, 166
  of denying the hypothesis, 165
False, 3, 8
Family tree, 505
Fast matrix multiplication, 314–315
Fast matrix multiplication algorithm, 314, 317
Fast multiplication algorithm, 313–314, 317
Fermat, Pierre de, 138
Fermat's last theorem, 197, 234
Fermat's little theorem, 139, 157
Fibonacci, 195
Fibonacci numbers, 195, 297–298, 309
  explicit formula for, 309
Fibonacci tree, rooted, 453
Field of a record, 352
Final assertion, 209, 216
Final state, 637, 639
Final value of a loop variable, A-7
Finite set, 37, 85
Finite-state automata, 637, 653
Finite-state machine, 626–653, 657
  with no output, 637–642
  with output, 628–632
First difference, 304
First forward difference, 338
Fleury's algorithm, 456, 465
Floor function, 63–64, 86
Floyd's algorithm, 475
Forest, 505–506, 571
  minimal spanning, 570
  spanning, 563
for loops, A-7
Form:
  Backus-Naur, 623
  conjunctive normal, 588
  disjunctive normal, 585
  infix, 536
  postfix, 538
  prefix, 536
  prenex normal, 32
Formal language, 615
Formula:
  Bayes', 292
  Euler's, 479
  well-formed, 199
Forward difference, 304, 338
Four-color theorem, 488, 497, B-2
Free variable, 25, 85
Frequency assignment, 492
Full adder, 594–595, 610
Full m-ary tree, 508, 514, 571
Full subtractor, 596
Function, 54–82, 86, B-0
  Ackermann's, 202
  Boolean, 578
  ceiling, 63–64, 86
  characteristic, 66
  codomain of, 55
  composition of, 61, 86
  computable, 74
  definition of, 54–66
  domain of, 55
  Euler-phi, 119
  exponential, A-0
  factorial, 73, 78, 194, 205–206, 212–213
  floor, 63–64, 86
  generating, A-9
  graph of, 62–63
  greatest integer, 64
  hashing, 115
  identity, 59
  injective, 59
  inverse, 60
  invertible, 60
  logarithmic, A-1
  McCarthy, 91, 219

mod, 112–113, 156
number of, 226
one-to-one, 57, 86
onto, 58, 86, 329–331, 336
partial, 66
partition, 202
product of, 56
propositional, 19, 20, 85
range of, 55
recursive, 194–198, B-1
as a relation, 342
strictly decreasing, 58
strictly increasing, 58
sum of, 56
surjective, 58
threshold, 613
transition, 628
Functional completeness, 18, 586
Functionally complete set of operators:
for Boolean functions, 586, 610
for propositions, 18
Fundamental theorem of arithmetic, 107, 157, 188
Fuzzy logic, 11–12, B-0
Fuzzy set, 53–54, B-0

Gate, logic, 588–589
AND, 588
NAND, 596
NOR, 596
OR, 588
threshold, 612
XOR, 612
Gauss, Karl Friedrich, 113
Generalized induction principle, 407
Generalized pigeonhole principle, 235–236, 287
Generating combinations, 282–284
Generating function, A-9–A-14
Generating permutations, 284–286
Geometric mean, 192
Geometric progression, 69
Goldbach's conjecture, 173
Graceful tree, 575
Grammar:
ambiguous, 658
context-free, 620, 656
context-sensitive, 620, 656
phrase-structure, 620, 656
regular, 620, 656
type 0, 620, 656
type 1, 620, 656
type 2, 620, 656
type 3, 620, 656
Graph:
bandwidth of, 502
bipartite, 420–422
coloring, 487, 557, 497
complementary, 429
complete, 420
complete bipartite, 422, 490
complete m-partite, 499
connected, 443
connected components of, 444
crossing number of, 486
diameter of, 502
directed, 362, 411
dual of a map, 487
edge of, 362, 409
of a function, 62–63
homeomorphic, 482
influence, 413
intersection, 415
isomorphism, 434
niche-overlap, 413
nonplanar, 477
n-regular, 428
orientable, 501
Petersen, 483
planar, 476
planar representation of, 477
precedence, 414
radius of, 502
regular, 428
self-complementary, 441
simple, 409
strongly connected directed, 445
thickness of, 486
underlying undirected, 419
undirected, 409–411
union of, 426
vertices of, 362, 409
weakly connected undirected, 445
weighted, 466
Graph theory, 408–503, B-2
Gray code, 461
Greater than, in a poset, 387
Greatest common divisor, 110–112, 120–122, 130–131, 156, 197–198
Greatest element of a poset, 393, 402
Greatest integer function, 64
Greatest lower bound, 395, 402
Greedy algorithm, 564, 571

Half adder, 593, 610
Half subtractor, 596
Hamilton, Sir William Rowan, 458
Hamilton circuit, 458–460, 497
Hamilton path, 458, 497
Hamilton's round the world puzzle, 458–459
Handle of an S_k-tree, 574
Handshaking theorem, 418
Harmonic number, 183, 192
Harmonic series, 183
Hashing function, 115
Hasse, Helmut, 391
Hasse diagram, 390–392
Hatcheck problem, 331–334
Height:
star, 659
of a tree, 449, 571
Hexadecimal expansion, 123, 156
Homeomorphic graphs, 483, 497
Hop, 424

Horner's method, 104
Horses, 191
Huffman coding, 525, B-2
Hummingbirds, 29
Hybrid topology for a local area network, 423
Hydrocarbons, 510
Hypothesis, 6
  inductive, 177
Hypothetical syllogism, 163

Idempotent laws:
  for Boolean variables, 581
  for propositions, 15
  for sets, 46
Identities for sets, 45–47
Identity function, 59
Identity laws:
  for Boolean functions, 581
  for propositions, 15
  for sets, 46
Identity matrix, 149, 157
If then statement, 7, A-5
Image, 55, 86
  of an element, 55
  of a set, 56
Implication, 6–8, 85, 166–170
Incidence matrix, 432, 496
Incident vertices, 417, 496
Incident with, 417
Inclusion-exclusion, 44, 229–230, 319–324, 336
Inclusive or, 5
Incomparable elements in a poset, 387, 402
In-degree, 419, 496
Independence number, 502
Independent events, 261, 287
  mutually, 291
Independent set of vertices, 502
Index of summation, 68
Index register, 493
Indirect proof, 168, 216
Indistinguishable objects, permutations of, 276–277
Induced subgraph, 499
Induction, mathematical, 175–190, B-1
  generalized principle, 407
  second principle, 188
Inductive definition, 194
Inductive hypothesis, 177, 188
Inductive step, 177, 188, 216
Inequalities, proof by mathematical induction, 180, 183, 185
Inequality:
  Bernoulli's, 190
  Bonferroni's, 269
  Boole's, 269
Infinite series, A-10–A-11
Infinite set, 37, 85
Infix expression, 540
Infix form, 536
Infix notation, 536, 571
Influence graph, 413
Inhomogeneous linear recurrence relation, 311
Initial assertion, 209, 216
Initial conditions for a recurrence relation, 296, 335
Initial state, 637
Initial value of a loop variable, A-7
Initial vertex of a directed edge, 418
Injection, 57
Injective function, 57
Inorder traversal, 530, 571
Inorder traversal algorithm, 535
Input alphabet for a finite-state automaton, 628, 637, 639
Insertion sort, 574
Intension of a data base, 353
Interconnection network, 423–425
Interior vertices of a path, 373
Internal vertex of a tree, 508, 514, 571
Intersection:
  of fuzzy sets, 54
  graph, 415
  of multisets, 53
  of sets, 42, 49, 86
Invariant for graph isomorphism, 434–435, 446–447, 496
Invariant, loop, 212–213
Inverse function of, 60, 62, 86
Inverse image of a set, 65
Inverse of a matrix, 155
Inverse modulo m, 133, 156
Inverse relation, 350, 366, 401
Inverter, 588, 610
Invertible function, 60
Invertible matrix, 155
Irrational number, 118
Irreflexive property, 350
Isolated vertex, 417, 496
Isomers, 446
Isomorphic simple graphs, 429, 434, 496
  invariant of, 434–435, 446–447, 496
Isomorphism of graphs, 429, 434–437
Iteration, 205, 216

Jigsaw puzzle, 219
Jobs, assignment of, 331
Join of n-ary relations, 356, 402
Join of zero-one matrices, 151, 157

Kaliningrad, USSR, 451
Karnaugh, Maurice, 597
Karnaugh map, 597–603, 610, B-3
Kempe, Alfred, 488
Key:
  composite, 354
  for hashing, 115
  primary, 353
al-Khowarizmi, Abu Ja'far Mohammed ibn Musa, 92
Kleene, Stephen, 636
Kleene closure, 636, 657
Kleene's theorem, 646, B-3
Kneiphof Island, 451
Knight's tour, 245, 466
  reentrant, 466
Knuth, D., B-0–B-1, B-3
Königsberg bridge problem, 451–452
Kruskal, Joseph, 567

Kruskal's algorithm, 567
k-tuple coloring, 495
Kuratowski, Kazimierz, 482
Kuratowski's theorem, 482, 487

Labeled tree, 453
Lamé, Gabriel, 197
Lamé's theorem, 197
Landau, Edmund, 76
Landau symbol, 76
Language, 615, 618, 656, B-3
  formal, 615
  generated by a grammar, 618
  of a grammar, 618
  natural, 615
  over a vocabulary, 616
  recognition, 638
  recognition by a finite-state automaton, 638, 657
  recognition by a nondeterministic finite-state automaton, 640
  recognized by an automaton, 638
  recognizer, 632
Laplace, Pierre Simon, 251
Law:
  absorption, 17, 582
  associative, 15, 16, 581
  commutative, 15, 46, 581
  complementation, 46
  De Morgan's, 13, 15, 46, 581
  of detachment, 162
  distributive, 15, 46, 581
  denomination, 15, 46, 581
  of the double complement, 46, 581
  double complementation, 46
  double negation, 15
  idempotent, 15, 46, 581
  identity, 15, 46, 581
Leaf of a tree, 508, 514, 571
Least common multiple, 111–112, 156
Least element of a poset, 393, 402
Least upper bound, 394, 402
Left child of a vertex in a binary tree, 509
Left subtree of a vertex in a binary tree, 509
Lemma, 162, 215
  pumping, 656
Length of a path in a directed graph, 367
Length of a path in a weighted graph, 467
Length of a string, 9, 68, 200
Less than, in a poset, 387
Level of a vertex in a tree, 515, 571
Level order traversal, 574
Lexicographic ordering, 388–390, 402
Light fixtures, 591–593
Light switches, 591–593
Limit, definition of, 29
Linear algebra, B-1
Linear bounded automata, 654
Linear combination, *gcd* as, 130, 156
Linear complexity, 102
Linear congruence, 133, 156
Linear congruential random number generator, 116
Linear equations, simultaneous, 155
Linear homogeneous recurrence relation, 305, 335, B-1
  with constant coefficients, solving, 306–310
Linear inhomogeneous recurrence relations, 311
Linearly ordered set, 388, 402
Linear ordering, 388, 402
Linear search, 95, 100, 101–102, 156, 204
Linear search algorithm, 95
Literal, 585, 610
Little-*o* notation, 84
Local area network, 422–423
Logarithm, A-0–A-1
Logarithmic complexity, 102
Logarithmic function, A-1, B-3
Logic, 1–32, B-0
  fuzzy, 11–12
  gate, 588
  predicate, 19–32
  propositional, 1–18
Logical connective, 4
Logical equivalence, 13, 85
Logically equivalent, 13
Logical operator, 4, 84, 514
Longest monotone subsequence, 237
Loop, 362, 402, 411, 496
  in a directed graph, 362
  in an undirected graph, 411
Loop in pseudocode, A-6–A-8
Loop invariant, 212–213, 216
Lottery, 252–253
Lovelace, Countess of (Ada Augusta), 13, 16
Lower bound in a poset, 394, 402
Lower limit of a summation, 68
Lucas number, 311
Łukasiewicz, Jan, 536

Machine:
  deterministic finite state, 637
  finite state,626, 637–653, 657
  Mealy, 632
  Moore, 632, 634
  nondeterministic finite-state, 639
  Turing, 654
  unit-delay, 630
  vending, 626–627
McCarthy 91 function, 219
McCluskey, Edward J., 604, B-3
Map, coloring of, 486–488
Map, Karnaugh, 597–603
m-ary tree, 508, 516–517, 571, 572
  complete, 518
  full, 508, 571
Mathematical foundations for computer
    science, B-0
Mathematical induction, 175–192, 216, B-1
  second principle, 188, 216
Matrix, 144–154, 157, B-1
  addition, 145, 157
  adjacency, 430–431
  diagonal, 154
  fast multiplication, 314
  identity, 149, 157

Matrix *(continued)*
  incidence, 432
  inverse, 155
  invertible, 155
  multiplication, 146, 157
  power of, 153
  product, 146
  representing relations, 358–362
  square, 144
  sum of, 145
  symmetric, 150
  transpose, 150, 157
  upper triangular, 159
  zero-one, 150
Maximal element of a poset, 392, 402
Maximal spanning tree, 570
Maxterm, 587
Maze, 562–563, B-3
Mealy, G.H., 632, B-3
Mealy machine, 632, B-3
Mean, 98
  arithmetic, 192
  geometric, 192
Median, 98
Meet of zero-one matrices, 151, 157
Member of a set, 34, 85
Membership table, 47, 86
Merge sort, 547–551
Merging lists, 550
Mesh network, 424
Method, Horner, 104
Michigan, 486
Minimal element of a poset, 393, 396, 402
Minimal spanning forest, 570
Minimal spanning tree, 564–569, 571, B-3
  algorithms for, 565–569
  definition of, 564
Minimization of circuits, 593–607, B-3
Minimum dominating set, 500
Minterm, 585, 610
mod function, 112–113, 156
Mode, 98
Modular arithmetic, 112–114
Modulus, 113
Modus ponens, 162, 163
Modus tollens, 163
Mohammed's scimitars, 456
Montmort, Pierre Raymond de, 332
Moore, E.F., 632, B-3
Moore machine, 632, 634, B-3
Multigraph, 410, 412, 452, 496
  directed, 411
Multinomial coefficient, 281
Multinomial theorem, 281
Multiple of an integer, 106
Multiple edges, 410, 412, 452, 496
Multiple output circuit, 593
Multiplication matrix, 146
Multiplicity of membership in a multiset, 53
Multiset, 53, B-0
Mutual friends or enemies, 238
Mutually independent events, 291
Mutually relatively prime, 159

Naive set theory, 33
NAND, 18, 596, 610
NAND gate, 596
n-ary relation, 351–357, 401
  definition of, 352
  degree of, 352
  domains of, 352
Natural language, 615
Naur, Peter, 623
n-cube, 420, 425, 477, 496
Necessary, 6
  and sufficient, 8
Necessary and sufficient conditions for Euler circuits, 454
Necessary and sufficient conditions for Euler paths, 457
Negation, 3, 84
Negation of a universal quantification, 29, 30
Negation operator, 4
Negative of an existential quantification, 29, 30
Neighbors in a graph, 417
Network:
  interconnection, 423–425
  local area, 422–423
  mesh, 424–425
Niche overlap graph, 413
n log *n* complexity, 102
Node, 362
Nonconstructive existence proof, 172
Nondeterministic finite-state automaton, 639, 657
Nonregular set, 652
Nonterminal symbol, 617
NOR, 18, 596, 610
NOR gate, 596
Notation:
  Arabic, 92
  big-O, 74
  little-o, 84
  Polish, 536, 571
  product, 73, 86
  reverse Polish, 538
  set builder, 35
  summation, 68, 86
  theta, 83
n-queens problem, 558
n-regular graph, 428
Null set, 35, 85
Null string, 616
Number:
  chromatic, 487
  crossing, 486
  edge chromatic, 495
  Erdös, 405
  Fibonacci, 195
  harmonic, 84, 183
  independence, 502
  irrational, 118
  Lucas, 311
  perfect, 118
  pseudorandom, 116

rational, 74
telephone, 226, 236
Ulam, 218
Numbering plan, telephone, 106–144, 226
Number of subsets of a set, 184, 227
Number theory, 105, B-0–B-1

Object, 33
One's complement expansion, 129
One-to-one correspondence, 59, 60, 86
One-to-one function, 57, 86
number of, 226
Onto function, 58, 86, 329–331, 336
number of, 329–331, 336
Operator:
logical, 4, 84
negation, 4
Optimal algorithm, 105
OR, 9
Or:
exclusive, 5, 85
inclusive, 5, 84
Order:
functions of the same, 76
level, 574
pecking, 502
Ordered n-tuple, 38
Ordered pair, 39, 41
Ordered rooted tree, 509, 571
Ordering:
lexicographic, 388–390, 402
linear, 386
partial, 386
total, 388
Organization, 512
OR gate, 588–589, 610
Orientable graph, 501
Orientation of a graph, 501
Out-degree of a vertex, 419, 496
Output alphabet, 628
Output function for a finite-state machine, 628

Pair:
ordered, 39
Pairwise relatively prime, 111
Palindrome, 233, 625
Paradox, 33, 85
Russell's, 41
Parallel edges of a graph, 410
Parallel processing, 103, 424–425, 512
Parallel processors, 424–425, 512, B-2
tree-connected, 512
Parent of a vertex of a tree, 507, 571
Parse tree, 621, 657
Parsing:
bottom-up, 623
top-down, 623
Partial correctness of a program, 209
Partial function, 66
Partially ordered set, 386
Partial ordering, 386–398, 402, B-2
Partition:
of an integer, 202
refinement of, 386
of a set, 382–384, 402
Pascal, 55, 94, 614, B-3
Pascal, Blaise, 244
Pascal's identity, 243, 287
Pascal's triangle, 244, 287
Pass through a vertex, 442
Password, 228
Path, 367, 402, 442–443, 447–448, 452, 497
between vertices, counting, 447–448
in a directed graph, 367, 402, 443, 497
Euler, 452
in a graph, 442–443, 497
Hamilton, 457
interior vertices of, 373
of length n, 442
in a relation, 368
in an undirected graph, 442, 497
Pecking order, 502
Pendant vertex, 417, 496
Perfect number, 118
Permutation, 240–241, 287
generating, 282–284
with indistinguishable objects, 276–277
with repetition, 271–272
Petersen graph, 465, 483
Petersen, Julius, 465
Phrase-structure grammar, 616–624, 656
Pigeonhole principle, 233–238, 287, B-1
generalized, 235, 287
Planar graph, 476–486, 497
definition of, 477
regions of, 478
representation, 477
Planarity, 476–486
Playoff, 230–231
Poker, 253–254
Polish notation, 536
reverse, 538
Polynomial complexity, 102
Poset, 386–398, 402
antichain in, 407
chain in, 407
comparable elements, 387
greatest element of, 393
incomparable elements, 387
least element of, 393
maximal element of, 392
minimal element of, 393
Postfix expression, 538, 540
Postfix form, 538
Postfix notation, 571
Postorder traversal, 532–534, 535, 538–539, 571
Postorder traversal algorithm, 535
Postulate, 162
Power, Boolean (of zero-one matrices), 153
Power of a relation, 349, 401
Power set, 38, 85, 394
Precedence graph, 414

Precedence rules for logical operators, 7
Predicate, 19
Predicate logic, 19–32
Prefix code, 524–525, 571, B-2
Prefix expression, 536, 540
Prefix form, 536
Prefix notation, 536–537, 571
Pregel River, 451
Pre-image, 55, 86
Premise, 6
Prenex normal form, 32
Preorder traversal, 527–530, 571
Preorder traversal algorithm, 534
President of the USSR, 369
Prim, Robert Clay, 565
Primality test:
  probabilistic, 139
Primary key, 353, 401
Prime, 107, 156
  relatively, 110
Prime factorization, 107–109
Prim's algorithm, 565
Principle of inclusion-exclusion, 44, 229–230, 319–324
Principle of mathematical induction, 175–192, 216, B-1
Private key cryptography, 139, 157
Probability, discrete, 250–271, 287
  conditional, 260
  definition of, 251, 259
  Laplace's definition of, 251, 259
Problem:
  hatcheck, 331
  Königsberg bridge, 451–452
  n-queens, 559
  searching, 95
  shortest path, 466
  utilities, 476
procedure statement, A-4
Processing, parallel, 103
Product:
  of Boolean function, 579
  Boolean, of zero-one matrices, 151
  Cartesian, 38–40
  of functions, 56
  of matrices, 146
  notation, 73
Product-of-sums expansion, 588
Production of a grammar, 617, 656
Product rule for counting, 225–228, 287
Program correctness, 209–214, 216
Program verification, 209–214
Progression:
  arithmetic, 218
  geometric, 69
Projection of an *n*-ary relation, 354, 402
Project, software, scheduling of, 406
Proof, 162, 166–173, 215, B-1
  by cases, 169, 216
  combinatorial, 243, 245, 246
  constructive existence, 171–172
  by contradiction, 168–169, 216
  definition of, 162
  direct, 167, 215
  indirect, 168, 216
  by mathematical induction, 176–190
  methods of, 162–173
  nonconstructive existence, 171, 172
  of program correctness, 208–214
  trivial, 167, 215
  vacuous, 167, 215
Proper subset, 36, 85
Properties of an algorithm, 94
Property, well-ordering, 176
Proposition, 2, 84
  compound, 3, 84
  negation of, 3
Propositional equivalences, 12–17
Propositional function, 19, 85
Propositional logic, 1–18
Pseudocode, 93, A-4–A-9
Pseudograph, 410, 412, 496
Pseudoprime, 138–139, 157
Pseudorandom numbers, 116
Public key cryptography, 139–142, 157
Pumping lemma, 652–653, 656
Pumping lemma for context-free languages, 660
Pure multiplicative random number generator, 116
Pushdown automata, 654, B-3
Puzzle:
  Hamilton's round the world, 458
  jigsaw, 219

Quantification, 20–32
  existential, 22, 85
  uniqueness, 32
  universal, 20, 85
Queen on a chessboard, 559
Quick sort, 552
Quine-McClusky method, 604–607, B-3
Quine, Willard V., 604
Quotient, 109

Rabbits, 297–298
Radius of a graph, 502
Ramsey, F. P., 238
Ramsey theorem, 238
Random variable, 264–265, 287
Range of a function, 55, 86
Rational number, 74
r-combination, 242, 287
Reachable state, 659
Reasoning, mathematical, 161–173
Recognized string, 638, 640
Recognized string by a nondeterministic finite-state automaton, 638
Recognizer, language, 632
Record in a data base, 352
Recurrence relation, 295–317, 335
  divide-and-conquer, 313–318
  initial conditions of, 296
  linear homogeneous, 305
  linear inhomogeneous, 311
  simultaneous, 312

solution of, 295
using generating function to solve, A-14–A-15
Recursion, 202–207
Recursive algorithm, 203, 216
Recursive definition, 193–200, 216
Recursively defined set, 198–200, 216
Recursive program, 203–207
Reentrant knight's tour, 466
Refinement of a partition, 386
Reflexive:
property, 344–345, 359, 363, 401
relation, 344, 401
Reflexive closure of a relation, 366, 403
Region, degree of, 480
Region of a representation of a planar graph, 478, 497
Regular expression, 645, 657
Regular grammar, 620, 645, 649, 651–652, 656
Regular graph, 428
Regular language, 620
Regular set, 645–646, 657
definition of, 645–646
finite-state automaton recognizing, 645–652
regular grammar for, 649–652
Related to, 341
Relation, 341–407, B-2
antisymmetric, 345
asymmetric, 350
binary, 341
circular, 404
closure of, 366
complementary, 350
composite of, 348
connectivity, 369
counting, 347
diagonal, 366
difference, 348
divide-and-conquer, 312
equivalence, 379
intersection, 348
inverse, 350
irreflexive, 350
n-ary, 352
path in, 368
power of, 349
properties of, 344–347
recurrence, 295, 335
reflexive, 344
representing relations using directed graphs, 362–365
representing relations using matrices, 358–362
on a set, 342
symmetric, 345
transitive, 346
union, 348
Relational data base model, 352–357, 401, B-2
Relatively prime integers, 110, 156
mutually, 159
pairwise, 111, 156
Remainder, 109
Repetition:
combination with, 272–276
permutation with, 271–272
Representation:
binary, 123, 156
of integers, 121–125
Representative of equivalence class, 380
Representing a Boolean function, 584–587
Representing relations using directed graphs, 362–364
Representing relations using matrices, 358–362
Reversal of a string, 201
Reverse Polish notation, 538, 571
Right child of a vertex in a tree, 509
Right subtree of a vertex in a tree, 509
Ring topology for a local area network, 423
Rooted Fibonacci tree, 519
Rooted spanning tree, 563
Rooted tree, 507, 571
ordered, 509
Root of a tree, 507
Round-robin tournament, 413–414
Row of a matrix, 145
r-permutation, 240, 287
RSA cryptosystem, 140–142
decryption in, 141–142
encryption in, 140–141
as a public key system, 142
Rule:
product, 225
sum, 223
Rule of inference, 162, 215
for program correctness, 210
Russell, Bertrand, 33
Russell's paradox, 41

Sample space, 251
Sampling with replacement, 272
Saturated hydrocarbons, 510
Scheduling:
exams, 491–492
software project, 406
tasks, 398
Scimitars, Mohammed's, 456
Search:
binary, 95
breadth-first, 556
depth-first, 555
linear, 95
sequential, 95
ternary, 98
tree, binary, 520–522
Searching algorithms, 95–97, 156, B-2
Searching problem, 95
Second principle of mathematical induction, 188–190
Selection sort, 551–552
Self-complementary graph, 441
Self-dual, 611
Semantics, 615
Sentence, 616
Sequence, 67, 86
Sequential search, 95
Sequential search algorithm, 95
Series:
infinite, A-11–A-12

Set, 33–54, 85, B-0
  builder notation, 35
  complement of, 44–45, 86
  computer representation of, 49–51
  countable, 71, 86
  cut, 574
  description of, 34–35
  difference of, 44, 86
  disjoint, 43
  dominating, 500
  element of, 34, 85
  empty, 35, 85
  equality, 34, 85
  finite, 37, 85
  fuzzy, 53, B-0
  identities, 46
  image of, 56
  infinite, 37, 85
  intersection of, 42, 49, 86
  inverse image, 65
  member of, 34, 85
  nonregular, 652
  null, 35, 85
  operations, 41–51
  partially ordered, 386
  partition of, 382
  power, 38, 85
  recursively defined, 198–200
  regular, 645–646
  relation on, 342
  successor of, 53
  symmetric difference, 52
  totally ordered, 388
  uncountable, 71, 86
  union of, 42, 48, 86
  universal, 35, 85
  well-ordered, 396
Shannon, Claude, 577
Shift cipher, 117
Shifting, 127
Shortest path problem, 467, 497
Sibling, 507, 571
Sieve of Eratosthenes, 328–329, 330
Signed integer, 624
Simple circuit, 442
Simple graph, 409, 412, 496
Simple path, 442, 497
Simplification rule of inference, 163
Socks, 238
Sollin's algorithm, 570, B-3
Solution of a recurrence relation, 295
Solving counting problems using generating functions, A-12–A-13
Solving recurrence relations using generating functions, A-14
Sort, 543–552
  bubble, 545–547
  insertion, 574
  merge, 547–551
  quick, 552
  selection, 551–552
Sorting, 543–542, 572, B-2
  topological, 396–397
Sorting algorithms, 544–551
Sorting problem, 543
Space complexity, 99, 156
Space, sampling, 251
Spanning forest, 563
Spanning tree, 552–560, 571
  maximal, 570
  minimal, 564–570
Square matrix, 144
Squarefree integer, 334
Squirrel, 413
Stairs, climbing, 303
Star height, 659
Starting state, 637
State:
  initial, 637
  final, 637
  start, 637
State diagram for a finite-state machine, 628
Statement:
  assignment, A-4–A-5
  begin, A-5
  blocks of, A-5–A-6
  end, A-5
  if then, A-6–A-7
  procedure, A-4
State of a finite-state machine, 637
  final, 637
  reachable, 659
  start, 637
  transient, 659
State table for a finite-state machine, 637
Step:
  basis, 177, 188, 216
  inductive, 177, 188, 216
S_k-tree, 574
Strictly decreasing function, 58
Strictly decreasing sequence, 237
Strictly increasing function, 58
Strictly increasing sequence, 237
String, 67–68, 86, 199–200, 635–636
  balanced string of parentheses, 220
  bit, 9
  concatenation of, 199, 635
  empty, 68, 616
  length of, 68, 200
  lexicographic ordering of, 388–390
  null, 68, 616
  reversal, 201
  ternary, 304
Strongly connected directed graph, 445
Subgraph, 425, 496
  induced, 499
Subsequence, 237
Subset, 36, 85
  number of, 227
  proper, 36, 85
Subtractor:
  full, 596

half, 596
Subtree, 508, 571
left, 509
right, 509
Successor of a set, 53
Sufficient, 6
necessary and, 8
Summation, 68–70
double, 70
Summation notation, 68
Sum of functions, 56
Sum of matrices, 145
Sum of multisets, 53
Sum-of-products expansion, 585–586, 610, B-3
Sum rule for counting, 223, 287
Sums of subsets of integers, 559–560
Surjection, 58, 86
Surjective function, 58
Switching circuits, 588–595, B-3
Symbol:
nonterminal, 617
start, 617
terminal, 617
Symmetric:
property, 345–347
relation, 345, 401
Symmetric closure of a relation, 366, 403
Symmetric difference, 52, 86
Symmetric matrix, 150, 157
Syntax, 615

Table:
membership, 47, 86
state, 627
truth, 4, 84
Table representation of a data base, 353
Tautology, 12, 85
Telephone numbering plan, 226
Telescoping series, 73
Terminal symbol, 617
Terminal vertex of a directed edge, 418
Term of a sequence, 67
Ternary search, 98, 304
Test, probabilistic primality, 139
Theorem, 162, 215
binomial, 246, 287
Chinese remainder, 134–136, 157
Fermat's little, 139, 157
four-color, 487
fundamental theorem of arithmetic, 107
handshaking, 418
Kleene's, 646
Kuratowski's, 482
Lamé's, 197
multinomial, 281
Wilson's, 143
Theta notation, 83
Thickness of a graph, 486, B-2
Threshold function, 613, B-3
Threshold gate, 612
Threshold value, 612
Time complexity, 99, 156
Top-down parsing, 623
Topological sorting, 396–398, 402
Topological sorting algorithm, 397
Topology:
ring, 423
star, 422
Torus, 486
Totally ordered set, 388, 402
Total ordering, 386, 402
compatible, 396
Tournament, 501
round-robin, 413–414
Tower of Hanoi, 298–300
Transient state, 659
Transition function for a finite-state machine, 628, 639
Transitive closure, 369–376, 403
Transitive relation, 346, 349, 401
Transpose of a matrix, 150, 157
Traversal of a tree, 526–543
inorder, 530–532, 534, 535
level order, 574
postorder, 532–534, 535
preorder, 527–530, 534
Traverse vertices in an undirected graph, 442
Tree, 504–570, 571, B-2–B-3
balanced, 516
binary, 508
binary search, 520
B-tree, 574
center of, 519
complete *m*-ary, 518
decision, 522–523
definition of, 505
degree-constrained spanning, 575
derivation, 621
diagram, 230–231, 287
family, 505
Fibonacci, rooted, 519
full *m*-ary, 508, 514, 571
graceful, 575
labeled, 453
leaf, 508, 514, 571
m-ary, 508
maximal spanning, 570
minimal spanning, 563–564
ordered rooted, 509
parse, 621
rooted, 507
rooted Fibonacci, 519
rooted spanning, 563
root of, 507
spanning, 552–560
S_k-tree, 574
Tree-connected network, 512
Tree-connected parallel processors, 512
Tree diagram, 230–231
Tree traversal, 571
Triangle, 244
Trivial proof, 167, 215
True, 3, 8

Truth table, 4, 84
Truth value, 3, 84
Tukey, John, 8
Turing, Alan, 655
Turing machines, 654, B-3
Type 0 grammar, 620, 656
Type 1 grammar, 620, 656
Type 2 grammar, 620, 656
Type 3 grammar, 620, 656
Types of finite-state machines, 632
Types of grammars, 620, 621

Ulam numbers, 218
Uncountable set, 71, 72, 86
  set of real numbers as an, 72
Underlying undirected graph, 419, 496
Undirected graph, 409–411, 496
  underlying, 419
Union of events, 255, 260
Union of fuzzy sets, 54
Union of graphs, 426, 496
Union of multisets, 53
Union of sets, 42, 48, 49, 86
Uniqueness quantification, 32
Unit-delay machine, 630
United States of America, 369
Universal address system, 526
Universal quantification, 20, 29, 30, 85, 172
  negation of, 29, 30
Universal set, 35, 85
Universe of discourse, 20, 85
Upper bound, 394, 402
Upper limit of a summation, 68
Upper triangular matrix, 159
USSR, 369
Utilities problem, 476

Vacuous proof, 167, 215
Valid argument, 164
Value:
  expected, 266
  truth, 3, 84
Vandermonde, Abnit-Théophile, 245
Vandermonde's identity, 245
Variable:
  Boolean, 8, 578, 610
  bound, 25
  free, 25, 85
  random, 265
Veitch, E. W., 598, B-3
Vending machine, 626–627
Venn diagram, 35, 85
Venn, John, 35
Vertex, 362, 409
  adjacent, 417
  adjacent from, 418
  adjacent to, 418
  basis, 450
  cut, 444
  degree of, 417
  distance between, 502
  eccentricity of, in a tree, 519
  end, 418
  in-degree of, 419
  independent set of, 502
  initial, of a directed edge, 362, 418
  internal, of a tree, 508
  isolated, 417
  level of, in a tree, 515
  out-degree of, 419
  pendant, 417
  terminal, of a directed edge, 362, 418
Vocabulary, 616, 617, 656

Warshall, Stephen, 373
Warshall's algorithm, 373–376, B-2
Weakly connected directed graph, 445
Weighted graph, 467, 497
Well-formed formulae, 199, 543
Well-ordered set, 219, 396, 402
Well-ordering property, 176, 216
Wheel, 420, 496
Wilson's theorem, 143
Woodpecker, 413
Word, 616
Worst-case time complexity, 100

XOR, 9
XOR gate, 612

Zero-one matrix, 150, 157
  Boolean powers, 153
  Boolean product, 151–152
Zodiac, signs of, 290

LIST OF SYMBOLS

TOPIC	SYMBOL	MEANING	PAGE
COUNTING	$P(n, r)$	number of r-permutations of a set with n elements	240
	$C(n, r)$	number of r-combinations of a set with n elements	242
	$\binom{n}{r}$	binomial coefficient n over r	243
	$p(E)$	probability of E	251
	$p(E \mid F)$	conditional probability of E given F	260
	$E(X)$	expected value of the random variable X	266
	$C(n; n_1, n_2, \ldots, n_m)$	multinomial coefficient	281
	$N(P_{i_1} \cdots P_{i_n})$	number of elements having properties P_{i_j}, $j = 1, \ldots, n$	326
	$N(P'_{i_1} \cdots P'_{i_n})$	number of elements not having properties P_{i_j}, $j = 1, \ldots, n$	326
	D_n	number of derangements of n objects	332
RELATIONS	$S \circ R$	composite of the relations R and S	348
	R^n	nth power of the relation R	349
	R^{-1}	inverse relation	350
	$P_{i_1, i_2, \ldots, i_m}$	projection	354
	$J_p(R, S)$	join	356
	Δ	diagonal relation	366
	R^*	connectivity relation of R	369
	$[a]_R$	equivalence class of a with respect to R	380
	$[a]_m$	congruence class modulo m	381
	(S, R)	poset consisting of the set S and partial ordering R	386
	$a \prec b$	a is less than b	387
	$a \preccurlyeq b$	a is less than or equal to b	387
	$a \succ b$	a is greater than b	387
	$a \succcurlyeq b$	a is greater than or equal to b	387
GRAPHS AND TREES	(u, v)	directed edge	362
	$\{u, v\}$	undirected edge	411
	$G = (V, E)$	graph with vertex set V and edge set E	409
	$\deg(v)$	degree of the vertex v	417
	$\deg^-(v)$	in-degree of the vertex v	419
	$\deg^+(v)$	out-degree of the vertex v	419
	K_n	complete graph on n vertices	420
	$K_{m,n}$	complete bipartite graph of size m, n	422
	C_n	cycle of size n	420
	W_n	wheel of size n	420
	Q_n	n-cube	420